HANDBOOK OF NUTRITION, DIET, AND THE EYE

HANDBOOK OF NUTRITION, DIET, AND THE EYE

Edited by

VICTOR R. PREEDY
King's College London,
London, UK

AMSTERDAM • BOSTON • HEIDELBERG • LONDON
NEW YORK • OXFORD • PARIS • SAN DIEGO
SAN FRANCISCO • SINGAPORE • SYDNEY • TOKYO
Academic Press is an imprint of Elsevier

Academic Press is an imprint of Elsevier
The Boulevard, Langford Lane, Kidlington, Oxford, OX5 1GB, UK
225 Wyman Street, Waltham, MA 02451, USA

First published 2014

British Library Cataloguing in Publication Data
A catalogue record for this book is available from the British Library

Library of Congress Cataloguing in Publication Data
A catalogue record for this book is available from the Library of Congress

ISBN: 978-0-12-401717-7

For information on all Academic Press publications
visit our website at **store.elsevier.com**

Printed and bound by CPI Group (UK) Ltd, Croydon, CR0 4YY

14 15 16 17   10 9 8 7 6 5 4 3 2 1

Working together
to grow libraries in
developing countries

www.elsevier.com • www.bookaid.org

Contents

III
GLAUCOMAS

IV
CATARACTS

VII

MACRONUTRIENTS

VIII
MICRONUTRIENTS

X

NUTRIGENOMICS AND MOLECULAR BIOLOGY OF EYE DISEASE

XI

ADVERSE EFFECTS AND REACTIONS

Contributors

Winsome Abbott-Johnson Princess Alexandra Hospital, Woolloongabba, Qld, Australia

Niyazi Acar Eye and Nutrition Research Group, University of Burgundy, Centre des Sciences du Goût et de l'Alimentation, Dijon, France

Vaishali Agte Agharkar Research Institute, Pune, India

Daniel Agudelo Department of Chemistry–Biology, University of Québec at Trois-Rivières, Trois-Rivières, Québec, Canada

Maria Antonietta Altea Department of Clinical and Experimental Medicine, University of Pisa, Pisa, Italy

R.A. Armstrong Vision Sciences, Aston University, Birmingham, UK

Tin Aung Singapore National Eye Centre, Singapore

Bahri Aydın Istanbul Medeniyet University Medical School, Istanbul, Turkey

Fereshteh Bahmani Kashan University of Medical Sciences, Kashan, Iran

D. Balmer IRO, Institute for Research in Ophthalmology, Sion, Switzerland

S. Zahra Bathaie Tarbiat Modares University, Tehran, Iran

Lynne Bell School of Psychology and Clinical Language Sciences, University of Reading, Reading, UK

Tos T.J.M. Berendschot University Eye Clinic Maastricht, Maastricht, The Netherlands

Paul S. Bernstein Moran Eye Center, University of Utah School of Medicine, Salt Lake City, UT

Brian M. Besch Moran Eye Center, University of Utah School of Medicine, Salt Lake City, UT

Philippe Bourassa Department of Chemistry–Biology, University of Québec at Trois-Rivières, Trois-Rivières, Québec, Canada

R.B. Bozard Department of Cellular Biology and Anatomy, Georgia Health Sciences University, Augusta, GA, USA; Department of Ophthalmology, Georgia Health Sciences University, Augusta, GA, USA

Lionel Bretillon Eye and Nutrition Research Group, University of Burgundy, Centre des Sciences du Goût et de l'Alimentation, Dijon, France

Alain M. Bron Eye and Nutrition Research Group, University of Burgundy, Centre des Sciences du Goût et de l'Alimentation, Dijon, France

Benjamin Buaud ITERG – Equipe Nutrition Métabolisme & Santé, Bordeaux, France

Gabriëlle H.S. Buitendijk Erasmus Medical Center, Rotterdam, The Netherlands

Laurie T. Butler School of Psychology and Clinical Language Sciences, University of Reading, Reading, UK

Aldo Caporossi Policlinico Universitario A. Gemelli, Rome, Italy

Stefano Caragiuli Azienda Ospedaliera Universitaria Senese, Siena, Italy

Chloé Cartier Département de psychologie, Université du Québec à Montréal, Montréal, Québec, Canada

Cristina Cartiglia University of Genoa, Genoa, Italy

Chi-Ming Chan School of Medicine, Fu Jen Catholic University, New Taipei City, Taiwan; Department of Ophthalmology, Cardinal Tien Hospital, New Taipei City, Taiwan

Min-Lee Chang Jean Mayer USDA Human Nutrition Research Center on Aging at Tufts University, Boston, MA

Bashira A. Charles Center for Research on Genomic and Global Health, National Human Genome Research Institute, Bethesda, Maryland, USA

Emily Y. Chew National Eye Institute, National Institutes of Health, Bethesda, Maryland, USA

Ching-Yu Cheng Singapore Eye Research Institute, Singapore; Yong Loo Lin School of Medicine, National University of Singapore

Chung-Jung Chiu Jean Mayer USDA Human Nutrition Research Center on Aging at Tufts University, Boston, MA

Deepika Chopra Government Medical College, Amritsar, Punjab, India

Patricia Coelho de Velasco Instituto de Biologia, Universidade Federal Fluminense, Niterói, Brazil

David Coman Department of Metabolic Medicine, The Royal Children's Hospital, Brisbane, Queensland, Australia

Nicole Combe ITERG – Equipe Nutrition Métabolisme & Santé, Bordeaux, France

Dolores Corella Genetic and Molecular Epidemiology Unit, Department of Preventive Medicine and Public Health, School of Medicine, University of Valencia, Valencia, Spain; CIBER Fisiopatología de la Obesidad y Nutrición, ISCIII, Valencia, Spain

M. Cossenza Program of Neurosciences, Fluminense Federal University, Niterói, Brazil; Department of Physiology and Pharmacology, Biomedical Institute, Fluminense Federal University, Niterói, Brazil

Simonetta Costa Division of Neonatology, Catholic University of Rome, Rome, Italy

Catherine P. Creuzot-Garcher Eye and Nutrition Research Group, University of Burgundy, Centre des Sciences du Goût et de l'Alimentation, Dijon, France

Maria Cristina de Oliveira Izar Cardiology Division, Department of Medicine, Federal University of São Paulo, São Paulo, Brazil

R.P. Cubbidge Vision Sciences, Aston University, Birmingham, UK

Alyssa Cwanger FM Kirby Center for Molecular Ophthalmology, Scheie Eye Institute, University of Pennsylvania, Philadelphia, Pennsylvania, USA

Cécile Delcourt Universite de Bordeaux, Bordeaux, France; Inserm, ISPED, Centre INSERM U897-Epidemiologie-Biostatistique, Bordeaux, France

Marie-Noëlle Delyfer Universite de Bordeaux, Bordeaux, France; Inserm, ISPED, Centre INSERM U897-Epidemiologie-Biostatistique, Bordeaux, France; Service d'Ophtalmologie, CHU de Bordeaux, Bordeaux, France

I.C.L. Domith Program of Neurosciences, Fluminense Federal University, Niterói, Brazil

David Dunaief Medical Compass, MD Private Practice, New York, USA

Joshua L. Dunaief FM Kirby Center for Molecular Ophthalmology, Scheie Eye Institute, University of Pennsylvania, Philadelphia, Pennsylvania, USA

Rajan Elanchezhian Department of Animal Science, School of Life Sciences, Bharathidasan University, Tiruchirappalli, Tamilnadu, India

Andrew W. Eller Retina Service, UPMC Eye Center, University of Pittsburgh School of Medicine, and The Eye and Ear Institute, Pittsburgh, PA

T.G. Encarnação Program of Neurosciences, Fluminense Federal University, Niterói, Brazil

Mesut Erdurmuş Abant Izzet Baysal University Medical School, Bolu, Turkey

Evangelina Espósito University Clinic Reina Fabiola, Catholic University of Córdoba, Córdoba, Argentina

Asghar Farajzadeh Tarbiat Modares University, Tehran, Iran

David T. Field School of Psychology and Clinical Language Sciences, University of Reading, Reading, UK

Silvia C. Finnemann Department of Biological Sciences, Center for Cancer, Genetic Diseases, and Gene Regulation, Fordham University, Bronx, New York, USA

Steven J. Fliesler Veterans Administration Western New York Healthcare System; University at Buffalo/State University of New York; and the SUNY Eye Institute, Buffalo, NY

Nicolas Froger INSERM, U968, Institut de la Vision, Paris, France; Sorbonne Universités, UPMC Univ Paris 06, UMR_S 968 Paris, France; CNRS, UMR 7210, Institut de la Vision, Paris, France

P.S. Ganapathy Department of Cellular Biology and Anatomy, Georgia Health Sciences University, Augusta, GA, USA; Department of Ophthalmology, Georgia Health Sciences University, Augusta, GA, USA

V. Ganapathy Vision Discovery Institute, School of Medicine, Georgia Health Sciences University, Augusta, GA, USA; Department of Biochemistry and Molecular Biology, Georgia Health Sciences University, Augusta, GA, USA

Jose J. Garcia-Medina Department of Ophthalmology, University General Hospital Reina Sofía, Murcia, Spain; Department of Ophthalmology, School of Medicine, University of Murcia, Murcia, Spain

Pitchairaj Geraldine Department of Animal Science, School of Life Sciences, Bharathidasan University, Tiruchirappalli, Tamilnadu, India

Carmen Giannantonio Division of Neonatology, Catholic University of Rome, Rome, Italy

C.R. Gibson Wyle Science, Technology and Engineering, Houston, Texas, USA, and Coastal Eye Associates, Webster, Texas, USA

Snehal Gite Agharkar Research Institute, Pune, India

N. Goldenberg-Cohen Pediatric Ophthalmology Unit, Schneider Children's Medical Center of Israel, Petach Tikva, Israel; The Krieger Eye Research Laboratory, Felsenstein Medical Research Center, Sackler School of Medicine, Tel Aviv University, Tel Aviv, Israel

Glen Gole Paediatrics and Child Health, University of Queensland, Queensland, Australia

Ian R. Gorovoy University of California, San Francisco, CA

Julia A. Haller Wills Eye Hospital, Philadelphia, PA

Lisa Hark Wills Eye Hospital, Philadelphia, PA

Rijo Hayashi Department of Ophthalmology, Koshigaya Hospital, Dokkyo University, School of Medicine, Koshigaya, Saitama, Japan

Hui He Department of Diagnostic Medicine/Pathobiology, Kansas State University, Manhattan, Kansas, USA

Tatiana Helfenstein Cardiology Division, Department of Medicine, Federal University of São Paulo, São Paulo, Brazil

Francisco Antonio Helfenstein Fonseca Cardiology Division, Department of Medicine, Federal University of São Paulo, São Paulo, Brazil

Ken-ichi Hosoya Department of Pharmaceutics, Graduate School of Medicine and Pharmaceutical Sciences, University of Toyama, Toyama, Japan

Yi-Ling Huang Jean Mayer USDA Human Nutrition Research Center on Aging at Tufts University, Boston, MA

Chi-Feng Hung School of Medicine, Fu Jen Catholic University, New Taipei City, Taiwan

M. Ibberson Vital-IT Group, Swiss Institute of Bioinformatics, Lausanne, Switzerland

Hiroto Izumi School of Medicine, University of Occupational and Environmental Health, Fukuoka, Japan

Alberto Izzotti University of Genoa, Genoa, Italy

Riikka L. Järvinen Finnsusp Ltd, Lieto, Finland

Hua Ji Department of Nutritional Sciences, Oklahoma State University, Stillwater, Oklahoma; Institute of Genetics and Physiology, Hebei Academy of Agriculture and Forestry Sciences, Shijiazhuang, PR China

Yao Jin Nanjing Medical University Eye Hospital, Nanjing, Jiangsu Province, China

Corinne Joffre INRA and University of Bordeaux, Nutrition and Integrative Neurobiology (NutriNeuro) Bordeaux, France

Choun-Ki Joo The Catholic University of Korea, Seoul, Korea

Sang Hoon Jung Functional Food Center, Korea Institute of Science and Technology (KIST), Gangneung Institute, Gangneung, Republic of Korea

Heikki P. Kallio Food Chemistry and Food Development, Department of Biochemistry, University of Turku, Finland

Paul Kerlin Wesley Medical Centre, Auchenflower, Qld, Australia

Eun Chul Kim The Catholic University of Korea, Seoul, Korea

Jin Sook Kim Herbal Medicine Research Division, Korea Institute of Oriental Medicine, Daejeon, Republic of Korea

Amar U. Kishan UC Davis Eye Center, Sacramento, CA

Caroline C.W. Klaver Erasmus Medical Center, Rotterdam, The Netherlands

Kimitoshi Kohno School of Medicine, University of Occupational and Environmental Health, Fukuoka, Japan

Jean-François Korobelnik Universite de Bordeaux, Bordeaux, France; Inserm, ISPED, Centre INSERM U897-Epidemiologie-Biostatistique, Bordeaux, France; Service d'Ophtalmologie, CHU de Bordeaux, Bordeaux, France

Yoshiyuki Kubo Department of Pharmaceutics, Graduate School of Medicine and Pharmaceutical Sciences, University of Toyama, Toyama, Japan

Petra S. Larmo Aromtech Ltd, Tornio, Finland

Ryszard Lauterbach Jagiellonian University Medical College, Kraków, Poland

Ling-Jun Li Saw Swee Hock School of Public Health, National University of Singapore, Singapore

Dingbo Lin Department of Nutritional Sciences, Oklahoma State University, Stillwater, Oklahoma

M.I. Lopez-Galvez Universidad de Valladolid, Valladolid, Spain

Yi Lu Eye and ENT Hospital, Fudan University, Shanghai, P.R. China

Zhi-Quan Lu Liaoning Medical University, Liaoning Province, P.R. China

F. Manco Lavado Hospital Clinico Valladolid, Valladolid, Spain

Claudio Marcocci Department of Clinical and Experimental Medicine, University of Pisa, Pisa, Italy

Shilpa Mathew UC Davis Eye Center, Sacramento, CA

Cosimo Mazzotta Azienda Ospedaliera Universitaria Senese, Siena, Italy

Francesca Menconi Department of Clinical and Experimental Medicine, University of Pisa, Pisa, Italy

Naoya Miyamoto School of Medicine, University of Occupational and Environmental Health, Fukuoka, Japan

Bobeck S. Modjtahedi UC Davis Eye Center, Sacramento, CA

Lawrence S. Morse UC Davis Eye Center, Sacramento, CA

Sarah W. Mount School of Psychology and Clinical Language Sciences, University of Reading, Reading, UK

Arumugam R. Muralidharan Department of Animal Science, School of Life Sciences, Bharathidasan University, Tiruchirappalli, Tamilnadu, India

Benjamin P. Nicholson National Eye Institute, National Institutes of Health, Bethesda, Maryland, USA

R. Paes-de-Carvalho Program of Neurosciences, Fluminense Federal University, Niterói, Brazil; Department of Neurobiology, Institute of Biology, Fluminense Federal University, Niterói, Brazil

J.C. Pastor IOBA: Universidad de Valladolid, Valladolid, Spain

Dorota Pawlik Jagiellonian University Medical College, Kraków, Poland

John F. Payne Palmetto Retina Center, LLC, West Columbia, SC

Daniel Petrovič Institute of Histology and Embryology, Medical Faculty Ljubljana, University of Ljubljana, Ljubljana, Slovenia

Serge Picaud INSERM, U968, Institut de la Vision, Paris, France; Sorbonne Universités, UPMC Univ Paris 06, UMR_S 968 Paris, France; CNRS, UMR 7210, Institut de la Vision, Paris, France; Fondation Ophtalmologique Adolphe de Rothschild, Paris, France

Maria D. Pinazo-Duran Ophthalmology Research Unit 'Santiago Grisolia', Dr. Peset University Hospital, Valencia, Spain; Department of Surgery, School of Medicine, University of Valencia, Valencia, Spain

Adela Mariana Pintea University of Agricultural Sciences and Veterinary Medicine, Cluj-Napoca, Romania

Jogchum Plat Department of Human Biology, Maastricht University Medical Center, Maastricht, The Netherlands

C.C. Portugal Program of Neurosciences, Fluminense Federal University, Niterói, Brazil

Ananda S. Prasad Department of Oncology, Wayne State University School of Medicine and Barbara Ann Karmanos Cancer Institute, Detroit, Michigan, USA

Victor R. Preedy Diabetes and Nutritional Sciences, School of Medicine, King's College London, London, UK

Jiang Qin Nanjing Medical University Eye Hospital, Nanjing, Jiangsu Province, China

R. Roduit IRO, Institute for Research in Ophthalmology, Sion, Switzerland; Department of Ophthalmology, University of Lausanne, Lausanne, Switzerland

Costantino Romagnoli Division of Neonatology, Catholic University of Rome, Rome, Italy

Marie-Bénédicte Rougier Service d'Ophtalmologie, CHU de Bordeaux, Bordeaux, France

Dumitriţa Olivia Rugină University of Agricultural Sciences and Veterinary Medicine, Cluj-Napoca, Romania

Charumathi Sabanayagam Singapore Eye Research Institute, Singapore; Yong Loo Lin School of Medicine, National University of Singapore; Office of Clinical Sciences, Duke-NUS Graduate Medical School, Singapore

Sergio Claudio Saccà St Martino Hospital, Ophthalmology Unit, Genoa, Italy

José-Alain Sahel INSERM, U968, Institut de la Vision, Paris, France; Sorbonne Universités, UPMC Univ Paris 06, UMR_S 968 Paris, France; CNRS, UMR 7210, Institut de la Vision, Paris, France; Centre Hospitalier National d'Ophtalmologie des Quinze-Vingts, Paris, France; Institute of Ophthalmology, University College of London, UK; Fondation Ophtalmologique Adolphe de Rothschild, Paris, France; French Academy of Sciences, Paris, France

Dave Saint-Amour Département de psychologie, Université du Québec à Montréal, Montréal, Québec, Canada; Département d'ophtalmologie, Université de Montréal, Montréal, Québec, Canada; Centre de recherche, Centre hospitalier universitaire Sainte-Justine, Montréal, Québec, Canada

Pedro Sanz-Solana Department of Ophthalmology, Dr. Peset University Hospital, Valencia, Spain

Seang Mei Saw Saw Swee Hock School of Public Health, National University of Singapore, Singapore

D.F. Schorderet IRO, Institute for Research in Ophthalmology, Sion, Switzerland; Department of Ophthalmology, University of Lausanne, Lausanne, Switzerland; Faculty of Life Sciences, Ecole Polytechnique Fédérale de Lausanne, Lausanne, Switzerland

Claudio Alberto Serfaty Instituto de Biologia, Universidade Federal Fluminense, Niterói, Brazil

Horacio M. Serra Department of Clinical Biochemistry, Faculty of Chemical Science, National University of Córdoba, Córdoba, Argentina

Saumil Sethna Department of Biological Sciences, Center for Cancer, Genetic Diseases, and Gene Regulation, Fordham University, Bronx, New York, USA

Jay Siak Singapore National Eye Centre, Singapore

Hüseyin Simavlı Bolu Izzet Baysal State Hospital, Bolu, Turkey

S.B. Smith Department of Cellular Biology and Anatomy, Georgia Health Sciences University, Augusta, GA, USA; Department of Ophthalmology, Georgia Health Sciences University, Augusta, GA, USA; Vision Discovery Institute, School of Medicine, Georgia Health Sciences University, Augusta, GA, USA

S.M. Smith Biomedical Research and Environmental Sciences Division (MC SK3), NASA Johnson Space Center, Houston, Texas, USA

R. Socodato Program of Neurosciences, Fluminense Federal University, Niterói, Brazil

Marco Spinazzi Laboratory for the Research of Neurodegenerative Diseases, Department of Human Genetics, University of Leuven, Belgium

K. Srinivasan Department of Biochemistry and Nutrition, CSIR – Central Food Technological Research Institute, Mysore, India

Philip Storey Wills Eye Hospital, Philadelphia, PA

María Fernanda Suárez Department of Clinical Biochemistry, Faculty of Chemical Science, National University of Córdoba, Córdoba, Argentina

H.A. Tajmir-Riahi Department of Chemistry–Biology, University of Québec at Trois-Rivières, Trois-Rivières, Québec, Canada

Gavin S. Tan Singapore National Eye Centre, Singapore

Vin Tangpricha Emory University, Atlanta, GA

Akihiko Tawara School of Medicine, University of Occupational and Environmental Health, Fukuoka, Japan

P. Archana Teresa Institute of Ophthalmology, Joseph Eye Hospital, Tiruchirappalli, Tamilnadu, India

Philip A. Thomas Institute of Ophthalmology, Joseph Eye Hospital, Tiruchirappalli, Tamilnadu, India

Mehmet Tosun Abant Izzet Baysal University Medical School, Bolu, Turkey

Julio A. Urrets-Zavalía University Clinic Reina Fabiola, Catholic University of Córdoba, Córdoba, Argentina

Preejith P. Vachali Moran Eye Center, University of Utah School of Medicine, Salt Lake City, UT

Carole Vaysse ITERG – Equipe Nutrition Métabolisme & Santé, Bordeaux, France

Sabrina Viau MacoPharma, Tourcoing, France

Claire M. Williams School of Psychology and Clinical Language Sciences, University of Reading, Reading, UK

Tien Y. Wong Singapore Eye Research Institute, Singapore; Yong Loo Lin School of Medicine, National University of Singapore

Chen Xi Nanjing Medical University Eye Hospital, Nanjing, Jiangsu Province, China

Ramazan Yağcı Pamukkale University Medical School, Denizli, Turkey

Jia Yan Liaoning Medical University, Liaoning Province, P.R. China

Baoru Yang Food Chemistry and Food Development, Department of Biochemistry, University of Turku, Finland

Ji Yong Department of Pathophysiology, Nanjing Medical University, Nanjing, Jiangsu Province, China

Vicente Zanon-Moreno Genetic and Molecular Epidemiology Unit, Department of Preventive Medicine and Public Health, School of Medicine, University of Valencia, Valencia, Spain; CIBER Fisiopatología de la Obesidad y Nutrición, ISCIII, Valencia, Spain

Xiangjia Zhu Eye and ENT Hospital, Fudan University, Shanghai, P.R. China

S.R. Zwart Division of Space Life Sciences, Universities Space Research Association, Houston, Texas, USA

Preface

The eye is perhaps one of the least studied organs in diet and nutrition, yet the consequences of vision loss can be devastating. Partial or whole, vision loss can affect not only the quality of life of the individual but the family unit as well. One of the biggest contributors to complete vision loss in the western hemisphere is diabetes, precipitated by metabolic syndrome. In some developing countries, micronutrient deficiencies are major contributing factors to impaired vision. However, there is a range of ocular defects that have their origin in nutritional deficiencies or excess, or have been shown to respond favorably to nutritional components in the experimental (i.e., preclinical) or clinical setting. The eye, from the cornea to the retina, may be affected by nutritional components. As these effects may be physiologic or molecular, there are a great many approaches to understanding how dietary and nutritional factors affect the eye. However, to date there has been no comprehensive book on the eye and vision in relation to diet and nutrition, especially with the new molecular sciences. The **Handbook of Nutrition, Diet, and the Eye** addresses this. It contains 11 sections:

1. **Introductions and Overviews**
2. **Macular Degeneration**
3. **Glaucomas**
4. **Cataracts**
5. **Other Eye Conditions**
6. **Obesity, Metabolic Syndrome, and Diabetes**
7. **Macronutrients**
8. **Micronutrients**
9. **Nutraceuticals**
10. **Nutrigenomics and Molecular Biology of Eye Disease**
11. **Adverse Effects and Reactions**

Coverage includes overviews of eye diseases and vision loss, age-related macular degeneration, cataracts, glaucoma, diabetic retinopathy, dry eye, development,

aging, metabolic syndrome, obesity and liver disease, iron overload, hypoglycemia, space flight, and other pathophysiologic conditions. The book also covers many micronutrients including trace elements, selenium, zinc, vitamins A, B2, B7 (biotin), B12, C, and E, folate, carotenoids, and lutein. There is also material on dietary carbohydrates, glycemic index, lipids (including polyunsaturated fatty acids, fish oils and hyperlipidemia, and cholesterol deficiency), antioxidant status, fruit and vegetables, other natural products, quercetin, resveratrol, acetyl-l-carnitine, taurine, green tea extract, lycopene, amino acids, antiglycating phytochemicals, flavonoids, polyphenols, plant stanol and sterol esters, wolfberry, sea buckthorn, and *Cornus officinalis*. Detailed descriptions include all areas of molecular biology and genetics, biochemistry, cell biology, pathology, epidemiology, and other fields within the confines and remit of diet and nutrition. For example, the molecular biology includes material on gene expression, adenosine A2A receptor genes, glycoprotein IIIa gene, haptoglobin genotypes, SLC23A2 gene variation, integrins, and the nitric oxide system. Of course, the Editor recognizes the difficulty in ascribing chapters to a specific section, as some chapters will be comfortably placed in two or more sections, but this is offset by the excellent indexing.

Contributors are authors of international and national standing, leaders in the field, and trendsetters. Emerging fields of vision science and important discoveries relating to diet and nutrition are also incorporated in **The Handbook of Nutrition, Diet, and the Eye.** This represents essential reading for nutritionists, dietitians, optometrists, ophthalmologists, opticians, health care professionals, research scientists, molecular or cellular biochemists, physicians, general practitioners, public health practitioners, as well as those interested in eye health and vision in general.

Professor Victor R. Preedy,
King's College London

INTRODUCTIONS AND OVERVIEWS

1

The Eye and Vision: An Overview

R.A. Armstrong, R.P. Cubbidge

Vision Sciences, Aston University, Birmingham, UK

INTRODUCTION

Vision is the sense that we rely on most to inform us of the state of the world. For this reason, more is known about the scientific basis of vision than any of our other senses.[1] The major organ of vision, the eye, is highly specialized for photoreception. It focuses light from an object onto the light-sensitive part of the eye, the retina. Changes in specialized neurons in the retina result in nerve action potentials that are relayed to the brain via the optic nerve. Visual processing by the brain results in 'visual perception', the construction of a sensory image that is then consciously appreciated as vision[2,3]. All other structures of the eye are subsidiary to this function, either by facilitating focusing of light rays or supporting the tissues of the eye. This chapter is an introduction to the different parts of the eye and their various functions in achieving a visual image.

DEVELOPMENT OF THE EYE

The eyes develop from outgrowths of the brain called the optic vesicles (Fig. 1.1).[4] Five weeks after conception, the optic vesicle has emerged from the neural ectoderm of the brain and begins to fold inward producing an inner and an outer layer separated by a cavity. The retina and smooth muscle of the iris will develop from this structure. The optic vesicle also induces the formation of the lens placode that develops from an invagination of surface ectoderm in front of the vesicle and ultimately will develop into the crystalline lens. In addition, the hyaloid artery ramifies on the back of the developing lens while the outer surface of the optic vesicle develops a network of blood vessels in the mesoderm eventually forming the choroid. Outside this, the mesoderm forms the sclera and the extraocular muscles.

Eight weeks after conception, the thicker inner layer of the optic cup has detached from the thinner outer layer thus illustrating the weak attachment that exists between inner and outer layers of the developing retina. Within the vitreous chamber, the hyaloid blood vessels have developed that nourish the developing vitreous and the crystalline lens. These vessels normally disintegrate before birth but remnants of them may persist into childhood. The crystalline lens has developed at this stage and the primary lens fibers have now filled the cavity of the lens vesicle. The cornea develops from the surface ectoderm and mesoderm and the anterior chamber is formed between the developing cornea and lens. The fused eyelids can be seen at this stage, their skin and glands developing from the surface ectoderm, while connective tissue and muscle develop from the mesoderm.

THE OCULAR ADNEXA

The ocular adnexa refer to the accessory or adjoining parts of the eye and comprise several structures including the eyebrows, eyelids, eyelashes, lacrimal gland, and orbit. Many of these structures are illustrated in the anterior view of the eye shown in Figure 1.2.

Eyebrows

Each eyebrow is a thickened area of skin with accompanying hairs that are directed both upward and toward the temporal side of the head. The hairs function to prevent sweat formed on the forehead from entering the eyes. In many cultures, the eyebrows are also involved in facial expression.

Eyelids

The eyelids are moveable folds of skin that function to protect the eyes from particulate matter in the air. They also reduce the amount of light entering the eyes and provide some of the constituents making up the tears.

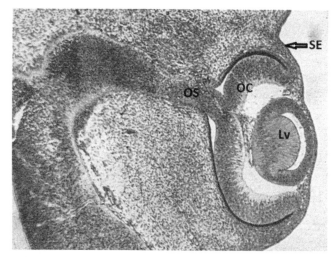

FIGURE 1.1 Development of the eye. The figure shows the lens vesicle, optic cup, and optic stalk, which will develop into the lens, retina, and optic nerve, respectively. Lv: lens vesicle, OC: optic cup; OS: optic stalk; SE: surface ectoderm. *Image courtesy RA Armstrong.*

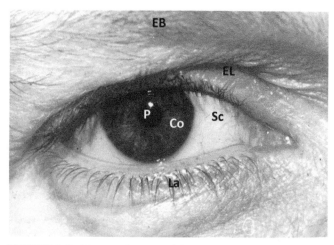

FIGURE 1.2 The anterior parts of the eye. The figure shows various parts of the anterior eye including the lids, cornea, sclera, and pupil. Co: cornea; EB: eyebrow; EL: eyelid; La: lashes; P: pupil; Sc: sclera. *Image courtesy RA Armstrong.*

The ocular surface can usually resist ocular infection both as a result of the mechanical action of the eyelids, which physically remove potential pathogens, and the washing effect of tears.[5]

Eyelashes

There are two or three rows of eyelashes located on the upper edge of the upper and lower lids, the lashes numbering approximately 150 on the upper lid and 75 on the lower lid. They function to protect the eye against small particles but are vulnerable to infection, especially by bacteria, a condition called 'blepharitis'.

Lacrimal Gland

The lacrimal gland lies inside the eye socket above the eye and functions to produce the tears. It is divided into a large 'orbital' and a smaller 'palpebral' portion connected by a canal. In the orbital part, ducts join with those from the palpebral portion entering the upper temporal part of the conjunctiva. Excess tears are drained via the canaliculi into the lacrimal sac and ultimately the nasal cavity. Tears contain lysozymes, lactoferrin, B-lysin, and immunoglobulin A, which are important in defense against infection.

Orbit

The orbit is the bony socket that contains the eye. The eye is situated in the anterior portion of the orbit closer to its lateral surface than the medial wall and nearer the roof of the orbit than the floor.[6] It comprises seven bones, including the ethmoid and sphenoid bones, and bones that make up the structure of the face such as the maxilla, frontal, and zygomatic. The purpose of the orbit is to protect the eye and to act as an anchorage point for the extraocular muscles and other ocular tissues.

Extraocular Muscles

There are six extraocular muscles that are attached to the eye by tendons at the sclera (the white outer coat of the eye). They function to move the eyes through 360° of gaze and are coordinated so that the two eyes move in unison, thus preventing double vision (diplopia). There are four rectus muscles: medial, lateral, superior, and inferior, which are attached to a common tendon ring at their posterior ends (the annulus of Zinn), which in turn, is attached to the posterior surface of the orbit. The primary action of the medial rectus is to pull the eye horizontally in the nasal direction, whereas the lateral rectus pulls the eye horizontally in the temporal direction. The primary action of the superior rectus is to pull the eye upward and the inferior rectus to pull the eye downward. The two remaining muscles, the superior and inferior oblique muscles are inserted more 'obliquely' into the upper and lower posterior temporal quadrants of the orbit. The inferior oblique, and superior, inferior, and medial recti muscles are controlled by the third cranial nerve (the oculomotor nerve) and the lateral rectus by the sixth cranial nerve (the abducens nerve). In addition, the superior oblique muscle is supplied by the fourth cranial nerve (the trochlear nerve). The primary actions of the superior and inferior oblique muscles are also to pull the eyes in an upward or downward direction, respectively. Nevertheless, only the primary muscle actions have been described, and several of the muscles act in concert to produce secondary and tertiary actions,

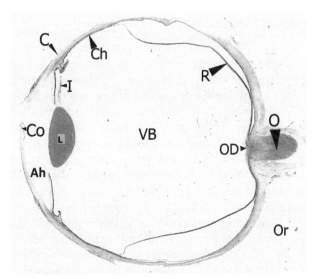

FIGURE 1.3 Regions of the eye as seen in a vertical section. The figure shows the various structures of the eye as seen in a vertical section. Ah: aqueous humor; C: conjunctiva; Ch: choroid; CO: cornea; I: iris; L: lens; O: optic nerve; OD: optic disc; Or: orbit; R: retina; VB: vitreous body. *Image courtesy RA Armstrong.*

which can move the eyes in more complex directions. The study of muscle action of the eyes and the coordination of eye movement is called 'binocular vision'.

The Anterior Structures of the Eye

The human eye is approximately spherical in shape, 25 mm in diameter, has a volume of 6.5 mL, while mean axial length of the globe is 24 mm (range 21–26 mm). It actually comprises the parts of two spheres, which are represented anteriorly by the cornea, which has a greater curvature than that represented by the curvature of the posterior sclera. Hence, the eye can be divided into two functionally distinct regions, the anterior eye and the posterior eye. The anterior segment comprises the cornea, iris, ciliary body, crystalline lens, aqueous humor, and the anterior part of the sclera (Fig. 1.3).

Conjunctiva

The conjunctiva is the outer membrane of the eye covering the white fibrous sclera. It is continuous with that of the transparent cornea and extends onto the surface of the upper and lower lids. It is a mucus membrane with a non-keratinized, stratified epithelium and subepithelial layers that are composed of adenoid and connective tissue and is a region especially vulnerable to infection ('conjunctivitis').

Cornea

The cornea is the most anterior structure of the eye and comprises one-sixth of the circumference of the globe. It

is a curved, transparent structure with a radius of 7.8 mm. The anterior surface is continually bathed by tears, while the posterior surface is bathed by aqueous humor. The surface of the cornea, together with the associated tear film, is responsible for most of the refractive power of the eye. Hence, the function of the cornea is to refract light rays so that they eventually come to a focus on the retina. The main thickness of the cornea is composed of regularly arranged collagen fibers, which together with the regular smooth epithelium and lack of blood vessels, is responsible for its transparency.

Sclera

With the exception of the cornea, the sclera forms the outermost layer of the eye. It is thickest in its posterior region and thinnest at the point of attachment of the tendons at the ends of the extraocular muscles. It comprises collagen fibers, which, unlike the cornea, are irregularly arranged resulting in an opaque appearance. The sclera is highly fibrous and provides protection, support, and anchorage for structures within and outside the eye such as the musculature, and also maintains the shape of the globe.

Iris

The iris is a 12-mm diameter structure that functions to regulate the amount of light entering the eye and also separates the eye into anterior and posterior chambers. It is analogous in action to the diaphragm of a camera. The pupil is an aperture in the center of the iris through which light rays pass on route to the retina. The iris also contains muscle that contracts in response to bright light, making the pupil smaller and reducing the amount of light entering the posterior segment. By contrast, dim light will cause the pupil to dilate thus increasing the light entering the posterior segment. The iris is controlled by branches of the autonomic nervous system. Hence, parasympathetic stimulation, supplied by the oculomotor nerve, will constrict the pupil while sympathetic stimulation, originating from the superior cervical ganglion, will act to dilate the pupil.

The posterior surface of the iris is covered in cells that contain the pigment melanin and that prevent light from entering the eye through the iris. The remaining part of the iris has varying amounts of pigment resulting in its characteristic color. Hence, an iris with relatively little pigment appears blue and progressively more pigment leads successively to green, hazel, and brown eyes. The amount of pigmentation present and therefore, the resultant eye color, is genetically determined. Albinism is a genetically determined condition that results in a lack of pigment in cells in the body. In humans, pigmentation on the back surface of the iris is never completely absent, so the eyes of humans with albinism often appear blue. In animals that are albino, however, no pigmentation is present and the eyes appear pink, for example, as in white mice.

Ciliary Body

The ciliary body is a 5–6-mm wide ring of tissue extending from the scleral spur anteriorly to the ora serrata posteriorly and is the anterior continuation of the choroid. It can be further divided into the pars plicata and pars plana. It consists of the ciliary muscle and a tissue that secretes the aqueous humor. In addition, it provides attachment to the zonular fibers, which attach to the peripheral region of the crystalline lens thus maintaining it in position. Contraction and relaxation of the ciliary muscle can change the thickness of the lens, which simultaneously alters its curvature thus enabling the eye to change power and focus at different distances, a process called 'accommodation'. The action of the ciliary body is also controlled by the oculomotor nerve.

Crystalline Lens

The lens consists of specialized surface ectoderm cells and is a highly elastic, circular, biconvex, transparent body lying immediately behind the pupil. It is suspended from the ciliary body by the zonular fibers and enclosed within a transparent capsule. The lens has less refractive power than the cornea and contraction of the ciliary muscle during accommodation relaxes the tension exerted by the zonular fibers on the lens, causing it to bulge, thus increasing its thickness and refractive power. This variable power enables us to focus from distant to near objects in the visual scene. The microscopic appearance of the lens is of long, thin cells that are regularly arranged in layers, essentially like an onion. These cells are continually produced through life as a result of mitotic cell division at the periphery of the lens [7], such that the thickness of the lens increases during a person's lifetime. Eventually the lens may become so thick that it is no longer able to support its own metabolism, leading to increased opacity and ultimately cataract.[7] The lens also becomes yellower with age as a result of pigment cells building up within its structure. These cells are thought to be a protective mechanism against ultraviolet light. As the lens becomes thicker, it also becomes less elastic, and this loss in elasticity and therefore, ability to focus, is thought to be responsible for a loss in reading ability in the middle years of life, a condition called 'presbyopia'.

Aqueous Humor

The aqueous humor is a transparent liquid produced by the ciliary body and which passes through the pupil, thus filling the anterior chamber. It ultimately passes through a fine meshwork at the cornea–sclera–iris junction called the trabecular meshwork into Schlemm canal, and then drains into the venous system. Aqueous humor has a consistency much like water and functions to provide nutrients to, and remove waste products of, metabolism from the transparent cornea and crystalline lens, both of which do not possess a blood supply. The continuous production of the aqueous humor and drainage through the trabecular meshwork results in a fluid pressure that has a range in normal individuals of 10–20 mm Hg. This pressure serves to maintain the shape of the eye. In some individuals, the trabecular meshwork can suddenly or more gradually become blocked leading to an increase in intraocular pressure (IOP) in the anterior chamber. This increase in IOP is transferred to the retina damaging its function, leading to a condition called glaucoma.[8]

THE POSTERIOR STRUCTURES OF THE EYE

The posterior part of the eye consists largely of the vitreous body, the choroid, retina, and optic disc (Fig. 1.3).

Vitreous Body

The vitreous cavity is the largest cavity in the eye comprising approximately two-thirds of its volume. It is bounded anteriorly by the lens and ciliary body and posteriorly by the retinal cup. It comprises two regions, a cortical zone of densely packed collagen fibers and a more liquid central region. Unlike the aqueous humor, the vitreous is a transparent, gel-like liquid with a relatively thick consistency. It consists largely of water (98%), with salts and collagen fibers making up the remainder. The vitreous fills the posterior eye and is only loosely attached to the retina, its function being mainly to maintain the shape of the eye. Vitreous is not continually produced, but remains fairly constant during life. Sometimes individuals complain of seeing particles or 'floaters' in their visual field. Floaters are a consequence of the aggregations of collagen fibers that circulate around the vitreous due to convection currents arising from body heat. When these particles are close to the retina, they cast a shadow that is perceived as floaters. In some older individuals, the loose attachment of the vitreous to the retina can break leading to a posterior vitreous detachment (PVD), a process that may cause a retinal tear.

The Choroid

The choroid lies between the sclera and retina and lines the posterior portion of the inner surface of the sclera. It is rich in blood vessels and a deep chocolate brown in color due to the dense pigmentation by melanin and serves to absorb stray light within the eye. The choroid also provides a vascular supply to structures of the anterior segment and provides nutrients and removes waste products from the retinal photoreceptor cells.

The Retina

The retina is the innermost layer of the posterior eye, lining approximately three-quarters of the eyeball. It is thickest at the back of the eye and thins out anteriorly ceasing just behind the ciliary body. The retina is the light-sensitive part of the eye and is responsible for converting the focused image into nerve action potentials, which are then relayed to the brain via the visual pathway. There are many layers of nerve cells within the retina that provide complex connections between the light-sensitive cells located toward the posterior parts of its surface.

The most posterior layer of the retina is the pigment epithelium (Fig. 1.4), which supplies metabolites and removes waste products from the light-sensitive cells embedded within it. The light-sensitive cells are of two types, rods and cones. Rods are highly specialized cells made up of an outer segment containing the photosensitive pigment rhodopsin, an inner segment containing large numbers of mitochondria, a nuclear region, and special synaptic structures. When exposed to light, rhodopsin causes a chemical cascade reaction that converts light energy into electrical energy. Rods are most active at low light intensities and, therefore, are responsible for vision in dim light and in night vision. Rods are only sensitive to the intensity of light and not to its wavelength, so they can only represent an image in black and white. The other type of light-sensitive cell, the cones, are much less numerous than rods but are capable of color detection. Hence, there are three types of cones sensitive to red, green, and blue light. Perception of a colored image is similar to that constructed on a television screen, where the red, green, and blue pixels interact to produce all of the colors of the visible spectrum.

When light enters the eye, it is brought to a focus by the cornea and crystalline lens on to the retina. This point of focus is called the macula lutea and located in its center is a depression called the fovea centralis. The density of cones is greatest at the macula while at the fovea centralis only cones are present. The density of cones is responsible for visual acuity, that is, the level of detail that an individual is capable of perceiving, which is analogous to the pixel resolution of a digital camera. Hence, the greater the number (and density) of pixels, the more detailed the image that the camera is able to resolve. In addition, at the fovea centralis, there is a layer of pigment that absorbs ultraviolet light and is, therefore, protective to the eye. This macular pigment can sometimes be observed with an ophthalmoscope as a tiny dot of yellow at the fovea centralis.

An image of the retina, termed the fundus, can be achieved by using an ophthalmoscope and several important features are visible. First, the retinal nerves,

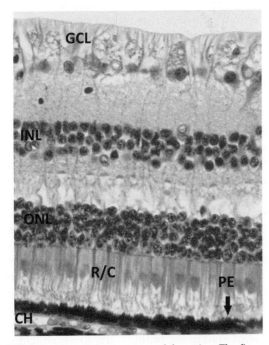

FIGURE 1.4 The cellular structure of the retina. The figure shows the various layers of the light-sensitive retina. CH: choroid; GCL: ganglion cell layer; INL: inner nuclear layer; ONL: outer nuclear layer; PE: pigment epithelium; R/C: rods and cones. *Image courtesy RA Armstrong.*

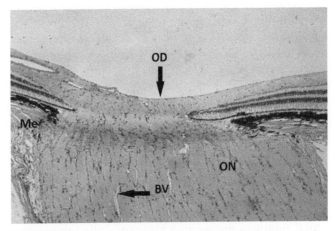

FIGURE 1.5 The optic nerve head. The figure shows the optic disc and nerve in section. BV: blood vessel; OD: optic disc; ON: optic nerve; Me: meninges. *Image courtesy RA Armstrong.*

which are generally too small to be seen individually, collect together, and leave the eye at the optic disc (Fig. 1.5). There are no rods or cones present within the optic nerve head and hence this region is known as the 'blind spot'. Second, the optic disc is the entry point for the major blood supply to the anterior surface of the retina. The eye is one of the few places in the body where the capillary network can be directly observed without an invasive procedure. Consequently, it is possible to observe structural changes within the blood vessels associated with disease such as aneurysm, embolism, and atherosclerosis.[9] In addition, the signs of malignant hypertension (high blood pressure) can be observed, which is a potentially life-threatening condition. The pattern of blood vessels across the retina is unique to each individual, much like fingerprints. Third, photoreceptor layer at the macula is thicker and has a greater and finer blood supply, resulting in a darker appearance. The increase in thickness is due to the greater packing density of the rods and cones.

The background color of the fundus is red–orange color, due to its blood supply. Because capillary vessels in the retina are particularly fine, they are also prone to damage from various diseases such as diabetes, which can cause leaking from the blood vessels. Other diseases result in degeneration of the photoreceptors as in age-related macular degeneration (AMD). In addition, the retina itself is a delicate structure and easily dislodged by trauma, a condition resulting in a detached retina.

VISUAL PATHWAY

The visual pathway describes the anatomical pathway by which electrical signals generated by the retina are sent to the brain (Fig. 1.6). The nerve fibers of the retina, representing the axons of the ganglion cells, collect together at the optic disc before passing out of the eye through the orbital bones and into the brain via the optic nerve (the second cranial nerve). The nerve fibers from different areas of the retina become more organized as they pass down the optic nerve. The optic nerves from each eye meet at the optic chiasm, a structure at the base of the brain. At this point, the nerve fibers that are associated with the nasal half of the retina from each eye cross over, so that on leaving the optic chiasm and passing into the optic tracts, the nerve fibers from the nasal retina of one eye travel down the optic tract with the nerve fibers originating in the temporal retina of the other eye. At the end of each optic tract, the retinal nerve fibers connect with other visual pathway nerves in a structure called the lateral geniculate nucleus (LGN) located in the midbrain. Some processing of the electrical signals occurs in the

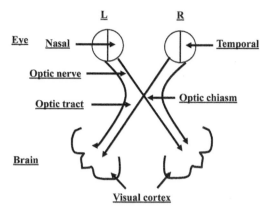

FIGURE 1.6 The visual pathway. The figure shows the pathway connecting the eyes to the brain; fibers form the temporal halves of each eye pass to the brain on the same side while those from the nasal halves of each eye cross over at the optic chiasm. L: left; R: right. *Image courtesy RA Armstrong.*

LGN before a series of radiating nerve fibers, the optic radiation, convey the information to the visual cortex in the posterior portion of the occipital lobe. Perception of sight ultimately derives from processing within this and adjacent areas of brain.

TAKE-HOME MESSAGES

- The eye is highly specialized for photoreception and focuses light from an object onto the retina.
- Visual processing by the brain results in 'visual perception', the construction of a sensory image in the brain that is then consciously appreciated as vision.
- The surface of the cornea, together with the associated tear film, is responsible for most of the refractive power of the eye.
- The lens has less refractive power than the cornea and contraction of the ciliary muscle during accommodation relaxes the tension exerted by the zonular fibers on the lens, causing it to bulge, thus increasing its thickness and refractive power.
- The retina is the light-sensitive part of the eye and is responsible for converting the focused image into electrical signals, which are then sent to the brain via the visual pathway.
- There are several important landmarks visible on the fundus of the eye including the optic disc, the capillary network of the retina, and the macula.
- The visual pathway describes the anatomical pathway by which electrical signals generated in the retina are sent to the brain.
- Perception of sight ultimately derives from processing within the visual cortex and adjacent areas of brain.

References

1. Smith CUM. *Biology of Sensory Systems.* New York: John Wiley & Sons, Ltd; 2000.

2. Hubel DH. *Eye, Brain, and Vision.* New York: Scientific American Library; 1998.

3. Valberg A. *Light, Vision, Color.* New York: John Wiley & Sons, Ltd; 2005.

4. Armstrong RA. Developmental anomalies of the eye: the genetic link. *Optom Today* 2006:29–32.

5. Armstrong RA. Fungal infections of the eye. *Microbiologist* 2013. in press.

6. Forrester J, Dick A, McMenamin P, Lee W. *The Eye: Basic Sciences in Practice.* London: W.B. Saunders Co. Ltd.; 1996.

7. Armstrong RA, Smith SN. The genetics of cataract. *Optom Today* 2000:33–5.

8. Armstrong RA, Smith SN. The genetics of glaucoma. *Optom Today* 2001:30–3.

9. Hamilton AMP, Gregson R, Fish GE. *Text Atlas of the Retina.* London: Martin Dunitz; 1998.

Age-Related Macular Degeneration: An Overview

Philip Storey, Lisa Hark, Julia A. Haller

Wills Eye Hospital, Philadelphia, PA

INTRODUCTION

Age-related macular degeneration (AMD) is the primary cause of legal blindness in the developed world and the third leading cause of blindness globally.[1] The two major forms of AMD are atrophic or non-neovascular ('dry') and neovascular ('wet'). The non-neovascular form of macular degeneration accounts for 85–90% of all cases and is characterized by thickening of the retinal pigment epithelium (RPE) and Bruch's membrane layer underlying the neurosensory retina, with formation of drusen, or yellowish deposits of extracellular material, and atrophy of the RPE and photoreceptors (Fig. 2.1).[2] The neovascular form of AMD is characterized by subretinal and sub-RPE proliferation of abnormal blood vessels that leak fluid, blood, and lipids, which leads to replacement of much of the central retina by fibrous scarring (Fig. 2.2). While much less prevalent, the neovascular form of the disease causes more than 80% of serious vision loss from AMD.[3]

Our understanding of macular degeneration and its treatment options have greatly evolved from just a decade ago when the disease was treatable only by thermal laser photocoagulation of neovascularization, in an effort to limit the size of the scarred central blind spot. A number of environmental and genetic risk factors have been identified and pharmacologic interventions have been developed to help treat the disease. Pharmaceutical research targeting vascular endothelial growth factor (VEGF) has yielded treatments to suppress development of abnormal blood vessels and the subsequent decline in visual function that historically made the neovascular form of AMD so debilitating. Finally, future research to expand methods to diagnose, manage, and treat AMD holds tremendous promise.

EPIDEMIOLOGY

AMD is more common among Caucasians and Asians. The Baltimore Eye Study found that Caucasians had 10 times the prevalence of AMD compared with African-Americans.[4] One cross-sectional study of Americans aged 40 years and older estimated AMD prevalence to be 6.5% and late AMD prevalence to be 0.8%.[5] The study also found that African-Americans were at lower risk of AMD than Caucasians with an AMD prevalence odds ratio of 0.37. An analysis of predominantly Caucasian populations in the US, Australia, and Europe reported a prevalence of less than 1% in people under 75 years old, increasing to 4.6% in those ages 75–84 years, and 13% for those ages 85 years and older.[6] One meta-analysis of population studies in Japan, China, Malaysia, India, and South Korea found that Asian people ages 40–79 years old had a combined prevalence rate of 6.8%, which was similar to corresponding prevalence rate of 8.8% among Caucasians.[7]

The prevalence of AMD is estimated at 9.1 million individuals within the US and 734,000 within the UK and is expected to double by 2050.[8,9] Within the US, cases of visual impairment and blindness from AMD are expected to increase from 620,000 to approximately 1 million by 2050 even with widespread use of vitamin prophylaxis at the earlier AMD stages and treatment for neovascular AMD.[8] The yearly economic cost of visual loss from AMD is estimated at more than USD575 million in direct medical costs in the US and approximately USD150 million in the UK.[9,10] Importantly, patients with AMD often have decreased quality of life and twice the rates of depression compared with age-matched community controls.[11] Treatment of AMD has also been shown to improve a patient's quality-adjusted life years.[11]

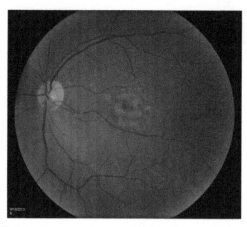

FIGURE 2.1 A fundus photograph of non-neovascular age-related macular degeneration. *Source: Wills Eye Hospital Diagnostic Testing Center. Reproduced with permission.* (See color plate section)

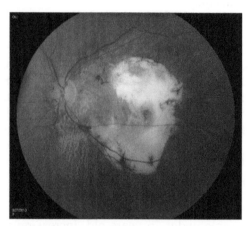

FIGURE 2.2 A fundus photo of neovascular age-related macular degeneration. *Source: Wills Eye Hospital Diagnostic Testing Center. Reproduced with permission.* (See color plate section)

RISK FACTORS

Epidemiologic studies have identified several risk factors with strong and consistent associations with AMD (Table 2.1). Older age is the strongest risk factor for AMD.[6,12] Cigarette smoking has been consistently associated with AMD and is the strongest modifiable risk factor.[13,14] A number of studies have also shown that Caucasians are at higher risk for the disease than African-American people.[4,15] Epidemiologic and twin studies have revealed heredity to influence AMD incidence, with the disease showing familial clustering and high rates of concordance among monozygotic twins.[16,17]

Risk factors showing moderate and consistent associations with AMD are obesity, hypertension, history of cardiovascular disease, higher plasma fibrinogen, and low intake of antioxidants and lutein.[13,18,19] Factors that have been weakly or inconsistently associated with

TABLE 2.1 Risk Factors for Age-Related Macular Degeneration (AMD). A Summary of Risk Factors Associated with AMD Stratified by Evidence Level

Risk factors with strong evidence for association
Age
Smoking
Caucasian ethnicity
Heredity
Risk factors with moderate evidence for association
Obesity
Hypertension
History of cardiovascular disease
Elevated plasma fibrinogen levels
Low intake of antioxidants and lutein
Risk factors with conflicting evidence for association
Female gender
Sunlight exposure
Iris color
Diabetes
History of cerebrovascular disease
Serum total cholesterol level
Serum total triglyceride level
History of cataract surgery in at least one eye

AMD are female gender, sunlight exposure, iris color, diabetes, history of cerebrovascular disease, and serum total cholesterol and triglyceride levels.[6,13,20] Two studies have shown cataract surgery to be associated with AMD.[21,22] However, this association is in doubt, as it was not confirmed in a prospective clinical trial.[23] Given our growing understanding of risk factors, it may be possible to prevent progression to advanced AMD by advocating dietary modifications, smoking cessation, maintenance of a healthy body weight, and blood pressure control. However, there are currently no prospective data on these proposed measures for AMD prevention – except for specific dietary interventions, which will be described in later chapters.

The mechanism by which some of these factors increase the risk of AMD appears to be related to chronic inflammation, at least in part via activation of the complement system, based on genetic evidence. The complement system is composed of a group of enzymes and regulatory proteins that balance activation of pathogen phagocytosis and inactivation for self-protection. Smoking, hypertension, and obesity have been shown to activate the complement system and individuals with these conditions typically present with increased plasma levels of C-reactive protein.[24,25] Twin studies have estimated that approximately one-third of the risk for advanced AMD is attributable to smoking, indicating that smoking, and potentially other environmental risk factors, may influence genetic predisposition.[26] Some studies confirmed that smokers with the Y402H mutation in gene LOC287715 were 10.2 times more likely to develop advanced AMD compared with smokers without the mutation, who were 5.1 times more likely to develop advanced AMD compared with non-smokers.[27]

Genetics have long been known to influence the development of AMD as demonstrated by clear evidence of heritability in twin studies and familial aggregation data.[28,29] The Human Genome Project and the HapMap, in which the genomes of 16 'normal' humans were sequenced, have allowed exploration of a number of diseases on the level of base sequences. Further analysis of the single nucleotide polymorphisms (SNPs) on these genetic maps have revealed a number of genes associated with AMD. Both predisposing and protective genes have been identified, all of which code for proteins involved in the inflammatory process, particularly the complement system. Variations in genes coding for complement factors H, B, C2, C3, and I, as well as a macrophage chemokine receptor, have been linked to AMD.[30,31] An SNP coding for a complement factor H gene, Tyr402His, shows Caucasians to be at a significantly higher risk for AMD.[32] A combined analysis of the several known genes associated with AMD found that variation in two loci can predict 74% of affected individuals.[33]

PATHOGENESIS

The most densely packed area of vision-producing nerve cells is located in the macula, which is the central part of the retina. Although the macula is only 5 mm in diameter, it accounts for the majority of our functional vision. The clinical hallmark of AMD is the appearance of drusen, visible excrescences like cobblestones on the thickened subretinal layer of RPE and Bruch's membrane.[34] In early stages, drusen appear as semi-translucent punctae and become obvious yellowish-white deposits as the RPE thins with AMD progression. Clinically, drusen are characterized as hard or soft based on the distinctness of their outer borders. Small drusen in one eye are not considered pathologic, as they are common in individuals over 50 years of age.[35] However, numerous intermediate-to large-sized drusen are an independent risk factor for visual loss in patients with AMD (Fig. 2.1).

In recent years, an understanding of the histopathology of drusen has helped to elucidate the underlying pathogenesis of AMD. Drusen are acellular debris and consist predominantly of cholesterol, vitronectin, and apolipoproteins B and E. Additionally, amyloid, multiple complement proteins, and immunoglobulin light chains have been identified in drusen.[36] Similar components are found in tissue deposits of systemic diseases including Alzheimer disease, atherosclerosis, amyloidosis, and membranoproliferative glomerulonephritis type II (MPGNII), suggesting that these systemic diseases share a common pathophysiologic mechanism with AMD.[37] Renal deposits in MPGNII have been shown to be identical in composition and structure to drusen, which is particularly notable as the disease consists of autoantibodies against a complement complex leading to uncontrolled activation of the alternative complement pathway.[38] The similarities between AMD and MPGNII highlight the central importance of the complement system and inflammation in AMD.

Geographic atrophy is a common sign of late-stage AMD and refers to areas of RPE cell death with atrophy of overlying photoreceptors often overlying drusen deposits.[34] Geographic atrophy is preceded by areas of increased autofluorescence, indicating that accumulation of lipofuscin in the RPE may also play a role in AMD pathogenesis.

Choroidal neovascularization (CNV) is growth of new blood vessels from the choroid underlying the RPE and is accompanied by subretinal or sub-RPE vascular leakage and hemorrhage. Additionally, new vessels can cause detachment of the RPE from the underlying choroid and fibrotic scars with severe vision loss.

The role of VEGF in AMD has been of particular interest as new treatments targeting the growth factor have been developed. Four VEGF isomers exist in humans and they are vital regulators of angiogenesis. VEGF inhibition leads to cessation of vascular growth as well as regression of new vessels.[39] VEGF has also been shown to cause increased capillary permeability leading to breakdown of the blood–retinal barrier and macular edema.[40]

CLASSIFICATION

While a number of classification schemes have been developed for diagnosis and classification of AMD, the Age-Related Eye Disease Study (AREDS) system has become the most widely used (Fig. 2.3).[42] According to AREDS, early AMD is characterized by a few (< 20) medium-size (< 63-μm) drusen or pigmentary abnormalities. Intermediate AMD consists of many intermediate-sized drusen (63–124 μm) or at least one larger drusen (> 124 μm) or geographic atrophy that does not extend into the central macula. Advanced AMD can be either non-neovascular, which consists of drusen and geographic atrophy extending into the central macula, or neovascular, which is characterized by proliferation of blood vessels and/or a disciform scar.

NATURAL HISTORY

In early AMD, vision loss is usually mild or even asymptomatic. Early symptoms can include blurred vision, a need for larger print, or decreased contrast sensitivity.[35] Patients with bilateral drusen and good vision have approximately an 8% annual rate of developing new atrophic or exudative lesions and a cumulative risk over 4 years of about 15% for developing CNV.[34]

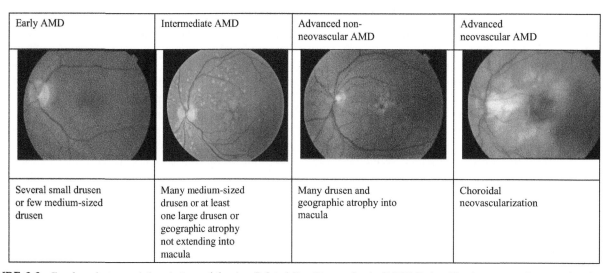

Early AMD	Intermediate AMD	Advanced non-neovascular AMD	Advanced neovascular AMD
Several small drusen or few medium-sized drusen	Many medium-sized drusen or at least one large drusen or geographic atrophy not extending into macula	Many drusen and geographic atrophy into macula	Choroidal neovascularization

FIGURE 2.3　Fundus photos and descriptions of the Age-Related Eye Disease Study (AREDS) classification system for age-related macular degeneration (AMD). *Sources: Coleman HR, Chan C, Ferris F, Chew E. Age-related macular degeneration. Lancet 2008;372:1835–45.[1] Stringham J, Hammond B, Nolan J, et al. The utility of using customized heterochromatic flicker photometry (cHFP) to measure macular pigment in patients with age-related macular degeneration. Exp Eye Res. 2008;87:445–53.[41] Reproduced with permission from Elsevier via the Copyright Clearance Centre.* (See color plate section)

FIGURE 2.4　Vision loss due to age-related macular degeneration. Upper photograph: normal vision; lower photograph: vision loss with age-related macular degeneration. *Source: National Eye Institute, National Institutes of Health.* (See color plate section)

For patients with atrophic ('dry') AMD, late symptoms of geographic atrophy usually include slow, progressive vision loss often over several years, leading to a central scotoma (Fig. 2.4). Geographic atrophy often develops in the parafoveal region, sparing the fovea until later stages of the disease. Historically, patients with geographic atrophy were not considered to be at risk for CNV. However, studies have shown that individuals with bilateral atrophy develop CNV at a rate of 2–4% over 2 years.[43] Geography atrophy is estimated to cause one-fifth of legal blindness from AMD.

For patients with neovascular ('wet') AMD, vision loss can be rapid, within days to weeks, due to CNV causing subretinal hemorrhage or macular edema. The natural history of CNV is quite poor. The Macular Photocoagulation Study showed that approximately two-thirds of patients with extrafoveal CNV or juxtafoveal CNV will have severe vision loss (loss of 6 or more lines of visual acuity) over a 5-year period.[44] While recent anti-VEGF therapies have greatly reduced severe vision loss from CNV, neovascular AMD is still responsible for the vast majority of vision loss from AMD.

RETINAL IMAGING FOR THE DIAGNOSIS AND MANAGEMENT OF AGE-RELATED MACULAR DEGENERATION

Advances in retinal imaging have improved the diagnosis and management of AMD. Fundus fluorescein angiography (FA) is the classic imaging modality for AMD and is a powerful technology for identifying the presence, size, (Fig. 2.5). In FA, a patient receives an intravenous

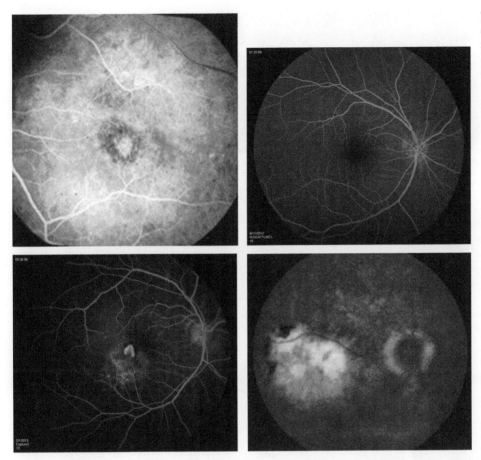

FIGURE 2.5 Fundus fluorescein angiography showing subretinal choroidal neovascularlization in AMD. *Source: Wills Eye Hospital Diagnostic Testing Center. Reproduced with permission.*

dye and sequential images of the retina are captured to evaluate retinal blood flow. Dye leakage due to increased immature capillary permeability caused in part by VEGF and inflammation is seen in neovascular AMD. Lesions can be classified by type – classic, occult, and mixed – and location – subfoveal, juxtafoveal, and extrafoveal. Classic lesions progress more rapidly and respond better to laser and photodynamic therapy than occult lesions. However, both lesion types have been shown to respond well to anti-VEGF therapy.

Optical coherence tomography (OCT) is a noninvasive imaging modality that uses near-infrared light and interferometric analysis to provide high-resolution images of the retina, vitreous, choroid, and optic nerve (Fig. 2.6).[45] OCT has evolved to supplement, and in some cases replace, traditional FA, allowing for topographic and cross-sectional analysis of choroidal neovascular lesions. OCT imaging is now widely used for early diagnosis and determining anti-VEGF retreatment criteria for neovascular lesions.

Fundus autofluorescence (FAF) imaging is another novel noninvasive modality (Fig. 2.7). Fluorophones are structures that possess fluorescent properties when exposed to light of a certain wavelength. Lipofuscin is the most commonly occurring fluorophone in the retina and is seen with phagocytosis of outer photoreceptors, which naturally occurs with aging. Additionally, excess lipofuscin granules in the RPE appears to represent a common downstream pathologic pathway of a number of complex diseases, including AMD. FAF imaging is commonly used to characterize progression of geographic atrophy, which is marked first by increased and later by decreased areas of autofluorescence.

MANAGEMENT

Prevention

A number of preventive strategies have been researched to reduce risk factors associated with AMD. In the initial AREDS, high doses of beta-carotene, zinc, and vitamins C and E were found to delay progression of AMD from intermediate to advanced states.[42] Omega-3 fatty acid, lutein, and zeaxanthin supplements were evaluated in the AREDS-2 and the recommended formulations were altered by omitting beta-carotene and adding lutein and zeaxanthine.[47] Given strong epidemiologic evidence of risk factors, a number of prevention strategies have been proposed. Smoking cessation, maintenance of a healthy weight, increased intake of antioxidants, and lower intake of cholesterol have all been proposed as lifestyle interventions that could potentially decrease incidence of AMD (see 'Macular Degeneration').

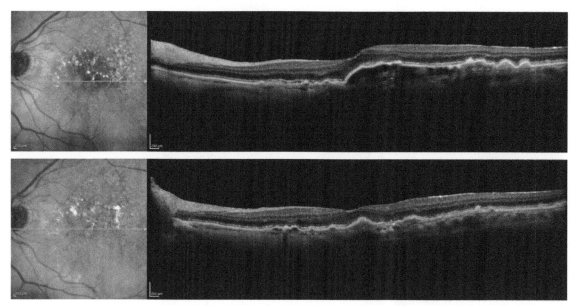

FIGURE 2.6 The effects of anti-vascular endothelial growth factor (VEGF) therapy for neovascularization. Optic coherence tomography of neovascular age-related macular degeneration prior to and post anti-VEGF therapy. *Source: Wills Eye Hospital Diagnostic Testing Center. Reproduced with permission.*

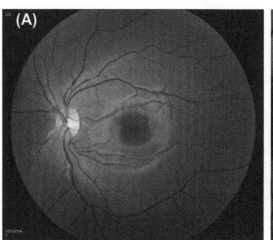
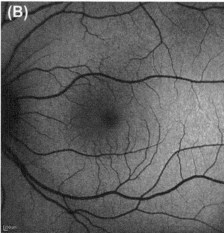

FIGURE 2.7 Fundus photo (A) and fundus autofluorescence image (B) in a normal subject. *Source: Wills Eye Hospital Retina Service. Reproduced with permission.* (See color plate section)

Laser Photocoagulation

In 1982, the Macular Photocoagulation Study Group published results of a randomized clinical trial evaluating argon laser photocoagulation for neovascularization outside the fovea.[48] After 18 months of follow-up, 25% of treated eyes had severe vision loss compared with 60% of untreated eyes. However, treated eyes did not gain vision and half of subjects had recurrence of neovascularization. Except for small lesions and/or those well away from the fovea, photocoagulation is rarely used today.

Photodynamic Therapy

In 1999, the results of the Treatment of Age-Related Macular Degeneration with Photodynamic Therapy study,

a 1-year randomized trial of verteporfin, were released.[49] In photodynamic therapy, verteporfin, a light-sensitive dye that preferentially concentrates in new blood vessels, is intravenously infused into a patient then activated by a laser focused on the retina, which leads to selective local thrombosis of neovascularization. In a 2-year randomized trial, 53% of verteporfin-treated patients lost fewer than three lines of vision compared with 38% of patients in the placebo group.[50] However, the verteporfin group still had a mean loss of 13 letters in visual acuity.

Anti-Vascular Endothelial Growth Factor Therapy

Since VEGF was recognized to play a role in subretinal and CNV, anti-VEGF therapy has become the first-line

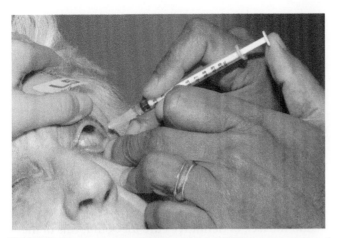

FIGURE 2.8 Intravitreal injection of anti-VEGF agent. A photo of a patient receiving an intravitreal injection of an anti-VEGF agent for neovascular age-related macular degeneration. *Source: Wills Eye Hospital Retina Service. Reproduced with permission.*

treatment for neovascular AMD. Three anti-VEGF medications, pegaptanib, ranibizumab, and aflibercept, are currently approved by the US Food and Drug Administration (FDA) for neovascular AMD. While bevacizumab is not FDA approved for AMD, it is the most widely used ocular anti-VEGF agent because of its relatively low cost.[51] All anti-VEGF agents are typically administered through intravitreal injection (Fig. 2.8).

Pegaptanib sodium (Macugen, Pfizer) was the first anti-VEGF agent to receive FDA approval for AMD treatment in 2004. Pegaptanib is a ribonucleic acid (RNA) aptamer that only binds one isoform: VEGF-165. One randomized controlled trial showed that pegaptanib significantly reduced severe vision loss in patients with neovascular AMD when administered every 6 weeks for 1 year; 70% of patients receiving pegaptanib lost fewer than 15 letters of visual acuity compared with 55% of sham-injected patients (P < 0.001).[52] However, pegaptanib is rarely used as other anti-VEGF medications that bind additional isoforms have been shown to be more effective.

Ranibizumab (Lucentis, Genentech) was the second anti-VEGF medication to receive FDA approval for the treatment of neovascular AMD. Ranibizumab is a recombinant antibody fragment that binds all VEGF-A isoforms and is derived from the same murine antibody as bevacizumab. In 2006, one landmark clinical trial showed that ranibizumab not only stabilized vision but resulted in substantial visual gain for the first time in the history of AMD treatment.[53] One-third of patients treated with ranibizumab gained 15 or more letters of visual acuity after monthly treatment for 2 years compared with less than 4% of patients treated with photodynamic therapy.

Bevacizumab (Avastin, Genentech) is a full-length antibody that binds all isoforms of VEGF and is FDA approved for a number of systemic malignancies. Despite lack of FDA approval for AMD, bevacizumab accounted for almost two-thirds of all anti-VEGF injections for neovascular AMD in 2008 among Medicare fee-for-service patients, which is attributed to apparent similar efficacy at a lower price.[51] The Comparisons of Age-Related Macular Degeneration Treatments Trial (CATT) showed that after 2 years of treatment, ranibizumab- and bevacizumab-treated patients had similar outcomes in visual acuity with similarly low rates of adverse events.[54]

Finally, aflibercept (Eylea, Regeneron) is a recombinant fusion protein consisting of portions of human VEGF receptors 1 and 2 fused to the Fc portion of a human antibody that binds all isoforms of VEGF-A and placental growth factor. It received FDA approval for neovascular AMD treatment in 2012. A potential benefit of interest is the possibility of less frequent injections required with aflibercept. Initial results from early trials showed that after an initial loading dose, aflibercept could be given every 2 months with similar outcomes in visual acuity as ranibizumab given monthly.[55]

While initial studies of ranibizumab and bevacizumab revealed few adverse events, safety is an important potential issue with anti-VEGF therapy. After ocular injection, anti-VEGF agents enter the systemic circulation, which in theory could lead to increased rates of vascular events, including cerebrovascular accident and hemorrhage. No trials have yet been powered to confirm this potential risk, although meta-analysis has suggested that while small, the risk may be real.[56]

FUTURE DIRECTIONS

The future of AMD treatment is evolving rapidly. A number of ongoing clinical trials are evaluating various combinations of current treatments, including nutrition therapies, anti-VEGF therapies, photodynamic therapy, and intravitreal steroids. In addition, novel treatment modalities built upon our understanding of the pathogenesis of AMD are being developed and include stem cell therapy, gene transfer, and optogenetics.

Stem Cell Therapy

Stem cell therapy for AMD has been investigated in animal models as well as in a few small human trials. Human embryonic stem cells have been used to generate RPE and were able to rescue photoreceptors and visual function in an animal model of retinal disease.[57] In this study, rats treated with RPE from stem cells improved 100% over untreated rats and maintained approximately 70% of visual function of a normal animal. Prospective trials have begun investigating RPE derived from stem cells transplanted into the subretinal space of patients with severely advanced AMD. Preliminary results of one trial 4 months after implantation of the RPE showed no serious adverse events and some improvement in visual function.[58] Additional studies are needed to determine safety, efficacy, and ideal surgical implantation method.

Gene Transfer

Investigators have begun using a novel treatment for AMD that involves transferring genes coding for antiangiogenic proteins to the human eye. Pigment epithelium-derived factor (PEDF) has been isolated from RPE cells and is a proteinase inhibitor that represses neovascularization and promotes photoreceptor survival.[59] In animal models, ocular injections of adenovirus expressing PEDF genes have led to regression of neovascularization, showing potential feasibility and efficacy of gene delivery systems.[60] A phase I clinical trial in 28 individuals with advanced neovascular AMD demonstrated that 70% of patients treated with adenoviral delivery of genes coding for PEDF maintained or improved their lesion and suffered no serious adverse events.[61] Currently, two clinical trials in patients with neovascular AMD are evaluating adenoviral delivery to express a VEGF-binding protein and lentiviral delivery to express endosatin and angiostatin, two proteins shown to regulate neovascularization in the retina.

Optogenetics

Optogenetics is a treatment approach that involves transferring light-sensitive transporters or ion channels to retinal neurons. In many retinal diseases, including AMD and retinitis pigmentosa, retinal neuron cells survive long after photoreceptors have degenerated. If an artificial light sensor can be linked to the neurons, light sensitivity can return to the retina. Animal studies have successfully restored vision in previously blind rodents by ectopic expression of melanopsin and microbial opsins in surviving cones or bipolar cells.[62] Artificially engineered light-gated ionotropic glutamate receptors have also been inserted into retinal ganglion cells of mice with retinal degeneration, achieving the return of light sensitivity.[63] While animal model evidence supports the efficacy of optogenetics for inherited ocular disease, clinical trials are needed to investigate its utility in humans.

TAKE-HOME MESSAGES

- AMD is a global disease without a definitive cure.
- AMD leads to substantial vision loss and decreased quality of life.
- Risk factors strongly associated with AMD include advancing age, cigarette smoking, Caucasian ethnicity, and heredity. Factors moderately associated with AMD include obesity, hypertension, history of cardiovascular disease, elevated plasma fibrinogen levels, and low dietary intake of antioxidants and lutein.
- Our understanding of the pathophysiology and genetics underlying the disease has greatly evolved.
- A number of interventions have been developed that increase visual acuity and delay visual decline in patients with AMD.
- Given the paucity of treatments for atrophic AMD and the lack of definitive cure for both forms of the disease, ongoing studies and future research are needed to continue to elucidate disease mechanisms and increase treatment options.

References

1. Pascolini D, Mariotti SP. Global estimates of visual impairment. *Br J Ophthalmol* 2010;**2012**(96):614–8.
2. Coleman HR, Chan C, Ferris F, Chew E. Age-related macular degeneration. *Lancet* 2008;**372**:1835–45.
3. Ferris 3rd FL, Fine SL, Hyman L. Age-related macular degeneration and blindness due to neovascular maculopathy. *Arch Ophthalmol* 1984;**102**:1640–2.
4. Friedman DS, Katz J, Bressler NM, Rahmani B, Tielsch JM. Racial differences in the prevalence of age-related macular degeneration: the Baltimore Eye Survey. *Ophthalmology* 1999;**106**:1049–55.
5. Klein R, Chou CF, Klein BE, Zhang X, Meuer SM, Saaddine JB. Prevalence of age-related macular degeneration in the US population. *Arch Ophthalmol* 2011;**129**:75–80.
6. Smith W, Assink J, Klein R, Mitchell P, Klaver CC, Klein BE, et al. Risk factors for age-related macular degeneration: pooled findings from three continents. *Ophthalmology* 2001;**108**:697–704.
7. Kawasaki R, Yasuda M, Song SJ, Chen SJ, Jonas JB, Wang JJ, et al. The prevalence of age-related macular degeneration in Asians: a systematic review and meta-analysis. *Ophthalmology* 2010;**117**:921–7.
8. Rein DB, Wittenborn JS, Zhang X, Honeycutt AA, Lesesne SB, Saaddine J. Forecasting age-related macular degeneration through the year 2050: the potential impact of new treatments. *Arch Ophthalmol* 2009;**127**:533–40.
9. Bonastre J, Le Pen C, Anderson P, Ganz A, Berto P, Berdeaux G. The epidemiology, economics and quality of life burden of age-related macular degeneration in France, Germany, Italy and the United Kingdom. *Eur J Health Econ* 2002;**3**:94–102.
10. Rein DB, Zhang P, Wirth KE, Lee PP, Hoerger TJ, McCall N, et al. The economic burden of major adult visual disorders in the United States. *Arch Ophthalmol* 2006;**124**:1754–60.
11. Slakter JS, Stur M. Quality of life in patients with age-related macular degeneration: impact of the condition and benefits of treatment. *Surv Ophthalmol* 2005;**50**:263–73.
12. Klein R, Cruickshanks KJ, Nash SD, Krantz EM, Javier Nieto F, Huang GH, et al. The prevalence of age-related macular degeneration and associated risk factors. *Arch Ophthalmol* 2010;**128**:750–8.
13. Chakravarthy U, Wong TY, Fletcher A, Piault E, Evans C, Zlateva G, et al. Clinical risk factors for age-related macular degeneration: a systematic review and meta-analysis. *BMC Ophthalmol* 2010;**10**:31.
14. Seddon JM, Willett WC, Speizer FE, Hankinson SE. A prospective study of cigarette smoking and age-related macular degeneration in women. *JAMA* 1996;**276**:1141–6.

15. Klein R, Klein BE, Knudtson MD, Wong TY, Cotch MF, Liu K, et al. Prevalence of age-related macular degeneration in 4 racial/ethnic groups in the multi-ethnic study of atherosclerosis. *Ophthalmology* 2006;**113**:373–80.

16. Smith W, Mitchell P. Family history and age-related maculopathy: the Blue Mountains Eye Study. *Aust N Z J Ophthalmol* 1998;**26**:203–6.

17. Gottfredsdottir MS, Sverrisson T, Musch DC, Stefansson E. Age related macular degeneration in monozygotic twins and their spouses in Iceland. *Acta Ophthalmol Scand* 1999;**77**:422–5.

18. Seddon JM, Cote J, Davis N, Rosner B. Progression of age-related macular degeneration: association with body mass index, waist circumference, and waist-hip ratio. *Arch Ophthalmol* 2003;**121**:785–92.

19. Snellen EL, Verbeek AL, Van Den Hoogen GW, Cruysberg JR, Hoyng CB. Neovascular age-related macular degeneration and its relationship to antioxidant intake. *Acta Ophthalmol Scand* 2002;**80**:368–71.

20. Mitchell P, Smith W, Wang JJ. Iris color, skin sun sensitivity, and age-related maculopathy. The Blue Mountains Eye Study. *Ophthalmology* 1998;**105**:1359–63.

21. Klein R, Klein BE, Jensen SC, Cruickshanks KJ. The relationship of ocular factors to the incidence and progression of age-related maculopathy. *Arch Ophthalmol* 1998;**116**:506–13.

22. Freeman EE, Munoz B, West SK, Tielsch JM, Schein OD. Is there an association between cataract surgery and age-related macular degeneration? Data from three population-based studies. *Am J Ophthalmol* 2003;**135**:849–56.

23. AREDS Study Group. Risk factors associated with age-related macular degeneration. A case-control study in the age-related eye disease study: Age-Related Eye Disease Study Report Number 3. *Ophthalmology* 2000;**107**:2224–32.

24. Seddon JM, Gensler G, Milton RC, Klein ML, Rifai N. Association between C-reactive protein and age-related macular degeneration. *JAMA* 2004;**291**:704–10.

25. Sung KC, Suh JY, Kim BS, Kang JH, Kim H, Lee MH, et al. High sensitivity C-reactive protein as an independent risk factor for essential hypertension. *Am J Hypertens* 2003;**16**:429–33.

26. Seddon JM, George S, Rosner B. Cigarette smoking, fish consumption, omega-3 fatty acid intake, and associations with age-related macular degeneration: the US Twin Study of Age-Related Macular Degeneration. *Arch Ophthalmol* 2006;**124**:995–1001.

27. Seddon JM, George S, Rosner B, Klein ML. CFH gene variant, Y402H, and smoking, body mass index, environmental associations with advanced age-related macular degeneration. *Hum Hered* 2006;**61**:157–65.

28. Hammond CJ, Webster AR, Snieder H, Bird AC, Gilbert CE, Spector TD. Genetic influence on early age-related maculopathy: a twin study. *Ophthalmology* 2002;**109**:730–6.

29. Seddon JM, Ajani UA, Mitchell BD. Familial aggregation of age-related maculopathy. *Am J Ophthalmol* 1997;**123**:199–206.

30. Chan CC, Tuo J, Bojanowski CM, Csaky KG, Green WR. Detection of CX3CR1 single nucleotide polymorphism and expression on archived eyes with age-related macular degeneration. *Histol Histopathol* 2005;**20**:857–63.

31. Fisher SA, Abecasis GR, Yashar BM, Zareparsi S, Swaroop A, Iyengar SK, et al. Meta-analysis of genome scans of age-related macular degeneration. *Hum Mol Genet* 2005;**14**:2257–64.

32. Edwards AO, Ritter 3rd R, Abel KJ, Manning A, Panhuysen C, Farrer LA. Complement factor H polymorphism and age-related macular degeneration. *Science* 2005;**308**:421–4.

33. Gold B, Merriam JE, Zernant J, Hancox LS, Taiber AJ, Gehrs K, et al. Variation in factor B (BF) and complement component 2 (C2) genes is associated with age-related macular degeneration. *Nat Genet* 2006;**38**:458–62.

34. Ambati J, Ambati BK, Yoo SH, Ianchulev S, Adamis AP. Age-related macular degeneration: etiology, pathogenesis, and therapeutic strategies. *Surv Ophthalmol* 2003;**48**:257–93.

35. Jager RD, Mieler WF, Miller JW. Age-related macular degeneration. *N Engl J Med* 2008;**358**:2606–17.

36. Zarbin MA. Current concepts in the pathogenesis of age-related macular degeneration. *Arch Ophthalmol* 2004;**122**:598–614.

37. Mullins RF, Russell SR, Anderson DH, Hageman GS. Drusen associated with aging and age-related macular degeneration contain proteins common to extracellular deposits associated with atherosclerosis, elastosis, amyloidosis, and dense deposit disease. *FASEB J* 2000;**14**:835–46.

38. Mullins RF, Aptsiauri N, Hageman GS. Structure and composition of drusen associated with glomerulonephritis: implications for the role of complement activation in drusen biogenesis. *Eye (Lond)* 2001;**15**:390–5.

39. Kondo S, Asano M, Suzuki H. Significance of vascular endothelial growth factor/vascular permeability factor for solid tumor growth, and its inhibition by the antibody. *Biochem Biophys Res Commun* 1993;**194**:1234–41.

40. Aiello LP, Bursell SE, Clermont A, Duh E, Ishii H, Takagi C, et al. Vascular endothelial growth factor-induced retinal permeability is mediated by protein kinase C *in vivo* and suppressed by an orally effective beta-isoform-selective inhibitor. *Diabetes* 1997;**46**:1473–80.

41. Stringham J, Hammond B, Nolan J, et al. The utility of using customized heterochromatic flicker photometry (cHFP) to measure macular pigment in patients with age-related macular degeneration. *Exp Eye Res* 2008;**87**:445–53.

42. AREDS Study Group. A randomized, placebo-controlled, clinical trial of high-dose supplementation with vitamins C and E and beta carotene for age-related cataract and vision loss: AREDS report no. 9. *Arch Ophthalmol* 2001;**119**:1439–52.

43. Holz FG, Wolfensberger TJ, Piguet B, Gross-Jendroska M, Wells JA, Minassian DC, et al. Bilateral macular drusen in age-related macular degeneration. Prognosis and risk factors. *Ophthalmology* 1994;**101**:1522–8.

44. Macular Photocoagulation Study Group. Argon laser photocoagulation for neovascular maculopathy. Five-year results from randomized clinical trials. *Arch Ophthalmol* 1991;**109**:1109–14.

45. Thomas D, Duguid G. Optical coherence tomography – a review of the principles and contemporary uses in retinal investigation. *Eye (Lond)* 2004;**18L**:561–70.

46. Gunther J, Altaweel M. Bevacizumab (Avastin) for the treatment of ocular disease. *Survey Ophthal* 2009;**54**:372–400.

47. Age-Related Eye Disease Study 2 Research Group. Lutein + zeaxanthin and omega-3 fatty acids for age-related macular degeneration: the Age-Related Eye Disease Study 2 (AREDS2) randomized clinical trial. *JAMA* 2013;**309**:2005–15.

48. Macular Photocoagulation Study Group. Argon laser photocoagulation for senile macular degeneration. Results of a randomized clinical trial. *Arch Ophthalmol* 1982;**100**:912–8.

49. Treatment of age-related macular degeneration with photodynamic therapy (TAP) Study Group. Photodynamic therapy of subfoveal choroidal neovascularization in age-related macular degeneration with verteporfin: one-year results of 2 randomized clinical trials – TAP report. *Arch Ophthalmol* 1999;**117**:1329–45.

50. Bressler NM. Photodynamic therapy of subfoveal choroidal neovascularization in age-related macular degeneration with verteporfin: two-year results of 2 randomized clinical trials-tap report 2. *Arch Ophthalmol* 2001;**119**:198–207.

51. Brechner RJ, Rosenfeld PJ, Babish JD, Caplan S. Pharmacotherapy for neovascular age-related macular degeneration: an analysis of the 100% 2008 Medicare fee-for-service part B claims file. *Am J Ophthalmol* 2011;**151**:887–95.

52. Gragoudas ES, Adamis AP, Cunningham Jr ET, Feinsod M, Guyer DR. Pegaptanib for neovascular age-related macular degeneration. *N Engl J Med* 2004;**351**:2805–16.

53. Rosenfeld PJ, Brown DM, Heier JS, Boyer DS, Kaiser PK, Chung CY, et al. Ranibizumab for neovascular age-related macular degeneration. *N Engl J Med* 2006;**355**:1419–31.

54. Martin DF, Maguire MG, Ying GS, Grunwald JE, Fine SL, Jaffe GJ. Ranibizumab and bevacizumab for neovascular age-related macular degeneration. *N Engl J Med* 2011;**364**:1897–908.

55. Dixon JA, Oliver SC, Olson JL, Mandava N. VEGF Trap-Eye for the treatment of neovascular age-related macular degeneration. *Exp Opin Investig Drugs* 2009;**18**:1573–80.

56. Schmucker C, Loke YK, Ehlken C, Agostini HT, Hansen LL, Antes G, et al. Intravitreal bevacizumab (Avastin) versus ranibizumab (Lucentis) for the treatment of age-related macular degeneration: a safety review. *Br J Ophthalmol* 2011;**95**:308–17.

57. Lund RD, Wang S, Klimanskaya I, Holmes T, Ramos-Kelsey R, Lu B, et al. Human embryonic stem cell-derived cells rescue visual function in dystrophic RCS rats. *Cloning Stem Cells* 2006;**8**:189–99.

58. Schwartz SD, Hubschman JP, Heilwell G, Franco-Cardenas V, Pan CK, Ostrick RM, et al. Embryonic stem cell trials for macular degeneration: a preliminary report. *Lancet* 2012;**379**:713–20.

59. Dawson DW, Volpert OV, Gillis P, Crawford SE, Xu H, Benedict W, et al. Pigment epithelium-derived factor: a potent inhibitor of angiogenesis. *Science* 1999;**285**:245–8.

60. Mori K, Duh E, Gehlbach P, Ando A, Takahashi K, Pearlman J, et al. Pigment epithelium-derived factor inhibits retinal and choroidal neovascularization. *J Cell Physiol* 2001;**188**:253–63.

61. Campochiaro PA, Nguyen QD, Shah SM, Klein ML, Holz E, Frank RN, et al. Adenoviral vector-delivered pigment epithelium-derived factor for neovascular age-related macular degeneration: results of a phase I clinical trial. *Hum Gene Ther* 2006;**17**:167–76.

62. Lin B, Koizumi A, Tanaka N, Panda S, Masland RH. Restoration of visual function in retinal degeneration mice by ectopic expression of melanopsin. *Proc Natl Acad Sci U S A* 2008;**105**:16009–14.

63. Caporale N, Kolstad KD, Lee T, Tochitsky I, Dalkara D, Trauner D, et al. LiGluR restores visual responses in rodent models of inherited blindness. *Mol Ther* 2011;**19**:1212–9.

3

Cataracts: An Overview

Mesut Erdurmuş[1], Hüseyin Simavlı[2], Bahri Aydın[3]

[1]Abant Izzet Baysal University Medical School, Bolu, Turkey, [2]Bolu Izzet Baysal State Hospital, Bolu, Turkey,
[3]Istanbul Medeniyet University Medical School, Istanbul, Turkey

INTRODUCTION

The crystalline human lens is a transparent, biconvex structure, whose fundamental functions are to maintain its own clarity, focus light on the retina, and provide accommodation. A cataract is any opacity within a lens that degrades the optical quality. While numerous factors, including trauma, inflammation, systemic diseases, and genetic mutations, can lead to the development of cataracts, age-related deterioration of the lens structure is the most common cause.

Cataracts pose a substantial economic and public health burden and are the leading cause of preventable blindness worldwide.[1] However, modern surgical techniques for cataract extraction and intraocular lens (IOL) implantation allow rapid visual rehabilitation; it is probably the most effective surgical procedure in all of medicine. This chapter focuses on cataracts, their pathogenesis, and well-known risk factors and treatment options.

ETYMOLOGY

The word 'cataract' is etymologically derived from the Latin 'cataracta', meaning 'waterfall', or from the Greek 'καταράκτης', meaning 'down-rushing'. As rapidly flowing water appears white, the term may later have been used figuratively to define the appearance of mature lens opacities.

EPIDEMIOLOGY

As people live longer due to advances in medical technology and healthier living, the overall prevalence of senile cataracts inevitably increases. According to the World Health Organization (WHO), cataracts are the most prominent cause of visual impairment and blindness throughout the world. Globally, the number of people of all ages visually impaired is estimated to be 285 million, of whom 39 million are blind.[2] Despite advances in surgical techniques, untreated age-related cataracts are responsible for approximately 51% of blindness in all areas of the world. In the US, senile lenticular changes have been reported in 42% of patients between the ages of 52 and 64 years, 60% of those between the ages of 65 and 74 years, and 91% of those between the ages of 75 and 85 years.[3,4] The Beaver Dam Eye Study results demonstrated that 38.8% of men and 45.9% of women older than 74 years of age had visually significant cataracts.[5] Another study by the same group reported that women were more likely than men to have nuclear cataracts, even after adjusting for age.[6]

The WHO projects that between 2000 and 2020, the number of cataract surgeries performed worldwide will need to triple. Because surgery is the only currently available treatment for visually significant lens opacities, in the near future, blindness due to cataracts will remain the leading cause of visual loss in many countries where surgical services are still inadequate. Therefore, the socioeconomic impact of the disease is enormous, in both developed and developing countries.

CLASSIFICATION, PATHOGENESIS, AND RISK FACTORS

The classification of cataracts basically involves the morphology, seen with slit-lamp biomicroscopy. Numerous classification schemes exist in the literature, including The Oxford Clinical Cataract Classification and Grading System.[7] and the Lens Opacities Classification System III (LOCS III).[8] These schemes are especially useful for epidemiologic studies, anticataract drug trials, and clinical studies where cataract hardness is an important issue.

Lens opacification may be partial or total, stationary or progressive, and hard or soft. Cataracts can be classified into the following major subheadings, according to etiology:

1. congenital cataracts: morphologic configurations include sutural, lamellar, polar, nuclear, zonular, membranous, and total patterns;
2. acquired cataracts:
   a. *age-related cataracts*: nuclear (Fig. 3.1a), cortical (Fig. 3.1b), and posterior subcapsular cataracts (PSCs) (Fig. 3.2);
   b. *secondary cataracts*: secondary to drugs (e.g., corticosteroids), ocular diseases (e.g., uveitis) (Fig. 3.3a), intraocular surgeries, metabolic diseases (e.g., diabetes mellitus), and other systemic diseases;
   c. *traumatic cataracts*: blunt and perforating trauma (Fig. 3.3b), ionizing radiation, infrared radiation, ultraviolet (UV) radiation, chemical trauma, and intraocular foreign bodies.

The etiology of senile cataract remains obscure; it has been considered to be multifactorial, with multiple genes and environmental factors influencing the phenotype. Advanced age is the most important risk factor for age-related cataracts. During cataract formation, post-translational modifications of lens proteins occur as a result of various chemical reactions, including oxidation, nonenzymatic glycosylation, Schiff base formation, proteolysis, transmidation, carbamylation, and phosphorylation. The most important pathogenic mechanisms leading to cataracts are:

1. photochemical insult and oxidative stress[9];
2. advanced glycation end products (AGEs)[10];
3. aggregation and insolubilization of lens proteins[11];
4. Na$^+$/K$^+$ ATPase underaction, leading to overhydration of the lens;
5. polyol pathway;
6. chromosomal and/or gene abnormalities that interfere with the normal functioning of the lens;
7. blunt and penetrating ocular trauma.

Photochemical insult, which is intraocular penetration of light and the consequent generation of reactive oxygen species, such as superoxide and singlet oxygen, and their derivatives, such as hydrogen peroxide and hydroxyl radical, induces damage to the epithelial cell deoxyribonucleic acid (DNA) of the lens, thus triggering a sequence of events leading to cataracts. Potential sources of oxidative stress to the lens include UV light, oxidants in the ocular fluids, endogenous oxidants produced in lens cells, and smoke constituents.[9,12,13]

Under hyperglycemic conditions, part of the excess glucose reacts nonenzymatically with proteins or other tissue or blood constituents, leading to the formation of AGE. Progressive accumulation of AGEs in the diabetic lens has been shown to contribute to the acceleration of cataractogenesis in hyperglycemic animals and diabetic humans.[10]

Nuclear cataracts are associated with increased light scattering resulting from the aggregation and insolubilization of lens proteins and their increased association with membranes. Lens proteins are known to undergo a wide variety of alterations with age, and many of these are accelerated in the presence of oxidative, osmotic, or other stresses.[14] In the lens crystallins, these modifications include proteolysis, an increase in disulfide bridges, deamidation of asparagines and glutamine residues, racemization of aspartic acid residues, phosphorylation, nonenzymatic glycosylation, and carbamylation. Thus, as the β- and γ-crystallins slowly accumulate damage over the lifetime of an individual, they lose the ability to participate in appropriate intermolecular

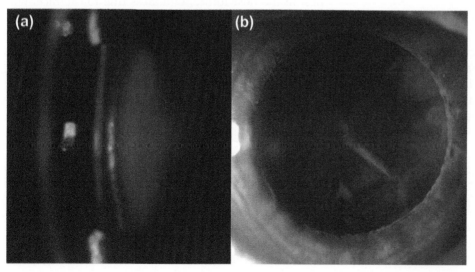

FIGURE 3.1 Appearance of age-related cataracts through a dilated pupil on slit-lamp biomicroscopy; a. nuclear sclerotic cataract; b. cortical cataract. (See color plate section)

interactions and even to remain in solution. As these crystallins begin to denature and precipitate, they are bound by the α-crystallins, which display a chaperone-like activity. The available α-crystallin is overwhelmed by increasing amounts of modified βγ-crystallin, and the complexes precipitate within the lens cell, forming lens opacities.[14] Nuclear cataracts are widely associated with a poor diet, low socioeconomic status, nonprofessional status, and low educational achievement. Smoking is a consistent risk factor for nuclear cataracts.[15] Patients who were treated daily for more than 1 year with hyperbaric oxygen therapy developed nuclear cataracts. In addition, postvitrectomy nuclear cataracts are associated with increased vitreous oxygen concentrations. Normally, the ascorbic acid (vitamin C) in the vitreous reacts with oxygen, decreasing the amount of oxygen that reaches the lens. Following vitrectomy, this protective mechanism is lost, leading to nuclear cataract formation. Studies of twins and examinations of familial associations have suggested that about one-third of the risk of nuclear cataracts is hereditary, although the

genes involved have not been identified. Twin studies and family associations suggest that 50–60% of the risk of developing cortical cataracts is hereditary.

Cortical cataracts involve the disruption of fiber cell membranes, followed by disintegration of the cytoplasmic contents of the damaged fiber cells. Age-dependent decrease in the activity of Na^+/K^+ ATPase leads to over-hydration, protein loss, and increased lenticular Na^+ and Ca^+ and decreased K^+ content. It seems possible that oxidative damage weakens or stresses the fiber cells, leading to their disruption, or that oxidation is not involved in causing cortical cataracts, but only occurs following the loss of membrane integrity. The superficial, nucleated cortical fibers of the lens are better able to resist oxidative damage than are nuclear fibers. High levels of sunlight exposure have been consistently associated with an increased risk of cortical cataracts. Several studies have shown that diabetics also have an increased risk of developing cortical cataracts and PSCs.

Histopathology of PSC has revealed posterior migration of lens epithelial cells from the lens equator to the

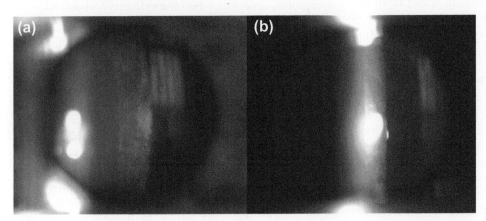

FIGURE 3.2 Appearance of posterior subcapsular cataract on slit-lamp biomicroscopy; a. subcapsular opacity is located in the cortex near the posterior capsule; b. the same eye under retroillumination. (See color plate section)

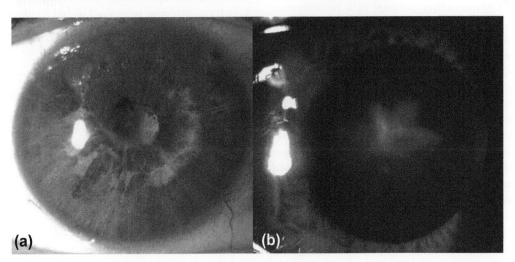

FIGURE 3.3 Secondary cataracts: a. cataract secondary to chronic Behçet uveitis. Posterior synechia and iris atrophy are noted. b. Cataract secondary to penetrating ocular trauma. Corneal perforation site can be seen on the upper left quadrant.

inner surface of the posterior capsule. The major risk factors associated with PSCs are high myopia, diabetes, exposure to therapeutic doses of steroids, intraocular inflammation, alcoholism, trauma, and ionizing radiation. One study, performed in rats, suggested that high-dose steroids might alter the expression of cell adhesion proteins, such as E- and N-cadherin, thereby interfering with fiber cell differentiation and migration.

The mechanism involved in the development of lenticular opacities due to diabetes is different from age-related cataracts.[16] The primary contributing factor is the accumulation of polyols within the lens substance. The crystalline lens does not require insulin for glucose and other simple sugars to enter into the lens through the capsule. In the case of diabetes, high concentrations of glucose in the aqueous humor can passively diffuse into the lens. The lens aldose reductase enzyme converts excess glucose to sorbitol or galactose to galactitol. These sugar alcohols (polyols – sorbitol or galactitol) cannot passively diffuse out of the lens, and they accumulate inside the lens. The accumulation of polyols inside the lens results in an osmotic gradient, which facilitates diffusion of water from the aqueous humor to the crystalline lens. The water drags sodium with it, and the lens swelling and electrolyte imbalances result in lens fiber disruption and cataract formation.

Congenital cataracts are visible at birth or during the first decade of life. This condition is one of the most frequent, treatable causes of visual impairment and blindness during infancy, with an estimated prevalence of one to six cases per 10,000 live births. Congenital cataracts can develop due to different causes, including metabolic disorders (e.g., galactosemia), infections during embryogenesis, gene defects, and chromosomal abnormalities. Cataracts may be an isolated anomaly, seen in association with another ocular developmental abnormality, or part of a multisystem syndrome, such as Down syndrome, Wilson disease, or myotonic dystrophy. Inherited cataracts account for 8–25% of congenital cataract cases, particularly bilateral cataracts; 27% of children with bilateral isolated congenital cataracts had a genetic basis, compared with 2% of unilateral cases.[17] Taking into account other genetic diseases causing congenital cataracts, approximately 50% of patients with congenital cataracts may have a genetic basis. Although all types of Mendelian inheritance have been documented for congenital cataracts, the autosomal-dominant transmission seems to be the most frequent type. Research on hereditary congenital cataracts has led to the identification of several classes of candidate genes that encode proteins, such as crystallins, lens-specific connexins, aquaporin, cytoskeletal structural proteins, and developmental regulators.

CLINICAL PRESENTATION

The main manifestation of cataracts is a worsening of vision, which is usually slowly progressive. Cataracts may also cause a variety of complaints and visual changes, including blurred vision, difficulty with glare, dulled color vision, increased nearsightedness accompanied by frequent changes in eyeglass prescription, and occasionally, double vision in one eye. PSC often occurs at a younger age than nuclear and cortical opacities do, and it causes greater difficulty with near vision, as it affects the central visual axis. When the pupil constricts in bright light, the involvement of the central visual axis causes difficulties during daytime activities. Cortical cataracts frequently cause glare and light scatter during activities such as driving at night. Nuclear cataracts can manifest as an increase in the refractive index of the lens, so that the refractive power of the eye changes in the direction of nearsightedness. Therefore, many patients with nuclear sclerotic cataracts have a greater loss of distance vision than of near vision. Cataracts are usually slowly progressive, and neither painful nor associated with any eye redness or other symptoms, unless they become extremely advanced.

DIAGNOSTIC EVALUATION

Clinicians can detect lens opacity by examining the anterior segments of the eye using slit-lamp biomicroscopy. Pupillary dilation may provide additional information about the site and extent of opacifications within the lens, and their relation to the optical axis of the eye can be determined. The patient should also undergo testing of best-corrected visual acuity, refraction, and contrast sensitivity; intraocular pressure measurement; examination of the other anterior segments' structures, including the iris and cornea; and fundus examination for possible retinal lesions, which can interfere with final visual acuity after surgery.

PREVENTION

Although there have been a vast number of trials designed to prevent the development of age-related cataracts, there is no universally accepted therapy that has been proven to prevent the formation or progression of age-related cataracts. Some remedies that have proven to be helpful are as follows:

1. wearing UV-filtering eyeglasses may slow the progression of cataracts;
2. a healthy lifestyle without smoking helps prevent cataracts, as well as other diseases in the body;
3. tight blood glucose control might inhibit cataract formation in patients with diabetes mellitus.

Several approaches to inhibiting cataract formation include antioxidant agents, vitamins[18,19], herbal drugs, aldose reductase inhibitors (ARIs)[20], nonsteroidal anti-inflammatory drugs (NSAIDs), and some other agents. Experimental studies in animal models have shown some proof of inhibition of cataract formation. However, human studies have shown no clear evidence of cataract inhibition.

Administration of pyruvate, one of several ARIs, prevented cataract development by inhibiting aldose reductase in diabetic rat lenses.[20] Similarly, ascorbic acid has shown potential as an ARI in both animal and clinical studies, indicating that it reduces lens sorbitol levels. Among ARIs, only sorbinil reached the advanced clinical trial stages in a cataract prevention program. However, the trial had to be discontinued, due to skin rashes. In spite of extensive research on the effects of sorbitol-lowering agents on cataract formation, none was able to produce convincing proof of their efficacy.

Cotlier and Sharma's study on aspirin use in patients with rheumatoid arthritis and diabetes was an inspiration for studies regarding the probable use of NSAIDs as prophylactic anticataract agents.[21] Extensively studied NSAIDs are aspirin, acetaminophen (paracetamol), ibuprofen, naproxen, sulindac, and bendazac.[22,23] The anticataract activity of these drugs is explained by different biochemical pathways, including acetylation, inhibition of glycosylation, and carbamylation of the lens crystallins. However, there is a need to explore their mechanisms of action in more detail in different experimental models.

Damage induced by reactive oxygen species is prevented by antioxidant enzymes (catalase, superoxide dismutase, and glutathione peroxidase). Antioxidants are of critical agents in preventing oxidation-related cataractogenesis. The most important function of glutathione (GSH) is to provide a reductive supply of glutathione peroxidase, to deactivate and render excess free radicals and keep them harmless.[24] The high concentration of GSH in the crystalline lens and the decreased concentration in most types of cataracts have led to many hypotheses on its role in the development of cataract formation. Carotenoids are natural, lipid-soluble antioxidants, and it has been reported that people with a high intake of carotenoids, such as α-carotene, β-carotene, lutein, lycopene, and cryptoxanthin, reduce their risk of cataracts. Numerous studies have proven stobadine, a novel synthetic pyrido indole, to be an efficient antioxidant. Stobadine has also been shown to delay the development of diabetic cataracts.

Herbal drugs, such as *Ginkgo biloba* extract, green tea (*Camellia sinensis*), herbal formulation Diabecon (used for diabetics, contains 25 herbal drugs), *Emblica officinalis*, (commonly known as *amla*), *Pterocarpus marsupium*, *Trigonella foenum-graecum*, and bilberry, have been reported to exert anticataract effects in various animal and clinical studies.[25]

In addition, miscellaneous agents, such as N-acetylcarnosine[26], angiotensin-converting enzyme (ACE) inhibitors[27], lipoic acid[28], pantethine, DL-penicillamine, deferoxamine, melatonin[29], and N-acetylcysteine[30], have been reported to be effective in the prevention of experimental cataract formation. Unfortunately, these drugs have not been evaluated clinically.

TREATMENT

The cataract is one of surgery's oldest topics, and there are numerous references to cataract surgery in the literature. The evolution of cataract surgery is summarized in Table 3.1. As there is no proven preventive or pharmacologic treatment for cataracts, the standard method of treatment is surgical removal of the opacified lens and implantation of an artificial IOL. Couching is the oldest traditional surgical procedure for cataract treatment. Sushruta, an ancient Indian surgeon, first described the procedure around 600 BC. Briefly, it typically involves the use of a blunt instrument to dislocate the cataractous lens posterior to the vitreous cavity. However, couching is an ineffective and hazardous method, and it often results in blindness or only partially restored vision. Today, in industrialized countries, cataract surgery is

TABLE 3.1 The Evolution of Cataract Surgery

- Surgeons in ancient India practiced *couching* as early as 600 BC.
- Celsus (25 BC–50 AD) drew the lens in the center of the eye.
- Ammar described the suction of a cataract through a needle in 1015.
- Fabricius ab Aquapendente illustrated the lens in its true anatomic position in 1600.
- Jacques Daviel (1696–1762) published the first account of cataract extraction through an inferior corneal incision.
- Samuel Sharp successfully performed intracapsular cataract extraction in 1753.
- von Graefe (1828–1870) improved Daviel's extracapsular cataract extraction technique using a superior corneal incision.
- Harold Ridley implanted the first artificial intraocular lens in 1949.
- Strampelli started to use anterior chamber intraocular lenses in Italy in 1951.
- Tadeusz Krwawicz developed a cryoprobe for safe intracapsular cataract extraction in 1961.
- Charles Kelman developed phacoemulsification in 1967.
- Mazzocco developed the first foldable intraocular lens in 1984.
- US FDA approved the first multifocal intraocular lens in 1997.
- US FDA approved the first toric intraocular lens in 1998.
- The first femtosecond laser system received FDA approval for cataract procedures in 2009.

FDA: Food and Drug Administration.

usually accomplished with ultrasound (phacoemulsification), followed by the implantation of an artificial IOL. There are several surgical options for removing a cataractous lens.

- *Intracapsular cataract extraction (ICCE)*: this is a large-incision cataract surgery in which the opacified lens is removed completely with its capsule, leaving no support for possible posterior chamber IOL implantation. In this procedure, IOLs can be either sutured to the sclera (transscleral fixation of IOL) or the iris in the plane of the pupil, or placed in the anterior chamber of the eye (anterior to the iris). ICCE is a form of cataract surgery developed in the early 1980s, but it is seldom used today, as better treatment options are available.
- *Extracapsular cataract extraction (ECCE)*: this is a large-incision cataract extraction in which the nucleus is expressed whole. This procedure leaves the posterior lens capsule intact, enabling an IOL to be positioned in its place. However, this procedure requires a large corneal or corneoscleral incision (8–10 mm), which can lead to corneoscleral sutures, wound-related problems, and high and irregular astigmatism.
- *Phacoemulsification*: today, the preferred standard method of cataract extraction is phacoemulsification. In this procedure, the cataractous lens is emulsified and aspirated away through a hollow needle that vibrates at a high ultrasonic frequency. A small corneal incision is made at the edge of the cornea. The corneal incision is commonly around 3 mm, but the latest developments allow removal through an incision of only 2 mm or even less. An ocular viscoelastic material is injected into the anterior chamber to maintain its volume and preserve the corneal endothelium during the cataract surgery. A round, continuous tear, approximately 5–6 mm in diameter, is created in the anterior capsule of the cataractous lens. The visibility of the anterior capsule may be enhanced by trypan blue dye in patients who have poor red reflex. The continuous curvilinear capsulorhexis allows access to the contents of the lens, which are removed by the phacoemulsification hand piece (Fig. 3.4). Once the cataract is removed and cortical cleaning is completed using irrigation/aspiration cannulas, more viscoelastic substance is injected to maintain the capsular bag space when inserting the foldable IOL. Clinical consensus is that phacoemulsification is less invasive, has fewer complications, and results in quicker and more stable visual rehabilitation than other cataract surgery techniques.
- *Small incisional manual cataract surgeries*: surgical removal of hard, mature cataracts by the

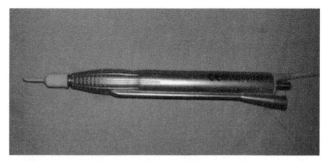

FIGURE 3.4 Hand piece of a conventional phacoemulsification device. It converts the electrical energy to mechanical vibration through a piezoelectric crystal.

phacoemulsification method can increase the rate of complications such as endothelial cell loss and posterior capsular rupture.[31,32] In contrast, performing ECCE on these eyes may cause problems such as wound healing and induced astigmatism.[33,34] These obstacles in the management of hard, mature cataracts have led to the development of small incisional manual cataract surgery techniques.[35] These techniques have aimed mainly to fragment the cataractous lens manually, without ultrasound energy, and to express it through a corneal or corneoscleral incision that is as small as possible.

Results of cataract surgery: numerous studies have been published regarding the outcomes of cataract surgery. However, the outcome measures employed are usually not comparable among the studies, as they are highly diverse and often subjective. It can be concluded that high-contrast visual acuity, which is the one of the most important visual functions and the one that is most easily measured through the method of successive approximation, enters an acceptable range after the implantation of current types of IOL. Minassian et al. performed a randomized clinical trial to compare the clinical outcomes of ECCE with small incision surgery by phacoemulsification.[33] The researchers found that the phacoemulsification technique is superior to ECCE in terms of intraoperative and postoperative surgical complications and visual acuity outcomes. Powe et al. reported in their systematic meta-analysis that 95.5% of eyes without any other ocular condition, and 89.7% of all eyes treated, reached an uncorrected postoperative visual acuity of 0.5 or better after ECCE or phacoemulsification surgery.[36] The Blue Mountains Eye Study reported the determinants of the postoperative visual outcome in older people who underwent cataract surgery as age and baseline cataract or presence of age-related maculopathy and baseline visual acuity.[37] The special optical designs of aspherical, toric, multifocal, and accommodative IOLs can further improve results in the particular situations for which

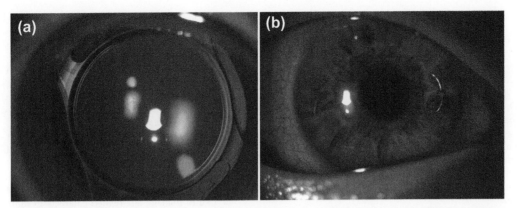

FIGURE 3.5 a. Appearance of four haptic posterior chamber intraocular lenses (Acreos, Bausch & Lomb) 6 months postoperative after uneventful phacoemulsification. b. iris claw anterior chamber intraocular lens in a patient with previous aphakia due to complicated cataract surgery. (See color plate section)

they are indicated. Figure 3.5 shows the appearance of two different models of IOL in the eye after successful implantation.

Complications of cataract surgery: today, most cataract surgeries are performed with great success, due to the development of modern surgical techniques. However, no surgery is free of risk, and occasional complications may occur. Cataract surgery complications can be divided into anesthesia-related complications, intraoperative complications, and postoperative complications. The anesthetic options for cataract surgery include general anesthesia, retrobulbar blockade, peribulbar blockade, sub-Tenon's anesthesia, and topical anesthesia. Clearly, the risk of most major complications associated with anesthesia are eliminated significantly, due to widespread use of topical and/or intracameral anesthesia. Intraoperative complications and their management options are beyond the scope of this article. Complications related to cataract surgery that might result in permanent visual loss are extremely rare. The major early and late complications of cataract surgery are infectious endophthalmitis, intraoperative suprachoroidal hemorrhage, cystoid macular edema, retinal detachment, corneal edema/bullous keratopathy, IOL dislocation, and posterior capsule opacification (Table 3.2). Although acute postoperative endophthalmitis after cataract surgery is rare, it continues to be a devastating complication. Over the years, the frequency of endophthalmitis has declined, due to various preoperative and perioperative measures that include the use of antiseptics and intracameral antibiotics, appropriate surgical draping techniques, and small incision cataract surgery techniques. Posterior capsule opacification, also called secondary cataract or aftercataract, is a late complication of extracapsular cataract surgery and IOL implantation. Despite advances in IOL designs and materials, there has been no noticeable decrease in the incidence of secondary cataracts. Posterior capsule

TABLE 3.2 Early and Late Postoperative Complications of Cataract Surgery

Early Postoperative Complications	Late Postoperative Complications
Endophthalmitis	Cystoid macular edema
Anterior chamber or vitreous hemorrhage	Retinal detachment
Wound gape/iris prolapse	Bullous keratopathy
İntraocular lens decentralization/dislocation	Uveitis
Increased intraocular pressure	Posterior capsule opacification

opacification is the most common complication of cataract surgery, and when it obscures a visual axis, neodymium-doped yttrium aluminum garnet laser posterior capsulotomy is necessary to restore visual function.

TAKE-HOME MESSAGES

- A cataract is an opacity of the lens of the eye.
- It is the most common cause of visual impairment and blindness in elderly people.
- Several risk factors have been identified in addition to aging, including genetic composition, UV light, smoking, diabetes, intraocular inflammation, and trauma.
- There exists no proven medical/pharmacologic treatment to prevent or reverse the development of age-related cataracts.
- The only effective management for senile cataract is surgical lens extraction.
- Cataract surgery using phacoemulsification allows small incisions and rapid visual rehabilitation.

References

1. Resnikoff S, Pascolini D, Etya'ale D, Kocur I, Pararajasegaram R, Pokharel GP, et al. Global data on visual impairment in the year 2002. *Bull World Health Organ* 2004;**82**:844–51.
2. Pascolini D, Mariotti SP. Global estimates of visual impairment. *Br J Ophthalmol* 2010;**2012**(96):614–8.
3. Sperduto RD, Seigel D. Senile lens and senile macular changes in a population-based sample. *Am J Ophthalmol* 1980;**90**:86–91.
4. Kahn HA, Leibowitz HM, Ganley JP, Kini MM, Colton T, Nickerson RS, et al. The Framingham Eye Study. I. Outline and major prevalence findings. *Am J Epidemiol* 1977;**106**:17–32.
5. Klein BE, Klein R, Linton KL. Prevalence of age-related lens opacities in a population. The Beaver Dam Eye Study. *Ophthalmology* 1992;**99**:546–52.
6. Klein BE, Klein R, Lee KE. Incidence of age-related cataract: the Beaver Dam Eye Study. *Arch Ophthalmol* 1998;**116**:219–25.
7. Sparrow JM, Bron AJ, Brown NAP, Ayliffe W, Hill AR. The Oxford clinical cataract classification and grading system. *Int Ophthalmol* 1986;**9**:207–25.
8. Chylack Jr LT, Wolfe JK, Singer DM, Leske MC, Bullimore MA, Bailey IL, et al. The Lens Opacities Classification System III. The Longitudinal Study of Cataract Study Group. *Arch Ophthalmol* 1993;**111**:831–6.
9. Spector A. Oxidative stress-induced cataract: mechanism of action. *FASEB J* 1995;**9**:1173–82.
10. Araki N, Ueno N, Chakrabati B, Morino Y, Horicuchi S. Immunological evidence for the presence of advanced glycation end products in human lens proteins and its positive correlation with ageing. *J Biol Chem* 1992;**267**:10211–4.
11. Benedek GB. Theory of transparency of the eye. *Appl Opt* 1971;**10**:459–73.
12. Taylor HR, West SK, Rosenthal FS, Munoz B, Newland HS, Abbey H, et al. Effect of ultraviolet radiation on cataract formation. *N Engl J Med* 1988;**319**:1429–33.
13. Micelli-Ferrari T, Vendemiale G, Grattogliango I, Boscia F, Arnese L, Altomare E, et al. Role of lipid peroxidation in the pathogenesis of myopic and senile cataract. *Br J Ophthalmol* 1996;**80**:840–3.
14. Hejtmancik JF, Kantorow M. Molecular genetics of age-related cataract. *Exp Eye Res* 2004;**79**:3–9.
15. Kelly SP, Thornton J, Edwards R, Sahu A, Harrison R. Smoking and cataract: review of causal association. *J Cataract Refract Surg* 2005;**31**:2395–404.
16. Monnier VM, Sell DR, Nagaraj RH, Miyata S, Grandhee S, Odetti P, et al. Maillard reaction-mediated molecular damage to extracellular matrix and other tissue proteins in diabetes, ageing and uremia. *Diabetes* 1992;**41**:36–41.
17. Santana A, Waiswo M. The genetic and molecular basis of congenital cataract. *Arq Bras Oftalmol* 2011;**74**:136–42.
18. Ayala MN, Söderberg PG. Vitamin C can protect against ultraviolet radiation induced cataract in albino rats. *Ophthalmic Res* 2004;**36**:264–9.
19. Labgle UW, Wolf A, Cordier A. Enhancement of SDZ ICT 322-induced cataracts and skin changes in rats following vitamin E- and selenium deficient diet. *Arch Toxicol* 1997;**71**:283–9.
20. Gupta SK, Joshi S. Relationship between aldose reductase inhibiting activity and anti-cataract action of various NSAIDs. *Dev Ophthalmol* 1991;**21**:151–6.
21. Cotlier E, Sharma YR. Aspirin and senile cataracts in rheumatoid arthritis. *Lancet* 1981;**1**:338–9.
22. Harding JJ. Pharmacological treatment strategies in age related cataracts. *Drugs Aging* 1992;**2**:287–300.
23. Harding JJ, Egerton M, Harding RS. Protection against cataract by aspirin, paracetamol, and ibuprofen. *Acta Ophthalmol (Copenh)* 1989;**67**:518–24.
24. Reddy VN, Giblin FJ. Metabolism and function of glutathione in the lens. *Ciba Found Symp* 1984;**106**:65–87.
25. Hiraoka T, Clark JI, Li XY, Thurston GM. Effect of selected anti-cataract agents of opacification in the selenite cataract model. *Exp Eye Res* 1996;**62**:11–9.
26. Babizhayey MA, Deyev AI, Yermakova VN, Brikman IV, Bours J. Lipid peroxidation and cataracts: N-acetylcarnosine as a therapeutic tool to manage age-related cataracts in human and in canine eyes. *Drugs R D* 2004;**53**:125–39.
27. Langade DG, Rao G, Girme RC, Patki PS, Bulakh PM. *In vitro* prevention by ACE inhibitors of cataract induced by glucose. *Indian J Pharmacol* 2006;**38**:107–10.
28. Maitra I, Serbinova E, Trischier H, Packer L. Alpha-lipoic acid prevents buthionine sulfoximine induced cataract formation in newborn rats. *Free Radic Biol Med* 1995;**18**:823–9.
29. Yağci R, Aydin B, Erdurmuş M, Karadağ R, Gürel A, Durmuş M, et al. Use of melatonin to prevent selenite-induced cataract formation in rat eyes. *Curr Eye Res* 2006;**31**:845–50.
30. Aydin B, Yagci R, Yilmaz FM, Erdurmus M, Karadağ R, Keskin U, et al. Prevention of selenite-induced cataractogenesis by N-acetylcysteine in rats. *Curr Eye Res* 2009;**34**:196–201.
31. Artzen D, Lundstrom M, Behndig A, Stenevi U, Lydahl E, Montan P. Capsule complication during cataract surgery: Case control study of preoperative and intraoperative risk factors: Swedish Capsule Study Rupture Study Group report 2. *J Cataract Refract Surg* 2009;**35**:1688–93.
32. Bourne RR, Minassian DC, Dart JK, Rosen P, Kaushal S, Wingate N. Effect of cataract surgery on the corneal endothelium: modern phacoemulsification compared with extracapsular cataract surgery. *Ophthalmology* 2004;**111**:679–85.
33. Minassian DC, Rosen P, Dart JK, Reidy A, Desai P, Sidhu M, et al. Extracapsular cataract extraction compared with small incision surgery by phacoemulsification: a randomised trial. *Br J Ophthalmol* 2001;**85**:822–9.
34. George R, Rupauliha P, Sripriya AV, Rajesh PS, Vahan PV, Praveen S. Comparison of endothelial cell loss and surgically induced astigmatism following conventional extracapsular cataract surgery, manual small-incision surgery and phacoemulsification. *Ophthalmic Epidemiol* 2005;**12**:293–7.
35. Keener GT. The nucleus division technique for small incision cataract extraction. In: Rozakis GW, editor. *Cataract Surgery: Alternative Small Incision Techniques.* 1st ed. New Delhi: Jaypee Brothers; 1995. pp. 163–91.
36. Powe NR, Schein OD, Gieser SC, Tielsch JM, Luthra R, Javitt J, et al. Synthesis of the literature on visual acuity and complications following cataract extraction with intraocular lens implantation. Cataract Patient Outcome Research Team. *Arch Ophthalmol* 1994;**112**:239–52.
37. Panchapakesan J, Rochtchina E, Mitchell P. Five-year change in visual acuity following cataract surgery in an older community: the Blue Mountains Eye Study. *Eye (Lond)* 2004;**18**:278–82.

4

Glaucoma: An Overview

Sergio Claudio Saccà[1], Cristina Cartiglia[2], Alberto Izzotti[2]

[1]St Martino Hospital, Ophthalmology Unit, Genoa, Italy, [2]University of Genoa, Genoa, Italy

INTRODUCTION

A 'Short History'

Glaucoma is a disease known since antiquity. The term 'glaucoma' appeared for the first time in the Aphorism of Hippocrates (460–375BC), and generally referred to the characteristic color assumed by the anterior segment of the affected eye and not to a specific disease. This denomination resulted in confusion among different diseases, especially with cataract. The situation was clarified in 1600, when Richard Banister, in the first English treaty of ophthalmology, pointed out that the glaucomatous bulb seemed more solid and hard than usual. Still nowadays in German language, the Banister terminology is used to distinguish the true cataract (Grauer Star) from glaucoma (Gruner Star). For the first time a visual field defect (a nasal loss) was described in 1722 by Charles Major Seminary of Saint-Yves, who devotes a chapter of his textbook on eye diseases to glaucoma. The description of an acute attack of glaucoma is due to Antonio Scarpa who, in 1818, described it as 'amaurosis'. James George Guthrie, founder of the Royal Westminster Infirmary for the Cure of Diseases of the Eye, still existing as part of the Moorfield Eye Hospital, London, UK, in 1823 described the ocular hypertension by associating it with an alteration of the vitreous humor. William Mackenzie (1791–1868) foresaw the importance of ocular hypertension in the glaucomatous disease. In the *"Practical treatise of the eye"*, he described the different panels of the eye hypertension, both acute and chronic, and all ophthalmologists of his era referred to it as a cult book. In 1850, Hermann Helmholtz (1821–1894) invented the ophthalmoscope thus enabling the *in vivo* exploration of the inner of the eye. Wilheilm Friederich Albrecht von Graefe (1828–1870) in the same historical period invented the iridectomy as a surgical solution of an acute attack of glaucoma. The concept of aqueous humor (AH) as a stagnant liquid was put in doubt since the early years of the 1900s by Theodore Leber (1840–1917). In 1923, Seidel, working with stains, concluded that there is a continuous flow of AH from the ocular anterior chamber (AC). In 1965, it was discovered that the AH passes through both the trabecular meshwork (TM) and uveoscleral pathways while there is a small loss of liquid through the limbar sclera.[1] The AH drainage is also influenced by selective uptake of certain substances by the iris.[2] Thus, it was discovered that AH flows through the iridocorneal angle and then reaches the general circulation through the Schlemm canal (SC) (Fig. 4.1). Alteration of this pathway determines hypertonus. Hjalmar August Schiotz (1850–1927) was the inventor of the indentation tonometer, while Hans Golmann (1899–1991) created the tonometer that is still used in the 2010s. Many others should be mentioned for their contributions to the study of glaucoma, such as Jannik Peterson Bjerrum (1851–1920) for his studies of the visual field, Adlof Weber (1829–1915) for his description of the excavation papillary, and Eight Barkan, pioneer of modern gonioscopy. We would like to conclude by quoting Jonas Stein Friedenwald (1897–1955) that ahead of his time wrote: "to follow-up the patient with glaucoma by the aid of a tonometer halo is a clinical dullness like following him with the determination of the single visual acuity".

DEFINITION OF GLAUCOMA AND ITS CLASSIFICATION

Glaucoma is a neurodegenerative syndrome characterized by a progressive optical atrophy that is the result of the retina ganglion cell death by apoptosis as a consequence of a combination of various factors such as oxidative stress, ischemia, mitochondrial damage, and proteome alterations. The target tissues of this complex disease are in the AC, the TM, in the central nervous system, the retinal ganglion cells (RGCs) of the optic nerve head, and the lateral geniculate nucleus (LGN). Increased intraocular pressure (IOP) is a major risk factor

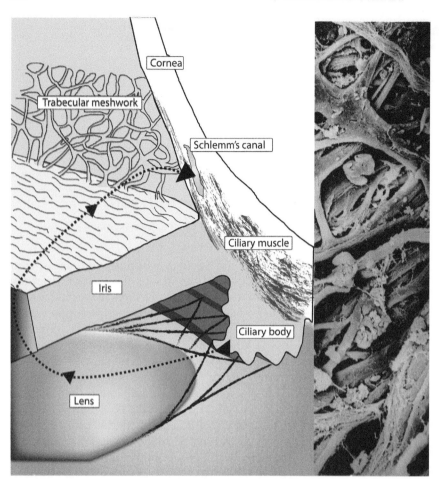

FIGURE 4.1 Conventional outflow pathway. Aqueous humor secreted by ciliary body flows through the iris into the anterior chamber then reaches circulation flowing through trabecular meshwork into the Schlemm canal. In the right panel scanning electron microscope photograph of the sclero-corneal trabecular meshwork (magnification x2500).

for glaucomatous damage. Nosologically, glaucomatous syndromes can be divided in two substantially different groups: normal-tension glaucoma (NTG) and high-tension glaucoma (HTG). This definition is essential but controversial. Indeed some authors would like to abolish this terminology: the definition of what is a high pressure remains arbitrary and many primary open-angle glaucoma (POAG) patients actually have IOP lower than 20 mm Hg.[3] IOP measurement is influenced by central corneal thickness and curvature as well as axial length of the eye and its biomechanical properties such as corneal hysteresis and its viscous damping properties. Indeed, measurements obtained with Goldmann-type tonometers can be used with confidence to monitor changes in the single patient's IOP, but it should not be relied on to determine the absolute manometric pressure within an eye or to compare the IOPs in eyes of different individuals.[4] Goldmann applanation tonometer, which is the current 'gold standard', is not sufficiently precise to measure the true IOP.[5] These considerations may generate confusion that tends to homogenize the various types of glaucoma drawing the erroneous feeling that glaucoma is a single nosologic entity. IOP increase is typical of HTG and its decrease up to physiologic levels is necessary to avoid RGC death. Conversely, in NTG, hypotonic drug therapy is ineffective.[6] Nevertheless, the damage of the visual fields are similar in both types of glaucoma (Fig. 4.2), even if the involved pathogenic factors are considerably different. Retinal arteriovenous passage times are significantly prolonged in NTG.[7] Vascular or perfusion abnormalities in NTG include the increased frequency of migraine headaches, Raynaud phenomenon, and sleep apnea.[8] The relationship between nocturnal blood pressure reduction and progression of the visual field defect suggests that in patients with NTG, the disturbance in the physiologic nocturnal blood pressure may be involved in the progression of glaucoma. The progression of visual field damage in NTG occurs when patients are lying flat during sleep.[9] Furthermore circadian mean ocular perfusion pressure fluctuation is a risk factor for NTG development, and vascular risk factors need to be considered when NTG patients have central visual field defects.[10]

NTG arises from IOP-independent mechanisms of RGCs, although the pathogenic mechanism is not yet understood. Underestimation of the IOP in patients with POAG who have thin corneas may lead to a misdiagnosis of NTG, while overestimation of the IOP in normal subjects who have thick corneas may lead to a misdiagnosis of ocular hypertension.[11] Various study

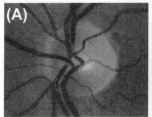 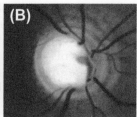 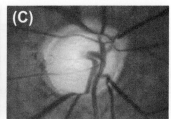

FIGURE 4.2 Clinical evolution of the damage to the optic nerve head during glaucoma course. Clinical outcome is similar in both high-tension and low-tension glaucoma. a: normal optic nerve head; b: vertical diameter of the excavation increased, which is reflected in the visual field defects; c: marked reduction in the papillary rim on 360° with tubular visual field.

groups have found no relationship between IOP and progression of the disease.[12] In NTG, RGC death is not determined by IOP increase, but by other factors bearing pathogenetic relevance. For instance, translaminar pressure gradient and low spinal fluid pressure may be important in NTG; indeed the optic nerve sheath diameter is significantly increased in NTG patients.[13] IOP, cerebrospinal fluid pressure, and arterial blood pressure are reciprocally correlated. The elevated retrolamina cribrosa pressure led to a normal translaminar pressure difference in eyes with elevated IOP, so that glaucomatous optic nerve damage did not develop.[14] Furthermore, patients with POAG exhibit low levels of circulating glutathione (GSH); conversely, in NTG patients, circulating total GSH levels were not different compared with those of normal subjects.[7] The serum total antioxidant status is significantly higher in NTG compared with healthy controls, while in POAG there is an antioxidant defense impairment.[15] Therefore, the first classification of glaucomas should be made distinguishing between high IOP and normal IOP and then dividing the glaucomas on the basis of angle closure (open or closed). POAG is, therefore, an optic nerve damage, in an eye that does not have evidence of angle closure on gonioscopy, and where there is no identifiable secondary cause. The definition of POAG was promulgated in 1996 by the American Academy of Ophthalmology defining it as a multifactorial optic neuropathy with a characteristic acquired loss of optic nerve fibers. Therefore, its severity is related to the visual field damage. POAG is the most frequent form of glaucoma since it represents about 60–70% of all the glaucomas, while the prevalence of primary angle-closure glaucoma (PACG) is 2–8%.[16] PACG is correlated to the size of the iridocorneal angle that can close for the affixing iris. This can determine the partial or total blockage of AH outflow and the consequent IOP increase. In this type of glaucoma, it is important to distinguish acute and chronic forms. The acute form occurs in only a minority of those with PACG patients, as indicated by population-based surveys in African and Asian settings, while the chronic, asymptomatic form of PACG is predominant.[17]. The severity of these glaucoma forms is not related to their symptoms, eyes with asymptomatic PACG often presenting severe to end-stage visual field loss. The glaucomas are defined secondary in the presence of other eye pathologic processes such as uveitis, trauma, neovascularization, or lens size-related defects. To this group belong pigmentary glaucoma and pseudoexfoliative glaucoma. The action of many factors is necessary for developing glaucoma, including aging, genetic predisposition, environmental, and endogenous factors, but in the HTG TM play a predominant pathogenetic role, TM malfunctioning leading to poor AH outflow and then IOP increase.

INTRAOCULAR PRESSURE AS A RISK FACTOR

Since elevated IOP is a major risk factor for glaucoma progression, at present the only possible therapeutic intervention in glaucoma is to lower the pressure. The therapeutic goal is, thus, to achieve an ideal IOP level in patients without interfering too much with their quality of life. IOP values in the normal population follow a normal distribution mostly being below the 20 mm Hg threshold. An exceedingly low IOP can cause refractive errors, alterations of the hemato-ophthalmic, lenticular opacities, macular folds, and optic disc edema. On the contrary, a high IOP can cause iris abnormalities, cataract, and glaucomatous optic atrophy. Therefore, IOP balance is essential for the maintenance of eyeball physiologic conditions.

The IOP is the result of three vector forces given by: a) the content of the eyeball, b) the elasticity of sclera, continuous with the cornea, and c) the peribulbar content orbit. The IOP is uniformly distributed across all intraocular structures.

An important quantitative relationship is provided below:

$$IOP = F/C + PV$$

where F = aqueous fluid formation rate, C = outflow rate, and PV = episcleral venous pressure.

Only one-tenth of arterial blood pressure variations are reflected on IOP because the change of the volume occurring in blood reaching the choroid determines compensatory changes in AH. Orbit content has importance for IOP value: in Graves disease muscle thickness can increase to compress the bulb, giving rise to a constant increase of pressure. Lowering IOP by medical treatment reduces the incidence of glaucomatous damage in ocular-hypertensive individuals.

However, results are sometimes controversial. Actually, Schulzer and coworkers [18] have found no differences between the treated group and the placebo. It should be emphasized that in the eyes with higher IOP, damage occurred earlier. Secondary glaucomas provide the most convincing evidence that IOP can be the cause of optic nerve damage. Patients undergoing post-traumatic angular recession develop a high IOP that, if not reduced, results in glaucomatous damage, while damage does not develop if the pressure is normalized. Different factors contribute to the glaucomatous optic neuropathy in different cases of NTG, interacting with IOP to different degrees and, thereby, affecting the magnitude of benefit of lowering IOP.[19] In addition, the results obtained after filtering surgery are not always consistent. In a prospective study on patients who had first trabeculectomies for primary open- or closed-angle glaucoma, IOP was well controlled by trabeculectomy. However, a steady decline in IOP control, visual acuity, and visual field occurred during follow-up.[20] In NTG patients, the favorable effect of IOP reduction on visual change progression was only found when the impact of cataracts on visual field progression was removed.[21] Anyway, it was demonstrated that progressive loss of RGC function in early glaucoma may be alleviated after IOP lowering.[22]

These conflicting opinions emphasize the variety of factors that can affect the health of the ganglion cells. Nowadays, the ophthalmologist's therapeutic approach only aims to decrease the IOP neglecting other potentially important factors. The relationship between IOP and glaucoma, although evident, must still be clarified. IOP value is not constant but varies during the 24 hours (Fig. 4.3). The greater fluctuations occur in patients affected by POAG and visual field deterioration occurs in eyes that have circadian variations of greater magnitude. AH volume is subject to circadian variations.

This rhythm was noticed for the first time by Sidler and Hugenin in 1899 and subsequently confirmed by many other researchers. The fact that these oscillations are present in all patients, even those undergoing treatment, leads to the assumption of a central or neurohormonal control. For example on light-dark entrained rabbits, IOP increases around the onset of dark due to the increased activities of ocular sympathetic nerves.[23]

Studies performed by fluorophotometry showed that the secretion of AH remains constant during the day while markedly decreasing during sleep. This led to the hypothesis that the night pressure could be lower than the diurnal values. However, the value of the IOP and the AH production are independent events. In fact, the pressure peaks encountered upon awakening does not reflect an increase of AH production. They are sudden, of about 6 mm Hg, and are probably related to awakening and more properly to the sleeping phases. This means that the night tonometry curves are not very reliable, because of bias by pressure increases not reflecting the true pressure performance. In our studies, we have identified six types of daily curves of the pressure in 50% of cases of POAG reflecting a typically concave circadian rhythm, with mean variations higher in the morning and lower in the early afternoon. The same trend was observed in approximately 35% of normal and NTG patients. In these two groups, pressures were stable in 45% of patients and in the stability of POAG pressure was only observed in 13.7% of patients. Our data also show that the higher the pressure, the more evident are the oscillations. In most cases, we found that the highest values were found in the morning (at 8 a.m.). In humans, the cycle followed by these oscillations is schematized with a sine curve where the concave part is the daytime, while the convex is night-time.[24] TM is a structure through which AH passes, made of endothelial cells organized in a meshwork to increase the overall filtering surface area. The interaction between these cells and free radicals in AH has a primary responsibility in causing glaucoma. From this point of view, the IOP may also be the indicator of TM health; that is, IOP increase is the symptom of a bad state of health of TM cells. We have demonstrated that both timolol (a beta-blocker) and dorzolamide (a carbonic anhydrase inhibitor), very popular drugs used to reduce the IOP, are truly antioxidant substances that are able to protect the endothelial cells and their mitochondria.[25] Therefore, in glaucoma therapy the ultimate aim should be not only to lower the IOP but also to protect TM cells. Anyway, there is still so much to discover about TM physiology. Probably the daily IOP fluctuations are related to the motility of the TM and more specifically the opening of TM intertrabecular spaces, whose widening or narrowing modification varies the number of endothelial cells exposed in the AH. In glaucomatous population, patients apparently undergoing

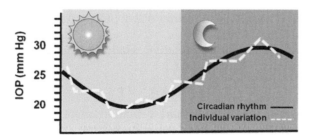

FIGURE 4.3 Circadian intraocular pressure (IOP) variation and individual variations during the day. Hypothetical graph indicating IOP throughout the day could be expressed by a sinusoidal wave that is governed by circadian rhythms. In most cases, the descending part of the wave corresponds to the morning IOP values, while the rising part of the curve corresponds to the IOP values during the night. Individual pressure variations can be superimposed on this wave to represent the ranges in IOP that occur at any given point along the 24-hour curve.

progressive visual field loss were found to have significantly more frequent IOP peaks than patients with stable visual fields. Statistically, in a population with a 30% prevalence of progressive visual field loss, 75% of the patients with peaks have progressive loss, and 75% of those without peaks do not have visual field progression. Thus, IOP peaks have an association with the progression of vision loss independently of the mean IOP.[26] These peaks are probably related to the dysfunction of the entire TM whose molecular bases will be described below.

PATHOGENESIS OF GLAUCOMA

High-pressure glaucoma differs from low-pressure glaucoma for the involvement of the TM not occurring in the latest.

In PACG, the iris tissue clogs up TM thus preventing AH outflow and causing the IOP increase. In POAG, on the contrary, there is a true degeneration of the TM impairing its normal function.

The TM is a key region of the conventional aqueous outflow pathway, and its malfunctioning represents the *conditio sine qua non* for the glaucoma onset. The human TM consists of interconnected channels lined by well-differentiated endothelial cell type, the iuxtacanalicular connective tissue is located between TM and SC that is the major site of resistance across the conventional aqueous outflow pathway. The loss of TM efficiency, that is, the fault of its endothelial cells, causes the dysfunction of the conventional outflow pathway thus starting glaucoma pathogenesis.[27]

In order to maintain a fluid barrier to prevent the passage of AH, the conventional outflow pathway is organized with a plumbing arrangement consisting of trabecular lamellae covered with TM cells, in front of a resistor, consisting of juxtacanalicular (JC) TM cells and the inner wall of SC (Fig. 4.1). The subendothelial region of SC does not form a continuous fluid system. The pathways through the connective tissue of the cribriform region are responsible for outflow facility and determine the filtration area of the inner wall of SC. The outermost PC or cribriform region has no collagenous beams, but rather several cell layers, which some authors claim to be immersed in loose extracellular material/matrix. TM pores contribute only to 10% of the aqueous outflow resistance. Overall, these data indicate that AH outflow is probably regulated through an active mechanism.[28] Thus, the tridimensional architecture of human trabecular meshwork (HTM) considerably increases the filtration surface. Its particular structure allows widening or narrowing of the intertrabecular spaces hence varying the quantity of TM cells involved in the outflow. Therefore, TM malfunction due to degeneration, resulting in

decay of HTM cellularity causes IOP increase and triggers glaucoma pathogenesis.[29]

The TM barrier and the SC barrier are both composed of endothelial cells supported from their matrix. Between these two barriers, there is the JC tissue that contains a loose extracellular matrix (ECM) through which AH flows. The TM cells release factors into the AH, and the ligands flow downstream from TM endothelial cells to bind and actively regulate the permeability properties of the SC endothelial cells. These factors upon binding to SC cells increase SC permeability, with a 400% enhancement in SC endothelium conductivity by means of the activation of TM endothelium genes.[27] In particular interleukin-1α and 1β and tumor necrosis factor (TNF)-α released by TM endothelium induce cell division and migration in those cells near Schwalbe's line, while inducing the release of matrix metalloproteinases (MMPs)[30] and an increase of fluid flow across ECM tissues near JC tissue.[27]

AC endothelial cells of the cornea, TM, and iris, are immersed into AH being, therefore, exposed to the free radicals. During POAG course, the most severe TM alterations occur in the anatomical layers in closest contact with the AC.[29] This finding led to the conclusion that toxic substances, identified as oxidative free radicals, contained in AH, contribute to the onset of pathogenetic alterations in TM, whose cells are exposed to relatively high H_2O_2 (hydrogen peroxide) concentrations. Indeed, in glaucomatous subjects, a total reactive antioxidant decrease occurs in AC and levels of 8-hydroxydeoxyguanosine (8-OHdG) are significantly higher in TM of glaucoma patients than in controls.[31] The decline of TM cellularity is linearly related to age.[29] It has been calculated that at 20 years of age the estimated TM cell number is 763,000 and that this number decreases to 403,000 by the age of 80 years, with a loss rate of 6000 TM cells per year. The progressive loss of TM cells in glaucomatous patients may be attributable to the long-term effect of oxidative free radical damage. Patients with glaucoma have lower levels of circulating GSH,[30] superoxide dismutase (SOD) 1 and 2, and glutathione-S-transferase (GST)1 enzymes than controls while the pro-oxidant enzymes NOS2 and glutamine synthetase (GS) are significantly higher in POAG patients than in controls.[32] The reduced expression of the antioxidant enzymes could aggravate the unbalance between both oxygen- and nitrogen-derived free radical production and detoxification. These findings suggest that IOP increase, which characterizes HTG, is related to oxidative degenerative processes affecting TM and specifically its endothelial cells. Much evidence indicates that in this region, reactive oxygen species (ROS) play a fundamental pathogenic role by reducing local antioxidant activities, inducing outflow resistance, and exacerbating the activities of SOD and glutathione peroxidase in glaucomatous eyes. Further, a loss of trabecular cells with age could result in a reduction in MMP activity in the TM, resulting in a reduced capacity of the TM to break down extracellular material. Blockage of the endogenous activity of the MMPs reduces outflow facility, probably because of ECM turnover, initiated by one or more MMPs, which appears to be essential to maintain IOP homeostasis.[33] Yet, outflow resistance increases in the presence of high levels of H_2O_2 in eyes with the GSH-depleted TM. The H_2O_2 effect on the adhesion of HTM cells to ECM proteins results in the rearrangement of cytoskeletal structures inducing decrease of TM cell adhesion, cell loss, and compromising HTM integrity.

In AH, there are many factors with a protective role for endothelial cells, among these GSH and vitamins. GSH within the cell maintains vitamins C and E in their reduced (active) forms while vitamin C helps to protect membrane lipids from peroxidation by recycling vitamin E. Among AC tissues, the TM is the most sensitive tissue to oxidative damage [17] (Fig. 4.4). It is likely that this different sensibility to oxidative damage can depend on the specific composition of each tissue. The cornea is rich in antioxidant enzymes such as SOD, catalase, glutathione peroxidase, and glutathione reductase, all involved in the removal of free radicals and ROS herein generated by constant absorption of ultraviolet light.[34] Antioxidant AH defenses are numerous too, and include vitamins, enzymes, and proteins such as albumins that have a protective role toward the TM.[35] In POAG, increased levels of lipid peroxidation products occur in the AH, TM, and SC.[36]

Currently, two theories exist explaining the origin of glaucoma: the vascular and the mechanic theories, both ascribing to oxidative stress a pathogenetic role. In the 'vascular theory', free radicals generated following ischemia are responsible for the oxidative damage to the axons. In the 'pressure theory', the oxidative deoxyribonucleic acid (DNA) damage would be the *primum*

movens of TM alterations. Oxidative phosphorylation takes place in the mitochondria and is one of the main endogenous sources of free radicals. Mitochondrial damage is involved in the pathogenesis of many chronic degenerative diseases. Mitochondria increase the production of ROS as byproducts of aerobic metabolism in aging tissues. Endothelial cell mitochondria have a central role in glaucoma pathogenesis. Mitochondria are equipped with circular molecules of DNA (mitochondrial deoxyribonucleic acid (mtDNA)). MtDNA damage is highlighted by a typical and common deletion of 4977 nucleotides. MtDNA common deletion is dramatically increased in TM of POAG patients as compared with controls. Furthermore, the amount of nuclear DNA per milligram of wet tissue and the mtDNA/nuclear DNA ratio are decreased too, thus confirming the severe mitochondrial damage occurring in the TM of POAG patients.[37] TM cells of POAG patients have low adenosine triphosphate (ATP) levels, because mitochondrial functionality is impaired by an intrinsic mitochondrial complex I defect resulting in a respiratory chain deficit. Exfoliation syndrome is an age-related ECM disorder characterized by the production and progressive accumulation of a fibrillar material not only in ocular tissues but also in skin and connective tissue portions of various visceral organs. This syndrome can be associated with glaucoma when the cellular pseudoexfoliative degeneration hits the TM. Oxidative DNA damage and formation, and accumulation of mtDNA deletions in TM of pseudoexfoliative glaucomatous patients, is greater than in POAG patients.[31,38] Pseudoexfoliative glaucoma is more aggressive than POAG, even if the clinical course is similar.

Mitochondrial dysfunction does not occur in other forms of open-angle glaucoma, the primary insult being different.[38] For instance, in a trauma with blood effusion in AC, glaucoma may develop both because of the direct TM trauma (angle recession glaucoma) or because of the blood injury that can destroy the endothelial cells (ghost cell glaucoma). Nonetheless, the decline in mitochondrial energy is manifested as tissue bioenergy mosaics, characterized by poor cytochrome c oxidase activity.[39] Cytochrome c oxidase is associated with the inner membrane of the mitochondrion and is in relationship with apoptosis activation in HTM cells of POAG patients. Oxidative stress causes activation of mitochondrial matrix caspase activity that is secondary to cytochrome c release.[40] Mitochondrial dysfunction leads to impaired oxidative phosphorylation and increases endogenous ROS production. Furthermore, severe mitochondrial damage results in intracellular calcium overload, triggering apoptosis through the intrinsic activation pathway. It is not clear if the primary damage is an excess of ROS in the AH or the mitochondrial impairment; however, mitochondrial damage contributes to the endothelial TM dysfunction. In

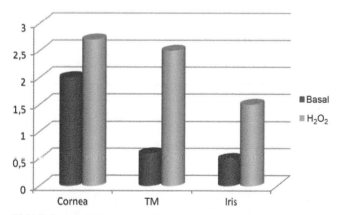

FIGURE 4.4 Different sensitivity to oxidative damage in various anterior chamber tissues. The trabecular meshwork (TM) is the most sensitive tissue to oxidative damage, as after exposure to hydrogen peroxide (H_2O_2) both markers of oxidative damage dramatically increased in the TM but not in the cornea and iris.

glaucoma, the following molecular events typically occur during TM endothelial dysfunction: decreased biosynthesis and/or bioavailability of nitric oxide (NO), excess superoxide and an excess of endothelin (ET) production. TNF-α regulates NOS expression and/or activity, which exert direct effects on NO production. TNF-α may increase inducible nitric oxide synthase (iNOS) expression by activating nuclear factor kappa-light-chain-enhancer of activated B cells (NF-κB). Increased TNF-α expression induces ROS production. TNF-α also activates NF-κB transcription, which regulates the expression of genes involved in inflammation, oxidative stress, and endothelial dysfunction (Fig. 4.5). In addition, there is an impaired equilibrium between ETs and NO that is at the base of the endothelial barrier. Vasoconstriction induced by ETs in the anterior part of the eye causes a decrease in ocular blood flow followed by pathologic changes in the retina and the optic nerve head, thus contributing to the degeneration of the RGCs. Trabecular outflow is modulated by TM contractility, which is affected by ET.[41] Conversely, the increased NO production by iNOS present in the TM of POAG patients contributes to the death of TM cells.[42] POAG patients have significantly impaired endothelial function,[43] which has been linked to an increased inflammatory status that is a causative mechanism of endothelial dysfunction. Furthermore, ET-1 has been linked to various other glaucoma-associated effects on the optic nerve and RGC including astrogliosis, ECM remodeling, and NO-induced damage.[44]

Upregulation of the RhoA/RhoA kinase (ROCK) signaling cascade, observed in different cardiovascular disorders, seems to have an impact on NO-signaling and vice versa: sustained Rho GTPase signaling activation in the AH outflow pathway, fundamentally increases the resistance to AH outflow through the trabecular pathway by influencing the actomyosin assembly, cell adhesive interactions, and expression of ECM proteins and cytokines in TM cells. Further, RhoA/ROCK regulation by Notch signaling in endothelial cells triggers a senescence phenotype that is associated with endothelial barrier dysfunction. This cellular dysfunction is difficult to appreciate *in vivo* but can be assayed by the analysis of the AH.

AH has 0.1–0.2% of the concentration of plasma protein and has higher concentrations of amino acids than plasma. A number of tissue growth factors have been detected in this fluid. AH promotes regulatory T-cell immunity and stimulates immune cell function. AH may play a primary role in POAG pathogenesis by facilitating the migration of cytokines that stimulate the activity of TM cells. The AH proteome profile undergoes dramatic changes in POAG patients as compared with matched controls: many proteins expressed at high levels in controls are reduced in POAG patients, while other proteins detected at low levels in controls are increased in POAG patients. AH proteome changes reflect molecular and cellular damage in POAG target tissues, that is, TM and optic nerve head (Table 4.1).

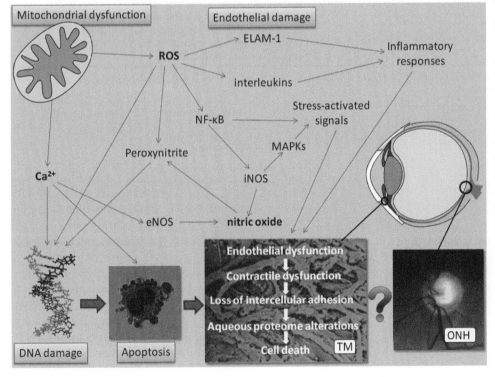

FIGURE 4.5 Multifactorial cascade involved in glaucoma pathogenesis. The oxidative damage plays a key role in the glaucoma pathogenesis. Mitochondria in turn contribute both to oxidative damage and the dysfunction of the TM endothelial cells. The TM malfunction finally determines the IOP increase. It is not clear yet what in turn determines the passage of information of the damage to the retina and the glia. However, it is likely that some cytokine is expressed by the damaged tissue of the trabecular meshwork. The involvement of the trabecular meshwork does not occur In normal pressure glaucomas. Probably other factors interact directly with the optic nerve head determining apoptosis of the retinal ganglion cells. *DNA: deoxyribonucleic acid; ELAM1: endothelial-leukocyte adhesion molecule 1; eNOS: endothelial nitric oxide synthase; iNOS: inducible nitric oxide synthase; MAPK: Mitogen-activated protein kinase; NF-κB: nuclear factor kappa-light-chain-enhancer of activated B cell; ONH: optic nerve hypoplasia; ROS: reactive oxygen species; TM: trabecular meshwork.*

TABLE 4.1 Proteome Analysis in Primary Open-Angle Glaucoma (POAG). Proteins Undergoing Significant (P < 0.05) Alteration in Aqueous Humor of POAG Patients as Compared with Controls Identified by k-Nearest Neighborhood Algorithm and Volcano-Plot Filtering.

Protein Name	Ontology	Function
TWO PROTEINS (5%) ARE INVOLVED IN CELLULAR HOMEOSTASIS:		
ATPase, Na⁺/K⁺ transporting, β 3 polypeptide**	Cell homeostasis	Noncatalytic component of the active enzyme, which catalyzes the hydrolysis of ATP coupled with the exchange of Na^+ and K^+ ions across the plasma membrane. Membrane protein
Glutamate-ammonia ligase (glutamine synthase)**	Glutamate detoxification Glutamine production Neural tissue homeostasis	Expressed in retina and neural tissues. Glutamate catabolism and glutamine production. Catalyzes glutamate transformation into glutamine by ATP-dependent NH_3 addiction. Glutamine is a main source of energy and is involved in cell proliferation, inhibition of apoptosis
FIVE PROTEINS (19%) ARE MITOCHONDRIAL PROTEINS INVOLVED IN ELECTRON TRANSPORT CHAIN, TRANSMEMBRANE TRANSPORT, PROTEIN REPAIR, MITOCHONDRIAL INTEGRITY MAINTENANCE:		
A kinase anchor protein 1, mitochondrial*	Mitochondria integrity	Anchors the cytoplasmic face of the mitochondrial outer membrane Located in mitochondrial outer membrane
Calcium-binding mitochondrial carrier protein Aralar1**	Mitochondrial transporter	Calcium-dependent mitochondrial aspartate and glutamate carrier. Located in mitochondria inner membrane
Mitochondrial heat shock 60-kDa protein 1**	Mitochondrial protein repair	Molecular chaperone
Translocase of inner mitochondrial membrane 23**	Mitochondrial transporter	Mitochondrial import inner membrane translocase
Cytochrome c**	Cell respiration Apoptosis	Mitochondrial electron-transport chain. Plays a role in apoptosis. Suppression of the antiapoptotic members or activation of the pro-apoptotic members of the Bcl-2 family leads to altered mitochondrial membrane permeability resulting in release of cytochrome c into the cytosol. Binding of cytochrome c to Apaf-1 triggers the activation of caspase-9, which then accelerates apoptosis by activating other caspases. Located in mitochondrial matrix
SIX PROTEINS (23%) ARE DIRECTLY INVOLVED IN APOPTOSIS INDUCTION MAINLY THROUGH THE INTRINSIC, I.E., MITOCHONDRIAL-DEPENDENT PATHWAY:		
BCL2-associated X protein (BAX)**	Apoptosis	Positive regulation of apoptosis. Located in mitochondria
BCL2-interacting killer (apoptosis-inducing) (BIK)**	Apoptosis	Activation of apoptosis through the intrinsic pathway. Located in mitochondria
Caspase 8**	Apoptosis	Apoptosis-related cysteine protease. Most upstream protease of the activation cascade of caspases responsible for the TNFRSF6/FAS mediated and TNFRSF1A induced cell death. Cleaves and activates CASP3, CASP4, CASP6, CASP7, CASP9, and CASP10
Caspase 9**	Apoptosis	Involved in the activation cascade of caspases responsible for apoptosis execution
TNF receptor-associated factor 2**	Apoptosis	Links members of the TNF receptor family to different signaling pathways. Mediates activation of NF-κ-B and JNK. Involved in apoptosis
Fas (TNFRSF6)-associated via death domain**	Apoptosis	Activation of apoptosis resulting from inflammation. Activation of apoptosis through the extrinsic pathway in response to inflammation and/or oxidative stress

TABLE 4.1 Proteome Analysis in Primary Open-Angle Glaucoma (POAG). Proteins Undergoing Significant (P < 0.05) Alteration in Aqueous Humor of POAG Patients as Compared with Controls Identified by k-Nearest Neighborhood Algorithm and Volcano-Plot Filtering.—cont'd

Protein Name	Ontology	Function
FIVE PROTEINS (19%) ARE COMPONENT OF INTERCELLULAR JUNCTION AND CONTRIBUTE TO TISSUE HOMEOSTASIS AND MAINTENANCE OF CELL ADHESION:		
Cadherin 3**	Tissue integrity	Idem
Cadherin 5**	Tissue integrity	Calcium-dependent cell adhesion proteins connecting cells. Control cohesion and organization of the intercellular junctions
Calnexin**	Protein repair	Calcium-binding protein playing a major role in the quality control apparatus of the endoplasmic reticulum by the retention of incorrectly folded proteins. Located in melanosomes
Catenin alpha**	Tissue integrity	Cadherin-associated protein. The association of catenins to cadherins produces a complex which is linked to the actin filament network, and which seems to be of primary importance for cadherin's cell-adhesion properties
Junction** plakoglobin	Tissue integrity Cell adhesion	Junctional plaque protein. Presence of plakoglobin in both the desmosomes and in the intermediate junctions suggests that it plays a central role in the structure and function of submembranous plaques
FOUR PROTEINS (15%) ARE TYPICALLY LOCATED IN NEURONS AND ARE INVOLVED IN VARIOUS NEURONAL FUNCTIONS INCLUDING NEURON SURVIVAL, NEUROGLYCANE PRODUCTION, NMDA RECEPTOR ACTIVATION, GLUTAMATE DETOXIFICATION AND PROTEIN REPAIR:		
Ankyrin 2, neuronal**	Neuron survival	Adapter protein in the postsynaptic density of excitatory synapses that interconnects receptors of the postsynaptic membrane including NMDA-type and metabotropic glutamate receptors, and the actin-based cytoskeleton
Chondroitin sulfate proteoglycan 5 (neuroglycan C)**	Neuron survival	Growth and differentiation factor involved in neuritogenesis and neuron survival. Expressed in neural tissue
Optineurin*	Neuron survival Apoptosis	Plays a neuroprotective role in the eye and optic nerve. Probably part of the TNF-α signaling pathway that can shift the equilibrium toward induction of cell death
Neural precursor cell expressed, developmentally downregulated 4 (NEDD 4)**	Protein repair in neurons	Molecular chaperone
NOS2**	Oxidative stress	Produces nitric oxide
SOD1 and 2*	Antioxidant defense	Destroys radicals that are normally produced within the cells and that are toxic to biologic systems
Microsomal glutathione S-transferase 1**	Antioxidant defense	Conjugation of reduced glutathione to a wide number of exogenous and endogenous electrophiles
Dynein, cytoplasmic, light polypeptide 1**	Tissue integrity Oxidative stress	Play a role in maintaining the spatial distribution of cytoskeletal structures. Binds and inhibits the catalytic activity of neuronal NOS

ATP: adenosine triphosphate; JNK: Jun N-terminal kinase; NF-κ-B: nuclear factor kappa-light-chain-enhancer of activated B cells; NMDA: N-methyl-D-aspartate; NOS: nitric oxide synthase; SOD: superoxide dismutase; TNF: tumor necrosis factor.
*P < 0.05.
**P < 0.01.

From an anatomic, physiologic, and pathologic point of view, the AC is a vessel and behaves as a vessel. Although the iris loses its endothelial lining at birth, as a matter of fact, during the course of glaucoma, all the early markers in the atherosclerotic plaque are significantly increased in the AH. The expression of these proteins in the AH of glaucomatous patients reflects the damage occurring in AC endothelia, mainly including

TABLE 4.2 Vascular Protein as Detected in Aqueous Humor (AH) of Primary Open-Angle Glaucoma (POAG) Patients by Antibody Microarray. The Presence of these Proteins in AH Indicates that POAG Course Damages the Trabecular Meshwork and in Particular its Endothelial and Cytoskeletal Components thus Altering Both TM Functionality and Motility.

Protein	Role	Target
ELAM1	Inflammatory responses	Endothelium and cytoskeleton
Apo B	Inflammatory responses	Endothelium
Apo E	Oxidative stress response	Endothelium
VASP	Cell adhesion and motility	Endothelium
Hsp60	Mitochondrial chaperonin	Endothelium
Hsp70	Immunoreactivity	Cytoskeleton
Myogenin	Muscle growth and regeneration	Cytoskeleton
Myogenic factor 3	Regulator of skeletal myogenesis	Cytoskeleton
Myotrophin	Myofibrillar growth pattern	Cytoskeleton
Ankyrin	Anchors cytoskeletal components	Cytoskeleton
Phospholipase C	PKC activator	Cytoskeleton
Ubiquitin	Regulation of endothelial NOS activity	Endothelium

NOS: nitric oxide synthase; PKC: protein kinase c.
Source: data from Saccà *et al., 2012.*[45]

the TM, in a state of cell suffering.[45] Changes occurring in the AC during the course of POAG include endothelial dysfunction, lipoprotein alteration, modification of smooth muscle cell functions, oxidative damage, inflammation, loss of intercellular adhesion, mitochondrial failure, and apoptosis (Table 4.2). POAG is associated with altered expression levels of adhesion molecules that are produced by endothelial cells, the recruitment of inflammatory cells, and the production of cytokines targeting endothelial cells, vascular smooth muscle cells, ECM, and mitochondria. These proteomic changes ultimately lead to apoptosis and TM degeneration. These molecules are part of the AH cytokine system designed to communicate signals from the anterior to the posterior segment of the eye and to the central nervous system, probably inducing ganglion cell death and apoptosis of retinal cells.

EPIDEMIOLOGY HINTS

About 12.3% of the worldwide population and the 21.8% of European adults (including 18% of those over 50 years of age) have been diagnosed with glaucoma.[46]

Racial factors play a prominent role. The Tajimi Study conducted in Japan on the prevalence of POAG has shown that the majority of Japanese POAG has a low or normal IOP.[47] POAG prevalence rate for Asian American people (6.52%) was similar to that of Latino people (6.40%) and higher than that of non-Hispanic white people (5.59%). The POAG and NTG prevalence rates were considerably higher among Asian American people (3.01% and 0.73%, respectively) relative to other races.[48] Worldwide, the highest prevalence of POAG occurs in Africans, and the lowest prevalence of POAG occurs in Inuit people. Evidence suggests that there is an important genetic contribution to POAG. As highlighted in studies performed in twins and relatives, it is possible that glaucoma segregates as a complex trait with multiple genes contributing to the phenotype along with unidentified environmental factors.[49] In conclusion, population projections for the years 2010 and 2020 indicate that open-angle glaucoma will become the most prevalent type of glaucoma in Europe, 60.5 million people with open-angle and angle-closure glaucoma are expected in 2010, increasing to 79.6 million by 2020, and of these, 74% will have open-angle glaucoma.

CONCLUSIONS

Currently the only form of therapy used to counteract the course of the glaucoma is to reduce IOP. Nowadays, the role of oxidative stress in the development and maintenance of the glaucomatous disease has been established. Therapeutic drugs such as timolol and dorzolamide have important antioxidant effects: indeed, timolol has an antioxidant effect on the entire endothelial cell, whereas dorzolamide exerts protective activity toward oxidative stress only in the presence of intact mitochondria.[25] From this point of view, the rise of IOP may be interpreted as an indicator of the 'health' state of the endothelial cells of the TM and reflect their function in AH outflow. Therefore, we think that it may be useful to add an additional therapy to antihypertensive therapy that is more selective to the pathogenesis of glaucoma. The manipulation of intracellular redox status by means of antioxidants might be a new therapeutic tool for preventing glaucomatous cell death. Better knowledge of the pathogenesis has opened up new therapeutic approaches that are often referred to as non-IOP-lowering treatment. These new agents, some of which are still under investigation, such as green tea or ginseng can improve the oxidative stress impact. In addition, substances such as the Ginkgo biloba extract have already been identified as useful in counteracting the development of glaucoma.[50] The question of glaucoma and nutrition is an intriguing one and deserves further investigation.

References

1. Bill A. Aqueous humor dynamics in monkeys (*Macaca irus* and *Cercopithecus ethiops*). *Exp Eye Res* 1971;**11**:195–206.
2. Raviola G, Butler JM. Asymmetric distribution of charged domains on the two fronts of the endothelium of iris blood vessels. *Invest Ophthalmol Vis Sci* 1985;**26**:597–608.
3. Sommer A. Ocular hypertension and normal-tension glaucoma: time for banishment and burial. *Arch Ophthalmol* 2011;**129**:785–7.
4. Whitacre MM, Stein R. Sources of error with use of Goldmann-type tonometers. *Surv Ophthalmol* 1993;**38**:1–30.
5. Chihara E. Assessment of true intraocular pressure: the gap between theory and practical data. *Surv Ophthalmol* 2008;**53**:203–18.
6. de Jong N, Greve EL, Hoyng PF, Geijssen HC. Results of a filtering procedure in low tension glaucoma. *Int Ophthalmol* 1989;**13**:131–8.
7. Park MH, Moon J. Circulating total glutathione in normal tension glaucoma patients: comparison with normal control subjects. *Korean J Ophthalmol* 2012;**26**:84–91.
8. Shields MB. Normal-tension glaucoma: is it different from primary open-angle glaucoma? *Curr Opin Ophthalmol* 2008;**19**:85–8.
9. Kiuchi T, Motoyama Y, Oshika T. Relationship of progression of visual field damage to postural changes in intraocular pressure in patients with normal-tension glaucoma. *Ophthalmology* 2006;**113**:2150–5.
10. Park HY, Jung KI, Na KS, Park SH, Park CK. Visual field characteristics in normal-tension glaucoma patients with autonomic dysfunction and abnormal peripheral microcirculation. *Am J Ophthalmol* 2012;**154**:466–75.
11. Doyle A, Bensaid A, Lachkar Y. Central corneal thickness and vascular risk factors in normal tension glaucoma. *Acta Ophthalmol Scand* 2005;**83**:191–5.
12. Greenfield DS, Liebmann JM, Ritch R, Krupin T. Low-pressure glaucoma study group. visual field and intraocular pressure asymmetry in the Low-Pressure Glaucoma Treatment Study. *Ophthalmology* 2007;**114**:460–5.
13. Jaggi G, Miller NR, Flammer J, Weinreb RN, Remonda L, Killer HE. Optic nerve sheath diameter in normal-tension glaucoma patients. *Br J Ophthalmol* 2012;**96**:53–6.
14. Ren R, Zhang X, Wang N, Li B, Tian G, Jonas JB. Cerebrospinal fluid pressure in ocular hypertension. *Acta Ophthalmol* 2011;**89**:e142–8.
15. Izzotti A, Bagnis A, Saccà SC. The role of oxidative stress in glaucoma. *Mutat Res* 2006;**612**:105–14.
16. Cedrone C, Mancino R, Cerulli A, Cesareo M, Nucci C. Epidemiology of primary glaucoma: prevalence, incidence, and blinding effects. *Prog Brain Res* 2008;**173**:3–14.
17. Izzotti A, Saccà SC, Longobardi M, Cartiglia C. Sensitivity of ocular anterior chamber tissues to oxidative damage and its relevance to the pathogenesis of glaucoma. *Invest Ophthalmol Vis Sci* 2009;**50**:5251–8.
18. Schulzer M, Drance SM, Douglas GR. A comparison of treated and untreated glaucoma suspects. *Ophthalmology* 1991;**98**:301–7.
19. Anderson DR, Drance SM, Schulzer M. Collaborative Normal-Tension Glaucoma Study Group. Factors that predict the benefit of lowering intraocular pressure in normal tension glaucoma. *Am J Ophthalmol* 2003;**136**:820–9.
20. Bevin TH, Molteno AC, Herbison P. Otago Glaucoma Surgery Outcome Study. Long-term results of 841 trabeculectomies. *Clin Exp Ophthalmol* 2008;**36**:731–7.
21. Collaborative Normal-Tension Glaucoma Study Group. The effectiveness of intraocular pressure reduction in the treatment of normal-tension glaucoma. *Am J Ophthalmol* 1998;**126**:498–505.
22. Ventura LM, Feuer WJ, Porciatti V. Progressive loss of retinal ganglion cell function is hindered with IOP-lowering treatment in early glaucoma. *Invest Ophthalmol Vis Sci* 2012;**53**:659–63.
23. Liu JH, Shieh BE, Alston CS. Short-wavelength light reduces circadian elevation of intraocular pressure in rabbits. *Neurosci Lett* 1994;**180**:96–100.
24. Saccà SC, Rolando M, Marletta A, Macrí A, Cerqueti P, Ciurlo G. Fluctuations of intraocular pressure during the day in open-angle glaucoma, normal-tension glaucoma and normal subjects. *Ophthalmologica* 1998;**212**:115–9.
25. Saccà SC, La Maestra S, Micale RT, Larghero P, Travaini G, Baluce B, et al. Ability of dorzolamide hydrochloride and timolol maleate to target mitochondria in glaucoma therapy. *Arch Ophthalmol* 2011;**129**:48–55.
26. Zeimer RC, Wilensky JT, Gieser DK, Viana MA. Association between intraocular pressure peaks and progression of visual field loss. *Ophthalmology* 1991;**98**:64–9.
27. Alvarado JA, Yeh RF, Franse-Carman L, Marcellino G, Brownstein MJ. Interactions between endothelia of the trabecular meshwork and of Schlemm's canal: a new insight into the regulation of aqueous outflow in the eye. *Trans Am Ophthalmol Soc* 2005;**103**:148–62.
28. Johnstone MA. The aqueous outflow system as a mechanical pump: evidence from examination of tissue and aqueous movement in human and non-human primates. *J Glaucoma* 2004;**13**:421–38.
29. Alvarado JA, Murphy C, Juster R. Trabecular meshwork cellularity in primary open-angle glaucoma and nonglaucomatous normals. *Ophthalmology* 1984;**91**:564–79.
30. Kelley MJ, Rose AY, Song K, Chen Y, Bradley JM, Rookhuizen D, et al. Synergism of TNF and IL-1 in the induction of matrix metalloproteinase-3 in trabecular meshwork. *Invest Ophthalmol Vis Sci* 2007;**48**:2634–43.
31. Izzotti A, Saccà SC, Cartiglia C, De Flora S. Oxidative deoxyribonucleic acid damage in the eyes of glaucoma patients. *Am J Med* 2003;**114**:638–46.
32. Bagnis A, Izzotti A, Centofanti M, Saccà SC. Aqueous humor oxidative stress proteomic levels in primary open angle glaucoma. *Exp Eye Res* 2012;**103**:55–62.
33. Bradley JM, Vranka J, Colvis CM, Conger DM, Alexander JP, Fisk AS, et al. Effect of matrix metalloproteinases activity on outflow in perfused human organ culture. *Invest Ophthalmol Vis Sci* 1998;**39**:2649–58.
34. Rao NA, Romero JL, Fernandez MA, Sevanian A, Marak Jr GE. Role of free radicals in uveitis. *Surv Ophthalmol* 1987;**32**:209–13.
35. Saccà SC, Izzotti A, Rossi P, Traverso C. Glaucomatous outflow pathway and oxidative stress. *Exp Eye Res* 2007;**84**:389–99.
36. Babizhayev MA, Bunin AY. Lipid peroxidation in open-angle glaucoma. *Acta Ophthalmol (Copenh)* 1989;**67**:371–7.
37. Izzotti A, Saccà SC, Longobardi M, Cartiglia C. Mitochondrial damage in the trabecular meshwork of patients with glaucoma. *Arch Ophthalmol* 2010;**128**:724–30.
38. Izzotti A, Longobardi M, Cartiglia C, Saccà SC. Mitochondrial damage in the trabecular meshwork occurs only in primary open-angle glaucoma and in pseudoexfoliative glaucoma. *PLoS One* 2011;**6**:e14567.
39. Linnane AW, Zhang C, Baumer A, Nagley P. Mitochondrial DNA mutation and the ageing process: bioenergy and pharmacological intervention. *Mutat Res* 1992;**275**:195–208.
40. Takahashi A, Masuda A, Sun M, Centonze VE, Herman B. Oxidative stress-induced apoptosis is associated with alterations in mitochondrial caspase activity and Bcl-2-dependent alterations in mitochondrial pH (pHm). *Brain Res Bull* 2004;**62**:497–504.
41. Rosenthal R, Fromm M. Endothelin antagonism as an active principle for glaucoma therapy. *Br J Pharmacol* 2011;**162**:806–16.
42. Fernández-Durango R, Fernández-Martínez A, García-Feijoo J, Castillo A, de la Casa JM, García-Bueno B, et al. Expression of nitrotyrosine and oxidative consequences in the trabecular meshwork of patients with primary open-angle glaucoma. *Invest Ophthalmol Vis Sci* 2008;**49**:2506–11.

43. Siasos G, Tousoulis D, Siasou G, Moschos MM, Oikonomou E, Zaromitidou M, et al. The association between glaucoma, vascular function and inflammatory process. *Int J Cardiol* 2011;**146**:113–5.

44. Good TJ, Kahook MY. The role of endothelin in the pathophysiology of glaucoma. *Exp Opin Ther Targets* 2010;**14**:647–54.

45. Saccà SC, Centofanti M, Izzotti A. New proteins as vascular biomarkers in primary open angle glaucomatous aqueous humor. *Invest Ophthalmol Vis Sci* 2012;**53**:4242–53.

46. Prokofyeva E, Zrenner E. Epidemiology of major eye diseases leading to blindness in Europe: a literature review. *Ophthalmic Res* 2012;**47**:171–88.

47. Iwase A, Suzuki Y, Araie M, Yamamoto T, Abe H, Shirato S, et al. The prevalence of primary open-angle glaucoma in Japanese: the Tajimi Study. *Ophthalmology* 2004;**111**:1641–8.

48. Stein JD, Kim DS, Niziol LM, Talwar N, Nan B, Musch DC, et al. Differences in rates of glaucoma among Asian Americans and other racial groups, and among various Asian ethnic groups. *Ophthalmology* 2011;**118**:1031–7.

49. Lichter PR. Genetics of the glaucomas. *J Glaucoma* 2001;**10**:S13–5.

50. Cybulska-Heinrich AK, Mozaffarieh M, Flammer J. Ginkgo biloba: an adjuvant therapy for progressive normal and high tension glaucoma. *Mol Vis* 2012;**18**:390–402.

Diabetic Retinopathy: An Overview

M.I. Lopez-Galvez[1], F. Manco Lavado[2], J.C. Pastor[3]

[1]Universidad de Valladolid, Valladolid, Spain; [2]Hospital Clinico Valladolid, Valladolid, Spain;
[3]IOBA: Universidad de Valladolid, Valladolid, Spain

INTRODUCTION

Diabetes mellitus (DM) is an important health problem affecting more than 350 million people worldwide, and its prevalence is still growing.[1] It has reached pandemic status and, as a serious chronic metabolic disease, it can bring about many types of complications that can severely impair people's quality of life.

DM is a group of metabolic diseases characterized by high blood sugar levels (hyperglycemia). There are two main classifications of this disease proposed by the American Diabetes Association (ADA) and the World Health Organization (WHO).

The ADA recognizes four DM forms: type 1, type 2, gestational diabetes, and other DM forms.

- Type 1 DM results from the body's failure to produce insulin due to autoimmune destruction of β cells of the islets of Langerhans. The diagnosis is often made around 25 years old but also affects older people. This form was previously referred to as *'insulin-dependent diabetes mellitus'* or *'juvenile diabetes'*.
- Type 2 DM is characterized by insulin resistance, a condition in which cells fail to use insulin properly. It is often observed in adults and is associated with obesity. This diabetes mellitus form was previously referred to as *'non-insulin-dependent diabetes mellitus'* or *'adult-onset diabetes'*.
- Gestational DM occurs when pregnant women develop a high blood glucose level without a previous diagnosis of diabetes. Rarely, it occurs after childbirth.
- Other specific types of DM include congenital diabetes, steroid diabetes induced by high doses of glucocorticoids, cystic fibrosis-related diabetes, and several forms of monogenic diabetes. Congenital diabetes is due to genetic defects of insulin secretion, and monogenic diabetes forms are caused by mutations in a single gene.

According to the WHO, DM is classified only into three groups: type 1, type 2, and gestational diabetes.

Diabetes without proper treatments may cause acute and chronic complications. Acute complications include hypoglycemia, diabetic ketoacidosis, and non-ketotic hyperosmolar coma. Serious long-term complications include macroangiopathic complications (ischemic heart disease, cerebrovascular disease, and peripheral vascular disease) and microangiopathic complications (retinopathy, neuropathy, and nephropathy). DR affects small blood vessels in the retina and may lead to visual symptoms, reduced vision, and potentially blindness (Box 5.1).

EPIDEMIOLOGY

DR is one of the most serious diseases affecting the microvasculature of the retina.[2] DR accounts for about 2.4 million cases of blindness globally, and a proportion of 4.8% of the global population has DR.

In 2012, Yau *et al.*[3] estimated the prevalence of global DR and its severe stages (proliferative diabetic retinopathy (PDR) and DME) using individual-level data from population-based studies worldwide. Based on the data obtained from 35 studies on more than 20,000 participants, the overall prevalence of any DR was 34.6%, PDR was 7%, DME was 6.8%, and vision-threatening DR (VTDR) was 10.2%. The prevalence of DR and VTDR was similar in men and women.

The number of people with diabetes in the world is predicted to grow to 429 million by 2030, owing to the rising frequency of obesity, increasing life span, and improved detection of the disease.[4] The DR prevalence rate is substantially higher in people with type 1 diabetes than in type 2 diabetes[3] and increases with the duration of diabetes, glycosylated-hemoglobin (HbA1c) values, blood pressure, and cholesterol. Total serum

---

BOX 5.1

Diabetes mellitus

Acute complications
- hypoglycemia
- diabetic ketoacidosis
- nonketotic hyperosmolar coma

Chronic complications
- macroangiopathic complications:
- ischemic heart disease
- cerebrovascular disease
- peripheral vascular disease

Microangiopathic complications
- diabetic retinopathy
- diabetic neuropathy
- diabetic nephropathy

BOX 5.2

The three major risk factors for diabetic retinopathy are:

1. diabetes duration
2. HbA1c
3. arterial blood pressure

---

cholesterol is also associated with a higher prevalence of DME.[3]

Epidemiologic studies have shown the effects of hyperglycemia, hypertension, and dyslipidemia on the incidence and progression of DR and clinically significant DME (CSME).

The three major risk factors for DR are diabetes duration,[3] HbA1c,[3,5] and high blood pressure.[3] Diabetes duration is the main risk for DR.[3] After 20 years of the disease, DR is present in more than 90% of type 1 diabetic patients and in nearly 60% of type 2 diabetic patients, although DR is higher in those patients taking insulin. Once retinopathy appears, duration of diabetes is less important than metabolic control for the further development and progression. The HbA1c level is the strongest risk factor for predicting the progression of DR, although it accounted for only 11% of the retinopathy risk in the DCCT.[6] Data from several studies suggest roles of other factors including sleep apnea[7]; nonalcoholic fatty liver disease[8]; and serum prolactin, adiponectin, and homocysteine levels,[9] as well as genetic factors including mutations in the erythropoietin gene promoter.[10] However, the relative contributions that these factors may have on the retinopathy risk in populations remains unknown.

DR is a diabetic complication that eventually develops to some degree in nearly all diabetic patients with long-standing disease. There are defects in the neurosensory retina before vascular damage is detected, but the earliest clinical manifestation is the presence of microaneurysms in the ocular fundus. Vascular permeability-related lesions (microaneurysms, hard exudates, hemorrhages) are first seen in diabetes,

but with time retinal capillary nonperfusion appears (cotton wool spots, intraretinal microvascular abnormalities (IRMA), venous beading). In the advanced stages of the disease, closure of the retinal vessels and new vessels are the main changes.

Diabetic macular edema may be present in all stages of DR. Clinically significant macular edema is the term used to describe the presence of retinal thickening and hard exudates in the center of the macular area, and usually it is considered a vision-threatening complication (Box 5.2).

NATURAL HISTORY

DR may progress from minimal changes to more severe stages. It is very important to identify vision-threatening stages in order to prevent blindness. Several clinical trials have demonstrated that it is possible to prevent severe vision loss in more than 90% of the cases when patients are managed correctly.

These clinical trials are the Diabetes Control and Complications Trial (DCCT),[6] the United Kingdom Prospective Diabetic Retinopathy Study (UKPDS),[11] the Diabetic Retinopathy Study (DRS),[12] the Early Treatment Diabetic Retinopathy Study (ETDRS),[12] and the Diabetic Retinopathy Vitrectomy Study (DRVS).[13]

There are two main classifications of DR: the classification proposed by the ETDRS and the Global Diabetic Retinopathy Project Group (GDRPG). The ETDRS classification is considered the gold standard in clinical trials. However, it is not used in clinical practice due to its complexity.[14] In 2002, The GDRPG proposed the International Scale Severity of Retinopathy. This new classification of diabetic retinopathy is based on the results of the ETDRS but is easier to use in practice (Table 5.1; Fig. 5.1; Fig. 5.2). A detailed correlation between both classifications can be found in http://www.rcophth.ac.uk/page.asp?section=451.[1]

The overall prevalence of nonproliferative DR is higher in type 2 diabetic patients.[15]

There are also several classifications for DME. The ETDRS classification has been the most commonly used worldwide.

TABLE 5.1 Classification of Diabetic Retinopathy adapted from the ETDRS*

DR Severity Level	Ocular Fundus Findings
Mild nonproliferative retinopathy	At least one microaneurysm; and definition not met for moderate nonproliferative retinopathy
Moderate nonproliferative retinopathy (Fig. 5.1)	Hemorrhages and/or microaneurysms ≥ standard photograph 2A*; and/or soft exudates, venous beading, or intraretinal microvascular abnormalities definitely present
Severe nonproliferative retinopathy	Soft exudates, venous beading, and intraretinal microvascular abnormalities in at least two fields; or two of the preceding three lesions present in at least two of fields four through seven and hemorrhages and microaneurysms present in these four fields, equaling or exceeding standard photograph 2A in at least one of them; or intraretinal microvascular abnormalities present in each of fields four through seven and equaling or exceeding standard photograph 8A in at least two of them
Early proliferative retinopathy	New vessels; and definition not met for high-risk proliferative retinopathy
High-risk proliferative retinopathy (Fig. 5.2)	New vessels on or within one disc diameter of the optic disc (NVD) ≥ standard photograph 10A* (about one-quarter to one-third disc area), with or without vitreous or preretinal hemorrhage; or vitreous and/or preretinal hemorrhage accompanied by new vessels, either NVD < standard photograph 10A or new vessels elsewhere (NVE) ≥ one-quarter disc area

DR: diabetic retinopathy.
*Early Treatment Diabetic Retinopathy Study Research Group.[35]

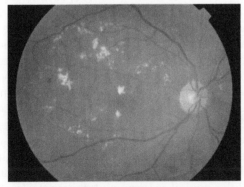

FIGURE 5.1 Nonproliferative diabetic retinopathy with diabetic macular edema. (See color plate section)

To diagnose CSME, one of the following characteristics must be present on clinical examination:

1. any retinal thickening within 500 μm of the center of the macula;

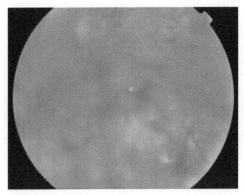

FIGURE 5.2 Proliferative diabetic retinopathy with vitreous hemorrhage. (See color plate section)

TABLE 5.2 Diabetic Macular Edema International Severity Scale.[14]

Diabetic Macular Edema (DME) Severity Level	Ocular Fundus Findings
DME NOT PRESENT	
Thickening apparently absent	Not thickening, no exudates in the posterior pole
DME PRESENT	
Thickening or hard exudates in the posterior pole	Thickening or hard exudates in the posterior pole
	Mild: some thickening or hard exudates distant from the center of the macula
	Moderate: some thickening or hard exudates approaching the center of the macula but no involvement
	Severe: some thickening or hard exudates involving the center of the macula

2. hard exudates within 500 μm of the center of the macula with adjacent retinal thickening;
3. retinal thickening at least one disc area in size, any part of which is within one disc diameter of the center of the macula.

Trying to improve communication, the GDRPG also proposed the use of the International Diabetic Macular Edema Severity Scale (Table 5.2).[14]

PATHOPHYSIOLOGY

The pathogenesis of DR is highly complex and involves multiple interlinked mechanisms leading to cellular damage and adaptive changes in the retina.[16]

Traditionally DR has been considered a microvascular disease, in which small retinal vessels are injured by chronic hyperglycemia and the induced endothelial damage is primarily responsible for the development of microangiopathy.[17]

However, DR is a multifactorial progressive disease with a very complex pathophysiology. The underlying mechanism is still poorly understood. Both neural and vascular damage seem to occur in this disease. DR is indeed a neurovascular disease in which neural dysfunction can be initiated in early-stage diabetes.[18]

Hyperglycemia has been considered as the initiator of retinal damage in DR, but this is a simple way to understand the development of this complication. Several metabolic pathways are activated. There are many studies suggesting that excess plasma glucose may not account for all the cellular and functional changes involved in DR.[19] In addition, it is well known that an intensive metabolic control reduces the risk of progression of DR but is not enough for its prevention.

This disease induces deregulated levels of metabolites such as glucose, lipids, or hormones that induce changes in the production of a number of mediators resulting in increased vascular permeability, apoptosis, and angiogenesis.[20] Many of the diabetic complications are clearly associated with oxidative stress, inflammation, mitochondrial dysfunction, and apoptosis.[21]

DR is a disease that progresses chronologically into two different phases. The first phase represents the passive suffering of the affected structures (neurons and vessels), while the second phase is a compensatory but aberrant response. In the first phase, biochemical changes, especially hyperglycemia, will deteriorate the retinal neural and microvascular system, leading to a hypoxic situation. The hypoxic retinal tissue triggers compensatory vasodilation and neovascularization to increase blood flow and tissue oxygenation. The reactive dilation of retinal vessels causes increased vascular permeability and subsequently edema (DME).

Several pathogenic mechanisms seem to be involved in the natural history of DR, and they can be grouped into biochemical, physiologic, hematologic, endocrinologic, and anatomical changes.[17,19] Biochemical alterations are most important in the early stages of DR and anatomic changes are more relevant in later stages of the disease.[17]

The cellular damage in the retina has been speculated to be caused by biochemical alterations, but many of these hypotheses have not yet been validated in human studies or clinical trials (Box 5.3).[16]

BOX 5.3

Hyperglycemia has been considered as the initiator of retinal damage in diabetic retinopathy

---

BIOCHEMICAL CHANGES

The major biochemical changes induced by hyperglycemia, which are implicated in the pathogenesis of DR, include[16–20]:

- increase of polyol pathway flux;
- increase of intracellular advanced glycation end product (AGE) formation;
- protein kinase C (PKC) activation;
- hexosamine pathway;
- polyadenosine diphosphate ribose polymerase (PARP) activation;
- increase of oxidative stress and reactive oxygen species (ROS) production;
- nuclear factor kappa-light-chain-enhancer of activated B cell (NF-κB) activation;
- Ras activation.

Other relevant biochemical changes are:

- high plasma homocysteine level;
- higher levels of branched chain amino acids in the diabetic retina;
- dysregulation of taurine level and its transporter;
- increase of adenosine level in the retina;
- dysregulation of nutrient levels such as α-lipoic acid, folic acid, vitamin C, vitamin E, and minerals.

Polyol or Sorbitol Pathway

In normal conditions, glucose is metabolized enzymatically by the glycolytic pathway and the pentose pathway. In chronic hyperglycemia, an alternative pathway, the sorbitol pathway, is activated.[17,22,23]

In this situation, glucose is metabolized by two enzymes: aldose reductase (AR) and sorbitol dehydrogenase (SDH). AR is an enzyme found in the endothelial cells, the ganglion cells, and the nerve fibers of the retina. It has an important role in the genesis of cataract, DR, and diabetic neuropathy.[17] This enzyme reduces aldehydes produced by oxygen free radicals to inactive alcohols and catalyze glucose to sorbitol using nicotinamide adenine dinucleotide phosphate (NADPH) as a cofactor (Fig. 5.3). Thus, the NADPH level is reduced, resulting in increased oxidative stress, a major factor in retinal damage.[22] NADPH is a cofactor required for glutathione reductase, a necessary enzyme that maintains the intracellular pool of reduced glutathione (GSH). GSH is an important scavenger protecting the endothelial cells against oxygen free radicals.

Sorbitol does not readily cross cell membranes. Then its intracellular concentration becomes higher and produces an increase of the osmotic pressure leading to water diffusion into the cell with the subsequent intracellular edema. The edema produces osmotic stress, which

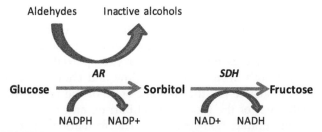

FIGURE 5.3 Polyol pathway. AR catalyzes the reduction of glucose to sorbitol and aldehydes to inactive alcohols. In this reaction NADPH is used, so its intracellular level decreases. SDH oxidizes sorbitol to fructose slowly. AR: aldose reductase. NAD: Nicotinamide adenine dinucleotide; NADH: reduced nicotinamide adenine dinucleotide; NADP+: oxidized form of nicotinamide adenine dinucleotide phosphate; NADPH: nicotinamide adenine dinucleotide phosphate; SDH: sorbitol dehydrogenase.[17,20]

TABLE 5.3 Actions Induced by Advanced Glycation End Product Overproduction

- Endothelial dysfunction
- Powerful free-radical generators
- Coagulation disorder that predisposes to the microthrombi formation
- PKC activation
- Blood–retinal barrier dysfunction
- Increase of nitrative stress in the retinal vascular cells
- Increase of VEGF, MCP-1, and ICAM-1 expressed in microvascular endothelial cells through intracellular ROS generation
- Retinal capillary cell apoptosis via activation of NF-κB and caspase-3
- Activation of NF-κB and NADPH oxidase with increase in ROS and apoptosis of pericytes and other retinal cells

ICAM-1: intercellular adhesion molecule-1; MCP-1: monocyte chemoattractant protein-1; NADPH: nicotinamide adenine dinucleotide phosphate; NF-κB: nuclear factor kappa-light-chain-enhancer of activated B cells; PKC: protein kinase C; ROS: reactive oxygen species; VEGF: vascular endothelial growth factor.[17,20,21]

alters the electrolytes' balance and the cell membrane's permeability, causing tissue hypoxia.[23]

Increased nitrotyrosine, increased lipid peroxidation products, and depletion of antioxidant enzymes are other biochemical consequences of polyol pathway activation.[23,24]

Intracellular Advanced Glycation End Product Formation

Biochemical changes observed in DR have also been associated with nonenzymatic glycosylation of proteins and the Maillard reaction between reducing sugars such as glucose and amino residues of proteins, lipids, or nucleic acids. The increase of nonenzymatic glycation leads to a high level of AGEs. This biochemical reaction is irreversible and depends on the glycemic level present in the body.[17,24]

Overproduction of AGEs is followed by intracellular and extracellular actions and by a breakdown of the blood-retinal barrier (Table 5.3).

TABLE 5.4 Actions Induced by Protein Kinase C Activation

- Increase of vascular permeability
- Decrease of nitric oxide production
- Increase of endothelin-1 activity
- Change in vascular smooth muscle contractility
- Increase of basement membrane protein synthesis and basal membrane thickening
- Stimulation of neovascularization, endothelial proliferation, and apoptosis
- Activation of cytokines and vasoactive factors such as VEGF, IGF-1, and TGF-β
- Increase of blood flow, extracellular matrix expansion, and leukocyte adhesion

IGF-1: insulin-like growth factor 1; TGF: transforming growth factor; VEGF: vascular endothelial growth factor.[16,17]

AGEs alter microvascular homeostasis and play a central role in inflammation, neurodegeneration, and microvascular dysfunction in DR.[24] The receptor for AGEs (RAGE) is ubiquitously expressed in various retinal cells and is upregulated in the retinas of diabetic patients, resulting in activation of pro-oxidant and proinflammatory signaling pathways.[24]

Protein Kinase C Activation

PKCs are a family of enzymes with at least 11 isoforms. Nine of them are activated by the second messenger diacylglycerol (DAG), which is the physiologic activator of PKC.[25]

In very early DR stages, hyperglycemia increases *de novo* synthesis of DAG from glucose via triose phosphate, which activates PKC. Hyperglycemia primarily activates the β and δ isoforms of PKC in vascular cells. The activation of β isoforms of PKC (PKC-β) has been implicated in the pathogenesis of early and late manifestations of RD (Table 5.4).[17,25]

PKC activation, hypoxia, and hyperglycemia collaborate in the stimulation of vascular endothelial growth factor (VEGF) expression.[17] The PKC-β activation is also essential for facilitating VEGF activity on vascular permeability and neovascularization.[25]

Hexosamine Pathway

Hexosamine content is increased in retinal tissues of humans and rats with diabetes and may mediate some of the toxic effects of high glucose and ROS concentrations into the cell.[26] In contrast, ROS may increase the hexosamine biosynthesis pathway by inhibition of glyceraldehyde 3-phosphate dehydrogenase (GAPDH), a multifunctional protein with diverse cytoplasmic membrane and nuclear activities.[17]

The GAPDH inhibition causes an increase in the production of uridine diphosphate N-acetylglucosamine

(UDP-GlcNAc) and may result in increased levels of the glycolytic metabolite glyceraldehyde 3-phosphate that can activate the AGE pathway by activating methylglyoxal, an intracellular AGE precursor.[27]

Polyadenosine Diphosphate Ribose Polymerase Activation

PARP is a family of enzymes found in the cell's nucleus and implicated in a number of cellular processes involving mainly deoxyribonucleic acid (DNA) repair and programmed cell death (apoptosis). It can also inhibit GAPDH activity and control the progression of DR. Zheng et al.[21] demonstrated that the activity of PARP is increased in the retina, endothelial cells, and pericites of diabetic rats and contributes to the induced cellular death of vascular cells in this disease.

Oxidative Stress

Retinal cellular homeostasis is maintained by a tight balance between the formation and elimination of ROS.[17] In diabetes, retinal blood flow is reduced, even at the early stages of the disease,[5] and, in more established stages, retinal oxygen supply will be reduced not only because of a reduced retinal blood flow but also because there is a decrease in choroidal oxygen partial pressure (PO_2).[28] Both decreased blood flow and choroidal PO_2 are responsible for the hypoxic state of the retina, the breakdown of the blood-retinal barrier, and the alteration in cellular homeostasis.[29] The formation of reactive nitrogen oxide species (RNOS) also stimulates the expression of growth factors in DR.[17]

The main biochemical pathways involved in the pathogenesis of DR (polyol pathway, intracellular AGE formation, PKC activation, and hexosamine pathway) seem to be interconnected (Fig. 5.4; Box 5.4).

Nuclear Factor-κB Activation

Nuclear Factor-κB (NF-κB) is a protein complex that controls the transcription of DNA. It is an inducible transcription factor and an important regulator of many genes involved in inflammatory (including adhesion molecules) and immune responses, proliferation, and apoptosis.[17] Activation of NF-κB leads to endothelial

FIGURE 5.4 Possible pathogenesis of diabetic retinopathy. Hyperglycemia ends in a common final pathway that involves the interaction between apoptosis, inflammation, and oxidative stress, leading to the diabetic retinopathy production. AGE: advanced glycation end product; DAG: diacylglycerol; GAPDH: glyceraldehyde 3-phosphate dehydrogenase; GSH: glutathione; NADPH: nicotinamide adenine dinucleotide phosphate; NF-κB: nuclear factor kappa-light-chain-enhancer of activated B cells; PARP: polyadenosine diphosphate ribose polymerase; PKC: protein kinase C; ROS: reactive oxygen species.[16,17,20,21,23,24,26,27,30,31]

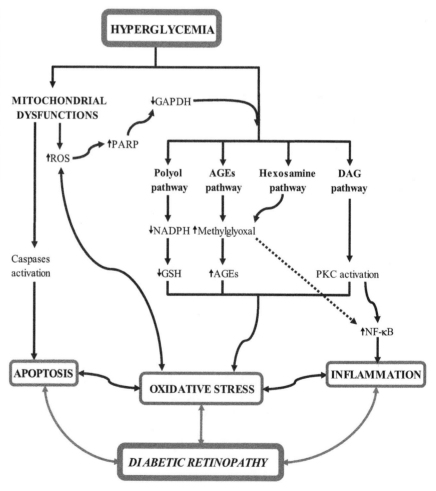

apoptosis via Bcl-2, caspase-3, and caspase-9 pathways.[30] NF-κB is also a critical regulator of antioxidant enzymes.

Oxidative stress can trigger a redox-sensitive NF-κB, and thus it activates a new pathway for the development of DR.

Several studies suggest that NF-κB plays an important role in the pathogenesis of early-stage DR. It has been found that inhibiting NF-κB or proteins whose expression is regulated by NF-κB (such as inducible nitric oxide synthase (iNOS) and intracellular adhesion molecule (ICAM)-1) may inhibit the development of the retinopathy (Box 5.5).

MITOCHONDRIAL DYSFUNCTION

Mitochondria are organelles involved in cell metabolism, energy production, and apoptosis. They are considered to be the main source of superoxide radicals.[17]

Hyperglycemia is followed by an increase in glucose oxidation at the mitochondrial level and an increase in the voltage gradient across the mitochondrial membrane. This leads to superoxide production and cell damage.[31]

All of the molecular mechanisms mentioned above ultimately may produce neuronal dysfunction, inflammation, and neovascularization in the late stages of the disease.

NEURONAL DYSFUNCTION AND INFLAMMATION

The neurosensory retina is also altered in diabetes. There is an impairment in the metabolism of glutamate (main neurotransmitter) and a loss in the synaptic activity and apoptosis in the ganglion cells and nuclear layers.[18]

The neuroprotective effect of the platelet-derived growth factor (PDGF) is also reduced in these eyes, and it contributes to the pericyte loss and to the impairment of the blood-retinal barrier. Nowadays most authors recognize the concept of the neurovascular unit. This neurovascular unit is altered as result of diabetes.[32]

Inflammation is also present in DR and plays an important role in its development. Inflammatory mediators such as interleukin-1β, tumor necrosis factor (TNF)-α, ICAM-1, and angiotensin II have been found elevated in DR. Microglial cells are also activated, and all these changes are implicated in the progressive retinal impairment.[32]

VASCULAR DAMAGE

As a result of all molecular changes there is an important increase in expression of intraocular VEGF levels and other cytokines. VEGF probably plays an important role in mediating the structural and functional changes in the retina.[33]

VEGF is a glycoprotein secreted by vascular endothelial cells, smooth muscle cells, and other cell types. There is an increase in the expression of VEGF and its receptors in DR and DME.[34]

VEFG belongs to a dimeric glycoprotein family that includes the placental growth factor (PIG) and several VEGF isoforms: VEGF-A, VEGF-B, VEGF-C, VEGF-D, VEGF-E, and VEGF-F. The isoform involved in DR and the most investigated is VEGF-A. It can act through the specific receptors VEGFR-1 and VEGFR-2.[33]

VEGF-165 is the main VEGF-A isoform. It is a 45-kDa homodimeric glycoprotein and it causes increased vascular permeability and neovascularization stimulation in both physiologic and pathologic processes.[34]

The expression of VEGFR-1 is upregulated by hypoxia and it binds VEGF with higher affinity than VEGFR-2. The latter is the main mediator of the mitogenic and angiogenic effects of VEGF and is responsible for the increased vascular permeability.[34]

Endothelial cells and pericytes are damaged in the early stages of the disease and microaneurysms appear in the ocular fundus. The blood-retinal barrier is broken and blood and hard exudates appear in the retina. In the advanced stages of the disease, hypoxia and vascular occlusion are present.

DIAGNOSIS AND PREVENTION

The appropriate care process for DR includes a medical history, an ophthalmic examination, and close follow-up.

TABLE 5.5 Actual Follow-Up Recommendations According to Diabetic Retinopathy Severity

Severity Level	Follow-Up Recommendation
No apparent DR	Annually*
Mild nonproliferative DR	Annually
Moderate nonproliferative DR	Every 6 months
Severe nonproliferative DR	Every 4–6 months
Low-risk proliferative DR	Consider treatment
High-risk proliferative DR	Treatment
CSME	Treatment

CSME: clinically significant diabetic macular edema; DR: diabetic retinopathy.
*Some authors consider every 2 years in type 2 diabetic patients who are well controlled due to the low risk.

The features of DR are detected by direct examination of the eye fundus (ophthalmoscopy). DR is asymptomatic in the early stages, and there is no visual dysfunction until the disease is well established. Patients with nonproliferative diabetic retinopathy (NPDR) may suffer some disturbances in their activities, but visual acuity only decreases when the macula is affected.

A comprehensive exam must be done in all diabetic patients. Dilated fundoscopy and biomicroscopy of the posterior are the best way to detect DR. Before inducing mydriasis, it is necessary to examine the iris to exclude the presence of iris neovascularization.

Undilated ophthalmoscopy has poor sensitivity and specificity.

Annual exams of the ocular fundus are recommended by the American Academy of Ophthalmology (AAO) in all diabetic patients, starting at the moment of diagnosis in type 2 diabetic patients and 3–5 years after diagnosis in type 1 diabetes. After DR has been diagnosed, the follow-up schedule depends on the stage of the disease. The actual recommendations are summarized in Table 5.5.

In type 2 diabetic patients without risk factors, annual exams can be replaced by biannual exams due to the slow progression of the disease.

The gold standard for detection and classification of DR is stereoscopic color fundus photographs in seven standard fields, as defined by the ETDRS group.[35] This procedure is time consuming and requires trained personnel. It is only used for clinical trials. In clinical practice, dilated ophthalmoscopy, biomicroscopy, and manual grading are considered the best methods.

However, in clinical practice, manual screening of all the diabetic population by trained ophthalmologists seems impossible due to the high prevalence of the disease. Even in the most developed countries, less than 50% of the diabetic patients are being followed correctly.

BOX 5.6

Laser photocoagulation remains the standard technique for the treatment of diabetic retinopathy

Several types of screening programs have been designed throughout the world to meet this problem. Screening exams or tests should aim for a sensitivity of at least 60%, though higher levels are usually achievable.[36] Specificity levels of 90–95% and technical failure rates of 5–10% are considered appropriate for both measures.[37]

The AAO recognizes that screening for DR using appropriately validated digital image technology is effective and sensitive for this purpose. These screening programs have great value in circumstances in which access to ophthalmic care is limited,[37] although they do not replace a comprehensive ophthalmic exam.

Patients must be aware of the importance of blood glucose and blood pressure control in both type 1 and type 2 diabetes. The benefits of metabolic control on the retinopathy progression have been well established in randomized clinical trials (DCCT, UKPDS).

The DCCT showed that the development and progression of DR in type 1 diabetic patients can be delayed if glucose concentrations are maintained in the near-normal range. This study showed how the risk of retinopathy was decreased by 53% in children aged 13–17 years without retinopathy at study entry, and the risk of retinopathy progression was decreased by 70% in those who had retinopathy at the beginning of the study. The benefits of intensive therapy continued to be evident 7 years after the end of the DCCT, as demonstrated in the Epidemiology of Diabetes Interventions and Complications study.[38] Evidence about the effects of controlling hyperglycemia in type 2 diabetic patients comes from observational data as well as the UKPDS (Box 5.6).

TREATMENT OF DIABETIC RETINOPATHY AND DIABETIC MACULAR EDEMA

DR is defined by progressive development of microvascular and neurovascular damage in the retina that leads to vision-threatening complications such as PDR and DME.

Retinal laser photocoagulation has been considered the standard technique for treating DR since the publication of the ETDRS results in the early 1980s. Laser photocoagulation techniques can be classified as panretinal, focal, or grid. Panretinal photocoagulation (PRP) should be performed whenever PDR is reached. Focal or grid

---

BOX 5.7

Intravitreal corticosteroids

Triamcinolone
Dexamethasone
Fluocinolone acetonide

Intravitreal antiangiogenics

Bevacizumab
Ranibizumab
Aflibercept

---

macular laser treatment should be considered for all eyes with CSME.

According to the ETDRS, scatter PRP significantly reduced the risk of severe vision loss (severe blindness) from PDR, and focal or grid laser photocoagulation can reduce the risk of moderate vision loss (doubling of the visual angle) from CSME. ETDRS results were achieved by rigorously applying laser recommendations and through close follow-up, with retreatment as needed. The regression of new vessels occurs often at three months, but in nearly 4.5% of patients the disease progresses to advanced disease even in the presence of adequate PRP.[39]

Recommendations for the type and pattern of laser treatment for DR have not changed since the 1980s.[40,41]

In patients with advanced DR and vitreous hemorrhages or tractional retinal detachment, laser is not useful and patients must undergo surgery (vitrectomy) (Box 5.7).

Since the 1990s, significant developments in pharmacotherapy have emerged, but they are still an adjunct to panretinal photocoagulation.

Corticosteroids and antiangiogenic drugs have shown promising results in preventing neovascularization and in the management of DME.

Intravitreal triamcinolone acetonide (IVTA) has been studied for its potent effects in DME. The Diabetic Retinopathy Clinical Research (DRCR) Network investigated the efficacy and safety of two doses (1 and 4 mg) of IVTA compared with the standard of care (focal/grid laser) in patients with DME. After 3 years of treatment, laser was associated with better visual acuity and fewer side effects (see http://drcrnet.jaeb.org).

Sustained drug delivery systems of corticosteroids are being tested in clinical trials in patients with DME. Ozurdex® (Allergan) is a biodegradable dexamethasone intravitreal implant approved by the Food and Drug Administration (FDA) and the European Medicines Agency (EMA) for treatment of macular edema secondary to retinal vein occlusion. Its efficacy in DME is now being investigated in randomized clinical trials.

Iluvien® (Alimera Sciences, Inc.) is a nonbiodegradable implant designed to release the drug fluocinolone acetonide for up to 3 years. It has been approved in Europe, but the FDA considers that this device cannot be approved in the USA in its present form because of the high prevalence of side effects, mainly cataracts and increased intraocular pressure.[42]

There are other sustained drug delivery systems under investigation such as the Cortiject® implant (NOVA63035, Novagali Pharma) and the Versiome® drug delivery system (Icon Bioscience, Inc.) but only in phase I and II clinical trials.[43]

In addition, antiangiogenic drugs have been found to be effective for treatment of PDR and DME. As has been mentioned, there is an overexpression of VEGF in DR. The standard average point for evaluating effectiveness of anti-VEGF drugs is 6 months, but only a few studies include this 6-month follow-up period after intravitreal injection.

Avastin® (bevacizumab, Genentech, Inc.) is a full-length humanized antibody that binds to all subtypes of VEGF and has shown promising results in DR. Avery et al.[44] reported transient iris and retinal neovascularization regression after intravitreal use of this drug. Shin et al.[45] have demonstrated that intravitreal bevacizumab reduced the VEGF expression to some degree in a limited number of PDR patients. It is used off label as a clinical adjunct to laser treatment in patients with PDR and in advanced disease 3–5 days previous to vitrectomy to reduce intraoperative and postoperative complications, mostly hemorrhages. In most studies, bevacizumab is administered 1 week preoperatively to avoid the occurrence of tractional retinal detachment.[46] The same authors have also suggested its use in cases of vitreous hemorrhage after vitrectomy.

Lucentis® (ranibizumab, Genentech Inc. and Novartis Pharma) is a recombinant humanized antibody fragment against VEGF approved by the FDA and the EMA for DME treatment. There are no final reports on the effect of ranibizumab in the treatment of PDR. The results of the clinical trial evaluating this drug by the DRCR Network in this stage of the disease are awaited by the scientific community.

Other anti-VEGF drugs are under investigation for this purpose, such as VEGF Trap-Eye® (aflibercept, Eylea, Bayer HealthCare and Regeneron Pharmaceuticals, Inc.).

A major limitation of anti-VEGF therapy is the recurrence of new vessels after 2 weeks of intravitreal injection. Some authors have suggested a period of reinjection of 3 months.[47] There is lack of information about long-term side effects of this therapy in PDR. Caution is warranted.

There is some evidence that the hyaloid and the posterior vitreous may influence the progression of DR and DME. It is accepted that the course of the disease seems more favorable in patients with posterior vitreous detachment.[43] Then, the posterior vitreous detachment induction may be beneficial in the treatment of DME and PDR. Intravitreal microsplamine (Ocriplasmin) has been recently approved by the FDA for this purpose. It is useful in resolving vitreomacular adhesion and it is safe and well-tolerated.[48]

Several attempts have been made with systemic therapy to prevent DR or its progression to PDR. Fenofibrate is a lipid-lowering drug used to treat dyslipidemia. Its main clinical effect is due to the inhibition of the actions of peroxisome proliferator-activated receptor alpha (PPAR-α). Some benefits in the prevention of DR and slowing the progression to advanced disease have been demonstrated with this drug in type 2 diabetic patients in the Fenofibrate Intervention and Event Lowering in Diabetes (the FIELD study), although the mechanism of this effect is unknown.[49]

In addition, some benefits have been shown with renin-angiotensin system inhibition in the DIabetic Retinopathy Candesartan Trial (DIRECT) study with candesartan[50] and with PKC inhibition that seemed to ameliorate vision loss without definitive conclusions.

There is still a long way to go in this field and a lot of work to do. New insights in the pathogenesis of this disease may open new possibilities for treatment. Many new compounds are currently in or entering clinical trials. The role of these new compounds in diabetic retinopathy must be defined.

References

1. Anon. The global challenge of diabetes. *Lancet* 2008;**371**:1723.
2. Idil A, Caliskan D, Ocaktan E. The prevalence of blindness and low vision in older onset diabetes mellitus and associated factors: a community-based study. *Eur J Ophthalmol* 2004;**14**:298–305.
3. Yau JW, Rogers SL, Kawasaki R, et al. Meta-Analysis for Eye Disease (META-EYE) Study Group. Global prevalence and major risk factors of diabetic retinopathy. *Diabetes Care* 2012;**35**(3):556–64.
4. Al-Rubeaan K. Type 2 diabetes mellitus red zone. *Int J Diabetes Mellitus* 2010;**2**(1):1–2.
5. Patel A, MacMahon S, Chalmers J, et al. ADVANCE Collaborative Group. Intensive blood glucose control and vascular out-comes in patients with type 2 diabetes. *N Engl J Med* 2008;**358**:2560–72.
6. Cundiff DK, Nigg CR. Diet and diabetic retinopathy: insights from the Diabetes Control and Complications Trial (DCCT). *Med Gen Med* 2005;**7**(1):3.
7. West SD, Groves DC, Lipinski HJ, et al. The prevalence of retinopathy in men with Type 2 diabetes and obstructive sleep apnoea. *Diabet Med* 2010;**27**:423–30.
8. Targher G, Bertolini L, Chonchol M, et al. Non-alcoholic fatty liver disease is independently associated with an increased prevalence of chronic kidney disease and retinopathy in type 1 diabetic patients. *Diabetologia* 2010;**53**:1341–8.
9. Arnold E, Rivera JC, Thebault S, et al. High levels of serum prolactin protect against diabetic retinopathy by increasing ocular vaso-inhibins. *Diabetes* 2010;**59**:3192–7.
10. Tong Z, Yang Z, Patel S, et al. Promoter polymorphism of the erythropoietin gene in severe diabetic eye and kidney complications. *Proc Natl Acad Sci U S A* 2008;**105**:6998–7003.
11. Kohner EM, Aldington SJ, Stratton IM, et al. United Kingdom Prospective Diabetes Study, 30: diabetic retinopathy at diagnosis of non-insulin-dependent diabetes mellitus and associated risk factors. *Arch Ophthalmol* 1998;**116**(3):297–303.
12. Cantrill HL. The diabetic retinopathy study and the early treatment diabetic retinopathy study. *Int Ophthalmol Clin* 1984;**24**(4):13–29.
13. Anon. Early vitrectomy for severe vitreous hemorrhage in diabetic retinopathy. Four-year results of a randomized trial: Diabetic Retinopathy Vitrectomy Study Report 5. *Arch Ophthalmol* 1990;**108**:958–64.
14. Wilkinson CP, Ferris III FL, Klein RE, et al. Global Diabetic Retinopathy Project Group. Proposed International Clinical Diabetic Retinopathy and diabetic macular edema. Disease Severity Scales. *Ophthalmology* 2003;**110**:1677–82.
15. Sparrow JM, McLeod BK, Smith TD, et al. The prevalence of diabetic retinopathy and maculopathy and their risk factors in the non-insulin-treated diabetic patients of an English town. *Eye (Lond)* 1993;**7**:158–63.
16. Frank RN. Diabetic retinopathy. *N Engl J Med* 2004;**350**(1):48–58.
17. Kowluru RA, Chan PS. Oxidative stress and diabetic retinopathy. *Exp Diabetes Res* 2007;**2007**:43603.
18. Barber AJ. A new view of diabetic retinopathy: a neurodegenerative disease of the eye. *Prog Neuropsychopharmacol Biol Psychiatry* 2003;**27**:283–90.
19. Anttoneti DA, Barber AJ, Bronson SK, et al. Diabetic retinopathy. Seeing beyond glucose-induced microvascular disease. *Diabetes* 2006;**55**:2401–11.
20. Ola MS, Nawaz MI, Siddiquei MM, et al. Recent advances in understanding the biochemical and molecular mechanism of diabetic retinopathy. *J Diabetes Complications* 2012;**26**(1):56–64.
21. Zheng L, Kern TS. Role of nitric oxide, superoxide, peroxynitrite and PARP in diabetic retinopathy. *Front Biosci* 2009;**14**:3974–87.
22. Obrosova IG, Pacher P, Szabó C, et al. Aldose reductase inhibition counteracts oxidative–nitrosative stress and poly(ADP-ribose) polymerase activation in tissue sites for diabetes complications. *Diabetes* 2005;**54**:234–42.
23. Barba I, Garcia-Ramírez M, Hernández C, et al. Metabolic fingerprints of proliferative diabetic retinopathy. An 1H-NMR-based metabonomic approach using vitreous humor. *Invest Ophthalmol Vis Sci* 2010;**51**:4416–21.
24. Zong H, Ward M, Stitt AW. AGEs, RAGE, and diabetic retinopathy. *Curr Diabetes Rep* 2011;**11**:244–52.
25. Galvez MI. Protein kinase C inhibitors in the treatment of diabetic retinopathy. *Rev Curr Pharm Biotechnol* 2011;**12**(3):386–91.
26. Giacco F, Brownlee M. Oxidative stress and diabetic complications. *Circ Res* 2010;**107**:1058–70.
27. Brownlee M. The pathobiology of diabetic complications: a unifying mechanism. *Diabetes* 2005;**54**:1615–25.
28. Padnick-Silver L, Linsenmeier RA. Effect of acute hyperglycemia on oxygen and oxidative metabolism in the intact cat retina. *Invest Ophthalmol Vis Sci* 2003;**44**:745–50.
29. Cao J, McLeod S, Merges CA, et al. Choriocapillaris degeneration and related pathologic changes in human diabetic eyes. *Arch Ophthalmol* 1998;**116**:589–97.
30. Cai L, Kang YJ. Oxidative stress and diabetic cardiomyopathy: a brief review. *Cardiovasc Toxicol* 2001;**1**(3):181–93.
31. Young TA, Cunningham CC, Bailey SM. Reactive oxygen species production by the mitochondrial respiratory chain in isolated rat hepatocytes and liver mitochondria: studies using myxothiazol. *Arch Biochem Biophys* 2002;**405**(1):65–72.

32. Antonetti DA, Klein R, Gardner TW. Diabetic retinopathy. *N Engl J Med* 2012;**366**(13):1227–39.

33. Wirostko B, Wong TY, Simó R. Vascular endothelial growth factor and diabetic complications. *Prog Retin Eye Res* 2008;**27**(6):608–21.

34. Valiatti FB, Crispim D, Benfica C, et al. The role of vascular endothelial growth factor in angiogenesis and diabetic retinopathy. *Arq Bras Endocrinol Metabol* 2011;**55**(2):106–13.

35. Early Treatment Diabetic Retinopathy Study Research Group. Grading diabetic retinopathy from stereoscopic color fundus photographs – an extension of the modified Airlie House classification. ETDRS report number 10. *Ophthalmology* 1991;**98**(Suppl. 5):786–806.

36. Heng LZ, Comyn O, Peto T, et al. Diabetic retinopathy: pathogenesis, clinical grading, management and future developments. *Diabet Med* 2013;**30**(6):640–50.

37. Fonda SJ, Bursell SE, Lewis DG, et al. The relationship of a diabetes telehealth eye care program to standard eye care and change in diabetes health outcomes. *Telemed J E Health* 2007;**13**(6):635–44.

38. White NH, Sun W, Cleary PA, et al. Effect of prior intensive therapy in type 1 diabetes on 10-year progression of retinopathy in the DCCT/EDIC: comparison of adults and adolescents. *Diabetes* 2010;**59**(5):1244–53.

39. Askew DA, Crossland L, Ware RS, et al. Diabetic retinopathy screening and monitoring of early stage disease in general practice: design and methods. *Contemp Clin Trials* 2012;**33**(5):969–75.

40. Giuliari GP. Diabetic retinopathy: current and new treatment options. *Curr Diabetes Rev* 2012;**8**(1):32–41.

41. Pareja-Ríos A, Serrano-García MA, Marrero-Saavedra MD, et al. Guidelines of clinical practice of the SERV (Spanish Retina and Vitreous Society): management of ocular complications of diabetes. Diabetic retinopathy and macular oedema. *Arch Soc Esp Oftalmol* 2009;**84**(9):429–50.

42. Schwartz SG, Flynn Jr HW. Fluocinolone acetonide implantable device for diabetic retinopathy. *Curr Pharm Biotechnol* 2011;**12**(3):347–51.

43. Kumar B, Gupta SK, Saxena R, et al. Current trends in the pharmacotherapy of diabetic retinopathy. *J Postgrad Med* 2012;**58**(2):132–9.

44. Avery RL, Pearlman J, Pieramici DJ, et al. Intravitreal bevacizumab (Avastin) in the treatment of proliferative diabetic retinopathy. *Ophthalmology* 2006;**113**(10):1695.

45. Shin YW, Lee YJ, Lee BR, et al. Effects of an intravitreal bevacizumab injection combined with panretinal photocoagulation on high-risk proliferative diabetic retinopathy. *Korean J Ophthalmol* 2009;**23**(4):266–72.

46. Arevalo JF, Sanchez JG, Lasave AF, et al. Intravitreal Bevacizumab (Avastin) for Diabetic Retinopathy: the 2010 GLADAOF Lecture. *J Ophthalmol* 2011:584238.

47. Moradian S, Ahmadieh H, Malihi M, et al. Intravitreal bevacizumab in active progressive proliferative diabetic retinopathy. *Graefes Arch Clin Exp Ophthalmol* 2008;**246**(12):1699–705.

48. Tsui I, Pan CK, Rahimy E, et al. Ocriplasmin for vitreoretinal diseases. *J Biomed Biotechnol* 2012:354979.

49. Keech AC, Mitchell P, Summanen PA, et al. Effect of fenofibrate on the need for laser treatment for diabetic retinopathy (FIELD study): a randomised controlled trial. *Lancet* 2007;**370**(9600):1687–97.

50. Chaturvedi N, Porta M, Klein R, et al. Effect of candesartan on prevention (DIRECT-Prevent 1) and progression (DIRECT-Protect 1) of retinopathy in type 1 diabetes: randomised, placebo-controlled trials. *Lancet* 2008;**372**(9647):1394–402.

1. INTRODUCTIONS AND OVERVIEWS

MACULAR DEGENERATION

6

Trace Elements, Vitamins, and Lipids and Age-Related Macular Degeneration: An Overview of the Current Concepts on Nutrients and AMD

Gabriëlle H.S. Buitendijk, Caroline C.W. Klaver

Erasmus Medical Center, Rotterdam, The Netherlands

INTRODUCTION

Dietary nutrients have been implicated in the development of age-related macular degeneration (AMD) for years. The first paper reporting a beneficial effect of a nutrient on AMD was published in 1988, in which the National Health and Nutritional Examination Survey (NHANES)-1 study investigated the dietary intake of vitamin A (P trend = 0.058) in AMD.[1] Since then, many more papers on this topic have been published. This chapter will provide an overview of nutrients investigated in relation to AMD.

CAROTENOIDS

There are many different carotenoids; over 600 are known to date. These can be split in two groups: carotenes, which refer to hydrocarbon carotenoids, and xanthophyll, a carotenoid with one or more oxygen groups. Carotenoids are pigments and can be found in chloroplasts and chromoplasts predominantly in plants and algae. Their function is to absorb blue light to protect the plants and algae from photodamage and absorb the light energy for use in photosynthesis. In the eye, lutein and zeaxanthin are xanthophylls that protect the macula from blue and ultraviolet (UV)-light damage. All dietary carotenoids have antioxidant function; α carotene, β carotene, γ carotene, and β cryptoxanthin also have vitamin A activity. These four carotenoids are converted to retinal in herbivores and omnivores.[2] These are reviewed extensively elsewhere in this book.

TRACE ELEMENTS

Trace elements are dietary minerals and are needed in small amounts for normal cell function. Often a trace element is the core of enzymes. However, in large amounts trace elements are toxic.

Zinc

The retina contains high amounts of zinc, suggesting a crucial role for this trace element.[3] Zinc has antioxidant functions and acts as a cofactor in several enzymes including retinol dehydrogenase, an important enzyme in the vitamin A visual cycle.

Dietary Zinc

Several observational studies have investigated the role of dietary zinc and early AMD. Two studies reported an inverse trend for dietary and supplementary intake of zinc approximately 10 years before ophthalmic exam and pigmentary abnormalities.[4,5] However, no association was found for early AMD (odds ratio (OR) 0.91; 95% confidence interval (CI) 0.74 to 1.11). A significant interaction was found between dietary zinc intake and the major susceptibility genes: *CFH* (Y402H) and *AMRS2* (A69S). Carriers of risk variants had a higher risk of AMD in the lower tertile of zinc intake, but risks lowered dramatically when intakes increased (Fig. 6.1).[6]

Supplementation of Zinc

Zinc supplementation had an inverse effect on developing advanced AMD in participants with signs of early

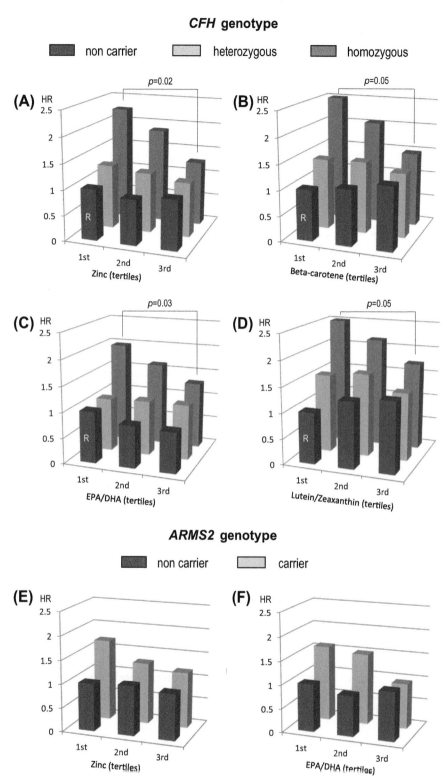

FIGURE 6.1 Gene-environment interactions in the Rotterdam Study. A–D. Joint effect of dietary nutrient intake and *CFH* Y402H genotype on the risk of early age-related macular degeneration (AMD); E–F. Joint effect of dietary nutrient intake and *LOC387715 (ARMS2)* A69S genotype on the risk of early AMD. R is the common reference group. Hazard ratios (HRs) are estimates of the relative risk of early AMD and represent the risk of disease (early AMD vs. no AMD) in the various genetic-environmental risk groups divided by the risk of disease (early AMD vs. no AMD) in the common reference group (R). HRs are estimated with Cox regression analyses and include age, sex, smoking status, and atherosclerosis. DHA: docosahexaenoic acid; EPA: eicosapentaenoic acid.

AMD in one meta-analysis (OR 0.73; 95% CI 0.58 to 0.93). Zinc was supplemented as zinc oxide (80 mg/day; together with cupric oxide, 2 mg/day) in the Age-related Eye Disease Study (AREDS) trial and as zinc sulfate (200 mg/day) in the other trials. These dosages are highly above the recommended upper level of zinc intake (Table 6.1).[7] Zinc sulfate is the most common zinc salt in the diet and supplements; zinc oxide has the longest history as a medicine, especially for skin irritations and wounds. Supplements versus AMD genotype appeared less significant than diet versus genotype. In the AREDS trial, noncarriers of *CFH*

TABLE 6.1 Estimated Average Requirements (EAR), Recommended Dietary Allowances (RDA), Adequate Intakes (AI), and Tolerable Upper Intake Levels

	EAR				RDA/AI				Tolerable Upper Intake			
	Males		Females		Males		Females		Males		Females	
Nutrient	51–70 yrs	> 70 yrs	51–70 yrs	> 70 yrs	51–70 yrs	> 70 yrs	51–70 yrs	> 70 yrs	51–70 yrs	> 70 yrs	51–70 yrs	> 70 yrs
Choline (mg/day)*	ND	ND	ND	ND	*550*	*550*	*425*	*425*	3500	3500	3500	3500
Lutein and zeaxanthin[†]	ND	ND	ND	ND	ND	ND	ND	ND	ND	ND	ND	ND
Lycopene**	ND	ND	ND	ND	ND	ND	ND	ND	ND	ND	ND	ND
Retinol activity equivalents (RAEs)[‡]	625	625	500	500	900	900	700	700	3000	3000	3000	3000
Omega-3 fatty acids[†]	ND	ND	ND	ND	ND	ND	ND	ND	ND	ND	ND	ND
Selenium (µg/day)	45	45	45	45	55	55	55	55	400	400	400	400
Vitamin C (mg/day)	75	75	60	60	90	90	75	75	2000	2000	2000	2000
Vitamin D (µg/day)[§]	10	10	10	10	15	20	15	20	100	100	100	100
Vitamin E (mg/day)[¶]	12	12	12	12	15	15	15	15	1000	1000	1000	1000
Zinc (mg/day)	9.4	9.4	6.8	6.8	11	11	8	8	40	40	40	40

NOTE: EAR is the average daily nutrient intake level estimated to satisfy the needs of 50% of the healthy individuals in a group.
RDA in ordinary type and AI in italic type; ND: no data; yrs: years.
*Only adequate intake for choline has been set. No data available for betaine of methionine.
[†]No RDA, AI, or upper limit has been set for these nutrients.
[‡]1 RAE = retinol 1 µg, β-carotene 12 µg, α-carotene 24 µg, β-cryptoxanthin 24 µg; upper intake was not determined for carotenoids.
[§]Vitamin D as cholecalciferol. Cholecalciferol 1 µg = vitamin D 40 IU; also under the assumption of minimal sunlight.
[¶]As α-tocopherol. α-tocopherol 1 mg = α-tocopherol 1.5 IU.
Taken from United States Department of Agriculture.

(Y402H) showed a more decreased risk after zinc supplementation than carriers, contradicting the findings with diet.[8]

Zinc also has the capability to interact with the complement cascade; it is known to downregulate complement activation. Complement activation is an established mechanism of AMD pathogenesis; hence, this explains the positive effect of high zinc intake. *ARMS2* (A69S) may influence mitochondrial function leading to increased complement activation, which again can be counteracted by zinc.[6] Zinc supplementation at high levels may lead to side effects and complications, including gastrointestinal symptoms, anemia, and more severe genitourinary causes; therefore, caution is warranted.[7,9]

Selenium

Selenium has antioxidant and inflammatory capacities and has, therefore, been investigated in AMD. No positive associations have been found.[10] Selenium is supplemented, but always in combination with other antioxidants, prohibiting the study of the single effect.

VITAMINS

Historically, all vitamins were thought to be amines; hence, in 1921, Kazimierz Funk put together the words 'vital' and 'amine', and composed 'vitamine'. Vitamins are needed only in limited amounts.

Vitamin A

This fat-soluble vitamin was discovered in the beginning of the 20th century in butterfat, and it appeared to be associated with yellow-plant pigments, the carotenoids. Later, vitamin A was found in the retinal tissue of rats and was named 'retinol' after the retina.[11] Total vitamin A includes carotenoids, originating from plants, as well as retinol, derived from animals. Foods containing high levels of retinol are liver and butter (Table 6.2).

The influence of vitamin A on ophthalmologic health has long been known; a vitamin A-deficient diet leads to diseases of the cornea (i.e., xerolphthalmia and keratomalacia) and also to diseases of the retina causing nyctalopia and hemeralopia. Retinal, the active form of retinol, bonds with 'opsine' to form 'rhodopsine', the photosensitive molecule of rod photoreceptors. Aside

TABLE 6.2　Nutrients in Foods

Nutrient	Foods
Astaxanthin	Salmon, trout, shrimp, crayfish
Betaine	Grain products, fish, spinach, sugar beets
α-Carotene	Dark-leafy vegetables (spinach, kale), yellow/orange vegetables (carrots, bell peppers)
β-Carotene	Dark-leafy vegetables (spinach, kale), yellow/orange vegetables (carrots, bell peppers)
β-Cryptoxanthin	Orange rind, egg yolk, papaya, apples
Lutein	Dark-leafy vegetables (spinach, kale), yellow/orange vegetables (carrots, bell peppers)
Lycopene	Red fruits and vegetables (tomatoes, bell peppers, watermelon)
Meso-zeaxanthin	Seafood
Methionine	Poultry, fish, dairy products
Omega-3 fatty acids – ALA	Vegetable oils (flaxseed, canola oil)
Omega-3 fatty acids – DHA/EPA	Oily fish (herring, salmon, sardines, trout)
Omega-6 fatty acids – LA	Vegetable oils (canola oil, safflower oil, corn oil)
Omega-6 fatty acids – AA	Poultry, meat
Resveratrol	Skin of red grapes, other fruits, red wine
Selenium	Shellfish and crustacea, egg yolk
Vitamin A	Liver, butter, cheddar cheese, milk
Vitamin C	Fruits and vegetables (kiwi, peppers, parsley, rose hips)
Vitamin D	Oily fish, dairy products, beef and fish liver*
Vitamin E†	Corn oil, soybean oil, margarine, dressings
Zinc	Fortified cereals, meats, dairy products, nuts, seeds
Zeaxanthin	Dark-leafy vegetables (spinach, kale), yellow/orange vegetables (carrots, bell peppers)

AA; arachidonic acid; ALA: α-linoleic acid; DHA: docosahexaenoic acid; EPA: eicosapentaenoic acid; LA: linoleic acid.
*Levels of vitamin D depend on country. Some foods (mostly dairy products) are fortified with vitamin D.
†There are many different forms of vitamin E; the most common is α-tocopherol.

from the visual function, vitamin A is also needed for growth, survival, and immunity. Supplementation of vitamin A has been investigated since the 1920s and has been associated with a reduction of mortality and morbidity in different infectious diseases.[12,13]

Total Vitamin A and Retinol

Goldberg *et al.* (1988) investigated the data from NHANES-I and suggested a negative association

between dietary vitamin A and AMD (P trend = 0.058). Since then, vitamin A has been investigated by other scientists. One meta-analysis included three prospective cohort studies and found no association of dietary vitamin A intake and incident early AMD (OR 0.98; 95% CI 0.81 to 1.18).[14] Nevertheless, total vitamin A intake including supplement use appeared to be associated with a decrease of pigmentary abnormalities (P trend = 0.01).[4]

For advanced AMD, a trend for total vitamin A intake without supplements was found in one case control study (P trend = 0.05).[15] Inclusion of supplement use increased the association (P trend = 0.02). In this study, retinol per se did not have a significant effect. In the AREDS study, dietary vitamin A intake including retinol had a beneficial effect on advanced AMD.[16]

Vitamin C

L-ascorbic acid, or vitamin C, is a powerful water-soluble antioxidant. Most mammals can synthesize vitamin C from glucose in their liver, except for some species, such as humans. These species lack the enzyme gulonolactone oxidase; therefore, diet is their only source of vitamin C. When diet does not include vitamin C, this can lead to scurvy, a lethal condition if not treated appropriately. Thus, vitamin C is needed to survive.[17]

Vitamin C may play a role in the pathogenesis of AMD. This potent antioxidant could potentially inhibit cellular damage from free radicals provoked by UV exposure in the retina. This hypothesis has been tested in many different study designs.

Dietary Vitamin C

One meta-analysis has pooled the point estimates from four large, high-quality cohort studies. The meta-analysis results, just like the single study results, showed no association of vitamin C with incident early AMD (OR 1.11; 95% CI 0.84 to 1.46). No meta-analysis was carried out for advanced AMD, since not every study had investigated advanced AMD. Those studies that did found no association of vitamin C with incident advanced AMD.[14] One case-control study found a trend for intake of vitamin C versus a lower risk of neovascular AMD (P trend = 0.03).[15]

Supplementation of Vitamin C

There are almost no studies that investigated vitamin C as the only supplement used. Most studies combined vitamin C with other antioxidants.[9] Findings from one randomized, double-masked, placebo-controlled trial showed that there was no association of 500 mg vitamin C supplementation with incident AMD (hazard ratio (HR) 0.99; 95% CI 0.75 to 1.31).[18] Supplementation of vitamin C mostly included a dose of 500 mg. Experiments have shown that plasma of subjects is saturated at doses of 400 mg daily.[17] Increasing the administered dose will probably provide the same results.

In conclusion, no strong associations with AMD have been found for vitamin C, either in the diet or as a supplement.

Vitamin D

Vitamin D plays a role in bone mineralization and it has anti-inflammatory and antiangiogenic properties.[19]

There are different forms of vitamin D. The two main forms are vitamin D2 (ergocalciferol) and vitamin D3 (cholecalciferol), which only differ from each other by one methyl group.[20,21] Vitamin D2 is a plant-derived form and can be produced through UV exposure of foods. A food source naturally rich in vitamin D3 is oily fish (Table 6.2).[20] However, diet only contributes 10% of total serum level of vitamin D.[22] The majority of vitamin D3 is from endogenous-produced vitamin D3, which is synthesized by the skin under the influence of the UV light of the sun. In the liver, vitamin D3 is hydroxylated into 25-hydroxyvitamin D3 (25(OH)D), and, in the kidney, it is hydroxylated to 1,25-dihydroxyvitamin D3 (1,25(OH)D), the active metabolite of vitamin D.[20]

Several studies investigated the role of vitamin D with any form of AMD, as well as early and advanced AMD, and studied the nutrient intake as sunlight exposure, serum levels of 25(OH)D, and intake of vitamin D via diet or supplements.

Vitamin D through Sunlight Exposure

The majority of vitamin D is produced through sunlight exposure of the skin. This environmental factor has been analyzed since the beginning of AMD research, with contradicting results. The combination of a potential harmful effect by deoxyribonucleic acid (DNA) damage as well as a beneficial effect through vitamin D may be an explanation.

Serum Levels of 25-Hydroxyvitamin D3

Parekh et al. (2007) were the first to report an association of serum levels of 25(OH)D with AMD in white non-Hispanics. The three highest quintiles of serum level of 25(OH)D were inversely associated with early AMD after adjustment for age and serum cotinine, a biomarker for exposure to tobacco (P trend = 0.003).[19] Among non-Hispanic black people and Mexican American people a similar, albeit nonsignificant, effect was found. This study also analyzed the different characteristics of early AMD, soft drusen, and pigmentary changes on the entire population. Soft drusen, but not pigmentary changes, were associated with 25(OH) D serum levels (P trend = 0.006 and 0.40, respectively). An association with advanced AMD was not found, most likely due to the small number of advanced cases.

An inverse association of 25(OH)D levels with early AMD, but not with advanced AMD, was found in postmenopausal women under 75 years of age (P trend = 0.01).[22]

A few studies have analyzed hypovitaminosis D as a marker for AMD. No associations within these analyses has been found for neovascular AMD,[23] but a significant association was reported for hypovitaminosis with any form of AMD.[24] Serum level of 25(OH)D lower than 50 nmol/L increased the risk of AMD three-fold (OR 3.03; 95% CI 1.04 to 8.80). Although interesting, these results need to be interpreted with caution since the number of cases and controls was small (n = 65), and the study has not yet been replicated.

Dietary Vitamin D

A positive association with decreased intake was found by study of twins, which selected 28 discordant twin pairs: one twin was diagnosed with advanced AMD while the other twin had no or only signs of early AMD. Vitamin D intake appeared significantly higher in twins with no or only early AMD (P = 0.048).[25] In some countries, dairy products are enriched with vitamin D, such as liquid milk in the US. One study showed that milk consumption was positively correlated with serum vitamin D levels (Pearson correlation coefficient 0.2; P < 0.001), and more than weekly milk consumption was associated with lower odds for early AMD. But only soft drusen were significantly inversely associated with the consumption of milk. No association of advanced AMD with milk consumption was found.[19] Oily fish is another food that is rich in vitamin D, and this may also add to the protective effect that has been found for oily fish.[26,27]

Supplementation of Vitamin D

Results of supplement use are inconclusive.[19,22] Most supplements contain vitamin D2, which is almost 10 times less active than vitamin D3.[21] This may explain the difficulty in finding a strong association. Parekh et al. (2007) found no association in the total population, but after excluding those with high intake due to milk consumption, they did. To disentangle the effects, it would have been more interesting if they analyzed the use of vitamin D supplements in the population not consuming any milk.

Vitamin E

Vitamin E refers to a group of tocopherols and tocotrienols, of which the former is the most profuse form in nature and has the highest biologic activity. Vitamin E deficiency, therefore, mainly refers to α-tocopherol. Vitamin E is a fat-soluble, chain-breaking antioxidant and may protect the retina from damage caused by free radicals. High levels of this nutrient can be found in corn and soybean oil (Table 6.2).[28]

Plasma Levels of α-Tocopherol

In the POLA (Pathologies Oculaires Liées à L'Age) study, no significant association was found between plasma levels of α-tocopherol and advanced AMD (P = 0.08). When α-tocopherol plasma levels were standardized to plasma lipids such as cholesterol and triglycerides, the ratio was negatively associated with advanced AMD (P = 0.004). The ratio was also associated with any signs of early AMD, drusen, or pigmentary changes in subjects free of advanced AMD (P = 0.04). All associations were adjusted for potential confounding factors.[29]

Dietary Vitamin E

Several prospective cohort studies have investigated the role of dietary vitamin E and the risk of developing AMD. Of these studies, three were of high quality and results were pooled quantitatively using meta-analytic methods. An almost significant association was found for vitamin E and incident early AMD (OR 0.83; 95% CI 0.69 to 1.01).[14] For advanced AMD, an association was found in the long-term follow-up of one study; higher intakes of total vitamin E (including supplement use) predicted advanced AMD.[30] This was not confirmed by the other studies.

Supplementation of Vitamin E

One meta-analysis has been published concerning three large trials supplementing vitamin E in healthy subjects.[31] Vitamin E had been supplemented in different dosages of α-tocopherol, 50–402 mg = 75–600 international units (IU) versus placebo for 4–10 years. The complete sample consisted of 40,887 participants of which 20,438 had received α-tocopherol supplementation; 20,449 had been assigned to placebo. In the supplemented group, 405 individuals developed early AMD and 42 advanced AMD. In the placebo group, 458 progressed to early AMD and 31 to advanced AMD. Thus, no associations were found (any type of AMD: RR 0.98; 95% CI 0.89 to 1.08; advanced AMD: RR 1.05; 95% CI 0.80 to 1.39).

Many studies have investigated the role of vitamin E in chronic diseases. High-dose vitamin E supplements (268 mg or greater (400 IU or greater)) have been linked to an increased risk of heart failure in people with vascular disease and diabetes. Supplements given in these trials were up to 15-fold higher than the maximum dietary intake in the cohort studies. Since a protective effect against AMD is dubious for vitamin E, extra care should be taken not to prescribe supplementation to those with cardiovascular risk factors.[32]

LIPIDS

Almost 20% of the dry weight of the retina is accountable to lipids. Over half of all the retinal fatty acids are unsaturated; the majority of these are polyunsaturated fatty acids (PUFAs). Docosahexaenoic acid (DHA) is a PUFA, which can be found in the photoreceptor outer segments and which has been shown to be a survival factor for photoreceptors. The high concentrations of this lipid in the retina and its anti-inflammatory properties[33] suggests a potential role for omega-3 fatty acids in retinal disease.

Dietary Intake of Lipids

Reported outcomes of dietary omega-3 fatty acids, fish consumption (rich in omega-3 fatty acids), and nut consumption have all been shown to be protective against early and late AMD.[33] One meta-analysis showed an inverse effect of high intake of omega-3 fatty acids and late AMD (OR 0.62; 95% CI 0.48 to 0.82).[34] In contrast to omega-3 fat, high levels of trans-fat may potentially increase the risk of AMD, but consistent evidence on this notion is lacking.[33]

One large, multicenter, case-control study found a link between the positive effect of omega-3 fatty acids and linolic acid, an omega-6 fatty acid. The beneficial effect of high intake of omega-3 fatty acids was particularly found in people with a low intake of linolic acid.[35] Similar results have been described for fish intake.[26] This indicates that the ratio of omega-6/omega-3 fatty acids needs to be of the right balance. The ideal ratio is 3:1 or 4:1, while in reality it is 10–50:1 for the average American.[36]

Interaction between AMD genes and intake of DHA and the other omega-3 lipid, eicosapentaenoic acid (EPA), was studied in the Rotterdam Study. Homozygous carriers of the CFH variant Y402H could lower their risk of developing early AMD with high intake of DHA/EPA (P trend = 0.03). The same effect was found for carriers of the risk variant ARMS2 A69S (P trend = 0.01).[6]

A similar effect was found for fish intake and CFH Y402H carriers. Weekly consumption of fish was associated with reduced risk of late AMD for carriers of the risk variant.[27]

Supplementation of Omega-3 Fatty Acids

Very little is known about the benefits of supplementation with omega-3 fatty acids as a single nutrient and AMD. One pilot trial was carried out in 38 patients with drusenoid pigment epithelial detachment in one eye, and the effect of oral supplementation with EPA 720 mg/day and DHA 480 mg/day was compared with no treatment (control group). After 6 months of supplementation, a significant increase of serum levels and red blood cell membranes of EPA and DHA was found, while no change was found in the control group. Since no exudative AMD occurred during the short follow-up time, no inferences on supplementation of EPA/DHA can be made.[37] Before long, the results of the AREDS2 trial are expected. In this trial, the supplementation of lutein/zeaxanthin and/or omega-3 fatty acids and the risk of advanced AMD are being studied.[38]

EPIGENETICS AND NUTRIENTS

Epigenetics refers to functional changes of the genome without a change in DNA nucleotide sequence. This could explain phenotypical differences in diseased monozygotic twins.[39] Various nutrients may cause epigenetic changes. Among those reported to have this capacity, betaine (a choline derivate) and methionine have been associated with advanced AMD.[25] In the US Twin Study of Age-Related Macular Degeneration, Seddon et al. (2011) found high dietary intake of betaine to be inversely associated with grade of AMD (P = 0.009; adjusted for age, smoking, and differences between twins) but not with drusen size, drusen area, or pigment area. Seddon found an inverse association for dietary methionine and drusen area (P = 0.033; but not with AMD grade, drusen size, or pigment area). There was no significant association for dietary choline intake with the macular phenotype.

Betaine, choline, methionine, and homocysteine are involved in the one-carbon metabolism pathway, which occurs in DNA methylation. Choline is oxidized to betaine, which, together with homocysteine, will produce methionine under the influence of vitamin B12. Methionine is an essential amino acid for DNA methylation.[39] Dietary choline, betaine, and methionine influence DNA methylation; higher intakes of choline and betaine showed lower plasma levels of homocysteine and a reduction of inflammatory markers in serum.[39,40] Higher levels of homocysteine or inflammatory markers were found to be risk factors for AMD.[25] Further studies are needed to confirm these associations with AMD.

RESVERATROL

Resveratrol is a natural phenol produced by plants, which may have anticancer and anti-inflammatory effects. It can be found in the skin of red grapes and is also present in red wine, although in very low concentrations. Experiments have indicated that resveratrol may protect retinal pigment epithelial cells from oxidative stress in culture.[41] This nutrient could have a beneficial effect on eye health and help to protect the macula

against AMD. We are not aware of any ongoing trials investigating resveratrol supplementation and AMD.

SUPPLEMENTATION WITH COMBINED NUTRIENTS

Many nutrients have been supplemented together, making it impossible to investigate the role of the individual nutrients and risk of AMD. Studies use different combinations and dosages of supplements, further hampering comparison. Nevertheless, it appears that supplementation of several nutrients combined can be beneficial; they could enforce each other. In the AREDS trial, a double-blind clinical trial, participants were randomly subscribed to oral use of (1) antioxidant (vitamin C 500 mg; vitamin E 400 IU; β-carotene 15 mg), (2) zinc (zinc oxide 80 mg; cupric oxide 2 mg), (3) antioxidants plus zinc, or (4) placebo. Estimates of RR show that those who were taking antioxidant and zinc had a 25% risk reduction of advanced AMD, while the groups that were taking antioxidant alone or zinc alone had a reduction of 17% and 21%, respectively.[9]

TABLE 6.3 Overview of Nutrients and Risk of Macular Degeneration

Nutrient	Early AMD	Late AMD	References
Astaxanthin	NA	NA	-
Betaine	NA	↓?	25
α-Carotene	↑?	↓?	14,42
β-Carotene	~	~	6,14,31
β-Cryptoxanthin	~	~	14,15
Lutein/zeaxanthin	↓↓	↓↓	43,44
Lycopene	↓↓	↓↓	4,45–47
Methionine	NA	↓?	25
Omega-3 fatty acids	↓↓	↓↓	33,34
Omega-6 fatty acids	NA	↑?	33
Resveratrol	NA	NA	-
Selenium	NA	NA	-
Vitamin A (total)	~	↓?	14–16
Vitamin A (retinol)	~	~	15,16
Vitamin C	~	↓?	14,15
Vitamin D	↓?	↓?	19,22,24,25
Vitamin E*	~	~	14,31
Zinc	↓?	↓↓	6,7,14

↓↓: lowering; ↓?: possible lowering; ~: questionable: ↑?: possible increase; AMD: age-related macular degeneration; NA: not able, no results available.
*There are many different forms of vitamin E; the most common is α-tocopherol.

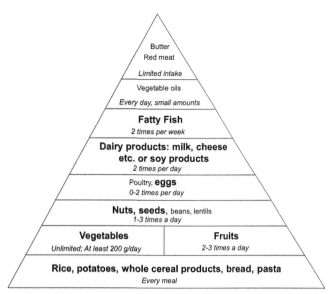

FIGURE 6.2 Nutritional advice for lowering risk on age-related macular degeneration (AMD). This food pyramid gives nutritional advice for lowering risk of AMD. All the foods in bold contain nutrients that have been associated with a lower risk of developing AMD.

However, 'the more the merrier' does not account for all nutrients. Nutrients that are alike mostly use the same uptake and transportation routes throughout the body. High levels of supplementation could then lead to competition between these nutrients, and this could even lead to deficiencies. An example of this is copper deficiency caused by increased zinc consumption.[7,9]

CONCLUSION

Nutrients and AMD have been widely studied. Not all prior questions have been answered, and new questions have already been launched. Nonetheless, from the current findings we conclude that a healthy diet, for instance, as recommended by the US Department of Agriculture in the food pyramid, may help lower the risk of AMD with a special focus on carotenoids (lutein, zeaxanthin, lycopene), zinc, and omega-3 fatty acids (Table 6.3; Fig. 6.2). Less firmly established, but also promising, is the apparent link between a high intake of vitamin D and nutrients influencing epigenetics. Beneficial effects may be particularly present in those carrying a genetic risk of AMD.

A word of concern is necessary. We need to keep in mind that the nutrients mentioned in this chapter are generally not consumed as single items but accompanied by other nutrients. Furthermore, a healthy diet is correlated with other lifestyle factors, which makes it difficult to interpret the positive effect of a healthy diet by itself.

TAKE-HOME MESSAGES

- Vitamins and minerals are essential for normal retinal physiology. Supplementation of these nutrients mostly has a dosage outside the recommended daily allowance and therefore should be prescribed with caution.
- We consume foods, not nutrients.
- Dietary intake of carotenoids (lutein, zeaxanthin, lycopene), zinc, and omega-3 fatty acids can help lower the risk of AMD.
- Supplementation of lutein, zeaxanthin, and zinc can help lower the risk of AMD.
- Vitamin D and nutrients influencing epigenetics are promising new topics for more in-depth research.
- Those at genetic risk should be made aware of their potential to lower the risk of AMD through diet.

References

1. Goldberg J, Flowerdew G, Smith E, Brody JA, Tso MO. Factors associated with age-related macular degeneration. An analysis of data from the first National Health and Nutrition Examination Survey. *Am J Epidemiol* 1988;**128**(4):700–10.
2. Armstrong GA, Hearst JE. Carotenoids 2: genetics and molecular biology of carotenoid pigment biosynthesis. *FASEB J* 1996;**10**(2):228–37.
3. Ugarte M, Osborne NN. Zinc in the retina. *Prog Neurobiol* 2001;**64**(3):219–49.
4. Morris MS, Jacques PF, Chylack LT, Hankinson SE, Willett WC, Hubbard LD, et al. Intake of zinc and antioxidant micronutrients and early age-related maculopathy lesions. *Ophthal Epidemiol* 2007;**14**(5):288–98.
5. VandenLangenberg GM, Mares-Perlman JA, Klein R, Klein BE, Brady WE, Palta M. Associations between antioxidant and zinc intake and the 5-year incidence of early age-related maculopathy in the Beaver Dam Eye Study. *Am J Epidemiol* 1998;**148**(2):204–14.
6. Ho L, van Leeuwen R, Witteman JC, van Duijn CM, Uitterlinden AG, Hofman A, et al. Reducing the genetic risk of age-related macular degeneration with dietary antioxidants, zinc, and omega-3 fatty acids: the Rotterdam study. *Arch Ophthalmol* 2011;**129**(6):758–66.
7. Evans J. Antioxidant supplements to prevent or slow down the progression of AMD: a systematic review and meta-analysis. *Eye (Lond)* 2008;**22**(6):751–60.
8. Klein ML, Francis PJ, Rosner B, Reynolds R, Hamon SC, Schultz DW, et al. CFH and LOC387715/ARMS2 genotypes and treatment with antioxidants and zinc for age-related macular degeneration. *Ophthalmology* 2008;**115**(6):1019–25.
9. Age-Related Eye Disease Study Research Group. A randomized, placebo-controlled, clinical trial of high-dose supplementation with vitamins C and E, beta carotene, and zinc for age-related macular degeneration and vision loss: AREDS report no. 8. *Arch Ophthalmol* 2001;**119**(10):1417–36.
10. Eye Disease Case-Control Study Group. Antioxidant status and neovascular age-related macular degeneration. *Arch Ophthalmol* 1993;**111**(1):104–9.
11. Yudkin AM. The presence of vitamin A in the retina. *Trans Am Ophthalmol Soc* 1931;**29**:263–72.
12. Semba RD. Vitamin A as "anti-infective" therapy, 1920–1940. *J Nutrition* 1999;**129**(4):783–91.

13. Semba RD. Vitamin A and immunity to viral, bacterial and protozoan infections. *Proc Nutrition Soc* 1999;**58**(3):719–27.

14. Chong EW, Wong TY, Kreis AJ, Simpson JA, Guymer RH. Dietary antioxidants and primary prevention of age related macular degeneration: systematic review and meta-analysis. *BMJ* 2007;**335**(7623):755.

15. Seddon JM, Ajani UA, Sperduto RD, Hiller R, Blair N, Burton TC, et al. Dietary carotenoids, vitamins A, C, and E, and advanced age-related macular degeneration. Eye Disease Case-Control Study Group. *JAMA* 1994;**272**(18):1413–20.

16. SanGiovanni JP, Chew EY, Clemons TE, Ferris 3rd FL, Gensler G, Lindblad AS, et al. The relationship of dietary carotenoid and vitamin A, E, and C intake with age-related macular degeneration in a case-control study: AREDS Report No. 22. *Arch Ophthalmol* 2007;**125**(9):1225–32.

17. Padayatty SJ, Katz A, Wang Y, Eck P, Kwon O, Lee JH, et al. Vitamin C as an antioxidant: evaluation of its role in disease prevention. *J Am Coll Nutr* 2003;**22**(1):18–35.

18. Christen WG, Glynn RJ, Sesso HD, Kurth T, Macfadyen J, Bubes V, et al. Vitamins E and C and medical record-confirmed age-related macular degeneration in a randomized trial of male physicians. *Ophthalmology* 2012;**119**(8):1642–9.

19. Parekh N, Chappell RJ, Millen AE, Albert DM, Mares JA. Association between vitamin D and age-related macular degeneration in the Third National Health and Nutrition Examination Survey, 1988 through 1994. *Arch Ophthalmol* 2007;**125**(5):661–9.

20. Lips P. Vitamin D physiology. *Prog Biophysics Mol Biol* 2006;**92**(1):4–8.

21. Houghton LA, Vieth R. The case against ergocalciferol (vitamin D2) as a vitamin supplement. *Am J Clin Nutrition* 2006;**84**(4):694–7.

22. Millen AE, Voland R, Sondel SA, Parekh N, Horst RL, Wallace RB, et al. Vitamin D status and early age-related macular degeneration in postmenopausal women. *Arch Ophthalmol* 2011;**129**(4):481–9.

23. Day S, Acquah K, Platt A, Lee PP, Mruthyunjaya P, Sloan FA. Association of vitamin d deficiency and age-related macular degeneration in medicare beneficiaries. *Arch Ophthalmol* 2012;**130**(8):1070–1.

24. Graffe A, Milea D, Annweiler C, Beauchet O, Mauget-Faysse M, Kodjikian L. Association between hypovitaminosis D and late stages of age-related macular degeneration: a case-control study. *J Am Geriatr Soc* 2012;**60**(7):1367–9.

25. Seddon JM, Reynolds R, Shah HR, Rosner B. Smoking, dietary betaine, methionine, and vitamin D in monozygotic twins with discordant macular degeneration: epigenetic implications. *Ophthalmology* 2011;**118**(7):1386–94.

26. Smith W, Mitchell P, Leeder SR. Dietary fat and fish intake and age-related maculopathy. *Arch Ophthalmol* 2000;**118**(3):401–4.

27. Wang JJ, Rochtchina E, Smith W, Klein R, Klein BE, Joshi T, et al. Combined effects of complement factor H genotypes, fish consumption, and inflammatory markers on long-term risk for age-related macular degeneration in a cohort. *Am J Epidemiol* 2009;**169**(5):633–41.

28. Brigelius-Flohe R, Traber MG, Vitamin E. function and metabolism. *FASEB J* 1999;**13**(10):1145–55.

29. Delcourt C, Cristol JP, Tessier F, Leger CL, Descomps B, Papoz L. Age-related macular degeneration and antioxidant status in the POLA study. POLA Study Group. Pathologies Oculaires Liees a l'Age. *Arch Ophthalmol* 1999;**117**(10):1384–90.

30. Tan JS, Wang JJ, Flood V, Rochtchina E, Smith W, Mitchell P. Dietary antioxidants and the long-term incidence of age-related macular degeneration: the Blue Mountains Eye Study. *Ophthalmology* 2008;**115**(2):334–41.

31. Evans JR, Lawrenson JG. Antioxidant vitamin and mineral supplements for preventing age-related macular degeneration. *Cochrane Database Syst Rev* 2012;**6**:CD000253.

32. Lonn E, Bosch J, Yusuf S, Sheridan P, Pogue J, Arnold JM, et al. Effects of long-term vitamin E supplementation on cardiovascular events and cancer: a randomized controlled trial. *JAMA* 2005;**293**(11):1338–47.

33. Kishan AU, Modjtahedi BS, Martins EN, Modjtahedi SP, Morse LS. Lipids and age-related macular degeneration. *Surv Ophthalmol* 2011;**56**(3):195–213.

34. Chong EW, Kreis AJ, Wong TY, Simpson JA, Guymer RH. Dietary omega-3 fatty acid and fish intake in the primary prevention of age-related macular degeneration: a systematic review and meta-analysis. *Arch Ophthalmol* 2008;**126**(6):826–33.

35. Seddon JM, Rosner B, Sperduto RD, Yannuzzi L, Haller JA, Blair NP, et al. Dietary fat and risk for advanced age-related macular degeneration. *Arch Ophthalmol* 2001;**119**(8):1191–9.

36. Seddon JM, George S, Rosner B. Cigarette smoking, fish consumption, omega-3 fatty acid intake, and associations with age-related macular degeneration: the US Twin Study of Age-Related Macular Degeneration. *Arch Ophthalmol* 2006;**124**(7):995–1001.

37. Querques G, Benlian P, Chanu B, Portal C, Coscas G, Soubrane G, et al. Nutritional AMD treatment phase I (NAT-1): feasibility of oral DHA supplementation in age-related macular degeneration. *Eur J Ophthalmol* 2009;**19**(1):100–6.

38. Chew EY, Clemons T, Sangiovanni JP, Danis R, Domalpally A, McBee W, et al. The Age-Related Eye Disease Study 2 (AREDS2): Study Design and Baseline Characteristics (AREDS2 Report Number 1). *Ophthalmology* 2012;**119**(11):2282–9.

39. Anderson OS, Sant KE, Dolinoy DC. Nutrition and epigenetics: an interplay of dietary methyl donors, one-carbon metabolism and DNA methylation. *J Nutr Biochem* 2012;**23**(8):853–9.

40. Rajaie S, Esmaillzadeh A. Dietary choline and betaine intakes and risk of cardiovascular diseases: review of epidemiological evidence. *ARYA Atheroscler* 2011;**7**(2):78–86.

41. Baur JA, Sinclair DA. Therapeutic potential of resveratrol: the *in vivo* evidence. *Nat Rev Drug Discov* 2006;**5**(6):493–506.

42. Zhou H, Zhao X, Johnson EJ, Lim A, Sun E, Yu J, et al. Serum carotenoids and risk of age-related macular degeneration in a Chinese population sample. *Invest Ophthalmol Vis Sci* 2011;**52**(7):4338–44.

43. Ma L, Dou HL, Wu YQ, Huang YM, Huang YB, Xu XR, et al. Lutein and zeaxanthin intake and the risk of age-related macular degeneration: a systematic review and meta-analysis. *Br J Nutr* 2012;**107**(3):350–9.

44. Sabour-Pickett S, Nolan JM, Loughman J, Beatty S. A review of the evidence germane to the putative protective role of the macular carotenoids for age-related macular degeneration. *Mol Nutr Food Res* 2012;**56**(2):270–86.

45. Mares-Perlman JA, Brady WE, Klein R, Klein BE, Bowen P, Stacewicz-Sapuntzakis M, et al. Serum antioxidants and age-related macular degeneration in a population-based case-control study. *Arch Ophthalmol* 1995;**113**(12):1518–23.

46. Simonelli F, Zarrilli F, Mazzeo S, Verde V, Romano N, Savoia M, et al. Serum oxidative and antioxidant parameters in a group of Italian patients with age-related maculopathy. *Clin Chim Acta* 2002;**320**(1–2):111–5.

47. Cardinault N, Abalain JH, Sairafi B, Coudray C, Grolier P, Rambeau M, et al. Lycopene but not lutein nor zeaxanthin decreases in serum and lipoproteins in age-related macular degeneration patients. *Clin Chim Acta* 2005;**357**(1):34–42.

The Role of Lipids and Lipid Metabolism in Age-Related Macular Degeneration

Bobeck S. Modjtahedi, Amar U. Kishan, Shilpa Mathew, Lawrence S. Morse

UC Davis Eye Center, Sacramento, CA

INTRODUCTION

Age-related macular degeneration (AMD) is a disease of significant public health importance because it is the leading cause of severe visual impairment in individuals over the age of 65 years in developed countries.[1] As the proportion of elderly individuals increases in the industrialized world, the burden of this disease will only be magnified. In countries such as the US, advanced AMD cases are projected to increase by 50% by the year 2020.[2,3] AMD is a multifactorial disease that results from non-modifiable factors, such as genetics and age, as well as modifiable factors, such as nutrition, smoking, and exercise.[4] Consequently, much effort has been directed toward discovering principles of pathogenesis in order to identify modifiable risk factors and to develop treatments and preventive measures. Though some modifiable risk factors have been identified such as smoking,[5] the exact pathogenesis of AMD remains unknown and treatment options are mostly limited to exudative AMD.[6,7] As such, there has been an ongoing need for further investigation into the pathogenesis of AMD and for the development of effective prevention measures and treatments.

Researchers have proffered several clues about the pathogenesis of AMD, but the potential vascular basis of AMD first suggested by Verhoeff and Grossman in 1937[6,8] has gained momentum with time and has become increasingly popular.[8–16] Although the vascular basis for AMD remains debated,[1,17–20] many researchers have suggested that choroidal vascular disease develops from an atherosclerosis-like process and leads to retinal dysfunction and the development of AMD.[1,21,22] Due to this proposed link, modern inquiry has focused on defining the roles of dietary fat and lipid metabolism as risk factors for the progression of AMD. This chapter will explore these novel investigations in both animal and human models and examine the relationship between AMD and lipids, with a particular emphasis on the benefit of omega-3 (n-3) fatty acid (FA) supplementation.

Understanding the relationship between AMD and lipids requires a basic working knowledge of the biochemistry of lipids. Accordingly, relevant biochemistry has been integrated into the discussion. The information provided is not exhaustive but will provide an adequate framework in which to understand the concepts discussed in this chapter. The chapter first begins with an overview of lipids in the retina and finishes with an examination of the current research involving AMD and lipids and its implications.

RETINAL LIPIDS

Elongation of Very Long Chain Fatty Acids-4 (ELOV4) and Lipids

The retina is highly enriched with very long chain polyunsaturated fatty acids (VLC-PUFAs) containing glycerophospholipids. Though little is known about the synthesis and function of these retinal lipids, studies have shown that elongation of very long chain fatty acids-4 (ELOV4), an FA elongase protein, is involved in the synthesis of VLC-PUFAs. Mutations in ELOV4 have been identified in the pathogenesis of Stargardt disease, a juvenile form of macular degeneration, and studies have shown that ELOV4 is essential for VLC-PUFA synthesis.[23] Additionally, studies have shown that pathogenic mutations in the ELOV4 gene in mice result in reduced levels of retinal C28–C36 acyl phosphatidylcholines (PCs)[24] and photoreceptor loss, disorganized inner and outer segments, and diminished electroretinogram (ERG) responses in pigs.[25] Though most research in this area has focused on the linkage to Stargardt disease, some of the observed retinal changes caused by the mutation in ELOV4 have been implicated in AMD and may eventually be linked to its pathogenesis.

Lipid Peroxidation

Oxidative stress has been proposed as a possible mechanism for AMD because the retina has high lipid content and suffers prolonged exposure to radiant energy.[26–28] Twenty percent of the retina's dry weight is composed of lipids, with 50% or more being unsaturated FAs. Approximately 60% of the unsaturated FAs in the retina are polyunsaturated fatty acids (PUFAs). The region of the retina called the photoreceptor outer segment (POS) contains more of the long-chain n-3 FA docosahexaenoic acid (DHA; Fig. 7.1), a type of PUFA, than any other mammalian cell membrane.[29,30]

Eukaryotic cells utilize oxidation–reduction reactions in the metabolism of fuels. Reduced nicotinamide adenine dinucleotide (NADH) and reduced flavin adenine dinucleotide (FADH$_2$) are used to carry electrons produced in these reactions to the mitochondria. In the mitochondria, these electrons are donated to the electron transport chain, and the process ultimately results in the production of a proton gradient that enables the production of adenosine triphosphate (ATP) to maintain the organism's energy requirements. This process is not 100% efficient and as a result, electrons can fall off the chain and form highly reactive oxygen species (ROS). The presence of ROS creates a domino effect by producing sequentially more free radicals by undergoing oxidation–reduction reactions in the quest for stability (Fig. 7.2). This mechanism is particularly important in tissues such as the retina, which have high oxygen use for two reasons. The first is that irradiation can generate ROS via photo-oxidative stress.[27,31] The second is that because the retina has a high concentration of PUFAs, it is more susceptible to the effects of oxidation (also referred to as lipid peroxidation).

PUFAs are particularly susceptible to lipid peroxidation because they are rich in double bonds, which act as electron donors in the oxidation reduction reactions (see Fig. 7.3). The lipid peroxidation reactions can also produce reactive aldehyde intermediates that can function as secondary toxic messengers.[32,33] For example, products such as maldonialdehyde (MDA), carboxyethylpyrrole (CEP; formed from the peroxidation of DHA), and 4-hydroxynonenal (HNE; formed from the peroxidation of linoleic acid, an n-6 FA) can form adducts with cellular proteins, creating advanced lipid peroxidation end products (ALEs) that may be connected to AMD pathogenesis.[32] Proteomic analysis has revealed that drusen proteins in AMD patients have an increased abundance of CEP adducts.[34] Lysosomal degradation *in vitro* has decreased as a result of modification of POS proteins by HNE,[35] which results in a collection of material that is ultimately transcytosed and may contribute to the formation of drusen.[36]

ALEs may also be involved in AMD by initiating an inflammatory response by acting as an antigen. Mice immunized with mouse serum albumin conjugated to CEP have developed increased drusen and other changes mimicking those in dry AMD,[37] and anti-CEP autoantibodies have been found in the serum of AMD patients.[38] Studies have also shown that molecules normally present in the inflammatory response, immunoglobulin G, and terminal C5b-9 complement complexes have been found in drusen.[39] Other research has shown that AMD has been associated with genetic variations in complement proteins,[40–44] thereby implicating a potential autoimmune mechanism for AMD by which proteins instigate an autoantibody response leading to complement fixation and local inflammatory damage that can lead to AMD. Shaw *et al.* recently demonstrated that the increased risk for AMD conferred by some complement factor H genotypes may be from interactions from oxidized phospholipids.[45]

Retinal Lipoproteins

It is important to understand the morphologic differences between retinal and retinal pigment epithelium (RPE) lesions. Basal laminar deposits (BLamDs) are deposits between RPE cells and their basement membranes, which when thin and homogeneous are age-related changes. However, when BLamDs thicken and

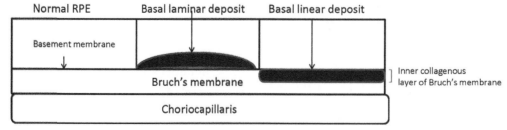

FIGURE 7.1   Diagram of basal laminar deposits (BLamDs) and basal linear deposits (BLinDs). BLamDs form between the retinal pigment epithelium (RPE) and its basement membrane whereas BLinDs form in the inner collagenous zone of Bruch's membrane.

accumulate heterogeneous debris, lipoproteins, inflammatory proteins, and 'long spacing collagen', they are associated histologically with AMD.[46] Conversely, basal linear deposits (BLinDs) are very strongly considered histologic markers for AMD and are accumulations of deposits within the inner collagenous layer of Bruch's membrane (BM).[46] Eventually large drusen may be seen when enough debris accumulates within the BLinDs to form a dome or mound[46] (see Fig. 7.1).

Studies have also shown a connection between retinal lipoproteins and AMD. RPE constitutively produces an apolipoprotein-B (ApoB) containing lipoprotein particles. It is hypothesized that as people age, deposition of these particles leads to the creation of a hydrophobic barrier, impairing nutrient exchange via the choriocapillary network. This process may initiate AMD.[47] Determining the source of lipids deposited in BM has been an area of expanding research. Some have suggested that the RPE deposits lipids derived from phagocytosed POS membrane components into BM.[48–50] Compositional analysis has demonstrated a relative paucity of DHA and abundance of linoleate in BM lipoproteins, which suggests that photoreceptor outer segments (which are rich in DHA) do not directly form these lipoproteins.[46,51] Therefore, if lipoproteins are derived from photoreceptor outer segments they must undergo significant

processing at the RPE level that allows them to take their final form. Similarly, differences between BM and plasma lipoproteins in cholesterol distribution, density profile, and morphology suggest that plasma lipoproteins are unlikely to be simply deposited in the BM.[52] Curcio et al. have suggested that plasma lipoproteins are significant upstream sources to an ApoB lipoprotein that originates from the RPE.[51] Curcio et al. have provided a model ('Oil spill') wherein the RPE secretes an unusual composition ApoB-lipoprotein, which accumulates with age and forms a lipid wall that is a precursor of basal linear deposits. These lesions then interact with ROS, forming peroxidized lipids that are pro-inflammatory and may elicit neovascularization.[51] The increasing knowledge of the origin of BM lipoproteins may help explain the imperfect associations with systemic hyperlipidemia and mixed results with statin therapy (see below). A better understanding of BM lipoprotein source may allow for development of superior animal models. It may also provide a better framework with which to tackle the problem of BM lipoprotein deposition therapeutically. Diet may play a role by limiting some upstream sources of lipoproteins. Novel approaches directed at removal of lipid from the BM and modulation of ApoB lipoprotein deposition from the RPE to BM may also be developed but are at the very early stages of development.[51]

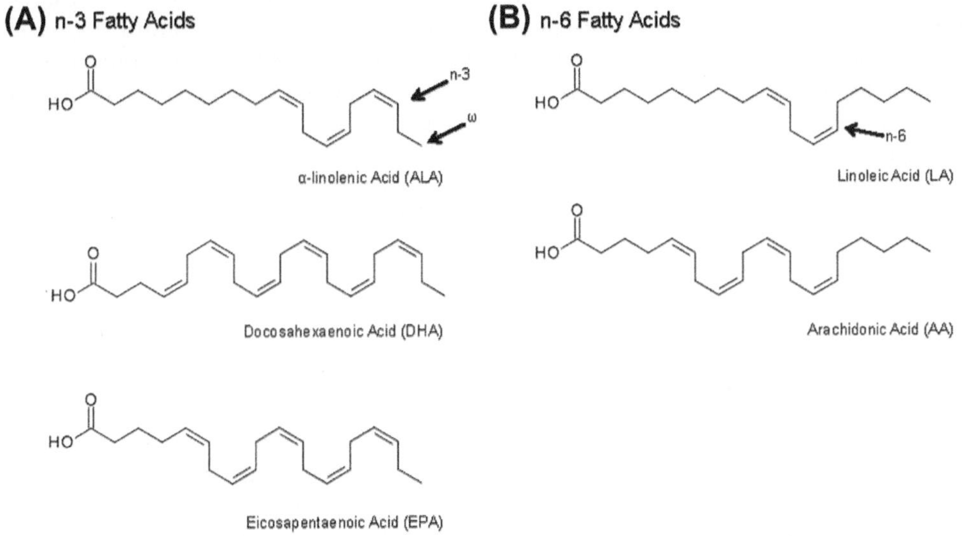

(A) n-3 Fatty Acids

α-linolenic Acid (ALA)

Docosahexaenoic Acid (DHA)

Eicosapentaenoic Acid (EPA)

(B) n-6 Fatty Acids

Linoleic Acid (LA)

Arachidonic Acid (AA)

FIGURE 7.2 n-3 and n-6 fatty acids: chemical structures of common n-3 (A) and n-6 (B) fatty acids. The omega carbon is the methyl end of the carbon backbone of the fatty acid, as depicted. n-3 fatty acids have their final double bond at the n-3 position, whereas n-6 fatty acids have their final double bond at the n-6 position. *Source: courtesy of Kishan* et al.[98] *reprinted with permission.*

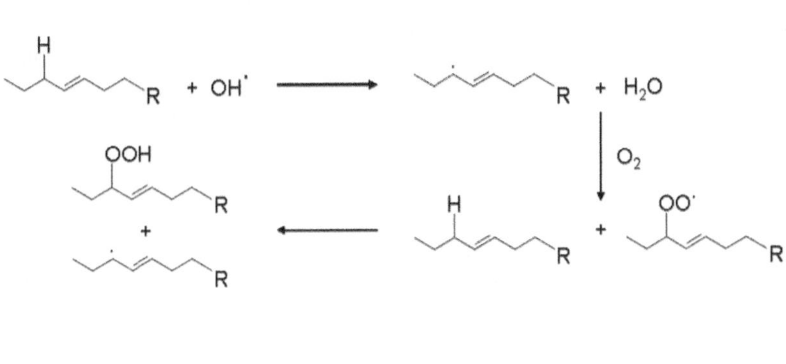

FIGURE 7.3 Lipid peroxidation schematic: lipid peroxidation begins with an initiation step, in which an unstable radical (here, the hydroxyl radical) abstracts an electron from an unsaturated lipid species. This generates a reactive lipid radical, which can react with molecular oxygen to form a lipid peroxyl radical. The lipid peroxyl radical can abstract an electron from another unsaturated lipid species, allowing for a chain reaction and the generation of numerous lipid peroxide species. *Source: courtesy of Kishan* et al.[98] *reprinted with permission.*

The clearance of retinal lipids may rely on reverse cholesterol transport (RCT). This hypothesis is supported by the presence of RCT components such as ATP binding cassette transporter AI,[53] apolipoprotein E (ApoE),[54] and scavenger receptor BI (SR-BI)[53,55] in the retina. High-density lipoprotein (HDL) has been shown to help lipids of POS origin exit from the basal surfaces of RPE cells in culture[56] by crossing the BM. There is a large reduction in the conductivity of the BM with age.[57] This results in a lipid buildup, which may contribute to the development of AMD. Methods of increasing RCT in the RPE and BM are currently under investigation.[58]

An enzyme involved in HDL metabolism, hepatic lipase gene (LIPC), has also been found to be associated with advanced AMD.[59] The T allele of the *LIPC* single nucleotide polymorphism, which is associated with increased HDL, was found to be protective against AMD. This association is unlikely to be the result of direct HDL modification because there was a trend between two HDL raising alleles and increased AMD risk. Additionally, studies examining the relationship between HDL and AMD have been contradictory.[59] As such, it has been hypothesized that *LIPC* may have a pleiotropic effect.

Dietary Fats and the Retina

The high content of DHA in the retina suggests that n-3 FAs play an essential role in retinal physiology. Other types of n-3 FAs include the long-chain eicosapentaenoic acid (EPA) and the short-chain α-linolenic acid (ALA). A related group of PUFAs, the omega-6 (n-6) FAs, includes the long-chain arachidonic acid (AA) and the short-chain linoleic acid (LA) (see Fig. 7.2 for chemical structures of all listed FAs). Many of these can be obtained via the diet; DHA and EPA can be obtained from marine oils and fish, ALA can be consumed in canola and flaxseed oil, LA can be found in oils such as safflower and corn oil, and AA can be obtained from meat. FAs are termed 'essential' if the delta-6-desaturase enzyme present in the body can act upon them to initiate synthesis of DHA/EPA or AA. Since ALA can result in DHA/EPA and LA can result in AA, these two are deemed 'essential'.

Evidence suggests that supplementation with n-3 FAs, particularly DHA and EPA, can benefit rheumatoid arthritis,[60] depression,[61,62] cardiovascular disease,[63,64] dementia and Alzheimer disease,[62] and neurologic development.[65,66] Research has shown that DHA and EPA breakdown yields anti-inflammatory intermediaries known as resolvins and protectins. Resolvins are synthesized from EPA and DHA via the action of cyclo-oxygenase-2. Protectins are derived from DHA via 15-lipoxygenase, and neuroprotectin D$_1$ has been shown to have an anti-apoptotic effect on RPE cells exposed to oxidative stress.[66,67]

DHA supplementation has been shown to be effective in the treatment of at least one human eye disease, Stargardt-like retinal dystrophy,[68] and it has also been shown to be a survival factor for photoreceptors.[69–72] An imbalanced n-6/n-3 ratio has been shown to lead to atypical ERG in animal models.[73,74] Additionally, the Age Related Eye Disease Study (AREDS) established the efficacy of dietary supplementation as a preventive measure for AMD by showing that high doses of vitamin C, vitamin E, beta-carotene, and zinc were protective against moderately severe early AMD to advanced AMD.[10]

Evidence also suggests that the *in vivo* derivatives of n-6 FAs are pro-inflammatory while *in vivo* derivatives of n-3 FAs have antioxidative, anti-inflammatory, and neuroprotective effects. Thus, assessing the n-3/n-6 ratio is important in determining risk of developing AMD. The data in these studies taken together imply that dietary modification and ingestion of n-3 FAs may be essential in the treatment of retinal diseases such as AMD.

MECHANISM AND PATHOGENESIS OF AGE-RELATED MACULAR DEGENERATION

Diet and Age-Related Macular Degeneration

The international obesity epidemic should be of great concern to the ophthalmic community as a growing number of studies have shown how dietary fat can induce AMD-like changes. In animal studies, mice fed a diet containing 15% fat over 30 weeks developed retinal changes including atrophy of the RPE and buildup of lipid-like droplets in the cytosol. Mice on a normal diet did not develop these changes even after 45 weeks.[75] The presence of a continuous layer of BLamDs is also indicated in early AMD, as the deposits can serve as a precursor to soft drusen. In late AMD, the BLamDs become amorphous and thick. Mice fed a high-fat diet for 8 months were more likely to have increased electron-lucent debris and a thicker BM.[76] This effect was magnified when the eyes of the mice were exposed to laser. Those on a high-fat diet accumulated significantly more BLamD-like material in BM. Age may only worsen risk, as aged mice on normal diets developed mild subretinal deposits, while many mice at 9 months or older on high-fat diets developed moderately severe BLamDs, especially if exposed to photo-oxidative stress.[77]

In human models, data also suggest that high total fat[78–80] and trans-fat[78,81–83] intake may increase risk of AMD while dietary n-3 FA intake,[78,81,82,84–94] fish consumption,[78,79,81,83,85–88,91,92,95,96] and nut consumption[86,92] are protective toward AMD. Study results concerning unsaturated and saturated fat intake are more variable.[80–85,87,92,95,97] Kishan *et al.* provide a table summary of statistically significant associations described in these studies.[98]

The first study in humans that evaluated the relationship between dietary fat and AMD showed that the high intake of total fat, saturated fat, and cholesterol was significantly associated with increased odds of early AMD. This study saw no correlation between intake of oleate, linoleate, and seafood (a proxy for n-3 FAs) and early or late AMD.[80] However, the retrospective nature of data collection and overall low intake of seafood in the study population may have hidden any relationship with n-3 FAs. Serum cholesterol and HDL levels were found not to be associated with AMD risk in India while dyslipidemia was found not to be associated with AMD in a Chinese cohort.[99,100] These findings underline the complex and multifactorial nature of AMD, as additional environmental/dietary and ethnic/genetic factors may be at play.

A subsequent cross-sectional study of the Blue Mountain Eye Study population showed that subjects with higher energy adjusted intakes of cholesterol were significantly more likely to have late AMD and those with higher intakes of monounsaturated fats were more likely to have early AMD. This study also found that higher fish consumption was associated with decreased odds of late AMD; however, they noted that there was a certain level of protective threshold after which higher levels of consumption did not confer greater benefit.[95] Another investigation had similar results, stressing the importance of the type of fat in assessing AMD risk.[78]

Recent investigations have sought to investigate the roles of specific types of fat more thoroughly, the interactions between those relevant types of fat, and the concept of protective threshold in assessing AMD risk. The role of the n-6/n-3 ratio has been investigated with implications that higher n-3 FA intake confers a lower risk of neovascular (NV) AMD among those with a low LA intake. Interestingly, neither animal fat nor saturated fat or cholesterol was found to be associated with an increased risk of AMD, while vegetable, monounsaturated, and polyunsaturated fats were associated.[82] The increased risk of AMD associated with polyunsaturated fat consumption was thought to be possibly derived from LA intake. Another study confirmed the importance of the n-6/n-3 ratio and found that fish intake was protective against AMD progression only among those with low LA intake.[85] The results further indicated that intake of higher total fat, vegetable fat (but not animal fat), saturated, monounsaturated, polyunsaturated, and trans-fat increased the risk of AMD progression and that nuts were protective against AMD while processed baked goods increased risk of AMD. Again, the effect of fish appeared to have a threshold effect, with protection manifesting at intake levels of more than two servings per week. One study that showed similar results noted high n-3 FA intake conferred a protective fraction of 22%.[86]

Others have re-examined the data from the Blue Mountain Eye Study and found that the higher intake of total n-3 FAs in general and of ALA in particular were associated with a decreased the risk of AMD. In contrast, participants with lower than normal intakes of either total or monounsaturated fat were at increased risk of developing AMD.[87] They also further quantified the threshold concept by stating that the intake of fish more than once per week was protective against early AMD, while intake greater than three times per week was protective against late AMD. One study found that eating oily fish at least once per week and high levels of DHA and EPA intake was protective toward NV AMD.[88]

Several analyses based on AREDS data have further elucidated a protective role for n-3 FA specifically. One study found that a higher intake of total dietary n-3 FAs, DHA, total fish, and broiled/baked fish were protective toward NV AMD, while high dietary AA intake resulted in higher risk. In this study, DHA was found to have a protective effect even at the highest quintiles of LA intake, but n-3 FA intake was still most protective at lower levels of AA intake.[84] Another study showed a reduced likelihood of progression to central geographic atrophy (CGA) but not NV AMD among study participants with the highest levels of EPA and EPA plus DHA consumption. However, this is the only report highlighting a differential protection for CGA and NV AMD,[90] and more recent studies have expanded upon these baselines and results by reporting that higher intake of EPA, DHA, and EPA plus DHA were protective against progression to either CGA or NV AMRD.[89]

Additional focus has been placed on the aggregate effects of nutrients from a composite diet rather than the effect of a single nutrient on the risk of AMD. Other nutritional factors, such as glycemic index (GI), are gaining increasing attention for their role in AMD specifically and inflammatory pathways generally.[101,102] Mice with lower GI diet demonstrate significantly reduced frequency and severity of age-related retinal lesions that precede AMD compared with mice with higher GI diet.[101,102] Additionally, high GI diets have been found to increase the risk of photoreceptor abnormalities.[101] Higher GI diets are felt to confer this risk by exerting glycative stress, which in turn results in the deposition of toxic advanced glycation end products in the retina.[101,102] One study examined AREDS participants and assigned a percentile rank score to vitamins C and E, zinc, lutein/zeaxanthin, DHA, EPA, and low-dietary GI and summed these together to create a compound score. Higher compound scores were associated with lower risk of both drusen and AMD.[91] The study also reported that a subgroup of AREDS participants with the highest intakes of DHA and

EPA was associated with a lower risk of progression to advanced AMD. In looking at composite diets, they noted there was an antagonistic relationship between the AREDS supplement of zinc, vitamin C, and vitamin E and dietary DHA intake with respect to early (but not late) AMD development. Consequently, the authors concluded that DHA supplementation by itself may be protective against early AMD while co-supplementation with AREDS supplements may be beneficial in preventing late AMD. One study showed that increased intake of n-3 FAs combined with lower intake of LA both improved the enrichment of n-3 FA-containing tissues such as the neurosensory retina and RPE and upregulated genes involved in lipid trafficking.[103] Connection between dietary GI and DHA and EPA consumption has also been investigated. Individuals with the lowest GI group and the highest DHA or EPA intake had half the risk of progression to advanced AMD compared with other individuals in the study. The authors explained that this effect might be because the compensatory hyperlipidemia following intake of a high GI meal plays a role in AMD pathogenesis and that n-3 FAs may modulate this process.[93]

Finally, some studies have looked at the role of lipids in the formation of macular pigment optical density (MPOD), which exhibits neuroprotective function against oxidative stress and inflammation, both of which have been identified in the pathogenesis of AMD. One such study indicated a positive correlation between MPOD and plasma n-3 long-chain PUFAs and a negative correlation with two types of n-6 long-chain PUFAs.[4] HDL has also been found to be significantly related to serum lutein level, serum zeaxanthin, and MPOD, which was also associated with total cholesterol.[104] Though it must be noted that some studies have disagreed with the results of the dietary-themed studies mentioned above,[105] it is possible that study methods and design may account for these differences.[97]

Though many studies have provided evidence for a protective role of n-3 supplementation in the context of AMD risk, the consensus is still to refrain from making any clinical recommendations. Pooled data from many of the previously mentioned studies were examined. The authors reported that dietary intake of n-3 FA was associated with a reduced risk of late AMD while fish intake at least twice a week was associated with a reduced risk of both early and late AMD. However, the authors concluded that there was insufficient evidence to support routine supplementation of n-3 FAs for AMD prevention.[81] An earlier review reached similar conclusions and stated that no clinical recommendations should be made.[106] AREDS2 examined supplementation of the original AREDS formula with several combinations of new nutrients, including DHA+EPA. Serum levels of DHA and EPA increased 30–40% and 90–120%,

respectively, during the study in those randomized to receive DHA+EPA; however, this did not affect visual acuity or confer a decreased risk of progression to advanced AMD. Participants in the AREDS2 trial had excellent nutritional status, which may have influenced the results. Eleven percent of subjects were taking DHA+EPA supplements on their own. Additionally, compared to participants in the National Health and Nutrition Examination Survey (NHANES), AREDS2 subjects had statistically significant higher serum levels of DHA and EPA (p<0.01). It is foreseeable that supplementation may not benefit those with high baseline intake in DHA+EPA and as such the results of AREDS2 may not be generalizable to those with the average western diet. Treatment dose, duration, type of PUFA, and DHA:EPA ratio may all play a role in the possible protective effect of dietary supplementation.[107]

Inflammation, n-3 Fatty Acids, and Age-Related Macular Degeneration

In animal studies, dietary n-3 FA supplementation has been shown to have protective anti-inflammatory effects. For example, mice with an engineered predisposition to develop AMD-like pathologies[108] had lesion regression when fed a high n-3 FA diet.[109] The regression was associated with increased levels of the anti-inflammatory intermediary prostaglandin $(PG)D_2$, lower amounts of the pro-inflammatory mediators leukotriene $(LT)B_4$ and PGE_2, and suppression of transcription of the pro-inflammatory molecules tumor necrosis factor (TNF)-α and interleukin (IL)-6. Additionally, mice exposed to laser photocoagulation and given a supplementation of 5% EPA showed reduced risk for choroidal neovascularization (CNV)[110] and not only had a noteworthy drop in the expression and production of inflammatory and angiogenic markers in the RPE, macrophages, and endothelial cells but also a lower serum IL-6 and C-reactive protein (CRP).

When considering oxygen induced retinopathy, carriage of the fat-1 gene (which encodes a desaturase enzyme that can change an n-6 FA to an n-3 Fas[111]) and n-3 FA supplementation have been shown to be protective.[112] These scientists have suggested that the protective effect of n-3 FAs is anti-inflammatory in nature because mice eating an n-6 FA-rich diet with either supplementation of small doses of the neuroprotectin D_1, resolvin D_1, resolvin E_1 or that receive an intraperitoneal or intraocular injection of the TNF-α-neutralizing agent, etanercept, have reduced risk of AMD-like symptoms.

The connection between retinal function and n-3 FA supplementation has also been explored. fat-1 mice with an n-6 FA-enriched diet showed abnormally elevated a- and b-wave amplitude on ERG evaluation.[113] Further, the photoreceptor proteins in these mice had 2-(ω-carboxyethyl)pyrrole (CEP), a harmful lipid peroxidation product, and the mice also had elevated levels of glial fibrillary acidic protein (GFAP), a marker of neurologic stress. Other studies showed that *fat-1* mice fed on an n-6 FA-enriched diet and exposed to photo-oxidative stress suffered decreased thickness of the outer nuclear layer, lower ERG amplitudes, increased photoreceptor apoptosis, and modification of retinal proteins by a per-oxidation product of DHA.[114] The concept of peroxida-tion products contributing to AMD pathogenesis was further explored.[37] Mice injected with CEP-modified mouse serum developed antibodies. The antibodies then produced a cascade of events: the deposition of comple-ment protein C3d in BM, a buildup of drusen, and the development of dry AMD-like lesions. When considered collectively, the evidence suggests that retinal function may be improved by supplementation with antioxidants and a decrease in the n-6/n-3 ratio.

Lipoprotein Metabolism and Age-Related Macular Degeneration

Because of the proposed vascular association with AMD, the role of lipoproteins and lipoprotein metabolism in AMD is of great intrigue. Studies have been conducted in both animal and human models. In murine studies, mice engineered to model human type III hyperlipopro-teinemia fed a high-cholesterol/high-fat diet showed BLamDs after 9 months. Only 33% of these mice showed BLamDs on a normal diet and 0% of knockout mice on either diet. As in humans, the mice BLamDs showed immunoreactivity toward anti-ApoE antibodies, imply-ing that dysfunctional ApoE may be necessary for BLamD accumulation[115]; however, ApoE-deficient mice on a high-cholesterol diet for 35 weeks showed altered ERG per-formance, as well as upregulation of pro-apoptotic genes in neuroretinal cells, suggesting that ApoE deficiency by itself can affect retinal function in the presence of high cho-lesterol.[116] Further, aged mice expressing the ApoE4 allele and maintained on a high-cholesterol diet were found to be especially prone to develop ARD-like changes, includ-ing patchy atrophy, diffuse sub-RPE deposits, and CNV.[117] Paradoxically, despite cholesterol levels being higher in ApoE2 mice, the ApoE4 mice had worse outcomes. This may be explained by human data, which suggests that ApoE4 is a protective allele while ApoE2 is a susceptibil-ity allele.[118]

There has also been inquiry into low-density lipo-protein (LDL) metabolism in the context of AMD. Mice expressing the human ApoB100 protein (ApoB100 mice) that were exposed to photo-oxidative stress were more susceptible to BLamD formation.[119] Another study found that 12-month-old ApoB100 mice developed BLamDs and highly AMD-specific basal linear-like deposits (BLinDs) even without exposure to photo-oxi-dative stress. Most interestingly, this study showed that ApoB100 status made these mice more vulnerable than eating a high-fat diet.[120] The study suggested the follow-ing mechanism for AMD: the RPE secretes ApoB100-rich lipoproteins to combat the effects of hyperlipidemia and avoid lipotoxicity, thereby causing aggregation of lipo-proteins and contributing to the pathogenesis of AMD.[121]

Studies have examined the LDL receptor in the context of AMD. Mice engineered to model human hyperchole-sterolemia developed membrane-bound translucent par-ticles in BM. Subsequently when they were fed a high-fat diet for 2 months, they developed even more particles and the membrane got thicker. Vascular endothelial growth factor (VEGF) staining was also higher in the mice on a high-fat diet, supporting the concept that impaired LDL clearance in a hyperlipidemic setting could accelerate the pathogenesis of AMD.[122] In addition, ApoB100 and mice lacking an LDL receptor suffered reduced photopic b-wave and scotopic amplitudes even though they ate a normal diet. The basement membrane of RPE had accu-mulated esterified cholesterol.[123] This research indicates that lipoprotein accumulation in concert with hyperlip-idemia may be related to AMD pathogenesis.

The SR-B1 has also been investigated because of its role in RCT, its significance in retinal vitamin E-transport,[124,125] and reports of its expression in RPE.[126] SR-B1 knockout mice that ate a high-fat/cholesterol diet for 30 weeks showed various retinal changes involving thickening of BM with sparse sub-RPE deposits and greater lipid inclu-sions.[127] The study suggested these changes were due to increased implicit time in a- and b-waves on scotopic ERGs. Thus, these conclusions indicate that the absence of SR-B1 and hypercholesterolemia leads to the changes resembling dry AMD, thereby suggesting that nutrient exchange in the retina is impeded by lipid deposition.

In human models, the research in the lipoprotein arena has often focused on the efficacy of statins (medi-cations used to lower cholesterol) in treating AMD. Because they may lower lipid levels in the body,[128] this may reduce the deposition of lipids in BM. Additionally, because statins have pleiotropic effects including antiox-idant and anti-inflammatory properties,[129,130] they may both beneficially improve lipid metabolism and mitigate inflammatory processes that damage the eye.

Studies investigating utility of statins have had mixed results. A protective effect of statins has been validated in certain studies,[20,131–138] a lack of a significant association has been noted in others,[139–143] and one study showed statin users to have increased risk of AMD.[144] It is possible that some of these mixed results may be explained because statins may have a delayed therapeutic advantage. This theory was supported in a study that revisited the Blue Mountains Eye Study Group after a 10-year period that showed statin use was protective for soft drusen, a precur-sor lesion for late AMD.[134] These findings from a 10-year follow-up period contradict the negative findings found in one study that had only a 5-year follow-up among the same study groups.[145] Renzi et al. examined the effect of statins on MPOD and found that MPOD decreased significantly with increased duration of statins, although globally MPOD was not lower among statin users compared to matched nonstatin users. Additionally, atorvastatin, but not rosuvastatin, was noted to reduce MPOD.[104]

Although some of the above-mentioned studies have provided some faint glimmers of hope in this therapeutic arena, two meta-analysis studies have concluded that the data did not convincingly support the use of lipid-lowering agents for AMD prevention.[146–148] Consequently, despite the theoretical appeal of using statins in the treatment of AMD, the consensus in the medical community is to make no clinical recommendation in support of statins as an AMD therapeutic until there is better, more consistent evidence.

CONCLUSION

There appears to be a relationship between lipid intake and metabolism and AMD; however, the ability to harness this information with concrete clinical guidelines is limited by the lack of appropriately designed interventional studies. The potential therapeutic benefit of n-3 FA supplementation is supported by the role of these FAs in normal retinal physiology and their anti-inflammatory properties and the putative role of inflammation in AMD. The susceptibility of n-3 FAs to oxidation may counteract this purported benefit. Consequently, adding antioxidants to the diet to alleviate this vulnerability has been suggested. The n-6/n-3 FA ratio and levels of cholesterol appear to play an important role as well. Increasing attention has been given to the role of the RPE in synthesizing BM lipoproteins, which have led some to postulate that the role of plasma lipoproteins in AMD may be limited.[46] A better understanding of where BM lipoproteins originate remains an active area of research and may allow for more targeted therapy.

The reader is directed elsewhere in this text for discussions of other dietary factors and AMD, including a more detailed discussion of fish intake and GI. Given the enormous public health problem inflicted by AMD, the identification of effective primary and secondary prevention measures is of critical importance. Future studies will likely provide more enlightenment.

TAKE-HOME MESSAGES

- AMD is a disease of significant public health burden and a leading cause of severe visual impairment among elderly people in the industrialized world.
- Lipid peroxidation plays an important role because irradiation can generate ROS via photo-oxidative stress. The retina's high concentration of PUFAs makes it particularly susceptible to oxidation.
- Evidence suggests that choroidal vascular disease develops from accumulation of lipid-rich particles in the sub-RPE space between the RPE basement membrane and the elastic layer in the BM in an atherosclerosis-like process and leads to retinal dysfunction and the development of AMD.
- The RPE likely plays an important role in the formation of BM lipoproteins.
- Both genetic and inflammatory processes contribute to pathogenesis of AMD.
- Research studies have shown mixed results on the efficacy of statin use in treating AMD.
- There may be a role of n-3 nutritional supplementation to reduce the risk of progression of AMD. The lack of significant impact of DHA+EPA supplementation in AREDS2 raises questions about which patients may benefit from supplementation in additional to dietary modification.

Acknowledgment

This chapter contains text, images, and figures originally from Kishan et al.[98] and has been republished with permission from the original publisher.

References

1. Friedman E. A hemodynamic model of the pathogenesis of age-related macular degeneration. *Am J Ophthalmol* 1997; **124**(5):677–82.
2. Brown MM, Brown GC, Sharma S, Stein JD, Roth Z, Campanella J, et al. The burden of age-related macular degeneration: a value-based analysis. *Curr Opin Ophthalmol* 2006;**17**(3):257–66.
3. Friedman DS, O'Colmain BJ, Munoz B, Tomany SC, McCarty C, de Jong PT, et al. Prevalence of age-related macular degeneration in the United States. *Arch Ophthalmol* 2004;**122**(4):564–72.
4. Delyfer MN, Buaud B, Korobelnik JF, Rougier MB, Schalch W, Etheve S, et al. Association of macular pigment density with plasma omega-3 fatty acids: the PIMAVOSA study. *Invest Ophthalmol Vis Sci* 2012;**53**(3):1204–10.
5. Thornton J, Edwards R, Mitchell P, Harrison RA, Buchan I, Kelly SP. Smoking and age-related macular degeneration: a review of association. *Eye (Lond)* 2005;**19**(9):935–44.
6. Verhoeff FH, Grossman HP. The pathogenesis of disciform degeneration of the macula. *Trans Am Ophthalmol Soc* 1937;**35**:262–94.
7. Webers CA, Beckers HJ, Nuijts RM, Schouten JS. Pharmacological management of primary open-angle glaucoma: second-line options and beyond. *Drugs Aging* 2008;**25**(9):729–59.
8. Hyman LG, Lilienfeld AM, Ferris 3rd FL, Fine SL. Senile macular degeneration: a case-control study. *Am J Epidemiol* 1983;**118**(2):213–27.
9. Evereklioglu C, Doganay S, Er H, Cekmen M, Ozerol E, Otlu B. Serum leptin concentrations are decreased and correlated with disease severity in age-related macular degeneration: a preliminary study. *Eye (Lond)* 2003;**17**(3):350–5.
10. Hyman L, Schachat AP, He Q, Leske MC. Hypertension, cardiovascular disease, and age-related macular degeneration. Age-Related Macular Degeneration Risk Factors Study Group. *Arch Ophthalmol* 2000;**118**(3):351–8.

11. Klein R, Klein BE, Tomany SC, Cruickshanks KJ. The association of cardiovascular disease with the long-term incidence of age-related maculopathy: the Beaver Dam Eye Study. *Ophthalmology* 2003;**110**(6):1273–80.

12. Seddon JM, Gensler G, Milton RC, Klein ML, Rifai N. Association between C-reactive protein and age-related macular degeneration. *JAMA* 2004;**291**(6):704–10.

13. TEDC-CS G. Risk factors for neovascular age-related macular degeneration. The Eye Disease Case-Control Study Group. *Arch Ophthalmol* 1992;**110**(12):1701–8.

14. van Leeuwen R, Ikram MK, Vingerling JR, Witteman JC, Hofman A, de Jong PT. Blood pressure, atherosclerosis, and the incidence of age-related maculopathy: the Rotterdam Study. *Invest Ophthalmol Vis Sci* 2003;**44**(9):3771–7.

15. van Leeuwen R, Klaver CC, Vingerling JR, Hofman A, van Duijn CM, Stricker BH, et al. Cholesterol and age-related macular degeneration: is there a link? *Am J Ophthalmol* 2004;**137**(4): 750–2.

16. Vingerling JR, Dielemans I, Bots ML, Hofman A, Grobbee DE, de Jong PT. Age-related macular degeneration is associated with atherosclerosis. The Rotterdam Study. *Am J Epidemiol* 1995;**142**(4):404–9.

17. Abalain JH, Carre JL, Leglise D, Robinet A, Legall F, Meskar A, et al. Is age-related macular degeneration associated with serum lipoprotein and lipoparticle levels? *Clin Chim Acta* 2002;**326**(1–2):97–104.

18. Blumenkranz MS, Russell SR, Robey MG, Kott-Blumenkranz R, Penneys N. Risk factors in age-related maculopathy complicated by choroidal neovascularization. *Ophthalmology* 1986; **93**(5):552–8.

19. Goldberg J, Flowerdew G, Smith E, Brody JA, Tso MO. Factors associated with age-related macular degeneration. An analysis of data from the first National Health and Nutrition Examination Survey. *Am J Epidemiol* 1988;**128**(4):700–10.

20. Klein R, Deng Y, Klein BE, Hyman L, Seddon J, Frank RN, et al. Cardiovascular disease, its risk factors and treatment, and age-related macular degeneration: Women's Health Initiative Sight Exam ancillary study. *Am J Ophthalmol* 2007;**143**(3):473–83.

21. Friedman E. The role of the atherosclerotic process in the pathogenesis of age-related macular degeneration. *Am J Ophthalmol* 2000;**130**(5):658–63.

22. Snow KK, Seddon JM. Do age-related macular degeneration and cardiovascular disease share common antecedents? *Ophthalmic Epidemiol* 1999;**6**(2):125–43.

23. Harkewicz R, Du H, Tong Z, Alkuraya H, Bedell M, Sun W, et al. Essential role of ELOVL4 protein in very long chain fatty acid synthesis and retinal function. *J Biol Chem* 2012;**287**(14):11469–80.

24. McMahon A, Butovich IA, Kedzierski W. Epidermal expression of an ELOVl4 transgene rescues neonatal lethality of homozygous Stargardt disease-3 mice. *J Lipid Res* 2011;**52**(6):1128–38.

25. Sommer JR, Estrada JL, Collins EB, Bedell M, Alexander CA, Yang Z, et al. Production of ELOVL4 transgenic pigs: a large animal model for Stargardt-like macular degeneration. *Br J Ophthalmol* 2011;**95**(12):1749–54.

26. Bazan NG. The metabolism of omega-3 polyunsaturated fatty acids in the eye: the possible role of docosahexaenoic acid and docosanoids in retinal physiology and ocular pathology. *Prog Clin Biol Res* 1989;**312**:95–112.

27. Beatty S, Koh H, Phil M, Henson D, Boulton M. The role of oxidative stress in the pathogenesis of age-related macular degeneration. *Surv Ophthalmol* 2000;**45**(2):115–34.

28. Fukuzumi K. Relationship between lipoperoxides and diseases. *J Environ Pathol Toxicol Oncol* 1986;**6**(3–4):25–56.

29. Anderson RE, Maude MB. Lipids of ocular tissues. 8. The effects of essential fatty acid deficiency on the phospholipids of the photoreceptor membranes of rat retina. *Arch Biochem Biophys* 1972;**151**(1):270–6.

30. Fliesler SJ, Schroepfer Jr GJ. Sterol composition of bovine retinal rod outer segment membranes and whole retinas. *Biochim Biophys Acta* 1982;**711**(1):138–48.

31. Wiegand RD, Giusto NM, Rapp LM, Anderson RE. Evidence for rod outer segment lipid peroxidation following constant illumination of the rat retina. *Invest Ophthalmol Vis Sci* 1983;**24**(10):1433–5.

32. Catalá A. An overview of lipid peroxidation with emphasis in outer segments of photoreceptors and the chemiluminescence assay. *Int J Biochem Cell Biol* 2006;**38**(9):1482–95.

33. Esterbauer H, Schaur RJ, Zollner H. Chemistry and biochemistry of 4-hydroxynonenal, malonaldehyde and related aldehydes. *Free Radic Biol Med* 1991;**11**(1):81–128.

34. Crabb JW, Miyagi M, Gu X, Shadrach K, West KA, Sakaguchi H, et al. Drusen proteome analysis: an approach to the etiology of age-related macular degeneration. *Proc Natl Acad Sci USA* 2002;**99**(23):14682–7.

35. Kaemmerer E, Schutt F, Krohne TU, Holz FG, Kopitz J. Effects of lipid peroxidation-related protein modifications on RPE lysosomal functions and POS phagocytosis. *Invest Ophthalmol Vis Sci* 2007;**48**(3):1342–7.

36. Krohne TU, Holz FG, Kopitz J. Apical-to-basolateral transcytosis of photoreceptor outer segments induced by lipid peroxidation products in human retinal pigment epithelial cells. *Invest Ophthalmol Vis Sci* 2010;**51**(1):553–60.

37. Hollyfield JG, Bonilha VL, Rayborn ME, Yang X, Shadrach KG, Lu L, et al. Oxidative damage-induced inflammation initiates age-related macular degeneration. *Nat Med* 2008;**14**(2):194–8.

38. Gu X, Meer SG, Miyagi M, Rayborn ME, Hollyfield JG, Crabb JW, et al. Carboxyethylpyrrole protein adducts and autoantibodies, biomarkers for age-related macular degeneration. *J Biol Chem* 2003;**278**(43):42027–35.

39. Johnson LV, Ozaki S, Staples MK, Erickson PA, Anderson DH. A potential role for immune complex pathogenesis in drusen formation. *Exp Eye Res* 2000;**70**(4):441–9.

40. Scholl HP, Fleckenstein M, Charbel Issa P, Keilhauer C, Holz FG, Weber BH. An update on the genetics of age-related macular degeneration. *Mol Vis* 2007;**13**:196–205.

41. Haines JL, Hauser MA, Schmidt S, Scott WK, Olson LM, Gallins P, et al. Complement factor H variant increases the risk of age-related macular degeneration. *Science* 2005;**308**(5720):419–21.

42. Edwards AO, Ritter 3rd R, Abel KJ, Manning A, Panhuysen C, Farrer LA. Complement factor H polymorphism and age-related macular degeneration. *Science* 2005;**308**(5720):421–4.

43. Donoso LA, Vrabec T, Kuivaniemi H. The role of complement Factor H in age-related macular degeneration: a review. *Surv Ophthalmol* 2010;**55**(3):227–46.

44. Klein RJ, Zeiss C, Chew EY, Tsai JY, Sackler RS, Haynes C, et al. Complement factor H polymorphism in age-related macular degeneration. *Science* 2005;**308**(5720):385–9.

45. Shaw PX, Zhang L, Zhang M, Du H, Zhao L, Lee C, et al. Complement factor H genotypes impact risk of age-related macular degeneration by interaction with oxidized phospholipids. *Proc Natl Acad Sci USA* 2012;**109**(34):13757–62.

46. Ebrahimi KB, Handa JT. Lipids, lipoproteins, and age-related macular degeneration. *J Lipids* 2011;**2011**:802059.

47. Wang L, Li CM, Rudolf M, Belyaeva OV, Chung BH, Messinger JD, et al. Lipoprotein particles of intraocular origin in human Bruch membrane: an unusual lipid profile. *Invest Ophthalmol Vis Sci* 2009;**50**(2):870–7.

48. Malek GLC, Guidry C, Medeiros NE, Curcio CA. Apolipoprotein B in cholesterol-containing drusen and basal deposits of human eyes with age-related maculopathy. *Am J Pathol* 2003;**162**(2):413–25.

49. Li CM, Zhang X, Dashti N, Chung BH, Medeiros NE, Guidry C, et al. Retina expresses microsomal triglyceride transfer protein: implications for age-related maculopathy. *J Lipid Res* 2005;**46**(4):628–40.

50. Holz FG, Pauleikhoff D, Bird AC. Analysis of lipid deposits extracted from human macular and peripheral Bruch's membrane. *Arch Ophthalmol* 1994;**112**(3):402–6.

51. Curcio CA, Johnson M, Rudolf M, Huang JD. The oil spill in ageing Bruch membrane. *Br J Ophthalmol* 2011;**95**(12):1638–45.

52. Li CM, Chung BH, Presley JB, Malek G, Zhang X, Dashti N, et al. Lipoprotein-like particles and cholesteryl esters in human Bruch's membrane: initial characterization. *Invest Ophthalmol Vis Sci* 2005;**46**(7):2576–86.

53. Duncan KG, Bailey KR, Yang H, Lowe RJ, Matthes MT, Kane JP, et al. Expression of reverse cholesterol transport proteins ATP-binding cassette A1 (ABCA1) and scavenger receptor BI (SR-BI) in the retina and retinal pigment epithelium. *Br J Ophthalmol* 2009;**93**(8):1116–20.

54. Ishida BY, Duncan KG, Chalkley RJ, Burlingame AL, Kane JP, Schwartz DM. Regulated expression of apolipoprotein E by human retinal pigment epithelial cells. *J Lipid Res* 2005;**45**(2):263–71.

55. Duncan KG, Kane JP, Schwartz DM. Human retinal pigment epithelial cells express scavenger receptors BI and BII. *Biochem Biophys Res Commun* 2002;**292**(4):1017–22.

56. Ishida BY, Bailey KR, Kane JP, Schwartz DM. High density lipoprotein mediated lipid efflux from retinal pigment epithelial cells in culture. *Br J Ophthalmol* 2006;**90**(5):616–20.

57. Moore DJ. The effect of age on the macromolecular permeability of human Bruch's membrane. *Invest Ophthalmol Vis Sci* 2001;**42**(12):2970–5.

58. Schwartz D. Methods to increase reverse cholesterol transport in the retinal pigment epithelium (RPE) and Bruch's membrane (BM). *US Patent US* 2008;**7470659**.

59. Neale BM, Fagerness J, Reynolds R, Sobrin L, Parker M, Raychaudhuri S, et al. Genome-wide association study of advanced age-related macular degeneration identifies a role of the hepatic lipase gene (LIPC). *Proc Natl Acad Sci USA* 2010;**107**(16):7395–400.

60. Calder PC. n-3 polyunsaturated fatty acids, inflammation, and inflammatory diseases. *Am J Clin Nutr* 2006;**83**(6 Suppl). 1505S–19S.

61. Parker G, Gibson NA, Brotchie H, Heruc G, Rees AM, Hadzi-Pavlovic D. Omega-3 fatty acids and mood disorders. *Am J Psychiatry* 2006;**163**(6):969–78.

62. Kotani S, Sakaguchi E, Warashina S, Matsukawa N, Ishikura Y, Kiso Y, et al. Dietary supplementation of arachidonic and docosahexaenoic acids improves cognitive dysfunction. *Neurosci Res* 2006;**56**(2):159–64.

63. Lavie CJ, Milani RV, Mehra MR, Ventura HO. Omega-3 polyunsaturated fatty acids and cardiovascular diseases. *J Am Coll Cardiol* 2009;**54**(7):585–94.

64. Russo GL. Dietary n-6 and n-3 polyunsaturated fatty acids: from biochemistry to clinical implications in cardiovascular prevention. *Biochem Pharmacol* 2009;**77**(6):937–46.

65. Innis SM. Omega-3 Fatty acids and neural development to 2 years of age: do we know enough for dietary recommendations? *J Pediatr Gastroenterol Nutr* 2009;**48**(Suppl 1):S16–24.

66. Mukherjee PK, Marcheselli VL, Barreiro S, Hu J, Bok D, Bazan NG. Neurotrophins enhance retinal pigment epithelial cell survival through neuroprotectin D1 signaling. *Proc Natl Acad Sci USA* 2007;**104**(32):13152–7.

67. Mukherjee PK, Marcheselli VL, de Rivero Vaccari JC, Gordon WC, Jackson FE, Bazan NG. Photoreceptor outer segment phagocytosis attenuates oxidative stress-induced apoptosis with concomitant neuroprotectin D1 synthesis. *Proc Natl Acad Sci USA* 2007;**104**(32):13158–63.

68. MacDonald IM, Hebert M, Yau RJ, Flynn S, Jumpsen J, Suh M, et al. Effect of docosahexaenoic acid supplementation on retinal function in a patient with autosomal dominant Stargardt-like retinal dystrophy. *Br J Ophthalmol* 2004;**88**(2):305–6.

69. Rotstein NP, Aveldano MI, Barrantes FJ, Politi LE. Docosahexaenoic acid is required for the survival of rat retinal photoreceptors. *in vitro J Neurochem* 1996;**66**(5):1851–9.

70. Rotstein NP, Aveldano MI, Barrantes FJ, Roccamo AM, Politi LE. Apoptosis of retinal photoreceptors during development in vitro: protective effect of docosahexaenoic acid. *J Neurochem* 1997;**69**(2):504–13.

71. Politi LE, Rotstein NP, Carri NG. Effect of GDNF on neuroblast proliferation and photoreceptor survival: additive protection with docosahexaenoic acid. *Invest Ophthalmol Vis Sci* 2001;**42**(12):3008–15.

72. Polit L, Rotstein N, Carri N. Effects of docosahexaenoic acid on retinal development: cellular and molecular aspects. *Lipids* 2001;**36**(9):927–35.

73. Neuringer M, Connor WE, Lin DS, Barstad L, Luck S. Biochemical and functional effects of prenatal and postnatal omega 3 fatty acid deficiency on retina and brain in rhesus monkeys. *Proc Natl Acad Sci USA* 1986;**83**(11):4021–5.

74. Pawlosky RJ, Denkins Y, Ward G, Salem Jr N. Retinal and brain accretion of long-chain polyunsaturated fatty acids in developing felines: the effects of corn oil-based maternal diets. *Am J Clin Nutr* 1997;**65**(2):465–72.

75. Miceli MV, Newsome DA, Tate Jr DJ, Sarphie TG. Pathologic changes in the retinal pigment epithelium and Bruch's membrane of fat-fed atherogenic mice. *Curr Eye Res* 2000;**20**(1):8–16.

76. Dithmar S, Sharara NA, Curcio CA, Le NA, Zhang Y, Brown S, et al. Murine high-fat diet and laser photochemical model of basal deposits in Bruch membrane. *Arch Ophthalmol* 2001;**119**(11):1643–9.

77. Cousins SW, Espinosa-Heidmann DG, Alexandridou A, Sall J, Dubovy S, Csaky K. The role of aging, high fat diet and blue light exposure in an experimental mouse model for basal laminar deposit formation. *Exp Eye Res* 2002;**75**(5):543–53.

78. Cho E, Hung S, Willett WC, Spiegelman D, Rimm EB, Seddon JM, et al. Prospective study of dietary fat and the risk of age-related macular degeneration. *Am J Clin Nutr* 2001;**73**(2):209–18.

79. Delcourt C, Carriere I, Cristol JP, Lacroux A, Gerber M. Dietary fat and the risk of age-related maculopathy: the POLANUT study. *Eur J Clin Nutr* 2007;**61**(11):1341–4.

80. Mares-Perlman JA, Brady WE, Klein R, VandenLangenberg GM, Klein BE, Palta M. Dietary fat and age-related maculopathy. *Arch Ophthalmol* 1995;**113**(6):743–8.

81. Chong EW, Kreis AJ, Wong TY, Simpson JA, Guymer RH. Dietary omega-3 fatty acid and fish intake in the primary prevention of age-related macular degeneration: a systematic review and meta-analysis. *Arch Ophthalmol* 2008;**126**(6):826–33.

82. Seddon JM, Rosner B, Sperduto RD, Yannuzzi L, Haller JA, Blair NP, et al. Dietary fat and risk for advanced age-related macular degeneration. *Arch Ophthalmol* 2001;**119**(8):1191–9.

83. Chong EW, Robman LD, Simpson JA, Hodge AM, Aung KZ, Dolphin TK, et al. Fat consumption and its association with age-related macular degeneration. *Arch Ophthalmol* 2009;**127**(5):674–80.

84. SanGiovanni JP, Chew EY, Clemons TE, Davis MD, Ferris 3rd FL, Gensler GR, et al. The relationship of dietary lipid intake and age-related macular degeneration in a case-control study: AREDS Report No. 20. *Arch Ophthalmol* 2007;**125**(5):671–9.

85. Seddon JM, Cote J, Rosner B. Progression of age-related macular degeneration: association with dietary fat, transunsaturated fat, nuts, and fish intake. *Arch Ophthalmol* 2003;**121**(12):1728–37.

86. Seddon JM, George S, Rosner B. Cigarette smoking, fish consumption, omega-3 fatty acid intake, and associations with age-related macular degeneration: the US Twin Study of Age-Related Macular Degeneration. *Arch Ophthalmol* 2006;**124**(7):995–1001.

87. Chua B, Flood V, Rochtchina E, Wang JJ, Smith W, Mitchell P. Dietary fatty acids and the 5-year incidence of age-related maculopathy. *Arch Ophthalmol* 2006;**124**(7):981–6.

88. Augood C, Chakravarthy U, Young I, Vioque J, de Jong PT, Bentham G, et al. Oily fish consumption, dietary docosahexaenoic acid and eicosapentaenoic acid intakes, and associations with neovascular age-related macular degeneration. *Am J Clin Nutr* 2008;**88**(2):398–406.

89. Sangiovanni JP, Agron E, Meleth AD, Reed GF, Sperduto RD, Clemons TE, et al. {omega}-3 Long-chain polyunsaturated fatty acid intake and 12-y incidence of neovascular age-related macular degeneration and central geographic atrophy: a prospective cohort study from the Age-Related Eye Disease Study. *Am J Clin Nutr* 2009;**90**(6):1601–7.

90. SanGiovanni JP, Chew EY, Agron E, Clemons TE, Ferris 3rd FL, Gensler G, et al. The relationship of dietary omega-3 long-chain polyunsaturated fatty acid intake with incident age-related macular degeneration: AREDS report no. 23. *Arch Ophthalmol* 2008;**126**(9):1274–9.

91. Chiu CJ, Milton RC, Klein R, Gensler G, Taylor A. Dietary compound score and risk of age-related macular degeneration in the age-related eye disease study. *Ophthalmology* 2009;**116**(5):939–46.

92. Tan JS, Wang JJ, Flood V, Mitchell P. Dietary fatty acids and the 10-year incidence of age-related macular degeneration: the Blue Mountains Eye Study. *Arch Ophthalmol* 2009;**127** (5):656–65.

93. Chiu CJ, Klein R, Milton RC, Gensler G, Taylor A. Does eating particular diets alter the risk of age-related macular degeneration in users of the Age-Related Eye Disease Study supplements? *Br J Ophthalmol* 2009;**93**(9):1241–6.

94. Merle B, Delyfer MN, Korobelnik JF, Rougier MB, Colin J, Malet F, et al. Dietary omega-3 fatty acids and the risk for age-related maculopathy: the Alienor Study. *Invest Ophthalmol Vis Sci* 2011;**52**(8):6004–11.

95. Smith W, Mitchell P, Leeder SR. Dietary fat and fish intake and age-related maculopathy. *Arch Ophthalmol* 2000;**118**(3):401–4.

96. Arnarsson A, Sverrisson T, Stefansson E, Sigurdsson H, Sasaki H, Sasaki K, et al. Risk factors for five-year incident age-related macular degeneration: the Reykjavik Eye Study. *Am J Ophthalmol* 2006;**142**(3):419–28.

97. Heuberger RA, Mares-Perlman JA, Klein R, Klein BE, Millen AE, Palta M. Relationship of dietary fat to age-related maculopathy in the Third National Health and Nutrition Examination Survey. *Arch Ophthalmol* 2001;**119**(12):1833–8.

98. Kishan AU, Modjtahedi BS, Martins EN, Modjtahedi SP, Morse LS. Lipids and age-related macular degeneration. *Surv Ophthalmol* 2011;**56**(3):195–213.

99. Jonas JB, Nangia V, Kulkarni M, Gupta R, Khare A. Associations of early age-related macular degeneration with ocular and general parameters. The Central India Eyes and Medical Study. *Acta Ophthalmol* 2012;**90**(3):e185–91.

100. Wang S, Xu L, Jonas JB, You QS, Wang YX, Yang H. Dyslipidemia and eye diseases in the adult Chinese population: the Beijing eye study. *PloS One* 2012;(3):7. e26871.

101. Weikel KA, Fitzgerald P, Shang F, Caceres MA, Bian Q, Handa JT, et al. Natural history of age-related retinal lesions that precede AMD in mice fed high or low glycemic index diets. *Invest Ophthalmol Vis Sci* 2012;**53**(2):622–32.

102. Uchiki T, Weikel KA, Jiao W, Shang F, Caceres A, Pawlak D, et al. Glycation-altered proteolysis as a pathobiologic mechanism that links dietary glycemic index, aging, and age-related disease (in nondiabetics). *Aging Cell* 2012;**11**(1):1–13.

103. Simon E, Bardet B, Gregoire S, Acar N, Bron AM, Creuzot-Garcher CP, et al. Decreasing dietary linoleic acid promotes long chain omega-3 fatty acid incorporation into rat retina and modifies gene expression. *Exp Eye Res* 2011;**93**(5):628–35.

104. Renzi LM, Hammond Jr BR, Dengler M, Roberts R. The relation between serum lipids and lutein and zeaxanthin in the serum and retina: results from cross-sectional, case-control and case study designs. *Lipids Health Dis* 2012;**11**:33.

105. Kabasawa S, Mori K, Horie-Inoue K, Gehlbach PL, Inoue S, Awata T, et al. Associations of cigarette smoking but not serum fatty acids with age-related macular degeneration in a Japanese population. *Ophthalmology* 2011;**118**(6):1082–8.

106. Hodge WG, Schachter HM, Barnes D, Pan Y, Lowcock EC, Zhang L, et al. Efficacy of omega-3 fatty acids in preventing age-related macular degeneration: a systematic review. *Ophthalmology* 2006;**113**(7):1165–72.

107. Age-Related Eye Disease Study 2 Research Group. Lutein + zeaxanthin and omega-3 fatty acids for age-related macular degeneration: the Age-Related Eye Disease Study 2 (AREDS2) randomized clinical trial. http://www.ncbi.nlm.nih.gov/pubmed/23644932. *JAMA.* 2013 May 15;**309**(19):2005–15.

108. Chan CC, Ross RJ, Shen D, Ding X, Majumdar Z, Bojanowski CM, et al. Ccl2/Cx3cr1-deficient mice: an animal model for age-related macular degeneration. *Ophthalmic Res* 2008;**40**(3–4):124–8.

109. Tuo J, Ross RJ, Herzlich AA, Shen D, Ding X, Zhou M, et al. A high omega-3 fatty acid diet reduces retinal lesions in a murine model of macular degeneration. *Am J Pathol* 2009;**175**(2):799–807.

110. Koto T, Nagai N, Mochimaru H, Kurihara T, Izumi-Nagai K, Satofuka S, et al. Eicosapentaenoic acid is anti-inflammatory in preventing choroidal neovascularization in mice. *Invest Ophthalmol Vis Sci* 2007;**48**(9):4328–34.

111. Kang JX, Wang J, Wu L, Kang ZB. Transgenic mice: fat-1 mice convert n-6 to n-3 fatty acids. *Nature* 2004;**427**(6974):504.

112. Connor KM, SanGiovanni JP, Lofqvist C, Aderman CM, Chen J, Higuchi A, et al. Increased dietary intake of omega-3-polyunsaturated fatty acids reduces pathological retinal angiogenesis. *Nat Med* 2007;**13**(7):868–73.

113. Suh M, Sauve Y, Merrells KJ, Kang JX, Ma DW. Supranormal electroretinogram in fat-1 mice with retinas enriched in docosahexaenoic acid and n-3 very long chain fatty acids (C24–C36). *Invest Ophthalmol Vis Sci* 2009;**50**(9):4394–401.

114. Tanito M, Brush RS, Elliott MH, Wicker LD, Henry KR, Anderson RE. High levels of retinal membrane docosahexaenoic acid increase susceptibility to stress-induced degeneration. *J Lipid Res* 2009;**50**(5):807–19.

115. Kliffen M, Lutgens E, Daemen MJ, de Muinck ED, Mooy CM, de Jong PT. The APO(*)E3-Leiden mouse as an animal model for basal laminar deposit. *Br J Ophthalmol* 2000;**84**(12):1415–9.

116. Ong JM, Zorapapel NC, Aoki AM, Brown DJ, Nesburn AB, Rich KA, et al. Impaired electroretinogram (ERG) response in apolipoprotein E-deficient mice. *Curr Eye Res* 2003;**27**(1):15–24.

117. Malek G, Johnson LV, Mace BE, Saloupis P, Schmechel DE, Rickman DW, et al. Apolipoprotein E allele-dependent pathogenesis: a model for age-related retinal degeneration. *Proc Natl Acad Sci USA* 2005;**102**(33):11900–5.

118. Fritsche LG, Freitag-Wolf S, Bettecken T, Meitinger T, Keilhauer CN, Krawczak M, et al. Age-related macular degeneration and functional promoter and coding variants of the apolipoprotein E gene. *Hum Mutat* 2009;**30**(7):1048–53.

119. Espinosa-Heidmann DG, Sall J, Hernandez EP, Cousins SW. Basal laminar deposit formation in APO B100 transgenic mice: complex interactions between dietary fat, blue light, and vitamin E.. *Invest Ophthalmol Vis Sci* 2004;**45**(1):260–6.

120. Fujihara M, Bartels E, Nielsen LB, Handa JT. A human apoB100 transgenic mouse expresses human apoB100 in the RPE and develops features of early AMD. *Exp Eye Res* 2009;**88**(6): 1115–23.

121. Curcio CA, Johnson M, Huang JD, Rudolf M. Apolipoprotein B-containing lipoproteins in retinal aging and age-related macular degeneration. *J Lipid Res* 2010;**51**(3):451–67.

122. Rudolf M, Winkler B, Aherrahou Z, Doehring LC, Kaczmarek P, Schmidt-Erfurth U. Increased expression of vascular endothelial growth factor associated with accumulation of lipids in Bruch's membrane of LDL receptor knockout mice. *Br J Ophthalmol* 2005;**89**(12):1627–30.

123. Bretillon L, Acar N, Seeliger MW, Santos M, Maire MA, Juaneda P, et al. ApoB100,LDLR–/– mice exhibit reduced electroretinographic response and cholesteryl esters deposits in the retina. *Invest Ophthalmol Vis Sci* 2008;**49**(4):1307–14.

124. Tachikawa M, Okayasu S, Hosoya K. Functional involvement of scavenger receptor class B, type I, in the uptake of alpha-tocopherol using cultured rat retinal capillary endothelial cells. *Mol Vis* 2007;**13**:2041–7.

125. Akanuma S, Yamamoto A, Okayasu S, Tachikawa M, Hosoya K. High-density lipoprotein-associated alpha-tocopherol uptake by human retinal pigment epithelial cells (ARPE-19 cells): the irrelevance of scavenger receptor class B, type I. *Biol Pharm Bull* 2009;**32**(6):1131–4.

126. Duncan KG, Bailey KR, Kane JP, Schwartz DM. Human retinal pigment epithelial cells express scavenger receptors BI and BII. *Biochem Biophys Res Commun* 2002;**292**(4):1017–22.

127. Provost AC, Vede L, Bigot K, Keller N, Tailleux A, Jais JP, et al. Morphologic and electroretinographic phenotype of SR-BI knockout mice after a long-term atherogenic diet. *Invest Ophthalmol Vis Sci* 2009;**50**(8):3931–42.

128. Watts GF. Treating patients with low high-density lipoprotein cholesterol: choices, issues and opportunities. *Curr Control Trials Cardiovasc Med* 2001;**2**(3):118–22.

129. Ludman A, Venugopal V, Yellon DM, Hausenloy DJ. Statins and cardioprotection – more than just lipid lowering? *Pharmacol Ther* 2009;**122**(1):30–43.

130. Montecucco F, Mach F. Update on statin-mediated anti-inflammatory activities in atherosclerosis. *Semin Immunopathol* 2009;**31**(1):127–42.

131. McGwin Jr G, Owsley C, Curcio CA, Crain RJ. The association between statin use and age related maculopathy. *Br J Ophthalmol* 2003;**87**(9):1121–5.

132. McGwin Jr G, Xie A, Owsley C. The use of cholesterol-lowering medications and age-related macular degeneration. *Ophthalmology* 2005;**112**(3):488–94.

133. Wilson HL, Schwartz DM, Bhatt HR, McCulloch CE, Duncan JL. Statin and aspirin therapy are associated with decreased rates of choroidal neovascularization among patients with age-related macular degeneration. *Am J Ophthalmol* 2004;**137**(4): 615–24.

134. Tan JS, Mitchell P, Rochtchina E, Wang JJ. Statins and the long-term risk of incident age-related macular degeneration: the Blue Mountains Eye Study. *Am J Ophthalmol* 2007;**143**(4):685–7.

135. Delcourt C, Michel F, Colvez A, Lacroux A, Delage M, Vernet MH. Associations of cardiovascular disease and its risk factors with age-related macular degeneration: the POLA study. *Ophthalmic Epidemiol* 2001;**8**(4):237–49.

136. Hall NF, Gale CR, Syddall H, Phillips DI, Martyn CN. Risk of macular degeneration in users of statins: cross sectional study. *BMJ* 2001;**323**(7309):375–6.

137. McCarty CA, Mukesh BN, Fu CL, Mitchell P, Wang JJ, Taylor HR. Risk factors for age-related maculopathy: the Visual Impairment Project. *Arch Ophthalmol* 2001;**119**(10):1455–62.

138. McCarty CA, Mukesh BN, Guymer RH, Baird PN, Taylor HR. Cholesterol-lowering medications reduce the risk of age-related maculopathy progression. *Med J Aust* 2001;**175**(6):340.

139. Smeeth L, Cook C, Chakravarthy U, Hubbard R, Fletcher AE. A case control study of age related macular degeneration and use of statins. *Br J Ophthalmol* 2005;**89**(9):1171–5.

140. Klein R, Knudtson MD, Klein BE. Statin use and the five-year incidence and progression of age-related macular degeneration. *Am J Ophthalmol* 2007;**144**(1):1–6.

141. Klein R, Klein BE, Tomany SC, Danforth LG, Cruickshanks KJ. Relation of statin use to the 5-year incidence and progression of age-related maculopathy. *Arch Ophthalmol* 2003;**121**(8):1151–5.

142. Smeeth L, Douglas I, Hall AJ, Hubbard R, Evans S. Effect of statins on a wide range of health outcomes: a cohort study validated by comparison with randomized trials. *Br J Clin Pharmacol* 2009;**67**(1):99–109.

143. Tomany SC, Wang JJ, Van Leeuwen R, Klein R, Mitchell P, Vingerling JR, et al. Risk factors for incident age-related macular degeneration: pooled findings from 3 continents. *Ophthalmology* 2004;**111**(7):1280–7.

144. McGwin Jr G, Modjarrad K, Hall TA, Xie A, Owsley C. 3-hydroxy-3-methylglutaryl coenzyme a reductase inhibitors and the presence of age-related macular degeneration in the Cardiovascular Health Study. *Arch Ophthalmol* 2006;**124**(1):33–7.

145. van Leeuwen R, Tomany SC, Wang JJ, Klein R, Mitchell P, Hofman A, et al. Is medication use associated with the incidence of early age-related maculopathy? Pooled findings from 3 continents. *Ophthalmology* 2004;**111**(6):1169–75.

146. Chuo JY, Wiens M, Etminan M, Maberley DA. Use of lipid-lowering agents for the prevention of age-related macular degeneration: a meta-analysis of observational studies. *Ophthalmic Epidemiol* 2007;**14**(6):367–74.

147. Martini E, Scorolli L, Burgagni MS, Fessehaie S. Valutazione degli effetti retinici della somministrazione di simvastatina in pazienti affetti da degenerazione maculare senile. *Ann Ottalmol Clin Oculistica* 1991;**117**(11):1121–6.

148. Gehlbach P, Li T, Hatef E. Statins for age-related macular degeneration. *Cochrane Database Syst Rev* 2009;(3):CD006927.

Carotenoids and Age-Related Macular Degeneration

Preejith P. Vachali, Brian M. Besch, Paul S. Bernstein

Moran Eye Center, University of Utah School of Medicine, Salt Lake City, UT

INTRODUCTION

Age-related macular degeneration (AMD) is an eye disease that usually affects older adults, generally over 50 years old, whose incidence rises dramatically in the elderly population greater than 70 years old.[1-4] There are two common forms of AMD: 'dry' (non-neovascular) and 'wet' (neovascular).[5] Around 80–90% of AMD cases are the dry form, characterized by the formation of drusen (small yellow deposits composed of oxidized lipids, proteins, and inflammatory material) under the macular retina.[5] These degenerative lesions can progress to a condition known as geographic atrophy, which can result in a gradual loss of visual acuity and legal blindness.[5] When neovascularization occurs in the macula due to upregulation of vascular endothelial growth factor (VEGF), dry AMD progresses to wet AMD. This results in hemorrhage, detachment of the retinal pigment epithelium (RPE) from Bruch's membrane, submacular swelling, and scarring of the retinal tissues.

Although the 'macula lutea' or 'yellow spot' was described in 1798 by Home,[6] its physiologic significance was not identified at the time. Bone and Landrum later identified the macular pigments as a 1:1:1 mixture of two dietary carotenoids: lutein [(3R,3′R,6′R)-β,ε-carotene-3,3′-diol] and zeaxanthin [(3R,3′R)-β,β-carotene-3,3′-diol], as well as the nondietary stereoisomer metabolite *meso*-zeaxanthin [(3R,3′S,*meso*)-β,β-carotene-3,3′-diol]. In nature, over 600 carotenoids have been identified, approximately 30–50 of which are found in what is considered a normal human diet. Furthermore, only 10–15 of these carotenoids are usually detected in the human bloodstream, and only lutein and zeaxanthin make their way into the human retina.[6] The remarkable selectivity of these pigments' uptake by the human retina can be attributed to specific binding proteins present in the human retina.[6] Our laboratory has identified the proteins responsible for the selective binding of lutein and zeaxanthin, and these will be discussed in more detail later in this chapter.[7,8]

Figure 8.1 presents the ophthalmoscopic view of the human retina and the major macular pigments.[6] The macular pigments are mainly associated with the Henle fiber layer, which consists of the axons of the foveal cones.[6] The major roles of these macular pigments are to function as a blue light filter,[5,6,9] to protect the retina from light-induced oxidative damage to polyunsaturated lipids and other oxidation-sensitive targets, and possibly to decrease ocular levels of A2E, a toxic metabolite of vitamin A that is a component of RPE lipofuscin.[5] Epidemiologic evidence suggests that intake of foods high in xanthophylls (specifically lutein and zeaxanthin) and their resultant increased levels in the blood are related to a reduced risk of advanced AMD, cataract, and other blinding disorders.[10] To support this claim further, an autopsy study revealed that maculae from patients with a history of AMD have lower concentrations of macular carotenoid pigments compared with control eyes from patients without a known history of AMD.[11] These findings suggest the importance of routine ophthalmic screening to monitor macular pigment optical density (MPOD) in aging adults and that increased dietary intake and supplementation of the macular pigments could delay the progression of these disease conditions.

CHEMISTRY OF MACULAR CAROTENOIDS

Lutein and its isomer zeaxanthin belong to the group of oxygenated derivatives of carotenoids called xanthophylls. They are both tetraterpenoids, each constructed from eight isoprene units, and they thus have a basic $C_{40}H_{56}$ carotene structure with an absorbance in the range of 300–600 nm. Their carbon backbone contains

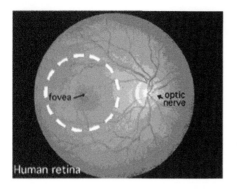

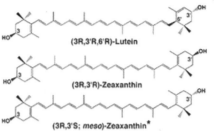

FIGURE 8.1 Ophthalmoscopic view of a human retina and the chemical structures of the major macular pigment carotenoids are shown on the right.[6] *Reproduced by permission of The Royal Society of Chemistry (RSC) on behalf of the European Society for Photobiology, the European Photochemistry Association, and RSC. (See color plate section)*

conjugated carbon–carbon double bonds, which allow relatively free electron movement and consequently explains absorbance of light in the blue–violet region of the visible spectrum. In general, these pigments are responsible for the red–yellow colors observed in nature. Because of their characteristic absorbance, they could act as efficient blue light filters.[6] Both pigments have two hydroxyl groups, which are believed to be responsible for the unique biologic roles these xanthophylls play.[12] Since xanthophylls are more polar compared with hydrocarbon carotenes, they could quench dissolved singlet oxygen more effectively than their nonpolar counterparts.[12] In 2010, Li *et al.* reported the direct correlation between singlet oxygen quenching and macular pigments using *in vitro* assays with electron spin resonance (ESR) spectroscopy in human donor eye tissues.[13] They concluded that light could induce singlet oxygen in RPE/choroid but not in human retina, which could be quenched by the outer photoreceptor macular pigment. It was also hypothesized that the more abundant inner retinal macular pigments could act as intrinsic filters to prevent the generation of singlet oxygen in RPE in the first place.

MEASUREMENT OF MACULAR CAROTENOIDS

Macular pigment density and risk of visual loss from AMD appear to be correlated, so noninvasive techniques for measuring the macular pigments could be very important tools for early identification of high-risk individuals, who could then be encouraged to modify diet, lifestyle, or other risk factors. The macular pigments can be measured using various psychophysical and optical techniques such as heterochromatic flicker photometry (HFP), autofluorescence imaging (AFI), reflectometry, or resonance Raman spectroscopy (RRS).[14] The major downside of HFP is its poor spatial resolution and the fact that its intrasubject variability can be more than 50% unless the subjects are rigorously trained.[15] As HFP is a psychophysical test, it cannot be correlated with the 'gold standard' high-performance liquid chromatography

(HPLC) method.[14] AFI is a noninvasive spectroscopic technique to measure macular pigment optical density (MPOD) and its spatial distribution.[16] This technique is based on measurement of attenuation of fundus background fluorescence of RPE lipofuscin by the macular pigment. Reflectometry measures macular pigment by assessment of attenuation by the macular carotenoids of light reflected from the sclera. This technique relies on complicated modeling of confounding factors such as lens opacities and vitreous scattering that increase with age. More recently, RRS has been developed to measure MPOD. It is a technique that is more rapid, specific, sensitive, and highly reproducible, making it an ideal tool for the efficient screening of large groups.[15] Bernstein *et al.* used resonance Raman measurement of macular carotenoids in elderly, normal, and AMD patients with supplementation and it was concluded that AMD patients not taking high-dose lutein supplements had macular carotenoid levels 32% lower than normal elderly eyes (P = 0.001, two-sided t-test).[17] The group of AMD patients regularly consuming high-dose lutein supplements had values that were significantly higher than their cohorts not using high-dose supplements (P = 0.038, two-sided t-test) and that were indistinguishable from normal subjects (P = 0.829, two-sided t-test).[17] In our recently published Age-Related Eye Disease Study 2 (AREDS2) ancillary study, we analyzed MPOD and macular pigment distribution using a dual-wavelength AFI system along with skin carotenoid levels by RRS.[18] Although the baseline MPODs of these subjects were unusually high relative to an age-matched control group, the serum and skin concentrations did not correlate with the MPOD. This could be attributed to many subjects' habitual lutein and zeaxanthin consumption prior to study entry. Once treatment arms are unmasked, our ancillary study will uniquely provide insights on the ocular carotenoid status of the AREDS2 participants throughout the entire 5-year study period. Figure 8.2 shows peak MPOD measured with the AFI on different age groups in the AREDS2 study population. It was concluded that the peak MPOD declined with age insignificantly in this previously supplemented cohort and was much higher than the mean

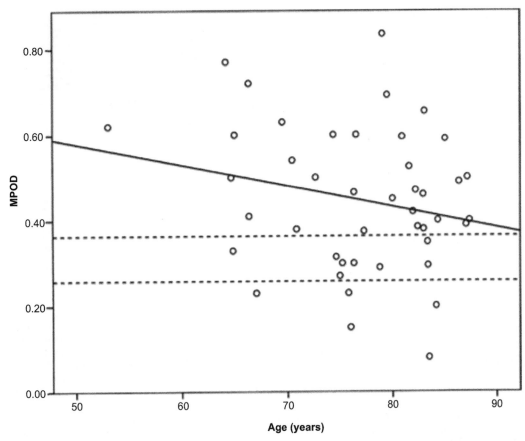

FIGURE 8.2 Peak macular pigment optical density (MPOD) versus age in a study population measured using autofluorescence imaging.[18,37] The solid line is the regression line of peak MPOD versus age (n = 44; r = −0.2268; P = 0.1388). The dashed lines denote the mean ± standard deviation of an age-matched unsupplemented control population. *Reproduced by permission of Association of Research in Vision and Ophthalmology (ARVO) on behalf of* Investigative Ophthalmology & Visual Science *journal.*

values for an unsupplemented age-matched control group.[18]

DIET AND MACULAR CAROTENOIDS

Diet plays an important role in the amount and type of carotenoids distributed among the various tissues in the body. The carotenoid content in several fruits and vegetables is summarized in Table 8.1. These pigments are mainly found in dark green leafy vegetables and fruits. In addition, egg yolk, fortified milk, cheese, cream, liver, kidney, cod, and halibut fish oil are also good dietary sources of carotenoids. Microbial sources of carotenoids such as algae are also getting attention as an alternative for supplementation.[19] Given their hydrophobic nature, there is evidence that consuming carotenoid-rich foods in the presence of oils or cholesterol may increase their uptake. This may explain some study results suggesting a higher bioavailability of lutein from lutein-enriched eggs versus leafy greens such as spinach or other forms

of supplementation.[20] The dietary intake of carotenoids varies largely between individuals and demographic factors; epidemiologic studies show that carotenoid intake across all age groups and ethnicities, as well as both sexes, has overall greater lutein than zeaxanthin consumption.[21] Lower zeaxanthin to lutein ratios have been observed in populations at increased risk for AMD, namely elderly people and women.[21] Most carotenoids are recognized as being generally safe for human consumption, which allows food manufacturers to use them as additives. In 2011, the European Food Safety Authority (EFSA) established an ADI (acceptable daily intake) of 1 mg/kg body weight/day for lutein preparations derived from *Tagetes erecta* containing at least 80% carotenoids.[22] Based on the available data, the EFSA concluded that an intake of 0.75 mg/kg body weight/day of synthetic zeaxanthin does not raise any safety concerns.[23] These values correspond to a daily intake of 53 mg of zeaxanthin and 70 mg of lutein for a person weighing 70 kg. These numbers are much higher than the earlier claims that 20 mg/day per person was safe in

TABLE 8.1　Carotenoid Contents of Various Fruits and Vegetables

Vegetable	β-Carotene	Lycopene	Lutein/ Zeaxanthin
Broccoli	779	-	2445
Brussels sprouts	450	-	1590
Cabbage	65	-	310
Carrots, baby	7275	-	358
Corn	30	-	-
Green beans	377	-	-
Grapefruit, red	603	1462	13
Kale, raw	9226	-	39,550
Leaf lettuce	1272	-	2635
Oranges	51	-	187
Papaya	276	-	75
Peas (green)	320	-	1350
Spinach	5597	-	11,938
Summer squash	410	-	2125
Tomatoes	393	3025	130
Winter squash	220	-	38

Units: μg/100 g. From U.S. Department of Agriculture (1998).
Adapted from Alves-Rodrigues and Shao (2004)[38] with the permission of Elsevier on behalf of the Toxicology Letters.

dietary supplements.[23] Mutagenic studies conducted by Kruger *et al.* revealed that lutein and zeaxanthin are safe for human consumption.[12,24]

ABSORPTION, BIOAVAILABILITY, AND METABOLISM OF MACULAR CAROTENOIDS

The absorption of carotenoids takes place in the gastrointestinal tract by intestinal absorptive cells. Figure 8.3 shows a schematic representation of a possible carotenoid uptake mechanism.[6] It is influenced by many factors, including the form, formulation, fat, and fiber content, as well as the type of food matrix in which it is present.[6] First, carotenoids are saponified by the intestinal cholesterol esterases and lipase and are taken up by the enterocytes in the form of micelles.[6,25] *In vitro* studies using Caco-2 intestinal cells suggest the roles of receptor proteins in the uptake of dietary carotenoids.[26] The scavenger receptor protein (SR-B1) is a possible nonspecific receptor involved in the uptake of carotenoids.[6,26] This receptor may play a critical role in the competitive inhibition of carotenoid uptake in the presence of beta-carotene or vitamin A.

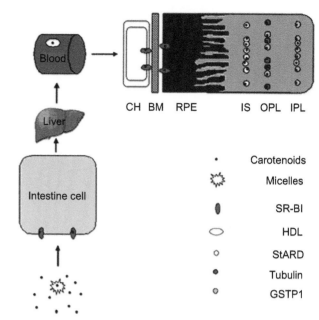

FIGURE 8.3　Possible pathway for macular pigment carotenoid uptake, transport, and accumulation in the human retina.[6] BM: Bruch's membrane; CH: choriocapillaris; HDL: high-density lipoprotein; IPL: inner plexiform layer; IS: inner segments; OPL: outer plexiform layer; RPE: retinal pigment epithelium; SR-BI: scavenger receptor class BI. *Reproduced by permission of The Royal Society of Chemistry (RSC) on behalf of the European Society for Photobiology, the European Photochemistry Association, and RSC.* (See color plate section)

The role of beta-carotene 15,15′-mono-oxygenase 1 enzyme (BCMO1) and a xanthophyll cleavage enzyme β,β-carotene-9′,10′-oxygenase (BCDO2) are responsible for the metabolic cleavage of carotenoids in mouse models.[27]

The carotenoids are further transported into the hepatic system by chylomicrons and to the rest of the body tissues by the general circulation.[6] In the blood, albumin, high-density lipoprotein (HDL), low-density lipoprotein (LDL), and very low-density lipoprotein (VLDL) play important roles as the carriers of carotenoids to the target tissues.[6] Connor *et al.* reported that chickens with genetic HDL defects showed reduced levels of retinal lutein, while the lutein levels in other tissues remained unaffected.[28] Further, Delyfer *et al.* found a positive correlation between omega-3 fatty acids and macular pigment densities.[29] This might be attributed to the increase in the level of HDL due to omega-3 fatty acids.[30] These studies offer some insight as to the role of HDL in the transport of macular pigments to the retinal cells. Bhosale *et al.* purified human macular xanthophyll-binding protein (XBP) and identified the major protein spot on two-dimensional gels as a Pi isoform of glutathione S-transferase (GSTP1). Pharmacologic and spectroscopic binding studies with human recombinant GSTP1 demonstrated that its interactions with zeaxanthin closely match those of XBP purified from human macula.[8] Li *et al.* provided evidence to identify StARD3 (also known as MLN64) as a human retinal lutein-binding protein based on database searches, Western blotting, and immunohistochemistry. An antibody to StARD3, N-62 StAR, localizes to all neurons of monkey macular retina and especially cone inner segments and axons. Further, recombinant StARD3 selectively binds lutein with high affinity (Kd = 0.45 μM) when assessed by surface plasmon resonance (SPR) binding assays.[7] Thus, high concentrations of macular pigments in the retina are attributed to GSTP1 for zeaxanthin[8] and StARD3 for lutein.[7]

PROTECTIVE ROLE OF MACULAR PIGMENTS AND AGE-RELATED MACULAR DEGENERATION

Even though some dietary intake studies have been inconsistent with blood serum levels in relation to AMD,[18] epidemiologic studies conducted by various researchers lend support for the protective role that xanthophylls play against the progression of AMD.[31] The first study conducted by the Eye Disease Case-Control study (EDCC) and a follow-up study done by Seddon *et al.* concluded that individuals with the highest blood levels and highest dietary intake of lutein and zeaxanthin had a 43% reduction of risk for AMD.[10,31] Two studies currently evaluating the role of supplementation and progression of AMD are CARMA and AREDS2. The Carotenoids in Age-Related Maculopathy (CARMA) study is investigating the efficacy of lutein and zeaxanthin supplementation (6 mg and 0.3 mg, respectively) for 36 months on distance visual acuity, contrast sensitivity, photopic interferometric acuity, and shape discrimination in 433 individuals with early AMD. In AREDS2, subjects are given the carotenoids lutein (10 mg/day) and zeaxanthin (2 mg/day), alone or in combination with the omega-3 fatty acids, docosahexaenoic acid (DHA) (350 mg/day) and eicosapentaenoic acid (EPA) (650 mg/day), to assess their influence on progression to advanced AMD in individuals at high risk for the disease with bilateral large, soft drusen and/or advanced AMD in one eye. In June 2008, 80 participating US centers recruited over 4000 AMD patients for the study, with each patient scheduled to receive their assigned treatment in a randomized, placebo-controlled, double-blind manner for 5 years.[31,32] Table 8.2 and Table 8.3 summarize some of the studies done by various researchers on the association of lutein, zeaxanthin, *meso*-zeaxanthin, and AMD. *In vitro* studies have shown prevention of lipid peroxidation by xanthophylls, which can be attributed to their blue light filtering and antioxidant capacity and subsequent protection of the photoreceptors from photo-oxidative damage.[33,34] The blue light filtering role

of macular pigments may also enhance visual function and reduce disability glare.[31,35] One epidemiologic study involving 828 Irish subjects found a significant correlation between dietary intake of lutein and zeaxanthin, serum concentration, and central MPOD.[36]

CONCLUSION

Overall, mounting evidence suggests dietary lutein and zeaxanthin play a unique biologic role in the prevention and mediation of several ocular maladies such as AMD and cataract. Several factors lend support for their importance in ocular health; first is their unique chemical structure, and their resultant ability to absorb light in the blue–violet spectrum and to quench singlet oxygen species. Second is the discovery of specific binding proteins responsible for concentrating these carotenoids in the macula. Continued research into the biochemical pathways of carotenoid uptake and metabolism will provide a basis for improving the efficiency and efficacy of increasing lutein and zeaxanthin via diet and supplementation. Furthermore, recent advancements in the noninvasive quantification of macular pigment density provide a means to gather relevant data, which may strengthen the correlation between carotenoids and ocular health. Ultimately, a better understanding of dietary carotenoids may reveal strategies to maintain healthy vision and quality of life with advanced age.

TAKE-HOME MESSAGES

- Lutein and zeaxanthin may play a protective role in the prevention and mediation of age-related macular degeneration (AMD) and cataract.
- Macular carotenoids have a chemical structure with a conjugated carbon backbone, the capacity to quench singlet oxygen species, and an absorbance in the blue–violet light spectrum.
- Several noninvasive techniques to measure macular carotenoids exist, including heterochromatic flicker photometry (HFP), autofluorescence imaging (AFI), reflectometry, or resonance Raman spectroscopy (RRS).
- Lutein and zeaxanthin must be obtained via diet, with fruits, dark leafy vegetables, and egg yolk all common sources; due to their hydrophobic nature, consumption in the presence of lipids may increase bioavailability.
- Macular carotenoid absorption and transportation to ocular tissues is a complex process, with GSTP and StARD3 binding proteins responsible for the high concentrations observed in the macula.

TABLE 8.2   Clinical Studies on the Association between Lutein and Zeaxanthin on Age-Related Macular Degeneration

Author (Year) [Ref]	Area of Study	Outcomes
Seddon et al. (1994)[10]	Dietary intake of carotenoids and AMD risk	Significant inverse relation between lutein intake and AMD risk
Curran-Celentano et al. (2001)[39]	Lutein intake and serum levels and MPD	Significant positive relation between lutein intake and serum lutein, respectively, and MPOD
Bone et al. (2001)[11]	Lutein concentration in the retina and AMD risk	Significant inverse relation between lutein concentration and central retina and AMD risk
Broekmans et al. (2002)[40]	Serum lutein and zeaxanthin, and adipose lutein and MPOD	Significant positive correlation between serum lutein and MPOD (in men)
Bernstein et al. (2002)[17]	Lutein supplementation and MPOD in AMD patients	MPOD significantly higher in AMD patients using a lutein supplement vs. those with no supplement
Goodrow et al. (2006)[41]	Lutein and zeaxanthin intake via egg yolk, MPOD, serum lipid levels	Significant positive relation between egg yolk consumption and raised serum lutein and zeaxanthin, MPD, with serum lipid levels constant
Vishwanathan et al. (2009)[42]	Lutein intake via egg yolk and MPOD	Significant positive relation between egg yolk consumption and lutein/zeaxanthin serum levels and MPD
Thurnham et al (2008)[43]	Supplementation of human subjects with lutein, zeaxanthin, and meso-zeaxanthin	Lower plasma response of zeaxanthin in presence of meso-zeaxanthin
Johnson et al. (2008)[44]	Effect of lutein and DHA supplementation on cognition in older women	Supplementation showed a positive relation with verbal fluency, memory, and learning efficiency
Connoly et al (2011)[45]	Serum and macular response to meso-zeaxanthin, lutein, and zeaxanthin supplementation in healthy humans	Subjects exhibited significant increases in serum concentrations of these carotenoids and a subsequent increase in central MPOD
Ma et al. (2012)[46]	Review and meta-analysis; lutein and zeaxanthin and risk of AMD	Xanthophyll consumption not associated with decreased risk of developing early AMD, but may benefit late AMD
Ma et al. (2012)[47]	Lutein and zeaxanthin supplementation in early AMD patients	Significant positive relation with lutein/zeaxanthin supplementation and increased MPOD, improved BCVA, CS
Chung et al. (2004)[20]	Bioavailability of lutein from several dietary and supplemental sources	In healthy men, lutein-enriched eggs showed greater bioavailability than spinach and supplements
Ma et al. (2012)[48]	Lutein supplementation in early AMD patients using mfERG	Early functional abnormalities of central retina in early AMD patients improved by lutein supplementation

AMD: age-related macular degeneration; BCVA: best-corrected visual acuity; CS: contrast sensitivity; DHA: docosahexaenoic acid; mfERG: multifocal electroretinogram; MPD: macular pigment density; MPOD: macular pigment optical density; ref: reference.

TABLE 8.3   Basic Research Studies on Macular Pigments

Author (Year) [Ref]	Area of Study	Outcomes
Bhosale et al. (2004)[8]	GSTP1 as a xanthophyll-binding protein	GSTP is a specific binding protein for zeaxanthin and meso-zeaxanthin
Bhosale et al. (2005)[49]	Quantification of 3′ oxolutein	3′ oxolutein quantification in human retina using HPLC
Bhosale et al. (2007)[50]	HPLC identification of a new carotenoid metabolite	Positive correlation between 3-methoxyzeaxanthin in human macula and age
Li et al. (2011)[7]	StARD3 as a lutein binding protein	StARD3 selectively binds lutein
Mamatha et al. (2011)[51]	Effect of dietary components in relation to intestinal lutein uptake in a rat model	Positive correlation between soybean oil/phospholipids and inverse correlation between dietary fiber and β-carotene on lutein absorption
Vachali et al. (2012)[52]	Surface plasmon resonance studies with carotenoids and their binding proteins	Quantification of lutein and zeaxanthin binding specificity against StARD3 and GSTP1, respectively

HPLC: high-performance liquid chromatography; ref: reference.

2. MACULAR DEGENERATION

- Epidemiologic evidence suggests a correlation with increased lutein and zeaxanthin consumption and a reduced risk for AMD; several large-scale studies, including CARMA and AREDS2, are currently assessing the effects of carotenoid supplementation.

References

1. Liutkeviciene R, Lesauskaite V, Asmoniene V, Zaliuniene D, Jasinskas V. Factors determining age-related macular degeneration: a current view. *Medicina (Kaunas)* 2010;46(2):89–94.

2. Congdon N, O'Colmain B, Klaver CC, Klein R, Munoz B, Friedman DS, et al. Causes and prevalence of visual impairment among adults in the United States. *Arch Ophthalmol* 2004;122(4):477–85.

3. Mitchell P, Foran S. Age-Related Eye Disease Study severity scale and simplified severity scale for age-related macular degeneration. *Arch Ophthalmol* 2005;123(11):1598–9.

4. Ferris FL, Davis MD, Clemons TE, Lee LY, Chew EY, Lindblad AS, et al. A simplified severity scale for age-related macular degeneration: AREDS Report No. 18. *Arch Ophthalmol* 2005;123(11):1570–4.

5. Trumbo PR, Ellwood KC. Lutein and zeaxanthin intakes and risk of age-related macular degeneration and cataracts: an evaluation using the Food and Drug Administration's evidence-based review system for health claims. *Am J Clin Nutr* 2006;84(5):971–4.

6. Li B, Vachali P, Bernstein PS. Human ocular carotenoid-binding proteins. *Photochem Photobiol Sci* 2010;9(11):1418–25.

7. Li B, Vachali P, Frederick JM, Bernstein PS. Identification of StARD3 as a lutein-binding protein in the macula of the primate retina. *Biochemistry* 2011;50(13):2541–9.

8. Bhosale P, Larson AJ, Frederick JM, Southwick K, Thulin CD, Bernstein PS. Identification and characterization of a Pi isoform of glutathione S-transferase (GSTP1) as a zeaxanthin-binding protein in the macula of the human eye. *J Biol Chem* 2004;279(47):49447–54.

9. Nolan JM. The role of the macular carotenoids as a blue light filter and an antioxidant. In: Stratton RD, Hauswirth WW, Gardner TW, editors. *Studies on Retinal and Choroidal Disorders*. New York: Humana Press; 2012. p. 595–611.

10. Seddon JM, Ajani UA, Sperduto RD, Hiller R, Blair N, Burton TC, et al. Dietary carotenoids, vitamins A, C, and E, and advanced age-related macular degeneration. Eye Disease Case-Control Study Group. *JAMA* 1994;272(18):1413–20.

11. Bone RA, Landrum JT, Mayne ST, Gomez CM, Tibor SE, Twaroska EE. Macular pigment in donor eyes with and without AMD: a case-control study. *Invest Ophthalmol Vis Sci* 2001;42(1):235–40.

12. Ma L, Lin XM. Effects of lutein and zeaxanthin on aspects of eye health. *J Sci Food Agric* 2010;90(1):2–12.

13. Li B, Ahmed F, Bernstein PS. Studies on the singlet oxygen scavenging mechanism of human macular pigment. *Arch Biochem Biophys* 2010;504(1):56–60.

14. Gellermann W, Bernstein PS. Noninvasive detection of macular pigments in the human eye. *J Biomed Opt* 2004;9(1):75–85.

15. Bernstein P, Gellermann W. Measurement of carotenoids in the living primate eye using resonance Raman spectroscopy. In: Amstrong D, editor. *Oxidants and Antioxidants: Methods in Molecular Biology*. New York: Humana Press; 2002. p. 321–9.

16. Delori FC, Gragoudas ES, Francisco R, Pruett RC. Monochromatic ophthalmoscopy and fundus photography. The normal fundus. *Arch Ophthalmol* 1977;95(5):861–8.

17. Bernstein PS, Zhao DY, Wintch SW, Ermakov IV, McClane RW, Gellermann W. Resonance Raman measurement of macular carotenoids in normal subjects and in age-related macular degeneration patients. *Ophthalmology* 2002;109(10):1780–7.

18. Bernstein PS, Ahmed F, Liu A, Allman S, Sheng X, Sharifzadeh M, et al. Macular pigment imaging in AREDS2 participants: an ancillary study of AREDS2 subjects enrolled at the Moran Eye Center. *Invest Ophthalmol Vis Sci* 2012;53(10):6178–86.

19. Vachali P, Bhosale P, Bernstein PS. Microbial carotenoids. *Methods Mol Biol* 2012;898:41–59.

20. Chung H-Y, Rasmussen HM, Johnson EJ. Lutein bioavailability is higher from lutein-enriched eggs than from supplements and spinach in men. *J Nutr* 2004;134(8):1887–93.

21. Johnson EJ, Maras JE, Rasmussen HM, Tucker KL. Intake of lutein and zeaxanthin differ with age, sex, and ethnicity. *J Am Diet Assoc* 2010;110(9):1357–62.

22. EFSA Panel on Food Additives and Nutrient Sources Added to Food (ANS). Scientific opinion on the re-evaluation of lutein preparations other than lutein with high concentrations of total saponified carotenoids at levels of at least 80%. *EFSA J* 2011;9(5):2144–69.

23. EFSA Panel on Dietetic Products Nutrition and Allergies (NDA). Statement on the safety of synthetic zeaxanthin as an ingredient in food supplements. *EFSA J* 2012;10(10):2891–905.

24. Kruger CL, Murphy M, DeFreitas Z, Pfannkuch F, Heimbach J. An innovative approach to the determination of safety for a dietary ingredient derived from a new source: case study using a crystalline lutein product. *Food Chem Toxicol* 2002;40(11):1535–49.

25. Kijlstra A, Tian Y, Kelly ER, Berendschot TT. Lutein: more than just a filter for blue light. *Prog Retin Eye Res* 2012;31(4):303–15.

26. During A, Doraiswamy S, Harrison EH. Xanthophylls are preferentially taken up compared with beta-carotene by retinal cells via a SRBI-dependent mechanism. *J Lipid Res* 2008;49(8):1715–24.

27. Lobo GP, Amengual J, Palczewski G, Babino D, von Lintig J. Mammalian carotenoid-oxygenases: key players for carotenoid function and homeostasis. *Biochim Biophys Acta* 2012;1821(1):78–87.

28. Connor WE, Duell PB, Kean R, Wang Y. The prime role of HDL to transport lutein into the retina: evidence from HDL-deficient WHAM chicks having a mutant ABCA1 transporter. *Invest Ophthalmol Vis Sci* 2007;48(9):4226–31.

29. Delyfer MN, Buaud B, Korobelnik JF, Rougier MB, Schalch W, Etheve S, et al. Association of macular pigment density with plasma omega-3 fatty acids: the PIMAVOSA study. *Invest Ophthalmol Vis Sci* 2012;53(3):1204–10.

30. Thomas TR, Smith BK, Donahue OM, Altena TS, James-Kracke M, Sun GY. Effects of omega-3 fatty acid supplementation and exercise on low-density lipoprotein and high-density lipoprotein subfractions. *Metabolism* 2004;53(6):749–54.

31. Bernstein PS, Delori FC, Richer S, van Kuijk FJ, Wenzel AJ. The value of measurement of macular carotenoid pigment optical densities and distributions in age-related macular degeneration and other retinal disorders. *Vision Res* 2010;50(7):716–28.

32. Chew EY, Clemons T, Sangiovanni JP, Danis R, Domalpally A, McBee W, et al. The Age-Related Eye Disease Study 2 (AREDS2): Study Design and Baseline Characteristics (AREDS2 Report Number 1). *Ophthalmology* 2012;119(11):2282–9.

33. van de Kraats J, van Norren D. Directional and nondirectional spectral reflection from the human fovea. *J Biomed Opt* 2008;13(2):024010.

34. Barker 2nd FM, Snodderly DM, Johnson EJ, Schalch W, Koepcke W, Gerss J, et al. Nutritional manipulation of primate retinas, V: effects of lutein, zeaxanthin, and n-3 fatty acids on retinal sensitivity to blue-light-induced damage. *Invest Ophthalmol Vis Sci* 2011;52(7):3934–42.

35. Wooten BR, Hammond BR. Macular pigment: influences on visual acuity and visibility. *Prog Retin Eye Res* 2002;21(2):225–40.

36. Nolan JM, Stack J, O'Donovan O, Loane E, Beatty S. Risk factors for age-related maculopathy are associated with a relative lack of macular pigment. *Exp Eye Res* 2007;84(1):61–74.

37. Sharifzadeh M, Bernstein PS, Gellermann W. Nonmydriatic fluorescence-based quantitative imaging of human macular pigment distributions. *J Opt Soc Am A Opt Image Sci Vis* 2006;23(10):2373–87.

38. Alves-Rodrigues A, Shao A. The science behind lutein. *Toxicol Lett* 2004;**150**(1):57–83.

39. Curran-Celentano J, Hammond BR, Ciulla TA, Cooper DA, Pratt LM, Danis RB. Relation between dietary intake, serum concentrations, and retinal concentrations of lutein and zeaxanthin in adults in a Midwest population. *Am J Clin Nutr* 2001;**74**(6):796–802.

40. Broekmans WM, Berendschot TT, Klopping-Ketelaars IA, de Vries AJ, Goldbohm RA, Tijburg LB, et al. Macular pigment density in relation to serum and adipose tissue concentrations of lutein and serum concentrations of zeaxanthin. *Am J Clin Nutr* 2002;**76**(3):595–603.

41. Goodrow EF, Wilson TA, Houde SC, Vishwanathan R, Scollin PA, Handelman G, et al. Consumption of one egg per day increases serum lutein and zeaxanthin concentrations in older adults without altering serum lipid and lipoprotein cholesterol concentrations. *J Nutr* 2006;**136**(10):2519–24.

42. Vishwanathan R, Goodrow-Kotyla EF, Wooten BR, Wilson TA, Nicolosi RJ. Consumption of 2 and 4 egg yolks/d for 5 wk increases macular pigment concentrations in older adults taking cholesterol-lowering statins. *Am J Clin Nutr* 2009;**90**(5):1272–9.

43. Thurnham DI, Tremel A, Howard AN. A supplementation study in human subjects with a combination of meso-zeaxanthin, (3R,3′R)-zeaxanthin and (3R,3′R,6′R)-lutein. *Br J Nutr* 2008;**100**(6):1307–14.

44. Johnson EJ, McDonald K, Caldarella SM, Chung HY, Troen AM, Snodderly DM. Cognitive findings of an exploratory trial of docosahexaenoic acid and lutein supplementation in older women. *Nutr Neurosci* 2008;**11**(2):75–83.

45. Connolly EE, Beatty S, Loughman J, Howard AN, Louw MS, Nolan JM. Supplementation with all three macular carotenoids: response, stability, and safety. *Invest Ophthalmol Vis Sci* 2011;**52**(12):9207–17.

46. Ma L, Dou HL, Wu YQ, Huang YM, Huang YB, Xu XR, et al. Lutein and zeaxanthin intake and the risk of age-related macular degeneration: a systematic review and meta-analysis. *Br J Nutr* 2012;**107**(3):350–9.

47. Ma L, Yan SF, Huang YM, Lu XR, Qian F, Pang HL, et al. Effect of lutein and zeaxanthin on macular pigment and visual function in patients with early age-related macular degeneration. *Ophthalmology* 2012;**119**(11):2290–7.

48. Ma L, Dou HL, Huang YM, Lu XR, Xu XR, Qian F, et al. Improvement of retinal function in early age-related macular degeneration after lutein and zeaxanthin supplementation: a randomized, double-masked, placebo-controlled trial. *Am J Ophthalmol* 2012;**154**(4):625–34.

49. Bhosale P, Bernstein PS. Quantitative measurement of 3′-oxolutein from human retina by normal-phase high-performance liquid chromatography coupled to atmospheric pressure chemical ionization mass spectrometry. *Anal Biochem* 2005;**345**(2):296–301.

50. Bhosale P, Zhao DY, Serban B, Bernstein PS. Identification of 3-methoxyzeaxanthin as a novel age-related carotenoid metabolite in the human macula. *Invest Ophthalmol Vis Sci* 2007;**48**(4):1435–40.

51. Mamatha BS, Baskaran V. Effect of micellar lipids, dietary fiber and beta-carotene on lutein bioavailability in aged rats with lutein deficiency. *Nutrition* 2011;**27**(9):960–6.

52. Vachali P, Li B, Nelson K, Bernstein PS. Surface plasmon resonance (SPR) studies on the interactions of carotenoids and their binding proteins. *Arch Biochem Biophys* 2012;**519**(1):32–7.

GLAUCOMAS

9

Glaucoma and Antioxidant Status

Mehmet Tosun[1], Ramazan Yağcı[2], Mesut Erdurmuş[1]

[1]Abant Izzet Baysal University Medical School, Bolu, Turkey, [2]Pamukkale University Medical School, Denizli, Turkey

INTRODUCTION

Glaucoma is a group of chronic eye diseases that share certain common features. It can irreversibly damage the retinal ganglion cells (RGCs) of the optic nerve and lead to severe vision loss and blindness if left untreated. Glaucoma is characterized by both typical structural glaucomatous optic nerve changes and visual field loss (Fig. 9.1), with or without intraocular pressure (IOP) elevation. The structural loss of axons, deepening and excavation of the cup, can be recognized by thinning of the nerve fiber layer surrounding the disk, either by clinical exam or by imaging methods, such as optical coherence tomography (Fig. 9.2), confocal scanning laser ophthalmoscopy, or scanning laser polarimetry.

Glaucoma is the second-leading cause of blindness worldwide, after age-related cataract, and it is one of the leading causes of preventable blindness. Globally, an estimated 60.5 million people had glaucoma in 2010. Of these, an estimated 44.7 million had primary open-angle glaucoma (POAG). The prevalence of glaucoma is increasing worldwide as the population ages, and it is expected to reach 79.6 million in 2020.[1]

There are many classification systems for glaucoma, according to etiologic, genetic, and anterior chamber angle characteristics. The most widely used classification system relies on the appearance of the drainage angle of the anterior chamber, which separates open-angle glaucoma from angle-closure glaucoma. There are many types, but the major forms of glaucoma include POAG, angle-closure glaucoma, and secondary open-angle glaucoma including pseudoexfoliation (PEX) glaucoma.[2] The precise pathophysiologic mechanisms that lead to glaucomatous optic neuropathy remain uncertain. Increasing evidence indicates that oxidative stress plays a critical role as an etiologic risk factor in the pathophysiology of glaucomatous RGC loss.[3-6] This chapter focuses on the role of oxidative stress in the pathogenesis of open-angle glaucoma.

OXIDATIVE STRESS

Oxidative stress can be described as elevated intracellular concentrations of reactive oxygen species (ROS) that include superoxide anion (O_2^-), hydrogen peroxide (H_2O_2), hydroxyl radical ($OH\cdot$), peroxyl radical, and singlet oxygen ($^1O_2$) over the range of physiologic values. In other words, it is an imbalance between ROS production and breakdown by endogenous antioxidants, which is associated with aging and a number of degenerative diseases. The eye is an exceptional organ that is more vulnerable to oxidative stress through light exposure, radiation, trauma, high level of O2 consumption, continuous exposure to environmental chemicals, and atmospheric oxygen (Fig. 9.3). The formation of multiple ROS can initiate and propagate free radicals. Under normal physiologic conditions, the resultant free radicals can be terminated by antioxidant defense systems; enzymatically through superoxide dismutase (SOD), catalase (CAT), glutathione peroxidase (GPx), and glutathione reductase; and nonenzymatically by glutathione (GSH), thioredoxin, and antioxidant vitamins (Fig.9.4).[7] An imbalance between oxidants and antioxidants in favor of the oxidants can lead to damage to lipids, proteins, and deoxyribonucleic acid (DNA), culminating in cell death (Fig. 9.3).

ROS, which are partially reduced metabolites of molecular oxygen (O_2), are generated intracellularly through a variety of processes, including normal aerobic metabolism and various signal transduction pathways. ROS can also be generated from exogenous sources as a consequence of the cell's exposure to some environmental insult.[8] Any electron-transferring protein or enzymatic system may produce ROS during cellular electron transfer reactions. The mitochondrial electron transport system consumes approximately 85–90% of O_2 in the cell; thus, it is the major source of the endogenous ROS production.

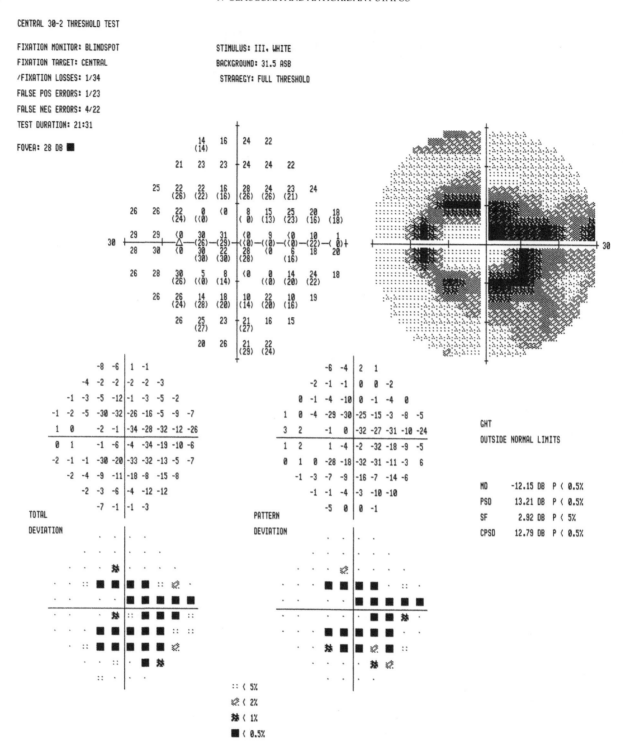

FIGURE 9.1 Humphrey visual field analysis in a patient with glaucoma. This patient had inferior and superior arcuate defects due to glaucoma.

In addition to the detrimental effects on cellular integrity through protein, lipid, and nucleic acid oxidation, high levels of ROS can further inhibit electron transport chain enzymes in the mitochondria and terminate mitochondrial energy production. In contrast, controlled production of ROS at low levels during normal physiologic conditions and regulated redox modifications of transcription factors or enzymes are required parts of signal transduction pathways, which serve important regulatory functions. Therefore, there is an obvious paradox regarding the role of ROS in the regulation of cellular functions.

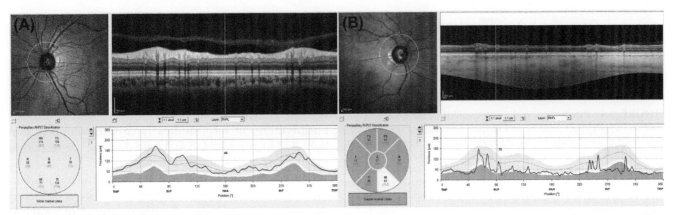

FIGURE 9.2 Spectral domain optical coherence tomography (OCT) images, thickness maps, and profiles of thickness of the circumpapillary retinal nerve fiber layer in a healthy subject (A) and a glaucoma patient (B). Note the loss of temporal retinal nerve fiber layer thickness in the glaucoma patient.

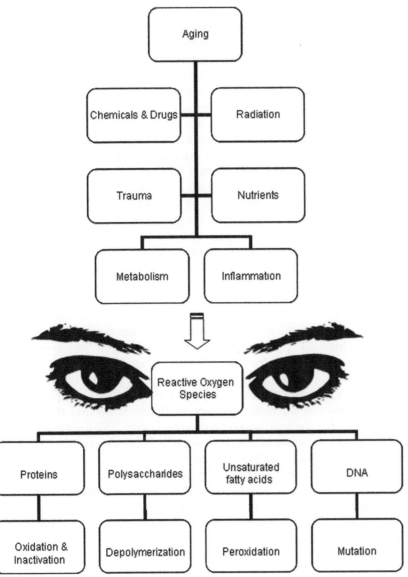

FIGURE 9.3 Sources of reactive oxygen species in the eye. Damage to cellular macromolecules (unsaturated fatty acids, proteins, polysaccharides, and deoxyribonucleic acid) may occur when cellular production of reactive oxygen species overwhelms its antioxidant capacity.

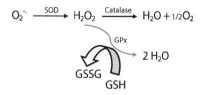

FIGURE 9.4 Biochemical pathways involved in the generation of reactive oxygen species (ROS) and in their scavenging. *The reduction of O^2 leads to the formation of the superoxide anion ($O^{2-}\cdot$) in the cell. $O^{2-}\cdot$ can react with H_2 to produce the hydroperoxide radical and then hydrogen peroxide (H_2O_2). As H_2O_2 is detrimental to cell integrity, catalase catalyzes its dismutation to O_2 and H_2O, and glutathione peroxidase (GPx) catalyzes its reduction to H_2O. GSH: glutathione; GSSG: glutathione disulfide; SOD: superoxide dismutase.*

The neural cells in the central nervous system exhibit a distinct susceptibility to oxidative stress, due to its enhanced metabolic rate with high levels of O_2 utilization, higher lipid content in neuron membranes, and limited cellular regeneration capacity. Besides the increased risk factors for the production of high levels of ROS, the brain might have a deficient defense system against oxidative stress. RGCs, like other central nervous system neurons, are particularly vulnerable to oxidative damage, due to their tremendous O2 consumption and high proportion of unsaturated fatty acid contents. Moreover, direct light exposure can detrimentally affect RGCs through oxidative stress, unlike cortical neurons.[8,9]

OXIDATIVE STRESS IN THE PATHOPHYSIOLOGY OF GLAUCOMA

Aqueous humor is an important fluid in the physiology of the human eye. It is secreted by the ciliary epithelium at a flow rate of 2–3 μL/minute, enters the posterior chamber, passes through the pupil into the anterior chamber, and then to the trabecular meshwork (TM) in the anterior chamber angle. It provides nutrients to anterior and posterior chamber structures and removes metabolic wastes. Aqueous humor also helps to maintain IOP in physiologic ranges. The TM, which drains the aqueous humor into the Schlemm canal, is composed of collagen fibers and elastic tissue covered by trabecular cells. These cells have phagocytic activity and may produce matrix metalloproteinase enzymes, extracellular matrix components, and certain growth factors.[10–12] The mechanism of elevated IOP in open-angle glaucoma is impaired outflow of aqueous humor resulting from increased resistance within the TM, particularly in the juxtacanalicular portion.

Oxidative stress may have a role in the pathogenesis of glaucoma, through either TM degeneration or RGC loss. TM degeneration by oxidative stress has been implicated as causing increased IOP, thus contributing to alterations in the aqueous outflow pathway.[13] The progressive loss of TM cells in patients with glaucoma may be ascribed to the long-term oxidative damage induced by ROS.[14,15] Treatment of human TM cells *in vitro* with hydrogen peroxide impairs TM cell adhesion and compromises cytoskeletal cell integrity. It has also been demonstrated that oxidative DNA damage is significantly greater in the TM cells of glaucoma patients, compared with controls.[16] Furthermore, *in vivo* studies have demonstrated that both IOP increase and visual field damage are related to the amount of oxidative DNA damage.[12,17]

The endothelium may be involved in the modulation of vascular permeability by the release of endothelins and nitric oxide (NO), which are important oxidants that can worsen TM metabolism.[18] Endothelins may also influence TM motility and IOP regulation.[19] Aqueous humor endothelin levels have been demonstrated to be higher in glaucoma patients than in unaffected controls.[20] As a result, various mechanisms can contribute to the production of oxidizing free radicals in the TM. However, a predominant role is played by the endogenous aerobic metabolism, which may be abetted by vascular dysregulation.[21]

Increasing evidence supports the idea that the pathogenic mechanism leading to glaucomatous RGC loss is oxidative stress. Oxidative stress induces apoptotic cell death by the activation of c-Jun *N*-terminal kinase (stress-activated protein kinase) and p38 mitogen-activated protein kinase, leading to caspase 3 activation.[22] It is hypothesized that fluctuations of ocular blood flow in patients with normal IOP can lead to ischemia reperfusion injury and result in oxidative tissue damage.[23] This interpretation is in agreement with the vascular pathogenic theory that suggests vessel dysregulation as the fundamental pathogenetic step in POAG.[24] Peripheral vascular insufficiency and restoration of blood flow induces a proinflammatory state reflected by enhanced O_2^- and H_2O_2 generation, which places organs at risk. These ROSs are usually derived from mitochondrial autoxidation, and they are also found within the RGCs, nerve fiber layer, outer plexiform layer, inner segments of photoreceptors, and the retinal pigment epithelium.[25] Increased ROS generation leads to RGC degeneration, glial malfunction, and activation of autoimmune response in glaucoma. However, many aspects of the relationship between oxidative stress and the neurodegenerative course remain obscure. During glaucomatous neuronal cell degeneration, ROS may be directly neurotoxic to RGCs, may cause secondary degeneration by stimulating glial dysfunction, and may also function as a second messenger and/or in regulating redox modifications of downstream effectors.[6]

PEX syndrome is characterized by the production and progressive deposition of fibrilogranular extracellular material in ocular tissues; it is most commonly seen on the pupillary border and anterior lens capsule.[26]

TABLE 9.1 Current Evidence Implicating Oxidative Stress in the Pathophysiology of Glaucoma

Tissue	Oxidative Parameter*	References
Aqueous humor	Antioxidant status/enzyme activity	3,7,20,28–31
	DNA damage	31
Trabecular meshwork	Antioxidant status/enzyme activity	4,18,32
	DNA damage	4,17
Retina/vitreous	Antioxidant status/enzyme activity	49
Serum	Antioxidant status/enzyme activity	29,31,33–38
	DNA damage	31,35,38

DNA: deoxyribonucleic acid.
These studies represent investigations on human subjects.

Accumulation of the PEX material or pigment particles in the anterior chamber angle can predispose the individual to both open-angle and angle-closure glaucoma.[12,27] Reported findings also suggest a role for oxidative stress in the pathogenesis and progression of PEX glaucoma.

Table 9.1 shows the current evidence implicating oxidative stress in the pathophysiology of glaucoma.

ANTIOXIDANTS

Cellular levels of ROS are controlled by antioxidants. An antioxidant is any substance that, when present in low concentrations compared with those of an oxidizable substrate, significantly delays or prevents oxidation of that substrate, such as DNA, lipids, and proteins.[39] In biologic systems, enzymatic and nonenzymatic antioxidant defense systems have evolved to protect against oxidative damage. The enzymatic antioxidant defense system includes SOD, CAT, and GPx. Nonenzymatic antioxidants include proteins (e.g., transferrins or haptoglobins), heat-shock proteins, and low-molecular-mass molecules (e.g., α-tocopherol, ascorbic acid, and GSH).

Enzymatic Antioxidants

Superoxide Dismutase

SOD is a key enzyme in the detoxification of free radicals. It removes superoxide anions derived from extracellular sources, including ionizing radiation and oxidative insult, together with those primarily generated within the mitochondria as byproducts of O_2 metabolism through the electron transport chain and prevents hydroxyl radical formation.[40] O_2^- formation is the first step in the cascade of univalent reductions of O_2, and it is the first

indicator of increased ROS production. Therefore, SOD may be an indicator of the antioxidant defense system.[33] SOD also plays a critical role in inhibiting the oxidative inactivation of NO, thereby preventing peroxynitrite formation. Three distinct isoforms of SOD have been identified and characterized in mammals: copper-zinc SOD, manganese SOD, and extracellular SOD.

Catalase

CAT is a ferriheme-containing enzyme that is responsible for the dismutation of hydrogen peroxide to water and O_2.[41] It is usually localized in peroxisomes and may also be found in other cell components, including cytoplasm and mitochondria. It has a minor role at low levels of hydrogen peroxide, and it becomes activated at higher levels of hydrogen peroxide production. CAT resembles GPx enzyme in terms of removing hydrogen peroxide.

Glutathione Peroxidase

GPx is an intracellular enzyme that reduces H_2O_2 or organic hydroperoxides to water or corresponding alcohols by oxidizing GSH. Re-reduction of the oxidized form of glutathione (GSSG) is then catalyzed by the enzyme glutathione reductase. In mammalian tissues, there are four major forms of selenium-dependent GPx. GPx 1 is known to localize primarily in the glial cells, where GP activity is 10-fold higher than in neurons.

Paraoxonase

Paraoxonases are a group of antioxidant enzymes involved in the hydrolysis of organophosphates; they are widely distributed among tissues. Paraoxonase 1, a high-density, lipoprotein-associated enzyme found in the eye, has a role in oxidative stress.[42] Paraoxonase 1 protects low-density lipoproteins from oxidative damage, both *in vivo* and *in vitro*.[43]

Nonenzymatic Antioxidants

Glutathione

GSH is provided by dietary intake, and it is the main antioxidant in the central nervous system. It consists of a tripeptide of glutamate, cysteine, and glycine, characterized by a reactive thiol group. Reduced GSH is the main intracellular endogenous antioxidant produced, and it can nonenzymatically act directly with free radicals, including superoxide radicals, hydroxyl radicals, NO, and carbon radicals for their removal. The strong antioxidative property of GSH is due to a sulfur atom in the sulfhydryl groups, which enables GSH to donate electrons readily. It maintains exogenous antioxidants such as vitamins C and E in their active forms.[13,44] GSH peroxidase and GSH reductase can act enzymatically to remove H_2O_2 and maintain GSH in a reduced state.

Vitamins

Vitamin E is a lipid-soluble molecule with antioxidant function that prevents mitochondrial production of ROS and apoptosis during O_2 reperfusion.[45] It appears to neutralize the effect of peroxide and prevent lipid peroxidation in membranes.

Ascorbic acid is thought to be a primary substrate in ocular protection because of its high concentration in the eye. Within the cell, vitamin C helps to protect membrane lipids from peroxidation by recycling vitamin E. One of presumed functions of ascorbate is to protect the lens and retina from the damaging effects of ultraviolet radiation.[46] A high level of ascorbic acid is necessary to maintain oxidative balance in the aqueous humor, while vitamin E deficiency increases H_2O_2 levels.[13,44,47]

Miscellaneous

Transition metals (e.g., copper and iron) are essential in most biologic reactions, including DNA, ribonucleic acid (RNA), and protein synthesis, as well as cofactors of numerous enzymes. However, the accumulation of redox-active transition metals in tissues in excess of the capacity of cellular proteins (catalytic, transport, storage) to bind to these metals is cytotoxic. Transferrins play a role in antioxidant defense by binding free iron or metal ions in forms that will not stimulate free-radical reactions.[48]

Independent of the type of stress induced, such as elevated temperatures, heavy metal exposure, or oxidative stress, cells produce heat-shock proteins. The biologic role of heat-shock proteins is to function as molecular chaperones, regulating the functions of other proteins by binding to them, as well as modulating their function, transport, and folding state.[48]

ANTIOXIDANT STATUS IN GLAUCOMA

Humans have effective defense mechanisms against free radicals. Ascorbic acid plays an important protective role in the eye, and high concentrations of ascorbic acid in the vitreous humor,[49] cornea,[50] lachrymal film,[51] central corneal epithelium,[46] and aqueous humor [7] have been described previously. Therefore, ascorbic acid is regarded as the main substrate involved in ocular protection. GSH is another antioxidant present in the eye; it is made up of three amino acids (L-cysteine, glycine, and glutamic acid). GSH protects eye tissues from damage caused by low concentrations of H_2O_2, while the SOD and CAT enzymes protect the eye from elevated H_2O_2 concentrations.[52] High concentrations of GSH are also found in the aqueous humor [7] and the TM.[32]

Our previous study demonstrated decreased antioxidant defense and increased oxidative stress in patients with POAG and PEX glaucoma compared with healthy controls.[34] We reported decreased total antioxidant capacity and increased SOD levels in the serum of patients with POAG and PEX glaucoma, and we found higher total oxidant status levels in the glaucoma groups compared with the controls.

Ferreirae et al. examined the antioxidant status of the aqueous humor of POAG patients.[3] The levels of water-soluble antioxidants (e.g., GSH and ascorbate) in the aqueous humor of glaucoma patients were found to be approximately half of those found in healthy control subjects. CAT activity was not different between the glaucoma and control groups, but SOD activity was approximately 60% higher in the glaucoma group compared with the healthy control eyes. GPx activity was nearly three times greater in the aqueous humor of glaucoma patients compared with controls.[53]

The enzymatic activities of SOD, GPx, and CAT were assessed, along with GSH, thiobarbituric acid-reactive substances (TBARS), and melatonin levels from retinal homogenates, in order to determine the potential for oxidative stress and the response of antioxidant defenses in glaucoma in rats.[54] SOD activity in elevated IOP eyes declined to half of that in the fellow eye. CAT activity declined to approximately 75% of control 3 weeks after IOP elevation. A compensatory 25% increase in GPx activity was observed in experimental eyes compared with controls. Despite the presumed oxidative burden caused by IOP elevation, GSH levels were only moderately reduced in the experimental samples. TBARS levels, a measure of lipid peroxidation, increased to approximately 160% of those of the controls. Melatonin is believed to increase the activity of GPx, raise SOD messenger RNA levels, and function as a direct antioxidant. Its levels progressively declined to a minimum of approximately 25% of those of the controls. The most startling result of this study was the lack of a decline in GSH levels with adequate IOP elevation. The authors suggest that this result might represent an adaptation to redox imbalance with compensatory activation of GSH synthesis pathways.[53]

Majsterek et al.[35] evaluated the oxidative stress markers in the pathogenesis of open-angle glaucoma, and they estimated the activity of antioxidant enzymes (CAT, SOD, and GPx), as well as the total antioxidant status (TAS). A significant decrease in antioxidant enzymes (CAT, SOD, and GPx) and a nonstatistical decrease in TAS were found in glaucoma patients compared with controls. The authors concluded that modulation of the pro-oxidant/antioxidant status might be a relevant target for glaucoma prevention and therapy.

Tanito et al.[36] recently reported that after adjustment for differences in age and sex among Japanese individuals, using multiple regression analysis, lower biologic antioxidant potential values correlated significantly with POAG and PEX glaucoma. The authors concluded that lower systemic antioxidant capacity, measured by ferric-reducing activity, is involved in the pathogenesis of POAG and PEX glaucoma. Another study provided evidence that TAS decreases in the plasma of PEX glaucoma patients, suggesting that TAS may have an important role in the pathogenesis of PEX glaucoma.[37]

Selenium (Se) is a component of the antioxidant enzymes GPx and thioredoxin reductase. Yilmaz et al. found reduced levels of Se in the aqueous humor, conjunctival specimens, and serum of patients with PEX syndrome, which may support the role of impairment in the antioxidant defense system in the pathogenesis of PEX syndrome.[55] Koliakos et al. [29] investigated the pro-oxidant/antioxidant balance in aqueous humor and serum samples of 20 cases of PEX syndrome, 20 of PEX glaucoma, and 20 age-matched controls. CAT activity was also measured in these samples. The authors found that a significant shift in the pro-oxidant/antioxidant balance in favor of oxidants was detected in the PEX glaucoma group, but not in the PEX syndrome group, compared with the controls. CAT activity in the aqueous humor of PEX syndrome and PEX glaucoma patients was significantly lower than that measured in normal aqueous humor. The authors suggested that in PEX syndrome, oxidative stress is counterbalanced in the aqueous humor, whereas the development of PEX glaucoma is accompanied by a disruption of this balance in favor of oxidants.

Bagnis et al. [30] investigated the expression of glutamine synthase (GS), nitric oxide synthase (NOS), SOD, and glutathione transferase (GST) in the aqueous humor of patients with POAG, as well as in controls. In that study, aqueous humor levels of SOD and GST were found to be significantly lower among POAG patients than among controls, and both NOS and GS expression were found to be significantly higher among POAG patients than among controls. The authors surmised that reduced expression of the antioxidant enzymes SOD and GST could aggravate the imbalance between O_2 and nitrogen and lead to free-radical production and detoxification.

Sorkhabi et al. [31] evaluated the DNA damage markers and the antioxidant status of the serum and aqueous humor of glaucoma patients. Aqueous levels of 8-hydroxy-2′-deoxyguanosine (8-OHdG) were higher in glaucoma patients than in the cataract group. Serum levels of 8-OHdG were also higher in glaucoma patients than in the cataract group. The TAS levels in the serum and aqueous humor in glaucoma patients were lower than in cataract patients. The authors concluded that the formation of ROS and/or a decrease in TAS might have an important role in the pathogenesis of glaucoma. In a similar study, Yuki et al. [38] investigated the amount of systemic oxidative stress-related DNA damage and serum TAS of patients with normal-tension glaucoma compared with healthy controls. Interestingly, the authors found increased serum TAS and decreased 8-OHdG levels, which may reflect compensatory alterations in response to increased systemic oxidative stress

in patients. Table 9.2 summarizes the levels of antioxidant enzymes/molecules in aqueous humor and serum of patients with glaucoma.

In conclusion, based on the results of this comprehensive chapter, we concluded that patients with open-angle glaucoma have low levels of serum antioxidant capacity, suggesting a general compromise of the defense system against oxidative stress.

POTENTIAL VALUE OF ANTIOXIDANTS FOR THE TREATMENT OF GLAUCOMA

Antioxidant supplementation may enhance TM function and preserve the RGCs in patients with glaucoma. Because the pathogenesis of glaucoma involves various factors, one of which may be oxidative stress, the possibility exists that natural dietary antioxidants or vitamin supplements may be beneficial. Table 9.3 shows the external sources of certain antioxidant molecules. If increasing dietary intake of antioxidants is shown to reduce optic nerve damage, this would be a promising means of primary prevention for glaucoma. Kang et al. [68] prospectively examined the role of specific carotenoids and vitamins E and C in relation to POAG risk among 116,484 participants followed for at least 10 years. They did not find any strong associations between antioxidant consumption and the risk of POAG. However, in that study, oral intake of antioxidants was determined using questionnaires. Participants' error in dietary reporting would result in biases toward null associations. Conversely, Giaconi et al. [69] reported higher intake of certain fruits and vegetables high in vitamins A and C and carotenoids may be associated with a decreased likelihood of glaucoma in older African-American women.

There is still much debate about the value of natural antioxidant or vitamin supplementation on glaucoma progression. There is a realistic hope that in the near future, we will be able to give more specific advice concerning diet to glaucoma patients.

TAKE-HOME MESSAGES

- Glaucoma is the second most common cause of visual impairment and blindness in elderly people.
- Increased IOP is the major risk factor.
- Oxidative stress may have a role in the pathogenesis of glaucoma.
- Increasing evidence supports the idea that the pathogenic mechanisms leading to glaucomatous RGC loss are increased oxidative stress and a lowered antioxidant defense system.
- Pro-oxidant/antioxidant status might be a relevant target for glaucoma prevention and therapy.

TABLE 9.2　The Levels of Antioxidant Enzymes/Molecules in Aqueous Humor and Serum of Patients with Glaucoma

Antioxidant Molecule	POAG	PEX Syndrome	PEX Glaucoma	NTG	PACG	References
Nitric oxide	↑ (AH, serum)	↑ (AH, serum)	↑ (AH, serum)	-	↑ (AH)	56–59
Nitric oxide markers	↓ (AH, serum)	-	-	-	-	
cGMP	↓ (AH, serum)	-	-	↓ (serum)	-	
Nitrite						
Total reactive antioxidant potential	↓ (AH)	-	↓ (AH)	-	-	3,60
Superoxide dismutase	↑ (AH, serum) ↓ (serum)	↓ (serum)	↑ (serum)	-	-	3,34,29
Catalase	No change (AH, serum) ↓ (serum)	↓ (serum)	-	-	-	3,35,61,62
Glutathione peroxidase	↑ (AH) ↓(serum)	-	↑ (AH)	-	-	3,35,60
Glutathione transferase	↑ (AH)	-	-	-	-	30
Iron-regulating proteins	Increased expression and protein levels	-	-	-	-	63
Myeloperoxidase	No change (serum)	-	-	-	-	61
Homocysteine	No change (serum)	↑ (serum)	↑ (serum)	-	-	64
Eicosapentaenoic	↓ (serum)	-	-	-	-	65
Docosahexaenoic	↓ (serum)	-	-	-	-	
Melatonin	No change (AH, serum)	-	-	-	-	66
Endothelin-1	↑ (AH)	-	-	↑ (serum)	↑ (AH)	58
Vitamin C	↓ (AH)	-	↓ (AH)	-	-	60
Vitamin E	-	-	-	No change (serum)	-	38
Uric acid	-	-	-	↑ (serum)	-	38
Selenium	-	↓ (AH, serum)	-	-	-	55
Total antioxidant status	↓ (serum)	↓ (serum)	↓ (serum)	-	-	34,38
Glutathione	-	-	-	No change (serum)	-	67

AH: aqueous humor; cGMP: cyclic guanosine monophosphate; NTG: normotensive glaucoma; PEX syndrome: pseudoexfoliation syndrome; PEX glaucoma: pseudo-exfoliative glaucoma; PACG: primary angle-closure glaucoma; POAG: primary open-angle glaucoma. ↑: represents increased levels; ↓: represents decreased levels.

TABLE 9.3　External Sources of Certain Antioxidant Molecules

Antioxidant Molecule	Source
Vitamin C (ascorbic acid)	Citrus fruits, green peppers, broccoli, green leafy vegetables, strawberries, raw cabbage, and potatoes
Vitamin E (α-tocopherol)	Wheat germ, nuts, seeds, whole grains, green leafy vegetables, vegetable oil, and fish-liver oil
Glutathione	Asparagus, avocado, grapefruit, squash, potato, cantaloupe, peach, zucchini, spinach, broccoli, watermelon, and strawberries. Fish, meat, and foods that yield sulfur containing amino acids (e.g., eggs) are the preferred sources for maintaining and increasing glutathione levels. Supplemental glutathione is only available in one active form, GSH
Beta-carotene	Carrots, squash, broccoli, sweet potatoes, tomatoes, kale, collards, cantaloupe, peaches, and apricots
Selenium	Fish, shellfish, red meat, grains, eggs, chicken and garlic; vegetables are a good source only if grown in selenium-rich soils
Flavonoids	Cranberries, kale, beets, berries, red and black grapes, oranges, lemons, grapefruits, and green tea
Capsaicin	Chili peppers
Uric acid	Fructose, which is found abundantly in fruits, significantly elevates uric acid levels

References

1. Quigley HA, Broman AT. The number of people with glaucoma worldwide in 2010 and 2020. *Br J Ophthalmol* 2006;**90**:262–7.

2. Lee DA, Higginbotham EJ. Glaucoma and its treatment: a review. *Am J Health Syst Pharm* 2005;**62**:691–9.

3. Ferreira SM, Lerner SF, Brunzini R, Evelson PA, Llesuy SF. Oxidative stress markers in aqueous humor of glaucoma patients. *Am J Ophthalmol* 2004;**137**:62–9.

4. Izzotti A, Saccà SC, Cartiglia C, De Flora S. Oxidative deoxyribonucleic acid damage in the eyes of glaucoma patients. *Am J Med* 2003;**114**:638–46.

5. Levin LA, Clark JA, Johns LK. Effect of lipid peroxidation inhibition on retinal ganglion cell death. *Invest Ophthalmol Vis Sci* 1996;**37**:2744–9.

6. Tezel G. Oxidative stress in glaucomatous neurodegeneration: mechanisms and consequences. *Prog Retin Eye Res* 2006;**25**:490–513.

7. Richer SP, Rose RC. Water soluble antioxidants in mammalian aqueous humor: interaction with UV B and hydrogen peroxide. *Vision Res* 1998;**38**:2881–8.

8. Mainster MA. Light and macular degeneration: a biophysical and clinical perspective. *Eye (Lond)* 1987;**1**:304–10.

9. Organisciak DT, Darrow RM, Barsalou L, Darrow RA, Kutty RK, Kutty G, et al. Light history and age-related changes in retinal light damage. *Invest Ophthalmol Vis Sci* 1998;**39**:1107–16.

10. Tripathi RC, Chan WF, Li J, Tripathi BJ. Trabecular cells express the TGF-beta 2 gene and secrete the cytokine. *Exp Eye Res* 1994;**58**: 523–8.

11. Yun AJ, Murphy CG, Polansky JR, Newsome DA, Alvarado JA. Proteins secreted by human trabecular cells. Glucocorticoid and other effects. *Invest Ophthalmol Vis Sci* 2012;**30**:2012–22.

12. Aslan M, Cort A, Yucel I. Oxidative and nitrative stress markers in glaucoma. *Free Radic Biol Med* 2008;**45**:367–76.

13. Sacca SC, Izzotti A, Rossi P, Traverso C. Glaucomatous outflow pathway and oxidative stress. *Exp Eye Res* 2007;**84**:389–99.

14. Alvarado JA, Murphy CG, Polansky JR, Juster R. Age-related changes in trabecular meshwork cellularity. *Invest Ophthalmol Vis Sci* 1981;**21**:714–27.

15. Alvarado JA, Murphy C, Juster R. Trabecular meshwork cellularity in primary open-angle glaucoma and nonglaucomatous normals. *Ophthalmology* 1984;**91**:564–79.

16. Izzotti A. DNA damage and alterations of gene expression in chronic-degenerative diseases. *Acta Biochim Pol* 2003;**50**:145–54.

17. Sacca SC, Pascotto A, Camicione P, Capris P, Izzotti A. Oxidative DNA damage in the human trabecular meshwork: clinical correlation in patients with primary open-angle glaucoma. *Arch Ophthalmol* 2005;**123**:458–63.

18. Tamm ER, Russell P, Johnson DH, Piatigorsky J. Human and monkey trabecular meshwork accumulate alpha B-crystallin in response to heat shock and oxidative stress. *Invest Ophthalmol Vis Sci* 1996;**37**:2402–13.

19. Haefliger IO, Dettmann E, Liu R, Meyer P, Prunte C, Messerli J, et al. Potential role of nitric oxide and endothelin in the pathogenesis of glaucoma. *Surv Ophthalmol* 1999;**43**:51–8.

20. Noske W, Hensen J, Wiederholt M. Endothelin-like immunoreactivity in aqueous humor of patients with primary open-angle glaucoma and cataract. *Graefes Arch Clin Exp Ophthalmol* 1997;**235**: 551–2.

21. Izzotti A, Bagnis A, Sacca SC. The role of oxidative stress in glaucoma. *Mutat Res* 2006;**612**:105–14.

22. Saeki K, Kobayashi N, Inazawa Y, Zhang H, Nishitoh H, Ichijo H, et al. Oxidation-triggered c-Jun N-terminal kinase (JNK) and p38 mitogen-activated protein (MAP) kinase pathways for apoptosis in human leukaemic cells stimulated by epigallocatechin-3-gallate (EGCG): a distinct pathway from those of chemically induced and receptor-mediated apoptosis. *Biochem J* 2002;**368**:705–20.

23. Mozaffarieh M, Grieshaber MC, Flammer J. Oxygen and blood flow: players in the pathogenesis of glaucoma. *Mol Vis* 2008;**14**: 224–33.

24. Flammer J, Haefliger IO, Orgül S, Resink T. Vascular dysregulation: a principal risk factor for glaucomatous damage? *J Glaucoma* 1999;**8**:212–9.

25. Andrews RM, Griffiths PG, Johnson MA, Turnbull DM. Histochemical localisation of mitochondrial enzyme activity in human optic nerve and retina. *Br J Ophthalmol* 1999;**83**:231–5.

26. Ritch R, Schlotzer-Schrehardt U. Exfoliation syndrome. *Surv Ophthalmol* 2001;**45**:265–315.

27. Ritch R, Schlotzer-Schrehardt U, Konstas AG. Why is glaucoma associated with exfoliation syndrome? *Prog Retin Eye Res* 2003;**22**:253–75.

28. Galassi F, Renieri G, Sodi A, Ucci F, Vannozzi L, Masini E. Nitric oxide proxies and ocular perfusion pressure in primary open angle glaucoma. *Br J Ophthalmol* 2004;**88**:757–60.

29. Koliakos GG, Befani CD, Mikropoulos D, Ziakas NG, Konstas AG. Prooxidant-antioxidant balance, peroxide and catalase activity in the aqueous humour and serum of patients with exfoliation syndrome or exfoliative glaucoma. *Graefes Arch Clin Exp Ophthalmol* 2008;**246**:1477–83.

30. Bagnis A, Izzotti A, Centofanti M, Saccà SC. Aqueous humor oxidative stress proteomic levels in primary open angle glaucoma. *Exp Eye Res* 2012;**103**:55–62.

31. Sorkhabi R, Ghorbanihaghjo A, Javadzadeh A, Rashtchizadeh N, Moharrery M. Oxidative DNA damage and total antioxidant status in glaucoma patients. *Mol Vis* 2011;**7**:41–6.

32. Kahn MG, Giblin FJ, Epstein DL. Glutathione in calf trabecular meshwork and its relation to aqueous humor outflow facility. *Invest Ophthalmol Vis Sci* 1983;**24**:1283–7.

33. Yagci R, Gürel A, Ersöz I, Keskin UC, Hepşen IF, Duman S, et al. Oxidative stress and protein oxidation in pseudoexfoliation syndrome. *Curr Eye Res* 2006;**31**:1029–32.

34. Erdurmuş M, Yağcı R, Atış Ö Karadağ R, Akbaş A, Hepşen IF. Antioxidant status and oxidative stress in primary open angle glaucoma and pseudoexfoliative glaucoma. *Curr Eye Res* 2011;**36**:713–8.

35. Majsterek I, Malinowska K, Stanczyk M, Kowalski M, Blaszczyk J, Kurowska AK, et al. Evaluation of oxidative stress markers in pathogenesis of primary open-angle glaucoma. *Exp Mol Pathol* 2011;**90**:231–7.

36. Tanito M, Kaidzu S, Takai Y, Ohira A. Status of systemic oxidative stresses in patients with primary open-angle glaucoma and pseudoexfoliation syndrome. *PLoS One* 2012;**7**:e49680.

37. Abu-Amero KK, Kondkar AA, Mousa A, Osman EA, Al-Obeidan SA. Decreased total antioxidants status in the plasma of patients with pseudoexfoliation glaucoma. *Mol Vis* 2011;**17**:2769–75.

38. Yuki K, Murat D, Kimura I, Tsubota K. Increased serum total antioxidant status and decreased urinary 8-hydroxy-2'-deoxyguanosine levels in patients with normal-tension glaucoma. *Acta Ophthalmol* 2010;**88**:259–64.

39. Halliwell B, Gutteridge JM. The definition and measurement of antioxidants in biological systems. *Free Radic Biol Med* 1995;**18**:125–6.

40. Enghild JJ, Thogersen IB, Oury TD, Valnickova Z, Hojrup P, Crapo JD. The heparin-binding domain of extracellular superoxide dismutase is proteolytically processed intracellularly during biosynthesis. *J Biol Chem* 1999;**274**:14818–22.

41. Qi X, Hauswirth WW, Guy J. Dual gene therapy with extracellular superoxide dismutase and catalase attenuates experimental optic neuritis. *Mol Vis* 2007;**13**:1–11.

42. Hashim Z, Ilyas A, Saleem A, Salim A, Zarina S. Expression and activity of paraoxonase 1 in human cataractous lens tissue. *Free Radic Biol Med* 2009;**46**:1089–95.

43. Baskol G, Demir H, Baskol M, Kilic E, Ates F, Kocer D, et al. Assessment of paraoxonase 1 activity and malondialdehyde levels in patients with rheumatoid arthritis. *Clin Biochem* 2005;**38**:951–5.

44. Özkaya D, Naziroğlu M, Armağan A, Demirel A, Köroglu BK, Çolakoğlu N, et al. Dietary vitamin C and E modulates oxidative stress induced-kidney and lens injury in diabetic aged male rats through modulating glucose homeostasis and antioxidant systems. *Cell Biochem Funct* 2011;**29**:287–93.

45. Southam E, Thomas PK, King RH, Goss-Sampson MA, Muller DP. Experimental vitamin E deficiency in rats. Morphological and functional evidence of abnormal axonal transport secondary to free radical damage. *Brain* 1991;**114**:915–36.

46. Ringvold A, Anderssen E, Kjonniksen I. Distribution of ascorbate in the anterior bovine eye. *Invest Ophthalmol Vis Sci* 2000;**41**:20–3.

47. Chow CK, Ibrahim W, Wei Z, Chan AC. Vitamin E regulates mitochondrial hydrogen peroxide generation. *Free Radic Biol Med* 1999;**27**:580–7.

48. Mozaffarieh M, Grieshaber MC, Orgül S, Flammer J. The potential value of natural antioxidative treatment in glaucoma. *Surv Ophthalmol* 2008;**53**:479–505.

49. Hanashima C, Namiki H. Reduced viability of vascular endothelial cells by high concentration of ascorbic acid in vitreous humor. *Cell Biol Int* 1999;**23**:287–98.

50. Brubaker RF, Bourne WM, Bachman LA, McLaren JW. Ascorbic acid content of human corneal epithelium. *Invest Ophthalmol Vis Sci* 2000;**41**:1681–3.

51. Dreyer R, Rose RC. Lachrymal gland uptake and metabolism of ascorbic acid. *Proc Soc Exp Biol Med* 1993;**202**:212–6.

52. Costarides AP, Riley MV, Green K. Roles of catalase and the glutathione redox cycle in the regulation of the anterior-chamber hydrogen peroxide. *Ophthal Res* 1991;**23**:284–94.

53. Kumar DM, Agarwal N. Oxidative stress in glaucoma: a burden of evidence. *J Glaucoma* 2007;**16**:334–43.

54. Moreno MC, Campanelli J, Sande P, Sánez DA, Keller Sarmiento MI, Rosenstein RE. Retinal oxidative stress induced by high intraocular pressure. *Free Radic Biol Med* 2004;**37**:803–12.

55. Yilmaz A, Ayaz L, Tamer L. Selenium and pseudoexfoliation syndrome. *Am J Ophthalmol* 2011;**151**:272–6.

56. Galassi F, Renieri G, Sodi A, Ucci F, Vannozzi L, Masini E. Nitric oxide proxies and ocular perfusion pressure in primary open angle glaucoma. *Br J Ophthalmol* 2004;**88**(6):757–60.

57. Gulaia NM, Zhaboedov GD, Petrenko OV, Kurilina EI, Kosiakova GV, Berdyshev AG. Changes of nitric oxide level at different stages of primary open-angle glaucoma. *Ukr Biokhim Zh* 2003;**75**(5):85–9.

58. Ghanem AA, Elewa AM, Arafa LF. Endothelin-1 and nitric oxide levels in patients with glaucoma. *Ophthalmic Res* 2011;**46**(2):98–102.

59. Borazan M, Karalezli A, Kucukerdonmez C, Bayraktar N, Kulaksizoglu S, Akman A, et al. Aqueous humor and plasma levels of vascular endothelial growth factor and nitric oxide in patients with pseudoexfoliation syndrome and pseudoexfoliation glaucoma. *J Glaucoma* 2010;**19**(3):207–11.

60. Ferreira SM, Lerner SF, Brunzini R, Evelson PA, Llesuy SF. Antioxidant status in the aqueous humour of patients with glaucoma associated with exfoliation syndrome. *Eye (Lond)* 2009;**23**(8): 1691–7.

61. Yildirim O, Ateş NA, Ercan B, Muşlu N, Unlü A, Tamer L, et al. Role of oxidative stress enzymes in open-angle glaucoma. *Eye (Lond)* 2005;**19**(5):580–3.

62. Zoric L, Miric D, Milenkovic S, Jovanovic P, Trajkovic G. Pseudoexfoliation syndrome and its antioxidative protection deficiency as risk factors for age-related cataract. *Eur J Ophthalmol* 2006;**16**(2):268–73.

63. Farkas RH, Chowers I, Hackam AS, Kageyama M, Nickells RW, Otteson DC, et al. Increased expression of iron-regulating genes in monkey and human glaucoma. *Invest Ophthalmol Vis Sci* 2004;**45**(5):1410–7.

64. Altintaş O, Maral H, Yüksel N, Karabaş VL, Dillioğlugil MO, Cağlar Y. Homocysteine and nitric oxide levels in plasma of patients with pseudoexfoliation syndrome, pseudoexfoliation glaucoma, and primary open-angle glaucoma. *Graefes Arch Clin Exp Ophthalmol* 2005;**243**(7):677–83.

65. Ren H, Magulike N, Ghebremeskel K, Crawford M. Primary open-angle glaucoma patients have reduced levels of blood docosahexaenoic and eicosapentaenoic acids. *Prostaglandins Leukot Essent Fatty Acids* 2006;**74**(3):157–63.

66. Chiquet C, Claustrat B, Thuret G, Brun J, Cooper HM, Denis P. Melatonin concentrations in aqueous humor of glaucoma patients. *Am J Ophthalmol* 2006;**142**(2):325–7.

67. Park MH, Moon J. Circulating total glutathione in normal tension glaucoma patients: comparison with normal control subjects. *Korean J Ophthalmol* 2012;**26**(2):84–91.

68. Kang JH, Pasquale LR, Willett W, Rosner B, Egan KM, Faberowski N, et al. Antioxidant intake and primary open-angle glaucoma: a prospective study. *Am J Epidemiol* 2003;**158**:337–46.

69. Giaconi JA, Yu F, Stone KL, Pedula KL, Ensrud KE, Cauley JA, et al. Study of Osteoporotic Fractures Research Group. The association of consumption of fruits/vegetables with decreased risk of glaucoma among older African-American women in the study of osteoporotic fractures. *Am J Ophthalmol* 2012;**154**:635–44.

10

Quercetin and Glaucoma

Naoya Miyamoto, Hiroto Izumi, Akihiko Tawara, Kimitoshi Kohno

School of Medicine, University of Occupational and Environmental Health, Fukuoka, Japan

INTRODUCTION

Glaucoma is a blinding optic neuropathy characterized by the progressive degeneration of retinal ganglion cells (RGCs) and visual field defects. It affects approximately 70 million people worldwide and is the leading cause of irreversible blindness. Elevated intraocular pressure (IOP), caused by a reduction in aqueous outflow, is a major risk factor in the development of glaucoma[1] and the progression of glaucomatous damage to the optic nerve.[2]

IOP elevation and visual field damage are reported to be proportional to the deoxyribonucleic acid (DNA) oxidative damage found in the human trabecular meshwork (TM).[3] This finding forms the basis for the role of oxidative stress in the pathogenesis of glaucoma and provides new insight into the molecular mechanisms involved.

Flavonoids such as quercetin (3,5,7,3′,4′-pentahydroxy flavone) can protect cells from oxidative stress.[4,5] Quercetin is one of the most widely distributed flavonoids.[6] It has been shown that certain flavonoids can induce antioxidant responsive element-dependent gene expression through the activation of nuclear factor (erythroid-derived 2)-like 2 (Nrf2).[7] Most importantly, oxidative stress plays an important role in the pathogenesis of multiple ocular diseases, including glaucoma.[8]

OXIDATIVE STRESS

Under physiologic conditions, there is a state of equilibrium between the endogenous production of free radicals and their neutralization capacity ('scavenging' activity). When damage ensues, this condition is known as oxidative stress.

More than 90% of the oxygen is consumed by mitochondria in aerobic organisms. Under normal physiologic conditions, about 1–5% of the oxygen consumed by mitochondria is converted to reactive oxygen species (ROS), including superoxide anions, hydrogen peroxide (H_2O_2), and hydroxyl radicals.[9] Mitochondrial respiratory function declines with age,[10] and this increases the production of ROS and free radicals in mitochondria. Consequently, ROS production essentially depends on mitochondrial function and on the levels of antioxidant defenses.[11]

Free radicals are neutralized by three major antioxidant systems in mammalian cells: superoxide dismutase/catalase, glutathione (GSH), and peroxiredoxin (PRDX) (Fig. 10.1).[12] Oxidative stresses are neutralized by numerous molecules that are either endogenously produced, such as GSH, or are a part of dietary consumption, such as flavonoids, vitamin C, and vitamin E. This mechanism is believed to be involved in the etiopathogenesis of many degenerative diseases (Fig. 10.2). Both vitamin C and GSH operate in fluids outside the cell and within the cell,[13] whereas vitamin E prevents endogenous mitochondrial production of ROS.[14] This may be important in maintaining cellular homeostasis, which is relevant to the etiology of primary open-angle glaucoma (POAG).[15]

OXIDATIVE STRESS AND TRANSCRIPTION

ROS, including H_2O_2, are toxic and potent inducers of oxidative damage, but they have been shown to also function as necessary second messengers in various signal transduction pathways. Knowledge of the molecular links between stress signaling pathways and transcription factors is essential for understanding the complexity of the genomic response. Many transcription factors are activated by oxidative stress, which induces the expression of target genes such as cellular antioxidant molecules for defense and survival (Fig. 10.3). The genomic response system is thought to decline during aging,

FIGURE 10.1 Scheme for intracellular generation of reactive oxygen species (ROS). O_2^- is dismutated to H_2O_2, which can generate •OH. Cell resistance to oxidative stress and repair depends in large part on removal of H_2O_2 and reduction of phospholipid hydroperoxides (there is no specific scavenger for •OH). H_2O_2 can be removed by multiple enzymes, including catalase, glutathione (GSH) peroxidases, and all peroxiredoxins (PRDXs). *GCS: glutamylcysteine synthetase; GPX: glutathione peroxidase; GR: glutathione reductase; GS: glutathione synthetase; GSSG: glutathione disulfide; GST: glutathione S-transferase; NADPH: nicotinamide adenine dinucleotide phosphate; oxTRX: oxidized thioredoxin; redTRX: reduced thioredoxin; RFK: riboflavin kinase; SOD: superoxide dismutase.*

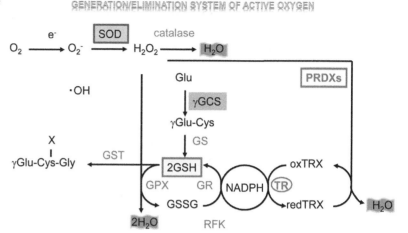

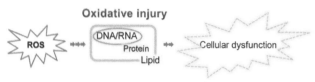

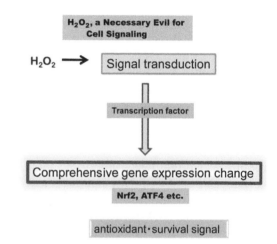

FIGURE 10.2 Systemic and ocular diseases due to oxidative stress. *ALS: amyotrophic lateral sclerosis; DNA: deoxyribonucleic acid; RA: rheumatoid arthritis; RNA: ribonucleic acid; ROP: retinopathy of prematurity; ROS: reactive oxygen species; SLE: systemic lupus erythematosus.*

FIGURE 10.3 Hydrogen peroxide is a metabolic byproduct or common mediator for signal transduction. *ATF4: activating transcription factor 4; Nrf2: nuclear respiratory factor 2.*

suggesting that the dysfunction of antioxidant systems induces various age-related diseases. It is well known that Nrf2 is a master transcription factor involved in stress responses, including oxidative stress. Many transcription factors activated by oxidative stress have been identified, including nuclear factor kappa-light-chain-enhancer of activated B cells (NF-κB), activating transcription factor 4 (ATF4), and nuclear respiratory factor 1 (Nrf1). Recently, Nrf2 regulatory factors and the interacting molecules have been identified.[16] Mechanistic details of the total genomic response are essential to facilitate the development of glaucoma treatments.

OXIDATIVE STRESS AND GLAUCOMA

The pathogenic role of oxidative stress in increasing IOP is supported by various studies. *In vitro* treatment of human TM cells with H_2O_2 alters cellular adhesion and integrity.[17] Perfusion of calf TM cells with H_2O_2 reduces aqueous humor drainage from the anterior chamber of the eye.[18]

In humans, DNA damage is significantly higher in the TM cells of glaucoma patients than in those of age-matched controls.[19] Further studies demonstrated abundant oxidative nucleotide modification (8-hydroxyguanosine (8-OH-dG)) levels in humans. The TM is significantly involved in the increase in IOP and visual field damage.[3,18] Glaucoma patients display a significant depletion of total antioxidant potential in their aqueous humor,[20] display an increase in serum antibodies against glutathione S-transferase,[21] and display a decrease in plasma GSH levels.[22]

Growing evidence supports the involvement of oxidative stress as a common component of neurodegenerative glaucoma in different subcellular compartments of RGCs.[23,24] In addition to the evidence of a direct cytotoxicity leading to RGC death, ROS may also act as a second messenger to modulate protein function by redox modifications of downstream enzymatic oxidation of specific amino acid residues.[25] Studies provide increasing evidence that supports the association of ROS with different aspects of the neurodegenerative process.[26,27] Oxidative protein modifications during neurodegenerative glaucoma increase neuronal susceptibility to damage and also lead to glial dysfunction.[28] Oxidative stress-induced dysfunctional glial cells may contribute to spreading neuronal damage by secondary degeneration.[29] Oxidative stress also promotes the accumulation of advanced glycation end products in glaucomatous tissues.[30]

OXIDATIVE STRESS AND TRABECULAR MESHWORK

It has been suggested that age- and disease-related loss of TM cells, followed by substitution with extracellular matrix, contributes to an increased resistance to aqueous outflow and to the subsequent increase in IOP found in POAG patients.[31,32] With age, resistance increases and alterations of the extracellular matrix in the juxtacanalicular region occur.[33] The loss or altered functionality of human TM cells may be the result of an increase in oxidative stress.[34] Resistance to aqueous humor outflow increases in the presence of high levels of H_2O_2 in eyes with GSH-depleted TM.[18] Moreover, the specific activity of superoxide dismutase demonstrates an age-dependent decline in normal human TM collected from cadavers.[34] The H_2O_2 effect on the adhesion of TM cells to the extracellular matrix proteins results in rearrangement of cytoskeletal structures that may induce a decrease in TM cell adhesion, cell loss, and compromised TM integrity.[17] Oxidative stress can also influence biologic reactions of human TM cells,[35] and may contribute to the changes observed in aging and in POAG.[35] These changes may include trabecular thickening and trabecular fusion.[36] Oxidative damage to the DNA of TM cells is significantly higher in affected patients than in age-matched control subjects, as demonstrated by analysis of 8-hydroxyguanosine (8-OH-dG), the most common oxidatively modified nucleotide.[19] Additional studies report a significant correlation among 8-OH-dG levels in the TM, increased IOP, and visual field damage.[3] The importance of oxidative damage in POAG has been further substantiated by the findings that glaucoma-affected patients have a significant depletion of total antioxidant potential in the aqueous humor,[20] an increase in serum antibodies against glutathione S-transferase,[21] a decrease in plasma GSH levels,[22] and an increase of lipid peroxidation products in the plasma[37] when compared with nonaffected individuals. These findings provide the basis for a possible role of oxidative stress in the pathogenesis of glaucoma and provide new

insight into the molecular mechanisms involved in this blinding disease.[8,38] This pathogenic mechanism plays a fundamental role in POAG, in which TM pathologic changes, mainly including oxidative DNA damage, trigger the 'glaucomatous cascade'.

QUERCETIN AND GLAUCOMA

Flavonoids comprise a large family of plant-derived polyphenolic compounds widely distributed in fruits and vegetables and, therefore, regularly consumed in the human diet.[39,40] They are particularly abundant in beverages derived from plants such as tea, cocoa, and red wine. Flavonoids are believed to exert protective as well as beneficial effects on multiple disease states, including cancer, cardiovascular disease, and neurodegenerative disorders.[39,40] The physiologic benefits of flavonoids are generally thought to be derived from their antioxidant and free-radical scavenging properties.[41] Accordingly, flavonoids may also have therapeutic potential in ocular diseases including glaucoma. Flavonoids such as quercetin, catechins, and kaempferol are better antioxidants than vitamin C and vitamin E.[42] Quercetin is one of the most widely distributed flavonoids, present in fruits, vegetables, and many other dietary sources (apples, onions, tomatoes) (Fig. 10.4). Among the flavonoids, quercetin has several pharmacologic effects such as suppression of cell proliferation, protection of low-density lipoprotein (LDL) oxidation, prevention of platelet aggregation, and induction of

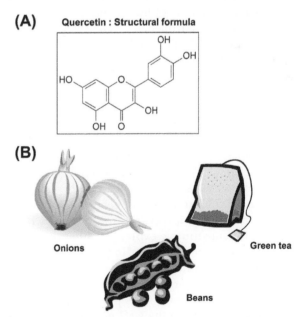

FIGURE 10.4 (A) Chemical structure of quercetin. (B) Quercetin is the main representative of the flavonol class and a polyphenolic antioxidant found in a variety of fruits and vegetables. It is highly concentrated in onions, green tea, apples, grapes (red wine), and soybeans.

apoptosis.[43,44] The preventive effects of quercetin on apoptosis have been reported in several cells such as fibroblasts and epithelial cells.[45,46] Quercetin was able to induce apoptosis in tumor cells through activation of the caspase 3 cascade and suppression of heat shock protein 70.[47]

Flavonoids can induce the expression of phase-2 proteins that function to enhance the cell's natural defenses against oxidative stress. Phase-2 proteins catalyze several different reactions that neutralize ROS and increase the intracellular concentrations of antioxidants such as GSH.[48] Among these phase-2 proteins are some of the key enzymes involved in GSH metabolism (glutathione S-transferase (GSH) and glutamate cysteine ligase) and other antioxidant enzymes, including heme oxygenase 1 (HO-1).[48] Overexpression of HO-1 in cells resulted in a marked reduction in injury and cytotoxicity induced by oxidative stress.[49] Quercetin prevented H_2O_2-induced apoptosis via antioxidant activity and HO-1 gene expression in macrophages.[4] It is reported that quercetin decreased oxidative stress, NF-κB activation, and inducible nitric oxide synthase (iNOS) overexpression in livers of streptozotocin-induced diabetic rats.[50]

Several transcription factors are activated under oxidative stress induced by H_2O_2 and inflammatory cytokines, such as tumor necrosis factor alpha (TNF-α) and interleukin 1 beta (IL-1β). Among them, both NF-κB and Nrf2 are well-known transcription factors related to oxidative stress.[51,52] PRDXs can eliminate H_2O_2 efficiently and participate in many physiologic processes such as signal transduction and apoptosis.[53] There are six distinct members of this family located in various subcellular compartments. PRDX1, 2, and 6 are in the cytoplasm, and PRDX3 is found in mitochondria. PRDX4 is in endoplasmic reticulum and is secreted. PRDX5 is found in various compartments (Table 10.1). It has been previously shown that oxidative stress induces *PRDX1* and *PRDX5* through the activation of Ets1.[54] Furthermore, *PRDX2* expression is regulated by the transcription factor Foxo3a via treatment with the antiglaucoma agents nipradilol and timolol.[55] Miyamoto *et al.* (2011) found that the Nrf2/NRF1 transcription pathway was also involved in the expression of both the *PRDX3* and *PRDX5* genes.[56] Nrf2, a basic leucine zipper transcription factor, is essential for the inducible and constitutive expression of several phase II detoxification proteins including those required for mitochondrial respiratory function.[57] NRF1 was found to act on many nuclear genes required for mitochondrial respiratory function.[58] This primary function was confirmed by disrupting the *Nrf1* gene in mice, resulting in a phenotype of peri-implantation lethality and a striking decrease in the mitochondrial DNA content of Nrf1-null blastocysts.[59]

TABLE 10.1 Cellular H_2O_2 Levels Are Controlled Sequentially by Peroxiredoxins

Human Gene	Amino Acid (aa)	Localization
PRDX1	199 aa	Cytoplasm Nucleus
PRDX2	198 aa	Cytoplasm Cellular membrane
PRDX3	256 aa (cleaved at 63–64 aa)	Mitochondria
PRDX4	271 aa (cleaved at 36–37 aa)	Cytoplasm Golgi body Secretion
PRDX5	214 aa (cleaved at 52–53 aa)	Mitochondria Peroxisome Cytoplasm
PRDX6	224 aa	Cytoplasm

Peroxiredoxins (PRDXs) are a family of small (22–27 kDa) nonselenium peroxidases currently thought to be composed of six mammalian isoforms. Although their individual roles in cellular redox regulation and antioxidant protection are quite distinct, they all catalyze peroxide reduction of H_2O_2, organic hydroperoxides, and peroxynitrite. They are found to be expressed ubiquitously and in high levels, suggesting that they are both an ancient and important enzyme family. PRDXs can be divided into three major subclasses: typical 2-cysteine (2-Cys) PRDXs (PRDX1–4), atypical 2-Cys PRDXs (PRDX5), and 1-Cys PRDXs (PRDX6).

One specific ROS, H_2O_2, is produced by mitochondria. Since PRDXs can eliminate H_2O_2 efficiently, mitochondrial PRDX3 may protect mitochondrial DNA from ROS damage.[60,61] We have previously reported that a member of the high mobility group protein family, mitochondrial transcription factor A (mtTFA), can recognize oxidatively damaged DNA.[62] Furthermore, it has been shown that mtTFA binds to mitochondrial deoxyribonucleic acid (mtDNA) in the same way that histones bind to nuclear DNA.[63] Because mtTFA is not protected by chromatin proteins like histones, it is highly sensitive to oxidative stress. mtTFA may protect mtDNA, acting as a preserver of mitochondrial function.[64] Miyamoto *et al.* (2011)[56] reported that quercetin induced mtTFA protein expression and protected against H_2O_2 toxicity. This indicates that quercetin may protect mitochondria from oxidative stress through the induction of both mtTFA and PRDX3 (Fig. 10.5). Quercetin inhibits the activation of caspase 3 and abolishes the H_2O_2-dependent induction of apoptosis-associated proteins such as Bcl2.[65] This also suggests that quercetin inhibits the mitochondrial apoptotic pathway induced by various stresses.

The endothelium plays a key role in maintenance of anterior chamber homeostasis and is also involved in glaucoma pathogenesis.[66] The expression of PRDX proteins was investigated in Fuchs' endothelial dystrophy, and the expression of PRDX2, 3, and 5 was significantly downregulated.[67] It has been reported that PRDX3 oxidation is found in TNF-α treated cells and is the early

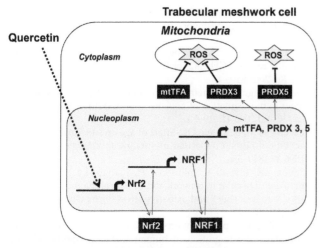

FIGURE 10.5 Scheme for protective effect of quercetin. Quercetin protects the trabecular meshwork (TM) cells by the modulation of an oxidative stress-protective pathway involving control of peroxiredoxin (PRDX)3, PRDX5, and mitochondrial transcription factor A (mtTFA) expression by the transcription factors nuclear respiratory factor 2 (Nrf2) and NRF1.

event of apoptosis. This leads to an increase of H_2O_2, which modulates the progression of apoptosis.[68] These data indicate that the expression of PRDXs in endothelial cells may also be related to glaucoma pathogenesis. PRDX6 reduces oxidative stress and transforming growth factor beta (TGF-β)-induced abnormalities of TM cells.[69] TGF-β is a fibrogenic cytokine that increases ROS production,[70] indicating that our study may be relevant to the physiology and pathophysiology of the outflow pathway in glaucoma.

CONCLUSIONS

Based on the previous reports, it remains to be resolved whether oxidative stress is a principal cause of ocular diseases including glaucoma. However, the involvement of oxidative stress has certainly been confirmed in a spectrum of damaging processes. Oxidative stress plays an important role in glaucoma pathogenesis, affecting the TM cells, RGCs, and the optic nerve head. As the pharmacologic properties of quercetin specifically target the factors involved in glaucomatous disease (oxidative stress and impairment of mitochondrial functions in TM cells), quercetin could theoretically be beneficial for glaucoma. This is certainly supported by solid data, but definitive proof is still required. Indeed, several factors are known to play a role in these processes, but many more remain to be elucidated. Further research in this area will help to understand the physiopathology of glaucoma and to develop new approaches for its prevention and treatment.

TAKE-HOME MESSAGES

- Oxidative stress plays an important role in glaucoma pathogenesis, affecting the TM cells, RGCs, and the optic nerve head.
- Many transcription factors are activated by oxidative stress, which induces the expression of target genes, such as cellular antioxidant molecules for defense and survival.
- Mechanistic details of the total genomic response are essential to facilitate the development of glaucoma treatments.
- The loss or altered functionality of human TM cells may be the result of an increase in oxidative stress.
- Flavonoids are believed to exert protective as well as beneficial effects on multiple disease states.
- Quercetin, a flavonoid, can protect cells from oxidative stress.
- Quercetin induces the expression of antioxidant proteins and protects TM cells from oxidative stress.

References

1. Gordon MO, Beiser JA, Brandt JD, Heuer DK, Higginbotham EJ, Johnson CA, et al. The Ocular Hypertension Treatment Study: baseline factors that predict the onset of primary open-angle glaucoma. *Arch Ophthalmol* 2002;**120**:714–20.
2. Leske MC, Heiji A, Hussein M, Bengtsson B, Hyman L, Komaroff E. Factors for glaucoma progression and effect of treatment: the early manifest glaucoma trial. *Arch Ophthalmol* 2003;**121**:6–48.
3. Saccà SC, Pascotto A, Camicione P, Capris P, Izzotti A. Oxidative DNA damage in the human trabecular meshwork: clinical correlation in patients with primary open-angle glaucoma. *Arch Ophthalmol* 2005;**123**:458–63.
4. Chow JM, Shen SC, Huan SK, Lin HY, Chen YC. Quercetin, but not rutin and quercetin, prevention of H_2O_2-induced apoptosis via anti-oxidant activity and heme oxygenase 1 gene expression in macrophages. *Biochem Pharmacol* 2005;**69**:1839–51.
5. Kook D, Wolf AH, Yu AL, Neubauer AS, Priglinger SG, Kampik A, et al. The protective effect of quercetin against oxidative stress in the human RPE *in vitro*. *Invest Ophthal Vis Sci* 2008;**49**:1712–20.
6. Pawlikowska-Pawlega B, Guszecki WI, Misiak LE, Gawron A. The study of the quercetin action on human erythrocyte membranes. *Biochem Pharmacol* 2003;**66**:605–12.
7. Johnson J, Maher P, Hanneken A. The flavonoid, eriodictyol, induces long-term protection in ARPE-19 cells through its effects on Nrf2 activation and phase II gene expression. *Invest Ophthalmol Vis Sci* 2009;**50**:2398–406.
8. Saccà SC, Izzotti A, Rossi P, Traverso C. Glaucomatous outflow pathway and oxidative stress. *Exp Eye Res* 2007;**84**:389–99.
9. Trounce I, Byrne E, Marzuki S. Decline in skeletal muscle mitochondrial respiratory chain function: possible factor in ageing. *Lancet* 1989;**1**:637–9.
10. Yen TC, Chen YS, King KL, Yeh SH, Wei YH. Liver mitochondrial respiratory functions decline with age. *Biochem Biophy Res Commun* 1989;**165**:994–1003.
11. Camougrand N, Rigoulet M. Aging and oxidative stress: studies of some genes involved both in aging and in response to oxidative stress. *Respir Physiol* 2001;**128**:393–401.

12. Rhee SG, Kang SW, Jeong W, Chang TS, Yang KS, Woo HA. Intracellular messenger function of hydrogen peroxide and its regulation by peroxiredoxins. *Curr Opin Cell Biol* 2005;**17**:183–9.

13. Cardoso SM, Pereira C, Oliveira CR. The protective effect of vitamin E, idebenone and reduced glutathione on free radical mediated injury in rat brain synaptosomes. *Biochem Biophys Res Commun* 1998;**246**:703–10.

14. Southam E, Thomas PK, King RH, Goss-Sampson MA, Muller DP. Experimental vitamin E deficiency in rats. Morphological and functional evidence of abnormal axonal transport secondary to free radical damage. *Brain* 1991;**114**:915–36.

15. Veach J. Functional dichotomy: glutathione and vitamin E in homeostasis relevant to primary open-angle glaucoma. *Br J Nutr* 2004;**91**:809–29.

16. Papp D, Lenti K, Módos D, Fazekas D, Dúl Z, Türei D, et al. The NRF2-related interactome and regulome contain multifunctional proteins and fine-tuned autoregulatory loops. *FEBS Lett.* 2012;**586**:1795–802.

17. Zhou L, Li Y, Yue BY. Oxidative stress affects cytoskeletal structure and cell-matrix interactions in cells from an ocular tissue: the trabecular meshwork. *J Cell Physiol* 1999;**180**:182–9.

18. Kahn MG, Giblin FJ, Epstein DL. Glutathione in calf trabecular meshwork and its relation to aqueous humor outflow facility. *Invest Ophthalmol Vis Sci* 1983;**24**:1283–7.

19. Izzotti A, Saccà SC, Cartiglia C, De Flora S. Oxidative deoxyribonucleic acid damage in the eyes of glaucoma patients. *Am J Med* 2003;**114**:638–46.

20. Ferreira SM, Lerner SF, Brunzini R, Evelson PA, Llesuy SF. Oxidative stress markers in aqueous humor of glaucoma patients. *Am J Ophthalmol* 2004;**137**:62–9.

21. Yang J, Tezel G, Patil RV, Romano C, Wax MB. Serum autoantibody against glutathione S-transferase in patients with glaucoma. *Invest Ophthalmol Vis Sci* 2001;**42**:1273–6.

22. Gherghel D, Griffiths HR, Hilton EJ, Cunliffe IA, Hosking SL. Systemic reduction in glutathione levels occurs in patients with primary open-angle glaucoma. *Invest Ophthalmol Vis Sci* 2005;**46**:877–83.

23. Ganapathy PS, White RE, Ha Y, Bozard BR, McNeil PL, Caldwell RW, et al. The role of N-methyl-D-aspartate receptor activation in homocysteine-induced death of retinal ganglion cells. *Invest Ophthalmol Vis Sci* 2011;**52**:5515–24.

24. Li GY, Fan B, Su GF. Acute energy reduction induces caspase-dependent apoptosis and activates p53 in retinal ganglion cells (RGC-5). *Exp Eye Res* 2009;**89**:581–9.

25. Williams D, Norman G, Khoury C, Metcalfe N, Briard J, Laporte A, et al. Evidence for a second messenger function of dUTP during Bax mediated apoptosis of yeast and mammalian cells. *Biochim Biophys Acta* 2011;**1813**:315–21.

26. Kaur H, Chauhan S, Sandhir R. Protective effect of lycopene on oxidative stress and cognitive decline in rotenone induced model of Parkinson's disease. *Neurochem Res* 2011;**36**:1435–43.

27. Wang F, Zhai H, Huang L, Li H, Xu Y, Qiao X, et al. Aspirin protects dopaminergic neurons against lipopolysaccharide-induced neurotoxicity in primary midbrain cultures. *J Mol Neurosci* 2012;**46**:153–61.

28. Allaman I, Gavillet M, Bélanger M, Laroche T, Viertl D, Lashuel HA, et al. Amyloid-beta aggregates cause alterations of astrocytic metabolic phenotype: impact on neuronal viability. *J Neurosci* 2010;**30**:3326–38.

29. Fitzgerald M, Bartlett CA, Payne SC, Hart NS, Rodger J, Harvey AR, et al. Near infrared light reduces oxidative stress and preserves function in CNS tissue vulnerable to secondary degeneration following partial transection of the optic nerve. *J Neurotrauma* 2010;**27**:2107–19.

30. Giacco F, Brownlee M. Oxidative stress and diabetic complications. *Circ Res* 2010;**107**:1058–70.

31. Alvarado J, Murphy C, Polansky J, Juster R. Age-related changes in trabecular meshwork cellularity. *Invest Ophthalmol Vis Sci* 1981;**21**:714–27.

32. Lütjen-Drecoll E. Morphological changes in glaucomatous eyes and the role of TGFbeta2 for the pathogenesis of the disease. *Exp Eye Res* 2005;**81**:1–4.

33. Tian B, Geiger B, Epstein DL, Kaufman PL. Cytoskeletal involvement in the regulation of aqueous humor outflow. *Invest Ophthalmol Vis Sci* 2000;**41**:619–23.

34. De La Paz MA, Epstein DL. Effect of age on superoxide dismutase activity of human trabecular meshwork. *Invest Ophthalmol Vis Sci* 1996;**37**:1849–53.

35. Tamm ER, Russell P, Johnson DH, Piatigorsky J. Human and monkey trabecular meshwork accumulate alpha B-crystallin in response to heat shock and oxidative stress. *Invest Ophthalmol Vis Sci* 1996;**37**:2402–13.

36. Hogg P, Calthorpe M, Batterbury M, Grierson I. Aqueous humor stimulates the migration of human trabecular meshwork cells. *in vitro. Invest Ophthalmol Vis Sci* 2000;**41**:1091–8.

37. Yildirim O, Ateş NA, Ercan B, Muşlu N, Unlü A, Tamer L, et al. Role of oxidative stress enzymes in open-angle glaucoma. *Eye* 2005;**19**:580–3.

38. Izzotti A, Bagnis A, Saccà SC. The role of oxidative stress in glaucoma. *Mutat Res* 2006;**612**:105–14.

39. Middleton E, Kandaswami C, Theoharides TC. The effects of plant flavonoids on mammalian cells: implications for inflammation, heart disease, and cancer. *Pharmacol Rev* 2000;**52**:673–751.

40. Ross JA, Kasum CM. Dietary flavonoids: bioavailability, metabolic effects, and safety. *Annu Rev Nutr* 2002;**22**:19–34.

41. Ishige K, Schubert D, Sagara Y. Flavonoids protect neuronal cells from oxidative stress by three distinct mechanisms. *Free Radic Biol Med* 2001;**30**:433–46.

42. Noroozi M, Angerson WJ, Lean ME. Effects of flavonoids and vitamin C on oxidative DNA damage to human lymphocytes. *Am J Clin Nutr* 1998;**67**:1210–8.

43. Mardla V, Kobzar G, Samel N. Potentiation of anti-aggregating effect of prostaglandins by alpha-tocopherol and quercetin. *Platelets* 2004;**15**:319–24.

44. Shen SC, Chen YC, Hsu FL, Lee WR. Differential apoptosis-inducing effect of quercetin and its glycosides in human promyeloleukemic HL-60 cells by alternative activation of the caspase 3 cascade. *J Cell Biochem* 2003;**89**:1044–55.

45. Park YH, Xu XR, Chiou GC. Structural requirements of flavonoids for increment of ocular blood flow in the rabbit and retinal function recovery in rat eyes. *J Ocul Pharmacol Ther* 2004;**20**:35–42.

46. Yoshizumi M, Tsuchiya K, Kirima K, Kyaw M, Suzaki Y, Tamaki T. Quercetin inhibits Shc- and phosphatidylinositol 3-kinase-mediated c-Jun N-terminal kinase activation by angiotensin II in cultured rat aortic smooth muscle cells. *Mol Pharmacol* 2001;**60**:656–65.

47. Wei YQ, Zhao X, Kariya Y, Fukata H, Teshigawara K, Uchida A. Induction of apoptosis by quercetin: involvement of heat shock protein. *Cancer Res* 1994;**54**:4952–7.

48. Talalay P. Chemoprotection against cancer by induction of phase 2 enzymes. *Biofactors* 2000;**12**:1–4.

49. Taillé C, El-Benna J, Lanone S, Dang MC, Ogier-Denis E, Aubier M, et al. Induction of heme oxygenase-1 inhibits NAD(P)H oxidase activity by down-regulating cytochrome b558 expression via the reduction of heme availability. *J Biol Chem* 2004;**279**:28681–8.

50. Dias AS, Porawski M, Alonso M, Marroni N, Collado PS, González-Gallego J. Quercetin decreases oxidative stress, NF-kappa B activation, and iNOS overexpression in liver of streptozotocin-induced diabetic rats. *J Nutr* 2005;**135**:2299–304.

51. Gloire G, Legrand-Poels S, Piette J. NF-kappaB activation by reactive oxygen species: fifteen years later. *Biochem Pharmacol* 2006;**72**:1493–505.

52. Nguyen T, Nioi P, Pickett CB. The Nrf2-antioxidant response element signaling pathway and its activation by oxidative stress. *J Biol Chem* 2009;**284**:13291–5.

53. Oláhová M, Taylor SR, Khazaipoul S, Wang J, Morgan BA, Matsumoto K, et al. A redox-sensitive peroxiredoxin that is important for longevity has tissue- and stress-specific roles in stress resistance. *Proc Natl Acad Sci USA* 2008;**105**:19839–44.

54. Shiota M, Izumi H, Miyamoto N, Onitsuka T, Kashiwagi E, Kidani A, et al. Ets regulates peroxiredoxin1 and 5 expressions through their interaction with the high-mobility group protein B1. *Cancer Sci* 2008;**99**:1950–9.

55. Miyamoto N, Izumi H, Miyamoto R, Kubota T, Tawara A, Sasaguri Y, et al. Nipradilol and timolol induce Foxo3a and peroxiredoxin 2 expression and protect trabecular meshwork cells from oxidative stress. *Invest Ophthalmol Vis Sci* 2008;**50**:2777–84.

56. Miyamoto N, Izumi H, Miyamoto R, Kondo H, Tawara A, Sasaguri Y, et al. Quercetin induces the expression of peroxiredoxins 3 and 5 via the Nrf2/NRF1 transcription pathway. *Invest Ophthalmol Vis Sci* 2011;**52**:1055–63.

57. Ishii T, Itoh K, Takahashi S, Sato H, Yanagawa T, Katoh Y, et al. Transcription factor Nrf2 coordinately regulates a group of oxidative stress-inducible genes in macrophages. *J Biol Chem* 2000;**275**:16023–9.

58. Scarpulla RC. Transcriptional activators and coactivators in the nuclear control of mitochondrial function in mammalian cells. *Gene* 2002;**286**:81–9.

59. Huo L, Scarpulla RC. Mitochondrial DNA instability and peri-implantation lethality associated with targeted disruption of nuclear respiratory factor 1 in mice. *Mol Cell Biol* 2001;**21**:644–54.

60. Noh YH, Baek JY, Jeong W, Rhee SG, Chang TS. Sulfiredoxin translocation into mitochondria plays a crucial role in reducing hyperoxidized peroxiredoxin III. *J Biol Chem* 2009;**284**:8470–7.

61. Wood ZA, Poole LB, Karplus PA. Peroxiredoxin evolution and the regulation of hydrogen peroxide signaling. *Science* 2003;**300**:650–3.

62. Yoshida Y, Izumi H, Torigoe T, Ishiguchi H, Itoh H, Kang D, et al. P53 physically interacts with mitochondrial transcription factor A and differentially regulates binding to damaged DNA. *Cancer Res* 2003;**63**:3729–34.

63. Kanki T, Ohgaki K, Gaspari M, Gustafsson CM, Fukuoh A, Sasaki N, et al. Architectural role of mitochondrial transcription factor A in maintenance of human mitochondrial DNA. *Mol Cell Biol* 2004;**24**:9823–34.

64. Larsson NG, Wang J, Wilhelmsson H, Oldfors A, Rustin P, Lewandoski M, et al. Mitochondrial transcription factor A is necessary for mtDNA maintenance and embryogenesis in mice. *Nat Genet* 1998;**18**:231–6.

65. Park C, So HS, Shin CH, Baek SH, Moon BS, Shin SH, et al. Quercetin protects the hydrogen peroxide-induced apoptosis via inhibition of mitochondrial dysfunction in H9c2 cardiomyoblast cells. *Biochem Pharmacol* 2003;**66**:1287–95.

66. Resch H, Garhofer G, Fuchsjäger-Mayrl G, Hommer A, Schmetterer L. Endothelial dysfunction in glaucoma. *Acta Ophthalmol* 2009;**87**:4–12.

67. Jurkunas UV, Rawe I, Bitar MS, Zhu C, Harris DL, Colby K, et al. Decreased expression of peroxiredoxins in Fuchs' endothelial dystrophy. *Invest Ophthalmol Vis Sci* 2008;**49**:2956–63.

68. Cox AG, Pullar JM, Hughes G, Ledgerwood EC, Hampton MB. Oxidation of mitochondrial peroxiredoxin 3 during the initiation of receptor-mediated apoptosis. *Free Radic Biol Med* 2008;**44**:1001–9.

69. Liu RM, Choi J, Wu JH, Gaston Pravia KA, Lewis KM, Brand JD, et al. Oxidative modification of nuclear mitogen-activated protein kinase phosphatase 1 is involved in transforming growth factor beta1-induced expression of plasminogen activator inhibitor 1 in fibroblasts. *J Biol Chem* 2010;**285**:16239–47.

70. Liu RM, Gaston Pravia KA. Oxidative stress and glutathione in TGF-beta-mediated fibrogenesis. *Free Radic Biol Med* 2010;**48**:1–15.

11

Diabetes Mellitus and Glaucoma

Jay Siak, Gavin S. Tan, Tin Aung

Singapore National Eye Centre, Singapore

INTRODUCTION

Glaucoma refers to a spectrum of diseases (Figure 11.1) associated with progressive optic disc cupping and irreversible loss of ganglion cell neurons, with corresponding functional visual loss. Although an elevated intraocular pressure (IOP) is the primary risk factor linked with disease onset and progression, its pathogenic mechanisms are multifactorial, and vascular factors such as optic nerve head ischemia, nocturnal hypotension, and systemic hypoxia from obstructive sleep apnea are also contributory factors. As diabetes mellitus is a major systemic cause of microvasculopathy, and the world population is facing an increasing epidemic of the condition, it is important to understand whether diabetic patients are facing an increased risk of blindness from glaucoma (besides diabetic retinopathy).

EPIDEMIOLOGY: DIABETES MELLITUS AND PRIMARY OPEN-ANGLE GLAUCOMA

As primary open-angle glaucoma (POAG) is the most common type of glaucoma worldwide, most reported epidemiologic studies have examined the relationship of POAG with diabetes mellitus. In general, many studies revealed diabetes mellitus to be associated with an increased IOP,[1–7] but its association with POAG remains unclear. The major population-based epidemiologic studies are summarized in Table 11.1 and Figure 11.2. A statistically significant association between diabetes and POAG was reported in the Beaver Dam study, the Rotterdam study, the Blue Mountains Eye study, the Los Angeles Latino Eye study, and the Nurse Health Study,[1,8–11] but several other studies did not concur with this finding.[3,6,12–18] In 2004, Bonovas *et al.* reported a meta-analysis of studies that evaluated diabetes mellitus as a risk factor for POAG.[19] This included seven cross-sectional and five case-control studies, and the authors concluded that there was a statistically significant relationship between diabetes mellitus and POAG (odds ratio (OR) 1.50; 95% confidence interval (CI) 1.16 to 1.93) with the assumption of a random-effects model. However, the results of the studies were not homogeneous (P = 0.023), and the association was not significant among the case-control studies (OR 1.45; 95% CI 0.85 to 2.45).

There are a number of reasons for the conflicting findings of these studies. Many of the epidemiologic studies used different diagnostic criteria to define their diabetic cohort, such as the use of self-reported diabetes history, random or fasting serum glucose, random glycated hemoglobin A1c level, and nonfasting glucose tolerance testing. The definition of POAG was also heterogeneous, although the majority of the population-based studies did define glaucoma subjects using glaucomatous optic disc changes with some form of corresponding visual field defects. Unfortunately, many of the studies were not specifically planned to examine the link between the two conditions. The studies were also performed among different ethnic groups, and there might be a difference in the relationship between diabetes and glaucoma among people of various ethnicity. Further large population studies using similar diabetes and glaucoma definitions are needed in order to arrive at a more conclusive understanding of the relationship between diabetes and POAG.

EPIDEMIOLOGY: DIABETES MELLITUS AND OTHER TYPES OF GLAUCOMA

In normal-tension glaucoma (NTG), compared with POAG, vascular risk factors such as migraine,[20] Raynaud phenomenon,[21] and nocturnal hypotension[22] are considered to be more important in its pathogenesis due to impaired autoregulation of optic nerve head vascular supply. There was only one study that specifically examined diabetes and NTG; Kim observed that

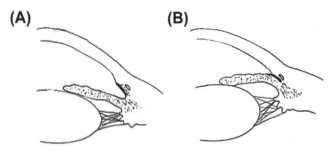

FIGURE 11.1   (A) Open-angle glaucoma; (B) angle-closure glaucoma.

diabetes mellitus increased the likelihood of bilateral involvement in NTG by 2.31 times (P = 0.004) compared with unilateral NTG.[23] Microvascular angiopathy and impaired vascular autoregulation due to autonomic neuropathy may potentiate the glaucomatous damage caused by nocturnal hypotension or large diurnal IOP fluctuations.

Pseudoexfoliation (PXF) syndrome is the most common identifiable cause of secondary open-angle glaucoma,[24] and it is an age-related disease of the extracellular

TABLE 11.1   Summary of Epidemiologic Studies on the Association between Diabetes Mellitus and Primary Open-Angle Glaucoma

Study	Study Location	Study Design	No. of Subjects	POAG Definition	Diabetes Definition	Odds Ratio	95% CI	Association
Klein 1994[8]	Beaver Dam, USA	Population cross-sectional	4926	VF, CDR, IOP ≥ 22, hx	DM history, HbA1c > 2 SD, random serum glucose > 11.1 mmol/L	1.84	1.09 to 3.11	Yes
Dielemans 1996[1]	Rotterdam, Netherlands	Population cross-sectional	4095	VF with CDR or IOP > 21	Random serum glucose or nonfasting glucose tolerance test > 11.1 mmol/L	3.11	1.12 to 8.66	Yes
Mitchell 1997[9]	Blue Mountains, Australia	Population cross-sectional	3642	VF, CDR	DM history, fasting serum glucose ≥ 7.8 mmol/L	2.12	1.18 to 3.79	Yes
Chopra 2008[11]	LALES, USA	Population cross-sectional	5894	VF, CDR	DM history, HbA1c ≥ 7%, random serum glucose ≥ 200 mg/dL	1.4	1.03 to 1.8	Yes
Pasquale 2006[10]	Nurse Health Study, USA	Cohort study of registered nurses	76,318	Glaucoma hx, VF, CDR	Self-reported DM history	RR 1.53	1.06 to 2.22	Yes
Tielsch 1995[3]	Baltimore, USA	Population cross-sectional	5308	VF, CDR	DM history	1.03	0.85 to 1.25	No
Quigley 2001[14]	Arizona, USA	Population cross-sectional	4774	VF, CDR	DM history, HbA1c ≥ 7%	1.24	0.79 to 1.94	No
Le 2003[15]	Melbourne, Australia	Population cross-sectional	3271	VF, CDR	DM history	-	-	No
Vijaya 2008[16]	Chennai, India	Population cross-sectional	3850	VF, CDR	DM history, random serum glucose > 200 mg/dL	1.1	0.8 to 1.3	No
Tan 2009[6]	Singapore	Population cross-sectional	3280	VF, CDR	DM history, random serum glucose ≥ 200 mg/dL	1.02	0.58 to 1.79	No
Xu 2009[17]	Beijing, China	Population cross-sectional	3251	CDR, FDP	DM history, fasting serum glucose ≥ 7.0 mmol/L	1.25	0.61 to 2.56	No
Ellis 2000[18]	Tayside, Scotland	Historical cohort population study	6631 diabetic, 166,144 nondiabetic	Glaucoma treatment, CDR, VF	DM history	RR 1.57	0.99 to 2.48	No
De Voogd 2006[12]	Rotterdam, Netherlands	Population cohort study	3837	VF, CDR	DM history, random serum glucose or nonfasting glucose tolerance test ≥ 11.1 mmol/L	RR 0.65	0.25 to 1.64	No

CI: confidence interval; DM: diabetes mellitus; DR: cup to disc ratio; FDP: frequency doubled perimetry; HbA1c: glycated hemoglobin; hx: history; IOP: intraocular pressure; POAG: primary open-angle glaucoma; RR: risk ratio; SD: standard deviation; VF: visual field.

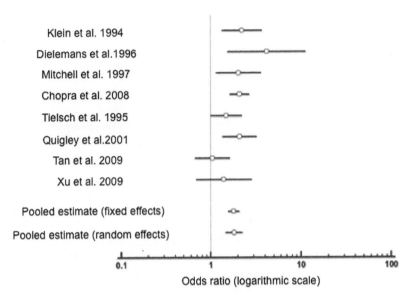

Klein et al. 1994
Dielemans et al.1996
Mitchell et al. 1997
Chopra et al. 2008
Tielsch et al. 1995
Quigley et al.2001
Tan et al. 2009
Xu et al. 2009

Pooled estimate (fixed effects)
Pooled estimate (random effects)

0.1   1   10   100

Odds ratio (logarithmic scale)

FIGURE 11.2 Analysis of eight population cross-sectional studies that examined the association between primary open-angle glaucoma and diabetes mellitus. Studies with insufficient data for analysis were excluded. The odds ratio and 95% confidence interval for each study as denoted by the first author and publication year are displayed on a logarithmic scale. The fixed effect model is appropriate as the test of heterogeneity is insignificant (P-value = 0.09).

matrix that was proposed to be associated with systemic vascular diseases such as aortic aneurysms.[25] However, in several large studies, PXF and PXF glaucoma were not found to be significantly associated with diabetes in multivariate analyses.[26–28]

Primary angle-closure glaucoma (PACG) is a disease that is particularly important in Asia, with a greater morbidity of blindness compared with open-angle glaucoma.[29] Previous studies have shown that the lens plays a role in the pathogenesis of primary angle closure (PAC), with increased lens thickness, lens vault,[30] and shallow anterior chamber depth[31] being significant risks. It was found that diabetic Chinese people have shallower mean anterior chamber depth (2.78 mm vs. 2.91 mm; P = 0.004) and thicker lenses (4.88 mm vs. 4.75 mm; P = 0.003) than non-diabetic Chinese people (Figure 11.3).[32] One population-based study conducted in India also observed an association between PACG and PAC with diabetes mellitus (OR 3.181; 95% CI 1.34 to 7.58; P = 0.001) in their multivariate analysis.[33] Sihota found that diabetes was more common among patients with asymptomatic chronic PACG than in patients with symptomatic chronic PACG (P < 0.01) (Table 11.2).[34] More studies are necessary to understand if the relationship between angle-closure glaucoma and diabetes is similar among the other populations.

Neovascular glaucoma can also develop with proliferative diabetic retinopathy or result from ocular ischemic syndrome, retinal vein occlusions, or diabetic complications such as tractional retinal detachment. Vitreous hemorrhage may also result in ghost cell glaucoma, and secondary glaucoma may develop after vitrectomy performed for the various complications of proliferative diabetic retinopathy.

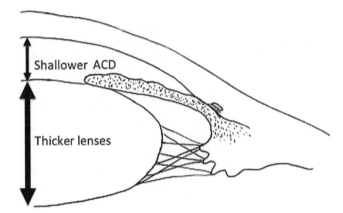

Shallower ACD

Thicker lenses

FIGURE 11.3 Diabetic subjects had shallower anterior chamber depth and thicker lenses, leading to increased risk of angle-closure disease.

PATHOPHYSIOLOGIC LINK BETWEEN DIABETES MELLITUS AND GLAUCOMA

How may glaucoma be related to diabetes mellitus? Although we still do not have a definitive explanation for this relationship, there is indirect evidence of similar ocular anatomical and physiologic changes that are common in diabetes and glaucoma (Table 11.3).

One of the key histologic features in glaucoma is the irreversible loss of ganglion cell neurons. It has been observed in streptozotocin (STZ)-induced diabetic rats that the ganglion cell layers are significantly thinner, whereas the outer retina layers are preserved.[35] Ganglion cell axon atrophy was also seen in type 1 diabetic BB/W-rats.[36] A significant reduction in retinal nerve fiber thickness was also observed in human diabetic eyes with no retinopathy using high-resolution optical

TABLE 11.2 Summary of Epidemiologic Studies on the Association between Diabetes Mellitus and Primary Angle-Closure Glaucoma

Study	Study Location	Study Design	No. of Subjects	Association
Saw 2007[32]	Singapore	Population cross-sectional	943	Diabetic Chinese people had shallower mean anterior chamber depth (2.78 mm vs. 2.91 mm; P = 0.004) and thicker lenses (4.88 mm vs. 4.75 mm; P = 0.003), which were associated with increased risk of angle-closure disease
Senthil 2010[33]	India	Population cross-sectional	3724	Diabetes mellitus was significantly associated with angle-closure disease (PACG and PAC) (OR 3.181; 95% CI 1.34 to 7.58)
Sihota 2000[34]	India	Clinic case-control	160	Diabetes was more common among patients with asymptomatic chronic PACG than those with symptomatic chronic PACG (P < 0.01)

CI: confidence interval; OR: odds ratio; PAC: primary angle closure; PACG: primary angle-closure glaucoma.

TABLE 11.3 Pathophysiologic Links between Diabetes Mellitus and Glaucoma

Changes in Diabetes	Changes in Glaucoma
• Ganglion cell axon atrophy • Reduction in retinal nerve fiber thickness	• Irreversible loss of ganglion cell neurons
• Altered vascular autoregulation • Abnormalities in retrobulbar blood flow and retinal microcirculation • Increased nitric oxide levels • Increased nitric oxide synthase activity	• Altered vascular autoregulation • Abnormalities in retrobulbar blood flow • Increased neuronal and inducible nitric oxide synthase activity in the astrocytes at lamina cribrosa
• Higher levels of endothelin-1	• Increased levels of endothelin-1 detected in POAG and NTG compared to normal controls
• Reduced retrograde axonal transport in retinal ganglion cells	• Inhibition of retrograde brain-derived neurotrophic factor transport in retinal ganglion cells
• Increased connective tissue growth factor mRNA expression • Higher connective tissue growth factor levels induced by advanced glycation end products, high glucose levels, and vascular endothelial growth factor	• Overexpression of connective tissue growth factor leads to modification of the trabecular meshwork actin cytoskeleton, causing increased IOP with optic nerve damage

mRNA: messenger ribonucleic acid; NTG: normal-tension glaucoma; POAG: primary open-angle glaucoma.

coherence tomography (OCT).[37] Retina studies on type 1 diabetic patients using OCT also confirmed a significant reduction in ganglion cell and retinal nerve fiber layer thickness.[38,39] In STZ-induced diabetic rats, there was an increase in apoptosis in the inner retina after the induction of chronic ocular hypertension (OHT) over 4 weeks by cauterization of episcleral veins, compared with control eyes.[40] The larger cell body changes were found to be at the level of the ganglion cell layer.

Pattern electroretinogram (PERG) is a well-established surrogate measure of retinal ganglion cell (RGC) function, and studies reported physiologic changes in the PERG in eyes with both glaucoma and diabetes. Over a period of up to 4 years, PERG waveforms were reduced significantly in patients with diabetes and OHT compared with normal control subjects with normal vision.[41] OHT subjects without diabetes did not have similar changes. The mean PERG amplitude of diabetic glaucoma subjects without retinopathy was significantly smaller (P < 0.05) than that of nondiabetic glaucoma subjects after adjustment for individual differences in age, visual acuity, visual field defect, and IOP.[42]

There are several ways in which diabetes may contribute toward neuronal damage in glaucoma. First, vascular autoregulation is altered in both diabetes[43] and glaucoma,[44] possibly related to autonomic dysfunction or local vasoactive agents such as nitric oxide. This may contribute toward ocular ischemia and reperfusion damage. Increased basal serum nitric oxide levels were found in diabetic subjects compared with nondiabetic controls (P = 0.0001).[45] Retina from postmortem eyes of subjects with diabetes and non-proliferative retinopathy revealed immunoreactivity for inducible nitric oxide synthase (iNOS) in retinal Müller glial cells compared with those of subjects without diabetes and ocular disease.[46] At high levels, nitric oxide has been shown to reduce blood flow at the optic nerve head,[47] possibly contributing to glaucomatous optic neuropathy. Increased neuronal and iNOS activity was also observed in supporting astrocyte cells at the lamina cribrosa of patients with glaucoma.[48]

Endothelin-1 (ET-1) is another vasoactive agent that mediates vasoconstriction, and increased levels have been reported in the aqueous humor of POAG patients compared with normal controls who underwent cataract surgery.[49] Plasma ET-1 levels were also significantly higher for 52 patients with NTG compared with normal controls.[50] Intravenous administration of

ET-1 also induced a reduction in the blood flow of the optic nerve head of rabbits. Higher levels of ET-1 were also observed in blood serum samples and the vitreous humor of patients with proliferative diabetic retinopathy compared with control patients,[51] and a higher ET-1 level was correlated with a higher glycated hemoglobin A1c level. Therefore, a higher ET-1 level in diabetic individuals may contribute to increased risk of optic nerve head ischemia and glaucomatous optic neuropathy.

RGCs require neurotrophic factors such as brain-derived neurotrophic factor (BDNF), and a substantial inhibition of retrograde transport of BDNF in RGC has been observed in rats after IOP elevation.[52] Diabetes has also been shown to reduce retrograde axonal transport progressively in RGCs in STZ-induced diabetic rat models.[53] Diabetes may accelerate the atrophy of ganglion cells in the presence of chronic OHT.

The trabecular meshwork aqueous humor outflow is influenced by transforming growth factor-β and its downstream mediator, connective tissue growth factor (CTFG). Overexpression of CTFG has been shown to increase IOP in mice with resultant optic nerve damage due to modification of the trabecular meshwork actin cytoskeleton.[54] CTFG messenger ribonucleic acid (mRNA) expression has been reported to be increased in diabetic eyes[55] and can be induced by advanced glycation end products,[56] high glucose levels[57] and vascular endothelial growth factor.[58]

CONCLUSIONS

Clinical and laboratory evidence suggest possible links between diabetes mellitus and various mechanisms of glaucomatous optic neuropathy. However, further studies are necessary to conclude its association with POAG, NTG, and angle-closure glaucoma.

TAKE-HOME MESSAGES

- Diabetes is associated with a higher IOP, but the evidence is conflicting regarding its association with POAG.
- Diabetes may be associated with bilateral NTG.
- Diabetes is not associated with PXF syndrome glaucoma.
- Diabetes is associated with angle-closure disease in Asian people.
- Diabetes and glaucoma share many ocular anatomical and electrophysiologic similarities.
- Mechanistic links may include impaired vascular autoregulation, increased RGC death from neurotrophic factor deficiency, and trabecular meshwork remodeling from increased CTFG levels.

References

1. Dielemans I, de Jong PT, Stolk R, Vingerling JR, Grobbee DE, Hofman A. Primary open-angle glaucoma, intraocular pressure, and diabetes mellitus in the general elderly population. The Rotterdam Study. *Ophthalmology* 1996;**103**(8):1271–5.
2. Klein BE, Klein R, Linton KL. Intraocular pressure in an American community. The Beaver Dam Eye Study. *Invest Ophthalmol Vis Sci* 1992;**33**(7):2224–8.
3. Tielsch JM, Katz J, Quigley HA, Javitt JC, Sommer A. Diabetes, intraocular pressure, and primary open-angle glaucoma in the Baltimore Eye Survey. *Ophthalmology* 1995;**102**(1):48–53.
4. Wu SY, Leske MC. Associations with intraocular pressure in the Barbados Eye Study. *Arch Ophthalmol* 1997;**115**(12):1572–6.
5. Matsuoka M, Ogata N, Matsuyama K, Yoshikawa T, Takahashi K. Intraocular pressure in Japanese diabetic patients. *Clin Ophthalmol* 2012;**6**:1005–9.
6. Tan GS, Wong TY, Fong CW, Aung T. Diabetes, metabolic abnormalities, and glaucoma. *Arch Ophthalmol* 2009;**127**(10):1354–61.
7. Hennis A, Wu SY, Nemesure B, Leske MC. Hypertension, diabetes, and longitudinal changes in intraocular pressure. *Ophthalmology* 2003;**110**(5):908–14.
8. Klein BE, Klein R, Jensen SC. Open-angle glaucoma and older-onset diabetes. The Beaver Dam Eye Study. *Ophthalmology* 1994;**101**(7):1173–7.
9. Mitchell P, Smith W, Chey T, Healey PR. Open-angle glaucoma and diabetes: the Blue Mountains eye study. *Aus Ophthalmol* 1997;**104**(4):712–8.
10. Pasquale LR, Kang JH, Manson JE, Willett WC, Rosner BA, Hankinson SE. Prospective study of type 2 diabetes mellitus and risk of primary open-angle glaucoma in women. *Ophthalmology* 2006;**113**(7):1081–6.
11. Chopra V, Varma R, Francis BA, Wu J, Torres M, Azen SP. Type 2 diabetes mellitus and the risk of open-angle glaucoma the Los Angeles Latino Eye Study. *Ophthalmology* 2008;**115**(2):227–32.
12. de Voogd S, Ikram MK, Wolfs RC, Jansonius NM, Witteman JC, Hofman A, et al. Is diabetes mellitus a risk factor for open-angle glaucoma? The Rotterdam Study. *Ophthalmology* 2006;**113**(10):1827–31.
13. Leske MC, Connell AM, Wu SY, Hyman LG, Schachat AP. Risk factors for open-angle glaucoma. The Barbados Eye Study. *Arch Ophthalmol* 1995;**113**(7):918–24.
14. Quigley HA, West SK, Rodriguez J, Munoz B, Klein R, Snyder R. The prevalence of glaucoma in a population-based study of Hispanic subjects: Proyecto VER. *Arch Ophthalmol* 2001;**119**(12):1819–26.
15. Le A, Mukesh BN, McCarty CA, Taylor HR. Risk factors associated with the incidence of open-angle glaucoma: the visual impairment project. *Invest Ophthalmol Vis Sci* 2003;**44**(9):3783–9.
16. Vijaya L, George R, Baskaran M, Arvind H, Raju P, Ramesh SV, et al. Prevalence of primary open-angle glaucoma in an urban south Indian population and comparison with a rural population. The Chennai Glaucoma Study. *Ophthalmology* 2008;**115**(4):648–54.
17. Xu L, Xie XW, Wang YX, Jonas JB. Ocular and systemic factors associated with diabetes mellitus in the adult population in rural and urban China. The Beijing Eye Study. *Eye (Lond)* 2009;**23**(3):676–82.
18. Ellis JD, Evans JM, Ruta DA, Baines PS, Leese G, MacDonald TM, Morris AD. Glaucoma incidence in an unselected cohort of diabetic patients: is diabetes mellitus a risk factor for glaucoma? DARTS/MEMO collaboration. Diabetes Audit and Research in Tayside Study. Medicines Monitoring Unit. *Br J Ophthalmol* 2000;**84**(11):1218–24.
19. Bonovas S, Peponis V, Filioussi K. Diabetes mellitus as a risk factor for primary open-angle glaucoma: a meta-analysis. *Diabet Med* 2004;**21**(6):609–14.

20. Phelps CD, Corbett JJ. Migraine and low-tension glaucoma. A case-control study. *Invest Ophthalmol Vis Sci* 1985;**26**(8):1105–8.

21. Broadway DC, Drance SM. Glaucoma and vasospasm. *Br J Ophthalmol* 1998;**82**(8):862–70.

22. Hayreh SS, Zimmerman MB, Podhajsky P, Alward WL. Nocturnal arterial hypotension and its role in optic nerve head and ocular ischemic disorders. *Am J Ophthalmol* 1994;**117**(5):603–24.

23. Kim C, Kim TW. Comparison of risk factors for bilateral and unilateral eye involvement in normal-tension glaucoma. *Invest Ophthalmol Vis Sci* 2009;**50**(3):1215–20.

24. Ritch R. Exfoliation syndrome - the most common identifiable cause of open-angle glaucoma. *J Glaucoma* 1994;**3**(2):176–7.

25. Schumacher S, Schlotzer-Schrehardt U, Martus P, Lang W, Naumann GO. Pseudoexfoliation syndrome and aneurysms of the abdominal aorta. *Lancet* 2001;**357**(9253):359–60.

26. Speckauskas M, Tamosiunas A, Jasinskas V. Association of ocular pseudoexfoliation syndrome with ischaemic heart disease, arterial hypertension and diabetes mellitus. *Acta Ophthalmol* 2012;**90**(6):e470–5.

27. Topouzis F, Wilson MR, Harris A, Founti P, Yu F, Anastasopoulos E, et al. Risk factors for primary open-angle glaucoma and pseudoexfoliative glaucoma in the Thessaloniki eye study. *Am J Ophthalmol* 2011;**152**(2):219–28.

28. Miyazaki M, Kubota T, Kubo M, Kiyohara Y, Iida M, Nose Y, et al. The prevalence of pseudoexfoliation syndrome in a Japanese population: the Hisayama study. *J Glaucoma* 2005;**14**(6):482–4.

29. Quigley HA, Broman AT. The number of people with glaucoma worldwide in 2010 and 2020. *Br J Ophthalmol* 2006;**90**(3):262–7.

30. Nongpiur ME, He M, Amerasinghe N, Friedman DS, Tay WT, Baskaran M, et al. Lens vault, thickness, and position in Chinese subjects with angle closure. *Ophthalmology* 2011;**118**(3):474–9.

31. Aung T, Nolan WP, Machin D, Seah SK, Baasanhu J, Khaw PT, et al. Anterior chamber depth and the risk of primary angle closure in 2 East Asian populations. *Arch Ophthalmol* 2005;**123**(4):527–32.

32. Saw SM, Wong TY, Ting S, Foong AW, Foster PJ. The relationship between anterior chamber depth and the presence of diabetes in the Tanjong Pagar Survey. *Am J Ophthalmol* 2007;**144**(2):325–6.

33. Senthil S, Garudadri C, Khanna RC, Sannapaneni K. Angle closure in the Andhra Pradesh Eye Disease Study. *Ophthalmology* 2010;**117**(9):1729–35.

34. Sihota R, Gupta V, Agarwal HC, Pandey RM, Deepak KK. Comparison of symptomatic and asymptomatic, chronic, primary angle-closure glaucoma, open-angle glaucoma, and controls. *J Glaucoma* 2000;**9**(3):208–13.

35. Barber AJ, Lieth E, Khin SA, Antonetti DA, Buchanan AG, Gardner TW. Neural apoptosis in the retina during experimental and human diabetes. Early onset and effect of insulin. *J Clin Invest* 1998;**102**(4):783–91.

36. Sima AA, Zhang WX, Cherian PV, Chakrabarti S. Impaired visual evoked potential and primary axonopathy of the optic nerve in the diabetic BB/W-rat. *Diabetologia* 1992;**35**(7):602–7.

37. Sugimoto M, Sasoh M, Ido M, Wakitani Y, Takahashi C, Uji Y. Detection of early diabetic change with optical coherence tomography in type 2 diabetes mellitus patients without retinopathy. *Ophthalmologica* 2005;**219**(6):379–85.

38. van Dijk HW, Kok PH, Garvin M, Sonka M, Devries JH, Michels RP, et al. Selective loss of inner retinal layer thickness in type 1 diabetic patients with minimal diabetic retinopathy. *Invest Ophthalmol Vis Sci* 2009;**50**(7):3404–9.

39. van Dijk HW, Verbraak FD, Kok PH, Garvin MK, Sonka M, Lee K, et al. Decreased retinal ganglion cell layer thickness in patients with type 1 diabetes. *Invest Ophthalmol Vis Sci* 2010;**51**(7):3660–5.

40. Kanamori ANM, Mukuno H, Maeda H, Negi A. Diabetes has an additive effect on neural apoptosis in rat retina with chronically elevated intraocular pressure. *Curr Eye Res* 2004;**28**(1):47–54.

41. Vesti ETG. Diabetes can alter the interpretation of visual dysfunction in ocular hypertension. *Ophthalmology* 1996;**103**(9):1419–25.

42. Ventura LM, Golubev I, Feuer WJ, Porciatti V. The PERG in diabetic glaucoma suspects with no evidence of retinopathy. *J Glaucoma* 2010;**19**(4):243–7.

43. Pemp B, Schmetterer L. Ocular blood flow in diabetes and age-related macular degeneration. *Can J Ophthalmol* 2008;**43**(3):295–301.

44. Flammer J, Orgul S, Costa VP, Orzalesi N, Krieglstein GK, Serra LM, et al. The impact of ocular blood flow in glaucoma. *Prog Ret Eye Res* 2002;**21**(4):359–93.

45. Ozden S, Tatlipinar S, Bicer N, Yaylali V, Yildirim C, Ozbay D, et al. Basal serum nitric oxide levels in patients with type 2 diabetes mellitus and different stages of retinopathy. *Can J Ophthalmol* 2003;**38**(5):393–6.

46. Abu El-Asrar AM, Desmet S, Meersschaert A, Dralands L, Missotten L, Geboes K. Expression of the inducible isoform of nitric oxide synthase in the retinas of human subjects with diabetes mellitus. *Am J Ophthalmol* 2001;**132**(4):551–6.

47. Nicolela MT, Hnik P, Drance SM. Scanning laser Doppler flowmeter study of retinal and optic disk blood flow in glaucomatous patients. *Am J Ophthalmol* 1996;**122**(6):775–83.

48. Neufeld AH, Hernandez MR, Gonzalez M. Nitric oxide synthase in the human glaucomatous optic nerve head. *Arch Ophthalmol* 1997;**115**(4):497–503.

49. Iwabe S, Lamas M. Vasquez Pelaez CG, Carrasco FG. Aqueous humor endothelin-1 (Et-1), vascular endothelial growth factor (VEGF) and cyclooxygenase-2 (COX-2) levels in Mexican glaucomatous patients. *Curr Eye Res* 2010;**35**(4):287–94.

50. Sugiyama T, Moriya S, Oku H, Azuma I. Association of endothelin-1 with normal tension glaucoma: clinical and fundamental studies. *Surv Ophthalmol* 1995;**39**(Suppl. 1):S49–56.

51. Adamiec-Mroczek J, Oficjalska-Mlynczak J, Misiuk-Hojlo M. Roles of endothelin-1 and selected proinflammatory cytokines in the pathogenesis of proliferative diabetic retinopathy: analysis of vitreous samples. *Cytokine* 2010;**49**(3):269–74.

52. Quigley HA, McKinnon SJ, Zack DJ, Pease ME, Kerrigan-Baumrind LA, Kerrigan DF, et al. Retrograde axonal transport of BDNF in retinal ganglion cells is blocked by acute IOP elevation in rats. *Invest Ophthalmol Vis Sci* 2000;**41**(11):3460–6.

53. Ino-Ue M, Zhang L, Naka H, Kuriyama H, Yamamoto M. Polyol metabolism of retrograde axonal transport in diabetic rat large optic nerve fiber. *Invest Ophthalmol Vis Sci* 2000;**41**(13):4055–8.

54. Junglas B, Kuespert S, Seleem AA, Struller T, Ullmann S, Bosl M, et al. Connective tissue growth factor causes glaucoma by modifying the actin cytoskeleton of the trabecular meshwork. *Am J Pathol* 2012;**180**(6):2386–403.

55. Tikellis C, Cooper ME, Twigg SM, Burns WC, Tolcos M. Connective tissue growth factor is up-regulated in the diabetic retina: amelioration by angiotensin-converting enzyme inhibition. *Endocrinology* 2004;**145**(2):860–6.

56. Twigg SM, Chen MM, Joly AH, Chakrapani SD, Tsubaki J, Kim HS, et al. Advanced glycosylation end products up-regulate connective tissue growth factor (insulin-like growth factor-binding protein-related protein 2) in human fibroblasts: a potential mechanism for expansion of extracellular matrix in diabetes mellitus. *Endocrinology* 2001;**142**(5):1760–9.

57. Paradis V, Perlemuter G, Bonvoust F, Dargere D, Parfait B, Vidaud M, et al. High glucose and hyperinsulinemia stimulate connective tissue growth factor expression: a potential mechanism involved in progression to fibrosis in nonalcoholic steatohepatitis. *Hepatology* 2001;**34**(4 Pt 1):738–44.

58. Suzuma K, Naruse K, Suzuma I, Takahara N, Ueki K, Aiello LP, et al. Vascular endothelial growth factor induces expression of connective tissue growth factor via KDR, Flt1, and phosphatidylinositol 3-kinase-akt-dependent pathways in retinal vascular cells. *J Biol Chem* 2000;**275**(52):40725–31.

Dietary Polyunsaturated Fatty Acids, Intraocular Pressure, and Glaucoma

Niyazi Acar, Catherine P. Creuzot-Garcher, Alain M. Bron, Lionel Bretillon

Eye and Nutrition Research Group, University of Burgundy, Centre des Sciences du Goût et de l'Alimentation, Dijon, France

INTRODUCTION

The glaucomas encompass several different clinical forms and are the second leading cause of blindness worldwide. Overall, about 80 million people will be suffering from glaucoma in 2020.[1] Although extensive research is carried out, the pathogenesis of glaucoma still remains unclear. The term 'glaucoma' covers a large variety of pathophysiologic processes, all sharing the common definition of an optic neuropathy with a characteristic optic nerve head remodeling and alteration of visual field. Primary open-angle glaucoma (POAG) represents the most common form of glaucomatous optic neuropathy. Elevated intraocular pressure (IOP) is recognized as a major risk factor for POAG.[2,3] As a consequence, the current strategies for preventing POAG are almost exclusively based on the use of ocular hypotensive drugs.[4,5] Among these agents, analogs of prostaglandin F2α, a derivative of the omega-6 polyunsaturated fatty acid (PUFA) arachidonic acid, are nowadays the first-line therapy and the most prescribed drugs in glaucoma and ocular hypertension.

Despite this successful therapeutic management of IOP, a substantial interest exists in other preventive approaches against the development of glaucomatous optic neuropathy. Indeed, several observations showing that an elevated IOP does not inevitably lead to glaucoma and that glaucoma can develop under normal IOP conditions[6,7] strongly suggest that factors other than IOP are likely to play a role in the pathogenesis of glaucoma.[8,9] The common objective of these approaches is to limit the damage to the optic nerve, and subsequently, the irreversible loss of visual fields through strategies other than IOP-lowering drugs. A number of epidemiologic and laboratory studies have shown, or at least partly shown, that modifiable lifestyle factors such as dietary fat may be associated with the occurrence of glaucoma.

EPIDEMIOLOGIC DATA

A significant number of studies on patients have identified modifiable determinants related to the prevalence of glaucoma. Some of the identified or at least suspected factors are cigarette smoking, dietary habits, exercise, and body mass index or obesity.[10,11] The influence of nutrition on the prevalence of glaucoma was almost exclusively evaluated though the assessment of dietary habits by food frequency questionnaires (FFQs). Intake of antioxidants and lipids has been found to be related to the prevalence of glaucoma. The involvement of oxidative stress in the pathophysiology of glaucoma has been extensively demonstrated in laboratory research (see Tezel for a review[12]). This relationship is less clear at an epidemiologic level even though a reduced risk of POAG in subjects having the highest consumption in vitamin A, vitamin E, and vitamin B_2 was suggested.[13,14] In one of these studies, namely the multicenter Study of Osteoporotic Fractures including 1155 women, no significant relationship was found between the total fat intake and the risk of POAG.[14] It might be hazardous to conclude that such a relationship does not exist since the connection between dietary lipids and glaucoma may be more subtle. Indeed, not a quantitative but a qualitative view of the dietary supply in lipids has to be considered. This qualitative consideration is based on the balance between the different families of PUFA rather than their absolute intake.

Dietary lipids are mainly composed of fatty acids that are esterified on triglycerides or phospholipids. When they are polyunsaturated, these fatty acids belong to omega-6 or omega-3 families. In the diet, the main represented omega-6 and omega-3 fatty acids are linoleic acid (C18:2n-6) and α-linolenic acid (C18:3n-3). After being ingested, these PUFAs can be the competitive substrate for endogenous enzymes, in order to be converted to longer-chain and more unsaturated PUFAs such as arachidonic acid (C20:4n-6) or docosahexaenoic acid (DHA, C22:6n-3) belonging to omega-6 and omega-3 families, respectively. It is now well established that the arachidonic acid derivative prostaglandin F2α is a very effective ocular hypotensive agent. Several prostaglandin analogs are currently used in glaucoma therapy.[4,5] However, in addition to influencing the ocular endogenous concentrations of prostaglandin F2α, dietary PUFA may interfere with several other cellular mechanisms such as oxidative stress, inflammation, or neurodegeneration.[15]

Only a few studies have evaluated the importance of dietary omega-6 and omega-3 PUFA on the incidence of POAG. Even though some studies did not reveal a relationship between glaucoma and dietary PUFA intake[16,17], some others have suggested several interesting associations. In the study of Kang and collaborators[18], the authors used subjects from the Nurses' Health Study and Health Professionals Follow-up Study, which investigated the roles of fat and antioxidant intake. Within this cohort, data from 474 self-reported glaucoma cases showed no relationship between the dietary intake of the major lipid classes (namely total fat, animal fat, vegetable fat, cholesterol, saturated fat, monounsaturated fat, polyunsaturated fat, *trans* unsaturated fat, total n-3 PUFA, and total n-6 PUFA). However, it was found that a higher dietary ratio of n-3 to n-6 PUFA was associated with a higher risk of high-tension POAG (Table 12.1). Indeed, when comparing the highest with the lowest quintile for dietary ratio of n-3 to n-6 PUFA, the association was weak but significant for high-tension POAG (relative risk of 1.68; 95% confidence interval 1.18 to 2.40; P = 0.009), whereas it was not significant for normal-tension glaucoma (NTG). These findings may be surprising since they would suggest a

beneficial effect of omega-6 to the detriment of omega-3 PUFAs, although these are known to display multiple useful cellular properties.[15] Nevertheless, the authors hypothesized that a lower dietary supply in omega-6 PUFA may reduce the tissue bioavailability of arachidonic acid, then leading to a lower endogenous production of omega-6-derived prostaglandins, including the hypotensive prostaglandin F2α. This hypothesis may be reinforced by the absence of such a relationship in subjects with normal-tension POAG.

Conversely, not only omega-6 but also omega-3 PUFA may be related to the pathophysiology of glaucoma. Several studies have revealed that patients suffering from POAG have lower circulating levels of omega-3 PUFA. An initial study of 10 POAG patients and eight healthy siblings has shown that the levels of eicosapentaenoic acid (C20:5n-3) and DHA were significantly reduced in patients with POAG.[19] These data were further confirmed by another study involving 31 POAG patients and 10 healthy controls.[20] As this latest study considered different subgroups with POAG, it has also revealed an additional loss of vinyl-ether phosphocholines (also named as plasmalogens) and different kinetics for the loss of omega-3 PUFA and vinyl-ether phosphocholines in erythrocytes (Fig. 12.1). In addition to suggesting the further involvement of vinyl-ether phosphocholines in the pathogenesis of POAG, the authors proposed that the mechanisms involved in the loss of vinyl-ether phospholines and omega-3 PUFA are not related. Indeed, linear regression analyses have shown that the reduction of vinyl-ether phosphocholine levels in red blood cells may start about 20 years prior to the first clinical changes whereas that of omega-3 PUFA might occur together with the alteration of visual field. Even if these two pilot biomarker studies were conducted in a reduced number of subjects, they underline a potential involvement of omega-3 PUFA but also that of other lipid entities in the risk for developing POAG. Of course, these modifications involve circulating molecules but recent data revealing the existence of associations between erythrocyte and retinal and optic nerve lipid levels may help

TABLE 12.1 Relative Risk for Normal-Tension Primary Open-Angle Glaucoma (POAG) and High-Tension POAG According to the Dietary n-3 Polyunsaturated Fatty Acid (PUFA) to n-6 PUFA Ratio

		Quintiles for Dietary n-3/n-6 Ratio				
		1	2	3	4	5
High-tension POAG	Relative risk	1.00	1.33	1.22	1.03	1.68
	95% CI		0.93 to 1.91	0.84 to 1.76	0.61 to 1.76	1.18 to 2.40
NPOAG	Relative risk	1.00	0.97	1.32	1.13	0.82
	95% CI		0.50 to 1.85	0.71 to 2.44	0.60 to 2.13	0.20 to 3.47

CI: confidence interval; NPOAG: nonprimary open-angle glaucoma; POAG: primary open-angle glaucoma.
Source: adapted from.[18]

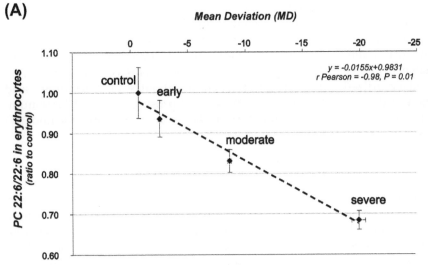

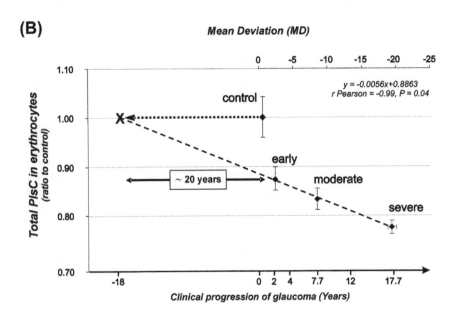

FIGURE 12.1 Kinetics of loss of omega-3 PUFA-rich phosphocholine and total vinyl-ether phosphocholines in erythrocytes of patients with POAG. Linear regression analyses showing the kinetics of diminution of phosphocholine esterified with omega-3 PUFA (PC22:6/22:6) (A) and total vinyl-ether phosphocholines (Total PlsC) (B) in erythrocytes of patients with POAG. *Reprinted from Experimental Eye Research, Vol 89/6, Niyazi Acar, Olivier Berdeaux, Pierre Juaneda, Stéphane Grégoire, Stéphanie Cabaret, Corinne Joffre, Catherine P. Creuzot-Garcher, Lionel Bretillon, Alain M. Bron, Red blood cell plasmalogens and docosahexaenoic acid are independently reduced in primary open-angle glaucoma, 840-853.[20] Copyright (2009), with permission from Elsevier.*

in extrapolating such changes to the eye.[21] Omega-3 PUFAs are known to be essential for the health of neuronal tissues including the retina, where they can influence multiple cellular processes involved in glaucoma, such as oxidative stress or neuroprotection.[15] The functions of vinyl-ether phospholipids are not yet fully elucidated but biochemical data showing their quantitative importance in the retina or in the optic nerve,[22,23] as well as the existence of optic nerve hypoplasia in vinyl-ether phospholipid-deficient animals,[24] reinforce their potential to preserve retinal and/or optic nerve structure during glaucomatous stress.

The mechanisms by which omega-3 PUFA are protective against POAG have not been elucidated; however, omega-6 PUFA may be involved in lowering IOP. This is suggested by data from epidemiologic studies carried out on the Japanese population. First, the Tajimi Study

investigated the prevalence of POAG in Japanese people.[25] The investigators found a very low proportion of high-tension POAG when compared with NTG patients (7.7% of the total subjects having POAG), thus confirming earlier observations on a high prevalence of NTG in Japan.[26] In parallel to the Tajimi study, another group has shown that the ratio of high-tension POAG to NTG is largely increased in Japanese Americans (16.9% of the total subjects having POAG),[27] suggesting that factors others than genetics may be involved in maintaining a physiologic IOP. In contrast to the Western diet, the traditional Japanese diet is characterized by its high content in omega-3-rich sea products; one hypothesis to explain this difference may be the impact of omega-3 PUFA on IOP. This hypothesis was further reinforced by data from animal experiments (see 'Laboratory research' section).

Nevertheless, the potential relationship between omega-3 PUFA and POAG may not be restricted to the regulation of IOP, as shown by a French study.[28] In this nationwide prospective study, ophthalmologists examined 339 cases of POAG and 339 age-matched people with ocular hypertension. Through a standardized questionnaire, the preferential consumption of the different types of fat, including omega-3-rich and omega-6-rich products was evaluated. In this study, POAG was significantly associated with a low consumption of oily fish and walnuts, two omega-3-rich food products. However, the molecular mechanisms of this protection are not yet clarified. These data are strongly reinforced by a genome-wide association study performed on more than 300 patients with NTG and 350 controls subjects.[29] By investigating more than 500,000 single-nucleotide polymorphisms, this study identified the *ELOVL5* gene (ELOngation of Very Long chain fatty acids family member 5) as a new susceptibility gene for NTG. The *ELOVL5* gene encodes a protein known to be one of the seven enzymes that catalyze the elongation of short-chain PUFA to longer-chain PUFA. Within the eight single-nucleotide polymorphisms identified in the *ELOVL5* region, almost all of them showed a strong disease association, and especially the rs735860 variant (Table 12.2).

LABORATORY RESEARCH

Implication of Polyunsaturated Fatty Acids in the Regulation of Intraocular Pressure

The control of IOP is dependent on the balance between aqueous humor production and outflow. While aqueous humor is produced in the epithelium of the ciliary body, the outflow pathway involves other structures such as the trabecular meshwork (conventional pathway) or the ciliary muscle (uveoscleral pathway). To date, the effects of PUFA on IOP were mainly reported on the aqueous humor outflow pathway. These data concern almost exclusively omega-6 PUFA and the oxygenated metabolites of arachidonic acid belonging to the series-2 prostaglandins. A large number of laboratory studies have investigated the underlying mechanisms.[4,5,30]

Prostaglandins are synthesized locally through the action of cyclooxygenase-2 (COX-2). The critical role of COX-2 as the primary source of prostaglandins was demonstrated by showing that COX-2 was naturally expressed in the nonpigmented secretory epithelium of the ciliary body, whereas its expression was completely lost in the nonpigmented secretory epithelium of patients at advanced stages of POAG.[31] Consistent with this finding, the authors also observed significantly lower levels of the arachidonic acid-derived series-2 prostaglandin in the aqueous humor of patients with POAG. The same group demonstrated the existence of a positive-feedback regulation of COX-2 expression by prostaglandin E_2 through the activation of p38 mitogen-activated protein kinase (MAPK) and p42/44 MAPK proteins.[32]

While other members of the series-2 prostaglandins such as prostaglandin E_2 have been studied, the majority of the data on the control of the aqueous humor outflow concern prostaglandin F2α or its pharmacologic analogs. The hypotensive effects of prostaglandin F2α and/or its analogs were attributed to their ability to strongly enhance uveoscleral outflow and to a lesser extent trabecular outflow, whereas little or no effect was observed on aqueous humor formation (see Cracknell and Grierson (2009),[4] Weinreb *et al.* (2002),[30] Civan and Macknight (2004)[33] for reviews). One of the proposed mechanisms of action is a remodeling of the ciliary muscle resulting

TABLE 12.2   Single-Nucleotide Polymorphisms Associated with Normal-Tension Glaucoma for the rs735860 Variant of the *ELOVL5* gene

SNP Identification	chr	Alleles	Allele/ Genotype	Nearest Gene	Frequency (%) Cases	Frequency (%) Controls	P value	OR (95% CI)
rs735860	6	T > C	C	ELOVL5	60.8	48.0	4.14×10^{-6}	1.69 (1.36 to 2.11)
			CC	ELOVL5	38.1	26.3	2.28×10^{-4}	1.73 (1.24 to 2.41)
			CT	ELOVL5	45.1	45.6		0.98 (0.72 to 1.34)
			TT	ELOVL5	16.8	28.2		0.51 (0.35 0.75)

chr: chromosome; CI: confidence interval; SNP: single-nucleotide polymorphism; OR: odds ratio.
Source: adapted from.[29]

in an increased space for outflow. Prostaglandins are thought to act through their binding to a prostaglandin FP receptor that is localized in many ocular tissues in monkeys, including the iris, ciliary muscle, and sclera.[34] The activation of the prostaglandin FP receptors is known to activate a G protein, protein kinase C, and to promote the release of intracellular calcium.[35,36] Another intracellular consequence of the exposure of ciliary muscle cells to prostaglandin F2α and its binding to the prostaglandin FP receptor is a significant induction of the nuclear factors c-fos and c-jun.[37] Following this induction, c-fos and c-jun are able to form heterodimers that bind the transcription regulatory element AP-1 in the promoter regions of a large number of genes.[38] One of the gene families whose promoters contain one or more copies of the AP-1 transcription element is encoding for matrix metalloproteinases (MMPs),[39] which are known to degrade many different components of the extracellular matrix. The ability of prostaglandin treatment to alter ciliary muscle expression of MMPs was confirmed *in vitro* and *in vivo* on cultures of human ciliary body smooth muscle cells as well as in monkeys.[40–42] These molecular events may explain the effects of prostaglandins at a cellular level. Indeed, it was shown that topical administration of prostaglandin F2α or its analogs may affect the ciliary muscle physiology by widening the connective tissue-filled spaces between ciliary muscle bundles through modifications of the extracellular matrix.[43,44]

To our knowledge, only one study was focused on the specific effects of dietary omega-3 PUFA on aqueous outflow.[45] The authors used two groups of Sprague-Dawley rats and placed one group in extreme conditions of dietary omega-3 PUFA deprivation and the second group in conditions of dietary omega-3 PUFA sufficiency (dietary omega-6 PUFA to omega-3 PUFA ratios of 6:1 and 334:1, respectively). The diets were delivered to the dams during gestation and lactation periods and to the pups until 40 weeks of age. When compared with the omega-3 PUFA-deficient group, IOP was significantly reduced in animals supplemented with omega-3 PUFA since 20 weeks of age (Fig. 12.2). Actually, when compared with basal IOP, the authors observed a significant increase of IOP in deficient animals whereas IOP was significantly decreased in animals supplemented with PUFA. The relative difference in IOP between the two groups of rats was 3.53 mm Hg at 40 weeks. By using a pulse-infusion methodology, the authors suggested that the difference observed in IOP may be the result of an improved aqueous outflow facility. They demonstrated that aqueous outflow facility was improved by 56% in rats supplemented with omega-3 PUFA when compared with omega-3-deficient animals. These physiologic changes were associated with biochemical modifications in tissue fatty acid composition since a 3.3-fold

increase in DHA concentration was observed in the ciliary body of supplemented animals. In this paper, the authors did not suppose any implication of a G-protein-derived change in aqueous humor production. Moreover, and since the concentrations of arachidonic acid in the ciliary body were unaffected by dietary treatments, they did not consider that the changes in prostaglandin production explained the divergence in IOP between supplemented and deficient animals. Based on previous work,[46] they considered there to be a more direct effect of omega-3 PUFA-derived metabolites on increasing aqueous outflow; however, the exact molecular mechanisms still remain unclear. Nevertheless, these results reinforce data from human studies reporting a lower prevalence of high-tension glaucoma in Japanese populations when compared with Western populations,[25] as well as those showing decreased circulating omega-3 PUFA levels in patients with POAG.[19,20]

Implication of Polyunsaturated Fatty Acids in Tissue Reaction to Experimental Intraocular Pressure Elevation

The potential role of omega-6 and omega-3 PUFA was also investigated in the retina. The aim was to use dietary lipids to influence the secretion of retinal prostaglandins in order to act on the retinal inflammation caused by an elevated IOP. Indeed, not only arachidonic acid but also di-homo-gamma-linolenic acid (C18:3n-6) and eicosapentaenoic acid (C20:5n-3) are precursors of prostaglandins. Whereas arachidonic acid-derived prostaglandins belong to series-2 prostaglandins, those

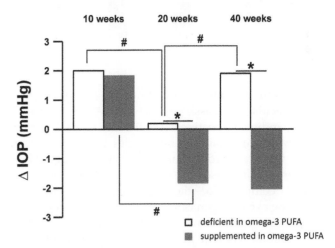

FIGURE 12.2 Effects of a supplementation and a deficiency in omega-3 polyunsaturated fatty acids (PUFAs) on intraocular pressure (IOP) in rats. IOP was significantly different between PUFA-supplemented and PUFA-deficient animals at 20 weeks and 40 weeks of age. #: P < 0.05 when compared with the same group at a previous time-point; *: P < 0.05 between the two groups at the same time point. Adapted from.[45]

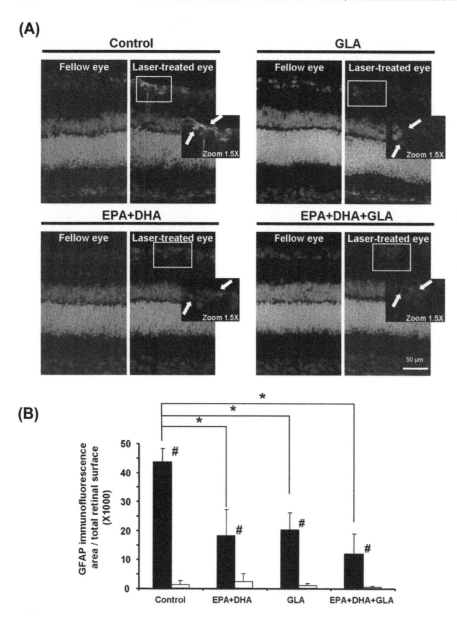

FIGURE 12.3 Effect of omega-6 and/or omega-3 polyunsaturated fatty acid (PUFA) supplementation on short-term retinal glial activation following intraocular pressure (IOP) elevation. Representative pictures (A) and semiquantitative analysis (B) of glial activation evaluated by glial fibrillary acidic protein (GFAP) immunohistochemistry in rats supplemented for 3 months with omega-3 PUFA (eicosapentaenoic acid (EPA) + docosahexaenoic acid (DHA)), omega-6 PUFA (gamma-linolenic acid (GLA)) or omega-3 and omega-6 PUFA (EPA+DHA+GLA) 24 hours after the induction of elevated IOP. *Reprinted from Nutrition Research, Vol 31, Coralie Schnebelen, Cynthia Fourgeux, Bruno Pasquis, Catherine P. Creuzot-Garcher, Alain M. Bron, Lionel Bretillon, Niyazi Acar, Dietary polyunsaturated fatty acids reduce retinal stress induced by an elevation of intraocular pressure in rats, 286-295.[49] Copyright (2011), with permission from Elsevier.*

formed from di-homo-gamma-linolenic and eicosapentaenoic acids are classified into series-1 and series-3 prostaglandins, respectively (see the article of Calder for a review[47]). Prostaglandins from series-1 and series-3 are known to be less proinflammatory than prostaglandins from series-2. Because PUFA from omega-6 and omega-3 families compete for the same metabolic enzymes (including cyclooxygenases that convert them into prostaglandins), a preliminary study aimed to characterize the retinal content in PUFA and prostaglandins in rats whose diet was supplemented with omega-6 and/or omega-3 PUFA.[48] In this study, the retinal fatty acid composition followed the lipid content of the diet since animals receiving omega-6 PUFA displayed increased levels of retinal omega-6 PUFA (particularly di-homo-gamma-linolenic and arachidonic acids), and animals receiving omega-3 PUFA had higher levels of omega-3

PUFA (namely eicosapentaenoic acid and DHA). The group supplemented with both omega-6 and omega-3 PUFA exhibited increased levels of di-homo-gamma-linolenic acid, eicosapentaenoic acid, and DHA, but a limited modification in the levels of arachidonic acid. These results would suggest a preferential production of the less proinflammatory series-1 and series-3 prostaglandins whereas that of series-2 would be limited. However, the modifications in the levels of retinal series-1 and series-2 prostaglandins did not reach statistical significance. The retinal concentrations of the series-3 prostaglandins were not evaluated in this study due to technical reasons.

Following this work, the same experimental design was used on an animal model of ocular hypertension.[49,50] After a 3-month dietary supplementation, the preventive effect of omega-6 and omega-3 PUFA was evaluated on early and

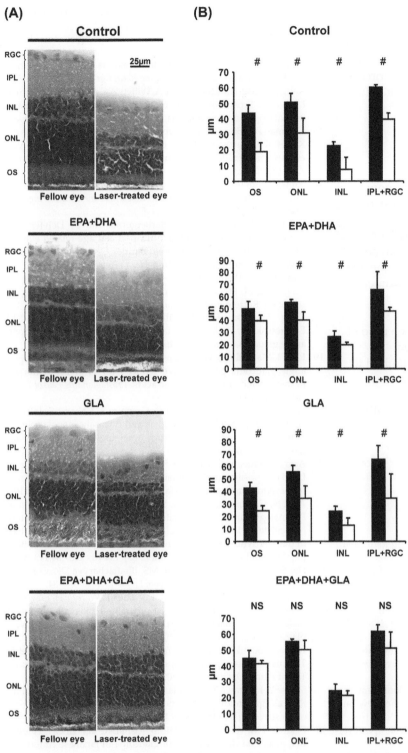

(A)

Control

RGC
IPL
INL
ONL
OS

Fellow eye   Laser-treated eye

25µm

EPA+DHA

RGC
IPL
INL
ONL
OS

Fellow eye   Laser-treated eye

GLA

RGC
IPL
INL
ONL
OS

Fellow eye   Laser-treated eye

EPA+DHA+GLA

RGC
IPL
INL
ONL
OS

Fellow eye   Laser-treated eye

(B)

Control

EPA+DHA

GLA

EPA+DHA+GLA

FIGURE 12.4 Morphometric analysis of the retinal cell layers in rats fed with diets differing in their content in omega-3 and omega-6 PUFAs 3 months after the laser photocoagulation. A) Representative pictures of retinal cryosections stained with hematoxylin and eosin. Retinal thickness is reduced in the laser-treated eyes compared to the fellow eyes, except for animals from the eicosapentaenoic acid (EPA) + docosahexaenoic acid (DHA) + gamma-linolenic acid (GLA) group. B) Retinal cell layer thickness was measured using the Nikon NIS Br software. A significant decrease was observed in all retinal cell layers in the laser-treated eyes (open bars) when compared to the fellow eyes (black bars) except for the retinas of animals from the EPA+DHA+GLA group. OS: outer segments; ONL: outer nuclear layer; INL: inner nuclear layer; IPL: inner plexiform layer; RGC: retinal ganglion cells. Values are mean ± standard deviation from three retinas in each group. #: statistically significant difference between laser-treated eye and fellow eye (Mann-Whitney test, P < 0.05); NS: non-significant. *Springer, Graefe's Archive for Clinical and Experimental Ophthalmology, 247, 2009, 1191-1203, A dietary combination of omega-3 and omega-6 polyunsaturated fatty acids is more efficient than single supplementations in the prevention of retinal damage induced by elevation of intraocular pressure in rats, Coralie Schnebelen, Bruno Pasquis, Manuel Salinas-Navarro, Corinne Joffre, Catherine P. Creuzot-Garcher, Manuel Vidal-Sanz, Alain M. Bron, Lionel Bretillon, Niyazi Acar, Figure 3, © Springer-Verlag.[50] With kind permission from Springer Science and Business Media.*

late stages of retinal injury. It was shown that retinal glial activation (a marker of retinal stress evaluated through glial fibrillary acidic protein (GFAP) immunoreactivity) was reduced in all animal groups supplemented with PUFA (omega-6 and/or omega-3) after the induction of ocular hypertension (Fig. 12.3). When late stages of retinal injury 3 months after the induction of ocular hypertension were considered, only the combination of dietary omega-6

and omega-3 PUFA was efficient in preventing glial activation. In this group of animals, not only was the activation of Müller cells prevented by dietary PUFA, but also the loss of retinal structure as evaluated by a morphometric analysis of retinal cell layers (Fig. 12.4). However, neither a modification of retinal concentrations of series-1 and series-2 prostaglandins, nor differences in the expression of retinal proinflammatory cytokines (such as interleukin-1beta, interleukin-6, and tumor necrosis factor-alfa) could explain the protection observed. As proposed by Nguyen and collaborators to explain the enhanced aqueous outflow facility, a potential involvement of omega-3 PUFA-derived docosanoids has been suggested.[45] Future experiments may help to elucidate the molecular mechanisms involved.

TAKE-HOME MESSAGES

- The beneficial properties of omega-6 PUFA-derived prostaglandins on lowering IOP are now well established, leading to the widespread use of prostaglandin analogs in humans.
- Independently from the action of omega-6 PUFA-derived prostaglandins on IOP, several human and animal studies suggest the existence of a relationship between the metabolism of omega-6 and/or omega-3 PUFA and the pathogenesis of glaucoma.
- More than the absolute dietary fat supply, a qualitative view of dietary lipids, such as the ratio of n-6 PUFA to n-3 PUFA, has to be considered.
- The beneficial effects of PUFA may not be limited to lowering IOP but might be extended to glaucoma subtypes displaying a normal IOP.
- Dietary PUFA can modulate the activation of retinal glia in stressful conditions.
- More work is needed to identify or clarify the mechanisms by which dietary omega-3 and omega-6 PUFAs may be beneficial in preventing glaucoma.

References

1. Quigley HA, Broman AT. The number of people with glaucoma worldwide in 2010 and 2020. *Br J Ophthalmol* 2006;**90**:262–7.
2. Leske MC, Wu SY, Hennis A, Honkanen R, Nemesure B. Risk factors for incident open-angle glaucoma: the Barbados Eye Studies. *Ophthalmology* 2008;**115**:85–93.
3. Anderson DR. Glaucoma: the damage caused by pressure. XLVI Edward Jackson memorial lecture. *Am J Ophthalmol* 1989;**108**:485–95.
4. Cracknell KP, Grierson I. Prostaglandin analogues in the anterior eye: their pressure lowering action and side effects. *Exp Eye Res* 2009;**88**:786–91.
5. Vetrugno M, Cantatore F, Ruggeri G, Ferreri P, Montepara A, Quinto A, et al. Primary open angle glaucoma: an overview on medical therapy. *Prog Brain Res* 2008;**173**:181–93.
6. Drance SM. Bowman Lecture. Glaucoma – changing concepts. *Eye* 1992;**6**(Pt 4):337–45.
7. Sommer A, Tielsch JM, Katz J, Quigley HA, Gottsch JD, Javitt J, et al. Relationship between intraocular pressure and primary open angle glaucoma among white and black Americans. The Baltimore Eye Survey. *Arch Ophthalmol* 1991;**109**:1090–5.
8. Shields MB. Normal-tension glaucoma: is it different from primary open-angle glaucoma? *Curr Opin Ophthalmol* 2008;**19**:85–8.
9. Pache M, Flammer J. A sick eye in a sick body? Systemic findings in patients with primary open-angle glaucoma. *Surv Ophthalmol* 2006;**51**:179–212.
10. Coleman AL, Kodjebacheva G. Risk factors for glaucoma needing more attention. *Open Ophthalmol J*. 200;3:38–42.
11. Pasquale LR, Kang JH. Lifestyle, nutrition, and glaucoma. *J Glaucoma* 2009;**18**:423–8.
12. Tezel G. Oxidative stress in glaucomatous neurodegeneration: mechanisms and consequences. *Prog Retin Eye Res* 2006;**25**:490–513.
13. Kang JH, Pasquale LR, Rosner BA, Willett WC, Egan KM, Faberowski N, et al. Prospective study of cigarette smoking and the risk of primary open-angle glaucoma. *Arch Ophthalmol* 2003;**121**:1762–8.
14. Coleman AL, Stone KL, Kodjebacheva G, Yu F, Pedula KL, Ensrud KE, et al. Glaucoma risk and the consumption of fruits and vegetables among older women in the study of osteoporotic fractures. *Am J Ophthalmol* 2008;**145**:1081–9.
15. SanGiovanni JP, Chew EY. The role of omega-3 long-chain polyunsaturated fatty acids in health and disease of the retina. *Prog Retin Eye Res* 2005;**24**:87–138.
16. Delcourt C, Korobelnik JF, Barberger-Gateau P, Delyfer MN, Rougier MB, Le Goff M, et al. Nutrition and age-related eye diseases: the Alienor (Antioxydants, Lipides Essentiels, Nutrition et maladies OculaiRes) Study. *J Nutr Health Aging* 2010;**14**:854–61.
17. Ramdas WD, Wolfs RC, Kiefte-de Jong JC, Hofman A, de Jong PT, Vingerling JR, et al. Nutrient intake and risk of open-angle glaucoma: the Rotterdam Study. *Eur J Epidemiol* 2012;**27**:385–93.
18. Kang JH, Pasquale LR, Willett WC, Rosner BA, Egan KM, Faberowski N, et al. Dietary fat consumption and primary open-angle glaucoma. *Am J Clin Nutr* 2004;**79**:755–64.
19. Ren H, Magulike N, Ghebremeskel K, Crawford M. Primary open-angle glaucoma patients have reduced levels of blood docosahexaenoic and eicosapentaenoic acids. *Prostaglandins Leukot Essent Fatty Acids* 2006;**74**:157–63.
20. Acar N, Berdeaux O, Juaneda P, Gregoire S, Cabaret S, Joffre C, et al. Red blood cell plasmalogens and docosahexaenoic acid are independently reduced in primary open-angle glaucoma. *Exp Eye Res* 2009;**89**:840–53.
21. Acar N, Berdeaux O, Gregoire S, Cabaret S, Martine L, Gain P, et al. Lipid composition of the human eye: are red blood cells a good mirror of retinal and optic nerve fatty acids?. *PLoS One* 2012;**7**: e35102.
22. Acar N, Gregoire S, Andre A, Juaneda P, Joffre C, Bron AM, et al. Plasmalogens in the retina: in situ hybridization of dihydroxyacetone phosphate acyltransferase (DHAP-AT) – the first enzyme involved in their biosynthesis – and comparative study of retinal and retinal pigment epithelial lipid composition. *Exp Eye Res* 2007;**84**:143–51.
23. Das SK, Steen ME, McCullough MS, Bhattacharyya DK. Composition of lipids of bovine optic nerve. *Lipids* 1978;**13**:679–84.
24. Rodemer C, Thai TP, Brugger B, Kaercher T, Werner H, Nave KA, Wieland F, et al. Inactivation of ether lipid biosynthesis causes male infertility, defects in eye development and optic nerve hypoplasia in mice. *Hum Mol Genet* 2003;**12**:1881–95.
25. Iwase A, Suzuki Y, Araie M, Yamamoto T, Abe H, Shirato S, et al. The prevalence of primary open-angle glaucoma in Japanese: the Tajimi Study. *Ophthalmology* 2004;**111**:1641–8.
26. Shiose Y, Kitazawa Y, Tsukahara S, Akamatsu T, Mizokami K, Futa R, et al. Epidemiology of glaucoma in Japan – a nationwide glaucoma survey. *Jpn J Ophthalmol* 1991;**35**:133–55.
27. Pekmezci M, Vo B, Lim AK, Hirabayashi DR, Tanaka GH, Weinreb RN, et al. The characteristics of glaucoma in Japanese Americans. *Arch Ophthalmol* 2009;**127**:167–71.
28. Renard JP, Rouland JF, Bron A, Sellem E, Nordmann JP, Baudouin C, et al. Nutritional, lifestyle and environmental factors in ocular hypertension and primary open-angle glaucoma: an exploratory case-control study. *Acta Ophthalmol* 2013;**91**:505–13.
29. Meguro A, Inoko H, Ota M, Mizuki N, Bahram S. Genome-wide association study of normal tension glaucoma: common variants in SRBD1 and ELOVL5 contribute to disease susceptibility. *Ophthalmology* 2010;**117**:1331–8.

30. Weinreb RN, Toris CB, Gabelt BT, Lindsey JD, Kaufman PL. Effects of prostaglandins on the aqueous humor outflow pathways. *Surv Ophthalmol* 2002;**47**(Suppl 1):S53–64.

31. Maihofner C, Schlotzer-Schrehardt U, Guhring H, Zeilhofer HU, Naumann GO, Pahl A, et al. Expression of cyclooxygenase-1 and -2 in normal and glaucomatous human eyes. *Invest Ophthalmol Vis Sci* 2001;**42**:2616–24.

32. Rosch S, Ramer R, Brune K, Hinz B. Prostaglandin E2 induces cyclooxygenase-2 expression in human non-pigmented ciliary epithelial cells through activation of p38 and p42/44 mitogen-activated protein kinases. *Biochem Biophys Res Commun* 2005;**338**:1171–8.

33. Civan MM, Macknight AD. The ins and outs of aqueous humour secretion. *Exp Eye Res* 2004;**78**:625–31.

34. Ocklind A, Lake S, Wentzel P, Nister M, Stjernschantz J. Localization of the prostaglandin F2 alpha receptor messenger RNA and protein in the cynomolgus monkey eye. *Invest Ophthalmol Vis Sci* 1996;**37**:716–26.

35. Abramovitz M, Boie Y, Nguyen T, Rushmore TH, Bayne MA, Metters KM, et al. Cloning and expression of a cDNA for the human prostanoid FP receptor. *J Biol Chem* 1994;**269**:2632–6.

36. Graves PE, Pierce KL, Bailey TJ, Rueda BR, Gil DW, Woodward DF, et al. Cloning of a receptor for prostaglandin F2 alpha from the ovine corpus luteum. *Endocrinology* 1995;**136**:3430–6.

37. Lindsey JD, To HD, Weinreb RN. Induction of c-fos by prostaglandin F2 alpha in human ciliary smooth muscle cells. *Invest Ophthalmol Vis Sci* 1994;**35**:242–50.

38. Meng Q, Xia Y. c-Jun, at the crossroad of the signaling network. *Protein Cell* 2011;**2**:889–98.

39. Woessner Jr JF. Matrix metalloproteinases and their inhibitors in connective tissue remodeling. *Faseb J* 1991;**5**:2145–54.

40. Oh DJ, Martin JL, Williams AJ, Peck RE, Pokorny C, Russell P, et al. Analysis of expression of matrix metalloproteinases and tissue inhibitors of metalloproteinases in human ciliary body after latanoprost. *Invest Ophthalmol Vis Sci* 2006;**47**:953–63.

41. Hinz B, Rosch S, Ramer R, Tamm ER, Brune K. Latanoprost induces matrix metalloproteinase-1 expression in human nonpigmented ciliary epithelial cells through a cyclooxygenase-2-dependent mechanism. *Faseb J* 2005;**19**:1929–31.

42. Gaton DD, Sagara T, Lindsey JD, Gabelt BT, Kaufman PL, Weinreb RN. Increased matrix metalloproteinases 1, 2, and 3 in the monkey uveoscleral outflow pathway after topical prostaglandin F(2 alpha)-isopropyl ester treatment. *Arch Ophthalmol* 2001;**119**: 1165–70.

43. Nilsson SF, Samuelsson M, Bill A, Stjernschantz J. Increased uveoscleral outflow as a possible mechanism of ocular hypotension caused by prostaglandin F2 alpha-1-isopropylester in the cynomolgus monkey. *Exp Eye Res* 1989;**48**:707–16.

44. Sagara T, Gaton DD, Lindsey JD, Gabelt BT, Kaufman PL, Weinreb RN. Topical prostaglandin F2alpha treatment reduces collagen types I, III, and IV in the monkey uveoscleral outflow pathway. *Arch Ophthalmol* 1999;**117**:794–801.

45. Nguyen CT, Bui BV, Sinclair AJ, Vingrys AJ. Dietary omega 3 fatty acids decrease intraocular pressure with age by increasing aqueous outflow. *Invest Ophthalmol Vis Sci* 2007;**48**:756–62.

46. Thieme H, Stumpff F, Ottlecz A, Percicot CL, Lambrou GN, Wiederholt M. Mechanisms of action of unoprostone on trabecular meshwork contractility. *Invest Ophthalmol Vis Sci* 2001;**42**: 3193–201.

47. Calder PC. Polyunsaturated fatty acids and inflammatory processes: new twists in an old tale. *Biochimie* 2009;**91**:791–5.

48. Schnebelen C, Gregoire S, Pasquis B, Joffre C, Creuzot-Garcher CP, Bron AM, et al. Dietary n-3 and n-6 PUFA enhance DHA incorporation in retinal phospholipids without affecting PGE(1) and PGE (2) levels. *Lipids* 2009;**44**:465–70.

49. Schnebelen C, Fourgeux C, Pasquis B, Creuzot-Garcher CP, Bron AM, Bretillon L, et al. Dietary polyunsaturated fatty acids reduce retinal stress induced by an elevation of intraocular pressure in rats. *Nutr Res* 2011;**31**:286–95.

50. Schnebelen C, Pasquis B, Salinas-Navarro M, Joffre C, Creuzot-Garcher CP, Vidal-Sanz M, et al. A dietary combination of omega-3 and omega-6 polyunsaturated fatty acids is more efficient than single supplementations in the prevention of retinal damage induced by elevation of intraocular pressure in rats. *Graefes Arch Clin Exp Ophthalmol* 2009;**247**:1191–203.

CATARACTS

13

Riboflavin and the Cornea and Implications for Cataracts

Cosimo Mazzotta[1], Stefano Caragiuli[1], Aldo Caporossi[2]

[1]Azienda Ospedaliera Universitaria Senese, Siena, Italy, [2]Policlinico Universitario A. Gemelli, Rome, Italy

INTRODUCTION

Riboflavin, also known as vitamin B_2, belongs to the class of water-soluble vitamins. The etymology of the word riboflavin, composed of ribo(se) and flavin, refers to its chemical composition. Riboflavin is composed of a nucleus of dimethyl isoalloxazine (isoalloxazine is the basic structure of all flavin molecules) and ribitol (the reduced form of the sugar ribose): 7,8-dimethyl-10-(1'-d-ribityl) isoalloxazine (Fig. 13.1).[1–3]

As suggested by its etymology (flavus in Latin means 'yellow'), riboflavin crystals have a yellow–orange color, whereas neutral solutions of riboflavin have a typical–green color. This is why they are used as the food coloring known as E101. Although riboflavin belongs to the group of water-soluble vitamins, it is in fact one of the least soluble in water (12 mg/100 mL at 25 °C) and even less soluble in alcohol. It is more soluble in acidic solutions up to a temperature of 120 °C, but unstable in alkaline media and when exposed to ultraviolet and visible light. The spectrum of riboflavin has four absorption peaks at wavelengths 223, 266, 373, and 445 nm. Riboflavin exhibits native fluorescence with an emission peak at 530 nm.[1–3]

HISTORY

Vitamin B_2 was first discovered in 1879 by the English chemist Alexander Wynter Blyth who called it 'lactoflavin' because he found it in cow's milk. Similar substances were called 'ovoflavin', 'hepatoflavin', and so forth, because they were discovered in eggs, liver, and other plant and animal tissues. It was not until 1934 that the Swiss chemist Paul Karrer and the Austrian-born German chemist Richard Kuhn, both Nobel laureates in chemistry, discovered that lacto-, ovo-, and hepatoflavin were the same substance, which they called riboflavin.[1,2]

DIETARY REQUIREMENTS

The vitamins are a heterogeneous group of compounds of relatively low molecular weight, indispensable, albeit in small quantities, for essential vital functions. Unlike microorganisms and plants, humans and higher animals are unable to synthesize them and must, therefore, receive them in food. Since its function makes it an essential constituent of all living cells, vitamin B_2 is widespread in nature. Brewer's yeast and calf and pig's liver are the richest sources of this vitamin, but are not frequent in the daily diet of most families. The main source of vitamin B_2 is therefore milk and milk products, eggs, meat, and green leafy vegetables. Because riboflavin is sensitive to light, milk should be sold in opaque containers. Traditional cooking methods cause loss of small quantities of riboflavin and boiling of vegetables in large quantities of water leaches out a considerable amount. The US Food and Nutritional Board established the following recommended daily allowances (RDA) for vitamin B_2: 1.3 mg/day for adult males and 1.1 mg/day for adult females.[4] The RDA is the quantity of a nutrient necessary to meet minimum daily requirements.

BIOLOGY

Riboflavin is not metabolically active, unlike its derivatives flavin mononucleotide (FMN) and flavin adenine dinucleotide (FAD). FMN is derived from phosphorylation of ribitol (which with isoalloxazine constitutes riboflavin). FAD is formed from FMN by binding with a molecule of adenosine monophosphate (AMP) from adenosine triphosphate (ATP) (Fig. 13.1).[5] FMN and FAD have coenzyme functions, specifically as electron transporters in biochemical reactions of oxidation–reduction (redox). In transferring high-energy electrons, FMN and FAD bind up to two electrons and two protons to their isoalloxazine group, undergoing reduction, while the molecule from which they obtain the electrons is oxidized. This process has an intermediate stage in which a semiquinone is formed (Fig. 13.2).[6]

The biochemical reactions in which it takes part are numerous (Fig 13.3).

The important aspect for our purposes is the role of the enzyme glutathione reductase as coenzyme in regeneration of the reduced form of glutathione (GSH).

GSH is a tripeptide with antioxidant properties. Its role is, therefore, to prevent damage to cellular and extracellular biomolecules by reactive oxygen species (ROS), which are oxygen free radicals. ROS are radicals, that is, very reactive chemical species, usually having a very short half-life.[7] They consist of an atom or molecule with an unpaired electron. This electron makes the radical extremely reactive and capable of binding to nearby molecules and capturing an electron. Major ROS include superoxide anion O_2^-, hydrogen peroxide H_2O_2, and the hydroxyl radical $\bullet OH$. These oxidants may damage cells and trigger chemical chain reactions, such as lipid peroxidation, or oxidize deoxyribonucleic acid (DNA) or proteins. It is easy to imagine the severe consequences of DNA damage such as genetic mutations, in the absence

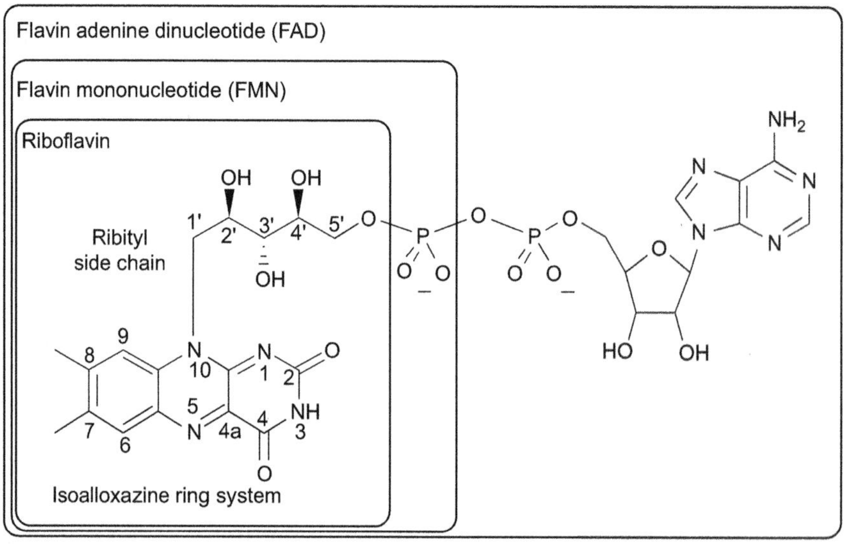

FIGURE 13.1 Biochemical formulas of flavin mononucleotide (FMN) and flavin adenine dinucleotide (FAD). The figure shows the formulas of the riboflavin, with its isoalloxazine ring, of the FMN and FAD.

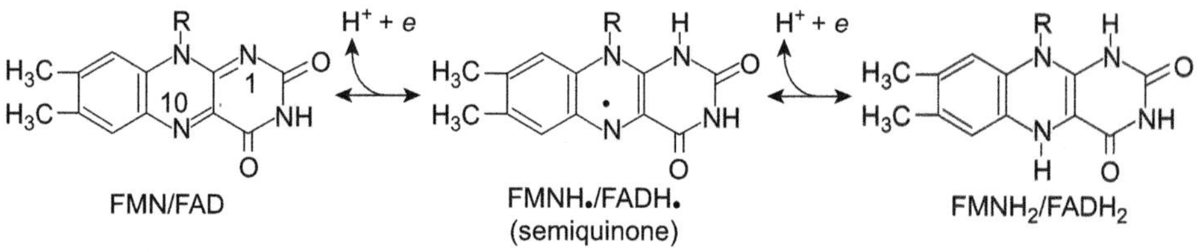

FIGURE 13.2 Oxidation–reduction of flavin mononucleotide (FMN) or flavin adenine dinucleotide (FAD). The figure shows the oxidation–reduction of the isoalloxazine ring of FMN or FAD via the intermediate step of semiquinone. R: side chain.

FIGURE 13.3 Biochemical reactions involving the flavoproteins as coenzymes. The figure shows reactions involving flavin adenine dinucleotide (FAD) as coenzymes and reactions involving FMN as coenzymes. Both of them are flavoproteins because of their chemical formulas.

Enzyme containing FAD as coenzyme	Enzyme containing FMN as coenzyme
Glutathione reductase	NADH dehydrogenase
Methylene-H$_4$folate reductase	Pyridoxine phosphate oxidase
Fatty acyl-CoA dehydrogenase	
Succinate dehydrogenase	
Xanthine dehydrogenase	
Choline dehydrogenase	
Monoamine oxidase (MAO)	

of specific repair mechanisms, whereas protein damage may include denaturing, breakdown, and inhibition of enzyme function. We illustrate the 'dire' concept of chain reaction with the following example. A peroxide radical (as in hydrogen peroxide) can attack fatty acids of the body, such as those of cell membranes, transforming them into organic free radicals and triggering a chain reaction of oxidation: at each step, the starting radical is regenerated and the reaction continues until total destruction of the substrate.

To prevent this, evolution has provided organisms with antioxidants, among which GSH plays a major role. In defending the body, two molecules of reduced GSH give up a hydrogen (H^+), which acts as electron acceptor from free radicals. The GSH molecules are thus oxidized to glutathione disulfide (GSSG), consisting of two molecules of oxidized glutathione joined by a disulfide bond. Once oxidized glutathione has deactivated the ROS, it must return to its reduced form to reacquire its antioxidant properties. This is carried out by an FAD-dependent enzyme known as glutathione reductase (Fig. 13.4). Although the names 'flavin mononucleotide' and 'flavin adenine dinucleotide' are widely used, both are inappropriate for FMN and FAD because neither is a true nucleotide.[5] The correct terms are 'riboflavin monophosphate' and 'riboflavin adenine diphosphate'. However, according to the Romans: "Error communis ius facit" or "Common error makes the law".

KERATOCONUS AND RIBOFLAVIN

The word keratoconus is derived from the Greek *kéras, kératos* 'cornea' plus *kônos* 'cone' (i.e., cone-shaped cornea). Keratoconus is a bilateral asymmetric degenerative disorder of the cornea characterized by corneal thinning. It is the main cause of primary ectasia.[8,9] Patients show thinning of the corneal stroma due to progressive reduction in the number of collagen lamellae, anomalous orientation of collagen fibrils (especially at the cone apex), and reduction in the quantity of proteoglycans between fibrils. The effect of this stromal thinning is bulging of the cornea under intraocular pressure, forming the corneal ectasia known as keratoconus. The cornea usually bulges in the inferior temporal paracentral region because this is physiologically the thinnest and weakest part of the cornea (Fig. 13.5; Fig. 13.6).[10,11]

Keratoconus typically manifests in the second decade (usually during puberty) and progresses until the fourth decade when it generally stabilizes due to a physiologic process of cross-linking of stromal collagen. This physiologic increase in cross-linking is associated with increasing corneal rigidity.[12] Diabetics seem exempt from progressive keratoconus, possibly due to increased cross-linking associated with hyperglycemia caused by the so-called Maillard reaction.[13,14]

Cross-linking of the collagen may be obtained by various physicochemical methods: glyceraldehyde, glutaraldehyde, and irradiation with ultraviolet light. Although glyceraldehyde and glutaraldehyde highly increase corneal rigidity, they are very toxic and clinically unacceptable. The one that has proven to have the best risk–benefit ratio in the corneal clinical application is the low-dose controlled irradiation with ultraviolet A light having a wavelength of 370 nm combined with riboflavin drops at a concentration of 0.1% (Fig. 13.7).

Riboflavin plays a fundamental role in the cornea acting as photosensitizer, increasing the production of free radicals that cause cross-linking of stromal collagen. These free radicals or ROS are thought to induce oxidative deamination of collagen leading to the formation of molecular bridges or cross-links between and within fibrils. Riboflavin also has another role, namely that of favoring absorption and concentration of ultraviolet light in the anterior half of the corneal stroma of corneas having a thickness of 400 μm or more, thus ensuring protection of the corneal endothelium and other vulnerable structures such as the lens and retina. It is worth

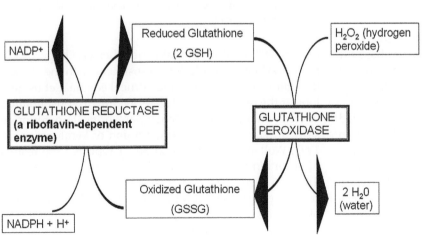

FIGURE 13.4 Ox-redox cycle of glutathione. The figure shows the cycle of oxidation–reduction of the glutathione. Note that the glutathione reductase needs flavin adenine dinucleotide (FAD) as coenzyme. FMN: flavin mononucleotide; NADH: nicotinamide adenine dinucleotide.

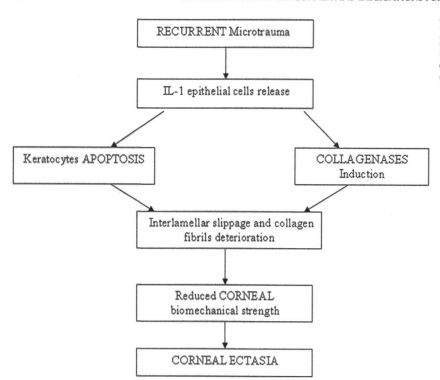

FIGURE 13.5 Hypothesis concerning the pathogenesis of keratoconus. The figure shows a flow chart that illustrates the current hypothesis concerning the pathogenesis of corneal ectasia underlying keratoconus. IL: interleukin.

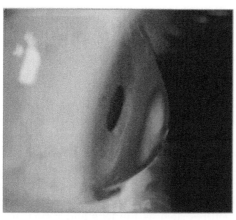

FIGURE 13.6 Keratoconic eye. The figure shows a keratoconic eye. Note the considerable bulging of the cornea.

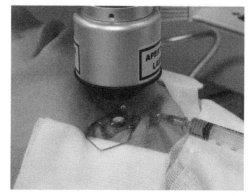

FIGURE 13.7 Corneal cross-linking. The figure shows riboflavin ultraviolet (UV) A corneal collagen cross-linking to stop keratoconus progression with the Siena CBM X linker, developed at Siena University by Caporossi, Baiocchi and Mazzotta, under the intellectual property of Siena University.

recalling that at birth, corneal endothelial cells lose their capacity to replicate and divide, remaining in phase G1 of the cell cycle. Further confirmation of the absence of side effects on the endothelium, lens and macula comes from prospective studies by Mazzotta (endothelium) [15] and Grewal (lens and macula).[16] Irradiation is conducted with ultraviolet light of 370 nm (in the spectral region in which riboflavin has an absorption peak (Fig. 13.2) at a power of 3 mW/cm². The instrument giving the best corneal irradiation is the VEGA C.B.M. X LINKER, developed by the Siena school. C.B.M. represents the names of the three researchers who developed it: Caporossi, Baiocchi, and Mazzotta. The riboflavin drops also

contain 20% dextran to make the otherwise hyperosmolar 0.1% riboflavin solution iso-osmolar with the cornea, thus preventing osmosis and corneal edema. Let us now look at the main historical steps leading to consolidation of cross-linking as the elective parasurgical operation to slow down or arrest the progression of keratoconus. In 1997, Wollensak, Spoerl, and Seiler reported the results of an *in vitro* study on cross-linking of pig corneas.[17,18] The first case of *in vivo* cross-linking in a patient with corneal ectasia was conducted in 1998. The first report of a series of 23 eyes treated with cross-linking and followed up for 48 months was published in 2003 by the eye school of Dresden (Germany).[19] In 2006, three researchers of the Siena eye school in Italy (Caporossi, Mazzotta *et al.*) published the results of a series of 10 eyes with keratoconus treated with cross-linking.[20,21] They reported arrested progression of keratoconus in treated eyes compared to progression in 37% of the untreated other eyes of the same patients. The results of confocal microscope examination of the same cohort of patients were reported by the same authors in 2007.[15] This article demonstrated depletion of corneal keratocytes in the anterior-intermediate stroma after cross-linking, with gradual repopulation by new keratocytes over a period of about 6 months. No endothelial changes were observed.

ARIBOFLAVINOSIS

Ariboflavinosis is the medical term used to indicate riboflavin deficiency. Pure riboflavin deficiency is practically nonexistent because it is often accompanied by a deficiency in other vitamins of the B group. Symptoms of riboflavin deficiency are essentially skin, mouth, and eye alterations.

- Skin symptoms include seborrheic dermatitis especially around the nose, mouth, scrotum, and vulva.
- Mouth symptoms include angular stomatitis with erythema and fissures at the corners of the mouth; cheilosis or fissuring and dry scaling of the vermilion surface of the lips in the absence of inflammation; glossitis with magenta tongue due to atrophy of filiform and hypertrophy of fungiform lingual papillae.
- Eye symptoms include photophobia, sight deterioration, corneal neovascularization, interstitial keratitis, corneal ulcers, conjunctivitis, and cataract.
- Other possible symptoms are normochromic normocytic anemia, characterized by erythroid hypoplasia with reticulocytopenia due to reduced mobilization of iron from intracellular deposits of ferritin or altered iron uptake by the intestine, and peripheral neuropathy of the extremities, characterized by hyperesthesia, sensations of cold and pain, and even decreased sensitivity to touch, temperature, vibration, and position.[5,22–24]

However, the above symptoms of riboflavin deficiency are rare in developed countries, where subclinical deficiency, characterized by altered biochemical indices (erythrocyte glutathione reductase function based on determination of activation coefficient after addition of FAD) without clinical symptoms, is more common.[4,25,26] Although the long-term effects of subclinical deficiency are unknown in adults, they lead to growth retardation in children.

The Italian Society for Human Nutrition (SINU) states that severe riboflavin deficiency does not exist in Italy. To assess intake of this vitamin it is possible to measure plasma levels and/or do the above erythrocyte glutathione reductase test. Epidemiologic studies in Italy show a prevalence of marginal deficiency of riboflavin (subclinical forms) in about 20% of elderly people.[26] Similar results are reported by the Food and Agriculture Organization in the US, where the percentages are 12% in the general population and 27% of black people.[27]

CATARACT

Cataract is the name used to describe any partial or complete loss of transparency of the lens. It is the major cause of blindness in the world, accounting for 51% of all cases of blindness according to the last report of the World Health Organization (2010).[28] A positive fact is that cataract causes reversible blindness, not through self-healing but through surgical replacement of the lens with an intraocular lens (IOL). Cataracts cause slow but progressive loss of visual capacity and are not painful.

Etymologically, the term *cataract* comes from the Greek word for *cascade*. In antiquity, it was thought that opacity of the eyes was caused by descent of a veil, in the same way that water falls (Fig. 13.8).

In developing countries, where riboflavin nutritional deficiency is frequent and becomes symptomatic, many studies have demonstrated a correlation between the deficiency and onset of nuclear cataract.[29–33] For example, a study conducted in Linxian (China) demonstrated a 44% decrease in the prevalence of nuclear cataract after vitamin supplementation.

The etiology of cataract due to riboflavin deficiency is presumably related to the fact that the vitamin is the coenzyme of glutathione reductase, which regenerates

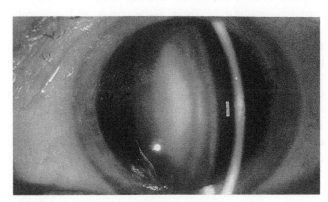

FIGURE 13.8 Nuclear senile cataract. The figure shows a slit-lamp photograph of an eye with an evident nuclear senile cataract.

reduced glutathione. The importance of the latter depends on three mechanisms.[34]

1. It maintains sulfydryl groups (SH⁻) of high-molecular-weight proteins, such as crystallins, in reduced state, preventing their alteration, aggregation, and precipitation, conditions leading to lens opacity.
2. It maintains sulfydryl groups (SH⁻) of ion channels in lenticular cells in reduced state, preventing cell imbibition, a condition that leads to lens opacity.
3. It counteracts denaturing of any type of molecule by oxygen free radicals, a condition leading to lens opacity.

A study by Shalka and Prchal on a middle-class population sample confirmed that only patients subject to severe riboflavin deficiency with clinical manifestations develop secondary cataract.[35] As already mentioned, cataract due to riboflavin deficiency only occurs in developing countries, and does not explain the high prevalence of cataract in developed countries. In developed countries, the high prevalence of cataract must therefore be explained without reference to riboflavin deficiency. First, let us recall that there are two types of cataract, classified in relation to site and/or etiology. The most frequent in developed countries is senile nuclear cataract. According to the Framingham Eye Study, it occurs in 65.5% of people over 75 years of age. Since the study was conducted at Framingham, a county in the state of Massachusetts, US, it provides statistics for developed countries, not developing countries.[36] GSH is involved in the etiopathogenesis of senile nuclear cataract and in riboflavin deficiency, but in the former case there is no dysfunction of glutathione reductase due to lack of the coenzyme riboflavin, but instead a decrease in lens nuclear concentrations of GSH, increasing vulnerability to ROS generated by absorption of ultraviolet light or as subproducts of endogenous cell metabolism. Let us now consider the question in more detail.

GSH is the main agent for detoxifying and preventing damage to the lens by oxygen free radicals. These radicals would otherwise oxidize and denature lens proteins, leading to their aggregation and precipitation, which causes opacity of the lens nucleus known as senile nuclear cataract. GSH is, therefore, important in preventing this type of cataract. It is worth recalling that the lens has no vasculature and that GSH is only synthesized in the lens by the anterior lenticular epithelium and by younger and more superficial lenticular fibers. These are the only lens cells containing organelles. However, an abundance of gap junctions between superficial and deep lenticular fibers enables movement of ions and small molecules among the fibers, permitting newly synthesized molecules of GSH to penetrate the lens nucleus from the surface. With increasing age, the concentration of GSH decreases together with its reducing power in

the lens nucleus.[37] Sweeney and Truscott proposed a fascinating revolutionary hypothesis, sustained by scientific evidence, to explain this fact.[38] Their experiment was based on incubation of human lenses in artificial aqueous humor containing labeled cysteine. Recall that GSH is a tripeptide composed of cysteine, glycine, and glutamate, so that incubation of the lens in this medium leads to incorporation of the labeled amino acid into the GSH molecule. After 48 hours of incubation, the lenses were sectioned and the distribution of ³⁵S-labeled cysteine detected by phosphorimaging using a fluorescent screen sensitive to the radiation emitted. The two Australian researchers observed that the marker was uniformly distributed in lenses from patients under 30 years of age, but in patients over 30 years of age very little labeled cysteine penetrated the lens nucleus. To explain this, they postulated an age-related barrier preventing diffusion of GSH from the lens surface into the nucleus. They also compared the dimensions of the area delineated by this barrier with the yellowish–brown sclerotic area of senile nuclear cataract and found that they coincided. Their first intuition was to ascribe the etiopathogenesis of senile nuclear cataract to the hypothesis of age-dependent decrease in GSH synthesizing and reducing enzymes. This decrease does in fact exist, but is not by itself sufficient to explain the onset of opacity.

Fourteen years later, this hypothesis has never been refuted and indeed new articles have been published explaining the nature of the proposed barrier. The barrier is not macroscopic, like a membrane separating the lens nucleus from the cortex and preventing GSH synthesized on the surface to penetrate the lens, but functional, due to age-related loss of gap junctions and aquaporins.[39,40] The aging lens can, therefore, be likened to a castle surrounded by a moat that can only be crossed by bridges. If these bridges are cut, exchange between the castle and its territory becomes impossible. The only preventive measure currently available to prevent age-related degeneration of the lens is for subjects exposed to high levels of sunlight to wear protective glasses that absorb ultraviolet rays.[41]

CONCLUSIONS

Riboflavin (vitamin B₂) is a substance with various implications for the eyes. The salient points of our discussion are as follows.

• Concentrations of GSH decrease with age, irrespective of riboflavin and its role as coenzyme of glutathione reductase. This decrease weakens defenses against ultraviolet radiation, which, through production of oxygen free radicals, denatures lens molecules, which in turn aggregate, precipitate, and cause lens opacity or cataract.

- Riboflavin deficiency increases the risk of eye pathologies, including cataract, which arises because glutathione reductase, the enzyme defending the lens against ROS, is FAD dependent. This only applies in developing countries, as already explained. Thus, in developing countries, riboflavin deficiency prevents the coenzyme function of the vitamin.
- Riboflavin causes photosensitization by absorbing ultraviolet light (wavelength 370 nm) and stimulating production of oxygen free radicals. This physicochemical property of riboflavin, not its coenzyme function, is exploited by ophthalmologists to stabilize or slow down keratoconus.

TAKE-HOME MESSAGES

- Riboflavin, also known as vitamin B$_2$, belongs to the class of water-soluble vitamins.
- The US Food and Nutritional Board established the following RDAs) for riboflavin: 1.3 mg/day for adult males and 1.1 mg/day for adult women.
- Riboflavin is not metabolically active, unlike its derivatives FMN and FAD.
- The biochemical reactions in which it takes part are numerous, the most important concern with the biochemical reactions of oxidation–reduction (redox) involving the glutathione reductase.
- GSH is a tripeptide with antioxidant properties.
- Riboflavin ultraviolet A-induced corneal collagen cross-linking is a new therapy to halt keratoconus and secondary corneal ectasia progression.
- Riboflavin acts as photosensitizer, increasing the production of free radicals that cause cross-linking of stromal collagen.
- Riboflavin also favors the absorption and concentration of ultraviolet light in the anterior half of the corneal stroma.
- Riboflavin deficiency plays an important role in the etiology of cataract due to its involvement in regeneration of reduced glutathione.
- Only patients subject to severe riboflavin deficiency with clinical manifestation develop secondary cataract.

References

1. Arienti G. *Le Basi Molecolari della Nutrizione*. 2nd edn. Padova: Piccin; 2003.
2. Cabras P, Martelli A. *Chimica Degli Alimenti*. Padova: Piccin; 2004.
3. Ball G. *Vitamins in Foods: Analysis, Bioavailability, and Stability*. Boca Raton: CRC Press; 2006.
4. The Institute of Medicine. *Dietary Reference Intakes for Thiamin, Riboflavin, Niacin, Vitamin B6, Folate, Vitamin B12, Pantothenic Acid, Biotin, and Choline*. Washington: The National Academies Press; 1998.
5. Combs GF. *The Vitamins*. San Diego: Elsevier Inc.; 2008.
6. Berg JM, Tymoczko JL, Stryer L. *Biochemistry*. 5th edn. New York: WH Freeman Publishers; 2002.
7. Benzi G. *Ossigeno, Energia e Radicali Liberi: il Ruolo Biologico Dell'ossigeno*. Milan: Edimac s.r.l.; 1988.
8. Rabinowitz Y. Keratoconus. *Surv Ophthalmol* 1998;**42**(4):97–319.
9. Romero-Jiménez M, Santodomingo-Rubido J, Wolffsohn JS. Keratoconus: a review. *Contact Lens Anterior Eye* 2010;**33**(4):157–66.
10. Sherwin T, Nigel N. Morphological changes in keratoconus: pathology or pathogenesis. *Clin Exp Ophthalmol* 2004;**32**:211–7.
11. Daxer A, Fratzl P. Collagen fibril orientation in the human corneal stroma and its implication in keratoconus. *Invest Ophthalmol Vis Sci* 1997;**38**(1):121–9.
12. Malik NS, Moss SJ, Ahmed N, Furth AJ, Wall RS, Meek KM. Ageing of the human corneal stroma: structural and biochemical changes. *Biochim Biophys Acta* 1992;**1138**(3):222–8.
13. Kuo IC, Broman A, Pirouzmanesh A, Melia M. Is there an association between diabetes and keratoconus? *Ophthalmology* 2006;**113**(2):184–90.
14. Sady C, Khosrof S, Nagaraj R. Advanced Maillard reaction and crosslinking of corneal collagen in diabetes. *Biochem Biophys Res Commun* 1995;**214**(3):793–7.
15. Mazzotta C, Balestrazzi A, Traversi C, Baiocchi S, Caporossi T, Tommasi C, et al. Treatment of progressive keratoconus by riboflavin-UVA-induced cross-linking of corneal collagen: ultrastructural analysis by Heidelberg Retinal Tomograph II *in vivo* confocal microscopy in humans. *Cornea* 2007;**26**:390–7.
16. Grewal DS, Brar GS, Jain R, Sood V, Singla M, Grewal SP. Corneal collagen crosslinking using riboflavin and ultraviolet-A light for keratoconus: one-year analysis using Scheimpflug imaging. *J Cataract Refract Surg* 2009;**35**:425–32.
17. Spoerl E, Huhle M, Seiler T. Induction of cross-links in corneal tissue. *Exp Eye Res* 1998;**66**(1):97–103.
18. Wollensak G, Spoerl E, Seiler T. Stress-strain measurements of human and porcine corneas after riboflavin-ultraviolet-A-induced cross-linking. *J Cat Refract Surg* 2003;**29**(9):1780–5.
19. Wollensak G, Spoerl E, Seiler T. Riboflavin/ultraviolet-A-induced collagen crosslinking for the treatment of keratoconus. *Am J Ophthalmol* 2003;**135**(5):620–7.
20. Caporossi A, Baiocchi S, Mazzotta C, Traversi C, Caporossi T. Parasurgical therapy of keratoconus by riboflavin-UVA-induced cross-linking of corneal collagen: preliminary refractive results in an Italian study. *J Cataract Refract Surg* 2006;**32**:837–45.
21. Caporossi A, Mazzotta C, Baiocchi S, Caporossi T. Long-term results of riboflavin ultraviolet a corneal collagen cross-linking for keratoconus in Italy: the Siena Eye Cross Study. *Am J Ophthalmol* 2010;**149**(4):585–93.
22. Powers HJ. Riboflavin (vitamin B-2) and health. *Am J Clin Nutr* 2003;**77**(6):1352–60.
23. Pirie A. The relation of riboflavin to the eye. A review article. *Br J Ophthalmol* 1943;**27**(7):291–301.
24. Robinson FA. *The Vitamin B Complex*. New York: Wiley; 1951.
25. Herrmann W, Obeid R. *Vitamins in the Prevention of Human Disease*. Berlin: De Gruyter; 2011.
26. Tomassi G, Leclercq C, Ticca M. *LARN (Livelli Raccomandati di Assunzione di Nutrienti)*. Florence: SINU; 1996.
27. Latham MC. *Human Nutrition in the Developing World*. Rome: FAO; 1997.
28. Pascolini D, Mariotti SP. Global estimates of visual impairment: 2010. *Br J Ophthalmol* 2012;**96**(5):614–8.
29. Sperduto RD, Hu TS, Milton RC, Zhao JL, Everett DF, Cheng QF, et al. The Linxian cataract studies. Two nutrition intervention trials. *Arch Ophthalmol* 1993;**111**(9):1246–53.
30. Leske MC, Wu SY, Hyman L, Sperduto R, Underwood B, Chylack LT, et al. Biochemical factors in the lens opacities. Case-control study. The Lens Opacities Case-Control Study Group. *Arch Ophthalmol* 1995;**113**(9):1113–9.

31. Hankinson SE, Stampfer MJ, Seddon JM, Colditz GA, Rosner B, Speizer FE, et al. Nutrient intake and cataract extraction in women: a prospective study. *BMJ* 1992;**305**(6849):335–9.

32. Jacques PF, Taylor A, Moeller S, Hankinson SE, Rogers G, Tung W, et al. Long-term nutrient intake and 5-year change in nuclear lens opacities. *Arch Ophthalmol* 2005;**123**(4):517–26.

33. Maraini G, Williams SL, Sperduto RD, Ferris FL, Milton RC, Clemons TE, et al. Effects of multivitamin/mineral supplementation on plasma levels of nutrients. Report No. 4 of the Italian-American clinical trial of nutritional supplements and age-related cataract. *Ann Ist Super Sanita* 2009;**45**(2):119–27.

34. Head KA. Natural therapies for ocular disorders, part two: cataracts and glaucoma. *Altern Med Rev* 2001;**6**(2):141–66.

35. Skalka HW, Prchal JT. Cataracts and riboflavin deficiency. *Am J Clin Nutr* 1981;**34**(5):861–3.

36. Sperduto RD, Hiller R. The prevalence of nuclear, cortical, and posterior subcapsular lens opacities in a general population sample. *Ophthalmology* 1984;**91**(7):815–8.

37. Harding JJ. Free and protein-bound glutathione in normal and cataractous human lenses. *Biochem J* 1970;**117**(5):957–60.

38. Sweeney MH, Truscott RJ. An impediment to glutathione diffusion in older normal human lenses: a possible precondition for nuclear cataract. *Exp Eye Res* 1998;**67**(5):587–95.

39. Berthoud VM, Beyer EC. Oxidative stress, lens gap junctions, and cataracts. *Antioxid Redox Signal* 2009;**11**(2):339–53.

40. Korlimbinis A, Berry Y, Thibault D, Schey KL, Truscott RJ. Protein aging: truncation of aquaporin 0 in human lens regions is a continuous age-dependent process. *Exp Eye Res* 2009;**88**(5):966–73.

41. Young RW. The family of sunlight-related eye diseases. *Optom Vis Sci* 1994;**71**(2):125–44.

14

Diabetic Cataract and Role of Antiglycating Phytochemicals

Vaishali Agte, Snehal Gite

Agharkar Research Institute, Pune, India

INTRODUCTION

Diabetes mellitus is one of the fastest growing metabolic diseases worldwide. As reported in the World Health Organization's (WHO) report,[1] the estimated prevalence of diabetes for all age groups worldwide was 2.8% in 2000 and has been predicted to be 4.4% by 2030. The projected diabetics are going to rise from 171 million in 2000 to 366 million in 2030. India has the world's largest population of diabetic patients having type 2 diabetes, and the predicted rise over 30 years in Indian diabetics is 40%. Since diabetes is a metabolic disorder, it is associated with secondary complications such as atherosclerosis, chronic renal insufficiency, nephropathy, neuropathy, and cataract. These complications may be significant with the increased level of blood sugar over prolonged periods of time in diabetes mellitus.[2] Although diabetes and cataract have some common and uncommon etiologies, our previous studies have shown that when diabetes and cataract are both present, particularly in women, the condition is associated with highest levels of oxidative stress and the most compromised micronutrient status.[3]

Cataract in diabetic patients is considered as a major cause of blindness in the developed and developing world. The pathogenesis of diabetic cataract development is yet to be fully understood.[4] Diabetic patients tend to accumulate glycosylated protein (advanced glycation end products (AGEs)) in their body tissues. The degree of accumulation of AGEs is related to the severity of diabetic complications.[5] The concentration of glycated hemoglobin (HbA1c), the first characterized Amadori product, which has the N-terminals of the b-chains (valine) linked to glucose, is proportionately increased with hyperglycemia in diabetic patients. Glycation of hemoglobin by fructose (fructation) also causes iron release from the heme protein and associated free radical reactions.[6] Fructose and its derivatives (fructose-1-phosphate and D-glyceraldehyde) are significantly more effective glycating agents than glucose in the Maillard reaction. The human plasma level of fructose is estimated in the range of few micromoles per liter; whereas the glucose concentration is 5 mmol/L. Diets high in fructose as well as caloric intake may affect the accumulation of AGEs, but it is tissue specific.[7]

The polyol pathway is the second mechanism of fructose accumulation and is activated by hyperglycemia in various tissues such as ocular lens, kidney, and peripheral nerves, but the concentration of fructose in these tissues is less. The polyol pathway may substantially contribute to intracellular AGE formation by the reaction of fructose-3-phosphate and 3-deoxyglucosone intermediates with proteins.[8] The concentrations of fructose are similar in magnitude to that of glucose and are strongly increased by hyperglycemia, making *in vivo* glycation by fructose a highly probable event.[7] Despite its high reactivity, the contribution of extracellular glycation of proteins by fructose is considerably less than D-glucose because of much lower levels of fructose in blood. Still, it might be possible that the higher reactivity of fructose and its subsequent metabolism may contribute to alterations of cellular proteins and dysfunction within the cells.

Aldose reductase (AR) is a monomeric, adenine dinucleotide phosphate (NADPH)-dependent enzyme belonging to the aldo-keto reductase super family. It has the ability to reduce a broad variety of aldehydes, ranging from membrane phospholipids to glucose. AR therefore plays multifarious roles in cellular metabolism and signaling. It was considered initially as a glucose-reducing enzyme. However, now the enzyme is assumed to be an important component of antioxidant defense involved in the depletion and detoxification of reactive aldehydes that are formed due to lipid peroxidation. Inhibition of AR is reported to prevent the development of diabetic complications through animal studies; however, a critical evaluation of the clinical efficacy of AR inhibitors awaits a clearer understanding of the role of AR in regulating inflammation and cell growth. More selective and effective inhibitors are needed to inhibit the cytotoxic role of AR in cell signaling specifically affecting its detoxification role. Such inhibitors are likely to be more effective in treating secondary diabetic complications by preventing inflammation due to chronic hyperglycemia.[9]

High blood glucose levels are considered to be responsible for the glycosylation of ε-amino groups of lysine residues in bovine and rat lens crystallins. *In vitro*, this glycosylation is considered to increase susceptibility of the crystallins to sulfhydryl oxidation. Disulfide cross-links can lead to the formation of high molecular weight aggregates and an increase in opacity of the crystallin solutions. Nevertheless, adding reducing agents may not only prevent but also reverse the formation of high molecular weight aggregates and the opacity of lens proteins. These concepts are a new interpretation of earlier results on cataractogenesis and offer a new approach for drugs to prevent lens opacity.[10]

The polyol pathway is a two-step metabolic pathway converting glucose to fructose via sorbitol. It is one of the most plausible mechanisms to partly explain the cellular toxicity of diabetic hyperglycemia because (i) it becomes active when glucose levels within cells are elevated, (ii) both AR and sorbitol dehydrogenase (SDH) are present in sites of diabetic complications, and (iii) the products that are formed in the pathway and the modified levels of cofactors generate the same types of cellular stress similar to diabetic complications. Inhibition of AR in the pathway reproducibly prevents diabetic retinopathy in diabetic rodent models. The endothelial cells of human retinal vessels show presence of AR. Specific polymorphisms in the promoter region of the AR gene were associated with susceptibility or development of diabetic retinopathy. This has created interest in a possible role of the polyol pathway in diabetic retinopathy and related complications.[11]

It has been postulated[12] that during the states of hyperglycemia, reduction of glucose to sorbitol by AR constitutes the initial and also the rate-limiting step of the polyol pathway, which converts glucose to fructose through the enzyme SDH. As a consequence, both NADPH and nicotinamide adenine dinucleotide (NAD$^+$) are used as cofactors for AR and SDH. Osmotic stress due to accumulation of sorbitol and oxidative stress due to changes in the ratio of NADPH/NADP$^+$ and reduced nicotinamide adenine dinucleotide (NADH)/NAD+ are

the major causes of various complications of secondary diabetes.

The fructose generated in the polyol pathway can convert to fructose-3-phosphate and be broken down to 3-deoxyglucosone, both compounds being potent glycosylating agents in the formation of AGEs. The participation of NADPH by AR may result in reduction of cofactor available for glutathione reductase, which is crucial to maintain the intracellular pool of reduced glutathione (GSH). This reduces the capacity of cells to manage the oxidative stress that produces a number of metabolic and signaling changes altering cell function. The surplus of NADH could serve as a substrate for NADH oxidase, and this could produce intracellular oxidant species. Thus, activation of the polyol pathway, generating AGE precursors, and exposing cells to oxidative stress through decreased antioxidant defenses and generation of oxidant species can start and proliferate mechanisms that are responsible for cellular damage.[11] Retinal ganglion cells, Muller glia, and vascular pericytes and endothelial cells contain AR in humans. Therefore, these cell types are exposed to polyol pathway activation during diabetes and exhibit noticeable changes or damage in diabetes. The biochemical consequences of polyol pathway activation are the accumulation of sorbitol and fructose and the generation or enhancement of oxidative stress in the whole retina of diabetic animals. These abnormalities are prevented by drugs that inhibit AR.[13]

INHIBITORY POTENTIAL OF FOODS

In India, there are certain plants that seem to provide protection from diabetes and diabetic complications. To study the chemical composition and functional activity of these natural inhibitors will be a first step in determining whether they may be useful as therapeutic agents. One of these plants is called amla or gooseberry. Curcumin is an active principle present in turmeric, a spice that is used in Indian cooking. Curcumin, which is known to protect against certain diabetic complications, is also a very potent and selective inhibitor of AR. Although amla, curcumin, or turmeric could not avert streptozotocin-induced hyperglycemia and insulin levels, observations from slit-lamp microscopy have indicated that these supplements delayed the progression and maturation of cataract.[14, 15] Curcumin, at the levels close to dietary consumption, has been also reported to prevent the loss of chaperone-like activity of α-crystallin with reference to cataract formation due to diabetes in rat lenses.[16]

Ingestion of high-isoflavone soy protein resulted in lowering of glucose levels along with reduced incidence of cataracts in diabetic rats. The advantageous effects of soy isoflavones might be due to increased insulin secretion, better glycemic control, and antioxidant protection.[17] Liu has proposed that a diet rich in fruits and vegetables is a complex mixture of phytochemicals, with their additive and synergistic effects responsible for their health benefit.[18] There is a need to explore these preventive benefits in such food materials through systematic studies.

INHIBITORY POTENTIAL OF SINGLE AND POLYHERBAL/AYURVEDIC DRUGS

The Indian traditional health system called Ayurveda has herb- and food-based treatments that are dependent upon the specific type of diabetic complication. Despite the fact that they might be safer and efficient, herbal medicines need to undergo properly designed, rigorous clinical trials in order to receive scientific merit over existing drugs. Traditional plant remedies have been used for centuries in the treatment of diabetes,[19–22] but only a few have been scientifically evaluated (Table 14.1; Table 14.2). To achieve a blockbuster status, clear evidence of an advantage over existing therapy is the most important requirement.[57]

The efficacy of *Momordica charantia* (MC), *Eugenia jambolana* (EJ), *Tinospora cordifolia* (TC), and *Mucuna pruriens* (MP) was assessed for 4 months in the prevention of diabetic cataract in a murine model given alloxan. MC and EJ prevented the development of cataract, while the protective effect was less with TC and MP along with a significant reduction of plasma glucose levels (P < 0.001).[58] *Scutellaria baicalensis* enhanced the antidiabetic effect of metformin in streptozotocin (STZ)-induced diabetic rats by improving the antioxidant status. It also increased pancreatic insulin content and the lipid profile in rats.[59] The effects of aerial parts of *Phlomis anisodonta* methanolic extract (PAE) on STZ-induced diabetic rats suggested that PAE was useful in the reduction of blood glucose, rising insulin levels, and combating oxidative stress by activation of hepatic antioxidant enzymes.[60]

Vanadate (SOV) at lowered doses administered in combination with Trigonella (TSP) was effective at controlling the altered glucose metabolism and antioxidant status in diabetic lenses, which reduced lens opacity.[61] Further, with combined treatment of SOV and TSP, there was a significant (P < 0.05) decrease in blood glucose, HbA1c levels, and polyol pathway enzymes AR and SDH than control levels.[62] Treatment with ginsenoside Re restored the levels of both GSH and malondialdehyde in the eye and kidney compared with those found in the control rats. The results indicated that ginsenoside Re could be used as an effective antidiabetic agent, particularly in the prevention of diabetic microvasculopathy.[63]

TABLE 14.1 Antidiabetic and Other Beneficial Effects in Traditional Medicine

Plant Name	Activity	Reference
Phaseolus vulgaris	Hypoglycemic, hypolipidemic, inhibit α-amylase activity, antioxidant; altered level of insulin receptor and GLUT-4 mRNA in skeletal muscle	Knott et al. (1992)[23] Tormo (2004)[24]
Salacia reticulata	Inhibitory activity against sucrase, α-glucosidase inhibitor	Yoshikawa (1998)[25]
Gymnema sylvestre	Antihyperglycemic effect, hypolipidemic	Chattopadhyay (1999)[26] Preuss et al. (1998)[27]
Emblica officinalis	Decreases lipid peroxidation, antioxidant, hypoglycemic	Bhattacharya (1999)[28] Kumar and Muller (1999)[29] Devasagayam et al. (1995)[30]
Punica granatum	Antiglycating and antioxidant activity of polysaccharides isolated from fruit extract	Kokila et al. (2010)[31]
Aloe vera	Maintains glucose homeostasis by controlling the carbohydrate metabolizing enzymes	Rajasekaran et al. (2004)[32] Okyar et al. (2001)[33]
Aegle marmelos	Leaf increases utilization of glucose; either by direct stimulation of glucose uptake or via the mediation of enhanced insulin secretion and has potent antioxidant activity, which may account for the hypoglycemic potential	Sachdewa et al. (2001)[34]
Eugenia jambolana	Seed powder exhibits normoglycemia and better glucose tolerance	Ravi et al. (2004)[35]
Enicostemma littorale	Its leaves decrease the levels of glycosylated hemoglobin and glucose-6-phosphatase	Srinivasan et al. (2005)[36] Maroo et al. (2002)[37]
Ocimum sanctum	Its powdered leaf has produced potent hypoglycemic and hypolipidemic effect in normal and diabetic rats	Vats et al. (2004)[38]
Scoparia dulcis	Leaves decrease the levels of glycosylated hemoglobin and increase the total hemoglobin, insulin-secretagog activity	Pari et al. (2002)[39]

TABLE 14.2　Polyhedral Formulations Available to Prevent Diabetes and Its Secondary Complications

Sr. No	Polyherbal Formulation	Components of Formulation	Benefits	Reference
1	PM021	*Mori folium* and *Aurantii fructus*	Distinct antidiabetic effects without any adverse effects or toxicities	Kim (2006)[40]
2	Okudiabet	*Stachytarpheta angustifolia, Alstonia congensis* bark, and *Xylopia acthiopica* fruits extract	Effective in decreasing plasma glucose levels in diabetic rats; the high LD50 value (16.5 g/kg) indicates that formulation could be safe for use	Ogbonnia *et al.* (2010)[41]
3	Karmin Plus	*Momordica charantia, Azadirachta indica, Picrorrhiza kurroa, Ocimum sanctum,* and *Zinziber officinale*	Antidiabetic activity	Om *et al.* (2009)[42]
4	Glyoherb	A polyherbal formulation	Antihyperglycemic, antihyperlipidemic, and antioxidant effects; improves kidney and liver functions	Thakkar *et al.* (2010)[43]
5	5EPHF	*Aegle marmelos, Murraya koenigii, Aloe vera, Pongamia pinnata,* and *Elaeodendron glaucum*	At 200 mg/kg to diabetic rats, significant reduction of serum glucose, glycosylated hemoglobin, total cholesterol, triglyceride, low-density lipoprotein, creatinine, and urea whereas significant increased level of insulin and high-density lipoprotein was observed along with improved antioxidant status	Sweety *et al.* (2011)[44]
6	EFPTT/09	Five ingredients of herbal origin that are used in medicine to treat diabetes	Hypoglycemic, antidiabetic, and antioxidant effect	Yoganandam and Bimlendu (2010)[45]
7	ESF/AY/500	*Aerva lanata, Aegle marmelos, Ficus benghalensis, Catharanthus roseus, Bambusa arundinaceae, Salacia reticulate, Szygium cumini,* and 'Eruca sativa'	Antioxidant activity	Sajeeth *et al.* (2010)[46]
8	Diabeta	A formulation containing *Gymnema sylvestre, Vinca rosea* (Periwinkle), *Curcumalonga* (Turmeric), *Azadirachta indica* (Neem), *Pterocarpus marsupium* (Kino Tree), *Momordica charantia* (Bitter Gourd), *Syzygiumcumini* (Black Plum), *Acacia arabica* (Black Babhul), *Tinospora cordifolia,* and *Zingiber officinale* (Ginger) available in the capsule form	Antidiabetic fortified with potent immunomodulators, antihyperlipidemics, antistress, and hepatoprotective of plant origin	Modak *et al.* (2007)[47]
9	Diabecure	*Juglans regia, Berberis vulgaris, Erytherea centaurium, Millefolium,* and *Taraxacum*	Effective in lowering the blood sugar level	Modak *et al.* (2007)[47]
10	Diabetes-Daily Care	α-lipoic acid, cinnamon 4% extract, chromax, vanadium, fenugreek 50% extract, *Gymnema sylvestre* 25% extract, *Momordica* 7% extract, licorice root 20% extract in a unique, natural formula	Effectively and safely improves sugar metabolism	Modak *et al.* (2007)[47]
11	Dia-Care	Sanjeevan Mool, Himej, Jambu beej, Kadu, Namejav, Neem chal	Effective for both type 1 and type 2 diabetes	Srivastava *et al.* (2012)[48]
12	Diabecon	*Gymnema sylvestre, Pterocarpus marsupium, Glycyrrhiza glabra, Casearia esculenta, Syzygium cumini, Asparagus racemosus, Boerhavia diffusa, Sphaeranthus indicus, Tinospora cordifolia, Swertia chirata, Tribulus terrestris, Phyllanthus amarus, Gmelina arborea, Gossypium herbaceum, Berberis aristata, Aloe vera,* Triphala, *Commiphora wightii,* shilajeet, *Momordica charantia, Piper nigrum, Ocimum sanctum, Abutilon indicum, Curcuma longa,* and *Rumex maritimus*	Reduces the glycated hemoglobin levels, normalizes the microalbuminurea, and modulates the lipid profile; it minimizes long-term diabetic complications, prevents complications such as retinopathy in diabetic patients; antioxidant and anti-inflammatory activity	Himalaya Health care[49]

TABLE 14.2   Polyhedral Formulations Available to Prevent Diabetes and Its Secondary Complications—cont'd

Sr. No	Polyherbal Formulation	Components of Formulation	Benefits	Reference
13	Diasulin	*Cassia auriculata, Coccinia indica, Curcuma longa, Emblica officinalis, Gymnema sylvestre, Momordica charantia, Scoparia dulcis, Syzygium cumini, Tinospora cordifolia,* and *Trigonella foenum graecum*	Antioxidant, controls the blood glucose level by increasing glycolysis and decreasing gluconeogenesis with a lower demand of pancreatic insulin	Ramalingam *et al.* (2005)[50]
14	Diakyur	*Cassia javanica, Cassia auriculata, Salacia reticulate, Gymnema sylvestre, Mucuna pruriens, Syzygium jambolaum,* and *Terminalia arjuna*	Hypoglycemic activity as well as antilipid peroxidative activity, treats diabetes as well as delays the late complications of diabetes	Chandra *et al.* (2007)[46] Joshi *et al.* (2007)[51]
15	Diashis	Composed of eight medicinal plants	Antioxidant and antidiabetic potential	Bera *et al.* (2010)[52]
16	DRF/ AY/5001	*Gymnema sylvestre, Syzygiumcumini, Pterocarpus marsupium, Momordica charantia, Emblica officinalis, Terminalia belirica, Terminalia chebula,* and *Shudh shilajit*	Antidiabetic activity	Mandlik *et al.* (2008)[53]
17	Diasol	*Eugenia jambolana, Foenum graceum, Terminalia chebula, Quercus infectoria, Cuminum cyminum, Taraxacum officinale, Emblica officinalis, Gymnea sylvestre, Phyllanthus nerui,* and *Enicostemma littorale*	Effective antidiabetic formulation	Babuji *et al.* (2010)[54]
18	Dihar	*Syzygium cumini, Momordica charantia, Emblica officinalis, Gymnema sylvestre, Enicostemma, Azadirachta indiaca, Tinospora cordifolia,* and *Curcuma longa*	Antihyperglycemic activity and antioxidant activity	Patel *et al.* (2009)[55]

LD50: lethal dose 50%.
Source: Srivastava et al. (2012)[56].

INHIBITORY POTENTIAL OF INDIVIDUAL PHYTOCHEMICALS

Diabetic cataract appears to be a complex process involving multiple mechanisms. The first and foremost might be the consequences of the overt oxidative stress in diabetes that results in toxic aldehydes generated as by-products of carbohydrate autoxidation and lipid peroxidation leading to modification (glycation) of essential proteins. Reported studies on use of phytochemicals are mainly on animal models by inducing diabetes through alloxan or streptozotocin. These have been done either using the individual molecules or their combinations or as adjuvant to insulin therapy. The type of molecule may be antioxidant, micronutrient, and enzyme inhibitor.

Carotenoids

Protective effect of lutein against the development of cataract has been reported wherein a combined treatment with lutein and insulin prevented the development of cataracts in STZ-induced diabetic rats.[64] Further, lutein and insulin, alone or in combination, significantly inhibited lipid peroxidation in the diabetic lens and the diabetes-induced decrease of GSH content. These are suggestions of normalization of the biochemical, histologic, and functional modifications induced by diabetes.[65] The effect of dietary antioxidants, such as astaxanthin and flavangenol, and their combination in combating oxidative stress in STZ-induced diabetes was investigated. The degree of cataract formation in the flavangenol and mixed groups were lower than the control group. These results indicate that the combination of astaxanthin with flavangenol has better protective effect on oxidative stress associated with STZ-induced diabetes than either agent used alone and may be beneficial in preventing the progression of diabetic complications.[66] Lycopene, another carotenoid when supplemented in the medium, significantly ($P < 0.001$) restored GSH and malondialdehyde levels. A significant delay in the onset and progression of galactose cataract was observed with oral feeding of lycopene.[67]

Vitamins

Uncontrolled hyperglycemia in the early phase of diabetes was associated with elevated plasma malondialdehyde adducts (MDA), hypertension, and proteinuria. Insulin therapy alone resulted in significant but incomplete reduction of plasma MDA and blood pressure. Antioxidant therapy (vitamin E-fortified food, tocopherol 5000 U/kg chow and vitamin C-fortified H_2O, 1000 mg/L) was ineffective when given alone. However, when combined with insulin treatment, it normalized plasma MDA, blood pressure, and reduced urinary protein excretion.[68] Long-term treatment of diabetic animals with stobadine (STB), vitamin E, or butylated hydroxytoluene (BHT) resulted in marked delay in the development of advanced stages of cataract.[69]

Lipid peroxidation increases in the lens and kidney of diabetic animals, and this could be due to decrease in antioxidant vitamins and enzymes. However, dietary vitamin C and E supplementation coupled with moderate exercise was shown to help the antioxidant defense system through the reduction of ROS and blood glucose levels.[70] In two more studies, combined treatment with either vitamin E or probucol and insulin was useful in preventing the development and progression of diabetic cataracts.[71,72] Effects of single or combined treatments with vitamin A (retinol acetate, 30 mg/kg/day, for 12 weeks) and insulin (8–10 IU/rat/day for the final 6 weeks) on vasomotor activity, oxidative stress, and retinol metabolism have been reported for 12 weeks in STZ diabetic rats.[73] Vitamin A together with insulin provided a better metabolic control and more benefits than use of insulin alone in the reduction of diabetes-induced vascular complications. The intake of some vitamins needed for health effects are given in Table 14.3.

Amino Acids

Nitric oxide (NO)-generating compounds viz. L-arginine (L-Arg) and sodium nitroprusside (SNP) improved most of the biochemical abnormalities and antioxidant levels in diabetic rats. The beneficial effects of NO-generating compounds can be attributed to the generation of NO and/or enhanced antioxidant enzyme activities.[76] Oral administration of glycine has significantly delayed the onset and the progression of diabetic cataract in rats. These effects were attributed to its antiglycating action and also the inhibition of oxidative stress and polyol pathway.

Other Molecules

Carnosine (CA) is a dipeptide of beta-alanine and histidine. After treatment with CA, aminoguanidine (AG), and aspirin (ASA) diabetic rats had a lower level of glycated lens protein compared with untreated diabetic rats ($P < 0.001$).[77] Taurine, or 2-aminoethanesulfonic acid, is an organic acid widely distributed in animal tissues, which is a major constituent of bile. Addition of taurine to the diet of diabetic animals resulted in a significant decrease of gamma-crystallin leakage into the vitreous but not the aqueous humor. Taurine had no effect on the lens adenosine triphosphate (ATP) levels.[78] Poly (adenosine diphosphate-ribose) polymerase (PARP) activation is implicated in the formation of diabetic cataract and in early retinal changes. PARP inhibitors delayed, but did not prevent, the formation of diabetic cataract. These findings provided a basis for the development of PARP inhibitors for the prevention of diabetic ocular complications.[79] Rats fed high dietary fructose were documented to form an acquired model of insulin resistance. Administration of L-carnitine (CA) resulted in significant decline in enzyme and nonenzyme antioxidants and an increase in lipid peroxidation products, protein oxidation, protein glycation, GSSG/GSH ratio, and aldehyde formation were observed in lens samples obtained from fructose-fed rats.[80]

Kametaka et al.[81] examined the effects of Ca^{2+}-channel blockers on sugar-related cataract formation in streptozotocin (65 mg/kg, intravenously)-induced diabetic rats that were given 5% D-glucose as drinking water. Nifedipine slowed the progression rate of diabetic cataracts without affecting the period of time required for the onset of this disease, whereas verapamil had no significant inhibitory effect on the diabetic cataract.

Treatment with resveratrol significantly reduced malondialdehyde (MDA), xanthine oxidase (XO), and NO production and increased GSH levels when compared with the STZ-induced diabetic-untreated group. This study demonstrated that resveratrol is a potent neuroprotective agent against diabetic oxidative damage.[82] Chen et al.[83] examined the ability of a pyridoxal-aminoguanidine adduct with both antiglycation and antioxidant activities in vitro to protect against neuropathy and cataract in

TABLE 14.3   Inhibitory Potential of Individual Phytochemicals

	RDA (mg)[1,2]	Intakes for Health Benefit	% Deficient from RDA
Vitamin C	90 (men)75 (women)+35 (smokers)	≥ 2.5 RDA	> 50
Vitamin E	15 mg (natural)30 mg (synthetic)	≥ 4 RDA	> 90
Lutein and zeaxanthin	Not defined	6 mg	

RDA: recommended daily allowance.
Source:
[1]Institute of Medicine, 2000.[74]
[2]Institute of Medicine, 2001.[75] Vitamin and mineral data were obtained from the Continuing Survey of Food Intakes by Individuals (CSFII) 1994–1996. Carotenoid data were gathered from National Health and Nutrition Examination Survey (NHANES III), 1988–1994.

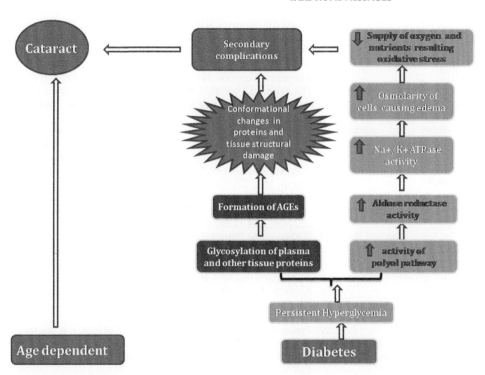

FIGURE 14.1 Pathogenesis of diabetic complications. (See color plate section)

streptozotocin-diabetic rats and compared the result with that of aminoguanidine.

Diabetic cataract is a disease that requires a biochemical/pharmacologic and a rather holistic approach for its better management. Thus, a good control of glucose metabolism will remain to be a first milestone to prevent lens opacification. However, some phytochemicals, the natural inhibitors of undesired oxidation and glycation reactions, show a promise to be next generation therapeutic agents.

TAKE-HOME MESSAGES

- Cataracts, the opacity of lens proteins, is one of the earliest secondary complications of diabetes mellitus, occurring due to chronic conditions of hyperglycemia (Fig. 14.1).
- AR, due to involvement in polyol pathway, has been a drug target because of its involvement in the development of secondary complications of diabetes including cataract.
- Antioxidant-rich foods such as gooseberry and spices such as turmeric (*Curcuma longa*) show potential in preventing diabetic cataract (Table 14.4).
- Traditionally used Ayurvedic herbs, such as Gudmar (*Gymnema sylvestre*), Tulsi (*Ocimum sanctum),* bael (*Aegle marmelos*), and jambhul (*Eugenia jambolana*), have been supported through animal experiments (Table 14.4).
- Antioxidant vitamins (A, C, and E), carotenoids, certain amino acids, and their peptides/metabolites show preventive action.

TABLE 14.4 Foods Recommended as a Source of Antioxidants for the Prevention of Cataractogenesis

Food	β-Carotene	Ascorbic Acid	Polyphenolics
Amaranth	Y	Y	Y
Agathi	Y	Y	Y
Coriander leaves	Y	Y	Y
Curry leaves	Y	Y	Y
Mint	Y	Y	Y
Mustard leaves	Y	Y	Y
Spinach (Bachali)	Y	Y	Y
Rape leaves	Y	Y	Y
Ponnaganti	Y	Y	Y
Fenugreek leaves	Y	Y	Y
Drumstick leaves	Y	Y	Y
Cluster beans	Y	Y	Y
Chilis (green)	Y	Y	Y
Carrots	Y	Y	Y
Amla	N	Y	Y
Lime	N	Y	Y
Tomato (ripe)	Y	Y	Y
Papaya (ripe)	Y	Y	Y
Mango (ripe)	Y	Y	Y
Guava (country)	N	Y	Y
Lemon	N	Y	Y

Source: Rao (2003).[84]

References

1. Wild S, Roglic G, Green A, Sicree R, King H. Global prevalence of diabetes estimates for the year 2000 and projections for 2030. *Diabetes Care* 2004;**27**:1047–53.

2. Roy A, Sil R, Chakraborti AS. Non-enzymatic glycation induces structural modifications of myoglobin. *Mol Cell Biochem* 2010;**338**:105–14.

3. Agte VV, Tarwadi KV. Combination of diabetes and cataract worsens the oxidative stress and micronutrient status in Indians. *Nutrition* 2008;**24**(7–8):617–24.

4. Andreas P, Ursula SE. Diabetic cataract – pathogenesis, epidemiology and treatment. *J Ophthalmol* 2010. 2010:608751.

5. Rosen P, Nawroth PP, King C, Moller W, Tritschler HJ, Packer L. The role of oxidative stress in the onset and progression of diabetes and its complications: a summary of Congress Series sponsored by UNESCO-MCBN, the American Diabetes Association and the German Diabetes Society. *Diabetes Metab Res Rev* 2001;**17**:189–212.

6. Bry L, Chen PC, Sacks DB. Effects of hemoglobin variants and chemically modified derivatives on assay for glycohemoglobin. *Clin Chem* 2001;**47**:153–63.

7. Schalkwijk CG, Stehouwer CDA, Hinsbergh VWM. Fructose-mediated non-enzymatic glycation: sweet coupling or bad modification. *Diabetes Metab Res Rev* 2004;**20**:369–82.

8. Avendano GF, Agarwal RK, Bashey RI. Effects of glucose intolerance on myocardial function and collagen-linked glycation. *Diabetes* 1999;**48**:1443–7.

9. Srivastava SK, Ramana KV, Bhatnagar A. Role of aldose reductase and oxidative damage in diabetes and the consequent potential for therapeutic options. *Endocr Rev* 2005;**26**(3):380–92.

10. Stevens VJ, Rouzer CA, Monnier VM, Cerami A. Diabetic cataract formation: potential role of glycosylation of lens crystallins. *Proc Natl Acad Sci USA* 1978;**75**:2918–22.

11. Lorenzi M. The polyol pathway as a mechanism for diabetic retinopathy: attractive, elusive, and resilient. *Exp Diabetes Res* 2007. 2007:61038.

12. Sivakumari K, Rathinabai AFMC, Kaleena PK, Jayaprakash P, Srikanth R. Molecular docking study of bark-derived components of *Cinnamomum cassia* on aldose reductase. *Indian J Sci Technol* 2010;**3**(8):1081–8.

13. Hotta N. New concepts and insights on pathogenesis and treatment of diabetic complications: polyol pathway and its inhibition. *Nagoya J Med Sci* 1997;**60**:89–100.

14. Suryanarayana P, Saraswat M, Petrash JM, Reddy GB. Emblica officinalis and its enriched tannoids delay streptozotocin-induced diabetic cataract in rats. *Mol Vis* 2007;**13**:1291–7.

15. Suryanarayana P, Saraswat M, Mrudula T, Krishna TP, Krishnaswamy K, Reddy GB. Curcumin and turmeric delay streptozotocin-induced diabetic cataract in rats. *Invest Ophthalmol Vis Sci* 2005;**46**(6):2092–9.

16. Kumar PA, Suryanarayana P, Reddy PY, Reddy GB. Modulation of alpha-crystallin chaperone activity in diabetic rat lens by curcumin. *Mol Vis* 2005;**11**:561–8.

17. Lu MP, Wang R, Song X, Chibbar R, Wang X, Wu L, Meng QH. Dietary soy isoflavones increase insulin secretion and prevent the development of diabetic cataracts in streptozotocin-induced diabetic rats. *Nutr Res* 2008;**28**(7):464–71.

18. Liu RH. Health benefits of fruit and vegetables are from additive and synergistic combinations of phytochemicals. *Am J Clin Nutr* 2003;**78**(Suppl):517S–20S.

19. Akhtar MS, Ali MR. Study of anti diabetic effect of a compound medicinal plant prescription in normal and diabetic rabbits. *J Pak Med Assoc* 1984;**34**(8):239–44.

20. Kesari AN, Gupta RK, Watal G. Hypoglycemic effects of *Murraya koenigii* on normal and alloxan-diabetic rabbits. *J Ethnopharmacol* 2005;**97**(2):247–51.

21. Kesari AN, Kesari S, Singh SK, Gupta RK, Watal G. Studies on the glycemic and lipidemic effect of *Murraya koenigii* in experimental animals. *J Ethnopharmacol* 2007;**112**:305–11.

22. Rai E, Sharma S, Koul A, Bhat AK, Bhanwer AJ, Bamezai RN. Interaction between the UCP2-866G/A, mtDNA 10398G/A and PGC1alpha p.Thr394Thr and p.Gly482Ser polymorphisms in type 2 diabetes susceptibility in North Indian population. *Hum Genet* 2007;**122**(5):535–40.

23. Knott RM, Grant G, Bardocz S, Pusztai A, de Carvalho AF, Hesketh JE. Alterations in the level of insulin receptor and GLUT-4 mRNA in skeletal muscle from rats fed a kidney bean (*Phaseolus vulgaris*) diet. *Int J Biochem* 1992;**24**:897–902.

24. Tormo MA, Gil-Exojo I, Romero de Tejada A, Campillo JE. Hypoglycemic and anorexigenic activities of an alpha-amylase inhibitor from white kidney beans (*Phaseolus vulgaris*) in Wistar rats. *Br J Nutr* 2004;**92**:785–90.

25. Yoshikawa M, Murakami T, Yashiro K, Matsuda H. Kotalanol, a potent α-glucosidase inhibitor with thiosugar sulfonium sulphate structure, from antidiabetic Ayurvedic medicine. *Salacia reticulata*. *Chem Pharm Bull* 1998;**46**:1339–40.

26. Chattopadhyay RR. A comparative evaluation of some blood sugar lowering agents of plant origin. *J Ethnopharmacol* 1999;**67**:367–72.

27. Preuss HG, Jarrell ST, Scheckenbach R, Lieberman S, Anderson RA. Comparative effects of chromium, vanadium and *Gymnema sylvestre* on sugar-induced blood pressure elevations in SHR. *J Am Coll Nutr* 1998;**17**:116–23.

28. Bhattacharya A, Chatterjee A, Ghosal S, Bhattacharya SK. Antioxidant activity of active tannoid principles of *Emblica officinalis* (amla). *Indian J. Exp. Biol* 1999;**37**:676–80.

29. Kumar KCS, Muller K. Medicinal plants from Nepal, II. Evaluation as inhibitors of lipid peroxidation in biological membranes. *J Ethnopharmacol* 1999;**64**:135–9.

30. Devasagayam TPA, Subramanian M, Singh BB, Ramanathan R, Das NP. Protection of plasmid pBR322 DNA by flavonoids against single-strand breaks induced by singlet molecular oxygen. *J Photochem Photobiol* 1995;**30**:97–103.

31. Kokila NR, Chethan Kumar M, Gangadhara NS, Harsha R, Dinesha R, Thammanna Gowda SS. Antiglycation and antioxidant activity of polysaccharides isolated from fruit extract of pomegranate (*Punica granatum*). *Pharmacol Online* 2010;**1**:821–9.

32. Rajasekaran S, Sivagnanam K, Ravi K, Subramanian S. Hypoglycemic effect of Aloe vera gel on streptozotocin-induced diabetes in experimental rats. *J Medic Food* 2004;**7**:61–6.

33. Okyar A, Can A, Akev N, Baktir G, Sutlupinar N. Effect of Aloe vera leaves on blood glucose level in type I and type II diabetic rat models. *Phytother Res* 2001;**1**:157–61.

34. Sachdewa A, Nigam R, Khemani LD. Hypoglycemic effect of *Hibiscus rosa sinensis* L. leaf extract in glucose and streptozotocin induced hyperglycemic rats. *Indian J Exp Biol* 2001;**39**:284–6.

35. Ravi K, Ramachandran B, Subramanian S. Protective effect of *Eugenia jambolana* seed kernel on tissue antioxidants in streptozotocin induced diabetic rats. *Biol Pharm Bull* 2004;**27**:1212–7.

36. Srinivasan M, Padmanabhan M, Prince PS. Effect of aqueous *Enicostemma littorale Blume* extract on key carbohydrate metabolic enzymes, lipid peroxides and antioxidants in alloxan-induced diabetic rats. *J Pharm Pharmacol* 2005;**57**:497–503.

37. Maroo J, Vasu VT, Aalinkeel R, Gupta S. Glucose lowering effect of aqueous extract of *Enicostemma littorale Blume* in diabetes: a possible mechanism of action. *J Ethnopharmacol* 2002;**81**:317–20.

38. Vats V, Yadav SP, Grover JK. Ethanolic extract of *Ocimum sanctum* leaves partially attenuates streptozotocin-induced alterations in glycogen content and carbohydrate metabolism in rats. *J Ethnopharmacol* 2004;**90**:155–60.

39. Pari L, Venkateswaran S. Hypoglycemic activity of *Scoparia dulcis* L. in alloxan induced hyperglycemic rats. *Phytother Res* 2002;**16**:662–4.

40. Kim JD, Kang SM, Seo BI, Choi HY, Choi HS, Ku SK. Antidiabetic activity of SMK001, a poly herbal formula in streptozotocin induced diabetic rats: therapeutic study. *Biol Pharm Bull* 2006;**29**:477–82.

41. Ogbonnia SO, Mbaka GO, Adekunle A, Anyika EN, Gbolade OE, Nwakakwa N. Effect of a poly-herbal formulation, Okudiabet, on alloxan-induced diabetic rats. *Agricult Biol J N Am* 2010;**1**:139–45.

42. Om PB, Edwin J, Asghar S, Ahmad S. Antidiabetic activity of polyherbal formulation (Karmin Plus). *Int J Green Pharm* 2009;**3**:211–4.

43. Thakkar NV, Jagruti AP. Pharmacological evaluation of "Glyoherb": a polyherbal formulation on streptozotocin-induced diabetic rats. *Int J Diab Dev Countries* 2010;**30**:1–7.

44. Sweety L, Debapriya G, Dheeraj A, Chand RA, Bharti A, Sanjay LK. Pharmacognostic standardization and hypoglycemic evaluations of novel polyherbal formulations. *Pharmacia Lett* 2011;**3**:319–33.

45. Prakash YG, Bimlendu JK. Effect of EFPTT/09, a herbal formulation, on blood sugar of normal and alloxan induced diabetic rats. *Res J Pharm Bio Curr Sci* 2010;**1**(4):987.

46. Sajeeth CI, Manna PK, Manavalan R, Jolly CI. Phytochemical investigation and antioxidant activity of a polyherbal formulation (ESF/AY/500) on streptozotocin induced oxidative stress in rats. *Pharma Chem* 2010;**2**:184–9.

47. Modak M, Dixit P, Londhe J, Ghaskadbi S, Devasagayam TPA. Indian herbs and herbal drugs used for the treatment of diabetes. *J Clin Biochem Nutr* 2007;**40**:163–73.

48. Srivastava S, Lal VK, Kumar K. Polyherbal formulations based on Indian medicinal plants as antidiabetic phytotherapeutics. *Plant Phytopharmacology* 2012;**2**(1):1–15.

49. http://www.himalayahealthcare.com (last accessed 17 November 2013).

50. Saravanan R, Pari L. Antihyperlipidemic and antiperoxidative effect of Diasulin, a polyherbal formulation in alloxan induced hyperglycemic rats. *BMC Complement Altern Med* 2005;**5**:14.

51. Joshi CS, Priya ES, Venkataraman S. Hypoglycemic and antilipid peroxidation of polyherbal formulation Diakyur in experimentally induced diabetes. *J Health Sci* 2007;**53**:734–9.

52. Bera TK, De D, Chatterjee K, Ali KM, Ghosh D. Effect of Diashis, a polyherbal formulation, in streptozotocin-induced diabetic male albino rats. *Int J Ayurveda Res* 2010;**1**:18–24.

53. Mandlik RV, Desai SK, Naik SR, Sharma G, Kohli RK. Antidiabetic activity of a polyherbal formulation (DRF/AY/5001). *Indian J Exp Biol* 2008;**46**:599–606.

54. Sant Sacha Baba Herbal Pharmacy, 2010. http://www.ssbherbs.com (last accessed 17 November 2013).

55. Patel SS, Shah RS, Goyal RK. Antihyperglycemic, antihyperlipidemic, antioxidant effect of Dihar, a polyherbal Ayurvedic formulation in streptozotocin induced diabetes rats. *Indian J Exp Biol* 2009;**47**:564–70.

56. Srivastava S, Lal VK, Pant KK. Polyherbal formulations based on Indian medicinal plants as antidiabetic phytotherapeutics. *Phytopharmacology* 2012;**2**(1):1–15.

57. Tiwari AK, Rao JM. Diabetes mellitus and multiple therapeutic approaches of phytochemicals: present status and future prospects. *Curr Sci* 2002;**83**:10.

58. Rathi SS, Grover JK, Vikrant V, Biswas NR. Prevention of experimental diabetic cataract by Indian Ayurvedic plant extracts. *Phytother. Res* 2002;**16**:774–7.

59. Waisundara VY, Hsu A, Huang D, Tan BK. *Scutellaria baicalensis* enhances the anti-diabetic activity of metformin in streptozotocin-induced diabetic Wistar rats. *Am J Chin Med* 2008;**36**(3):517–40.

60. Sarkhail P, Rahmanipour S, Fadyevatan S, Mohammadirad A, Dehghan G, Amin G, Shafiee A, et al. Antidiabetic effect of *Phlomis anisodonta*: effects on hepatic cells lipid peroxidation and antioxidant enzymes in experimental diabetes. *Pharmacol Res* 2007;**56**(3):261–6.

61. Preet A, Gupta BL, Yadava PK, Baquer NZ. Efficacy of lower doses of vanadium in restoring altered glucose metabolism and antioxidant status in diabetic rat lenses. *J Biosci* 2005;**30**(2):221–30.

62. Preet A, Siddiqui MR, Taha A, Badhai J, Hussain ME, Yadava PK, et al. Long-term effect of *Trigonella foenum graecum* and its combination with sodium orthovanadate in preventing histopathological and biochemical abnormalities in diabetic rat ocular tissues. *Mol Cell Biochem* 2006;**289**(1–2):137–47.

63. Cho WC, Chung WS, Lee SK, Leung AW, Cheng CH, Yue KK. Ginsenoside Re of Panax ginseng possesses significant antioxidant and antihyperlipidemic efficacies in streptozotocin-induced diabetic rats. *Eur J Pharmacol* 2006;**550**(1–3):173–9.

64. Arnal E, Miranda M, Almansa I, Muriach M, Barcia JM, Romero FJ, et al. Lutein prevents cataract development and progression in diabetic rats. *Graefes Arch Clin Exp Ophthalmol* 2009;**247**(1):115–20.

65. Arnal E, Miranda M, Johnsen-Soriano S, Alvarez-Nölting R, Díaz-Llopis M, Araiz J, et al. Beneficial effect of docosahexanoic acid and lutein on retinal structural, metabolic, and functional abnormalities in diabetic rats. *Curr Eye Res* 2009;**34**(11):928–38.

66. Nakano M, Orimo N, Katagiri N, Tsubata M, Takahashi J, Van CN. Inhibitory effect of astraxanthin combined with Flavangenol on oxidative stress biomarkers in streptozotocin-induced diabetic rats. *Int J Vitam Nutr Res* 2008;**78**(4–5):175–82.

67. Gupta SK, Trivedi D, Srivastava S, Joshi S, Halder N, Verma SD. Lycopene attenuates oxidative stress induced experimental cataract development: an *in vitro* and *in vivo* study. *Nutrition* 2003;**19**(9):794–9.

68. Koo JR, Ni Z, Oviesi F, Vaziri ND. Antioxidant therapy potentiates antihypertensive action of insulin in diabetic rats. *Clin Exp Hypertens* 2002;**24**(5):333–44.

69. Kyselova Z, Gajdosik A, Gajdosikova A, Ulicna O, Mihalova D, Karasu C, et al. Effect of the pyridoindole antioxidant stobadine on development of experimental diabetic cataract and on lens protein oxidation in rats: comparison with vitamin E and BHT. *Mol Vis* 2005;**11**:56–65.

70. Kutlu M, Naziroğlu M, Simşek H, Yilmaz T, Sahap KA. Moderate exercise combined with dietary vitamins C and E counteracts oxidative stress in the kidney and lens of streptozotocin-induced diabetic-rat. *Int J Vitam Nutr Res* 2005;**75**(1):71–80.

71. Yoshida M, Kimura H, Kyuki K, Ito M. Combined effect of vitamin E and insulin on cataracts of diabetic rats fed a high cholesterol diet. *Biol Pharm Bull* 2004;**27**(3):338–44.

72. Yoshida M, Kimura H, Kyuki K, Ito M. Combined effect of probucol and insulin on cataracts of diabetic rats fed a high cholesterol diet. *Eur J Pharmacol* 2005;**513**(1–2):159–68.

73. Zobali F, Besler T, Ari N, Karasu C. Hydrogen peroxide-induced inhibition of vasomotor activity: evaluation of single and combined treatments with vitamin A and insulin in streptozotocin-diabetic rats. *Int J Exp Diabetes Res* 2002;**3**(2):119–30.

74. Institute of Medicine. *Dietary reference intakes for vitamin C, vitamin E, selenium, and carotenoids.* ; 2000. Available online at http://www.iom.edu/reports/2000/dietary-reference-intakes-for-vitamin-c-vitamin-e-selenium-and-carotenoids.aspx (last accessed 17.11.13).

75. Institute of Medicine. Dietary reference intakes for vitamin A, vitamin K, arsenic, boron, chromium, copper, iodine, iron, manganese, molybdenum, nickel, silicon, vanadium, and zinc. Available online at http://www.iom.edu/reports/2001/dietary-reference-intakes-for-vitamin-a-vitamin-k-arsenic-boron-chromium-copper-iodine-iron-manganese-molybdenum-nickel-silicon-vanadium-and-zinc.aspx (last accessed 17.11.13).

76. Mohamadin AM, Hammad LN, El-Bab MF, Gawad HS. Can nitric oxide-generating compounds improve the oxidative stress response in experimentally diabetic rats? *Clin Exp Pharmacol Physiol* 2007;**34**(7):586–93.

77. Yan H, Guo Y, Zhang J, Ding Z, Ha W, Harding JJ. Effect of carnosine, aminoguanidine, and aspirin drops on the prevention of cataracts in diabetic rats. *Mol Vis* 2008;**14**:2282–91.

78. Mitton KP, Linklater HA, Dzialoszynski T, Sanford SE, Starkey K, Trevithick JR. Modelling cortical cataractogenesis 21: in diabetic rat lenses taurine supplementation partially reduces damage resulting from osmotic compensation leading to osmolyte loss and antioxidant depletion. *Exp Eye Res* 1999;**69**(3):279–89.

79. Drel VR, Xu W, Zhang J, Kador PF, Ali TK, Shin J, et al. Poly(ADP-ribose)polymerase inhibition counteracts cataract formation and early retinal changes in streptozotocin-diabetic rats. *Invest Ophthalmol Vis Sci* 2009;**50**(4):1778–90.

80. Balasaraswathi K, Rajasekar P, Anuradha CV. Changes in redox ratio and protein glycation in precataractous lens from fructose-fed rats: effects of exogenous L-carnitine. *Clin Exp Pharmacol Physiol* 2008;**35**(2):168–73.

81. Kametaka S, Kasahara T, Ueo M, Takenaka M, Saito M, Sakamoto K, et al. Effect of nifedipine on severe experimental cataract in diabetic rats. *J Pharmacol Sci* 2008;**106**(4):651–8.

82. Ates O, Cayli SR, Yucel N, Altinoz E, Kocak A, Durak MA, et al. Central nervous system protection by resveratrol in streptozotocin-induced diabetic rats. *J Clin Neurosci* 2007;**14**(3):256–60.

83. Chen AS, Taguchi T, Sugiura M, Wakasugi Y, Kamei A, Wang MW, et al. Pyridoxal-aminoguanidine adduct is more effective than aminoguanidine in preventing neuropathy and cataract in diabetic rats. *Horm Metab Res* 2004;**36**(3):183–7.

84. Rao BN. Review article. Bioactive phytochemicals in Indian foods and their potential in health promotion and disease prevention. *Asia Pacific J Clin Nutr* 2003;**12**(1):9–22.

Role of Amino Acids on Prevention of Nonenzymatic Glycation of Lens Proteins in Senile and Diabetic Cataract

S. Zahra Bathaie[1], Fereshteh Bahmani[2], Asghar Farajzadeh[1]

[1]Tarbiat Modares University, Tehran, Iran, [2]Kashan University of Medical Sciences, Kashan, Iran

INTRODUCTION

Recently, there has been more attention on complement therapy of diabetes to prevent its harmful complications. Amino acid therapies have attracted more attention because they are known by body and easily metabolized. This chapter reviews the process of glycation and its inhibition by amino acids in comparison with other compounds, with emphasis on reaction mechanism especially in the lens.

HISTORY AND OVERVIEW OF THE FORMATION OF ADVANCED GLYCATION END-PRODUCTS

Louis-Camille Maillard (February 4, 1878–May 12, 1936), the French scientist, undertook studies of the reaction between amino acids and sugars, and his first paper was published in 1912 (*Compt. Rend.* 1912;154:66. Cited in: http://cen.acs.org/articles/90/i40/Maillard-Reaction-Turns-100.html). Then he explained the principles of the browning phenomenon that occurs in meat following long-term expose to air. This cascade of reactions has been named "the Maillard reaction". Further studies on this pathway showed that the Maillard reaction is an important phenomenon in food, beverages, paper, textile, biopharmaceutical, and even soil. After the 1970s, substantial attention has been given to the Maillard reaction in *in vivo* conditions.[1,2] Then extensive studies on its chemistry under physiologic conditions were carried out.

The Maillard reaction, or nonenzymatic glycation of amino acids and proteins (Fig. 15.1), initiates by a nucleophilic addition reaction between a free amino group (e.g., lysine or arginine side chain) and a carbonyl group from a reducing sugar to form a Schiff base. This reaction occurs rapidly in a reversible process, the rate of which depends on the concentration of available sugar, so that by lowering the sugar concentration, the unstable product is degraded. A Schiff base can undergo further rearrangement to form a more stable Amadori product, which is an irreversible process for ketoamine production. Under suitable conditions, such as high sugar concentration, Amadori products accumulate over time and can undergo additional complex rearrangements giving rise to different types of advanced glycation end-products (AGEs) (Fig. 15.2).

GLYCATING AGENTS

In reality, all reducing sugars, such as fructose, and certain sugar derivatives, such as ascorbic acid, can initiate the Maillard reaction *in vivo*. However, because of the slow rate of the reaction of glucose (Glc) with proteins and its high extracellular concentration in diabetic patients, it has been thought that AGEs only form at long-lived and/or extracellular proteins. Nevertheless, later studies showed that the short-lived proteins, intracellular, and even the nuclear proteins also can be a target for glycation.[3,4]

The type of sugar, in addition to the sugar concentration and half-life of biomacromolecules, influences the

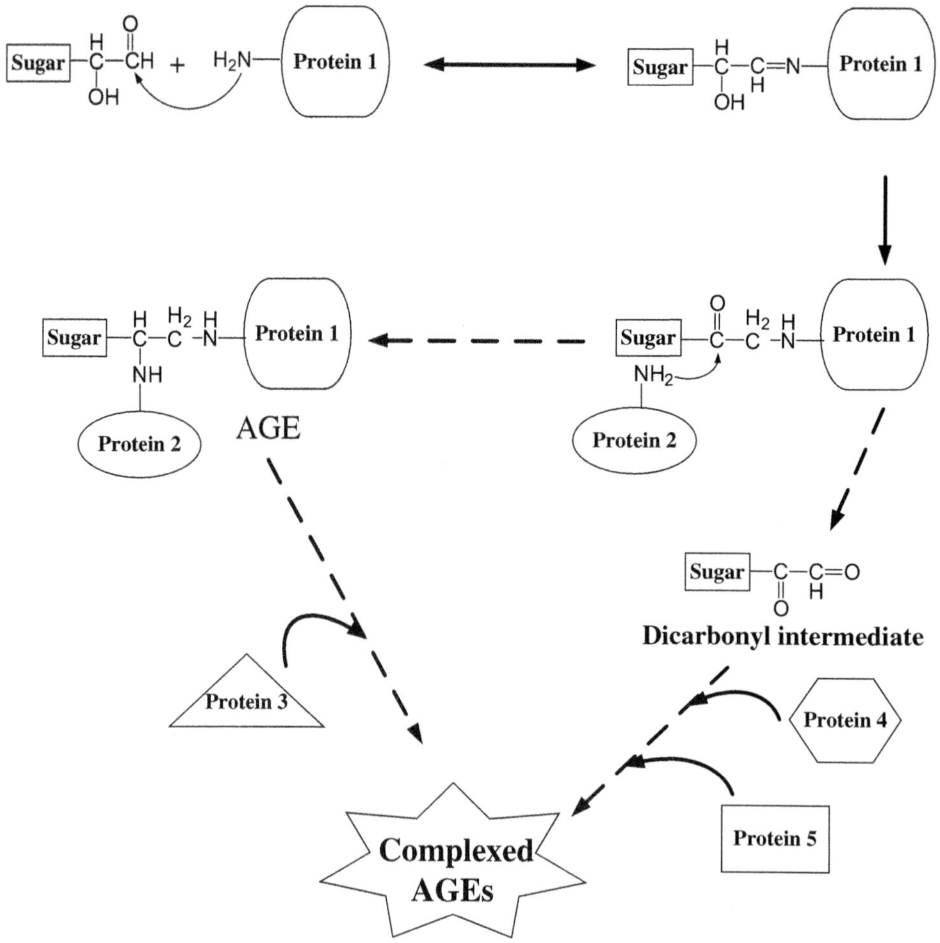

FIGURE 15.1 Glycation of a protein by reducing sugars like glucose and the subsequent formation of AGEs. The initial interaction between a reducing sugar (such as glucose) and free amino groups in the proteins form a reversible Schiff base that rearranges to a stable ketoamine or Amadori product. With time, these Amadori products directly or via dicarbonyl intermediates (such as 3-deoxyglucosone (3-DG)) interact with other protein molecules to form advanced glycation end-products (AGEs)/more complex AGEs.

process of glycation. The rate of glycation is also directly proportional to the percentage of sugar in the open-chain form.[5] It has been shown that Glc in the solution is 0.002% in the open-chain form at 20 °C, but glyceraldehyde-3-phosphate (an intermediate in the glycolysis) is 100% in the open-chain form.[6] Therefore, the latter produces over 200-fold more glycated hemoglobin (HbA1c) than the first, after 72 hours of incubation.[7] In addition, the rate of human serum albumin glycation by fructose was 10-fold more than Glc in vitro.[8] Thus, precursors other than Glc, such as Glc-6-phosphate, fructose, glyceraldehyde, dihydroxy-acetone-phosphate, glyceraldehyde-3-phosphate, and the dicarbonyl compounds like glyoxal (GO), methylglyoxal (MGO), and 3-deoxyglucosone (3-DG) are of great importance for the intracellular Maillard reaction. Since accumulation of these reactive glycolytic intermediates is high at some metabolic situation in the cell, they have an important role in the in vivo formation of AGEs.[7–9]

In some organs, such as the ocular lens, fructose is produced from sorbitol in an oxidation reaction catalyzed by sorbitol dehydrogenase (SDH). This pathway that is also named the polyol pathway increases the concentrations of fructose in the same order of magnitude

as that of Glc and is strongly increased by hyperglycemia, making in vivo glycation by fructose a highly probable event.[9,10] In addition, due to the increased levels of fructose by dietary intake, the fructose content of the lens and some other tissues is markedly elevated during hyperglycemia. Therefore, up to 23-fold increase in fructose concentration has been reported in the lenses of diabetic patients, which is twice that of Glc. Previous in vivo studies on the endogenous glycation reaction by fructose have shown that 10–20% of the sugar moieties binds to human ocular lens proteins via carbon-2.[10] It has also been reported that AGE production can be facilitated by some exogenous compounds derived from foods, tobacco, and stress.[11] Therefore, formation of AGEs progressively increases with age, even in the absence of diseases like diabetes.

TYPES OF ADVANCED GLYCATION END-PRODUCT

As mentioned above, AGEs are complex, heterogeneous molecules that are produced by cross-linking

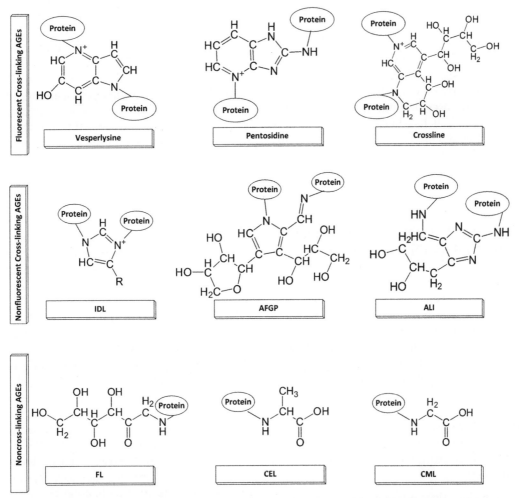

FIGURE 15.2 Chemical structure of different types of AGEs. (Top) Fluorescent cross-linking advanced glycation end-products (AGEs) such as vesperlysine, pentosidine, and crossline. (Middle) Nonfluorescent cross-linking AGEs such as IDL: imadazolium dilysine cross-links; AFGP: alkyl formyl glycosyl pyrroles; and ALI: arginine–lysine imidazole cross-links. (Lower) Noncross-linking AGEs such as FL: N-fructosyl-lysine; CEL: N-carboxyethyl-lysine; and CML: N-carboxymethyllysine.

of proteins and result in browning. Various types of AGEs have been identified (Fig. 15.2) and can be divided into three main categories based on their chemical structure and ability to emit fluorescence. These are:

1) The cross-linking AGEs with fluorescence emission, such as pentosidine, crossline, vesper lysine A, glyoxal lysine dimmer (GOLD), methylglyoxal lysine dimmer (MOLD), and glucosepane.
2) The nonfluorescent cross-linking AGEs, such as imidazolium dilysine (IDL), alkyl formyl glycosyl pyrrole (AFGP), and arginine lysine imidazole (ALI) cross-links.
3) Noncross-linking and nontoxic AGEs such as pyrraline, N-carboxymethyllysine (CML), N-carboxyethyl-lysine (CEL), and N-fructosyl-lysine (FL).[12]

AGEs derived from glycolaldehyde or glyceraldehyde are toxic.[13] Thus, AGEs can be divided into the toxic and nontoxic compounds.

As Figure 15.2 depicts, the majority of AGEs are formed on the free amines of Lys and Arg residues of proteins (e.g., pentosidine, glucosepane, MOLD, GOLD, and crossline); however, the N-terminal of proteins are also candidates for glycation. Such an event happens in the N-terminal valine of hemoglobin β-chain and results in the formation of HbA1c.[14] Cysteine is also a candidate for glycation in some proteins.[15]

The role of different types of AGEs in the pathogenesis of diabetic complications has been extensively studied. However, some contradictions exist in the reported data that result in confusion in understanding the mechanisms involved. For example, it has been reported that serum levels of CML increase in diabetic patients with retinopathy but not with nephropathy, whereas the levels of pentosidine increased in both groups.[16]

In addition to aging, diabetes and the known diabetic complications such as cataracts, diabetic nephropathy, and atherosclerosis, AGEs have been introduced as markers and causative factors for pathogenesis of neurodegenerative diseases, including Alzheimer and Parkinson disease.[12,17,18] It has been shown that AGE formation under physiologic conditions is significantly related to the half-life of proteins so that AGEs accumulate prominently in the long-lived proteins such as lens crystallins.[19]

ADVANCED GLYCATION END-PRODUCT FORMATION IN THE LENS

The pathophysiology behind senile cataracts is complex and depends on several factors. Normal aging is accompanied by a progressive increase of AGEs in lens proteins. Advanced glycation occurs during normal aging but to a greater degree in diabetes. Thus, diabetes is considered as a major risk factor for the development of cataract, not only for the nonenzymatic glycation of lens proteins but also for oxidative stress and activated polyol pathway.[20]

The extent of protein glycation in lens fiber cells has been estimated to be up to 10-fold higher than other tissues with usual protein turnover.[21] Some of the most important reasons are:

1) Glc uptake in the lens does not depend on insulin, thus it is constantly exposed to high concentrations of this sugar, especially in diabetes.
2) Lens crystallins are structural proteins that have little or no turnover; therefore, they are particularly vulnerable to glycation.[22]
3) Those substantial modifications of lens proteins may stimulate further glycation, oxidation, and consequently formation of water insoluble and high-molecular weight aggregates.

Nowadays, the process of lens opacification is well understood and it has been found that coloration of the human lens in certain types of cataracts is also related to the formation of AGEs and accumulation of various Amadori products on lens proteins.[22–24] Various structurally characterized AGEs in the lenses are shown in Figure 15.3. It has been shown that cataractous lenses contain significantly higher levels of GOLD, MOLD, and methylglyoxal hydroimidazolones compared with the normal lenses from age-matched subjects. The increase in the serum pentosidine, argpyrimidine, GOLD, and MOLD concentrations has been also reported in diabetic patients.[25]

EFFECTS OF ADVANCED GLYCATION END-PRODUCTS ON THE FUNCTION OF PROTEINS

Cross-linked species accumulate in different tissues such as the cataractous lens, cornea, skin, smooth muscle, and vascular collagen both as a consequence of diabetes and normal aging, thus affecting the whole function of the organ. Alterations in the organ function is induced both directly, due to the changes in the structure and function of proteins, and indirectly, due to the binding of AGEs to AGE receptors (RAGEs) on the cell surface.

Direct Effect

As mentioned above, increased nonenzymatic glycation of proteins in the presence of reducing sugar and buildup of AGEs alter the proteins' structure, which results in the changes in their function. Proteins have different functions, including: enzymatic activity, ligand binding, transport of other proteins or ligands, deoxyribonucleic acid (DNA) binding activity, and so on. Any changes in the protein structure may affect their activities (Fig. 15.4A), modify protein half-life, and alter immunogenicity. The results of our recent studies have indicated the named changes in the structure and function of various proteins from different locations. For example, serum proteins such as albumin [26] and HSP70,[27] plasma proteins such as fibrinogen,[28] extracellular proteins such as lysozyme,[29] cytosolic proteins such as α-crystallin,[30] and even nuclear proteins such as histone H1[4] are the substrates for glycation reaction and AGE formation.

As a result of glycation, many proteins lose their activity[4,26,27,30], but activity of a few proteins like fibrinogen is increased.[28] The increased activity of fibrinogen also results in an increase in clot formation, which is the reason for higher chance of atherosclerosis in diabetic patients.[28] Loss of the activity of molecular chaperones (HSP70 or α-crystallin) results in the misfolding and alteration in the function of other proteins, thus affecting many proteins and systems in the body. Another example of the effect of glycation on decreasing the protein activity is Na^+-K^+ ATPase, which results in alteration of the intracellular ion concentration and subsequent water movement via osmosis, after *in vitro* glycation.[31] Such an effect *in vivo* may contribute toward cataract formation in diabetes. The impaired esterase activity of MGO-modified serum albumin compared with unmodified albumin is another example. Moreover, cysteine proteases such as cathepsins are inhibited by MGO modification at the active site cysteines. The glycation of low-density lipoprotein (LDL) reduces their uptake by their normal receptors and glycation of superoxide dismutase increases reactive oxygen species generation and amplifies the oxidative stress.[12] Glycated histone H1 that was

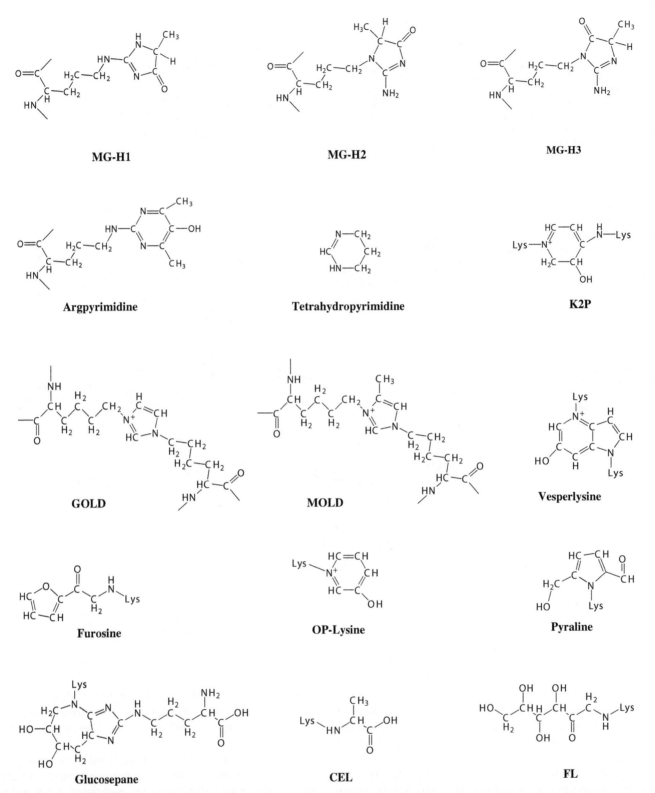

FIGURE 15.3 Advanced glycation end-products in diabetic and aging lenses. Various AGEs found in aging and diabetic lenses. For detail see the text. CEL: N-carboxyethyl-lysine; FL: N-fructosyl-lysine; GOLD: glyoxal lysine dimmer; K2P: 1-(5-amino-5-carboxypentyl)-4-(5-amino-5-carboxypentyl-amino)-3-hydroxy-2, 3-dihydropyridinium; MG-H1, MG-H2, and MG-H3: Hydroimidazolones isomers 1, 2, and 3, respectively; MOLD: methylglyoxal lysine dimmer; OP-Lys: 2-Ammonio-6-(3-oxidopyridinium-1-yl)hexanoate.

FIGURE 15.4 Overview of cellular advanced glycation end-product (AGE) interactions and functions. (A) Changing in protein function due to the glycation modification. Enzyme (e)-substrate (s) interaction in normal condition (upper) is disrupted after enzyme glycation (lower). (B) AGE formation due to the cross-linking of proteins after glycation, leading to tissue rigidity. (C) Interaction of AGE with its membrane bound receptor (RAGE), which results in the activation of some signaling pathways and induction of the expression of some genes to produce proteins such as tumor necrosis factor (TNF)-α and interleukin (IL)-6, as the inflammatory mediators (AP-1: activator protein-1; MAP: mitogen-activated protein; NF-κB: nuclear factor kappa B; PI3K: phosphoinositide 3-kinase; ROS: reactive oxygen species.) (D) Inducing the formation of free radicals; production of reducing equivalents in the mitochondrial matrix during normal catabolism of Glc that was followed by the appearance of superoxide anion (O2-°) and its subsequent elimination by the action of superoxide dismutase (SOD) and catalase (CAT) as the antioxidant defense system. In diabetic conditions, due to the glycation and inactivation of the named enzymes and alteration in the activity of complex III, free radicals are accumulated and induce cell damage.

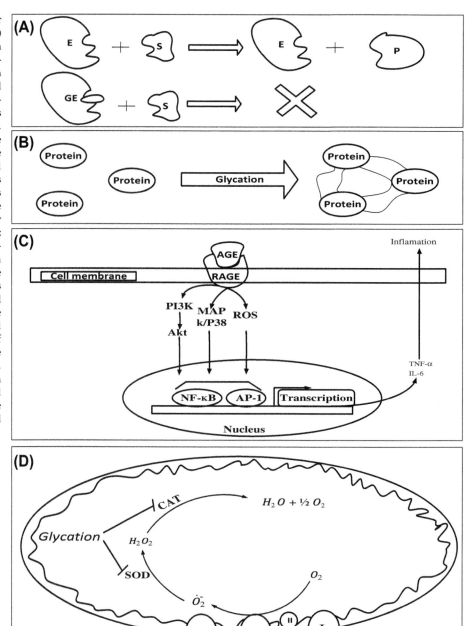

extracted and purified from the liver of diabetic rats has also shown a lower binding affinity to DNA, compared with the histone H1 separated from normal rat liver.[4] This alteration led to the changes in the expression of some genes.

Glycoxidatively modified proteins, such as fibronectin, can form large aggregates in the extracellular space (Fig. 15.4B). Cross-links formed between extracellular matrix components are known to affect cell–matrix interactions, impair matrix assembly, reduce protein turnover (possibly due to a decreased proteolytic digestibility of glycated proteins or altered proteolytic enzyme activity), and increase arterial and myocardial stiffness.[32,33] The

reduced activity of glycated lysozyme in this compartment can also reduce its function as the innate immune system and result in inflammation in the extracellular fluid.[29]

As mentioned above, the long-lived structural proteins such as lens crystallins are the most important target for nonenzymatic glycation associated with aging and complications of diabetes. Crystallins largely determine the transparency of the eye lens. The extent of glycation of a variety of proteins, including lens crystallins and lens capsule, is approximately two times more in diabetes and the aging process. It has been reported that the extent of glycation, AGE formation, and appearance of yellow color aggregates increases with the severity of diabetic complications.[34] α-crystallin, a member of the small heat-shock protein family, is a major structural element in the protein matrix of the vertebrate eye lens. The chaperoning ability of α-crystallin is believed to be essential for the maintenance of transparency of the lens, thus preventing the formation of cataract.[35] The *in vitro* glycation of α-crystallin disrupts its structural stability, resulting in decreased chaperone activity, similar to that seen in the *in vivo* studies.[30,36]

Indirect Effect

The AGE–RAGE interaction (Fig. 15.4C) initiates the cascade of activation reactions and signal transduction pathways. Other major features of the AGE–RAGE complex formations relate to their endocytosis, degradation, pro-oxidant, and proinflammatory events. Recent studies suggest that interaction of AGEs with RAGEs alters all the named processes, including: intracellular signaling, expression of some genes, release of free radicals, and pro-inflammatory molecules that contribute toward the pathology of diabetic complications.[37–41]

RAGEs are members of the immunoglobulin receptor family and bind several ligands in addition to AGEs, such as HMG-1, S-100 proteins, and β-amyloid peptide. Binding of ligands to RAGEs results in activation of NADPH oxidase that leads to an increased production of reactive oxygen species (ROS). Furthermore, activation of signaling pathways including ERK, p38/MAPK, JAK/STAT-pathway, rho-GTPases, and phosphoinositol 3-kinase (PI3K) are linked to RAGE activation. One major downstream target of RAGE is the proinflammatory NF-κB pathway, which in turn leads to elevated RAGE expression and perpetuation of the cellular inflammatory state. In addition to RAGEs, other binding proteins have also been described, including oligosaccharyl transferase (OST48, AGE-R1), 80K-H phosphoprotein (AGE-R2), galectin-3 (AGE-R3), CD36, and scavenger receptors II-a and II-b. Taken together, several AGE-binding molecules are involved in binding, signaling, and degradation of AGE. The AGE-mediated effects will, therefore, depend on the occurrence of these receptors on the individual cell type.[37–43]

Moreover, exogenous dietary AGEs that were formed from cooked food can enter blood circulation; however, their contribution to pathophysiologic processes is under discussion.[44]

Production of free radicals and oxidative stress is another consequence of AGE formation (Fig. 15.4D). Glycation-derived free radicals can cause protein fragmentation and oxidation of nucleic acids and lipids,[12] which cause some complications including cell apoptosis.

BIOLOGICAL DETOXIFICATION OF ADVANCED GLYCATION END-PRODUCTS

Available data suggest that AGEs are eliminated from the blood mainly by scavenger receptor-mediated uptake in Kupffer cells and liver sinusoidal endothelial cells.[45] The liver enzymes such as α-ketogluteraldehyde dehydrogenase are capable of inactivating 3-DG and preventing AGE formation. An important role has also been suggested for kidneys to eliminate AGEs.[45] Macrophages possess receptors enabling them to recognize and remove harmful AGE-proteins by endocytosis associated with the role of lysozyme in the extracellular fluid.

A variety of plasma amines, such as spermine, may also react with sugar and Amadori carbonyl groups to reduce AGEs.[46] Antioxidants can also protect against glycation-derived free radicals.[47]

The 'detoxifying' enzymes such as glutathione-dependent glyoxalase complex (formed from glyoxalase I and glyoxalase II components) act as an effective detoxification system for GO and MGO.[48] The enzyme complex catalyses conversion of MGO to s-D-lactoyl-glutathione, then it is subsequently converted to D-lactate by glyoxalase II. In oxidative situations associated with low glutathione concentrations, the antiglycation defense mediated by the glyoxalase system is insufficient; this establishes links between oxidative stress and glycation. Cells that overexpress this enzyme show less accumulation of MGO-derived AGEs. Glyoxalase-I enzymatic activity becomes progressively impaired with the aging process; this contributes to an accelerated AGE accumulation with aging.[49]

PREVENTION/INHIBITION OF ADVANCED GLYCATION END-PRODUCT FORMATION

A major goal of therapy in diabetic patients, especially in type 2 diabetes, is to reduce hyperglycemia by modification of the diet. However, dietary compliance is often difficult, and an alternative is to use pharmacologic compounds that can reduce AGEs.[50] Complement therapy using natural products is another method, which was used not only by ancient cultures but also by the modern world.

The main reason for diabetic complications is glycation of biomacromolecules, especially proteins. Therefore, protection of proteins against glycation either by prevention or by dealing with the consequences of glycation has been considered.

Various steps of the protein glycation reaction and AGE formation, emphasizing the inhibitor(s) used in each stage, are outlined in Figure 15.5. Since glycation

FIGURE 15.5 Schematic representation of the sites of action of antidiabetic compounds to inhibit protein glycation and advanced glycation end-product (AGE) production. Each compound affects the process in one or more steps according to the mechanism of its action and prevents glycation or AGE formation. Some compounds can break the AGEs or interfere with its interaction with its receptor (RAGE). Inhibition of production of reducing sugars is also another strategy for diabetes treatment and prevention. sRAGE: soluble RAGE.

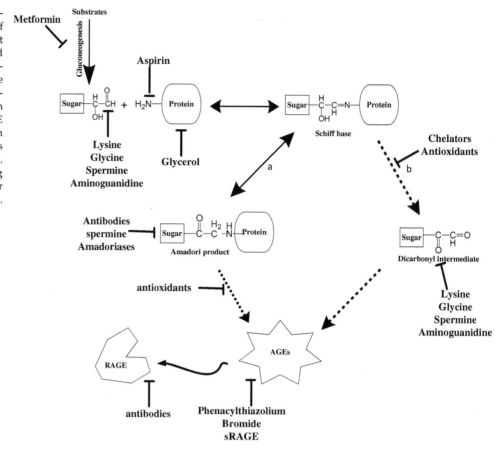

and AGE formation is a spontaneous process with no need of enzymes (nonenzymatic), the suggested inhibitors should have long half-life with no toxicity and no immune reaction, preferably familiar with the physiologic system, so that they can be used for long periods of time. Obviously these criteria are difficult to meet, but Figure 15.5 represents some of the compounds that have been considered and shows the distinct effects of these compounds as explained below. Figure 15.6 also shows the structure of some of the mentioned compounds in this section.

1) Inhibition of glucose production. Metformin (or *N,N*-dimethylimidodicarbonimidic diamide, a biguanine compound) activates AMP-activated protein kinase (AMPK), which is an enzyme that plays an important role in insulin signaling, whole body energy balance, and metabolism of glucose and fats; therefore, it inhibits the production of glucose by liver cells.[51]

2) Compounds that block the free amino groups on proteins, preventing their glycation by free reducing sugars. One of the famous examples of this group of materials is aspirin (acetylsalicylic acid), which can acetylate the free amine groups specially epsilon amine in Lys residue of proteins. Both *in vivo* and *in vitro* results have shown the inhibition of protein (such as α-crystallin and albumin) glycation by aspirin.[47,52–54] This mechanism is different from the antidiabetic effect of other anti-inflammatory drugs.

3) Blocking the carbonyl groups on reducing sugars, Amadori products, and dicarbonyl intermediates (3-DG, MGO, etc.) effectively reduce glycation and AGE formation. Compounds such as aminoguanidine or pimagedine reduce the levels of AGEs through interacting with 3-DG and thus make them unavailable to proteins. Aminoguanidine slows down the progression of lens opacification in moderately diabetic rats.[55] This drug is capable of protecting skin proteins (elastin and collagen), nerves, eye lenses, and kidney proteins from cross-linking; however, its toxic effect on higher doses should also be considered. Carnosine (β-alanyl-L-histidine), which can react with the sugars and prevent them from reacting with protein,[26,56] can also be included in this group of antiglycating compounds. Free amino acids (such as Lys and Gly) and other amino-containing compounds (such as polyamines) are also including in this category of materials, which will be discussed in detail in the following section.

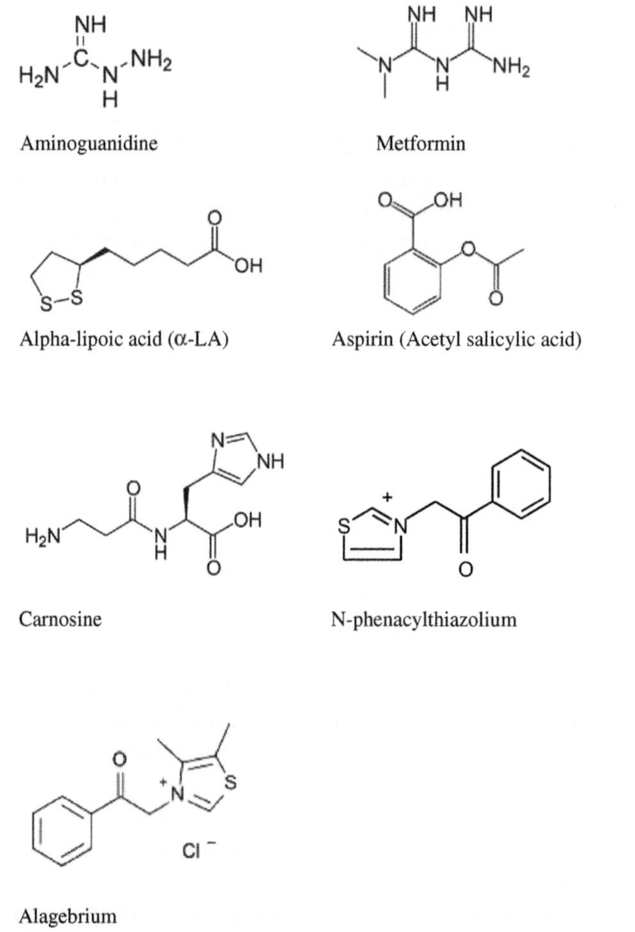

Aminoguanidine

Metformin

Alpha-lipoic acid (α-LA)

Aspirin (Acetyl salicylic acid)

Carnosine

N-phenacylthiazolium

Alagebrium

FIGURE 15.6 Chemical structure of some of the named antiglycating compounds. Chemical structures of some of the named antiglycating compounds in Figure 15.5 are presented here to compare their structure.

4) Chemical chaperones from the polyols family (such as glycerol and inositol), which stabilize protein structure and prevent their modification and conformational change, are another group of antiglycating compounds.

5) Antioxidants can protect cells and proteins against free radicals derived via autoxidative glycation, glycoxidation, and AGE compounds. Amino acids such as Lys,[26] dipeptides such as carnosine or N-acetyl carnosine,[57] and some drugs such as aspirin[47] can also act through this mechanism.
Carnosine has attracted much attention as a natural antioxidant and transition-metal ion-sequestering agent. In addition, carnosine can nonenzymatically react with detrimental hexoses, pentoses, and trioses and protect proteins, such as α-crystallin against glycation. Therefore, it can decrease cross-linking induced by sugars, diminish the modification of α-crystallin, and disaggregate glycated α-crystallin.[58]

Some studies have indicated α-lipoic acid (α-LA), a naturally occurring dithiol compound, is an antioxidant that acts through direct radical scavenging and metal chelating, interacting with other antioxidants, and increasing the reduced intracellular glutathione and vitamin C levels. Studies have showed that α-LA facilitates nonoxidative and oxidative Glc metabolism and increases Glc uptake leading to improved Glc utilization both *in vitro* and *in vivo*. Other reports have also indicated that α-LA may prevent protein glycation in cultured lenses under hyperglycemic conditions.[59] The ability of α-LA to prevent cataractogenesis has also been demonstrated *in vitro* in rat lens cell cultures exposed to high concentrations of Glc, and also *in vivo* in an oxidative stress model of cataract, as well as in nutritionally induced type 2 diabetic rats.[60] Its biosynthesis decreases as people age and is reduced in people with compromised health. These powerful hypoglycemic and antioxidant effects led to the use of α-LA supplementation in treatment of diabetic cataract and neuropathy.[61]

6) Amadoriases have been known as enzymes that deglycate Amadori products or inactivate intermediates such as 3-DG.[62] Some amadoriases function as fructosamine oxidases, which convert fructosyllysine to Lys, hydrogen peroxide, and glucosone. Another class of enzymes that destabilize the glycated proteins is fructosamine 3-kinase, which phosphorylates the Amadori compounds at C3, producing a fructosamine 3-phosphate residue on the protein, which then decomposes spontaneously to release phosphate and 3-DG.[63]

7) AGE cross-link breakers such as alagebrium and N-phenacylthiazolium bromide offer the potential of reversing diabetic complications, especially cardiovascular disease. It was postulated that alagebrium reverses the stiffening of blood vessel walls, thus it is effective in reducing blood pressure and providing therapeutic effect for patients with diastolic heart failure.[64]

8) Regulation of RAGEs to prevent or inhibit progression of diabetic complications is the other strategy to control diabetes.[65] RAGE blockers could prevent interaction of AGEs with RAGEs to suppress the cellular and inflammatory changes associated with the development of diabetic complications. It has also been shown that after cleavage of RAGE from membrane the soluble RAGE (sRAGE) is formed, which can bind to AGEs and work as decoy receptors against ligand-RAGE interaction.[39]

9) Chelation of transition metals by compounds such as ethylenediaminetetraacetic acid (EDTA) to reduce the glycation-derived free radicals. But in fact, their complete removal may be undesirable and cause distortion of the reactions needing trace elements. In addition, application of these compounds in the *in vivo* condition is under question.

10) Antibodies may be used to block Amadori products or RAGEs. The specificity of this approach is more than other methods that recognize carbonyl groups; however, antibody therapy is accompanied with some limitations that affect its ordinary use.

INHIBITORY EFFECTS OF AMINO ACIDS ON ADVANCED GLYCATION END-PRODUCT FORMATION AND CATARACT

For the first time, in 1961, Reddy et al. showed that some amino acids can be actively transported from plasma into the posterior aqueous humor and from there into the lens.[66] Therefore, amino acids have a role in the lens, and any distortion in the transfer system results in some eye diseases. In contrast, disorders due to a deficiency in these transfer systems can be compensated for by administration of amino acids. Using these hypotheses, several studies have been done on the distribution of amino acids in the eye and the beneficial effect of their free form on eye diseases such as cataract.[67–69] Both *in vitro* and *in vivo* studies show antiglycating effects of free amino acids on lens proteins. High solubility of some amino acids, especially Gly, in comparison with other anticataract agents is a very important aspect of their application.[70] Some examples of these studies are presented here.

In Vitro Studies

Antiglycating effects of several amino acids including Lys, Gly, Ala, Glu, and Asp on lens proteins have been reported.[30,69,71,72] It has been shown that varying concentrations of Lys and Gly can significantly reduce the extent of glycation of lens proteins in Glc-treated homogenates of normal human and goat lenses.[73] In addition, Ala, Asp, and Glu, which present in relatively larger amounts in the lens, have been found to undergo nonenzymatic glycation and significantly reduce the extent of *in vitro* glycation of lens proteins in the Glc-treated homogenates of normal human lenses.[71] Incubation of lens homogenate with galactose has also increased glycation of proteins; however, addition of the above-named amino acids decreased the glycation of lens proteins.[69]

Another study has shown reduced nonenzymatic glycation of lens proteins by acidic amino acids. In this study lens homogenates of the goat, rat, and human cataractous lens proteins were incubated with different concentrations of Glc and the effects of Asp and Glu were studied individually. The results indicated that these two amino acids can protect lens proteins from glycation.[72]

Our *in vitro* study using α-crystallin, purified protein from cow lens, showed that incubation of this protein with Glc in the presence or absence of Lys or Gly prevented its glycation and conformational change.[30] The inhibitory role of these amino acids on glycation of other proteins such as humans or bovine serum albumin was also shown by us.[26]

In Vivo Studies

Antiglycating effects of Gly and Lys were also investigated in animal models of diabetes and diabetic patients. Besides being a stimulator of insulin secretion, Gly and Lys showed antidiabetic effects by other mechanisms.[28,70,74,75] Diabetic rats treated with Gly showed less enlargement of the glomerular basal membrane than controls. These rats showed a diminution in the microaneurysms in the eyes. In addition, the isolated peripheral blood mononuclear cells from Gly-treated diabetic rats showed a better proliferative response to mitogens phytohemogluttinin or concavalin A compared with those obtained from nontreated diabetic rats. Gly-treated rats had less-intense corporal weight loss compared with nontreated animals.[70] In another study, Gly diminished hyperglycemia, hypercholesterolemia, and HbA1c concentrations in diabetic rats. It has been suggested that the effect of Gly may be secondary to its higher solubility, which prevents the formation and precipitation of AGE products.[76]

Lys has also been known as an inhibitor of protein glycation; however, its long-term use in diabetes treatment considering different aspects of diabetic complications is limited in the literature. In our previous study, the effect of Lys, as a chemical chaperone, on both an animal model of diabetes and type 2 diabetic patients was investigated. Results showed improvement of some metabolic and lipid parameters' alteration, after 5 months' treatment with 0.1% Lys in rats[26] and after 3 months of treatment with coadministration of 3 mg/day of Lys with glibenclamide and metformin in diabetic patients.[28] This treatment decreased fasting serum Glc, increased serum insulin, and improved antioxidant defense system and lipid profile in both rat and human.

In one recent study, we also investigated the inhibitory effect of two amino acids Lys and Gly, separately, on the progression of eye lens opacification in streptozotocin (STZ)-induced diabetic rats. Our results indicated the beneficial effect of both amino acids.[30] Table 15.1 shows

TABLE 15.1 Lens Parameters, in the Three Main Pathways of This Tissue, in the Normal and Streptozotocin-Induced Diabetic Rats, with or without Receiving L-Glycine.

Lens Parameters		N	D	DG	NG
Polyol pathway	AR activity (μmol NADPH oxidized/h/100 mg pr)	30.8 ± 3.1	40.2 ± 3.0 [a,#]	32.8 ± 2.3 [b,*]	29.9 ± 4.9 [b,#]
	SDH activity (μmol NADH oxidized/h/100 mg pr)	4.63 ± 0.54	5.12 ± 0.84	5.14 ± 0.59	4.58 ± 0.90
Glycation reaction	Lens AGE (FI%)	18.5 ± 3.1	82.6 ± 5.9 [a,†]	23.7 ± 4.9 [b,†]	19.3 ± 3.7 [b,†]
	Glycated proteins (μmol HMF/mg lens pr)	0.52 ± 0.06	1.87 ± 0.30 [a,†]	0.86 ± 0.15 [b,†]	0.50 ± 0.14 [b,†]
	Total protein (mg/lens)	17.68 ± 0.72	10.19 ± 3.10 [a,#]	16.69 ± 2.02 [b,#]	17.17 ± 1.47 [b,#]
	Soluble protein (mg/lens)	14.50 ± 0.66	4.79 ± 2.28 [a,#]	8.25 ± 0.84 [a,b,#]	13.97 ± 1.04 [b,#]
Antioxidant capacity	GSH (mg/g lens)	178.0 ± 36.9	51.0 ± 14.8 [a,†]	126.9 ± 25.5 [b,*]	175.0 ± 23.6 [b,†]
	CAT activity (μmol/g lens pr)	59.8 ± 8.6	31.7 ± 11.5 [a,#]	39.1 ± 10.7	55.7 ± 13.2 [b,*]
	SOD activity (units/min/mg lens pr)	6.20 ± 0.20	15.39 ± 5.34 [a,†]	11.09 ± 1.72 [a,b,*]	6.50 ± 0.45 [b,†]

AGE: advanced glycation end-product; AR: aldose reductase; CAT: catalase; D: diabetic; DG: diabetic group receiving L-glycine; FI: fluorescence intensity; GSH: glutathione; h: hour; HMF: hydroxymethylfurfural; min: minute; N: normal; NADH: nicotinamide adenine dinucleotide; NADPH: nicotinamide adenine dinucleotide phosphate; NG: normal group receiving L-glycine; pr: protein; SDH: sorbitol dehydrogenase; SOD: superoxide dismutase.

[a]indicates the significance of the data that compares group N versus all groups

[b]indicates the significance of the data that compares group D versus groups of Glycine-treated rats

*P < 0.05

#P < 0.01

†P < 0.001

Table created from Bahmani et al. (2012), Tables 2–4.[74]

and compares various parameters in the lenses of normal and diabetic rats with or without receiving Gly.[74] As can be seen, all three important pathways in the lenses were changed in the diabetic rats (D) compared with the normal group (N), but due to Gly administration (DG) they were returned or were close to the normal values. Gly had no effect on the normal group (NG). Figure 15.7 shows the incidence of cataract in the rats under experimental conditions. A specialist determined this biweekly using a handheld ophthalmoscope that was equipped with a slit lamp. As it is seen, there are significant differences between the percentage of cataract (from clear to completely opaque or grade 4) in the lenses of diabetic animals that received Gly in comparison with the untreated group.[74]

Several mechanisms have been proposed for the inhibitory effect of amino acids on protein glycation, especially in the lenses. Some of them include:

- **Interaction of free amino acids with sugars.** Lys and Gly, like some other amino acids, were found to react with Glc at physiologic pH and temperature and undergo nonenzymatic glycation. Formation of the glycated Lys has been shown by paper and thin-layer chromatography and high-performance liquid chromatography. Confirmation was made by studies on incorporation of U-[14C] Glc into Lys and Gly. These amino acids also formed adducts after incubation with galactose at physiologic pH and temperature. Studies on Lys interaction showed that the extent of glycation of the free amino acid increased with time.[69] Direct interaction of amino acids with sugars results in decreased free sugar concentration in the medium; thus availability of these toxic agents to form protein adduct was reduced.

- **Preventing the loss of protein content in diabetic lenses.** It has been shown that the total and soluble protein content of the cataractous lenses is significantly reduced. Formation of high molecular weight aggregates of proteins may be the reason for the observed decrease in protein concentration. Administration of the mentioned amino acids compensate it and make both soluble and total protein content of lenses close to the normal value.[74,77,78]

FIGURE 15.7   Effect of glycine (Gly) therapy on the opacity of the lenses in streptozotocin-diabetic rats. (A) Photographs of lenses from each group at the end of 12 weeks. (B) Progression of cataract in diabetic rats throughout the experimental period. Cataract formation was scored biweekly according to the following classification: clear normal lens (O), peripheral vesicles (I), peripheral vesicles and cortical opacities (II), diffuse central opacities (III), and mature cataract (IV). The scores of cataracts in each group were meaned at the given time and the mean score of the cataract was plotted as a function of time. There was a significant difference (P < 0.001) between the mean score of the cataract of groups D and DG from 6 weeks to the end of the study. (C) Maturation of the cataract in diabetic rats after 12 weeks. Cataract development in rats was observed on week 12 of the study and the number of lenses that developed opacity against the total number of lenses was considered for calculating the percentage of incidence of cataract in each group. D: diabetic rats; DG: diabetic rats with Gly; N: normal rats; NG: normal rats with Gly. *Source: Bahmani et al. (2012), Figure 3C.*[74]

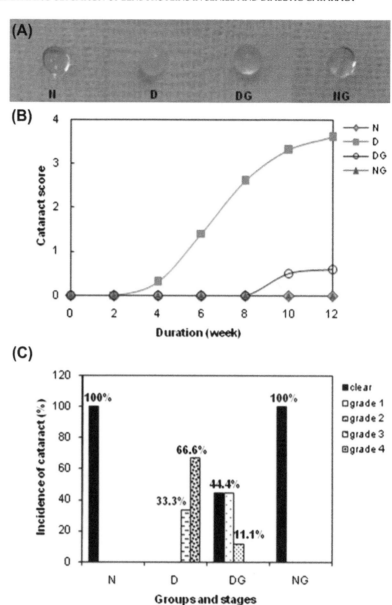

- **Potentiate the antioxidant defense system.** Oxidative damage to the constituents of the eye lens has been considered as another important mechanism in the development of cataract. The decreased glutathione (GSH) and the altered activities of the antioxidant enzymes, including catalase (CAT) and superoxide dismutase (SOD), are due to the increased oxidative stress in diabetic conditions. Such alteration was also seen in various types of cataract.[79] Lys and Gly significantly increased the antioxidant capacity in diabetic rat lenses.[30]

- **Effect on polyol pathway.** The dicarbonyl intermediates of the Maillard reaction, such as GO, MGO, and the amount of AGEs, increase during hyperglycemia. The dicarbonyl intermediates are the substrates of aldose reductase (AR) and the AGEs are also inducers of it. Thus, when these compounds increase due to hyperglycemia, the activity of the polyol pathway is also increased. As our results indicated, the activity of the AR was significantly increased in diabetic rats and the treatment of these rats with Lys or Gly caused a significant decrease in the activity of this enzyme. However, SDH activity was not significantly altered in the diabetic group.[30]

- **Specific transporters for some amino acids in the cortex and nuclear proteins of lenses.**[80,81] In rat lenses, Gly uptake is mediated by a family of Na/Cl-dependent neurotransmitter transporter proteins. These transporters were named as Gly transporters 1 and 2 (GLYT1 and GLYT2). Although GLYT1 and GLYT2 are likely to mediate Gly uptake in cortical fiber cells, GLYT2 alone appears to be responsible for the accumulation of Gly in the center of the lens. Since the lens core lacks the capacity to synthesize GSH, the role of Gly in this region is unclear. Enhancing the delivery of Gly to the core via the sutures may represent a pathway to protect the lens against the protein modifications associated with cataract and aging.[80]

- **Hypoglycemic effect by metabolic control.** The serum Glc level of diabetic rats was reduced significantly after treatment with Lys and Gly, not only through direct interaction with sugars but also for their metabolic effect. For example, catabolism of L-Lys in the liver was down by using 2 mole of α-ketoglutarate (α-KG) as acceptor and cosubstrate. Thus, catabolism of more Lys caused entrance of more Glc to the liver (insulin-independent entry) to produce more α-KG to catabolize excess L-Lys, which in turn caused a decrease in blood Glc.[26] In humans, it has been found that Gly administration increases the response to insulin.[82] Oral Gly stimulates the secretion of a gut hormone that potentiates the effect of insulin on Glc removal from the circulation.[83]

- **Enhancing both secretion of insulin and activity of insulin receptor.** It has been reported that Lys administration (1 g/day) with antidiabetic tablets (glyciphage or chlorformine) in type 2 diabetes mellitus can enhance insulin receptor tyrosine kinase activity in monocytes.[84] Our results have also indicated the increase in insulin secretion in both diabetic humans and rats after Lys administration.[26,28]

- **The nature of amino acids** that are physiologically available, water soluble, and nontoxic for tissues. In several studies including ours, many parameters were also measured in normal rats and humans treated with Lys/Gly. The results indicated there are no differences between the determined parameters in the normal subjects who were treated with these amino acids when compared with those not treated. Therefore, these amino acids are nontoxic and have no harmful effect on healthy subjects. In addition, their administration even decreased the mortality of diabetic animals.[26,28,30,70]

In conclusion, a variety of antiglycating agents have recently been introduced; some of them are nonphysiologic and toxic and may cause several side effects. According to the results of the mentioned research, as well as other references that limitations did not allow us to include, oral administration of amino acids, especially Lys and Gly, can significantly delay the onset and progression of diabetic cataract in model animals and humans. Amino acids, being highly soluble in water, are rapidly absorbed and transported by mechanisms that do not require energy. These naturally occurring metabolites are less toxic and may prove to be suitable for preventing nonenzymatic glycation and hence senile and diabetic cataract. Scavenging intracellular sugar and thereby protecting the lens proteins from excessive glycation appears to be the most important mechanism by which amino acids, especially Lys and Gly, show their beneficial effect on cataract.

TAKE-HOME MESSAGES

- The spontaneous or nonenzymatic reaction between sugar and proteins is named glycation.
- Glycation plays an important role in several complications related to aging and diabetes, for example in senile and diabetic cataract.
- The antiglycating effect of some amino acids has been shown in *in vivo* and *in vitro* studies.
- Some amino acids inhibit binding of sugar with proteins, which is the first step in the glycation pathway.
- Free amino acids mitigate the glycation of lens proteins and delay cataract. Thus, they can be used as effective complement therapy to treat or prevent cataract.
- Some amino acids with free-radical scavenging/ antioxidant activities are able to preserve cells in the direct or indirect process, through regenerating other antioxidants such as glutathione, and vitamins C and E.
- Some amino acids improve Glc utilization through improvement of cellular uptake and/or interfere with Glc metabolic pathway.
- Some amino acids can block the inflammatory processes (the root cause of diabetic complications, cardiovascular disease, arthritis, and some cancers) at the cellular level.
- Some amino acids can chelate metals that may be implicated in disease promotion through the creation of particularly aggressive free radicals.

References

1. Sgarbieri VC, Amaya J, Tanaka M, Chichester CO. Nutritional consequences of the Maillard reaction. Amino acid availability from fructose-leucine and fructose-tryptophan in the rat. *J Nutr* 1973;**103**(5):657–63.
2. Eble AS, Thorpe SR, Baynes JW. Nonenzymatic glucosylation and glucose-dependent cross-linking of protein. *J Biol Chem* 1983;**258**(15):9406–12.
3. Giardino I, Edelstein D, Brownlee M. Nonenzymatic glycosylation *in vitro* and in bovine endothelial cells alters basic fibroblast growth factor activity: a model for intracellular glycosylation in diabetes. *J Clin Invest* 1994;**94**:110–7.
4. Rahmanpour R, Bathaie SZ. Histone H1 structural changes and its interaction with DNA in the presence of high glucose concentration *in vivo* and in vitro. *J Biomol Struct Dyn* 2011;**28**(4):575–86.

5. Bunn HF, Higgins PJ. Reaction of monosaccharides with proteins: possible evolutionary significance. *Science* 1981;**213**:222–4.

6. Angyal SJ. The composition of reducing sugars in solution. In: Tipson RS, Derek H, editors. *Advances in Carbohydrate Chemistry and Biochemistry*. Academic Press; 1984. pp. 15–68.

7. Stevens VJ, Vlassara H, Abati A, Cerami A. Nonenzymatic glycosylation of hemoglobin. *J Biol Chem* 1977;**252**(9):2998–3002.

8. Suarez G, Rajaram R, Oronsky AL, Gawinowicz MA. Nonenzymatic glycation of bovine serum albumin by fructose (fructation). Comparison with the Maillard reaction initiated by glucose. *J Biol Chem* 1989;**264**(7):3674–9.

9. Lal S, Szwergold BS, Taylor AH, Randall WC, Kappler F, Wells-Knecht K, et al. Metabolism of fructose-3-phosphate in the diabetic rat lens. *Arch Biochem Biophys* 1995;**318**:191–9.

10. McPherson JD, Shilton BH, Walton DJ. Role of fructose in glycation and cross-linking of proteins. *Biochemistry* 1988;**27**(6):1901–7.

11. Cerami C, Founds H, Nicholl I, Mitsuhashi T, Giordano D, Vanpatten S, et al. Tobacco smoke is a source of toxic reactive glycation products. *Proc Natl Acad Sci USA* 1997;**94**:13915–20.

12. Ahmed N. Advanced glycation endproducts: role in pathology of diabetic complications. *Diabetes Res Clin Pract* 2005;**67**:3–21.

13. Sato T, Iwaki M, Shimogaito N, Wu X, Yamagishi S, Takeuchi M. TAGE (toxic AGEs) theory in diabetic complications. *Curr Mol Med* 2006;**6**:351–8.

14. Bunn HF, Shapiro R, McManus M, Garrick L, McDonald MJ, Gallop PM, et al. Structural heterogeneity of human hemoglobin A due to nonenzymatic glycosylation. *J Biol Chem* 1979;**254**(10):3892–8.

15. Zeng J, Davies MJ. Protein and low molecular mass thiols as targets and inhibitors of glycation reactions. *Chem Res Toxicol* 2006;**19**:1668–76.

16. Miura J, Yamagishib SI, Uchigataa Y, Takeuchic M, Yamamotod H, Makitab Z, et al. Serum levels of non-carboxymethyllysine advanced glycation endproducts are correlated to severity of microvascular complications in patients with Type 1 diabetes. *J Diabetes Complications* 2003;**17**:16–21.

17. Sayre LM, Smith MA, Perry G. Chemistry and biochemistry of oxidative stress in neurodegenerative disease. *Curr Med Chem* 2001;**8**:721–38.

18. Stitt AW. The Maillard reaction in eye diseases. *Ann N Y Acad Sci* 2005;**1043**:582–97.

19. Bunn HF. Modification of hemoglobin and other proteins by non-enzymatic glycosylation. *Prog Clin Biol Res* 1981;**51**:223–39.

20. Gula A, Rahmanb MA, Salimc A, Simjeeb SU. Advanced glycation end products in senile diabetic and nondiabetic patients with cataract. *J Diabetes Complications* 2009;**23**:343–8.

21. Ahmed N, Thornalley PJ, Dawczynski J, Franke S, Strobel J, Stein G, et al. Methylglyoxal-derived hydroimidazolone advanced glycation end-products of human lens proteins. *Invest Ophthalmol Vis Sci* 2003;**44**:5287–92.

22. Franke S, Dawczynski J, Strobel J, Niwa T, Stahl P, Stein G. Increased levels of advanced glycation end products in human cataractous lenses. *J Cataract Refract Surg* 2003;**29**:998–1004.

23. Nagaraj RH, Sady C. The presence of a glucose-derived Maillard reaction product in the human lens. *FEBS Lett* 1996;**382**:234–8.

24. Zarina S, Zhao HR, Abraham EC. Advanced glycation end products in human senile and diabetic cataractous lenses. *Mol Cell Biochem* 2000;**210**:29–34.

25. Wilker SC, Chellan P, Arnold BM, Nagaraj RH. Chromatographic quantification of argpyrimidine, a methylglyoxal-derived product in tissue proteins: comparison with pentosidine. *Anal Biochem* 2001;**290**:353–8.

26. Jafarnejad A, Bathaie SZ, Nakhjavani M, Hassan A, Banasadegh S. The improvement effect of L-Lys as a chemical chaperone on STZ-induced diabetic rats, protein structure and function. *Diabetes Metab Res Rev* 2008;**24**:64–73.

27. Bathaie SZ, Jafarnejad A, Hosseinkhani S, Nakhjavani M. The effect of hot-tub therapy on serum Hsp70 level and its benefit on diabetic rats: a preliminary report. *Int J Hyperthermia* 2010;**26**(6):577–85.

28. Mirmiranpour H, Bathaie SZ, Khaghani S, Nakhjavani M, Kebriaeezadeh A. Investigation of the mechanism(s) involved in decreasing increased fibrinogen activity in hyperglycemic conditions using L-lysine supplementation. *Thromb Res* 2012;**130**(3):e13–9.

29. Bathaie SZ, Nobakht BB, Mirmiranpour H, Jafarnejad A, Moosavi-Nejad SZ. Effect of chemical chaperones on glucose-induced lysozyme modifications. *Protein J* 2011;**30**(7):480–9.

30. Bahmani F. *Mechanisms of the Effect of Chemical Chaperones on Prevention of Cataract in Diabetic Rats and Their Inhibitory Effects on Glycation of Related Proteins*. Tehran: Tarbiat Modares University; 2012.

31. Stevens A. The contribution of glycation to cataract formation in diabetes. *J Am Optom Assoc* 1998;**69**:519–30.

32. Tarsio JF, Wigness B, Rhode TD, Rupp WM, Buchwald H, Furcht LT. Non-enzymatic glycation of fibronectin and alterations in the molecular association of cell matrix and basement membrane components in diabetes mellitus. *Diabetes* 1985;**34**:477–84.

33. Aronson D. Cross-linking of glycated collagen in the pathogenesis of arterial and myocardial stiffening of aging and diabetes. *J Hypertens* 2003;**21**:3–12.

34. Harding JJ. Non-enzymatic post-translational modification of proteins in vivo. *Adv Protein Chem* 1985;**37**:247–334.

35. Reddy GB, Kumar PA, Kumar MS. Chaperone-like activity and hydrophobicity of a-crystallin. *IUBMB Life* 2006;**58**(11):632–41.

36. Kumar PA, Kumar MS, Reddy GB. Effect of glycation on α-crystallin structure and chaperone-like function. *Biochem J* 2007;**408**:251–8.

37. Schmidt AM, Yan SD, Yan SD, Stern DM. The biology of the receptor for advanced glycation end products and its ligands. *Biochim Biophys Acta* 2000;**1498**:99–111.

38. Vlassara H. The AGE-receptor in the pathogenesis of diabetic complications. *Diabetes Metab Res Rev* 2001;**17**:436–43.

39. Bierhaus A, Humpert PM, Stern DM, Arnold B, Nawroth PP. Advanced glycation end product receptor-mediated cellular dysfunction. *Ann NY Acad Sci* 2005;**1043**:676–80.

40. Kim W, Hudson BI, Moser B, Guo J, Rong LL, Lu Y, et al. Receptor for advanced glycation end products and its ligands: a journey from the complications of diabetes to its pathogenesis. *Ann NY Acad Sci* 2005;**1043**:553–61.

41. Yan SF, Ramasamy R, Schmidt AM. The receptor for advanced glycation endproducts (RAGE) and cardiovascular disease. *Exp Rev Mol Med* 2009;**11**:1–13.

42. Stern DM, Yan SD, Yan SF, Schmidt AM. Receptor for advanced glycation endproducts (RAGE) and the complications of diabetes. *Ageing Res Rev* 2002;**1**:1–15.

43. Bierhaus A, Humpert PM, Morcos M, Wendt T, Chavakis T, Arnold B, et al. Understanding RAGE, the receptor for advanced glycation end products. *J Mol Med* 2005;**83**:876–86.

44. Nass N, Bartling B, Navarrete Santos A, Scheubel RJ, Börgermann J, Silber RE, et al. Advanced glycation end products, diabetes and ageing. *Z Gerontol Geriat* 2007;**40**:349–56.

45. Svistounov D, Smedsrød B. Hepatic clearance of advanced glycation end products (AGEs) – myth or truth? *J Hepatol* 2004;**41**:1038–40.

46. Jafarnejad A, Bathaie SZ, Nakhjavani M, Hassan MZ. Effect of spermine on lipid profile and HDL functionality in the streptozotocin-induced diabetic rat model. *Life Sci* 2008;**82**(5–6):301–7.

47. Jafarnejad A, Bathaie SZ, Nakhjavani M, Hassan MZ. Investigation of the mechanisms involved in the high-dose and long-term acetyl salicylic acid therapy of type I diabetic rats. *JPET* 2008;**324**:850–7.

48. Thornalley PJ. Glutathione-dependent detoxification of alpha-oxoaldehydes by the glyoxalase system: involvement in disease mechanisms and antiproliferative activity of glyoxalase I inhibitors. *Chem Biol Interact* 1998;**111–112**:137–51.

49. Bolton WK, Cattran DC, Williams ME, Adler SG, Appel GB, Cartwright K, et al. Randomized trial of an inhibitor of formation of advanced glycation end products in diabetic nephropathy. *Am J Nephrol* 2004;**24**:32–40.

50. Shirali S, Zahra Bathaie S, Nakhjavani M. Effect of crocin on the insulin resistance and lipid profile of streptozotocin-induced diabetic rats. *Phytother Res* 2013;**27**(7):1042–7.

51. Peyrouxa J, Sternberga M. Advanced glycation endproducts (AGEs): pharmacological inhibition in diabetes. *Pathol Biol* 2006;**54**:405–19.

52. Swamy MS, Abraham EC. Inhibition of lens crystallin glycation and high molecular weight aggregate formation by aspirin *in vitro* and in vivo. *Invest Ophthalmol Vis Sci* 1989;**30**(6):1120–6.

53. Blakytny R, Harding JJ. Prevention of cataract in diabetic rats by aspirin, paracetamol (acetaminophen) and ibuprofen. *Exp Eye Res* 1992;**54**(4):509–18.

54. Lin PP, Barry RC, Smith DL, Smith JB. *In vivo* acetylation identified at lysine 70 of human lens α-crystallin. *Protein Sci* 1998;**7**:71451–7.

55. Swamy-Mruthinti S, Green K, Abraham EC. Inhibition of cataracts in moderately diabetic rats by aminoguanidine. *Exp Eye Res* 1996;**62**:505–12.

56. Yan H, Guo Y, Zhang J, Ding Z, Ha W, Harding JJ. Effect of carnosine, aminoguanidine, and aspirin drops on the prevention of cataracts in diabetic rats. *Mol Vis* 2008;**14**:2282–91.

57. Babizhayeva MA, Deyeva AI, Yermakovab VN, Semiletova YA, Davydovab NG, Kuryshevab NI, et al. N-Acetylcarnosine, a natural histidine-containing dipeptide, as a potent ophthalmic drug in treatment of human cataracts. *Peptides* 2001;**22**:979–94.

58. Hipkiss AR, Brownson C, Carrier MJ. Carnosine, the anti-ageing, anti-oxidant dipeptide, may react with protein carbonyl groups. *Mech Ageing Dev* 2001;**122**:1431–45.

59. Suzuki YJ, Tschiya M, Packer L. Lipoate prevents glucose-induced protein modifications. *Free Radic Res Commun* 1992;**17**:211–7.

60. Borenshtein D, Ofri R, Werman M, Stark A, Tritschler HJ, Moeller W, et al. Cataract development in diabetic sand rats treated with α-lipoic acid and its c-linolenic acid conjugate. *Diabetes Metab Res Rev* 2001;**17**:44–50.

61. Singh U, Jialal I. Alpha-lipoic acid supplementation and diabetes. *Nutr Rev* 2008;**66**(11):646–57.

62. Brown SM, Smith DM, Alt N, Thorpe SR, Baynes JW. Tissue-specific variation in glycation of proteins in diabetes: evidence for a functional role of amadoriase enzymes. *Ann NY Acad Sci* 2005;**1043**:817–23.

63. Delpierrre G, Vertommen D, Communi D, Rider MH, Van Schaftingen E. Identification of fructosamine residues deglycated by fructosamine-3-kinase in human hemoglobin. *J Biol Chem* 2004;**279**(26):27613–20.

64. Coughlan MT, Forbes JM, Cooper ME. Role of the AGE crosslink breaker, alagebrium, as a renoprotective agent in diabetes. *Kidney Int Suppl* 2007;**106**:S54–60.

65. Win MT, Yamamoto Y, Munesue S, Saito H, Han D, Motoyoshi S, et al. Regulation of RAGE for attenuating progression of diabetic vascular complications. *Exp Diabetes Res* 2012;**2012**:894605.

66. Reddy DVN. Ammo acid transport in the lens in relation to sugar cataracts. *Invest Ophthalmol* 1965;**4**(4):700–8.

67. Kinoshita JH, Barber GW, Merola LO, Tung B. Changes in the levels of free amino acids and myo-inositol in the galactose-exposed lens. *Invest Ophthalmol* 1969;**8**(6):625–32.

68. Durham DG. Distribution of free amino acids in human intraocular fluids. *Trans Am Ophthalmol Soc* 1970;**68**:462–500.

69. Ramakrishnan S, Sulochana KN, Punitham R. Free lysine, glycine, alanine, glutamic acid and aspartic acid reduce the glycation of human lens proteins by galactose. *Indian J Biochem Biophys* 1997;**34**(6):518–23.

70. Alvarado-Vásquez N, Lascurain R, Cerón E, Vanda B, Carvajal-Sandovala G, Tapia A, et al. Oral glycine administration attenuates diabetic complications in streptozotocin-induced diabetic rats. *Life Sci* 2006;**79**:225–32.

71. Ramakrishnan S, Sulochana KN, Punitham R, Arunagiri K. Free alanine, aspartic acid, or glutamic acid reduce the glycation of human lens proteins. *Glycoconj J* 1996;**13**(4):519–23.

72. Sivakumar R, Raj DG, Ramakrishnan S, Shyamala DCS. Influence of aspartic acid and glutamic acid on non-enzymatic glycation of lens proteins in vitro. *Indian J Pharmacol* 1995;**27**(3): 197–200.

73. Ramakrishnan S, Sulochana KN. Decrease in glycation of lens proteins by lysine and glycine by scavenging of glucose and possible mitigation of cataractogenesis. *Exp Eye Res* 1993;**57**(5):623–8.

74. Bahmani F, Bathaie SZ, Aldavood SJ, Ghahghaei A. Glycine therapy inhibits the progression of cataract in streptozotocin-induced diabetic rats. *Mol Vis* 2012;**18**:439–48.

75. Jafarnejad A, Bathaie SZ, Nakhjavani M, Hassan MZ, Banasadegh S. The improvement effect of L-Lys as a chemical chaperone on STZ-induced diabetic rats, protein structure and function. *Diabetes Metab Res Rev* 2008;**24**(1):64–73.

76. Alvarado-Vasquez N, Zamudio P, Ceron E, Vanda B, Zenteno E, Carvajal-Sandovala G. Effect of glycine in streptozotocin-induced diabetic rats. *Comp Biochem Physiol Part C* 2003;**134**:521–7.

77. Luthra M, Balasubramanian D. Nonenzymatic glycation alters protein structure and stability: a study of two eye lens crystallins. *J Biol Chem* 1993;**268**(24):18119–27.

78. Suryanarayana P, Saraswat M, Mrudula T, Krishna P, Krishnaswamy K, Reddy GB. Curcumin and turmeric delay streptozotocin-induced diabetic cataract in rats. *Invest Ophthalmol Vis Sci* 2005;**46**(6): 2092–9.

79. Zhang S, Chai FY, Yan H, Guo Y, Harding JJ. Effects of N-acetylcysteine and glutathione ethyl ester drops on streptozotocin-induced diabetic cataract in rats. *Mol Vis* 2008;**14**:862–70.

80. Lim J, Li L, Jacobs MD, Kistler J, Donaldson PJ. Mapping of glutathione and its precursor amino acids reveals a role for GLYT2 in glycine uptake in the lens core. *Invest Ophthalmol Vis Sci* 2007;**48**(11):5142–51.

81. Lim J, Lorentzen KA, Kistler J, Donaldson PJ. Molecular identification and characterisation of the glycine transporter (GLYT1) and the glutamine/glutamate transporter (ASCT2) in the rat lens. *Exp Eye Res* 2006;**83**:447–55.

82. Gonzalez-Ortiz M, Medina-Santillan R, Martinez-Abundis E, von Drateln CR. Effect of glycine on insulin secretion and action in healthy first-degree relatives of type 2 diabetes mellitus patients. *Horm Metab Res* 2001;**33**(6):358–60.

83. Gannon MC, Nuttall JA, Nuttall FQ. The metabolic response to ingested glycine. *Am J Clin Nutr* 2002;**76**:1302–7.

84. Sulochana KN, Rajesh M, Ramakrishnan S. Insulin receptor tyrosine kinase activity in monocytes of type 2 diabetes mellitus patients receiving oral L-lysine. *Indian J Biochem Biophys* 2001;**38**(5):331–4.

16

Selenium Supplementation and Cataract

Xiangjia Zhu, Yi Lu

Eye and ENT Hospital, Fudan University, Shanghai, P.R. China

INTRODUCTION

Recently, the number of investigations surrounding the protective effect of selenium in cataract development have increased. Proper dosage with selenium may help to strengthen the defensive system of lens against oxidation by modulating certain enzymes and protein kinase pathways and preventing deoxyribonucleic acid (DNA) methylation. This chapter gives us an overview of research findings on the relationship between selenium supplementation and cataract.

DOSE OF SELENIUM RECOMMENDED FOR SUPPLEMENTATION

Selenium is an essential trace element for mammals. It is required by humans in microgram amounts and has a relatively narrow margin of safety. Its recommended dietary allowance (RDA) for adults is 70–350 µg/day according to the World Health Organization (WHO). For adults it is 55 µg/day, for pregnant females 60 µg/day, and for lactating females 70 µg/day (USA) (Table 16.1). A daily intake should not exceed 400 µg/day of selenium, which the National Academy of Sciences/Institute of Medicine (NAS/IOM) report concluded is the tolerable upper intake level (UL) from foods and supplements likely to pose no risk of adverse health effects on almost all people. The same NAS/IOM report indicated that selenium 800 µg/day is the no-observed-adverse-effect level (NOAEL).

PREVIOUS STUDIES ON SELENIUM SUPPLEMENTATION

Selenium has been found to lower the risk of cardiovascular diseases.[1,2] Taking a daily supplement containing 200 µg of selenium is sufficient to inhibit the development of certain cancers.[3] Some researchers have found that selenium can also decrease the activity of brain monoamine oxidase B (MAO-B) in adult rats and slow down the aging process.[4] Studies showed that in patients with senile cataract, selenium levels in the aqueous humor and serum of those patients decreased significantly, which may reflect defective antioxidative defense systems that may lead to the formation of cataract.[5] Based on these findings, ophthalmic investigators believe that selenium may slow the progress of age-related diseases, and age-related cataract may be one of these.

However, the anticipated protective roles of small doses of selenium on the lens remains unclear, since previous studies concerning the relationship between cataract and selenium have been mainly focused on selenite cataract. It has been proved that a high dose of sodium selenite (0.70 mg/kg body weight (BW) per day for over 5 days) injection to suckling rats can induce selenite cataract. However, such injections could no longer cause permanent cataract in rats after 18 days' postpartum.[6]

SELENIUM DEFICIENCY AND CATARACT

There is evidence of the associations between occurrence of cataract and deficiency of selenium. Normal human eyes have a relatively higher volume of selenium than other tissues, and it is especially rich in the uveal membrane, retina, and lens. According to previous studies, the content of selenium in the lens of cataract patients is far decreased compared with people with no cataract, while the content of hydrogen peroxide is two to three times higher.[7] This indicates that selenium may closely relate to the occurrence of cataract. Studies that are more intensive are presently investigating the correlation between them.

The bioactivity of selenium is mostly measured by glutathione peroxidases (GPx), and the classical glutathione

TABLE 16.1 Recommended Dose for Adult Selenium Supplementation (USA, UK, WHO, China)

For Adult	USA (µg/day) (RDA)	UK (µg/day) (RDA)	WHO (µg/day) (RDA)	China (µg/day) (Chinese Nutrition Society)
Recommended dose	55	Male: 75 Female: 60	70–350	50–250
UL	400	400	400	400

RDA: recommended dietary allowances; UL: tolerable upper intake level; WHO: World Health Organization.

peroxidase, GPx-1, and GPx-3, the extracellular variant, decline most readily in selenium deficiency and recover slowly upon re-supplementation. Cai *et al.* have done a series of tests on selenium-deficient and control rats and found a significant decrease of GPx activity in the lens of the selenium-deficient group,[8] where GPx works insufficiently due to the lack of selenium, which can lower the enzyme activity by 83.3%. In previous experiments by Lawrence *et al.*,[9] such inactivity of antioxidant enzymes in the lens of cataract patients gradually got worse with a decrease of glutathione (GSH) content, leaving less effective protection against the progression of cataract. The selenium-deficient diet has also been found to accelerate the progress of SDZ ICT 322 (a selective hydroxytryptamine antagonist)-induced cataract.[10] With an insufficiency of selenium, it is possible to trigger the pathologic process of accumulation of free radicals and lipid peroxidation, destroy the construction and function of crystallins, and then affect the intraocular metabolism, which is the primary reason for the lens changes.

However, there is another theory of selenium shortage-mediated oxidative stress in cataract development. Recent data revealed an early onset of nuclear cataract without severe cellular oxidative stress in 15-kDa selenoprotein knockout mice. Some researchers suggested the possibility of improper folding of crystallins with the genetic dysfunction of selenoprotein.[11]

It is noteworthy that excessive intake of selenium will in turn lead to cataract, which is known as the 'selenium cataract'. However, one recent study has provided preliminary evidence of the appropriate dose of selenium application in cataract prevention of Sprague-Dawley (SD) rat models.[12] It proves the protective effect of selenium to the lens from the opposite aspect to some extent.

In summary, it is likely that insufficient consumption of selenium will weaken the ability of various enzymes to fight against harmful peroxides and then damage the visual function of the lens. For early-stage cataract patients or those who are not suitable for immediate

surgery, appropriate intake of a selenium-rich diet might be appropriate. A convincing explanation of selenium deficiency and cataract, however, needs more investigation.

PREVENTIVE ROLES OF SELENIUM AGAINST CATARACT

Scientists now have begun to study the protective effects of this important micronutrient against cataract.

Investigations carried out in some Chinese villages with very low incidence of cataract have shown that selenium was more abundant in the soil and water of those areas, compared with that in other areas.[13] Meanwhile, coadministration of selenium (1.5 mg/kg BW intraperitoneally) and high doses of vitamin E could protect the lens against cisplatin-induced oxidative damages.[14] Although antioxidants may be a part of the body's defense system to prevent age-related cataract, environmental contaminants may also contribute to cataractogenesis. Interestingly, one epidemiologic study has shown that in fish-eating populations of the lower Tapajós region, elevated exposure to mercury (Hg) has been found; however, the cataractogenic effects of Hg may be offset by selenium.[15]

Naphthalene cataract is considered as an ideal model for age-related cataract. It has been shown that orally administered naphthalene is transformed into naphthalene dihydrodiol in the liver, which then passes through the blood-aqueous barrier of the eye and is finally converted into 1,2-naphthoquinone (NQ) and hydrogen peroxide in the aqueous humor, exerting oxidative stress on the lens.[16]

In one study, selenium supplementation has been shown to slow the development of naphthalene cataract in rats, possibly by attenuating the oxidative stress in the lens. Adult rats were fed sodium selenite at human equivalent doses of 100, 200, and 400 µg/day (i.e., 0.0104, 0.0208, and 0.0416 mg/kg BW, respectively). Meanwhile, all the intervention groups (selenium-supplemented groups and naphthalene control group) were administered 10% naphthalene solution orally for 5 weeks.

In the selenium-supplemented groups, the general conditions of the rats were better. Rats supplemented with human equivalent dose of 200 µg/day of selenium had the shiniest fur, followed by human equivalent dose of 100 µg/day, and 400 µg/day groups. The naphthalene controls had the dullest fur.[12]

Lens density was evaluated through an anterior segment photograph system. It is also interesting to see that both nuclear and cortical densities of lenses in the selenium-supplemented groups were on average lower than those in the naphthalene group, suggesting that

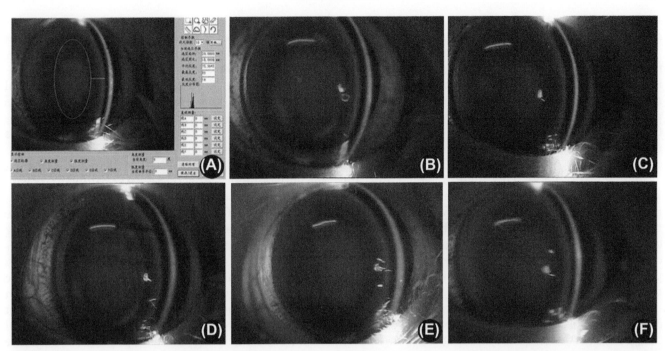

FIGURE 16.1 Lens images on day 37. Selected elliptical area for lens density analysis (A). Anterior segment images of naphthalene control (B), normal control (C), selenium group I (D), selenium group II (E), and selenium group III (F). Selenium groups I, II, and III were orally administrated with sodium selenite at human equivalent doses of 100, 200, and 400 µg/day (i.e., 0.0104, 0.0208, and 0.0416 mg/kg body weight). In addition, all the intervention groups (selenium-supplemented groups and naphthalene control) were orally administered with 10% naphthalene solution for 5 weeks. *Source: adapted from: Zhu and Lu (2012), Figure 1.*[12] *With permission from Current Eye Research.*

proper dose of selenium supplementation is beneficial for the lens. However, it is also notable that, despite that, selenium was constantly given throughout the experiment, and lens densities in the selenium-supplemented groups on day 29 were on average lower than those observed 1 week later (Fig. 16.1; Fig. 16.2). This suggests that since oxidative insults exist throughout a person's lifetime, selenium supplementation may be more of a generally preventive measure than a cure.[12] However, from another point of view, since the whole world is spending millions of dollars on cataract treatment each year, if the development of cataract can be postponed, the overall benefits would be enormous.

In selenium-supplemented groups, GPx activities in the blood and lens increased, as did selenium levels in aqueous humor. Selenium supplementation also ameliorated the decrease in GSH levels and increases in malondialdehyde (MDA) and hydroxyl radical levels in the lenses of rats treated with naphthalene.[12]

Thus, we can draw some interesting conclusions from above studies. It seems that whether selenium is cataractogenic depends on the dosage of the agent, and lower doses of selenium are more preventive for the lens than higher doses. Meanwhile, animal ages may be another important factor of the occurrence of cataract. Both of them deserve further studies in the future.

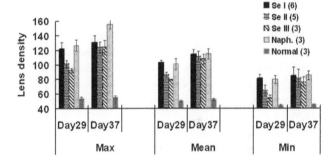

FIGURE 16.2 Lens densities on days 29 and 37. Maximum, mean, and minimum densities of lenses in each group on day 29 and day 37 (mean ± standard deviation (SD)). Selenium groups I, II, and III were orally administrated with sodium selenite at human equivalent doses of 100, 200, and 400 µg/day (i.e., 0.0104, 0.0208, and 0.0416 mg/kg body weight). In addition, all the intervention groups (selenium-supplemented groups and naphthalene control) were orally administered with 10% naphthalene solution for 5 weeks. *Source: adapted from: Zhu and Lu (2012), Figure 3.*[12] *With permission from Current Eye Research.*

SELENIUM AND GPx

The lens has a unique defensive system against oxidative insults. It uses primary defenses, including nonenzymatic (e.g., glutathione, carotenoids, vitamin E, etc.) as well as enzymatic (e.g., superoxide dismutase, GPx, and catalase, etc.) systems[17] to neutralize reactive oxygen species (ROS) and to repair, recover, or degrade those damaged molecules.

Among them, GPx is the general name of an enzyme family with peroxidase activity whose main biologic role is to protect the organism from oxidative damage. Its major biochemical functions are to reduce lipid hydroperoxides to their corresponding alcohols and to reduce free hydrogen peroxide to water by oxidizing GSH, a major free-radical scavenger in the lens, to glutathione disulfide (GSSG).

In mammals, GPx1, GPx2, GPx3, and GPx4 are known to be selenium-containing enzymes. Among them, GPx1 and GPx3 are the only two selenoproteins that have currently been shown to be present in the eye. GPx1 is the most abundant version found in the cytoplasm, while GPx3 is the major extracellular component. Both of them are very sensitive to selenium deficiency.[18]

One GPx1 gene-knockout study conducted by Reddy *et al.* has supported the view that GPx1 is the principal defense system that protects the lens against free radical-induced oxidative damage.[19] However, in the aqueous humor, GPx3, though it exists in a very tiny amount, is the major selenoprotein that may serve to eliminate the surrounding free radicals.[18] Previous studies have also shown that low levels of H_2O_2 are normally present in the aqueous humor,[20] and the principal mechanism for its removal is the GPx system,[21] possibly in the form of GPx3. Measuring selenium levels using atomic absorption spectrometry may be a feasible way to evaluate GPx3 content in the aqueous humor. Consequently, the elevated selenium levels in aqueous humor observed in Zhu's study may be responsible for alleviating the oxidative stress in the surrounding environment of the lens.[12]

In addition, in the selenium-supplemented study, GPx activities in different tissues were also studied to understand the underlying mechanisms of selenium-mediated protection. They showed that GPx activity in the plasma and red blood cells, as well as the lens, was increased in the supplemented groups, which indicated that the general defensive system against oxidation was strengthened in those groups.[12]

SELENIUM AND PHOSPHATIDYLINOSITOL 3-KINASE (PI3-K)/PROTEIN KINASE B (AKT) PATHWAY

Selenium is not an antioxidant, either in the chemical meaning of this term or in a biologic sense. Selenium does play an important role in the metabolism of H_2O_2 and other hydroperoxides by acting as a functional heteroatom in GPx, which contributes to the prevention of

free radical formation and related tissues damage. However, it functions as a protective agent for the lens in some different ways as well.

Scientists have found that by inhibiting apoptosis through the activation of phosphatidylinositol 3-kinase (PI3-K)/protein kinase B (Akt) pathway, selenium can also protect certain cell lines from damage induced by free radicals.[22–24] Sodium selenite is found to maintain survival of HT1080 cells (fibrosarcoma) and 3T3-L1 (rat adipocyte) cells by activating the antiapoptotic signal (PI3-K/Akt pathway), as well as blocking those apoptotic signals.[24] To explain, Akt, a downstream effecter of PI3-K, is an important serine threonine kinase. It plays a critical role in regulating survival responses of many cells in its phosphorylated form. As selenium also prevented secondary pathologic events in an animal model of traumatic brain injury by blocking apoptotic neuron cell death through phosphorylation of Akt,[23] is it possible that sodium selenite can protect the human lens epithelial cells (hLECs) from age-related apoptosis through the same pathway?

We know that the anterior surface of human lens is covered by lens epithelial cells, of which those in the central region serve to protect the underlying fibers from damage induced by oxidative insults. However, that damage can accumulate in these epithelial cells throughout the lifetime, causing eventual apoptosis of lens epithelial cells. Hence, the protective effect of central epithelium over the underlying fibers will be diminished.[25] Charakidas *et al.* also mentioned that the accumulation of small-scale apoptotic epithelial losses during a lifetime will finally alter the formation and homeostasis of lens fibers, leading to the opacity of the lens.[26] One recent study has summarized that apoptosis of the lens epithelial cells is an important cellular basis for noncongenital cataract formation.[27] Based on these findings, it would be very useful to find an effective method to inhibit the apoptosis of hLECs.

Recently, Zhu *et al.* have found that sodium selenite can increase hLECs viabilities (Fig. 16.3) and prevent 1,2-dihydroxynaphthalene (1,2-DHN),[28] the metabolite of naphthalene in the lens, which can be subsequently auto-oxidized to 1,2-naphthaquinone and H_2O_2, inducing hLECs apoptosis through phosphorylation and nuclear translocation of Akt (Fig. 16.3; Fig. 16.4).[29,30]

Since the effect of selenium is dose-dependent, scientists focused on cancer prevention and chemotherapy are interested in the induction of apoptosis by toxic concentrations of selenium,[31] while others have concentrated on the protective effect of selenium at low doses.[32]

The beneficial dosage of sodium selenite for the hLECs is also low, 1–16 ng/mL according to Zhu *et al.*[28] Within this range, the viabilities of hLECs were

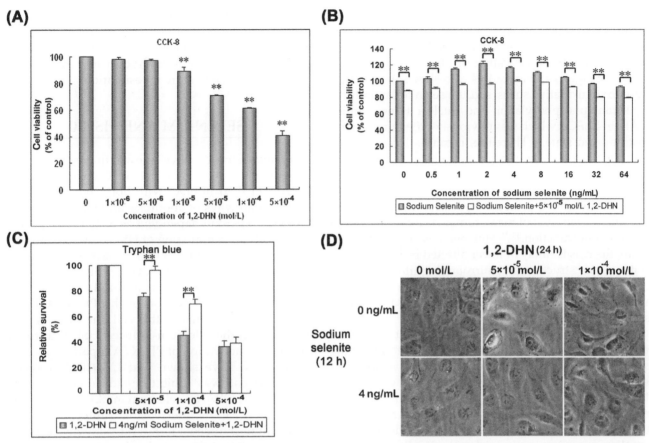

FIGURE 16.3 Sodium selenite increased cell viability and protected human lens epithelial cells (hLECs) from 1,2-dihydroxynaphthalene (1,2-DHN)-induced cell death. (A) Cultured hLECs were treated with 1,2-DHN at different concentrations for 24 h. Cell viabilities were assayed. (B) Cultured hLECs were either treated with sodium selenite alone at different concentrations for 12 h or given a combined treatment with 5×10^{-5} M 1,2-DHN for 24 h. Cell viabilities were assayed. (C) Cultured hLECs were treated with 1,2-DHN alone at different concentrations for 24 h or given a pretreatment with 4 ng/mL sodium selenite for 12 h before the 1,2-DHN treatment. The relative survival rate was estimated by tryphan blue exclusion assay. (D) Inverted microscope images of hLECs treated with 1,2-DHN at two concentrations with or without 4 ng/mL sodium selenite pretreatment (*P < 0.05, **P < 0.01). Each experiment was performed in triplicate. 1,2-DHN: 1,2-dihydroxynaphthalene; hLECs: human lens epithelial cells. *Source: adapted from Zhu* et al. *(2011), Figure 1,*[28] *with permission* (http://creativecommons.org/licenses/by-nc-nd/3.0/).

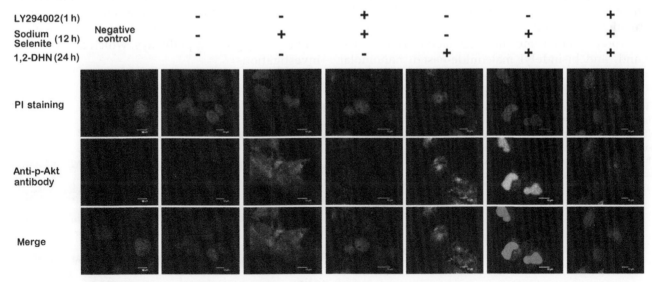

FIGURE 16.4 Immunofluorescence staining results also showed that sodium selenite induced Akt phosphorylation in a phosphatidylinositol 3-kinase (PI3-K)-dependent manner either in the normal or 1,2-dihydroxynaphthalene (1,2-DHN) stimulated state. *Source: adapted from Zhu* et al. *(2011),*[28] *with permission* (http://creativecommons.org/licenses/by-nc-nd/3.0/).

significantly increased whether in the stimulated or unstimulated state, and apoptosis induced by the strong oxidant, 1,2-DHN, was effectively inhibited by sodium selenite through the activation of PI3-K/Akt pathway (Fig. 16.3; Fig. 16.4). Figure 16.4 also shows that, in the presence of oxidative stress, sodium selenite can induce significant phosphorylation and nuclear translocation of Akt in hLECs, which was obviously different from its unstimulated state.[28] Therefore, it can be inferred that with the correct dose, sodium selenite could alleviate the oxidative stress in the lens also through its antiapoptotic effect.

It is noteworthy that the PI3-K/Akt pathway may also promote the survival of many tumor cells through such antiapoptotic function.[33,34] However, as an interesting organ, the lens is rarely or never affected by tumors due to its unique structure and surrounding immune environment, which suggests that the activation of PI3-K/Akt pathway by selenium in the lens may be safer than it is in any other cell lines.

The effect of nuclear translocation of Akt in hLECs remains unclear. It may further enhance the antiapoptotic effect of selenium against oxidative stress.[28] Studies have shown that insulin-like growth factor (IGF)-1-mediated protection against high glucose-induced apoptosis might be an important reason for nuclear redistribution of active Akt.[35] In dorsal root ganglion neurons, such nuclear redistribution promotes nuclear activation of survival transcription factors cyclic adenosine monophosphate response element binding protein (CREB) as well as nuclear exclusion of Foxo-1. In cardiomyocytes, scientists have also noticed that phosphorylation and nuclear translocation of Akt can accelerate the nuclear exclusion of Foxo-3a and Foxo-1.[36] These are two members of Forkhead box that promote cell death through activation of several Foxo-responsive genes, such as the pro-apoptotic tumor necrosis factor (TNF)-α, Fas ligand, and bisindolyl maleimide-based, nanomolar

protein kinase C inhibitors (Bim).[37–39] However, as the majority of studies on PI3-K/Akt pathway have been focused on the cytoplasmic region, the downstream signaling mechanism of the Akt nuclear translocation in hLECs still needs more investigation.

SELENIUM AND HSP70

Previous studies have shown that selenium can upregulate the expression of HSP70 in some cell lines. HSP70 can then interact with both the intrinsic and extrinsic pathways of apoptosis and inhibit cell death through chaperone-dependent as well as independent activities.[40] Such induction of HSP70 by selenium was also observed in hLECs at both messenger ribonucleic acid (mRNA) and protein levels.[28] Since it has been shown that in hLECs, selenium-induced protection was mediated by activation of the PI3-K/Akt pathway, it is possible that in this cell line there is also a link between the selenium-induced HSP70 upregulation and the activation of PI3-K/Akt pathway.

Zhu *et al.* have shown that LY294002, a specific PI3-K/Akt pathway inhibitor, had no effects on this induction of HSP70 by selenium in hLECs (Fig. 16.5).[28] Therefore, it can be inferred that protection of hLEC by selenium-induced PI3-K/Akt pathway activation was not mediated through the downstream upregulation of HSP70. In rat cerebellar granule neurons, the neuroprotective effect of Akt activation induced by thermal preconditioning was not mediated through downstream induction of HSP70 expression either.[41] However, in renal carcinoma cells (RCC4), the increased HSP70 expression after heat treatment was largely suppressed by the LY294002.[42] Contradictory evidence concerning a link between Akt activity and HSP70 expression may be due to the cell-specific differences in signaling pathways, which deserve future investigations.

FIGURE 16.5 Selenium and heat-shock protein 70 (HSP70). LY294002 did not affect the upregulated HSP70 expression induced by sodium selenite. bp: base pair; RT-PCR: reverse transcription-polymerase chain reaction. *Source: adapted from Zhu* et al. *(2011),*[28] *with permission* (http://creativecommons.org/licenses/by-nc-nd/3.0/).

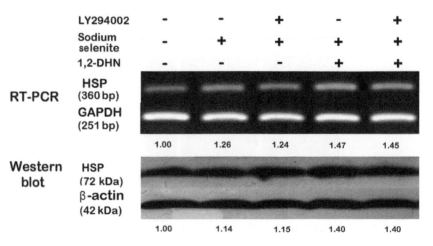

SELENIUM AND DEOXYRIBONUCLEIC ACID METHYLATION

Continuous oxidative stress is an important cause of cataract formation.[43] This oxidative damage to the lens is thought to be mediated through ROS, which are highly reactive compounds that can alter the lens configuration by DNA damage as well as lipid peroxidation and protein modifications.[44]

As a common epigenetic alteration, DNA hypermethylation is thought to contribute to the development of many diseases. It is an important biochemical process involved in DNA damage repair. In some prostate cancer studies, those changes are found in glutathione-S-transferase genes, which provide protection for DNA against oxidative stress.[45]

Recent data suggested that some dietary factors may be involved in epigenetic modifications to regulate cellular functions.[46] Selenium, although required in microgram amounts, is a key micronutrient essential for human health. In some experiments, researchers tested the effect of sodium selenite on various human cancer cells. The results show that it would modify the expression of certain hypermethylated genes, leading to decreasing of general DNA methylation and increasing of partial promoter demethylation.[9] For example, selenium has been shown to inhibit the expression of DNA methyltransferase (DNMT),[47] whose activation or malfunction can lead to hypermethylation. This can be a possible pathway to regulate such an epigenetic process.

The occurrence of many human diseases is closely related to DNA damage, including methylation. Cataract is one of them, which remains the primary cause of blindness in the whole world. One of the most important causes for cataract formation is continuous oxidative stress. Reactive ROS are highly responsible for the oxygen damage by lipid peroxidation, DNA damage, and protein modifications.[12] With those changes in the configuration of essential proteins, a cataract may develop. Methylation is also a probable reason for DNA damage in lenses except for oxidative stress. In tissue culture studies, Ferguson et al. reported that moderate levels of certain selenium compounds would be protective additives against DNA damage,[48] indicating the important role of selenium in the maintenance of genomic stability. The same results happened in some later experiments where rats were fed with selenium-deficient foods. After taking supranutritional levels of selenium, the global methylation in hepatocytes of rats showed significant reduction.[48] However, many cases with lens opacities are found to have a lower level of selenium in blood, aqueous humor, and lens than normal. That comes to the question: can selenium be a potential preventive agent for cataract patients through modulation of DNA methylation in hLECs? The answer

is possible, but the underlying mechanisms and implications of aforementioned observations are unclear and warrant further investigation.

In addition, considering the dual effect of selenium, some related studies suggest that it should be applied within safety range because it is of high possibility that it only maintains its epigenetic effects at nontoxic levels.[45]

In conclusion, selenium can protect the lens through known as well as unknown mechanisms, and all of those unanswered questions still need more exploration. Recent studies investigating the relationship between selenium and cataract might have presented a promising future for cataract prevention through special dietary supplementation.

TAKE-HOME MESSAGES

- Selenium may slow the progress of age-related diseases.
- The bioactivity of selenium is mostly measured by GPx, which decline most readily in selenium deficiency and recover slowly upon resupplementation.
- Insufficiency of selenium may trigger the pathologic process of the accumulation and lipid peroxidation of free radicals and then destroy the construction and function of crystallins.
- Selenium supplementation may be more of a generally preventive measure than cure, as oxidative insults exist throughout life. Whether selenium is cataractogenic depends on the dose of the agent, and lower doses of selenium are more preventive for the lens than the high doses.
- Selenium can also protect hLECs against damages from oxidative stress by suppressing apoptosis through the activation of PI3-K/Akt pathway.
- Under certain circumstances, selenium can upregulate the expression of HSP70 and inhibit cell death through chaperone-dependent as well as independent activities.
- Moderate levels of certain selenium compounds would be protective additives against DNA damage in lens by modulating the methylation process.

References

1. Bleys J, Navas-Acien A, Guallar E. Serum selenium levels and all-cause, cancer, and cardiovascular mortality among US adults. *Arch Intern Med* 2008;**168**:404–10.
2. Liu S, Zhang H, Zhu L, et al. Kruppel-like factor 4 is a novel mediator of selenium in growth inhibition. *Mol Cancer Res* 2008;**6**:306–13.
3. Alwahaibi NY, Budin SB, Mohamed J, et al. Nuclear factor-kappa B as a promising target for selenium chemoprevention in rat hepatocarcinogenesis. *J Gastroenterol Hepatol* 2010;**25**:786–91.

4. Tang YL, Wang SW, Lin SM. Both inorganic and organic selenium supplements can decrease brain monoamine oxidase B enzyme activity in adult rats. *Br J Nutr* 2008;**100**:660–5.

5. Karakucuk S, Ertugrul MG, Faruk EO, et al. Selenium concentrations in serum, lens and aqueous humour of patients with senile cataract. *Acta Ophthalmol Scand* 1995;**73**:329–32.

6. Shearer TR, Anderson RS, Britton JL. Influence of selenite and fourteen trace elements on cataractogenesis in the rat. *Invest Ophthalmol Visual Sci* 1983;**24**:417–23.

7. Swanson AA, Truesdale AW. Elemental analysis in normal and cataractous human lens tissue. *Biochem Biophys Res Commun* 1971;**45**:1488–96.

8. Cai QY, Chen XS, Zhu LZ, et al. Biochemical and morphological changes in the lenses of selenium and/or vitamin E deficient rats. *Biomed Environ Sci* 1994;**7**:109–15.

9. Lawrence RA, Sunde RA, Schwartz GL, et al. Glutathione peroxidase activity in rat lens and other tissues in relation to dietary selenium intake. *Exp Eye Res* 1974;**18**:563–9.

10. Langle UW, Wolf A, Cordier A. Enhancement of SDZ ICT 322-induced cataracts and skin changes in rats following vitamin E- and selenium-deficient diet. *Arch Toxicol* 1997;**71**:283–9.

11. Kasaikina MV, Fomenko DE, Labunskyy VM, et al. Roles of the 15-kDa selenoprotein (Sep15) in redox homeostasis and cataract development revealed by the analysis of Sep 15 knockout mice. *J Biol Chem* 2011;**286**:33203–12.

12. Zhu X, Lu Y. Selenium supplementation can slow the development of naphthalene cataract. *Curr Eye Res* 2012;**37**:163–9.

13. Huang B, Zhao Y, Sun W, et al. Relationships between distributions of longevous population and trace elements in the agricultural ecosystem of Rugao County, Jiangsu, China. *Environ Geochem Health* 2009;**31**:379–90.

14. Naziroglu M, Karaoglu A, Aksoy AO. Selenium and high dose vitamin E administration protects cisplatin-induced oxidative damage to renal, liver and lens tissues in rats. *Toxicology* 2004;**195**:221–30.

15. Lemire M, Fillion M, Frenette B, et al. Selenium and mercury in the Brazilian Amazon: opposing influences on age-related cataracts. *Environ Health Perspect* 2010;**118**:1584–9.

16. Lou MF, Xu GT, Zigler SJ, et al. Inhibition of naphthalene cataract in rats by aldose reductase inhibitors. *Curr Eye Res* 1996;**15**:423–32.

17. Lou MF. Redox regulation in the lens. *Prog Retinal Eye Res* 2003;**22**:657–82.

18. Flohe L. Selenium, selenoproteins and vision. *Dev Ophthalmol* 2005;**38**:89–102.

19. Reddy VN, Giblin FJ, Lin LR, et al. Glutathione peroxidase-1 deficiency leads to increased nuclear light scattering, membrane damage, and cataract formation in gene-knockout mice. *Invest Ophthalmol Visual Sci* 2001;**42**:3247–55.

20. Giblin FJ, McCready JP, Kodama T, et al. A direct correlation between the levels of ascorbic acid and H2O2 in aqueous humor. *Exp Eye Res* 1984;**38**:87–93.

21. Giblin FJ, Reddan JR, Schrimscher L, et al. The relative roles of the glutathione redox cycle and catalase in the detoxication of H_2O_2 by cultured rabbit lens epithelial cells. *Exp Eye Res* 1990;**50**:795–804.

22. Yeo JE, Kim JH, Kang SK. Selenium attenuates ROS-mediated apoptotic cell death of injured spinal cord through prevention of mitochondria dysfunction: *in vitro* and *in vivo* study. *Cell Physiol Biochem* 2008;**21**:225–34.

23. Yeo JE, Kang SK. Selenium effectively inhibits ROS-mediated apoptotic neural precursor cell death *in vitro* and *in vivo* in traumatic brain injury. *Biochim Biophys Acta* 2007;**1772**:1199–210.

24. Yoon SO, Kim MM, Park SJ, et al. Selenite suppresses hydrogen peroxide-induced cell apoptosis through inhibition of ASK1/JNK and activation of PI3-K/Akt pathways. *FASEB J* 2002;**16**:111–3.

25. Harocopos GJ, Alvares KM, Kolker AE, et al. Human age-related cataract and lens epithelial cell death. *Invest Ophthalmol Visual Sci* 1998;**39**:2696–706.

26. Charakidas A, Kalogeraki A, Tsilimbaris M, et al. Lens epithelial apoptosis and cell proliferation in human age-related cortical cataract. *Eur J Ophthalmol* 2005;**15**:213–20.

27. Li WC, Kuszak JR, Dunn K, et al. Lens epithelial cell apoptosis appears to be a common cellular basis for non-congenital cataract development in humans and animals. *J Cell Biol* 1995;**130**:169–81.

28. Zhu X, Guo K, Lu Y. Selenium effectively inhibits 1,2-dihydroxynaphthalene-induced apoptosis in human lens epithelial cells through activation of PI3-K/Akt pathway. *Mol Vision* 2011;**17**:2019–27.

29. Martynkina LP, Qian W, Shichi H. Naphthoquinone cataract in mice: mitochondrial change and protection by superoxide dismutase. *J Ocul Pharmacol Ther.* 2002;**18**:231–9.

30. van Heyningen R. Experimental studies on cataract. *Invest Ophthalmol Visual Sci* 1976;**15**:685–97.

31. Kato MA, Finley DJ, Lubitz CC, et al. Selenium decreases thyroid cancer cell growth by increasing expression of GADD153 and GADD34. *Nutr Cancer* 2010;**62**:66–73.

32. Roy S, Dontamalla SK, Mondru AK, et al. Downregulation of apoptosis and modulation of TGF-beta1 by sodium selenate prevents streptozotocin-induced diabetic rat renal impairment. *Biol Trace Elem Res* 2011;**139**:55–71.

33. Kawauchi K, Ogasawara T, Yasuyama M, et al. The PI3K/Akt pathway as a target in the treatment of hematologic malignancies. *Anti-cancer Agents Med Chem* 2009;**9**:550–9.

34. Lee DH, Szczepanski MJ, Lee YJ. Magnolol induces apoptosis via inhibiting the EGFR/PI3K/Akt signaling pathway in human prostate cancer cells. *J Cell Biochem* 2009;**106**:1113–22.

35. Leinninger GM, Backus C, Uhler MD, et al. Phosphatidylinositol 3-kinase and Akt effectors mediate insulin-like growth factor-I neuroprotection in dorsal root ganglia neurons. *FASEB J* 2004;**18**:1544–6.

36. Caporali A, Sala-Newby GB, Meloni M, et al. Identification of the prosurvival activity of nerve growth factor on cardiac myocytes. *Cell Death Differ* 2008;**15**:299–311.

37. Vogt PK, Jiang H, Aoki M. Triple layer control: phosphorylation, acetylation and ubiquitination of FOXO proteins. *Cell Cycle* 2005;**4**:908–13.

38. Sunters A, Fernandez DMS, Stahl M, et al. FoxO3a transcriptional regulation of Bim controls apoptosis in paclitaxel-treated breast cancer cell lines. *J Biol Chem* 2003;**278**:49795–805.

39. Modur V, Nagarajan R, Evers BM, et al. FOXO proteins regulate tumor necrosis factor-related apoptosis inducing ligand expression. Implications for PTEN mutation in prostate cancer. *J Biol Chem* 2002;**277**:47928–37.

40. Arya R, Mallik M, Lakhotia SC. Heat shock genes – integrating cell survival and death. *J Biosci* 2007;**32**:595–610.

41. Cao L, Cao DX, Su XW, et al. Activation of PI3-K/Akt pathway for thermal preconditioning to protect cultured cerebellar granule neurons against low potassium-induced apoptosis. *Acta Pharmacol Sin* 2007;**28**:173–9.

42. Zhou J, Schmid T, Frank R, et al. PI3K/Akt is required for heat shock proteins to protect hypoxia-inducible factor 1alpha from pVHL-independent degradation. *J Biol Chem* 2004;**279**:13506–13.

43. Marsili S, Salganik RI, Albright CD, et al. Cataract formation in a strain of rats selected for high oxidative stress. *Exp Eye Res* 2004;**79**:595–612.

44. Babizhayev MA, Deyev AI, Yermakova VN, et al. Lipid peroxidation and cataracts: N-acetylcarnosine as a therapeutic tool to manage age-related cataracts in human and in canine eyes. *Drugs R & D* 2004;**5**:125–39.

45. Xiang N, Zhao R, Song G, et al. Selenite reactivates silenced genes by modifying DNA methylation and histones in prostate cancer cells. *Carcinogenesis* 2008;**29**:2175–81.

46. Davis CD, Uthus EO. DNA methylation, cancer susceptibility, and nutrient interactions. *Exp Biol Med (Maywood)* 2004;**229**:988–95.

47. Barrera LN, Cassidy A, Johnson IT, et al. Epigenetic and antioxidant effects of dietary isothiocyanates and selenium: potential implications for cancer chemoprevention. *Proc Nutr Soc* 2012;**71**:237–45.

48. Ferguson LR, Karunasinghe N, Zhu S, et al. Selenium and its role in the maintenance of genomic stability. *Mutat Res* 2012;**733**:100–10.

OTHER EYE CONDITIONS

Vitamin A with Cyclosporine for Dry Eye Syndrome

Eun Chul Kim, Choun-Ki Joo
The Catholic University of Korea, Seoul, Korea

INTRODUCTION

Dry eye disease is "a disorder of the tear film due to tear deficiency or excessive evaporation that causes damage to the interpalpebral ocular surface and is associated with symptoms of discomfort".[1] Traditional treatments for dry eye disease, such as punctal occlusion and artificial tears, are palliative measures that attempt to increase the volume of the tear film by either decreasing drainage or supplementing it with an aqueous solution.[2] However, new insights into the inflammatory nature of this disease have led to a shift in the therapeutic approach to dry eye disease.[3] The realization that inflammation plays a role in dry eye syndrome led to the use of anti-inflammatory medications, such as short-term corticosteroids and long-term cyclosporine.[4,5]

Despite the plurality of underlying causes of dry eye disease, there are some common histopathologic manifestations to the ocular surface epithelia: loss of conjunctival goblet cells, abnormal enlargement of epithelial cells, increase in cellular stratification, and keratinization.[6] Normal secretory conjunctival mucosa gradually develops into a nonsecretory keratinized epithelium, a process referred to as squamous metaplasia.[7]

In this chapter, we will describe the efficacy of aqueous ophthalmic retinyl palmitate solution and that of cyclosporine A (CsA) 0.05% ophthalmic emulsion in treating dry eye disease.

DRY EYE

Dry eye disease is characterized by a variety of signs and symptoms, including decreased tear production, an altered and unstable tear film, epithelial damage evidenced by fluorescein staining of the cornea (Fig. 17.1), and rose bengal or lissamine green staining of the conjunctiva.[2] Patients report blurred vision, dryness, and ocular surface burning or stinging, which can be exacerbated by desiccating conditions or prolonged visual concentration. Dry eye affects quality of life by decreasing functional vision for reading, using a computer, and driving.[8]

DRY EYE AND INFLAMMATION

There is increasing evidence that dry eye is an inflammatory disease. Disease or dysfunction of the lacrimal glands leads to changes in tear composition, such as hyperosmolarity, which stimulates the production of inflammatory mediators on the ocular surface.[9] Increased production and activation of proinflammatory cytokines such as interleukin (IL)-1, IL-6, tumor necrosis factor (TNF)-α, and matrix metalloproteases (MMPs) by stressed ocular surface and glandular epithelial cells, as well as by the inflammatory cells and T lymphocytes that infiltrate these tissues, have been reported in dry eye.[10–12]

In Sjögren syndrome, significantly increased levels of IL-1α, IL-6, IL-8, TNF-α, and transforming growth factor (TGF)-β1 ribonucleic acid (RNA) transcripts have been found in the conjunctival epithelium.[13]

Tear MMP-9 activity levels also are positively associated with corneal fluorescein staining scores and with low contrast visual acuity.[14] In keratoconjunctivitis sicca (KCS), the increased MMP-9 activity is associated with deranged corneal epithelial barrier function (increased fluorescein permeability), increased corneal epithelial

Handbook of Nutrition, Diet, and the Eye
http://dx.doi.org/10.1016/B978-0-12-401717-7.00017-4

desquamation (punctate epithelial erosions), and corneal surface irregularity.[14]

There is increased evidence that CD4+ T cells are involved in the pathogenesis of dry eye. In an animal model, increased infiltration of CD4+ T cells in the goblet cell-rich area resulted in increased expression of interferon (IFN)γ, goblet cell loss, and conjunctival metaplasia (Fig. 17.2).[15]

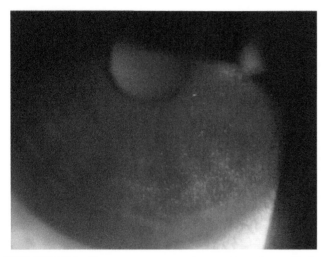

FIGURE 17.1 Fluorescent staining of cornea in patient with dry eye syndrome. Corneal epithelial damage is observed by fluorescein staining.

CONVENTIONAL TREATMENT OF DRY EYE

Dry eye treatment is directed at minimizing the symptoms and improving the quality of life for the patients. One of the first treatment options involves making simple environmental modifications to reduce evaporation of the tear film. These measures include increasing ambient humidity with humidifiers, avoiding air currents, and taking frequent breaks from visually demanding tasks.[16] The traditional approach to treating dry eye is lubricating the ocular surface. However, it does not directly address the ocular surface inflammation. Anti-inflammatory therapies should be considered for patients with moderate-to-severe disease with symptoms or evidence of corneal disease refractory to treatment.[17]

ANTI-INFLAMMATORY TREATMENT OF DRY EYE

Anti-inflammatory therapies inhibit inflammatory mediators and reduce the signs and symptoms of KCS.[18]

Corticosteroids are potent anti-inflammatory agents that are routinely used to control inflammation in many organs. They decrease the production of a number of inflammatory cytokines (IL-1, IL-6, IL-8, TNF-α, granulocyte-macrophage colony-stimulating

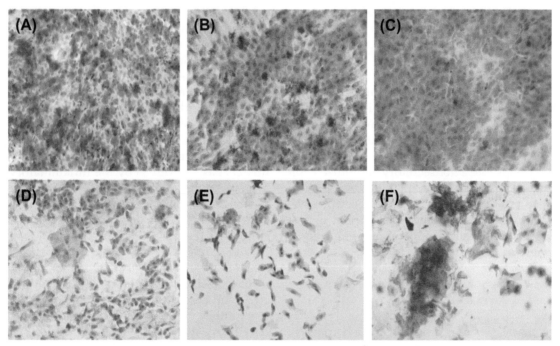

FIGURE 17.2 Photomicrographs showing each stage of impression cytologic analysis results in patients with dry eye syndrome. A: Stage 0 – normal, B: Stage 1 – early loss of goblet cells, C: Stage 2 – total loss of goblet cells, D: Stage 3 – early keratinization, E: Stage 4 – moderate keratinization, F: Stage 5 – advanced keratinization (100 ×, Periodic Acid Schiff (PAS)-hematoxylin stain).

factor (GM-CSF)) and MMP-9 by the corneal epithelium.[18] Topical administration of a 1% solution of non-preserved methylprednisolone, given three of four times daily for 2 weeks to patients with Sjögren syndrome KCS, provided moderate-to-complete relief of symptoms in all patients.[19]

Autologous serum contains inhibitors of inflammatory cytokines (e.g., interleukin-1 receptor antagonist (IL-1 RA) and soluble TNF-α receptors) and MMP (e.g., tissue inhibitors of metalloproteinases (TIMPs)) inhibitors.[20] Therefore, it may inhibit soluble mediators of the ocular surface inflammatory cascade of dry eye. Autologous serum drops improve ocular irritation symptoms and conjunctival and corneal dye staining in Sjögren syndrome-associated KCS.[21]

Tetracyclines are compounds that have antibiotic as well as anti-inflammatory properties that may make them useful for the management of chronic inflammatory diseases.[18] These agents decrease the activity of collagenase, phospholipase A2, nitric oxide production, and several MMPs.[18] They also decrease the production of IL-1α and TNF-α in the human corneal epithelium.[22] Doxycycline, the semi-synthetic tetracycline, has been reported to improve irritation symptoms, increase tear film stability, and decrease the severity of ocular surface disease in patients with ocular rosacea.[23] Doxycycline has also been reported to be effective for treating recurrent corneal epithelial erosions and phlyctenular keratoconjunctivitis.[24,25]

SUPPLEMENT TREATMENTS IN DRY EYE

Essential fatty acids, such as 18-carbon omega-6 and omega-3 fatty acids, are necessary for complete health, and they cannot be synthesized by vertebrates and must be obtained from dietary sources.[26] Omega-3 fatty acids inhibit the synthesis of these lipid mediators as well as block production of IL-1 and TNF-α.[26]

VITAMIN A

Vitamin A is essential for maintaining the health of epithelial cells throughout the body; it affects cellular regulation and differentiation.[27] Dietary deficiency results in keratinization of epithelial cells in the gastrointestinal tract, respiratory tract, and the ocular surface.[28] Severe vitamin A deficiency results in xerophthalmia and keratomalacia.[29] Vitamin A, as retinol, is the main constituent of the visual pigments.[30] It is also important for ocular surface integrity through its role in epithelial cell RNA synthesis.[30] Thus, vitamin A deficiency has many ocular manifestations, including night blindness, xerophthalmia, and loss of vision.[30]

Vitamin A in Corneal Wound Healing

Vitamin A is necessary for the normal growth and differentiation of corneal epithelium,[27] and vitamin A deficiency causes the loss of goblet cells, leading to increased epidermal keratinization and squamous metaplasia of mucus membranes, including the cornea and conjunctiva.[31]

The vitamin A-deficient rat cornea has significantly reduced frequency and size of hemidesmosome.[32] This is accompanied by redistribution of bullous pemphigoid antigen (BPA) to the cell cytoplasm and changes in the staining pattern for laminin.[32] Structural abnormalities of the epithelial basement membrane complex are responsible for the loose epithelial adhesion as well as other known abnormalities of healing in the vitamin A-deficient rat cornea.[32]

Vitamin A modulates the expression of thrombospondin-1 (TSP1) to accelerate the re-epithelialization of wounded corneas,[33] and TSP1 can also inhibit vascular endothelial growth factor (VEGF)-A signaling by directly binding to the protein and by competing with VEGF-A for binding to cell surface heparin sulfate proteoglycans.[34] Retinyl palmitate eye drops can inhibit VEGF-A and activate TSP2 and improve conjunctival impression cytologic findings.[35] Furthermore, retinyl palmitate eye drops were found to promote corneal healing after an alkali burn.[35]

Vitamin A and Inflammation

The functions of vitamin A in regulation of differentiation and function of various immune cells and tissue cells suggest that vitamin A intake and levels of retinol and retinoids in the body have significant impacts on tissue inflammation.[36] Vitamin A deficiency can exacerbate or suppress tissue inflammation.[36]

Vitamin A deficiency decreases intestinal inflammation through the suppression of the recruiting of inflammatory T cells into the gut.[37] This is because T cells require retinoic acid for expression of two gut-homing receptors, CCR9 and α4β7.[38] However, tissue inflammation can be worsened in vitamin A deficiency because of decreased retinoic acid-dependent generation of tolerogenic dendritic cells and macrophages.[39] Retinoic acid is important for both formation of immunity and prevention of inflammatory diseases.[36] Promotion of immunity by retinoic acid is through terminal differentiation of phagocytes, generation of gut-homing T cells, and induction of immunoglobulin (Ig)A-producing B cells.[36] Prevention of inflammation and autoimmunity by retinoic acid is mediated through generation of FoxP3+ regulatory T cells and tolerogenic antigen-presenting cells or myeloid cells.[36]

Vitamin A in Dry Eye

Vitamin A may exist in three forms: retinol, retinal, and retinoic acid.[8] Many tissues requiring vitamin A store the vitamin as an ester of retinal.[8] Retinyl palmitate is found in cells of the lacrimal gland.[40] Retinol is found in the tears of rabbits and humans.[41] Its presence in tears provides the rationale for treating dry eye disease with vitamin A.[41]

Topical all-trans retinoic acid ointment was effective in the treatment of four severe cases of the following ocular surface diseases: KCS, Stevens-Johnson syndrome, drug-induced pseudopemphigoid, and surgery-induced dry eye.[41] Impression cytologic analysis, dry eye symptoms, visual acuity, keratopathy, and the Schirmer test results all improved after the use of topical all-trans retinoic acid ointment.[42]

CYCLOSPORINE A

CsA is a neutral, hydrophobic, cyclic metabolite of the fungi *Tolypocladium inflatum* and *Beauveria nivea*.[43] One of the mechanisms of action of CsA is the inhibition of calcineurin (a serine/threonine phosphatase) with subsequent restriction in the expression of certain genes involved in T-cell activation (IL-2, IL-4, IL-12p40).[43] CsA also can bind cyclophiline D, and the complex formed can prevent the opening of the mitochondrial permeability transition pore in response to stress stimuli, thus preventing apoptosis.[43] There is evidence that CsA blocks JNK and p38 signaling pathways as well, and it has been shown to induce the synthesis of TGF-β *in vivo* and *in vitro*.[43]

CYCLOSPORINE A IN DRY EYE

A licensed topical CsA (0.2% ointment, Optimmune, Schering-Plough) has been available in veterinary ophthalmology for the treatment of canine KCS.[44] Cyclosporine 0.05% ophthalmic emulsion (Restasis, Allergan Inc., Irvine, CA) is the first commercially available therapy for dry eye disease that actually increases production of natural tears.[5]

Two independent phase III clinical trials compared twice-daily treatment with CsA 0.05% or 0.1% or vehicle in 877 patients with moderate-to-severe dry eye disease.[4,5] Patients treated with CsA, 0.05% or 0.1%, showed significantly greater improvement in two objective signs of dry eye disease (corneal fluorescein staining and anesthetized Schirmer test values) than those treated with vehicle (P ≤ 0.05). An increased Schirmer test score was observed in 59% of patients treated with CsA, with 15% of patients having an increase of 10 mm or more. In contrast, only 4% of vehicle-treated patients had this magnitude of change in their Schirmer test scores (P < 0.0001). CsA 0.05% treatment also produced significantly greater improvements in three subjective measures of dry eye disease (blurred vision symptoms, need for concomitant artificial tears, and the global response to treatment) (P < 0.05).[4,5]

In the original phase III trials, topical administration of cyclosporine 0.05% or 0.1% ophthalmic emulsions resulted in very low plasma levels of cyclosporine: cyclosporine was detectable (lower limit of quantitation, 0.1 ng/mL) in only seven of 310 blood samples, all of which were taken from patients administered cyclosporine 0.1%.[45] Therapy of chronic dry eye disease with cyclosporine 0.1% ophthalmic emulsion for 1–3 years was safe, well tolerated, and not associated with systemic side effects.[45]

CYCLOSPORINE A VERSUS VITAMIN A

Kim *et al.* compared the efficacy of vitamin A (retinyl palmitate) and CsA 0.05% eye drops in treating patients with dry eye disease.[46] A total of 150 patients with defined dry eye disease participated (50 in each treatment group). In three identical clinical trials, patients were treated twice daily with CsA 0.05%, or four times daily with retinyl palmitate 0.05%, or with neither CsA nor retinyl palmitate. The decrease from baseline in blurred vision was statistically significant in the vitamin A group at month two (from 1.35 ± 0.23 to 0.91 ± 0.19) and month three (from 1.35 ± 0.23 to 1.02 ± 0.21) and in the CsA 0.05% group at month three (from 1.21 ± 0.25 to 0.95 ± 0.19; P < 0.05) (Fig. 17.3). At month two, there

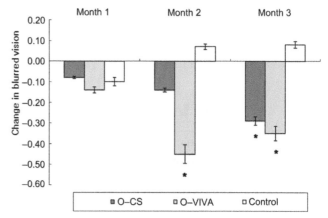

FIGURE 17.3 Graph showing the change in blurred vision from baseline after topical cyclosporine 0.05%, topical retinyl palmitate 0.05% (Refresh Plus), or artificial lubricants alone treatment in patients with dry eye syndrome. The decrease from baseline in blurred vision was statistically significant in the vitamin A group at 2 months and in the cyclosporine and vitamin A treatment groups at 3 months (*P < 0.05, analysis of variance (ANOVA)). Control: Refresh Plus alone; O-CS: cyclosporine 0.05%; O-Viva: retinyl palmitate 0.05%.

was a statistically significant improvement in tear film breakup time (BUT) in both the CsA 0.05% group (from 3.43 ± 1.32 seconds to 5.93 ± 2.02 seconds) and vitamin A group (from 3.57 ± 1.56 seconds to 5.65 ± 2.39 seconds; P < 0.05) (Fig. 17.4). At month two, there was a statistically significant improvement in Schirmer values in the vitamin A group (P < 0.05). At month three, there was a statistically significant improvement in both the CsA 0.05% group (from 3.93 ± 1.21 mm to 7.30 ± 1.79 mm) and vitamin A group (from 3.67 ± 1.63 mm to 8.13 ± 1.64 mm; P < 0.05) (Fig. 17.5). The decrease from baseline in the impression cytologic grade was statistically significant in both the CsA 0.05% group (from 2.26 ± 0.58 to 1.75 ± 0.56) and vitamin A group (from 2.03 ± 0.25 to 1.38 ± 0.63) at month three (P < 0.05). At month three, there was a statistically significant increase in the goblet cell density in both the CsA 0.05% group (from 113.25 ± 45.26 cells/mm^2 to 203.85 ± 52.39 cells/mm^2) and vitamin A group (from 109.59 ± 39.58 cells/mm^2 to 192.57 ± 45.27 cells/mm^2; P < 0.05). At month two, there was a statistically significant decrease in the corneal staining score in the CsA 0.05% group. At month three, there was a statistically significant decrease in both the CsA 0.05% group (from 2.23 ± 0.88 mm to 1.89 ± 0.24 mm) and vitamin A group (from 2.18 ± 0.74 mm to 1.95 ± 0.35 mm; P < 0.05) (Fig. 17.6).

In this study, blurred vision improved in the vitamin A group faster than in the CsA 0.05% group because the Schirmer score increased in the vitamin A group at month two. The corneal staining score improved faster in the CsA group than in the vitamin A group.

Therefore, CsA may be a more effective treatment for the ocular surface inflammation caused by dry eye in the relatively early treatment period.

CONCLUSION

Both vitamin A eye drops and topical CsA 0.05% treatments can improve symptoms of blurred vision, tear film BUT, Schirmer I score, and impression cytologic findings in patients with severe dry eye syndrome. Therefore,

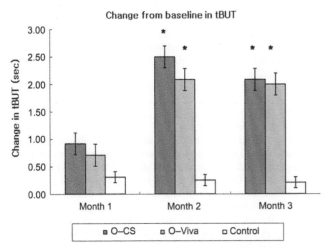

FIGURE 17.4 Graph showing the change in tear film break-up time (tBUT) from baseline after treatment with topical cyclosporine 0.05%, topical retinyl palmitate 0.05% (Refresh Plus), or artificial lubricants alone in patients with dry eye syndrome (mean value ± standard error). The increase in tear film tBUT was statistically significant in the cyclosporine and vitamin A treatment groups at 2 and 3 months (*P < 0.05 by analysis of variance (ANOVA)). Control: Refresh Plus alone; O-CS: cyclosporine 0.05%; O-Viva: retinyl palmitate 0.05%.

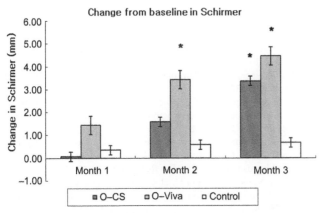

FIGURE 17.5 Bar graph showing the change in Schirmer I test results from baseline after treatment with topical cyclosporine 0.05%, topical retinyl palmitate 0.05% (Refresh Plus), or artificial lubricants alone in patients with dry eye syndrome (mean value ± standard error). The increase in Schirmer test value (mm) was statistically significant in the cyclosporine group at 3 months and in the vitamin A group at 2 and 3 months (*P < 0.05 by analysis of variance (ANOVA)). Control: Refresh Plus alone; O-CS: cyclosporine 0.05%; O-Viva: retinyl palmitate 0.05%.

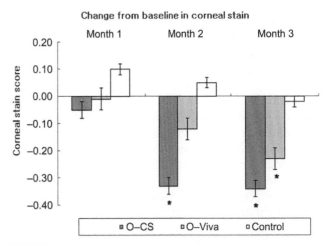

FIGURE 17.6 Bar graph showing the change in corneal fluorescein staining score from baseline after treatment with topical cyclosporine 0.05%, topical retinyl palmitate 0.05%, or artificial lubricants alone in patients with dry eye syndrome. The decrease in corneal staining score was statistically significant in the cyclosporine group at 2 and 3 months and in the vitamin A group at 3 months (*P < 0.05 by Kruskal-Wallis test). Values are mean ± standard errors (SE) and were graded on a scale from 0 through 4. Control: Refresh Plus alone; O-CS: cyclosporine 0.05%; O-Viva: retinyl palmitate 0.05%.

vitamin A eye drops as well as CsA can be used as an adjunct therapy with artificial lubricants for dry eye syndrome.

TAKE-HOME MESSAGES

- Dry eye is a dysfunction of the lacrimal glands leading to changes in tear composition including hyperosmolarity that stimulates the production of inflammatory mediators on the ocular surface.
- Anti-inflammatory therapies such as corticosteroids, autologous serum, and tetracyclines are commonly used for dry eye syndrome.
- Vitamin A is essential for maintaining the health of epithelial cells throughout the body; it affects cellular regulation and differentiation.
- Retinoic acid is important for both formation of immunity and prevention of inflammatory diseases.
- Topical all-trans retinoic acid ointment was effective in the treatment of four severe cases of the following ocular surface diseases: KCS, Stevens-Johnson syndrome, drug-induced pseudopemphigoid, and surgery-induced dry eye.
- CsA inhibits calcineurin (a serine/threonine phosphatase), with subsequent restriction in the expression of certain genes involved in T-cell activation (IL-2, IL-4, IL-12p40).
- Patients treated with CsA, 0.05% or 0.1%, showed significantly ($P \leq 0.05$) greater improvement in corneal fluorescein staining, anesthetized Schirmer test, blurred vision symptoms, need for concomitant artificial tears, and the global response to treatment.
- Blurred vision and the Schirmer score without anesthesia improved in the vitamin A group faster than in the CsA 0.05% group. The corneal staining score improved faster in the CsA group than in the vitamin A group.
- Both vitamin A eye drops and topical CsA 0.05% treatments can improve symptoms of blurred vision, tear film BUT, Schirmer I score, and impression cytologic findings in patients with severe dry eye syndrome.
- Vitamin A eye drops as well as CsA can be used as an adjunct therapy with artificial lubricants for dry eye syndrome.

References

1. Lemp MA. Report of the National Eye Institute/Industry workshop on Clinical Trials in Dry Eyes. *CLAO J* 1995;**21**:221–32.
2. Pflugfelder SC, Tseng SC, Sanabria O, et al. Evaluation of subjective assessments and objective diagnostic tests for diagnosing tear-film disorders known to cause ocular irritation. *Cornea* 1998;**17**:38–56.
3. Wilson SE. Inflammation: a unifying theory for the origin of dry eye syndrome. *Manag Care* 2003;**12**(Suppl):14–9.
4. Stevenson D, Tauber J, Reis BL. Cyclosporine Phase 2 Study Group. Efficacy and safety of cyclosporine ophthalmic emulsion in the treatment of moderate-to-severe dry eye disease: a dose-ranging, randomized trial. *Ophthalmology* 2000;**107**:967–74.
5. Sall K, Stevenson OD, Mundorf TK, et al. Two multicenter, randomized studies of the efficacy and safety of cyclosporine ophthalmic emulsion in moderate to severe dry eye disease. *Ophthalmology* 2000;**107**:631–9.
6. Nelson JD, Havener VR, Cameron JD. Cellulose acetate impressions of the ocular surface. Dry eye states. *Arch Ophthalmol* 1983;**101**:1869–72.
7. Tseng SC, Hirst LW, Maumenee AE, et al. Possible mechanisms for the loss of goblet cells in mucin-deficient disorders. *Ophthalmology* 1984;**91**:545–52.
8. Barber LD, Pflugfelder SC, Tauber J, et al. Phase III safety evaluation of cyclosporine 0.1% ophthalmic emulsion administered twice daily to dry eye disease patients for up to 3 years. *Ophthalmology* 2005;**112**:1790–4.
9. Luo L, Li DQ, Doshi A, et al. Experimental dry eye stimulates production of inflammatory cytokines and MMP-9 and activates MAPK signaling pathways on the ocular surface. *Invest Ophthalmol Vis Sci* 2004;**45**:4293–301.
10. Zhu X, Topouzis S, Liang LF, et al. Myostatin signaling through Smad2, Smad3 and Smad4 is regulated by the inhibitory Smad7 by a negative feedback mechanism. *Cytokine* 2004;**26**:262–72.
11. López Bernal D, Ubels JL. Quantitative evaluation of the corneal epithelial barrier: effect of artificial tears and preservatives. *Curr Eye Res* 1991;**10**:645–56.
12. López Bernal D, Ubels JL. Artificial tear composition and promotion of recovery of the damaged corneal epithelium. *Cornea* 1993;**12**:115–20.
13. Pflugfelder SC, Jones D, Ji Z, et al. Altered cytokine balance in the tear fluid and conjunctiva of patients with Sjögren's syndrome keratoconjunctivitis sicca. *Curr Eye Res* 1999;**19**:201–11.
14. Pflugfelder SC, Farley W, Luo L, et al. Matrix metalloproteinase-9 knockout confers resistance to corneal epithelial barrier disruption in experimental dry eye. *Am J Pathol* 2005;**166**:61–71.
15. De Paiva CS, Villarreal AL, Corrales RM, et al. Dry eye-induced conjunctival epithelial squamous metaplasia is modulated by interferon-gamma. *Invest Ophthalmol Vis Sci* 2007;**48**: 2553–60.
16. Yao W, Davidson RS, Durairaj VD, et al. Dry eye syndrome: an update in office management. *Am J Med* 2011;**124**:1016–8.
17. Utine CA, Stern M, Akpek EK. Clinical review: topical ophthalmic use of cyclosporin A. *Ocul Immunol Inflamm* 2010;**18**:352–61.
18. Pflugfelder SC. Antiinflammatory therapy for dry eye. *Am J Ophthalmol* 2004;**137**:337–42.
19. Marsh P, Pflugfelder SC. Topical non-preserved methylprednisolone therapy of keratoconjunctivitis sicca in Sjögren's syndrome. *Ophthalmology* 1999;**106**:811–6.
20. Liou LB. Serum and *in vitro* production of IL-1 receptor antagonist correlate with C-reactive protein levels in newly diagnosed, untreated lupus patients. *Clin Exp Rheumatol* 2001;**19**:515–23.
21. Tsubota K, Goto E, Fujita H, et al. Treatment of dry eye by autologous serum application in Sjögren's syndrome. *Br J Ophthalmol* 1999;**83**:390–5.
22. Solomon A, Rosenblatt M, Li DQ, et al. Doxycycline inhibition of interleukin-1 in the corneal epithelium. *Invest Ophthalmol Vis Sci* 2000;**41**:2544–57.
23. Frucht-Pery J, Sagi E, Hemo I, et al. Efficacy of doxycycline and tetracycline in ocular rosacea. *Am J Ophthalmol* 1993;**116**:88–92.
24. Hope-Ross MW, Chell PB, Kervick GN, et al. Oral tetracycline in the treatment of recurrent corneal erosions. *Eye* 1994;**8**:384–8.

25. Culbertson WW, Huang AJ, Mandelbaum SH, et al. Effective treatment of phlyctenular keratoconjunctivitis with oral tetracycline. *Ophthalmology* 1993;**100**:1358–66.

26. Endres S, Ghorbani R, Kelley VE, et al. The effect of dietary supplementation with n-3 polyunsaturated fatty acids on the synthesis of interleukin-1 and tumor necrosis factor by mononuclear cells. *N Engl J Med* 1989;**320**:265–71.

27. Kobayashi TK, Tsubota K, Takamura E, et al. Effect of retinol palmitate as a treatment for dry eye: a cytological evaluation. *Ophthalmologica* 1997;**211**:358–61.

28. Moore T, Holmes PD. The production of experimental vitamin A deficiency in rats and mice. *Lab Anivi* 1971;**5**:239–50.

29. Sommer A. *Nutritional Blindness: Xerophthalmia and Keratomalacia.* New York: Oxford University Press; 1982.

30. Sommer A. Effects of vitamin A deficiency on the ocular surface. *Ophthalmology* 1983;**90**:592–600.

31. Hatchell DL, Sommer A. Detection of ocular surface abnormalities in experimental vitamin A deficiency. *Arch Ophthalmol* 1984;**102**:1389–93.

32. Shams NB, Hanninen LA, Chaves HV, et al. Effect of vitamin A deficiency on the adhesion of rat corneal epithelium and the basement membrane complex. *Invest Ophthalmol Vis Sci* 1993;**34**:2646–54.

33. Uno K, Kuroki M, Hayashi H, et al. Impairment of thrombospondin-1 expression during epithelial wound healing in corneas of vitamin A-deficient mice. *Histol Histopathol* 2005;**20**:493–9.

34. Gupta K, Gupta P, Wild R, et al. Binding and displacement of vascular endothelial growth factor (VEGF) by thrombospondin: effect on human microvascular endothelial cell proliferation and angiogenesis. *Angiogenesis* 1999;**3**:147–58.

35. Kim EC, Kim TK, Park SH, et al. The wound healing effects of vitamin A eye drops after a corneal alkali burn in rats. *Acta Ophthalmol* 2012;**90**:e540–6.

36. Kim CH. Retinoic acid, immunity, and inflammation. *Vitam Horm* 2011;**86**:83–101.

37. Kang SG, Wang C, Matsumoto S, et al. High and low vitamin A therapies induce distinct FoxP3+ T-cell subsets and effectively control intestinal inflammation. *Gastroenterology* 2009;**137**:1391–402.

38. Iwata M, Hirakiyama A, Eshima Y, et al. Retinoic acid imprints gut-homing specificity on T cells. *Immunity* 2004;**21**:527–38.

39. Saurer L, McCullough KC, Summerfield A. *In vitro* induction of mucosa-type dendritic cells by all-trans retinoic acid. *J Immunol* 2007;**179**:3504–14.

40. Ubels JL, Osgood TB, Foley KM. Vitamin A is stored as fatty acyl esters of retinol in the lacrimal gland. *Curr Eye Res* 1988;**7**:1009–16.

41. Ubels JL, MacRae SM. Vitamin A is present as retinol in the tears of humans and rabbits. *Curr Eye Res* 1984;**3**:815–22.

42. Tseng SC. Topical retinoid treatment for dry eye disorders. *Trans Ophthalmol Soc UK* 1985;**104**:489–95.

43. Matsuda S, Koyasu S. Mechanisms of action of cyclosporine. *Immunopharmacology* 2000;**47**:119–25.

44. Kaswan RL, Salisbury MA, Ward DA. Spontaneous canine keratoconjunctivitis sicca. A useful model for human keratoconjunctivitis sicca: treatment with cyclosporine eye drops. *Arch Ophthalmol* 1989;**107**:1210–6.

45. Small DS, Acheampong A, Reis B, et al. Blood concentrations of cyclosporin A during long-term treatment with cyclosporine A ophthalmic emulsions in patients with moderate to severe dry eye disease. *J Ocul Pharmacol Ther* 2002;**18**:411–8.

46. Kim EC, Choi JS, Joo CK. A comparison of vitamin A and cyclosporine A 0.05% eye drops for treatment of dry eye syndrome. *Am J Ophthalmol* 2009;**147**:206–13.

Dietary N-3 Polyunsaturated Fatty Acids and Dry Eye

Corinne Joffre[1], Sabrina Viau[2]

[1]INRA and University of Bordeaux, Nutrition and Integrative Neurobiology (NutriNeuro) Bordeaux, France,
[2]MacoPharma, Tourcoing, France

INTRODUCTION

Dry eye syndrome is a growing public health problem. It can have considerable impact on visual acuity and daily activities such as reading, computer use, driving and watching television, and professional work.[1] Current therapies display side effects and provide only temporary improvement in symptoms. The underlying inflammation occurring in case of dry eye and the knowledge of the anti-inflammatory effects of food components such as n-3 polyunsaturated fatty acids (PUFA) provide a rationale for their use in dry eye therapy. In comparison, a diet deficient or unbalanced in n-3 PUFA may exacerbate the inflammation. The rapid changes in our diet and lifestyle, particularly since the 1910s, have led to an increased consumption of n-6 PUFA at the expense of n-3 PUFA. This western diet resulted in an unbalanced ratio between n-6 and n-3 PUFA of 15–16 instead of the four that was recommended.[2] We will see below the importance of the n-3 PUFA supply in dry eye syndrome.

DRY EYE: AN INFLAMMATORY PATHOLOGY

Dry Eye: Definition and Prevalence

In 1995, members of the Dry Eye WorkShop (DEWS) Definition and Classification Subcommittee adopted a definition for describing dry eye that was, until then, only considered as a set of symptoms and clinical signs of diverse origin and giving rise to unstable tear film: "Dry eye is a disorder of the tear film due to tear deficiency or excessive evaporation, which causes damage to the interpalpebral ocular surface and is associated with symptoms of ocular discomfort".[3]

In 2007, after advances in the field of dry eye describing a more complex situation, the definition was updated by the DEWS as follows: "Dry eye is a multifactorial disease of the tears and ocular surface that results in symptoms of discomfort, visual disturbance, and tear film instability with potential damage to the ocular surface, accompanied by increased osmolarity of the tear film and inflammation of the ocular surface".[4]

Inflammation is a common feature in all forms of dry eye syndrome, and this definition emphasizes its role in ocular surface, maintaining a chronic state that may persist even when the cause of dry eye has been removed.[5]

The main epidemiologic studies on the prevalence of dry eye were compared in the review of Schaumberg and colleagues[6] and in the report of the DEWS[7]: studies in Salisbury (Salisbury Eye Evaluation),[8] Melbourne (Melbourne Visual Impairment Project),[9] Beaver Dam (Beaver Dam Eye Study),[10] and the cohort study of women's program on the evaluation of the benefits and risks of aspirin and vitamin E in the primary prevention of cardiovascular disease and cancer (Women's Health Study).[11] The comparison of these studies led to the prevalence of dry eye of 5–10% in people aged 45–60 years, 5–15% in people aged 60–70 years, and 10–20% for people aged more than 75 years. These studies focused on female populations. Men appear to be less exposed to dry eye, with a prevalence 5–6% in men aged 70–74 years and 6–8% for men aged more than 75 years.[12]

Dry Eye: Multiple Symptoms Underlying Chronic Inflammation

The mechanisms involved in dry eye are complex and still poorly understood. The most studied are the instability of the lacrimal film caused by a change in the quality or the quantity of the aqueous or lipidic phase, the injury of the ocular surface (characterized by an impairment of the conjunctival and corneal epithelial cells), and the inflammation of the lacrimal glands. The common denominator of all forms of dry eye syndrome is the inflammation of the ocular surface and/or lacrimal glands whether dry eye is caused by a disease (blepharitis, Sjögren syndrome) or a combination of predisposing factors (wear of contact lenses, prolonged visual tasking) associated with a trigger event (menopause, bacterial infection). The hyperosmolarity of the tear film is consistently observed and considered as a central mechanism.[4] Whether due to increased evaporation of tears and/or to insufficient secretion, this hyperosmolarity is responsible for tissue damage in the cornea and conjunctiva. The mucus cells of the conjunctiva are particularly damaged, leading to the production of mucus of lower quality and/or quantity. Concerning the cornea, epithelial cells undergo apoptosis, leading to a repeated nervous stimulation of the lacrimal glands. The injury caused in the conjunctiva triggers inflammatory activity in epithelial cells leading to the release of inflammatory mediators in the tears, increasing instability of the tear film. Once engaged, inflammation escapes to the regulation system and a vicious circle of chronic inflammation sustains independently of the original causes.[4,5] Indeed, the corneal epithelium is no longer protected from the environment due to the instability of the tear film and becomes the target of inflammatory mediators present in tears. Moreover, the injury of corneal and conjunctival epithelia leads to a breakdown in epithelial junctions, thereby promoting bacterial growth. The presence of bacterial flora contributes to the instability of the lipid phase of the tear film[13] and maintains inflammation at the ocular surface. The instability of the lacrimal film contributes to the increased evaporation of the tears. It has been shown that the lipid composition of the tear film can also be modified. In blepharitis patients, a decrease in triglycerides and cholesterol has been observed,[14] as well as a decrease in saturated and monounsaturated fatty acids [15,16] and an increase in branched-chain fatty acids.[17] These modifications may have important consequences on the properties of the tear film. It has not been demonstrated whether they are cause or consequences of the inflammation of the ocular surface. We hypothesize that the increase in the branched-chain fatty acids and the decrease in saturated fatty acids may be an adaptive response since branched-chain fatty acids may enhance lipid fluidity and offer a greater resistance to evaporation, thereby counterbalancing the lack of tears (Fig. 18.1).

In dry eye syndrome, the ocular surface is injured leading to variable degrees of inflammation characterized by an enhanced production of proinflammatory cytokines and chemokines in the tears such as interleukin (IL)-6, IL-1, and tumor necrosis factor (TNF)-α (reviewed in[18]) and by an increased production and activation of proinflammatory cytokines in glandular epithelial cells and inflammatory cells that infiltrate this tissue.[19] In experimental animal models, dry eye induced an increase in the expression of IL-1, IL-6, and TNF-α in rat and mice.[20,21] In humans, these cytokines were also increased in the conjunctival epithelium in Sjögren syndrome patients, the autoimmune and most severe type of dry eye.[22] The marker of immune activation, HLA-DR (human leukocyte antigen-D related) was also increased in the conjunctival epithelium of dry eye patients [5,23,24]

FIGURE 18.1 The 'vicious circle' of the mechanisms maintaining a chronic inflammation during dry eye syndrome. Whether this is an evaporative or a hyposecretive dry eye, tear film instability and hyperosmolarity occur. They can be cause and consequence of tissue damage in the cornea and the conjunctiva (loss of mucus cell) and ocular surface inflammation. These two phenomena maintain each other. The tissue damage leads to a continuous nervous stimulation of the reflex lacrimal loop, generating neurogenic inflammation, thus increasing the hyposecretion and the ocular surface inflammation.

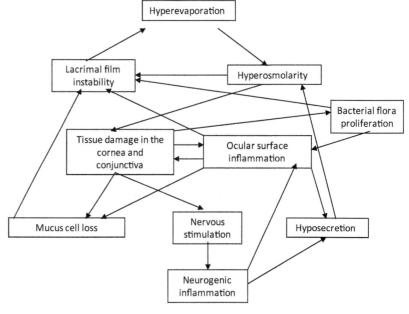

and its equivalent, major histocompatibility complex, (MHC) II, in a rat model of dry eye.[20]

DIETARY N-3 POLYUNSATURATED FATTY ACIDS: THEIR ROLES IN INFLAMMATION

Lipids represent 35–40% of the energy intake in France.[25] They are essentially found (90–95%) in the form of triglycerides, a structure consisting of a glycerol backbone and three fatty acids. They are also found in the form of phospholipids, in which the fatty acid in the 3-position on the glycerol is replaced by a phosphory-lated alcohol function. The structure of a triglyceride and a phospholipid is shown schematically in Figure 18.2. Fatty acids have many physiologic roles. They are the primary source of energy for the tissues and, as components of membrane phospholipids, play a structural role. Fatty acids of phospholipids are also mobilized by the cells as precursors of lipid mediators involved in

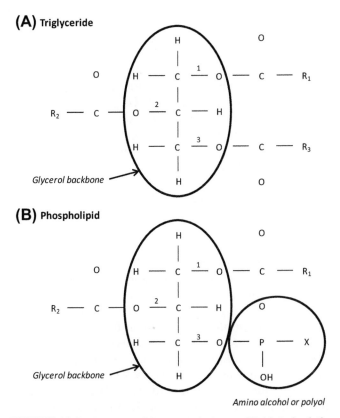

(A) Triglyceride

Glycerol backbone

(B) Phospholipid

Glycerol backbone

Amino alcohol or polyol

FIGURE 18.2 Structure of the two main forms of lipids in food, the triglyceride (A) and phospholipid (B). Triglycerides and phospholipids contain a glycerol backbone on which fatty acids are esterified (R1, R2, and R3 in position 1, 2, and 3 for the triglyceride and R1 and R2 for the phospholipid). At position 3 of the phospholipid, a phosphate group associated to a radical was esterified, which may be an amino alcohol or a polyol.

many physiologic processes, such as inflammation or neuroprotection.

Anti-inflammatory Actions of Polyunsaturated Fatty Acids

PUFA, fatty acids comprising more than one double bond on their carbon chain, can be classified into two categories, the n-6 PUFA and the n-3 PUFA. Linoleic acid (18:2 n-6; LA) and α-linolenic acid (18:3 n-3; ALA) are respectively the precursors of these two series. They are called essential fatty acids because they cannot be synthesized by mammals. However, once absorbed, they are metabolized *in vivo* by elongation, desaturation, and β-oxidation in fatty acids containing more unsaturations and/or more carbon atoms (Fig. 18.3). The main metabolites are arachidonic acid (20:4 n-6; AA) and docosapentaenoic acid (22:5 n-6) for the n-6 family and eicosapentaenoic acid (20:5 n-3; EPA) and docosahexaenoic acid (22:6 n-3; DHA) for the n-3 family. These two families share the same enzymatic equipment for the biosynthesis of the long-chain PUFA derivatives and can thus compete.[26] N-3 PUFA exert many physiologic roles in the cells among which is a benefit role in inflammation.

In pathologic conditions, n-6 and n-3 PUFA play a role in the expression of adhesion molecules and the production of cytokines and reactive oxygen species.[27] A large number of studies have demonstrated the anti-inflammatory effects of the long-chain n-3 PUFA, EPA, and DHA, *in vitro*, *in vivo*, and in humans.[27] *In vitro*, it has been shown that EPA and/or DHA inhibit the overexpression of intercellular adhesion molecule (ICAM)-1 and HLA-DR and the production of proinflammatory cytokines (IL-1, IL-2, IL-6, TNF-α) in stimulated immune cells (monocytes, macrophages).[28–34] *In vivo*, EPA and/or DHA inhibit the expression of inflammatory markers (including ICAM-1) in healthy mice or in a murine model of choroidal neovascularization, as well as the production of inflammatory mediators (monocyte chemotactic protein (MCP)-1 and IL-6) in the choroid.[30,35] In humans, it has been shown that a combined intake of EPA and DHA decreases the expression of HLA-DR and adhesion molecules in healthy subjects and the production of proinflammatory cytokines (IL-1, TNF-α) in healthy subjects and in patients with rheumatoid arthritis.[36–39] DHA alone is sufficient to decrease the production of IL-1 and TNF-α by peripheral blood mononuclear cells in healthy subjects.[40]

Involvement of Lipid Mediators Synthesized from PUFA during Inflammatory Processes

Much of the pro- or anti-inflammatory actions of PUFA are attributed to the lipid mediators synthesized

FIGURE 18.3 Elongation and desaturation of n-6 and n-3 polyunsaturated fatty acids (PUFA). PUFA are synthesized from linoleic acid (18:2 n-6) and α-linolenic acid (18:3 n-3) following elongation (addition of carbon atoms) and desaturation (addition of double bonds) steps in endoplasmic reticulum. C24 PUFA is converted to C22 PUFA by β-oxidation in peroxisomes. These elongation and desaturation steps lead to the formation of docosapentaenoic acid (22:5 n-6) for the n-6 series and docosahexaenoic acid (22:6 n-3) for the n-3 series.

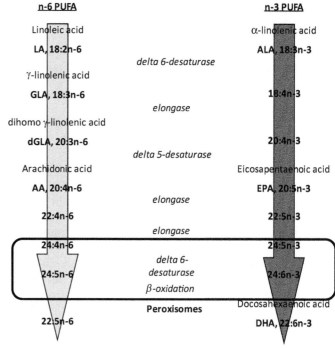

by oxygenation, hydroxylation, and peroxidation of AA and EPA, called eicosanoids.[41,42] AA-derived mediators include 2-series prostaglandins (PG), 4-series leukotrienes (LT), 2-series thromboxanes (TX), and 5-hydroxyeicosatetraenoic acids (5-HETE). The main EPA-derived mediators include 3-series PG, 5-series LT, and 3-series TX (inactive) (Fig. 18.4). DHA is also converted to 3-series PG. N-3 PUFA are thought to act via several mechanisms. They can serve as an alternative substrate for cyclo-oxygenase (COX) or lipoxygenase (LOX), thereby limiting the metabolism of AA into eicosanoids and resulting in the production of less potent products.[43] They can also be converted to potent anti-inflammatory and protective mediators. E-series resolvins (Rvs) are produced from EPA, and D-series Rvs, neuroprotectin D1 (NPD1), and maresins are produced from DHA (reviewed in[44]) (Fig. 18.4).

The eicosanoid receptors (G-protein coupled) are present on many cell types, including smooth muscle, gastric mucosa, adipose tissue, central nervous system tissues, blood platelets, and the cells of the immune system. Eicosanoids synthesized from EPA and those from AA share the same receptors, thus they act in competition. However, eicosanoids from EPA and especially 3-series PG and TX and 5-series LT are biologically less active than eicosanoids derived from AA.[45] For example, the PGE_3 is much less effective than PGE_2 to induce the synthesis of IL-6 and COX-2 in macrophages.[46] Thus, eicosanoids derived from EPA antagonize those synthesized from AA. Moreover, the EPA has the property to inhibit the production of

eicosanoids from AA, as a competitive inhibitor for the enzymes involved.[45] First, EPA has the ability to inhibit the activity of delta 5-desaturase, converting dihomo γ-linolenic acid (dGLA) to AA. It can also inhibit the activity of phospholipase A2 (PLA2) preventing the release of AA, the activity of COX-2 generating the PG, prostacyclins, and TX[47,48] and the activity of 5-LOX that generate the LT.[49] Thus, EPA reduces both the proportion of AA and the production of proinflammatory eicosanoids derived from AA.

DHA, as well as EPA, is able to inhibit the production of eicosanoids from AA. A dietary supply in DHA in healthy subjects reduces the production of PGE_2 and LTB_4 by stimulated peripheral blood mononuclear cells.[40]

The docosanoids (Rvs and NPD1) synthesized from EPA and DHA act in physiologic dose ranges (picomolar–nanomolar) and possess anti-inflammatory and pro-resolving properties. They stimulate the clearance of apoptotic cells and inflammatory debris by macrophages, inhibit the expression of the proinflammatory cytokines, and block neutrophil infiltration.[42,50–53] They act via cell-surface G-protein coupled receptors such as GPR32 and ALX/FPRL2 for RvD1 and chemoattractant receptor 23 for RvE1.[52] NPD1 exhibits potent anti-inflammatory activity and can activate the resolution.[44] It inhibits leukocyte infiltration in the brain and the production of IL-1 by stimulated glial cells.[54,55] It sustains neuronal function and protects synapses and circuits in the brain.[56] It protects the photoreceptors from death by promoting the

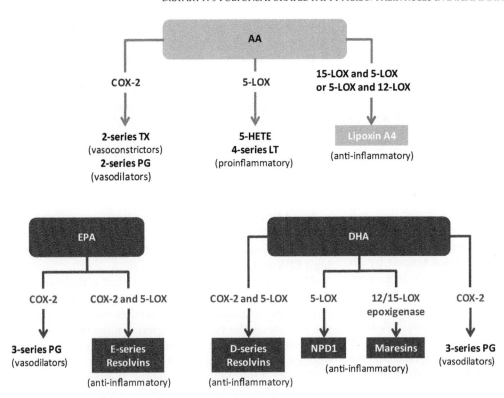

FIGURE 18.4 Pathways in active eicosanoid metabolism from arachidonic acid (AA, n-6 polyunsaturated fatty acids (PUFA) in gray) and eicosapentaenoic acid and docosahexaenoic acid (EPA and DHA, n-3 PUFA in dark gray). AA, EPA, and DHA are hydrolyzed from the phospholipids by phospholipases A2. The actions of cyclo-oxygenase (COX) and lipoxygenase (LOX) lead to the formation of pro- and anti-inflammatory lipid mediators. The main anti-inflammatory mediators described are lipoxin A4 from AA, E-series resolvins from EPA and D-series resolvins, NPD1, and maresins from DHA. 5-HETE: 5-hydroxyeicosatetraenoic acid; LT: leukotrienes; NPD1: neuroprotectin D1; PG: prostaglandins; TX: thromboxanes. *Adapted from de Roos* et al. *(2009).*[42]

Modulation of Gene Expression by n-3 Polyunsaturated Fatty Acids

The regulation of inflammation by n-3 PUFA may be mediated by transcription factors activated by PUFA themselves or by the lipid mediators synthesized.

The PPAR (peroxisome proliferator activated receptors) are nuclear hormone receptors, described in three forms: α, β/δ, and γ.[60]

PPARα was the first PPAR identified. It is involved in the oxidation of fatty acids and lipoprotein metabolism.[60] It has been shown that PPARα is expressed in monocytes/macrophages[61,62] and endothelial cells[63] and participates in these cells in the regulation of the inflammatory response.[60,64] The activation of PPARα in macrophages induces their apoptosis when activated, inhibits the production of metalloproteinase (MMP)-9 and TNF-α, and induces the activity of iNOS (inducible nitric oxide synthase), a reactive oxygen species-producing enzyme.[61,62,65,66] The activation of PPARα in the endothelial cells inhibits the activation of COX-2, the expression of VCAM-1 (vascular cell adhesion molecule-1), and the production of IL-6.[63,67,68] The precursor ALA, and also the long-chain PUFA EPA and DHA, are ligands of PPARα.[69,70]

PPARγ, initially described as a key regulator of the differentiation of adipocytes and involved in lipid metabolism, is also linked to the regulation of inflammation.[60] Its expression has been shown in macrophages, in which it inhibits the activation and induces apoptosis.[61] Specifically, its activation represses transcription of genes encoding proinflammatory cytokines (TNF-α, IL-1, IL-6), adhesion molecule VCAM-1, MMP-9, and iNOS.[62,65,66,71,72] The binding of ALA to PPARγ has been demonstrated,[70] but few data are currently available regarding the potential of n-3 PUFA to activate PPARγ.

The nuclear factor-kappaB (NF-κB) is a ubiquitous heterodimeric transcription factor inducing many inflammatory genes, including COX-2, iNOS, adhesion molecules ICAM-1 and VCAM-1, cytokines TNF-α, IL-1, IL-6 and IFN-γ, and MMP.[73] Under basal conditions, NF-κB is associated in the cytoplasm with its inhibitor I-κB. NF-κB is activated by the phosphorylation of I-κB, leading to the release of NF-κB. The latter then undergoes a translocation to the nucleus where it binds to its target genes, while I-κB is degraded by the proteasome. It has been shown that EPA, by blocking the phosphorylation of I-κB and thus the activation of NF-κB, inhibits the production of TNF-α induced by the lipopolysaccharide in human monocytes.[29] The transcription factor PPARα inhibits the activation of NF-κB.[74]

expression of antiapoptotic proteins at the expense of proapoptotic ones.[57,58] It significantly inhibits choroidal neovascularization and induces regeneration of corneal nerves.[59]

N-3 POLYUNSATURATED FATTY ACIDS AND DRY EYE

Dietary Habits versus Recommended Food Intake

As stated before, human metabolism is not able to synthesize PUFA but only to produce them from their precursors, LA and ALA. Therefore, the dietary intake of PUFA is crucial to maintain the balance between n-6 and n-3 PUFAs. However, in the western diet, there is an imbalance between n-6 and n-3 PUFA leading to an n-3 PUFA consumption 12–20 times lower than n-6 PUFA consumption.[2,26] This is due to increased industrialization in developed nations accompanied by changes in dietary habits. It is particularly characterized by an increase in LA, abundant in many corn oils (e.g., sunflower oil), and a decrease in ALA, found in some green vegetables, soybeans, and nuts, and EPA and DHA found in oily fish (salmon, mackerel, tuna).[75] A high intake of LA associated with a low intake of ALA leads to the accumulation of n-6 PUFA, including AA. Dietary recommendations state a ratio LA/ALA close to 4–5 and a ~500 mg/day supply in EPA and DHA sufficient to meet all the needs of the body of DHA and to protect against cardiovascular disease.[76,77] Indeed, although mammals have the necessary enzymes to make the long-chain PUFA from the precursors, less than 1% LA, ~5% ALA, and less than 0.5% ALA are converted to AA, EPA, and DHA, respectively.[78,79] Hence, a supply in EPA and DHA is needed.

Natural n-3 PUFA seem to be a promising therapy since they exert many anti-inflammatory effects as described in the previous paragraph. In fact, clinical studies on PUFA and dry eye consist of, on one hand, epidemiologic studies that focused on the relationship between food intake and dry eye (with a particular focus on the n-3 PUFA) and, on the other hand, interventional studies that primarily evaluated the efficacy of n-6 PUFA in forms of GLA and LA.

N-3 Polyunsaturated Fatty Acids Deficiency: Consequences on Dry Eye

Observational studies that have examined the relationship between PUFA and dry eye syndrome emphasized the negative impact of a low dietary intake of n-3 PUFA and an imbalance between n-6 PUFA and n-3 PUFA on dry eye syndrome.

The study of Cermak et al. showed that dietary n-3 PUFA, especially EPA and DHA, are lower in female patients with Sjögren syndrome in comparison with healthy individuals.[80] The Women's Health Study showed that the prevalence of dry eye is negatively correlated with the dietary n-3 PUFA level and tuna consumption and positively correlated with the ratio of dietary n-6 PUFA to n-3 PUFA (68% lower prevalence of dry eye in women eating more than five or six tuna servings per week as compared with less than one).[81] The study of Oxholm et al. showed that plasma levels of dGLA, EPA, and DHA, as well as the ratio of n-3 PUFA to n-6 PUFA, are inversely correlated with the severity of the damage of the exocrine glands in patients with Sjögren syndrome.[82] The same study showed that the presence of circulating autoantibodies is positively correlated with AA rate and negatively with the ratio of n-3 PUFA to n-6 PUFA.

In addition, lipid profile of meibomius secretions in patients with Sjögren syndrome with a low dietary EPA and DHA level is different from that of patients with a high EPA and DHA diet.[83]

Thus, these results highlighted that a deficiency in dietary n-3 PUFA seems to have an impact on the prevalence of dry eye syndrome, as well as on the severity of the disease in the case of Sjögren syndrome. Moreover, these studies pointed out the importance of the dietary balance between n-6 and n-3 PUFA.

To our knowledge, we were the only group to perform a study aiming at evaluating the impact of a severe n-3 PUFA deficiency on the severity of dry eye in an animal model. We found restricted effects of an n-3 PUFA-deficient diet on clinical and biochemical parameters in a rat model of moderate dry eye (Fig. 18.5; Fig. 18.6).[84] Hence, we hypothesize that a n-3 PUFA deficiency had minor effects on the severity of the dry eye but instead had a greater effect on its incidence as reported above.

N-3 Polyunsaturated Fatty Acids Supplementation and Dry Eye

The benefits of PUFA from the n-6 and n-3 series were documented in dry eye syndrome, which results in inflammation of the components of the lacrimal functional unit.[85] Interventional studies have suggested the potential benefit of dietary n-6 or n-3 PUFA alone or a combination of both on the clinical signs of ocular dryness.[86–90] Two studies were conducted in vivo in mice and rats. The study of Rashid and collaborators (2008) on mice showed that topical application of ALA led to a significant reduction of dry eye signs and inflammatory changes.[91] Corneal epithelial damage as evaluated by fluorescein staining was reversed. This effect could be mediated by NPD1 since NPD1 has been reported in the murine cornea.[92] The corneal and conjunctival expression of IL-1α and TNF-α, which are important mediators of dry eye syndrome, decreased following the application of ALA. These results are encouraging. However, more studies are needed to describe how a topical application of a fatty acid can influence the

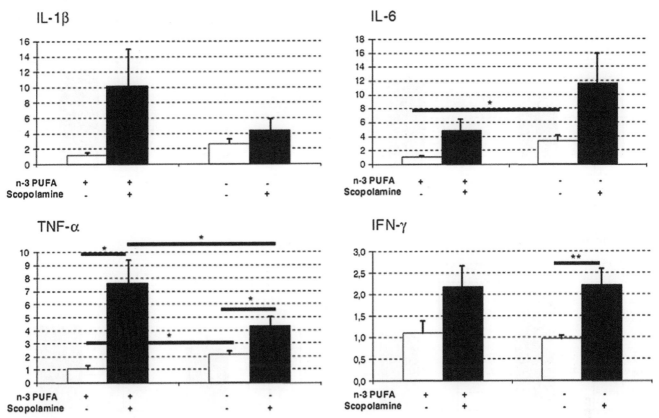

FIGURE 18.5 Expression of cytokines in the exorbital lacrimal glands of rats fed with the balanced diet or the n-3 polyunsaturated fatty acids (PUFA)-deficient diet for three generations and treated or not with continuous scopolamine (12.5 mg/day) for 10 days. The induction of dry eye induced a significant increase in the transcript levels of tumor necrosis factor (TNF)-α (P = 0.03) in the exorbital lacrimal gland of the balanced diet rats and of TNF-α (P = 0.03) and interferon (IFN)-γ (P = 0.009) in the exorbital lacrimal glands of the n-3 PUFA-deficient rats. The increase of the transcript levels of interleukin (IL)-1β (P = 0.08) in the exorbital lacrimal glands of the balanced diet rats and of IL-6 (P = 0.08) in the exorbital lacrimal glands of the n-3 PUFA-deficient rats failed to reach the threshold of significance (P = 0.08). Results were given as relative level of transcript (means ± standard error). In the balanced diet group, n = 3 and n = 4 for control and dry eye (DE) animals, respectively. In the n-3 PUFA-deficient group, n = 5 for both. * P < 0.05; ** P < 0.01. *Reprinted from Viau* et al. *(2011),[84] with permission.*

inflammation. Our study on rats first demonstrated the efficacy of n-3 PUFA in corneal alterations.[93] The n-3 PUFA-enriched diet improved the clinical signs of dry eye, although at a moderate level, prevented the occurrence of corneal keratitis, and partially protected from a mucin decrease.

The rationale of the potential benefit of n-3 PUFA supplementation in dry eye syndrome is supported by the few data available on the incorporation of PUFA in the lacrimal functional unit and especially in the lacrimal glands and conjunctival cells as target tissues (Table 18.1).

Data in lacrimal glands showed that dietary supplementation in EPA and DHA induced an increase in n-6 and/or n-3 PUFA in rats.[93,94] Concerning the conjunctiva, one *in vitro* study demonstrated the possibility to modulate the precursors of PG, dGLA, AA, and EPA in conjunctional epithelial cells (Table 18.1). The supplementations with GLA and/or EPA in the culture medium induced the incorporation of these fatty acids and their metabolites in the neutral lipids and phospholipids of

the conjunctival cells.[95] This result demonstrated the possibility to modulate the content of the precursors of PG, dGLA, AA, and EPA in conjunctival epithelial cells. It is particularly interesting because conjunctiva has been described to have the ability to synthesize PGE_2 and PGE_3.[96] It was also shown that the composition of tear lipids changed with the lipidic dietary intake.[83] It seems that a nutritional modulation of the lipid mediator synthesis is possible.

Dietary PUFA are of great interest since they can modulate the composition of the cell membrane due to their incorporation into a variety of tissues including the lacrimal gland, the retina, and the brain.[93,94,97,98] PG synthesis is affected by altering the fatty acid composition of the cell membrane and can be modulated by manipulation of PUFA intake. This was already shown in several tissues including the retina[99] and lacrimal glands,[93] but also in tears.[86] However, neither in the lacrimal glands *in vivo* nor in the conjunctival cells *in vitro* was the PG content associated to the n-3 PUFA amount present in the diet or in the culture medium. This means that the

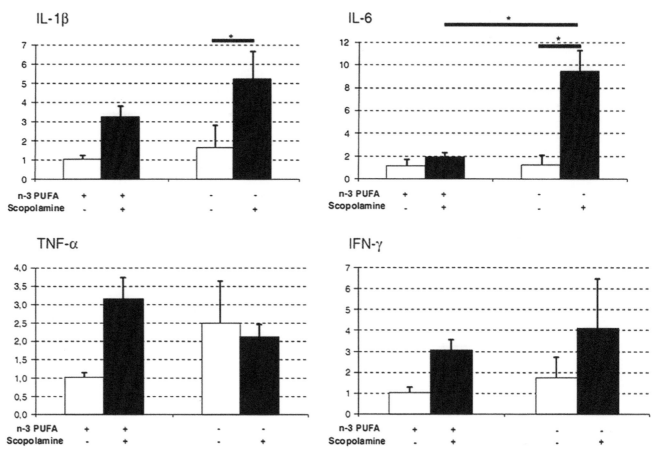

FIGURE 18.6 Expression of cytokines in the conjunctiva of rats fed with the balanced diet or the n-3 polyunsaturated fatty acids (PUFA)-deficient diet for three generations and treated or not with continuous scopolamine (12.5 mg/day) for 10 days. The n-3 PUFA-deficient diet did not increase the levels of the proinflammatory cytokine transcripts as compared to the balanced diet in control animals. The induction of dry eye in the balanced diet rats did not significantly modify the transcript levels of inflammatory cytokines in the conjunctiva. On the contrary, it significantly increased the expression of interleukin (IL)-1β (P = 0.05) and IL-6 (P = 0.014) in the conjunctiva of n-3 PUFA-deficient rats. The n-3 PUFA-deficient diet did not affect the severity of dry eye. Results were given as relative level of transcript (means ± standard error). In the balanced diet group, n = 3 and n = 4 for control and dry eye (DE) animals, respectively. In the n-3 PUFA-deficient group, n = 5 for both. IFN: interferon; TNF: tumor necrosis factor. *P < 0.05; **P < 0.01. *Reprinted from Viau et al. (2011),[84] with permission.*

benefits observed cannot be restricted to the production of fatty acid-derived PGs. Other anti-inflammatory actions of dietary PUFA may be suggested, such as the inhibition by EPA of the proinflammatory transcription factor NF-κB, or the activation by EPA and DHA of the PPAR, key elements in the regulation of the inflammatory response, or the involvement of Rv and NPD1 synthesized from EPA and DHA.

Even if the benefits of dietary n-3 PUFA were moderate, PUFA may influence ocular surface alterations and then can be used in combination with other therapies. In humans, therapeutic strategies, including nutritional components such as PUFA, usually consist of associating dietary recommendations or dietary supplements with classical therapies such as artificial tears and/or eyelid care. In this context, PUFA intake is interesting in the prevention and/or treatment of inflammation in dry eye syndrome. In practice, such contributions could help to

limit the use of heavier treatments, often poorly tolerated and responsible of side effects.

TAKE-HOME MESSAGES

- Dry eye disease is characterized by inflammation of the ocular surface.
- PUFA are promising molecules due to their anti-inflammatory effect via the synthesis of lipid mediators or the modulation of gene expression.
- The imbalance between dietary n-6 and n-3 PUFA led to lower n-3 PUFA consumption, which has a negative impact on dry eye, mainly on its incidence rather than the severity.
- n-3 PUFA supplementation improved the symptoms of dry eye.

TABLE 18.1 Modification in the Fatty Acid Content of the Phospholipids of the Exorbital Lacrimal Glands of Animals after Induction of Dry Eye and of the Chang Conjunctival Cells Treated with Interferon (IFN)-γ (% of Total Fatty acids)

	In Vivo Lacrimal Glands of Animals					In Vitro Chang Conjunctival Cells				
	C/Balanced	DE/Balanced	DE/GLA	DE/EPA+DHA	DE/GLA+EPA+DHA	C	IFN	IFN/GLA25	IFN/EPA25	IFN/GLA25+EPA25
dGLA	3.3c	3.2c	4.8a	1.6d	4.4b	0.2c	0.2c	1.7b	0.2c	4.7a
AA	18.6c	22.5b	27.1a	9.5e	14.4d	3.8b	3.7b	11.7a	2.9c	11.3a
EPA	0.4c	0.3cd	0.1d	5.6a	3.2b	0.5c	0.4c	0.04d	6.4a	5.4b
DHA	4.6c	4.6c	2.6d	5.3b	5.8a	1.7a	1.6a	1.0b	0.6c	0.5c
n-6/n-3	5.8c	6.8b	14.2a	1.8e	2.8d	1.9b	2.0b	11.4a	0.3d	1.5c

Animals were randomly distributed into four diet groups: control, GLA (n-6), EPA+DHA (n-3), and GLA+EPA+DHA (n-6+n-3). After 2 months of feeding with the specific diets, dry eye was induced with scopolamine treatment (12.5 mg/day). The treatment lasted 28 days, during which the specific diets were maintained. Chang conjunctival cells were incubated with 25 µg/mL GLA and/or 25 µg/mL EPA for 72 hours and then stimulated with IFN-gamma (300 µ/mL) for 48 hours. Balanced: balanced diet; C: control (no induction of dry eye or no stimulation); DE: induction of dry eye; DHA: docosahexaenoic acid; EPA: eicosapentaenoic acid; EPA+DHA: diet containing 10% EPA + 5% DHA; EPA25: supplementation of the culture medium with 25 µg/mL EPA; GLA: γ-linolenic acid; GLA: diet containing 10% GLA; GLA+EPA+DHA: diet containing 10% GLA + 10% EPA + 5% DHA; GLA25: supplementation of the culture medium with 25 µg/mL GLA; GLA25+EPA25: supplementation of the culture medium with 25 µg/mL GLA and 25 µg EPA; IFN: interferon.
Adapted from Viau et al. (2009)[84] and Viau et al. (2012).[95]

- n-3 PUFA supplementation induced the incorporation of PUFA at least in conjunctiva and lacrimal glands, showing that a nutritional modulation of the lipid mediators is possible.

References

1. Miljanovic B, Dana R, Sullivan DA, et al. Impact of dry eye syndrome on vision-related quality of life. *Am J Ophthalmol* 2007;**143**:409–15.
2. Simopoulos AP. The importance of the ratio of omega-6/omega-3 essential fatty acids. *Biomed Pharmacother* 2002;**56**:365–79.
3. Lemp MA. Report of the National Eye Institute/Industry workshop on clinical trials in dry eyes. *Clao J* 1995;**21**:221–32.
4. Lemp MA, Baudouin C, Baum J, et al. The definition and classification of dry eye disease: Report of the Definition and Classification Subcommittee of the International Dry Eye WorkShop (2007). *Ocul Surf.* 2007;(5):75–92.
5. Baudouin C. A new approach for better comprehension of diseases of the ocular surface. *J Fr Ophtalmol* 2007;**30**:239–46.
6. Schaumberg DA, Sullivan DA, Dana MR. Epidemiology of dry eye syndrome. *Adv Exp Med Biol* 2002;**506**:989–98.
7. Smith JA, Albeitz J, Begley C, et al. The epidemiology of dry eye disease: Report of the Epidemiology Subcommittee of the International Dry Eye WorkShop. *Ocul Surf.* 2007;**2007**(5):93–107.
8. Schein OD, Miunoz B, Tielsch JM, et al. Prevalence of dry eye among the elderly. *Am J Ophthalmol* 2007;**124**:723–8.
9. McCarty CA, Bansal AK, Livingston PM, et al. The epidemiology of dry eye in Melbourne, Australia. *Ophthalmology* 1998;**105**:1114–9.
10. Moss SE, Klein R, Klein BE. Prevalence of and risk factors for dry eye syndrome. *Arch Ophthalmol* 2000;**118**:1264–8.
11. Schaumberg DA, Sullivan DA, Buring JE, et al. Prevalence of dry eye syndrome among US women. *Am J Ophthalmol* 2003;**136**:318–26.
12. Schaumberg DA, Dana R, Buring JE, et al. Prevalence of dry eye disease among US men: estimates from the Physicians' Health Studies. *Arch Ophthalmol* 2009;**127**:763–8.
13. Driver PJ, Lemp MA. Meibomian gland dysfunction. *Surv Ophthalmol* 1996;**40**:343–67.
14. Mathers WD, Lane JA. Meibomian gland lipids, evaporation, and tear film stability. In: Sullivan DA, Dartt DA, Meneray MA, editors. *Lacrimal Gland, Tear Film, and Dry Eye Syndromes 2.* New York: Plenum Press; 1998. pp. 349–60.
15. Dougherty JM, Osgood JK, McCulley JP. The role of wax and sterol ester fatty acids in chronic blepharitis. *Invest Ophthalmol Vis Sci* 1991;**32**:1932–7.
16. Shine WE, McCulley JP. Association of meibum oleic acid with meibomian seborrhea. *Cornea* 2000;**19**:72–4.
17. Joffre C, Souchier M, Gregoire S, et al. Differences in meibomian fatty acid composition in patients with meibomian gland dysfunction and aqueous-deficient dry eye. *Br J Ophthalmol* 2008;**92**:116–9.
18. Barabino S, Chen Y, Chauhan S, et al. Ocular surface immunity: homeostatic mechanisms and their disruption in dry eye disease. *Prog Retin Eye Res* 2012;**31**:271–85.
19. De Paiva CS, Corrales RM, Villarreal AL, et al. Apical corneal barrier disruption in experimental murine dry eye is abrogated by methylprednisolone and doxycycline. *Invest Ophthalmol Vis Sci* 2006;**47**:2847–56.
20. Viau S, Maire MA, Pasquis B, et al. Time course of ocular surface and lacrimal gland changes in a new scopolamine-induced dry eye model. *Graefes Arch Clin Exp Ophthalmol* 2008;**246**:857–67.
21. Corrales RM, Villarreal A, Farley W, et al. Strain-related cytokine profiles on the murine ocular surface in response to desiccating stress. *Cornea* 2007;**26**:579–84.
22. Pflugfelder SC, Jones D, Ji Z, et al. Altered cytokine balance in the tear fluid and conjunctiva of patients with Sjogren's syndrome keratoconjunctivitis sicca. *Curr Eye Res* 1999;**19**:201–11.
23. Tsubota K, Fukagawa K, Fujihara T, et al. Regulation of human leukocyte antigen expression in human conjunctival epithelium. *Invest Ophthalmol Vis Sci* 1999;**40**:28–34.
24. De Saint Jean M, Brignole F, Feldmann G, et al. Interferon-gamma induces apoptosis and expression of inflammation-related proteins in Chang conjunctival cells. *Invest Ophthalmol Vis Sci* 1999;**40**:2199–212.
25. Malvy D, Preziosi P, Galan P, et al. La consommation de lipides en France: contribution à l'équilibre nutritionnel (données préliminaires de l'étude SU.VI.Max. *Oleagineux, Corps gras, Lipides* 1999;**6**:21–5.

26. Simopoulos AP. Evolutionary aspects of diet: the omega-6/omega-3 ratio and the brain. *Mol Neurobiol* 2011;**44**:203–15.
27. Calder PC. n-3 polyunsaturated fatty acids, inflammation, and inflammatory diseases. *Am J Clin Nutr* 2006;**83**:1505S–19S.
28. Hughes DA, Southon S, Pinder AC. (n-3) Polyunsaturated fatty acids modulate the expression of functionally associated molecules on human monocytes *in vitro*. *J Nutr* 1996;**126**:603–10.
29. Zhao Y, Joshi-Barve S, Barve S, et al. Eicosapentaenoic acid prevents LPS-induced TNF-alpha expression by preventing NF-kappaB activation. *J Am Coll Nutr* 2004;**23**:71–8.
30. Koto T, Nagai N, Mochimaru H, et al. Eicosapentaenoic acid is anti-inflammatory in preventing choroidal neovascularization in mice. *Invest Ophthalmol Vis Sci* 2007;**48**:4328–34.
31. Zurier RB. Prostaglandins. Their potential in clinical medicine. *Postgrad Med* 1980;**68**:70–81.
32. Santoli D, Phillips PD, Colt TL, et al. Suppression of interleukin 2-dependent human T cell growth *in vitro* by prostaglandin E (PGE) and their precursor fatty acids. Evidence for a PGE-independent mechanism of inhibition by the fatty acids. *J Clin Invest* 1990;**85**:424–32.
33. Purasiri P, McKechnie A, Heys SD, et al. Modulation *in vitro* of human natural cytotoxicity, lymphocyte proliferative response to mitogens and cytokine production by essential fatty acids. *Immunology* 1997;**92**:166–72.
34. Lo CJ, Chiu KC, Fu M, et al. Fish oil decreases macrophage tumor necrosis factor gene transcription by altering the NF kappa B activity. *J Surg Res* 1999;**82**:216–21.
35. Miles EA, Wallace FA, Calder PC. Dietary fish oil reduces intercellular adhesion molecule 1 and scavenger receptor expression on murine macrophages. *Atherosclerosis* 2000;**152**:43–50.
36. Espersen GT, Grunnet N, Lervang HH, et al. Decreased interleukin-1 beta levels in plasma from rheumatoid arthritis patients after dietary supplementation with n-3 polyunsaturated fatty acids. *Clin Rheumatol* 1992;**11**:393–5.
37. Caughey GE, Mantzioris E, Gibson RA, et al. The effect on human tumor necrosis factor alpha and interleukin 1 beta production of diets enriched in n-3 fatty acids from vegetable oil or fish oil. *Am J Clin Nutr* 1996;**63**:116–22.
38. Hughes DA, Pinder AC, Piper Z, et al. Fish oil supplementation inhibits the expression of major histocompatibility complex class II molecules and adhesion molecules on human monocytes. *Am J Clin Nutr* 1996;**63**:267–72.
39. Miles EA, Thies F, Wallace FA, et al. Influence of age and dietary fish oil on plasma soluble adhesion molecule concentrations. *Clin Sci Lond* 2001;**100**:91–100.
40. Kelley DS, Taylor PC, Nelson GJ, et al. Docosahexaenoic acid ingestion inhibits natural killer cell activity and production of inflammatory mediators in young healthy men. *Lipids* 1999;**34**:317–24.
41. Arita M. Mediator lipidomics in acute inflammation and resolution. *J Biochem* 2012;**152**:313–9.
42. de Roos B, Mavrommatis Y, Brouwer IA. Long-chain n-3 polyunsaturated fatty acids: new insights into mechanisms relating to inflammation and coronary heart disease. *Br J Pharmacol* 2009;**158**:413–28.
43. Schmitz G, Ecker J. The opposing effects of n-3 and n-6 fatty acids. *Prog Lipid Res* 2008;**47**:147–55.
44. Bannenberg G, Serhan CN. Specialized pro-resolving lipid mediators in the inflammatory response: an update. *Biochim Biophys Acta* 2010;**1801**:1260–73.
45. Calder PC. Dietary modification of inflammation with lipids. *Proc Nutr Soc* 2002;**61**:345–58.
46. Bagga D, Wang L, Farias-Eisner R, et al. Differential effects of prostaglandin derived from omega-6 and omega-3 polyunsaturated fatty acids on COX-2 expression and IL-6 secretion. *Proc Natl Acad Sci U S A* 2003;**100**:1751–6.
47. Needleman P, Raz A, Minkes MS, et al. Triene prostaglandins: prostacyclin and thromboxane biosynthesis and unique biological properties. *Proc Natl Acad Sci U S A* 1979;**76**:944–8.
48. Obata T, Nagakura T, Masaki T, et al. Eicosapentaenoic acid inhibits prostaglandin D2 generation by inhibiting cyclo-oxygenase-2 in cultured human mast cells. *Clin Exp Allergy* 1999;**29**:1129–35.
49. Sperling RI, Benincaso AI, Knoell CT, et al. Dietary omega-3 polyunsaturated fatty acids inhibit phosphoinositide formation and chemotaxis in neutrophils. *J Clin Invest* 1993;**91**:651–60.
50. Arita M, Yoshida M, Hong S, et al. Resolvin E1, an endogenous lipid mediator derived from omega-3 eicosapentaenoic acid, protects against 2,4,6-trinitrobenzene sulfonic acid-induced colitis. *Proc Natl Acad Sci U S A* 2005;**102**:7671–6.
51. Schwab JM, Chiang N, Arita M, et al. Resolvin E1 and protectin D1 activate inflammation-resolution programmes. *Nature* 2007;**447**:869–74.
52. Fredman G, Serhan CN. Specialized proresolving mediator targets for RvE1 and RvD1 in peripheral blood and mechanisms of resolution. *Biochem J* 2011;**437**:185–97.
53. Zhang MJ, Spite M. Resolvins: anti-inflammatory and proresolving mediators derived from omega-3 polyunsaturated fatty acids. *Annu Rev Nutr* 2012;**32**:203–27.
54. Hong S, Gronert K, Devchand PR, et al. Novel docosatrienes and 17S-resolvins generated from docosahexaenoic acid in murine brain, human blood, and glial cells. Autacoids in anti-inflammation. *J Biol Chem* 2003;**278**:14677–87.
55. Marcheselli VL, Hong S, Lukiw WJ, et al. Novel docosanoids inhibit brain ischemia-reperfusion-mediated leukocyte infiltration and pro-inflammatory gene expression. *J Biol Chem* 2003;**278**:43807–17.
56. Bazan NG, Musto AE, Knott EJ. Endogenous signaling by omega-3 docosahexaenoic acid-derived mediators sustains homeostatic synaptic and circuitry integrity. *Mol Neurobiol* 2011;**44**:216–22.
57. Lukiw WJ, Cui JG, Marcheselli VL, et al. A role for docosahexaenoic acid-derived neuroprotectin D1 in neural cell survival and Alzheimer disease. *J Clin Invest* 2005;**115**:2774–83.
58. Bazan NG, Calandria JM, Serhan CN. Rescue and repair during photoreceptor cell renewal mediated by docosahexaenoic acid-derived neuroprotectin D1. *J Lipid Res* 2010;**51**:2018–31.
59. Bazan NG, Molina MF, Gordon WC. Docosahexaenoic acid signalolipidomics in nutrition: significance in aging, neuroinflammation, macular degeneration, Alzheimer's, and other neurodegenerative diseases. *Annu Rev Nutr* 2011;**31**:321–51.
60. Clark RB. The role of PPARs in inflammation and immunity. *J Leukoc Biol* 2002;**71**:388–400.
61. Chinetti G, Griglio S, Antonucci M, et al. Activation of proliferator-activated receptors alpha and gamma induces apoptosis of human monocyte-derived macrophages. *J Biol Chem* 1998;**273**:25573–80.
62. Shu H, Wong B, Zhou G, et al. Activation of PPARalpha or gamma reduces secretion of matrix metalloproteinase 9 but not interleukin 8 from human monocytic THP-1 cells. *Biochem Biophys Res Commun* 2000;**267**:345–9.
63. Marx N, Bourcier T, Sukhova GK, et al. PPARgamma activation in human endothelial cells increases plasminogen activator inhibitor type-1 expression: PPARgamma as a potential mediator in vascular disease. *Arterioscler Thromb Vasc Biol* 1999;**19**:546–51.
64. Devchand PR, Keller H, Peters JM, et al. The PPARalpha-leukotriene B4 pathway to inflammation control. *Nature* 1996;**384**:39–43.
65. Colville-Nash PR, Qureshi SS, Willis D, et al. Inhibition of inducible nitric oxide synthase by peroxisome proliferator-activated receptor agonists: correlation with induction of heme oxygenase 1. *J Immunol* 1998;**161**:978–84.
66. Hill MR, Clarke S, Rodgers K, et al. Effect of peroxisome proliferator-activated receptor alpha activators on tumor necrosis factor expression in mice during endotoxemia. *Infect Immun* 1999;**67**:3488–93.

67. Staels B, Koenig W, Habib A, et al. Activation of human aortic smooth-muscle cells is inhibited by PPARalpha but not by PPAR-gamma activators. *Nature* 1998;**393**:790–3.

68. Delerive P, De Bosscher K, Besnard S, et al. Peroxisome proliferator-activated receptor alpha negatively regulates the vascular inflammatory gene response by negative cross-talk with transcription factors NF-kappaB and AP-1. *J Biol Chem* 1999;**274**:32048–54.

69. Forman BM, Chen J, Evans RM. Hypolipidemic drugs, polyunsaturated fatty acids, and eicosanoids are ligands for peroxisome proliferator-activated receptors alpha and delta. *Proc Natl Acad Sci U S A* 1997;**94**:4312–7.

70. Kliewer SA, Sundseth SS, Jones SA, et al. Fatty acids and eicosanoids regulate gene expression through direct interactions with peroxisome proliferator-activated receptors alpha and gamma. *Proc Natl Acad Sci U S A* 1997;**94**:4318–23.

71. Jiang C, Ting AT, Seed B. PPAR-gamma agonists inhibit production of monocyte inflammatory cytokines. *Nature* 1998;**391**:82–6.

72. Ricote M, Li AC, Willson TM, et al. The peroxisome proliferator-activated receptor-gamma is a negative regulator of macrophage activation. *Nature* 1998;**391**:79–82.

73. Christman JW, Lancaster LH, Blackwell TS. Nuclear factor kappa B: a pivotal role in the systemic inflammatory response syndrome and new target for therapy. *Intensive Care Med* 1998;**24**:1131–8.

74. Poynter ME, Daynes RA. Peroxisome proliferator-activated receptor alpha activation modulates cellular redox status, represses nuclear factor-kappaB signaling, and reduces inflammatory cytokine production in aging. *J Biol Chem* 1998;**273**:32833–41.

75. James MJ, Gibson RA, Cleland LG. Dietary polyunsaturated fatty acids and inflammatory mediator production. *Am J Clin Nutr* 2000;**71**:343S–8S.

76. Burdge G. Alpha-linolenic acid metabolism in men and women: nutritional and biological implications. *Curr Opin Clin Nutr Metab Care* 2004;**7**:137–44.

77. Lucas M, Asselin G, Merette C, et al. Validation of an FFQ for evaluation of EPA and DHA intake. *Public Health Nutr* 2009;**12**:1783–90.

78. Plourde M, Cunnane SC. Extremely limited synthesis of long chain polyunsaturates in adults: implications for their dietary essentiality and use as supplements. *Appl Physiol Nutr Metab* 2007;**32**:619–34.

79. Cunnane SC, Guesnet P. Linoleic acid recommendations – a house of cards. *Prostaglandins Leukot Essent Fatty Acids* 2011;**85**:399–402.

80. Cermak JM, Papas AS, Sullivan RM, et al. Nutrient intake in women with primary and secondary Sjogren's syndrome. *Eur J Clin Nutr* 2003;**57**:328–34.

81. Miljanovic B, Trivedi KA, Dana MR, et al. Relation between dietary n-3 and n-6 fatty acids and clinically diagnosed dry eye syndrome in women. *Am J Clin Nutr* 2005;**82**:887–93.

82. Oxholm P, Asmussen K, Wiik A, et al. Essential fatty acid status in cell membranes and plasma of patients with primary Sjogren's syndrome. Correlations to clinical and immunologic variables using a new model for classification and assessment of disease manifestations. *Prostaglandins Leukot Essent Fatty Acids* 1998;**59**:239–45.

83. Sullivan BD, Cermak JM, Sullivan RM, et al. Correlations between nutrient intake and the polar lipid profiles of meibomian gland secretions in women with Sjogren's syndrome. *Adv Exp Med Biol* 2002;**506**:441–7.

84. Viau S, Pasquis B, Maire MA, et al. No consequences of dietary n-3 polyunsaturated fatty acid deficiency on the severity of scopolamine-induced dry eye. *Graefes Arch Clin Exp Ophthalmol* 2011;**249**:547–57.

85. Stern ME, Pflugfelder SC. Inflammation in dry eye. *Ocul Surf* 2004;**2**:124–30.

86. Aragona P, Bucolo C, Spinella R, et al. Systemic omega-6 essential fatty acid treatment and pge1 tear content in Sjogren's syndrome patients. *Invest Ophthalmol Vis Sci* 2005;**46**:4474–9.

87. Barabino S, Rolando M, Camicione P, et al. Systemic linoleic and gamma-linolenic acid therapy in dry eye syndrome with an inflammatory component. *Cornea* 2003;**22**:97–101.

88. Creuzot-Garcher C. Lacrimal film and the ocular surface. *J Fr Ophtalmol* 2006;**29**:1053–9.

89. Macri A, Giuffrida S, Amico V, et al. Effect of linoleic acid and gamma-linolenic acid on tear production, tear clearance and on the ocular surface after photorefractive keratectomy. *Graefes Arch Clin Exp Ophthalmol* 2003;**241**:561–6.

90. Wojtowicz JC, Butovich I, Uchiyama E, et al. Pilot, prospective, randomized, double-masked, placebo-controlled clinical trial of an omega-3 supplement for dry eye. *Cornea* 2011;**30**:308–14.

91. Rashid S, Jin Y, Ecoiffier T, et al. Topical omega-3 and omega-6 fatty acids for treatment of dry eye. *Arch Ophthalmol* 2008;**126**:219–25.

92. Gronert K. Lipoxins in the eye and their role in wound healing. *Prostaglandins Leukot Essent Fatty Acids* 2005;**73**:221–9.

93. Viau S, Maire MA, Pasquis B, et al. Efficacy of a 2-month dietary supplementation with polyunsaturated fatty acids in dry eye induced by scopolamine in a rat model. *Graefes Arch Clin Exp Ophthalmol* 2009;**247**:1039–50.

94. Schnebelen C, Viau S, Gregoire S, et al. Nutrition for the eye: different susceptibility of the retina and the lacrimal gland to dietary omega-6 and omega-3 polyunsaturated fatty acid incorporation. *Ophthalmic Res* 2009;**41**:216–24.

95. Viau S, Leclere L, Buteau B, et al. Polyunsaturated fatty acids induce modification in the lipid composition and the prostaglandin production of the conjunctival epithelium cells. *Graefes Arch Clin Exp Ophthalmol* 2012;**250**:211–22.

96. Kulkarni PS, Srinivasan BD. Cyclooxygenase and lipoxygenase pathways in anterior uvea and conjunctiva. *Prog Clin Biol Res* 1989;**312**:39–52.

97. Schnebelen C, Pasquis B, Salinas-Navarro M, et al. A dietary combination of omega-3 and omega-6 polyunsaturated fatty acids is more efficient than single supplementations in the prevention of retinal damage induced by elevation of intraocular pressure in rats. *Graefes Arch Clin Exp Ophthalmol* 2009;**247**:1191–203.

98. Bourre JM. Effects of nutrients (in food) on the structure and function of the nervous system: update on dietary requirements for brain. Part 2: macronutrients. *J Nutr Health Aging* 2006;**10**:386–99.

99. Schnebelen C, Gregoire S, Pasquis B, et al. Dietary n-3 and n-6 PUFA enhance DHA incorporation in retinal phospholipids without affecting PGE(1) and PGE (2) levels. *Lipids* 2009;**44**:465–70.

OBESITY, METABOLIC SYNDROME, AND DIABETES

Metabolic Syndrome and Cataract

Charumathi Sabanayagam[1,2,3], Ching-Yu Cheng[1,2], Tien Y. Wong[1,2]

[1]Singapore Eye Research Institute, Singapore, [2]Yong Loo Lin School of Medicine, National University of Singapore, [3]Office of Clinical Sciences, Duke-NUS Graduate Medical School, Singapore

INTRODUCTION

Metabolic syndrome, characterized by a cluster of metabolic abnormalities including hyperglycemia, hypertension, dyslipidemia, and obesity,[1] is an emerging public health problem affecting 100 million people worldwide. The prevalence of metabolic syndrome is expected to increase further, contributed by the rising prevalence of obesity[2] associated with changing dietary behavior and sedentary lifestyle. Metabolic syndrome increases the risk of cardiovascular disease by two-fold[3] and the risk of type 2 diabetes by up to five-fold.[4] Further, clustering of multiple components of metabolic syndrome confers greater risk than the sum of the risks associated with each individual risk factor.[5] Cataract is the leading cause of blindness in the elderly population, accounting for an estimated 17 million cases of blindness.[6] Age-related cataract has been shown to be related to higher risk of mortality and decreased rate of survival in population-based studies.[7,8] Although surgical extraction of the cataract is a well-recognized and effective treatment for cataract, in many places and countries surgical resources are not readily available. It has been shown that a 50% reduction in cataract surgery could be achieved by delaying the onset of cataract by 10 years.[9] Experimental evidence shows that control of risk factors with therapeutic interventions[10,11] or lifestyle modifications[12] potentially delay the onset of cataract.

There is some evidence from several clinical and epidemiologic studies suggesting that metabolic syndrome[13–17] and its components are associated with age-related cataract[18–28]. In this chapter, we summarize current evidence supporting an association between metabolic syndrome and age-related cataract from observational studies and current developments in understanding the pathophysiologic mechanisms underlying this association.

DEFINITIONS OF THE METABOLIC SYNDROME

Metabolic syndrome is known by different names, such as the syndrome X,[29] insulin resistance syndrome, dysmetabolic syndrome, cardiometabolic syndrome, and the deadly quartet.[30] The concept of the metabolic syndrome originated several decades ago. In 1923, Kylin, a Swedish physician, first described the clustering of hypertension, hyperglycemia, and gout as a syndrome.[31] Later in 1947, the Marseilles physician Dr. Jean Vague observed that visceral obesity is associated with type 2 diabetes and cardiovascular disease.[32] In 1988, Dr. Gerald Reaven, in his landmark Banting lecture, first introduced the term 'syndrome X' to describe the clustering of metabolic abnormalities centering on insulin resistance.[29] Since then, many different definitions of metabolic syndrome have been proposed. The first formal definition was proposed by the World Health Organization (WHO) in 1998.[33] This one emphasized insulin resistance and required insulin resistance as one of the main criteria apart from additional criteria including obesity, hypertension, dyslipidemia, or microalbuminuria. In 2001, the National Cholesterol Education Program Adult Treatment Panel III (NCEP-ATP III) published a new set of criteria based on common clinical measurements, and it did not require insulin resistance as a necessary criterion.[34] In 2006, the International Diabetes Federation (IDF) proposed a new definition placing a greater focus on central obesity defined by ethnicity-specific cutoffs.[35] All these definitions agree on the core components including hyperglycemia, obesity, dyslipidemia, and hypertension but differ in the cutoffs and criteria (Table 19.1). Currently, the NCEP-ATP III and the IDF criteria are largely used in epidemiologic studies.

Handbook of Nutrition, Diet, and the Eye
http://dx.doi.org/10.1016/B978-0-12-401717-7.00019-8

TABLE 19.1   Definitions of Metabolic Syndrome

WHO, 1998[33]

Diabetes, or impaired fasting glycemia or impaired glucose tolerance or insulin resistance, PLUS any two of the following criteria:

Blood pressure ≥140/90 mm Hg

Dyslipidemia: raised plasma triglycerides (≥ 150 mg/dL [≥ 1.7 mmol/L] or low HDL cholesterol (< 35 mg/dL [< 0.9 mmol/L] in men; < 40 mg/dL [< 1.0 mmol/L] in women)

Obesity (waist-to-hip ratio > 0.9 in men and > 0.85 in women and/or BMI > 30 kg/m²)

Microalbuminuria (urinary albumin excretion >20 μg/minute or ACR ≥ 30 mg/g)

ATP III, 2001[34]

Any three or more of the following criteria:

Abdominal obesity: waist circumference > 102 cm in men and > 88 cm in women

Raised triglycerides: ≥ 150 mg/dL (≥ 1.7 mmol/L)

Low HDL-cholesterol: < 40 mg/dL (< 1.0 mmol/L) in men; < 50 mg/dL (< 1.3 mmol/L) in women

Raised blood pressure: ≥ 130/85 mm Hg or antihypertensive medication

Fasting plasma glucose: ≥ 110 mg/dL (≥ 6.1 mmol/L)

IDF, 2006[35]

Central obesity as defined by ethnicity-specific waist circumference values, PLUS any two of the following criteria:

Raised triglycerides: ≥ 150 mg/dL (≥ 1.7 mmol/L) or specific treatment for this lipid abnormality

Low HDL-cholesterol: < 40 mg/dL (< 1.0 mmol/L) in men; < 50 mg/dL (< 1.3 mmol/L) in women or specific treatment for this lipid abnormality

Raised blood pressure: ≥ 130/85 mm Hg or antihypertensive medication

Fasting plasma glucose: ≥ 100 mg/dL (≥ 5.6 mmol/L) or previously diagnosed type 2 diabetes

ACR: albumin:creatinine ratio; ATP III: Adult Treatment Panel III; BMI: body mass index; HDL, high-density lipoprotein; IDF: International Diabetes Federation; WHO: World Health Organization.

PATHOPHYSIOLOGY OF METABOLIC SYNDROME

The core mechanisms underlying the metabolic syndrome are insulin resistance, central or visceral adiposity, and atherogenic dyslipidemia.[1,30] Insulin resistance and associated compensated hyperinsulinemia impair insulin signaling at the beta cell, resulting in the development of impaired glucose tolerance and diabetes mellitus. Central adiposity contributes to excessive free fatty acid mobilization into the liver associated

with triglyceride (TG) accumulation. In the presence of hypertriglyceridemia, TG enrichment of high-density lipoprotein (HDL)-cholesterol coupled with lipolysis of HDL by hepatic lipase leads to enhanced metabolic clearance of HDL resulting in lower levels of HDL-cholesterol. Insulin resistance and adiposity both lead to oxidative stress, endothelial dysfunction,[36] activation of renin-angiotensin system (RAS), and abnormal expression of adipocytokines including tumor necrosis factor (TNF)-α, interleukin (IL)-6, resistin, C-reactive protein, and plasminogen activator inhibitor-1 (PAI-1). In addition, excessive fat accumulation is associated with a decreased production of adiponectin,[37] an adipose-tissue-specific protein with insulin sensitizing and anti-inflammatory effects, and an increased production of leptin, an appetite-suppressing hormone associated with obesity.[38] Mechanisms including insulin resistance, inflammation, oxidative stress, and RAS activation in the metabolic syndrome induce sympathetic overactivity,[39] salt sensitivity, vasoconstriction, intravascular fluid accumulation, impaired vasodilation, and development of hypertension.[40,41]

ASSOCIATION BETWEEN METABOLIC SYNDROME, COMPONENTS OF THE METABOLIC SYNDROME, AND CATARACT

Metabolic Syndrome and Cataract

Metabolic syndrome has been shown to be associated with an increased risk of cataract in several observational studies (Table 19.2). Paunksnis *et al.* reported an association between metabolic syndrome and cataract among middle-aged European men and women.[16] In the Blue Mountains Eye Study (BMES), metabolic syndrome was associated with an increased risk of all subtypes of cataract including cortical, nuclear, and posterior subcapsular cataract (PSC) among elderly Australians.[17] In a population of Malay adults in Singapore, a significant association between metabolic syndrome and cataract was also found.[13] A dose–response relationship was also observed between an increasing number of metabolic syndrome components and cataract. Among the subtypes, cortical cataract showed a positive association with metabolic syndrome.[13] Lindblad *et al.* examined a large, population-based cohort of Swedish women who participated in the Swedish Mammography Cohort and found that a combination of three components of metabolic syndrome, including raised waist circumference, diabetes, and hypertension, increased the risk of cataract extraction by 68% compared to those without any of these components.[15] In addition, metabolic syndrome increased the risk of cataract extraction by

TABLE 19.2 Association between Metabolic Syndrome and Cataract in Observational Studies

Author, Year	Study Design	Study Population	Criteria Used	Results
Paunksnis et al. 2007[16]	Population-based cross-sectional study	1282 people ages 35–64 years in Lithuania	ATP III	Metabolic syndrome was associated with cataract in men ages 55–64 years (OR 1.59) and in women ages 45–64 years (OR 1.60)
Tan et al. 2008[17]	Population-based prospective cohort study Follow-up 10 years	3564 people ages ≥ 49 years in Australia	WHO	Metabolic syndrome was associated with cataract subtypes; OR 1.72 (95% CI 1.11 to 2.67), 2.03 (95% CI 1.11 to 3.71), and 1.66 (95% CI 1.03 to 2.67) for cortical, PSC, and nuclear cataract
Lindblad et al. 2008[15]	Population-based prospective cohort study Follow-up 8 years	35,369 women ages 49–83 years in Sweden	IDF	Metabolic syndrome was associated with cataract extraction: OR 1.68 (95% CI 1.40 to 2.03)
Galeone et al. 2010[14]	Hospital-based case-control study	761 patients with cataract extraction and 1522 controls in Italy	IDF	Metabolic syndrome was associated with cataract extraction: OR 2.11 (95% CI 1.32 to 3.37)
Sabanayagam et al. 2011[13]	Population-based cross-sectional study	2794 Malay adults ages 40–80 years in Singapore	ATP III	Metabolic syndrome was associated with cataract: OR 1.27 (95% CI 1.04 to 1.55)

ATP III: Adult Treatment Panel III; CI: confidence interval; IDF: International Diabetes Federation; OR: odds ratio; PSC: posterior subcapsular cataract; WHO: World Health Organization.

approximately three-fold among women aged less than 65 years. Galeone et al. found that metabolic syndrome was associated with a two-fold increased risk of cataract extraction in a clinic-based study in Italy.[14] Further, a significant linear trend in risk was also reported with an increasing number of metabolic syndrome components.

Individual Components of the Metabolic Syndrome and Cataract

Several studies have also assessed the association between individual metabolic syndrome components and the risk of cataract (Table 19.3; Table 19.4; Table 19.5; Table 19.6). Diabetes has long been associated with age-related cataract in several cross-sectional and prospective studies[17,25,42–48] (Table 19.3). Among the specific types of cataract, a majority of the studies have reported association of diabetes with cortical and PSC.[13,19,47,49,50] However, in the BMES[17] and in the Los Angeles Latino Eye Study (LALES),[51] diabetes showed an association with nuclear cataract. Cataract occurs at an earlier age and progresses rapidly in diabetic people compared with those without diabetes. Evidence support that even impaired glucose tolerance,[52] impaired fasting glycemia,[17] or glycosylated hemoglobin (HbA1c),[51] a measure of chronic glycemic exposure, increase the risk of cataract. In addition, studies showed poor glycemic control among those with diabetes to be an independent risk factor for cataract.[19,49,51,53,54]

Several lines of evidence suggest that obesity (Table 19.4) may be an important factor in the development of cataract. In the Physicians' Health Study (PHS), a prospective cohort study of a large cohort of physicians in

the US, higher body mass index (BMI) levels and higher waist-to-hip ratio (WHR) significantly increased the risk of cataract. Similarly, in a hospital-based case-control study in China, both overweight and obesity were associated with cataract surgery.[55] Among specific types of cataract, cortical and PSC were shown to be associated with obesity in several studies. In the Framingham Heart Study, a population-based cohort study in the US, higher BMI was associated with both cortical and PSC.[27] In the Beaver Dam Eye Study (BDES), a prospective cohort study of whites in Wisconsin, US, each unit increase in BMI increased the risk of PSC by 5% among those without diabetes and that of cortical cataract by 8% among those with diabetes.[56] In the Age-Related Eye Disease Study (AREDS), a prospective cohort study of participants enrolled in a clinical trial of antioxidant vitamins and minerals, weight change of 24 kg and above increased the risk of both cortical and PSC.[54] In the BMES, BMI increased the risk of PSC alone.[17] Evidence from cross-sectional studies including the Barbados Eye Study,[57] Shihpai Eye Study,[58] Singapore Malay Eye Study,[13] and a case-control study in Italy[48] showed an association of obesity with cortical cataract. In the Nurses' Health Study (NHS), cross-sectional evaluation showed an association of PSC with obesity measured by both BMI and waist circumference.[59]

A growing body of evidence suggests that lipid abnormalities might play a role in the development of cataract[13,16,26,60] (Table 19.5). In the Framingham cohort, elevated levels of TGs (≥ 5.52 mg/dL) increased the risk of PSC three-fold among men.[26] In one case-control study of cataract extraction among women in Italy, dyslipidemia showed a positive association with

TABLE 19.3 Association between Metabolic Syndrome Component, Diabetes/Hyperglycemia, and Cataract in Observational Studies

Author, Year	Study Design, Location	Study Population	Component	Cataract Type	Results
Ederer et al. 1981[42]	Framingham Eye Study and the HANES Cross-sectional, US		Diabetes	Any	Excess prevalence of cataract among people < 65 year old; RR 4.02 in the Framingham and 2.97 in the HANES
Karasik et al. 1984[52]	The Israel GOH Study	930 adults ages 40–70 years	IGT	Any	In women, OR 6.1 (95% CI 3.3 to 11.1, P < 0.001) for IGT; in men, no association was found
Tavani et al. 1995[48]	Hospital-based case-control study, Italy	287 women who had cataract extraction and 1227 controls	Diabetes	Cataract surgery	OR 2.2 (95% CI 1.4 to 3.4)
Klein et al. 1998[19]	Beaver Dam Eye Study Population-based prospective cohort study, US, follow-up 5 years	4926 adults ages 43–86 years	Diabetes	Cortical, nuclear, PSC	Diabetes was associated with incidence of cortical and PSC (P ≤ 0.001 for both)
Leske et al. 1999[57]	Barbados Eye Study Cross-sectional, Barbados	4341 black participants ages 40–84 years	Diabetes	Any, cortical	For all cataract, OR 2.23 (95% CI 1.63 to 3.24) for ages < 60 years and 1.63 (95% CI 1.22 to 2.17) for ages ≥ 60 years
Delcourt et al. 2000[49]	POLA Study Cross-sectional, France	2584 adults ages 60–95 years	Diabetes	Cortical, nuclear, PSC, mixed, cataract surgery	Known diabetes with duration ≥ 10 years was associated with PSC, cortical, mixed, and cataract surgery
Rowe et al. 2000[25]	Blue Mountains Eye Study, Australia Cross-sectional	3564 adults ages ≥ 49 years	Diabetes	PSC, cataract surgery	OR 1.8 (95% CI 1.0 to 3.1) for PSC and 2.5 (95% CI 1.5 to 4.2) for cataract surgery
Jacques et al. 2003[59]	Nurses' Health Study Cross-sectional, US	466 women ages 53–73 years	Diabetes	Cortical, nuclear, PSC	OR 4.1 (95% CI 1.8 to 9.4) for PSC; no significant association with cortical or nuclear cataract
Foster et al. 2003[47]	Tanjong Pagar survey Cross-sectional, Singapore	1206 Chinese adults ages 40–81 years	Diabetes	Any, cortical, nuclear, PSC Cataract surgery	OR 3.1 (95% CI 1.6 to 6.1), 2.2 (95% CI 1.2 to 4.1), and 2.3 (95% CI 1.3 to 4.1) for cortical, PSC, and cataract surgery
Hennis et al. 2004[50]	Barbados Eye Study Population-based prospective cohort study, Barbados Follow-up 4 years	2040 black participants ages 40–84 years	Diabetes HbA1c	Cortical and PSC	In diabetes, RR 2.4 (95% CI 1.8 to 3.2) for cortical and 2.9 (95% CI 1.9 to 4.5) for PSC; each unit increase in HbA1c increases the risk of cortical by 16% and PSC by 23%
Saxena et al. 2004[45]	Blue Mountains Eye Study, Australia Follow-up 5 years	2335 adults ages ≥ 49 years	Impaired fasting glycemia	Cortical	OR 2.2 (95% CI 1.1 to 4.1)
Mukesh et al. 2006[85]	The Visual Impairment Project, Australia Population-based prospective cohort study, follow-up 5 years	3721 adults ages ≥ 40 years	Diabetes	Cortical, nuclear, PSC	OR 2.9 (95% CI 1.7 to 5.1) for PSC; no significant association with nuclear or cortical
Tan et al. 2008[17]	Blue Mountains Eye Study, Australia Population-based prospective cohort study, follow-up 10 years	3564 adults ages ≥ 49 years	Diabetes IFG	Cortical, nuclear, PSC Cataract surgery	Diabetes: OR 1.57 (95% CI 1.10 to 2.24) and 1.64 (95% CI 1.02 to 2.64) for cortical and nuclear; IFG: 2.01 (95% CI 1.20 to 3.36) for cortical
Lindblad et al. 2008[15]	Swedish Mammography Cohort, Population-based prospective cohort study, follow-up 8 years	35,369 women ages 49–83 years	Diabetes	Cataract surgery	OR 1.43 (95% CI 1.10 to 1.86)

(Continued)

TABLE 19.3 Association between Metabolic Syndrome Component, Diabetes/Hyperglycemia, and Cataract in Observational Studies—cont'd

Author, Year	Study Design, Location	Study Population	Component	Cataract Type	Results
Chang et al. 2011[54]	AREDS, US Clinic-based cohort study	4425 adults ages 55–80 years	Diabetes	Cortical, nuclear, PSC, cataract surgery	OR 1.45 (95% CI 1.19 to 1.75), 1.31 (95% CI 1.04 to 1.66), and 1.71 (95% CI 1.29 to 2.27) for surgery, cortical, and PSC
Sabanayagam et al. 2011[13]	SiMES Population-based cross-sectional	2794 Malay adults ages 40–80 years	Diabetes	Any, cortical, nuclear, PSC	OR 1.77 (95% CI 1.38 to 2.25) for any; 2.28 (95% CI 1.83 to 2.83) and 1.39 (1.09 to 1.77) for cortical and PSC; no significant association with nuclear
Richter et al. 2012[51]	LALES, US Population-based prospective cohort study, follow-up 4 years	4658 adults ages ≥ 40 years	Diabetes	Nuclear, cortical PSC, mixed	OR 1.11 (95% CI 1.01 to 1.22), 2.32 (95% CI 1.58 to 3.41), and 4.73 (95% CI 2.86 to 7.82) for nuclear, cortical, and mixed cataracts; no significant association with PSC

AREDS: Age-Related Eye Disease Study; CI, confidence interval; IFG, impaired fasting glycemia; HANES: Health and Nutrition Examination Survey; HbA1c: glycosylated hemoglobin; IGT, impaired glucose tolerance; OR, odds ratio; LALES: Los Angeles Latino Eye Study; POLA: Pathologies Oculaires Liées à l'Age; PSC, posterior subcapsular; RR: risk ratio; SiMES: Singapore Malay Eye Study.

TABLE 19.4 Association between Metabolic Syndrome Component, Obesity, and Cataract in Observational Studies

Author, Year	Study Design, Location	Study Population	Measure of Obesity	Cataract Type	Results
Hiller et al. 1998[27]	Framingham Heart Study Cohort Population-based prospective cohort study	714 adults ages 52–80 years	BMI	Cortical, nuclear, PSC	Increasing BMI was associated with higher risk of developing cortical and PSC (P = 0.002 for both)
Klein et al. 1998[19]	Beaver Dam Eye Study Population-based prospective cohort study, US, follow-up 5 years	4926 adults ages 43–86 years	BMI	Cortical, nuclear, and PSC	OR (95% CI) = 1.05 (1.02-1.09) per unit increase in BMI for PSC among those without diabetes; 1.08 (1.01-1.16) per unit increase in BMI for cortical among those with diabetes
Tavani et al. 1995[48]	Hospital-based case-control study, Italy	287 women who had cataract extraction and 1227 controls	BMI	Cataract surgery	OR 2.2 (95% CI 1.2 to 3.8) for BMI ≥ 30 versus < 20
Leske et al. 1999[57]	Barbados Eye Study Population-based cross-sectional, Barbados	4341 black participants ages 40–84 years	WHR	Cortical	OR 1.36 (95% CI 1.00 to 1.84)
Schaumberg et al. 2001[21]	Physicians' Health Study Population-based cohort, US, follow-up 14 years	20,271 men	BMI WHR	Cataract and cataract surgery	RR 1.25 (95% CI 1.03 to 1.51) and 1.36 (95% CI 1.06 to 1.75) for cataract and cataract surgery comparing BMI ≥ 27.8 to < 22; 1.31 (95% CI 1.10 to 1.55) for cataract comparing WHR ≥ 0.99 vs. < 0.9
Jacques et al. 2003[59]	Nurses' Health Study Population-based cross-sectional, US	466 women ages 53–73 years	BMI ≥ 30 kg/m² WC ≥ 89 cm	Cortical, nuclear, PSC	PSC: OR 2.5 (95% CI 1.2 to 5.2) for BMI ≥ 30 and 2.3 (95% CI 1.0 to 5.2) for WC ≥ 89 cm
Foster et al. 2003[47]	Tanjong Pagar Survey Population-based cross-sectional, Singapore	1206 Chinese adults ages 40–81 years	BMI	Any, cortical, nuclear, PSC Cataract surgery	Lower BMI (lowest vs. highest quintile) was associated with cortical cataract, OR 2.3 (95% CI 1.3 to 4.0) and 1.8 (95% CI 1.1 to 2.9) for any and cortical

(Continued)

TABLE 19.4 Association between Metabolic Syndrome Component, Obesity, and Cataract in Observational Studies—cont'd

Author, Year	Study Design, Location	Study Population	Measure of Obesity	Cataract Type	Results
Kuang *et al.* 2005[58]	Shihpai Eye Study Population-based cross-sectional, Taiwan	1361 adults ages ≥ 65 years	BMI	Cortical, nuclear, PSC	OR (95% CI) per unit increase in BMI = 1.52 (1.04-2.34) for cortical and 0.73 (0.54-0.98) for nuclear; no significant association with PSC
Paunksnis *et al.* 2007[16]	Population-based cross-sectional, Lithuania	1282 adults ages 35–64 years	Central obesity	Any	OR 1.80 (95% CI 1.01 to 3.20) and 1.54 (95% CI 1.01 to 2.35) in men and women ages 45–64 years
Tan *et al.* 2008[17]	Blue Mountains Eye Study, Australia, follow-up 10 years	3564 adults ages ≥ 49 years	BMI	Cortical, nuclear, PSC, cataract surgery	RR per SD of BMI 1.20 (95% CI 1.03 to 1.41) for PSC
Sabanayagam *et al.* 2011[13]	Singapore Malay Eye Study Population-based cross-sectional, Singapore	2794 Malay adults ages 40–80 years	BMI ≥ 25 kg/m^2	Any cortical, nuclear, PSC	OR 1.22 (95% CI 1.01 to 1.48) and 0.65 (95% CI 0.52 to 0.81) for cortical and nuclear; no significant association with any cataract or PSC
Chang *et al.* 2011[54]	AREDS, clinic-based cohort study, US	4425 adults ages 55–80 years	Weight change	Cortical, nuclear, PSC, cataract surgery	Comparing weight change ≥ 24 vs. ≤ 4.5 kg, OR 1.35 (95% CI 1.09 to 1.67), 1.48 (95% CI 1.10 to 1.97) for cortical and PSC
Lu *et al.* 2012[55]	Hospital-based case-control study, China	362 men with cataract surgery and 362 controls ages 45–85 years	Overweight and obesity defined by BMI	Cataract surgery	OR 1.55 (95% CI 1.02 to 1.98) and 1.71 (95% CI 1.32 to 2.39) for overweight and obesity

ARDES: Age-Related Eye Disease Study; BMI: body mass index; CI: confidence interval; OR: odds ratio; PSC: posterior subcapsular; RR: risk ratio; WC: waist circumference; WHR: waist-to-hip ratio.

TABLE 19.5 Association between Metabolic Syndrome Component, Lipids, and Cataract in Observational Studies

Author, Year	Study Design, Location	Study Population	Component of Lipid	Cataract Type	Results
Tavani A *et al.* 1995[48]	Hospital-based case-control study, Italy	287 women with cataract extraction and 1227 controls	Dyslipidemia	Cataract surgery	OR 1.8 (95% CI 1.2 to 2.7)
Klein BE *et al.* 1997[60]	Beaver Dam Eye Study Population-based cross-sectional, US	4926 adults ages 43–86 years	HDL and total cholesterol to HDL ratio	Cortical, nuclear, PSC	Higher HDL was associated with lower risk of cortical cataract in women; higher ratio of total to HDL was associated with increased risk of PSC in men
Hiller R *et al.* 2003[26]	Framingham Offspring Heart Study Cohort, nested case-control study, US	1869 adults ages ≥ 49 years	High triglycerides	PSC	OR 3.1 (95% CI 1.3 to 7.4) for PSC in men; no significant association in women
Meyer D *et al.* 2003[28]	Hospital-based cross-sectional study, South Africa	115 dyslipidemic subjects	HDL and LDL:HDL ratio	Any	OR 7.33 (95% CI 2.06 to 26.1) for low HDL cholesterol (< 1.5 mmol/L) and 2.35 (95% CI 1.09 to 5.04) for high LDL:HDL ratio (> 5)
Paunksnis *et al.* 2007[16]	Population-based cross-sectional, Lithuania	1282 adults ages 35–64 years	Higher serum triglycerides	Any	OR 1.86 (95% CI 1.20 to 2.90) in women ages 45–64 years
Sabanayagam *et al.* 2011[13]	Population-based cross-sectional Singapore	2794 Malay adults ages 40–80 years	Low HDL	Any, cortical, nuclear, PSC	OR 1.34 (95% CI 1.09 to 1.65) for cortical; no significant association with other cataracts

CI: confidence interval; HDL: high-density lipoprotein; LDL: low-density lipoprotein; OR: odds ratio; PSC: posterior subcapsular.

TABLE 19.6 Association between Metabolic Syndrome Component, Blood Pressure, and Cataract in Observational Studies

Author, Year	Study Design, Location	Study Population	Component of Blood Pressure	Cataract Type	Results
Klein et al. 1995[64]	Beaver Dam Eye Study Population-based cross-sectional, US	4926 adults ages 43–84 years	Hypertension	PSC	OR 1.39 (95% CI 1.05 to 1.84)
Tavani et al. 1995[48]	Hospital-based case-control study, Italy	287 women who had cataract extraction and 1227 controls	Hypertension	Cataract surgery	OR 1.5 (95% CI 1.1 to 2.0)
Leske et al. 1999[57]	Barbados Eye Study Cross-sectional, Barbados	4341 black participants ages 40–84 years	High DBP	Cortical	OR 1.49 (95% CI 1.00 to 2.23) for age < 60 years
Schaumberg et al. 2001[65]	Physicians' Health Study Population-based cohort, US, follow-up 12 years	20,271 men	SBP, DBP, hypertension	Cataract	RR 1.31 (95% CI 1.04 to 1.66) for SBP comparing ≥ 150 vs. ≤ 120 mm Hg
Paunksnis et al. 2007[16]	Population-based cross-sectional, Lithuania	1282 adults ages 35–64 years	High blood pressure	Any	OR 1.98 (95% CI 1.21 to 3.25) in women ages 45–64 years
Tan et al. 2008[17]	Blue Mountains Eye Study, Australia, follow-up 10 years	3564 adults ages ≥ 49 years	SBP, DBP, and PP	Cataract and cataract surgery	RR (95% CI) of PSC was 0.82 (0.68-0.99) per unit increase in DBP; RR (95% CI) of nuclear was 1.16 (1.01-1.33) per unit increase in PP
Sabanayagam et al. 2011[13]	Singapore Malay Eye Study Population-based cross-sectional, Singapore	2794 Malay adults ages 40–80 years	High blood pressure	Any, cortical, nuclear, PSC	OR 1.82 (95% CI 1.38 to 2.41) for any; significant association with cortical, nuclear, and PSC

CI: confidence interval; DBP: diastolic blood pressure; OR: odds ratio; PP: pulse pressure; PSC: posterior subcapsular; RR: risk ratio; SBP: systolic blood pressure.

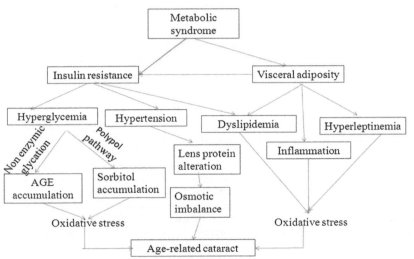

FIGURE 19.1 Potential pathophysiological mechanisms underlying the association between metabolic syndrome and cataract.

cataract extraction. Further, the association was stronger in women aged less than 60 years compared with those aged 60 years or over.[48] Cross-sectional evaluation of the BDES cohort showed that higher levels of total cholesterol:HDL-cholesterol ratio were associated with PSC and a higher levels of HDL was protective against cortical cataract.[60] Further evidence from cross-sectional studies conducted in South Africa,[28] Lithuania,[16] and Singapore[13] supports increased risk of cataract associated with low HDL cholesterol[13,28] and high TGs.[16] Few studies have examined the relationship between dietary fat intake and risk of cataract.[61–63] Tavani et al. reported a higher risk of cataract extraction associated with a higher intake of dietary fat in Italian patients.[62] This is further supported by the NHS, where total fat intake was associated with a higher risk of cataract extraction. However, higher intake of long-chain omega-3 fatty acid was associated with a 12% lower risk of cataract extraction.[61] In the BMES, higher intake of polyunsaturated fat was associated with a lower risk of cortical cataract.[63]

Finally, there are also studies supporting an association between hypertension and cataract[13,16,57,64,65] (Table 19.6). In the PHS, higher levels of systolic blood pressure (BP) increased the risk of cataract in men.[65] In the BMES, elevated pulse pressure, an indicator of arterial compliance, was associated with an increased risk of nuclear cataract.[17] An association between high BP and cataract was also reported among Italian women and Malay adults in Singapore.[13] In the BDES, hypertension was associated with PSC,[64] and in the Barbados Eye Study, high diastolic BP was associated with cortical cataract.[57]

Potential Biological Mechanism Underlying the Association between Metabolic Syndrome and Cataract

The principal mechanisms causing lens damage and cataract formation are oxidative stress, osmotic imbalance, and nonenzymatic protein glycation.[66] Metabolic syndrome affects all of these physiologic mechanisms.[40] The link between hyperglycemia and cataract is activated by nonenzymatic glycation of lens proteins,[67] activation of polypol pathway with sorbitol accumulation, and increased oxidative stress.[68,69] The rate of leukocyte telomere shortening, a marker for cataractogenesis in aging, has been shown to be mediated by oxidative stress.[70] Treatment with oral[71] or topical[72] antioxidants has been shown to delay the progression of cataract in humans[72] and animal models.[71,72] Diet rich in antioxidants including lutein and zeaxanthin was associated with a decreased risk of cataract in large prospective studies.[73] The link between obesity and cataract may be mediated by oxidative stress, inflammation, and endothelial dysfunction.[74] Moderate caloric restriction delayed the onset of age-related cataracts in mouse models.[75] In human studies, inflammatory markers including C-reactive protein (CRP),[76] IL-6, and intracellular adhesion molecule (ICAM)[77] have been shown to be associated with cataract. Hyperleptinemia and leptin resistance due to excessive adiposity also promote cataract formation.[78] The contribution of serum lipids to cataract formation may be more complex, mediated by oxidative stress.[79] Studies have shown a positive correlation between elevated TGs, low HDL, and oxidative stress.[80] Diet rich in cholesterol accelerated the onset of cataract in streptozin-induced diabetic rats.[81] The mechanism linking hypertension and cataract is not clear. In animal studies, genetic models of salt-sensitive hypertensive rats had up

to 50% increased risk of, and reversal of, early cataract lesions with sodium restriction, suggesting that alterations in the protein structures of lens capsule,[82] defective ion transport at the lens epithelium and ciliary body,[83] and extracellular fluid volume state may contribute to cataractogenesis.[84]

TAKE-HOME MESSAGES

- Metabolic syndrome is a well-recognized condition affecting millions of people worldwide.
- Increasing evidence suggests an association between metabolic syndrome and cataract.
- Among metabolic syndrome components, diabetes, obesity, lipids, and hypertension are shown to be associated with cataract.
- Understanding the pathophysiologic mechanisms linking metabolic syndrome and cataract may identify possible preventative measures and novel therapeutic targets for intervention.
- Early identification of metabolic risk factors and lifestyle modifications may help delay the onset of cataract and reduce the burden of blindness in the growing elderly population.

References

1. Cornier MA, Dabelea D, Hernandez TL, Lindstrom RC, Steig AJ, Stob NR, et al. The metabolic syndrome. *Endocr Rev* 2008;**29**(7): 777–822.
2. Grundy SM. Obesity, metabolic syndrome, and cardiovascular disease. *J Clin Endocrinol Metab* 2004;**89**(6):2595–600.
3. Mottillo S, Filion KB, Genest J, Joseph L, Pilote L, Poirier P, et al. The metabolic syndrome and cardiovascular risk a systematic review and meta-analysis. *J Am Coll Cardiol* 2010;**56**(14):1113–32.
4. Grundy SM, Brewer Jr HB, Cleeman JI, Smith Jr SC, Lenfant C. Definition of metabolic syndrome: report of the National Heart, Lung, and Blood Institute/American Heart Association conference on scientific issues related to definition. *Circulation* 2004;**109**(3):433–8.
5. Ninomiya JK, L'Italien G, Criqui MH, Whyte JL, Gamst A, Chen RS. Association of the metabolic syndrome with history of myocardial infarction and stroke in the Third National Health and Nutrition Examination Survey. *Circulation* 2004;**109**(1):42–6.
6. Congdon NG, Friedman DS, Lietman T. Important causes of visual impairment in the world today. *JAMA* 2003;**290**(15):2057–60.
7. Williams SL, Ferrigno L, Mora P, Rosmini F, Maraini G. Baseline cataract type and 10-year mortality in the Italian-American Case-Control Study of age-related cataract. *Am J Epidemiol* 2002;**156**(2):127–31.
8. Thompson JR, Sparrow JM, Gibson JM, Rosenthal AR. Cataract and survival in an elderly nondiabetic population. *Arch Ophthalmol* 1993;**111**(5):675–9.
9. Kupfer C. Bowman lecture. The conquest of cataract: a global challenge. *Trans Ophthalmol Soc UK* 1985;**104**(Pt 1):1–10.
10. Kametaka S, Kasahara T, Ueo M, Takenaka M, Saito M, Sakamoto K, et al. Effect of nifedipine on severe experimental cataract in diabetic rats. *J Pharmacol Sci* 2008;**106**(4):651–8.

11. Kanthan GL, Wang JJ, Rochtchina E, Mitchell P. Use of antihypertensive medications and topical beta-blockers and the long-term incidence of cataract and cataract surgery. *Br J Ophthalmol* 2009;**93**(9):1210–4.

12. Fernandez MM, Afshari NA. Nutrition and the prevention of cataracts. *Curr Opin Ophthalmol* 2008;**19**(1):66–70.

13. Sabanayagam C, Wang JJ, Mitchell P, Tan AG, Tai ES, Aung T, et al. Metabolic syndrome components and age-related cataract: the Singapore Malay Eye Study. *Invest Ophthalmol Vis Sci* 2011;**52**(5):2397–404.

14. Galeone C, Petracci E, Pelucchi C, Zucchetto A, La VC, Tavani A. Metabolic syndrome, its components and risk of age-related cataract extraction: a case-control study in Italy. *Ann Epidemiol* 2010;**20**(5):380–4.

15. Lindblad BE, Hakansson N, Philipson B, Wolk A. Metabolic syndrome components in relation to risk of cataract extraction: a prospective cohort study of women. *Ophthalmology* 2008;**115**(10):1687–92.

16. Paunksnis A, Bojarskiene F, Cimbalas A, Cerniauskiene LR, Luksiene DI, Tamosiunas A. Relation between cataract and metabolic syndrome and its components. *Eur J Ophthalmol* 2007;**17**(4):605–14.

17. Tan JS, Wang JJ, Mitchell P. Influence of diabetes and cardiovascular disease on the long-term incidence of cataract: the Blue Mountains Eye Study. *Ophthalmic Epidemiol* 2008;**15**(5):317–27.

18. Weintraub JM, Willett WC, Rosner B, Colditz GA, Seddon JM, Hankinson SE. A prospective study of the relationship between body mass index and cataract extraction among US women and men. *Int J Obes Relat Metab Disord* 2002;**26**(12):1588–95.

19. Klein BE, Klein R, Lee KE. Diabetes, cardiovascular disease, selected cardiovascular disease risk factors, and the 5-year incidence of age-related cataract and progression of lens opacities: the Beaver Dam Eye Study. *Am J Ophthalmol* 1998;**126**(6):782–90.

20. Lim LS, Tai ES, Aung T, Tay WT, Saw SM, Seielstad M, et al. Relation of age-related cataract with obesity and obesity genes in an Asian population. *Am J Epidemiol* 2009;**169**(10):1267–74.

21. Schaumberg DA, Glynn RJ, Christen WG, Hankinson SE, Hennekens CH. Relations of body fat distribution and height with cataract in men. *Am J Clin Nutr* 2000;**72**(6):1495–502.

22. Anon. Does obesity increase the risk for cataracts? *J Am Optom Assoc* 1997;**68**(7):463.

23. Harding JJ. Recent studies of risk factors and protective factors for cataract. *Curr Opin Ophthalmol* 1997;**8**(1):46–9.

24. Glynn RJ, Christen WG, Manson JE, Bernheimer J, Hennekens CH. Body mass index. An independent predictor of cataract. *Arch Ophthalmol* 1995;**113**(9):1131–7.

25. Rowe NG, Mitchell PG, Cumming RG, Wans JJ. Diabetes, fasting blood glucose and age-related cataract: the Blue Mountains Eye Study. *Ophthalmic Epidemiol* 2000;**7**(2):103–14.

26. Hiller R, Sperduto RD, Reed GF, D'Agostino RB, Wilson PW. Serum lipids and age-related lens opacities: a longitudinal investigation: the Framingham Studies. *Ophthalmology* 2003;**110**(3):578–83.

27. Hiller R, Podgor MJ, Sperduto RD, Nowroozi L, Wilson PW, D'Agostino RB, et al. A longitudinal study of body mass index and lens opacities. The Framingham Studies. *Ophthalmology* 1998;**105**(7):1244–50.

28. Meyer D, Parkin D, Maritz FJ, Liebenberg PH. Abnormal serum lipoprotein levels as a risk factor for the development of human lenticular opacities. *Cardiovasc J S Afr* 2003;**14**(2):60–4.

29. Reaven GM. Banting lecture 1988. Role of insulin resistance in human disease. *Diabetes* 1988;**37**(12):1595–607.

30. Kassi E, Pervanidou P, Kaltsas G, Chrousos G. Metabolic syndrome: definitions and controversies. *BMC Med* 2011;**9**:48.

31. Kylin E. Studien uber das Hypertonie-Hyperglyka "mie-Hyperurika" miesyndrom. *Zentral Innere Med* 1923;**44**:105–27.

32. Vague J. *Presse Med* 1947;**55**(30):339.

33. Alberti KG, Zimmet PZ. Definition, diagnosis and classification of diabetes mellitus and its complications. Part 1: diagnosis and classification of diabetes mellitus provisional report of a WHO consultation. *Diabet Med* 1998;**15**(7):539–53.

34. Executive Summary of The Third Report of The National Cholesterol Education Program (NCEP) Expert Panel on Detection. Evaluation, and Treatment of High Blood Cholesterol in Adults (Adult Treatment Panel III). *JAMA* 2001;**285**(19):2486–97.

35. Alberti KG, Zimmet P, Shaw J. Metabolic syndrome – a new worldwide definition. A Consensus Statement from the International Diabetes Federation. *Diabet Med* 2006;**23**(5):469–80.

36. Hansel B, Giral P, Nobecourt E, Chantepie S, Bruckert E, Chapman MJ, et al. Metabolic syndrome is associated with elevated oxidative stress and dysfunctional dense high-density lipoprotein particles displaying impaired antioxidative activity. *J Clin Endocrinol Metab* 2004;**89**(10):4963–71.

37. Pittas AG, Joseph NA, Greenberg AS. Adipocytokines and insulin resistance. *J Clin Endocrinol Metab* 2004;**89**(2):447–52.

38. Mantzoros CS. The role of leptin in human obesity and disease: a review of current evidence. *Ann Intern Med* 1999;**130**(8):671–80.

39. Mancia G, Bousquet P, Elghozi JL, Esler M, Grassi G, Julius S, et al. The sympathetic nervous system and the metabolic syndrome. *J Hypertens* 2007;**25**(5):909–20.

40. Reaven GM. Insulin resistance, the insulin resistance syndrome, and cardiovascular disease. *Panminerva Med* 2005;**47**(4):201–10.

41. Yanai H, Tomono Y, Ito K, Furutani N, Yoshida H, Tada N. The underlying mechanisms for development of hypertension in the metabolic syndrome. *Nutr J* 2008;**7**:10.

42. Ederer F, Hiller R, Taylor HR. Senile lens changes and diabetes in two population studies. *Am J Ophthalmol* 1981;**91**(3):381–95.

43. Nielsen NV, Vinding T. The prevalence of cataract in insulin-dependent and non-insulin-dependent-diabetes mellitus. *Acta Ophthalmol (Copenh)* 1984;**62**(4):595–602.

44. Klein R, Klein BE, Moss SE. Visual impairment in diabetes. *Ophthalmology* 1984;**91**(1):1–9.

45. Saxena S, Mitchell P, Rochtchina E. Five-year incidence of cataract in older persons with diabetes and pre-diabetes. *Ophthalmic Epidemiol* 2004;**11**(4):271–7.

46. West SK, Valmadrid CT. Epidemiology of risk factors for age-related cataract. *Surv Ophthalmol* 1995;**39**(4):323–34.

47. Foster PJ, Wong TY, Machin D, Johnson GJ, Seah SK. Risk factors for nuclear, cortical and posterior subcapsular cataracts in the Chinese population of Singapore: the Tanjong Pagar Survey. *Br J Ophthalmol* 2003;**87**(9):1112–20.

48. Tavani A, Negri E, La VC. Selected diseases and risk of cataract in women. A case-control study from northern Italy. *Ann Epidemiol* 1995;**5**(3):234–8.

49. Delcourt C, Cristol JP, Tessier F, Leger CL, Michel F, Papoz L. Risk factors for cortical, nuclear, and posterior subcapsular cataracts: the POLA study. Pathologies Oculaires Liees a l'Age. *Am J Epidemiol* 2000;**151**(5):497–504.

50. Hennis A, Wu SY, Nemesure B, Leske MC. Risk factors for incident cortical and posterior subcapsular lens opacities in the Barbados Eye Studies. *Arch Ophthalmol* 2004;**122**(4):525–30.

51. Richter GM, Torres M, Choudhury F, Azen SP, Varma R. Risk factors for cortical, nuclear, posterior subcapsular, and mixed lens opacities: the Los Angeles Latino Eye Study. *Ophthalmology* 2012;**119**(3):547–54.

52. Karasik A, Modan M, Halkin H, Treister G, Fuchs Z, Lusky A. Senile cataract and glucose intolerance: the Israel Study of glucose Intolerance Obesity and Hypertension (The Israel GOH Study). *Diabetes Care* 1984;**7**(1):52–6.

53. Olafsdottir E, Andersson DK, Stefansson E. The prevalence of cataract in a population with and without type 2 diabetes mellitus. *Acta Ophthalmol* 2012;**90**(4):334–40.

54. Chang JR, Koo E, Agron E, Hallak J, Clemons T, Azar D, et al. Risk factors associated with incident cataracts and cataract surgery in the Age-Related Eye Disease Study (AREDS): AREDS report number 32. *Ophthalmology* 2011;**118**(11):2113–9.

55. Lu ZQ, Sun WH, Yan J, Jiang TX, Zhai SN, Li Y. Cigarette smoking, body mass index associated with the risks of age-related cataract in male patients in northeast China. *Int J Ophthalmol* 2012;**5**(3): 317–22.

56. Klein BE, Klein R, Moss SE. Incidence of cataract surgery in the Wisconsin Epidemiologic Study of Diabetic Retinopathy. *Am J Ophthalmol* 1995;**119**(3):295–300.

57. Leske MC, Wu SY, Hennis A, Connell AM, Hyman L, Schachat A. Diabetes, hypertension, and central obesity as cataract risk factors in a black population. The Barbados Eye Study. *Ophthalmology* 1999;**106**(1):35–41.

58. Kuang TM, Tsai SY, Hsu WM, Cheng CY, Liu JH, Chou P. Body mass index and age-related cataract: the Shihpai Eye Study. *Arch Ophthalmol* 2005;**123**(8):1109–14.

59. Jacques PF, Moeller SM, Hankinson SE, Chylack Jr LT, Rogers G, Tung W, et al. Weight status, abdominal adiposity, diabetes, and early age-related lens opacities. *Am J Clin Nutr* 2003;**78**(3):400–5.

60. Klein BE, Klein R, Lee KE. Cardiovascular disease, selected cardiovascular disease risk factors, and age-related cataracts: the Beaver Dam Eye Study. *Am J Ophthalmol* 1997;**123**(3):338–46.

61. Lu M, Cho E, Taylor A, Hankinson SE, Willett WC, Jacques PF. Prospective study of dietary fat and risk of cataract extraction among US women. *Am J Epidemiol* 2005;**161**(10):948–59.

62. Tavani A, Negri E, La VC. Food and nutrient intake and risk of cataract. *Ann Epidemiol* 1996;**6**(1):41–6.

63. Cumming RG, Mitchell P, Smith W. Diet and cataract: the Blue Mountains Eye Study. *Ophthalmology* 2000;**107**(3):450–6.

64. Klein BE, Klein R, Jensen SC, Linton KL. Hypertension and lens opacities from the Beaver Dam Eye Study. *Am J Ophthalmol* 1995;**119**(5):640–6.

65. Schaumberg DA, Glynn RJ, Christen WG, Ajani UA, Sturmer T, Hennekens CH. A prospective study of blood pressure and risk of cataract in men. *Ann Epidemiol* 2001;**11**(2):104–10.

66. Harding J. *Cataract: Biochemistry, Epidemiology, and Pharmacology.* 1st edn. London: Chapman and Hall; 1991.

67. Perejda AJ, Uitto J. Nonenzymatic glycosylation of collagen and other proteins: relationship to development of diabetic complications. *Coll Relat Res* 1982;**2**(1):81–8.

68. Reddy PY, Giridharan NV, Reddy GB. Activation of sorbitol pathway in metabolic syndrome and increased susceptibility to cataract in Wistar-Obese rats. *Mol Vis* 2012;**18**:495–503.

69. Suryanarayana P, Patil MA, Reddy GB. Insulin resistance mediated biochemical alterations in eye lens of neonatal streptozotocin-induced diabetic rat. *Indian J Exp Biol* 2011;**49**(10):749–55.

70. Sanders JL, Iannaccone A, Boudreau RM, Conley YP, Opresko PL, Hsueh WC, et al. The association of cataract with leukocyte telomere length in older adults: defining a new marker of aging. *J Gerontol A Biol Sci Med Sci* 2011;**66**(6):639–45.

71. Randazzo J, Zhang P, Makita J, Blessing K, Kador PF. Orally active multi-functional antioxidants delay cataract formation in streptozotocin (type 1) diabetic and gamma-irradiated rats. *PLoS One* 2011;(4):6. e18980.

72. Babizhayev MA, Deyev AI, Yermakova VN, Brikman IV, Bours J. Lipid peroxidation and cataracts: N-acetylcarnosine as a therapeutic tool to manage age-related cataracts in human and in canine eyes. *Drugs R D* 2004;**5**(3):125–39.

73. Moeller SM, Voland R, Tinker L, Blodi BA, Klein ML, Gehrs KM, et al. Associations between age-related nuclear cataract and lutein and zeaxanthin in the diet and serum in the Carotenoids in the Age-Related Eye Disease Study, an Ancillary Study of the Women's Health Initiative. *Arch Ophthalmol* 2008;**126**(3):354–64.

74. Abraham AG, Condon NG, West GE. The new epidemiology of cataract. *Ophthalmol Clin North Am* 2006;**19**(4):415–25.

75. Taylor A, Zuliani AM, Hopkins RE, Dallal GE, Treglia P, Kuck JF, et al. Moderate caloric restriction delays cataract formation in the Emory mouse. *FASEB J* 1989;**3**(6):1741–6.

76. Schaumberg DA, Ridker PM, Glynn RJ, Christen WG, Dana MR, Hennekens CH. High levels of plasma C-reactive protein and future risk of age-related cataract. *Ann Epidemiol* 1999;**9**(3):166–71.

77. Klein BE, Klein R, Lee KE, Knudtson MD, Tsai MY. Markers of inflammation, vascular endothelial dysfunction, and age-related cataract. *Am J Ophthalmol* 2006;**141**(1):116–22.

78. Gomez-Ambrosi J, Salvador J, Fruhbeck G. Is hyperleptinemia involved in the development of age-related lens opacities? *Am J Clin Nutr* 2004;**79**(5):888–9.

79. Hashim Z, Zarina S. Assessment of paraoxonase activity and lipid peroxidation levels in diabetic and senile subjects suffering from cataract. *Clin Biochem* 2007;**40**(9–10):705–9.

80. Katsuki A, Sumida Y, Urakawa H, Gabazza EC, Murashima S, Nakatani K, et al. Increased oxidative stress is associated with serum levels of triglyceride, insulin resistance, and hyperinsulinemia in Japanese metabolically obese, normal-weight men. *Diabetes Care* 2004;**27**(2):631–2.

81. Tsutsumi K, Inoue Y, Yoshida C. Acceleration of development of diabetic cataract by hyperlipidemia and low high-density lipoprotein in rats. *Biol Pharm Bull* 1999;**22**(1):37–41.

82. Lee SM, Lin SY, Li MJ, Liang RC. Possible mechanism of exacerbating cataract formation in cataractous human lens capsules induced by systemic hypertension or glaucoma. *Ophthalmic Res* 1997;**29**(2):83–90.

83. Rodriguez-Sargent C, Cangiano JL, Berrios CG, Marrero E, Martinez-Maldonado M. Cataracts and hypertension in salt-sensitive rats. A possible ion transport defect. *Hypertension* 1987;**9**(3):304–8.

84. Rodriguez-Sargent C, Berrios G, Irrizarry JE, Estape ES, Cangiano JL, Martinez-Maldonado M. Prevention and reversal of cataracts in genetically hypertensive rats through sodium restriction. *Invest Ophthalmol Vis Sci* 1989;**30**(11):2356–60.

85. Mukesh BN, Le A, Dimitrov PN, Ahmed S, Taylor HR, McCarty CA. Development of cataract and associated risk factors: the Visual Impairment Project. *Arch Ophthalmol* 2006;**124**(1):79–85.

20

Childhood Obesity, Body Fatness Indices, and Retinal Vasculature

Ling-Jun Li[1], Tien Y. Wong[2], Seang Mei Saw[1]

[1]Saw Swee Hock School of Public Health, National University of Singapore, Singapore,
[2]Singapore Eye Research Institute, Singapore National Eye Center, Singapore

INTRODUCTION

Obesity is a condition of abnormal accumulation of excessive adipose tissue that causes negative effects to health.[1] Since the 1980s and 1990s, the worldwide prevalence of childhood overweight and obesity has risen steeply, not only in most industrialized countries but also in several low-income countries such as China, Chile, and Brazil.[2–4] In children, obesity has become the most common cause of insulin resistance,[5] which can adversely affect almost every organ system, such as pubertal advancement[6] and asthma,[7] and is associated with type 2 diabetes[8] and long-term vascular complications later in life.[9–11]

Overweight and obesity should ideally be defined by the amount of total body fat. In research settings, better measurement techniques are reported using dual-energy X-ray absorptiometry (DEXA),[12] underwater weighting (densitometry),[13] rapid and noninvasive methods such as bioelectrical impedance analysis (BIA),[14] computed tomography (CT/CAT scan),[15] and magnetic resonance imaging (MRI/NMR).[16] However, these measurement techniques are complex, costly, and impractical for clinical assessment, especially among children. Thus, in clinical settings, body mass index (BMI), waist circumference, body surface area (BSA), and subcutaneous skinfold thickness measurements are common, inexpensive, and practical body fatness indices used in children.

The retinal microcirculation is part of the microcirculation *in vivo* and with advances in retinal photographic techniques and retinal vasculature can now be visualized and graded noninvasively.[17–23] Based on its special architectural anatomy, which consists of retinal arteriole and venules that contain neither internal elastic lamina nor a continuous muscular coat,[20,24,25] changes on retinal vasculature including retinal vascular caliber, tortuosity, branching angle, and fractal dimension can be reflected promptly and sensitively. Therefore, relative alterations of retinal vascular parameters have been quantified in the past 10 years[26] and were able to detect earliest signs of systemic diseases (hypertension (HTN) and diabetes),[26–32] cardiovascular diseases (CVD), and cerebrovascular diseases by numerous population-based epidemiologic studies.[33–38]

Since childhood obesity is associated with a series of adverse metabolic and vascular complications where the underlying mechanisms remain unknown, the relationship of body fatness indices and retinal vasculature among children may be an important entry point to investigate such pathophysiologic mechanisms. This chapter aims to 1) summarize major findings from recent population-based epidemiologic studies (from 2000 onwards) regarding the associations of childhood obesity, body fatness indices, and retinal vasculature among children; 2) explore possible pathophysiologic mechanisms from these associations; and 3) suggest potential clinical implications in future retinal vasculature study on childhood obesity.

CHILDHOOD OBESITY

Childhood Obesity and Metabolic and Cardiovascular Consequence

There are approximately 22 million children diagnosed as overweight or obese under the age of 5 years,[39] and at least two-thirds of obese school-aged children will become obese adults.[40,41] Due to the possibility that metabolic syndrome or an abnormality that links insulin resistance and type 2 diabetes or atherosclerotic CVD in adulthood may originate *in utero*[5] and early life,[9,10] childhood obesity has been widely studied in recent years in order to investigate the mechanisms better as well as prevention strategies.

The prevalence of pediatric obesity parallels the increased incidence of type 2 diabetes.[42] Since type 2 diabetes does not display itself until adulthood after developing over many years, impaired glucose tolerance and insulin resistance play the role of intermediate stages in type 2 diabetes development.[43] Thus, obese children whose metabolic process is accelerated may have a shorter transition between impaired glucose tolerance and type 2 diabetes than children with normal weight.

Furthermore, obese children have significant increased levels of low-density lipoprotein-cholesterol (LDL-C), vascular adhesion molecules, tumor necrosis factor (TNF)-α, C-reactive protein (CRP), interleukin (IL)-6, and decreased levels of high-density lipoprotein-cholesterol (HDL-C),[44] triggering a series of proinflammatory and proatherogenic changes resulting in vascular dysfunction such as reduced endothelial function and arterial compliance[44] and even long-term vascular complications such as HTN and CVD.[10,11,43]

Strong evidence from autopsy and epidemiologic studies has indicated such relationships between CVD risk factors caused by obesity early in life and vascular signs of atherosclerosis later in life. Two postmortem studies in children and youth autopsy – the Bogalusa Heart Study and the Pathobiological Determinants of Atherosclerosis in Young (PDAY) – reported associations between early atherosclerotic lesions in the aorta and coronary arteries and a range of CVD risk factors such as LDL-C and triglycerides (TG).[45,46] Three longitudinal studies on young and middle-aged adults – the Bogalusa Heart Study, the Cardiovascular Risk in Young Finns Study, and the Muscatine Study – also reported similar results that current higher carotid intima-media thickness (IMT) could be traced back to higher levels of CVD risk factors in childhood and adolescence.[47–49]

In conclusion, although childhood obesity is known to be at higher risk for early development of impaired glucose level and vascular pathology, which will further turn into type 2 diabetes and vascular complications later in life, the underlying pathophysiologic mechanisms are still unclear. Therefore, assessments on macro- and microvascular function can provide evidence of predisposition of vascular pathology changes long before overt clinical manifestations.

Body Fatness Indices

BMI is the most commonly used measure of adiposity status in children. In addition, for children, especially preschool children, the definition of obesity and overweight varies from country to country. Two well-known and commonly used definitions include the US Centers for Disease Control and Prevention (CDC) guideline and the International Obesity Task Force (IOTF) guideline. The CDC guideline defines overweight and obesity among children based on nationally representative data, which is arbitrary and not appropriate for application to other populations or countries. According to data collected from six countries (US, UK, Hong Kong, the Netherlands, Brazil, and Singapore), the IOTF guideline defines percentile curves for overweight and obesity that intersect the adult cutoff points of a BMI of $25 \, \text{kg}/\text{m}^2$ and $30 \, \text{kg}/\text{m}^2$ at age of 18 years, respectively. Therefore, the cutoff points proposed by the IOTF guideline are age and gender specific. The threshold of obesity from the IOTF guideline is higher than that of the CDC guideline at most ages.[43] However, two recent studies showed similar sensitivity and specificity in applying the CDC and IOTF guidelines in predicting CVD risk factors in adulthood.[50,51]

Considering the limited accuracy of BMI in distinguishing adiposity from muscularity, body fatness indices such as waist circumference, BSA, and skinfold thickness are used in conjunction with BMI measurements, yet these parameters still carry limitations such as poor reliability and reproducibility. Therefore, combining the use of all body fatness indices may compensate for mutual measurement errors and provide a better assessment of body adiposity.

ASSESSMENTS OF RETINAL VASCULATURE

Retinal microcirculation can be assessed invasively, and the technology in examining retinal vascular characteristics, such as caliber, tortuosity, branching angle, and fractal dimension, has been widely validated and recognized.

Recent population-based studies have used computer-assisted programs to measure individual arterioles and venules and to combine them according to formulas developed firstly by Parr and Spear,[52,53] subsequently modified by Hubbard *et al*,[54] and further improved by Knudtson *et al*.[55] The use of computer-assisted programs differs in all population-based epidemiologic studies. For example, Computer Assisted Image Analysis of the Retina program (CAIAR) and Retinal Image MultiScale Analysis were used in the UK in adult studies.[56,57] Retinal Imaging Software Fractal (IRIS-Fractal) was used in an Australian study of children,[58–61] Non-mydriatic Vessel Analyser (SVA-T) was used in a German study of children,[62] Interactive Vessel Analysis (IVAN) was widely used in US studies[21,63,64] and Asian studies,[29,65] while the Singapore I Vessel Assessment (SIVA) was newly developed and applied in recent studies in Singapore.[26,32]

Regardless of the various models of computer-assisted programs, the calibration of all these programs works similarly. After determining the true size of the optic disc and locating retinal vessels with minimum detectable width varied by different computer-assisted programs, calibration of the computer-assisted program will generate three fundamental variables, which are projected caliber size of the central retinal arteriole equivalent (CRAE), the projected caliber size of central retinal venular equivalent (CRVE), and the ratio of the two variables (arteriole-to-venule ratio (AVR)). With the substantial reproducibility demonstrated in recent studies (intra- and intergrader correlation coefficient ranged from 0.67 to 0.99), computed-assisted programs have been proved to be precise and reliable for assessing structural retinal vascular caliber changes among general populations. The outputs of IVAN and SIVA computer programs are shown in Figure 20.1, Figure 20.2, and Figure 20.3.

CHILDHOOD OBESITY, BODY FATNESS INDICES, AND RETINAL VASCULATURE

The Relationships between Childhood Obesity, Body Fatness Indices, and Retinal Vasculature

The association between childhood obesity and retinal vasculature has been repeatedly investigated by a few school-based or population-based and cross-sectional studies in Asian and white children in the past five years. Childhood obesity was assessed by a series of body fatness indices such as BMI, BSA, waist circumference, and triceps skinfold thickness (TSF). Table 20.1, Table 20.2, and Table 20.3 shows the relationship between childhood obesity, body fatness indices, and retinal vasculature. In five population-based/school-based and cross-sectional studies in Germany,[62] Australia,[27,66] and Singapore,[67,68] children with greater values of body fatness indices tended to have retinal arteriolar narrowing and/or retinal venular widening.

Among these five cross-sectional studies, four were school-based studies[27,62,66,67] while one was population-based study.[68] Furthermore, three studies had

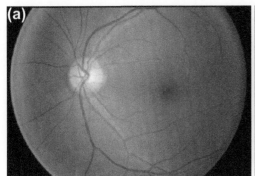

FIGURE 20.1 Retina allows for a noninvasive visualization of human microcirculation. Computer-assisted program for the measurement of retinal vascular caliber to quantify structural vascular microcirculatory changes. Zone B is marked in IVAN software by 0.5–1.0 optic disc diameter away from the margin of optic disc. The biggest eight retinal vascular arterioles and venules were located and assessed within zone B. (See color plate section)

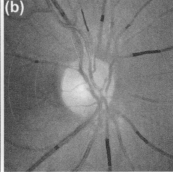

FIGURE 20.2 Vessels are assessed by clicking on the seed point (the proximal end of the vessel trace) on the image display or by clicking on the vessel number button on the data table. Retinal arteriolar analysis control (a) and retinal venular analysis control (b) show the visual vessel tracing and the identified data (angle, width, Sigma) for each vessel. (See color plate section)

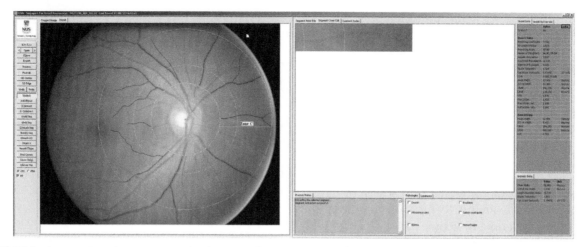

FIGURE 20.3 A screenshot of a computer-assisted program for measurement of new geometrical retinal vascular parameters from a retinal fundus photograph. Zone B and zone C are marked in IVAN software by 0.5–1.0 and 0.5–2.0 optic disc diameter away from the margin of optic disc, respectively. All retinal arterioles and venules larger than 25 μm are marked and assessed within zone B and zone C. (See color plate section)

TABLE 20.1 Association between Body Mass Index (BMI) and Retinal Vasculature among Children

Study	Study Population and Study Design	Sample Size	Mean/Range of Age (Years)	Childhood BMI	Retinal Arteriolar Caliber Narrowing	Retinal Venular Caliber Widening
Hanssen 2012[62] JuvenTUM3	School-based cross-sectional	578	11.1	Each 1 kg/m² increase	0.37 μm (P = 0.04)	0.37 μm (P = 0.05)
Li 2011[65] STARS Family	Population-based cross-sectional	136	10.7	Each SD (3.52 kg/m²) increase	–	3.40 μm (P < 0.01)
Gopinath 2011[27] SCES	School-based cross-sectional	2353	12.7	4th vs. 1st quartile	2.8 μm (P < 0.001)	4.2 μm (P < 0.01)
Taylor 2007[66] SCES	School-based cross-sectional	1608	6	Each SD increase	0.76 μm (P < 0.01)	1.13 μm (P < 0.001)
Cheung 2007[67] SCORM	School-based cross-sectional	768	7–9	Each SD increase	–	2.55 μm (P < 0.001)

SCES: the Sydney Children Eye Study; SCORM: the Singapore Cohort Study of the Risk Factors of Myopia SD: standard deviation; STARS: Family Study, the Strabismus, Amblyopia and Refractive Error Study in Singaporean Chinese Preschoolers Family Study.

multiple ethnicities involved while two studies had only one ethnicity involved. JuvenTUM3 included a total of 578 fifth-grade European children ages 10–13 years from 15 secondary general and intermediate schools in the greater Munich area, Germany.[62] Similarly to the German study design, the Sydney Children Eye Study (SCES) included 1740 multiethnic children age 6 years from a random cluster sample of 34 primary schools[66] and 2353 multiethnic children age 12.7 years from another random cluster sample of 21 high schools[27] in the Sydney metropolitan area, Australia. The Singapore Cohort Study of Risk Factors for Myopia (SCORM) recruited 768 multiethnic children ages 7–9 years who were randomly and systematically sampled from three primary schools in Singapore.[67] As a population-based study, the Strabismus, Amblyopia and Refractive Error Study in

Singapore Chinese Preschoolers (STARS) Family study was a population-based study that recruited 136 Chinese children ages 6–16 years who resided in South-Western Singapore.[68]

Table 20.1 shows the association between BMI and abnormal retinal vessel signs (e.g., retinal arteriolar narrowing, retinal venular widening, and reduction in arteriole-to-venule ratio) among children in multiple linear regression models. In the JuvenTUM 3 study on 578 children with mean age of 11.1 years, obese children had a smaller arteriole-to-venular ratio than those of normal weight peers (0.85 vs. 0.89, P < 0.05).[62] Among Singaporean children, each standard deviation (SD) increase in BMI was reported to be associated with a 3.40 μm (P < 0.01) widening in retinal venular caliber in the STARS Family Study[68] and a 2.55 μm (P < 0.001) widening

TABLE 20.2 Association between Childhood Overweight/Obesity and Retinal Vasculature

Study	Study Population and Study Design	Sample Size	Mean/ Range of Age	Childhood Overweight/Obesity	Retinal Arteriolar Caliber Narrowing	Retinal Venular Caliber Widening	AVR
Hanssen 2012[62] JuvenTUM3	School-based cross-sectional	578	11.1	Self-definition: Obese vs. normal weight Overweight vs. normal weight	–	–	0.85 vs. 0.89 (P < 0.001) 0.87 vs. 0.89 (P = 0.03)
Li 2011[65] STARS Family	Population-based cross-sectional	136	10.7	CDC definition: Above vs. below BMI threshold	–	9.33 μm (P = 0.02)	–
Gopinath 2011[27] SCES	School-based cross-sectional	2353	12.7	IOTF definition: Above vs. below BMI threshold	2.8 μm (P = 0.01)	4.5 μm (P = 0.01)	–
Taylor 2007[66] SCES	School-based cross-sectional	1608	6	IOTF definition: Above vs. below BMI threshold	1.7 μm (P < 0.01)	2.7 μm (P < 0.001)	–

AVR: arteriole-to-venule ratio; BMI: body mass index; CDC: Centers for Disease Control and Prevention; IOTF: International Obesity Task Force; SCES: the Sydney Children Eye Study; STARS Family Study: the Strabismus, Amblyopia and Refractive Error Study in Singaporean Chinese Preschoolers Family Study.

TABLE 20.3 Association between Other Body Fatness Indices and Retinal Vasculature among Children

Study	Study Population and Study Design	Sample Size	Mean/Range of Age (Years)	Childhood Body Fatness Indices	Retinal Arteriolar Caliber Narrowing	Retinal Venular Caliber Widening	AVR
Hanssen 2012[62] JuvenTUM3	School-based cross-sectional	578	11.1	Waist circum: Each 10 cm increase	–	–	0.01 (P < 0.001)
				% body fat: Each 10% increase	–	–	0.01 (P < 0.01)
Li 2011[65] STARS Family	Population-based cross-sectional	136	10.7	Triceps skinfold: Each SD increase	–	2.94 μm (P = 0.01)	–
Taylor 2007[66] SCES	School-based cross-sectional	1608	6	Waist circum: Each SD increase	–	0.99 μm (P < 0.01)	–
				Body surface area: Each SD increase	–	1.97 μm (P < 0.0001)	–

AVR: arteriole-to-venule ratio; SCES, the Sydney Children Eye Study; SD: standard deviation; STARS Family Study, the Strabismus, Amblyopia and Refractive Error Study in Singaporean Chinese Preschoolers Family Study; Waist circum: waist circumference.

in retinal venular caliber in the SCORM study,[67] respectively. Similarly in Australian obese children, each SD increase in BMI was associated with a 0.76 μm (P = 0.001) narrowing in retinal arteriolar caliber and a 1.13 μm (P = 0.01) widening in retinal venular caliber compared with normal weight peers, respectively.[27]

As for the other body fatness indices shown in Table 20.2, TSF, waist circumference, and BSA were consistently associated with retinal vasculature. Each SD increase in TSF, waist circumference, and body surface was accordingly associated with a 2.94 μm (P = 0.01), a 0.99 μm (P < 0.05), and a 1.97 μm (P < 0.05) widening in retinal venular caliber in Singaporean children, Chinese children,[68] and Australian children.[66]

By using the CDC guideline and the IOTF guideline for classifying children into overweight and obesity, researchers also show a significant association between childhood overweight/obesity and retinal vasculature changes (Table 20.3). By using these two systems, children whose BMI was equal to and above 85th percentile of age- and gender-specific CDC/IOTF BMI chart were considered as overweight to obese subjects. Compared with children below the threshold, overweight to obese Singaporean children above the CDC threshold had a 9.33 μm (P = 0.02) wider retinal venular caliber,[68] while overweight to obese Australian children above IOTF threshold had a 1.7–2.8 μm (P < 0.01) narrower retinal arteriolar caliber and a 2.7–4.5 μm (P < 0.01) wider retinal venular caliber.[27,66]

The Pathological Mechanism

Obesity has been suggested to upregulate systemic inflammatory status and to be associated with a range of cardiovascular risk factors such as white blood cell counts, CRP, IL-6, serum amyloid A (SAA), fibrinogen, TG, VDL-C, and HLD-C both in children and adults.[69–72] However, microcirculation in obese subjects is still an unexplored domain. The demonstrations of structural and/or functional abnormalities in microvascular physiology very early in life may provide solid evidence for future metabolic disease development.

It is hypothesized that retinal vasculature abnormality is caused by a series of pathophysiologic changes and the sequential biomolecular regulations *in vivo*, among which nitric oxide (NO)-dependent endothelial dysfunction is the most plausibly postulated mechanism underlying retinal arteriolar narrowing and retinal venular widening.[20,68,73] A number of animal models (mainly in rats) have shown that vasorelaxation could be evoked by leptin, a hormone found to be elevated in obese subjects that could modulate endovascular nitric oxide syntheses without modifying peripheral blood pressure.[69,70,72] Also, administration of lipid hydroperoxide in rats could stimulate leukocyte–endothelium interaction and retinal venular dilation in retinal microcirculation.[71]

A number of epidemiologic studies have reported that retinal vascular caliber changes were associated with a range of metabolic and cardiovascular risk factors such as CRP, IL-6, LDL-C, and HDL-C in adults[21,28–30,74–76] and even in children,[57,62] which favor endothelial dysfunction. The study in retinal vasculature in obese children may provide epidemiologic basis in such pathophysiologic mechanism.

Clinical Recommendations and Limitation

Since obese children are described to have impairments of endothelial function and increased concentrations of inflammatory markers, it is reasonable to explain the association between obesity and retinal vasculature from the aspect of endothelial dysfunction. Thus, studies on childhood obesity, body fatness indices, and retinal vasculature can provide further pathophysiologic mechanisms of metabolic and vascular regulations among children.

Moreover, a body of evidence has showed that both childhood obesity and abnormal signs of retinal vasculature are associated with metabolic diseases and CVD in later life. It is likely that development of obesity early in life may be sustained into adulthood, and this process can be observed and monitored by the changes of retinal vasculature. Future studies should focus on the retinal vasculature of normal-weight children who have risk factors such as low-fiber and high-fat dietary, decreased physical activity, increased television viewing time, elevated blood pressure, and elevated serum lipid levels in order to assess different ranks of risk in developing obesity and to indicate early behavioral prevention and pharmacologic intervention.

The five studies on childhood obesity, body fatness indices, and retinal vasculature were population-based/school-based, had large sample sizes and multiple ethnicities involved, and followed standard protocols, of which the epidemiologic findings were substantial and reliable. However, there were some inevitable limitations. First, these were all cross-sectional studies. The temporal sequence between childhood obesity and retinal vascular changes were still unclear. Second, due to the variation in population, study design, retinal exam and vessel assessment, childhood obesity classification system, and multiple linear regression model application, the magnitude of retinal vascular signs differed from study to study. Since most of the studies were school-based, there might be selection bias. Third, the body fatness indices and their relationships with retinal vasculature were studied separately. These findings can only estimate but not represent the relationship between integral body fatness (including subcutaneous and visceral fat) and retinal vasculature.

FUTURE STUDIES

Childhood obesity might be an early predisposition for systemic diseases occurring later in life. Microcirculation (arterioles and venules) is a key compartment in translating fat accumulation into cardiovascular risks. Future longitudinal study should concentrate on exploring the true relationship between childhood obesity and retinal vasculature. Once the association is established, predictive value of retinal vasculature on future metabolic disease development can be further explored.

TAKE-HOME MESSAGES

- Childhood obesity has become a critical public health problem both in developed and developing countries.
- The association between pediatric obesity and development of cardiovascular risk factors such as type 2 diabetes, HTN, and coronary heart disease has been widely reported, but very little is known about the process of vascular pathology in childhood long before any overt clinical manifestations show up in adulthood.
- The technology improvement in retinal exam has granted an invasive and convenient method in assessing retinal vasculature characteristics.

- A body of evidence has shown that abnormal retinal vascular signs such as retinal arteriolar narrowing and retinal venular widening are associated with vascular risk and vascular disease and even incidence of CVD.
- The association between childhood obesity and retinal vasculature in children helps to provide deleterious changes for early predisposition of vascular dysfunction and also help to understand better the underlying pathophysiologic mechanism in childhood obesity from the impact of retinal microcirculation.
- Future longitudinal study should be performed on exploring the true relationship between childhood obesity and retinal vasculature. Once the association is established, predictive value of retinal vasculature on future metabolic disease development can be further explored.

References

1. World Health Organization. Obesity: preventing and managing the global epidemic. Report of a WHO consultation. *World Health Organ Tech Rep Ser* 2000;**894**:1–253.
2. Ogden CL, Carroll MD, Kit BK, Flegal KM. Prevalence of obesity and trends in body mass index among US children and adolescents, 1999–2010. *JAMA* 2012;**307**(5):483–90.
3. Han JC, Lawlor DA, Kimm SY. Childhood obesity. *Lancet* 2010;**375**(9727):1737–48.
4. Wang Y, Mi J, Shan XY, Wang QJ, Ge KY. Is China facing an obesity epidemic and the consequences? The trends in obesity and chronic disease in China. *Int J Obes (Lond)* 2007;**31**(1):177–88.
5. Weiss R, Dziura J, Burgert TS, Tamborlane WV, Taksali SE, Yeckel CW, et al. Obesity and the metabolic syndrome in children and adolescents. *N Engl J Med* 2004;**350**(23):2362–74.
6. Mamun AA, Hayatbakhsh MR, O'Callaghan M, Williams G, Najman J. Early overweight and pubertal maturation – pathways of association with young adults' overweight: a longitudinal study. *Int J Obes (Lond)* 2009;**33**(1):14–20.
7. Thomsen SF, Ulrik CS, Kyvik KO, Sorensen TI, Posthuma D, Skadhauge LR, et al. Association between obesity and asthma in a twin cohort. *Allergy* 2007;**62**(10):1199–204.
8. Arslanian S. Type 2 diabetes in children: clinical aspects and risk factors. *Horm Res* 2002;**57**(Suppl 1):19–28.
9. Berenson GS, Srinivasan SR, Bao W, Newman 3rd WP, Tracy RE, Wattigney WA. Association between multiple cardiovascular risk factors and atherosclerosis in children and young adults. The Bogalusa Heart Study. *N Engl J Med* 1998;**338**(23):1650–6.
10. Must A, Jacques PF, Dallal GE, Bajema CJ, Dietz WH. Long-term morbidity and mortality of overweight adolescents. A follow-up of the Harvard Growth Study of 1922 to 1935. *N Engl J Med* 1992;**327**(19):1350–5.
11. Steinberger J, Daniels SR. Obesity, insulin resistance, diabetes, and cardiovascular risk in children: an American Heart Association scientific statement from the Atherosclerosis, Hypertension, and Obesity in the Young Committee (Council on Cardiovascular Disease in the Young) and the Diabetes Committee (Council on Nutrition, Physical Activity, and Metabolism). *Circulation* 2003;**107**(10):1448–53.
12. Hodge S, Bunting BP, Carr E, Strain JJ, Stewart-Knox BJ. Obesity, whole blood serotonin and sex differences in healthy volunteers. *Obes Facts* 2012;**5**(3):399–407.
13. Probst M, Goris M, Vandereycken W, Van Coppenolle H. Body composition of anorexia nervosa patients assessed by underwater weighing and skinfold-thickness measurements before and after weight gain. *Am J Clin Nutr* 2001;**73**(2):190–7.
14. Kyle UG, Bosaeus I, De Lorenzo AD, Deurenberg P, Elia M, Gomez JM, et al. Bioelectrical impedance analysis – part I: review of principles and methods. *Clin Nutr* 2004;**23**(5):1226–43.
15. Tokunaga K, Matsuzawa Y, Ishikawa K, Tarui S. A novel technique for the determination of body fat by computed tomography. *Int J Obes* 1983;**7**(5):437–45.
16. Thomas EL, Saeed N, Hajnal JV, Brynes A, Goldstone AP, Frost G, et al. Magnetic resonance imaging of total body fat. *J Appl Physiol* 1998;**85**(5):1778–85.
17. Wong TY. Is retinal photography useful in the measurement of stroke risk? *Lancet Neurol* 2004;**3**(3):179–83.
18. Wong TY, Mitchell P. The eye in hypertension. *Lancet* 2007;**369**(9559):425–35.
19. Gopinath B, Baur LA, Wang JJ, Teber E, Liew G, Cheung N, et al. Blood pressure is associated with retinal vessel signs in preadolescent children. *J Hypertens* 2010;**28**(7):1406–12.
20. Sun C, Wang JJ, Mackey DA, Wong TY. Retinal vascular caliber: systemic, environmental, and genetic associations. *Surv Ophthalmol* 2009;**54**(1):74–95.
21. Liew G, Sharrett AR, Wang JJ, Klein R, Klein BE, Mitchell P, et al. Relative importance of systemic determinants of retinal arteriolar and venular caliber: the atherosclerosis risk in communities study. *Arch Ophthalmol* 2008;**126**(10):1404–10.
22. Wang JJ, Liew G, Klein R, Rochtchina E, Knudtson MD, Klein BE, et al. Retinal vessel diameter and cardiovascular mortality: pooled data analysis from two older populations. *Eur Heart J* 2007;**28**(16):1984–92.
23. Ikram MK, de Jong FJ, Bos MJ, Vingerling JR, Hofman A, Koudstaal PJ, et al. Retinal vessel diameters and risk of stroke: the Rotterdam Study. *Neurology* 2006;**66**(9):1339–43.
24. Hayreh SS. Hypertensive retinopathy. *Introduction Ophthalmologica* 1989;**198**(4):173–7.
25. Scheie HG. Evaluation of ophthalmoscopic changes of hypertension and arteriolar sclerosis. *AMA Arch Ophthalmol* 1953;**49**(2):117–38.
26. Cheung CY, Tay WT, Mitchell P, Wang JJ, Hsu W, Lee ML, et al. Quantitative and qualitative retinal microvascular characteristics and blood pressure. *J Hypertens* 2011;**29**(7):1380–91.
27. Gopinath B, Baur LA, Teber E, Liew G, Wong TY, Mitchell P. Effect of obesity on retinal vascular structure in pre-adolescent children. *Int J Pediatr Obes* 2011;**6**(2–2):e353–9.
28. Klein R, Klein BE, Moss SE, Wong TY, Sharrett AR. Retinal vascular caliber in persons with type 2 diabetes: the Wisconsin Epidemiological Study of Diabetic Retinopathy: XX. *Ophthalmology* 2006;**113**(9):1488–98.
29. Sun C, Liew G, Wang JJ, Mitchell P, Saw SM, Aung T, et al. Retinal vascular caliber, blood pressure, and cardiovascular risk factors in an Asian population: the Singapore Malay Eye Study. *Invest Ophthalmol Vis Sci* 2008;**49**(5):1784–90.
30. Wong TY, Islam FM, Klein R, Klein BE, Cotch MF, Castro C, et al. Retinal vascular caliber, cardiovascular risk factors, and inflammation: the multi-ethnic study of atherosclerosis (MESA). *Invest Ophthalmol Vis Sci* 2006;**47**(6):2341–50.
31. Wong TY, Knudtson MD, Klein R, Klein BE, Meuer SM, Hubbard LD. Computer-assisted measurement of retinal vessel diameters in the Beaver Dam Eye Study: methodology, correlation between eyes, and effect of refractive errors. *Ophthalmology* 2004;**111**(6):1183–90.
32. Cheung CY, Zheng Y, Hsu W, Lee ML, Lau QP, Mitchell P, et al. Retinal vascular tortuosity, blood pressure, and cardiovascular risk factors. *Ophthalmology* 2011;**118**(5):812–8.

33. Cheung N, Liew G, Lindley RI, Liu EY, Wang JJ, Hand P, et al. Retinal fractals and acute lacunar stroke. *Ann Neurol* 2010;**68**(1):107–11.

34. Cooper LS, Wong TY, Klein R, Sharrett AR, Bryan RN, Hubbard LD, et al. Retinal microvascular abnormalities and MRI-defined subclinical cerebral infarction: the Atherosclerosis Risk in Communities Study. *Stroke* 2006;**37**(1):82–6.

35. Sabanayagam C, Shankar A, Koh D, Chia KS, Saw SM, Lim SC, et al. Retinal microvascular caliber and chronic kidney disease in an Asian population. *Am J Epidemiol* 2009;**169**(5):625–32.

36. Sasongko MB, Wong TY, Donaghue KC, Cheung N, Jenkins AJ, Benitez-Aguirre P, et al. Retinal arteriolar tortuosity is associated with retinopathy and early kidney dysfunction in type 1 diabetes. *Am J Ophthalmol* 2012;**153**(1):176–83.

37. Tikellis G, Arnett DK, Skelton TN, Taylor HW, Klein R, Couper DJ, et al. Retinal arteriolar narrowing and left ventricular hypertrophy in African Americans: the Atherosclerosis Risk in Communities (ARIC) study. *Am J Hypertens* 2008;**21**(3):352–9.

38. Ikram MK, Witteman JC, Vingerling JR, Breteler MM, Hofman A, de Jong PT. Retinal vessel diameters and risk of hypertension: the Rotterdam Study. *Hypertension* 2006;**47**(2):189–94.

39. Deitel M. The International Obesity Task Force and "globesity". *Obes Surg* 2002;**12**(5):613–4.

40. Must A. Does overweight in childhood have an impact on adult health? *Nutr Rev* 2003;**61**(4):139–42.

41. Magarey AM, Daniels LA, Boulton TJ, Cockington RA. Predicting obesity in early adulthood from childhood and parental obesity. *Int J Obes Relat Metab Disord* 2003;**27**(4):505–13.

42. Miller J, Rosenbloom A, Silverstein J. Childhood obesity. *J Clin Endocrinol Metab* 2004;**89**(9):4211–8.

43. Jolliffe CJ, Janssen I. Vascular risks and management of obesity in children and adolescents. *Vasc Health Risk Manag* 2006;**2**(2):171–87.

44. Short KR, Blackett PR, Gardner AW, Copeland KC. Vascular health in children and adolescents: effects of obesity and diabetes. *Vasc Health Risk Manag* 2009;**5**:973–90.

45. Tracy RE, Newman 3rd WP, Wattigney WA, Berenson GS. Risk factors and atherosclerosis in youth autopsy findings of the Bogalusa Heart Study. *Am J Med Sci* 1995;**310**(Suppl 1):S37–41.

46. Zieske AW, Malcom GT, Strong JP. Natural history and risk factors of atherosclerosis in children and youth: the PDAY study. *Pediatr Pathol Mol Med* 2002;**21**(2):213–37.

47. Davis PH, Dawson JD, Riley WA, Lauer RM. Carotid intimal-medial thickness is related to cardiovascular risk factors measured from childhood through middle age: the Muscatine Study. *Circulation* 2001;**104**(23):2815–9.

48. Li S, Chen W, Srinivasan SR, Bond MG, Tang R, Urbina EM, et al. Childhood cardiovascular risk factors and carotid vascular changes in adulthood: the Bogalusa Heart Study. *JAMA* 2003;**290**(17):2271–6.

49. Raitakari OT, Juonala M, Kahonen M, Taittonen L, Laitinen T, Maki-Torkko N, et al. Cardiovascular risk factors in childhood and carotid artery intima-media thickness in adulthood: the Cardiovascular Risk in Young Finns Study. *JAMA* 2003;**290**(17):2277–83.

50. Janssen I, Katzmarzyk PT, Srinivasan SR, Chen W, Malina RM, Bouchard C, et al. Utility of childhood BMI in the prediction of adulthood disease: comparison of national and international references. *Obes Res* 2005;**13**(6):1106–15.

51. Katzmarzyk PT, Tremblay A, Perusse L, Despres JP, Bouchard C. The utility of the international child and adolescent overweight guidelines for predicting coronary heart disease risk factors. *J Clin Epidemiol* 2003;**56**(5):456–62.

52. Parr JC, Spears GF. Mathematic relationships between the width of a retinal artery and the widths of its branches. *Am J Ophthalmol* 1974;**77**(4):478–83.

53. Parr JC, Spears GF. General caliber of the retinal arteries expressed as the equivalent width of the central retinal artery. *Am J Ophthalmol* 1974;**77**(4):472–7.

54. Hubbard LD, Brothers RJ, King WN, Clegg LX, Klein R, Cooper LS, et al. Methods for evaluation of retinal microvascular abnormalities associated with hypertension/sclerosis in the Atherosclerosis Risk in Communities Study. *Ophthalmology* 1999;**106**(12):2269–80.

55. Knudtson MD, Lee KE, Hubbard LD, Wong TY, Klein R, Klein BE. Revised formulas for summarizing retinal vessel diameters. *Curr Eye Res* 2003;**27**(3):143–9.

56. Mahal S, Strain WD, Martinez-Perez ME, Thom SA, Chaturvedi N, Hughes AD. Comparison of the retinal microvasculature in European and African-Caribbean people with diabetes. *Clin Sci (Lond)* 2009;**117**(6):229–36.

57. Owen CG, Rudnicka AR, Nightingale CM, Mullen R, Barman SA, Sattar N, et al. Retinal arteriolar tortuosity and cardiovascular risk factors in a multi-ethnic population study of 10-year-old children; the Child Heart and Health Study in England (CHASE). *Arterioscler Thromb Vasc Biol* 2011;**31**(8):1933–8.

58. Gopinath B, Baur LA, Pfund N, Burlutsky G, Mitchell P. Differences in association between birth parameters and blood pressure in children from preschool to high school. *J Hum Hypertens* 2013;**27**(2):79–84.

59. Gopinath B, Flood VM, Rochtchina E, Baur LA, Smith W, Mitchell P. Influence of high glycemic index and glycemic load diets on blood pressure during adolescence. *Hypertension* 2012;**59**(6):1272–7.

60. Gopinath B, Flood VM, Wang JJ, Smith W, Rochtchina E, Louie JC, et al. Carbohydrate nutrition is associated with changes in the retinal vascular structure and branching pattern in children. *Am J Clin Nutr* 2012;**95**(5):1215–22.

61. Gopinath B, Schneider J, Hickson L, McMahon CM, Burlutsky G, Leeder SR, et al. Hearing handicap, rather than measured hearing impairment, predicts poorer quality of life over 10 years in older adults. *Maturitas* 2012;**72**(2):146–51.

62. Hanssen H, Siegrist M, Neidig M, Renner A, Birzele P, Siclovan A, et al. Retinal vessel diameter, obesity and metabolic risk factors in school children (JuvenTUM 3). *Atherosclerosis* 2012;**221**(1):242–8.

63. Wong TY, Klein R, Klein BE, Meuer SM, Hubbard LD. Retinal vessel diameters and their associations with age and blood pressure. *Invest Ophthalmol Vis Sci* 2003;**44**(11):4644–50.

64. Wong TY, Klein R, Sharrett AR, Schmidt MI, Pankow JS, Couper DJ, et al. Retinal arteriolar narrowing and risk of diabetes mellitus in middle-aged persons. *JAMA* 2002;**287**(19):2528–33.

65. Li LJ, Cheung CY, Liu Y, Chia A, Selvaraj P, Lin XY, et al. Influence of blood pressure on retinal vascular caliber in young children. *Ophthalmology* 2011;**118**(7):1459–65.

66. Taylor B, Rochtchina E, Wang JJ, Wong TY, Heikal S, Saw SM, et al. Body mass index and its effects on retinal vessel diameter in 6-year-old children. *Int J Obes (Lond)* 2007;**31**(10):1527–33.

67. Cheung N, Saw SM, Islam FM, Rogers SL, Shankar A, de Haseth K, et al. BMI and retinal vascular caliber in children. *Obesity (Silver Spring)* 2007;**15**(1):209–15.

68. Li LJ, Cheung CY, Chia A, Selvaraj P, Lin XY, Mitchell P, et al. The relationship of body fatness indices and retinal vascular caliber in children. *Int J Pediatr Obes* 2011;**6**(3–4):267–74.

69. Fernandez-Alfonso MS. Regulation of vascular tone: the fat connection. *Hypertension* 2004;**44**(3):255–6.

70. Lembo G, Vecchione C, Fratta L, Marino G, Trimarco V, d'Amati G, et al. Leptin induces direct vasodilation through distinct endothelial mechanisms. *Diabetes* 2000;**49**(2):293–7.

71. Tamai K, Matsubara A, Tomida K, Matsuda Y, Morita H, Armstrong D, et al. Lipid hydroperoxide stimulates leukocyte-endothelium interaction in the retinal microcirculation. *Exp Eye Res* 2002;**75**(1):69–75.

72. Vecchione C, Aretini A, Maffei A, Marino G, Selvetella G, Poulet R, et al. Cooperation between insulin and leptin in the modulation of vascular tone. *Hypertension* 2003;**42**(2):166–70.

73. Nguyen TT, Wong TY. Retinal vascular manifestations of metabolic disorders. *Trends Endocrinol Metab* 2006;**17**(7):262–8.

74. Yim-Lui Cheung C, Wong TY, Lamoureux EL, Sabanayagam C, Li J, Lee J, et al. C-reactive protein and retinal microvascular caliber in a multiethnic Asian population. *Am J Epidemiol* 2010;**171**(2):206–13.

75. Stettler C, Witt N, Tapp RJ, Thom S, Allemann S, Tillin T, et al. Serum amyloid A, C-reactive protein, and retinal microvascular changes in hypertensive diabetic and nondiabetic individuals: an Anglo-Scandinavian Cardiac Outcomes Trial (ASCOT) substudy. *Diabetes Care* 2009;**32**(6):1098–100.

76. Ikram MK, de Jong FJ, Vingerling JR, Witteman JC, Hofman A, Breteler MM, et al. Are retinal arteriolar or venular diameters associated with markers for cardiovascular disorders? The Rotterdam Study. *Invest Ophthalmol Vis Sci* 2004;**45**(7):2129–34.

Visual Evoked Potentials and Type-2 Diabetes Mellitus

Deepika Chopra

Government Medical College, Amritsar, Punjab, India

INTRODUCTION

The word 'diabetes', meaning a 'siphon' or 'running through', was used by Areteaus the Cappadocian in the second century AD to describe polyuria; he also noted thirst and emaciation as features of this fatal disease. Sushuruta in India is said to have referred to diabetes mellitus (DM) as "honey urine" in the fifth century AD. Willis in 1679 wrote "those laboring with the disease piss a great deal more then they drink" and went on to say that "urine is wonderfully sweet as if it were imbedded with honey or sugar". In 1815, the famous French chemist Chevreul discovered that the sugar in the urine was glucose.[1]

The incidence of diabetes is increasing day by day due to increasing prevalence of obesity and lack of physical activity. According to data, 171 million people in the world were suffering from diabetes in the year 2000, and this may increase to 366 million by 2030.[2,3]

The metabolic dysregulation associated with DM causes secondary pathophysiologic changes in multiorgan systems that impose a tremendous burden on the individual with diabetes and on the healthcare system.[4]

Normally, the hormone insulin, made by the pancreas, regulates blood sugar levels. The basic defect is an absolute or relative lack of insulin, which leads to abnormalities of metabolism.

The two broad categories of DM are designated as type 1 and type 2.

- Type 1 DM (also known as insulin-dependent diabetes mellitus (IDDM) or juvenile diabetes) is characterized by beta-cell destruction caused by an autoimmune process, usually leading to absolute insulin deficiency.

- Type 2 DM (also known as noninsulin dependent diabetes mellitus (NIDDM) or adult onset) is characterized by insulin resistance in peripheral tissues and an insulin secretory defect of beta cells of the pancreas. This is highly associated with a family history of diabetes, older age, obesity, and lack of exercise.[5–7]

In both types of DM, metabolism of all the main foodstuffs is altered not only in carbohydrates but also in proteins and fats. Most of the body cells are not able to utilize the available glucose efficiently either due to absolute lack of insulin or insulin resistance. As a result, utilization of glucose falls tremendously and utilization of fats and proteins increases, which leads to a catabolic state. The catabolic state also includes inadequate synthesis and breakdown of proteins and a complex hormone imbalance. The result is poor tissue growth and repair.[8,9]

It has been observed that there is a clear relationship between the degree of hyperglycemia and risk of developing diabetic retinopathy, diabetic nephropathy, and diabetic neuropathy. The risk for microvascular and neuropathic complication is related to both duration of diabetes and the severity of hyperglycemia.[10]

Diabetic neuropathy is the most common clinical picture of nervous system disorder caused by DM. It is well known that patients with DM develop peripheral and autonomic neuropathy. Recent reviews have suggested that they may also have central neuropathy or degeneration of the higher nervous system, which frequently gets involved with complications of diabetes as the duration of diabetes increases. Though impairment of the central nervous system (CNS) is a frequent complication, its clinical importance is still underestimated.[11–15]

For many years, it was believed that visual dysfunction that develops in diabetic patients was due to diabetic retinopathy, which in turn had a vascular pathology. However, now it has been emphasized that neuronal cells of the retina also are affected by diabetes, resulting in dysfunction and even degeneration of some neuronal cells.[16]

The concept of 'diabetic encephalopathy' was introduced by De Jong in a case report in 1950. He observed diffuse histologic abnormalities throughout the CNS. He pointed to clinical and pathologic evidence that the brain parenchyma might be affected in DM. The information about these changes provided by electroencephalography (EEG) was limited, particularly in the assessment of deeper brain structures as, for many years, EEG was the only technique available to study the electrophysiologic activity of the brain.

Since the 1990s, the advent of newer electroneurophysiologic techniques to assess retinal and cerebral function such as electroretinography (ERG) and measurement of electrical-evoked potentials such as visual evoked potential (VEP) has increased our understanding of normal visual function and possible effects that diabetes may exert.[17]

VISUAL EVOKED POTENTIALS

VEPs are visually evoked electrophysiologic signals extracted from the electroencephalographic activity in the visual cortex recorded from the overlying scalp in response to the stimulation of visual receptors. VEPs depend on the functional integrity of central vision at any level of the visual pathway including the eye, retina, optic nerve, optic radiations, and occipital cortex.

The International Society for Clinical Electrophysiology of Vision (ISCEV) has selected a subset of stimuli and recording conditions, which provide core clinical information that can be performed by most clinical electrophysiology laboratories throughout the world. These are:

1. Pattern-reversal VEP is elicited by checkerboard stimuli with large 1° (60 min of arc) and small 0.25° (15 min) checks.
2. Pattern onset/offset VEP is elicited by checkerboard stimuli with large 1° (60 min) and small 0.25° (15 min) checks.
3. Flash VEP is elicited by a brief luminance increment, a flash, which subtends a visual field of at least 20°.

Pattern reversal is the preferred stimulus for most clinical purposes. Pattern-reversal VEPs are less variable in waveform and timing than the VEPs elicited by other stimuli. The pattern-reversal stimulus consists of black

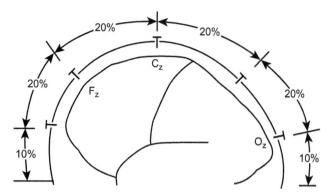

FIGURE 21.1 Position of active and reference electrodes for standard responses in visual evoked potentials.

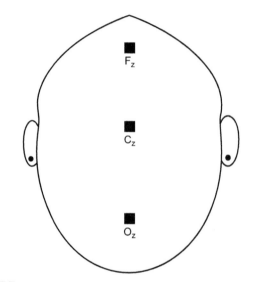

FIGURE 21.2 O_z - active, F_z- reference electrode, C_z – ground electrode (The subscript z indicates a midline position)

and white checks that change phase (i.e., black to white and white to black) abruptly and repeatedly at a specified number of reversals per second.

Skin electrodes such as sintered silver–silver chloride, standard silver–silver chloride, or gold disk electrodes are recommended for recording VEPs. The skin should be prepared by cleaning and a suitable paste or gel used to ensure good, stable electrical connection. Figure 21.1 shows the position of scalp electrodes, which should be placed relative to bony landmarks, in proportion to the size of the head, according to the International 10/20 system. These electrode positions are used to ensure reproducible electrode placement in serial studies. The anterior/posterior midline measurements are based on the distance between the nasion and the inion over the vertex. The active electrode is placed on the scalp over the visual cortex at O_z with the reference electrode at F_z (Fig. 21.2). A separate electrode should be attached to a

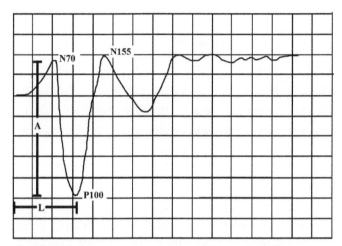

FIGURE 21.3 Visual evoked potentials in normal individual. A: amplitude; L: latency.

relatively indifferent point and connected to the ground; commonly used ground electrode positions include the forehead, vertex (C_z), mastoid, and earlobes.[18,19]

VEP is primarily a reflection of activity originating in the central 3–6° of the visual field, which is relayed to the surface of the occipital lobe.[20]

VEP consist of series of waveforms of opposite polarity, the negative waveform (denoted as N) and a positive waveform (denoted as P), which are followed by the approximate latency in milliseconds (Fig. 21.3). The P100 waveform is generated in the striate and peristriate occipital cortex due to activation of the primary visual cortex and also due to thalamocortical fiber discharge. N70 reflects activity of the fovea and primary visual cortex while N155 reflects the activity of the visual association area.[21]

VEP is a simple, sensitive, and noninvasive method for detecting early alterations in central optic pathways in diabetics. It has no side effects, and nerve tract damage increases the latency and reduces the amplitude of the response. Its use has occupied the electrophysiologic mainstream for evaluating visual functions in humans. Pattern VEPs have become a routine diagnostic method in neurology setups, because its intraindividual stability allows the detection of several neurologic diseases affecting the visual pathway.[22–24]

REVIEW OF LITERATURE

Neurobehavioral and electrophysiologic studies were carried out by Ryu (1995) to determine the effect of DM on brain function. The aim of this study was to determine the change of brain function in diabetics and to evaluate the correlation between brain function and clinical factors. The results showed that latencies of brainstem auditory evoked potentials (BAEP), sensory evoked potentials (SEP) wave N6 P100 N125 P160 of VEP, and the interpeak latency of I-V of BAEP were prolonged significantly compared with the controls. These results demonstrate that brain dysfunction is present in type-2 DM.[25]

Imam (2009) suggested that in DM visual defects are usually of retina, optic nerve, and further visual pathways up to the striate area. The detection of early CNS damage is only possible through electrophysiologic and psychomotor tests.[26]

Comi (1997) revealed abnormalities of central afferent and efferent pathways by evoked potentials in diabetic patients. He suggested that evoked potentials can be abnormal even in patients without neuropathy.[27]

Goldenberg et al. (2004) found that, in spite of missing clinical signs of CNS lesion in type 1 DM patients, there is a statistically significant prolongation of central motor conduction time (CMCT) compared to the controls.[13]

The exact pathophysiology of the central nervous dysfunction is not clear, but it seems to be due to various factors, involving metabolic and vascular factors, similar to the pathogenesis of diabetic peripheral neuropathy.

The delay in central transmission time in diabetics may be related to diffuse neuropathologic changes that have been found in the optic nerves, periventricular regions, brainstem, and spinal cord in postmortem pathologic studies. Similar changes have been found in animals with experimental diabetes.[28]

It has been suggested that hyperglycemia-induced metabolic and enzymatic changes lead to production of toxic metabolites. Ischemia, reduced protein synthesis, depleted myoinositol, and high sorbitol levels have been demonstrated in patients with diabetes, which alters the functional and structural properties of Schwann cells (SC) and axons and may result in nerve fiber loss in optic nerves.[29,30]

A study was conducted to obtain ample electrophysiologic documentation of possible neurologic abnormalities in 25 diabetic patients (both type 1 and type 2 DM patients) with a short duration of disease and without overt complications, taking into account metabolic control. It was found that neurophysiologic abnormalities were present in type 1 and type 2 DM patients only a few years after clinical diagnosis but before the appearance of overt complications, and these abnormalities seem to be correlated with metabolic control status.[31]

To evaluate central optic pathway involvement in diabetics, VEP, and in particular the latency of positive peak (P100), was studied in 35 diabetic patients without

TABLE 21.1 Visual Evoked Potential Parameters of Left and Right Eye

Parameters		Group 1 Mean ± SD		Group 2 Mean ± SD		Group 3 Mean ± SD		Group 4 Mean ± SD	
		Left	Right	Left	Right	Left	Right	Left	Right
Latency (ms)	N70	65.59 ± 3.35	65.72 ± 3.30	68.78 ± 3.07	68.11 ± 2.54	68.58 ± 3.54	68.14 ± 3.63	69.57 ± 6.84	69.34 ± 7.25
	P100	95.28 ± 3.16	95.03 ± 3.33	97.54 ± 2.45	97.34 ± 2.81	98.37 ± 2.77	98.46 ± 2.73	101.70 ± 5.29	101.57 ± 5.61
Amplitude (µv)	N70-P100	5.42 ± 2.00	5.33 ± 1.73	5.28 ± 1.97	5.34 ± 2.25	4.38 ± 1.38	4.53 ± 1.53	4.21 ± 1.41	4.00 ± 1.35
	P100-N155	7.62 ± 2.73	7.65 ± 2.59	7.42 ± 2.23	7.66 ± 2.68	5.90 ± 1.76	6.12 ± 1.60	5.86 ± 2.02	6.02 ± 1.76

ms: millisecond; SD: standard deviation; µv: microvolt.

retinopathy and 35 normal controls using pattern-reversal stimulation. A positive correlation was found between latencies of VEP and duration of disease.[32]

One group of researchers examined flash ERG in a group of recently diagnosed type 1 DM patients and compared the results with changes in VEP. They found no statistically significant differences in flash ERG, but P100 latency was markedly increased in the diabetic group compared to the non-diabetic control group.[17]

A study conducted among 51 diabetic patients with type 2 DM to evaluate central neuropathy by multimodel evoked potentials concluded that central and peripheral neuropathies and DM are related to duration of disease and not to degree of hyperglycemia and metabolic control.[33]

A comparison study was conducted among three groups (30 patients each) of type 2 DM patients with different durations of disease and 30 age- and sex-matched healthy controls. The groups were divided as:

- Group 1 → 30 controls, age- and sex-matched healthy individuals.
- Group 2 → 30 patients with type 2 DM of duration less than 10 years.
- Group 3 → 30 patients with type 2 DM of duration 10–15 years.
- Group 4 → 30 patients with type 2 DM of duration greater than 15 years.

Subjects having a history of any disorder that could influence the interpretation of results, like a demyelinating disorder such as multiple sclerosis, retinopathy, cataract, glaucoma, vitreous opacities, or having any evidence of optic atrophy and visual acuity less than 6/18 even with corrective lenses, were excluded from the study. Sensory neuropathy was excluded by monofilament test.

Parameters recorded were: the latencies of waves N70 and P100 (in milliseconds) and peak-to-peak amplitudes of waves N70-P100 and P100-N155 (in microvolts).

Mean value ± standard deviations (SD) of VEP parameters of left and right eye in the control group as well as in the diabetic patients of variable duration (groups 2, 3, and 4) are shown in Table 21.1. VEP parameters of both the eyes in all the study groups were statistically

analyzed and no significant difference was found between parameters of left and right eye (Table 21.2). So, for convenience, further intergroup analysis interpretations of results were confined to left eye parameters only (Table 21.3).

Figure 21.4 shows that the absolute latency of N70 has increased in all the diabetic groups as compared with normal healthy subjects, indicating early central neuropathic changes in DM.

Figure 21.5 shows that there is prolongation of P100 latency in all diabetic groups as compared to healthy subjects. The P100 is a prominent peak that shows relatively little variation between subjects, minimal within-subject interocular difference, and minimal variation with repeated measurements over time. Delay in P100 latency is highly significant in diabetic groups 3 (DM 10–15 years) and 4 (DM greater than 15 years), whereas significant prolongation is seen in group 2 (DM less than 10 years) patients as compared with that of normal healthy controls (group 1), indicating that diabetic patients with more than 10 years of disease have undergone more damage to central neurons as compared with diabetic patients with a disease duration less than 10 years. This shows that center degeneration caused by DM is positively correlated with the duration of the disease.

Figure 21.6 shows that N70-P100 and P100-N155 amplitudes are significantly decreased in all patients of disease duration greater than 10 years as compared with healthy controls. These findings suggest that there is definite neurologic deficit in type 2 DM, which can involve CNS at the much earlier stage and does increase with increased duration of disease.

TABLE 21.2 Statistical Analyses of Visual Evoked Potential Parameters of Right and Left Eye

Parameters		Group 1	Group 2	Group 3	Group 4
Latency P value	N70	0.666[NS]	0.091[NS]	0.188[NS]	0.538[NS]
	P100	0.666[NS]	0.552[NS]	0.830[NS]	0.713[NS]
Amplitude P value	N70-P100	0.680[NS]	0.754[NS]	0.483[NS]	0.133[NS]
	P100-N155	0.887[NS]	0.272[NS]	0.265[NS]	0.406[NS]

NS: not significant (P > 0.05).

TABLE 21.3 Intergroup Statistical Analysis of Visual Evoked Potentials Parameters of Left Eyes

Parameters		Gp 2 vs. Gp 1	Gp 3 vs. Gp 1	Gp 4 vs. Gp 1	Gp 3 vs. Gp 2	Gp 4 vs. Gp 2	Gp 4 vs. Gp 3
Latency	N70	< 0.001[s*]	0.001[s]	0.006[s]	0.821[NS]	0.564[NS]	0.484[NS]
	P100	0.003[s]	< 0.001[s*]	< 0.001[s*]	0.222	< 0.001[s*]	0.003[s]
Amplitude	N70-P100	0.794[NS]	0.023[s]	0.009[s]	0.045[s]	0.018	0.630[NS]
	P100-N155	0.760[NS]	0.005[s]	0.006[s]	0.005[s]	0.006[s]	0.942[NS]

Gp: group; NS: not significant (P > 0.05); S: significant (P < 0.05); S*: highly significant (P < 0.001).

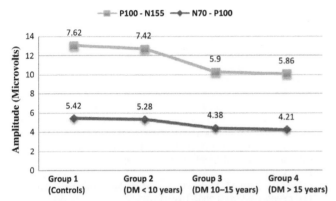

FIGURE 21.4 Graphical analysis of mean value of N70 latency in various study groups. The mean latency of N70 has increased in all the diabetic groups as compared with normal healthy subjects, indicating early central neuropathic changes in diabetes mellitus (DM).

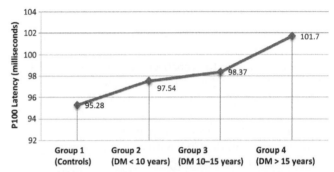

FIGURE 21.6 Graphical analysis of mean value of P100-N70 and N70-P100 amplitude in various study groups. N70-P100 and P100-N155 amplitudes are significantly decreased in all the patients of disease duration greater than 10 years as compared with healthy controls. These findings suggest that there is definite neurologic deficit in type 2 diabetes mellitus (DM), which can involve the central nervous system at a much earlier stage and increases with increased duration of disease.

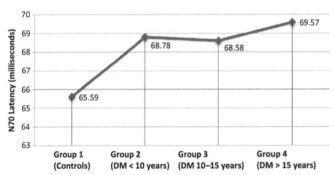

FIGURE 21.5 Graphical analysis of mean value of P100 latency in various study groups. P100 latency is prolonged in all diabetic groups as compared with healthy subjects. Delay in P100 latency is highly significant in diabetic groups 3 (DM 10–15 years) and 4 (DM > 15 years), whereas just significant prolongation is seen in group 2 (DM < 10 years) patients as compared with that of normal healthy controls (group 1), indicating that diabetic patients with more than 10 years of disease have undergone more damage to central neurons as compared with diabetic patients of disease duration less than 10 years. This shows that center degeneration caused by diabetes mellitus is positively correlated with the duration of the disease.

Neuropoietic cytokines, including interleukin (IL)-1, IL-6, leukemia inhibitory factor (LIF), ciliary neurotrophic factor (CNTF), tumor necrosis factor (TNF)-α, and transforming growth factor (TGF)-β, exhibit pleiotropic effects on homeostasis of the glia and neurons in the central, peripheral, and autonomic nervous systems. These cytokines are produced locally by resident and infiltrating macrophages, lymphocytes, mast cells, SC, fibroblasts, and sensory neurons.[29,30]

Accumulation of neurodegenerative mediators probably delays conduction in the visual pathway, which can be the probable cause of delay in latencies found in diabetics as compared with healthy controls. As the duration of diabetes increases, these mediators accumulate further, which can further delay latencies in diabetics with more duration of disease as compared with diabetics with lesser duration of disease.

Thus, it can be concluded that changes in VEP response do occur in diabetic patients much earlier than the development of overt retinopathy or clinically apparent sensory neuropathy, and these changes are positively correlated with duration of disease.

Therefore, VEP measurement, which is a highly sensitive, reliable, noninvasive, and reproducible method for detecting early alterations in central optic pathways in diabetics, should be recommended when possible and must be added to the list of screening tools for a more complete and early assessment of the neurologic involvement of diabetic patients to offer an early opportunity for early and proper management of the disease.

TAKE-HOME MESSAGES

- Type 2 DM is becoming one of the most serious challenges to healthcare primarily because of the increase in prevalence of sedentary lifestyle and obesity. Neuropathies are common complications of type 2 DM.
- Electrophysiology can play a very important role in determining peripheral and central neuropathy in diabetic patients.
- VEPs are visually evoked electrophysiologic signals extracted from the electroencephalographic activity in the visual cortex recorded from the overlying scalp in response to the stimulation of visual receptors.
- VEPs depend on the functional integrity of central vision at any level visual pathway (i.e., retina, optic nerve, optic radiations, and occipital cortex). Neural damage can delay the latency and reduce the amplitude. VEPs can be used to evaluate disturbances in the CNS with a simple, sensitive, and noninvasive methodology.
- Studies show that changes in VEP response do occur in diabetic patients much earlier than the development of overt retinopathy or clinically apparent sensory neuropathy.
- VEP abnormalities are positively correlated with duration of disease. So VEP measurements should be recommended when possible and must be added to the list of tools for a more complete and early assessment of the neurologic involvement of diabetic patients to offer an early opportunity for proper management.

References

1. Wright S. The endocrine functions of the pancreas. In: Keele CA, Neil E, Joels N, editors. *Samson Wright's Applied Physiology*. 13th ed. New Delhi: Oxford University Press; 2006. p. 508.
2. Wild S, Roglic G, Green A, et al. Global prevalence of diabetes, estimates for the year 2000 and predictions for the 2030. *Diab Care* 2004;27(5):1047–53.
3. Hogan P, Dall T, Nikolov P. Economic costs of diabetes in the US in 2002, American Diabetes Association. *Diab Care* 2003;26:917–32.
4. Powers CA. Endocrinology and metabolism. In: Fausi AS, Braunwald E, Kasper DL, et al., editors. 17th ed. *Harrison's Principal of Internal Medicine*, Vol. II. New York: McGraw Hill Companies; 2008. p. 2275.
5. Achenbach P, Bonifacio E, Koczwara K, et al. Natural history of type 1 diabetes. *Diabetes* 2005;54(2):25–31.
6. Mayfield J. Diagnosis and classification of diabetes mellitus: new criteria. *Am Fam Physician* 1998;58(6):1355–70.
7. American diabetes association. Diagnosis and classification of diabetes mellitus. *Diab Care* 2004;27(1):S5–S10.
8. Guyton AC, Hall JE. Insulin, glucagon and diabetes mellitus. In: Guyton AC, Hall JE, editors. *Textbook of Medical Physiology*. 11th ed. New Delhi: Elsevier Indis Private Limited; 2006. pp. 972–6.
9. Hoffman JM. Clinical and laboratory findings. In: Beigelman PM, Kumar D, editors. *Diabetes Mellitus for House Officer*. 1st ed. Baltimore: Williams and Wilkins; 1986. p. 11.
10. Byron JH. Review, complication of diabetes mellitus. *Int J Diab Dev Countries* 2005;25:63–9.
11. Raman PG, Sodani A, George B. A study of visual evoked potential changes in diabetes mellitus. *Int J Diab Dev Countries* 1997;17:69–73.
12. Varkonyi TT, Peto T, Degi R, et al. Impairment of visual evoked potentials. An early central manifestation of diabetic neuropathy. *Diab Care* 2002;25:1661–2.
13. Goldenberg Z, Kucera P, Brezinova M, et al. Clinically unapparent central motor pathways lesion in patients with type I diabetes mellitus. A transcranial magnetic stimulation study. *Bratisl Lek Listy* 2004;105:400–3.
14. Suzuki C, Ozaki I, Tanosaki M, et al. Peripheral and central conduction abnormalities in diabetes mellitus. *Neurology* 2000;4:1932–7.
15. Uzon N, Uludoz D, Mikla S, et al. Evolution of asymptomatic central neuropathy in type 1 diabetes mellitus. *Electromyoer Clin Neurophysiol.* 2006;463:131–7.
16. Kern TS, Barber AJ. Retinal ganglion cells in diabetes. *J Physiol.* 2008;586:4401–8.
17. Ewing FME, Deary IJ, Strachan MWJ, et al. Seeing beyond retinopathy in diabetes: electrophysiological and psychophysical abnormalities and alterations in vision. *Endocr Rev.* 1998;194:462–76.
18. Odom JV, Bach M, Brigell M, et al. ISCEV standard for clinical visual evoked potentials (2009 update). *Doc Ophthalmol.* 2010;120:11–9.
19. Odom JV, Bach M, Barber C. Visual evoked potentials standard. *Doc Ophthalmol.* 2004;2004(108):115–23.
20. Misra UK, Kalita J. Visual evoked potential. In: Misra UK, Kalita J, editors. *Clinical Neurophysiology*. 2nd ed. New Delhi: Elsevier; 2006. pp. 309–24.
21. Jain AK. Neuro-electro diagnostic techniques. In: Jain AK, editor. *Manual of Practical Physiology For MBBS*. New Delhi: Arya Publication; 2003. pp. 261–4.
22. Azal O, Ozkardes A, Onde ME, et al. Visual evoked potentials in diabetic patients. *Tr J Med Sci.* 1998;28:139–42.
23. Mincewicz MW, Dulska HT, Szajewska BE, et al. Co-existence of abnormalities in the peripheral nervous system and in the auditory and visual evoked potentials in children with type 1 diabetes. *Diabetol Doswiadczalna Liniczna* 2007;71:44–9.
24. Antal A, Kincses TZ, Nitsche MA, et al. Excitability changes induced in the human primary visual cortex by transcranial direct current stimulation: direct electrophysiological evidence. *Invest Ophthalmol Vis Sci.* 2004;452:702–7.
25. Ryu N, Chin K. Decreased brain function in patients with non insulin-dependent diabetes mellitus. *No To Shinkei* 1995;476:543–8.
26. Imam M, Shehata OH. Subclinical central neuropathy in type 2 diabetes mellitus. *Bull Alex Fac Med.* 2009;2006(451):65–73.
27. Comi G. Evoked potentials in diabetes mellitus. *Clin Neurosci* 1997;463:74–9.
28. Fawi GH, Khalifa GA, Kasim MA. Central and peripheral conduction abnormalities in diabetes mellitus. *Egypt J Neurol Psychiat Neurosurg* 2005;421:209–21.
29. Myers AR. Endocrine and metabolic disease. In: Myers AR, editor. *National Medical Series for Independent Study, Medicine*. 3rd ed. New Delhi: B.I. Waverly Private Ltd; 1997. 1997;475–84.
30. Pan CH, Chen SS. Pattern shift visual evoked potentials in diabetes mellitus. *Gaoxiong Yi Xue Ke Xue Za Zhi* 1992;8(7):374–83.
31. Pozzessre G, Rizzo PA, Valle E, et al. A longitudinal study of multimodel evoked potential in diabetes mellitus. *Diabetes Res.* 1989;101:17–20.
32. Moreo G, Mariani E, Pizzamiglio G, et al. Visual evoked potentials in NIDDM: a longitudinal study. *Diabetologia* 1995;1995(385):573–6.
33. Dolu H, Ulas VH, Bolu E, et al. Evaluation of central neuropathy in type II diabetes mellitus by multimodel evoked potentials. *Acta Neurol Belg* 2003;103:2006–11.

MACRONUTRIENTS

Glycemic Index and Age-Related Macular Degeneration

Yi-Ling Huang, Min-Lee Chang, Chung-Jung Chiu

Jean Mayer USDA Human Nutrition Research Center on Aging at Tufts University, Boston, MA

INTRODUCTION

Glucose homeostasis and carbohydrate nutrition play a crucial role in human aging as well as in disease pathogenesis. Many pathophysiologic effects follow postprandial hyperglycemia after eating a high-glycemic index (GI) meal (Fig. 22.1). It may manifest different diseases, such as diabetes, diabetic retinopathy (DR), age-related macular degeneration (AMD), and cardiovascular disease (CVD). However, most of the studies on hyperglycemia have focused on diabetes or diabetic complications instead of age-related disorders. Studies also suggest that older people with diabetes are more susceptible to age-related diseases, including AMD, than people without diabetes. Therefore, this review will focus on how dietary hyperglycemia may increase the risk for metabolic retinal disease and the plausible mechanisms.

GLYCEMIC INDEX

The GI, proposed by Dr David J Jenkins and colleagues in 1981,[1] is a measure of the effects of carbohydrate-containing foods on postprandial glycemia. It is defined as the percentage of the area under 2-hour blood glucose curve (AUC) following the ingestion of a tested food versus a standard food (Fig. 22.2). Foods that break down quickly during digestion and result in higher levels of blood glucose have a high GI. Foods that break down more slowly, releasing glucose more gradually into the bloodstream, have a low GI. The most updated formal publication of GI values for almost 2500 food items is available online at http://dx.doi.org/10.2337/dc08-1239.

GI values have been measured in variety of subjects, including both diabetic and nondiabetic healthy people.

However, the published GI values for apparently similar foods may vary from study to study. This is because the GI of a food is determined by several factors that affect gastric emptying and rate of intestinal digestion, including amylase, fiber, moisture content, cooking time, ingredients or processing methods, etc.[2,3]

The glycemic response to mixed meals can be predicted with reasonable accuracy from the GI of constituent foods when standard methods are used. The overall GI for a person's diet (dietary GI) can be calculated as the weighted mean of the GI scores for each food item, with the amount of carbohydrate consumed from each food item as the weight $(S(GI_i \times W_i)/W)$,[4] where GI_i is the GI of an individual food, W is the weight of total carbohydrate, and W_i is the weight of available carbohydrate of individual food (i.e., the fiber content is subtracted from the carbohydrate content).

Another measure of carbohydrate nutrition, glycemic load (GL), was defined to summarize the combined effects of quantity and quality of carbohydrate foods. It is calculated as the product of the GI and the carbohydrate amount (in grams) of the food item divided by 100.

Glycemic Index to Metabolism and Maintaining Physical Function

The human body has an obligatory requirement for glucose, approaching 200 g/day.[5] The blood glucose concentration is tightly regulated by homeostatic regulatory systems and maintained between 40 mg/dL (2.2 mmol/L) and 180 mg/dL (10.0 mmol/L). Hypoglycemia below the lower limit may result in coma, seizure, or even death. Hyperglycemia, exceeding the upper limit, is associated with immediate (glycosuria and calorie loss) and long-term (retinopathy, atherosclerosis, renal failure, etc.) consequences.

Handbook of Nutrition, Diet, and the Eye
http://dx.doi.org/10.1016/B978-0-12-401717-7.00022-8

A low-GI food results in a better postprandial glycemia because it raises blood glucose gradually. Gradual increases in blood glucose reduce the postprandial levels of gut hormones (e.g., incretins) and insulin. This will suppress the free fatty acid concentrations and the counter-regulatory responses. Under this condition, the respiratory quotient is raised (i.e., sustained tissue insulinization) and glucose is withdrawn from the circulation at a faster rate (i.e., better glucose clearance).

In contrast, the rapid absorption of glucose elicits a sequence of hormonal events that challenge glucose homoeostasis after high-GI meals. Compared with a low-GI meal, a high-GI meal induces a significant excursion

and fluctuation of blood glucose over the whole postprandial period. This results in a high insulin to glucagon ratio during early postprandial stage (0–2 hours), hypoglycemia, suppressed free fatty acid concentration during middle postprandial stage (2–4 hours), and counter-regulatory hormone responses and a compensatory increase in free fatty acid concentration during late postprandial stage (4–6 hours).

GLYCEMIC INDEX AND HUMAN DISEASES

The quantity (amount) and glycemic quality (GI) of carbohydrate foods have been related to the risk for diabetes, CVD, and AMD in human studies. Many epidemiologic observations and both long-term and short-term interventional trials have focused on the amount of carbohydrate intake to risk of diabetes. However, the conclusions of the studies are inconsistent and difficult to interpret because of the differences in the degrees of weight loss between the diet groups and in the duration of follow-up. However, overall, GI offers a better measure of carbohydrate foods than quantity for the associations with these diseases.

Glycemic Index to Diabetes

Type 2 diabetes is characterized by insulin resistance and reduced responsiveness of the pancreatic islet cells to glucose, ultimately leading to hyperglycemia and the development of clinical diabetes. Animal models have shown that hyperglycemia contributes

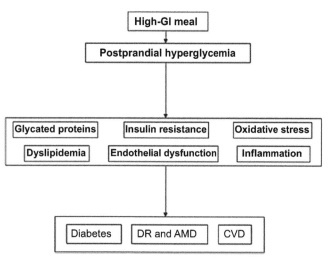

FIGURE 22.1 Adverse metabolic events relating high-glycemic index (GI) diets to diabetes and cardiovascular disease (CVD). AMD: age-related macular degeneration; DR: diabetic retinopathy.

FIGURE 22.2 Glycemic responses demonstrate the definition of glycemic index (GI). AUC: area under the curve; h: hour.

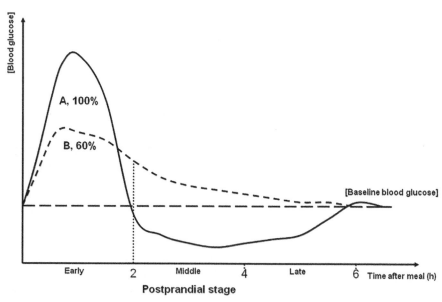

Curve A: Glucose (Reference food): $GI_{Glucose} = 100$

Curve B: Test food: $GI_{Test\ food} = (AUC_B / AUC_A) \times 100 = 60$

to insulin resistance and defects in insulin secretion.[6,7] Thus, dietary factors that decrease plasma glucose and insulin responses could plausibly decrease the risk of type 2 diabetes. Over the past three decades, studies have demonstrated that, independent of the effect of fat intake, consuming a low-GI diet may improve long-term blood glucose control and blood lipids in diabetic and, probably, nondiabetic people. This may help reduce the risk for obesity, insulin resistance, diabetes, CVD, etc. Overall, prospective epidemiologic studies support a protective effect of low-GI diets against diabetes.

Positive associations were obtained in seven of the 10 prospective epidemiologic studies that examined the relation between GI and risk of type 2 diabetes.[8-16] GL was also positively associated with diabetes, and this finding was confirmed based on 20 years of follow-up.[17] Methodological difficulties might explain the three studies with null findings.[10,11,14] A possible reason for the lack of association in the Iowa Women's Study is that the diagnosis of diabetes was made only on self-report without confirmation.[10] Stevens *et al.* used an abbreviated food questionnaire that deliberately focused on dietary fat rather than carbohydrate.[14] Sahyoun *et al.* assessed only 99 cases of diabetes.[11]

In one meta-analysis of studies of GI and GL in relation to risk of type 2 diabetes, Barclay calculated 40% and 27% higher summary risk ratios (RR) when comparing the highest with lowest quintiles of GI (95% confidence interval (CI) 1.23 to 1.59; $P < 0.0001$) and for GL (95% CI 1.12 to 1.45; $P < 0.0001$), respectively.[18] Additional data from the Black Women's Health Study[9] and the Shanghai Women's Health Study[15] provided valuable evidence

that the adverse effects of GI and GL also apply to non-Caucasian ethnic groups. In summary, although not every study found positive associations between GI and GL and risk of type 2 diabetes, the overall epidemiologic evidence strongly supports a positive relationship.

Glycemic Index to Age-Related Macular Degeneration

Some epidemiologic studies have consistently found positive relationships between GI and AMD in nondiabetic people, and the associations are independent of fiber intake (Fig. 22.3).[19-22] In the first study published in 2006, women in the third tertile of dietary GI compared with those in the first tertile had an approximately 2.7-fold increased risk for early AMD, mainly pigment abnormality, in a case-control study of the Nutrition and Vision Project (NVP) of the Nurses' Health Study (NHS).[23] The findings were replicated in a much larger American cohort, the Age-Related Eye Disease Study (AREDS).[20] In that case-control study, a diet in the highest quintile of dietary GI compared with a diet in the lowest quintile was associated with an over 40% increased risk for large drusen. When comparing the upper 50% with the lower 50% of the dietary GI, an almost 50% increased risk for advanced AMD was noted.

The positive relationship between GI and AMD was further strengthened in a prospective study that followed the AREDS subjects for up to 8 years (mean 5.4 years).[21] Overall the multivariate-adjusted risk of progression was significantly higher (hazard ratio 1.10; 95% CI 1.00 to 1.20; $P = 0.047$) in the upper 50% of the dietary GI than in the lower 50%. Furthermore, the more

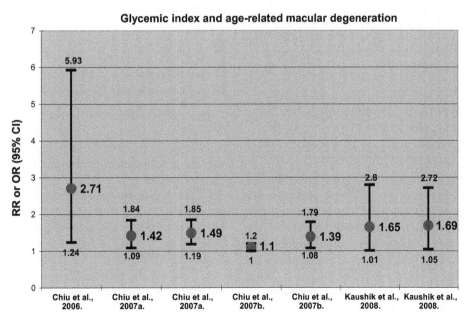

FIGURE 22.3 Studies relating glycemic index (GI) to age-related macular degeneration (AMD) indicate that consuming a low-GI diet is associated with lower risk for both early and advanced AMD. CI: confidence interval; OR: odds ratio; RR: risk ratio.

advanced the grade of AMD at baseline, the higher the increase of risk for progression during the follow-up period (P for trend < 0.001). Importantly, consuming lower GI diets appears to provide ophthalmic benefit in addition to that gained from currently known dietary factors. Analysis of a compound score summarizing dietary intakes of antioxidants (including vitamins C and E and lutein/zeaxanthin), zinc, omega-3 fatty acids (including docosahexaenoic acid (DHA) and eicosapentaenoic acid (EPA)), and GI suggested that the associations between the compound score and risk for drusen and advanced AMD are largely driven by dietary GI. A prospective analysis of the AREDS AMD trial indicated that consuming a low-GI diet augmented the protective effects of the AREDS formula (antioxidants plus zinc) and of DHA/EPA against progression to advanced AMD.[24]

The GI-AMD relationship was further confirmed in a 10-year follow-up in the Blue Mountains Eye Study (BMES).[22] After multivariate adjustment, a higher dietary GI was associated with a 77% increased risk of early AMD comparing the fourth with the first quartiles of dietary GI (95% CI 1.13 to 2.78; P for trend = 0.03), and further including cereal fiber in the model did not change the association. Conversely, greater consumption of cereal fiber (in a comparison of quartiles 1 and 4, RR 0.68; 95% CI 0.44 to 1.04; P for trend = 0.05) was associated with a reduced risk of incident early AMD.

In one prospective study, it is shown that GI played a more important role in individuals with bilateral AMD progression (i.e., those who are more susceptible to AMD progression) than those with unilateral AMD progression, especially in the later stages of AMD.[21] This finding implies that the interaction between AMD susceptibility and GI affects the risk for AMD progression and that the interaction plays a more important role in the later stages. It is possible that genetic susceptibility represents a major component of the underlying relationship between GI and AMD.

MECHANISM FOR HYPERGLYCEMIA TO DIABETIC RETINOPATHY AND AGE-RELATED MACULAR DEGENERATION

In the following, potential underlying mechanisms for the GI-disease associations on DR and AMD are discussed. Beginning with a brief review of the pathologies for DR and AMD, we then continue discussing the effects of hyperglycemia and the molecular mechanisms. Five well-developed hyperglycemic mechanisms are described, including four glycolysis-associated pathways and one mitochondria-associated pathway. In addition, we propose a novel hyperglycemic, hypoxia-inducible factor (HIF) pathway to complement the current theories regarding hyperglycemic pathogenesis.

Finally, inter-relationships among the six pathways are discussed.

Hyperglycemic Pathology of Diabetic Retinopathy and Age-Related Macular Degeneration

Hyperglycemia is recognized as a critical factor in the development of DR. DR is a retinal microvascular complication of diabetes that primarily affects the retinal circulation. A main pathologic feature of very early-stage DR is hyperglycemia-associated iBRB (inner blood–retinal barrier) breakdown. The iBRB breakdown begins with the loss of tight junctions between adjacent microvascular endothelial cells. This could interfere with the ability of the basement membranes to bind various growth factors, and if the disease becomes more severe, it may progress to retinopathy or even blindness.

The first indication of AMD is observed in the outer retina, primarily involving the retinal pigment epithelium (RPE) and associated tissues.[25] The RPE lays on a basal lamina, known as Bruch's membrane, and together they form the outer blood retinal barrier (oBRB). The oBRB separates the retina from the choroidal plexus. The RPE also provides a major transport pathway for the exchange of metabolites and ions. All of these functions indicate that the RPE plays a central role in the health of the outer retina.

With the function of RPE compromised during aging, drusen, the early stage of maculopathy, and more advanced lesions can begin to develop. In addition to drusen, the accumulation of deposits within the RPE-Bruch membrane-choriocapillaris complex includes extracellular basal lamina deposits (BLDs). Such abnormalities are thought to be important in the development of AMD.

Hypothesized Mechanisms Relating Dietary Hyperglycemia to Age-Related Macular Degeneration and Diabetic Retinopathy

In aerobic cellular respiration, glucose is metabolized through three steps: (1) glycolytic pathway, (2) tricarboxylic acid (TCA) cycle (also known as citric acid cycle or Kreb cycle), and (3) electron transport chain (ETC). Under normal glucose level (euglycemia, left panel in Fig. 22.4a), glucose generates energy (i.e., adenosine triphosphate (ATP)) for normal physiologic needs without inducing deleterious side reactions. However, under hyperglycemic conditions that exceed the physiologic needs (right panel in Fig. 22.4a), the glycolytic pathway may induce four adverse side pathways to relieve the influx of excess glucose. The four glycolysis-related hyperglycemic pathways include: (1) intracellular production of AGE (advanced glycation end product) precursors, (2) increased flux through the polyol pathway,

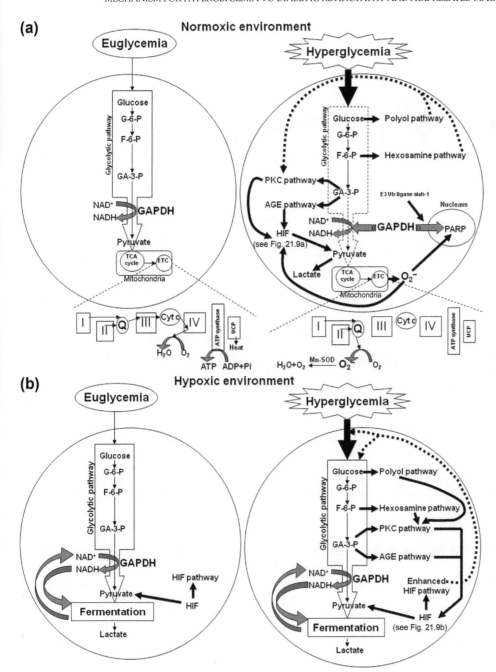

FIGURE 22.4 Cellular responses to euglycemia (normal glycemia) and hyperglycemia under normoxia (a) and hypoxia (b). (a) Glucose metabolism in euglycemia vs. hyperglycemia under normoxic conditions. Compared with euglycemia, hyperglycemia induces mitochondria-derived superoxide (O_2^-) and four glycolysis-related pathways (see Figs. 22.5–22.8), including polyol, hexosamine, advanced glycation end products (AGE), and protein kinase C (PKC) pathways, and excess cytosolic hypoxia-inducible factor (HIF). The left panel demonstrates normal aerobic respiration in a euglycemic condition. After glycolysis, the glucose metabolite, pyruvate, is produced. Pyruvate enters the mitochondria to generate adenosine triphosphate (ATP) and water (H_2O). The right panel demonstrates that hyperglycemia drives glycolysis to generate the four adverse side pathways noted above (also see Figs. 22.5–22.8). (b) Glucose metabolism in euglycemia vs. hyperglycemia under hypoxic conditions. In euglycemia, HIF pathway is turned on by hypoxia-activated HIF. Under hyperglycemic conditions, the HIF pathway is enhanced by hyperglycemia-induced AGE and PKC pathways. ADP: adenosine diphosphate; Cyt c: cytochrome-c; ETC: electron transport change; F-6-P: fructose-6 phosphate; G-6-P: glucose-6 phosphate; GA-3-P: glyceraldehyde 3-phosphate; GAPDH: glyceraldehyde 3-phosphate dehydrogenase; Mn-SOD: manganese superoxide dismutase; NADH: reduced nicotinamide adenine dinucleotide; PARP: poly ADP ribose polymerase; Pi: inorganic phosphate; TCA: tricarboxylic acid cycle; UCP: uncoupling protein.

(3) protein kinase C (PKC) activation, and (4) increased hexosamine pathway activity. Each of these will be discussed below. Under normoxic conditions (normal oxygen tension), the TCA cycle will induce an abnormally high mitochondrial membrane potential. This will induce the ETC to reduce O_2 into superoxide (O_2^-), which in turn will generate intracellular and even extracellular oxidative stress.

We hypothesized that the hyperglycemic pathogenesis consists of six pathways, including four glycolysis-related pathways, a mitochondria-derived reactive oxygen species (ROS) pathway, and an HIF (hypoxia-inducible factor) pathway. The four glycolysis-related pathways are applicable in both normoxic and hypoxic conditions. The mitochondria-derived ROS pathway participates during normoxia and the HIF pathway dominates the hyperglycemic pathogenesis in hypoxia. In the following sections, we will discuss these six molecular mechanisms regarding how dietary hyperglycemia results in cellular dysfunctions, which are associated with tissue damage and clinical manifestations of AMD and DR.

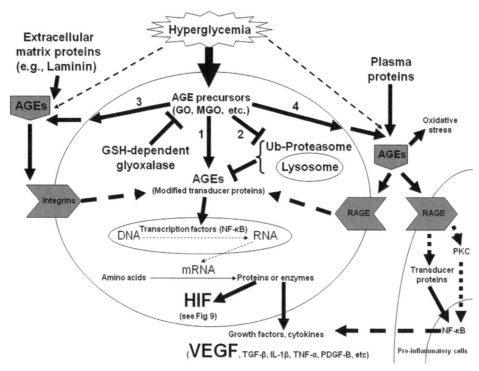

FIGURE 22.5 Hyperglycemic advanced glycation end products (AGE) pathway. The hyperglycemia-induced intracellular AGE precursors, such as MGO, induce pathologic consequences in four routes: 1) direct intracellular glycation of proteins, including proteins involved in the regulation of gene transcription, such as nuclear factor kappa-light-chain-enhancer of activated B cells (NF-κB), 2) inhibiting enzymes responsible for protein degradation, such as proteasomal (including ubiquitin) and lysosomal systems, 3) the intracellular AGEs precursors diffusing out of the cell and modifying nearby cells (even the same cell itself) and extracellular matrix, such as Bruch membrane and choroidal capillary membranes, and 4) the intracellular AGEs precursors diffusing out of the cell to modify circulating proteins in the blood, which in turn activate AGE receptors (RAGE) on proinflammatory cells or circulating endothelial cells (CEC), thereby causing the production of inflammatory cytokines and/or growth factors. DNA: deoxyribonucleic acid; GO: glyoxal; GSH: reduced glutathione; IL: interleukin; MGO: methylglyoxal; mRNA: messenger ribonucleic acid; PDGF: platelet-derived growth factor; TGF: transforming growth factor; TNF: tumor necrosis factor; Ub: ubiquitin; VEGF: vascular endothelial growth factor.

Hyperglycemic Advanced Glycation End Products Pathway

Figure 22.5 summarizes the plausible mechanism of hyperglycemic AGE pathway. First, AGE accumulation in RPE can appear as free AGE adducts in the cytoplasm and as AGE-modified proteins in lipofuscin granules. While some cytoplasm AGE-modified transducer proteins can affect the activity of downstream transcriptional factors, others are transported through a receptor-mediated transportation to the lysosomal compartment for degradation. Therefore, the AGE precursors trigger different pathological outcomes through (1) directly inducing intracellular glycation of proteins, (2) inhibiting the protein degradation, (3) modifying the nearby cells by diffusing out of cell, and (4) modifying plasma proteins that induce inflammatory cytokines, growth factors, and oxidative stress.

HYPERGLYCEMIC ADVANCED GLYCATION END PRODUCTS PATHWAY AND DIABETIC RETINOPATHY

Clinical studies have shown that the levels of AGEs in serum, skin, and cornea correlate with the onset or grade of DR. Importantly, AGEs are significantly increased in diabetic prepubescent children and adolescents with early or pre-proliferative retinopathy compared to both healthy and diabetic controls who are free from clinical signs of retinopathy.

Exposure to AGEs results in several deleterious effects on the retinal vessels, including increasing vasopermeability, neovascularization, evoking proinflammatory pathways, etc. For example, in vitro and in vivo studies showed that exposure to AGEs causes significant upregulation of vascular endothelial growth factor (VEGF),[26–29] which can also be induced by a variety of stimuli, such as PKC and HIF, to increase vascular permeability and induce DR-related neovascularization.

The importance of AGEs in DR can also be seen from several novel therapeutics of DR. These include peroxisome proliferator-activated receptor (PPAR) agonist, blockade of the renin-angiotensin system with an angiotensin converting enzyme inhibitor or by using angiotensin II type 1-receptor blockers, and intravitreal anti-VEGF antibody administration, all of which have been at least indirectly related to AGEs.

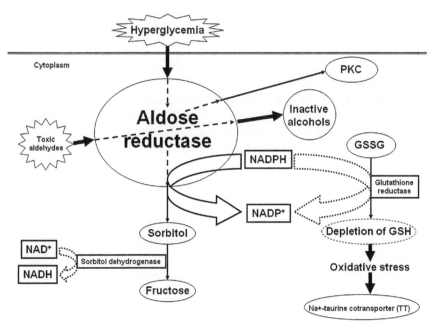

FIGURE 22.6 Hyperglycemic polyol pathway. Under hyperglycemia, aldose reductase (AR) reduces glucose to sorbitol (a polyol or sugar alcohol), which is later oxidized to fructose. In this process, the AR consumes cofactor nicotinamide adenine dinucleotide phosphate (NADPH). Therefore, the hyperglycemic polyol pathway consumes NADPH and hence results in the depletion of reduced glutathione (GSH). This increases intracellular oxidative stress. GSSG: glutathione disulfide; PKC: protein kinase C.

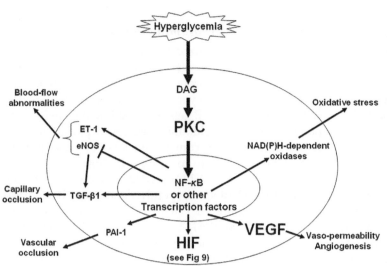

FIGURE 22.7 Hyperglycemic protein kinase C (PKC) pathway. The pathogenic consequences of hyperglycemic PKC through activating transcription factors for a wide range of proteins, including cytokines. Many transcription factors, such as nuclear factor kappa-light-chain-enhancer of activated B cells (NF-κB), are activated through hyperglycemia-induced PKC activation, resulting in oxidative stress, increased vasopermeability, angiogenesis, vascular occlusion, capillary occlusion, abnormal blood flow, etc. DAG: diacylglycerol; eNOS: endothelial nitric oxide synthase; ET: endothelin; HIF: hypoxia-inducible factor; PAI: plasminogen activator inhibitor; TGF: transforming growth factor; VEGF: vascular endothelial growth factor.

HYPERGLYCEMIC ADVANCED GLYCATION END PRODUCTS (AGE) PATHWAY AND AGE-RELATED MACULAR DEGENERATION

Although many of the adverse effects caused by AGEs are mediated via AGE receptors, these receptors play a critical role in AGE-related pathobiology associated with diabetes and aging disorders. Activation of the receptor for advanced glycation end products (RAGE) evokes downstream proinflammatory responses that could play a critical role in aging, such as skin aging, and age-related diseases, such as Alzheimer disease, atherosclerosis, dysfunction of cardiomyocytes, and retinal diseases.

Studies focusing on the outer retina showed that RAGE is expressed on RPE and that RAGE levels are significantly increased in AMD (in postmortem tissue), especially on cells adjacent to drusen. Activation of the RAGE axis in RPE cells upregulates the expression and secretion of VEGF.

Hyperglycemic Polyol Pathway

Dietary hyperglycemia may manifest age-related or diabetic disorders through increase in the polyol pathway (Fig. 22.6), which in turn leads to intracellular accumulation of sorbitol and oxidative stress. The polyol pathway is primarily controlled by the enzyme aldose reductase (AR). Under euglycemia, AR can reduce toxic aldehydes in the cell to inactive alcohols, but when the glucose concentration in the cell becomes too high, AR also reduces that glucose to sorbitol (a polyol or sugar alcohol), which is later oxidized to fructose. In the process of reducing high intracellular glucose to sorbitol, the AR consumes the cofactor nicotinamide adenine dinucleotide phosphate (NADPH). However, as shown in Figure 22.6, NADPH is also the essential cofactor for regenerating a critical intracellular antioxidant, reduced glutathione (GSH). By competing NADPH with glutathione reductase and hence resulting in a reduced amount of GSH, the polyol pathway increases susceptibility to intracellular oxidative stress.

HYPERGLYCEMIC POLYOL PATHWAY AND DIABETIC RETINOPATHY

In diabetes, elevated expression of AR may impair antioxidant defense, which may determine tissue susceptibility to chronic diabetic complications. Thus, increased expression of AR has been implicated as the critical link between chronic glucose toxicity and tissue damage.

HYPERGLYCEMIC POLYOL PATHWAY AND AGE-RELATED MACULAR DEGENERATION

The precise pathophysiologic mechanism linking polyol pathway to AMD remains uncertain, but depletion of the osmolyte and antioxidant taurine have been invoked through an inhibitory effect on its Nat-taurine cotransporter (TT).[30,31] Indeed, it is found that in RPE cells the TT is regulated by oxidative stress and that overexpression of AR and hyperglycemia impair this response (Fig. 22.6).

Hyperglycemic Protein Kinase C Pathway

Hyperglycemia-induced PKC activation may lead to blood–retinal barrier (BRB) breakdown. PKC also affects the expression of gap junction proteins, connexins, which are critical for intercellular communication between RPE cells. Many correlations between gap junctional intercellular communication and cellular processes, such as cellular growth control, cell differentiation, regulation of development, tissue homeostasis, etc., have been described.[32]

As shown in Figure 22.7, intracellular hyperglycemia increases the synthesis of diacyglycerol (DAG), a critical activating cofactor for the isoforms of PKC, -b, -d, and -a. Moreover, the PKC activates the downstream transcriptional factors, such as NF-κB (nuclear factor kappa-light-chain-enhancer of activated B cells), transforming growth factor (TGF)-β. Furthermore, hyperglycemia-induced PKC activation has also been implicated in the activation of membrane-associated NAD(P)H-dependent oxidase, which may increase the generation of ROS. In addition to DAG-related pathway, hyperglycemia may also activate the PKC pathway indirectly through ligation of AGE receptors (Fig. 22.5),[33] increased flux of

the polyol pathway (Fig. 22.6),[34] and increased flux of the hexosamine pathway (Fig. 22.8).[35]

HYPERGLYCEMIC PROTEIN KINASE C PATHWAY AND DIABETIC RETINOPATHY

In retinal endothelial cells and pericytes, high glucose causes activation of PKC-b and consequent expression of VEGF, contributing to the progression of DR.[36–38] It is also suggested that RPE cells may contribute to the pathogenesis of DR caused by hyperglycemia (Fig. 22.7) and hypoxia (Fig. 22.4b) through the PKC-mediated expression of VEGF.[39]

HYPERGLYCEMIC PROTEIN KINASE C (PKC) PATHWAY AND AGE-RELATED MACULAR DEGENERATION

In RPE cells, VEGF is expressed in response to mechanical stretch, hypoxia, and high glucose and may be mediated by PKC activation. This is corroborated by the observation that inhibition of the PKC pathway using a mixture of ethanol extracts from herbal medicines inhibits high glucose or AGE-induced VEGF expression in human RPE.

Hyperglycemic Hexosamine Pathway

Glucose is metabolized through glycolysis, going first to glucose-6 phosphate (G-6-P), then fructose-6 phosphate (F-6-P), and then on through the rest of the glycolytic pathway (Fig. 22.8). However, when glucose is high inside a cell, some of that F-6-P gets diverted into the hexosamine pathway in which glutamine:fructose-6 phosphate amidotransferase (GFAT) converts the F-6-P to glucosamine-6 phosphate and finally to uridine diphosphate N-acetyl glucosamine (UDPGlcNAc). Studies showed that inhibition of GFAT blocks hyperglycemia-induced increases in the transcription of TGF-α, TGF-β1, and plasminogen activator inhibitor (PAI)-1. It is suggested that, in the hexosamine pathway, the hyperglycemia-induced increases in gene transcription may be through glycosylation of the transcription factor, Sp1, by UDPGlcNAc. Furthermore, the glycosylated form of Sp1 seems to be more transcriptionally active than the deglycosylated form.

HYPERGLYCEMIC HEXOSAMINE PATHWAY AND DIABETIC RETINOPATHY

It is known that during the early stages of DR, there is significant death of the retinal microvascular pericytes. For example, it is shown that gangliosides in retinal pericytes are increased in response to the increase flux of the hexosamine pathway and are involved in the antiproliferative effect of glucosamine. In addition, AGEs can increase a-series ganglioside (GM3, GM2, GM1, GD1a) levels to inhibit bovine retinal pericyte cell proliferation. The possible mechanism could involve an increase in GM3 synthase activity (Fig. 22.8).[40]

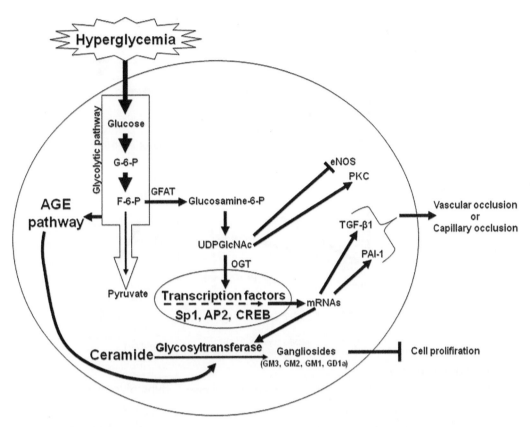

FIGURE 22.8 Hyperglycemic hexosamine pathway. The hyperglycemic hexosamine pathway starting from the glycolytic intermediate, fructose-6 phosphate (F-6-P), which is converted by glutamine:fructose-6 phosphate amidotransferase (GFAT) to glucosamine-6-phosphate and eventually to uridine diphosphate N-acetyl glucosamine (UDPGlcNAc), an O-linked N-acetylglucosamine (GlcNAc). Intracellular glycosylation, by adding GlcNAc moieties to serine and threonine residues of proteins (e.g., transcription factors), is catalyzed by O-GlcNAc transferase (OGT). Increased glycosylation of transcription factors, such as specificity protein (Sp)1, activating protein (AP)2, and CREB (cAMP response element-binding protein), often at phosphorylation sites, increases the expression of cytokines and enzymes, including transforming growth factor (TGF)-β1, plasminogen activator inhibitor (PAI)-1, and glycosyl transferase. In addition, advanced glycation end products (AGEs) can exert cellular effects by increasing a-series ganglioside levels to inhibit retinal pericyte cell proliferation. G-6-P: glucose-6 phosphate; eNOS: endothelial nitric oxide synthase; mRNA: messenger ribonucleic acid; PKC: protein kinase C; UDPGIcNAc: uridine diphosphate N-acetyl glucosamine.

HYPERGLYCEMIC HEXOSAMINE PATHWAY AND AGE-RELATED MACULAR DEGENERATION

In macula-derived RPE cells taken from fresh human donors, there are age-related decreases in the activity of N-acetyl-beta-glucosaminidase, an enzyme that is responsible for the degradation of N-acetyl glucosamine (GlcNAc). Because GlcNAc is the major carbohydrate monomer of the oligosaccharide chains of human rhodopsin, defects in its degradation may lead to the accumulation of undigested residual material in the RPE.

Hyperglycemic Mitochondria-Derived Reactive Oxygen Species

There are four protein complexes in the mitochondrial ETC, including complexes I, II, III, and IV. Under euglycemic and normoxic conditions (left panel of Fig. 22.4a), after pyruvate is metabolized through the TCA cycle, it generates electrons that are passed to coenzyme Q

through complexes I and II and then transferred to complex III, cytochrome-c (Cyt c), complex IV, and finally to molecular oxygen (O_2), which is then reduced to water. While electrons are transported from left to right in the left panel of Figure 22.4a, some of the energy of those electrons is used to pump protons across the membrane at complexes I, III, and IV to generate a voltage potential across the mitochondrial membrane. The energy from this voltage gradient drives the synthesis of ATP by adenosine triphosphate (ATP) synthase. Regulation of the rate of ATP generation is achieved in part by uncoupling proteins (UCPs) that can dissipate the voltage gradient to generate heat.

However, under hyperglycemic normoxic conditions (right panel of Fig. 22.4a), more glucose is oxidized. This pushes more electron donors into the ETC. When the voltage gradient across the mitochondrial membrane increases to a critical threshold, transfer in complex III is blocked,[41]

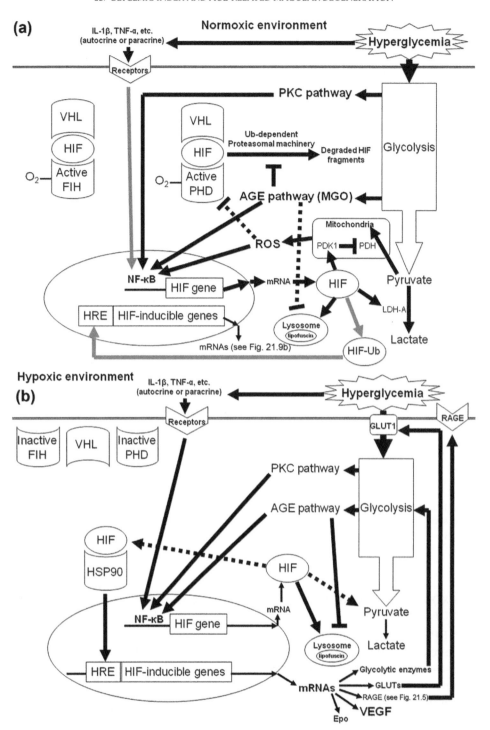

FIGURE 22.9 Hyperglycemic HIF pathway in both normoxic (a) and hypoxic conditions (b). (a) Normoxic, hyperglycemic HIF pathway. In normoxia, hyperglycemic PKC activation, AGE formation, mitochondrial ROS, and proinflammatory cytokines (e.g., IL-1b and TNF-a) decrease the degradation (through impairing the proteasomal system) and/or increase the expression of HIF (through activating NF-kB). The elevated cytoplasmic HIF proteins may switch glucose metabolism from aerobic respiration to fermentation, giving rise to lactate accumulation (also see Fig. 22.4a). The hyperglycemia-induced excess cytosolic HIF proteins may also lead to increased autophagy, while the lysosomal proteases are impaired by hyperglycemia. The combination of the two effects may also results in the accumulation of lysosomal lipofuscin. In addition, the excess cytosolic HIF proteins, such as a ubiquitinated form of HIF-1a induced by TNF-a, can also transactivate HIF-inducible genes (also see (b)). However, under hypoxia the hyperglycemia-induced HIF protein is more stable. (b) Hypoxic, hyperglycemic HIF pathway. In hypoxia, the trans-activation activity of HIF is turned on because both oxygen sensors, PHD and FIH, become inactive. This eliminates proteasomal degradation of HIF proteins. The HIF proteins are further stabilized by hypoxia-induced HSP90 and bind to the HREs in the promoter or enhancer region of HIF-inducible genes to transactivate the transcriptions of the genes. The HIF pathway consists of many HIF-inducible genes, which encode a wide range of proteins, including RAGE, glycolytic enzymes, GLUTs, Epo, and VEGF. The hyperglycemic, hypoxic HIF pathway may be enhanced by

causing the electrons to accumulate in coenzyme Q. This allows coenzyme Q to donate the electrons one at a time to O_2 thereby generating superoxide (O_2-). The cell defends itself against this ROS using the mitochondrial isoform of superoxide dismutase (Mn-SOD). This enzyme degrades the oxygen free radical to hydrogen peroxide, which is then converted to H_2O and O_2 by catalase.

HYPERGLYCEMIC MITOCHONDRIA-DERIVED REACTIVE OXYGEN SPECIES AND DIABETIC RETINOPATHY

The cellular antioxidant response element is important for the amelioration of oxidative stress. It responds to hyperglycemia and can be used to evaluate the complications of diabetes mellitus. It has been shown that in retinal endothelial cells the induction of mitochondrial oxidative stress is more sensitive to hyperglycemia than the induction of the antioxidant response element. In other words, it appears that endothelial cells are particularly vulnerable to hyperglycemia-induced mitochondrial ROS.

HYPERGLYCEMIC MITOCHONDRIA-DERIVED REACTIVE OXYGEN SPECIES AND AGE-RELATED MACULAR DEGENERATION

Age-related pathology, including AMD, has been related to mitochondrial genomic instability. For example, an increased level of the mitochondrial superoxide dismutase (SOD2) (Fig. 22.4a) has been shown to decrease the disruption of mitochondrial transmembrane potential and the release of Cyt c, and thus to prevent apoptotic cell death in mouse RPE.

Furthermore, it is suggested that in RPE mitochondria are the main target of oxidative injury, that the mitochondrial genome is a weak link in the antioxidant defenses, and that deficits in mitochondrial DNA repair pathways are important contributors to the pathogenesis of AMD.[42,43] These data suggest that oxidative stress-induced mitochondrial genomic instability will result in loss of cell function and greater susceptibility to stress. This mitochondrial overproduction of ROS may play a significant role in AMD pathogenesis.

Hyperglycemic Hypoxia-Inducible Factor Pathway

HIF was first characterized from human hepatoma cells in which it influenced the transcription of erythropoietin (Epo) gene. Roles for HIF in the etiologies of DR and AMD have also been proposed.[44,45]

Although HIF is expressed under physiologic conditions, the expression and degradation of the HIF protein are kept in balance (Fig. 22.9). Under normoxic conditions (Fig. 22.9a),[46] the first mechanism is through the Ub-dependent proteasomal degradation pathway. The conserved proline residues (402 and 564) of HIF are first hydroxylated by proline hydroxylase (PHD) enzymes. Next, the hydroxylated HIF interacts with a Ub-protein ligase, von Hippel-Lindau (VHL) protein, becomes ubiquitinated, and is degraded via the proteasomal pathway. The degradation is dependent on the protein motifs found in the carboxyl terminus of HIF proteins, termed hypoxia-responsive domains. Thus, HIF is kept at a low level in the cytosol.

In the other mechanism, HIF is prevented from binding to the HREs in the promoter or enhancer regions of HIF-inducible genes. Specifically, in normoxia, HIF binding to DNA is inhibited through asparagine hydroxylation by an oxygen-dependent factor inhibiting HIF (FIH) (Fig. 22.9a). In this situation, transcription of HIF-inducible genes, such as VEGF, is limited.

Under hypoxia both oxygen sensors, praline hydroxylase (PHD) and factor inhibiting HIF (FIH), become inactive and unable to hydroxylate HIF-1a. The stabilized HIF binds to HIF-inducible genes (Fig. 22.9b). In addition, hypoxia also induces the expression of heat shock proteins (HSPs). HSPs are molecular chaperones required for the stability and function of a number of conditionally activated and/or expressed protein kinases and transcription factors, such as HIF. Under hypoxia, the transcription activities of HIF-inducible genes are turned on by HIF. As noted earlier, these HIF-inducible genes are responsible for a wide range of hyperglycemic pathogenesis, such as VEGF-induced angiogenesis (Fig. 22.4b; Fig. 22.9b).

HYPERGLYCEMIA AND HYPOXIA-INDUCIBLE FACTOR

Accumulating evidence implies that there is an undescribed HIF-related mechanism that results in pathology under hyperglycemic, hypoxic conditions (Fig. 22.9b). Hyperglycemia can enhance the activity of HIF through glycolysis-derived hyperglycemic AGE and PKC pathways (Fig. 22.9b). First, hypoxia has been shown to result in the activation of NF-kB,

---

hyperglycemic AGE formation and PKC activation, which can activate RAGE signaling cascades and increase HIF expression. Furthermore, in adaptation of lower efficiency of ATP generation from fermentation, the activation of GLUTs and glycolytic enzymes increases glucose uptake and upregulates the glycolysis pathway (also see Fig. 22.4b). Therefore, in hyperglycemic, hypoxic conditions the HIF pathway may further deteriorate the four glycolysis-associated pathways and lactate accumulation in the cytoplasm. Furthermore, the hyperglycemia-induced excess cytosolic HIF proteins may also enhance autophage, while the lysosomal system is impaired by hyperglycemia. Together, they may also result in the accumulation of lysosomal lipofuscin. In addition, hyperglycemia-induced proinflammatory cytokines (e.g., IL-1b and TNF-a) can further enhance the HIF pathway. Accumulation of lysosomal lipofuscin. In addition, hyperglycemia-induced proinflammatory cytokines (e.g., IL-1b and TNF-a) can further enhance the HIF pathway.

which can bind to the HIF promoter in response to hypoxia. As described previously, NF-kB can also be activated by hyperglycemia-induced AGE formation (Fig. 22.5) and PKC activation (Fig. 22.7). Thus, coincidental hyperglycemia and hypoxia lead to enhanced expression of HIF and HIF-inducible genes (Fig. 22.9b). It has been shown that when hyperglycemia coincides with hypoxia, the secretion of VEGF is enhanced. This is partially mediated via activation of PKC. Remarkably, PKC can be also activated through hyperglycemic polyol pathway (Fig. 22.6) and hyperglycemic hexosamine pathway (Fig. 22.8). Second, the hyperglycemic HIF pathway may amplify the deleterious effects of hyperglycemic AGE formation (Fig. 22.5) by HIF-inducible RAGE expression (Fig. 22.9b). Actually, under both normoxic and hypoxic conditions (Fig. 22.9a and b), hyperglycemia can lead to HIF accumulation, resulting in the formation of intracellular deposits and expression of HIF-inducible genes.

Hyperglycemia Induces Inflammation and Apoptosis

Both DR and AMD have been characterized as chronic inflammatory diseases leading to cell death in the retina. This is consistent with many molecular and epidemiological observations, reviewed above, that hyperglycemia results in increased production of proinflammatory cytokines and apoptosis of the cells. Caspases, a family of cysteine proteases, are known to be critically involved in both activation of proinflammatory cytokines and the initiation and execution of apoptosis. Two caspase pathways have been described by which cells undergo apoptosis. The extrinsic (receptor-mediated) pathway is triggered via cell surface receptors, which are represented by tumor necrosis factor (TNF)-α family receptors, leading to activation of caspase-8 and caspase-3 proteolytic enzymes. The intrinsic (mitochondrial) pathway involves the mitochondrial Cyt c release and activation of caspase-9, with subsequent activation of caspase-3. The intrinsic pathway can be activated by agents that directly target the mitochondria or indirectly via the extrinsic pathway through caspase-8-mediated cleavage of the inactive cytosolic protein BID. Once activated, BID translocates to the mitochondria, where it stimulates Cyt c release.

In the human RPE, the activation of caspase-8 pathway (i.e., extrinsic pathway), but not the mitochondrial caspase-9 pathway (i.e., intrinsic pathway), was shown to involve the 7-ketocholesterol (an oxidative stressor)-induced apoptosis. However, in TNF-α-induced apoptosis in the human RPE cells it was shown that the mitochondrial caspase-9 could be used to amplify the death signal mediated by caspase-8.

As discussed in the four glycolysis-related pathways above, hyperglycemia can induce proinflammatory cytokines (e.g., Fig. 22.5), such as interleukin (IL)-1β and TNF-α. Interestingly, IL-1β and TNF-α can prolong the activation of HIF-1a protein under conditions of inflammation via enhancing the translation of HIF-1a messenger ribonucleic acid (mRNA), further leading to increased expression of HIF-inducible genes. Importantly, this can happen under both normoxic (Fig. 22.9a) and hypoxic (Fig. 22.9b) conditions. This gives additional support to the idea that hyperglycemia is able to induce some effects of HIF. In conclusion, HIF is not only a hypoxia-inducible factor: it can also be described as a "hyperglycemia-inducible factor". Through this mechanism, the hyperglycemic HIF pathway can affect oxidative stress responses, inflammation, proteolytic mechanisms.

TAKE-HOME MESSAGES

- Glucose homeostasis and carbohydrate nutrition are involved in human aging as well as in disease pathogenesis, including DR and AMD. In this context, these diseases can be considered as metabolic retinal diseases.
- GI is an indicator measuring how fast a carbohydrate-containing food raises blood glucose. Moreover, consuming high-GI foods is associated with diabetes, CVD, and AMD.
- Under hyperglycemic conditions, the glycolytic pathway may induce four adverse side pathways to relieve the influx of excess glucose. The four glycolysis-related hyperglycemic pathways include: (1) intracellular production of AGE precursors, (2) increased flux through the polyol pathway, (3) PKC activation, and (4) increased hexosamine pathway activity.
- Under normoxic conditions, the TCA cycle will induce an abnormally high mitochondrial membrane potential, and this will induce the ETC to reduce O_2 into superoxide (O_2-), which in turn will generate intracellular and even extracellular oxidative stress.

Acknowledgments

Any opinions, findings, conclusions, or recommendations expressed in this publication are those of the authors and do not necessarily reflect the views or policies of the U.S. Department of Agriculture, nor does mention of trade names, commercial products, or organizations imply endorsement by the U.S. Government.

The figures in this book chapter were adapted from Chiu CJ, Taylor A. Dietary hyperglycemia, glycemic index and metabolic retinal diseases. *Prog Retin Eye Res.* 2011;30:18–53.

References

1. Jenkins DJ, Wolever TM, Taylor RH, Barker H, Fielden H, Baldwin JM, et al. Glycemic index of foods: a physiological basis for carbohydrate exchange. *Am J Clin Nutr* 1981;**34**:362–6.
2. Foster-Powell K, Holt SH, Brand-Miller JC. International table of glycemic index and glycemic load values. *Am J Clin Nutr* 2002;**2002**(76):5–56.
3. Riccardi G, Rivellese AA, Giacco R. Role of glycemic index and glycemic load in the healthy state, in prediabetes, and in diabetes. *Am J Clin Nutr* 2008;**87**:269S–74S.
4. Wolever TM, Nguyen PM, Chiasson JL, Hunt JA, Josse RG, Palmason C, et al. Determinants of diet glycemic index calculated retrospectively from diet records of 342 individuals with non-insulin-dependent diabetes mellitus. *Am J Clin Nutr* 1994;**59**:1265–9.
5. Cahill Jr GF. Starvation in man. *New Engl J Med* 1970;**282**:668–75.
6. DeFronzo RA, Bonadonna RC, Ferrannini E. Pathogenesis of NIDDM. A balanced overview. *Diabetes Care* 1992;**15**:318–68.
7. Leahy JL, Bonner-Weir S, Weir GC. Beta-cell dysfunction induced by chronic hyperglycemia. Current ideas on mechanism of impaired glucose-induced insulin secretion. *Diabetes Care* 1992;**15**:442–55.
8. Hodge AM, English DR, O'Dea K, Giles GG. Glycemic index and dietary fiber and the risk of type 2 diabetes. *Diabetes Care* 2004;**27**:2701–6.
9. Krishnan S, Rosenberg L, Singer M, Hu FB, Djousse L, Cupples LA, et al. Glycemic index, glycemic load, and cereal fiber intake and risk of type 2 diabetes in US black women. *Arch Intern Med* 2007;**167**:2304–9.
10. Meyer KA, Kushi LH, Jacobs Jr DR, Slavin J, Sellers TA, Folsom AR. Carbohydrates, dietary fiber, and incident type 2 diabetes in older women. *Am J Clin Nutr* 2000;**71**,:921–30.
11. Sahyoun NR, Anderson AL, Tylavsky FA, Lee JS, Sellmeyer DE, Harris TB. Dietary glycemic index and glycemic load and the risk of type 2 diabetes in older adults. *Am J Clin Nutr* 2008;**87**,:126–31.
12. Salmeron J, Ascherio A, Rimm EB, Colditz GA, Spiegelman D, Jenkins DJ, et al. Dietary fiber, glycemic load, and risk of NIDDM in men. *Diabetes Care* 1997;**20**:545–50.
13. Schulze MB, Liu S, Rimm EB, Manson JE, Willett WC, Hu FB. Glycemic index, glycemic load, and dietary fiber intake and incidence of type 2 diabetes in younger and middle-aged women. *Am J Clin Nutr* 2004;**80**:348–56.
14. Stevens J, Ahn K, Juhaeri Houston D, Steffan L, Couper D. Dietary fiber intake and glycemic index and incidence of diabetes in African-American and white adults: the ARIC study. *Diabetes Care* 2002;**25**:1715–21.
15. Villegas R, Liu S, Gao YT, Yang G, Li H, Zheng W, et al. Prospective study of dietary carbohydrates, glycemic index, glycemic load, and incidence of type 2 diabetes mellitus in middle-aged Chinese women. *Arch Intern Med* 2007;**167**:2310–6.
16. Zhang C, Liu S, Solomon CG, Hu FB. Dietary fiber intake, dietary glycemic load, and the risk for gestational diabetes mellitus. *Diabetes Care* 2006;**29**:2223–30.
17. Halton TL, Liu S, Manson JE, Hu FB. Low-carbohydrate-diet score and risk of type 2 diabetes in women. *Am J Clin Nutr* 2008;**87**:339–46.
18. Barclay AW, Petocz P, McMillan-Price J, Flood VM, Prvan T, Mitchell P, et al. Glycemic index, glycemic load, and chronic disease risk – a meta-analysis of observational studies. *Am J Clin Nutr* 2008;**87**:627–37.
19. Chiu CJ, Milton RC, Gensler G, Taylor A. Dietary carbohydrate intake and glycemic index in relation to cortical and nuclear lens opacities in the Age-Related Eye Disease Study. *Am J Clin Nutr* 2006;**83**:1177–84.
20. Chiu CJ, Milton RC, Gensler G, Taylor A. Association between dietary glycemic index and age-related macular degeneration in nondiabetic participants in the Age-Related Eye Disease Study. *Am J Clin Nutr* 2007a;**86**:180–8.
21. Chiu CJ, Milton RC, Klein R, Gensler G, Taylor A. Dietary carbohydrate and the progression of age-related macular degeneration: a prospective study from the Age-Related Eye Disease Study. *Am J Clin Nutr* 2007b;**86**:1210–8.
22. Kaushik S, Wang JJ, Flood V, Tan JS, Barclay AW, Wong TY, et al. Dietary glycemic index and the risk of age-related macular degeneration. *Am J Clin Nutr* 2008;**88**:1104–10.
23. Chiu CJ, Hubbard LD, Armstrong J, Rogers G, Jacques PF, Chylack Jr LT, et al. Dietary glycemic index and carbohydrate in relation to early age-related macular degeneration. *Am J Clin Nutr* 2006;**83**:880–6.
24. Chiu CJ, Klein R, Milton RC, Gensler G, Taylor A. Does eating particular diets alter the risk of age-related macular degeneration in users of the Age-Related Eye Disease Study supplements? *Br J Ophthalmol.* 2009b;**93**:1241–6.
25. Glenn JV, Stitt AW. The role of advanced glycation end products in retinal ageing and disease. *Biochim Biophys Acta* 2009b;**1790**:1109–16.
26. Lu M, Kuroki M, Amano S, Tolentino M, Keough K, Kim I, et al. Advanced glycation end products increase retinal vascular endothelial growth factor expression. *J Clin Invest.* 1998;**101**:1219–24.
27. Stitt AW, Bhaduri T, McMullen CB, Gardiner TA, Archer DB. Advanced glycation end products induce blood-retinal barrier dysfunction in normoglycemic rats. *Mol Cell Biol Res Commun* 2000;**3**:380–8.
28. Treins C, Giorgetti-Peraldi S, Murdaca J, Van Obberghen E. Regulation of vascular endothelial growth factor expression by advanced glycation end products. *J Biol Chem.* 2001;**276**:43836–41.
29. Yamagishi S, Amano S, Inagaki Y, Okamoto T, Koga K, Sasaki N, et al. Advanced glycation end products-induced apoptosis and overexpression of vascular endothelial growth factor in bovine retinal pericytes. *Biochem Biophys Res Commun* 2002;**290**:973–8.
30. Nakashima E, Pop-Busui R, Towns R, Thomas TP, Hosaka Y, Nakamura J, et al. Regulation of the human taurine transporter by oxidative stress in retinal pigment epithelial cells stably transformed to overexpress aldose reductase. *Antioxid Redox Signal* 2005;**7**:1530–42.
31. Stevens MJ, Hosaka Y, Masterson JA, Jones SM, Thomas TP, Larkin DD. Downregulation of the human taurine transporter by glucose in cultured retinal pigment epithelial cells. *Am J Physiol* 1999;**277**:E760–71.
32. Goodenough DA, Goliger JA, Paul DL. Connexins, connexons, and intercellular communication. *Ann Rev Biochem* 1996;**65**:475–502.
33. Portilla D, Dai G, Peters JM, Gonzalez FJ, Crew MD, Proia AD. Etomoxir-induced PPARalpha-modulated enzymes protect during acute renal failure. *Am J Physiol Renal Physiol* 2000;**2000**(278):F667–75.
34. Keogh RJ, Dunlop ME, Larkins RG. Effect of inhibition of aldose reductase on glucose flux, diacylglycerol formation, protein kinase C, and phospholipase A2 activation. *Metab Clin Exp.* 1997;**46**:41–7.
35. Goldberg HJ, Whiteside CI, Fantus IG. The hexosamine pathway regulates the plasminogen activator inhibitor-1 gene promoter and Sp1 transcriptional activation through protein kinase C-beta I and -delta. *J Biol Chem.* 2002;**277**:33833–41.
36. Clarke M, Dodson PM. PKC inhibition and diabetic microvascular complications. *Best Pract Res Clin Endocrinol Metab* 2007;**21**:573–86.
37. Enaida H, Kabuyama Y, Oshima Y, Sakamoto T, Kato K, Kochi H, et al. VEGF-dependent signaling in retinal microvascular endothelial cells. *Fukushima J Med Sci* 1999;**45**:77–91.

38. Hata Y, Rook SL, Aiello LP. Basic fibroblast growth factor induces expression of VEGF receptor KDR through a protein kinase C and p44/p42 mitogen-activated protein kinase-dependent pathway. *Diabetes* 1999;**48**:1145–55.

39. Young TA, Wang H, Munk S, Hammoudi DS, Young DS, Mandelcorn MS, et al. Vascular endothelial growth factor expression and secretion by retinal pigment epithelial cells in high glucose and hypoxia is protein kinase C-dependent. *Exp Eye Res* 2005;**80**:651–62.

40. Masson E, Troncy L, Ruggiero D, Wiernsperger N, Lagarde M, El Bawab S. Series gangliosides mediate the effects of advanced glycation end products on pericyte and mesangial cell proliferation: a common mediator for retinal and renal microangiopathy? *Diabetes* 2005;**54**:220–7.

41. Korshunov SS, Skulachev VP, Starkov AA. High protonic potential actuates a mechanism of production of reactive oxygen species in mitochondria. *FEBS Lett* 1997;**416**:15–8.

42. Cai J, Nelson KC, Wu M, Sternberg Jr P, Jones DP. Oxidative damage and protection of the RPE. *Prog Retin Eye Res* 2000;**19**:205–21.

43. Jarrett SG, Lin H, Godley BF, Boulton ME. Mitochondrial DNA damage and its potential role in retinal degeneration. *Prog Retin Eye Res* 2008;**27**:596–607.

44. Arjamaa O, Nikinmaa M. Oxygen-dependent diseases in the retina: role of hypoxia-inducible factors. *Exp Eye Res* 2006;**83**:473–83.

45. Arjamaa O, Nikinmaa M, Salminen A, Kaarniranta K. Regulatory role of HIF-1alpha in the pathogenesis of age-related macular degeneration (AMD). *Ageing Res Rev* 2009;**8**:349–58.

46. Iyer NV, Leung SW, Semenza GL. The human hypoxia-inducible factor 1alpha gene: HIF1A structure and evolutionary conservation. *Genomics* 1998;**52**:159–65.

Fish-Oil Fat Emulsion and Retinopathy in Very Low Birth Weight Infants

Ryszard Lauterbach, Dorota Pawlik

Jagiellonian University Medical College, Kraków, Poland

INTRODUCTION

Two families of essential fatty acids (EFAs), known as long-chain polyunsaturated fatty acids (LC-PUFAs), exist in nature: ω-3 and ω-6. They are structurally classified by the number of carbons, double bonds, and proximity of the first double bond to the methyl (omega) terminal of the fatty acid acyl chain. Fatty acids of the ω-3 family contain a double bond at the third carbon, whereas those of the ω-6 family contain a double bond at the sixth carbon. Omega-3 LC-PUFAs demonstrate the capacity to modulate production, activation, and potency of bioactive molecules.[1,2] Docosahexaenoic acid (DHA), a major dietary ω-3 LC-PUFA, is also a major structural lipid in sensory and vascular retina. DHA and its substrate, eicosapentaenoic acid (EPA), influence eicosanoid metabolism by reducing ω-6 LC-PUFA levels (mainly arachidonic acid (AA)) and competing for enzymes (cyclooxygenase (COX)) used to produce AA-based angiogenic and proinflammatory series 2- and 4-eicosanoids.[3] Omega-3 LC-PUFAs present antivasoproliferative and neuroprotective actions and influence the factors and processes implicated in the pathogenesis of proliferative retinal diseases. The main sources of DHA are breast milk, algal oils, and fish oil. Infants aged between 1 and 6 months of life, fed with breast milk, consume about 315 mg of DHA daily.[4]

PRETERM INFANT DEFICIT OF LC-PUFAS: THE RATIONALE FOR ADMINISTRATION

During the last trimester of pregnancy, the placenta provides the fetus with DHA. The placental transfer involves a multistep process of uptake and translocation facilitated by placental fatty acid binding and transport proteins (FATP – fatty acid transport protein) that favor n-3 fatty acids.[5,6,7] It has been established from autopsy data that the fetal accretion of DHA during the last trimester is approximately 45 mg/kg/day.[8] Prematurely delivered infants, having missed the important period of intrauterine nutrient accretion, accumulate only limited energy and fat reserves. Moreover, they usually do not tolerate enteral feeding or it is insufficient to meet the requirements during the first weeks of life. This is particularly evident in neonates under 1500 g birth weight. Therefore, parenteral nutrition should be introduced in those infants shortly after birth. Intravenous lipid emulsions are an integral part of established parenteral nutrition practices. A well-balanced fatty acid supply during the first weeks of life plays a potentially significant role in the modulation of developmental processes affecting short- and long-term health outcomes related to growth, visual and cognitive development, immune and allergic responses, and the prevalence of nutrition-related chronic disease.[9–11] However, most commonly used lipid emulsions are comprised of soybean oils, with or without olive oils or medium-chain triglyceride oil, and thus are rich in linoleic acid (LA; C18:2 ω-6) but contain relatively low amounts of α-linolenic acid (α-LNA; C18: ω-3) and AA (C20:4 ω-6) and no ω-3 LC-PUFAs such as EPA (C20:5 ω-3) and DHA (C22:6 ω-3). Actually, there are only two lipid emulsions that contain fish oil and thus the supply of EPA and DHA: 10% Omegaven (Fresenius Kabi, Bad Homburg, Germany) and 20% SMOFlipid (Fresenius Kabi, Bad Homburg, Germany).

The recommended dietary intake of energy, protein, lipids, and carbohydrates for preterm infants is frequently based on the assumption that optimal postnatal growth should approximate that of the fetus *in utero* at the same gestational age.[12] These recommendations did not take into account the possible needs to compensate for an early DHA deficiency. When the total energy intake is greater than 100 kcal/kg/day, the absolute quantity of DHA synthesized from α-LNA in preterm infants that are fed a formula containing LC-PUFAs reaches 12.6 mg/kg/day. However, when the energy intake is less than 100 kcal/kg/day, only one-third of this amount is synthesized.[13] Lapillone and colleagues found that the cumulative DHA deficiency in infants born at less than 28 weeks of pregnancy increased significantly every week and on average reached its maximum extent at 661 ± 100 mg/kg at the end of the fourth week of life (minimum − 539 mg/kg; maximum − 914 mg/kg).[14] This finding was highly correlated with the birth weight, demonstrating that the smaller the infant, the larger was the DHA deficit. When compared with values measured in term infants, the DHA content of erythrocyte phospholipids of preterm infants at term gestation were 31% and 14% lower when neonates were breastfed compared with nutrition with a currently available DHA-supplemented formula.

In term infants, DHA is stored in the adipose tissue, which becomes a reservoir of LC-PUFAs that can be mobilized and consumed if the diet is lacking DHA. It is known that percentage of adipose tissue in newborns delivered at term is around 18%, whereas in very-low-birth weight infants delivered prematurely it reaches only 1% of total body mass. Consequently, apart from the significantly shortened period of DHA placental transfer, the storage of DHA is markedly reduced in prematurely born infants. The comparison between full-term newborns and preterm infants' plasma DHA concentrations, measured in blood samples that were obtained within the first hour after birth, were markedly higher and statistically significant. DHA concentrations were five times higher in plasma and three times higher in erythrocytes when measured in neonates delivered at term.[15] Conversely, plasma EPA levels did not differ significantly between preterm and term infants (author's unpublished observation).

POTENTIAL CONSEQUENCES OF DHA DEFICIENCY IN THE VISUAL SYSTEM IN PRETERM INFANTS

It is well known that between the 23rd and 40th week of pregnancy, the weight of the brain increases from approximately 75 to 400 g. DHA is a nutrient absolutely required for the development of sensory, perceptual, cognitive, and motor neural system during this rapid brain growth.[16] The retina is functionally an extension of the brain. The rods and cones have membranes highly enriched with DHA. This LC-PUFA is the major fatty acid in structural lipids of retinal photoreceptor outer segment disc membranes. The outer segment disc contains rhodopsin, the photopigment necessary for initiating visual sensation;

DHA is efficiently incorporated and selectively retained in disc membranes. Highest body concentrations of DHA per unit area are found in the disc membranes, where the overall percentage of DHA is 50% greater than in the next most concentrated tissue.[17] The biophysical and biochemical properties of DHA affect membrane function by altering permeability, fluidity, thickness, lipid phase properties, and the activation of membrane-bound proteins. DHA-rich membranes impart properties to outer segment discs that influence the dynamics of inter- and intracellular communication.[18] The stereochemical structure of DHA with its 22 carbons and six double bonds allows an efficient conformational change of the trans-membrane protein rhodopsin during light absorption. A more fluid membrane allows a faster response to stimulation. It was established that DHA comprises 20% of total fatty acid content of the infant retina. DHA operates as a trophic molecule in photoreceptor development, differentiation, and growth. DHA increases opsin expression in developing rat photoreceptors *in vitro*. DHA also prolongs survival of rat photoreceptors *in vitro*.

DHA deficiency is associated with structural and functional abnormalities in the visual system. Rodents and primates with experimentally induced DHA deficiency show abnormal retinal structure, reduced visual acuity, and diminished cognitive performance. In contrast, a higher dose of DHA in the neonatal period improves visual acuity of preterm infants.[19,20]

RETINOPATHY OF PREMATURITY – POSSIBLE ROLE OF ω-3 LC-PUFAs IN PREVENTION OF DISEASE

Retinopathy of prematurity (ROP) is the leading cause of blindness and visual impairment during infancy and is associated with extreme preterm birth. It is a vascular disorder that leads to retinal destruction, but lesser stages of ROP can also result in functional visual deficits in early childhood. In infants with ROP, severe retinopathy is an independent risk factor for myopia. Although peripheral retinal disease associated with ROP can cause functional visual disorders, the most pronounced deficits arise from disruptions in the central part of the retina and are manifested by defective cone and rod functions.[21–23] ROP exhibits aspects of both reactive neovascularization and cellular degeneration that may be modulated by ω-3 LC-PUFAs. Incidence is the highest among extremely low-birth weight infants, and it is directly proportional to the degree of a significantly shortened gestation. Approximately 90% of infants weighing less than 750 g at birth develop some form of ROP, and many may reach the threshold for treatment. Laser photocoagulation is the most frequent method used for treatment of ROP (Fig. 23.1).

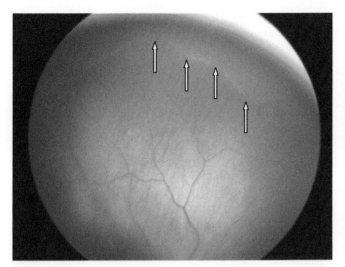

FIGURE 23.1   Retinopathy of prematurity (ROP) – vascular ridge with peripheral avascular zone. The second stage of retinopathy of prematurity: arrows indicate vascular ridge and the beginning of avascular zone. *Source: author's collection.* (See color plate section)

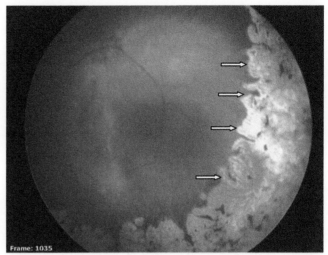

FIGURE 23.2   Retina – the effect of laser photocoagulation. The arrows indicate scars of the retina – the effect of laser photocoagulation. *Source: author's collection.*

The long-term benefits from diode laser photocoagulation include preservation of 1) near acuity, 2) contrast sensitivity, and 3) refraction-related color vision. However, the distance visual acuity in treated patients is significantly more frequently disturbed when compared with data observed in preterm infants with spontaneous regression. Moreover, about 50% of the laser-treated eyes are myopic, with an overall mean spherical equivalent of –2.10D.[24,25]

Retinal blood vessels initiate development in the human fetus at around 16 weeks of gestation. This is central to peripheral wave of growth at a rate of about 0.1 mm/day.[26] The nasal retina is fully vascularized by 32 weeks of gestation and the temporal retina is vascularized around 40 weeks. Premature infants are exposed to higher intra-arterial oxygen tensions after birth compared with *in utero* measurements and that leads to a downregulation of the hypoxia-triggered VEGF (vascular endothelial growth factor) production and results in apoptotic obliteration of the developing and existing retinal microvasculature (Fig. 23.2). This is one of the first events in the development of ROP. As neural tissue develops in the avascular retina, there is a concomitant increase in metabolic load that results in hypoxia at higher than physiologic levels. This condition leads to ischemia and is associated with activation of hypoxia-inducible factor-1 (HIF-1). It modulates transcription of VEGF and other genes for growth factors, proteinases, and cell-surface molecules that stimulate reactive neovascularization. In the infant rat model of ROP, disorganized and shortened rod outer segments were observed. Relative to animals housed in room air, attenuated photo-responses were measured in oxygen-exposed animals and resulted in variable to heightened

alterations in rhodopsin absorbance. It was shown in a mouse model of ischemia-induced ROP that astrocytes can sustain a period of hyperoxia and may be a key factor in recovery from hypoxic insult. Evidence supporting the role of astrocytes in vascular development is strengthened on the basis that astrocyte-free areas are avascular. These cells also have the capacity to supply DHA from α-LNA precursors, and DHA was shown to be a neurotrophic and survival factor in model systems of developing photoreceptors.[27,28] Also, mouse pups fed n-3 LC-PUFAs as supplements starting on postnatal day 1 had decreased oxygen-induced vaso-obliteration and neovascularization.[29,30] There is rapidly increasing evidence that suggests n-3 LC-PUFAs exert antiangiogenic and vasculogenic properties through modulation of processes involved in intracellular signaling, activation of transcription factors, and production of inflammatory mediators.

Inflammatory processes interfere with normal retinal vascularization in the most vulnerable retinas. The consequence of inflammation is oxidative stress that has been implicated in ROP etiology. The role of inflammation in ROP has been poorly investigated; however, some evidence in humans points to its contribution in the disease process. Late-onset sepsis, pneumonia, necrotizing enterocolitis, and *Candida* bloodstream infections are associated with more frequent and severe ROP. Sood and colleagues found that in patients who later developed ROP, there were higher systemic levels of interleukin (IL)-6 and C-reactive protein compared with controls at the third postnatal day.[31] They also observed higher levels of IL-18 during the second and third postnatal weeks in neonates who subsequently developed ROP. Sato and associates found higher levels of several cytokines and growth factors in the eyes of infants with ROP.[32] These biomolecules included IL-6, fibroblast growth factor, GCSF (granulocyte colony-stimulating factor), GMCSF (granulocyte-macrophage colony-stimulating factor), and particularly VEGF. Studies performed by Rivera and colleagues also showed that COX-2, an early gene product of inflammation, contributes to neovascularization in the vitreous of ischemic retinopathies through generation of prostaglandin (PG) E_2, which in turn acts on its EP3 receptor.[33] It is well known from experimental studies that modulation of tumor necrosis factor (TNF)-α production influences vascular growth,[34,35] and mice lacking TNF-α have less oxygen-induced retinopathy.[36] In contrast, DHA or EPA incubation of human adult saphenous vein endothelial cell cultures reduced endothelial expression of vascular cell adhesion protein (VCAM)-1, E-selectin, intercellular adhesion molecule (ICAM)-1, IL-6, and IL-8 after challenge with IL-1, TNF-α, or bacterial endotoxin. It was also shown experimentally that supplementation with n-3 LC-PUFAs led to increased formation of cytoprotective and anti-inflammatory metabolites, most notably neuroprotectinD1, resolvinD1, and resolvinE1, which exert their effects partially by suppressing TNF-α. These bioactive ω-3-LC-PUFA-derived mediators at physiologic levels reduce pathologic neovascularization through enhanced vessel regrowth after vascular loss and injury.[37] Moreover, Sapieha and associates found that a 5-lipoxygenase metabolite of DHA, 4-hydroxy-docosahexaenoic acid (4-HDHA), is a mediator of the anti-angiogenic effect of ω-3 LC-PUFAs and is independent from VEGF formation.[37] These investigators found that 4-HDHA directly inhibited angiogenesis and that it did so by activating a nuclear factor, peroxisome proliferator-activated receptor (PPAR)-γ.

On the grounds of consistent evidence the general conclusion is that ω-3 LC-PUFAs act in a protective role against pathologic factors that influence growth of the immature retina, namely ischemia, hyperoxia or hypoxia, and inflammatory agents.

SAFETY OF PARENTERAL SUPPLEMENTATION OF FISH-OIL-BASED FAT EMULSION IN PRETERM INFANTS

Premature neonates have limited muscle and fat mass and thus decreased hydrolytic capacity of the enzyme lipoprotein lipase.[38] As a consequence, very low birth weight infants are at higher risk for parenteral nutrition-associated hypertriglyceridemia than term infants. The monitoring of serum triglyceride concentrations has been recommended in premature infants to avoid hypertriglyceridemia. The recent recommendations of the European Society for Clinical Nutrition and Metabolism/European Society of Paediatric Gastroenterology, Hepatology and Nutrition (ESPEN/ESPGHAN) consider a concentration of serum triglycerides exceeding 2.82 mmol/L (250 mg/dL).[9] Therefore, all studies evaluating the safety and efficacy of parenteral fat emulsions should include the monitoring of serum triglyceride levels.

In 2008, Gura and associates demonstrated that parenteral fish-oil-based fat emulsion was safe and effective in the treatment of parenteral nutrition-associated liver disease.[39] In that study, 18 infants with parenteral nutrition-associated liver disease received fish-oil-based fat emulsion (10% Omegaven, Fresenius Kabi AG). For the first 2 days of treatment, patients received fish-oil emulsion at 0.5 g/kg/day and then were progressed to a maintenance dosage of 1 g/kg/day over 12 hours. The provision of fish-oil-based fat emulsion was not associated with EFA deficiency, hypertriglyceridemia, coagulopathy, infections, or growth delay. Thereafter, Tomsits and colleagues showed in a group of 30 premature infants the safety and efficacy of an intravenous lipid emulsion containing a mixture of soybean oil, medium-chain triglycerides, olive oil, and fish oil (20% SMOFlipid Fresenius Kabi Germany).[40] A comparison study of the safety, tolerability, and efficacy of SMOFlipid has been performed using the generally accepted fat emulsion, namely 20% Intralipid (Fresenius Kabi Germany). The compositions of 20% SMOFlipid and 20% Intralipid are presented in Table 23.1. Infusion of SMOFlipid was safe and well tolerated and showed a potential benefit by reducing cholestasis and elevating ω-3 LC-PUFAs and vitamin E levels in premature infants requiring parenteral nutrition. Recently, D'Ascenzo and associates analyzed concentrations of plasma lipid classes, plasma fatty acids,

TABLE 23.1   Composition of 20% SMOFlipid and 20% Intralipid

	20% SMOFlipid	20% Intralipid
Soybean oil (g/L)	60	200
Medium-chain triglycerides (g/L)	60	–
Olive oil (g/L)	50	–
Fish oil (g/L)	30	–
Glycerol (g/L)	25	22.5
Egg phospholipids (g/L)	12	12
Vitamin E (mg α-tocopherol/L)	Approx 200	38
Water for injection	ad 1000 mL	ad 1000 mL
pH value	7.5–8.8	7–8
Osmolarity (mosmol/L)	273	265

The comparison of these two fat emulsions is performed to show that 20% Intralipid emulsion does not contain docosahexaenoic acid (DHA). SMOFlipid contains 350 mg DHA in 100 mL of solution.

and erythrocyte fatty acids in preterm infants weighing less than 1250 g birth weight.[41] On day 7, the infants receiving 10% fish oil lipid emulsion had significantly lower plasma phospholipids, cholesterol esters, and free cholesterol but similar triglyceride concentrations. They also had significantly higher plasma DHA and EPA concentrations but lower AA levels when compared with those infants given soybean oil emulsion. Similar differences in fatty acid content were found in erythrocytes of the two groups. It is important to emphasize that plasma DHA concentrations in infants receiving fish-oil-containing emulsion were similar to those found in preterm infants fed on their mother's milk. It might be explained that the lower plasma cholesterol concentrations (free cholesterol and cholesterol esters) observed in the studied infants were a reflection of an increased clearance of exogenous and endogenous lipids or could be caused by a reduced endogenous lipogenesis. Short-term use of parenteral nutrition with a lipid emulsion containing a mixture of soybean oil, olive oil, medium-chain triglycerides, and fish oil (SMOFlipid Fresenius Kabi Germany) was also evaluated in a randomized double-blind study performed in preterm infants by Rayyan and colleagues.[42] The plasma triglyceride concentrations increased similarly in both groups receiving SMOFlipid or Intralipid. Throughout the study, mean serum triglyceride levels remained below the upper limit of normal range in both groups. A significant and more rigorous decrease in total and direct bilirubin compared with baseline was seen in the group of infants receiving fat-emulsion-containing fish oil compared with that of the control group. Other safety evaluations, namely hematologic profiles, a range of clinical laboratory tests, and vital signs, found that both lipid emulsions were equally safe and well tolerated. Moreover, both plasma and erythrocyte phospholipids, EPA, and DHA were higher and the ω-6/ω-3 fatty acid ratio was lower in the group of infants receiving lipid emulsion-containing fish oil. If all studies are taken together, the parenteral administration of fish-oil-enriched emulsions was safe and well tolerated by preterm infants and there was a beneficial fatty acid profile.

The effects of parenteral nutritional support with fish-oil emulsion (10% Omegaven Fresenius Kabi Germany) have also been evaluated in experimental studies. Tural-Emon and associates analyzed the effects of parenteral nutrition with fish-oil emulsion on spinal cord recovery in rats after traumatic spinal cord injury.[43] The researchers observed less neuronal injury and cord edema in the group of animals receiving fish-oil fat emulsion. The ω-3 LC-PUFAs had a marked neurite-promoting potential in adult and aged animals.[44] The anti-inflammatory effects of fish-oil also decreased edema and prevented neuronal injury.

FISH-OIL FAT EMULSION SUPPLEMENTATION MAY REDUCE THE RISK OF SEVERE RETINOPATHY IN VLBW INFANTS

One report showed that a fish-oil lipid emulsion (10% Omegaven, Fresenius Kabi Germany) given intravenously to infants weighing less than 1250 g at birth reduced the incidence of ROP.[45] This trial was designed to replace the use of a soybean and olive oil emulsion as part of the total daily lipid requirement. Preterm infants received a fish-oil emulsion that fulfilled one-third of their total daily intravenous lipid intake. Two-thirds of daily intravenous intake was the 20% soybean and olive oil emulsion (20% Clinoleic, Baxter SA, Norfolk, UK). The comparison of the two parenteral fat emulsions, specifically 10% Omegaven and 20% Clinoleic, is presented in Table 23.2. The occurrence of ROP in infants fed parenterally with the fish-oil fat emulsion was compared with historic data obtained from preterm neonates that were nourished exclusively with 20% Clinoleic. Patients in both groups had similar prenatal steroid use, birth weight, gestational age, clinical status after birth (Apgar score, CRIB score), and frequency of prenatal steroid administration. The monitoring of oxygen saturation using pulse oximetry revealed both groups remained within the ranges of target oxygenation, specifically 88% and 93%. Although the number of infants with ROP in the two groups was similar (13 vs. 16 infants), 10 of 13 patients with ROP that received fish-oil supplement had spontaneous regression of disease. Three patients needed laser therapy in the fish-oil

TABLE 23.2 Comparison of Parenteral Fat Emulsions: 10% Omegaven and 20% Clinoleic (g/100 mL)

Variable	Omegaven	Clinoleic
Oil		
Soybean	–	4
Olive	–	16
Fish	10	–
Glycerol	2.5	2.25
Egg phospholipids	1.2	1.2
Linoleic	0.1–0.7	50
α-Linoleic	< 0.2	8
DHA	1.4–3.1	–
EPA	1.3–2.8	–
Arachidonic acid	0.1–0.4	–

The absolute quantity of DHA synthesized from α-linoleic acid in preterm infants is dependent on energy intake. When it is less than 100 kcal/kg/day, only one-third of total amount is synthesized.
DHA: docosahexaenoic acid; EPA: eicosapentaenoic acid.

group while 12 of 16 infants receiving intravenous soybean and olive oil fat emulsion had to undergo laser photocoagulation of the retina (P = 0.023). Follow-up visits assessed visual acuity in 85% of infants that developed ROP in treatment groups. Only 15% of these infants tested had acuity below the normal range. There were no differences regarding the integrity of the visual pathways (i.e., neural connections) between the two groups of infants. No patient in either group developed any visible side effects from intravenous lipid administration. Mean platelets counts, catheter-related infections, and overall infection rates were comparable between the two groups of infants given intravenous lipid emulsions. The total intravenous intake of DHA in infants fed with a fish-oil emulsion supplement ranged approximately from 2.0 to 4.0 g per infant. Since the concentration of DHA in breast milk was not evaluated, it was impossible to calculate the total dose of DHA that an infant received in minimizing the risk of laser therapy for ROP. The same investigators in a subsequent study prospectively measured the fluctuations in plasma DHA concentrations among the preterm infants that either were treated or were not treated with a fish oil emulsion over the first 4 weeks of life.[15] The plasma DHA concentrations measured immediately after birth, before parenteral administration of lipid emulsion was introduced, showed no differences between both groups of infants. Thereafter, a statistically significant increase in mean plasma DHA concentration was noted on the seventh day of the study in the fish-oil emulsion-related group. However, the highest value of mean plasma DHA seen on the seventh day of life was still significantly lower than that observed in full-term neonates right after birth (P = 0.00001). On the 14th day of life, there was a decrease in plasma DHA levels, and these concentrations remained similar on the 21st and 28th day of study (Fig. 23.3). Alternatively, plasma DHA concentrations quantified in preterm infants fed parenterally with a traditional fat emulsion (without fish-oil supplement) showed a slight decrease, with the lowest value observed on the 14th day of the study (P = 0.43). Afterward, plasma DHA concentration rose on the 21st day of life, but this rise may be explained by diminished parenteral nutrition and increased enteral nutrition with breast milk or a formula enriched with DHA. The increased enteral intake of either breast milk or DHA-enriched formula in these infants was still probably insufficient to reach or even come close to the values of plasma DHA concentration observed in infants given fish-oil emulsion. The mean values of plasma DHA concentration measured in infants receiving traditional lipid emulsion were significantly lower at all stages of the study and by comparison with respective data obtained in patients supplemented with fat emulsion enriched with fish oil.

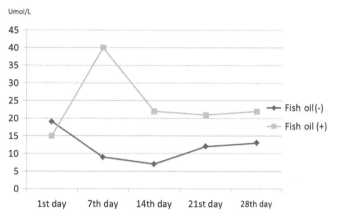

FIGURE 23.3 The fluctuations of mean values of plasma DHA concentration (μmol/L). Fish oil (−): the group of infants receiving lipids intravenously without addition of fish-oil-based fat emulsion. Fish oil (+): the group of infants receiving fish-oil-based fat emulsion. 1st, 7th, 14th, 21st, 28th day of life: data of measurement of plasma DHA concentration. Statistically significant increase in mean plasma DHA concentration was noted on the 7th day of the study in the fish-oil emulsion-related group. However, that value was still significantly lower than that observed in full-term neonates immediately after birth. The mean values of plasma DHA concentration measured in infants receiving traditional lipid emulsion were significantly lower at all stages of the study by comparison with respective data obtained in patients supplemented with fat emulsion enriched with fish oil.

In a retrospective analysis, the same researchers analyzed the frequency of ophthalmologic intervention (laser photocoagulation) in 337 very-low-birth weight infants. The preterm infants were divided into two groups: 1) those given either a fish-oil-based emulsion or 2) a nonfish-oil emulsion.[46] Patients in the two groups had similar demographics at and after birth. In a group of 152 preterm infants supplemented with fish-oil lipid emulsion, laser therapy was required in nine cases, whereas in a group of 185 infants treated with a fish-oil-free fat emulsion, laser photocoagulation was performed in 23 patients (P = 0.06). Combining the statistical analysis from the observational investigation[45] and from the retrospective study[46] showed a significantly reduced risk of severe ROP in the group of preterm infants supplemented with fish-oil lipid emulsion (see Table 23.3).

The results presented above are promising. Recently, we confirmed the association between the risk of severe ROP development and the beneficial effect of fish-oil emulsion in parenteral nutrition administered from the first day of life in a randomized prospective pilot study (author's personal communication). However, the effects of this modification of parenteral nutrition should be assessed in randomized, blinded, multicenter studies. The question of whether the dosage of fish-oil-based lipid emulsion as a component of total parenteral nutrition should be increased so that preterm infants at the 40th week of postconceptional age achieve plasma DHA levels similar to those seen in term

TABLE 23.3 Demographic and Clinical Characteristics of Infants Receiving or Not Receiving Fish-Oil-Based Fat Emulsion

	Infants Receiving Fish-Oil-Based Fat Emulsion n = 192	Infants Not Receiving Fish-Oil-Based Fat Emulsion n = 229	P
Birth weight (g): median (range)	960/570–1500	970/670–1500	NS
Gestational age (weeks): median (range)	28.3 (25–32)	28 (24–32)	NS
Gender (female) n (%)	109 (57.5%)	144 (63.5%)	NS
APGAR score, median (range)	7.0 (1–9)	7.0 (1–9)	NS
SGA infant n (%)	38 (20%)	40 (17%)	NS
ROP – laser therapy	12	35	0.0032*

Data shows a significantly reduced risk of severe ROP in the group of preterm infants supplemented with fish-oil emulsion.
NS: not significant; ROP: retinopathy of prematurity; SGA: small for gestational age.
Two-tailed Fisher exact test.

infants remains undefined. In other words, dose ranging studies are required in the context of future randomized clinical trials.

TAKE-HOME MESSAGES

- DHA is a ω-3 LC-PUFA and is considered the major structural lipid in sensory and vascular retina. Highest body concentrations of DHA per unit area are found in the disc membranes, and the overall percentage of DHA is 50% higher in the retina than in the next most concentrated tissue.
- Prematurely delivered infants miss an important period of intrauterine nutrient accretion that creates a cumulative DHA deficit and that correlates with earlier gestational ages and lower birth weights.
- DHA deficiency is associated with structural and functional abnormalities in the visual system. Rodents and primates with experimentally induced DHA deficiency show abnormalities in retinal structure, hindered visual acuity, and reduced cognitive performance. Then again, higher blood concentrations of DHA in preterm infants during the neonatal period are associated with improved visual acuity.
- A substantial body of evidence indicates that parenteral administration with fat emulsions containing fish-oil lipids are safe and well tolerated and result in a beneficial fatty acid profile in preterm infants.

- The association between reduced incidences of severe ROP following nutrition of preterm infants with intravenous fish-oil containing emulsion is encouraging. However, a large, multicentered, and randomized clinical trial that compares a fish oil-versus a nonfish-oil-based lipid emulsion in very preterm infants is the next step that will delineate a reduction in severe ROP while showing no adverse effects.

References

1. Niu SL, Mitchell DC, Litman BJ. Optimization of receptor-G protein coupling by bilayer lipid composition II: formation of meta-rhodopsin II-transducin complex. *J Biol Chem* 2001;**276**:42807–11.
2. Mitchell DC, Niu SL, Litman BJ. Optimization of receptor-G protein coupling by bilayer lipid composition I: kinetics of rhodopsin-transducin binding. *J Biol Chem* 2001;**276**:42801–6.
3. Ringbom T, Huss U, Stenholm A, Flock S, Skattebol L, Perera P, et al. COX-2 inhibitory effects of naturally occurring and modified fatty acids. *J Nat Prod* 2001;**64**:747–9.
4. Lien EL. Toxicology and safety of DHA. *Prostaglandins Leukot Essent Fatty Acids* 2009;**81**:125–32.
5. Innis H. Essential fatty acid transfer and fetal development. *Placenta* 2005;**26**(Suppl A.):S70–5.
6. Haggarty P. Effect of placental function on fatty acid requirements during pregnancy. *Eur J Clin Nutr* 2004;**58**:1559–70.
7. Larque E, Krauss-Etschmann S, Campoy C, Hartl D, Linde J, Klingler M, et al. Docosahexaenoic acid supply in pregnancy affects placental expression of fatty acid transport protein. *Am J Clin Nutr* 2006;**84**:853–61.
8. Lapillonne A, Jensen CL. Reevaluation of the DHA requirement for the premature infant. *Prostaglandins Leukot Essent Fatty Acids* 2009;**81**:143–50.
9. Koletzko B, Goulet O, Hunt J, Krohn K, Shamir R. Guidelines on Paediatric Parenteral Nutrition of the European Society of Paediatric Gastroenterology, Hepatology and Nutrition (ESPGHAN) and the European Society for Clinical Nutrition and Metabolism (ESPEN), supported by the European Society of Paediatric Research (ESPR). *J Pediatr Gastroenterol Nutr* 2005;**41**(Suppl 2.):S1–87.
10. Fleith M, Clandinin MT. Dietary PUFA for preterm and term infants: review of clinical studies. *Crit Rev Food Sci Nutr* 2005;**45**:205–29.
11. Innis SM. Fatty acids and early human development. *Early Hum Dev* 2007;**83**:761–6.
12. Klein CJ. Nutrient requirements for preterm infant formulas. *J Nutr* 2002;**132**:1395S–577S.
13. Carnielli VP, Simonato M, Verlato G, Luijendijk I, De Curtis M, Sauer PJ, et al. Synthesis of long-chain polyunsaturated fatty acids in preterm newborns fed formula with long-chain polyunsaturated fatty acids. *Am J Clin Nutr* 2007;**86**:1323–30.
14. Lapillone A, Eleni dit Trolli S, Kermorvant-Duchemin E. Postnatal docosahexaenoic acid deficiency is an inevitable consequence of current recommendations and practice in preterm infants. *Neonatology* 2010;**98**:397–403.
15. Pawlik D, Lauterbach R, Walczak M, Hurkała J. Docosahexaenoic acid (DHA) concentration in very low birth weight newborns receiving a fish-oil based fat emulsion from the first day of life. Preliminary clinical observation. *Med Wieku Rozwoj* 2011;**15**:312–7.
16. McCann JC, Ames BN. Is docosahexaenoic acid, an n-3 long-chain polyunsaturated fatty acid, required for development of normal brain function? An overview of evidence from cognitive and behavioral tests in human and animals. *Am J Clin Nutr* 2005;**82**:281–95.

17. Neuringer M, Jeffrey BG. Visual development: neural basis and new assessment methods. *J Pediatr* 2003;**143**:S87–95.

18. Niu SL, Mitchell DC, Litman BJ. Manipulation of cholesterol levels in rod disk membranes by methyl-beta-cyclodextrin: effects on receptor activation. *J Biol Chem* 2002;**277**:20139–45.

19. Greenberg JA, Bell SJ, Van Ausdal W. Omega-3 fatty acid supplementation during pregnancy. *Rev Obstet Gynecol* 2008;**1**:162–9.

20. Smithers LG, Gibson RA, McPhee A, Makrides M. Higher dose of docosahexaenoic acid in the neonatal period improves visual acuity of preterm infants: results of a randomized controlled trial. *Am J Clin Nutr* 2008;**88**:1049–56.

21. Pollan C. Retinopathy of prematurity: an eye toward better outcomes. *Neonatal Netw* 2009;**28**:93–101.

22. Siatkowski RM, Dobson V, Quinn GE, Summers CG, Palmer EA, Tung B. Severe visual impairment in children with mild or moderate retinal residua following regressed threshold retinopathy of prematurity. *J AAPOS* 2007;**11**:148–52.

23. Fulton AB, Hansen RM, Petersen RA, Vanderveen DK. The rod photoreceptors in retinopathy of prematurity: an electroretinographic study. *Arch Ophthalmol* 2001;**119**:499–505.

24. McLoone E, O'Keefe M, McLoone S, Lanigan B. Long term functional and structural outcomes of laser therapy for retinopathy of prematurity. *Br J Ophthalmol* 2006;**90**:754–9.

25. O'Connor AR, Stephenson TJ, Johnson A, Tobin MJ, Ratib S, Moseley M, et al. Visual function in low birth weight children. *Br J Ophthalmol* 2004;**88**:1149–53.

26. Ashton N. Retinal angiogenesis in the human embryo. *Br Med Bull* 1970;**26**:103–6.

27. Bernoud N, Fenart L, Benistant C, Pageaux JF, Dehouck MP, Moliere P, et al. Astrocytes are mainly responsible for the polyunsaturated fatty acid enrichment in blood-brain barrier endothelial cells *in vitro*. *J Lipid Res* 1998;**39**:1816–24.

28. Williard DE, Harmon SD, Kaduce TL, Preuss M, Moore SA, Robbins ME, et al. Docosahexaenoic acid synthesis from n-3 polyunsaturated fatty acids in differentiated rat brain astrocytes. *J Lipid Res* 2001;**42**:1368–76.

29. Connor KM, SanGiovani JP, Lofqvist C, Aderman CM, Chen J, Higuchi A, et al. Increased dietary intake of omega-3-polyunsaturated fatty acids reduces pathological retinal angiogenesis. *Nat Med* 2007;**13**:868–73.

30. SanGiovani JP, Berkey CS, Dwyer JT, Colditz GA. Dietary essential fatty acids and visual resolution acuity in healthy full-term infants: a systematic review. *Early Hum Dev* 2000;**57**:165–88.

31. Sood BG, Madan A, Saha S, Schendel D, Thorsen P, Skogstrand K, et al. NICHD Neonatal Research Network: Perinatal systemic inflammatory response syndrome and retinopathy of prematurity. *Pediatr Res* 2010;**67**:394–400.

32. Sato T, Kusaka S, Shimojo H, Fujikado T. Simultaneous analyses of vitreous levels of 27 cytokines in eyes with retinopathy of prematurity. *Ophthalmology* 2009;**116**:2165–9.

33. Rivera JC, Sapieha P, Joyal JS, Duhamel F, Shaou Z, Sitaras N, et al. Understanding retinopathy of prematurity: update on pathogenesis. *Neonatology* 2011;**100**:343–53.

34. Goukassian DA, Dolan C, Murayama T, Silver M, Curry C, Eaton E, et al. Tumor necrosis factor-α receptor p75 is required in ischemia-induced neovascularization. *Circulation* 2007;**115**:752–62.

35. Kishore R, Qin G, Luedemann C, Bord E, Hanley A, Silver M, et al. The cytoskeletal protein ezrin regulates EC proliferation and angiogenesis via TNF-α-induced transcriptional repression of cyclin A. *Am J Clin Invest* 2005;**115**:1785–96.

36. Gardiner TA, Gibson DS, de Gooyer TE, de la Cruz VF, McDonald DM, Stitt AW. Inhibition of tumor necrosis factor-α improves physiological angiogenesis and reduces pathological neovascularization in ischemic retinopathy. *Am J Pathol* 2005;**166**:637–44.

37. Sapieha P, Stahl A, Chen J, Seaward MR, Willett KL, Krah NM, et al. 5-Lipoxygenase metabolite 4-HDHA is a mediator of the anti-angiogenic effect of ω-3 polyunsaturated fatty acids. *Sci Transl Med* 2011;**3**:69.

38. Ziegler EE, O'Donnell AM, Nelson SE, Fomon SJ. Body composition of the reference fetus. *Growth* 1976;**40**:329–41.

39. Gura KM, Lee S, Valim C, Zhou J, Kim S, Modi BP, et al. Safety and efficacy of a fish-oil based fat emulsion in the treatment of parenteral nutrition associated liver disease. *Pediatrics* 2008;**121**:678–86.

40. Tomsits E, Pataki M, Tolgyesi A, Fekete G, Rischak K, Szollar L. Safety and efficacy of a lipid emulsion containing a mixture of soybean oil, medium-chain triglycerides, olive oil and fish oil: a randomized, double-blind clinical trial in premature infants requiring parenteral nutrition. *J Pediatr Gastroenterol Nutr* 2010;**51**:514–21.

41. D'Ascenzo R, D'Egidio S, Angelini L, Bellagamba MP, Manna M, Pompilio A, et al. Parenteral nutrition of preterm infants with a lipid emulsion containing 10% fish oil: effect on plasma lipids and long-chain polyunsaturated fatty acids. *J Pediatr* 2011;**159**:33–8.

42. Rayyan M, Devlieger H, Jochum F, Allegaert K. Short-term use of parenteral nutrition with a lipid emulsion containing a mixture of soybean oil, olive oil, medium-chain triglycerides, and fish oil: a randomized double-blind study in preterm infants. *J Parented Enteral Nutr* 2012;**36**:81S–94S.

43. Tural-Emon S, Gercek-Irban A, Uyar-Bozkurt S, Akakin D, Konya D, Ozgen S. Effects of parenteral nutritional support with fish-oil emulsion on spinal cord recovery in rats with traumatic spinal cord injury. *Turk Neurosurg* 2011;**21**:197–202.

44. Michael-Titus AT. Omega-3 fatty acids and neurological injury. *Prostaglandins Leukot Essent Fatty Acids* 2007;**77**:295–300.

45. Pawlik D, Lauterbach R, Turyk E. Fish-oil fat emulsion supplementation may reduce the risk of severe retinopathy in VLBW infants. *Pediatrics* 2011;**127**:223–8.

46. Pawlik D, Lauterbach R, Hurkała J. The efficacy of fish-oil based fat emulsion administered from the first day of life in very low birth weight newborns. *Med Wieku Rozwoj* 2011;**15**:306–11.

The Impact of Low Omega-3 Fatty Acids Diet on the Development of the Visual System

Patricia Coelho de Velasco, Claudio Alberto Serfaty

Instituto de Biologia, Universidade Federal Fluminense, Niterói, Brazil

DEVELOPMENT OF THE VISUAL SYSTEM

During the development of visual as well as other brain areas, massive amounts of neurons and glial cells are generated. After migration to appropriate positions and differentiation into specific cell types, neurons start dendritic and axonal outgrowth, target selection, and synaptogenesis. During this process, axons and their growth cones navigate through selective pathways guided by repulsive and attractive molecular gradients.[1] On reaching appropriate neuronal targets, the development of synapses takes place, mostly during postnatal life (Fig. 24.1).

Development of the Retina and Retinofugal Connections

Studies in rodents have shown that during early development, common progenitor cells give rise to all neuronal cell types that will populate the retina as well as Muller glial cells. In rats, retinal neurogenesis starts at embryonic day (E) 10 and is completed by postnatal day (PND) 12. Retinal ganglion cells are the first cell type to be produced, followed by horizontal cells, cones, amacrine cells, rods, bipolar cells, and Muller glia cells[2] in a central-peripheral gradient of cell proliferation across the retina. To accomplish such a task, several transcription factors have been shown to be expressed by subsets of retinal progenitor cells or their postmitotic progeny at particular periods during retinal development.[3] Retinal ganglion cells extend their axons toward the optic nerve head, a process that is directed by several molecular markers including chondroitin sulfate proteoglycans (CSPGs), Netrins, and Semaphorins.[4]

In rodents, a partial decussation of retinal axons is observed at the optic chiasm: axons from ganglion cells lying at the periphery of the temporal retina (about 5% of the ganglion cell population) do not cross the midline, forming an ipsilateral pathway along the optic tract. Conversely, axons from ganglion cells lying in the nasal retina, and comprising 95% of the ganglion cell population, cross the midline, joining the contralateral optic tract. This pattern of axonal direction involves EprinB-2 molecules expressed in glial cells at the chiasmatic midline, which selectively repeals EphB1-expressing axons from the temporal retina out of the midline.[4]

Development of Visual Topographical Maps

Since the pioneering studies of Roger Sperry,[5] synaptic specification and the importance of topographical maps have been recognized as major features of brain organization and neural processing. So far, the concept that sensory as well as motor and cognitive abilities depend in a fundamental way on the correct patterning of connections has become part of our current knowledge about brain function.[6] In this way, a keystone feature of visual processing relies on the correct patterning of visual connections that allow the proper development of various visual attributes of form, color, and motion, which directly influence visual acuity.[7]

The development of organized and specific connections found among mammalian species depends on two general strategies: an initial overproduction of neurons and synaptic contacts followed by the death of excess neurons and elimination of misplaced ineffective axons/synapses.[8,9] About the time that neuronal contacts begin to emerge (starting during the last trimester in human gestation), about 50% of the immature neuronal population undergoes a process of natural neuronal death. However, natural neuronal death seems to play only a marginal role in topographical maps development.[4,10] More importantly, in rodents during early postnatal development and in humans during the last gestational

Human brain development

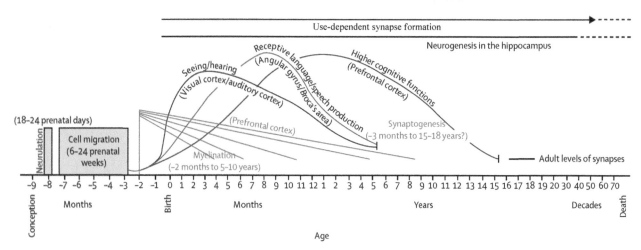

FIGURE 24.1 Time course of major events of the human brain development. The diagram displays the major steps of sensory, motor, and cognitive development. *Reproduced with kind permission from Springer Science+Business Media: from Handbook of Behavior, Food and Nutrition: Tryptophan intake and the influence of serotonin on development and plasticity of sensory circuits, Serfaty, C.A. fig1, 2011. Reprinted from Grantham-McGregor et al.,[114] with permission from Elsevier. Adapted from Thompson and Nelson (2001)[115] and reproduced with personal permission of authors.*

trimester, axonal arbors are expanded over the target territories and, as a result, immature neurons establish a set of transitory synaptic contacts.[7,11] Hence, during postnatal development, transitory axons are selectively eliminated and synapses from adjacent retinal neurons converge into appropriate postsynaptic sites, giving rise to precisely organized circuits. Those regressive events of neurogenesis are among the most important mechanisms that govern the selection of interconnecting neuronal populations in the brain (Fig. 24.1).[12]

As a model system for the study of organized brain circuits, the connections between retinal ganglion cells and their target nuclei, the superior colliculus (SC) and the dorsal lateral geniculate nucleus (dLGN) – the so-called retinofugal connections – have been extensively studied in rodents.[13–15] It has been shown that the rodent retinocollicular pathway develops within the first 3 postnatal weeks, producing highly specific patterns of axonal connections (Fig. 24.2). This form of developmental plasticity occurs mainly through axonal elimination and synaptic growth at appropriate territories.[16]

The initial development of retinocollicular topography is strongly influenced by repulsive/attractive molecules between retinal axons and target neurons. Retinal ganglion cell axons and target cells in the SC express Ephrins and Eph receptors in complementary gradients that vary along the main retinal axis (dorsal to ventral/temporal to nasal).[17] A later and complementary step of topographical refinement is achieved by activity-dependent mechanisms that are required to ensure the fine tuning of the correct representation of retinal axons over their targets, and thus circuitry maturation.[18] This includes both spontaneous and evoked activity of retinal ganglion cells.[19]

CRITICAL PERIODS FOR BRAIN DEVELOPMENT

The use-dependent development of functional, organized central connections in the mammalian brain occurs during a time window known as the critical period. The duration of the critical period is highly variable between mammalian species and is inversely related to the species longevity: rodents display a 3-week critical period while humans develop protracted critical periods that extend up to 5–12 years and possibly beyond.[20,21]

In a broader conception the critical period corresponds to a developmental stage in which environmental cues provide rapid plasticity of neuronal circuits necessary for the acquisition of appropriate sensory, motor, as well as cognitive skills. Indeed critical periods have been described in many brain systems and in a large variety of species: song learning in birds, auditory localization in barn owls, and, in humans, the development of sensory acuity, motor, and language skills.[22] The end of the critical period considerably affects plasticity of primary sensory areas of the brain, and as a result a considerable slowdown in use-dependent modifications has been described in visual, auditory, and somatosensory cortical[21,23] and subcortical areas such as the SC.[16] Although the end of the critical period affects plasticity in cortical and subcortical primary sensory areas, it does not limit plasticity in other associative regions, which are still able to undergo use-dependent modifications even in adulthood.[24–26]

One important question is how use-dependent activation of primary sensory areas during early postnatal development impacts development and plasticity of other brain areas later on. It has been shown that early

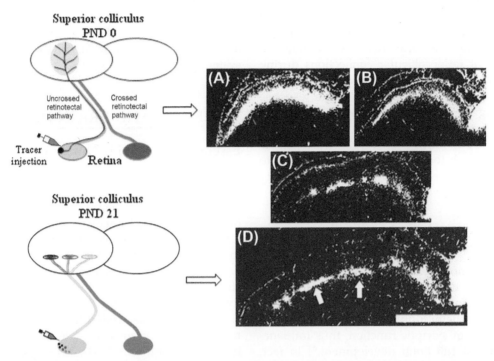

FIGURE 24.2 Development of the retinocollicular visual pathway. Injection of a neuroanatomical tracer into the vitreous chamber reveals the pattern of axonal labeling of retinal ganglion cell axons in the visual layers of the ipsilateral superior colliculus (bright labeling in dark-field photomicrographs – coronal sections of the superior colliculus). The postnatal development starts with expanded axons over the collicular visual layers found soon after birth (diagram in upper left; photomicrographs A, B). By the second and third postnatal weeks, misplaced retinal axons have been selectively eliminated and topographically correct axons concentrate at specific termination zones at the ventral aspect of the collicular visual layers (diagram lower left; photomicrographs C, D, arrows). PND: postnatal day. Scale bar: 500 μm. *Reproduced with kind permission from Springer Science+Business Media: from Handbook of Behavior, Food and Nutrition: Tryptophan intake and the influence of serotonin on development and plasticity of sensory circuits, Serfaty, C.A. fig2, 2011. Reprinted from Serfaty* et al.,[16] *with permission from Elsevier.*

visual experience directly influences plasticity in the adult visual cortex,[27] and early musical training (before the age of 7 years) influences sensory motor integration and, thus, performance of musicians.[28] Therefore, the experiences acquired during this developmental window may directly affect the emergence of several other brain abilities and influence the outcome of our individuality. Importantly, those use-dependent modifications in neuronal connectivity ultimately result in certain behaviors or capabilities, which would not be revealed/developed otherwise.[11,20,22,28]

The influence of critical periods on the development of sensory brain connections was originally defined after the experiments made by Wiesel and Hubel (1963). Those investigators showed that a temporary visual deprivation of one eye causes a dramatic change in the ocular dominance distribution, in favor of the open eye, in kittens between the fourth and the eighth postnatal weeks.[29,30] In humans, neonatal strabismus can also result in similar loss of visual acuity: eye misalignment, if not appropriately treated until the age of 5 years, produces a permanent loss of visual acuity also known as amblyopia.[31] This acuity loss results from the weakening of synapses originating from the nonaligned eye. Furthermore, cats raised under visually biased environments

(e.g., exposed to visually stereotyped patterns of horizontal or vertical lines) do not develop accurate discrimination of visual stimuli (except for those horizontal or vertical stimuli), as well as the proper binocular representation of the visual field.[32]

In the visual cortex, the critical period has been correlated with mechanisms involving neurotrophin signaling, especially brain-derived neurotrophic factor (BDNF), which aids the differentiation of inhibitory gamma-aminobutyric acid (GABA) circuits.[22] The development of GABAergic innervation seems to be crucial for the onset of the critical period,[33] and inhibitory circuits are under control of both visual experience and BDNF.[34] Also, insulin-like growth factor 1 (IGF-1) has been shown to facilitate the development of inhibitory innervation and increase visual acuity.[35]

The closure of the critical period in the primary visual cortex involves extracellular matrix (ECM) molecules that develop an environment that inhibits axon and dendritic remodeling. Such molecules include CSPGs,[36] tissue plasminogen activator (tPA),[37] and growth inhibitory proteins such as Nogo, MAG, and OMgp.[38]

In visual subcortical nuclei such as the SC, the critical period overlaps with the period of fine tuning of topographical maps.[9,39] Lesion studies, either monocular enucleation or restricted retinal lesions, have been used to induce reorganization of axons originating from the intact eye.[16,40,41] Those experiments revealed that the plastic capacity of retinocollicular connections during the critical period is characterized by a rapid reactive growth of axons from the nonlesioned eye in response to lesions during the first 3 postnatal weeks. After the third postnatal week, a single retinal lesion was still able to elicit a certain amount of reorganization of the intact pathway, which took, however, several weeks to develop. Therefore, a second slow stage of plasticity does occur even after the end of the critical period.[16]

Mechanisms for Use-Dependent Plasticity During the Visual System Critical Period

Use-dependent modifications that occur at a rapid rate during the critical period occur at specific loci called dendritic spines. Those specialized dendritic structures concentrate the biochemical machinery for use-dependent modifications in synaptic function, thus influencing the outcome of full brain development.[42] In fact, disorders related to mental retardation are often correlated to structural and functional modifications in those structures, with an enormous impact on cognitive and sensorimotor development.[43]

Use-dependent modifications of developing synapses relate to the expression of α-amino-3-hydroxy-5-methyl-4-isoxazolepropionic acid (AMPA) and N-methyl-D-aspartate (NMDA) glutamate receptors (AMPAr/NMDAr) that play a critical role in synaptic plasticity.[44] For instance, during early postnatal development NMDAr display a composition with NR1/NR2B subunits, a configuration that allows an increased capacity for processing simultaneous inputs (extended synaptic integration), as would be expected for immature synapses. By the time of eye opening, however, NMDAr switch their subunit composition to NR1/NR2A, a configuration that requires strict temporal co-activation times, necessary for the development of highly specific connections.[45] In accordance with a role for NMDAr in synaptic development and stabilization, it has been shown that the pharmacologic blockade of NMDAr disrupt the normal connectivity pattern and, as a result, increase the sprouting of the uncrossed pathway following the partial deafferentation of the contralateral converging retinotectal projection.[46]

Evidence also shows that the expression of AMPAr, and their GLUR1 and GLUR2 subunits, regulate the size and shape of neuronal dendritic spines, affecting the number and type of synaptic inputs, as well as the complexity of the retinotectal circuits in xenopus.[47] Activity and NMDA-induced calcium influx can trigger either the membrane insertion or the internalization of AMPAr at synaptic membranes, resulting in long-term potentiation (LTP) or long-term depression (LTD) of synaptic activity.[48] Both LTP and LTD have been associated with the development of synaptic specificity in the rodent visual system.[49,50]

OMEGA-3 AND BRAIN DEVELOPMENT

Essential Fatty Acids Nutritional Status

Several studies have extensively demonstrated that the development and maturation of the central nervous system (CNS) is directly influenced by diet.[51–53] Indeed, nutrition has been recognized among the main factors that influence CNS development, especially during critical periods of formation and differentiation of neural circuitry.[54]

Nutritional deficiencies are related to changes in structural, biochemical, electrophysiologic, and cognitive aspects of brain functioning, thus influencing brain size, neurogenesis, cell differentiation, synthesis and release of neurotransmitters, levels of neuronal excitation, and inhibition.[55,56] The consequences of malnutrition will depend on its nature and severity, including the type of nutritional deficiency, level of malnutrition, and exposure time.[57]

Essential nutrients are exclusively acquired through diet. Therefore, social and cultural factors involved in nutritional needs could lead to functional changes in neurochemical signaling pathways, with serious consequences for neural circuitry maturation.[53,58] Lipids represents about 50–60% of brain dry weight of an adult, 35% of which is long-chain polyunsaturated fatty acids (LC-PUFA). α-Linolenic acid (ALA; omega-3 fatty acid) and linoleic acid (omega-6 fatty acid) are considered essential fatty acids (EFAs) as they cannot be synthetized endogenously.[59,60] Docosahexaenoic acid (DHA) and arachidonic acid (AA), derived from their respective precursors, omega-3/omega-6, are the main LC-PUFA found in the brain.[61,62]

Adequate supplies of EFAs are required during development and in adulthood in order to allow normal brain function.[63–65] Numerous studies on the deficiency of omega-3 fatty acids mention the initial stages of development as a critical period of brain modeling.[13,66] Several factors can interfere with the conversion of EFAs into their main products including: intake of saturated fatty acids and hydrogenated lipids, deficiency of vitamins and minerals that act as cofactors (mainly zinc deficiency), excessive alcohol consumption, and stress-related hormones. These potential blockers of EFA metabolism indicate that even under an adequate intake of dietary EFAs, deficiencies of AA and DHA may occur. Moreover, differences in the metabolism of constitutional EFAs are being increasingly recognized as possible risk factors for neurodevelopmental disorders.[67]

Indeed, DHA levels in the brain seem to be strictly controlled, since any disturbance leads to severe impairment in brain development and maturation.[68–70] Most of DHA incorporation in the brain and retina occurs throughout the last trimester of gestation and continues up to 4 years of age in humans.[71] This time course overlaps with major landmarks of visual system development, from neurogenesis to axonal elimination and its underlying critical period.

During prenatal development, DHA transfer takes place from mother to offspring through placental transport. After birth, omega-3 incorporation is achieved by breast milk consumption.[63,72] Thus, fetuses and newborns are completely dependent on the maternal supply of EFAs. One major issue of EFA transfer is that the composition of fatty acids in breast milk has been shown to be highly variable and strictly dependent on the mother's diet composition.[61,73,74]

The availability of DHA during early life has been shown to affect the development of cognitive and visual functions.[75] Randomized clinical studies have found significant associations between red blood cell or plasma levels of DHA and improvements in visual status and pointed out the importance of the introduction of omega-3 fatty acids in infant formulas as a demand to reach children's needs during growth and development.[74,76] A study reported a persistent effect of DHA on visual acuity during the first year of life in children fed with breast milk and with a DHA-supplemented formula as opposed to children fed with a nonsupplemented formula.[77] The results also showed similar lipid composition in red blood cells from the breast milk group and supplemented formula groups, indicating that blood levels of DHA are directly related to changes in visual acuity.[71] Birch and colleagues demonstrated, in a clinical trial study, that infants at 12 months of age that received supplemented formulas with different levels of DHA increased visual evoked potentials and visual acuity when compared with a control formula containing 0% of DHA.[78]

The balance of omega-3/omega-6 (ω-3/ω-6) fatty acids is an important determinant of homeostasis maintenance and normal brain development.[79,80] An adequate ratio of ω-3/ω-6 has changed in the modern Western human diets.[81] A decreased ratio in ω-3/ω-6 fatty acids and the increased intake of saturated fat has been implicated in DHA deficiency even in breastfed infants.[61] The main food sources of ALA are flaxseed and some nuts, but DHA is obtained primarily from fish and other seafood.[82,83]

Incorporation of DHA in neuronal membranes occurs primarily by the accumulation of DHA from circulating plasma or by biosynthesis that occurs in the liver. Local synthesis may still occur in brain tissue.[84] Omega-3 fatty acids are found esterified in membrane glycerophospholipids containing phosphatidylserine and phosphatidylethanolamine and are released to participate directly or indirectly in the regulation of physiologic and metabolic processes.[63] About 20% of the fatty acids in human retina are DHA enriched in phospholipids integrating membrane discs in the outer segment of rods and cones.[63] DHA also supports the efficiency in phototransduction and proper regulation of the visual pathway.[75] Moreover, DHA promotes survival and inhibition of apoptosis of photoreceptors.[85]

Omega-3, Docosahexaenoic Acid, and Synaptic Plasticity

Omega-3 fatty acids are necessary for brain maturation and synaptic plasticity, regulating synaptic strength, and increasing the fluidity and function of neuronal membranes.[86,87] As an integral component of neural membranes, DHA modulates the activity of membrane-bound enzymes and receptors. DHA facilitates exocytosis of neurotransmitter-containing vesicles, indicating an important role in neurotransmission.[88] Moreover, it not only modulates the physical properties of neuronal membranes[89] but also promotes the formation of second messengers that activate signaling pathways.[90] DHA has also been shown to modulate the levels of neuronal synaptic proteins.[13,91] Indeed, Cansev (2009) reported in rats that DHA and uridine supplementation during the gestational period until PND21 increased the levels of synapsin-1, mGluR1, and PSD-95.[92] DHA also increases the number of dendritic spines and excitatory synapses in hippocampal neurons,[93] suggesting its involvement in synaptic plasticity.

As a free fatty acid, DHA can regulate the activity and insertion of glutamate receptors such as NMDAr and AMPAr.[66] It also modulates dopaminergic[94] and serotonergic neurotransmission.[95] DHA can also restore LTP attenuated by blocking the activity of phospholipase A2 (PLA2),[96] since PLA2 activity is related to the insertion of AMPAr in postsynaptic densities.[97] DHA as a free form, or by way of bioactive derivatives such as neuroprotectin D1 (NPD1), can regulate and interact with multiple signaling cascades, contributing to the development, differentiation, synaptic function, protection, and repair of cells in the nervous system.[98]

DHA has also been seen as a trophic molecule that is supported by an increased expression of genes related to signal transduction mechanisms, synaptic plasticity, energy metabolism, and traffic through membrane transporters.[99] DHA has an important role as an endogenous ligand for retinoid X receptors (RXR).[100] The receptors RXRα and RXRβ together as peroxisome proliferator-activated receptor (PPAR) gamma can be activated by DHA, leading to dimerization of these receptors and their insertion into the nucleus, where they can act as transcription factors related to protein differentiation and synaptic stability.[84] Studies have shown that DHA can regulate the expression of BDNF messenger ribonucleic acid (mRNA), a neurotrophic factor directly related to processes of neuronal survival and synaptic strengthening.[101–103]

ROLE OF OMEGA-3 ON DEVELOPMENT OF CENTRAL VISUAL CONNECTIONS

DHA has been shown to exert several roles in the visual system from photoreceptor differentiation to synaptic plasticity in a series of events leading to a direct influence on visual acuity.[74,104] As the development of topographical maps, especially the fine tuning of visual connections onto primary subcortical and cortical targets, is strongly influenced by patterns of spontaneous and visually driven activity,[22] we tested the roles of EFAs during development of the rat retinofugal pathways.

AA has been recognized as a retrograde messenger that stabilizes axonal arbors in the optic tectum.[105] Schmidt and colleagues[106] showed that the pharmacologic blockade of AA release increased the dynamics of developing retinotectal axonal arbors in goldfish. In the mammalian visual system it was shown that the blockade of PLA2 activity between PND21 and PND28 is able to induce a similar increase in retinocollicular terminal field size, which is expanded in comparison to control animals even beyond the critical period closure (Fig. 24.3).[107] These data corroborate a role for PLA2 metabolites in synaptic maintenance. Since pharmacologic blockers of PLA2 activity such as quinacrine are able to block not only AA but also DHA release from cellular membranes, it seems likely that those previous results could be due, in part, to a reduction of both free AA as well as DHA content at the level of retinocollicular synapses.

In order to address a more specific role for DHA in the developing visual system, we used a nutritional approach in which female rats were given an isocaloric diet containing coconut oil as a lipid source.[13] This diet protocol started 5 weeks before mating in order to deplete omega-3 fatty acids. Females were then kept under this nutritional restriction during mating, pregnancy, and after delivery until the litters reached PND42 (Fig. 24.4). Lipid levels were measured in samples from the collicular visual layers of rats at PND28. Those samples revealed a 53% reduction in the levels of DHA without any changes in AA content.[13] The topographical distribution of uncrossed retinocollicular terminal fields was determined by the anterograde transport of horseradish peroxidase (HRP) as an anterograde tracer. It was shown that this chronic form of malnutrition was able to disrupt the topographical development as early as the second postnatal week (PND13), when terminal fields displayed a two-fold increase in label density when compared to control, soy oil-fed animals (Fig. 24.5A, B). Coconut-fed litters also revealed topographically expanded terminal fields at PND28 and PND42 (Fig. 24.5C), strongly suggesting that an omega-3 restriction, and the subsequent reduction of DHA levels, produced abnormal connections in the rodent visual system. This could be due either to a slowdown in axonal elimination of transitory synapses or to unspecific sprouting as a result of a decrease in DHA-induced synaptic stabilization mechanisms. The latter mechanism was in part confirmed by a decrease in phospho-GAP43 (pGAP-43) content observed in the visual layers of the SC 13. The phosphorylated form of GAP-43 (pGAP-43) protein has been involved in hippocampal synaptic plasticity and in the stabilization of developing synapses.[40,108]

The disturbance of visual system development induced by omega-3/DHA reduction was not confined

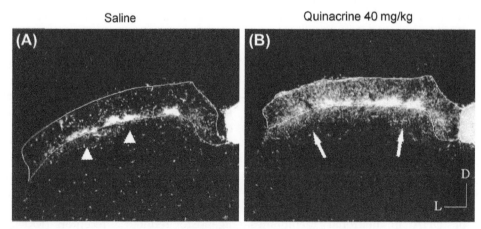

FIGURE 24.3 Role of essential fatty acids (EFAs) in the maintenance of retinocollicular topography. Coronal sections of the visual layers of superior colliculus showing the control group (A) with a well-defined pattern of discrete clusters of terminal labeling located at the ventral border of collicular visual layers. Pharmacologic blockade of PLA2 disrupts the topographical distribution of retinal axons, with a resulting dispersion of axonal terminals throughout the visual layers of the superior colliculus (B), indicating the role for the EFA derivatives, arachidonic acid and docosahexaenoic acid, in the maintenance of visual circuits. *Reprinted from Campello-Costa* et al.,[107] *with permission from Elsevier.*

to the ipsilateral retinocollicular pathway. It was shown that both crossed and uncrossed retinogeniculate terminal fields overlapped in an incorrect way in rats fed with a coconut-based diet.[13] At PND28, when most of the development of retinogeniculate segregation has finished,[109] the ipsilateral terminal zones were still expanded in relation to control animals, suggesting that the disturbances in the fine tuning of the visual system topography were a common finding in all major subcortical visual nuclei.[13]

Since DHA is related to several mechanism that influence neurogenesis, apoptosis, and cell differentiation, we asked whether the disturbances in retinal connectivity onto visual nuclei would be reversed by a DHA supplementation protocol. This was achieved by the oral administration of fish oil, a source of DHA and eicosapentaenoic acid (EPA). As shown in Figure 24.5D, fish oil supplementation for 3 weeks (from PND 7/28) was able to reverse the effects of a gestational/neonatal DHA

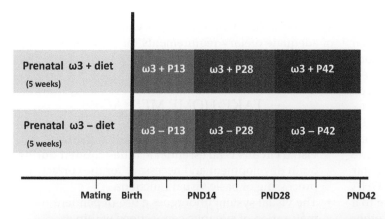

FIGURE 24.4 Representative scheme of diet administration schedules and experimental groups. Females were fed *ad libitum* with either control (ω-3+ soy oil) or experimental (ω-3− coconut oil) diets starting at 5 weeks before mating. After delivery, litters were fed through their mothers, and after weaning, young rats received diets directly. PND: postnatal day. *Reprinted from Velasco et al.,[13] with permission from Elsevier.*

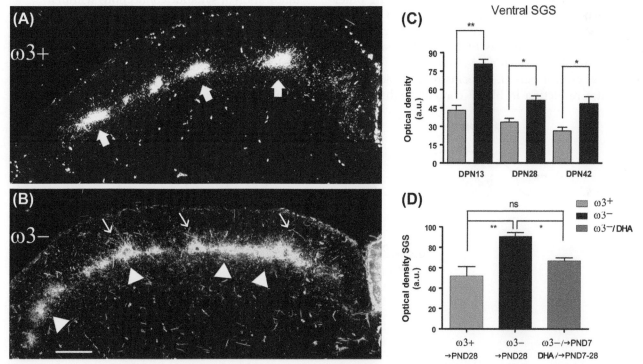

FIGURE 24.5 Omega-3 restricted diet disrupts the topographical distribution of uncrossed retinocollicular axons during early and late postnatal development. Dark field photomicrographs of coronal sections of the visual layers of superior colliculus at postnatal day (PND)13. (A) Omega-3+ control rats presented a normal pattern of terminal labeling with major terminal fields located at the ventral border of the collicular visual layers (large arrows). (B) In contrast, omega-3− restriction group presented an increased optical density throughout the same layers (small arrows and arrowheads), which denotes a loss of topographical fine tuning of visual connections. (C) Data quantification showed that, in omega-3 restricted rats, the density of terminal labeling was abnormally high at PND13 but still above control levels at PND28 and PND42, long after the period when axonal elimination and topographical refinement should have occurred. (D) Docosahexaenoic acid (DHA) supplementation (with fish oil) for 3 weeks (from PND 7/28) was able to reverse completely the effects of perinatal omega-3 deprivation on the density of retinal innervation across the superior colliculus. ns: not significant; SGS: stratrum griseum superficiale. *Adapted and reprinted from Velasco et al.,[13] with permission from Elsevier.*

deprivation completely, resulting in normal densities of terminal fields across the SC.[13] Thus, the results clearly show that disturbances in visual connectivity can be restored by an adequate amount of DHA intake during the critical period of development.

Thus, the results described by de Velasco and colleagues are consistent with a general delay in development, which results in errors in topographical fine tuning of retinal connections.[13] To address whether other aspects of development could also exhibit a similar delay directly, we made a series of retinal lesion experiments that are suitable to access the critical period limits (Fig. 24.6). As described earlier,[16] restricted retinal lesions to one eye induce a rapid sprouting of axons from the intact eye that converge to the same aspect of the SC contralateral to the lesioned eye. It has been shown that after the third postnatal week, a slow plasticity is observed only within weeks or months 16. Under normal conditions no sprouting of intact axons can be detected one week after a retinal lesion made at PND21, which characterizes the end of the collicular critical period.[107,110] However, animals depleted of DHA surprisingly displayed a vigorous plastic response to a retinal lesion (Fig. 24.6), suggesting that among other things, DHA restriction altered the duration of the critical period.[13] Since DHA is involved in transcription of neurotrophic factors such as BDNF[111] and that neurotrophin reduction as well as dark rearing[112,113] delays the cortical critical period, it seems reasonable to suppose that a reduction in BDNF content might explain the developmental delay of retinofugal connections found in this study.

In conclusion, omega-3 nutritional restriction directly impacts DHA availability within visual nuclei and dramatically alters the time course of topographical refinement and critical period windows. The consequences of those influences on such a precisely regulated timecourse may explain the dysfunctions observed in DHA-deficient children, who display reduced visual acuity[77] and impaired cognitive performance.[74,93] Thus, an improved understanding of the role of EFAs in visual system development is mandatory for the establishment of adequate dietary requirements for these essential lipids during early postnatal life.

TAKE-HOME MESSAGES

- The development of topographically organized visual circuits is an essential feature acquired during brain development and necessary for the correct processing of basic visual functions.
- The visual system undergoes a use-dependent maturation of synaptic connections within a developmental critical period.
- Omega-3/DHA nutritional restriction induces an abnormal development of topography in the visual system either through a delay in axonal elimination or through the lack of appropriate synaptic stabilization mechanisms.

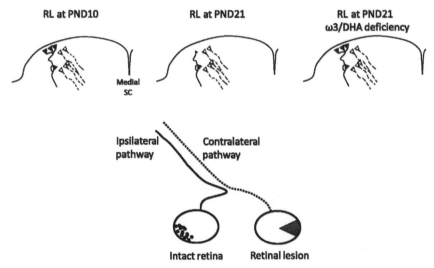

FIGURE 24.6 Schematic representation of retinal lesion (RL)-induced plasticity protocol.[16] The diagram displays a lesion to the temporal periphery of the retina and the ipsilateral pathway from the intact eye (bottom). As a result, the corresponding pathways from each retina present a lesioned contralateral pathway (dotted lines) and an intact ipsilateral pathway. Therefore, the visual layers of the superior colliculus (SC) are partially denervated by the lesion in the contralateral retina and the ipsilateral pathway is able to fill in the denervated territories depending on the age at which the lesion was inflicted (top). Lesions at postnatal day (PND)10 (during the critical period) result in a rapid reorganization of the intact ipsilateral retinocollicular projection, mostly toward the surface of the superior colliculus (upper, left). At the end of the critical period, at PND21, an RL is no longer able to induce a rapid growth of retinal axons from the intact eye (upper, center). However, in omega-3 restricted rats, the critical period closure is delayed so that a similar RL at PND21 induces a conspicuous growth of retinal axons from the intact retina (upper, right). DHA: docosahexaenoic acid.

- Omega-3/DHA nutritional restriction delays the closure of the critical period, thus altering the development of connections to subcortical visual nuclei.
- Omega-3/DHA nutritional deficiency leads to deficits in visual acuity that may be related to abnormal wiring patterns and use-dependent plasticity.

References

1. Charron F, Tessier-Lavigne M. Novel brain wiring functions for classical morphogens: a role as graded positional cues in axon guidance. *Development* 2005;**132**(10):2251–62.
2. Rapaport DH, Wong LL, Wood ED, Yasumura D, LaVail MM. Timing and topography of cell genesis in the rat retina. *J Comp Neurol* 2004;**474**(2):304–24.
3. Ohsawa R, Kageyama R. Regulation of retinal cell fate specification by multiple transcription factors. *Brain Res* 2008;**1192**:90–8.
4. Reese BE. Development of the retina and optic pathway. *Vision Res* 2011;**51**(7):613–32.
5. Sperry RW. Chemoaffinity in the orderly growth of nerve fiber patterns and connections. *Proc Natl Acad Sci U S A* 1963;**50**:703–10.
6. Meyer RL. Roger Sperry and his chemoaffinity hypothesis. *Neuropsychologia* 1998;**36**(10):957–80.
7. Graven SN. Early neurosensory visual development of the fetus and newborn. *Clin Perinatol* 2004;**31**(2):199–216.
8. Linden R, Perry VH. Ganglion cell death within the developing retina: a regulatory role for retinal dendrites? *Neuroscience* 1982;**7**(11):2813–27.
9. Serfaty CA, Linden R. Development of abnormal lamination and binocular segregation in the retinotectal pathways of the rat. *Brain Res Dev Brain Res* 1994;**82**(1–2):35–44.
10. Serfaty CA, Reese BE, Linden R. Cell death and interocular interactions among retinofugal axons: lack of binocularly matched specificity. *Brain Res Dev Brain Res.* 1990;**56**(2):198–204.
11. Serfaty CA, Oliveira-Silva P, Faria Melibeu Ada C, Campello-Costa P. Nutritional tryptophan restriction and the role of serotonin in development and plasticity of central visual connections. *Neuroimmunomodulation* 2008;**15**(3):170–5.
12. Cowan WM, Fawcett JW, O'Leary DD, Stanfield BB. Regressive events in neurogenesis. *Science* 1984;**225**(4668):1258–65.
13. de Velasco PC, Mendonca HR, Borba JM, Andrade da Costa BL, Guedes RC, Navarro DM, et al. Nutritional restriction of omega-3 fatty acids alters topographical fine tuning and leads to a delay in the critical period in the rodent visual system. *Exp Neurol* 2012;**234**(1):220–9.
14. Kano M, Hashimoto K. Synapse elimination in the central nervous system. *Curr Opin Neurobiol* 2009;**19**(2):154–61.
15. Huberman AD, Feller MB, Chapman B. Mechanisms underlying development of visual maps and receptive fields. *Annu Rev Neurosci* 2008;**31**:479–509.
16. Serfaty CA, Campello-Costa P, Linden R. Rapid and long-term plasticity in the neonatal and adult retinotectal pathways following a retinal lesion. *Brain Res Bull* 2005;**66**(2):128–34.
17. Cang J, Wang L, Stryker MP, Feldheim DA. Roles of ephrin-as and structured activity in the development of functional maps in the superior colliculus. *J Neurosci* 2008;**28**(43):11015–23.
18. O'Leary DD, McLaughlin T. Mechanisms of retinotopic map development: ephs, ephrins, and spontaneous correlated retinal activity. *Prog Brain Res* 2005;**147**:43–65.
19. Torborg CL, Feller MB. Spontaneous patterned retinal activity and the refinement of retinal projections. *Prog Neurobiol* 2005;**76**(4):213–35.
20. Berardi N, Pizzorusso T, Maffei L. Critical periods during sensory development. *Curr Opin Neurobiol* 2000;**10**(1):138–45.
21. Hensch TK. Critical period mechanisms in developing visual cortex. *Curr Topics Dev Biol* 2005;**69**:215–37.
22. Levelt CN, Hubener M. Critical-period plasticity in the visual cortex. *Annu Rev Neurosci* 2012;**35**:309–30.
23. Erzurumlu RS, Gaspar P. Development and critical period plasticity of the barrel cortex. *Eur J Neurosci* 2012;**35**(10):1540–53.
24. Majewska AK, Newton JR, Sur M. Remodeling of synaptic structure in sensory cortical areas *in vivo*. *J Neurosci* 2006;**26**(11):3021–9.
25. Tailby C, Wright LL, Metha AB, Calford MB. Activity-dependent maintenance and growth of dendrites in adult cortex. *Proc Natl Acad Sci U S A* 2005;**102**(12):4631–6.
26. Barnes SJ, Finnerty GT. Sensory experience and cortical rewiring. *Neuroscientist* 2010;**16**(2):186–98.
27. Hofer SB, Mrsic-Flogel TD, Bonhoeffer T, Hubener M. Prior experience enhances plasticity in adult visual cortex. *Nat Neurosci* 2006;**9**(1):127–32.
28. Penhune VB. Sensitive periods in human development: evidence from musical training. *Cortex* 2011;**47**(9):1126–37.
29. Wiesel TN, Hubel DH. Single-cell responses in striate cortex of kittens deprived of vision in one eye. *J Neurophysiol* 1963;**26**:1003–17.
30. Hubel DH, Wiesel TN. The period of susceptibility to the physiological effects of unilateral eye closure in kittens. *J Physiol* 1970;**206**(2):419–36.
31. Kanonidou E. Amblyopia: a mini review of the literature. *Int Ophthalmol* 2011;**31**(3):249–56.
32. Crair MC, Gillespie DC, Stryker MP. The role of visual experience in the development of columns in cat visual cortex. *Science* 1998;**279**(5350):566–70.
33. Hensch TK, Fagiolini M, Mataga N, Stryker MP, Baekkeskov S, Kash SF. Local GABA circuit control of experience-dependent plasticity in developing visual cortex. *Science* 1998;**282**(5393):1504–8.
34. Hanover JL, Huang ZJ, Tonegawa S, Stryker MP. Brain-derived neurotrophic factor overexpression induces precocious critical period in mouse visual cortex. *J Neurosci* 1999;**19**(22). RC40.
35. Ciucci F, Putignano E, Baroncelli L, Landi S, Berardi N, Maffei L. Insulin-like growth factor 1 (IGF-1) mediates the effects of enriched environment (EE) on visual cortical development. *PLoS ONE* 2007;**2**(5):e475.
36. Pizzorusso T, Medini P, Berardi N, Chierzi S, Fawcett JW, Maffei L. Reactivation of ocular dominance plasticity in the adult visual cortex. *Science* 2002;**298**(5596):1248–51.
37. Mataga N, Nagai N, Hensch TK. Permissive proteolytic activity for visual cortical plasticity. *Proc Natl Acad Sci U S A* 2002;**99**(11):7717–21.
38. McGee AW, Yang Y, Fischer QS, Daw NW, Strittmatter SM. Experience-driven plasticity of visual cortex limited by myelin and Nogo receptor. *Science* 2005;**309**(5744):2222–6.
39. Simon DK, O'Leary DD. Limited topographic specificity in the targeting and branching of mammalian retinal axons. *Dev Biol* 1990;**137**(1):125–34.
40. Mendonca HR, Araujo SE, Gomes AL, Sholl-Franco A, da Cunha Faria Melibeu A, Serfaty CA, et al. Expression of GAP-43 during development and after monocular enucleation in the rat superior colliculus. *Neurosci Lett* 2010;**477**(1):23–7.
41. Thompson ID, Cordery P, Holt CE. Postnatal changes in the uncrossed retinal projection of pigmented and albino Syrian hamsters and the effects of monocular enucleation. *J Comp Neurol* 1995;**357**(2):181–203.
42. Yuste R. Dendritic spines and distributed circuits. *Neuron* 2011;**71**(5):772–81.
43. Kulkarni VA, Firestein BL. The dendritic tree and brain disorders. *Mol Cell Neurosci* 2012;**50**(1):10–20.

44. Luscher C, Malenka RC. NMDA receptor-dependent long-term potentiation and long-term depression (LTP/LTD). *Cold Spring Harb Perspect Biol* 2012;**4**(6).

45. Lu W, Constantine-Paton M. Eye opening rapidly induces synaptic potentiation and refinement. *Neuron* 2004;**43**(2):237–49.

46. Colonnese MT, Constantine-Paton M. Chronic NMDA receptor blockade from birth increases the sprouting capacity of ipsilateral retinocollicular axons without disrupting their early segregation. *J Neurosci* 2001;**21**(5):1557–68.

47. Haas K, Li J, Cline HT. AMPA receptors regulate experience-dependent dendritic arbor growth *in vivo*. *Proc Natl Acad Sci U S A* 2006;**103**(32):12127–31.

48. Terashima A, Pelkey KA, Rah JC, Suh YH, Roche KW, Collingridge GL, et al. An essential role for PICK1 in NMDA receptor-dependent bidirectional synaptic plasticity. *Neuron* 2008;**57**(6):872–82.

49. Lo FS, Mize RR. Properties of LTD and LTP of retinocollicular synaptic transmission in the developing rat superior colliculus. *Eur J Neurosci* 2002;**15**(9):1421–32.

50. Du JL, Wei HP, Wang ZR, Wong ST, Poo MM. Long-range retrograde spread of LTP and LTD from optic tectum to retina. *Proc Natl Acad Sci U S A* 2009;**106**(45):18890–6.

51. Labrousse VF, Nadjar A, Joffre C, Costes L, Aubert A, Gregoire S, et al. Short-term long chain omega3 diet protects from neuroinflammatory processes and memory impairment in aged mice. *PLoS One* 2012;**7**(5):e36861.

52. Boitard C, Etchamendy N, Sauvant J, Aubert A, Tronel S, Marighetto A, et al. Juvenile, but not adult exposure to high-fat diet impairs relational memory and hippocampal neurogenesis in mice. *Hippocampus* 2012;**22**(11):2095–100.

53. Gonzalez EM, Penedo LA, Oliveira-Silva P, Campello-Costa P, Guedes RC, Serfaty CA. Neonatal tryptophan dietary restriction alters development of retinotectal projections in rats. *Exp Neurol* 2008;**211**(2):441–8.

54. Morgane PJ, Austin-LaFrance R, Bronzino J, Tonkiss J, Diaz-Cintra S, Cintra L, et al. Prenatal malnutrition and development of the brain. *Neurosci Biobehav Rev* 1993;**17**(1):91–128.

55. Ranade SC, Rose A, Rao M, Gallego J, Gressens P, Mani S. Different types of nutritional deficiencies affect different domains of spatial memory function checked in a radial arm maze. *Neuroscience* 2008;**152**(4):859–66.

56. Dauncey MJ. New insights into nutrition and cognitive neuroscience. *Proc Nutr Soc* 2009;**68**(4):408–15.

57. Joseph J, Cole G, Head E, Ingram D. Nutrition, brain aging, and neurodegeneration. *J Neurosci* 2009;**29**(41):12795–801.

58. Ballabriga A. Malnutrition and the central nervous system. In: Suskind RM, editor. *The Malnourished Child*. Raven Press; 1990. pp. 177–95.

59. Wainwright PE. Dietary essential fatty acids and brain function: a developmental perspective on mechanisms. *Proc Nutr Soc* 2002;**61**(1):61–9.

60. Heird WC, Lapillonne A. The role of essential fatty acids in development. *Annu Rev Nutr* 2005;**25**:549–71.

61. Innis SM. Dietary omega 3 fatty acids and the developing brain. *Brain Res* 2008;**1237**:35–43.

62. Yehuda S, Rabinovitz S, Mostofsky DI. Essential fatty acids and the brain: from infancy to aging. *Neurobiol Aging* 2005;**26**(Suppl 1):98–102.

63. Innis SM. The developing brain and dietary omega-3 fatty acids. In: Preedy VR, editor. *Handbook of Behavior, Food and Nutrition*. Springer Science+Business Media; 2011. pp. 2069–87.

64. Bhatia HS, Agrawal R, Sharma S, Huo YX, Ying Z, Gomez-Pinilla F. Omega-3 fatty acid deficiency during brain maturation reduces neuronal and behavioral plasticity in adulthood. *PLoS One* 2011;**6**(12):e28451.

65. Luchtman DW, Song C. Cognitive enhancement by omega-3 fatty acids from child-hood to old age: findings from animal and clinical studies. *Neuropharmacology* 2013;**64**:550–65.

66. Moreira JD, Knorr L, Ganzella M, Thomazi AP, de Souza CG, de Souza DG, et al. Omega-3 fatty acids deprivation affects ontogeny of glutamatergic synapses in rats: relevance for behavior alterations. *Neurochem Int* 2010;**56**(6–7):753–9.

67. Richardson AJ, Puri BK. The potential role of fatty acids in attention-deficit/hyperactivity disorder. *Prostaglandins Leukot Essent Fatty Acids* 2000;**63**(1–2):79–87.

68. Fedorova I, Salem Jr N. Omega-3 fatty acids and rodent behavior. *Prostaglandins Leukot Essent Fatty Acids* 2006;**75**(4–5):271–89.

69. Kim W, McMurray DN, Chapkin RS. n-3 polyunsaturated fatty acids – physiological relevance of dose. *Prostaglandins Leukot Essent Fatty Acids* 2010;**82**(4–6):155–8.

70. Kitajka K, Puskas LG, Zvara A, Hackler Jr L, Barcelo-Coblijn G, Yeo YK, et al. The role of n-3 polyunsaturated fatty acids in brain: modulation of rat brain gene expression by dietary n-3 fatty acids. *Proc Natl Acad Sci U S A* 2002;**99**(5):2619–24.

71. Carlson SE. Docosahexaenoic acid and arachidonic acid in infant development. *Semin Neonatol* 2001;**6**(5):437–49.

72. Koletzko B, Lien E, Agostoni C, Bohles H, Campoy C, Cetin I, et al. The roles of long-chain polyunsaturated fatty acids in pregnancy, lactation and infancy: review of current knowledge and consensus recommendations. *J Perinat Med* 2008;**36**(1):5–14.

73. Ozias MK, Carlson SE, Levant B. Maternal parity and diet (n-3) polyunsaturated fatty acid concentration influence accretion of brain phospholipid docosahexaenoic acid in developing rats. *J Nutr* 2007;**137**(1):125–9.

74. Hoffman DR, Boettcher JA, Diersen-Schade DA. Toward optimizing vision and cognition in term infants by dietary docosahexaenoic and arachidonic acid supplementation: a review of randomized controlled trials. *Prostaglandins Leukot Essent Fatty Acids* 2009;**81**(2–3):151–8.

75. Lien EL, Hammond BR. Nutritional influences on visual development and function. *Prog Retin Eye Res* 2011;**30**(3):188–203.

76. Uauy R, Castillo C. Lipid requirements of infants: implications for nutrient composition of fortified complementary foods. *J Nutr* 2003;**133**(9):2962S–72S.

77. Birch EE, Hoffman DR, Uauy R, Birch DG, Prestidge C. Visual acuity and the essentiality of docosahexaenoic acid and arachidonic acid in the diet of term infants. *Pediatr Res* 1998;**44**(2):201–9.

78. Birch EE, Carlson SE, Hoffman DR, Fitzgerald-Gustafson KM, Fu VL, Drover JR, et al. The DIAMOND (DHA Intake And Measurement Of Neural Development) study: a double-masked, randomized controlled clinical trial of the maturation of infant visual acuity as a function of the dietary level of docosahexaenoic acid. *Am J Clin Nutr* 2010;**91**(4):848–59.

79. Gomez-Pinilla F. Brain foods: the effects of nutrients on brain function. *Nat Rev Neurosci* 2008;**9**(7):568–78.

80. Simopoulos AP. Evolutionary aspects of diet: the omega-6/omega-3 ratio and the brain. *Mol Neurobiol* 2011;**44**(2):203–15.

81. Simopoulos AP. Evolutionary aspects of diet, the omega-6/omega-3 ratio and genetic variation: nutritional implications for chronic diseases. *Biomed Pharmacother* 2006;**60**(9):502–7.

82. Marszalek JR, Lodish HF. Docosahexaenoic acid, fatty acid-interacting proteins, and neuronal function: breastmilk and fish are good for you. *Annu Rev Cell Dev Biol* 2005;**21**:633–57.

83. Siriwardhana N, Kalupahana NS, Moustaid-Moussa N. Health benefits of n-3 polyunsaturated fatty acids: eicosapentaenoic acid and docosahexaenoic acid. *Adv Food Nutr Res* 2012;**65**:211–22.

84. Kim HY. Novel metabolism of docosahexaenoic acid in neural cells. *J Biol Chem* 2007;**282**(26):18661–5.

85. German OL, Insua MF, Gentili C, Rotstein NP, Politi LE. Docosahexaenoic acid prevents apoptosis of retina photoreceptors by activating the ERK/MAPK pathway. *J Neurochem* 2006;**98**(5):1507–20.

86. Bazan NG. Lipid signaling in neural plasticity, brain repair, and neuroprotection. *Mol Neurobiol* 2005;**32**(1):89–103.

87. Chen C, Bazan NG. Lipid signaling: sleep, synaptic plasticity, and neuroprotection. *Prostaglandins Other Lipid Mediat* 2005;**77**(1–4):65–76.

88. Pongrac JL, Slack PJ, Innis SM. Dietary polyunsaturated fat that is low in (n-3) and high in (n-6) fatty acids alters the SNARE protein complex and nitrosylation in rat hippocampus. *J Nutr* 2007;**137**(8):1852–6.

89. Tanabe Y, Hashimoto M, Sugioka K, Maruyama M, Fujii Y, Hagiwara R, et al. Improvement of spatial cognition with dietary docosahexaenoic acid is associated with an increase in Fos expression in rat CA1 hippocampus. *Clin Exp Pharmacol Physiol* 2004;**31**(10):700–3.

90. Phillis JW, Horrocks LA, Farooqui AA. Cyclooxygenases, lipoxygenases, and epoxygenases in CNS: their role and involvement in neurological disorders. *Brain Res Rev* 2006;**52**(2):201–43.

91. Mazelova J, Ransom N, Astuto-Gribble L, Wilson MC, Deretic D. Syntaxin 3 and SNAP-25 pairing, regulated by omega-3 docosahexaenoic acid, controls the delivery of rhodopsin for the biogenesis of cilia-derived sensory organelles, the rod outer segments. *J Cell Sci* 2009;**122**(Pt 12):2003–13.

92. Cansev M, Marzloff G, Sakamoto T, Ulus IH, Wurtman RJ. Giving uridine and/or docosahexaenoic acid orally to rat dams during gestation and nursing increases synaptic elements in brains of weanling pups. *Dev Neurosci* 2009;**31**(3):181–92.

93. Wurtman RJ. Synapse formation and cognitive brain development: effect of docosahexaenoic acid and other dietary constituents. *Metab Clin Exp* 2008;**57**(Suppl 2):S6–10.

94. Kuperstein F, Eilam R, Yavin E. Altered expression of key dopaminergic regulatory proteins in the postnatal brain following perinatal n-3 fatty acid dietary deficiency. *J Neurochem* 2008;**106**(2):662–71.

95. Chalon S. Omega-3 fatty acids and monoamine neurotransmission. *Prostaglandins Leukot Essent Fatty Acids* 2006;**75**(4–5):259–69.

96. Fujita S, Ikegaya Y, Nishikawa M, Nishiyama N, Matsuki N. Docosahexaenoic acid improves long-term potentiation attenuated by phospholipase A(2) inhibitor in rat hippocampal slices. *Br J Pharmacol* 2001;**132**(7):1417–22.

97. Martel MA, Patenaude C, Menard C, Alaux S, Cummings BS, Massicotte G. A novel role for calcium-independent phospholipase A in alpha-amino-3-hydroxy-5-methylisoxazole-propionate receptor regulation during long-term potentiation. *Eur J Neurosci* 2006;**23**(2):505–13.

98. Bazan NG. Cellular and molecular events mediated by docosahexaenoic acid-derived neuroprotectin D1 signaling in photoreceptor cell survival and brain protection. *Prostaglandins Leukot Essent Fatty Acids* 2009;**81**(2–3):205–11.

99. Kitajka K, Sinclair AJ, Weisinger RS, Weisinger HS, Mathai M, Jayasooriya AP, et al. Effects of dietary omega-3 polyunsaturated fatty acids on brain gene expression. *Proc Natl Acad Sci U S A* 2004;**101**(30):10931–6.

100. de Urquiza AM, Liu S, Sjoberg M, Zetterstrom RH, Griffiths W, Sjovall J, et al. Docosahexaenoic acid, a ligand for the retinoid X receptor in mouse brain. *Science* 2000;**290**(5499):2140–4.

101. Bousquet M, Gibrat C, Saint-Pierre M, Julien C, Calon F, Cicchetti F. Modulation of brain-derived neurotrophic factor as a potential neuroprotective mechanism of action of omega-3 fatty acids in a parkinsonian animal model. *Prog Neuropsychopharmacol Biol Psychiatry* 2009;**33**(8):1401–8.

102. Katsuki H, Kurimoto E, Takemori S, Kurauchi Y, Hisatsune A, Isohama Y, et al. Retinoic acid receptor stimulation protects midbrain dopaminergic neurons from inflammatory degeneration via BDNF-mediated signaling. *J Neurochem* 2009;**110**(2):707–18.

103. Venna VR, Deplanque D, Allet C, Belarbi K, Hamdane M, Bordet R. PUFA induce antidepressant-like effects in parallel to structural and molecular changes in the hippocampus. *Psychoneuroendocrinology* 2009;**34**(2):199–211.

104. Jeffrey BG, Weisinger HS, Neuringer M, Mitchell DC. The role of docosahexaenoic acid in retinal function. *Lipids* 2001;**36**(9):859–71.

105. Schmidt JT. Activity-driven sharpening of the retinotectal projection: the search for retrograde synaptic signaling pathways. *J Neurobiol* 2004;**59**(1):114–33.

106. Schmidt JT, Fleming MR, Leu B. Presynaptic protein kinase C controls maturation and branch dynamics of developing retinotectal arbors: possible role in activity-driven sharpening. *J Neurobiol* 2004;**58**(3):328–40.

107. Campello-Costa P, Fosse Jr AM, Oliveira-Silva P, Serfaty CA. Blockade of arachidonic acid pathway induces sprouting in the adult but not in the neonatal uncrossed retinotectal projection. *Neuroscience* 2006;**139**(3):979–89.

108. Schaechter JD, Benowitz LI. Activation of protein kinase C by arachidonic acid selectively enhances the phosphorylation of GAP-43 in nerve terminal membranes. *J Neurosci* 1993;**13**(10):4361–71.

109. Guido W. Refinement of the retinogeniculate pathway. *J Physiol* 2008;**586**(Pt 18):4357–62.

110. Campello-Costa P, Fosse Jr AM, Ribeiro JC, Paes-De-Carvalho R, Serfaty CA. Acute blockade of nitric oxide synthesis induces disorganization and amplifies lesion-induced plasticity in the rat retinotectal projection. *J Neurobiol* 2000;**44**(4):371–81.

111. Wu A, Ying Z, Gomez-Pinilla F. The interplay between oxidative stress and brain-derived neurotrophic factor modulates the outcome of a saturated fat diet on synaptic plasticity and cognition. *Eur J Neurosci* 2004;**19**(7):1699–707.

112. Fagiolini M, Pizzorusso T, Berardi N, Domenici L, Maffei L. Functional postnatal development of the rat primary visual cortex and the role of visual experience: dark rearing and monocular deprivation. *Vision Res* 1994;**34**(6):709–20.

113. Viegi A, Cotrufo T, Berardi N, Mascia L, Maffei L. Effects of dark rearing on phosphorylation of neurotrophin Trk receptors. *Eur J Neurosci* 2002;**16**(10):1925–30.

114. Grantham-McGregor S, Cheung YB, Cueto S, Glewwe P, Richter L, Strupp B. Developmental potential in the first 5 years for children in developing countries. *Lancet* 2007;**369**(9555):60–70.

115. Thompson RA, Nelson CA. Developmental science and the media. Early brain development. *Am Psychol* 2001;**56**(1):5–15.

Prenatal Omega-3 Fatty Acid Intake and Visual Function

Chloé Cartier[1], Dave Saint-Amour[1, 2, 3]

[1]Département de psychologie, Université du Québec à Montréal, Montréal, Québec, Canada, [2]Département d'ophtalmologie, Université de Montréal, Montréal, Québec, Canada, [3]Centre de recherche, Centre hospitalier universitaire Sainte-Justine, Montréal, Québec, Canada

INTRODUCTION

Lipids influence neuronal function by modifying the characteristics of the membrane, gene expression, and cell signaling, all of which play a critical role in neurotransmission, metabolism, growth, and cell differentiation.[1] A considerable accumulation of long-chain polyunsaturated fatty acids (PUFAs) occurs in neural and retinal membranes during gestation, and it continues throughout the first postnatal years, a period that is associated with a substantial growth of the brain[2, 3] (Fig. 25.1). Inadequate intake of PUFAs can perturb cerebral maturational processes and thus have functional consequences on sensory and cognitive neurodevelopment at birth or later in life.

Docosahexaenoic acid (DHA) is the principal omega-3 (n-3) PUFA found in the mammalian central nervous system. It is highly concentrated in the gray matter of the cerebral cortex as well as in phospholipids of the photoreceptor outer segment membranes of the retina.[5] Although DHA can be synthesized in the liver from its precursor, α-linolenic acid (ALA, 18:3n-3), only small amounts can be produced in humans. It is estimated that healthy adults can convert about 0.1% of ALA from their diet to DHA.[6] Because of the low human capacity to synthesize DHA from its precursors, in particular early in life,[7] the source of DHA in the fetus and breastfed infants is highly dependent on maternal-to-fetal transfer before birth and breast milk postnatally. Indeed, there is a positive correlation between the mother's circulating DHA level intakes during pregnancy and lactation and infant DHA status at birth or in the few months postpartum. Maternal dietary intake of DHA is crucial for adequate DHA supply to the fetus and infant.[8–10] Pregnant women need to have a diet with enough DHA-containing products for their own health and to supply sufficient DHA to their developing infant. Because fetal DHA demands deplete maternal DHA stores during pregnancy, several factors may influence maternal availability to the fetus, such as the period (prenatal or postnatal), number of infants per pregnancy, and number of months since the last delivery.[11] Worldwide variations in dietary patterns as well as regional and personal dietary habits lead to important differences in DHA breast milk content between women and raise the question of whether DHA intakes may be sufficient in some populations with a Western diet to provide enough DHA to the developing baby. For example, women with higher dietary consumption of fish or seafood, found in Japan, the Philippines, and Chile, were reported to have two- to five-fold higher DHA milk contents than women with low fish product consumption[12] (in Canada and the USA).

In recent decades, a growing body of evidence from epidemiologic and experimental studies has linked DHA prenatal exposure and DHA formula supplementation with visual acuity in preterm or term infants. As described in the following sections, these studies have also provided a better understanding of the implications of DHA in visual processes and highlighted the necessity of an adequate DHA supply for optimal visual system development.

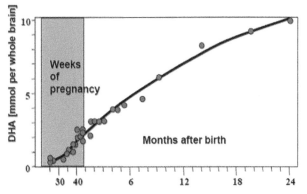

FIGURE 25.1 Docosahexaenoic acid (DHA) accumulation in the brain from pregnancy to the first years of life. DHA brain levels increase rapidly in the last trimester of pregnancy and during the first months postpartum. *Source: Morse NL. Benefits of docosahexaenoic acid, folic acid, vitamin D and iodine on foetal and infant brain development and function following maternal supplementation during pregnancy and lactation. Nutrients 2012;4:799–840.[4] with permission*

FIRST EVIDENCE OF OMEGA-3 FATTY ACID EFFECTS ON VISUAL FUNCTION

The idea that PUFAs, regardless of the carbon–carbon bond position (n-3 or n-6), may be involved in visual functioning emerged from clinical cases. In the early 1960s the first indicators of the role of PUFAs in healthy development came from infants with PUFA deficiencies due to skim-milk-based formula or lipid-free parenteral nutrition.[13] Abnormal skin manifestations and a delay in growth were reported in those infants lacking linoleic acid (LA), the n-6 precursor. Since LA supplementation reversed the symptoms, this provided evidence of PUFAs' role in normal infant development. The more specific role for n-3 PUFAs in vision was only discovered later, in 1982. Holman *et al.*[14] described a clinical case of blurred vision in a 6-year-old girl under long-term ALA-free parenteral nutrition that was associated with depletion in n-3 PUFA serum concentration. As for the LA deficiency mentioned above, n-3 PUFA supply led to a restoration of visual function, indicating their involvement in human visual processing, although no explicit mechanism of action was suggested in that study.

Experimental studies in animals, particularly in nonhuman primates, have established the beneficial role of n-3 PUFAs in visual development. An n-3 fatty acid-deficient diet was given to healthy rhesus monkeys before conception and during pregnancy[15] (see Chapter 7, 'The Role of Lipids and Lipid Metabolism in Age-Related Macular Degeneration'). Infants were fed with a diet very low in ALA content from birth to 22 months of age. Prenatal and early postnatal months were preferentially chosen for DHA depletion since, as mentioned in the first section of this chapter, they represent the key periods for DHA accumulation in the brain and retina. n-3 PUFA depletion resulted in a lower mother DHA blood level and lower infant DHA plasma levels at birth compared with the control groups, confirming the infant's dependence on the mother's diet for intrauterine DHA supply. Furthermore, a diet depleted in n-3 PUFA during pregnancy and during the first months postpartum was correlated with a reduced DHA accretion rate in cerebral cortex and retinal structures.[16] At 22 months of age, DHA concentration in the occipital cortex of the depleted group was onesixth of that in the control group. A similar pattern was observed in retinal DHA content. In both groups, visual acuity assessed with preferential looking at 4, 8, and 12 weeks of life improved with age as a result of visual acuity maturation. However, at each assessment time-point, visual acuity was lower in the depleted group. Prenatal and postnatal DHA depletion also induce impairments in retinal activity as measured by electroretinography (ERG).

A more recent experimental study investigated whether DHA supply after birth may reverse lower visual acuity caused by prenatal n-3 PUFA deficiency.[17] Prenatal n-3 PUFA depletion was produced by giving an ALA-deficient diet to pregnant rhesus monkeys. The control group included pregnant monkeys with a normal diet (i.e., without ALA depletion). Infants from both groups were given a diet high in ALA from birth to 3 years of age. At 15 weeks of life infants from the experimental group showed blood and cerebral DHA levels similar to those of the control group. No differences were detected in visual acuity between the two groups. Nonetheless, infants with prenatal ALA depletion showed altered electrophysiologic activity of cone and rod photoreceptors at 3–4 months of age compared with the nondepleted group and had a reduced level of DHA in the retina at 3 years of age. Thus, although visual acuity was preserved in the prenatal DHA-depleted group following postnatal DHA supplementation, subclinical retina dysfunction was detectable early in life. These results suggest that prenatal life may constitute a window of vulnerability for the action of DHA on visual system maturation and in particular on retina development.

The findings of the aforementioned clinical and experimental studies have motivated researchers to conduct studies related to n-3 long-chain PUFAs in visual acuity development in humans. The next sections will review the actual knowledge accumulated over the past two decades on the prenatal and early postnatal benefits of DHA in the development of visual acuity.

BENEFICIAL EFFECTS ON HUMAN RETINAL FUNCTION

The beneficial effects of DHA supplementation on retina maturation were first shown in preterm infants. Preterm infants were primarily targeted for experimental DHA formula supplementation feeding because they were hypothesized to have a higher risk of DHA deficiency than term infants, owing to their shorter intrauterine life and immature central nervous system. Very low birth weight healthy preterm newborns were assigned to receive different formulas with or without DHA supplementation, from 10 days to 57 weeks postconception.[17–19] A breastfed group of preterm infants was also studied. Cone and rod photoreceptor activity was evaluated at 36 and 57 weeks postconception using ERG. Infants without supplementation presented a delay in retinal maturation relative to the two other groups. Furthermore, a higher luminance was necessary to elicit rod activation in the no-supplementation group and amplitude responses remained lower than in infants in the DHA-supplemented and breastfed groups. Rod activity was similar in these last two groups. In all cases, n-3 PUFA plasma and erythrocyte levels were significantly associated with improved electrophysiologic rod response. No differences in ERG between groups were seen at 57 weeks postconception, however, suggesting that the retina accumulated sufficient DHA to achieve normal function with time, even on very low ALA intakes.

Preterm infants are known to be a population at risk of developing retinopathy of prematurity, an eye disease characterized by abnormal retinal vessel growth (see Chapter 7, 'The Role of Lipids and Lipid Metabolism in Age-Related Macular Degeneration'). Infants may recover spontaneously from the pathology or it can lead to important vision complications. Pawlik et al.[20] conducted a clinical trial on preterm newborns under parenteral nutrition in which one group was given a classic fat emulsion and another group was given a fish oil emulsion containing DHA. While infants from both groups developed retinopathy in a similar way, a significantly better spontaneous recovery was associated with fish oil administration in the first days of life. Recent evidence suggests that the protective mechanism of action of the n-3 PUFA effect may be mediated by blood vessel growth. Indeed, 4-hydroxydocosahexaenoic acid (4-HDHA), a DHA metabolite, has been found to have antiangiogenic properties and to protect against abnormal retinal vessel proliferation.[21]

Retinal sensitivity has also been studied in the first days of life using ERG in healthy term infants from mothers supplemented with fish oil from 15 weeks of gestation until delivery.[22] In contrast to studies conducted on preterm infants, no differences in retinal activity for either rod or cone responses were detected in these infants compared with placebo, evidence that adequate DHA for retinal development is accumulated in utero. However, subtle differences may still exist; for example, in both groups, DHA in cord blood correlated positively with retinal responses, reaffirming the necessity of an adequate DHA dietary intake during pregnancy for optimal retina development. In addition to their benefits in infants, n-3 PUFAs also constitute an essential nutrient for maintaining retinal functioning throughout life. Indeed, DHA dietary intakes have been found to provide protection against the risk of developing age-related macular degeneration,[23] the main cause of vision loss in people over 65 years of age (see Chapter 2, 'Age-Related Macular Degeneration: An Overview').

Different mechanisms have been proposed to explain why DHA deficiency may affect retinal function. It is likely that n-3 PUFA deficiency in the retina is mediated by an alteration of the photoreceptors per se, considering the important role of n-3 fatty acids in membrane fluidity and phototransduction of rhodopsin in rods.[24] Although the specific mechanisms of n-3 PUFA action on the retina remain to be defined, these studies provide evidence to promote n-3 PUFA intake, particularly from fish, to protect and optimize the eye structure and retinal function, especially in young and elderly populations.

OMEGA-3 FATTY ACID EXPOSURE AND INFANT VISUAL ACUITY

Acuity is the visual function most studied clinically in relation to n-3 PUFA exposure. Because visual acuity depends on the maturation and integrity of the retina, the thalamocortical pathway, and the primary visual cortex, it provides an excellent probe to assess the visual system.

Observational studies comparing infants fed human milk and formula and using both electrophysiologic and behavioral measurements of visual acuity have found improved performances in breastfed infants.[25–27] Because human milk is known to contain DHA[12] and since comparisons were made with non-DHA-supplemented formula, it was suggested that this beneficial effect was, at least in part, a result of the action of DHA on visual pathways. To confirm this hypothesis, randomized clinical trials were conducted in term infants with n-3 PUFA postnatal supplementation. Using visual evoked potentials (VEPs), scalp recording, and the Teller Acuity Card behavioral procedure (Fig. 25.2), several studies have shown that postnatal DHA supplementation improves visual acuity during the first months of life[27,29–31] and at 12 months,[32] while others have failed to demonstrate such an effect.[27,33] Differential results may be due to heterogeneity between studies regarding sample sizes, confounding factors such as family stimulation, DHA sources, duration, and dose, as well as the methods used to assess visual function.

In contrast, studies conducted on preterm infants have provided more consistent results and nearly all have concluded that there is a beneficial effect of DHA postnatal supplementation on visual acuity development.[18,25,34–36] In these studies, preterm newborns were

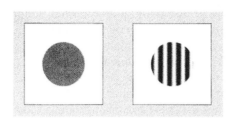  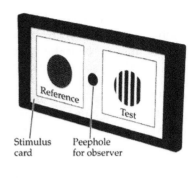

FIGURE 25.2 Teller Acuity Card test. This test is based on the preferential looking measure. The infant is presented with two items: one with contrasting stripes and one without stripes. The infant will prefer to look at the more complex one (the card with stripes); he or she will look equally at both cards if stripes cannot be discriminated. *Source: Adapted from Wolfe JM, Kluender KR, Levi DM, Bartoshuk LM, Herz RS, Klatzky RL, et al.* Sensation and Perception, *2nd edn. Sinauer Associates, Sunderland, MA, 2009.[28] Copyright © 2012, with permission from Sinauer Associates.*

randomly assigned to formula enriched with n-3 PUFAs or not. Visual acuity was measured between birth and 1 year of age with behavioral and electrophysiologic assessments. Using a behavioral assessment, Carlson and colleagues[34,35] showed higher visual acuity in the preterm group supplemented with DHA from fish oil at 2 and 4 months of age, although this difference was not observed in the interval from 6.5 to 12 months. In O'Connor's study,[36] the researchers did not notice a significant effect of n-3 PUFA supplementation on visual acuity at 6 months of age using the same Teller Acuity Card procedure, but they found better visual acuity in supplemented infants when assessed with VEPs, evidence that the electrophysiologic measurement is more sensitive in detecting DHA-related beneficial effects on visual function.

As discussed above, the beneficial effects of n-3 PUFAs on visual acuity have mostly been detected before 6 months of age. Morale et al.[37] reported that, in a large group of DHA-supplemented children (n = 243), the improved visual acuity observed with VEP recording during the first months of life was also present at 1 year. Thus, the fact that clear enhancements of visual acuity are not commonly detected after 6 months of age does not necessarily mean that they do not occur. First, as illustrated above, the method chosen to measure visual function (i.e., from behavior or brain activity) can make a difference. Second, since visual acuity develops rapidly during the first 6 months of life, it is possible that this function reaches a plateau, making it difficult to highlight differences between groups later in life because of ceiling effects.[38]

The fact that more robust DHA beneficial effects were reported with preterm infants suggests that the last months of gestation are a critical time window for DHA to affect visual acuity development. Accordingly, observational human studies have reported a significant relationship between prenatal n-3 PUFA exposure

and visual acuity. Jacobson et al.[39] studied DHA prenatal exposure on visual acuity in term infants in an Inuit cohort from Nunavik, the northernmost region of Quebec. Inuit individuals are chronically exposed to high amounts of n-3 PUFA because of their fish and marine mammal diet. Using the Teller Acuity Card, the authors found that cord DHA phospholipids were associated with better visual acuity at 6 months of life with no remaining effects detected at 11 months. To the present authors' knowledge, three clinical trials have evaluated the effect of DHA supplementation during pregnancy on visual acuity development in infants using the same protocol, in which pregnant women were given DHA or placebo from midpregnancy until delivery.[40–42] Behavioral assessments were conducted at birth and at 2, 4, and 6 months of age to evaluate the impact of DHA prenatal exposure on visual acuity development. Innis and collaborators[40] reported better visual acuity in the supplemented group at 2 months of age compared with the placebo group. Judge et al.[41] found the same positive effect of DHA at 4 months but no associations between DHA and visual acuity 2 months later. In Malcolm's study,[42] no difference in visual acuity performance was detected between the DHA-supplemented and the control group at birth or at 50 and 66 weeks postconception. Nonetheless, for all infants, either in the placebo or in the DHA group, a significant positive association was found between the DHA level measured at birth and postnatal visual acuity performance.

In conclusion, while the beneficial effect of DHA on visual acuity maturation seems to be clearly demonstrated in preterm infants, the results obtained with prenatal or postnatal DHA exposure in term populations are less consistent. Although some studies find enhanced visual acuity related to DHA prenatal exposure only in the first few months of life, this may be due to the fact that the behavioral assessment of acuity is not sensitive to longer term effects on visual acuity.

LONG-TERM BENEFITS OF DEVELOPMENTAL EXPOSURE TO OMEGA-3 POLYUNSATURATED FATTY ACIDS

Although the role of n-3 PUFAs in visual acuity has been widely investigated, most of the studies focused only on early postnatal life, and it has not been established whether DHA exposure impacts other visual and cognitive functions.

In 2007, Birch et al.[26] conducted a randomized clinical trial with 84 healthy term infants to evaluate the long-term beneficial effect of formula supplementation with DHA on visual acuity development. Infants were breastfed, formula fed with DHA, or formula fed with no supplementation from birth to 17 weeks of age. At 4 years of age, visual acuity measured in the nonsupplemented group was found to be lower than in the supplemented and the breastfed groups. Although the difference was subtle and only detectable in the right eye, these results suggest that early postnatal DHA administration has long-term benefits for visual maturation.

In addition to standard visual acuity, some authors have included measurements of stereoacuity to evaluate the effect of DHA exposure on visual cortex maturation.[29,43,44] Stereoacuity, or stereoscopic acuity, is the smallest detectable depth difference induced by binocular disparity. In these studies, stereoacuity performance was compared between infants formula fed with DHA supplementation and infants formula fed without supplementation.[29] Comparisons were also made between formula-fed and breastfed infants, and maternal intake of DHA during pregnancy was taken into account.[43,44] At 4 months of age better stereoacuity was detected in DHA-supplemented infants compared with the nonsupplemented group.[29] At 3.5 years of age, stereoacuity was higher in breastfed children than in the formula-fed group.[29] It is worth noting that better stereoacuity was observed at 3.5 years in children whose mothers had high fish oil intake during pregnancy compared with children whose mothers had low DHA intake. These results suggest a beneficial action of not only postnatal but also prenatal exposure to DHA on visual brain maturation.

More recently, a VEP study was conducted in 136 school-aged (11-year-old) Inuit children in Nunavik to investigate the long-term beneficial effects of prenatal intake of n-3 PUFAs on visual functions.[45] Because fish and marine mammals represent an important part of the Inuit diet, n-3 PUFA intake is substantially greater in this population than in southern Canada,[46] although this population is concomitantly exposed to environmental contaminants such as methylmercury, polychlorinated biphenyls, and lead.[47] In this study, several variables including contaminants were controlled statistically to isolate the effects of n-3 PUFAs on children's visual

processing. DHA amounts were measured in umbilical cord blood samples taken at delivery as well as at the time of testing to control current exposure. Different VEP paradigms were used to assess parvocellular and magnocellular brain responses, two pathways that carry different types of visual information.[48] The magnocellular pathway is optimally sensitive to low-to-medium spatial frequencies, low achromatic contrasts, and high temporal frequencies, while the parvocellular pathway is optimally sensitive to medium-to-high spatial frequencies, high contrast, and low temporal frequency. As a consequence, the magnocellular pathway is more sensitive to motion, whereas the parvocellular pathway plays a major role in processing stimulus detail and chromatic analysis. Considering that the parvocellular system mediates visual acuity, the authors hypothesized that the beneficial effects of prenatal n-3 PUFA intake on acuity observed in infancy[39] continue to be evident in childhood, as revealed by parvocellular-related VEPs. Accordingly, the results showed a beneficial impact of prenatal exposure to n-3 PUFAs in school-aged children. Indeed, after adjustment for confounders, cord plasma DHA was associated with shorter latencies of the N1 and P1 components of the isoluminant pattern-reversal VEPs, whereas no effects were found for low contrasted motion-onset VEPs (Fig. 25.3). These findings support the notion that the beneficial effects of fetal DHA intake on visual development may persist into late childhood and suggest that this effect is specific to parvocellular function. This effect was subtle and subclinical, as it was not significantly related to behavioral measurement of visual acuity. As a result, the VEP findings in this study may be difficult to interpret in terms of clinical significance, but they are clearly non-negligible in terms of optimal visual function.

The aforementioned studies in Nunavik[39,45] showed beneficial effects on visual function in relation to cord but not child plasma DHA. This suggests that DHA intake during the prenatal period plays a critical role in the early development of the visual system that may still be detectable at school age and during adulthood, although data are missing to support the latter hypothesis. Although DHA concentrations in the cord phospholipid plasma in Arctic Quebec are about three times higher than in southern Quebec,[46] cord DHA levels in Nunavik are similar to those reported in several other Western countries, notably in Europe[49] and in Massachusetts in the USA.[50]

Recently, it has been proposed that the effects of n-3 PUFAs on visual cortical processing may preferentially involve the dorsal stream,[51] which is often contrasted with the so-called ventral visual streams as the two systems are anatomically and functionally segregated (Fig. 25.4). The ventral system, which relays primary visual cortex input to temporal areas, refers to the

FIGURE 25.3 Effects of docosahexaenoic acid (DHA) on visual evoked potential (VEP). (A) Color (Oz site) and motion (T5–T6 sites) VEP grand mean average from valid subjects for 134 and 70 children, respectively; (B) color VEP N1 and P1 latency as a function of cord plasma phospholipid DHA concentration. *Source: Reprinted from Jacques C, Levy E, Muckle G, Jacobson SW, Bastien C, Dewailly E, et al. Long-term effects of prenatal omega-3 fatty acid intake on visual function in school-age children. J Pediatr. 2011;158:83–90,e1.[45] Copyright © 2012, with permission from Elsevier.*

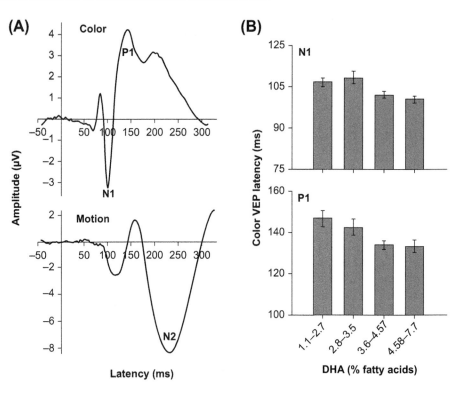

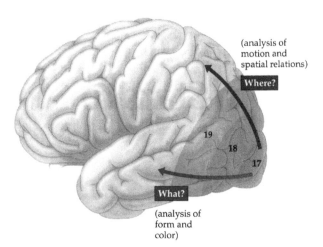

FIGURE 25.4 Dorsal and ventral visual streams. The dorsal stream deals with information about spatial localization of objects whereas the ventral stream ensures object identification. *Source: Adapted from Wolfe JM, Kluender KR, Levi DM, Bartoshuk LM, Herz RS, Klatzky RL, et al. Sensation and Perception, 2nd edn. Sinauer Associates, Sunderland, MA, 2009.[28] Copyright © 2012, with permission from Sinauer Associates.*

'what' because it ensures object identification. The dorsal system, which projects from the primary visual cortex to the parietal cortex, is called the 'where' and the 'how' as it deals with the spatial localization of objects and actions in relation to them (e.g., how to move one's hand to pick up a spoon from a table). Dysfunction of the dorsal stream has been reported in association with dyslexia,[52] and n-3 PUFA deficiency has been detected in children and adults with dyslexia.[53,54] Although indirect,

this finding supports a relationship between DHA and dorsal stream processing. Moreover, a clinical trial has revealed improved reading abilities in dyslexic children following DHA intake for 5 months.[55] Although such findings suggest that DHA is related to dyslexia, an abnormal ratio of n-6 to n-3 may be an important factor in the neurophysiopathology of reading ability in people with dyslexia.[56]

The hypothesis of the action of n-3 PUFAs on the dorsal stream is supported by Dunstan and colleagues' studies on visuomotor coordination.[57,58] Australian women were randomly assigned to receive four capsules per day of fish oil containing DHA or four capsules per day of olive oil from midpregnancy until delivery. Infants were submitted to Griffiths Mental Development Scales at 2.5 years postpartum. Maternal milk samples taken at 6 weeks postpartum revealed a significantly higher DHA concentration in milk from mothers of the fish oil group in comparison with the control group. At 2.5 years of age, children from the fish oil group displayed better eye and hand coordination performance when compared with the olive oil group. The authors found a positive association between DHA concentrations in breast milk at 3 days postdelivery and enhanced eye and hand coordination at 2.5 years of age. These studies provide further evidence of the long-term beneficial effects of n-3 PUFA exposure and demonstrate the role of n-3 PUFAs in improving eye and hand coordination, which is known to involve dorsal stream processing.

Animal and human studies have also explored the effect of DHA on visual attention.[59–62] In humans, the Fagan Test of Infant Intelligence is commonly used to assess visual attention and recognition in infants. This test measures the infant's natural tendency to spend more time looking at a novel stimulus than at a familiar stimulus. The infant is presented with two identical faces during a predetermined period. Then, the same items are presented a second time but each one is paired with a novel unfamiliar item. The total time the infant spends looking at each item is recorded. Using this test, two studies have reported shorter total looking duration in infants supplemented with DHA formula with no effects on visual recognition.[59,61] According to the authors, this may reflect 'more rapid visual information processes and a more mature attention',[59] including a better capacity to disengage or shift attention. Kannass et al.[63] used a multiple object free-play task, in which children aged 12 and 18 months were presented with objects to explore, in order to assess attention in relation to DHA status at birth. Children were presented different toys at the same time. The time for which the children looked at each toy, the number of looks at the toys, and episodes of inattention (i.e., not looking at the toys) were evaluated. While a shorter look duration is usually considered to reflect efficient visual processing during the first year of life, look duration typically increases after 1 year because of the time-course of attention development in order for the child to become able to maintain his or her attention on a task and to resist being distracted (sustained attention). Kannass et al. found a positive association between the mother's DHA red blood cell phospholipid concentration at delivery and the total duration of looking at the toys. They also found fewer episodes of inattention in children with high maternal DHA status at birth.[63] These results show a positive effect of DHA exposure on visual attention development beyond the first year of life. The neural basis of the beneficial impact of DHA on attention is unknown, but the frontal cortex may be involved.[64]

All together, the studies discussed in this section suggest that n-3 PUFA preferentially affects the parvocellular system as far as subcortical and low-level visual structures are concerned and the dorsal stream at the cortical level. Of the two cortical visual streams, the dorsal one has been considered more 'vulnerable',[65] and this may explain why the impact of n-3 PUFAs is more evident for dorsal visual function. However, the dorsal and ventral streams do interact, and normal visual perception is intrinsically dependent on both streams. It would be rather provocative to claim that n-3 PUFAs do not influence ventral visual stream processing. Moreover, other cortical areas, such as the frontal cortex, are likely to be involved.[64] There is a lack of data on the mechanisms underlying how n-3 PUFAs improve visual function for both prenatal and postnatal exposure. Further studies are therefore needed.

TAKE-HOME MESSAGES

- Maternal dietary omega-3 (n-3) long-chain polyunsaturated fatty acid (PUFA) intake influences fetus and breastfed infant n-3 status.
- The n-3 long-chain polyunsaturated fatty acids are highly concentrated in the gray matter of the brain and in photoreceptors of the retina.
- This localization has been studied to determine whether n-3 long-chain PUFAs are required for optimal visual acuity and cognitive development.
- Clinical cases and depletion experimental studies have highlighted the benefits of n-3 long-chain PUFAs for vision.
- Docosahexaenoic acid (DHA) enhances retinal function in humans and may protect against age-related eye disease.
- Supplementation with n-3 long-chain PUFAs has a positive effect on visual acuity development in preterm infants, and results are found more consistently than in term infants.
- A few studies have also found long-term beneficial effects of n-3 long-chain PUFAs on visual acuity and suggested that n-3 long-chain PUFAs act on specific visual systems (i.e., parvocellular pathway and dorsal pathway).
- Effects of n-3 long-chain PUFAs on visual attention and eye/hand coordination while requiring normal visual acuity are probably related to early mental and motor development, reflected in higher brain DHA accumulation.

Acknowledgments

We are very grateful to Susan Carlson for her comments in revising the draft of this chapter. This research was supported by the Canadian Institutes of Health Research and the Vision Health Research Network of the Fonds de recherche du Québec, awarded to Dave Saint-Amour.

References

1. Jump DB, Clarke SD. Regulation of gene expression by dietary fat. Annu Rev Nutr 1999;19:63–90.
2. Clandinin MT, Chappell JE, Leong S, Heim T, Swyer PR, Chance GW. Extrauterine fatty acid accretion in infant brain: implications for fatty acid requirements. Early Hum Dev 1980;4:131–8.
3. Martinez M. Tissue levels of polyunsaturated fatty acids during early human development. J Pediatr 1992;120(4 Pt 2):S129–38.
4. Morse NL. Benefits of docosahexaenoic acid, folic acid, vitamin D and iodine on foetal and infant brain development and function following maternal supplementation during pregnancy and lactation. Nutrients 2012;4:799–840.
5. Giusto NM, Pasquare SJ, Salvador GA, Castagnet PI, Roque ME, Ilincheta de Boschero MG. Lipid metabolism in vertebrate retinal rod outer segments. Prog Lipid Res 2000;39:315–91.
6. Plourde M, Cunnane SC. Extremely limited synthesis of long chain polyunsaturates in adults: implications for their dietary essentiality and use as supplements. Appl Physiol Nutr Metab 2007;32:619–34.

7. Salem Jr N, Wegher B, Mena P, Uauy R. Arachidonic and docosa-hexaenoic acids are biosynthesized from their 18-carbon precursors in human infants. *Proc Natl Acad Sci USA* 1996;**93**:49–54.

8. Dunstan JA, Mori TA, Barden A, Beilin LJ, Holt PG, Calder PC, et al. Effects of n-3 polyunsaturated fatty acid supplementation in pregnancy on maternal and fetal erythrocyte fatty acid composition. *Eur J Clin Nutr* 2004;**58**:429–37.

9. Helland IB, Saugstad OD, Saarem K, Van Houwelingen AC, Nylander G, Drevon CA. Supplementation of n-3 fatty acids during pregnancy and lactation reduces maternal plasma lipid levels and provides DHA to the infants. *J Matern Fetal Neonatal Med* 2006;**19**:397–406.

10. Marc I, Plourde M, Lucas M, Sterescu A, Piedboeuf B, Dufresne A, et al. Early docosahexaenoic acid supplementation of mothers during lactation leads to high plasma concentrations in very preterm infants. *J Nutr* 2011;**141**:231–6.

11. Al MD, van Houwelingen AC, Hornstra G. Relation between birth order and the maternal and neonatal docosahexaenoic acid status. *Eur J Clin Nutr* 1997;**51**:548–53.

12. Yuhas R, Pramuk K, Lien EL. Human milk fatty acid composition from nine countries varies most in DHA. *Lipids* 2006;**41**:851–8.

13. Caldwell MD, Jonsson HT, Othersen Jr HB. Essential fatty acid deficiency in an infant receiving prolonged parenteral alimentation. *J Pediatr* 1972;**81**:894–8.

14. Holman RT, Johnson SB, Hatch TF. A case of human linolenic acid deficiency involving neurological abnormalities. *Am J Clin Nutr* 1982;**35**:617–23.

15. Neuringer M, Connor WE, Van Petten C, Barstad L. Dietary omega-3 fatty acid deficiency and visual loss in infant rhesus monkeys. *J Clin Invest* 1984;**73**:272–6.

16. Anderson GJ, Neuringer M, Lin DS, Connor WE. Can prenatal n-3 fatty acid deficiency be completely reversed after birth? Effects on retinal and brain biochemistry and visual function in rhesus monkeys. *Pediatr Res* 2005;**58**:865–72.

17. Birch DG, Birch EE, Hoffman DR, Uauy RD. Retinal development in very-low-birth-weight infants fed diets differing in omega-3 fatty acids. *Invest Ophthalmol Vis Sci* 1992;**33**:2365–76.

18. Hoffman DR, Birch EE, Birch DG, Uauy RD. Effects of supplementation with omega 3 long-chain polyunsaturated fatty acids on retinal and cortical development in premature infants. *Am J Clin Nutr* 1993;**57**(Suppl. 5). 807–12S.

19. Uauy RD, Birch DG, Birch EE, Tyson JE, Hoffman DR. Effect of dietary omega-3 fatty acids on retinal function of very-low-birth-weight neonates. *Pediatr Res* 1990;**28**:485–92.

20. Pawlik D, Lauterbach R, Turyk E. Fish-oil fat emulsion supplementation may reduce the risk of severe retinopathy in VLBW infants. *Pediatrics* 2011;**127**:223–8.

21. Sapieha P, Stahl A, Chen J, Seaward MR, Willett KL, Krah NM, et al. 5-Lipoxygenase metabolite 4-HDHA is a mediator of the anti-angiogenic effect of omega-3 polyunsaturated fatty acids. *Sci Transl Med* 2011;**3**:69ra12.

22. Malcolm CA, Hamilton R, McCulloch DL, Montgomery C, Weaver LT. Scotopic electroretinogram in term infants born of mothers supplemented with docosahexaenoic acid during pregnancy. *Invest Ophthalmol Vis Sci* 2003;**44**:3685–91.

23. Ho L, van Leeuwen R, Witteman JC, van Duijn CM, Uitterlinden AG, Hofman A, et al. Reducing the genetic risk of age-related macular degeneration with dietary antioxidants, zinc, and omega-3 fatty acids: the Rotterdam study. *Arch Ophthalmol* 2011;**129**:758–66.

24. Kurlak LO, Stephenson TJ. Plausible explanations for effects of long chain polyunsaturated fatty acids (LCPUFA) on neonates. *Arch Dis Child Fetal Neonatal Ed* 1999;**80**:F148–54.

25. Birch EE, Birch DG, Hoffman DR, Uauy R. Dietary essential fatty acid supply and visual acuity development. *Invest Ophthalmol Vis Sci* 1992;**33**:3242–53.

26. Birch EE, Garfield S, Castaneda Y, Hughbanks-Wheaton D, Uauy R, Hoffman D. Visual acuity and cognitive outcomes at 4 years of age in a double-blind, randomized trial of long-chain polyunsaturated fatty acid-supplemented infant formula. *Early Hum Dev* 2007;**83**:279–84.

27. Carlson SE, Ford AJ, Werkman SH, Peeples JM, Koo WW. Visual acuity and fatty acid status of term infants fed human milk and formulas with and without docosahexaenoate and arachidonate from egg yolk lecithin. *Pediatr Res* 1996;**39**:882–8.

28. Wolfe JM, Kluender KR, Levi DM, Bartoshuk LM, Herz RS, Klatzky RL, et al. *Sensation and Perception*. 2nd ed. Sunderland, MA: Sinauer Associates; 2009.

29. Birch EE, Hoffman DR, Castaneda YS, Fawcett SL, Birch DG, Uauy RD. A randomized controlled trial of long-chain polyunsaturated fatty acid supplementation of formula in term infants after weaning at 6 wk of age. *Am J Clin Nutr* 2002;**75**:570–80.

30. Birch EE, Castaneda YS, Wheaton DH, Birch DG, Uauy RD, Hoffman DR. Visual maturation of term infants fed long-chain polyunsaturated fatty acid-supplemented or control formula for 12 mo. *Am J Clin Nutr* 2005;**81**:871–9.

31. Hoffman DR, Birch EE, Birch DG, Uauy R, Castaneda YS, Lapus MG, Wheaton DH. Impact of early dietary intake and blood lipid composition of long-chain polyunsaturated fatty acids on later visual development. *J Pediatr Gastroenterol Nutr* 2000;**31**: 540–53.

32. Birch EE, Carlson SE, Hoffman DR, Fitzgerald-Gustafson KM, Fu VL, Drover JR, et al. The DIAMOND (DHA Intake And Measurement Of Neural Development) Study: a double-masked, randomized controlled clinical trial of the maturation of infant visual acuity as a function of the dietary level of docosahexaenoic acid. *Am J Clin Nutr* 2010;**91**:848–59.

33. Auestad N, Scott DT, Janowsky JS, Jacobsen C, Carroll RE, Montalto MB, et al. Visual, cognitive, and language assessments at 39 months: a follow-up study of children fed formulas containing long-chain polyunsaturated fatty acids to 1 year of age. *Pediatrics* 2003;**112**(3 Pt 1):e177–83.

34. Carlson SE, Werkman SH, Rhodes PG, Tolley EA. Visual-acuity development in healthy preterm infants: effect of marine-oil supplementation. *Am J Clin Nutr* 1993;**58**:35–42.

35. Carlson SE, Werkman SH, Tolley EA. Effect of long-chain n-3 fatty acid supplementation on visual acuity and growth of preterm infants with and without bronchopulmonary dysplasia. *Am J Clin Nutr* 1996;**63**:687–97.

36. O'Connor DL, Hall R, Adamkin D, Auestad N, Castillo M, Connor WE, et al. Growth and development in preterm infants fed long-chain polyunsaturated fatty acids: a prospective, randomized controlled trial. *Pediatrics* 2001;**108**:359–71.

37. Morale SE, Hoffman DR, Castaneda YS, Wheaton DH, Burns RA, Birch EE. Duration of long-chain polyunsaturated fatty acids availability in the diet and visual acuity. *Early Hum Dev* 2005;**81**: 197–203.

38. Cheatham CL, Colombo J, Carlson SE. n-3 fatty acids and cognitive and visual acuity development: methodologic and conceptual considerations. *Am J Clin Nutr* 2006;**83**(Suppl. 6): 1458–1466S.

39. Jacobson JL, Jacobson SW, Muckle G, Kaplan-Estrin M, Ayotte P, Dewailly E. Beneficial effects of a polyunsaturated fatty acid on infant development: evidence from the Inuit of Arctic Quebec. *J Pediatr* 2008;**152**:356–64.

40. Innis SM, Friesen RW. Essential n-3 fatty acids in pregnant women and early visual acuity maturation in term infants. *Am J Clin Nutr* 2008;**87**:548–57.

41. Judge MP, Harel O, Lammi-Keefe CJ. A docosahexaenoic acid-functional food during pregnancy benefits infant visual acuity at four but not six months of age. *Lipids* 2007;**42**:117–22.

42. Malcolm CA, McCulloch DL, Montgomery C, Shepherd A, Weaver LT. Maternal docosahexaenoic acid supplementation during pregnancy and visual evoked potential development in term infants: a double blind, prospective, randomised trial. *Arch Dis Child Fetal Neonatal Ed* 2003;**88**:F383–90.

43. Birch E, Birch D, Hoffman D, Hale L, Everett M, Uauy R. Breast-feeding and optimal visual development. *J Pediatr Ophthalmol Strabismus* 1993;**30**:33–8.

44. Williams C, Birch EE, Emmett PM, Northstone K. Stereoacuity at age 3.5 y in children born full-term is associated with prenatal and postnatal dietary factors: a report from a population-based cohort study. *Am J Clin Nutr* 2001;**73**:316–22.

45. Jacques C, Levy E, Muckle G, Jacobson SW, Bastien C, Dewailly E, et al. Long-term effects of prenatal omega-3 fatty acid intake on visual function in school-age children. *J Pediatr* 2011;**158**:83–90, e1.

46. Lucas M, Dewailly E, Muckle G, Ayotte P, Bruneau S, Gingras S, et al. Gestational age and birth weight in relation to n-3 fatty acids among Inuit (Canada). *Lipids* 2004;**39**:617–26.

47. Muckle G, Ayotte P, Dewailly E, Jacobson SW, Jacobson JL. Determinants of polychlorinated biphenyls and methylmercury exposure in Inuit women of childbearing age. *Environ Health Perspect* 2001;**109**:957–63.

48. Shapley R. Visual sensitivity and parallel retinocortical channels. *Annu Rev Psychol* 1990;**41**:635–58.

49. Krauss-Etschmann S, Shadid R, Campoy C, Hoster E, Demmelmair H, Jimenez M, et al. Effects of fish-oil and folate supplementation of pregnant women on maternal and fetal plasma concentrations of docosahexaenoic acid and eicosapentaenoic acid: a European randomized multicenter trial. *Am J Clin Nutr* 2007;**85**:1392–400.

50. Donahue SM, Rifas-Shiman SL, Olsen SF, Gold DR, Gillman MW, Oken E. Associations of maternal prenatal dietary intake of n-3 and n-6 fatty acids with maternal and umbilical cord blood levels. *Prostaglandins Leukot Essent Fatty Acids* 2009;**80**:289–96.

51. Molloy C, Doyle LW, Makrides M, Anderson PJ. Docosahexaenoic acid and visual functioning in preterm infants: a review. *Neuropsychol Rev* 2012;**22**:425–37.

52. Jednorog K, Marchewka A, Tacikowski P, Heim S, Grabowska A. Electrophysiological evidence for the magnocellular–dorsal pathway deficit in dyslexia. *Dev Sci* 2011;**14**:873–80.

53. Richardson AJ, Calvin CM, Clisby C, Schoenheimer DR, Montgomery P, Hall JA, et al. Fatty acid deficiency signs predict the severity of reading and related difficulties in dyslexic children. *Prostaglandins Leukot Essent Fatty Acids* 2000;**63**:69–74.

54. Taylor KE, Higgins CJ, Calvin CM, Hall JA, Easton T, McDaid AM, Richardson AJ. Dyslexia in adults is associated with clinical signs of fatty acid deficiency. *Prostaglandins Leukot Essent Fatty Acids* 2000;**63**:75–8.

55. Lindmark L, Clough PA. 5-month open study with long-chain polyunsaturated fatty acids in dyslexia. *J Med Food* 2007;**10**:662–6.

56. Cyhlarova E, Bell JG, Dick JR, Mackinlay EE, Stein JF, Richardson AJ. Membrane fatty acids, reading and spelling in dyslexic and non-dyslexic adults. *Eur Neuropsychopharmacol* 2007;**17**:116–21.

57. Dunstan JA, Mitoulas LR, Dixon G, Doherty DA, Hartmann PE, Simmer K, Prescott SL. The effects of fish oil supplementation in pregnancy on breast milk fatty acid composition over the course of lactation: a randomized controlled trial. *Pediatr Res* 2007;**62**:689–94.

58. Dunstan JA, Simmer K, Dixon G, Prescott SL. Cognitive assessment of children at age 2(1/2) years after maternal fish oil supplementation in pregnancy: a randomised controlled trial. *Arch Dis Child Fetal Neonatal Ed* 2008;**93**:F45–50.

59. Carlson SE, Werkman SH. A randomized trial of visual attention of preterm infants fed docosahexaenoic acid until two months. *Lipids* 1996;**31**:85–90.

60. Reisbick S, Neuringer M, Gohl E, Wald R, Anderson GJ. Visual attention in infant monkeys: effects of dietary fatty acids and age. *Dev Psychol* 1997;**33**:387–95.

61. Werkman SH, Carlson SE. A randomized trial of visual attention of preterm infants fed docosahexaenoic acid until nine months. *Lipids* 1996;**31**:91–7.

62. Colombo J, Kannass KN, Shaddy DJ, Kundurthi S, Maikranz JM, Anderson CJ, et al. Maternal DHA and the development of attention in infancy and toddlerhood. *Child Dev* 2004;**75**:1254–67.

63. Kannass KN, Colombo J, Carlson SE. Maternal DHA levels and toddler free-play attention. *Dev Neuropsychol* 2009;**34**:159–74.

64. Makrides M, Neumann MA, Byard RW, Simmer K, Gibson RA. Fatty acid composition of brain, retina, and erythrocytes in breast- and formula-fed infants. *Am J Clin Nutr* 1994;**60**:189–94.

65. Braddick O, Atkinson J, Wattam-Bell J. Normal and anomalous development of visual motion processing: motion coherence and 'dorsal-stream vulnerability'. *Neuropsychologia* 2003;**41**:1769–84.

Omega-3 and Macular Pigment Accumulation: Results from the Pimavosa Study

Marie-Noëlle Delyfer[1,2,3], Benjamin Buaud[4], Jean-François Korobelnik[1,2,3], Marie-Bénédicte Rougier[3], Carole Vaysse[4], Nicole Combe[4], Cécile Delcourt[1,2]

[1]Universite de Bordeaux, Bordeaux, France, [2]Inserm, ISPED, Centre INSERM U897-Epidemiologie-Biostatistique, Bordeaux, France, [3]Service d'Ophtalmologie, CHU de Bordeaux, Bordeaux, France, [4]ITERG – Equipe Nutrition Métabolisme & Santé, Bordeaux, France

The characterization of macular pigment (MP) has taken more than 200 years. Yet, its precise pathophysiologic properties remain elusive. The existence of a macular central yellow spot was first observed on postmortem histologic sections by Buzzi in 1782.[1] Decades later, *in vivo* observation of the macular area using Hermann von Helmholtz's ophthalmoscope (1851) enabled scientists to determine that this yellow spot was due to the presence of a macular yellow pigment. In 1945, Wald demonstrated that MP absorbs wavelengths between 430 and 490 nm, with a maximum absorption at 465 nm, and that MP contains carotenoids belonging to the xanthophyll family.[2,3] Xanthophylls are a class of oxygen-containing carotenoid pigments,[4] responsible for the color of many of the yellow, orange, and red hues of flowers, fruits, vegetables (corn, pepper, etc.), egg yolks, and feathers, shells, or flesh of many animal species (flamingo, canary, shrimp, lobster, chicken, or salmonids).[5] In plants, they are involved in photosynthesis with chlorophyll and are responsible for the red, yellow, and/or brown colors of autumn foliage as the chlorophyll levels decline. Approximately 600 different carotenoids have been characterized to date.[6] However, in 1985, Bone demonstrated that, in the retina, macular xanthophylls that compose MP are specifically restricted to two – lutein (L) and its structural isomer zeaxanthin (Z).[7] Anatomically, L and Z are concentrated in the Henle fiber layer of the macular area[8] and display a particular spatial distribution, Z being clearly dominant in the center of the fovea with a Z:L ratio decreasing peripherally.[7]

In humans, L and Z exclusively derive from dietary intake.[9,10] They are lipid-soluble and their metabolism is, therefore, strongly interlinked with lipids.[11,12] High dietary intake of L and Z, or their oral supplementation, is known to result in an increase of their plasma concentrations and, consequently, in their specific accumulation within the macula, in which they form the MP.[13–16] The precise mechanism of MP accumulation remains to be determined. It is suspected that it is based on a regulated active transport mechanism, thus explaining a 10,000-fold higher concentration in the macula than in the blood. As proposed recently, it may rely on specific carotenoid-binding proteins.[17,18] The macular over-concentration of L and Z further implies organ-specific biologic roles. The two main biologic properties suggested include, first, a distinctive blue light-absorbing capability due to the presence of the long chromophore of conjugated double bonds (polyene chain) and, second, an antioxidant capability through reactive oxygen species scavenging (e.g., superoxide anion and hydroxyl radical).[11] This last role underpins the growing interest for MP in aging macular diseases, especially in age-related macular degeneration (AMD). Indeed, AMD is the leading cause of blindness in developed countries and represents an increasing burden for public health systems.[19–21] AMD is a multifactorial disease that results from the combination of nonmodifiable factors (i.e., genetics, sex, age) and identified modifiable factors (i.e., nutritional and/or smoking status).[22] Controlling these modifiable factors may therefore be a way of preventing a significant percentage of AMD cases. Evidence from both epidemiologic and laboratory studies has demonstrated an inverse association between dietary intake of xanthophyll carotenoids –L and Z – and risk of advanced AMD.[22–24]

In contrast, long-chain polyunsaturated fatty acids (omega-3 LC-PUFAs) – notably docosahexaenoic acid (DHA) – are abundant in the human retina, in which they exert some identified structural, functional, and neuroprotective roles.[26,27] DHA reaches its highest concentration in the membranes of photoreceptors and is important in photoreceptor differentiation and survival, as well as in retinal function.[26] Omega-3 LC-PUFAs furthermore exhibit anti-inflammatory properties,[26] which are of particular interest in AMD, since inflammation appears to play a pivotal role in the pathogenesis of the disease.[28] The neuroprotective role of omega-3 LC-PUFAs in AMD has been demonstrated by a number of epidemiologic studies that observed a decreased risk for AMD in subjects with high intakes of omega-3 LC-PUFAs.[26,27,29–31] Among other potential mechanisms supporting neuroprotection, it has been suggested that dietary intake of omega-3 LC-PUFAs may favor the retinal accumulation of L and Z and thereby increase MP density.[12,13] The PIMAVOSA (PIgment MAculaire chez le VOlontaire SAin, i.e., MP in the healthy volunteer) study is an observational study that was hence initiated to evaluate the inter-relations between MP, plasma L and Z, and plasma omega-3 LC-PUFAs (as well as other fatty acids) in a homogeneous population of healthy volunteers.[32]

The characteristics of the studied population are detailed in Table 26.1. All of the recruited participants were healthy adults ages 20–60 years and born in a restricted geographic area (i.e., the southwest of France) in order to optimize the homogeneity of nutritional habits. None had any previous ocular history (all being phakic with a visual acuity of 20/24 or greater, without chronic diseases with significant ocular consequences or myopia exceeding 4 diopters). The use of vitamins and/or supplements was systematically checked and recorded (only seven subjects out of the 107 recruited were supplemented). Participants underwent bilateral eye exam including measurement of best-corrected visual acuity, refraction, retinal photographs after pupil dilation, fundus autofluorescence imaging, macular pigment optical density (MPOD) measurements, and fasting blood tests on the same day.[32]

The choice of the technique to be used for our MPOD measurements was difficult since several techniques can measure MP *in vivo*. We wanted a method that would be the most i) objective, ii) reproducible, and iii) easy to perform both for participants and technicians. Heterochromatic flicker photometry is classically considered as the 'gold standard'; however, although rather easy to perform and reliable, it requires cooperation from the subject. Raman spectroscopy is also highly reproducible but again requires active participation of the subject for the initial setting up of the device. In contrast, imaging techniques offer a means of measuring the spatial distribution of MPOD objectively and easily and require less subject compliance.[33,34] They are based on retinal excitation by an incident light and the analysis of the reflected signal obtained (Fig. 26.1). However, absence of nuclear opacities seems important to avoid any bias.[35] This was the technique we finally chose for our MPOD measurements, first because it appeared to be the most objective and, second, because MPOD mapping offered the possibility of studying the correlations between MPOD at different degrees of eccentricity from the center of the fovea and our other variables.

Quantification of our nutritional variables (i.e., L, Z, and fatty acids) could either have been estimated from food intakes or directly measured from plasma samples. Because of the multiple difficulties of dietary assessment (including memory bias, difficulties in taking into account high complexity and day-to-day variability of the human diet, imprecisions in estimations of quantities of ingested foods and of nutrient content of foods, etc.) and of interindividual variability in nutrient absorption, we deemed it more objective – and potentially more effective – to use plasma measurements. Plasma L and Z measurements were determined by reversed-phase high-performance liquid chromatography. Plasma phospholipid fatty acid measurements not only focused on omega-3 (alfa-linolenic acid and omega-3 LC-PUFAs) but also on omega-6 (linoleic acid (LA) and omega-6 LC-PUFAs), monounsaturated, and saturated fatty acids.

The first step of our work was a type of 'internal validation test' of our measurements in the studied population. Relationships between MP and xanthophylls, indeed, are now well established. MP is composed of two xanthophyll carotenoids, L and Z. Supplementation or high dietary intake of L and Z increase L and Z plasma levels and, in turn, MP density.[13–16]

We hence checked the association of plasma L and Z with MPOD in our sample. This was confirmed within 1° of eccentricity and beyond, plasma L and Z being considered separately and/or together (Table 26.2, Fig. 26.2A).[32] The result was not affected by the exclusion of the seven subjects declaring use of dietary supplements (data not shown). However, MPOD within 0.5° was not found to be significantly correlated with plasma macular xanthophylls. Interindividual variations in the spatial distribution of MPOD at the very center of the macula may explain – at least in part – this lack of correlation.[34,36]

The identification of the mechanisms underlying the specific macular accumulation of xanthophylls represents a key step toward a better understanding of macular physiology and disease. Among the different determinants of MP concentration under focus, omega-3 LC-PUFAs have been proposed as key factors.[12,13]

TABLE 26.1   Characteristics of the Studied Population

	Total (n = 107)	20–39 Years (n = 53)	40–60 Years (n = 54)	P Value
Age (years)	38.9 ± 12.1	28.2 ± 5.9	49.3 ± 5.5	< 0.0001
Gender (men)	43	23	20	0.50
Best corrected visual acuity (LogMAR units)	−0.10 ± 0.1	−0.10 ± 0.09	−0.09 ± 0.1	0.48
MPOD (within 6° of eccentricity, optical density units)	0.2 ± 0.1	0.2 ± 0.0	0.2 ± 0.1	0.11
PLASMA PHOSPHOLIPID OMEGA-3 PUFAs (% OF TOTAL FATTY ACIDS)				
Total	6.9 ± 1.9	6.5 ± 1.8	7.2 ± 1.9	0.05
ALA	0.2 ± 0.1	0.2 ± 0.1	0.2 ± 0.1	0.96
Omega-3 LC-PUFAs				
Total	6.7 ± 1.9	6.3 ± 1.8	7.0 ± 1.9	0.05
EPA	1.2 ± 0.7	1.1 ± 0.7	1.4 ± 0.7	0.02
DPA	0.9 ± 0.2	0.9 ± 0.3	1.0 ± 0.2	0.08
DHA	4.5 ± 1.2	4.3 ± 1.3	4.7 ± 1.2	0.18
PLASMA PHOSPHOLIPID OMEGA-6 PUFAs (% OF TOTAL FATTY ACIDS)				
Total	34.8 ± 2.4	34.9 ± 2.2	34.7 ± 2.5	0.60
Linoleic acid	18.7 ± 2.4	18.5 ± 2.6	18.9 ± 2.2	0.40
Omega-6 LC-PUFAs				
Total	15.3 ± 2.1	15.6 ± 1.9	15.0 ± 2.2	0.19
Eicosadienoic acid	0.3 ± 0.1	0.3 ± 0.1	0.3 ± 0.1	0.47
Dihomo-γ-linolenic acid	3.0 ± 0.7	3.1 ± 0.7	2.9 ± 0.7	0.11
Arachidonic acid	12.0 ± 2.0	12.1 ± 2.0	11.8 ± 2.0	0.43
Plasma phospholipid saturated fatty acids (% of total fatty acids)	44.4 ± 1.2	44.3 ± 1.3	44.5 ± 1.1	0.35
Plasma phospholipid monounsaturated fatty acids (% of total fatty acids)	13.0 ± 1.4	13.4 ± 1.3	12.7 ± 1.4	0.01
*PLASMA XANTHOPHYLLS**				
Plasma lutein* (µg/L)	150.1 ± 58.9	137.8 ± 48.4	161.8 ± 65.8	0.04
Plasma zeaxanthin* (µg/L)	40.9 ± 20.2	40.4 ± 17.5	41.3 ± 22.6	0.83
Plasma lutein + zeaxanthin* (µg/L)	191.1 ± 75.4	178.2 ± 62.4	203.1 ± 84.8	0.10

Data are the mean ± standard deviation; n = 107 unless specified otherwise. ALA: alfa-linolenic acid; DHA: docosahexaenoic acid; DPA: docosapentaenoic acid; EPA: eicosapentaenoic acid; LC-PUFA: long-chain polyunsaturated fatty acid; MPOD: macular pigment optical density; PUFA: polyunsaturated fatty acid; * = due to technical failure, L and Z plasma measurements were available only in 99 subjects.
Reprinted from Delyfer et al., Invest Ophthalmol Vis Sci. 53:1204–1210, with permission.

Omega-3 PUFAs include a precursor (ALA) and three long-chain derivatives (eicosapentaenoic acid (EPA, 20:5 omega-3), docosapentaenoic acid (DPA; 22:5 omega-3), and DHA (22:6 omega-3)]. ALA is an essential nutrient, since humans cannot synthesize it *de novo* and, therefore, rely on diet as its sole source (mainly from vegetables and vegetable oils). Synthesis of the long-chain derivatives is very limited in humans,[37] who must, therefore, also rely

on their dietary supply, mainly through fish and seafood. In the PIMAVOSA study, the analysis of the fatty acid composition of total plasma phospholipids was used as a valid biomarker of LC-PUFA dietary intakes.[38] As shown in Table 26.3, high plasma levels of total omega-3 PUFAs were associated with high MPOD (Table 26.3).[32] This was even more marked when considering total omega-3 LC-PUFAs only (Table 26.3, Fig. 26.2B). The mechanisms

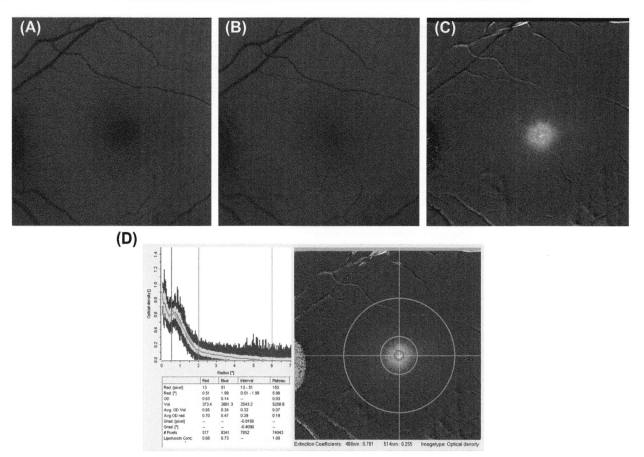

FIGURE 26.1 Macular pigment optical density (MPOD) measurements with the modified confocal scanning laser ophthalmoscope (mpHRA; Heidelberg Engineering, Heidelberg, Germany) using autofluorescence images obtained at two wavelengths: 488 nm (A) and 514 nm (B), with a high-pass filter transmitting at a wavelength greater than 530 nm. MPOD was quantified by calculating a MPOD map (C) and comparing foveal and parafoveal autofluorescence at 488 nm and 514 nm. Density maps were processed to estimate MPOD within a circle centered on the fovea at different degrees of eccentricities (0.5°, 1°, 2°, and 6°), using the software provided by the manufacturer of the device (D).

TABLE 26.2   Correlation of Macular Pigment Optical Density (MPOD) with Plasma Lutein and Zeaxanthin Levels (n = 99)

	MPOD within 0.5°	MPOD within 1°	MPOD within 2°	MPOD within 6°
Lutein + zeaxanthin	0.16 (0.1)	0.26 (0.01)*	0.33 (0.001)*	0.36 (0.0005)*
Lutein	0.16 (0.1)	0.24 (0.01)*	0.32 (0.001)*	0.35 (0.0006)*
Zeaxanthin	0.15 (0.1)	0.24 (0.02)*	0.29 (0.005)*	0.30 (0.003)*

Results are expressed as 'r (p)', r being the correlation coefficient and p the P value. *indicates significance. MPOD: macular pigment optical density.
Reprinted from Delyfer et al., Invest Ophthalmol Vis Sci. *53:1204–10, with permission.*

through which omega-3 LC-PUFAs correlate with MPOD remain to be determined. A first hypothesis could be that omega-3 LC-PUFAs may act as a modulator of L and Z gastrointestinal uptake. However, such a mechanism would have implied a correlation between plasma levels of omega-3 LC-PUFAs and xanthophyll carotenoids, which was not observed (Table 26.4).[32] A second hypothesis could be that omega-3 LC-PUFAs influence L and Z carriage by lipoproteins.[39] An increase of high-density lipoprotein (HDL) and low-density lipoprotein (LDL) subfractions has indeed been observed after omega-3

LC-PUFA supplementation.[40–42] Finally, omega-3 LC-PUFAs may favor L and Z concentration within the macular area through an influence on xanthophyll-binding proteins.[17]

When the different omega-3 LC-PUFAs were detailed, we observed that EPA and DPA both correlated significantly with MPOD, while, surprisingly, DHA did not (Table 26.3).[32] DHA is the major LC-PUFA in structural lipids of the human retina, accounting for about 30% of total retinal fatty acids. DHA is an essential structural component of retinal membranes and exhibits several

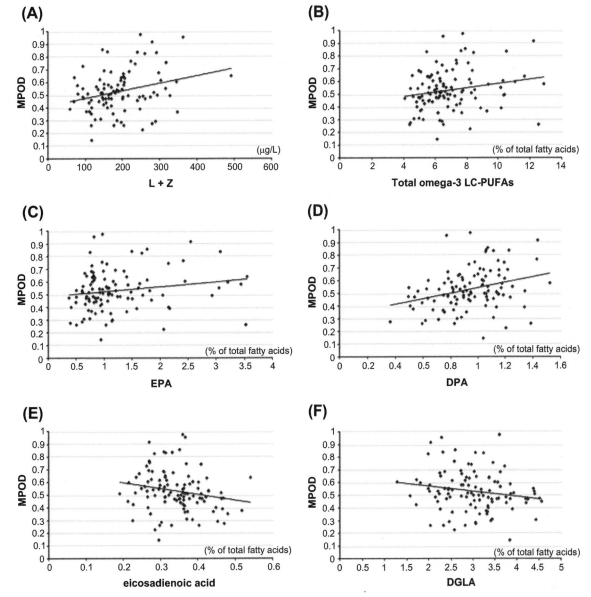

FIGURE 26.2 Scatter plots depicting the correlations between macular pigment optical density (MPOD) at 1° of eccentricity with (A) lutein and zeaxanthin (L + Z); with (B) total omega-3 long-chain polyunsaturated fatty acids (LC-PUFAs); with (C) EPA (eicosapentaenoic acid); with (D) DPA (docosapentaenoic acid); with (E) eicosadienoic acid; and with (F) DGLA (dihomo-γ-linolenic acid). Note that the associations are not driven by isolated cases. *Reprinted from Delyfer et al., Invest Ophthalmol Vis Sci. 53:1204–10, with permission.*[32]

essential neuroprotective properties.[26] The lack of correlation we observed between plasma DHA level and MPOD is not sufficient to assess the importance of DHA in MP accumulation. Nevertheless, the absence of correlation between DHA and MPOD seems corroborated by a previous study showing that DHA supplementation did not induce a significant increase of total MPOD values,[12] although it may influence MPOD distribution. EPA, the other major dietary omega-3 LC-PUFA present in plasma, is poorly accreted to the retina, as it is quickly converted to DHA or eicosanoid biosynthesis. EPA undergoes oxidative metabolism by cyclo-oxygenases and lipoxygenases to produce eicosanoids

with vasoregulatory and anti-inflammatory properties.[26] Contrary to DHA, a significant positive relationship was observed between plasma EPA level and MPOD (Table 26.3, Fig. 26.2C).[32] The positive correlation was even more significant with DPA (Table 26.3, Fig. 26.2D).[32] DPA is a metabolic intermediary between EPA and DHA and the second most abundant omega-3 LC-PUFA found within the retina after DHA, its endogenous level being around one-tenth that of DHA in the retinal lipids.[43] The precise functions of DPA have not been identified yet. DPA appears to be the precursor of omega-3 very long-chain polyunsaturated fatty acids (VLC-PUFAs) not found in normal human diet, and

TABLE 26.3 Correlation of Macular Pigment Optical Density (MPOD) with Plasma Phospholipid Omega-3 PUFA Levels and Other Plasma Phospholipid Fatty Acids (n = 107)

	MPOD within 0.5°	MPOD within 1°	MPOD within 2°	MPOD within 6°
OMEGA-3 PUFAs				
Total	0.19 (0.04)*	0.21 (0.03)*	0.20 (0.04)*	0.22 (0.02)*
ALA	0.0035 (0.97)	-0.0011 (0.99)	-0.00074 (0.99)	0.0016 (0.98)
OMEGA-3 LC-PUFAs				
Total	0.20 (0.04)*	0.22 (0.02)*	0.20 (0.04)*	0.22 (0.02)*
EPA	0.18 (0.06)	0.21 (0.04)*	0.20 (0.04)*	0.21 (0.03)*
DPA	0.33 (0.0006)*	0.32 (0.0007)*	0.30 (0.002)*	0.31 (0.001)*
DHA	0.13 (0.18)	0.14 (0.16)	0.12 (0.23)	0.14 (0.14)
OTHER FATTY ACIDS				
Saturated fatty acids	-0.15 (0.1)	-0.11 (0.2)	-0.11 (0.3)	-0.12 (0.2)
Monounsaturated fatty acids	-0.04 (0.7)	-0.09 (0.3)	-0.08 (0.4)	-0.06 (0.5)
OMEGA-6 PUFAs				
Total	-0.07 (0.5)	-0.064 (0.5)	-0.076 (0.5)	-0.092 (0.3)
Linoleic acid	0.02 (0.8)	0.014 (0.9)	0.0088 (0.9)	-0.027 (0.8)
Omega-6 LC-PUFAs				
Total	-0.07 (0.5)	-0.064 (0.5)	-0.076 (0.5)	-0.092 (0.3)
Eicosadienoic acid	-0.30 (0.001)*	-0.24 (0.008)*	-0.22 (0.02)*	-0.21 (0.03)*
Dihomo-γ-linolenic acid	-0.21 (0.03)*	-0.20 (0.04)*	-0.19 (0.05)*	-0.19 (0.05)*
Arachidonic acid	0.02 (0.8)	0.022 (0.8)	0.0083 (0.9)	0.0075 (0.9)

Results are expressed as 'r (p)', r being the correlation coefficient and p the P value. *Indicates significance. ALA: alfa-linolenic acid; DHA: docosahexaenoic acid; DPA: docosapentaenoic acid; EPA: eicosapentaenoic acid; LC-PUFAs: long-chain polyunsaturated fatty acids; PUFA: polyunsaturated fatty acid.
Reprinted from Delyfer et al., Invest Ophthalmol Vis Sci. *53:1204–10, with permission.*

especially of 24:5 omega-3, the most abundant omega-3 VLC-PUFA in the retina.[43] These omega-3 VLC-PUFAs are present in restricted mammalian organs, that is, retina, brain, testes, and thymus. Although identified early, their precise role has not been elucidated due to their great length and minor abundance, which makes them very difficult to analyze. The synthesis of 24:5 omega-3 VLC-PUFA, however, is an important metabolic step in the retina because 24:5 omega-3 plays a central role as a metabolic precursor in the synthesis of other omega-3 VLC-PUFAs and is an obligatory intermediate in the synthesis of DHA.[44] Moreover, alterations in omega-3 VLC-PUFA biosynthesis have been shown to result in macular alteration. In particular, defect in the elongation of the very long-chain fatty acids 4 (*ELOVL4*) gene is associated with dominant Stargardt macular dystrophy.[45] Recently, a decrease of DPA, DHA, and some omega-3 VLC-PUFAs (notably 24:5 omega-3) was observed in early and intermediate AMD retinas as compared with age-matched controls,[43] suggesting a retinal vulnerability associated with a decreased level of omega-3 LC-PUFAs and VLC-PUFAs.

Analysis of correlations between MPOD and fatty acids was further continued with plasma phospholipid monounsaturated, saturated, and omega-6 fatty acids (Table 26.3).[32] Neither saturated nor monounsaturated fatty acids were found to be associated with MPOD. Interest in omega-6 fatty acids has increased, since imbalance between omega-6 and omega-3 LC-PUFAs has emerged as a potential risk factor of AMD.[43,46] Our results concerning omega-6 were heterogeneous. Total plasma phospholipid omega-6 LC-PUFAs did not display any association with MPOD measurements (Table 26.3). However, when the different omega-6 were detailed, we observed that two minor omega-6 LC-PUFAs (i.e., eicosadienoic acid (20:2 omega-6) and dihomo-γ-linolenic acid (DGLA; 20:3 omega-6)), exhibited a negative relationship with MPOD (Table 26.3, Fig. 26.2E and 2F). In plasma, DGLA is almost exclusively localized in phospholipids and represents about 20% of omega-6 LC-PUFAs. In the retina, DGLA accounts for less than 2.5% of total retinal fatty acids. DGLA has been reported to have notable anti-inflammatory properties.[47] Regarding eicosadienoic acid (20:2

TABLE 26.4 Correlation of Plasma Carotenoids with Plasma Phospholipid Omega-3 PUFAs (n = 99)

	Lutein + Zeaxanthin	Lutein	Zeaxanthin
Total omega-3 PUFAs	0.14 (0.2)	0.15 (0.1)	0.07 (0.5)
ALA	0.03 (0.8)	0.008 (0.9)	0.05 (0.6)
OMEGA-3 LC-PUFAS			
Total	0.14 (0.2)	0.15 (0.1)	0.06 (0.5)
EPA	0.15 (0.1)	0.16 (0.1)	0.09 (0.4)
DPA	0.08 (0.4)	0.12 (0.2)	-0.06 (0.6)
DHA	0.08 (0.4)	0.09 (0.4)	0.02 (0.8)

Results are expressed as 'r (p)', r being the correlation coefficient and p the P value. ALA: alfa-linolenic acid; DHA: docosahexaenoic acid; DPA: docosapentaenoic acid; EPA: eicosapentaenoic acid; LC-PUFA: long-chain polyunsaturated fatty acid; PUFA: polyunsaturated fatty acids.
Reprinted from Delyfer et al., Invest Ophthalmol Vis Sci. *53:1204–10, with permission.*

omega-6), its physiologic role remains unknown. It is a relatively minor metabolite of LA (18:2 omega-6) found in human plasma and erythrocytes.[48] Lately, eicosadienoic acid has been shown to modulate the inflammatory response *in vitro*.[49] Since inflammation is postulated to be involved in AMD pathogenesis, the negative relationship of both plasma DGLA and eicosadienoic acid with MPOD we observed may suggest a reduced risk of AMD by a metabolic utilization of these anti-inflammatory omega-6 LC-PUFAs.

The PIMAVOSA study hence demonstrates that MP density is associated not only with plasma L and Z but also with plasma phospholipid omega-3 LC-PUFAs, and more particularly with DPA and EPA.[32] MP is further correlated – but negatively – with two omega-6 LC-PUFAs, which may have anti-inflammatory properties. The mechanisms underlying the associations of omega-3 or omega-6 LC-PUFAs with MP remain to be determined.

Still, our results suggest that macular xanthophylls and omega-3 LC-PUFAs may act synergistically in the constitution of MP, thereby providing a rationale for combined supplementation in patients at high risk for AMD.

TAKE-HOME MESSAGES

- MPOD correlates with plasma levels of L and Z.
- MPOD is associated positively with total plasma omega-3 polyunsaturated fatty acids (PUFAs).
- These results suggest that xanthophylls and omega-3 PUFAs may act synergistically in the constitution of MP.

- Among the different omega-3 PUFAs, DPA has the highest correlation with MPOD, while correlation with EPA is moderate and did not reach statistical significance for DHA.
- In contrast, MPOD is not significantly associated with plasma saturated fatty acids or monounsaturated fatty acids or total omega-6 PUFAs. However, MPOD exhibits a negative relationship with two minor omega-6 PUFAs: eicosadienoic acid and dihomo-γ-linolenic acid.

Acknowledgment

This study received financial support from Laboratoires Théa (Clermont-Ferrand, France). This sponsor participated in the design of the study but not in the collection, management, statistical analysis, and interpretation of the data, neither in the preparation, review, and approval of the present manuscript. This study received a grant from the Institut Carnot Lisa (Lipids for Industry and Health).

References

1. Buzzi F. Nuove sperienze fatte sull' occhio umano. *Opuscoli Scetti Sulle Scienze e Arti* 1782;**5**:87.
2. Wald G. Human vision and the spectrum. *Science* 1945;**101**:653–8.
3. Brown PK, Wald G. Visual pigments in human and monkey retinas. *Nature* 1963;**200**:37–43.
4. Goodwin TW. *The Biochemistry of the Carotenoids*. London: Chapman & Hall; 1980.
5. Mangels AR, Holden JM, Beecher GR, Forman MR, Lanza E. Carotenoid content of fruits and vegetables: an evaluation of analytic data. *J Am Diet Assoc* 1993;**93**:284–96.
6. Pfander H. *Key to Carotenoids*. Basel: Birkhauser; 1987.
7. Bone RA, Landrum JT, Tarsis SL. Preliminary identification of the human macular pigment. *Vision Res* 1985;**24**:1531–5.
8. Snodderly DM, Auran JD, Delori FC. The macular pigment, II: spatial distribution in primate retinas. *Invest Ophthalmol Vis Sci* 1984;**24**:674–85.
9. Snodderly DM, Brown PK, Delori FC, Auran JD. The macular pigment. I. Absorbance spectra, localization, and discrimination from other yellow pigments in primate retinas. *Invest Ophthalmol Vis Sci* 1984;**24**:660–73.
10. Bernstein PS, Khachik F, Carvalho LS, Muir GJ, Zhao DY, Katz NB. Identification and quantitation of carotenoids and their metabolites in the tissues of the human eye. *Exp Eye Res* 2001;**72**:215–23.
11. Whitehead AJ, Mares JA, Danis RP. Macular pigment: a review of current knowledge. *Arch Ophthalmol* 2006;**124**:1038–45.
12. Johnson EJ, Chung HY, Caldarella SM, Snodderly DM. The influence of supplemental lutein and docosahexaenoic acid on serum, lipoproteins, and macular pigmentation. *Am J Clin Nutr* 2008;**87**:1521–9.
13. Mares JA, LaRowe TL, Snodderly DM, Moeller SM, Gruber MJ, Klein ML, et al. Predictors of optical density of lutein and zeaxanthin in retinas of older women in the Carotenoids in Age-Related Eye Disease Study, an ancillary study of the Women's Health Initiative. *Am J Clin Nutr* 2006;**84**:1107–22.
14. Kvansakul J, Rodriguez-Carmona M, Edgar DF, Barker FM, Köpcke W, Schalch W, et al. Supplementation with the carotenoids lutein or zeaxanthin improves human visual performance. *Ophthalmic Physiol Opt* 2006;**26**:362–71.
15. Richer S, Devenport J, LAST II Lang JC. Differential temporal responses of macular pigment optical density in patients with atrophic

age-related macular degeneration to dietary supplementation with xanthophylls. *Optometry* 2007;**78**:213–9.

16. Trieschmann M, Beatty S, Nolan JM, Hense HW, Heimes B, Austermann U, et al. Changes in macular pigment optical density and serum concentrations of its constituent carotenoids following supplemental lutein and zeaxanthin: the LUNA study. *Exp Eye Res* 2007;**84**:718–28.

17. Li B, Vachali P, Bernstein PS. Human ocular carotenoid-binding proteins. *Photochem Photobiol Sci* 2010;**9**:1418–24.

18. Li B, Vachali P, Frederick JM, Bernstein PS. Identification of StARD3 as a lutein-binding protein in the macula of the primate retina. *Biochemistry* 2011;**50**:2441–9.

19. Klein R, Klein BE, Linton KL. Prevalence of age-related maculopathy: the Beaver Dam Eye Study. *Ophthalmology* 1992;**99**:933–43.

20. Mitchell P, Smith W, Attebo K, Wang JJ. Prevalence of age-related maculopathy in Australia: the Blue Mountains Eye Study. *Ophthalmology* 1995;**102**:1450–60.

21. Vingerling JR, Dielemans I, Hofman A, Grobbee DE, Hijmering M, Kramer CF, et al. The prevalence of age-related maculopathy in the Rotterdam Study. *Ophthalmology* 1995;**102**:205–10.

22. Jager RD, Mieler WF, Miller JW. Age-related macular degeneration. *N Engl J Med* 2008;**358**:2606–17.

23. Eye Disease Case-Control Study Group. Antioxidant status and neovascular age-related macular degeneration. *Arch Ophthalmol* 1993;**111**:104–9.

24. SanGiovanni JP, Chew EY, Clemons TE, Ferris 3rd FL, Gensler G, Lindblad AS, et al. The relationship of dietary carotenoid and vitamin A, E, and C intake with age-related macular degeneration in a case-control study: AREDS Report No. 22. *Arch Ophthalmol* 2007;**124**:1224–32.

25. Mares-Perlman JA, Fisher AI, Klein R, Palta M, Block G, Millen AE, et al. Lutein and zeaxanthin in the diet and serum and their relation to age-related maculopathy in the third national health and nutrition examination survey. *Am J Epidemiol* 2001;**153**:424–32.

26. SanGiovanni JP, Chew EY. The role of omega-3 long-chain polyunsaturated fatty acids in health and disease of the retina. *Prog Retin Eye Res* 2005;**24**:87–138.

27. Kishan AU, Modjtahedi BS, Martins EN, Modjtahedi SP, Morse LS. Lipids and age-related macular degeneration. *Surv Ophthalmol* 2011;**56**:195–213.

28. Donoso LA, Kim D, Frost A, Callahan A, Hageman G. The role of inflammation in the pathogenesis of age-related macular degeneration. *Surv Ophthalmol* 2006;**51**:137–52.

29. SanGiovanni JP, Chew EY, Agrón E, Clemons TE, Ferris 3rd FL, Gensler G, et al. The relationship of dietary omega-3 long-chain polyunsaturated fatty acid intake with incident age-related macular degeneration: AREDS report no. 23. *Arch Ophthalmol* 2008;**126**:1274–9.

30. Swenor BK, Bressler S, Caulfield L, West SK. The impact of fish and shellfish consumption on age-related macular degeneration. *Ophthalmology* 2010;**117**:2395–401.

31. Merle B, Delyfer MN, Korobelnik JF, Rougier MB, Colin J, Malet F, et al. Dietary omega3 fatty acids and the risk for age-related maculopathy: the Alienor Study. *Invest Ophthalmol Vis Sci* 2011;**52**:6004–11.

32. Delyfer MN, Buaud B, Korobelnik JF, Rougier MB, Schalch W, Etheve S, et al. Association of macular pigment density with plasma ω-3 fatty acids: the PIMAVOSA study. *Invest Ophthalmol Vis Sci* 2012;**53**:1204–10.

33. Delori FC, Goger DG, Hammond BR, Snodderly DM, Burns SA. Macular pigment density measured by autofluorescence spectrometry: comparison with reflectometry and heterochromatic flicker photometry. *J Opt Soc Am* 2001;**18**:1212–30.

34. Wolf-Schnurrbusch UE, Röösli N, Weyermann E, Heldner MR, Höhne K, Wolf S. Ethnic differences in macular pigment density and distribution. *Invest Ophthalmol Vis Sci* 2007;**48**:3783–7.

35. Sasamoto Y, Gomi F, Sawa M, Sakaguchi H, Tsujikawa M, Nishida K. Effect of cataract in evaluation of macular pigment optical density by autofluorescence spectrometry. *Invest Ophthalmol Vis Sci* 2011;**52**:927–32.

36. Davies NP, Morland AB. Macular pigments: their characteristics and putative role. *Prog Retin Eye Res* 2004;**23**:533–59.

37. Arterburn LM, Hall EB, Oken H. Distribution, interconversion, and dose response of n-3 fatty acids in humans. *Am J Clin Nutr* 2006;**83**:1467S–76S.

38. Hodson L, Murray Skeaff C, Fielding BA. Fatty acid composition of adipose tissue and blood in humans and its use as a biomarker of dietary intake. *Prog Lipid Res* 2008;**47**:348–80.

39. Parker RS. Absorption, metabolism, and transport of carotenoids. *FASEB J* 1996;**10**:542–51.

40. Nelson GJ, Schmidt PC, Bartolini GL, Kelley DS, Kyle D. The effect of dietary docosahexaenoic acid on plasma lipoproteins and tissue fatty acid composition in humans. *Lipids* 1997;**32**:1137–46.

41. Foulon T, Richard MJ, Payen N, Bourrain JL, Beani JC, Laporte F, et al. Effects of fish oil fatty acids on plasma lipids and lipoproteins and oxidant-antioxidant imbalance in healthy subject. *Scand J Clin Lab Invest* 1999;**59**:239–48.

42. Thomas TR, Smith BK, Donahue OM, Altena TS, James-Kracke M, Sun GY. Effects of omega-3 fatty acid supplementation and exercise on low-density lipoprotein and high-density lipoprotein subfractions. *Metabolism* 2004;**53**:749–54.

43. Liu A, Chang J, Lin Y, Shen Z, Bernstein PS. Long-chain and very long-chain polyunsaturated fatty acids in ocular aging and age-related macular degeneration. *J Lipid Res* 2010;**51**:3217–29.

44. Rotstein NP, Pennacchiotti GL, Sprecher H, Aveldanno MI. Active synthesis of C24:5 n-3 fatty acid in retina. *Biochem J* 1996;**316**:859–64.

45. Zhang K, Kniazeva M, Han M, Li W, Yu Z, Yang Z, et al. A 5-bp deletion in ELOVL4 is associated with two related forms of autosomal dominant macular dystrophy. *Nat Genet* 2001;**27**:89–93.

46. Seddon JM, Cote J, Rosner B. Progression of age-related macular degeneration: association with dietary fat, transunsaturated fat, nuts, and fish intake. *Arch Ophthalmol* 2003;**121**:1728–37.

47. Umeda-Sawada R, Fujiwara Y, Ushiyama I, Sagawa S, Morimitsu Y, Kawashima H, et al. Distribution and metabolism of dihomo-γ-linolenic acid (DGLA, 20:3n-6) by oral supplementation in rats. *Biosci Biotechnol Biochem* 2006;**70**:2121–30.

48. Park WJ, Kothapalli KSD, Lawrence P, Tyburczy C, Brenna JT. An alternate pathway to long-chain polyunsaturates: the FADS2 gene product D8-desaturates 20:2n-6 and 20:3n-3. *J Lipid Res* 2009;**50**:1195–202.

49. Huang YS, Huang WC, Li CW, Chuang LT. Eicosadienoic acid differentially modulates production of pro-inflammatory modulators in murine macrophages. *Mol Cell Biochem* 2011;**358**:85–94.

Dietary Carbohydrate and Age-Related Cataract

Zhi-Quan Lu, Jia Yan

Liaoning Medical University, Liaoning Province, P.R. China

INTRODUCTION

Age-related cataract (ARC) is the leading cause of blindness worldwide. In less-developed countries, cataract often results in blindness.[1,2] Although in developed countries cataract surgery is usually available and effective, it imposes a heavy personal and societal economic burden.[2] Measures to delay or prevent the onset of cataract could have major economic and health effects. Hyperglycemia is thought to be a risk factor for cataract development, as considerable evidence has linked aberrant glucose metabolism or diabetes to cataract risk, and *in vitro* and *in vivo* animal studies also suggest that carbohydrates play an important role in cataractogenesis. Different carbohydrates, as a result of their individual physical and chemical characteristics, induce distinct plasma glucose responses, which can be quantified by the glycemic index (GI).[3,4] Dietary glycemic load (GL) is the product of the GI of each specific food and its carbohydrate content, summed over all foods, and thus it represents the quantity and quality of carbohydrates in the overall diet and their interaction. Only a few epidemiologic studies have investigated the association between carbohydrate nutrition and cataract, or cataract surgery, and the findings to date have been inconsistent.[5-11] A role for carbohydrate nutrition in cataractogenesis is biologically plausible. Available mechanisms have been put forward.

DEFINITION OF DIETARY GLYCEMIC INDEX AND DIETARY GLYCEMIC LOAD

Dietary Glycemic Index

We used the dietary GI to assess carbohydrate quality. Jenkins *et al.*[3] introduced the concept of GI for individual foods to facilitate the identification of potentially clinically useful foods that result in relatively low glycemic responses. The dietary GI for each subject was calculated as the weighted mean of the GI scores for each food item, with the amount of carbohydrate consumed from each food item as the weight.

The GI has proven to be a more useful nutritional concept than is the chemical classification of carbohydrates (as simple or complex, as sugars or starches, or as available or unavailable), permitting new insights into the relation between the physiologic effects of carbohydrate-rich foods and health.[12] The GI is a numerical system of measuring how much of a rise in circulating blood sugar a carbohydrate triggers – the higher the number, the greater the blood sugar response. Peak blood glucose concentrations and mean blood glucose concentrations are higher after high GI foods. Metabolic studies show that after consuming equivalent quantities of high compared with low GI foods, it takes a longer time to return to baseline blood glucose concentrations.[13] A GI of 70 or more is high, a GI of 56 to 69 inclusive is medium, and a GI of 55 or less is low.

Dietary Glycemic Load

The GL is a relatively new way to assess the impact of carbohydrate consumption that takes the GI into account but gives a fuller picture than does GI alone. A GI value indicates only how rapidly a particular carbohydrate turns into sugar. It does not indicate how much of that carbohydrate is in a serving of a particular food. You need to know both of these to understand a food's effect on blood sugar. That is where GL comes in. Dietary GL is calculated as the sum of the product of the GI for each foodstuff and its available carbohydrate content divided by 100; thus it represents the quantity and quality of carbohydrates in the overall diet and their interaction.[14] Each unit of dietary GL represents the equivalent of 1 g carbohydrate from white bread. For example, the GL of one serving of cooked potatoes was determined to be 38 because the carbohydrate content of one serving of potatoes is 37 g and the GI of potatoes (with white bread as the reference) is 102 (i.e., 102*37/100 = 38). We then multiply this dietary GL score by the frequency of consumption and sum the products over all food items to produce the dietary GL.

A GL of 20 or more is high, a GL of 11 to 19 inclusive is medium, and a GL of 10 or less is low. Foods that have a low GL almost always have a low GI. Foods with an intermediate or high GL range from very low to very high GI. The carbohydrate in watermelon, for example, has a high GI. However, there is not a great quantity of it, so the GL of watermelon is relatively low. Table 27.1 shows the glycemic index by glycemic load of certain common foods.

EPIDEMIOLOGIC STUDIES

In recent years, more and more epidemiologic studies have shown that the long-term diet with a high GL is independently associated with an increased risk of developing type 2 diabetes,[15,16] cardiovascular disease,[17–19] and other chronic conditions.[20] However, epidemiologic studies that have examined the relation between dietary carbohydrate nutrition and three principal types of cataract or cataract surgery were limited. Moreover, the few studies have produced inconsistent results.[5–11] Epidemiologic studies that have examined the relation between dietary carbohydrate nutrition and cataract are shown in Table 27.2. Only four previous studies have examined the association between GI and cataract.[7–10] Figure 27.1 presents the risk of different ARCs among (the highest vs. the lowest of GI) these studies. In an Australian population of The Blue Mountains Eye Study (BMES),[9] after many confounding factors were controlled for, GI significantly predicted incident cortical cataract, but no association was found between GI and nuclear or posterior subcapsular cataract (PSC) cataract. In contrast, the Age-Related Eye Disease Study (AREDS) found that higher GI was associated with nuclear but not cortical cataract.[8] A cross-sectional study reported that the risk of nuclear cataract comparing the third quartile of dietary glycemic index (dGI) with the first quartile significantly increased, but there was not a consistent dose–response association.[10] The Nutrition and Vision Project (NVP), which examined cortical and nuclear opacities, found no association with GI.[7]

Figure 27.2 shows the risk of different ARCs among (the highest vs. the lowest) studies that have evaluated the association between dietary carbohydrate intake and ARC. Consistent with the findings of BMES in a cross-sectional report,[5] the NVP, the Melbourne Visual

TABLE 27.1 Glycemic Index by Glycemic Load of Common Food Items*§

	Low GI	Medium GI	High GI
Low GL	All-bran cereal (8, 42)	Beets (5, 64)	Popcorn (8, 72)
	Apples (6, 38)	Cantaloupe (4, 65)	Watermelon (4, 72)
	Carrots (3, 47)	Pineapple (7, 59)	Whole wheat flour bread (9, 71)
	Chana dal (3, 8)	Sucrose (table sugar) (7, 68)	White wheat flour bread (10, 70)
	Chick peas (8, 28)		
	Grapes (8, 46)		
	Green peas (3, 48)		
	Kidney beans (7, 28)		
	Nopal (0, 7)		
	Oranges (5, 42)		
	Peaches (5, 42)		
	Peanuts (1, 14)		
	Pears (4, 38)		
	Pinto beans (10, 39)		
	Red lentils (5, 26)		
	Strawberries (1, 40)		
	Sweet corn (9, 54)		
Medium GL	Apple juice (11, 40)	Life cereal (16, 66)	Cheerios (15, 74)
	Bananas (12, 52)	New potatoes (12, 57)	Shredded wheat (15, 75)
	Buckwheat (16, 54)	Sweet potatoes (17, 61)	
	Fettucine (18, 40)	Wild rice (18, 57)	
	Navy beans (12, 38)		
	Orange juice (12, 50)		
	Parboiled rice (17, 47)		
	Pearled barley (11, 25)		
	Sourdough wheat bread (15, 54)		
High GL	Linguine (23, 52)	Couscous (23, 65)	Baked russet potatoes (26, 85)
	Macaroni (23, 47)	White rice (23, 64)	Cornflakes (21, 81)
	Spaghetti (20, 42)		

*First number in parentheses is glycemic load (GL)
§the second number is glycemic index (GI).
By Mendosa David, which originally appeared on Mendosa.com on 23 April 2003.

TABLE 27.2 Epidemiologic Studies That Have Examined the Relation between Dietary Carbohydrate Nutrition and Cataract

Study	Design	Author	Sample Size	Outcome	Comparison	OR/RR (95% CI)
BMES	Cross-sectional	Cumming et al. (2000)[5]	2900	The three main types of cataract	Highest vs. lowest quintiles of carbohydrate intake	Cortical cataract: 1.4 (1.0 to 1.9)
NHS HPFS	Prospective	Schaumberg et al. (2004)[6]	71,919 (female) 39,926 (male)	extraction of nuclear cataract and PSC	Highest vs. lowest quintiles of GL	Nuclear cataract: 1.01 (0.83 to 1.22) PSC: 0.96 (0.79 to 1.16)
NHS	Prospective	Chiu et al. (2005)[7]	417 (female)	The three main types of cataract	Highest vs. lowest tertile of carbohydrate intake	Cortical cataract: 2.46 (1.30 to 4.64)
AREDS	Cross-sectional	Chiu et al. (2006)[8]	3377	Cortical and nuclear opacities	Highest 25% vs. lowest 25% of GI	Nuclear opacities: 1.29 (1.04 to 1.59)
BMES	Prospective	Tan et al. (2007)[9]	3654	The three main types of cataract	Highest vs. lowest quartile of GI	Cortical cataract: 1.77 (1.13 to 2.78)
MVIP	Cross-sectional	Chiu et al. (2010)[10]	1609	Cortical and nuclear opacities	Highest vs. lowest quartile of carbohydrate intake	Cortical cataract: 3.19 (1.10 to 9.27)
	Case-control	Yan et al. (2012)[11]	360 cases 360 controls	Cataract extraction	Highest vs. lowest quartile of carbohydrate intake	Cortical cataract: 2.47 (1.35 to 6.04)

AREDS: Age-Related Eye Disease Study; BMES: Blue Mountains Eye Study; CI: confidence interval; GI: glycemic index; GL: glycemic load; HPFS: Health Professionals Follow-up Study; MVIP: Melbourne Visual Impairment Project; NHS: Nurses' Health Study; OR: odds ratio; PSC: posterior subcapsular cataract; RR: risk ratio.

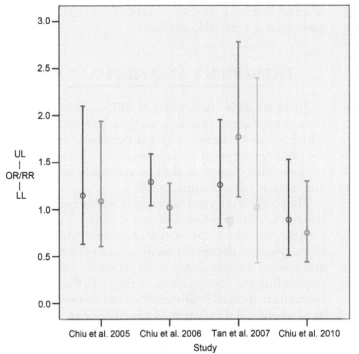

FIGURE 27.1 Compared with participants of the lowest GI, the odds ratio (OR)/risk ratio (RR) and corresponding 95% confidence interval of different ARC for subjects in the highest GI among (the highest vs. the lowest of glycemic index (GI)) studies that have evaluated the association between dietary GI.

LL: lowest limit; PSC: posterior subcapsular cataract; UL: upper limit

Impairment Project (MVIP),[10] and a case-control study in China[11] also reported an association between dietary carbohydrate quantity or GL as another aspect of carbohydrate nutrition and increased risk of cortical but not nuclear cataract. Despite the differences across studies in the period covered by dietary questionnaires and the cutoff points for defining cortical opacities, similar findings have been reported in US as well as Australian

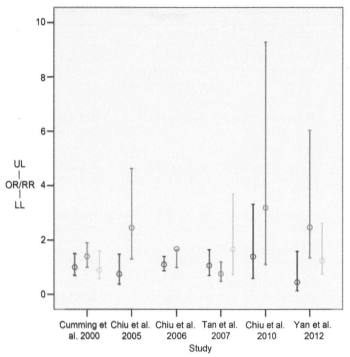

FIGURE 27.2 Compared with participants of the lowest quantity of carbohydrate, the odds ratio (OR)/risk ratio (RR) and corresponding 95% confidence interval of different ARC for subjects in the highest carbohydrate intake among studies that have evaluated the association between dietary carbohydrate intake and ARC.

LL: lowest limit; PSC: posterior subcapsular cataract; UL: upper limit.

and Chinese populations. The baseline BMES[5] and the MVIP study evaluated the past year's diet, whereas the NVP study reported the influence of long-term diet (previous 14 years) on cataract. The baseline BMES and the MVIP study used cutoff points for cortical opacities that were comparable to the cutoff point used for moderate cortical opacities in the AREDS, whereas in the MNVP a cutoff point for cortical opacities was used that was similar to that for mild cortical opacities in the AREDS.

Figure 27.3 shows the risk of different ARCs among (the highest vs. the lowest of GL) studies that have evaluated the association between dietary GL and ARC. Results of the Nurses' Health Study (NHS) and Health Professionals Follow-up Study (HPFS), which examined cataract extractions and nuclear and PSC subtypes,[6] and 10-year follow-up of BMES, which examined three types of cataract,[9] showed that there was no significant association between GL and cataract. Differences in the sample size, study design, and participants' ethnic backgrounds between studies may explain the difference in findings.

In the baseline BMES, the risk of cortical cataract was 40% higher for participants in the highest quintile of daily carbohydrate intake than for participants in the first quintile. However, after 10-year follow-up of the same study sample, poorer dietary carbohydrate quality (high dGI, which was not examined in the previous cross-sectional report), but not quantity, predicted incident cortical cataract. Although this inconsistency needs further clarification, results at baseline and in the 10-year follow-up in the BMES indicate that dietary carbohydrate affects the risk of cortical cataract. In view that carbohydrate foods represent the main energy

source for humans and the present studies exploring dietary carbohydrate and cataract are limited and the results are inconsistent, understanding the potentially harmful effects of a high carbohydrate diet on the lens is important and worthy of further longitudinal, and randomized, controlled studies.

PATHOGENY AND MECHANISMS

There are three main types of ARCs defined by their clinical appearance: nuclear, cortical, and PSC cataract, which affect the center, adjacent peripheral tissue, and the posterior aspect of the lens, respectively. The etiology of age-related changes in the lens is not fully understood and is likely to be multifactorial.[21]

The possible mechanisms of relationship between carbohydrate nutrition and ARC are shown in Figure 27.4. Despite the rapid, facilitated uptake of glucose and its subsequent decline to basal amounts in some cells and tissues, glucose is taken up relatively slowly from plasma into the aqueous humor, the fluid that provides nutrients to the lens.[10] Glucose then passes freely into the lens, where it is slowly turned over. Prolonged exposure of the lens proteins to elevated glucose was observed to cause accumulation of polyol and extensive glycation, the consequences of which may include oxidation, cross-linking, aggregation, and precipitation of the modified lens proteins.[22–24] The glycation mechanism has also been linked to the etiology of other age-related chronic diseases, such as diabetes, coronary heart disease, and cancers.[25]

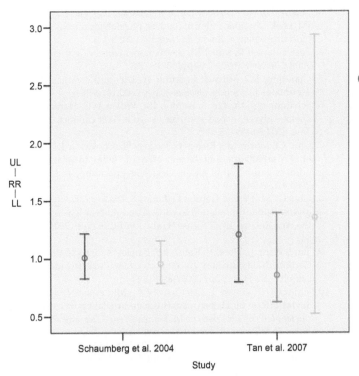

FIGURE 27.3 Compared with participants of the lowest GL, the risk ratio (RR) and corresponding 95% confidence interval of different ARC for subjects in the highest GL among studies that have evaluated the association between dietary GL and ARC.

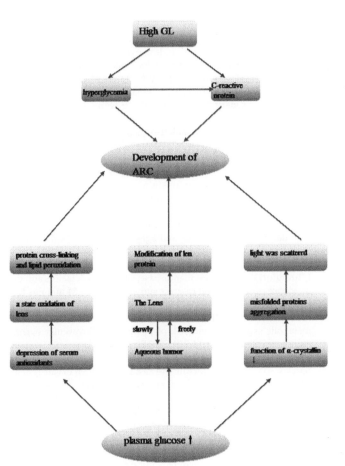

FIGURE 27.4 The possible mechanisms of relationship between carbohydrate nutrition and age-related cataract (ARC). GL: glycemic load.

A potential role for oxidative damage in the association between high GI foods and cataract is further supported by studies showing that higher postprandial glucose concentrations result in greater depression of serum antioxidants.[13,26] A protective association between higher nutritional intake and serum concentrations of antioxidants and age-related cataract has been indicated in previous studies.[27,28] When antioxidant concentrations are inadequate and the redox balance in the lens is shifted toward a state of oxidation, sulfhydryl groups are oxidized, resulting in protein cross-linking and lipid peroxidation. Protein aggregates form within the lens, and these theoretically begin to scatter light when they become larger to some extent.[29]

Diets with a high GL are associated with hyperglycemia and hyperinsulinemia,[30] and diabetes mellitus has long been recognized as a risk factor for cataract.[31–33] Research on the health effects of dietary GL has shed doubt on the healthfulness of a high GL diet. Previous studies also showed that higher GL was linked with higher plasma concentrations of the inflammatory marker C-reactive protein,[34] which in turn has been associated with an increased risk of cataract in at least one study.[35]

However, a higher dietary GL is not a risk factor for the development of cataract in several prospective studies.[6,9] It is possible that there is a threshold for hyperglycemia that must be reached before the risk of cataract is affected, and laboratory data in animals suggest that this threshold may be well within the diabetic range.[36]

Whereas the risk of cortical opacity has been related to total carbohydrate intake, there has been no report of a significant association between total carbohydrate intake and nuclear cataract. Studies also suggest that higher plasma glucose concentrations may have a greater effect on the lens cortex than on the nucleus.[7] Nuclear magnetic resonance studies suggest that glucose concentrations remain higher in the cortex than in the nucleus.[37] Thus, it is possible that higher carbohydrate intakes and plasma concentrations of glucose will result in chronically enhanced exposure of lens cortex proteins to glucose.

In addition, enzyme activities involved in the metabolism of glucose decrease toward the center of the lens, consistent with higher concentrations of glucose and glucose transporters in the lens cortex than in the nucleus.[7,38] Because of the extreme long life of lens proteins and the limited opportunities for repair or removal and replacement, these effects may cause enhanced cortical protein modification, precipitation, and cataract.[39] *In vivo* evidence indicates that the insolubilization of glycated lens proteins in diabetic cataractogenesis is initiated in the cortical region.[40] Epidemiologic studies also support a more consistent link between diabetes and cortical rather than nuclear cataract.[41–43]

TAKE-HOME MESSAGES

- The dietary GI is a numerical system of measuring how much of a rise in circulating blood sugar a carbohydrate triggers – the higher the number, the greater the blood sugar response.
- Dietary GL represents the quantity and quality of carbohydrates in the overall diet and their interaction.
- The limited studies that have examined the relation between dietary carbohydrate nutrition and three principal types of cataract or cataract surgery have produced inconsistent results.
- A potential role for oxidative damage in the association between high GI foods and cataract is further supported by studies.
- Diets with a high GL are associated with hyperglycemia and hyperinsulinemia, and diabetes mellitus has long been recognized as a risk factor for cataract.
- Epidemiologic studies also support a more consistent link between diabetes and cortical rather than nuclear cataract.

References

1. Congdon NG, Friedman DS, Lietman T. Important causes of visual impairment in the world today. *JAMA* 2003;**290**:2057–60.
2. Asbell PA, Dualan I, Mindel J, Brocks D, Ahmad M, Epstein S. Age-related cataract. *Lancet* 2005;**365**:599–609.
3. Jenkins DJ, Wolever TM, Taylor RH, Barker H, Fielden H, Baldwin JM, et al. Glycemic index of foods: a physiological basis for carbohydrate exchange. *Am J Clin Nutr* 1981;**34**:362–6.
4. Foster-Powell K, Miller JB. International tables of glycemic index. *Am J Clin Nutr* 1995;**62**(Suppl):871S–90S.
5. Cumming RG, Mitchell P, Smith W. Diet and cataract: the Blue Mountains Eye Study. *Ophthalmology* 2000;**107**:450–6.
6. Schaumberg DA, Liu S, Seddon JM, Willett WC, Hankinson SE. Dietary glycemic load and risk of age-related cataract. *Am J Clin Nutr* 2004;**80**:489–95.
7. Chiu CJ, Morris MS, Rogers G, Jacques PF, Chylack Jr LT, Tung W, et al. Carbohydrate intake and glycemic index in relation to the odds of early cortical and nuclear lens opacities. *Am J Clin Nutr* 2005;**81**:1411–6.
8. Chiu CJ, Milton RC, Gensler G, Taylor A. Dietary carbohydrate and glycemic index in relation to cortical and nuclear lens opacities in the Age-Related Eye Disease Study. *Am J Clin Nutr* 2006;**83**:1177–84.
9. Tan J, Wang JJ, Flood V, Kaushik S, Barclay A, Brand-Miller J, et al. Carbohydrate nutrition, glycemic index, and the 10-y incidence of cataract. *Am J Clin Nutr* 2007;**86**:1502–8.
10. Chiu CJ, Robman L, McCarty CA, Mukesh BN, Hodge A, Taylor HR, et al. Dietary carbohydrate in relation to cortical and nuclear lens opacities in the Melbourne Visual Impairment Project. *Invest Ophthalmol Vis Sci* 2010;**51**:2897–905.
11. Yan J, Sun WH, Zhang D, Jiang TX, Zhai SN, Li Y, et al. Association between dietary macronutrient intake and age-related cataract. *Chin Gen Pract* 2012;**15**:160–3.
12. Foster-Powell K, Holt SHA, Brand-Miller JC. International table of glycemic index and glycemic load values. *Am J Clin Nutr* 2002;**2002**(76):55–6.
13. Jenkins DJ, Kendall CW, Augustin LS, Franceschi S, Hamidi M, Marchie A, et al. Glycemic index: overview of implications in health and disease. *Am J Clin Nutr* 2002;**76**(Suppl):266S–73S.
14. Liu S, Willett WC, Stampfer MJ, Hu FB, Franz M, Sampson L, et al. A prospective study of dietary glycemic load, carbohydrate intake, and risk of coronary heart disease in US women. *Am J Clin Nutr* 2000;**71**:1455–61.
15. Jacques PF, Moeller SM, Hankinson SE, Chylack Jr LT, Rogers G, Tung W, et al. Weight status, abdominal adiposity, diabetes, and early age-related lens opacities. *Am J Clin Nutr* 2003;**78**:400–5.
16. Lindblad BE, Hakansson N, Philipson B, Wolk A. Metabolic syndrome components in relation to risk of cataract extraction: a prospective cohort study of women. *Ophthalmology* 2008;**115**:1687–92.
17. Klein BE, Klein R, Lee KE. Cardiovascular disease, selected cardiovascular disease risk factors, and age-related cataracts: the Beaver Dam Eye Study. *Am J Ophthalmol* 1997;**123**:338–46.
18. Younan C, Mitchell P, Cumming R, Rochtchina E, Panchapakesan J, Tumuluri K. Cardiovascular disease, vascular risk factors and the incidence of cataract and cataract surgery: the Blue Mountains Eye Study. *Ophthal Epidemiol* 2003;**10**:227–40.
19. Ridker PM, Buring JE, Cook NR, Rifai N. C-reactive protein, the metabolic syndrome, and risk of incident cardiovascular events: an 8-year follow-up of 14 719 initially healthy American women. *Circulation* 2003;**107**:391–7.
20. Barclay AW, Petocz P, McMillan-Price J, Flood VM, Prvan T, et al. Glycemic index, glycemic load, and chronic disease risk – a meta-analysis of observational studies. *Am J Clin Nutr* 2008;**87**:627–37.
21. Streeten BW. Pathology of the lens. In: Albert DM, Jakobiec FA, editors. *Principles and Practice of Ophthalmology*. 2nd ed. Philadelphia: WB Saunders; 2000. pp. 3685–749.
22. Tessier F, Obrenovich M, Monnier VM. Structure and mechanism of formation of human lens fluorophore LM-1. Relationship to vesper-lysine A and the advanced Maillard reaction in aging, diabetes, and cataractogenesis. *J Biol Chem*. 1999;**274**:20796–804.

23. Stitt AW. Advanced glycation: an important pathological event in diabetic and age related ocular disease. *Br J Ophthalmol* 2001;**85**:746–53.

24. Lee AY, Chung SK, Chung SS. Demonstration that polyol accumulation is responsible for diabetic cataract by the use of transgenic mice expressing the aldose reductase gene in the lens. *Proc Natl Acad Sci USA* 1995;**92**:2780–4.

25. Augustin LS, Franceschi S, Jenkins DJ, Kendall CW, La Vecchia C. Glycemic index in chronic disease: a review. *Eur J Clin Nutr* 2002;**56**:1049–71.

26. Schaumberg DA, Ridker PM, Glynn RJ, Christen WG, Dana MR, Henneken CH. High levels of plasma C-reactive protein and future risk of age-related cataract. *Ann Epidemiol* 1999;**9**:166–71.

27. Gritz DC, Srinivasan M, Smith SD, Kim U, Lietman TM, Wilkins JH, et al. The Antioxidants in Prevention of Cataracts Study: effects of antioxidant supplements on cataract progression in South India. *Br J Ophthalmol* 2006;**90**:847–51.

28. Klein BE, Knudtson MD, Lee KE, Reinke JO, Danforth LG, Wealti AM, et al. Supplements and age-related eye conditions: the Beaver Dam Eye Study. *Ophthalmology* 2008;**115**:1203–8.

29. Lou MF. Redox regulation in the lens. *Prog Retin Eye Res.* 2003;**22**:657–82.

30. Miller JC. Importance of glycemic index in diabetes. *Am J Clin Nutr* 1994;**59**(Suppl):747S–52S.

31. Weintraub JM, Willett WC, Rosner B, Colditz GA, Seddon JM, Hankinson SE. A prospective study of the relationship between body mass index and cataract extraction among US women and men. *Int J Obes Relat Metab Disord* 2002;**26**:1588–95.

32. Leske MC, Wu SY, Nemesure B, Hennis A. Risk factors for incident nuclear opacities. *Ophthalmology* 2002;**109**:1303–8.

33. Kato S, Shiokawa A, Fukushima H, Numaga J, Kitano S, Hori S, et al. Glycemic control and lens transparency in patients with type 1 diabetes mellitus. *Am J Ophthalmol* 2001;**131**:301–4.

34. Liu S, Manson JE, Buring JE, Stampfer MJ, Willett WC, Ridker PM. Relation between a diet with a high glycemic load and plasma concentrations of high-sensitivity C-reactive protein in middle-aged women. *Am J Clin Nutr* 2002;**75**:492–8.

35. Klein BE, Klein R, Lee KE, Knudtson MD, Tsai MY. Markers of inflammation, vascular endothelial dysfunction, and age-related cataract. *Am J Ophthalmol* 2006;**141**:116–22.

36. Swamy-Mruthinti S, Shaw SM, Zhao HR, Green K, Abraham EC. Evidence of a glycemic threshold for the development of cataracts in diabetic rats. *Curr Eye Res* 1999;**18**:423–9.

37. Sawada T, Nakamura J, Nishida Y, Kani K, Okamoto R, Morikawa S, et al. Imaging of (13)C-labeled glucose and sorbitol in bovine lens by (1)H-detected (13)C nuclear magnetic resonance spectroscopy. *Magn Reson Imaging* 2003;**21**:1029–31.

38. Donaldson P, Kistler J, Mathias RT. Molecular solutions to mammalian lens transparency. *News Physiol Sci* 2001;**16**:118–23.

39. Taylor A. Nutritional and environmental influences on risk for cataract. In: Taylor A, editor. *Nutritional and Environmental Influences on the Eye.* Boca Raton, FL: CRC Press; 1999. pp. 53–93.

40. Mota MC, Carvalho P, Ramalho JS, Cardoso E, Gaspar AM, et al. Protein glycation and *in vivo* distribution of human lens fluorescence. *Int Ophthalmol* 1994–1995;**18**:187–93.

41. Klein BE, Klein R, Lee KE. Diabetes, cardiovascular disease, selected cardiovascular disease risk factors, and the 5-year incidence of age-related cataract and progression of lens opacities: the Beaver Dam Eye Study. *Am J Ophthalmol* 1998;**126**:782–90.

42. Saxena S, Mitchell P, Rochtchina E. Five-year incidence of cataract in older persons with diabetes and pre-diabetes. *Ophthal Epidemiol* 2004;**11**:271–7.

43. Leske MC, Wu SY, Hennis A, Connell AM, Hyman L, Schachat A. Diabetes, hypertension, and central obesity as cataract risk factors in a black population. The Barbados Eye Study. *Ophthalmology* 1999;**106**:35–41.

Fruit and Vegetable Intake and Age-Related Cataract

Zhi-Quan Lu, Jia Yan

Liaoning Medical University, Liaoning Province, P.R. China

INTRODUCTION

Age-related cataract (ARC) is one chronic disease that may be influenced by diet. It represents a considerable public health burden as the leading cause of visual disability in many developed and developing countries.[1] Basic science and animal research studies support a role for oxidative mechanisms in the etiology of cataract.[2–6] Thus, observational epidemiologic studies and randomized controlled trials (RCT) of dietary factors in cataract have focused on the amount of antioxidant nutrients contained in the diet.[7–11] An alternative approach is to examine the development of cataract in relation to specific foods or food groups, such as total fruit and vegetable intake.[12] This approach enables an assessment of the combined effects of antioxidant nutrients together with the effects of other components of the diet, such as other micronutrients, photochemicals, and fiber. However, to our knowledge, only a few studies have examined ARC in relation to total fruit and vegetable intake, which generally showed that a higher intake of fruit or vegetables was associated with a reduced risk of cataract. In the meantime, the possible beneficial effects of fruit and vegetables on the risk of many chronic diseases, including cataract, have a strong biologic basis and warrant the continued recommendation to increase total intakes of fruit and vegetables.

FRUIT AND VEGETABLES AND NUTRIENTS

Fruit and vegetables are important components of a healthy diet and essential parts of our daily routine intake, which gives us the most vital supplies without which our body system would not function properly. Accumulating evidence suggests that they could help prevent major diseases such as cardiovascular diseases[13,14] and certain cancers principally of the digestive system.[13] There are several mechanisms by which these protective effects may be mediated involving antioxidants and other micronutrients, such as flavonoids, carotenoids, vitamin C, and folic acid, as well as dietary fiber. These and other substances block or suppress the action of carcinogens and, as antioxidants, prevent oxidative deoxyribonucleic acid (DNA) damage.

Fruit and Nutrients

A fruit is a ripened ovary and an edible part of a plant. Some nonsweet fruits are used for cooking. These are called culinary fruits. There are basically three types of fruits: simple fruits, aggregate fruits, and multiple fruits. Some seedless varieties of fruits are also available. Many fruits such as apples can be eaten either as fresh fruits or preserved varieties as jams and juices. Fruits are rich in water, fiber, and vitamins. The nutritional value of fruits will vary significantly depending on the variety, size, and edible methods. The contents of vitamin C, vitamin E, lutein, and zeaxanthin of common fruits are shown in Figure 28.1, Figure 28.2, and Figure 28.3.

Vegetables and Nutrients

A vegetable is an edible part of a plant that is rich in nutrients. Vegetables are used in the diet in a variety of ways, as a part of meal or as a snack item. All vegetables are packed with protein, fat, vitamins, minerals, carbohydrates, and fiber. Some vegetables have antioxidant, antifungal, antibacterial, and anticarcinogenic properties. The words fruits and vegetables are mutually used sometimes. Some fruits are

Handbook of Nutrition, Diet, and the Eye
http://dx.doi.org/10.1016/B978-0-12-401717-7.00028-9

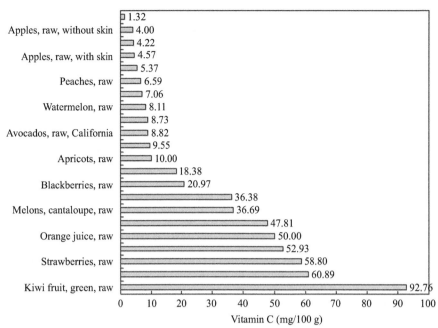

FIGURE 28.1 The contents of vitamin C in milligrams of 100 g of common fruits (unpublished data from United States Department of Agriculture (USDA) National Nutrient Database for Standard Reference, Release 24).

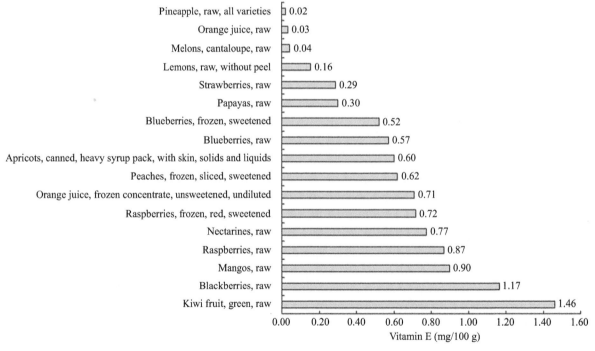

FIGURE 28.2 The contents of vitamin E in milligrams of 100 g of common fruits (unpublished data from United States Department of Agriculture (USDA) National Nutrient Database for Standard Reference, Release 24).

even classified as vegetables. The nutritional value of vegetables will differ substantially depending on the variety, size, and cooking methods. The contents of vitamin C, vitamin E, lutein, and zeaxanthin of common vegetables are shown in Figure 28.4, Figure 28.5, and Figure 28.6.

NUTRITION AND THE PREVENTION OF CATARACT

Many basic science and animal studies have suggested oxidative stress is a major cause of cataract development and have demonstrated a protective effect

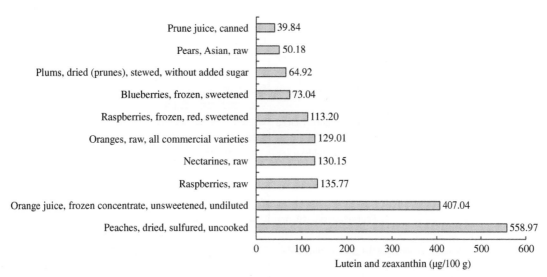

FIGURE 28.3 The contents of lutein and zeaxanthin in micrograms of 100 g of common fruits (unpublished data from United States Department of Agriculture (USDA) National Nutrient Database for Standard Reference, Release 24).

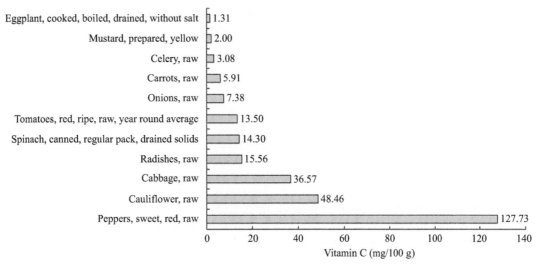

FIGURE 28.4 The contents of vitamin C in milligrams of 100 g of common vegetables (unpublished data from United States Department of Agriculture (USDA) National Nutrient Database for Standard Reference, Release 24).

of antioxidants on lens tissue.[2–6] The lens is subject to oxidative stress from molecular oxygen and free radicals generated by sunlight, cigarette smoke, and aging. Superoxide radicals, peroxide, hydroxyl radicals, and single oxygen are major reactive oxygen species that may be produced in the lens.[15] The lens also contains a high concentration of glutathione compared with other tissues, which is considered the first line of defense against reactive oxygen species. The concentration of glutathione in the lens of patients with severe lens opacity decreased dramatically.[3]

Small molecules that aid in the defense against reactive oxygen species include vitamin C, vitamin E, and the carotenoids. Vitamin C is an important antioxidant that may protect the lens against oxidative stress, which can be actively transported from the blood into the aqueous humor. *In vitro* studies[16–18] have shown that vitamin C

mediates glycation of lens proteins and generation of superoxide anions.[19]

One observational study correlated 10 years or more of vitamin C supplementation with a 28% lower incidence of cataract extraction in women.[20] This agreed with the 2005 report by Ferrigno *et al.*, of the Italian–American Clinical Trial of Nutrition Supplements, that higher plasma levels of vitamin C were associated with a lower prevalence of nuclear and posterior subcapsular cataracts.[21] The National Health and Nutrition Examination Survey II[22] also correlated a 1 mg/dL increase in plasma ascorbate with a 26% decreased risk of cataracts in 62–70-year-old Americans. However, there is not strong enough evidence to warrant vitamin C supplement use, particularly in higher dose and for longer duration in the general population.[23]

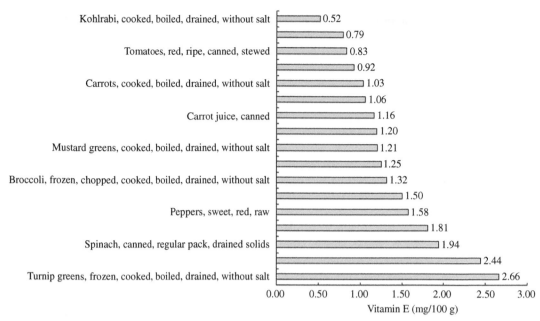

FIGURE 28.5 The contents of vitamin E in milligrams of 100 g of common vegetables (unpublished data from United States Department of Agriculture (USDA) National Nutrient Database for Standard Reference, Release 24).

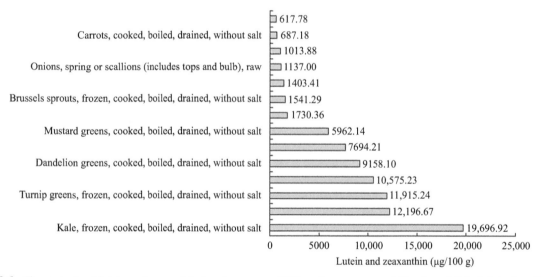

FIGURE 28.6 The contents of lutein and zeaxanthin in micrograms of 100 g of common vegetables (unpublished data from United States Department of Agriculture (USDA) National Nutrient Database for Standard Reference, Release 24).

Other nutrients of interest are the xanthophylls lutein and zeaxanthin. Lutein is believed to act primarily as a filter of blue light[24] but is also a free-radical scavenger because of the presence of an easily oxidizable hydroxyl group.[25] Lutein and zeaxanthin are also the only carotenoids that occur naturally in the lens and macula.[26] Dietary lutein is present in dark, leafy vegetables such as spinach and kale as well as yellow-colored foods such as corn and egg yolks.[27]

EPIDEMIOLOGIC STUDIES CORRELATING FRUIT AND VEGETABLE INTAKE WITH THE RISK OF CATARACT

To date, epidemiology studies that have examined the association between the categories of all fruit and vegetable intake and the risk of cataract are limited and the results were inconclusive. Table 28.1 shows these studies. A small, clinic-based, case-control study with

TABLE 28.1 Epidemiologic Studies That Have Examined the Relation between Fruit and Vegetable Intake and Cataract

Study	Design	Author	Sample Size	Outcome	Comparison	OR/RR (95% CI)
	Case-control	Jacques et al. (1991)[28]	77 cases 35 controls	Any cataract	≥ 3.5 servings vs. < 3.5 of total fruit and vegetable intake	5.7 (P < 0.05)
	Case-control	Lu et al. (2012)[29]	360 cases 360 controls	Cataract extraction	Highest vs. lowest quartile	Total fruit: 0.81 (0.67 to 0.97); Total vegetable: 0.81 (0.69 to 0.94)
NHS	Cross-sectional	Moeller et al. (2004)[30]	479 (female)	Nuclear cataract	Highest vs. lowest quartile	Fruit: 0.58 (0.32 to 1.05)
BDES	Cross-sectional	Mares-Perlman et al. (1995)[31]	1919	Nuclear cataract	Highest vs. lowest quintile of total fruit and vegetable intake	0.68 (0.43 to 1.09)
BDES	Prospective	Lyle et al. (1999)[7]	1919	Incident nuclear cataract	Highest vs. lowest quintile	Total fruit: 1.8 (1.0 to 3.3); Total vegetable: 0.7 (0.4 to 1.2)
WHS	Prospective	Christen et al. (2005)[12]	35,724 (female)	Incident cataract	Highest vs. lowest quintile of total fruit and vegetable intake	Total fruit and vegetable: 0.83 (0.70 to 0.99)

BDES: Beaver Dam Eye Study; CI: confidence interval; NHS: Nurse's Health Study; OR: odds ratio; RR: risk ratio; WHS: Women's Health Study.

77 cataract patients and 35 control subjects showed that those subjects with a mean daily intake of fruit and vegetables above the 20th percentile had nearly a half decreased risk of any type of cataract, compared with those below this frequency, after adjustment for multiple factors.[28]

Three hundred sixty cases ages 45–85 years for cataract extraction and 360 controls participated in another case-control study, which was conducted in a middle-aged and elderly population in northeast China. Fruit and vegetable intake was assessed with the use of a validated, semi-quantitative food-frequency questionnaire (FFQ). After adjusting for multiple potential confounders, intake of fruit, vegetables, and total fruit and vegetables were inversely associated with ARC. Compared with controls in the lowest quartile, the risk of ARC for cases in the highest quartile of intake decreased for 19% (odds ratio (OR) 0.81, 95% confidence interval (CI) 0.67 to 0.97), 19% (OR 0.81, 95% CI 0.69 to 0.94), and 29% (OR 0.71, 95% CI 0.60 to 0.93), respectively. The risk of ARC decreased with higher intake of fruit (P = 0.014), vegetables (P < 0.001), and total fruit and vegetables (P = 0.023).[29]

Cross-sectional data for 479 Nurse's Health Study (NHS) participants ages 52–73 years indicated a moderately reduced risk of prevalent nuclear opacity for those reporting a high intake of fruit but not vegetables. The OR of nuclear opacity for participants in the highest quartile was 0.58 (95% CI 0.32 to 1.05; P = 0.31) compared with those in the lowest quartile

of total fruit intake.[30] The Beaver Dam Eye Study (BDES), another cross-sectional study with 1919 adults ages 43–84 years, showed that a higher intake of fruit and vegetables was associated with a reduced risk of prevalent nuclear sclerosis in men rather than women. Compared with men in the lowest quintile, the risk of nuclear cataract for those in the highest quintile decreased by 32% for total intake of fruit and vegetables (OR 0.68, 95% CI 0.43 to 1.09).[31] After 5-year follow-up in the same population of BDES, 246 incident nuclear cataracts were identified, and no significant association between higher intake of vegetables and cataract was found (OR 0.7, 95% CI 0.4 to 1.2; P = 0.39, for the comparison of high and low quintiles). Meanwhile, a possible elevated risk of cataract for those with high intake of fruit (OR 1.8, 95% CI 1.0 to 3.3; P = 0.07) was indicated.[7] Separate results for men and women were not provided in that analysis.

In the Women's Health Study (WHS), only a small overall benefit was suggested, which was the largest to report on the category of all fruit and vegetable intake.[12] A total of 2067 cataracts and 1315 cataract extractions were confirmed after a mean of 10 years of follow-up of this prospective study. Compared with women in the lowest quintile of fruit and vegetable intake, women with higher intakes had a modest 10–15% reduced risk of cataract (P < 0.05). No significant negative relationship between fruit and vegetable intake and cataract extraction was observed (P = 0.12).

ASSESSMENT OF FRUIT AND VEGETABLE INTAKE IN EPIDEMIOLOGIC STUDIES

Almost all observational epidemiologic studies assessed the fruit and vegetable intake via FFQ. Participants were requested to complete an FFQ that included fruit items and vegetable items. For each food item, a standard unit or portion size was specified and participants were asked how often, on average, during the previous year they had consumed that amount. Nine responses were possible, ranging from 'almost never' to '≥ 6 times/day'. The amounts of fruit (grams/day) and vegetables (grams/day) consumed were calculated from the responses, and the mean daily intakes of individual fruit and vegetables were summed to compute total fruit and vegetable intake.[32] The high validity of FFQ for assessment of fruit and vegetable consumption was also shown in certain studies.[13,33]

TAKE-HOME MESSAGES

- Fruit and vegetables are important components of a healthy diet and essential parts of our daily routine intake, which gives us the most vital supplies without which our body system would not function properly.
- Fruits are rich in water, fiber, and vitamins. The nutritional value of fruits will vary significantly depending on the variety, size, and edible methods.
- All vegetables are packed with protein, fat, vitamins, minerals, carbohydrates, and fiber. Some vegetables have antioxidant, antifungal, antibacterial, and anticarcinogenic properties.
- Many basic science and animal studies have suggested oxidative stress is a major cause of cataract development and have demonstrated a protective effect of antioxidants on lens tissue.
- Fruit and vegetables may play a protective role in the prevention and mediation of ARC.

References

1. Congdon NG, Friedman DS, Lietman T. Important causes of visual impairment in the world today. *JAMA* 2003;**290**:2057–60.
2. Lou MF. Redox regulation in the lens. *Prog Retin Eye Res* 2003;**22**:657–82.
3. Truscott RJ. Age-related nuclear cataract: oxidation is the key. *Exp Eye Res* 2005;**80**:709–25.
4. Lin D, Barnett M, Grauer L, Robben J, Jewell A, Takemoto L, et al. Expression of superoxide dismutase in whole lens prevents cataract formation. *Mol Vis* 2005;**11**:853–8.
5. Bayer A, Evereklioglu C, Demirkaya E, Altun S, Karslioglu Y, Sobaci G. Doxorubicin-induced cataract formation in rats and the inhibitory effects of hazelnut, a natural antioxidant: a histopathological study. *Med Sci Monit* 2005;**11**:BR300–4.
6. Kumar PA, Suryanarayana P, Reddy PY, Reddy GB. Modulation of alphacrystallin chaperone activity in diabetic rat lens by curcumin. *Mol Vis* 2005;**11**:561–8.
7. Lyle BJ, Mares-Perlman JA, Klein BE, Klein R, Greger JL. Antioxidant intake and risk of incident age-related nuclear cataracts in the Beaver Dam Eye Study. *Am J Epidemiol* 1999;**149**:801–9.
8. Tan AG, Mitchell P, Flood VM, Burlutsky G, Rochtchina E, Cumming RG, et al. Antioxidant nutrient intake and the long-term incidence of age-related cataract: the Blue Mountains Eye Study. *Am J Clin Nutr* 2008;**87**:1899–905.
9. Chasan-Taber L, Willett WC, Seddon JM, Stampfer MJ, Rosner B, Colditz GA, et al. A prospective study of vitamin supplement intake and cataract extraction among U.S. women. *Epidemiology* 1999;**10**:679–84.
10. Brown L, Rimm EB, Seddon JM, Giovannucci EL, Chasan-Taber L, Spiegelman D, et al. A prospective study of carotenoid intake and risk of cataract extraction in US men. *Am J Clin Nutr* 1999;**70**:517–24.
11. Age-Related Eye Disease Study Research Group. A randomized, placebo-controlled, clinical trial of high-dose supplementation with vitamins c and e and beta carotene for age-related cataract and vision loss: AREDS report No. 9. *Arch Ophthalmol* 2001;**119**:1439–52.
12. Christen WG, Liu S, Schaumberg DA, Buring JE. Fruit and vegetable intake and the risk of cataract in women. *Am J Clin Nutr* 2005;**81**:1417–22.
13. Takachi R, Inoue M, Ishihara J, Kurahashi N, Iwasaki M, Sasazuki S, et al. Fruit and vegetable intake and risk of total cancer and cardiovascular disease: Japan Public Health Center-based Prospective Study. *Am J Epidemiol* 2008;**167**:59–70.
14. Bazzano LA, He J, Ogden LG, Loria CM, Vupputuri S, Myers L, et al. Fruit and vegetable intake and risk of cardiovascular disease in US adults: the first National Health and Nutrition Examination Survey Epidemiologic Follow-up Study. *Am J Clin Nutr* 2002;**76**:93–9.
15. Spector A. Oxidative stress-induced cataract: mechanism of action. *FASEB J* 1995;**9**:1173–82.
16. Linetsky M, Shipova E, Cheng R, Ortwerth BJ. Glycation by ascorbic acid oxidation products leads to the aggregation of lens proteins. *Biochim Biophys Acta* 2008;**1782**:22–34.
17. Cheng R, Lin B, Lee KW, Ortwerth BJ. Similarity of the yellow chromophores isolated from human cataracts with those from ascorbic acid-modified calf lens proteins: evidence for ascorbic acid glycation during cataract formation. *Biochim Biophys Acta* 2001;**1537**:14–26.
18. Cheng R, Feng Q, Ortwerth BJ. LC-MS display of the total modified amino acids in cataract lens proteins and in lens proteins glycated by ascorbic acid *in vitro*. *Biochim Biophys Acta* 2006;**1762**:533–43.
19. Linetsky M, James HL, Ortwerth BJ. Spontaneous generation of superoxide anion by human lens proteins and by calf lens proteins ascorbylated *in vitro*. *Exp Eye Res* 1999;**69**:239–48.
20. Chasan-Taber L, Willett WC, Seddon JM, Stampfer MJ, Rosner B, Colditz GA, et al. A prospective study of carotenoid and vitamin A intakes and risk of cataract extraction in US women. *Am J Clin Nutr* 1999;**70**:509–16.
21. Ferrigno L, Aldigeri R, Rosmini F, Sperduto RD, Maraini G. Italian-American Cataract Study Group. Associations between plasma levels of vitamins and cataract in the Italian–American Clinical Trial of Nutritional Supplements and Age-Related Cataract (CTNS): CTNS Report #2. *Ophthalmic Epidemiol* 2005;**12**:71–80.
22. Simon JA, Hudes ES. Serum ascorbic acid and other correlates of self-reported cataracts among older Americans. *J Clin Epidemiol* 1999;**52**:1207–11.
23. Rautiainen S, Lindblad BE, Morgenstern R, Wolk A. Vitamin C supplements and the risk of age-related cataract: a population-based prospective cohort study in women. *Am J Clin Nutr* 2010;**91**:487–93.
24. Alves-Rodrigues A, Shao A. The science behind lutein. *Toxicol Lett* 2004;**150**:57–83.

25. Trevithick-Sutton CC, Foote CS, Collins M, Trevithick JR. The retinal carotenoids zeaxanthin and lutein scavenge superoxide and hydroxyl radicals: a chemiluminescence and ESR study. *Mol Vis* 2006;**12**:1127–35.

26. Yeum KJ, Taylor A, Tang G, Russell RM. Measurement of carotenoids, retinoids, and tocopherols in human lenses. *Invest Ophthalmol Vis Sci* 1995;**36**:2756–61.

27. Sommerburg O, Keunen JE, Bird AC, van Kuijk FJ. Fruits and vegetables that are sources for lutein and zeaxanthin: the macular pigment in human eyes. *Br J Ophthalmol* 1998;**82**: 907–10.

28. Jacques PF, Chylack Jr LT. Epidemiologic evidence of a role for the antioxidant vitamins and carotenoids in cataract prevention. *Am J Clin Nutr* 1991;**53**(Suppl):352S–5S.

29. Lu ZQ, Yan J, Jiang TX, Zhang D, Sun WH, Zhai SN, et al. Influence of fruit and vegetable intake on age-related cataract. *Int Eye Sci* 2012;**12**:58–61.

30. Moeller SM, Taylor A, Tucker KL, McCullough ML, Chylack Jr LT, Hankinson SE, et al. Overall adherence to the dietary guidelines for Americans is associated with reduced prevalence of early age-related nuclear lens opacities in women. *J Nutr* 2004;**134**:1812–9.

31. Mares-Perlman JA, Brady WE, Klein BE, Klein R, Haus GJ, Palta M, et al. Diet and nuclear lens opacities. *Am J Epidemiol* 1995;**141**:322–34.

32. Liu S, Manson JE, Lee IM, Cole SR, Hennekens CH, Willett WC, et al. Fruit and vegetable intake and risk of cardiovascular disease: the Women's Health Study. *Am J Clin Nutr* 2000;**72**:922–8.

33. Willett WC. *Nutritional Epidemiology*. 2nd ed. New York, NY: Oxford University Press; 1998.

Retinal Degeneration and Cholesterol Deficiency

Steven J. Fliesler

Veterans Administration Western New York Healthcare System; University at Buffalo/State University of New York; and the SUNY Eye Institute, Buffalo, NY

INTRODUCTION

Most human diseases that involve inborn errors of metabolism are catabolic diseases, that is, they are due to failure to break down certain molecules efficiently; the resulting excessive accumulation of such molecules results in pathology. Yet, there also exists a family of related diseases where cholesterol *deficiency*, rather than cholesterol *excess*, is at the core of the pathology.[1,2] The first discovered such disease was the Smith-Lemli-Opitz syndrome (SLOS) – a recessive genetic disorder caused by mutations in the gene (*DHCR7*) that encodes the enzyme 3β-hydroxysterol-Δ7-reductase (EC1.3.1.21; DHCR7; OMIM# 602758), the penultimate enzyme in mammalian cholesterol biosynthesis – which in turn causes inefficient conversion of 7-dehydrocholesterol (7DHC, the immediate precursor of cholesterol) to cholesterol.[3–5] As might be predicted, the result of such enzymatic defects is that affected individuals exhibit abnormally low levels of cholesterol and excessive levels of 7DHC in all bodily tissues and fluids. In addition, these individuals have multiple dysmorphologies, a range of cognitive and neurologic impairments, including visual dysfunction, failure to thrive, and premature death. While the estimated frequency of SLOS is approximately 1 in 20,000 to 1 in 60,000 live births, the carrier frequency is estimated to be 1 in 30, which predicts a much higher disease incidence (i.e., 1 in 1,590 to 1 in 13,500).[6,7] Hence, SLOS is estimated to be the fourth most prevalent recessive human disease, after cystic fibrosis, phenyketonuria, and hemochromatosis.

The fact that diseases such as SLOS occur in humans highlights the necessity of producing and maintaining specific levels of cholesterol, particularly during critical stages of embryologic development, in order to achieve normal development, structure, and function of mammalian cells, tissues, organs, and, indeed, the entire organism. In addition, such diseases indicate that cholesterol per se – and not any of the myriad of other sterols that occur in nature, including those that are biogenic intermediates in the biosynthesis of cholesterol – is requisite for achieving normal mammalian biology. As shown in Figure 29.1, cholesterol (cholest-5-en-3β-ol) and its immediate precursor, 7DHC (cholesta-5,7-dien-3β-ol), are almost identical in size, chemical structure, and molecular form, differing only by one extra double bond (between carbons 7 and 8 in ring B of the sterol nucleus in 7DHC). At first glance, this difference would appear to be trivial; yet, replacement of cholesterol with 7DHC, as occurs in SLOS, clearly can result in devastating, even lethal, consequences! Why is this so? The answer to this question is not intuitively obvious, but elucidating that answer ultimately may lead not only to a better understanding of the biologic functions of cholesterol in mammals but also to more effective therapeutic interventions into SLOS and allied disorders of cholesterol anabolic metabolism.

The present review will focus in particular on the consequences of defective cholesterol biosynthesis with respect to achieving and maintaining the normal morphologic structure and electrophysiologic function of the retina. Some aspects of this work have been reviewed previously.[8,9]

THE AY9944 RAT MODEL OF SMITH-LEMLI-OPITZ SYNDROME AND RETINAL DEGENERATION

Studies performed largely in the author's laboratory have provided insights into the role of cholesterol in the retina, particularly with regard to the impact of

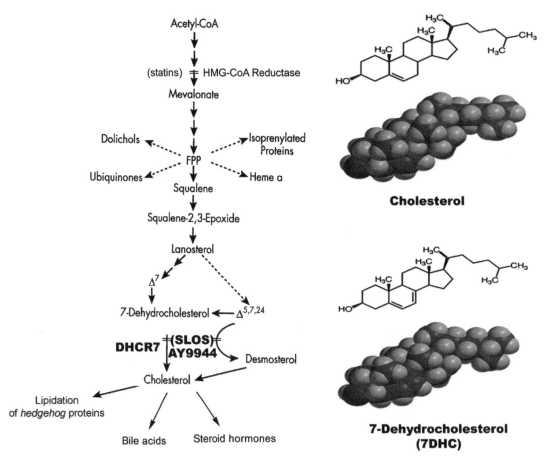

FIGURE 29.1 Simplified schematic of the *de novo* cholesterol biosynthetic pathway. Smith-Lemli-Opitz Syndrome (SLOS) is caused by mutations in the gene that encodes the enzyme DHCR7 (3β-hydroxysterol-Δ7-reductase). The drug AY9944 is a relatively selective inhibitor of DHCR7 and, hence, blocks cholesterol biosynthesis at this step, resulting in the abnormal accumulation of 7-dehydrocholesterol (7DHC) and decreased levels of cholesterol in bodily tissues and fluids. The chemical structures and space-filling representations of cholesterol and 7DHC are shown, illustrating that the only difference between the two is the presence of an additional double bond between carbons 7 and 8 in ring B of the sterol nucleus in 7DHC. Abbreviations: FPP, farnesylpyrophosphate; Δ7, lathosterol; Δ5,7,24, 7-dehydrodesmosterol. *Reproduced from Fliesler 2010,[8] with permission.*

disrupted cholesterol biosynthesis on the structure and function of the retina. Those studies have employed a pharmacologically induced animal model of SLOS originally developed by Roux and coworkers[10,11] and modified in the author's laboratory[12–14] to afford extended postnatal viability over a duration sufficient to study the retinal degeneration that ensues when cholesterol biosynthesis is interrupted as occurs in SLOS. An abbreviated schematic of the *de novo* cholesterol biosynthesis pathway, indicating the enzymatic step that is genetically defective in SLOS, is shown in Figure 29.1. This same enzymatic step, catalyzed by DHCR7, can be inhibited relatively selectively by the experimental drug, AY9944 (*trans*-1,4-bis(2-chlorobenzaminomethyl)-cyclohexane dihydrochloride).[10,11,15,16] AY9944 was developed initially as a cholesterol-lowering drug for potential use in preventing atherosclerosis and cardiovascular disease (a forerunner of statins) but never approved for human use due to its severe teratogenic effects.[10] It is interesting to note that the original report describing the unusual

constellation of phenotypic birth defects associated with SLOS[3] was published around the same time as the description of the chemistry and biologic properties of AY9944,[10,15,16] yet the connection between the two was not made until nearly three decades later, when the biochemical defect,[17,18] followed by the molecular (gene) defect,[19–21] were elucidated. The schematic illustrates why blocking cholesterol biosynthesis at the level of DHCR7 results in the observed increases in the steady-state levels of 7DHC and diminution in the levels of cholesterol in bodily tissues and fluids (serum, plasma), both in SLOS and in the AY9944 rat model of SLOS.

Using the conditions employed by Roux and coworkers[10, 11] did not afford sufficient postnatal viability to allow for full maturation of retinal structure and function, which takes at least 4 postnatal weeks to achieve in the rat.[22,23] Empirically, conditions were established by Fliesler and coworkers[12] that afforded postnatal survival of Sprague-Dawley rats exposed to AY9944 *in utero* when pregnant dams were fed the drug *ad lib* in their

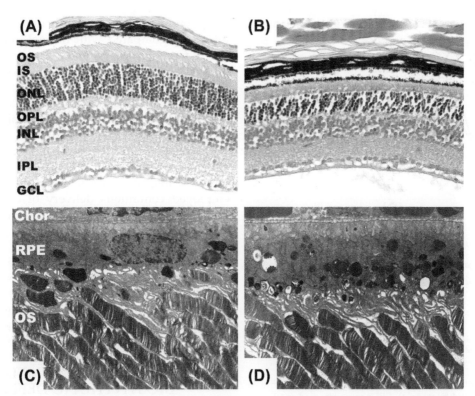

FIGURE 29.2 Morphologic and ultrastructural changes in the retina in the AY9944 rat model of Smith-Lemli-Opitz syndrome. Upper panels: histology of the retina in 10-week-old control (A) and an age-matched, AY9944-treated Sprague-Dawley rat (B). Overall thickness of the retina, particularly the outer nuclear layer (ONL) and the photoreceptor outer segment layer (OS), are reduced in the treated rat, compared with the control. Paraffin embedment; hematoxylin and eosin stain. Lower panels: electron micrographs of retinas from a companion, age-matched control (C) and an AY9944-treated rat (D). The retinal pigment epithelium (RPE) in the treated animal is hypertrophic and engorged with phagosomes and other membranous inclusions, but otherwise the cells have maintained normal polarity. Epon embedment, osmium tetroxide stain, lead citrate counterstain. Chor: choroid; IS: photoreceptor inner segment layer; OPL: outer plexiform layer; INL: inner nuclear layer; IPL: inner plexiform layer; GCL: ganglion cell layer. *Reproduced (in grayscale) from Fliesler and Bretillon (2010),*[9] *with permission. [Originally published in the* Journal of Lipid Research: *Fliesler, S. J. and Bretillon, L., 2010, Vol. 51:pp.3399-3413. © the American Society for Biochemistry and Molecular Biology.]*

chow (at 1 mg/100 g of cholesterol-free chow) during the last 2 gestational weeks, followed by subcutaneous injection of pups postnatally with the drug in an olive oil emulsion every other day (at a dose of 25–30 mg/kg). Despite dramatic changes (compared with age-matched, untreated control rats, where cholesterol is virtually the only detectable sterol) in the sterol profiles of retina, as well as serum, liver, and brain, by postnatal day 27 (P27), with retina 7DHC/cholesterol mole ratios in the range of 4:1, there was no apparent effect on retinal development, histologic or ultrastructural organization, or electrophysiologic (phototransduction) competence.[12]

However, if the treatment period was extended to 10 postnatal weeks, dramatic effects were produced both in retinal structure and function in Sprague-Dawley rats.[13] As shown in Figure 29.2, the thickness of the retina in AY9944-treated rats became dramatically reduced (Fig. 29.2B), compared with age-matched controls (Fig. 29.2A), primarily due to loss of photoreceptor cells, as evidenced by the thinning of the outer nuclear layer (ONL), which is comprised of the nuclei of the rod and cone photoreceptors. Histologic and quantitative morphometric analyses demonstrated that rod outer segments were reduced in length by nearly one-third, compared with age-matched untreated controls. However, electron microscopy (EM) revealed that the remnant rod outer segments in retinas of the AY9944-treated rats (Fig. 29.2D) had ultrastructural organization comparable to that of rod outer segments in retinas from untreated controls (Fig. 29.2C). By contrast, the retinal pigment epithelium (RPE) of treated rats (Fig. 29.2D) was clearly pathologic, being engorged with phagosomes (the ingested tips of rod outer segments that had been shed by the adjacent photoreceptor cells and taken up by the RPE cells via phagocytosis)[23] as well as other membranous and lipid inclusions. Despite this, however, the RPE cells exhibited normal polarization, with well-extended apical microvilli enveloping the adjacent rod outer segments and their mitochondria lined up just underneath the basal infoldings of basal plasmalemma, comparable to control RPE cells (Fig. 29.2C). As shown in Figure 29.3, electroretinographic (ERG) analysis demonstrated that response amplitudes of both the a- and b-wave components of the scotopic (dark-adapted,

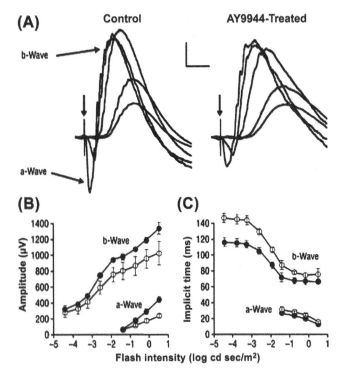

FIGURE 29.3 Scotopic (dark-adapted) electroretinograms (ERGs) from control and AY9944-treated rats. ERGs were recorded at 10 postnatal weeks, and wave amplitudes and implicit times were measured as a function of strobe flash intensity. (A) Representative ERGs from a control (left) and an AY9944-treated rat (right). ERG a- and b-waves are indicated; vertical arrow denotes stimulus onset. Calibration mark: 200 μV (Y-axis) and 50 ms (X-axis). (B) Flash intensity–response functions for a- and b-wave amplitudes and (C) implicit time values obtained from control (filled circles; n = 4) and AY9944-treated (open circles; n = 5) rats. Each point represents the mean value; error bars represent standard deviation. cd: candela. *Reproduced, in modified form, from Fliesler et al. (2004),[13] with permission.*

rod-driven) ERGs were substantially reduced in 10-week old, AY9944-treated Sprague-Dawley rats compared with age-matched controls (compare Fig. 29.3A, right- vs. left-hand tracings, respectively, and Fig. 29.3B). In addition, the 'implicit times' (response time from initial flash stimulus to maximum response amplitude) were greater from eyes of treated animals compared with controls (Fig. 29.3C). Hence, the responsiveness of rod photoreceptors in the SLOS rat model to photic stimulation was dramatically less robust and more sluggish than normal. The response amplitude deficits can be explained, in part, by the fact that the rod outer segments were reduced in length by 30–33% in the treated rats, compared with age-matched controls (see discussion above, and Fig. 29.2). Subsequent 'leading edge' analysis[24,25] of the scotopic ERG a-wave further revealed that the sensitivity (S-value, or gain) and maximal response parameter (R_{mP3}) were reduced by two-fold in treated animals, compared with age-matched controls. These results, in total, suggest that the efficiency of phototransduction in

rod photoreceptors is markedly diminished in the SLOS rat model compared with normal, untreated rats and that synaptic transmission is also negatively impacted by AY9944 treatment. Parallel analysis of photopic (cone-driven) electrophysiology revealed similar trends,[12] indicating that cone function also was significantly affected by disruption of cholesterol biosynthesis in this rat model of SLOS.

Feeding the drug to the pregnant dams potentially allows for too much variability in drug dosage during gestation, so subsequent studies (discussed below) employed continuous delivery of drug (dissolved in phosphate-buffered saline (PBS)) to pregnant rats via a subcutaneous Alzet® osmotic pump, placed under the dorsal skin.[14] In addition, the vehicle for systemic administration of AY9944 in neonatal and postnatal rats was switched from olive oil emulsion (given subcutaneously under the dorsal skin every other day) to a PBS solution of the drug administered by intraperitoneal injection (30 mg/kg).[14]

BEYOND THE IMMEDIATE BIOSYNTHETIC DEFECT IN SMITH-LEMLI-OPITZ SYNDROME

It should be noted that, with the extended AY9944 treatment protocol described above, the 7DHC/cholesterol mole ratio in retina increased from approximately 4:1 at 4 weeks to more than 5:1 by 10 weeks, indicating that 7DHC substituted for more than 80% of the cholesterol molecules normally present in the retina in AY9944-treated animals. Conceivably, this could significantly impact the formation of 'lipid rafts' – highly ordered membrane microdomains enriched in cholesterol and sphingolipids, which serve as transient platforms for the organization of signal transduction complexes and which support the activity of various ion channels, membrane receptors, and enzymes.[26,27] Such alterations could have dramatic, global effects on both cellular and systemic physiology. However, studies from at least two different laboratories have demonstrated that 7DHC can support lipid raft formation as well as, if not better than, cholesterol.[28,29] Furthermore, model membrane studies, using Langmuir monolayer films have demonstrated that the packing density and organization of 7DHC in membranes is almost identical to that of cholesterol.[30–32] Hence, the observed effects of AY9944 treatment on the retina in the rat model of SLOS cannot be ascribed to defects in membrane structure induced by replacing cholesterol with 7DHC, per se.

Further investigations of the lipid composition of whole neural retinas[33] and purified rod outer segment membranes[34] derived from AY9944-treated and age-matched control rats have demonstrated that the defect in cholesterol biosynthesis is the *initial*, but not the only, aspect of lipid metabolism affected in the SLOS rat model. Compared with controls, retinas in the SLOS rat model are markedly deficient in docosahexaenoic acid (DHA, or 22:6n3),[33] the dominant fatty acid in mammalian retinas and rod outer segment membranes.[35] Detailed lipidomic analysis demonstrated that the levels of DHA-containing molecular species of the major glycerophospholipid classes of the rat retina (phosphatidylcholine (PC), phosphatidylethanolamine (PE), and phosphatidylserine (PS)) were all dramatically lower in treated versus control rats.[33] PC molecular species containing palmitate plus DHA (16:0-22:6n3) and stearate plus DHA (18:0-22:6n3) were reduced in amount by more than 50% and more than 33%, respectively, in rats treated for 2 or 3 months with AY9944. Reductions in the levels of di-22:6n3 PE (> 60 mol%) and 18:0-22:6n3 PE (> 15 mol%) molecular species, as well as di-22:6n3 PS (> 80 mol%), also were observed in retinas from treated rats, compared with age-matched controls. Notably, these marked alterations in the steady-state glycerophospholipid profile of the retina occurred in the absence of n3 fatty acid deficiency in plasma or liver. This suggests that these global changes in the retinal 'lipidome' are due to local processes in the retina per se, such as altered uptake of DHA or DHA-containing glycerophospholipid molecular species from the blood, or possibly metabolic cross-talk between *de novo* cholesterol biosynthesis and glycerophospholipid and/or fatty acid metabolism in the retina.[8,9]

Given these findings, it was not surprising that the DHA content of retinal rod outer segment membranes also was dramatically reduced (by > 40%), relative to control retinas, by 3 postnatal months of AY9944 treatment, resulting in marked reduction in membrane fluidity (measured spectroscopically, using two different fluorescent probes).[34] These changes were progressive, as the DHA content of rod outer segment membranes was not altered significantly by 1 postnatal month of AY9944 treatment (although there were minor alterations in other fatty acid species), compared with age-matched controls, nor was membrane fluidity affected (consistent with the lack of histologic, ultrastructural, or electrophysiologic changes observed by 1 month of AY9944, as described above). Normalized to total lipid phosphorus content, the values obtained for total steady-state sterol content of rod outer segment membranes and whole retinas were not altered as a function of AY9944 treatment, nor were the protein/phospholipid or opsin/phospholipid ratios; only the 7DHC/cholesterol mole ratios were changed. This demonstrates that cholesterol was replaced, mole for mole, by 7DHC in the retina in the SLOS rat model and that the reduction in rod outer segment membrane fluidity was due predominantly to the marked decrease in DHA content.

Indeed, the retinal lipidome is not the only aspect of tissue composition or metabolism that is globally altered in the SLOS rat model. Affymetrix gene chip (microarray) analysis has shown that the retinal transcriptome also is dramatically and progressively changed by AY9944 treatment.[36,37,39] By 1 postnatal month (i.e., where there is no overt evidence of retinal degeneration or electrophysiologic deficits), only a relatively small cohort (24) of genes are differentially expressed, compared with normal controls, whereas by 2 months of treatment (where marked pyknosis in the ONL and some thinning of the retina and diminution of rod outer segment length are observed) nearly 1200 genes exhibit differential expression. The affected genes are implicated in a myriad of cellular processes, including (but not limited to) lipid metabolism, oxidative stress, cell death/survival, membrane trafficking, ion transport, and energy metabolism – consistent with the kinds of changes one would expect to be associated with the observed retinal degeneration. Some of these changes parallel the kinds of transcriptome alterations observed in brains from the embryonic *Dhcr7*-knockout mouse model of SLOS.[38] Consistent with such alterations in gene expression, it also has been shown, using an ion current-based, nano-LC/MS, gel-free method, that there are marked changes in the retinal proteome in the SLOS rat model compared with the normal retinal proteome ([39] and manuscript in preparation). The expression levels of 41 proteins were found to be elevated, while 61 proteins were reduced, in retinas of AY9944-treated rats, relative to controls, by 2 postnatal months. These included proteins involved in multiple biologic processes, such as visual perception, vesicular transport, lipid metabolism, oxidative stress response, and cell death/survival, again consistent with the observed retinal degeneration in the SLOS rat model. Using more conventional, two-dimensional (2D) isoelectric focusing polyacrylamide gel electrophoresis (IEF-PAGE) methods, proteomic alterations had been observed previously not only in the retina,[40] but in the brain[28] of the SLOS rat model, compared with untreated, age-matched control rats. More recently, using similar methods, multiple proteomic alterations also have been reported for embryonic day 18.5 mouse brain from the homozygous *Dhcr7*-knockout mouse, as well as the lathosterol-Δ5-desaturase (SC5D) knockout mouse.[41]

In sum, the above findings support the conclusion that while inhibition of cholesterol biosynthesis at the level of DHCR7 is the *initial* defect in SLOS, as well as in the AY9944 SLOS rat model, the mechanism of the disease goes well beyond cholesterol metabolism, affecting multiple other metabolic pathways, with resultant changes in the transcriptome, lipidome, and proteome of multiple tissues.

THERAPEUTIC INTERVENTION WITH DIETARY CHOLESTEROL SUPPLEMENTATION

Since the discovery of the central metabolic defect in SLOS at the level of DHCR7, therapeutic approaches to the clinical management of SLOS patients have centered upon cholesterol supplementation therapy (reviewed in[42,43]). While this approach has been relatively successful in promoting increased steady-state levels of cholesterol and decreased levels of 7DHC in serum of affected individuals, measurable improvements in behavioral and cognitive deficits, in addition to improvements in general quality of life and longevity, have been minimal and quite variable. Dietary cholesterol supplementation therapy was tested using the AY9944 SLOS rat model, monitoring the effects on the associated retinal degeneration and visual function deficits.[14] Feeding SLOS model rats a high-cholesterol (2%, by weight) diet, versus a cholesterol-free chow, starting at weaning (P27) and continuing through postnatal week 10, nearly normalized the cholesterol levels in serum, increasing the cholesterol content 32-fold while also diminishing the levels of 7DHC by nearly two-fold. Even more remarkably, the steady-state levels of cholesterol in the retina increased approximately three-fold, while levels of 7DHC were reduced by nearly one-third, with dietary cholesterol supplementation. Hence, unlike the case for the brain, where uptake of cholesterol from the circulation is prevented by the blood-brain barrier,[44] the retina is not only capable of taking up diet-derived, blood-borne cholesterol, but this uptake mechanism actually can dramatically alter the steady-state sterol profile of the retina. Improvements to scotopic (rod-driven) visual function, as assessed by dark-adapted ERGs, were modest, at best; a-wave amplitudes were 36 ± 7% larger in the cholesterol-supplemented group than those from SLOS rats on a cholesterol-free diet, while b-wave amplitudes only showed a modest (but not statistically significant) trend in improvement. Implicit times for rod a- and b-waves were decreased (improved) with cholesterol supplementation, but they still were not normal. However, marked improvements in light-adapted (cone-driven) ERGs were observed in AY9944-treated rats fed the high-cholesterol diet, compared with those fed cholesterol-free chow (Fig. 29.4), with nearly a two-fold increase in photopic ERG amplitudes and substantially reduced (by > 8 ms) implicit times, approaching normal values for these parameters. Despite these improvements in retinal function, there was no sparing of histologic damage to the retina afforded by dietary cholesterol supplementation: quantitative morphometric analysis along the

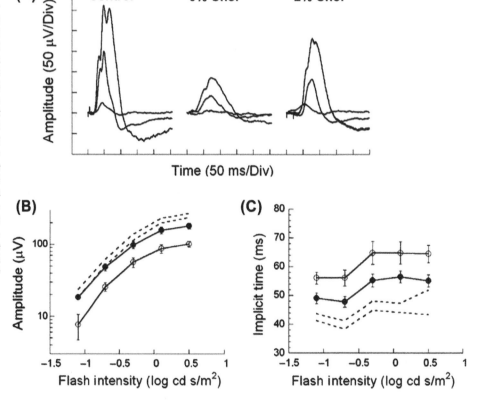

FIGURE 29.4 Dietary cholesterol supplementation improves cone function in the SLOS rat model. Photopic (light-adapted) electroretinograms (ERGs) measured at 10 postnatal weeks. (A) Representative ERG tracings from an untreated control rat (left), compared with rats treated with AY9944 and fed either a cholesterol-free (middle) or a high-cholesterol (2% by weight) diet (right). A total of five responses obtained to strobe flash stimuli from –1.1 to 0.5 log cd s/m² are overlaid per analysis. Intensity–response functions for the amplitude (B) and implicit time (C) of the cone ERG are shown. Data points represent mean values (± standard error of the mean, n = 6). Open circles (○), cholesterol-free diet; filled circles (●), high-cholesterol diet. Dashed lines indicate the 95% confidence interval for response values from control rats. *Reproduced from Fliesler et al. (2007),[14] with permission.*

vertical meridian of eyes from supplemented versus nonsupplemented SLOS model rats demonstrated on average about 21% reduction in ONL thickness, compared with age-matched controls. That said, while SLOS model rats exhibited about 20-fold more pyknotic ONL nuclei than observed in age-matched control rats, there was a significant reduction (by approximately 30%, $P < 0.007$) in the number of pyknotic ONL nuclei in the cholesterol-supplemented group compared with those on the cholesterol-free diet. Hence, dietary cholesterol supplementation reduced photoreceptor cell death somewhat, but this was not sufficient to prevent ONL thickness reduction or other histologic aspects of the AY9944-induced retinal degeneration.

Preliminary studies also have shown that dietary cholesterol supplementation can nearly normalize the transcriptome profile of the retina,[45] as well as substantially reduce the extent of oxidative, post-translational modification of retinal proteins (e.g., 4-hydroxynonenal adducts)[46] in the AY9944 SLOS rat model. However, as indicated above, this approach is insufficient to prevent the retinal degeneration. In addition, initial results from clinical trials employing combined dietary cholesterol supplementation plus simvastatin treatment, despite a fairly compelling rationale, do not appear promising as an improved therapeutic intervention in SLOS (F.D. Porter and R.D. Steiner, personal communications). Given the multiple alterations in tissue transcriptomes, proteomes, and lipidomes observed both in humans with SLOS as well as in animal models of SLOS, the disease is far more complex than 'conventional wisdom' might imply. Hence, cholesterol supplementation appears to be an overly simplistic and poorly efficacious approach to therapeutic intervention in this disease. Clearly, something important has been overlooked in the disease mechanism that could be informative for improving treatment strategies. Essential insights into this problem have come from animal studies utilizing the AY9944 rat model of SLOS.

OXIDATION OF 7-DEHYDROCHOLESTEROL: POTENTIAL KEY TO THE PATHOBIOLOGY AND THE TREATMENT OF SMITH-LEMLI-OPITZ SYNDROME

Vaughan and coworkers have shown that the retina in the AY9944 rat model of SLOS is extraordinarily susceptible to *visible* light-induced damage, compared with normal albino (Sprague-Dawley) rats.[47] In that study, rats were treated with AY9944 for 4 postnatal weeks and then were randomized into two groups: one group (control) was kept in normal, dim cyclic light (20–40 lux, 12-hour dark/12-hour light cycle), while the other (experimental) group was exposed for 24 hours to continuous, intense (1700 lux) green light (490–580 nm, corresponding to the wavelength of light maximally absorbed by the rod visual pigment rhodopsin) and then returned to normal, dim cyclic light. After 2 weeks, rats from both groups were examined by ERG and then sacrificed and tissues harvested for biochemical and morphologic analysis. The control group rats (by then 6 weeks old) exhibited only moderately diminished ERG amplitudes and minimally delayed implicit times, but their retinas were histologically normal. By contrast, rats in the experimental group had massively reduced a- and b-wave amplitudes and significantly delayed implicit times (especially the b-waves) and also exhibited extensive histologic damage to the retina that was far more severe in magnitude and geographic extent than what was produced by the same intense light exposure conditions in otherwise untreated (normal) albino rats. Critically, if the AY9944-treated rats were treated prior to intense light exposure by injection with dimethylthiourea (DMTU; a hydroxyl radical scavenger and antioxidant), there was marked protection against histologic damage as well as diminution of electrophysiologic abnormalities. Quantitative morphometric analysis of ONL thickness of retinas from treated and control groups, with and without DMTU treatment, versus age-matched controls, made this case even more compelling (see Fig. 29.5). As shown in Figure 29.5, whereas the reduction in ONL thickness (signifying photoreceptor degeneration and drop-out) following exposure to intense, constant green light is relatively restricted to the superior central zone in otherwise untreated rats, there was extensive and panretinal loss of photoreceptors in AY9944-treated, light-exposed rats, and this loss was largely prevented by pretreatment with the systemically administered antioxidant, DMTU. In a companion study,[48] using the same light exposure conditions (but harvesting retinas immediately following intense light exposure, as opposed to allowing a 2-week recovery period), it was shown that while normal, untreated albino rats had a low steady-state level of lipid hydroperoxides (LPO) in their retinas (about 1.5 nmol/retina), retinas from the SLOS rat model (not exposed to intense light) had about two-fold greater LPO levels (about 3.1 nmol/retina), comparable to the levels found in normal albino rats that have been subjected to the intense green light exposure paradigm (about 2.7 nmol/retina). However, exposure of AY9944-treated rats to intense, constant green light for 24 hours produced a three-fold increase in the steady-state levels of LPO in the retina (about 10.7 nmol/retina), and this correlated with massive photoreceptor pyknosis, death and drop-out, and even retinal gliosis, even after only 24 hours of intense light exposure.[48]

The above findings suggest that, somehow, the presence of abnormal levels of 7DHC in the retina of the SLOS

FIGURE 29.5 Photoreceptor survival in the Smith-Lemli-Opitz syndrome rat model is markedly reduced by exposure to intense green light and protected against photodamage by systemic antioxidant treatment. Quantitative morphometric analysis of retinal outer nuclear layer (ONL) thickness as a function of AY9944 treatment and intense light exposure (1700 lux, 24 hours, 490–580 nm), with and without systemic dimethylthiourea treatment. C: controls; A: AY9944-treated; U: not exposed to intense light; E: exposed to intense light; D: treated with dimethylthiourea prior to light exposure. Note the significant potentiation of visible light-induced damage by AY9944 administration (AE), compared with both nonlight-exposed groups (AU, CU) and to the control light-exposed group (CE), and the substantial protection from light-induced damage afforded by dimethylthiourea pretreatment (AED). *Reproduced in modified form, from Vaughan* et al. *(2006),[47] with permission.*

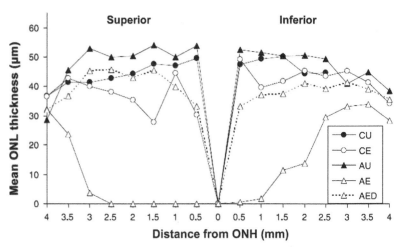

rat model potentiates light-induced damage of photoreceptor cells. This is curious, since 7DHC absorbs light in the ultraviolet (UV), not the visible, range; so it is highly unlikely that 7DHC is acting as a photosensitizing agent. The only known molecule that absorbs in the green wavelength range in the albino rat retina is the visual pigment rhodopsin. The nature of the lipid hydroperoxides formed in the retina cannot be ascertained from the experiments described above. However, the fact that the retina in the SLOS rat model contains steady-state levels of LPO comparable to those found in retinas from photodamaged albino rats, and that pretreatment of rats with a systemic antioxidant/free radical scavenger substantially protects the retina from photodamage, suggests that the formation of oxidized lipids is toxic to the retina (above a certain threshold level) and that perhaps 7DHC may be a substrate for such peroxidation. Indeed, 7DHC has been shown to be the most highly oxidizable lipid known[49] – nearly 200 times more readily oxidizable than cholesterol, and about seven times more susceptible to oxidation than DHA (which has six double bonds, compared with only two found in 7DHC). Perhaps photoactivated rhodopsin is able to transfer energy to 7DHC, thereby promoting its oxidation? This remains to be determined.

So, what (if any) is the link between oxidation of 7DHC and the pathobiology of SLOS or of the retinal degeneration observed in the AY9944 rat model of SLOS? For one, oxidation of 7DHC, either via photo-oxidation or by free radical mechanism, yields more than a dozen 'oxysterol' (oxidized sterol) products, some of which are benign, while others have been found to be extremely toxic to cells in culture.[50,51] For example, the 7DHC-derived epimeric oxysterols 6α-hydroxy- and 6β-hydroxy- 5,9-endoperoxy-cholest-7-en-3β-ol (referred to as compounds 2a and 2b in references[50,51]) kill 50% of Neuro2a cells in culture within 48 hours at a concentration of 5 μM, which is about

10-fold more toxic than 7-ketocholesterol, a 'benchmark' cytotoxic oxysterol implicated in the pathobiology of atherosclerosis, cardiovascular disease, neuronal degenerative diseases, and age-related macular degeneration.[52,53] Another 7DHC-derived oxysterol, abbreviated 'DHCEO' (3β,5α-dihydroxy-cholest-7-en-6-one; or compound 10 in references[50,51]), an α,β-unsaturated ketone, is about twice as toxic as is 7-ketocholesterol in this same culture system and recently has been suggested as a novel oxysterol biomarker for 7DHC oxidation in tissue specimens from SLOS patients and from SLOS animal models, as well as being implicated in the pathobiology of the disease.[54,55] If such oxysterols have any connection to the retinal degeneration observed in the AY9944 SLOS rat model, then they should be found in the retinas of those animals but not in untreated normal control rats. This is, indeed, the case. In fact, three novel, exclusively 7DHC-derived oxsterols – 4α- and 4β-hydroxy-7DHC and 24-hydroxy-7DHC – in addition to DHCEO and 7-ketocholesterol were isolated for the first time from tissues (including retina, brain, liver, and serum) obtained from AY9944-treated albino rats.[56,57] Whether any or all of these oxysterols plays a direct role in the observed retinal degeneration, as opposed to being merely an oxidized lipid 'by-stander' or sequela of the degeneration process, remains to be determined.

If 7DHC-derived oxysterols are causative, and not merely biomarkers, in the pathobiology of SLOS or its genetic or pharmacologically induced animal models, then this would explain why merely supplying exogenous cholesterol is not an effective therapy. One must prevent excessive accumulation of 7DHC, as well as block formation of its highly toxic oxysterol byproducts, in addition to providing exogenous cholesterol to meet the cellular and systemic physiologic needs of the body. In addition, one must consider the involvement of other oxidation products or oxidatively modified molecules, including those derived from nonsterol lipids and proteins. As stated in a recent review[8]: "Given the evidence for oxidation of both lipids and proteins in the SLOS rat model, in addition to the expected sterol pathway modifications, we suggest that future development of therapeutic interventions for clinical management of SLOS patients should include antioxidants in addition to cholesterol supplementation." Based, in part, on the translational findings obtained with the AY9944 rat model of SLOS described above, a limited clinical trial, employing combined cholesterol supplementation plus the multivitamin AquADEKs® (http://yasooproducts .com/aquadeks/; last accessed 27 November 2013), which contains both lipid-soluble and water-soluble antioxidants, is currently underway to evaluate the efficacy of this combined therapeutic approach to improving visual function outcomes in SLOS patients. Preliminary results are encouraging[58]: "On cholesterol alone, these [ERG] abnormalities continued to worsen over time. However, when antioxidants were added, statistically significant improvement in ERG function (increased amplitude and diminished implicit time) were seen."

HYPOTHESIS CONCERNING THE MECHANISM OF RETINAL DEGENERATION IN THE SMITH-LEMLI-OPITZ SYNDROME RAT MODEL

The experimental findings described above provide the basis for a hypothesis to explain why retinal degeneration occurs in the AY9944 SLOS rat model, as shown in Figure 29.6. AY9944 blocks formation of cholesterol at the level of DHCR7, leading to abnormal and excessive buildup of 7DHC. Some of the 7DHC then becomes oxidized *in situ* to form oxysterols, which then go on to exert a variety of biologic effects, including: perturbation of gene expression via cognate oxysterol-binding proteins (OSBPs), receptors (*e.g.*, liver X receptor/retinoid X receptor (LXR/RXR)), or sterol response elements (SREs) (provoking stress responses, upregulating proapoptotic genes, and suppressing survival factors); direct perturbation of membrane structure and function; and stimulation of mitochondria to form reactive oxygen species (ROS*) and reactive nitrogen species (RNS*), resulting in oxidation of lipids and proteins (which, in turn, can affect membrane structure and function), as well as nucleic acids. Some of the protein oxidative modifications may result from oxidative degradation of polyunsaturated fatty acids (PUFAs), leading to the formation of active aldehydes such as 4-hydroxynonenal (HNE), 4-hydroxyhexenal (HHE), and carboxyethylpyrrole (CEP), while others may derive from nitrotyrosine (NTyr) formation. Any or all of these biologic effects can promote cell dysfunction

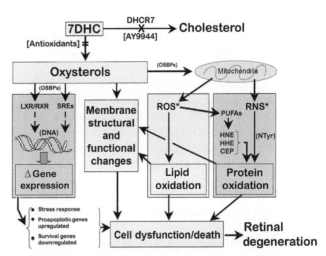

FIGURE 29.6 Hypothetical scheme to explain retinal degeneration in the Smith-Lemli-Opitz syndrome rat model. Blocking cholesterol biosynthesis at the level of 3β-hydroxysterol-Δ7-reductase (DHCR7) with AY9944 promotes 7-dehydrocholesterol (7DHC) formation, which in turn results in cytotoxic oxysterol formation. This leads to multiple biologic sequelae, as shown, resulting in retinal cell dysfunction and death and eventual degeneration. CEP: carboxyethylpyrrole; DNA: deoxyribonucleic acid; HHE: 4-hydroxyhexenal; HNE: 4-hydroxynonenal; LXR: liver X receptor; NTyr: nitrotyrosine; OSBP: oxysterol-binding protein; PUFA: polyunsaturated fatty acid; RNS*: reactive nitrogen species; ROS*: reactive oxygen species; RXR: retinoid X receptor; SRE: sterol response element.

and ultimately death (particularly of photoreceptor cells), resulting in retinal degeneration.

Several elements of the above hypothesis remain to be tested and either verified or negated, empirically. If the hypothesis is correct, then the retinal degeneration in the rat model (and, by inference, in the human disease) can be attenuated or prevented by blocking the formation of cytotoxic 7DHC-derived oxysterols, that is, with antioxidants, in addition to providing exogenous cholesterol to meet cellular and systemic needs. Hence, this 'bench-to-bedside' translational approach heralds the development of more effective therapeutic interventions for SLOS and allied disorders.[59]

TAKE-HOME MESSAGES

- SLOS involves genetic defects in the penultimate enzyme in cholesterol biosynthesis, 3β-hydroxysterol-Δ7-reductase (DHCR7), which catalyzes the conversion of 7-dehydrocholesterol to cholesterol.
- A rat model of SLOS has been developed, using a relatively selective inhibitor (AY9944) of DHCR7, which exhibits a progressive retinal degeneration and attendant visual function defects.
- Results obtained with this animal model indicate that the pathobiology mechanism of SLOS is more

complex than initially thought and likely involves multiple biochemical pathways, as well as protein and lipid oxidation, in addition to the initial defect in cholesterol biosynthesis.

- Formation and accumulation of cytotoxic, 7DHC-derived oxysterols likely play a key role in the pathobiology of SLOS, as well as in the retinal degeneration in the rat model of this disease.
- Cholesterol supplementation therapy alone is relatively ineffective and variable in treatment of SLOS but affords some sparing of visual function deficits (particularly those involving cones) and decreased photoreceptor cell death in the rat model of this disease.
- Combined antioxidant plus cholesterol supplementation holds promise for providing an improved therapy for the treatment of SLOS patients.

Acknowledgments

Studies performed in the author's laboratory, described herein, were supported, in part, by U.S.P.H.S. (NEI/NIH) grant EY007361, by departmental Unrestricted Grants from Research to Prevent Blindness (RPB), by a Senior Scientific Investigator Award from RPB, by a grant from the March of Dimes, and by resources and facilities provided by the Veteran Affairs Western New York Healthcare System (VAWNYHS). The author wishes to thank the following collaborators who contributed significantly to the studies summarized briefly in this review: Robert E. Anderson, Kathleen Boesze-Battaglia, R. Steven Brush, Deborah A. Ferrington, David A. Ford, Rebecca J. Kapphahn, R. Kennedy Keller, Drake C. Mitchell, Barbara A. Nagel, Neal S. Peachey, Bruce A. Pfeffer, F. Denny Porter, Ned A. Porter, Jun Qu, Michael J. Richards, Lowell Sheflin, Akbar Siddiqui, Chengjian Tu, Dana K. Vaughan, Christopher Wassif, and Libin Xu. Thanks to Christopher Goulah and Bruce A. Pfeffer for helpful comments and to Lisa Prince-Baker for clerical assistance in the preparation of this chapter. The opinions expressed herein do not necessarily reflect those of the Veterans Administration or the U.S. Government.

References

1. Nwokoro NA, Wassif CA, Porter FD. Genetic disorders of cholesterol biosynthesis in mice and humans. *Mol Genet Metab* 2001;**74**:105–19.

2. Porter FD, Herman GE. Malformation syndromes caused by disorders of cholesterol synthesis. *J Lipid Res* 2011;**52**:6–34.

3. Smith DW, Lemli L, Opitz JM. A newly recognized syndrome of multiple congenital anomalies. *J Pediatr* 1964;**64**:210–7.

4. Correa-Cerro LS, Porter FD. 3beta-hydroxysterol Delta7-reductase and the Smith-Lemli-Opitz syndrome. *Mol Genet Metab* 2005;**84**:112–26.

5. DeBarber AE, Eroglu Y, Merkens LS, Pappu AS, Steiner RD. Smith-Lemli-Opitz syndrome. *Expert Rev Mol Med* 2011;**13**:e24.

6. Battaile KP, Battaile BC, Merkens LS, Masten CL, Steiner RD. Carrier frequency of the common mutation IVS8-1G>C in DHCR7 and estimate of the expected incidence of Smith-Lemli-Opitz syndrome. *Mol Genet Metab* 2001;**72**:67–71.

7. Nowaczyk MJ, Waye JS, Douketis JD. DHCR7 mutation carrier rates and prevalence of the RSH/Smith-Lemli-Opitz syndrome: where are the patients? *Am J Med Genet A* 2006;**140**:2057–62.

8. Fliesler SJ. Retinal degeneration in a rat model of Smith-Lemli-Opitz syndrome: thinking beyond cholesterol deficiency. *Adv Exp Med Biol* 2010;**664**:481–9.

9. Fliesler SJ, Bretillon L. The ins and outs of cholesterol in the vertebrate retina. *J Lipid Res* 2010;**51**:3399–413.

10. Roux C, Aubry M. Teratogenic action in the rat of an inhibitor of cholesterol synthesis, AY 9944. *CR Seances Soc Biol Fil* 1966;**160**:1353–7.

11. Kolf-Clauw M, Chevy F, Wolf C, Siliart B, Citadelle D, Roux C. Inhibition of 7-dehydrocholesterol reductase by the teratogen AY9944: a rat model for Smith-Lemli-Opitz syndrome. *Teratology* 1996;**54**:115–25.

12. Fliesler SJ, Richards MJ, Miller C, Peachey NS. Marked alteration of sterol metabolism and composition without compromising retinal development or function. *Invest Ophthalmol Vis Sci* 1999;**40**: 1792–801.

13. Fliesler SJ, Peachey NS, Richards MJ, Nagel BA, Vaughan DK. Retinal degeneration in a rodent model of Smith-Lemli-Opitz syndrome: electrophysiologic, biochemical, and morphologic features. *Arch Ophthalmol* 2004;**122**:1190–200.

14. Fliesler SJ, Vaughan DK, Jenewein EC, Richards MJ, Nagel BA, Peachey NS. Partial rescue of retinal function and sterol steady-state in a rat model of Smith-Lemli-Opitz syndrome. *Pediatr Res* 2007;**61**:273–8.

15. Dvornik D, Kraml M, Dubuc J, Givner M, Guadry R. A novel mode of inhibition of cholesterol biosynthesis. *J Am Chem Soc* 1963;**85**:3309.

16. Kraml M, Bagli JF, Dvornik D. Inhibition of the conversion of 7-dehydrocholesterol to cholesterol by AY-9944. *Biochem Biophys Res Commun* 1964;**15**:455–7.

17. Irons M, Elias ER, Salen G, Tint GS, Batta AK. Defective cholesterol biosynthesis in Smith-Lemli-Opitz syndrome. *Lancet* 1993;**341**:1414.

18. Tint GS, Irons M, Elias ER, Batta AK, Frieden R, Chen TS, et al. Defective cholesterol biosynthesis associated with the Smith-Lemli-Opitz syndrome. *N Engl J Med* 1994;**330**:107–13.

19. Fitzky BU, Witsch-Baumgartner M, Erdel M, Lee JN, Paik YK, Glossmann H, et al. Mutations in the Delta7-sterol reductase gene in patients with the Smith-Lemli-Opitz syndrome. *Proc Natl Acad Sci U S A* 1998;**95**:8181–6.

20. Moebius FF, Fitzky BU, Lee JN, Paik YK, Glossmann H. Molecular cloning and expression of the human delta7-sterol reductase. *Proc Natl Acad Sci U S A* 1998;**95**:1899–902.

21. Wassif CA, Maslen C, Kachilele-Linjewile S, Lin D, Linck LM, Connor WE, et al. Mutations in the human sterol delta7-reductase gene at 11q12-13 cause Smith-Lemli-Opitz syndrome. *Am J Hum Genet* 1998;**63**:55–62.

22. Weidman TA, Kuwabara T. Development of the rat retina. *Invest Ophthalmol* 1969;**8**:60–9.

23. Braekevelt CR, Hollenberg MJ. The development of the retina of the albino rat. *Am J Anat* 1970;**127**:271–301.

24. Lamb TD, Pugh Jr EN. A quantitative account of the activation steps involved in phototransduction in amphibian photoreceptors. *J Physiol* 1992;**449**:719–58.

25. Pugh Jr EN, Lamb TD. Amplification and kinetics of the activation steps in phototransduction. *Biochim Biophys Acta* 1993;**1141**:111–49.

26. Owen DM, Magenau A, Williamson D, Gaus K. The lipid raft hypothesis revisited – new insights on raft composition and function from super-resolution fluorescence microscopy. *Bioessays* 2012;**34**:739–47.

27. Simons K, Sampaio JL. Membrane organization and lipid rafts. *Cold Spring Harb Perspect Biol* 2011;**3**. a004697.

28. Keller RK, Arnold TP, Fliesler SJ. Formation of 7-dehydrocholesterol-containing membrane rafts *in vitro* and *in vivo*, with relevance to the Smith-Lemli-Opitz syndrome. *J Lipid Res* 2004;**45**:347–55.

29. Wang J, Megha, London E. Relationship between sterol/steroid structure and participation in ordered lipid domains (lipid rafts): implications for lipid raft structure and function. *Biochemistry* 2004;**43**:1010–8.

30. Serfis AB, Brancato S, Fliesler SJ. Comparative behavior of sterols in phosphatidylcholine-sterol monolayer films. *Biochim Biophys Acta* 2001;**1511**:341–8.

31. Berring EE, Borrenpohl K, Fliesler SJ, Serfis AB. A comparison of the behavior of cholesterol and selected derivatives in mixed sterol-phospholipid Langmuir monolayers: a fluorescence microscopy study. *Chem Phys Lipids* 2005;**136**:1–12.

32. Lintker KB, Kpere-Daibo P, Fliesler SJ, Serfis AB. A comparison of the packing behavior of egg phosphatidylcholine with cholesterol and biogenically related sterols in Langmuir monolayer films. *Chem Phys Lipids* 2009;**161**:22–31.

33. Ford DA, Monda JK, Brush RS, Anderson RE, Richards MJ, Fliesler SJ. Lipidomic analysis of the retina in a rat model of Smith-Lemli-Opitz syndrome: alterations in docosahexaenoic acid content of phospholipid molecular species. *J Neurochem* 2008;**105**:1032–47.

34. Boesze-Battaglia K, Damek-Poprawa M, Mitchell DC, Greeley L, Brush RS, Anderson RE, et al. Alteration of retinal rod outer segment membrane fluidity in a rat model of Smith-Lemli-Opitz syndrome. *J Lipid Res* 2008;**49**:1488–99.

35. Fliesler SJ, Anderson RE. Chemistry and metabolism of lipids in the vertebrate retina. *Prog Lipid Res* 1983;**22**:79–131.

36. Siddiqui AM, Richards MJ, Fliesler SJ. *Global differential and temporal transcriptional profiling of retinas from AY9944-treated (SLOS) vs. control rats*. Abstract 2498, Annual Meeting. Lauderdale, FL: Association for Research in Vision and Ophthalmology (ARVO), Ft; May 2007. [on CD-ROM].

37. Siddiqui AM, Wassif CA, Porter FD, Richards MJ, Fliesler SJ. *A systems-level approach to temporal transcriptional profiling of retinas in a rat model of Smith-Lemli-Opitz syndrome*. Abstract 3067, Annual Meeting. Lauderdale, FL: Association for Research in Vision and Ophthalmology (ARVO), Ft; May 2009. [on CD-ROM].

38. Waage-Baudet H, Dunty Jr WC, Dehart DB, Hiller S, Sulik KK. Immunohistochemical and microarray analyses of a mouse model for the Smith-Lemli-Opitz syndrome. *Dev Neurosci* 2005;**27**:378–96.

39. Tu C, Li J, Jiang X, Sheflin LG, Pfeffer BA, Behringer M, Fliesler SJ, Qu J. Ion current-based proteomic profiling of the retina in a rat model of Smith-Lemli-Opitz syndrome. *Mol Cell Proteomics* 2013;**12**:3583–98.

40. Fliesler SJ, Kapphahn RJ, Ferrington DA. *Proteome alteration and enhanced oxidative modification of retinal proteins in a rat model of Smith-Lemli-Opitz Syndrome*. Abstract 447, Annual Meeting. Lauderdale, FL: Association for Research in Vision and Ophthalmology (ARVO), Ft; May 2009. [on CD-ROM].

41. Jiang XS, Backlund PS, Wassif CA, Yergey AL, Porter FD. Quantitative proteomics analysis of inborn errors of cholesterol synthesis: identification of altered metabolic pathways in DHCR7 and SC5D deficiency. *Mol Cell Proteomics* 2010;**9**:1461–75.

42. Porter FD. Smith-Lemli-Opitz syndrome: pathogenesis, diagnosis and management. *Eur J Hum Genet* 2008;**16**:535–41.

43. Svoboda MD, Christie JM, Eroglu Y, Freeman KA, Steiner RD. Treatment of Smith-Lemli-Opitz syndrome and other sterol disorders. *Am J Med Genet C Semin Med Genet* 2012;**160C**:275–94.

44. Dietschy JM. Central nervous system: cholesterol turnover, brain development and neurodegeneration. *Biol Chem* 2009;**390**:277–93.

45. Siddiqui AM, Richards MJ, Fliesler SJ. *Effect of cholesterol supplementation on the retinal gene transcriptome in a rat model of Smith-Lemli-Opitz syndrome*. Abstract 4152, Annual Meeting. Lauderdale, FL: Association for Research in Vision and Ophthalmology (ARVO), Ft; May 2009. [on CD-ROM].

46. Fliesler SJ, Kapphahn RJ, Ferrington DA. *Retinal proteome alterations in a rat model of Smith-Lemli-Opitz Syndrome, and the effects of a high-cholesterol diet*. Abstract, ARVO-ISOCB Biennial Meeting. Portugal: Ericeira; September 2009.

47. Vaughan DK, Peachey NS, Richards MJ, Buchan B, Fliesler SJ. Light-induced exacerbation of retinal degeneration in a rat model of Smith-Lemli-Opitz syndrome. *Exp Eye Res* 2006;**82**:496–504.

48. Richards MJ, Nagel BA, Fliesler SJ. Lipid hydroperoxide formation in the retina: correlation with retinal degeneration and light damage in a rat model of Smith-Lemli-Opitz syndrome. *Exp Eye Res* 2006;**82**:538–41.

49. Xu L, Davis TA, Porter NA. Rate constants for peroxidation of polyunsaturated fatty acids and sterols in solution and in liposomes. *J Am Chem Soc* 2009;**131**:13037–44.

50. Xu L, Korade Z, Porter NA. Oxysterols from free radical chain oxidation of 7-dehydrocholesterol: product and mechanistic studies. *J Am Chem Soc* 2010;**132**:2222–32.

51. Korade Z, Xu L, Shelton R, Porter NA. Biological activities of 7-dehydrocholesterol-derived oxysterols: implications for Smith-Lemli-Opitz syndrome. *J Lipid Res* 2010;**51**:3259–69.

52. Brown AJ, Jessup W. Oxysterols: sources, cellular storage and metabolism, and new insights into their roles in cholesterol homeostasis. *Mol Aspects Med* 2009;**30**:111–22.

53. Rodríguez IR, Larrayoz IM. Cholesterol oxidation in the retina: implications of 7KCh formation in chronic inflammation and age-related macular degeneration. *J Lipid Res* 2010;**51**:2747–62.

54. Xu L, Korade Z, Rosado Jr DA, Liu W, Lamberson CR, Porter NA. An oxysterol biomarker for 7-dehydrocholesterol oxidation in cell/mouse models for Smith-Lemli-Opitz syndrome. *J Lipid Res* 2011;**52**:1222–33.

55. Xu L, Mirnics K, Bowman AB, Liu W, Da J, Porter NA, Korade Z. DHCEO accumulation is a critical mediator of pathophysiology in a Smith-Lemli-Opitz syndrome model. *Neurobiol Dis* 2012;**45**:923–9.

56. Xu L, Liu W, Sheflin LG, Fliesler SJ, Porter NA. Novel oxysterols observed in tissues and fluids of AY9944-treated rats: a model for Smith-Lemli-Opitz syndrome. *J Lipid Res* 2011;**52**:1810–20.

57. Xu L, Sheflin LG, Porter NA, Fliesler SJ. 7-Dehydrocholesterol-derived oxysterols and retinal degeneration in a rat model of Smith-Lemli-Opitz syndrome. *Biochim Biophys Acta* 2012;**1821**:877–83.

58. Elias E, Braverman R, Tong S. *Beyond cholesterol: antioxidant treatment for patients with Smith-Lemli-Opitz syndrome*. Abstract 238, Annual Meeting. San Francisco, CA: American Society for Human Genetics; November 2012.

59. Fliesler SJ. Antioxidants: the missing key to improved therapeutic intervention in Smith-Lemli-Opitz syndrome? *Hereditary Genet* 2013;**2**:119.

MICRONUTRIENTS

Vitamin A, Zinc, Dark Adaptation, and Liver Disease

Winsome Abbott-Johnson[1], Paul Kerlin[2]

[1]Princess Alexandra Hospital, Woolloongabba, Qld, Australia, [2]Wesley Medical Centre, Auchenflower, Qld, Australia

HISTORY OF VITAMIN A AND NIGHT BLINDNESS

Although vitamin A itself was discovered by McCollum and Davis in 1912, observations relating to vitamin A-containing foods and vision extend into antiquity. In a collection from the school of Hippocrates (300 BC) entitled 'Concerning Vision', it was recommended that children with night vision problems should eat liver.[1] In the 19th century, many sailors had night blindness due to inadequate rations during long sea voyages.[2] In the US civil war, the prevalence of night blindness in black soldiers was higher than for white soldiers. This probably reflected poorer rations for the black soldiers.[2] A seasonal variation in night blindness was also observed, with visual problems being more common in spring and summer than after the harvest.[2]

McCollum and Davis (1912) found that 'fat-soluble factor A' was essential for growth in rats.[1] The important connection between vitamin A deficiency and night blindness however was made by Frederica and Holm in 1925, who observed slower generation of visual purple in light-adapted vitamin A-deficient rats than for normal rats when put into the dark.[1]

A relationship between night blindness and cirrhosis was reported by Haig *et al.* in 1938[3] and Patek and Haig in 1939.[4] It was thought that these patients may be deficient in vitamin A and the deficiency state was not thought to be attributable to inadequate intake of the vitamin in their food. Impairments of dark adaptation (DA) included delayed rod cone break (time when rods become more sensitive to light than cones), higher intensity of light seen at 20 minutes (postexposure to a bright light), and higher intensity of light seen at final reading (elevated final rod thresholds). Nineteen of 24 patients demonstrated night blindness but none was aware of this on direct questioning.

VITAMIN A ABSORPTION AND METABOLISM

Vitamin A is ingested either as retinyl esters or as β carotene (provitamin A). Food sources of vitamin A are liver, cod liver oil, eggs, and milk fat (e.g., in whole milk, butter, cheese). Sources of β carotene are orange-colored fruit and vegetables (e.g., carrots, pumpkin) and also some green vegetables in which the color is masked (e.g., broccoli and spinach). Retinyl ester is hydrolyzed to retinol within the intestine. Within the intestinal mucosa, β carotene is converted to retinal, then to retinol, and into retinyl palmitate.

Retinyl palmitate travels within the chylomicron to the liver, where it is hydrolyzed to retinol. Within the liver, retinol is converted into retinyl palmitate (reversible reaction) and stored in Ito cells. Prior to release into the bloodstream, retinyl palmitate is hydrolyzed to retinol and combines with retinol binding protein (RBP) to form *holo*-retinol binding protein. After release from the liver, *holo*-retinol binding protein combines with prealbumin, forming a ternary complex, and is transported in the bloodstream in this form. Prior to attachment to the retina, the prealbumin is released, leaving the *holo*-retinol binding protein as the form that attaches to the retina (Fig. 30.1).[5]

FIGURE 30.1 Vitamin A absorption, metabolism, and delivery to the eye. This figure shows the mode of absorption of vitamin A and β carotene, the role of the liver, and transport of the vitamin A complex to the retina. ARAT: acyl coenzyme A retinol acyl transferase; AREH: acid retinyl ester hydrolase; CEL: carboxyl ester lipase; HL: hepatic lipase; LPL: lipoprotein lipase; LRAT: lecithin retinol acyl transferase; NREH: neutral retinyl ester hydrolase; RBP: retinol-binding protein. *With permission from Abbott-Johnson et al. Dark adaptation in vitamin A-deficient adults awaiting liver transplantation: improvement with intramuscular vitamin A treatment. Br J Ophthamol. 2011;95(4):544–8, BMJ Publishing Group Ltd.*[33]

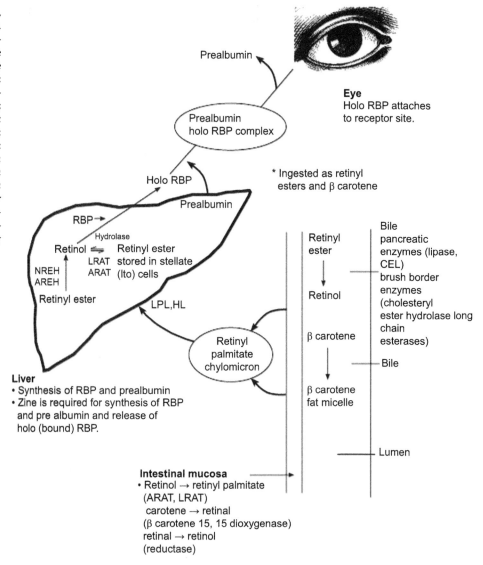

ALTERATIONS IN VITAMIN A STATUS IN LIVER DISEASE

Reasons for Reduction in Vitamin A Status

Factors associated with reduced vitamin A status include 1) reduced oral intake, 2) malabsorption (in cholestatic forms of liver disease), and 3) hepatic effects involving reduced synthesis of carrier proteins (i.e., RBP and prealbumin). Reduced oral intake of vitamin A can be associated with decreased appetite or displacement of oral intake by alcohol ingestion. McClain *et al.* found reduced oral intake of both vitamin A and zinc in a cohort of alcoholic cirrhotic patients.[6] Malabsorption is expected in cholestatic forms of liver disease since both vitamin A and β carotene are fat soluble and require bile for micelle formation.

Both RBP and prealbumin are synthesized in the liver. Factors that influence availability of these proteins in liver disease include malnutrition, inflammation, and zinc status.

Malnutrition is associated with reduced circulating levels of RBP and prealbumin.[7] Malnutrition is common in end-stage liver disease and worsens with severity of illness.[8] It is likely that the malnutrition that is present in end-stage liver disease impacts on levels of RBP and prealbumin.

Both RBP[9] and prealbumin are acute phase proteins that are depressed in the presence of inflammation. Inflammation is a feature of early liver disease across a range of etiologies including nonalcoholic steatohepatitis and hepatitis C.[1] Inflammation also leads to zincuria, which reduces zinc status.

The interrelationship between reduced zinc status and low vitamin A levels was first shown in rats.[11] Low zinc status is common in liver disease.[12] Since zinc is required for protein synthesis, this could lead to reduced hepatic synthesis of RBP and prealbumin. This is supported by Bates and McClain, who found significantly depressed prealbumin, albumin, and transferrin in a group of nonliver disease patients who had depressed serum zinc. Following zinc supplementation, the levels of all three proteins increased.[13]

Storage of Vitamin A

Vitamin A is stored in the liver in stellate cells (also called Ito cells). This is the major cell type involved in fibrosis. When fibrosis occurs, vitamin A is shed from the cell. Vitamin A stores are, therefore, very low in cirrhosis.[14]

Release of Vitamin A from the Liver

Mobilization of vitamin A from the liver is a highly regulated process that depends on both the production of RBP in the liver and secretion of RBP from the liver.[8] Zinc is also required for the secretion of holo-RBP (bound-RBP) from the liver. Smith *et al.* reported in zinc-deficient rats that if zinc was given in addition to vitamin A, there was a greater rise in plasma vitamin A than if vitamin A was given alone.[15]

Levels of Plasma Vitamin A in Liver Disease

Abbott-Johnson *et al.* studied 107 patients who were assessed for liver transplantation. The range of disease categories for these patients is given in Table 30.1. Seventy-five per cent of these patients were found to have plasma retinol of 1.0 μmol/L or less,[16] which is a level below which ocular impairments can occur (Fig. 30.2).[17] The high prevalence of low plasma vitamin A levels across a range of disease etiologies implies the importance of the metabolic abnormalities that are common in these diseases.

Although it has been commonly believed that blood levels of vitamin A are lower in cholestatic liver disease than in other forms of disease, Abbott-Johnson *et al.* found lower levels of vitamin A in the hepatocellular disease group than in cholestatic disease (Fig. 30.3). It was noted that the hepatocellular group had more severe disease (Child-Pugh score 9[5,6–14]) compared with the cholestatic disease group (Child-Pugh score 7[5,6–11]) (P < 0.0001). Subsequent analysis of the whole group (excluding familial amyloid polyneuropathy) showed inverse relationships between plasma retinol levels and both the Child-Pugh score and the Model of End Stage Liver Disease (MELD) score (Fig. 30.4).

DARK ADAPTATION STUDIES IN LIVER DISEASE

DA has been mainly studied in patients with alcoholic liver disease (ALD) and patients with primary biliary cirrhosis. Table 30.2 is a summary of studies of DA in patients with ALD. Note that serum retinol was not a good predictor of DA in the studies by Morrison

TABLE 30.1 Diagnostic Classification of 107 Patients Assessed for Liver Transplantation

Hepatocellular (n = 74)	ALD	32	ALD	22
			ALD with HCC	2
			ALD with hepatitis C	7
			ALD with hepatitis B	1
	Chronic hepatitis	34	Autoimmune chronic active hepatitis	6
			Hepatitis B	5
			Hepatitis B with HCC	2
			Hepatitis C	16
			Hepatitis C with HCC	4
			Hepatitis C with cholangiocarcinoma	1
	Cryptogenic cirrhosis	7	Cryptogenic cirrhosis	6
			Cryptogenic cirrhosis with HCC	1
	Other hepatocellular	1	Budd-Chiari syndrome	1
Cholestatic (n = 22)	Primary sclerosing cholangitis (PSC)	16	PSC	12
			PSC with cholangiocarcinoma	2
			PSC with pancreatitis	1
			PSC with autoimmune chronic active hepatitis	1
	PBC	4		
	Inflammatory pseudotumor	1		
	Cholestatic biliary fibrosis	1		
Metabolic (n = 8)	FAP	4		
	Other metabolic liver disease	4	Wilson disease	1
			α-1 antitrypsin deficiency	1
			Porphyrin synthesis disorder	1
			Multisystem amyloidosis	1
Miscellaneous (n = 3)	Congenital hepatic fibrosis	3		

ALD: alcoholic liver disease; FAP: familial amyloid polyneuropathy; HCC: hepatocellular carcinoma; PSC: primary sclerosing cholangitis.

With permission from Abbott-Johnson et al. *Relationships between blood levels of fat soluble vitamins and disease severity in adults awaiting liver transplantation.* J Gastroenterol Hepatol. *2011;26:1403. Wiley & Sons.[18]*

FIGURE 30.2 Plasma retinol levels and expected physiologic effects in patients awaiting liver transplantation. There was a high prevalence of low vitamin A levels, n = 107. *With permission, adapted from Abbott-Johnson et al. Relationships between blood levels of fat soluble vitamins and disease severity in adults awaiting liver transplantation. J Gastroenterol Hepatol. 2011;26:1405, Wiley & Sons.[18]*

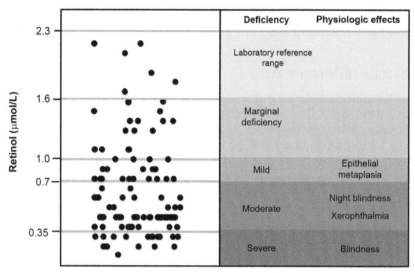

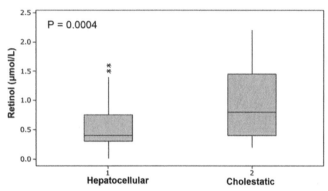

FIGURE 30.3 Comparison of plasma retinol concentrations for hepatocellular and cholestatic diseases in patients awaiting liver transplantation. Retinol levels were lower in hepatocellular disease than in cholestatic disease. Hepatocellular disease, n = 74; cholestatic disease, n = 22. Median, lower box: quartile 1 (Q1), upper box: quartile 3 (Q3), upper whisker: largest data that is ≤ 1.5 (Q3–Q1) above Q3, lower whisker: smallest data that is ≥ 1.5 (Q3–Q1) below Q1, * outliers. *With permission from Abbott-Johnson et al. Relationships between blood levels of fat soluble vitamins and disease severity in adults awaiting liver transplantation. J Gastroenterol Hepatol. 2011;26:1405, Wiley & Sons.[18]*

et al. (1976),[23] Dutta et al. (1979),[21] and Mobarhan et al. (1981).[24] Table 30.3 is a summary of studies of DA in patients with primary biliary cirrhosis. Note that serum retinol was not predictive of DA in the studies of Barber et al. (1989),[28] Herlong et al. (1981),[29] and Hussaini et al. (1998).[30] Little work has been done on DA in other etiologies of liver disease. One case study has been reported of grossly impaired DA in hepatitis C,[31] and impaired DA was present in a subset of eight patients with hepatitis in a study by Vahlquist (1978).[32] In a more recent study by Abbott-Johnson et al. using the SST-1 dark adaptometer, eight of 20 patients with low vitamin A had impaired DA. Five of these had ALD, two had ALD with coexistent hepatitis C, and one had hepatitis C.[33]

Zinc and Dark Adaptation

Zinc status has been shown to play an independent role in DA in some studies (McClain et al.,[6] Herlong et al.,[29] Russell et al.,[34] Morrison et al.[35]). In the study by McClain et al. (1979), 28 people with cirrhosis underwent DA testing. Sixteen had delayed rod cone break and 12 had elevated final thresholds. Six patients did not correct their final threshold with vitamin A supplementation but did with zinc therapy (Fig. 30.5).[6,36] In the study by Russell et al. (1978), 14 of 26 patients with mild-to-moderate alcohol-associated cirrhosis had DA abnormalities. Patients with elevated final DA thresholds were treated with supplemental doses of 3300 µg of vitamin A for 2–4 weeks. Four of these continued to have elevated final thresholds and two (who had low serum zinc) were treated with oral zinc supplements. The final DA thresholds in these patients returned to normal in 1–2 weeks.[34] In the study by Herlong et al. (1981), 11 patients with primary biliary cirrhosis were surveyed for evidence of vitamin A and zinc deficiency using serum vitamin A and zinc. Nine of the patients had abnormal DA and vitamin A levels. Seven of these received vitamin A therapy for 4–12 weeks. One patient who received vitamin A had a persistently elevated DA threshold in spite of normal serum vitamin A but was deficient in zinc. This patient achieved a normal DA threshold following oral zinc supplementation.[29] Morrison et al. (1978) studied six stable cirrhotic patients with low serum zinc. Two of these were initially given vitamin A and one showed improvement (but not normalization) in DA threshold. After zinc supplementation, DA threshold normalized in both patients. Three patients were given zinc only. DA normalized in two and improved in the third. Patient six was given vitamin A and zinc and DA normalized.[35]

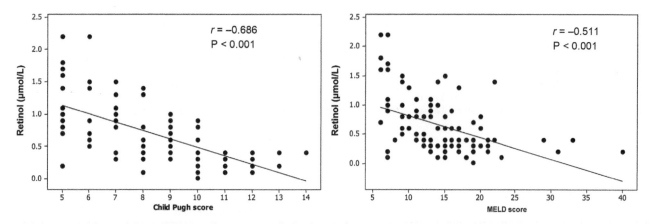

FIGURE 30.4 Comparison of plasma retinol with Child-Pugh score and Model of End Stage Liver Disease (MELD) score in patients awaiting liver transplantation. Lower retinol levels were associated with greater severity of illness as shown by both the Child-Pugh Score and the MELD score. Analysis for the whole group (excluding familial amyloid polyneuropathy), n = 103. *r*: Spearman rank order correlation. *With permission from Abbott-Johnson et al. Relationships between blood levels of fat soluble vitamins and disease severity in adults awaiting liver transplantation.* J Gastroenterol Hepatol. 2011;26:1407, Wiley & Sons.[18]

TABLE 30.2 Dark Adaptation Studies in Alcoholic Liver Disease

Reference	Subjects	Vitamin A Assessment	Findings	Instrument Used
Haig (1938)[3]	14 subjects with ALD	Observations of signs of VAD	13/14 had impaired DA Both cone and rod thresholds elevated	Dark adaptometer reported by Hecht 1938[19]
Patek (1939)[4]	24 cirrhotic patients (ages 38–64 years) with history of alcoholism Most patients malnourished 15 normal (ages 25–40 years)	Signs of vitamin deficiencies including vitamin A	None aware of night blindness on direct questioning 19/24 night blind Delayed speed of adaptation → delayed RC break This can occur alone or with elevated final threshold	Dark adaptometer reported by Hecht 1938[19]
Morrison (1976)[23]	21 cirrhotic patients with history of alcoholism DA performed at least 4 days after withdrawal 8 normal	Serum vitamin A Fecal fat (assess malabsorption)	13/21 abnormally high DA thresholds Serum vitamin A low in four people Fat malabsorption in three people	Not stated
Sandberg (1977)[20]	Two male subjects with chronic alcoholism (ages 55 and 38 years)	Serum vitamin A	Elevated final threshold Nondetectable vitamin A levels	Goldmann-Weekers
Russell (1978)[34]	26 male alcoholic cirrhotic patients (mean age 51 years) 21 controls	Serum vitamin A RBP prealbumin	14/26 had DA impairment with higher final thresholds than controls Higher retinol, RBP, and prealbumin in cirrhotics without DA impairment than in those with impairment but NS	Goldmann-Weekers
McClain (1979)[6]	28 alcoholic cirrhotic patients (mean age 51 years)	Serum vitamin A, RBP, and zinc done in patients who had elevated final thresholds	16/28 delayed RC breaks 12/28 elevated final threshold	Goldmann-Weekers
Dutta (1979) (abstract)[21]	One alcoholic cirrhotic, malnourished	Serum vitamin A, zinc, RBP, prealbumin, albumin	Elevated final DA threshold Normal serum vitamin A and zinc Low RBP, prealbumin, and albumin	Not stated

(Continued)

TABLE 30.2 Dark Adaptation Studies in Alcoholic Liver Disease—cont'd

Reference	Subjects	Vitamin A Assessment	Findings	Instrument Used
Carney (1980)[22]	67 patients with liver disease, gastrointestinal disease, or chronic alcoholism (27 alcoholic cirrhotic, 10 chronic alcoholism without liver disease)	Serum vitamin A serum zinc	Impaired rod thresholds The prevalence of impaired DA increased with lower levels of serum vitamin A Dark adaptation is a better measure of vitamin A sufficiency than serum vitamin A	Goldmann-Weekers
Mobarhan (1981)[24]	Eight ALD subjects with impaired DA 13 ALD subjects without impaired DA	Serum vitamin A, RBP, TTR, RDR	Subjects with impaired DA had impaired RDR test and lower TTR No difference between serum retinol, RBP, or zinc for subjects with or without DA impairment	Goldmann-Weekers

ALD: alcoholic liver disease; DA: dark adaptation; NS: not significant; RBP: retinol binding protein; RC: rod cone; RDR: relative dose response; TTR: transthyretin (prealbumin); VAD: vitamin A deficiency. W Abbott-Johnson, unpublished.

TABLE 30.3 Dark Adaptation Studies in Primary Biliary Cirrhosis

References	Subjects	Vitamin A Assessment	Findings	Instrument Used
Walt (1984)[25]	3 cases (female) with late-stage PBC	Serum vitamin A Cases 1 and 2: < 0.17 μmol/L Case 3: 0.2 μmol/L 1 week before	Symptomatic problems in 2/3 cases. Case 1, DA absent Case 2, rod adaptation grossly slowed Case 3, no rod function	Goldmann perimeter to measure photopic function Automated perimeter adaptometer to measure cone and rod function
Kemp (1988)[26]	Two female cases receiving monthly IM 100,000 IU vitamin A Subject P2 aged 58 years Subject P3 aged 55 years	Serum vitamin A Subject P2: < 5 μg/dL (0.17 μmol/L (RR > 30 μg/dL) Subject P3: 8 μg/dL (0.3 μmol/L)	Delayed cone adaptation No measurable rod function	Automated static perimeter
Shepherd (1984)[27]	9 patients with low serum vitamin A	Serum vitamin A 0.4–1.3 μmol/L (RR 1–3 μmol/L) RBP 1.5–2.8 μg/dL (0.7–1.3 μmol/L), RR 3–5 μg/dL (1.4–2.4 μmol/L)	DA within the normal range for the investigators	Friedmann analyzer
Herlong (1981)[29]	11 patients	Serum vitamin A 9 patients had low serum vitamin A < 40 μg/dL (< 1.4 μmol/L) RBP Prealbumin	Abnormal DA threshold No significant correlation between vitamin A levels and dark adaptation	Goldmann-Weekers dark adaptometer
Barber (1989)[28]	18 patients (age 41–79 years) with no visual symptoms 18 age-matched controls	Serum vitamin A 2 patients had vitamin A < 0.7 μmol/L (RR 0.7–3.3 μmol/L)	Impaired DA of patients compared with controls at 25 minutes Serum vitamin A abnormally low in 2 patients who had abnormally high DA thresholds Numerous cases of elevated DA in the face of normal vitamin A levels	Stimulus and adapting light as described in the article

(Continued)

TABLE 30.3 Dark Adaptation Studies in Primary Biliary Cirrhosis—cont'd

References	Subjects	Vitamin A Assessment	Findings	Instrument Used
Hussaini (1998)[30]	10 females (ages 45–74 years) regularly receiving 100,000 IU IM vitamin A/month Eight control subjects (ages 31–64 years)	Vitamin A, β carotene, RBP	Impaired DA in patients compared with controls after 6 minutes Time to RC break longer in PBC versus controls Low vitamin A, RBP, and β carotene did not predict patients with impaired DA Age and bilirubin were significantly related to final DA threshold (stepwise multiple regression)	Friedmann visual field analyzer

DA: dark adaptation; IM: intramuscular; IU: international unit; PBC: primary biliary cirrhosis; RC: rod cone; RR: reference range. W Abbott-Johnson, unpublished.

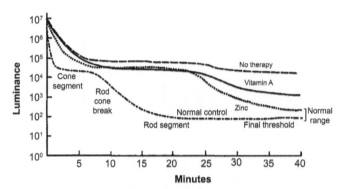

FIGURE 30.5 Dark adaptation curves for a patient on 'no therapy', 'vitamin A', and 'zinc supplementation' compared with a control subject. Dark adaptation for the patient improved on vitamin A and normalized with zinc supplementation. *With permission from McClain. Alterations in zinc, vitamin A, and retinol-binding protein in chronic alcoholics: a possible mechanism for night blindness and hypogonadism. Alcohol Clin Exp Res. 1979;3(2):138. Wiley & Sons.[36]*

It has been suggested that DA improvement with zinc is due to zinc being a coenzyme in alcohol dehydrogenase, which is required for the conversion of 11-cis retinol to 11-cis retinal in the visual cycle.[6,37] Severe zinc deficiency leads to photoreceptor degeneration in rats,[38] and a zinc finger protein has been found within the human retina.[39]

Lack of Perception of Impaired Dark Adaptation

Lack of awareness of night blindness has long been known. Patek and Haig (1939) reported night blindness in 19 of 24 cirrhotic patients mainly of alcoholic origin. None of these was aware of night blindness on direct questioning.[4] Similarly, a lack of awareness of night blindness was reported by Herlong *et al.* (1981)[29] and Hussaini *et al.* (1998)[30] in primary biliary cirrhosis. Abbott-Johnson *et al.* (2011) found six of eight patients with night blindness were unaware of this problem.[33] This lack of awareness is a risk for night driving and falls.

Interventions (with Vitamin A and/or Zinc) to Improve Dark Adaptation

Oral interventions and DA outcomes are given in Table 30.4. Vitamin A has generally been given as 10,000 international units (IU). It is important to avoid the risk of toxicity by using low doses of vitamin A, especially in ALD.[14]

Abbott-Johnson *et al.* (2011) gave an intramuscular injection of 50,000 IU of aqueous retinyl palmitate (G Streuli & Co. AG, Uznach, Germany) to 13 patients who had low plasma vitamin A on previous testing. Follow-up DA testing using the SST-1 dark adaptometer was possible in eight patients and was performed 30 days after the injection (Fig. 30.6). There was a significant improvement, with light of half the previous lowest intensity being seen at 30 days.[33] Aqueous retinyl palmitate given intramuscularly has been shown to have good bioavailability in healthy subjects, whereas oil-based retinyl palmitate is reported to have negligible bioavailability.[40]

TAKE-HOME MESSAGES

- Low serum levels of vitamin A are common and exist in all etiologies of liver disease.
- DA testing has been reported mainly in patients with ALD and in patients with primary biliary cirrhosis (cholestatic disease).
- Zinc is independently involved with DA.
- Impairment of DA can be under-perceived.
- Although DA impairment is more common in lower levels of vitamin A, blood levels of vitamin A can lack sensitivity in prediction of impairment of DA.
- Avoid giving high doses of vitamin A in ALD in view of reported toxicity.
- Aqueous retinyl palmitate given intramuscularly has better bioavailability than the oil-based product.

TABLE 30.4 Oral Intervention Studies in Patients with Chronic Liver Disease

References	Subjects	Product/Dosing	Response
Morrison (1976) (abstract)[23]	13 of 21 patients with cirrhosis and history of alcoholism who had impaired DA	2–10 weeks 10,000 IU/day Product not stated	Improved DA but not normalized
Sandberg (1977)[21]	Two ALD cases Case 1: 55-year-old male Case 2: 38-year-old male continued to drink during treatment	Aquasol 10,000 units daily for 4 weeks + 100,000 units days 7, 8, and 15 10,000 units daily + 100,000 units day 119	Case 1 DA improvement within 7 days Normalization by day 15 Case 2 Day 121 rod function normal
Russell (1978)[34]	12 subjects with mild-to-moderate alcoholic cirrhosis and impaired DA	Vitamin A-water miscible 3300 μg (11,000 IU)	8/12 normalized in 2–4 weeks on 3300 μg Two normalized 1 week after starting 220 mg ZnSO$_4$ Two noncompliant
Morrison (1978)[35]	Six stable alcoholic patients with cirrhosis with low serum zinc; vitamin A 15–37 μg/100 mL (RR 30–65)	Three subjects: zinc replacement only, 220 mg/day ZnSO$_4$ For 1–2 weeks Two subjects: vitamin A 10,000 IU daily for 2–4 weeks followed by ZnSO$_4$ 220 mg/day One subject: zinc + vitamin A concurrently	DA normalized in two and improved in one Improved DA following vitamin A in one DA normalized in both after zinc given to normal DA
Carney (1980)[26]	Nine alcoholic patients with cirrhosis with no Zn deficiency Part of a larger cohort	10,000 IU Product not stated. Dose continued for 1–4 weeks or until DA became normal Weekly DA testing	DA threshold improved in eight of nine people
McClain (1979)[6]	12/28 patients with alcoholic cirrhosis who had impaired DA Zn, RBP, and vitamin A depressed compared with normal people Given house diet for 1 week before intervention	Given either protocol A or B Protocol A: 50,000 IU for 10 days Protocol B: 10,000 IU for 4 weeks Product not stated If DA did not improve given 220 mg ZnSO$_4$ 10 days if on protocol A or 4 weeks if protocol B	DA improved on vitamin A in all but six, but these did improve with Zn Some of these still had delayed rod cone breaks
Mobarhan (1981)[24]	Eight ALD subjects with impaired DA had relative dose response tests	10,000 μg (33,000 IU) retinol daily for 4 weeks	Improved DA No change in serum retinol, serum zinc
Herlong (1981)[29]	Seven subjects with abnormal DA and low serum vitamin A	25,000–50,000 IU vitamin A for 4–12 weeks	Normalization of vitamin A levels In three patients, DA did not correct These had decreased zinc levels DA was corrected with oral zinc supplements

ALD: alcoholic liver disease, DA: dark adaptation, IU: international unit; RBP: retinol binding protein. W Abbott-Johnson, unpublished.

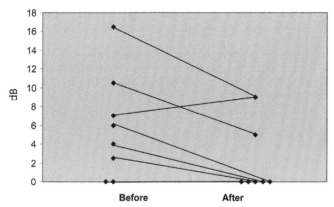

FIGURE 30.6 Final dark adaptation thresholds before and after vitamin A administration in patients with chronic liver disease. This shows improvement in dark adaptation after a single dose of vitamin A. Data for each patient represents the average of two eyes. Seven patients had alcoholic liver disease, two with coexistent hepatitis C. One patient had hepatitis C. Dark adaptation improved in five of six patients whose initial dark adaptation was nonzero. Two patients were zero dB before and after intervention (floor effect of instrument). *With permission from Abbott-Johnson et al, Dark adaptation in vitamin A- deficient adults awaiting liver transplantation: improvement with intramuscular vitamin A treatment.* Br J Ophthamol. 2011;95(4):544–8, BMJ Publishing Group Ltd.[33]

References

1. Wolf G. A history of vitamin A and retinoids. *FASEB* 1996;**10**:1102–7.
2. Semba RD. The vitamin A story, lifting the shadow of death. In: Koletzko B, editor. *World Review of Nutrition and Dietetics.* Basel: Karger; 2012, page 55.
3. Haig C, Hecht S, Patek A. Vitamin A and rod-cone dark adaptation in cirrhosis of the liver. *Science* 1938;**87**(2267):534–6.
4. Patek A, Haig C. The occurrence of abnormal dark adaptation and its relation to vitamin A metabolism in patients with cirrhosis of the liver. *J Clin Invest* 1939;**18**:609–16.
5. Bavik C, Levy F, Hellman U, Wernstedt C, Eriksson U. The retinal pigment epithelial membrane receptor for plasma retinol-binding protein. *J Biol Chem.* 1993;**268**(27):20540–6.
6. McClain C, Van Thiel D, Parker S, Badzin L, Gilbert H. Alterations in zinc, vitamin A, and retinol-binding protein in chronic alcoholics: a possible mechanism for night blindness and hypogonadism. *Alcohol Clin Exp Res.* 1979;**3**(2):135–41.
7. Goodman DS. Plasma retinol binding protein. *Ann N Y Acad Sci.* 1980:378–90.
8. Abbott W, Thomson A, Steadman C, Gatton M, Bothwell C, Kerlin P, et al. Child-Pugh class, nutritional indicators and early liver transplant outcomes. *Hepato-Gastroenterology* 2001;**48**:823–7.
9. Fex G, Felding P. Factors affecting the concentration of free holo retinol-binding protein in human plasma. *Eur J Clin Invest* 1984;**14**:146–9.
10. The Cleveland Clinic. Available at:. *Hepatology* 2010. (last accessed 27 November 2013). http://www.clevelandclinicmeded.com/medicalpubs/diseasemanagement/hepatology/.
11. Smith JC. The vitamin A-zinc connection: a review. *Ann N Y Acad Sci.* 1980:62–75.
12. McClain C, Marsano L, Burk L, Bacon B. Trace metals in liver disease. *Semin Liver Dis* 1991;**11**(4):321–39.
13. Bates J, McClain C. The effect of severe zinc deficiency on serum levels of albumin, transferrin, and prealbumin in man. *Am J Clin Nutr* 1981;**34**:1655–60.
14. Leo M, Lieber C. Alcohol, vitamin A, and B carotene: adverse interactions, including hepatotoxicity and carcinogenicity. *Am J Clin Nutr* 1999;**69**:1071–85.
15. Smith J, McDaniel E, Halstead J. Zinc: a trace element essential in vitamin A metabolism. *Science* 1973;**181**:954–5.
16. Abbott-Johnson W, Kerlin P, Clague A, Cuneo R. Relationships between blood levels of fat soluble vitamins and disease aetiology and severity in adults awaiting liver transplantation. *J Gastroenterol Hepatol* 2011;**26**:1402–10.
17. Sommer A. *Vitamin A Deficiency and Its Consequences. A Field Guide to Detection and Control.* 3rd ed. Geneva: World Health Organization; 1995.
18. Abbott-Johnson W, Kerlin P, Clague A, Johnson H, Cuneo R. Relationships between blood levels of fat soluble vitamins and disease severity in adults awaiting liver transplantation. *J Gastroenterol Hepatol* 2011;**26**:1403.
19. Hecht S, Shlaer S. An adaptometer for measuring human dark adaptation. *J Optic Soc Am.* 1938;**28**:269.
20. Sandberg M, Rosen J, Berson E. Cone and rod function in vitamin A deficiency with chronic alcoholism and retinitis pigmentosa. *Am J Ophthalmol* 1977;**84**:658–65.
21. Dutta SK, Russell RM, Lakhanpal V, Jacobs RA. Night blindness due to deficiency of vitamin A transport proteins in treated pancreatic insufficiency and stable cirrhosis (abstract). *Am J Clin Nutr* 1979;**32**:945.
22. Carney E, Russell R. Correlation of dark adaptation test results with serum vitamin A levels in diseased adults. *J Nutr* 1980;**110**:552–7.
23. Morrison S, Russell R, Carney E, Oaks E. Failure of cirrhotics with hypovitamosis A to achieve normal dark adaptation performance on vitamin A replacement. *Gastroenterology* 1976;**71**. A-30/922.
24. Mobarhan S, Russell R, Underwood B, Wallingford J, Mathieson R, Al-Midani H. Evaluation of the relative dose response test for vitamin A nutriture in cirrhotics. *Am J Clin Nutr* 1981;**34**:2264–70.
25. Walt RP, Kemp CM, Lyness L, Bird AC, Sherlock S. Vitamin A treatment for night blindness in primary biliary cirrhosis. *BMJ* 1984;**288**:1030–1.
26. Kemp C, Jacobson S, Faulkner D, Walt R. Visual function and rhodopsin levels in humans with vitamin A deficiency. *Exp Eye Res.* 1988;**46**:185–97.
27. Shepherd AN, Bedford GJ, Hill A, Bouchier AID. Primary biliary cirrhosis, dark adaptometry, electro-oculography and vitamin A state. *BMJ* 1984;**289**:1484–5.
28. Barber C, Brimlow G, Galloway N, Toghill P, Walt R. Dark adaptation compared with electrooculography in primary biliary cirrhosis. *Doc Ophthalmol* 1989;**71**:397–402.
29. Herlong H, Russell R, Maddrey W. Vitamin A and zinc therapy in primary biliary cirrhosis. *Hepatology* 1981;**1**(4):348–51.
30. Hussaini S, Henderson T, Morrell A, Losowsky M. Dark adaptation in early primary biliary cirrhosis. *Eye* 1998;**12**:419–26.
31. Elison J, Friedman A, Brodie S. Acquired subretinal flecks secondary to hypovitaminosis A in a patient with hepatitis C. *Doc Ophthalmol* 2004;**109**:279–81.
32. Vahlquist A, Sjolund K, Norden A, Peterson P, Stigmar G, Johansson B. Plasma vitamin A transport and visual dark adaptation in diseases of the intestine and liver. *Scand J Clin Lab Invest* 1978;**38**(4):301–8.
33. Abbott-Johnson W, Kerlin P, Abiad G, Clague A, Cuneo R. Dark adaptation in vitamin A-deficient adults awaiting liver transplantation: improvement with intramuscular vitamin A treatment. *Br J Ophthalmol* 2011;**95**(4):544–8.
34. Russell R, Morrison S, Smith F. Vitamin-A reversal of abnormal dark adaptation in cirrhosis. *Ann Intern Med* 1978;**88**:622–6.
35. Morrison A, Russell R, Carney E, Oaks E. Zinc deficiency: a cause of abnormal dark adaptation in cirrhotics. *Am J Clin Nutr* 1978;**31**:276–81.

36. McClain C. Alterations in zinc, vitamin A, and retinol-binding protein in chronic alcoholics: a possible mechanism for night blindness and hypogonadism. *Alcohol Clin Exp Res.* 1979;**3**(2):138.

37. Huber A, Gershoff S. Effects of zinc deficiency on the oxidation of retinol and ethanol in rats. *J Nutr* 1975;**105**:1486–90.

38. Leure-duPree A, McClain C. The effect of severe zinc deficiency on the morphology of the rat retinal pigment epithelium. *Invest Ophthalmol Vis Sci.* 1982;**23**:425–34.

39. Sharma S, Dimasi D, Higginson K, Della N. RZF, a zinc finger protein in the photoreceptors of human retina. *Gene* 2004;**342**: 219–29.

40. Hartmann D, Gysel D, Dubach U, Forgo I. Pharmokinetic modelling of the plasma concentration-time profile of the vitamin retinyl palmitate following intramuscular administration. *Biopharm Drug Dispos* 1990;**11**:689–700.

Vitamin C Functions in the Cornea: Ultrastructural Features in Ascorbate Deficiency

Horacio M. Serra[1], María Fernanda Suárez[1], Evangelina Espósito[2],
Julio A. Urrets-Zavalía[2]

[1]Department of Clinical Biochemistry, Faculty of Chemical Science, National University of Córdoba, Córdoba, Argentina, [2]University Clinic Reina Fabiola, Catholic University of Córdoba, Córdoba, Argentina

INTRODUCTION

The purpose of this chapter is to highlight the functions that vitamin C (ascorbic acid; AA) – an important antioxidant nutrient – has in a normal cornea. First, we will briefly describe its synthesis, its degradation, and its general properties. Second, we will summarize the latest advances concerning the cornea's structure, and finally, we will specifically focus on how this vitamin reaches the cornea and the multiple roles it plays in its different components.

VITAMIN C

Vitamin C/AA is synthesized by most vertebrate and invertebrate species. During evolution, the organ responsible for its synthesis in vertebrates changed twice from the kidneys to the liver. There exist many different pathways for vitamin C biosynthesis in the animal and vegetable kingdoms. Animals with this capacity begin AA synthesis from D-glucose. The last step in the vitamin C synthesis pathway is the oxidation of L-gulonolactone to L-AA by L-gulono-γ-lactone oxidase (GLO), an enzyme associated with the microsomal fractions of liver homogenates, particularly the endoplasmic reticulum.[1] Figure 31.1 shows in detail the animal vitamin C biosynthesis pathway starting from D-glucose-1-P. Glucuronate is also converted to D-xylulose in the pentose pathway.

A number of species, such as teleost fishes, passeriform birds, bats, guinea-pigs, and anthropoid primates (including humans) have lost their ability to synthesize vitamin C due to the absence of the GLO enzyme as

a result of mutations in its gene (Table 31.1).[2] Consequently, vitamin C must be incorporated in these animals' diet.

This nutrient has two major forms in a diet: L-AA and L-dehydroascorbic acid (DHA). Both are absorbed along the entire length of the human intestine in the brush border, after which DHA is reduced intracellularly into ascorbate mono anion. Vitamin C levels are tightly controlled when taken orally and its plasma levels are not only limited by absorption but also by reabsorption and excretion in the kidneys. If the ascorbate intake is low, it is not excreted, whereas ascorbate excess is excreted harmlessly in the urine.

In humans, maximum body pools are limited to about 2 g, and a total body pool below 300 mg is associated with scurvy, a severe vitamin C deficiency disease characterized by symptoms related to connective tissue defects. About 5% of the total vitamin C body content has to be replaced daily. This fact can only be explained by the presence of an efficient system for the recycling and intracellular rescue of vitamin C in which DHA reductases and DHA and AA transporters play a central role.

AA cellular transport is mediated by two different transporter families. One family consists of the low-affinity and high-capacity facilitative glucose transporters (GLUT) that mediate the DHA transport. The other family includes the high-affinity and low-capacity sodium-dependent vitamin C transporters (SVCT1 and SVCT2) that transport the AA. The two human isoforms, SVCT1 and SVCT2, were cloned for the first time in 1999. SVCT1 is encoded by the gene *SLC23A1* in the chromosome locus 5q31.2-31.3, which encodes a predicted protein of 598 amino acids. SVCT2 is the product of the gene *SLC23A2* in the chromosomal locus 20p12.2-12.3, which encodes a predicted protein of 650 amino acids.[5] Both of

Handbook of Nutrition, Diet, and the Eye
http://dx.doi.org/10.1016/B978-0-12-401717-7.00031-9

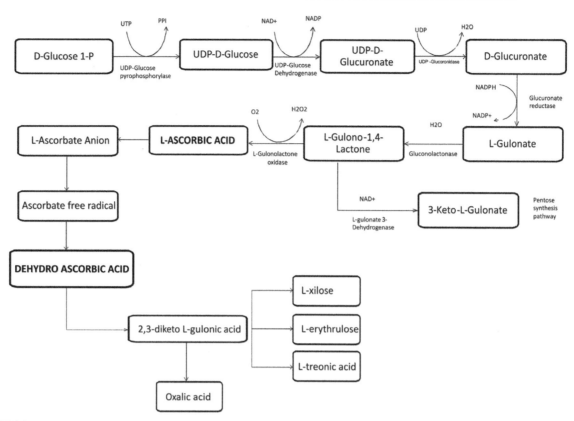

FIGURE 31.1   Vitamin C biosynthesis and degradation pathways. NAD: nicotinamide adenine dinucleotide; NADP: nicotinamide adenine dinucleotide phosphate; NADPH: reduced nicotinamide adenine dinucleotide phosphate; PPI: inorganic pyrophosphate; UDP: uridine diphosphate.

them, the SVCT1 and the SVCT2, are functionally similar, have a close sequence homology, show distinct tissue distributions, and are subject to regulation under different physiologic and pathophysiologic conditions at the transcriptional and post-transcriptional levels.[6]

The findings showing that different GLUT also act as DHA transporters have caused profound implications in our current understanding of vitamin C physiology in humans.[7]

Since every cell in the human body contains AA, it seems likely that the cell's redox state can influence the receptors' expressions, and, in this way, the resulting response will regulate the transport and intracellular ascorbate level.[8] High levels are maintained in the pituitary and adrenal glands, brain, leukocytes, and eye tissues, while low levels are found in plasma and saliva.

The biologic functions of AAs are based on its ability to provide reduced equivalents for a variety of biochemical reactions. Vitamin C is known to be an electron donor for different human enzymes. Vitamin C-dependent proline hydroxylation also plays a role in gene transcription mediated by the hypoxia inducible factor-1. Due to its reducing power, vitamin C can scavenge most physiologically relevant reactive oxygen species (ROS) and reactive nitrogen species (RNS) as well as singlet oxygen and hypochlorite. Ascorbate also provides antioxidant protection by regenerating other antioxidants (such as

glutathione and α-tocopherol) back to their active state. Vitamin C also plays a role in the endothelial nitric oxide synthase (eNOS) function by recycling tetrahydrobiopterin, which is relevant to blood pressure regulation. These, and other vitamin C functions, as well as the molecular mechanisms that make it essential for humans have been extensively described in a recent outstanding review by Traver et al. (2011).[9]

A specific pathway for the degradation of vitamin C has only been fully described in bacteria. Degradation of vitamin C in mammals is initiated by the hydrolysis of DHA. AA is oxidized to DHA in several enzymatic and nonenzymatic reactions: a catalyzed oxidative pathway, an uncatalyzed pathway, and a pathway under anaerobic conditions.[1] DHA can be reduced back to ascorbate, but it can also be easily hydrolyzed at neutral pH to 2,3-diketo-L-gulonate (2,3-DKG) with the consequent loss of nutritional value. 2,3-DKG is spontaneously degraded to oxalate, CO_2, L-xilose, L-threonate, and L-erythrulose.[10] A summary of vitamin C catabolism can be seen in Figure 31.1.

CORNEAL ULTRASTRUCTURE

For many years, cornea studies were limited to the use of light microscopy. As Figure 31.2 shows, our substantial increased understanding of the cornea's different

TABLE 31.1 Ability to Synthesize Vitamin C in Different Mammals

Animal	Vitamin C Biosynthesis	GLO Mutation/Defect
Elephant	Ok	-
Dog	Ok	-
Cat	Ok	-
Cow	Ok	-
Sheep	Ok	-
Pig	Ok	-
Horse	Ok	-
Bat	No	Eight amino acid changes[2]
Rabbit	Ok	
Mouse	Ok	
Rat	Ok	
Guinea-pig	No	Complete loss of exon I and V and partial loss of VI exon[2-4]
Squirrel	Ok	
Galago	Ok	
Lemur	Ok	
Owl monkey	No	Loss of seven out of twelve functional exons found in GLO functional genes[2]
Marmoset	No	
Macaque	No	
Gibbon	No	
Orangutan	No	
Gorilla	No	
Human	No	In humans, deletion of VIII and XI exons[4]

GLO: L-gulono-γ-lactone oxidase.

structures has derived from the use of electronic microscopy. More recently, laser scanner confocal, X-ray diffraction, second harmonic generation multiphoton imaging (SHG), and the atomic force microscope (AFM) have added new methods to study its ultrastructure. The analysis of the diffraction patterns obtained by X-rays passed through the corneal stroma provides quantitative information about the orientation and distribution of collagen molecules, fibrils, and lamellae. The way X-ray diffraction was used to map the distribution and orientation of collagen in the cornea was reported by Meek et al. (2001).[11]

The cornea is a transparent aspheric, avascular vascular tissue that plays important roles in the eye. It maintains the ocular pressure, protects against different aggressions and forces, and constitutes the primary lens in many animals' eyes. In order to fulfill these functions, the cornea requires sophisticated structures. For its nutrition, the cornea depends on glucose and oxygen that diffuse from the aqueous humor, the limbal circulation, and through the tear film. The plentiful sensory nerve fibers are extensions of the long ciliary nerves and form a subepithelial plexus. The cornea has one of the body's highest densities of nerve endings.

There is an epithelial-cell coating derived from the surface ectoderm beneath the covering tear film, composed of stratified, squamous, nonkeratinized epithelial cells, as well as histiocytes, macrophages, lymphocytes, and melanocytes. This layer represents about 5% of the total corneal thickness. The outer epithelium layer has microvilli, and together with the tear film, they form an optically smooth surface. The continuous proliferation of perilimbal basal epithelial cells gives rise to the other layers, which will later differentiate into superficial cells and desquamate into tears. Bowman's membrane – under the epithelium – is formed by randomly dispersed collagen fibrils. The cornea's bulk is the stroma, where keratocytes with numerous cytoplasmic lamellapodia are responsible for the production of proteoglycans (PGs) and collagens, very important components of the extracellular matrix (ECM). PGs chains – proteins attached to glycosaminoglycan (GAG) – are classified into two families: the chondroitin sulfates/dermatan sulfates (CS/DS) and the keratan sulfates (KS), with filaments 70 nm and 40 nm long, respectively.[12,13]

Types I and V collagen molecules form small diameter hybrid fibrils with a regular separation distance, which lie parallel to each other within layers.[14]

Different scientists have studied the PGs–collagen interactions, and their roles in cornea development have been reviewed by Andrew et al. (2008).[15] More recently, Keith et al. (2009) published a complete description of the corneal collagen arrangement and its participation in the maintenance of the cornea's form and transparency.[16]

Another excellent review written by Lewis et al. (2010) describes the three-dimensional interactions between collagen fibrils and PGs.[17] They proposed a new model of interactions between fibril and PGs in which the fibrils associate with sulfated PGs that appear as extended, variable-length linear structures. The PG network appears to tether two or more collagen fibrils (Fig. 31.3).

Transparency also depends on keeping the corneal stroma water content at 78%. Corneal hydration is largely controlled by intact epithelial and endothelial barriers and by the endothelial pump functioning, which is linked to an ion-transport system controlled by temperature-dependent enzymes such as Na^+,K^+-ATPase. The endothelium is composed of a single layer made up of closely interdigitated cells arranged in a hexagonal-shaped mosaic pattern firmly attached to Descemet's membrane.

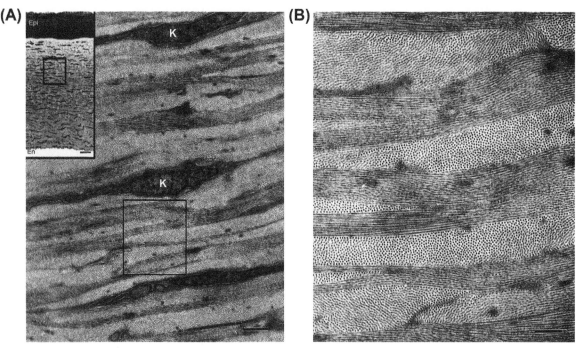

FIGURE 31.2   Transmission electron micrographs of cornea structure. The cornea is limited by an outer epithelium and an inner endothelium (low magnification inset in A). (A) The stroma makes up more than 90% of the corneal thickness and contains keratocytes (K) orientated parallel to the corneal surface and found between the stromal lamellae. (B) Enlargement of area in rectangle in A. The lamellae are composed of small diameter collagen fibrils with regular packing. Adjacent layers are at approximately right angles to one another, forming an orthogonal lattice. *Figure and legend reproduced from Hassell* et al. *(2010),[36] with permission of the copyright holder.*

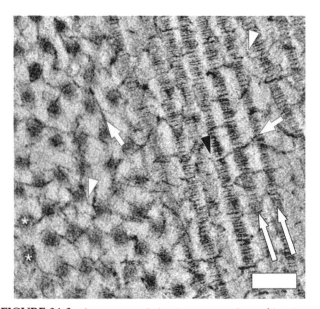

FIGURE 31.3   Appearance of a bovine cornea as imaged in a transmission electron microscope. The imaged region is at the interface between two lamellae and shows a transverse view on the left-hand side, where some of the collagen fibrils are marked by asterisks. On the right-hand side there is a longitudinal view where some collagen fibrils are highlighted by long arrows. Short arrows indicate large, stained proteoglycan filaments that connect two or more adjacent collagen fibrils. A white arrowhead points to a smaller proteoglycan between fibrils. A black arrowhead indicates proteoglycans running axially, not bridging adjacent fibrils. Scale bar represents 100 nm. *Figure and legend reproduced from Lewis* et al. *(2010),[17] with permission of the copyright holder.*

VITAMIN C IN THE CORNEA

Vitamin C Entry and Distribution into the Cornea Structures

There is no doubt that high AA concentrations are found in various ocular fluids and tissues, including the cornea.[18] One of the first reports appeared in 1945, when Pirie *et al.* demonstrated that there is a high concentration of AA in oxen and rabbit corneas and that the greatest concentration is in the corneal epithelium. Corneal stroma and endothelium present approximately the same AA concentration.[19]

Vitamin C uptake in the cornea is an active SVCT 2-mediated carrier mechanism from the plasma into the aqueous humor and across the iris-ciliary body. As this carrier-mediated pathway is specific for AA, this acid is actively accumulated in the endothelium at higher concentrations than those found in the aqueous humor, to finally reach the stroma and the epithelium (Fig. 31.4). However, these concentrations vary among the corneas of those animals that are capable of synthesizing AA versus other animals that need to incorporate it in their diet.[18,20,21] Guinea-pigs maintain a 10-fold higher AA concentration in their corneas than rats.

In the human corneal epithelium, the AA is concentrated by means of a selective active transport from the plasma into the aqueous humor and then through the endothelium into the stroma up to the epithelium.[22]

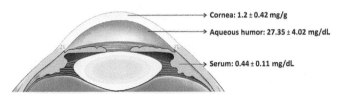

→ Cornea: 1.2 ± 0.42 mg/g

→ Aqueous humor: 27.35 ± 4.02 mg/dL

→ Serum: 0.44 ± 0.11 mg/dL

FIGURE 31.4 Distribution of ascorbate in the different compartments of bovine anterior eye segments. Vitamin C in the cornea is an active transport carrier mechanism from plasma into aqueous humor across the iris-ciliary body. Subsequently the ascorbic acid is actively accumulated in the endothelium, at higher concentrations than those found in aqueous humor, to finally reach the stroma and the epithelium.

Vitamin C distribution in the bovine's anterior eye was completely described by Ringvold et al.[23] They observed a peak AA concentration in the central corneal epithelium covering the pupillary area, and the ascorbate level in the central region of the epithelium was 12–23% higher than in the periphery. In the corneal stroma, Descemet's membrane and the endothelium, the vitamin C concentration was similar to the one found in the aqueous humor.

In 2006, a novel uptake mechanism was postulated by Talluri et al. in addition to the existing one. They hypothesized that lacrimal glands can uptake AA from plasma and then secrete it into tears. Vitamin C in the tear fluid may be incorporated into the corneal epithelium through SVCT 2.[20]

All these studies support the idea that AA concentration in the corneal epithelium is the highest in the eye region and higher than in any other tissue in the body.

Roles of Vitamin C in Corneal Epithelium

Since the idea that ultraviolet radiation (UVR) is a potential source of eye tissue damage has been established,[24] numerous studies have reported the correlation between AA eye concentration and its protective role against UVR exposure.[24–28] In 1980, Ringvold pointed out that AA levels are higher in the aqueous humor of diurnal animals than in the animals active in the dark.[25] Later, work performed in two closely related species of spiny mice (diurnal and nocturnal) showed that the vitamin C concentration in the aqueous humor of diurnal species is 35 times higher than in nocturnal species.[27]

Similar studies have been carried out in diurnal and nocturnal mammal species' corneal epitheliums. It was found that the AA level varies and adjusts accordingly to the ambient radiation dose, just as in the aqueous humor. However, the ascorbate concentration is higher in the epithelium than in the aqueous humor.[29] This is a logical finding considering that the corneal epithelium is the outermost tissue and that it has to deal with high doses of UVR. In another study carried out on different mammalian corneal epitheliums and aqueous humors, it was shown that ambient radiation is needed to maintain high AA concentrations in the epithelium

and that the corneal epithelial thickness and the cell number are susceptible to seasonal fluctuations regulated by UVR. In contrast, the AA content in the aqueous humor was not affected by environmental changes. These results suggest a seasonal adaptation of the mammalian corneal epithelium in response to the ambient UVR variation.[30]

The AA concentration in the human corneal epithelium is approximately 1.2 mg/g wet weight (14 times more concentrated than in the aqueous humor) (Fig. 31.4).[22] The purpose of such a high concentration is to protect the deeper layers of the cornea and the posterior ocular structures from the deleterious UVR effects.[27]

A study of two groups of guinea-pigs fed a normal or a low AA diet was performed to clarify the effect caused in ocular structures by a long-term (7 months) exposure to 8.2 J/cm² of ultraviolet B (UVB) radiation.[31] The authors reported that both groups presented corneal epithelial edema, neovascularization in stroma, increase in corneal thickness, and decrease in endothelial density and lens epithelial damage but no differences with respect to their diets. A different result was obtained by us when guinea-pigs fed on a normal diet and those fed on a low AA diet were exposed during 9 months to a more physiologic dose of UVB (0.12 J/cm²/day). The group fed on a low AA diet had more corneal epithelial defects (unpublished data).

Suh et al. carried out another study to evaluate the vitamin C protective effect on the corneas of rabbits exposed to UVB radiation.[32] Rabbits were given intravenous AA doses before being exposed to the radiation. Similarly to our study and opposed to the report of Wu et al. (2004), their experimental study showed that more corneal damage was found in the irradiated rabbit groups that did not get AA. We think that the different results shown by these studies could be due to the number of UVB doses used and the way the AA was administrated.

In the case of humans, it has been reported that corneal haze following photorefractive keratectomy has only occurred in Norway when the sun is visible 24 hours/day[33] but that pre- and postoperative vitamin-C supplementation reduced its incidence.[34] We have also found that patients with climatic droplet keratopathy in Patagonia (Argentina), who have normally experienced constant intense winds, high solar radiation, and low humidity during their whole lives,[35] show abnormally low ascorbate blood levels as a result of a low vitamin C diet (unpublished data).

Roles of Vitamin C in Corneal Stroma

As it was previously described, the stroma is a highly organized structure needed to produce a minimal light scattering, contributing to corneal transparency.[36] Abnormalities in the collagen organization result in a loss of cornea transparency.[37,38]

Due to its importance in the structure and transparency of the cornea, the collagen in AA metabolism and its participation in its synthesis have been extensively studied for many years. In addition to its role as an antioxidant, AA is also a cofactor in collagen synthesis since it is needed for the proline and lysine hydroxylations,[39,40] which allow the proper formation of stable collagen fibrils and the association with specific GAGs (Fig. 31.5).[41] For a comprehensive review see Peterkofsky (1991).[42]

There has been different evidence that AA participates directly or indirectly in the collagen biosynthesis. Using an animal scurvy as a model, Chojkier et al. (1983) observed that after a 2-week AA-deficient diet, the collagen biosynthesis significantly decreased whereas the proline hydroxylation was not affected. In vitro studies of normal bones revealed that adding a proline hydroxylation inhibitor did not affect the collagen biosynthesis.

In contrast, the proline hydroxylation was restored after adding AA to a cell culture made from scorbutic bones, while the collagen synthesis was not. They have postulated that the collagen biosynthesis decrease during scurvy is associated with the food intake decrease and weight loss induced by vitamin C deficiency, whereas proline hydroxylation constitutes an independent event.[43]

In a different publication, human fibroblasts exposed to scorbutic guinea-pig serum showed a reduced collagen synthesis.[44] This inhibition could be reversed with the addition of an insulin-like growth factor I (IGF-I), which is known to stimulate collagen and PGs syntheses.[45] The collagen synthesis inhibition seems to be related to the induction of a low concentration IGF-I inhibitor into the normal serum, with a 10-fold increase in the scorbutic or starving guinea-pigs' serum.[46] These results suggest that the collagen synthesis decrease is due to the

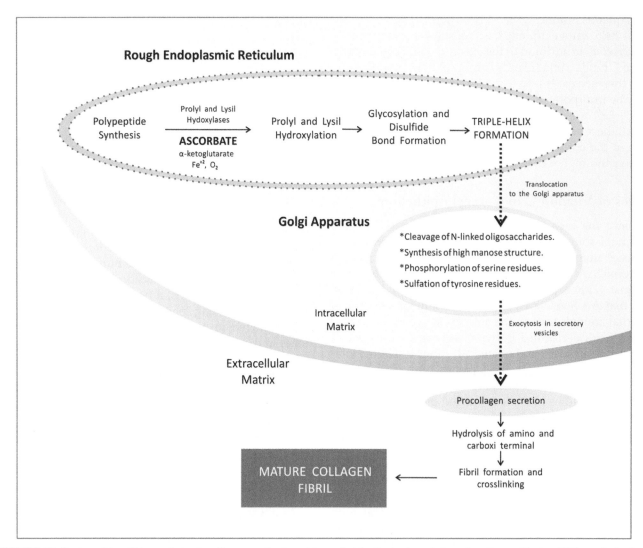

FIGURE 31.5 Ascorbic acid is a cofactor in collagen synthesis and is needed for the hydroxylation of proline and lysine. Without the hydroxylation of these amino acids, procollagen is unable to crosslink properly to form stable collagen fibrils.

vitamin C deficiency induced by fast and that it is not related to AA participation in proline hydroxylation.

Procollagen secretion takes place after the hydroxylation and triple helix formation. It is known to occur in two steps: most of the procollagen is secreted in a first-order reaction (with an approximately 30-minute mean life). The rest is slowly secreted in a first-order reaction too, but with a mean life of approximately 120 minutes. In primary avian tendon cultures, it was demonstrated that the addition of AA induced procollagen secretion and affected its rate, accelerating this event.[47] Pacifici et al. (1988) studied the AA role in the rough endoplasmic reticulum (RER) remodeling during procollagen secretion.[48] They discovered that after adding AA to a scorbutic culture of chondrocytes for 9 days, the accumulated under-hydroxylated procollagen type II is hydroxylated in about 1 hour, starting its secretion afterward. After a 24-hour ascorbate addition, the procollagen is totally secreted. They also found that during the AA treatment, the RER undergoes a remodeling with large cisternae formations like sacs and many flat cisternae (RER without treatment consisting of unique oval cisternae). After the procollagen redistribution takes place, large and flat cisternae establish communication and the secretion takes place.

In 1992 and 1993, Saika found that AA and ascorbic acid 2-phosphate (Asc-2P) – its oxidation-resistant derivative – enhanced the proliferation of cultured keratocytes and the formation of multilayered ones. At higher concentrations, only AA exerted a cytotoxic effect.[49,50] Taking advantage of Asc 2-P properties and using aligned scaffolding material to create an artificial cell matrix, Phu et al. (2009) studied their effect on rabbit corneal fibroblasts. It was found that although ascorbate induced cell stratification – suggesting that AA stimulates ECM production – the rabbit corneal fibroblasts cannot organize the matrix into an arrangement with minimal light scattering. For that reason, they used an aligned scaffold to guide the ECM deposition.[51]

In a recent study, we have investigated the effect of ascorbate deficiency and/or UVB exposure on the guinea-pig corneal stromal ultrastructure using four groups of animals (A, B, C, and D). For almost 4 months, groups A and C were fed on an ascorbate-rich diet and the other two groups on an ascorbate-deficient diet. During the last 3 months, groups C and D were exposed to a normal UVB intensity. After the treatment, the guinea-pigs were euthanized. Cornea electron microscopy images were taken and a small-angle X-ray scattering was used to obtain the collagen fibril separation distance and the fibril diameter. Even though there was no evidence of UVB-induced collagen aggregation, collagen fibrils were found to be more closely packed in animals fed on an ascorbate-deficient diet and/or exposed to UVB. The changes in corneal ultrastructure were most pronounced

in animals fed on an ascorbate-deficient diet.[52] This suggests that AA plays a vital role in the organization of the stroma and in the protection from the UVB harmful effects.

Roles of Vitamin C in Corneal Endothelium

As it has already been mentioned, the AA first enters the aqueous humor and then the cornea through the endothelium, and the corneal endothelial cells are responsible for maintaining corneal transparency by regulating the corneal hydration.[53]

As far as we know, there are not too many studies about AA roles in corneal endothelial cells. In 1991 Bode et al.[54] evaluated the corneal endothelial cells' transport capacity and the AA and DHA uptake in bovine endothelial cells. They found that DHA enters the cornea across the posterior or apical surface of corneal endothelium more rapidly than AA. The molecule reduction occurs immediately after it is taken up. It is worthy to mention that AA is normally the more abundant form in the aqueous humor, and consequently, it is attributed most of the uptakes.

ROS is generated – among others – by the exposure to sunlight, atmospheric oxygen, and metabolic activity, which makes the eye susceptible to oxidative damage.[55] The effects of vitamin C and other antioxidant vitamins (A, E) on the corneal endothelial cells, as well as its protective role against oxidative stress induced by iron, were studied in murine corneal endothelial cells.[56] It was observed that increased levels of iron promoted the apoptosis and lipid peroxidation and that vitamins C, A, and E have the ability to protect the endothelial cells against oxidative stress, suggesting that the three vitamins have common action targets.

Human corneal endothelial cells (HCECs) derive from the neural crest, and the absolute number of endothelial cells appears to decrease with increasing age.[57] It has been demonstrated that HCECs in vivo remain at the G1 phase of the cell cycle rather than in terminal differentiation.[58] It has also been shown that these cells have the ability to proliferate in vitro and that samples taken from elderly donors have greater heterogeneity and a shorter lifespan in long-term cultures than the HCEC taken from young donors.[59] Apart from the importance of the donor's age, a successful culture requires effective isolation, preservation, and expansion of HCEC, a complex combination of mitogens, an appropriate ECM, and culture supplements, such as AA.[60] Knowing the participation of AA in the proline hydroxylation during collagen synthesis, this vitamin is added to the culture media in order to facilitate this process and the basement membrane deposition.[61]

Many years ago, Yue et al. evaluated the effect of AA daily additions to rabbit corneal endothelial cell cultures and found a detrimental effect.[62] Cells were elongated, with large size and vacuoles; they had a shorter lifespan and did not reach confluence in the culture. They suggested that this unexpected result was due to the hydrogen peroxide, a product of oxidative reactions in the cells treated with AA. The authors remarked that corneal endothelial cells are bathed in an AA-rich aqueous humor, but as cells are not dividing, the vitamin's negative effect is produced when cells are undergoing mitosis.

More recently, Shima et al. evaluated these factors' effect on HCEC growth using Asc-2P, various mitogens, and ECMs.[63] They found that Asc-2P, basic fibroblast growth factor (bFGF), and atellocollagen were the perfect combination for primary culture and subculture of HCECs, achieving an increase in the proliferation and the replicative lifespan of HCECs from donors of a wide range of ages and owing these results to a protecting effect of Asc-2P against oxidative deoxyribonucleic acid (DNA) damage.

One year later, the same group elucidated the mechanism by which Asc-2P increased the HCECs' proliferation. They evaluated HCEC growth and collagen synthesis in the presence of various factors: Asc-2P, vitamin A, reduced and oxidized glutathione, carnosine, and a water-soluble vitamin E derivative. Only Asc-2P promoted the HCECs' proliferation but did not promote the collagen synthesis, and it was capable of stimulating hepatocyte growth factor (HGF) by HCECs and vice versa. These results suggested that Asc-2P stimulates HCEC growth by means of HGF production regulation.[64]

TAKE-HOME MESSAGES

- AA is synthesized by most vertebrate and invertebrate species.
- The ability to synthesize vitamin C has been lost in a number of species, such as teleost fishes, passeriform birds, bats, guinea-pigs, and anthropoid primates including humans, due to the absence of the enzyme L-gulono-γ-lactone oxidase. Consequently, vitamin C must be incorporated in these animals' diets.
- The cornea is a transparent aspheric, avascular vascular tissue that plays important roles in the eye. It consists of three layers: epithelium, a middle stroma, and endothelium.
- The cornea has high concentrations of AA.
- Vitamin C in the cornea is an active carrier mechanism from plasma into aqueous humor across the iris-ciliary body. Subsequently, the AA is actively accumulated in the endothelium at higher concentrations than those found in the aqueous humor, to finally reach the stroma and the epithelium.

- The concentration of this powerful antioxidant agent in human corneal epithelium is extremely high. The purpose of such high levels is to protect the deeper layers of the cornea and the posterior ocular structures from the deleterious effects of UVR.
- In addition to its role as an antioxidant, AA is also a cofactor in collagen synthesis since it is needed for the hydroxylation of proline and lysine, which allow the proper formation of stable collagen fibrils and their association with specific GAGs.
- AA combines with other molecules promoted the proliferation of human corneal endothelial cells in vitro.
- Low levels of AA seriously compromise the functions and ultrastructure of different corneal components.

Acknowledgments

HMS work is supported by grants from Argentina: SECYT UNC, and CONICET PIP: 112-200801-01455. The authors gratefully acknowledge Nicolás Crim (MD) for his help in the preparation of the figures and tables.

References

1. Linster CL, Van Schaftingen E, Vitamin C. Biosynthesis, recycling and degradation in mammals. FEBS J 2007;274:1–22.
2. Drouin G, Godin JR, Pagé B. The genetics of vitamin C loss in vertebrates. Curr Genomics 2011;12:371–8.
3. Nishikimi M, Kawai T, Yagi K. Guinea pigs possess a highly mutated gene for L-gulono-gamma-lactone oxidase, the key enzyme for L-ascorbic acid biosynthesis missing in this species. J Biol Chem 1992;267:21967–72.
4. Nishikimi M, Fukuyama R, Minoshima S, Shimizu N, Yagi K. Cloning and chromosomal mapping of the human nonfunctional gene for L-gulono-gamma-lactone oxidase, the enzyme for L-ascorbic acid biosynthesis missing in man. J Biol Chem 1994;269:13685–8.
5. Tsukaguchi H, Tokui T, Mackenzie B, Berger UV, Chen XZ, Wang Y, et al. A family of mammalian Na⁺-dependent L-ascorbic acid transporters. Nature 1999;399:70–5.
6. Liang WJ, Johnson D, Jarvis SM. Vitamin C transport systems of mammalian cells. Mol Membr Biol 2001;18:87–95.
7. Rumsey SC, Kwon O, Xu GW, Burant CF, Simpson I, Levine M. Glucose transporter isoforms GLUT1 and GLUT3 transport dehydroascorbic acid. J Biol Chem 1997;272:18982–9.
8. Mandl J, Szarka A, Bánhegyi G. Vitamin C: update on physiology and pharmacology. Br J Pharmacol 2009;157:1097–110.
9. Traber MG, Stevens JF. Vitamins C and E: beneficial effects from a mechanistic perspective. Free Radic Biol Med 2011;51:1000–13.
10. Bowman BA, Russell RM. Vitamina C. In: Ziegler EE, Filer Jr LJ, editors. Conocimientos Actuales Sobre Nutrición. 8th ed. Washington, DC: Organización Panamericana de la Salud, Oficina Regional de la Organización Mundial de la Salud; 2003. p. 193.
11. Meek KM, Quantock AJ. The use of X-ray scattering techniques to determine corneal ultrastructure. Prog Retin Eye Res 2001;20:95–137.
12. Laurent TC, Anseth A. Studies on corneal polysaccharides. II. Characterization. Exp Eye Res 1961;1:99–105.
13. Axelsson I, Heinegård D. Characterization of the keratan sulphate proteoglycans from bovine corneal stroma. Biochem J 1978;169:517–30.
14. Gerecke DR, Meng X, Liu B, Birk DE. Complete primary structure and genomic organization of the mouse Col14a1 gene. Matrix Biol 2004;22:595–601.

15. Quantock AJ, Young RD. Development of the corneal stroma, and the collagen-proteoglycan associations that help define its structure and function. *Dev Dyn* 2008;**237**:2607–21.

16. Meek KM. Corneal collagen – its role in maintaining corneal shape and transparency. *Biophys Rev* 2009;**1**:83–93.

17. Lewis PN, Pinali C, Young RD, Meek KM, Quantock AJ, Knupp C. Structural interactions between collagen and proteoglycans are elucidated by three-dimensional electron tomography of bovine cornea. *Structure* 2010;**18**:239–45.

18. DiMattio J. Ascorbic acid entry into cornea of rat and guinea pig. *Cornea* 1992;**11**:53–65.

19. Pirie A. Ascorbic acid content of cornea. *Biochem J* 1946;**40**:96–100.

20. Talluri RS, Katragadda S, Pal D, Mitra AK. Mechanism of L-ascorbic acid uptake by rabbit corneal epithelial cells: evidence for the involvement of sodium-dependent vitamin C transporter 2. *Curr Eye Res* 2006;**31**:481–9.

21. Socci RR, Delamere NA. Characteristics of ascorbate transport in the rabbit iris-ciliary body. *Exp Eye Res* 1988;**46**:853–61.

22. Brubaker RF, Bourne WM, Bachman LA, McLaren JW. Ascorbic acid content of human corneal epithelium. *Invest Ophthalmol Vis Sci* 2000;**41**:1681–3.

23. Ringvold A, Anderssen E, Kjønniksen I. Distribution of ascorbate in the anterior bovine eye. *Invest Ophthalmol Vis Sci.* 2000;**41**:20–3.

24. Ringvold A. Quenching of UV-induced fluorescence by ascorbic acid in the aqueous humour. *Acta Ophthalmol Scand* 1995;**73**:529–33.

25. Ringvold A. Aqueous humour and ultraviolet radiation. *Acta Ophthalmol (Copenh)* 1980;**58**:69–82.

26. Reiss GR, Werness PG, Zollman PE, Brubaker RF. Ascorbic acid levels in the aqueous humor of nocturnal and diurnal mammals. *Arch Ophthalmol* 1986;**104**:753–5.

27. Koskela TK, Reiss GR, Brubaker RF, Ellefson RD. Is the high concentration of ascorbic acid in the eye an adaptation to intense solar irradiation? *Invest Ophthalmol Vis Sci* 1989;**30**:2265–7.

28. Ringvold A. The significance of ascorbate in the aqueous humour protection against UV-A and UV-B. *Exp Eye Res.* 1996;**62**:261–4.

29. Ringvold A, Anderssen E, Kjønniksen I. Ascorbate in the corneal epithelium of diurnal and nocturnal species. *Invest Ophthalmol Vis Sci* 1998;**39**:2774–7.

30. Ringvold A, Anderssen E, Kjønniksen I. Impact of the environment on the mammalian corneal epithelium. *Invest Ophthalmol Vis Sci* 2003;**44**:10–5.

31. Wu K, Kojima M, Shui YB, Sasaki H, Sasaki K. Ultraviolet B-induced corneal and lens damage in guinea pigs on low-ascorbic acid diet. *Ophthalmic Res* 2004;**36**:277–83.

32. Suh MH, Kwon JW, Wee WR, Han YK, Kim JH, Lee JH. Protective effect of ascorbic acid against corneal damage by ultraviolet B irradiation: a pilot study. *Cornea* 2008;**27**:916–22.

33. Stojanovic A, Nitter TA. Correlation between ultraviolet radiation level and the incidence of late-onset corneal haze after photorefractive keratectomy. *J Cataract Refract Surg* 2001;**27**:404–10.

34. Stojanovic A, Ringvold A, Nitter T. Ascorbate prophylaxis for corneal haze after photorefractive keratectomy. *J Refract Surg* 2003;**19**:338–43.

35. Urrets-Zavalía JA, Maccio JP, Knoll EG, Cafaro T, Urrets-Zavalia EA, Serra HM. Surface alterations, corneal hypoesthesia, and iris atrophy in patients with climatic droplet keratopathy. *Cornea* 2007;**26**:800–4.

36. Hassell JR, Birk DE. The molecular basis of corneal transparency. *Exp Eye Res* 2010;**91**:326–35.

37. Maurice DM. The structure and transparency of the cornea. *J Physiol* 1957;**136**:263–86.

38. Meek KM, Leonard DW, Connon CJ, Dennis S, Khan S. Transparency, swelling and scarring in the corneal stroma. *Eye (Lond)* 2003;**8**:927–36.

39. Kivirikko KI, Prockop DJ. Hydroxylation of proline in synthetic polypeptides with purified protocollagen hydroxylase. *J Biol Chem* 1967;**242**:4007–12.

40. Ramachandran GN, Bansal M, Bhatnagar RS. A hypothesis on the role of hydroxyproline in stabilizing collagen structure. *Biochim Biophys Acta* 1973;**322**:166–71.

41. Pinnell SR. Abnormal collagens in connective tissue diseases. *Birth Defects Orig Artic Ser* 1975;**11**:23–30.

42. Peterkofsky B. Ascorbate requirement for hydroxylation and secretion of procollagen: relationship to inhibition of collagen synthesis in scurvy. *Am J Clin Nutr* 1991;**54**:1135–40.

43. Chojkier M, Spanheimer R, Peterkofsky B. Specifically decreased collagen biosynthesis in scurvy dissociated from an effect on proline hydroxylation and correlated with body weight loss. *In vitro* studies in guinea pig calvarial bones. *J Clin Invest* 1983;**72**:826–35.

44. Oyamada I, Bird TA, Peterkofsky B. Decreased extracellular matrix production in scurvy involves a humoral factor other than ascorbate. *Biochem Biophys Res Commun* 1988;**152**:1490–6.

45. Oyamada I, Palka J, Schalk EM, Takeda K, Peterkofsky B. Scorbutic and fasted guinea pig sera contain an insulin-like growth factor I-reversible inhibitor of proteoglycan and collagen synthesis in chick embryo chondrocytes and adult human skin fibroblasts. *Arch Biochem Biophys* 1990;**276**:85–93.

46. Peterkofsky B, Palka J, Wilson S, Takeda K, Shah V. Elevated activity of low molecular weight insulin-like growth factor-binding proteins in sera of vitamin C-deficient and fasted guinea pigs. *Endocrinology* 1991;**128**:1769–79.

47. Schwarz RI. Procollagen secretion meets the minimum requirements for the rate-controlling step in the ascorbate induction of procollagen synthesis. *J Biol Chem* 1985;**260**:3045–9.

48. Pacifici M, Iozzo RV. Remodeling of the rough endoplasmic reticulum during stimulation of procollagen secretion by ascorbic acid in cultured chondrocytes. A biochemical and morphological study. *J Biol Chem* 1988;**263**:2483–92.

49. Saika S. Ultrastructural effect of L-ascorbic acid 2-phosphate on cultured keratocytes. *Cornea* 1992;**11**:439–45.

50. Saika S. Ascorbic acid and proliferation of cultured rabbit keratocytes. *Cornea* 1993;**12**:191–8.

51. Phu D, Orwin EJ. Characterizing the effects of aligned collagen fibers and ascorbic acid derivatives on behavior of rabbit corneal fibroblasts. *Conf Proc IEEE Eng Med Biol Soc* 2009:4242–5.

52. Hayes S, Cafaro TA, Boguslawska PJ, Kamma-Lorger CS, Boote C, Harris J, et al. The effect of vitamin C deficiency and chronic ultraviolet-B exposure on corneal ultrastructure: a preliminary investigation. *Mol Vis* 2011;**17**:3107–15.

53. Kaufman HE, Capella JA, Robbins JE. The human corneal endothelium. *Am J Ophthalmol* 1966;**61**:835–41.

54. Bode AM, Vanderpool SS, Carlson EC, Meyer DA, Rose RC. Ascorbic acid uptake and metabolism by corneal endothelium. *Invest Ophthalmol Vis Sci* 1991;**32**:2266–71.

55. Shoham A, Hadziahmetovic M, Dunaief JL, Mydlarski MB, Schipper HM. Oxidative stress in diseases of the human cornea. *Free Radic Biol Med* 2008;**45**:1047–55.

56. Serbecic N, Beutelspacher SC. Vitamins inhibit oxidant-induced apoptosis of corneal endothelial cells. *Jpn J Ophthalmol* 2005;**49**:355–62.

57. Kaufman HE. The corneal endothelium in intraocular surgery. *J R Soc Med* 1980;**73**:165–71.

58. Joyce NC, Meklir B, Joyce SJ, Zieske JD. Cell cycle protein expression and proliferative status in human corneal cells. *Invest Ophthalmol Vis Sci* 1996;**37**:645–55.

59. Miyata K, Drake J, Osakabe Y, Hosokawa Y, Hwang D, Soya K, et al. Effect of donor age on morphologic variation of cultured human corneal endothelial cells. *Cornea* 2001;**20**:59–63.

60. Li W, Sabater AL, Chen YT, Hayashida Y, Chen SY, He H, et al. A novel method of isolation, preservation, and expansion of human corneal endothelial cells. *Invest Ophthalmol Vis Sci* 2007;**48**:614–20.

61. Perlman M, Baum JL, Kaye GI. Fine structure and collagen synthesis activity of monolayer cultures of rabbit corneal endothelium. *J Cell Biol* 1974;**63**:306–11.

62. Yue BY, Niedra R, Baum JL. Effects of ascorbic acid on cultured rabbit corneal endothelial cells. *Invest Ophthalmol Vis Sci* 1980;**19**:1471–6.

63. Shima N, Kimoto M, Yamaguchi M, Yamagami S. Increased proliferation and replicative lifespan of isolated human corneal endothelial cells with L-ascorbic acid 2-phosphate. *Invest Ophthalmol Vis Sci* 2011;**52**:8711–7.

64. Kimoto M, Shima N, Yamaguchi M, Amano S, Yamagami S. Role of hepatocyte growth factor in promoting the growth of human corneal endothelial cells stimulated by L-ascorbic acid 2-phosphate. *Invest Ophthalmol Vis Sci* 2012;**53**:7583–9.

Vitamin Transport Across the Blood–Retinal Barrier: Focus on Vitamins C, E, and Biotin

Ken-ichi Hosoya, Yoshiyuki Kubo

Department of Pharmaceutics, Graduate School of Medicine and Pharmaceutical Sciences, University of Toyama, Toyama, Japan

INTRODUCTION

The retina forms the anatomic and physiologic basis for vision and has a unique position with regard to solute transport in that the blood–retinal barrier (BRB) separates the neural retina from the circulating blood. The BRB, which is formed by complex tight junctions of retinal capillary endothelial cells (inner BRB) and retinal pigment epithelial cells (outer BRB), restricts nonspecific transport between the circulating blood and retina (Fig. 32.1).[1,2] The inner BRB is responsible for nourishment of the inner two-thirds of the retina whereas the outer BRB is responsible for nourishment of the remaining one-third of the retina.[2] Recent progress in BRB research has revealed that retinal capillary endothelial cells and retinal pigment epithelial cells express a variety of unique transporters, which play a pivotal role in the influx transport of essential molecules.[3]

For proper functioning, the retina needs to obtain macronutrients, such as D-glucose and amino acids, and micronutrients, such as vitamins and metals, from the circulating blood. Neuronal cells including photoreceptor cells require a large amount of metabolic energy for phototransduction and neurotransduction. Facilitative glucose transporter 1 (GLUT1), solute carrier 2A1 (SLC2A1), L (leucine-preferring)-type amino acid transporter 1 (LAT1), SLC7A5, and other amino acid transporters are expressed at the inner and outer BRB and supply D-glucose and amino acids to the retina to satisfy the great demand for metabolic energy and neurotransduction.[3] Because of consumption of a large amount of metabolic energy, oxygen consumption by the retina is much greater than that by any other tissue. In addition, the retina is subject to high levels of cumulative irradiation, which causes the production of reactive oxygen species, and it contains an abundance of chromophores

such as rhodopsin, melanin, and lipofuscin. Therefore, the retina needs antioxidants for protection against light-induced oxidative stress. Vitamin C, vitamin E, and glutathione (GSH) are important antioxidants that act as a defense against oxidative attack on the retina. Although GSH is synthesized from L-glutamate, L-cysteine, and glycine in the retina, vitamins need to be transported from the circulating blood across the BRB. This chapter presents an overview of solute transport across the BRB and focuses on how vitamins C, E, and biotin are transferred to the retina across the BRB.

TRANSPORT SYSTEM AT THE BLOOD–RETINAL BARRIER

Cellular transport can be classified into passive diffusion and carrier-mediated transport. However, all nonspecific transport across the BRB, namely, paracellular diffusion, is restricted by the tight junctions in the retinal capillary endothelial cells (inner BRB) and the retinal pigment epithelial cells (outer BRB). Thus, the physiologic role of transcellular transport is very important in the uptake of essential nutrients and elimination of toxic compounds and metabolites across the BRB.[3] Transcellular transport at the BRB is performed by transcellular diffusion and carrier-mediated transport (Fig. 32.2). Carrier-mediated transport depends on the functional properties of transporter or receptor proteins, such as substrate affinity and capacity, while transcellular diffusion depends on the physicochemical properties of compounds, such as lipophilicity and ionic nature.

Regarding carrier-mediated transport, receptor-mediated transport is known to act as the transport system for high-molecular weight compounds across the tissue barrier, and the involvement of scavenger receptors class

Handbook of Nutrition, Diet, and the Eye
http://dx.doi.org/10.1016/B978-0-12-401717-7.00032-0

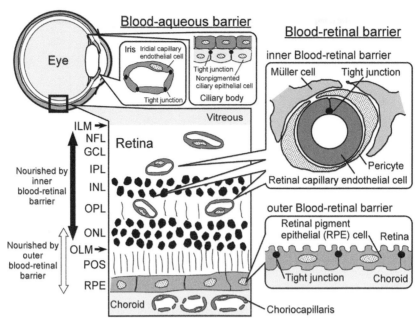

FIGURE 32.1 Schematic diagram of the blood–retinal barrier (BRB). The retinal cell layers seen histologically consist of: RPE: retinal pigment epithelium; POS: photoreceptor outer segments; OLM: outer limiting 'membrane'; ONL: outer nuclear layer; OPL: outer plexiform layer; INL: inner nuclear layer; IPL: inner plexiform layer; GCL: ganglion cell layer; NFL: nerve fiber layer; ILM: inner limiting 'membrane'. *Source: Hosoya and Tomi (2005).[2] Reproduced with kind permission from The Pharmaceutical Society of Japan.*

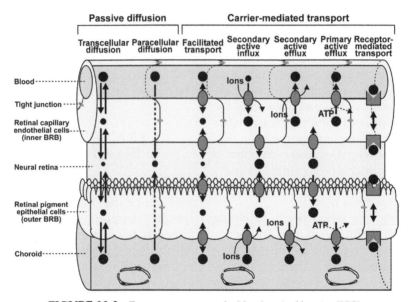

FIGURE 32.2 Transport system at the blood–retinal barrier (BRB).

B, type I (SR-BI), insulin receptors, and transferrin receptors has been reported.[4–6] In receptor-mediated transport, ligands, such as insulin, bind to their specific receptors on the cell surface to undergo endocytosis while transporter-mediated transport is largely involved in the transport of low-molecular weight compounds, such as nutrients, across the BRB. Transporter-mediated transport can be subdivided into facilitated transport, secondary active influx and efflux transport, and primary active efflux transport, and the molecules responsible for this are transporter proteins that are generally 12 membrane-spanning proteins.

To date, approximately 400 transporter genes have been identified in humans. They can be classified into adenosine triphosphate-binding cassette (ABC) transporters and SLC transporters, and the expression of various transporters at the BRB has been reported (Table 32.1).[3] ABC transporters are involved in the primary active efflux to eliminate toxic and waste compounds by coupling with ATP hydrolysis. P-Glycoprotein (MDR1/ABCB1) is a representative one at the BRB.[7,8] SLC transporters are involved in facilitated transport and secondary active influx and efflux transport by coupling with concentration or ion gradients between

TABLE 32.1 Membrane Transporters Expressed at the Inner Blood–Retinal Barrier

Transporters	Alias	Transport Substrates	Transport Direction (Influx or Efflux)
SLC1A4/ SLC1A5	ASCT	D/L-Serine	Influx
SLC2A1	GLUT1	D-Glucose	Influx
		DHAA	Influx
SLC5A6	SMVT	Biotin	Influx
SLC6A6	TauT	Taurine	Influx
		GABA	Influx
SLC6A8	CRT	Creatine	Influx
SLC6A9	GlyT1	Glycine	Influx
SLC7A1	CAT1	L-Arginine	Influx
SLC7A5	LAT1	L-Leucine	Influx
SLC7A11	xCT	L-Cystine	Influx
		L-Glutamate	Influx
SLC16A1	MCT1	L-Lactate	Influx
SLC19A1	RFC1	MTF	Influx
SLC22A5	OCTN2	L-Carnitine	Influx
SLC22A8	OAT3	Organic anions	Efflux
SLC29A2	ENT2	Nucleosides	Influx
SLC38A2	ATA2	L-Proline	Efflux
	SNAT2	L-Alanine	Efflux
SLCO1A4	OATP1A4	Organic anions	Efflux
	oatp2		
ABCB1	MDR1	Lipophilic drugs	Efflux
	P-gp	Organic cations	Efflux
ABCC4	MRP4	Organic anions	Efflux
ABCG2	BCRP	Organic anions	Efflux
	MTX		

DHAA: dehydroascorbic acid; GABA: γ-aminobutyric acid; MTF: methyltetrahydrofolate.

intracellular and extracellular spaces. At the BRB, the uptake of essential nutrients across the BRB is carried out by SLC transporters, and facilitated transport of D-glucose mediated by GLUT1 is a representative example of this in retinal capillary endothelial cells.[9]

In research involving pharmacokinetics and biologic chemistry, membrane transport can be monitored by means of various *in vivo* and *in vitro* methods. As major experimental approaches in the BRB research, integration plot and retinal uptake index methods have been adopted in *in vivo* analysis,[10] and primary-cultured cells and established cell lines of retinal capillary endothelial cells and retinal pigment epithelial cells are used in *in vitro* uptake analysis. In particular, a conditionally immortalized rat retinal endothelial cell line (TR-iBRB2 cells) is an *in vitro* model of the inner BRB,[11] with uptake properties that are significantly correlated with *in vivo* BRB permeability.[12] In addition to these established *in vivo* and *in vitro* methods, new approaches based on genomics and proteomics will be helpful to understand the detailed features of transport across the BRB since the mechanism for the transport of some nutrients and drugs remains to be discovered.[13–15]

VITAMIN C IN THE RETINA

L-Ascorbic acid (AA), also known as vitamin C, functions as a cofactor involved in the enzymatic biosynthesis of collagen, catecholamines, and peptide neurohormones and antioxidant and free radical scavengers to detoxify free radicals in the retina as well as the brain.[16] It is provided exogenously from the diet and transported intracellularly because it is not biosynthesized in the human body.[16] Even though rodents such as mice and rats can synthesize AA from D-glucose in the liver, AA needs to be transported from the circulating blood to the retina across the BRB. At physiologic pH (around pH 7.4), AA exists as a monovalent anion with pKa_1 and pKa_2 values of 4.2 and 11.8, respectively, and it does not easily cross the lipid bilayer. AA can be oxidized, enzymatically or nonenzymatically, with the removal of two hydrogen atoms to form dehydro-L-ascorbic acid (DHAA). Although DHAA is a weak acid, it is not ionized under physiologic conditions and is present in solution as a hydrate, with a structure that is very similar to that of D-glucose. The concentration of AA in plasma ranges from 50 to 100 μM, whereas DHAA is present at lower levels (about 10 μM).[17,18] In contrast, the intracellular concentration of AA in the neural retina (about 1.6 mM in the rat and guinea pig retina) is about 15 times higher than in plasma,[18] supporting the theory that vitamin C is transported to the retina from the circulating blood across the BRB.

Two different transporters have been identified at a molecular level for the transport of AA in mammalian cells. Sodium (Na+)-dependent vitamin C transporter (SVCT) 1 and 2, which are members of the SLC23A family, mediate concentrative high-affinity AA transport that is driven by an Na+ electrochemical gradient,[19] while GLUT1, GLUT3 (SLC2A3), and GLUT4 (SLC2A4) mediate equilibrative and relatively low-affinity DHAA transport.[20]

VITAMIN C TRANSPORT ACROSS THE BLOOD–RETINAL BARRIER

GLUT1, which transports D-glucose as well as DHAA, is localized in both the luminal and abluminal membranes of the inner BRB and in both the brush-border

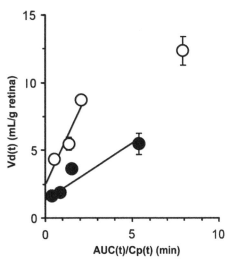

FIGURE 32.3 Integration plot of the initial uptake of [^{14}C]dehydro-L-ascorbic acid (DHAA) by the retina after intravenous administration. [^{14}C]DHAA (5 µCi/rat) was injected into streptozotocin-induced diabetic rats (closed circle) and normal rats (open circle) via the femoral vein. Each point represents the mean ± SEM (n = 3–5). The apparent blood-to-retina influx clearance (CL_{retina}) of [^{14}C]DHAA (µL/min/g retina) was determined by integration plot analysis. As an index of the tissue distribution characteristics of [^{14}C]DHAA, the apparent retina-to-plasma concentration ratio (Vd) as a function of time is used. This ratio ($Vd(t)$) (mL/g retina) is defined as the amount of [^{14}C] per gram retina divided by that per milliliter plasma, calculated over the period of the experiment. The CL_{retina} can be described by the following relationship: $Vd(t) = CL_{retina} \times AUC(t)/Cp(t) + V_i$, where $Cp(t)$ (dpm/mL) and V_i (mL/g tissue) represent the plasma concentration at time t and the rapidly equilibrated distribution volume of [^{14}C]DHAA, respectively; $AUC(t)$ (dpm·min/mL) is the area under the plasma concentration time curve of [^{14}C]DHAA from time 0 to t. When $AUC(t)/Cp(t)$ (min) is plotted versus $Vd(t)$, as shown in Fig. 3, the slope represents CL_{retina} (µL/min/g retina). *Source: Minamizono* et al. *(2006).[26] Reproduced with kind permission from The Pharmaceutical Society of Japan.*

and basolateral membranes of the outer BRB.[21] There is no evidence in the literature for the expression of SVCT1 and SVCT2 at the inner and outer BRB, although SVCT2 messenger RNA (mRNA) is expressed in TR-iBRB2 cells used as an *in vitro* model of the inner BRB, and SVCT1 and SVCT2 mRNAs are expressed in ARPE-19 cells used as an *in vitro* model of the outer BRB.[22,23] Notably, [^{14}C] AA uptake by TR-iBRB2 cells and primary-cultured human retinal pigment epithelial cells is negligible compared with the uptake of [^{14}C]DHAA.[22,24]

An *in vivo* intravenous administration study using rats demonstrated that the apparent blood-to-retina influx clearance of [^{14}C]DHAA across the BRB was about 38-fold greater than that of [^{14}C]AA, and most of the [^{14}C]DHAA injected into the femoral vein was converted to [^{14}C]AA in the retina.[22] It is suggested that vitamin C is mainly transported as the DHAA form across the BRB and accumulates as the AA form in the retina. The initial uptake rate of [^{14}C]DHAA in TR-iBRB2 cells was also found to be about 38-fold greater than that of [^{14}C] AA. [^{14}C]DHAA uptake by TR-iBRB2 cells occurred in

an Na$^+$-independent and concentration-dependent manner with a Michaelis–Menten constant (K_m) of 93 µM and was inhibited by substrates and inhibitors of GLUT.[22] In the light of these findings, GLUT1 at the BRB plays an important role in supplying vitamin C, as well as D-glucose, to the neural retina. Although the affinity of DHAA for GLUT1 (K_m = 93 µM) is greater than that of D-glucose (the K_m estimated for D-glucose transport across the BRB is 7.8 mM),[22,25] DHAA uptake through GLUT1 is competitively inhibited by D-glucose, and the normal D-glucose concentration in most mammals is about 5 mM. D-Glucose inhibited [^{14}C]DHAA uptake by TR-iBRB2 cells in a concentration-dependent manner with a 50% inhibition constant (IC$_{50}$) of 5.6 mM.[22] Therefore, DHAA transport by GLUT1 at the BRB does not exhibit complete inhibition under normal conditions. However, in diabetes mellitus, it is expected that hyperglycemia restricts the supply of vitamin C to the retina owing to inhibition of GLUT1. Indeed, the apparent blood-to-retina influx clearance of [^{14}C]DHAA across the BRB was reduced by 66% in streptozotocin-induced diabetic rats compared with normal rats (Fig. 32.3).[26] Accordingly, diabetic patients may experience enhanced oxidative stress in the retina following a reduction in the influx transport of DHAA.

VITAMIN C TRANSPORT IN MÜLLER CELLS

Müller cells are astrocyte-like glial cells found in the vertebrate retina, which serve as support cells for retinal neurons and the inner BRB.[27] Müller cell end-feet cover virtually all the capillary walls in the retina and express GLUT1.[9] There is no evidence in the literature for expression of SVCT1 and SVCT2 in Müller cells, although SVCT2 mRNA is expressed in a conditionally immortalized rat Müller cell line (TR-MUL5 cells) used as an *in vitro* model of Müller cells.[28,29] Notably, [^{14}C]AA uptake by TR-MUL5 cells is much lower than the uptake of [^{14}C]DHAA and [^{14}C]AA uptake is inhibited in the presence of D-glucose, suggesting that [^{14}C] AA in the medium is converted to [^{14}C]DHAA. [^{14}C] DHAA uptake by TR-MUL5 cells takes place in an Na$^+$-independent and concentration-dependent manner with a K_m of 198 µM and is inhibited by substrates and inhibitors of GLUT. TR-MUL5 cells and rat retina express GLUT1 with a molecular weight of 45 kDa. High-performance liquid chromatography analysis supports that DHAA is converted to AA in TR-MUL5 cells.[29] Although GLUT1 is not a concentrative transporter, DHAA is rapidly reduced to AA and accumulates in TR-MUL5 cells. TR-MUL5 cells can synthesize GSH through L-cystine uptake via a cystine/glutamate transporter (system xc$^-$/xCT/SLC7A11).[28] DHAA in Müller cells is

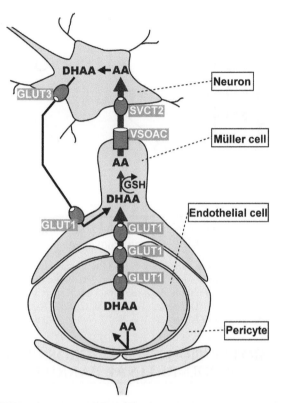

FIGURE 32.4 Vitamin C transport in the retina. DHAA: dehydro-ascorbic acid; AA: ascorbic acid; GSH: glutathione; VSOC: volume-sensitive organic osmolyte and anion channel; GLUT1: facilitative glucose transporter 1; GLUT3: facilitative glucose transporter 3; SVCT2: Na$^+$-dependent vitamin C transporter 2.

probably reduced directly by GSH and also by NADPH-dependent and GSH-dependent enzymes.[30,31] Müller cell end-feet are in close contact with retinal capillaries and neuronal cells.[26] This supported the hypothesis that DHAA crosses from blood via GLUT1 in the inner BRB and is then taken up by Müller cells. Müller cells also take up DHAA generated in the neuronal cells by oxidation of AA. Immunohistochemical studies have shown that a large portion of retinal AA is present in Müller cells in the guinea pig retina,[32] and GSH is preferentially distributed in Müller cells of the rabbit retina.[33] Neuronal cells cannot synthesize AA *de novo* from D-glucose but instead take up AA from the extracellular fluid. In an *in situ* hybridization, SVCT2 labeling was observed exclusively in the inner nuclear layer (bipolar, amacrine, and horizontal-cell bodies).[19] TR-MUL5 cells release AA when TR-MUL5 cell swelling is caused by glutamate and under hypotonic conditions. AA efflux by TR-MUL5 cells under hypotonic conditions was inhibited in the presence of volume-sensitive organic osmolyte and anion channel (VSOAC) inhibitors.[29,34] Thus, AA efflux by Müller cells may involve swelling-induced opening of VSOAC and occur just in time and in the right place to reach active neurons. It is proposed that AA is recycled between Müller cells and neurons in a process involving

GLUT1, GLUT3, and SVCT2.[35] Neurons express GLUT3 and probably also express SVCT2, whereas Müller cells express GLUT1.[36] Neurons take up AA from the extracellular space via SVCT2 for use as an antioxidant. When AA participates in the detoxification of free radicals, it is converted to DHAA, which undergoes efflux from neurons via GLUT3. GLUT3 is a facilitative transporter with no driving force, and the direction of transport is dependent on the gradient of the substrate concentration. DHAA is taken up via GLUT1 of Müller cells and reduced to AA by GSH and/or NADPH-dependent and GSH-dependent enzymes. AA is subsequently released, probably via VSOAC, into the extracellular space to be reused by neurons (Fig. 32.4).

VITAMIN E IN THE RETINA

Although naturally occurring vitamin E exists in eight chemical forms (α-, β-, γ-, and δ-tocopherol and α-, β-, γ-, and δ-tocotrienol), α-tocopherol is the major constituent found in mammalian tissues and exhibits the highest biologic activity.[37,38] Vitamin E is regarded as an effective antioxidant that is capable of modulating peroxidation activity.[38] Systemic vitamin E administration has been proposed to have preventive and therapeutic effects in human retinopathies.[39] In contrast, prolonged vitamin E deficiency may lead to retinal degeneration.[40] Weaned rats fed a diet deficient in vitamin E for 15 weeks exhibited about an eight-fold reduction in the concentration of vitamin E in the neural retina (10 µg vitamin E/g dry weight retina), compared with those fed a regular diet (78 µg vitamin E/g dry weight retina).[41] Therefore, vitamin E, mostly α-tocopherol, needs to be transported to the retina from the circulating blood to protect the neural retina against oxidative stress because the retina is the only tissue in which light is directly focused on cells and causes free radical oxidation.

α-Tocopherol is exclusively associated with lipoproteins, such as high-density lipoprotein (HDL) and low-density lipoprotein (LDL), in the circulating blood. Quantitatively, 80–95% of total plasma α-tocopherol is associated with HDL among mouse plasma lipoproteins.[42] Voegele *et al.* reported that human serum albumin does not bind to D-α-tocopherol.[43] Since HDL-associated α-tocopherol does not cross the BRB by passive diffusion, a transport system is necessary to cross the BRB.

Several HDL binding proteins have been identified and their roles in HDL metabolism have been elucidated. HBP (vigilin), which lacks a transmembrane domain, is responsive to cell cholesterol levels. Another candidate HDL receptor, HB2, one of a pair of liver HDL binding proteins, has a high sequence homology with adhesion molecules, particularly the activated leukocyte-cell adhesion molecule,[44] although their physiologic significance remains unknown. The cellular uptake of HDL-associated lipids is reported to be mediated by SR-BI.[45] The level of SR-BI expression correlates with both the selective transfer of cholesteryl ester into cells and cholesterol efflux from cells, and the transfers are probably mediated after the docking of HDL at the cell surface.[46]

VITAMIN E TRANSPORT ACROSS THE BLOOD–RETINAL BARRIER

High-density lipoprotein-associated [^{14}C]α-tocopherol ([^{14}C]α-tocopherol-HDL) uptake by TR-iBRB2 cells exhibits a time-dependent increase at 37°C and is reduced by 90% at 4°C compared with that at 37°C, suggesting involvement of energy-dependent carrier-mediated process(es).[6] BLT-1, a specific inhibitor of SR-BI-mediated lipid transfer,[47] inhibits [^{14}C]α-tocopherol-HDL uptake with an IC_{50} of 23.2 nM. Notably, immunoblot analysis revealed that SR-BI protein is expressed in TR-iBRB2 cells and immunohistochemical analysis showed that SR-BI protein is localized in rat retinal capillary endothelial cells. Moreover, an RNA interference study of SR-BI in TR-iBRB2 cells showed that the reduction in SR-BI protein expression is related to a reduction in [^{14}C]α-tocopherol-HDL uptake, supporting SR-BI-mediated α-tocopherol transport at the inner BRB.[6] Tserentsoodol et al. used immunohistochemical analysis to show that SR-BI proteins are found in retinal pigment epithelium/choriocapillaris regions, ganglion cells, and Müller cells as well as the photoreceptors in the monkey retina.[48] They proposed a mechanism of lipid transport including HDL via SR-BI in the retina. HDL in the circulating blood enters the retina via SR-BI in the retinal pigment epithelial cells. Retinal pigment epithelial cells take up lipoproteins and transfer lipid to neuronal cells including photoreceptors. Therefore, α-tocopherol-HDL may be taken up via SR-BI in the inner and outer BRB and transfer α-tocopherol-HDL to neuronal cells via SR-BI.

BIOTIN IN THE RETINA

Biotin is an essential cofactor for five carboxylases catalyzing essential steps in the fatty acid biosynthesis, glyconeogenesis, and catabolism of several branched-chain amino acids and odd-chain fatty acids.[49] Recently, biotin has also been implicated in glutamine synthetase activity and glutamate decarboxylase activity as a 42 kDa biotin-coupled protein in the chick retina.[50] Although there is no information about the biotin concentration in the mammalian retina, biotin in the retina needs to be supplied from the circulating blood because biotin is a water-soluble vitamin and is not synthesized in the retina.

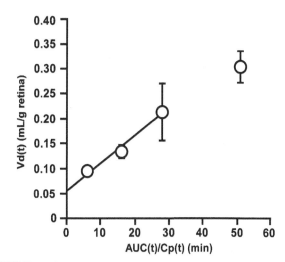

FIGURE 32.5 Integration plot of the initial uptake of [^{3}H]biotin by the retina after intravenous administration. [^{3}H]Biotin (3 μCi/rat) was injected into the normal rat via the femoral vein. Each point represents the mean ± SEM (n = 4). *Source: Ohkura et al. (2010).[10] Reproduced with kind permission from Elsevier.*

BIOTIN TRANSPORT ACROSS THE BLOOD–RETINAL BARRIER

An *in vivo* intravenous administration study using rats demonstrated that the apparent blood-to-retina influx clearance of [^{3}H]biotin across the BRB was 5.6 μL/min/g retina (Fig. 32.5).[10] This figure is nine-fold greater than that of D-mannitol, which is a paracellular diffusion marker (0.63 μL/min/g retina).[51] Using a carotid artery injection (retinal uptake index) method, [^{3}H]biotin uptake by the retina across the BRB was inhibited by 5 mM biotin and pantothenic acid, but not by ascorbic acid, supporting carrier-mediated transport of biotin from the circulating blood to the retina across the BRB. [^{3}H]Biotin uptake by TR-iBRB2 cells takes place in an Na$^+$-, temperature-, and concentration-dependent manner with a K_m of 146 μM. This process is inhibited in the presence of pantothenic acid, lipoic acid, and desthiobiotin.[10] The characteristics of [^{3}H]biotin uptake by ARPE-19 cells are the similar to those of TR-iBRB2 cells.[52] Sodium (Na$^+$)-dependent multivitamin transporter (SMVT/SLC5A6) mRNA is expressed in TR-iBRB2 cells, isolated rat retinal endothelial cells, and ARPE-19 cells.[10,52] In the light of these findings, SMVT at the inner and outer BRB plays a key role in supplying biotin as well as pantothenic acid to the retina. SMVT transports pantothenic acid as a substrate.[53]

CONCLUSIONS

Vitamins C, E, and biotin are essential for the maintenance of retinal health. The neural retina obtains vitamins across the inner and outer BRB, and the transfer processes are mediated by transporters and receptors. Oxidized ascorbic acid (DHAA) in the blood is transported

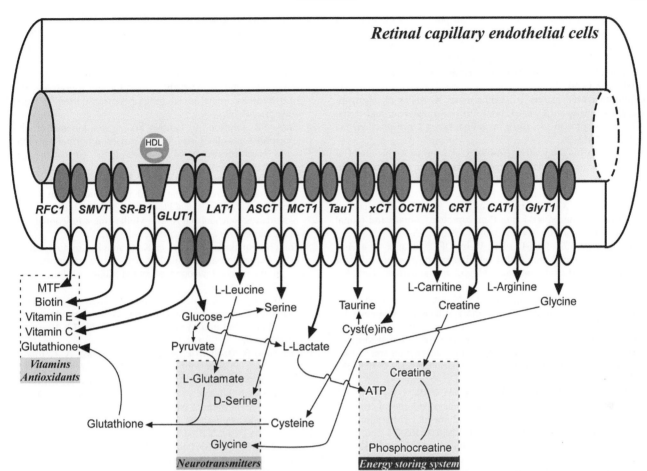

FIGURE 32.6 Hypothetical localization and physiologic function transporters and receptors at the inner blood–retinal barrier (BRB). Transporters and receptors at the inner BRB play an essential role in supplying vitamins and nutrients to the retina. MTF: methyltetrahydrofolate; HDL: high-density lipoprotein; RFC1: reduced folate carrier 1; SMVT: Na$^+$-dependent multivitamin transporter; SR-BI: scavenger receptor class B, type I; GLUT1: facilitative glucose transporter 1; LAT1: L (leucine-preferring)-type amino acid transporter 1; ASCT: alanine–serine–cysteine transporter; MCT1: monocarboxylate transporter 1; TauT: taurine transporter; xCT: cystine/glutamate transporter; OCTN2: organic cation transporter novel 2/carnitine transporter; CRT: creatine transporter; CAT1: cationic amino acid transporter 1; GlyT1: glycine transporter 1.

via GLUT1 of the inner and outer BRB to the retina and is taken up via GLUT1 of Müller cells and reduced to AA. AA is subsequently released, probably via VSOAC, into the extracellular space for use by neurons. It is likely that the transfer of DHAA via this mechanism is inhibited to a significant degree in diabetes. α-Tocopherol is exclusively associated with HDL. α-Tocopherol-HDL is taken up via SR-BI of the inner and outer BRB and transfers α-tocopherol to the neural retina. Biotin is transported to the retina via SMVT of the inner and outer BRB. In addition to the transport processes covered in this chapter, there are several transporters at the inner and outer BRB, which carry essential molecules from the circulating blood to the neural retina (Fig. 32.6).

TAKE-HOME MESSAGES

- The blood–retinal barrier (BRB) is composed of retinal capillary endothelial cells (inner BRB) and retinal pigment epithelial cells (outer BRB).

- The inner and outer BRB are equipped with a variety of transporters and receptors and regulate nutrient flux between the circulating blood and the retina.
- Oxidized ascorbic acid (DHAA) in the blood is transported via GLUT1 of the inner and outer BRB to the retina.
- DHAA is taken up via GLUT1 of Müller cells and reduced to ascorbic acid.
- α-Tocopherol high-density lipoprotein is taken up via scavenger receptor class B, type I of the inner and outer BRB and transfers α-tocopherol to the neural retina.
- Biotin is transported to the retina via the Na$^+$-dependent multivitamin transporter of the inner and outer BRB.

References

1. Cunha-Vaz JG. The blood–retinal barriers system. Basic concepts and clinical evaluation. *Exp Eye Res* 2004;**78**:715–21.
2. Hosoya K, Tomi M. Advances in the cell biology of transport via the inner blood–retinal barrier: establishment of cell lines and transport functions. *Biol Pharm Bull* 2005;**28**:1–8.

3. Hosoya K, Tomi M, Tachikawa M. Strategies for therapy of retinal diseases using systemic drug delivery: relevance of transporters at the blood–retinal barrier. *Expert Opin Drug Deliv* 2011;**8**:1571–87.

4. Naeser P. Insulin receptors in human ocular tissues. Immunohistochemical demonstration in normal and diabetic eyes. *Ups J Med Sci* 1997;**102**:35–40.

5. Yefimova MG, Jeanny JC, Guillonneau X, Keller N, Nguyen-Legros J, Sergeant C, et al. Iron, ferritin, transferrin, and transferrin receptor in the adult rat retina. *Invest Ophthalmol Vis Sci* 2000;**41**:2343–51.

6. Tachikawa M, Okayasu S, Hosoya K. Functional involvement of scavenger receptor class B, type I, in the uptake of alpha-tocopherol using cultured rat retinal capillary endothelial cells. *Mol Vis* 2007;**13**:2041–7.

7. Greenwood J. Characterization of a rat retinal endothelial cell culture and the expression of P-glycoprotein in brain and retinal endothelium. in vitro. *J Neuroimmunol* 1992;**39**:123–32.

8. Shen J, Cross ST, Tang-Liu DD, Welty DF. Evaluation of an immortalized retinal endothelial cell line as an *in vitro* model for drug transport studies across the blood–retinal barrier. *Pharm Res* 2003;**20**:1357–63.

9. Kumagai AK, Glasgow BJ, Pardridge WM. GLUT1 glucose transporter expression in the diabetic and nondiabetic human eye. *Invest Ophthalmol Vis Sci* 1994;**35**:2887–94.

10. Ohkura Y, Akanuma S, Tachikawa M, Hosoya K. Blood-to-retina transport of biotin via Na$^+$-dependent multivitamin transporter (SMVT) at the inner blood–retinal barrier. *Exp Eye Res* 2010;**91**:387–92.

11. Hosoya K, Tomi M, Ohtsuki S, Takanaga H, Ueda M, Yanai N, et al. Conditionally immortalized retinal capillary endothelial cell lines (TR-iBRB) expressing differentiated endothelial cell functions derived from a transgenic rat. *Exp Eye Res* 2001;**72**:163–72.

12. Kubo Y, Fukui E, Akanuma S, Tachikawa M, Hosoya K. Application of membrane permeability evaluated in *in vitro* analyses to estimate blood–retinal barrier permeability. *J Pharm Sci* 2012;**101**:2596–605.

13. Han YH, Sweet DH, Hu DN, Pritchard JB. Characterization of a novel cationic drug transporter in human retinal pigment epithelial cells. *J Pharmacol Exp Ther* 2001;**296**:450–7.

14. Zhang N, Kannan R, Okamoto CT, Ryan SJ, Lee VH, Hinton DR. Characterization of brimonidine transport in retinal pigment epithelium. *Invest Ophthalmol Vis Sci* 2006;**47**:287–94.

15. Hosoya K, Yamamoto A, Akanuma S, Tachikawa M. Lipophilicity and transporter influence on blood–retinal barrier permeability: a comparison with blood–brain barrier permeability. *Pharm Res* 2010;**27**:2715–24.

16. Friedman PA, Zeidel ML. Victory at C. *Nat Med* 1999;**5**:620–1.

17. Nakayama H, Akiyama S, Inagaki M, Gotoh Y, Oguchi K. Dehydroascorbic acid and oxidative stress in haemodialysis patients. *Nephrol Dial Transplant* 2001;**16**:574–9.

18. Nielsen JC, Naash MI, Anderson RE. The regional distribution of vitamin E and C in mature and premature human retinas. *Invest Ophthalmol Vis Sci* 1988;**29**:22–6.

19. Tsukaguchi H, Tokui T, Mackenzie B, Berger UV, Chen XZ, Wang Y, et al. A family of mammalian Na$^+$-dependent L-ascorbic acid transporters. *Nature* 1999;**399**:70–5.

20. Vera JC, Rivas CI, Fischbarg J, Golde DW. Mammalian facilitative hexose transporters mediate the transport of dehydroascorbic acid. *Nature* 1993;**364**:79–82.

21. Takata K, Kasahara T, Kasahara M, Ezaki O, Hirano H. Ultracytochemical localization of the erythrocyte/HepG2-type glucose transporter (GLUT1) in cells of the blood–retinal barrier in the rat. *Invest Ophthalmol Vis Sci* 1992;**33**:377–83.

22. Hosoya K, Minamizono A, Katayama K, Terasaki T, Tomi M. Vitamin C transport in oxidized form across the rat blood–retinal barrier. *Invest Ophthalmol Vis Sci* 2004;**45**:1232–9.

23. Ganapathy G, Ananth S, Smith SB, Martin PM. Vitamin C transporters in the retina. In: Tombran-Tink J, Barnstable CJ, editors. *Ocular Transporters in Ophthalmic Diseases and Drug Delivery*. Totowa, NJ: Humana Press; 2008. pp. 437–50.

24. Root-Bernstein R, Busik JV, Henry DN. Are diabetic neuropathy, retinopathy and nephropathy caused by hypoglycemic exclusion of dehydroascorbate uptake by glucose transporters? *J Theor Biol* 2002;**216**:345–59.

25. Ennis SR, Johnson JE, Pautler EL. *In situ* kinetics of glucose transport across the blood–retinal barrier in normal rats and rats with streptozocin-induced diabetes. *Invest Ophthalmol Vis Sci* 1982;**23**:447–56.

26. Minamizono A, Tomi M, Hosoya K. Inhibition of dehydroascorbic acid transport across the rat blood–retinal and –brain barriers in experimental diabetes. *Biol Pharm Bull.* 2006;**29**:2148–50.

27. Newman E, Reichenbach A. The Müller cell: a functional element of the retina. *Trends Neurosci* 1996;**19**:307–12.

28. Tomi M, Funaki T, Abukawa H, Katayama K, Kondo T, Ohtsuki S, et al. Expression and regulation of L-cystine transporter, system xc⁻, in the newly developed rat retinal Müller cell line (TR-MUL). *Glia* 2003;**43**:208–17.

29. Hosoya K, Nakamura G, Akanuma S, Tomi M, Tachikawa M. Dehydroascorbic acid uptake and intracellular ascorbic acid accumulation in cultured Müller glial cells (TR-MUL). *Neurochem Int* 2008;**52**:1351–7.

30. Del Bello B, Maellaro E, Sugherini L, Santucci A, Comporti M, Casini AF. Purification of NADPH-dependent dehydroascorbate reductase from rat liver and its identification with 3 alpha-hydroxysteroid dehydrogenase. *Biochem J* 1994;**304**:385–90.

31. May JM, Mendiratta S, Hill KE, Burk RF. Reduction of dehydroascorbate to ascorbate by the selenoenzyme thioredoxin reductase. *J Biol Chem* 1997;**272**:22607–10.

32. Woodford BJ, Tso MO, Lam KW. Reduced and oxidized ascorbates in guinea pig retina under normal and light-exposed conditions. *Invest Ophthalmol Vis Sci* 1983;**24**:862–7.

33. Pow DV, Crook DK. Immunocytochemical evidence for the presence of high levels of reduced glutathione in radial glial cells and horizontal cells in the rabbit retina. *Neurosci Lett* 1995;**193**:25–8.

34. Ando D, Kubo Y, Akanuma S, Yoneyama D, Tachikawa M, Hosoya K. Function and regulation of taurine transport in Müller cells under osmotic stress. *Neurochem Int* 2012;**60**:597–604.

35. Wilson JX. The physiological role of dehydroascorbic acid. *FEBS Lett* 2002;**527**:5–9.

36. Watanabe T, Nagamatsu S, Matsushima S, Kirino T, Uchimura H. Colocalization of GLUT3 and choline acetyltransferase immunoreactivity in the rat retina. *Biochem Biophys Res Commun* 1999;**256**:505–11.

37. Brigelius-Flohe R, Traber MG, Vitamin E: function and metabolism. *FASEB J* 1999;**13**:1145–55.

38. Burton GW, Traber MG, Vitamin E: antioxidant activity, biokinetics, and bioavailability. *Annu Rev Nutr* 1990;**10**:357–82.

39. Bursell SE, Clermont AC, Aiello LP, Aiello LM, Schlossman DK, Feener EP, et al. High-dose vitamin E supplementation normalizes retinal blood flow and creatinine clearance in patients with type 1 diabetes. *Diabetes Care* 1999;**22**:1245–51.

40. Parks E, Traber MG. Mechanisms of vitamin E regulation: research over the past decade and focus on the future. *Antioxid Redox Signal* 2000;**2**:405–12.

41. Stephens RJ, Negi DS, Short SM, van Kuijk FJ, Dratz EA, Thomas DW. Vitamin E distribution in ocular tissues following long-term dietary depletion and supplementation as determined by microdissection and gas chromatography–mass spectrometry. *Exp Eye Res* 1988;**47**:237–45.

42. Sattler W, Levak-Frank S, Radner H, Kostner GM, Zechner R. Muscle-specific overexpression of lipoprotein lipase in transgenic mice results in increased alpha-tocopherol levels in skeletal muscle. *Biochem J* 1996;**318**:15–9.

43. Voegele AF, Jerkovic L, Wellenzohn B, Eller P, Kronenberg F, Liedl KR, Dieplinger H. Characterization of the vitamin E-binding properties of human plasma afamin. *Biochemistry* 2002;**41**:14532–8.

44. Fidge NH. High density lipoprotein receptors, binding proteins, and ligands. *J Lipid Res* 1999;**40**:187–201.

45. Krieger M. Charting the fate of the 'good cholesterol': identification and characterization of the high-density lipoprotein receptor SR-BI. *Annu Rev Biochem* 1999;**68**:523–58.

46. Mardones P, Pilon A, Bouly M, Duran D, Nishimoto T, Arai H, et al. Fibrates down-regulate hepatic scavenger receptor class B type I protein expression in mice. *J Biol Chem* 2003;**278**:7884–90.

47. Nieland TJ, Penman M, Dori L, Krieger M, Kirchhausen T. Discovery of chemical inhibitors of the selective transfer of lipids mediated by the HDL receptor SR-BI. *Proc Natl Acad Sci USA* 2002;**99**:15422–7.

48. Tserentsoodol N, Gordiyenko NV, Pascual I, Lee JW, Fliesler SJ, Rodriguez IR. Intraretinal lipid transport is dependent on high density lipoprotein-like particles and class B scavenger receptors. *Mol Vis* 2006;**12**:1319–33.

49. Pacheco-Alvarez D, Solórzano-Vargas RS, Del Río AL. Biotin in metabolism and its relationship to human disease. *Arch Med Res* 2002;**33**:439–47.

50. Arunchaipong K, Sattayasai N, Sattayasai J, Svasti J, Rimlumduan T. A biotin-coupled bifunctional enzyme exhibiting both glutamine synthetase activity and glutamate decarboxylase activity. *Curr Eye Res* 2009;**34**:809–18.

51. Tachikawa M, Takeda Y, Tomi M, Hosoya K. Involvement of OCTN2 in the transport of acetyl-L-carnitine across the inner blood–retinal barrier. *Invest Ophthalmol Vis Sci* 2010;**51**:430–6.

52. Janoria KG, Boddu SH, Wang Z, Paturi DK, Samanta S, Pal D, Mitra AK. Vitreal pharmacokinetics of biotinylated ganciclovir: role of sodium-dependent multivitamin transporter expressed on retina. *J Ocul Pharmacol Ther* 2009;**25**:39–49.

53. Prasad PD, Wang H, Huang W, Fei YJ, Leibach FH, Devoe LD, Ganapathy V. Molecular and functional characterization of the intestinal Na^+-dependent multivitamin transporter. *Arch Biochem Biophys* 1999;**366**:95–106.

33

Vitamin D and Diabetic Retinopathy

John F. Payne[1], Vin Tangpricha[2]

[1]Palmetto Retina Center, LLC, West Columbia, SC; [2]Emory University, Atlanta, GA

OVERVIEW OF VITAMIN D INSUFFICIENCY/DEFICIENCY

Vitamin D, a secosteroid prohormone and fat-soluble vitamin, is initially produced in the skin after ultraviolet light exposure.[1] Only a small amount (less than 30%) of vitamin D can be obtained from the diet, since relatively few foods contain it naturally.[2] The vitamin D formed in skin or obtained from the diet then undergoes two hydroxylations to form the active hormone, 1,25-dihydroxy-vitamin D $(1,25(OH)_2D_3)$.[1] The first hydroxylation takes place in the liver and the second conversion primarily takes place in the kidneys. Many tissues in the body, including the eyes, heart, stomach, pancreas, brain, skin, gonads, and activated lymphocytes, have nuclear receptors for $1,25(OH)_2D_3$.[1] A full schematic outlining the synthesis and activation of vitamin D is shown in Figure 33.1.[1] While originally implicated in calcium and phosphorus homeostasis, emerging evidence has shown that vitamin D is essential for a vast number of physiologic processes.

The best overall measure of vitamin D status is the metabolite, 25-hydroxyvitamin D (25(OH)D).[1] Serum 25(OH)D is preferred over $1,25(OH)_2D_3$ for several reasons including (1) its long circulating half-life (~3 weeks versus ~8 hours); (2) the concentration of 25(OH)D is 1000 times higher in circulation compared with $1,25(OH)_2D_3$ (ng/mL versus pg/mL); and (3) the production of $1,25(OH)_2D_3$ is mainly under the influence of parathyroid hormone, which tightly regulates calcium levels.[3] Concentrations of 25(OH)D can be reported in units of nanograms per milliliter or nanomoles per liter. When converting the units from nanomoles per liter to nanograms per milliliter, one should divide the concentration by 2.496. While the definition of optimal vitamin D can be debated, most researchers agree that the most advantageous target concentration of serum 25(OH)D for multiple clinical endpoints begins at 75 nmol/L (30 ng/mL).[4] Thus, people with serum concentrations of 25(OH)D of 75 nmol/L or greater are typically labeled as

vitamin D sufficient. People with levels between 50 and 75 nmol/L are usually designated as vitamin D insufficient, and people below 50 nmol/L are considered vitamin D deficient.[4,5] The risk of vitamin D intoxication can occur when serum concentrations of 25(OH)D rise above 375 nmol/L and manifests as severe hypercalcemia, hyperphosphatemia, and ultimately renal failure.[5] Risk factors for vitamin D deficiency include older age, darker skin pigmentation, obesity, chronic kidney disease, malabsorption, and living in northern latitudes where winters may be prolonged.[4,5] The recommendations for ideal vitamin D intake continue to be debated. The Institute of Medicine recommended that patients maintain 25(OH)D concentrations above 50 nmol/L, while the Endocrine Society recommends a goal of at least 75 nmol/L. A summary of the most recent intake recommendations from the Institute of Medicine and The Endocrine Society are listed in Table 33.1.[6]

It has been estimated that vitamin D insufficiency has reached pandemic proportions, with more than half the world's population being at risk.[1] Vitamin D insufficiency has now been implicated in the development of diabetes and correlated with an elevated risk of cardiovascular disease, cancer, autoimmune diseases, and mortality.[1,3,7,8] Additionally, vitamin D insufficiency has been associated with neurologic conditions, such as multiple sclerosis and Parkinson disease.[9,10] The pathogenetic connection between vitamin D and these conditions has not been fully established, but there is mounting research to support the notion that a relationship nonetheless exists. The focus of this chapter is on the relationship between vitamin D and diabetic retinopathy.

OVERVIEW OF DIABETIC RETINOPATHY

Diabetic retinopathy is the leading cause of blindness among working-age adults in the world.[11,12] During the first two decades with the disease, nearly all of those with type 1 diabetes and more than 60% of those with type 2 diabetes have some degree of retinopathy.[11] One worldwide meta-analysis reported the overall prevalence of any diabetic retinopathy and proliferative

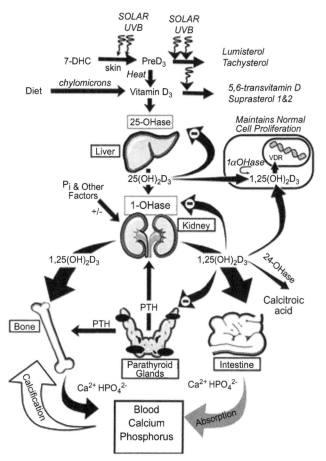

FIGURE 33.1 Schematic representation of the synthesis and metabolism of vitamin D.[1] 1,25(OH)$_2$D$_3$: 1,25-dihydroxy-vitamin D; 25(OH)D: 25-hydroxyvitamin D; 7-DHC: 7-dehydrocholesterol; PTH: parathyroid hormone; UVB: ultraviolet B; VDR: vitamin D receptor.

diabetic retinopathy (PDR) as 34.6% and 6.96%, respectively.[12] Furthermore, the prevalence of vision-threatening diabetic retinopathy, defined as PDR and/or diabetic macular edema (DME), was found to be 10.2%.[12] Interventions that could slow or even halt or reverse the disease course could potentially prevent blindness for millions of people worldwide.

Diabetic retinopathy typically progresses from mild nonproliferative abnormalities, characterized by microaneurysm formation and increased vascular permeability, to moderate and severe nonproliferative diabetic retinopathy (NPDR), which is often associated with retinal capillary nonperfusion and DME.[11] With further progression, PDR develops, which is characterized by extraretinal neovascularization onto the posterior surface of the vitreous. With advanced fibrovascular proliferation, tractional retinal detachments can occur and may lead to severe vision loss. Central vision loss is most often associated with DME and/or macular ischemia.[11] The mechanisms of pathogenesis of diabetic retinopathy are exquisitely complex and beyond the scope of this chapter. Figure 33.2 provides a simplified schematic of the pathogenesis of diabetic retinopathy.[13] Figure 33.3 displays fundus photographs of the progression of diabetic retinopathy from normal (top) to NPDR with DME (middle) and advanced PDR with tractional retinal detachment (bottom).

Many studies have examined various risk factors for diabetic retinopathy progression. The three most established risk factors are duration of diabetes, glycemic control, and blood pressure control.[11,12] The 4-year incidence of developing PDR in the Wisconsin Epidemiologic Study of Diabetic Retinopathy younger-onset group increased from 0% in the first 5 years to nearly 28% after 13–14 years of diabetes.[14] With regards to glycemic control, two landmark trials showed the benefits of tight glucose control on diabetic retinopathy. The Diabetes Control and

TABLE 33.1   Vitamin D Intake Recommendations of the Institute of Medicine and the Endocrine Society[6]

Life Stage Group	Institute of Medicine Recommendations		The Endocrine Society Recommendations	
	Recommended Daily Allowance (IU/day)	Tolerable Upper Intake Level (IU/day)	Recommended Daily Allowance (IU/day)	Tolerable Upper Intake Level (IU/day)
Infants (0–6 months)	400	1000	400–1000	2000
Infants (6–12 months)	400	1500	400–1000	2000
Children (1–3 years)	600	2500	600–1000	4000
Children (4–8 years)	600	3000	600–1000	4000
Children and adults (9–70 years)	600	4000	600–2000	4000–10,000
Adults (> 70 years)	800	4000	1500–2000	10,000
Pregnant/lactating females	600	4000	600–2000	4000–10,000

IU: International Units.

Complications Trial showed in a cohort of type 1 diabetic subjects that intensive insulin therapy reduced the risk of retinopathy progression by 54%.[15] The UK Prospective Diabetes Study (UKPDS), which examined the effects of glycemic control in type 2 diabetes, demonstrated a reduction in the overall rate of microvascular complications by 25% in those people receiving intensive insulin therapy compared with those receiving conventional therapy.[16–18] Furthermore, the UKPDS showed that those people assigned to tight blood pressure control had a 34% reduction in retinopathy progression.[16–18] While these three risk factors are well established, the exact pathophysiology of retinopathy progression remains unclear. Interestingly, several studies have shown that African-American subjects have a higher incidence of diabetic retinopathy and vision-threatening diabetic retinopathy.[12] While it is not clear why this occurs, it is possible that lower vitamin D levels in those with darker skin tones may be related. The mechanisms by which vitamin D may play a role in the development of diabetic retinopathy will be discussed in further detail below.

in turn leads to decreased insulin production. In contrast, type 2 diabetes is thought to be caused by insulin resistance and altered insulin secretion.[2] Despite the differences in pathophysiology, vitamin D has been implicated in both forms of diabetes. Several studies in rats and humans have shown that low serum 25(OH)D concentrations cause reduced insulin secretion,[19,20] and that $1,25(OH)_2D_3$ improves β-cell function and consequently glucose intolerance.[21,22] Additionally, vitamin D treatment has been shown to improve, and even prevent, type 1 diabetes in both human[23,24] and animal models.[25–27] Vitamin D not only influences insulin secretion through plasma calcium levels but also directly on the β-cells via the vitamin D receptor.[2] Similar to other steroid hormones, vitamin D has a range of effects that are thought not to involve gene expression, such as a rise in intracellular calcium and cyclic guanosine monophosphate (cGMP) levels and activation of protein kinase C.[2] Allelic differences in the vitamin D receptor gene may contribute to the genetic predisposition to certain diseases, such as type 2 diabetes.[2]

BASIC SCIENCE RESEARCH ON VITAMIN D AND DIABETIC RETINOPATHY

Type 1 versus Type 2 Diabetes Mellitus

Type 1 diabetes mellitus is considered to be an autoimmune disease whereby pancreatic β-cells are damaged via a T-cell-mediated immune response, which

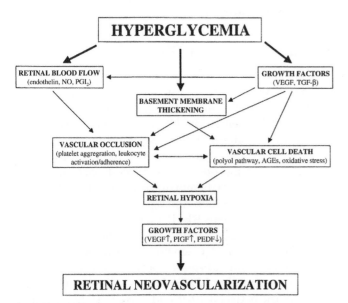

FIGURE 33.2 Schematic diagram of the pathogenesis of diabetic retinopathy.[13] AGE: advanced glycation end product; NO: nitric oxide; PEDF: pigment epithelium-derived factor; PGI_2: prostaglandin I_2; PlGF: placental growth factor; TGF-β: transforming growth factor beta; VEGF: vascular endothelial growth factor.

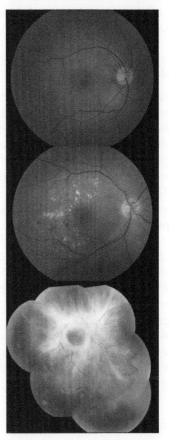

FIGURE 33.3 Fundus photographs depicting the progression of diabetic retinopathy: Normal (top); multiple microaneurysms, intraretinal hemorrhages, and hard exudates consistent with moderate-to-severe nonproliferative diabetic retinopathy with diabetic macular edema (middle); advanced proliferative diabetic retinopathy with tractional retinal detachment (bottom).

Inflammation

Inflammatory factors have often been associated with insulin resistance and β-cell failure, and there is evidence to suggest that vitamin D may play a role in the pathogenesis of diabetic retinopathy through its effects on the immune system. Inflammatory cytokines, such as tumor necrosis factor (TNF)-α, TNF-β, interleukin (IL)-6, C-reactive protein, and plasminogen activator inhibitor-1 are upregulated in patients with type 2 diabetes,[2,28] and it has been shown that vitamin D decreases the production of several proinflammatory cytokines, such as IL-2, IL-6, IL-8, IL-12, and TNF-α.[29,30] Vitamin D also exerts an anti-inflammatory effect by decreasing the proliferation of helper T-cells, cytotoxic T-cells, and natural killer cells.[2] One study found that vitamin D deficiency was associated with vascular endothelial dysfunction in middle-aged and older adults.[31] The authors concluded that this dysfunction was related to increased vascular endothelial cell expression of the proinflammatory transcription factor, nuclear factor κB. Because many of the inflammatory proteins involved in the pathogenesis of diabetic retinopathy are regulated by nuclear factor κB,[32] it is certainly plausible that vitamin D exerts an anti-inflammatory effect via this mechanism.

Angiogenesis

Vitamin D may also contribute to diabetic retinopathy via mechanisms of angiogenesis. Several studies have shown that the active metabolite of vitamin D, calcitriol, can directly affect the activity of endothelial cells and can inhibit their proliferation and budding.[33–35] It should be noted that these studies did not assess retinal endothelial cell activity. The effect of vitamin D on retinal neovascularization was not assessed until Albert and colleagues found that calcitriol was a potent inhibitor of retinal neovascularization *in vivo*.[36] This study also found that calcitriol inhibits retinal endothelial cell capillary morphogenesis *in vitro*.[36] Furthermore, calcitriol downregulates hypoxia-inducible factor-1 (HIF-1) transcriptional activity, as well as HIF-1 target genes, such as vascular endothelial growth factor (VEGF).[37] As several of the complications in diabetic retinopathy, such as macular edema and neovascularization, are driven by VEGF production,[38,39] vitamin D could exert its positive effect via calcitriol-mediated VEGF reduction.

Glycemic and Blood Pressure Control

Vitamin D may also play a protective role through its effects on glycemic control and hypertension, both significant risk factors for the development and progression of diabetic retinopathy.[16–18] There is evidence to suggest that vitamin D plays a role in normal insulin secretion in response to glucose and for maintenance of glucose tolerance.[2,40,41] Additionally, vitamin D and calcium citrate supplementation in patients with impaired fasting glucose leads to a reduction in insulin resistance and an attenuation of the rise in fasting glucose levels.[42] Vitamin D deficiency has also been implicated in the development of hypertension, as several studies have shown an inverse association between vitamin D levels and blood pressure.[43,44] One large prospective cohort study found that vitamin D deficiency was associated with 6.1-fold and 2.7-fold increased relative risk of developing hypertension in men and women, respectively.[45] It is possible that treating vitamin D insufficiency may lead to an improvement in blood sugar and blood pressure control, which could ultimately slow the progression of retinopathy.

CLINICAL STUDIES ASSESSING VITAMIN D AND DIABETIC RETINOPATHY

The scientific evidence supporting the role of vitamin D in the pathogenesis of various diseases continues to mount. Interestingly, there have only been a few clinical studies reporting the association between vitamin D and diabetic retinopathy. There have been two clinical studies assessing the relationship between vitamin D and type 1 diabetes mellitus[46,47] and three studies examining vitamin D and type 2 diabetes.[48–50] It should be noted that there currently have been no prospective, randomized clinical trials assessing vitamin D treatment and diabetic retinopathy.

Vitamin D and Type 1 Diabetes Mellitus

Kaur and colleagues examined serum 25(OH)D levels in a cross-sectional study of 517 children and adolescents ages 8–20 years with type 1 diabetes mellitus in Australia.[46] In their study, 16% of the subjects were considered to be vitamin D deficient, which was defined as serum 25(OH)D less than 50 nmol/L. Retinopathy, which was defined as the presence of at least one microaneurysm or intraretinal hemorrhage on fundus photography, was more common among people with vitamin D deficiency (18% versus 9%; P = 0.02). In the multivariate regression analysis, vitamin D status, duration of diabetes, and hemoglobin A_1c remained statistically significant.

Joergensen and colleagues performed a prospective, observational study of 220 children with type 1 diabetes in Denmark between 1979 and 1984.[47] They separated their cohort of patients into two groups: children with plasma 25(OH)D levels 15.5 nmol/L or less and those with levels greater than 15.5 nmol/L. Interestingly, they found an increased rate of overall mortality in the group

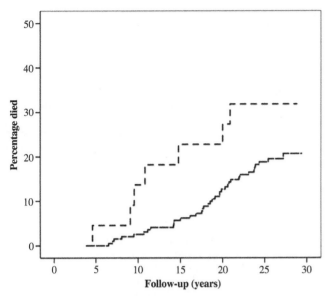

FIGURE 33.4 Kaplan-Meier curves of all-cause mortality in 220 type 1 diabetic subjects in accordance with a vitamin D level equal to and below or above the 10% percentile, 25(OH)D = 15.5 nmol/L.[47] Solid line represents the group with vitamin D levels ≤ 15.5 nmol/L and dotted line represents the group with vitamin D levels > 15.5 nmol/L (P = 0.06).

with lower 25(OH)D levels (32% versus 19%; P = 0.06). Figure 33.4 displays the Kaplan-Meier curve for mortality for the two groups. With regards to retinopathy, vitamin D levels did not predict the development of background or proliferative diabetic retinopathy. This finding is in contrast to that of Kaur *et al.*[46] and may be related to the different definitions of vitamin D deficiency used for the two studies.

Vitamin D and Type 2 Diabetes Mellitus

Aksoy and colleagues performed the first study assessing vitamin D and retinopathy in subjects with type 2 diabetes.[48] They measured serum concentrations of 1,25(OH)$_2$D$_3$, 25(OH)D, and parathyroid hormone among 66 diabetic and 20 control subjects. The reported mean serum 25(OH)D levels were higher in the nondiabetic patients than those with diabetes (60.6 versus 31.0 nmol/L; P < 0.001). Furthermore, they concluded that there was a significant inverse relationship between the severity of retinopathy and the concentrations of 1,25(OH)$_2$D$_3$. It should be noted that they did not include demographic variables, such as season of blood draw, body mass index, and multivitamin usage, which could affect the serum concentrations of vitamin D.

Payne *et al.* performed a large, cross-sectional analysis of vitamin D levels on diabetic retinopathy in 221 subjects.[49] Advantages of this study include the large sample size, minimized seasonal bias, and the inclusion of demographic variables, such as body mass index,

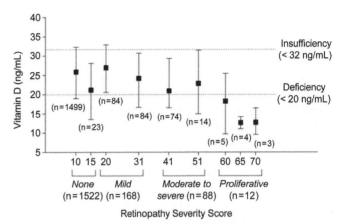

FIGURE 33.5 Serum 25(OH)D and the severity of diabetic retinopathy in 1790 study participants.[50]

renal function, and multivitamin use. Concurrent with the Aksoy study,[48] they found that diabetic subjects had lower serum 25(OH)D concentrations compared with nondiabetic subjects (57.2 versus 75.4 nmol/L). Additionally, there was a significant inverse association between vitamin D concentrations and the severity of retinopathy. However, in the multivariate analysis, only multivitamin usage remained significant. The authors concluded that lower vitamin D status was lower in subjects with type 2 diabetes, especially those with PDR, compared with subjects without diabetes.

Patrick and colleagues performed a cross-sectional analysis of individuals older than 40 years with diabetes mellitus who participated in an interview and medical examination as part of the Third National Health and Examination Survey (NHANES III) between 1988 and 1994.[50] Of the 1790 adults who met the inclusion criteria, there was a positive association between the prevalence of vitamin D deficiency and the severity of diabetic retinopathy. However, a regression analysis showed no significant relationship between retinopathy severity and serum 25(OH)D levels (P = 0.07). Figure 33.5 graphically displays the serum 25(OH)D concentrations in relation to the severity of retinopathy. Nonetheless, the authors concluded that the association between vitamin D and diabetic retinopathy warrants further study. While these clinical studies implicate vitamin D in the pathogenesis of diabetic retinopathy, it remains unclear whether treating vitamin D deficiency can slow or even prevent the onset or progression of diabetic retinopathy. Ideally, a prospective, randomized clinical trial will be designed to answer this question.

CONCLUSIONS

Vitamin D deficiency has now been implicated in numerous systemic conditions, and research suggests that it plays a role in the development of diabetic retinopathy, the leading cause of blindness in working-age adults. There is evidence to suggest that low vitamin D levels may contribute to the development of diabetic retinopathy through its effects on inflammation and angiogenesis, as well as glucose and blood pressure homeostasis. Clinical studies confirm an association between vitamin D deficiency and diabetic retinopathy, although more research is needed to determine the pathogenetic connections and to assess whether treating vitamin D deficiency will slow the progression of diabetic retinopathy.

TAKE-HOME MESSAGES

- Vitamin D is a fat-soluble prohormone that is primarily synthesized in the skin upon ultraviolet light exposure and plays a role in a vast array of physiologic processes including many involved in the pathogenesis of retinopathy.
- The best biomarker for vitamin D status is serum 25(OH)D mainly due to its longer circulating half-life and higher serum concentrations compared with the active metabolite, $1,25(OH)_2D_3$.
- Diabetic retinopathy is the leading cause of blindness among working-age adults in the world and has an estimated worldwide prevalence approaching 35%.
- The most established risk factors for progression of diabetic retinopathy are longer duration of diabetes, as well as poor blood pressure and glycemic control.
- While the mechanisms of diabetic retinopathy progression remain unclear, vitamin D may play a role through its effects on inflammation, insulin secretion, glucose tolerance, blood pressure control, and angiogenesis.
- While several large observational studies have shown an inverse association between vitamin D status and the severity of diabetic retinopathy, further prospective clinical trials are needed to discern whether treating vitamin D deficiency can slow or prevent the development or progression of retinopathy.

References

1. Holick MF. Vitamin D: importance in the prevention of cancers, type 1 diabetes, heart disease, and osteoporosis. *Am J Clin Nutr* 2004;**79**:362–71.
2. Palomer X, Gonzalez-Clemente JM, Blanco-Vaca F, et al. Role of vitamin D in the pathogenesis of type 2 diabetes mellitus. *Diabetes Obes Metab* 2008;**10**:185–97.
3. Judd SE, Tangpricha V. Vitamin D deficiency and risk for cardiovascular disease. *Am J Med Sci* 2009;**338**:40–4.
4. Bischoff-Ferrari HA, Giovanucci E, Willett WC, et al. Estimation of optimal serum 25-hydroxyvitamin D for multiple health outcomes. *Am J Clin Nutr* 2006;**84**:18–28.
5. Nemerovski CW, Dorsch MP, Simpson RU, et al. Vitamin D and cardiovascular disease. *Pharmacotherapy* 2009;**29**:691–708.
6. Holick MF, Binkley NC, Bischoff-Ferrari HA, et al. Evaluation, treatment, and prevention of vitamin D deficiency: an Endocrine Society clinical practice guideline. *J Clin Endocrinol Metab* 2011;**96**:1911–30.
7. Baz-Hecht M, Goldfine AB. The impact of vitamin D deficiency on diabetes and cardiovascular risk. *Curr Opin Endocrin Diab Obes* 2010;**17**:113–9.
8. Mitri J, Muraru MD, Pittas AG. Vitamin D and type 2 diabetes: a systematic review. *Eur J Clin Nutr* 2011;**64**:1005–15.
9. Simpson Jr S, Taylor B, Blizzard L, et al. Higher 25-hydroxyvitamin D is associated with lower relapse risk in multiple sclerosis. *Ann Neurol* 2010;**68**:193–203.
10. Evatt ML, Delong MR, Khazai N, et al. Prevalence of vitamin D insufficiency in patients with Parkinson disease and Alzheimer disease. *Arch Neurol* 2008;**65**:1348–52.
11. Fong DS, Aiello L, Gardner TW, et al. Diabetic retinopathy. *Diabetes Care* 2003;**26**(Suppl. 1):S99–102.
12. Yau JWY, Rogers SL, Kawasaki R, et al. Global prevalence and major risk factors for diabetic retinopathy. *Diabetes Care* 2012;**35**:556–64.
13. Cai J, Boulton M. The pathogenesis of diabetic retinopathy: old concepts and new questions. *Eye* 2002;**16**:242–60.
14. Klein R, Klein BE, Moss SE, et al. The Wisconsin Epidemiologic Study of Diabetic Retinopathy. II. Prevalence and risk of diabetic retinopathy when age at diagnosis is less than 30 years. *Arch Ophthalmol* 1984;**102**:520–6.
15. Diabetes Control and Complications Trial Research Group. The effect of intensive treatment of diabetes on the development and progression of long-term complications in insulin-dependent diabetes mellitus. *N Engl J Med* 1993;**329**:977–86.
16. UK Prospective Diabetes Study Group. Intensive blood-glucose control with sulphonylureas or insulin compared with conventional treatment and risk of complications in patients with type 2 diabetes (UKPDS 33). *Lancet* 1998;**352**:837–53.
17. UK Prospective Diabetes Study Group. Effect of intensive blood glucose control with metformin on complications in overweight patients with type 2 diabetes (UKPDS 34). *Lancet* 1998;**352**:854–65.
18. UK Prospective Diabetes Study Group. Tight blood pressure control and risk of macrovascular and microvascular complications in patients with type 2 diabetes: UKPDS 38. *BMJ* 1998;**317**:708–13.
19. Norman AW, Frankel BJ, Heldt AM, et al. Vitamin D deficiency inhibits pancreatic secretion of insulin. *Science* 1980;**209**:823–5.
20. Boucher BJ, Mannan N, Noonan K, et al. Glucose intolerance and impairment of insulin secretion in relation to vitamin D deficiency in East London Asians. *Diabetologia* 1995;**38**:1239–45.
21. Kumar S, Davies M, Zakaria Y, et al. Improvement in glucose tolerance and β-cell function in a patient with vitamin D deficiency during treatment with vitamin D. *Postgrad Med J* 1994;**70**:440–3.
22. Luong K, Nguyen LTH, Nguyen DNP. The role of vitamin D in protecting type 1 diabetes mellitus. *Diabetes Metab Res Rev* 2005;**21**:338–46.
23. The EURODIAB Substudy. 2 Study Group. Vitamin supplement in early childhood and risk for type 1 (insulin-dependent) diabetes mellitus. *Diabetologia* 1999;**42**:51–4.
24. Stene LC, Ulriksen J, Magnus P, et al. Use of cod liver oil in pregnancy is associated with lower risk of type 1 diabetes in the offspring. *Diabetologia* 2000;**43**:1093–8.
25. Mathieu C, Laureys J, Sobis H, et al. 1,25-dihydroxyvitamin D3 prevents insulitis in NOD mice. *Diabetes* 1992;**41**:1491–5.
26. Mathieu C, Waer M, Laureys J, et al. Prevention of autoimmune diabetes in NOD mice by 1,25-dihydroxyvitamin D3. *Diabetologia* 1994;**37**:552–8.
27. Gregori S, Giarratana N, Smiroldo S, et al. A 1alpha,25-dihydroxyvitamin D(3) analog enhances regulatory T-cells and arrests autoimmune diabetes in NOD mice. *Diabetes* 2002;**51**:1367–74.
28. Kolb H, Mandrup-Poulsen T. An immune origin of type 2 diabetes? *Diabetologia* 2005;**48**:1038–50.

29. Mauricio D, Mandrup-Poulsen T, Nerup J. Vitamin D analogues in insulin-dependent diabetes mellitus and other autoimmune disease: a therapeutic perspective. *Diabetes Metab Res Rev* 1996;**12**:57–68.

30. Lemire JM. Immunomodulatory actions of 1,25-dihydroxyvitamin D3. *J Steroid Biochem Mol Biol* 1995;**53**:599–602.

31. Jablonski KL, Chonchol M, Pierce GL, et al. 25-hydroxyvitamin D deficiency is associated with inflammation-linked vascular endothelial dysfunction in middle-aged and older adults. *Hypertension* 2011;**57**:63–9.

32. Kern TS. Contributions of inflammatory processes to the development of early stages diabetic retinopathy. *Exp Diabetes Res* 2007;**2007**:95103.

33. Suzuki T, Sano Y, Kinoshita S. Effects of 1alpha,25-dihydroxyvitamin D3 on Langerhans cell migration and corneal neovascularization in mice. *Invest Ophthalmol Vis Sci* 2000;**41**:154–8.

34. Mantell DJ, Owens PE, Bundred NJ, et al. 1alpha,25-dihydroxyvitamin D(3) inhibits angiogenesis *in vitro* and *in vivo*. *Circ Res* 2000;**87**:214–20.

35. Bernardi RJ, Johnson CS, Modzelewski RA, et al. Antiproliferative effects of 1alpha,25-dihydroxyvitamin D(3) and vitamin D analogs on tumor-derived endothelial cells. *Endocrinology* 2002;**143**:2508–14.

36. Albert DM, Scheef EA, Wang S, et al. Calcitriol is a potent inhibitor of retinal neovascularization. *Invest Ophthalmol Vis Sci* 2007;**48**:2327–34.

37. Ben-Shoshan M, Amir S, Dang DT, et al. 1alpha,25-dihydroxyvitamin D3 (calcitriol) inhibits hypoxia-inducible factor-1/vascular endothelial growth factor pathway in human cancer cells. *Mol Cancer Ther* 2007;**6**:1433–9.

38. Wang J, Xu X, Elliott MH, et al. Muller cell-derived VEGF is essential for diabetes-induced retinal inflammation and vascular leakage. *Diabetes* 2010;**59**:2297–305.

39. Wang X, Wang G, Wang Y. Intravitreous vascular endothelial growth factor and hypoxia-inducible factor 1a in patients with proliferative diabetic retinopathy. *Am J Ophthalmol* 2009;**148**:883–9.

40. Kadowaki S, Norman AW. Dietary vitamin D is essential for normal insulin secretion from the perfused rat pancreas. *J Clin Invest* 1984;**73**:759–66.

41. Gedik A, Akalin S. Effects of vitamin D deficiency and repletion on insulin and glucagon secretion in man. *Diabetologia* 1986;**29**:142–5.

42. Pittas AG, Harris SS, Stark PC, et al. The effects of calcium and vitamin D supplementation on blood glucose and markers of inflammation in nondiabetic adults. *Diabetes Care* 2007;**30**:980–6.

43. Kristal-Boneh E, Froom P, Harari G, Ribak J. Association of calcitriol and blood pressure in normotensive men. *Hypertension* 1997;**30**:1289–94.

44. Judd SE, Nanes MS, Ziegler TR, et al. Optimal vitamin D status attenuates the age-associated increase in systolic blood pressure in white Americans: results from the third National Health and Nutrition Examination Survey. *Am J Clin Nutr* 2008;**87**:136–41.

45. Forman JP, Giovannucci E, Homes MD, et al. Plasma 25-hydroxyvitamin D levels and risk of incident hypertension. *Hypertension* 2007;**49**:1063–9.

46. Kaur H, Donaghue KC, Chan AK, et al. Vitamin D deficiency is associated with retinopathy in children and adolescents with type 1 diabetes. *Diabetes Care* 2011;**34**:1400–2.

47. Joergensen C, Hovind P, Schmedes A, et al. Vitamin D levels, microvascular complications, and mortality in type 1 diabetes. *Diabetes Care* 2011;**34**:1081–5.

48. Aksoy H, Akcay F, Kurtul N, et al. Serum 1,25 dihydroxy vitamin D ($1,25(OH)_2D_3$), 25 hydroxy vitamin D (25(OH)D) and parathormone levels in diabetic retinopathy. *Clin Biochem* 2000;**33**:47–51.

49. Payne JF, Ray R, Watson DG, et al. Vitamin D insufficiency in diabetic retinopathy. *Endocr Pract* 2012;**18**:185–93.

50. Patrick PA, Visintainer PF, Shi Q, et al. Vitamin D and retinopathy in adults with diabetes mellitus. *Arch Ophthalmol* 2012;**130**:756–60.

Vitamin D and Age-Related Macular Degeneration

Yao Jin[1], Chen Xi[1], Jiang Qin[1], Victor R. Preedy[2], Ji Yong[3]

[1]Nanjing Medical University Eye Hospital, Nanjing, Jiangsu Province, China, [2]Diabetes and Nutritional Sciences, School of Medicine, Kings College London, London, UK, [3]Department of Pathophysiology, Nanjing Medical University, Nanjing, PR China

INTRODUCTION

The role of vitamins in the etiology of eye disease and vision is reviewed elsewhere in this book. This chapter focuses exclusively on vitamin D and age-related macular degeneration (AMD).

VITAMIN D

Vitamin D is a group of fat-soluble secosteroids; these are structurally similar but not identical to steroids. In humans, vitamin D (endogenous cholecalciferol; vitamin D_3 is the bioactive form) is mainly synthesized from cholesterol when sun exposure is sufficient. Vitamin D (ergocalciferol; vitamin D_2) can also be obtained from foods such as liver, eggs, fish, and milk, which is fortified in some countries. Vitamin D_2 is sourced from the ultraviolet (UV) irradiation of ergosterol, while vitamin D_3 is produced by the UV irradiation of its precursor 7-dehydrocholesterol to previtamin D_3 in the skin of humans, with a further thermal isomerization step to form vitamin D_3.[1] Excessive exposure to sunlight can degrade both previtamin D_3 and vitamin D_3 into inactive photoproducts. Vitamin D_2 or vitamin D_3 produced in the skin or ingested in the diet can be stored in and then released from fat cells when necessary.[2] Vitamin D is converted to calcidiol, 25-hydroxycholecalciferol ($25(OH)D_3$) by the enzyme D-25-hydroxylase in the liver. Calcidiol is a specific vitamin D metabolite, which can be measured in serum to determine vitamin D status. Calcitriol ($1,25(OH)2D_3$), the biologically active form of vitamin D, is converted from calcidiol in the kidneys. Calcitriol is a hormone, which acts on target organs such as the intestines, kidneys, and bones. Calcitriol can also be converted from calcidiol, which can influence the proliferation, differentiation, and apoptosis of cells. Calcitriol is involved in neuromuscular function and inflammation. Both calcitriol and 25-hydroxycholecalciferol can be autoregulated through feedback systems. Vitamins D_2 and D_3 are metabolized through the same pathway (Fig. 34.1). Vitamin D_3 is regulated by various factors including blood phosphate and calcium concentrations, parathyroid hormone (PTH), and calcitonin.

Vitamin D mainly plays a role in regulating, at the cellular and molecular levels, phosphate and blood calcium concentration,[3] the immune response, growth, and differentiation.[4] Some molecular events are mediated via the nuclear vitamin D receptor (VDR). Vitamin D_3 regulates hundreds of genes including those coding for proteins involved in cell differentiation and cell proliferation, as well as calcium and phosphate homeostasis.[5]

Three cytochrome P450 (CYP) hydroxylases are responsible for the synthesis and degradation of vitamin D, namely vitamin D-25-hydroxylase (25-OHase) in the liver, 25(OH)D-1α-hydroxylase (1α-OHase) or CYP27B1, and 25(OH)D-24-hydroxylase (24-OHase) or CYP24A1 in the kidneys.[6] Craig *et al.*[7] found that VDR is expressed in many organs in the zebra fish, such as the brain and spinal cord. Importantly in the context of this chapter, VDR is also expressed in the retina.

Recently, it has been demonstrated that there are measurable concentrations of vitamin D in tear fluid and aqueous and vitreous humor of rabbits.[8] Furthermore, oral supplementation of vitamin D can affect the concentration of vitamin D in the anterior segment of the rabbit's eye.[8] This study suggests that there is an association between vitamin D and the eye.

In recent decades, there has been an accumulating body of evidence to show that vitamin D deficiency is

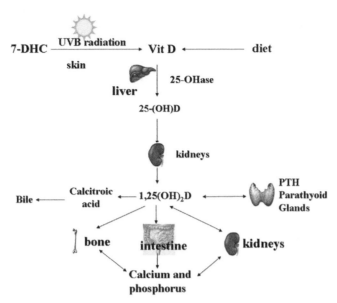

FIGURE 34.1 Schematic representation of the synthesis and metabolism of vitamin D for regulating calcium, phosphorus, and bone metabolism. Vitamin D_3 is produced by the ultraviolet B (UVB) irradiation of its precursor 7-dehydrocholesterol (7-DHC) to previtamin D_3 in the skin of humans; then, previtamin D_3 converts to vitamin D_3. Vitamin D can also be obtained through the diet. Vitamin D is converted to calcidiol, 25-hydroxycholecalciferol (25(OH)D_3) by the enzyme D-25-hydroxylase in the liver. The biologically active form of vitamin D, calcitriol (1,25(OH)2D_3), is converted from calcidiol in the kidneys. Calcitriol is a hormone, which acts on target organs such as the intestines, kidneys, and bones. Vitamin D_3 is regulated by various factors including the blood phosphate and calcium concentrations, parathyroid hormone (PTH), and calcitonin. 1,25(OH)2D is catabolized to the water-soluble, biologically inactive calcitroic acid and excreted in the bile.

related to numerous diseases, such as diabetes, hypertension, cancer, schizophrenia, cardiovascular disease, infection, and multiple sclerosis.[9–11] Some of these are considered as *diseases of aging*. In this regard, the ability to synthesize vitamin D declines during aging and supplemental vitamin D ameliorates the onset of disease, including AMD.[12]

AGE-RELATED MACULAR DEGENERATION

Age-related macular degeneration (AMD or ARMD), which is also known as senile macular degeneration (SMD), is the principal cause worldwide of registered legal blindness among those aged over 65. Almost two-thirds of the population over 80 years old will have some signs of AMD. The retina is considered an end organ of the central nervous system. The macula, which is at the center of retina, is particular prone to age-related degenerative changes.[13] AMD results in a loss of vision in the center of the visual field (the macula) because of damage to the retina. The definition of this disease varies,

but the condition of AMD is generally characterized by extensive drusen, often associated with pigmentary abnormalities.[14]

AMD can be classified into early-stage AMD, intermediate-stage AMD, and advanced-stage AMD. In the advanced stage of AMD, geographic atrophy and/or neovascular changes can be seen in the fundus of the patient's eye.[14] AMD can be classified into two phenotypes: atrophic (dry) and neovascular (wet) AMD according to the changes seen in the fundus. Wet AMD can be treated by antivascular endothelial growth factor (anti-VEGF) drugs, photodynamic therapy (PDT), or a combination of these two therapies. There are no effective methods for the treatment of dry AMD. Other methods to treat AMD include laser treatment, surgery, radiation, and other pharmacologic agents, but none of them can completely cure AMD[14] (Figs. 34.2 and 34.3).

The etiology and pathogenesis of AMD, especially the precise etiology of the molecular mechanism, are still unknown. The pathogenesis of AMD includes the following theories: genetics, oxidative stress, hydrodynamic changes of Bruch's membrane (BM), retinal pigment epithelium (RPE) senescence, hemodynamic changes of choroidal blood, choroidal neovascularization (CNV) and angiogenesis, and subclinical inflammation.[15]

The pathology of AMD is characterized by degeneration involving the retinal photoreceptors, RPE, BM, and alterations in choroidal capillaries in some cases.[16] RPE cell senescence is thought by some to play a key role in AMD.[17] Both human anatomopathologic findings and experimental models have provided large amounts of evidence to suggest that inflammation and the immune system contribute to the pathogenesis and progression of AMD. However, some consider that AMD is not a classical inflammatory disease.[18]

Autophagy (the process by which cells recycle cytoplasm and dispose of excess or defective organelles) is the main mechanism that eliminates damaged components in response to age-related cellular modifications.[19] As autophagy is related to some forms of neurodegenerative disease; it is conceivable that it may also be related to the pathogenesis of AMD.[20]

It has been suggested that the histopathology of early AMD is very important for the understanding of AMD.[21] In this regard, the formation of deposits between the RPE and BM are the two most obvious histopathologic changes. These histopathologic changes could be either focal or diffuse. When the change is focal, it is called drusen. Drusen can be divided into hard drusen and soft drusen. Hard drusen are small and usually have clear margins, while soft drusen are generally larger without well-defined margins, tending to merge and to enlarge histologically.[18]

Complement factor H (*CFH*) and age-related maculopathy susceptibility 2 (*ARMS2*)/high-temperature

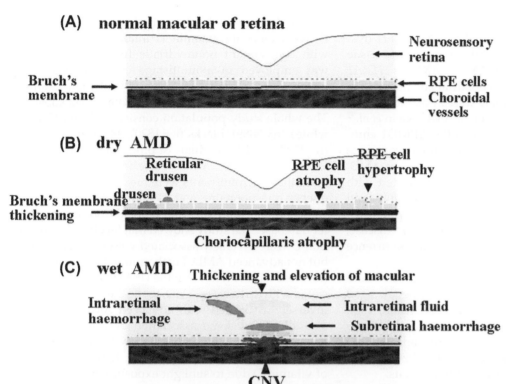

FIGURE 34.2 Normal macula of retina, dry age-related macular degeneration (AMD) and wet AMD. (A) Normal macula of retina; (B) drusen, retinal pigment epithelium (RPE) cell atrophy and hypertrophy, Bruch's membrane thickening, and choriocapillaris atrophy in dry AMD; (C) thickening and elevation of macula, intraretinal and subretinal hemorrhage, intraretinal fluid, and choroidal neovascularization (CNV) in wet AMD.

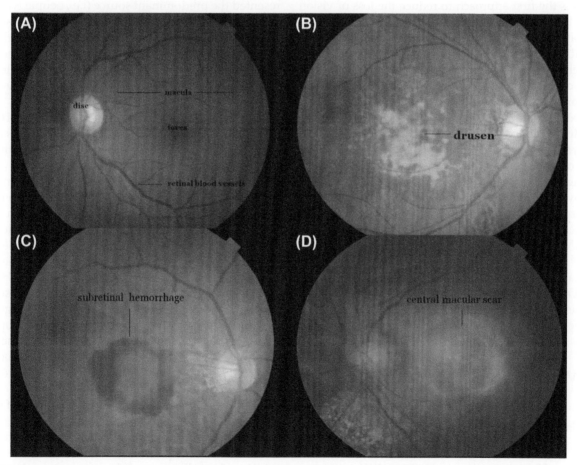

FIGURE 34.3 Retinal photographs: (A) normal; (B) dry age-related macular degeneration (AMD); (C) wet AMD; (D) end-stage macular scar. *Source: Nanjing Medical University Affiliated Eye Hospital.* (See color plate section)

requirement factor A1 (*HTRA1*) are the two genes already identified as being related to AMD. *CFH* dysfunction may lead to excessive inflammation and tissue damage in the pathogenesis of AMD.[21] *ARMS2/HTRA1* may affect AMD through mitochondrial alterations.[16] In a large family unit in which there were 21 members with the predominantly dry phenotype of AMD, Klein *et al.*[22] showed that AMD was localized to the ARMD1 chromosome 1q25-q31 as a dominant trait. Other groups are investigating whether other genes are related to AMD. Many studies have shown that genes associated with inflammation, oxidative stress, angiogenesis, and lipid metabolism are responsible for AMD susceptibility.[23] Overall, AMD is thought to be a polygenetic disease, with genes playing an important role in its occurrence and/or development.

The risk factors related to AMD are diverse and mainly include old age, female gender (for people aged over 75), cardiovascular diseases, hypertension, high body mass index, family history of the disease, cigarette smoking history, excessive UV exposure, and diet. The last factor includes low intakes of lutein and zeaxanthin or other dietary antioxidants or diets high in fats.

Since there is no effective method to cure AMD, prevention is the first approach to reduce the loss of vision caused by AMD.[14] Control of the risk factors of AMD listed above may reduce the damage encompassed by AMD. In 2001, the Age-Related Eye Disease Study Research Group prescribed that patients with the following conditions should consider taking a supplement of antioxidants (such as vitamin C and vitamin E) plus zinc: extensive intermediate-size drusen, at least one large druse, noncentral geographic atrophy in one or both eyes, or advanced AMD or AMD-induced vision loss in one eye, and without contraindications (such as smoking).

Many previous studies suggested that beta-carotene, lutein, zeaxanthin, vitamin C, vitamin E, and micronutrients such as zinc can reverse the development of early-stage AMD and/or reduce the development of advanced AMD.[25] However, some recent studies have suggested that intake of sufficient vitamin D can also lower the risk of AMD.[17] The following section describes in detail the relationship between vitamin D and AMD and presents putative mechanisms.

VITAMIN D AND AGE-RELATED MACULAR DEGENERATION

Associations between Vitamin D and Age-Related Macular Degeneration

Usually, 25-hydroxyvitamin D (25(OH)D) in serum, the preferred biomarker, is measured to determine vitamin D status as it reflects both endogenous vitamin D and exogenous vitamin D intake.[26] This approach, with concomitant nonmydriatic fundus photographs, was employed in a multistage cross-sectional study in the Third National Health and Nutrition Examination Survey of the USA, which studied 7752 subjects.[27] The whole study population consisted of non-Hispanic whites (n=3889), blacks (n=1820), Mexican Americans (n=1742), and individuals of other ethnicities (n=301). Of these, 11% had AMD.[27] The intakes of vitamin D and other micronutrients were measured with food frequency questionnaires and 24-hour dietary recalls using compositional data from the National Coordinating Center database. The study showed that levels of serum vitamin D were inversely associated with early-stage AMD but not advanced AMD. Furthermore, the intake of milk (fortified with vitamin D in the USA) was inversely associated with early-stage AMD, while the intake of fish was inversely associated with advanced AMD. The results also suggested some association between the drusen and pigmentary abnormalities and vitamin D levels, but this was not statistically significant. To remove the effect of vitamin D due to sunlight exposure, the authors also identified a cohort for whom endogenous vitamin D represented the predominant source (i.e., people who drank milk daily and had regular supplemental vitamin D). The odds for early-stage AMD were significantly lower among people in the highest compared with the lowest quintile for serum vitamin D level in this sample.[27] This study therefore suggests that vitamin D may protect against AMD.

Another survey drew the same conclusion.[28] The association between serum 25(OH)D concentrations and the prevalence of early-stage AMD among postmenopausal American women was investigated in this study. Serum 25(OH)D concentrations were analyzed according to the different seasons (in general, sunlight exposure is lower in the fall and winter seasons). This investigation showed that high levels of serum 25(OH)D concentrations protected against early-stage AMD for women less than 75 years old, and the participants had a 48% decreased odds of early-stage AMD when the serum 25(OH)D concentration was higher than 38 nmol/L. This association was consistent across the subtypes of early-stage AMD. In addition, the authors proposed that as people age, a greater proportion of early-stage AMD-susceptible, compared with AMD-unsusceptible, individuals with low 25(OH)D concentrations die from other chronic diseases before developing early-stage AMD.[28]

Convincing evidence for the protective role of vitamin D in AMD is found in a study among American caucasian male twin pairs.[29] This showed that higher dietary intake of vitamin D was present in the twins with less severe AMD and smaller size of drusen compared with their co-twins, adjusted for smoking and age.[29]

These studies suggest that in addition to genetic susceptibility, intake of vitamin D is associated with the appearance and development of AMD.

There are also differing opinions about the role of vitamin D. Golan *et al.*[30] evaluated serum vitamin D levels of members of the Maccabi Healthcare Services (one of the four largest Israeli Health Maintenance Organizations) who were aged over 60 years between 2002 and 2008. In this study 1045 had AMD and 8124 did not. However, Golan *et al.* could not detect any obvious relationship between serum vitamin D levels and the appearance of AMD.[30]

As the above study is negative (but not contradictory in that a negative correlation was not observed), more advanced and larger systematic studies need to be performed to identify confounding factors, if any. There is also a conceptual argument that excessive light exposure, especially UV rays, may damage the macula. Nevertheless, it would be encouraging if supplemental intake of vitamin D could prevent AMD, as this would offer practical opportunities.

Mechanisms

The precise mechanism whereby vitamin D protects against AMD is still unknown. One needs to consider that vitamin D plays an important role in both the inflammatory and immune systems. Vitamin D participates in the regulation of immune systems by inhibiting production of interleukin-2 (IL-2) and immunoglobulin, and it can also block lymphocyte proliferation.[31] Furthermore, vitamin D can suppress IL-1β-induced IL-8 production in human whole blood.[32] By activating macrophages and dendrites, vitamin D_3 can inhibit the production of IL-12; that is, it acts in an immunosuppressive capacity. The relevance of this relates to the fact that IL-12 is involved in the pathogenesis of chronic inflammatory autoimmune disorders.[33] It has been found that there is an association between immune events and drusen biogenesis. As mentioned above, drusen is the hallmark of AMD. Immune and/or inflammatory pathways are thus a possible mechanism by which vitamin D protects against AMD. Parekh *et al.*[27] speculated that vitamin D acts protectively in AMD by decreasing proliferation of T helper cells, T cytotoxic cells, and natural killer cells, enhancing T suppressor cell activity, decreasing the production of proinflammatory agents such as IL-2, IL-6, IL-8, and IL-12, and reducing C-reactive protein (CRP), a marker of systemic inflammation. Furthermore, they suggested that vitamin D might protect against AMD by virtue of its antiangiogenic properties. This may be relevant with respect to wet advanced-stage AMD, on account of its characteristic of growth. In addition, drusen contains a variety of molecular constituents including activation products and negative regulators of complement and oxidative fragments such as carboxyethylpyrrole protein (CEP) adducts. Activation of the complement system may contribute to the appearance of AMD.[34] As an oxidative protein modification, when elevated in the BM of AMD, CEP could stimulate neovascularization *in vivo*.[35] Other studies have shown that mice immunized with CEP modifications develop an AMD-like phenotype. This may also be one of the pathways by which vitamin D acts in a protective manner, by reducing drusen, and thus CEP, and thus CEP–protein adducts that initiate AMD.[34]

Vitamin D_3 is one of the dominant regulators of the immune system and may protect against the process of aging. There is growing evidence that vitamin D_3 can influence inflammation by promoting the differentiation of monocytes into macrophages and enhancing phagocytosis.[36] Vitamin D_3 can not only diminish macrophage production including proinflammatory cytokines and chemokines such as interferon (IFN) in chronic inflammation but also promote neural protection via some factors such as nerve growth factor (NGF).[37] Richards *et al.*[38] showed that vitamin D_3 delayed age-related diseases in a population-based cohort of women (n = 2160) across a wide age spectrum (18–80 years, average 49.4 years) in the UK. They demonstrated that the length of the leukocyte telomere is a predictor of age-related disease and chronic inflammation. The length of the leukocyte telomere is shorter after each cell cycle. Using this approach, Richards *et al.* showed those people with relatively high levels of vitamin D_3 have longer leukocyte telomeres, and this may delay the development of aging and age-related diseases.[38] Vitamin D supplement users were also found to have longer leukocyte telomere lengths than nonusers, which provides support for this hypothesis. Leukocyte telomere length serves as a cumulative index of an individual's lifelong burden of oxidative stress, inflammation, and some other factors. Some of the latter factors are genetic, while others impact after birth. For instance, environment, cigarette smoking, obesity, and sedentary lifestyle are all associated with shortened leukocyte telomere lengths. As AMD is closely related to aging, this may be one of the mechanisms through which vitamin D delays the appearance of AMD. They also found that the vitamin D concentration levels were negatively correlated with levels of CRP. As CRP is an inflammation factor, they suggested that vitamin D may reduce aging-induced damage via by a reduction in systemic inflammation.

The metabolic demand of the outer retina is one the highest in the body. In aging, the permeability of the interface between the outer retina and its blood supply declines owing to protein deposits, including amyloid-beta (Aβ). This is accompanied by an increase in the number of macrophages, an elevation of chronic inflammation, a reduction in photoreceptor numbers, and a decline in visual acuity. Lee *et al.*[39] confirmed that vitamin D_3 was important to the aging process of mice after conducting a short, 6-week experiment. Fourteen 12-month-old mice were divided into two groups, namely treated and controls. Visual function was recorded by electroretinography, the amounts and morphology of the macrophages were examined, and the contents of Aβ and C3d determined. In this regard, Aβ was used as a marker of aging while C3d was the marker of chronic inflammatory reactions. As is already known, retinal macrophages are responsible for phagocytosis, and their morphology can change to become dendritic during retinal inflammation. In the vitamin D_3-treated group, macrophages became more mobile and phagocytic, but the number declined compared with the control group. Significantly less C3d expression was found along with BM in vitamin D_3-treated animals than in the controls. Apart from this, the Aβ expression on BM in the treated group was less than that in controls. The magnitude of the a-wave and stimulation luminance in vitamin D_3-treated mice increased significantly above controls. This experiment indicated that vitamin D_3 could improve retinal function by inhibiting inflammation, reducing the number of activated macrophages, and enhancing the clearance of Aβ deposits.

Morrison *et al.*[40] found that ultraviolet irradiance could protect the development of neovascular AMD when the other risk factors of AMD, including polymorphisms of the genes encoding CFH and ARMS2/HTRA1, gender, age and smoking history, were controlled. They employed a candidate gene approach, by which they evaluated the common variations in key genes of vitamin D pathways. They found that single point variants in the gene called cytochrome P450, family 24, subfamily A, polypeptide 1 (*CYP24A1*) could influence the risk of AMD both separately and, more importantly, in a meta-analysis. These findings were supported by expression data from human donor eyes and human retinal cell lines. *CYP24A1* is the cytochrome P450 component of the 25-hydroxyvitamin D_3-24-hydroxylase enzyme that catalyzes the conversion of 25-hydroxyvitamin D_3 (25(OH)D_3) and 1,25-dihydroxyvitamin D_3 (1,25(OH)2D_3) into 24-hydroxylated products, which constitute the degradation of the vitamin D molecule.[6] Many scientists have found that increased expression of *CYP24A1* can lead to a deficiency of vitamin D status. However, there is a fundamental need to further study the inhibition of *CYP24A1* expression and its association with the appearance and/or development of AMD in depth. This has potential therapeutic implications for gene targeting AMD patients with variations in *CYP24A1*. In any event, this study shows that there is an interrelationship between the level of vitamin D, AMD, and *CYP24A1*, as the latter play an important role in the catabolism of vitamin D.

In addition, as in many other diseases, gene variations may influence the appearance and/or development of AMD. Genes that affect the metabolism of vitamin D, for instance, *CYP24A1*, have already been demonstrated to be related to AMD. As the human body is a complicated entity regulated by a series of interrelated factors, concerted efforts are needed to understand the complex mechanisms whereby vitamin D protects against AMD.

HOW TO SUPPLY VITAMIN D

In the past, vitamin D deficiency was defined as serum 25(OH)D concentrations less than 10ng/mL (25nmol/L), as these 25(OH)D levels were associated with rickets and osteomalacia, the classical markers of vitamin D deficiency. The reference level of 200 IU/day of vitamin D was introduced to maintain skeletal health for children and people up to 50 years old.[2] However, as recent studies show that the deficiency of vitamin D may contribute to many other diseases, including cardiovascular disease, selective cancers, type 2 diabetes, infectious disease, autoimmune disorders, and other pathologies,[11] the definition of *subclinical vitamin D deficiency* or *'vitamin D insufficiency'* was proposed.[2]

The US Institute of Medicine (IOM) recommended dietary allowance (RDA) of vitamin D is 600IU/day in children aged more than 1 year old and adults up to 70 years old and 800IU/day for adults aged more than 70 years old; the US IOM report concludes that serum 25(OH)D of 20ng/mL will cover the requirements of at least 97.5% of the population. This is the guidance for healthy populations.

Another institute has introduced separate guidelines for patients with specific diseases who are at risk of vitamin D deficiency. The Clinical Practice Guideline from the US Endocrine Society suggests that children aged more than 1 year old require 600–1000 IU/day of vitamin D and adults aged 19 years old or more require 1500–2000 IU/day of vitamin D to maintain serum 25(OH)D above the optimal level of 30ng/mL (Table 34.1).[41]

Recommended strategies for treatment of vitamin D deficiency should be instituted according to both the age and comorbidities of patients.[2] There are many strategies for the treatment of vitamin D deficiency. One strategy is to treat vitamin D deficiency with 50,000 IU of vitamin D_2 weekly for 8 weeks and to prevent the recurrence of vitamin D deficiency by maintaining the patient on 50,000 IU of vitamin D_2 twice a month. This is effective for up to as long as 6 years without evidence of toxicity.[2] However, for people who do not have vitamin D deficiency and just take vitamin D to prevent AMD, there must be an upper intake level or safe line. According to the data from the Endocrine Society clinical practice guideline published in 2011, the tolerable upper intake

TABLE 34.1 Vitamin D Intakes Recommended by the US Institute of Medicine (IOM) and the Endocrine Practice Guidelines Committee

		0–6 months	6–12 months	1–3 years	4–8 years	9–18 years	19–70 years	>70 years	Pregnancy and Lactation 14–18 years	19–50 years
US IOM recommendations (IU)	Adequate intake	400	400							
	Estimated average requirement			400	400	400	400	400	400	400
	Recommended dietary allowance			600	600	600	600	800	600	600
	Tolerable upper intake level	1000	1500	2500	3000	4000	4000	4000	4000	4000
Committee recommendations for patients at risk for vitamin D deficiency (IU)	Daily requirement	400–1000	400–1000	600–1000	600–1000	600–1000	1500–2000	1500–2000	600–1000	1500–2000
	Tolerable upper intake level	2000	2000	4000	4000	4000	10,000	10,000	4000	10,000

TABLE 34.2 Clinical Definitions of 25-Hydroxyvitamin D (25(OH)D) Levels

Vitamin D		Severe Deficiency	Deficiency	Insufficiency	Normal	Excess	Intoxication
25(OH)D	mmol/L	12.5	37.5	37.5–50	50–250	250	375
	ng/L	5	15	15–20	20–100	100	150

level of vitamin D is 4000 IU for people less than 18 years old and 10,000 IU for people over 18 years old, both men and women, but 2000 IU for infants (<12 months).[41] The treatment levels of vitamin D for diseases such as hypertension and obesity are listed in the guideline, but there is no guidance to protect against AMD.

It is now understood that vitamin D intake during childhood will impact on the health of people when they are aging. This supposition is founded on studies on the bone mass of elderly people. Elderly women with lower vitamin D intake during childhood had more health problems related to vitamin D than those with higher vitamin D intake when they were young.[42] Thus, there is an argument to suggest that supplementation of vitamin D should commence in children to prevent AMD, as AMD is an age-related disease.

However, a cautionary note is needed as, with vitamin D, the notion that 'the more, the better' may not be true. The half-life of vitamin D in fatty tissue is approximately 2 months, whereas it is 2 weeks in the circulation. Thus, there is a risk of toxicity if excessive vitamin D is consumed

in a short time. Vitamin D intoxication (VDI), also called vitamin D toxicity (VDT), as a result of supplementation is rare, but it has been reported more frequently in recent years. This is attributable to an increase in vitamin D supplement intake due to an understanding that vitamin D 25(OH)D is involved in the pathogenesis of several diseases.[43] Serum 25(OH)D levels above 150 ng/mL are considered as vitamin D toxicity (Table 34.2). The common symptoms of vitamin D intoxication or so-called hypervitaminosis D include vomiting, weight loss, thirst, anorexia, polyuria, and constipation.[43]

Sources of vitamin D include sun exposure, diet, and supplements.[26] A lack of light exposure is the main cause of vitamin D deficiency worldwide.[44] The synthesis of vitamin D via skin by light exposure is influenced by a number of factors including season of the year, skin pigmentation, latitude, use of sunscreens, clothing, and amount of skin exposed. Age is also a factor, in that synthesis of vitamin D declines with increasing age, owing in part to a fall in 7-dehydrocholesterol levels and in part to alterations in skin morphology.[44] Thus, while it is recommended for people to intake enough vitamin D via adequate light exposure, too much sun exposure can lead to sunburn, photoaging, and skin cancer. In addition, the amount of sunlight sufficient to achieve optimal vitamin D status varies depending on the host of factors mentioned above.[45] Supplemental intake of vitamin D may be an optional method to delay the appearance and/or the development of AMD. Vitamin D could be supplied through both intake of food or drugs and sunlight exposure. To reiterate the point above, sunlight is the main way to supply vitamin D for humans.[45] For fair skin types sunlight exposure can be achieved by exposure to 5–15 minutes of sunlight in summer (depending on latitude), just outside the most damaging period for ultraviolet B (UVB) exposure (10 a.m.–2 p.m., standard time), with longer exposures required in winter and in those with more pigmented skin.[42] However, several studies have reported AMD to have a possible association with UV exposure and considered excessive light exposure to be one risk factor for AMD.[46] Wearing of sunglasses and hats is a good way to solve this problem.

It is generally considered that the major foods containing vitamin D (either D_2 or D_3), either in their natural form or as a fortified food, are milk, fish (especially fatty fish), fish liver oil, yogurts, cheeses, orange juices, and cereals.[2,45] However, not all countries fortify foods with vitamin D in the same way. Vitamin D_3 is present in oil-rich fish such as salmon, mackerel, and herring.[2] Only a few natural foods, notably egg yolk and fatty fish, contain nutritionally significant quantities of vitamin D. Therefore, Liu[47] concluded that dietary vitamin D will only be a component of vitamin D supply, with UV exposure and/or oral supplements continuing to be the main sources of vitamin D to maintain adequate status, on the basis that diet alone is unlikely to achieve the recommended vitamin D values proposed by the US IOM. Reduced skin vitamin D synthesis is particularly a problem for elderly people, and if they dress modestly for cultural or religious reasons, the synthesis is reduced.[42]

Armas et al.[48] found that vitamin D_2 potency is less than one-third that of vitamin D_3 by measuring the concentration and duration of serum 25(OH)D. They suggested that physicians who resort to using vitamin D_2 should be aware of its markedly lower potency and shorter duration of action relative to that of vitamin D_3. Besides, through a meta-analysis it was found that although vitamins D_2 and D_3 are metabolized through the same pathway, vitamin D_3 is more efficacious at raising the serum 25(OH)D concentrations than is vitamin D_2, and therefore vitamin D_3 could potentially become the preferred choice for supplementation.[1] However, there is a different opinion suggesting that vitamin D_2 is as effective as vitamin D_3 in treating and preventing the recurrence of vitamin D deficiency. Until now, vitamin D_2 is the only pharmaceutical form of vitamin D approved by the US Food and Drug Administration

(FDA).[2] Vitamin D can also be obtained via the intramuscular route to increase serum 25(OH)D in people with obvious vitamin D deficiency.

To summarize, vitamin D intake can be via daily food such as fatty fish, milk (if fortified), egg yolk, and fish liver oil. The most convenient and comfortable way to supply vitamin D is from the sun, although protective sunglasses and hats are needed to protect the eyes at the same time. Supplemental vitamin D preparations are needed for those who do not receive enough sunshine. For those with insufficient sun exposure, there are increasing recommendations for much higher doses of supplementation than previously considered: around 1000 IU/day.[42] It is also recommended to supply vitamin D via a combination of the methods above. However, one needs to be cautious about vitamin D toxicity, especially in those who persistently take pharmacologic preparations of vitamins.[45]

PROSPECTIVE

AMD will affect more and more people with the global demographic increases in the aging population. So far, there is no effective method to treat AMD, even in developed countries. Treatments with PDT or anti-VEGF are expensive, while increasing vitamin D status is cheaper and more convenient. Understanding the mechanism whereby vitamin D protects against AMD will provide a firm foundation upon which to build other strategies for diagnosis, treatment, and prevention.

TAKE-HOME MESSAGES

- Vitamin D (i.e., D_2 or D_3) is a group of fat-soluble secosteroids, which can be synthesized via sun exposure or gained from food.
- Vitamin D_3 has an important role in maintaining calcium and phosphate homeostasis and also plays a part in the human immune response, cell growth, and differentiation.
- Age-related macular degeneration (AMD) is the principal cause of registered legal blindness among those aged over 65 worldwide. There is no effective cure for this disease.
- Recent studies have suggested that vitamin D could protect against AMD.
- Vitamin D may protect against AMD through regulating the immune system and reducing inflammation.
- Vitamin D could be supplied through intake of food or drugs and light exposure. Light exposure is the main way to supply vitamin D for humans.
- Dietary supplements of vitamin D may be helpful.

References

1. Tripkovic L, Lambert H, Hart K, Smith CP, Bucca G, Penson S, et al. Comparison of vitamin D_2 and vitamin D_3 supplementation in raising serum 25-hydroxyvitamin D status: a systematic review and meta-analysis. *Am J Clin Nutr* 2012;**95**:64–1357.

2. Pramyothin P, Holick MF. Vitamin D supplementation: guidelines and evidence for subclinical deficiency. *Curr Opin Gastroenterol* 2012;**28**:50–139.

3. Schulman RC, Weiss AJ, Mechanick JI. Nutrition, bone, and aging: an integrative physiology approach. *Curr Osteoporos Rep* 2011;**9**:1–184.

4. Becker S, Cordes T, Diesing D, Diedrich K, Friedrich M. Expression of 25 hydroxyvitamin D_3-1alpha-hydroxylase in human endometrial tissue. *J Steroid Biochem Mol Biol* 2007;**103**:5–771.

5. Haussler MR, Jurutka PW, Mizwicki M, Norman AW. Vitamin D receptor (VDR)-mediated actions of 1alpha,25(OH)(2)vitamin D(3): genomic and non-genomic mechanisms. *Best Pract Res* 2011;**25**:59–43.

6. Jones G, Prosser DE, Kaufmann M. 25-Hydroxyvitamin D-24-hydroxylase (CYP24A1): its important role in the degradation of vitamin D. *Arch Biochem Biophys* 2012;**523**:9–18.

7. Craig TA, Sommer S, Sussman CR, Grande JP, Kumar R. Expression and regulation of the vitamin D receptor in the zebrafish. *Danio rerio. J Bone Miner Res* 2008;**23**:96–1486.

8. Lin Y, Ubels JL, Schotanus MP, Yin Z, Pintea V, Hammock BD, et al. Enhancement of vitamin D metabolites in the eye following vitamin D_3 supplementation and UV-B irradiation. *Curr Eye Res* 2012;**37**:8–871.

9. Vaidya A, Forman JP. Vitamin D and vascular disease: the current and future status of vitamin D therapy in hypertension and kidney disease. *Curr Hypertens Rep* 2012;**14**:9–111.

10. McGrath JJ, Burne TH, Feron F, Mackay-Sim A, Eyles DW. Developmental vitamin D deficiency and risk of schizophrenia: a 10-year update. *Schizophr Bull* 2010;**36**:8–1073.

11. Khadilkar VV, Khadilkar AV. Use of vitamin D in various disorders. *Indian J Pediatr* 2013;**80**:8–215.

12. Thomas DR. Vitamins in aging, health, and longevity. *Clin Interv Aging* 2006;**1**:81–91.

13. Khandhadia S, Cherry J, Lotery AJ. Age-related macular degeneration. *Adv Exp Med Biol* 2012;**724**:15–36.

14. Coleman HR, Chan CC, Ferris III FL, Chew EY. Age-related macular degeneration. *Lancet* 2008;**372**:45–1835.

15. Ambati J, Ambati BK, Yoo SH, Ianchulev S, Adamis AP. Age-related macular degeneration: etiology, pathogenesis, and therapeutic strategies. *Surv Ophthalmol* 2003;**48**:93–257.

16. Ding X, Patel M, Chan CC. Molecular pathology of age-related macular degeneration. *Prog Retin Eye Res* 2009;**28**:1–18.

17. Kozlowski MR. RPE cell senescence: a key contributor to age-related macular degeneration. *Med Hypotheses* 2012;**78**:10–505.

18. Buschini E, Piras A, Nuzzi R, Vercelli A. Age related macular degeneration and drusen: neuroinflammation in the retina. *Prog Neurobiol* 2011;**95**:14–25.

19. Shintani T, Klionsky DJ. Autophagy in health and disease: a double-edged sword. *Science NY* 2004;**306**:5–990.

20. Ambati J, Fowler BJ. Mechanisms of age-related macular degeneration. *Neuron* 2012;**75**:26–39.

21. Johnson PT, Betts KE, Radeke MJ, Hageman GS, Anderson DH, Johnson LV. Individuals homozygous for the age-related macular degeneration risk-conferring variant of complement factor H have elevated levels of CRP in the choroid. *Proc Natl Acad Sci USA* 2006;**103**:61–17456.

22. Klein ML, Schultz DW, Edwards A, Matise TC, Rust K, Berselli CB, et al. Age-related macular degeneration. Clinical features in a large family and linkage to chromosome 1q. *Arch Ophthalmol* 1998;**116**:8–1082.

23. Tanaka K, Nakayama T, Yuzawa M, Wang Z, Kawamura A, Mori R, et al. Analysis of candidate genes for age-related macular degeneration subtypes in the Japanese population. *Mol Vision* 2011;**17**:8–2751.

24. Group A- REDSR. A randomized, placebo-controlled, clinical trial of high-dose supplementation with vitamins C and E, beta carotene, and zinc for age-related macular degeneration and vision loss: AREDS report no. 8. *Arch Ophthalmol* 2001;**119**:36–1417.

25. Sabour-Pickett S, Nolan JM, Loughman J, Beatty S. A review of the evidence germane to the putative protective role of the macular carotenoids for age-related macular degeneration. *Mol Nutr Food Res* 2012;**56**:86–270.

26. Holick MF. Vitamin D deficiency. *N Engl J Med* 2007;**357**:81–266.

27. Parekh N, Chappell RJ, Millen AE, Albert DM, Mares JA. Association between vitamin D and age-related macular degeneration in the Third National Health and Nutrition Examination Survey, 1988 through 1994. *Arch Ophthalmol* 2007;**125**:9–661.

28. Millen AE, Voland R, Sondel SA, Parekh N, Horst RL, Wallace RB, et al. Vitamin D status and early age-related macular degeneration in postmenopausal women. *Arch Ophthalmol* 2011;**129**:9–481.

29. Seddon JM, Reynolds R, Shah HR, Rosner B. Smoking, dietary betaine, methionine, and vitamin D in monozygotic twins with discordant macular degeneration: epigenetic implications. *Ophthalmology* 2011;**118**:94–1386.

30. Golan S, Shalev V, Treister G, Chodick G, Loewenstein A. Reconsidering the connection between vitamin D levels and age-related macular degeneration. *Eye (Lond)* 2011;**25**:9–1122.

31. Muller K, Gram J, Bollerslev J, Diamant M, Barington T, Hansen MB, et al. Down-regulation of monocyte functions by treatment of healthy adults with 1 alpha,25 dihydroxyvitamin D_3. *Int J Immunopharmacol* 1991;**13**:30–525.

32. Takahashi K, Horiuchi H, Ohta T, Komoriya K, Ohmori H, Kamimura T. 1Alpha,25-dihydroxyvitamin D_3 suppresses interleukin-1beta-induced interleukin-8 production in human whole blood: an involvement of erythrocytes in the inhibition. *Immunopharmacol Immunotoxicol* 2002;**24**:1–15.

33. D'Ambrosio D, Cippitelli M, Cocciolo MG, Mazzeo D, Di Lucia P, Lang R, et al. Inhibition of IL-12 production by 1,25-dihydroxyvitamin D_3. Involvement of NF-kappaB downregulation in transcriptional repression of the p40 gene. *J Clin Invest* 1998;**101**:62–252.

34. Hajrasouliha AR, Kaplan HJ. Light and ocular immunity. *Curr Opin Allergy Clin Immunol* 2012;**12**:9–504.

35. Ni J, Yuan X, Gu J, Yue X, Gu X, Nagaraj RH, et al. Plasma protein pentosidine and carboxymethyllysine, biomarkers for age-related macular degeneration. *Mol Cell Proteomics* 2009;**8**:33–1921.

36. Kamen DL, Tangpricha V. Vitamin D and molecular actions on the immune system: modulation of innate and autoimmunity. *J Mol Med (Berl)* 2010;**88**:50–441.

37. Aparna R, Subhashini J, Roy KR, Reddy GS, Robinson M, Uskokovic MR, et al. Selective inhibition of cyclooxygenase-2 (COX-2) by 1alpha,25-dihydroxy-16-ene-23-yne-vitamin D_3, a less calcemic vitamin D analog. *J Cell Biochem* 2008;**104**:42–1832.

38. Richards JB, Valdes AM, Gardner JP, Paximadas D, Kimura M, Nessa A, et al. Higher serum vitamin D concentrations are associated with longer leukocyte telomere length in women. *Am J Clin Nutr* 2007;**86**:5–1420.

39. Lee V, Rekhi E, Kam JH, Jeffery G. Vitamin D rejuvenates aging eyes by reducing inflammation, clearing amyloid beta and improving visual function. *Neurobiol Aging* 2012;**33**:9–2382.

40. Morrison MA, Silveira AC, Huynh N, Jun G, Smith SE, Zacharaki F, et al. Systems biology-based analysis implicates a novel role for vitamin D metabolism in the pathogenesis of age-related macular degeneration. *Hum Genomics* 2011;**5**:68–538.

41. Holick MF, Binkley NC, Bischoff-Ferrari HA, Gordon CM, Hanley DA, Heaney RP, et al. Evaluation, treatment, and prevention of vitamin D deficiency: an Endocrine Society clinical practice guideline. *J Clin Endocrinol Metab* 2011;**96**:30–1911.

42. Everitt AV, Hilmer SN, Brand-Miller JC, Jamieson HA, Truswell AS, Sharma AP, et al. Dietary approaches that delay age-related diseases. *Clin Interv Aging* 2006;**1**:11–31.

43. Ozkan B, Hatun S, Bereket A. Vitamin D intoxication. *Turk J Pediatr* 2012;**54**:8–93.

44. Institute of Medicine (US). Committee to Review Dietary Reference Intakes for Vitamin D and Calcium; Ross AC. In: Taylor CL, Yaktine AL, Del Valle HB, editors. *Dietary Reference Intakes for Calcium and Vitamin D. National Academies Press*. Washington, DC: National Academies Press; 2011.

45. Wagner CL, Taylor SN, Dawodu A, Johnson DD, Hollis BW. Vitamin D and its role during pregnancy in attaining optimal health of mother and fetus. *Nutrients* 2012;**4**:30–208.

46. Cruickshanks KJ, Klein R, Klein BE, Nondahl DM. Sunlight and the 5-year incidence of early age-related maculopathy: the Beaver Dam Eye Study. *Arch Ophthalmol* 2001;**119**:50–246.

47. Liu J. Vitamin D content of food and its contribution to vitamin D status: a brief overview and Australian focus. *Photochem Photobiol Sci* 2012;**11**:7–1802.

48. Armas LA, Hollis BW, Heaney RP. Vitamin D_2 is much less effective than vitamin D_3 in humans. *J Clin Endocrinol Metab* 2004;**89**:91–5387.

Folate Transport in Retina and Consequences on Retinal Structure and Function of Hyperhomocysteinemia

S.B. Smith[1, 2, 3], P.S. Ganapathy[1, 2], R.B. Bozard[1, 2], V. Ganapathy[3, 4]

[1]Department of Cellular Biology and Anatomy, Georgia Health Sciences University, Augusta, GA, USA, [2]Department of Ophthalmology, Georgia Health Sciences University, Augusta, GA, USA, [3]Vision Discovery Institute, School of Medicine, Georgia Health Sciences University, Augusta, GA, USA, [4]Department of Biochemistry and Molecular Biology, Georgia Health Sciences University, Augusta, GA, USA

INTRODUCTION

Folate is an essential nutrient; no cell can survive without it. Folate deficiency has direct deleterious consequences on the retina and is implicated in nutritional amblyopia and methanol-induced retinal toxicity.[1] In nutritional amblyopia, which may occur in the presence of other vitamin deficiencies[1,2] or in isolated folate deficiency,[3,4] an optic neuropathy develops characterized by selective degeneration of the smallest fibers of retinal ganglion cells (papillomacular bundle) of the optic nerve, leading to loss of central vision.[4-7] There have been reports in the literature of severe cases of anorexia nervosa in which patients manifest very poor near- and distance-visual acuity. The rapid improvement of vision upon repletion of folic acid, vitamin B_{12}, and thiamine implicates deficiency of these micronutrients in the cause of the visual dysfunction.[8] In methanol-induced ocular toxicity, formate, a highly toxic by-product of methanol metabolism, damages Müller cells, leading to blindness or serious visual impairment.[9]

It is well established that decreased levels of folate lead to an accumulation of homocysteine, termed hyperhomocysteinemia (HHcy), which is implicated in a number of visual diseases. This chapter reviews the mechanisms by which cells of the retina acquire folate and the consequences on retinal structure and function in the presence of HHcy.

FOLIC ACID (FOLATE)

Folate (folic acid, vitamin B_9) is a water-soluble vitamin that plays a critical role in one-carbon metabolism in a variety of mammalian biosynthetic processes. The folate molecule consists of a pteridine moiety linked via a methylene bridge to p-aminobenzoic acid, which is in turn linked to glutamate (Fig. 35.1). Mammals cannot synthesize folate and must obtain it from the diet; green, leafy vegetables are a rich source of folate. The parent folic acid is not found in nature but is commonly used in food fortification. It is chemically stable and highly bioavailable because it is readily converted to tetrahydrofolate, the active coenzyme form. The various forms of reduced folate serve as enzyme cofactors in one-carbon metabolism, acting as one-carbon donors and acceptors in metabolic reactions. Folates are metabolized to polyglutamate forms for biologic activity as they have greater affinity for folate-dependent enzymes than monoglutamate forms.[10-12]

ROLE OF FOLATE IN METABOLISM

Folates serve as coenzymes in many biologic reactions, including (1) the synthesis of nucleotides for DNA and RNA synthesis, (2) the interconversion of serine and glycine, (3) the generation of methionine, and (4) the methylation of a vast array of biologic molecules. To explain these briefly: first, in thymidylate synthesis, deoxyuridine monophosphate is converted to thymidine monophosphate by thymidylate synthase with the transfer of a methyl group from 5,10-methylene-tetrahydrofolate. For *de novo* purine biosynthesis, 10-formyltetrahydrofolate is used in reactions performed by glycinamide ribonucleotide transformylase and 5-amino-4-imidazolecarboxamide ribonucleotide transformylase to incorporate formate into the C8 and C2 positions of the purine ring. Second, in the interconversion of serine and glycine, serine hydroxymethyltransferase regenerates 5,10-methylene-tetrahydrofolate from tetrahydrofolate. The main provider of one-carbon units in folate-dependent one-carbon metabolism is serine. Third, 5-methyltetrahydrofolate, the predominant form of folate found in blood, is used as a methyl donor by methionine synthase to convert homocysteine to methionine. Fourth, methionine can be further metabolized by methionine adenosyl transferase to S-adenosylmethionine, a methyl donor in many different reactions. S-Adenosylmethionine is utilized in the methylation of histones, proteins, DNA, phospholipids, and neurotransmitters.[10–13]

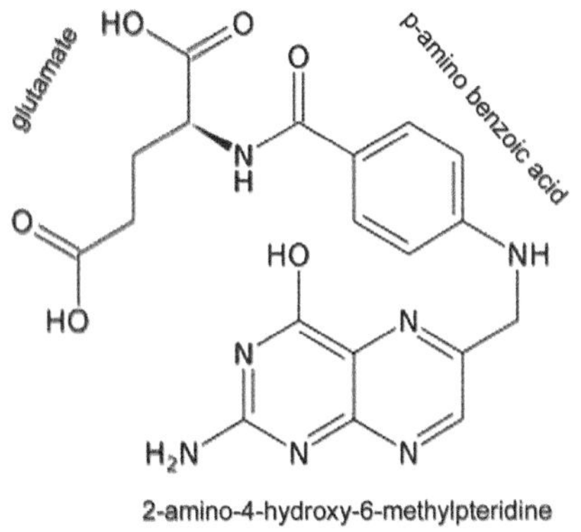

2-amino-4-hydroxy-6-methylpteridine

FIGURE 35.1 Structure of folic acid. Folic acid is comprised of a pteridine moiety containing two rings, which is linked by a methylene bridge to p-aminobenzoic acid, which is in turn linked to glutamate.

OVERVIEW OF THE RETINA

Owing to folate's key role in DNA and RNA synthesis, no cell, including those of the retina, can survive without adequate levels of this vitamin. Before describing the mechanisms for folate transport in the retina, it is necessary to understand the organization of this complex tissue, which is responsible for processing visual light stimulation and converting it to electrical signals to be computed by the brain.[14] The retina comprises the innermost tunic of the eyeball. Microscopically, the mammalian retina is composed of an outer pigmented layer, the retinal pigment epithelium, and an inner neurosensory layer, termed the neural retina. The retina is organized into 10 histologic layers (Fig. 35.2). Beginning at the innermost layer, the histologic layers of the retina are: the inner limiting membrane (ILM), the nerve fiber layer (NFL), the ganglion cell layer (GCL), the inner plexiform layer (IPL), the inner nuclear layer (INL), the outer plexiform layer (OPL), the outer nuclear layer (ONL), the outer limiting membrane (OLM), the inner and outer segments (IS/OS) of the photoreceptors, and the retinal pigment epithelium (RPE).[15]

The cells of the retina function in a highly coordinated fashion to achieve vision. Light that enters the eye passes through the entire thickness of the retina before reaching and activating the photoreceptors, neurons whose nuclei make up the ONL. Photoreceptors are the first order neurons of the visual pathway. There are two types of photoreceptor: rods and cones. Rod photoreceptors are responsible for dim vision, brightness, and

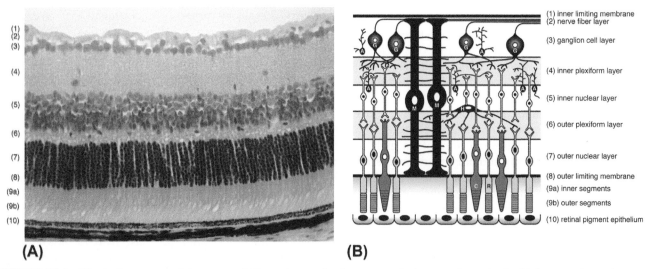

FIGURE 35.2 Histologic structure of the retina. (A) Photomicrograph of a mouse retinal cryosection stained with hematoxylin and eosin to visualize the 10 histologic layers of the retina. From the innermost to the outermost layer, they are: (1) inner limiting membrane (ILM), (2) nerve fiber layer (NFL), (3) ganglion cell layer (GCL), (4) inner plexiform layer (IPL), (5) inner nuclear layer (INL), (6) outer plexiform layer (OPL), (7) outer nuclear layer (ONL), (8) outer limiting membrane (OLM), (9a) inner segments (IS) and (9b) outer segments (OS) of the photoreceptors, and (10) retinal pigment epithelium (RPE). (B) Artist's rendition of the cell types within the various histologic layers, including the amacrine cell (A), bipolar cell (B), cone photoreceptor (C), rod photoreceptor cell (R), ganglion cell (G), horizontal cell (H), and Müller cell (M). *Source: Retinal diagram courtesy of Mrs. Laura McKie, Georgia Health Sciences University.*

contrast (scotopic vision), whereas cone photoreceptors are responsible for color vision and fine resolution (photopic vision).[16] The cone photoreceptor synapses with a single cone bipolar cell, the second order neuron whose cell bodies are located in the INL, and the cone bipolar cell synapses directly with the ganglion cell in the innermost retina, the axons of which all bundle together to form the optic nerve that exits the eye to the brain. Regarding the rod pathway, multiple rod photoreceptors synapse with a rod bipolar cell that indirectly synapses with ganglion cells via amacrine cells, whose nuclei are located in either the INL or the GCL. Horizontal cells, whose nuclei are found also in the INL, play an important role in maximizing spatial resolution and sharpening contrast via synaptic connections with photoreceptor and bipolar cells.[17] Two cell types serve as sustentacular cells within the retina: RPE cells and Müller cells. The RPE cells, which are critical in maintenance of the very metabolically active photoreceptors, form the outermost layer of the retina and consist of a single layer of low cuboidal epithelium. The Müller glial cell spans the entire neural retina, performing many supportive functions for adjacent neurons.[18,19] It is these two cell types, Müller and RPE, that have been most extensively studied with respect to transport of folate.

MECHANISMS OF FOLATE UPTAKE IN THE RETINA

Folates are small ($M_r \approx 500\,\mathrm{Da}$), hydrophilic, anionic molecules that require specific mechanisms to traverse the hydrophobic barrier of the plasma membrane. Three cellular mechanisms have been identified that facilitate folate acquisition in cells: folate receptor (FR), reduced folate carrier (RFC), and proton-coupled folate transporter (PCFT) (Fig. 35.3). All three transport mechanisms have been identified in the retina.[20–23]

Folate Receptor

FR is anchored to the plasma membrane by glycosylphosphatidylinositol (GPI) (Fig. 35.3A).[24–27] It acquires folate by receptor-mediated endocytosis. FR binds folate to form a receptor–folate complex; the plasma membrane invaginates to generate an endosome and subsequently internalizes folate by a nonclassical endocytic mechanism. GPI-anchored proteins (GPI-APs) are internalized via a nonclathrin-mediated endocytic route and delivered to peripheral tubular–vesicular endosomes called GPI-AP-enriched endosomal compartments (GEEC).[28] GPI anchoring leads to cholesterol-dependent and sphingolipid-dependent retention of endocytosed proteins in the recycling endosomal compartment.[29,30] Upon endosomal acidification, folate dissociates from the receptor

and is exported across the endosomal membrane into the cell by a mechanism that may involve PCFT, at least in retinal Müller glial cells as described below.[20]

There are four human isoforms of FR (α, β, γ, and δ). In mice, a very commonly used species in retinal cell biology,[31] the FR protein is referred to as folate binding protein (Folbp) and there are three isoforms (Folbp 1, 2, and 3). These are analogous to the α, β, and δ forms in humans. Depending on the isoform, FRs contain approximately 240–260 amino acids and have a molecular mass in the range of around 28–40kDa, reflecting the extent of glycosylation.[25,32] FRs function optimally at pH 7–7.4 and have a much greater affinity for folic acid than for

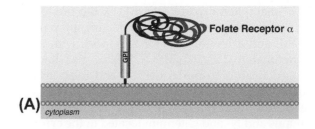

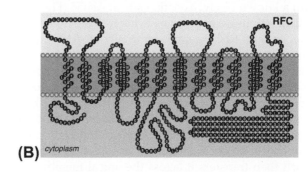

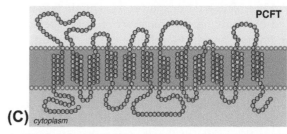

FIGURE 35.3 Proteins involved in the uptake of folate in the retina. (A) Folate receptor-α (FRα). A glycosylphosphatidylinositol (GPI) moiety anchors FRα to the plasma membrane. The core structure of a GPI is composed of an ethanolamine group, three mannose groups, a glucosamine group, an inositol group, and two phosphate groups. (B) Membrane topology of reduced folate carrier (RFC). RFC has 12 transmembrane domains and N- and C-termini located in the cytoplasm. An N-linked glycosylation site is present at Asn58 (N58) between the first and second transmembrane domains.[12] (C) Membrane topology of proton-coupled folate transporter (PCFT). PCFT is a typical transmembrane transporter that spans the membrane 12 times. N- and C-termini are oriented to the cytoplasm. N-linked glycosylation sites at asparagine residues 58 and 68 are located between the first and second transmembrane domains.[12] *Source: Figure courtesy of Mrs. Laura McKie, Georgia Health Sciences University.*

reduced folates. FRs have a high affinity for folic acid in the subnanomolar range ($K_d \approx 1$–10 nM) compared with folate transporters, which have a low affinity for folate in the micromolar range.[12] FRα is the only isoform of FR that has been studied in retina, and it is abundantly expressed in the GCL and INL cells, especially Müller cells and in photoreceptor cell inner segment layers.[20,22] The RPE expresses FRα on its basolateral surface.[21] Studies showing that FRα works coordinately with RFC in the RPE are described below.

Reduced Folate Carrier

RFC is a 57–65 kDa integral transmembrane and energy-dependent protein that exhibits a high affinity for N^5-methyltetrahydrofolate (MTF), the predominant form of folate in blood, with a K_m of approximately 2–7 µM, and a low affinity for folic acid, with a K_m of approximately 150–200 µM.[12,33,34] RFC, also known as reduced folate transporter (RFT-1) and folate transport protein (FOLT), is a member of the SLC19 family of solute carriers (SLC19A1).[34] RFC has 12 transmembrane domains and N- and C-termini located in the cytoplasm, with a known N-glycosylation site at Asn58 (Fig. 35.3B). RFC functions as an anion exchanger that is Na⁺ and H⁺ independent. RFC operates optimally at pH 7.4, and its activity and folate-concentrating ability decrease as pH decreases.[12] In the intact retina, RFC is associated primarily with the RPE. Laser scanning confocal microscopy experiments have shown that RFC is localized to the apical membrane of the RPE, while FRα is on the basolateral membrane. It was hypothesized that FRα takes up folate from choroidal circulation and transfers it into the RPE, which then transfers it across the apical membrane via RFC, a bidirectional transporter. This prediction was validated by studying the transport of radiolabeled folate in an immortalized RPE cell line (ARPE-19) cultured on permeable membranes.[35] Membrane vesicles were prepared from bovine RPE and the operational mechanism and energetics of RFC were studied.[21,34] It was found that RFC is energized by a transmembrane pH gradient ($pH_{out} < pH_{in}$), suggesting that the transport mechanism involves the influx of the anionic substrate coupled to either symport of H⁺ into cells or antiport of OH⁻ out of the cells. Studies of the intact retina limit RFC to the RPE; however, there have been reports of a functional role for RFC in the inner blood–retinal barrier (BRB).[36] In those studies, the rat retinal capillary endothelial cell line (TR-iBRB2) was used as an *in vitro* model of the inner BRB. Under these *in vitro* conditions RFC messenger RNA (mRNA) expression was significant. RFC in non-RPE cells has also been explored; it was found that primary mouse Müller cells, which had been allowed to proliferate in culture, also expressed RFC, although the protein was not detectable in freshly isolated primary cells.[20]

This may suggest that under conditions of proliferation, RFC expression is upregulated and the protein becomes active. To understand the role of RFC in retinal development, *in situ* hybridization and immunohistochemical studies have been performed in mouse embryos. During early development, RFC is expressed throughout the eye, but by embryonic day 12.5, RFC protein becomes localized to the RPE.[37]

Proton-Coupled Folate Transporter

PCFT is a folate transport protein that is the product of SLC46A1.[38] Originally identified as heme carrier protein 1 (HCP1),[39] PCFT mediates H⁺-coupled electrogenic transport of folate and its derivatives with similar affinity for the oxidized and reduced forms of folic acid.[38] PCFT has a molecular weight of 50–65 kDa depending upon the extent of glycosylation; it contains two N-linked glycosylation sites (Asn58 and Asn68).[40] Like RFC, PCFT is an integral membrane transporter protein and spans the membrane 12 times (Fig. 35.3C).[41] Folate transport by PCFT involves the influx of one H⁺ per transport cycle. Since the transport process is electrogenic, this stoichiometry suggests that the zwitterionic form of folate is recognized by the transporter as the substrate, and involvement of one H⁺ indicates that folate is accepted as a substrate only in its electroneutral form.[42]

In retina, PCFT is present in the GCL, OPL, OLM, and the inner segments of photoreceptor cells. It is present in the RPE and retinal Müller glial cells. In addition to detection in intact retina, PCFT is present in primary cultures of RPE, ganglion, and Müller cells.[42] Primary cultures of Müller cells have been used to investigate whether FRα colocalized with PCFT in endosomes.[20,43] Kamen and colleagues proposed that folate internalization by FRα is coupled to a putative carrier protein by four steps, termed receptor-coupled transmembrane transport.[43] This novel method of endocytosis is thought to begin with FRα binding 5-MTF. Subsequently, the ligand–receptor complex is internalized by receptor-mediated endocytosis. Liberation of folate from the receptor occurs upon acidification of the endosomal compartment, followed by export across the endosomal membrane by an anion carrier. Finally, polyglutamic acid residues are added to MTF, a critical step in maintenance of the intracellular pool of folate. To test this hypothesis, MA401 cells were used to measure folate accumulation in cytosolic versus vesicular membrane fractions, following [³H]MTF uptake. The investigators reported an increase in the fraction of total MTF exported from the endosome into the cytosol, which was significantly inhibited by the addition of probenecid, an inhibitor of organic anion transporters. After the work was reported, others obtained evidence

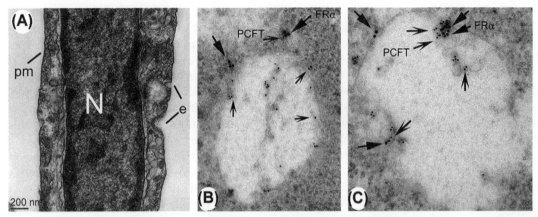

FIGURE 35.4 Electron microscopic immunolocalization of folate receptor-α (FRα) and proton-coupled folate transporter (PCFT) in Müller cells. Müller cells were fixed and postembedding electron-microscopic (EM) immunolocalization was employed to detect PCFT (10 nm gold particle, indented arrows) and FRα (18 nm gold particle, flat arrows). (A) EM photomicrograph (lower magnification) of a Müller cell. The nucleus (N) is prominent in the cell, 'e' denotes endosomes forming along the plasma membrane, and 'pm' denotes plasma membrane. (B) PCFT immunolabeling on the plasma membrane. (C) FRα and PCFT immunolabeling along the nuclear membrane. *Source: Figure adapted from Bozard BR, Ganapathy PS, Duplantier J, Mysona B, Ha Y, Roon P, et al. Molecular and biochemical characterization of folate transporter proteins in retinal Müller cells. Invest Ophthalmol Vis Sci. 2010;51:3226–35.[20] © Association for Research in Vision and Ophthalmology.*

supporting the hypothesis[34,44,45]; however, there was no electron microscopic evidence to support the proposed model. Very recently, the present authors obtained exciting data in retinal Müller cells using immuno-electron microscopic methods and demonstrated that PCFT and FRα colocalized in the endosomal compartment of Müller cells (Fig. 35.4). Figure 35.4(A) shows an electron-photomicrograph of the Müller cell with forming endosomes labeled with arrows. Figure 35.4(B, C) shows the immunogold labeling of FRα and PCFT that are clearly colocalized within endosomes. The data provide compelling evidence that PCFT and FRα function coordinately to mediate folate uptake into the Müller cell and that PCFT probably functions as an endosomal transporter for folate.

To summarize, FRα and PCFT are expressed abundantly in the retina, whereas RFC is limited (in the intact retina) to the RPE and possibly the inner blood–retinal barrier. FRα works coordinately with both of the folate transport proteins in retina. Specifically in the RPE, FRα sits at the basolateral surface of the cell to take folate from the choroidal circulation and transfer it into the cell by receptor-mediated endocytosis. Once inside the cell, folate can be transferred across the apical membrane to the subretinal space via the activity of RFC. FRα also works with PCFT. This has been studied in isolated primary cultures of retinal Müller cells, where it has been shown that folate resides on the plasma membrane as well as other organelle membranes (nuclear membrane, endosomal membrane, etc.). PCFT has also been localized to endosomal membranes of the Müller cells, providing the ideal mechanism for folate to be transferred from within the endosome into the cytoplasm.

REGULATION OF THE EXPRESSION AND ACTIVITY OF FOLATE TRANSPORT PROTEINS IN RETINAL CELLS

Reduced Folate Carrier

In addition to establishing mechanisms of folate transport in retinal cells, this laboratory has studied the regulation of these proteins extensively, particularly focusing on the sustentacular cells responsible for maintaining the retina, namely RPE and Müller cells. These investigations frequently asked whether conditions that are known to occur in retinal disease (e.g., oxidative stress, hyperglycemia, nitric oxide (NO), and HHcy) would alter the activities of the folate transporter proteins. RFC was investigated using RPE cells, and it was discovered that exposure of these cells to NO donors downregulated RFC activity significantly.[46] Elevation of NO is implicated in several retinal diseases including diabetic retinopathy. The consequences of hyperglycemia on RFC activity in RPE cells and its expression in retinas of diabetic mice were investigated. There was a marked reduction in RFC activity and expression in diabetic mouse retinas.[47] Functional studies showed that the uptake of radiolabeled MTF in RPE of diabetic mice was reduced by approximately 20% compared with that in nondiabetic, age-matched control animals. Semiquantitative reverse transcription–polymerase chain reaction (RT-PCR) demonstrated that the mRNA encoding RFC was decreased significantly in these mice as well. These data suggest that diabetes could markedly affect the availability of folate to cells of the retina. It is established that decreased availability of folate leads to elevation of homocysteine (Hcy), an amino acid involved in metabolism of methionine. Exposure of RPE to high levels of homocysteine (50 μM) attenuated RFC activity and downregulated its expression at the mRNA and protein levels.[48] Although homocysteine altered the activity of RFC, it did not perturb the function of several other nutrient transporters in RPE. The researchers also asked whether endogenous levels of folate itself regulate the RFC activity and demonstrated that indeed high levels of folate decreased the activity and expression of RFC.[49] When RPE cells were treated for several days with 2.26 μM MTF, they showed a 40% decrease in the uptake of MTF compared with cells exposed to much lower levels of MTF. The folate-induced attenuation of RFT-1 activity was associated with decreased maximal velocity of the transporter, while substrate affinity remained constant. Steady-state levels of RFT-1 mRNA and protein decreased significantly in the presence of excess folate.

Proton-Coupled Folate Transporter

The regulation of PCFT was investigated using primary cultures of mouse Müller cells to determine whether factors known to be associated with retinal disease regulated folate uptake in Müller cells *in vitro*. Stressors that had clearly regulated RFC (NO, reactive oxygen species, advanced glycation endproducts) did not alter PCFT activity in these cells.[50] This may reflect the highly proliferative state of these isolated cells, and indeed Müller cell proliferation (reactive gliosis) is a hallmark of retinal disease.[51,52] This led the researchers to investigate whether fumaric acid esters, compounds that have been used to treat the proliferative skin disease psoriasis, would alter PCFT activity in Müller cells.

It was found that monomethylfumarate inhibited PCFT activity and did so by decreasing the maximal velocity of the transporter with no change in its binding affinity for radiolabeled methyltetrahydrofolate.[53] The data suggest that this fumaric acid ester attenuates folate uptake in retinal Müller cells and may have potential therapeutic value in retinal disease.

These studies have enlightened us significantly on the role of folate transport in the retina, and it is clear that the mechanisms to take up this critical vitamin must function properly to ensure retinal health. As mentioned, one of the consequences of either dysfunctional folate uptake or a dietary deficiency of folate is that the amino acid homocysteine can accumulate. The consequences of excess homocysteine on retinal function are described below.

HYPERHOMOCYSTEINEMIA AND RETINAL HEALTH

Homocysteine, a sulfhydryl-containing, nonproteinogenic amino acid, is a key intermediate in methionine metabolism (Fig. 35.5). Homocysteine sits at the intersection of two metabolic pathways, termed the remethylation and transsulfuration pathways. HHcy can be caused by vitamin deficiencies (folate, B$_{12}$, or B$_6$) and by genetic defects involving enzymes: cystathionine-β-synthase (CBS), methylenetetrahydrofolate reductase (MTHFR), or methionine synthase. Normal fasting plasma levels (total homocysteine) are 5–15 μmol/L; higher levels are considered HHcy. Levels of HHcy are termed moderate (16–30 μmol/L), intermediate (31–100 μmol/L), and severe (>100 μmol/L).[54] Modest HHcy is found in 5–7% of the population.

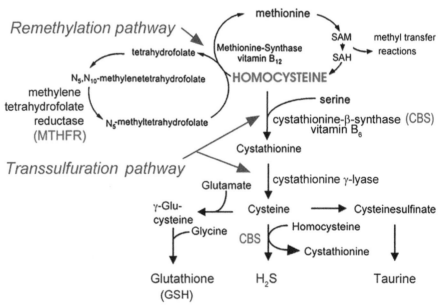

FIGURE 35.5 Metabolism of homocysteine (Hcy). Hcy is a nonprotein-forming sulfur amino acid whose metabolism lies at the intersection of two metabolic pathways: remethylation and transsulfuration. In remethylation, Hcy acquires a methyl group from the reduced form of folate, N^5-methyltetrahydrofolate (MTF), to form methionine in a vitamin B$_{12}$-dependent reaction catalyzed by methionine synthase. Insufficient levels of MTF lead to accumulation of Hcy. In the transsulfuration pathway, Hcy condenses with serine to form cystathionine in a reaction catalyzed by cystathionine-β-synthase (CBS) and is further converted to cysteine via the enzyme cystathionine-γ-lyase. Cysteine is required for synthesis of glutathione, hydrogen sulfide (H$_2$S), and taurine, all of which are known for their antioxidant properties.

HHcy is known to induce endoplasmic reticulum (ER) stress, excitotoxicity, oxidative stress, and DNA damage.[55] Homocysteine has gained considerable attention over the past 40 years because when elevated, even moderately, it is an independent risk factor in cardiovascular diseases (stroke, venous thrombosis, peripheral arterial occlusive disease)[55–58] and neurodegenerative diseases (Alzheimer's and Parkinson's diseases).[59,60] Given that the retina is a neurovascular tissue, it is not surprising that many clinical studies have investigated levels of homocysteine in retinopathies.[61] HHcy has been identified as a risk factor in retinal venous and arterial occlusions,[62] including central retinal vein occlusion.[63–67] It has been linked to pseudoexfoliation glaucoma,[68–70] a major form of glaucoma worldwide.[71] Elevated plasma homocysteine has been reported in macular degeneration[72–74] and diabetic retinopathy.[75–77]

Recently, very important clinical findings have emerged from studies of visual function during and after long-term spaceflight. After living aboard the international space station for 6 months, seven astronauts had ophthalmic findings, including optic disc edema, choroidal folds, cotton wool spots, and thickening of the NFL by optical coherence tomography.[78] The cause of the ophthalmic changes is not known, but the initial report suggested that perhaps microgravity-induced cephalad fluid shifts or localized intraorbital changes were involved. More recently, however, analysis of the blood samples drawn from these astronauts before, during, and after spaceflight revealed significantly higher concentrations of serum homocysteine and cystathionine in those individuals with ophthalmic changes versus those without.[79] The biochemical differences observed in crew members with visual deficits strongly suggest that folate- and vitamin B_{12}-dependent one-carbon transfer metabolism was affected before and during flight.

IN VITRO AND IN VIVO STUDIES OF HYERHOMOCYSTEINEMIA AND THE RETINA

To understand the role of HHcy in retinal disease various laboratory investigations have been conducted over the past few years. This laboratory has been actively engaged in these studies. The initial *in vitro* studies used a retinal neuronal cell line[80] and subsequently primary ganglion cells harvested from neonatal mouse retina. Using the primary cells it was established that micromolar concentrations of homocysteine were sufficient to induce rapid and significant ganglion cell death.[81] The effects of homocysteine in the retina have also been examined in RPE cells. Indeed, high levels of homocysteine in ARPE-19 cells increased steady-state vascular endothelial growth factor (VEGF) mRNA levels by as

much as five- to eight-fold depending upon the form of homocysteine used in the experiments.[82] The increase in VEGF mRNA levels was due to increased transcription, not mRNA stabilization. The investigators provided evidence of alterations in ER stress genes including GRP78, eIF2α, and ATF4 in the homocysteine-treated RPE cells. Alterations of RPE VEGF expression induced by elevated homocysteine are important with respect to the role of homocysteine in retinal vascular disease.

In vivo studies of the role of homocysteine in retina used high-dosage intravitreal injection of D,L-Hcy-thiolactone and demonstrated marked ganglion cell loss and inner retinal disruption within 5 days of exposure,[83,84] significant loss of photoreceptor cells (within 15 days), and ablation of the outer nuclear layer within 90 days following prolonged HHcy exposure. These studies provide clear evidence of neuronal cell loss in retina due to exogenous administration of excess homocysteine.

The availability of mouse models with endogenous elevation of homocysteine has provided a very powerful tool to examine effects of moderate/severe HHcy on retina. The most comprehensive studies have been performed in mice deficient or completely lacking CBS, a critical enzyme in the transsulfuration pathway (Fig. 35.5). The cbs mutant mice were developed in Dr. N. Maeda's laboratory at the University of North Carolina, Chapel Hill.[85] Homozygous ($cbs^{-/-}$) mice are a model for severe HHcy. They have plasma homocysteine levels around 40-fold greater than normal (203.6 ± 65.3 nmol/mL vs. 6.1 ± 0.8 nmol/mL), suffer from severe growth retardation, and die by 3–5 weeks. Heterozygous ($cbs^{+/-}$) mutants have around 50% reduction in *cbs* mRNA and enzyme activity in the liver and have two to four times the normal plasma homocysteine levels (13.5 ± 3.2 nmol/mL). The $cbs^{+/-}$ mutants are a promising model for studying the *in vivo* role of moderate HHcy in the etiology of retinal diseases over an extended lifespan. Homocysteine levels in retinas of $cbs^{-/-}$ and $cbs^{+/-}$ mice, determined by high-performance liquid chromatography, are seven- and two-fold, respectively, greater than WT ($cbs^{+/+}$).[86]

Using the *cbs* mutant mice, profound visual function loss,[87] marked disruption of retinal neuronal layers,[86] and florid vasculopathy were observed in homozygous ($cbs^{-/-}$) mutant mice.[88] Figure 35.6(A) shows the marked loss of a- and b-wave amplitude in the $cbs^{-/-}$ mouse compared with wild-type and heterozygous mice and a marked decrease in luminance response function (Fig. 35.6B). In the *cbs* heterozygous mice ($cbs^{+/-}$) the phenotype is much milder. Over a period of several months, cells are lost in the GCL,[86] and more modest declines in the a- and b-wave changes are detected by ERG (Fig. 35.6B–F). The histologic appearance of the $cbs^{+/+}$, $cbs^{+/-}$, and $cbs^{-/-}$ mice is shown in Figure 35.6(G–K). Preliminary studies suggest that the $cbs^{+/-}$ mice have milder vascular alterations than $cbs^{-/-}$ mice. The retinal

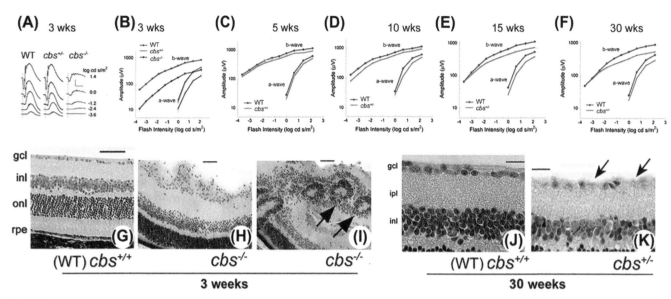

FIGURE 35.6 Morphologic and electrophysiologic analyses of cbs$^{+/-}$ and cbs$^{-/-}$ mice. (A) Dark-adapted electroretinography (ERG) showing a significant reduction in a- and b-wave ERG amplitudes in 3 week homozygous cbs$^{-/-}$ mice. (B) Luminance response functions are significantly decreased in cbs$^{-/-}$ mice compared with cbs$^{+/+}$ mice. (C–F) Luminance response functions in cbs$^{+/-}$ mice show modest decreases with advancing age (5, 10, 15, and 30 weeks) compared with cbs$^{+/+}$ mice. (G–I) Representative photomicrographs of retinal histology of 3 week cbs$^{+/+}$, cbs$^{+/-}$, and cbs$^{-/-}$ mice. The cbs$^{-/-}$ mice have profound disruption of the retina, with cellular dropout (H) and rosette formation (I, arrows). (J, K) Representative photomicrographs of retinal histology of 30 week cbs$^{+/+}$ and cbs$^{+/-}$ mice. Analysis of heterozygous cbs$^{+/-}$ retinas reveals a milder phenotype characterized by modest, gradual loss in the ganglion cell layer (gcl) and decreased thickness of the inner plexiform (IPL) and nuclear (INL) layers, which is evident only at older ages. The observed morphology is consistent with the electrophysiologic evidence of gradual retinal dysfunction. Scale bar = 50 μm (G–I), 20 μm (J, K). *Source: Figure adapted from Yu M, Sturgill-Short G, Ganapathy P, Tawfik A, Peachey NS, Smith SB. Age-related changes in visual function in cystathionine-beta-synthase mutant mice, a model of hyperhomocysteinemia. Exp Eye Res. 2012;96:124–31.*[87] © *Elsevier.*

ganglion cell loss in the *cbs*$^{+/-}$ mice is accompanied by alterations in the mitochondria of the NFL detectable at the ultrastructural level and by increased expression of two mitochondrial proteins, Opa1 and Fis1.[89] The study suggested that increased mitochondrial fission is a novel mechanism of homocysteine toxicity to neurons. The findings are particularly relevant to glaucoma because mitochondrial fission has been implicated in autosomal dominant optic atrophy, a progressive disease characterized by preferential loss of ganglion cells.[90]

The data obtained using the *cbs* mutant mice suggest that (1) HHcy compromises retinal structure and function, the severity of which correlates with the level of HHcy, and (2) HHcy alters retinal vasculature and triggers death of retinal neurons. The phenotype may be due to homocysteine-induced pathology (ER stress, oxidative stress, excitotoxicity). Studies are now underway to investigate the consequences of HHcy on retina in a mouse with deficiency/lack of MTHFR.[91] MTHFR converts N_5,N_{10}-methylenetetrahydrofolate into N_5-methyltetrahydrofolate, the predominant circulating form of folate in the body. Mutations of MTHFR are fairly common in the human population, especially the 677C→T mutation (with a frequency of approximately 30% in many ethnic groups).[92] *mthfr* mice have a defect in the remethylation pathway (Fig. 35.5) with concomitant HHcy; however, their transsulfuration pathway is intact.

Using these mice, it will be possible to dissect the role of HHcy in the retina. One of the directions of the research in this area will be to explore the effects of supplementation of the diet with folate/B$_{12}$/B$_6$ on HHcy-induced retinopathy.

In vivo studies have used dietary manipulation to explore the effects of HHcy on the retina. Studies of rats fed a high homocysteine diet showed an increase in vimentin, glial fibrillary acidic protein (GFAP), and VEGF immunoreactivity in the retina compared with controls.[93] The increase in vimentin immunoreactivity in the hyperhomocysteinemic rats was correlated with changes in GFAP immunoreactivity in astrocytes within the GCL. The investigators found that short-term HHcy-induced oxidative stress activated retinal glial cells and increased VEGF expression in the retina. Thus, these data provide additional *in vivo* evidence that sustained elevation of homocysteine has deleterious consequences on retinal health.

These studies in animal models have relevance to the human population. For example, the 7-year Women's Antioxidant and Folic Acid Cardiovascular Study (WAFACS), with an enrollment of more than 8000 subjects who did not have age-related macular degeneration (AMD) at the time of recruitment to the study, reported fewer new cases of AMD, fewer cases of visual loss, and decreased plasma homocysteine in patients receiving daily supplementation with folic acid/B_6/B_{12} compared with placebo.[94] Recommendations from the study urged vitamin supplementation to reduce the incidence of AMD. Wright and colleagues reviewed the clinical literature relevant to homocysteine. The measurement of plasma homocysteine is suggested and reduction of elevated homocysteine with folic acid for secondary prevention of retinal arterial and venous occlusion is recommended.[61]

In summary, studies of folic acid and its role in retinal health represent a new direction for research in nutrition and the eye. The clinical evidence is clear that lack of folate can compromise retinal function. In situations, either nutritional or genetic, in which homocysteine is elevated, the effects on retinal vasculature and neuronal function are likely to be disrupted. Further clinical studies documenting levels of homocysteine in various retinal pathologies, coupled with laboratory studies carefully assessing the consequences of HHcy, will elucidate mechanisms by which homocysteine affects retinal structure and function. Carefully executed studies of nutritional supplementation will provide more information on the benefits of folate/vitamin B_6/B_{12} supplementation for these retinal diseases.

TAKE-HOME MESSAGES

- The retina utilizes all three known folate transport proteins to acquire folate: folate receptor-α (FRα) and proton-coupled folate transporter (PCFT) are expressed abundantly throughout the retina, whereas reduced folate carrier (RFC) is limited (in the intact retina) to the retinal pigment epithelium (RPE) and possibly the inner blood–retinal barrier.
- FRα works coordinately with both of the folate transport proteins in retina.
- In the RPE, FRα, located on the basolateral surface, takes folate from choroidal circulation and transfers it into the cell via endocytosis and appears to release folate to the subretinal space via the RFC.
- In retinal Müller glial cells, FRα and PCFT have been localized to endosomes, suggesting that they work coordinately to transfer folate from within the endosome into the cytoplasm.
- One of the consequences of deficiency of folate (either dietary or through impaired transport mechanisms) is an elevation of homocysteine. Hyperhomocysteinemia can also be caused by genetic mutations that affect the remethylation and transsulfuration metabolic pathways.
- Retinal diseases that have been associated with hyperhomocysteinemia include central retinal vein occlusion, pseudoexfoliation glaucoma, macular degeneration, and diabetic retinopathy.

- Experimental models of hyperhomocysteinemia due to deficiency or lack of the enzymes cystathionine-β-synthase and methylene tetrahydrofolate reductase have provided excellent tools to understand the role of hyperhomocysteinemia in retinal neuronal and vascular function.

Acknowledgments

The authors thank Mrs. Laura McKie, Division of Visual and Instructional Design, Georgia Health Sciences University, for the illustrations used in Figures 35.2 and 35.3. This work was supported by NIH R01 EY012830 and EY014560.

References

1. Knox DL, Chen MF, Guilarte TR, Dang CV, Burnette J. Nutritional amblyopia: folic acid, vitamin B-12, and other vitamins. *Retina* 1982;**2**:288–93.
2. Smiddy WE, Green WR. Nutritional amblyopia. A histopathologic study with retrospective clinical correlation. *Graefe's Arch Clin Exp Ophthalmol* 1987;**225**:321–4.
3. Schaible ER, Golnik KC. Optic neuropathy associated with folate deficiency. *Invest Ophthal Vis Sci* 1993;**34**: S1215.
4. Golnik KC, Schaible ER. Folate-responsive optic neuropathy. *J Neurophthalmol* 1994;**14**:163–9.
5. Miller NR. The optic nerve. *Curr Opin Neurol* 1996;**9**:5–15.
6. Sadun AA. Metabolic optic neuropathies. *Semin Ophthalmol* 2002;**17**:29–32.
7. Sadun A, Rubin R. Residual psychophysical deficits following recovery from the Cuban epidemic of optic neuropathy. In: Lakshminarayan V, editor. *Basic and Clinical Applications of Vision Science.* Dordrecht: Kluwer; 1997. pp. 231–4.
8. Mroczkowski MM, Redgrave GW, Miller NR, McCoy AN, Guarda AS. Reversible vision loss secondary to malnutrition in a woman with severe anorexia nervosa, purging type, and alcohol abuse. *Int J Eat Disord* 2011;**44**:281–3.
9. Garner CD, Lee EW, Louis-Ferdinand RT. Müller cell involvement in methanol-induced retinal toxicity. *Toxicol Appl Pharmacol* 1995;**130**:101–7.
10. Shane B. Folate and vitamin B_{12} metabolism: overview and interaction with riboflavin, vitamin B_6, and polymorphisms. *Food Nutr Bull* 2008;**29**:S5–19.
11. Fox JT, Stover PJ. Folate-mediated one-carbon metabolism. *Vitam Horm* 2008;**79**:1–44.
12. Zhao R, Matherly LH, Goldman ID. Membrane transporters and folate homeostasis: intestinal absorption and transport into systemic compartments and tissues. *Expert Rev Mol Med* 2009;**28**:e4.
13. Hamid A, Wani NA, Kaur J. New perspectives on folate transport in relation to alcoholism-induced folate malabsorption – association with epigenome stability and cancer development. *FEBS J* 2009;**276**:2175–91.
14. Luo DG, Xue T, Yau KW. How vision begins: an odyssey. *Proc Natl Acad Sci USA* 2008;**105**:9855–62.
15. Dowling JE, Boycott BB. Organization of the primate retina: electron microscopy. *Proc R Soc Lond B Biol Sci* 1966;**166**:80–111.
16. Osterberg GA. Topography of the layers of the rods and cones in the human retina. *Acta Ophthalmol Suppl* 1935;**6**:1–103.
17. Wässle H, Boycott BB. Functional architecture of the mammalian retina. *Physiol Rev* 1991;**71**:447–80.
18. Bok D. The retinal pigment epithelium: a versatile partner in vision. *J Cell Sci* 1993;(Suppl. 17):189–95.

19. Sarthy V, Ripps H. *The Retinal Müller Cell, Structure and Function*. New York: Kluwer Academic/Plenum Publishers; 2001.

20. Bozard BR, Ganapathy PS, Duplantier J, Mysona B, Ha Y, Roon P, et al. Molecular and biochemical characterization of folate transporter proteins in retinal Müller cells. *Invest Ophthalmol Vis Sci* 2010;**51**:3226–35.

21. Chancy CD, Kekuda R, Huang W, Prasad PD, Kuhnel JM, Sirotnak FM, et al. Expression and differential polarization of the reduced-folate transporter-1 and the folate receptor α in mammalian retinal pigment epithelium. *J Biol Chem* 2000;**275**:20676–84.

22. Smith SB, Kekuda R, Gu X, Chancy C, Conway SJ, Ganapathy V. Expression of folate receptor alpha in the mammalian retinal pigmented epithelium and retina. *Invest Ophthalmol Vis Sci* 1999;**40**:840–8.

23. Huang W, Prasad PD, Kekuda R, Leibach FH, Ganapathy V. Characterization of N^5-methyltetrahydrofolate uptake in cultured human retinal pigment epithelial cells. *Invest Ophthalmol Vis Sci* 1997;**38**:1578–87.

24. Spiegelstein O, Eudy JD, Finnell RH. Identification of two putative novel folate receptor genes in humans and mouse. *Gene* 2000;**258**:117–25.

25. Antony AC. Folate receptors. *Annu Rev Nutr* 1996;**16**:501–21.

26. Lee HC, Shoda R, Krall JA, Foster JD, Selhub J, Rosenberry TL. Folate binding protein from kidney brush border membranes contains components characteristic of a glycoinositol phospholipid anchor. *Biochemistry* 1992;**31**:3236–43.

27. Shen F, Wu M, Ross JF, Miller D, Ratnam M. Folate receptor type gamma is primarily a secretory protein due to lack of an efficient signal for glycosylphosphatidylinositol modification: protein characterization and cell type specificity. *Biochemistry* 1995;**34**:5660–5.

28. Sabharanjak S, Mayor S. Folate receptor endocytosis and trafficking. *Adv Drug Deliv Rev* 2004;**56**:1099–109.

29. Mayor S, Sabharanjak S, Maxfield FR. Cholesterol-dependent retention of GPI-anchored proteins in endosomes. *EMBO J* 1998;**17**:4626–38.

30. Chatterjee S, Smith ER, Hanada K, Stevens VL, Mayor S. GPI anchoring leads to sphingolipid-dependent retention of endocytosed proteins in the recycling endosomal compartment. *EMBO J* 2001;**20**:1583–92.

31. Chalupa LM, Williams RW. *Eye, Retina and Visual System of the Mouse*. Cambridge, MA: MIT Press; 2008.

32. Salazar MD, Ratnam M. The folate receptor: what does it promise in tissue-targeted therapeutics? *Cancer Metastasis Rev* 2007;**26**:141–52.

33. Sirotnak FM, Tolner B. Carrier-mediated membrane transport of folates in mammalian cells. *Annu Rev Nutr* 1999;**19**:91–122.

34. Ganapathy V, Smith SB, Prasad PD. SLC19: the folate/thiamine transporter family. *Pflugers Arch* 2004;**447**:641–6.

35. Bridges CC, El-Sherbeny A, Ola MS, Ganapathy V, Smith SB. Transcellular transfer of folate across the retinal pigment epithelium. *Curr Eye Res* 2002;**24**:129–38.

36. Hosoya K, Fujita K, Tachikawa M. Involvement of reduced folate carrier 1 in the inner blood–retinal barrier transport of methyltetrahydrofolate. *Drug Metab Pharmacokinet* 2008;**23**:285–92.

37. Maddox DM, Manlapat A, Roon P, Prasad P, Ganapathy V, Smith SB. Reduced-folate transporter (RFT) is expressed in placenta and yolk sac, as well as in cells of the developing forebrain, hindbrain, neural tube, craniofacial region, eye, limb buds and heart. *BMC Dev Biol* 2003;**3**:6.

38. Qiu A, Jansen M, Sakaris A, Min SH, Chattopadhyay S, Tsai E, et al. Identification of an intestinal folate transporter and the molecular basis for hereditary folate malabsorption. *Cell* 2006;**127**:917–28.

39. Shayeghi M, Latunde-Dada GO, Oakhill JS, Laftah AH, Takeuchi K, Halliday N, et al. Identification of an intestinal heme transporter. *Cell* 2005;**122**:789–801.

40. Unal ES, Zhao R, Qiu A, Goldman ID. N-linked glycosylation and its impact on the electrophoretic mobility and function of the human proton-coupled folate transporter (HsPCFT). *Biochim Biophys Acta* 2008;**1778**:1407–14.

41. Zhao R, Goldman ID. The molecular identity and characterization of a proton-coupled folate transporter – PCFT; biological ramifications and impact on the activity of pemetrexed. *Cancer Metastasis Rev* 2007;**26**:129–39.

42. Umapathy NS, Gnana-Prakasam JP, Martin PM, Mysona B, Dun Y, Smith SB, et al. Cloning and functional characterization of the proton-coupled electrogenic folate transporter and analysis of its expression in retinal cell types. *Invest Ophthalmol Vis Sci* 2007;**48**:5299–305.

43. Kamen BA, Smith AK, Anderson RG. The folate receptor works in tandem with a probenecid-sensitive carrier in MA104 cells. *in vitro J Clin Invest* 1991;**87**:1442–9.

44. Wollack JB, Makori B, Ahlawat S, Koneru R, Picinich SC, Smith A, et al. Characterization of folate uptake by choroid plexus epithelial cells in a rat primary culture model. *J Neurochem* 2008;**104**:1494–503.

45. Zhao R, Min SH, Wang Y, Campanella E, Low PS, Goldman ID. A role for the proton-coupled folate transporter (PCFT-SLC46A1) in folate receptor-mediated endocytosis. *J Biol Chem* 2009;**284**:4267–74.

46. Smith SB, Huang W, Chancy C, Ganapathy V. Regulation of the reduced folate transporter by nitric oxide in cultured human retinal pigment epithelial cells. *Biochem Biophys Res Commun* 1999;**257**:279–83.

47. Naggar H, Ola MS, Moore P, Huang W, Bridges CC, Ganapathy V, Smith SB. Downregulation of reduced-folate transporter by glucose in cultured RPE cells and in RPE of diabetic mice. *Invest Ophthalmol Vis Sci* 2002;**43**:556–63.

48. Naggar H, Fei YJ, Ganapathy V, Smith S. Regulation of reduced-folate transporter-1 (RFT-1) by homocysteine and identity of transport systems for homocysteine uptake in retinal pigment epithelial (RPE) cells. *Exp Eye Res* 2003;**77**:687–97.

49. Naggar H, Van Ells TK, Ganapathy V, Smith SB. Regulation of reduced-folate transporter-1 in retinal pigment epithelial cells by folate. *Curr Eye Res* 2005;**30**:35–44.

50. Bozard BR. *Molecular and biochemical characterization and regulation of folate transport proteins in retinal Müller cells*. PhD thesis. Augusta, Georgia: Georgia Health Sciences University, College of Graduate Studies; 2011.

51. Bringmann A, Iandiev I, Pannicke T, Wurm A, Hollborn M, Wiedemann P, et al. Cellular signaling and factors involved in Müller cells gliosis: neuroprotective and detrimental effects. *Prog Ret Eye Res* 2009;**28**:423–51.

52. Bringmann A, Pannicke T, Biedermann B, Francke M, Iandiev I, Grosche J, et al. Role of retinal glial cells in neurotransmitter uptake and metabolism. *Neurochem Int* 2009;**54**:43–160.

53. Bozard BR, Chothe PP, Tawfik A, Williams C, Fulzele S, Prasad PD, et al. Regulation of proton-coupled folate transporter in retinal Müller cells by the antipsoriatic drug monomethylfumarate. *Glia* 2012;**60**:333–42.

54. Hankey GJ, Eikelboom JW. Homocysteine and vascular disease. *Lancet* 1999;**354**:407–13.

55. Perła-Kaján J, Twardowski T, Jakubowski H. Mechanisms of homocysteine toxicity in humans. *Amino Acids* 2007;**32**:561–72.

56. Maron BA, Loscalzo J. The treatment of hyperhomocysteinemia. *Annu Rev Med* 2009;**60**:39–54.

57. Wald DS, Law M, Morris JK. Homocysteine and cardiovascular disease: evidence on causality from a meta-analysis. *BMJ* 2002;**325**:1202.

58. Homocysteine Studies Collaboration. Homocysteine and risk of ischemic heart disease and stroke: a meta-analysis. *JAMA* 2002;**288**:2015–22.

59. Agnati LF, Genedani S, Rasio G, Galantucci M, Saltini S, Filaferro M, et al. Studies on homocysteine plasma levels in Alzheimer's patients. Relevance for neurodegeneration. *J Neural Transm* 2005;**112**:163–9.

60. Hogervorst E, Ribeiro HM, Molyneux A, Budge M, Smith AD. Plasma homocysteine levels, cerebrovascular risk factors, and cerebral white matter changes (leukoaraiosis) in patients with Alzheimer disease. *Arch Neurol* 2002;**59**:787–93.

61. Wright AD, Martin N, Dodson PM. Homocysteine, folates, and the eye. *Eye* 2008;**22**:989–93.

62. Cahill MT, Stinnett SS, Fekrat S. Meta-analysis of plasma homocysteine, serum folate, serum vitamin B(12), and thermolabile MTHFR genotype as risk factors for retinal vascular occlusive disease. *Am J Ophthalmol* 2003;**136**:1136–50.

63. Jaksic V, Markovic V, Milenkovic S, Stefanovic I, Jakovic N, Knezevic M. MTHFR C677T homozygous mutation in a patient with pigmentary glaucoma and central retinal vein occlusion. *Ophthal Res* 2010;**43**:193–6.

64. Gao L, Pulido JS, Hatfield RM, Dundervill III RF, McCannel CA, Shippy SA. Capillary electrophoretic assay for nitrate levels in the vitreous of proliferative diabetic retinopathy. *J Chromatogr B Anal Technol Biomed Life Sci* 2007;**847**:300–4.

65. Gumus K, Kadayifcilar S, Eldem B, Saracbasi O, Ozcebe O, Dundar S, Kirazli S. Is elevated level of soluble endothelial protein C receptor a new risk factor for retinal vein occlusion? *Clin Exp Ophthalmol* 2006;**34**:305–11.

66. Abu El-Asrar AM, Al-Obeidan SA, Abdel Gader AG. Retinal periphlebitis resembling frosted branch angiitis with nonperfused central retinal vein occlusion. *Eur J Ophthalmol* 2003;**13**:807–12.

67. Lahey JM, Kearney JJ, Tunc M. Hypercoagulable states and central retinal vein occlusion. *Curr Opin Pulm Med* 2003;**9**:385–92.

68. Tranchina L, Centofanti M, Oddone F, Tanga L, Roberti G, Liberatoscioli L, et al. Levels of plasma homocysteine in pseudoexfoliation glaucoma. *Graefes Arch Clin Exp Ophthalmol* 2011;**249**:443–8.

69. Turgut B, Kaya M, Arslan S, Demir T, Güler M, Kaya MK. Levels of circulating homocysteine, vitamin B_6, vitamin B_{12}, and folate in different types of open-angle glaucoma. *Clin Interv Aging* 2010;**5**:133–9.

70. Roedl JB, Bleich S, Reulbach U, Rejdak R, Naumann GO, Kruse FE, et al. Vitamin deficiency and hyperhomocysteinemia in pseudoexfoliation glaucoma. *J Neural Transm* 2007;**114**:571–5.

71. Schlötzer-Schrehardt U. Oxidative stress and pseudoexfoliation glaucoma. *Klin Monbl Augenheilkd* 2010;**227**:108–13.

72. Javadzadeh A, Ghorbanihaghjo A, Bahreini E, Rashtchizadeh N, Argani H, Alizadeh S. Serum paraoxonase phenotype distribution in exudative age-related macular degeneration and its relationship to homocysteine and oxidized low-density lipoprotein. *Retina* 2012;**32**:658–66.

73. Javadzadeh A, Ghorbanihaghjo A, Bahreini E, Rashtchizadeh N, Argani H, Alizadeh S. Plasma oxidized LDL and thiol-containing molecules in patients with exudative age-related macular degeneration. *Mol Vis* 2010;**16**:2578–84.

74. Carrillo-Carrasco N, Venditti CP. Combined methylmalonic acidemia and homocystinuria, cblC type. II. Complications, pathophysiology, and outcomes. *J Inherit Metab Dis* 2012;**35**:103–14.

75. Lim CP, Loo AV, Khaw KW, Sthaneshwar P, Khang TF, Hassan M, Subrayan V. Plasma, aqueous and vitreous homocysteine levels in proliferative diabetic retinopathy. *Br J Ophthalmol* 2012;**96**:704–7.

76. Cho HC. The relationship among homocysteine, bilirubin, and diabetic retinopathy. *Diabetes Metab J* 2011;**35**:595–601.

77. Satyanarayana A, Balakrishna N, Pitla S, Reddy PY, Mudili S, Lopamudra P, et al. Status of B-vitamins and homocysteine in diabetic retinopathy: association with vitamin-B_{12} deficiency and hyperhomocysteinemia. *PLoS ONE* 2011;**6**: e26747.

78. Mader TH, Gibson CR, Pass AF, Kramer LA, Lee AG, Fogarty J, et al. Optic disc edema, globe flattening, choroidal folds, and hyperopic shifts observed in astronauts after long-duration space flight. *Ophthalmology* 2011;**118**:2058–69.

79. Zwart SR, Gibson CR, Mader TH, Ericson K, Ploutz-Snyder R, Heer M, Smith SM. Vision changes after spaceflight are related to

alterations in folate- and vitamin B-12-dependent one-carbon metabolism. *J Nutr* 2012;**142**:427–31.

80. Martin PM, Ola MS, Agarwal N, Ganapathy V, Smith SB. The sigma receptor (σR) ligand (+)-pentazocine prevents retinal ganglion cell death induced *in vitro* by homocysteine and glutamate. *Brain Res Mol Brain Res* 2004;**123**:66–75.

81. Dun Y, Thangaraju M, Prasad P, Ganapathy V, Smith SB. Prevention of excitotoxicity in primary retinal ganglion cells by (+)-pentazocine, a sigma receptor-1 specific ligand. *Invest Ophthalmol Vis Sci* 2007;**48**:4785–94.

82. Roybal CN, Yang S, Sun CW, Hurtado D, Vander Jagt DL, Townes TM, Abcouwer SF. Homocysteine increases the expression of vascular endothelial growth factor by a mechanism involving endoplasmic reticulum stress and transcription factor ATF4. *J Biol Chem* 2004;**279**:14844–52.

83. Moore P, El-sherbeny A, Roon P, Schoenlein PV, Ganapathy V, Smith SB. Apoptotic cell death in the mouse retinal ganglion cell layer is induced *in vivo* by the excitatory amino acid homocysteine. *Exp Eye Res* 2001;**73**:45–57.

84. Chang HH, Lin DP, Chen YS, Liu HJ, Lin W, Tsao ZJ, et al. Intravitreal homocysteine-thiolactone injection leads to the degeneration of multiple retinal cells, including photoreceptors. *Mol Vis* 2011;**17**:1946–56.

85. Watanabe M, Osada J, Aratani Y, Kluckman K, Reddick R, Malinow MR. Mice deficient in cystathionine beta-synthase: animal models for mild and severe homocysteinemia. *Proc Natl Acad Sci USA* 1995;**92**:1585–9.

86. Ganapathy PS, Moister B, Roon P, Mysona B, Duplantier J, Dun Y, et al. Endogenous elevation of homocysteine induces retinal ganglion cell death in the cystathionine-β-synthase mutant mouse. *Invest Ophthalmol Vis Sci* 2009;**50**:4460–70.

87. Yu M, Sturgill-Short G, Ganapathy P, Tawfik A, Peachey NS, Smith SB. Age-related changes in visual function in cystathionine-beta-synthase mutant mice, a model of hyperhomocysteinemia. *Exp Eye Res* 2012;**96**:124–31.

88. Tawfik AM, Sonne S, Williams C, Al-Shabrawey M, Ganapathy V, Smith SB. Altered retinal vasculature in hyperhomocysteinemic mice. *Invest Ophthalmol Vis Sci* 2012: Abstract #2557.

89. Ganapathy PS, Perry RL, Tawfik A, Smith RM, Perry E, Roon P, et al. σHomocysteine-mediated modulation of mitochondrial dynamics in retinal ganglion cells. *Invest Ophthalmol Vis Sci* 2011;**52**:5551–8.

90. Williams PA, Morgan JE, Votruba M. Opa1 deficiency in a mouse model of dominant optic atrophy leads to retinal ganglion cell dendropathy. *Brain* 2010;**133**:2942–51.

91. Lawrance AK, Racine J, Deng L, Wang X, Lachapelle P, Rozen R. Complete deficiency of methylenetetrahydrofolate reductase in mice is associated with impaired retinal function and variable mortality, hematological profiles, and reproductive outcomes. *J Inherit Metab Dis* 2011;**34**:147–57.

92. Wilcken B, Bamforth F, Li Z, Zhu H, Ritvanen A, Renlund M, et al. Geographical and ethnic variation of the 677C>T allele of 5,10 methylenetetrahydrofolate reductase (MTHFR): findings from over 7000 newborns from 16 areas world wide. *J Med Genet* 2003;**40**:619–25.

93. Lee I, Lee H, Kim JM, Chae EH, Kim SJ, Chang N. Short-term hyperhomocysteinemia-induced oxidative stress activates retinal glial cells and increases vascular endothelial growth factor expression in rat retina. *Biosci Biotechnol Biochem* 2007;**71**:1203–10.

94. Christen WG, Glynn RJ, Chew EY, Albert CM, Manson JE. Folic acid, pyridoxine, and cyanocobalamin combination treatment and age-related macular degeneration in women: the Women's Antioxidant and Folic Acid Cardiovascular Study. *Arch Intern Med* 2009;**169**:335–41.

36

Selenium and Graves' Orbitopathy

Claudio Marcocci, Maria Antonietta Altea, Francesca Menconi

Department of Clinical and Experimental Medicine, University of Pisa, Pisa, Italy

INTRODUCTION

A cellular redox state is necessary for the maintenance of cellular homeostasis within certain boundaries. Intracellular reactive oxygen species (ROS) are the products of the oxidative cell metabolisms.[1,2] They include partially reduced forms of oxygen, such as hydrogen peroxide (H_2O_2), hydroxyl radicals ($OH°$) and superoxide anions (O_2^-), and lipid peroxides. ROS act as oxidizing agents, perturb intracellular reactions, and damage cell structures, including membranes and cellular proteins, lipids, and nucleic acids. Under physiologic conditions, an antioxidant system, which includes enzymes such as superoxide dismutase (SOD), catalase (CT), glutathione peroxidase (GPx), and glutathione reductase (GR), as well as small non-enzymatic molecules such as glutathione (GSH) and vitamins (ascorbic acid and tocopherol), defends the cells from ROS-induced damage. An increase in ROS production or a decrease in their scavenging ability will compromise the oxidative stability of the cell, resulting in a state of oxidative stress.[3]

Increased ROS generation is involved in the development of several diseases (cardiovascular, neurodegenerative, neoplastic, and endocrine),[3] in the control of the immune system, and in the pathogenesis of autoimmune disease. Recently, there has been growing interest in the possibility that antioxidant administration may counteract the disease-promoting effects of oxidative stress.

Selenium is a trace mineral, which is incorporated into several selenoproteins and enzymes, where selenium acts as a redox center and functions as an antioxidant.

SELENIUM

The distribution of selenium in nature varies widely depending on soils and geographic areas. Thus the intake of selenium varies from deficient (7 μg/day) to toxic concentration (4990 μg/day).[4] The selenium content in soil is poor in central and southern European countries, while is rich in most North American countries. Selenium reaches the human food chain through ingestion of edible plants, so diet is the most important source of selenium.[5]

The mean values of daily selenium intake in Europe are 40 μg, while in the USA they are 93 μg in women and 134 μg in men. The daily recommended selenium intake is 53 μg in women and 60 μg in men.[4]

Shellfish, crabs, kidney, liver, and Brazil nuts are the main dietary sources of selenium. The bioavailability of selenium in different foods depends on the content of different selenium forms.[5] Selenomethionine and selenocysteine, the organic forms of selenium, are selenium analogues of human amino acids, which are directly incorporated into body proteins. Selenomethionine is mostly found in plants (cereals) and selenocysteine in food of animal origin.[4] Selenate, the inorganic form of selenium, is mostly found in plants, in which it is directly translocated from the soil, and fish. These compounds are metabolized via different pathways and used for the biosynthesis of selenoproteins, and they affect the total selenium status in different ways.

Selenium status is assessed by measuring circulating selenoproteins or total selenium concentration. Two soluble selenoproteins, glutathione peroxidase-3 and selenoprotein P, account for blood selenium content.[5]

Individual selenium status varies widely in different parts of the world according to selenium intake.[5] Epidemiologic and intervention studies, looking at the effect of selenium supplementation on the risk of cancer, type 2 diabetes, and mortality, suggest that people whose serum or plasma selenium concentration is higher than 122 μg/L not only will not benefit from selenium supplementation but may be at higher risk. Conversely, various health benefits and no risks are associated with selenium supplementation in people with a serum or plasma selenium concentration lower than $122\,\mu g/L$.[5] Thus, baseline selenium status represents the most important predictor of either benefit or risk of selenium supplementation.

Selenomethionine or sodium selenite is usually present in selenium-containing supplements. It makes no difference whether selenomethionine or selenite is used as the supplement, until the circulating selenoproteins are saturated, because selenium is mainly used for the biosynthesis of selenoproteins. Once selenoproteins are saturated, only selenomethionine will be able to further increase serum selenium through its unregulated incorporation into proteins, whereas selenite will be no longer effectively be used for selenoprotein synthesis and will be excreted.

Therefore, when taking selenomethionine as a supplement, the blood selenium concentration will increase even in selenium-sufficient people, whereas the effect of selenite will depend on individual selenium status.

OXIDATIVE STRESS AND GRAVES' ORBITOPATHY

Graves' orbitopathy (GO) is an autoimmune disorder. Endogenous and environmental factors are involved in its pathogenesis.[6] A widely accepted hypothesis suggests that, after recognition of antigen(s) shared by the orbit and the thyroid, intraorbital activated T lymphocytes initiate a cascade of events leading to increased production of cytokines, growth factor, and ROS.[7] Increased proliferation of adipocytes and orbital fibroblasts and overproduction of hydrophilic glycosaminoglycans will increase the volume of orbital structures (fibroadipose tissue and extraocular muscles).[6]

In vitro and *in vivo* studies support a role of ROS in the pathogenesis of GO.[8] Moreover, cigarette smoking, the most important environmental factor involved in the development of GO, has been shown to enhance *in vitro* the generation of ROS and decrease the antioxidant defenses.[8]

Several *in vitro* studies have been performed using cultured orbital fibroblasts obtained from patients with GO undergoing orbital decompression.[9–14] As indicated in Table 36.1, evidence of increased oxidizing activities, decreased antioxidant defenses, and increased lipid peroxidation and DNA damage has been demonstrated.

HUMAN STUDIES

The hyperthyroid state of Graves' disease is characterized by increased ROS production, which may have a role in oxidative stress, as postulated in the

TABLE 36.1 *In Vitro* Experiments Proving an Involvement of Reactive Oxygen Species (ROS) in the Pathogenesis of Graves' Orbitopathy (GO)

Study	Experimental Data
Burch *et al.* (1997)[9]	Superoxide radicals increase orbital fibroblast proliferation and glycosaminoglycan production
Heufelder *et al.* (1992)[10]	H_2O_2 induces the expression of HLA-DR and heat shock protein-72, which is involved in antigen recognition and T-lymphocyte recruitment
Heufelder *et al.* (1992)[10]	Antioxidants (nicotinamide and allopurinol) decrease superoxide- and hydrogen peroxide-induced fibroblast proliferation, glycosaminoglycan production, and HLA-DR and heat shock protein-72 expression in GO orbital fibroblasts, probably by increasing the ROS scavenging activity
Lu *et al.* (1999)[11]	IL-1β increases ROS production in GO and control orbital fibroblasts and is associated with an increase in intracellular superoxide dismutase activity
Hondur *et al.* (2008)[12]; Tsai *et al.* (2010)[13]; Tsai *et al.* (2011)[14]	Increased oxidative DNA damage, lipid peroxidation, and H_2O_2 and decreased glutathione peroxidase activity in orbital fibroblasts from GO patients compared with orbital fibroblast from normal controls

H_2O_2: hydrogen peroxide; HLA: human leukocyte antigen; IL-1β: interleukin-1β.

pathogenesis of Graves' disease,[15] hyperthyroidism-induced damage,[16,17] and GO.[14]

Bednarek *et al.*, in a study that included hyperthyroid Graves' patients with and without active GO, showed that the increased peripheral ROS and antioxidant activities were normalized by antithyroid drug (ATD) therapy only in patients without GO, suggesting that orbital inflammation partially accounted for the increased circulating markers of the oxidative status.[18] Tsai *et al.* found that urinary levels of 8-hydroxy-2'-deoxyguanosine (8-OHdG), a marker of oxidative DNA damage, were still higher in GO patients rendered euthyroid by 6 months of treatment with ATD.[19] At this stage, oral glucocorticoid administration was followed by a significant decrease in urinary 8-OHdG compared with baseline, which paralleled changes in the clinical activity score and ophthalmopathy index. Relapse of GO during glucocorticoid therapy or after its withdrawal was associated with an increase in urinary 8-OHdG. The same authors also showed that mean urinary 8-OHdG levels were higher in GO patients compared with controls, and, among GO patients, in smokers compared with never-smokers, and in those with active GO compared with those with inactive GO.[20] Similar data have recently been reported by Akarsu *et al.*, who measured serum levels of malondialdehyde (MDA), a marker of the oxidative status.[21] Taken together, these data suggest that the orbital inflammatory process sustains a status of oxidative stress in patients with euthyroid Graves' disease and active GO and that parameters of oxidative status may become useful markers in the follow-up of patients with GO.

The involvement of oxidative stress in the pathogenesis of GO provided the rationale for using antioxidants in the management of GO. A small study suggested that antioxidant treatment may be beneficial in patients with GO. In this study, patients with moderate to severe GO were treated with a 3-month course of allopurinol (300 mg/day) and nicotinamide (300 mg/day) or placebo.[22] Improvement of GO occurred in nine of 11 antioxidant-treated patients (82%) and in only three of 11 patients who were given placebo (27%); pain and diplopia were reduced, while exophthalmos was poorly responsive.

SELENIUM AND GRAVES' ORBITOPATHY

Approximately half of patients with Graves' disease have ocular signs and symptoms of GO (also frequently referred to as ophthalmopathy).[23] Moderately severe and active forms of GO can benefit from treatment with glucocorticoids and/or orbital irradiation.[24,25] Milder forms of GO may improve spontaneously; patients usually do not receive any treatment and local measures (artificial tears, ointments, prisms) may be sufficient to control symptoms.

The 'wait-and-see' strategy can be challenged because many patients with even mild GO suffer from a significant decrease in their quality of life (QoL) and also because in natural history studies of mild GO, spontaneous improvement occurred in only 20% of patients, while eye disease remained static in 65% of cases and progressed in 15%.[26] Thus, only one-fifth of patients may actually benefit from an expectant strategy. Accordingly, intervention would be justified. Treatment should be affordable, safe, well tolerated, and widely available. One agent that may potentially inhibit the pathogenic mechanisms involved in GO is selenium.

Increased generation of free oxygen radicals seems to play a pathogenic role in GO. Moreover, selenium has an important effect on the immune system.[27,28] Selenium administration has been shown to be beneficial in Hashimoto's thyroiditis[29] and Graves' disease.[30] Recently, the European Group on Graves' Orbitopathy (EUGOGO; http://www.eugogo.eu) completed a multicenter, randomized, double-blind, placebo-controlled clinical trial, which showed that selenium, compared with placebo, may be beneficial in patients with mild GO.[31]

The study included patients with stable euthyroidism and at least one sign of mild GO (e.g., soft tissue swelling NO SPECS class 2a and 2b[32]), proptosis 22 mm or smaller, and a disease duration of less than 18 months. Patients with more severe eye disease (soft tissue swelling NO SPECS class 2c, proptosis larger than 22 mm, diplopia in primary or reading position, and sings or symptoms of optic neuropathy) were excluded.

Selenium, given as sodium selenite (100 μg), and placebo were administered twice daily. The intervention lasted for 6 months and was followed by a 6-month period of follow-up. All patients were investigated by the endocrinologist at baseline and after 3, 6, and 12 months.

Eye examination was performed by a blinded ophthalmologist using a modified EUGOGO case record form. The following parameters were evaluated: eyelid width (measured in millimeters), soft tissue involvement (according to the Color Atlas), proptosis (measured in millimeters using the same exophthalmometer), eye muscle involvement (assessed by the extent of ductions in degrees), and visual acuity (measured in decimals using the Snellen chart). In addition, the Clinical Activity Score (CAS) and Gorman's diplopia score were recorded. The CAS consists of a binary evaluation (yes/no) of seven items (spontaneous retrobulbar pain, pain on attempted up or down gaze, redness of the conjunctiva, redness of the eyelids, chemosis, swelling of the caruncle, and swelling of the eyelids), and the final score is the sum of all items present.[33] Gorman's score[34] includes four categories: (a) no diplopia (absent); (b) diplopia when tired or when first awakening (intermittent); (c) diplopia at extremes of gaze (inconstant); and (d) continuous diplopia in the primary or reading position (constant).

The QoL was evaluated using a disease-specific questionnaire (GO-QoL), which has been previously validated[35,36] and is available in several languages (http://www.eugogo.eu).

The primary endpoints of the study were comparisons of outcome rates at 6 months of overall ophthalmic assessment by a blinded ophthalmologist and the subjective disease-specific GO-QoL filled out by the patient, between patients assigned to active treatment and those receiving placebo (Table 36.2). Secondary outcome measurements were the changes in the diplopia score and the seven-item CAS. The observation at 12 months was scheduled to determine whether the treatment outcomes at 6 months persisted.

At the end of the active treatment period (6 months), the overall outcome was significantly better in the selenium group than in the placebo group (P=0.01) (Fig. 36.1). In particular, GO improved in 18 of 50 patients (36%) in the placebo group and 33 of 54 (61%) in the selenium group and worsened in 13 of 50 patients (26%) in the placebo

TABLE 36.2 Primary Outcomes Evaluation in Patients with Mild Graves' Orbitopathy (GO) Randomly Assigned to Treatment with Selenium or Placebo

OVERALL OPHTHALMIC OUTCOME WAS DEFINED AS FOLLOWS:

1. Improvement, when at least one of the following outcome measures improved in one eye, without worsening in any of these measures in either eye:
   (a) reduction in eyelid aperture by at least 2mm
   (b) reduction in any of the class 2 signs by at least one grade in the NO SPECS classification[32]
   (c) reduction in exophthalmos by at least 2mm
   (d) reduction in ocular motility by at least 8° in any duction.

2. Deterioration, if any of the following occurred:
   (a) worsening by at least one grade in any of the NO SPECS classes[32]
   (b) appearance of a new NO SPECS class[32]
   (c) decrease in visual acuity due to optic nerve involvement by at least one line on the Snellen chart, or other evidence (e.g., impaired color vision) or suspicion of optic nerve compression.

GO-SPECIFIC QUALITY OF LIFE (GO-QOL) QUESTIONNAIRE*

1. Improvement, if there was an increase of 6 or more points on either one (or both) of the GO-QoL scales (functioning and appearance).[†]
2. Deterioration, if there was a decrease of 6 points on either of the GO-QoL scales.

*The Graves' orbitopathy-specific quality of life (GO-QoL) questionnaire measures health-related quality of life in patients with Graves' orbitopathy.[34] It measures limitations in visual functioning as a consequence of diplopia and/or decreased visual acuity (eight questions) and limitations in psychosocial functioning as a consequence of a changed appearance (eight questions).

[†]A change of at least 6 points on one or both GO-QoL subscales was considered as an important change in daily functioning for GO patients.[35] This cut-off value was derived from a previous study in which the minimal clinically important difference (MCID), defined as the smallest difference in score on the domain of interest that patients perceive as benefit and that would mandate, in the absence of troublesome side effects and costs, a change in the patient's management in either GO-QoL subscale, appeared to be 6 points.

group and four of 54 (7%) in the selenium group. The positive effect of selenium compared with placebo persisted 6 months after withdrawal of therapy.

The beneficial effects of selenium were mostly due to an amelioration of soft tissue changes and a decrease in eyelid aperture, with a statistically significant improvement compared with placebo (Figs. 36.2 and 36.3). No changes were observed in exophthalmos and eye motility; however, very few patients had motility dysfunction at baseline.

GO-QoL scores at baseline showed a mild to moderate impairment in GO-QoL. Compared with baseline, selenium treatment was associated with a significant increase in both scores at 6 months, and this beneficial effect persisted at 12 months. As shown in Figure 36.4, a significantly greater proportion of patients in the selenium group showed an improvement in the GO-QoL at 6 months compared with the placebo group, and this difference was maintained after withdrawal of therapy

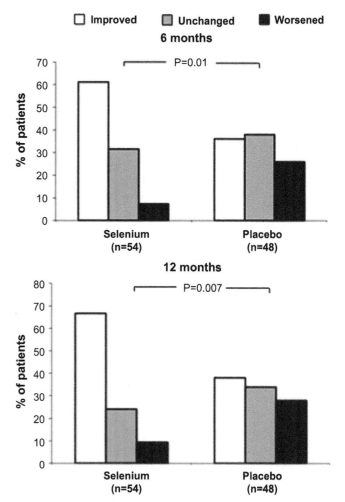

FIGURE 36.1 Changes in overall eye evaluation at 6 and 12 months in patients with mild Graves' orbitopathy (GO) randomly assigned to treatment with selenium (100 μg twice daily) or placebo (twice daily). Patients were treated with selenium or placebo for 6 months. The 12-month observation was scheduled as exploratory, for the sole purpose of determining whether the treatment effects at 6 months had been maintained. Changes in overall eye evaluation were scored as improved, unchanged, or worsened according to predefined criteria (Table 36.2). The data indicate that a greater proportion of patients treated with selenium compared with those treated with placebo showed an improvement in overall eye evaluation at the 6- and 12-month observations. P values were calculated with the use of the chi-squared test. *Source: Derived from data of Marcocci et al. (2011)[31]; partially reproduced from Marcocci et al. (2012),[7] with permission from S Karger AG, Basel, Switzerland.*

(12 months). Most of the patients who had improvement in soft tissue changes and eyelid aperture also had an improvement in the appearance subscale of the GO-QoL (Fig. 36.5).

Extraocular muscle dysfunction, assessed by the diplopia score, did not significantly change during the study in any group (data not shown). The CAS values at 6 and 12 months were significantly lower in the selenium group than in the placebo group (Fig. 36.6).

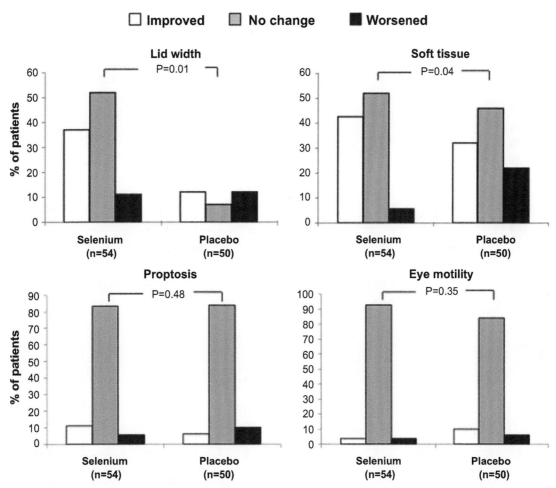

FIGURE 36.2 Changes in individual ophthalmic parameters at 6-month observation. Patients were treated with selenium (100 µg twice daily) or placebo (twice daily) for 6 months. Changes were graded according to predefined criteria as improved, unchanged, or worsened, according to predefined criteria (Table 36.2). The data indicate that treatment with selenium was associated with a statistically significantly better outcome compared with placebo for lid width and soft tissue changes but not for proptosis and eye motility. P values were calculated with the use of the chi-squared test. *Source: Derived from data of Marcocci* et al. *(2011).[31]*

Selenium treatment was well tolerated and no drug-related adverse events were observed.[31]

This study demonstrates than an intervention aimed at improving the imbalance of the antioxidant/oxidant status could be of help in patients with GO. A small study suggested that antioxidant treatment may be beneficial in patients with GO. In this study patients with moderate to severe GO were treated with a 3-month course of allopurinol (300 mg/day) and nicotinamide (300 mg/day) or placebo.[22] Improvement of GO occurred in nine of 11 antioxidant-treated patients (82%) and in only three of 11 patients who were given placebo (27%); pain and diplopia were reduced, while exophthalmos was poorly responsive.

This study has two potential limitations. First, there are no data on the changes in serum selenium concentrations following selenium selenite administration. Second, most of the patients included in the study came from areas of marginal selenium deficiency,[27,37,38] where, as reported in other studies, the beneficial effect of selenium may be favored. Similar studies performed in selenium-sufficient areas would be helpful in addressing this issue. Despite these limitations, the authors believe that the data clearly show that a 6-month selenium supplementation modified and improved the natural course of GO and the GO-related impairment of QoL. Therefore, a 6-month course of selenium in addition to symptomatic local measures should be offered to patients with mild GO.[31]

There have been some concerns about the potential risk of developing type 2 diabetes in patients given long-term selenium administration, particularly in those who already are selenium replete. An increased prevalence of type 2 diabetes, compared with placebo, was reported in the Nutritional Prevention of Cancer trial in the USA, where participants were given 200 µg selenium daily

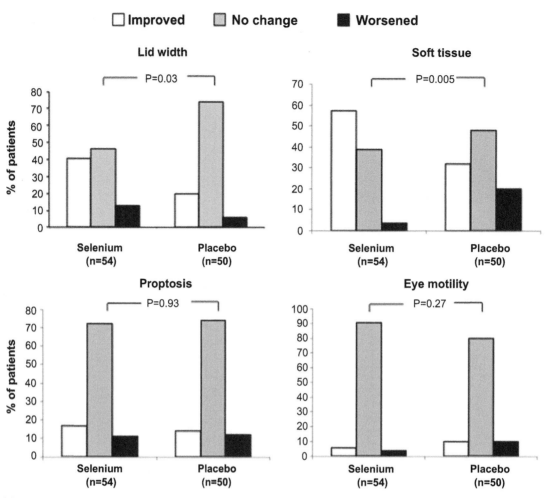

FIGURE 36.3 Changes in individual ophthalmic parameters at 12-month observation. Patients were treated with selenium (100 μg twice daily) or placebo (twice daily) for 6 months. The 12-month observation was scheduled as exploratory, for the sole purpose of determining whether the treatment effects at 6 months had been maintained. Changes were graded according to predefined criteria as improved, unchanged, or worsened, according to predefined criteria (Table 36.2). The data indicate that the beneficial effect of selenium treatment persisted at 12 months, i.e., 6 months after withdrawal of selenium. P values were calculated with the use of the chi-squared test. *Source: Derived from data of Marcocci et al. (2011).[31]*

for up to 12 years.[39] Population studies have provided conflicting results, showing either an association between selenium levels and diabetes, particularly in men,[40] or a sex-specific protective effect of higher selenium status at baseline on the later occurrence of dysglycemia.[41] A link between selenium supplementation and an increased incidence of glaucoma has also been suggested.[42]

To what extent are these concerns relevant to this trial? Patients included in the study came from areas in which the population is moderately selenium deficient; the daily dose of 200 μg sodium selenite contains 91 μg of selenium and patients were supplemented for only 6 months. Blood glucose was not routinely measured, but no changes compared with baseline were observed in a subgroup of patients enrolled in Greece. No cases of glaucoma were observed during the entire 12-month observation period. Therefore, it is reasonable to conclude that a 6-month course of low-dose selenium

is safe and should be offered to patients with mild GO, even if they are selenium replete.[31]

TAKE-HOME MESSAGES

- Oxidative stress is an important step in the pathogenesis of Graves' orbitopathy (GO).
- Selenium is a trace mineral that acts as an antioxidant and immune modulator.
- Mild GO can improve in 20% of patients, remain stable in 65%, and worsen in 15%.
- Mild GO affects the quality of life of patients.
- To date, no treatments are available for mild GO.
- Six months of selenium supplementation in patients with mild GO from marginally selenium-deficient areas improves ocular involvement and quality of life and reduces the progression of eye disease.

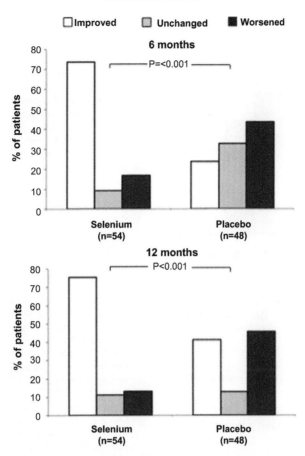

FIGURE 36.4 Changes in Graves' orbitopathy (GO)-specific quality of life questionnaire at 6 and 12 months in patients with mild GO randomly assigned to treatment with selenium (100 μg twice daily) or placebo (twice daily). Patients were treated with selenium or placebo for 6 months. The 12-month observation was scheduled as exploratory, for the sole purpose of determining whether the treatment effects at 6 months had been maintained. Changes in quality of life were scored as improved, unchanged, or worsened according to predefined criteria (Table 36.2). The data indicate that a greater proportion of patients treated with selenium compared with those treated with placebo showed an improvement in GO-specific quality of life scores at 6- and 12-month observation. P values were calculated with the use of the chi-squared test. *Source: Partially reproduced from Marcocci et al. (2012),[7] with permission from S Karger AG, Basel, Switzerland.*

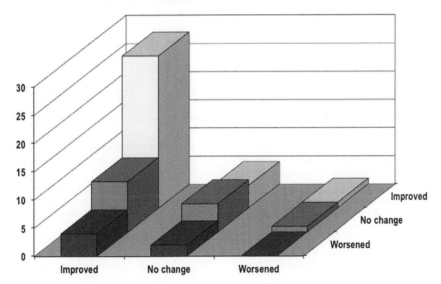

FIGURE 36.5 Changes in the appearance subscale of Graves' orbitopathy quality of life (GO-QoL) according to changes in eyelid aperture, soft tissue changes, or both at 6-month evaluation. Changes were graded according to predefined criteria as improved, unchanged, or worsened, as reported in Table 36.2. Most of the patients who had improvement in soft tissue changes and eyelid aperture also had an improvement in the appearance subscale of the GO-QoL questionnaire. *Source: Derived from data of Marcocci et al. (2011).[31]*

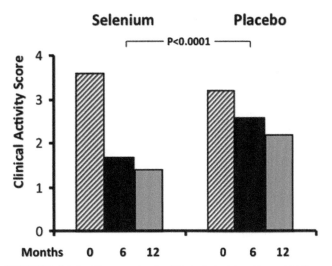

FIGURE 36.6 Changes in the Clinical Activity Score (CAS) at 6- and 12-month evaluation in patients treated with selenium or placebo. The CAS values at 6 and 12 months were significantly lower in the selenium group than in the placebo group. The P value was calculated using the Mann–Whitney test. *Source: Derived from data of Marcocci et al. (2011).*[31]

- Selenium is an affordable, safe, and well-tolerated treatment.
- Future study is required to establish whether selenium supplementation may have a beneficial effect in moderate GO.

References

1. Halliwell B, Gutteridge JM. Lipid peroxidation, oxygen radicals, cell damage, and antioxidant therapy. *Lancet* 1984;i:1396–7.
2. Valko M, Leibfritz D, Moncol J, Cronin MT, Mazur M, Telser J. Free radicals and antioxidants in normal physiological functions and human disease. *Int J Biochem Cell Biol* 2007;39:44–84.
3. Halliwell B, Gutteridge JMC. Cellular responses to oxidative stress: adaptation, damage, repair, senescence and death. In: *Free Radicals in Biology and Medicine*. 4th ed. New York: Oxford University Press; 2007. pp. 187–267.
4. Rayman MP. Selenium and human health. *Lancet* 2012;379:1256–68.
5. Duntas LH. Selenium and the thyroid: a close-knit connection. *J Clin Endocrinol Metab* 2010;95:5180–8.
6. Bahn RS. Graves' ophthalmopathy. *N Engl J Med* 2010;362:726–38.
7. Marcocci C, Leo M, Altea MA. Oxidative stress in Graves' disease. *Eur Thyroid J* 2012;2:80–7.
8. Bartalena L, Tanda ML, Piantanida E, Lai A. Oxidative stress and Graves' ophthalmopathy: *in vitro* studies and therapeutic implications. *Biofactors* 2003;19:155–63.
9. Burch HB, Lahiri S, Bahn RS, Barnes S. Superoxide radical production stimulates retroocular fibroblast proliferation in Graves' ophthalmopathy. *Exp Eye Res* 1997;65:311–6.
10. Heufelder AE, Wenzel BE, Bahn RS. Methimazole and propylthiouracil inhibit the oxygen free radical-induced expression of a 72 kilodalton heat shock protein in Graves' retroocular fibroblasts. *J Clin Endocrinol Metab* 1992;74:737–42.
11. Lu R, Wang P, Wartofsky L, Sutton BD, Zweier JL, Bahn RS, et al. Oxygen free radicals in interleukin-1beta-induced glycosaminoglycan production by retro-ocular fibroblasts from normal subjects and Graves' ophthalmopathy patients. *Thyroid* 1999;9:297–303.
12. Hondur A, Konuk O, Dincel AS, Bilgihan A, Unal M, Hasanreisoglu B. Oxidative stress and antioxidant activity in orbital fibroadipose tissue in Graves' ophthalmopathy. *Curr Eye Res* 2008;33:421–7.
13. Tsai CC, Wu SB, Cheng CY, Kao SC, Kau HC, Chiou SH, et al. Increased oxidative DNA damage, lipid peroxidation, and reactive oxygen species in cultured orbital fibroblasts from patients with Graves' ophthalmopathy: evidence that oxidative stress has a role in this disorder. *Eye (Lond)* 2010;24:1520–5.
14. Tsai CC, Wu SB, Cheng CY, Kao SC, Kau HC, Lee SM, Wei YH. Increased response to oxidative stress challenge in Graves' ophthalmopathy orbital fibroblasts. *Mol Vis* 2011;17:2782–8.
15. Zarkovic M. The role of oxidative stress on the pathogenesis of Graves' disease. *J Thyroid Res* 2012. article ID 302537, 1–5.
16. Asayama K, Dobashi H, Hayashibe H, Megata Y, Kato K. Lipid peroxidation and free radical scavengers in thyroid dysfunction in the rat: a possible mechanism of injury to heart and skeletal muscle in hyperthyroidism. *Endocrinology* 1987;121:2112–8.
17. Yamada T, Mishima T, Sakamoto M, Sugiyama M, Matsunaga S, Wada M. Oxidation of myosin heavy chain and reduction in force production in hyperthyroid rat soleus. *J Appl Physiol* 2006;100:1520–6.
18. Bednarek J, Wysocki H, Sowiński J. Oxidative stress peripheral parameters in Graves' disease: the effect of methimazole treatment in patients with and without infiltrative ophthalmopathy. *Clin Biochem* 2005;38:13–8.
19. Tsai CC, Kao SC, Cheng CY, Kau HC, Hsu WM, Lee CF, Wei YH. Oxidative stress change by systemic corticosteroid treatment among patients having active Graves ophthalmopathy. *Arch Ophthalmol* 2007;125:1652–6.
20. Tsai CC, Cheng CY, Liu CY, Kao SC, Kau HC, Hsu WM, Wei YH. Oxidative stress in patients with Graves' ophthalmopathy: relationship between oxidative DNA damage and clinical evolution. *Eye (Lond)* 2009;23:1725–30.
21. Akarsu E, Buyukhatipoglu H, Aktaran S, Kurtul N. Effects of pulse methylprednisolone and oral methylprednisolone treatments on serum levels of oxidative stress markers in Graves' ophthalmopathy. *Clin Endocrinol (Oxf)* 2011;74:118–24.
22. Bouzas EA, Karadimas P, Mastorakos G, Koutras DA. Antioxidant agents in the treatment of Graves' ophthalmopathy. *Am J Ophthalmol* 2000;129:618–22.
23. Bartalena L, Pinchera A, Marcocci C. Management of Graves' ophthalmopathy: reality and perspectives. *Endocr Rev* 2000;21:168–99.
24. Bartalena L, Baldeschi L, Dickinson A, Eckstein A, Kendall-Taylor P, Marcocci C, et al. European Group on Graves' Orbitopathy (EU-GOGO). Consensus statement of the European Group on Graves' orbitopathy (EUGOGO) on management of GO. *Eur J Endocrinol* 2008;158:273–85.
25. Marcocci C, Marinò M. Treatment of mild, moderate-to-severe and very severe Graves' orbitopathy. *Best Pract Res Clin Endocrinol Metab* 2012;26:325–37.
26. Perros P, Crombie AL, Kendall-Taylor P. Natural history of thyroid associated ophthalmopathy. *Clin Endocrinol (Oxf)* 1995;42:45–50.
27. Rayman MP. The importance of selenium to human health. *Lancet* 2000;356:233–41.
28. Hoffmann PR, Berry MJ. The influence of selenium on immune responses. *Mol Nutr Food Res* 2008;52:1273–80.
29. Negro R. Selenium and thyroid autoimmunity. *Biologics* 2008;2:265–73.
30. Vrca VB, Skreb F, Cepelak I, Romic Z, Mayer L. Supplementation with antioxidants in the treatment of Graves' disease: the effect on glutathione peroxidase activity and concentration of selenium. *Clin Chim Acta* 2004;341:55–63.
31. Marcocci C, Kahaly GJ, Krassas GE, Bartalena L, Prummel M, Stahl M, et al. European Group on Graves' Orbitopathy. Selenium and the course of mild Graves' orbitopathy. *N Engl J Med* 2011;364:1920–31.

32. Werner SC. Modification of the classification of the eye changes of Graves' disease: recommendation of the Ad Hoc Committee of the American Thyroid Association. *J Clin Endocrinol Metab* 1977;**44**:203–4.

33. Mourits MP, Prummel MF, Wiersinga WM, Koornneef L. Clinical activity score as a guide in the management of patients with Graves' ophthalmopathy. *Clin Endocrinol (Oxf)* 1997;**47**:9–14.

34. Bahn RS, Gorman CA. Choice of therapy and criteria for assessing treatment outcome in thyroid-associated ophthalmopathy. *Endocrinol Metab Clin North Am* 1987;**16**:391–407.

35. Terwee CB, Gerding MN, Dekker FW, Prummel MF, Wiersinga WM. Development of a disease specific quality of life questionnaire for patients with Graves' ophthalmopathy: the GO-QOL. *Br J Ophthalmol* 1998;**82**:773–9.

36. Terwee CB, Dekker FW, Mourits MP. Interpretation and validity of changes in scores on the Graves' ophthalmopathy quality of life questionnaire (GO–QOL) after different treatments. *Clin Endocrinol (Oxf)* 2001;**54**:391–8.

37. Rayman MP. Food-chain selenium and human health: emphasis on intake. *Br J Nutr* 2008;**100**:254–68.

38. Burney P, Potts J, Makowska J, Kowalski M, Phillips J, Gnatiuc L, et al. A case–control study of the relation between plasma selenium and asthma in European populations: a GAL2EN project. *Allergy* 2008;**63**:865–71.

39. Stranges S, Marshall JR, Natarajan R, Donahue RP, Trevisan M, Combs GF, et al. Effects of long-term selenium supplementation on the incidence of type 2 diabetes: a randomized trial. *Ann Intern Med* 2007;**147**:217–23.

40. Bleys J, Navas-Acien A, Guallar E. Serum selenium and diabetes in US adults. *Diabetes Care* 2007;**30**:829–34.

41. Akbaraly TN, Arnaud J, Rayman MP, Hininger-Favier I, Roussel AM, Berr C, Fontbonne A. Plasma selenium and risk of dysglycemia in an elderly French population: results from the prospective Epidemiology of Vascular Ageing Study. *Nutr Metab (Lond)* 2010;**7**:21.

42. Sheck L, Davies J, Wilson G. Selenium and ocular health in New Zealand. *N Z Med J* 2010;**123**:85–94.

Zinc Deficiency and the Eye

Benjamin P. Nicholson[1], Ananda S. Prasad[2], Emily Y. Chew[1]

[1]National Eye Institute, National Institutes of Health, Bethesda, Maryland, USA; [2]Department of Oncology, Wayne
State University School of Medicine and Barbara Ann Karmanos Cancer Institute, Detroit, Michigan, USA

INTRODUCTION

Zinc is an essential micronutrient, and it is the second most abundant trace element in the human body.[1] After some early observations in Iranian villagers,[2] zinc was first shown to be an essential micronutrient in 1963.[3] A zinc-responsive syndrome of iron-deficiency anemia, hepatosplenomegaly, dwarfism, and hypogonadism was described in a group of Egyptian subjects. They had decreased zinc levels in plasma, red blood cell, hair, and urine samples. Zn-65 studies demonstrated increased plasma zinc turnover, a smaller 24 hour exchangeable pool, and decreased urine and stool Zn-65 excretion.

Zinc is absorbed in the proximal small intestine and excreted primarily via the gastrointestinal tract. Most zinc in the human body is intracellular, and intracellular levels are generally unaltered in the event of nutritional deprivation. Important dietary sources include shellfish, beef, fortified cereals, pork, beans, and chicken.[4] Deficiency tends to occur in developing nations in populations that consume large quantities of unleavened grain products that contain phytates, which chelate zinc. Deficiency is rare in North America, but it may occur in the context of alcoholism, total parenteral nutrition, or gastrointestinal disease such as Crohn's disease, or with certain medications including penicillamine. Clinical zinc deficiency can manifest as growth retardation, immune deficiency, hypogonadism, and cognitive impairment.[1]

Several studies have shown that zinc supplementation has antioxidant effects and reduces expression of inflammatory cytokines. In a small, randomized trial of supplementation of 45 mg zinc gluconate, subjects had lower levels of lipid peroxidation products and DNA adducts.[5] Mononuclear cells from supplemented patients expressed less tumor necrosis factor-α and interleukin-1β (IL-1β) messenger RNA (mRNA), possibly as a result of activation of the A20 negative feedback loop.[6] A second randomized trial of zinc

supplementation (45 mg) in elderly subjects supported these findings and demonstrated other immune changes including decreased plasma C-reactive protein and IL-6 levels.[7] Zinc supplementation augmented a variety of immune parameters in a randomized trial of 36 sickle cell patients.[8] Patients in the treatment group had fewer infections during the study period. This reduction in infection rate was also shown in a randomized trial of zinc supplementation in 50 elderly subjects.[9]

Zinc is a cofactor or structural component of hundreds of metalloenzymes and transcription factors in addition to having a role in cell membrane integrity.[1] Metalloenzymes in the retina and retinal pigment epithelium (RPE), including catalase and Cu,Zn-superoxide dismutase, require zinc as a cofactor. Some have proposed that zinc supplementation could benefit the eye by augmenting these functions.

Zinc has a prominent place in modern clinical ophthalmology as a component of the Age-Related Eye Disease Study (AREDS) supplement for age-related macular degeneration (AMD). Zinc is found in relatively high concentrations in ocular tissues, particularly retina and choroid. One study reported choroidal concentrations of 472 µg zinc/g tissue compared with 20–30 µg/g in most other human tissues.[10] For these and other reasons, zinc is of great interest in both basic and clinical ophthalmic research.

ZINC DEFICIENCY IN ANIMAL MODELS

Animal models of zinc deficiency have demonstrated a variety of ocular pathologies. In the rat, zinc deficiency during gestation results in failure of optic cup invagination, coloboma, retinal dysplasia, and even anophthalmia.[1] Newborn of zinc-deficient adult rats showed mild ocular manifestations including delayed eyelid opening, keratopathy, and electron micrographic signs of

Handbook of Nutrition, Diet, and the Eye
http://dx.doi.org/10.1016/B978-0-12-401717-7.00037-X

photoreceptor degeneration.[1] Zinc-deficient cats have been reported to have electroretinogram (ERG) abnormalities, and zinc-deficient pigs have swollen outer segments and other RPE and retinal ultrastructural abnormalities on electron microscopy.[1]

The remainder of this chapter addresses the role of zinc in human ophthalmic disease and pathophysiology.

ZINC DEFICIENCY AND RETINAL PHYSIOLOGIC MANIFESTATIONS

Night blindness and impaired dark adaptation have been reported in numerous studies of zinc deficiency,[11–13] although most human reports are complicated by the presence of multiple nutritional deficiencies. A study of six patients with alcoholic cirrhosis, abnormal dark adaptation, and zinc deficiency showed that dark adaptation normalized after 1–2 weeks of zinc supplementation.[14] Four of the six patients were also vitamin A deficient, but improvement in dark adaptation correlated temporally with zinc supplementation rather than vitamin A supplementation. In a study of 13 sickle cell patients, who tend to be susceptible to zinc deficiency, six had poor dark adaptation.[15] Neutrophil zinc content correlated with the dark adaptation thresholds.

Another patient with combined zinc and vitamin A deficiency due to cirrhosis in primary sclerosing cholangitis first presented with acquired nyctalopia.[16] His severely reduced ERG amplitudes (rod more so than cone) prompted a comprehensive medical evaluation that disclosed the underlying liver disease. Serum zinc levels responded to treatment of his liver disease and vitamin A levels did not. As serum zinc levels improved, ERG parameters and symptoms also improved.

Several other reports support the association of zinc deficiency and night blindness. Low serum and scalp zinc levels were associated with subjective nyctalopia in a cohort of children in Pakistan.[17] Zinc supplementation alone led to resolution of nyctalopia in a patient with cystic fibrosis with combined zinc and vitamin A deficiency.[18] ERG rod responses improved dramatically after zinc supplementation. ERG characteristics of zinc deficiency include diminished scotopic A- and B-wave amplitudes; B-wave amplitudes may be affected to a greater degree.[1,16]

The effects of zinc deficiency on the retina remain poorly understood. Zinc has numerous roles in the retina, and each is a theoretical mechanism for retinal dysfunction in zinc deficiency. Zinc interacts with vitamin A and taurine in the retina, plays a role in plasma membrane integrity in outer segments, affects the light–rhodopsin reaction and downstream neural transmission, and has antioxidant effects that may be important in the eye.

Plasma levels of retinol binding protein are reduced in zinc deficiency,[19] and this may prevent mobilization of liver vitamin A stores. Zinc-deficient rats have a marked reduction in rhodopsin regeneration in the dark, although this may be related to impaired opsin synthesis rather than vitamin A metabolism.[20] Animal models of taurine deficiency have demonstrated reduced zinc content in rod photoreceptors.[21] Combined zinc and taurine deficiency results in a synergistic ERG depression.[1] Human taurine deficiency is sometimes seen in the context of a vegan diet or parenteral nutrition.

The retina contains relatively high concentrations of zinc, and histochemical studies show that most loosely bound zinc in retina localizes to photoreceptor outer segments.[22] Zinc plays a role in plasma membrane integrity in outer segments through effects on membrane conformation and protein–protein interactions.[1] Given the high rate of plasma membrane turnover, photoreceptor outer segments may be particularly sensitive to zinc intake. Zinc concentrations in outer segments are decreased in zinc deficiency.[1] Outer segment degeneration in a rat model of zinc deficiency and swelling of outer segments in a pig model suggest a particular stress on outer segments in zinc deficiency.[1]

Zinc is important for neuronal signal transduction in the retina, and zinc deficiency affects inner retinal function as evidenced by reduced B-wave amplitudes. Zinc ions are present in photoreceptor terminals.[23] They are released from the photoreceptor synaptic terminal at the time of depolarization and probably affect the neural signal. Animal studies have shown that zinc staining patterns in the retina are different in dark- and light-adapted conditions, further suggesting a role for zinc in phototransduction.[24]

Another theory of retinal dysfunction in zinc deficiency involves decreased activity of zinc-dependent metalloenzymes. Zinc is generally tightly bound to such enzymes, and most zinc metalloenzymes do not have reduced activity in zinc deficiency.[1,24] Nonetheless, several enzymes in the visual cycle are zinc metalloenzymes and their individual sensitivities to zinc deficiency are not known. The activity of α-mannosidase, an enzyme involved in digestion of disc membranes in RPE cells, is induced by zinc in RPE cell culture.[25] Dysfunction of α-mannosidase could lead to accumulation of rod outer segments or drusen formation. RPE zinc concentrations decline in deficient states before retinal concentrations; this may indicate that the RPE buffers zinc availability in the retina.[1]

Zinc status may also affect retinal function via zinc's role as an antioxidant. As discussed above, zinc supplementation is associated with decreased expression of inflammatory cytokines and improvements in inflammatory markers such as C-reactive protein. The retina is particularly sensitive to oxidative damage,[26] and *in vitro* studies of zinc have shown that it can protect outer segments from peroxidative damage.[27] The primary clinical evidence for a salutary effect of zinc supplementation in eye disease comes from AREDS and other studies that have found a benefit in AMD (discussed below). These effects are thought to be mediated via zinc's antioxidant properties, especially given the growing body of evidence that the pathogenesis of AMD involves oxidative stress.

Zinc levels in the neural retina and choroid have been shown to vary with age.[28] Zinc levels decline with age in the neural retina and increase with age in the choroid. High concentrations of zinc are found in sub-RPE deposits, particularly in eyes with AMD.[29] Zinc may therefore play a role in age-related eye disease.

ZINC AND AGE-RELATED MACULAR DEGENERATION

Zinc is a component of the protective AREDS supplement for AMD. AREDS was a multicenter, randomized, placebo-controlled trial designed to study the natural history of AMD and age-related cataract and to assess the impact of antioxidant vitamins and zinc supplementation on these conditions.[30] The intervention incorporated both antioxidants and zinc for two main reasons.[30] First, several epidemiologic studies and clinical trials at that time had suggested a role for antioxidants in reducing the risks of cancer, cardiovascular disease, and eye disease. A small trial had also suggested that pharmacologic doses of zinc reduced the risk of vision loss in AMD.[31] The second reason was the growing use of commercially available antioxidant and zinc supplements among AMD patients, despite a paucity of clinical evidence. A large, randomized trial was needed to evaluate these supplements for AMD.

The AREDS trial randomized 3640 participants with AMD to antioxidant supplements, zinc, combined antioxidants and zinc, or placebo (Table 37.1). The combined AREDS supplement contained 15 mg β-carotene, 500 mg vitamin C, 400 international units (IU) vitamin E, 80 mg zinc oxide, and 2 mg of copper as cupric oxide. Participants were stratified into four categories of AMD by clinical appearance:

- Category 1: No drusen to few drusen; 0.44% developed advanced AMD by year 5.

TABLE 37.1 Age-Related Eye Disease Study (AREDS) Treatment Groups

Formulations	β-Carotene	Vitamin C	Vitamin E	Zinc Oxide	Cupric Oxide
Placebo	–	–	–	–	–
Antioxidants	15 mg	500 mg	400 IU	–	–
Zinc	–	–	–	80 mg	2 mg
Antioxidants + zinc	15 mg	500 mg	400 IU	80 mg	2 mg

- Category 2: Extensive small drusen, pigment abnormalities, or at least one intermediate druse in at least one eye; 1.3% probability of progression to advanced AMD by year 5 (Fig. 37.1).
- Category 3: Extensive intermediate drusen, large drusen, or noncentral geographic atrophy (GA) in at least one eye; 18% probability of progression to advanced AMD by year 5. Patients within category 3 who had bilateral large drusen or noncentral GA in at least one eye at enrollment were four times more likely to progress to advanced AMD than the remaining participants in category 3 (27% vs. 6% at 5 years) (Fig. 37.2).
- Category 4: Advanced AMD or vision loss due to nonadvanced AMD in one eye; 43% probability of progression to advanced AMD in 5 years (Fig. 37.3).

The interventional AMD study results were published in 2001.[32] The combination of antioxidant vitamins with zinc was protective against the development of advanced AMD for category 3 and 4 participants (odds ratio (OR) 0.66, 95% confidence interval (CI) 0.47 to 0.91, P=0.001). Those with category 1 or 2 AMD had a very low risk of progression to advanced AMD, and a much larger sample size and longer follow-up would be required to evaluate for a treatment effect for the AREDS formulation for these patients.

The zinc without antioxidants treatment group in AREDS had a suggestive, but not statistically significant, reduction in risk of progression to advanced disease (OR 0.75, 99% CI 0.55 to 1.03). When analysis was restricted

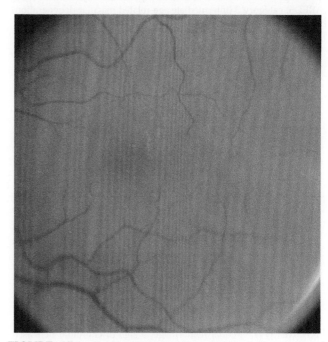

FIGURE 37.1 Small and intermediate drusen (yellow deposits) in the worse eye of a patient who meets criteria for Age-Related Eye Disease Study (AREDS) category 2 (early age-related macular degeneration). (See color plate section)

to category 3 and 4 participants, there was a significant reduction in progression to advanced AMD (OR 0.71, 99% CI 0.52 to 0.99). For zinc alone, there was no significant reduction in rates of moderate vision loss. Secondary analyses of the AREDS cohort revealed that participants who took zinc had a significantly lower mortality (mean follow-up 6.5 years, relative risk 0.73, 95% CI 0.61 to 0.89).

The first clinical evidence of a beneficial effect of zinc supplementation in AMD came in 1988 from a small, randomized, controlled trial of 100 mg of oral zinc sulfate for

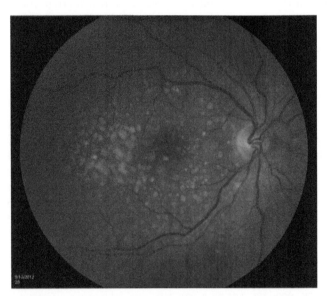

FIGURE 37.2 Extensive large drusen (yellow deposits) in the worse eye of a patient who meets criteria for Age-Related Eye Disease Study (AREDS) category 3 (intermediate age-related macular degeneration). (See color plate section)

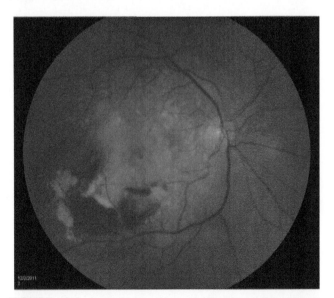

FIGURE 37.3 Retinal hemorrhage, a sign of choroidal neovascularization and advanced age-related macular degeneration (AMD), and large drusen in the worse eye of a patient who meets criteria for Age-Related Eye Disease Study (AREDS) category 4 (advanced AMD). (See color plate section)

a wide range of AMD patients. Since 1988, AREDS and several other studies have provided further evidence of a relationship between zinc status and AMD. In the Rotterdam Study, above median intake of vitamin E, vitamin C, β-carotene, and zinc was associated with a 35% reduced risk of incident AMD.[33] Furthermore, analysis of dietary zinc intake in the Rotterdam cohort showed a protective effect for zinc on incident AMD. Analyses from the Blue Mountains Eye Study also found higher dietary zinc intake to have a favorable effect on incident AMD.[34] In the Beaver Dam Eye Study, dietary zinc had a moderate favorable effect on incident pigmentary changes but not on overall incident AMD.[35]

There are also several studies that have not detected a relationship between zinc status and AMD. Stur and colleagues performed a randomized, controlled trial of 200 mg of oral zinc sulfate over 2 years in patients with unilateral exudative AMD (n = 112).[36] Just 14 patients developed new neovascularization during the 2-year study, and there was no apparent treatment benefit. The Nurses' Health Study and Health Professionals Follow-up Study found no relationship between dietary zinc and incident AMD,[37] and the Eye Disease Case Control Study showed no relationship between serum zinc levels and prevalent neovascular AMD.[38]

Oral zinc supplements are generally well tolerated, but toxicities can occur. Zinc can interfere with absorption of dietary copper and iron.[30] Zinc chloride is a gastrointestinal irritant.[32] In AREDS, patients taking zinc were hospitalized more often than controls for genitourinary complaints (7.5% vs. 4.9%, P = 0.001).[32] In AREDS2 (Tables 37.2 and 37.3), a lower zinc dose (25 mg) will be

TABLE 37.2 Treatment Groups in the Primary Randomization in Age-Related Eye Disease Study 2 (AREDS2)

Supplement	Daily Dose
Placebo	–
Lutein/zeaxanthin	10 mg/2 mg
DHA/EPA	350 mg/650 mg
Lutein/zeaxanthin + DHA/EPA	10 mg/2 mg + 350 mg/650 mg

DHA: docosahexaenoic acid; EPA: eicosapentaenoic acid.

TABLE 37.3 Treatment Groups in the Secondary Randomization in Age-Related Eye Disease Study 2 (AREDS2)

Formulation	Vitamin C	Vitamin E	β-Carotene	Zinc Oxide	Cupric Oxide
1	500 mg	400 IU	15 mg	80 mg	2 mg
2	500 mg	400 IU	0 mg	80 mg	2 mg
3	500 mg	400 IU	15 mg	25 mg	2 mg
4	500 mg	400 IU	0 mg	25 mg	2 mg

tested because 80 mg doses are now thought to overload the gastrointestinal tract's capacity for zinc absorption.

ZINC AND CATARACT

Zinc nutrition has been loosely linked to cataract. In a study from India, 140 patients with senile cataract had reduced plasma levels of zinc, among other micronutrients, compared with age-matched controls.[39] Zinc deprivation causes cataract in fish.[40] Cataract has been reported in acrodermatitis enteropathica, a rare, autosomal recessive disease that causes impaired zinc absorption.[41] Zinc levels seem to be elevated in cataractous human lens tissue[1] and greater in cataracts extracted from diabetic patients.[42]

Zinc deficiency could lead to lens opacity via impairment of the molecular chaperone activity of α-crystallins. Aggregation of γ-crystallins in the lens causes lens opacity, and this aggregation is prevented by α-crystallins.[43] α-Crystallin activity improves with Cu^{2+} and Zn^{2+} binding. Oxidative stress is implicated in the pathogenesis of age-related cataract,[44] and zinc's antioxidant properties via inhibition of NADPH oxidase, promotion of free radical scavenger production, and other pathways may promote lens clarity.

Zinc supplementation did not have an effect on the development of cataract in AREDS.[45] Similarly, no benefit was detected for a combined zinc and retinol supplement in the Linxian (China) cataract studies, although the odds ratio was suggestive for nuclear cataract (OR 0.77, 95% CI 0.58 to 1.02).[46]

ZINC AND OCULAR SURFACE DISEASE

Severe zinc deficiency appears to cause ocular surface disease. Animal models of zinc deficiency demonstrate conjunctivitis, corneal opacity, and corneal neovascularization.[47,48] In humans, ocular surface disease has been described in zinc deficiency, predominantly in acrodermatitis enteropathica. Patients may have conjunctivitis, keratomalacia, blepharitis, punctual stenosis, corneal edema, punctate corneal opacity, corneal ulcer, and corneal scarring.[49] Corneal disease responds to zinc therapy in acrodermatitis enteropathica.[10] A study from India suggests that recalcitrant corneal ulcers may be associated with zinc deficiency.[50]

TAKE-HOME MESSAGES

- Zinc deficiency causes a syndrome of iron-deficiency anemia, hepatosplenomegaly, dwarfism, and hypogonadism.

- The ocular manifestations of zinc deficiency in humans include impaired dark adaptation and nyctalopia.
- The mechanism of retinal dysfunction in zinc deficiency is not well understood. Proposed mechanisms include interaction with vitamin A and taurine metabolism, loss of plasma membrane integrity in outer segments, perturbation of the light–rhodopsin reaction and downstream neural transmission, and loss of protective antioxidant effects of zinc.
- The results of the Age-Related Eye Disease Study (AREDS) demonstrate a role for the specific AREDS antioxidant with zinc formulation for the prevention of advanced age-related macular degeneration. The AREDS formulation is recommended as a treatment for nonsmokers with extensive intermediate drusen, large drusen, noncentral geographic atrophy, or unilateral advanced age-related macular degeneration.
- AREDS2 is an ongoing study of supplements for patients at risk of vision loss due to age-related macular degeneration, and it will compare a lower zinc dose (25 mg) with the current AREDS dose (80 mg) in addition to investigating long-chain polyunsaturated fatty acids and lutein.
- Severe zinc deficiency, particularly in the form of acrodermatitis enteropathica, is associated with ocular surface disease.
- Randomized trials of zinc supplementation have not shown a clear benefit for cataract.

References

1. Grahn BH, Paterson PG, Gottschall-Pass KT, Zhang Z. Zinc and the eye. *J Am Coll Nutr* 2001;**20**(2 Suppl.):106–18.
2. Prasad AS, Halsted JA, Nadimi M. Syndrome of iron deficiency anemia, hepatosplenomegaly, hypogonadism, dwarfism and geophagia. *Am J Med* 1961;**31**:532–46.
3. Prasad AS, Miale Jr A, Farid Z, Sandstead HH, Schulert AR. Zinc metabolism in patients with the syndrome of iron deficiency anemia, hepatosplenomegaly, dwarfism, and hypognadism. *J Lab Clin Med* 1963;**61**:537–49.
4. Clemons TE, Kurinij N, Sperduto RD. Associations of mortality with ocular disorders and an intervention of high-dose antioxidants and zinc in the Age-Related Eye Disease Study: AREDS report no. 13. *Arch Ophthalmol* 2004;**122**:716–26.
5. Prasad AS, Bao B, Beck FW, Kucuk O, Sarkar FH. Antioxidant effect of zinc in humans. *Free Radic Biol Med* 2004;**37**:1182–90.
6. Prasad AS, Bao B, Beck FW, Sarkar FH. Zinc-suppressed inflammatory cytokines by induction of A20-mediated inhibition of nuclear factor-kappaB. *Nutrition* 2011;**27**:816–23.
7. Bao B, Prasad AS, Beck FW, Fitzgerald JT, Snell D, Bao GW, et al. Zinc decreases C-reactive protein, lipid peroxidation, and inflammatory cytokines in elderly subjects: a potential implication of zinc as an atheroprotective agent. *Am J Clin Nutr* 2010;**91**:1634–41.
8. Bao B, Prasad AS, Beck FW, Snell D, Suneja A, Sarkar FH, et al. Zinc supplementation decreases oxidative stress, incidence of infection, and generation of inflammatory cytokines in sickle cell disease patients. *Transl Res* 2008;**152**:67–80.

9. Prasad AS, Beck FW, Bao B, Fitzgerald JT, Snell DC, Steinberg JD, et al. Zinc supplementation decreases incidence of infections in the elderly: effect of zinc on generation of cytokines and oxidative stress. *Am J Clin Nutr* 2007;**85**:837–44.

10. Karcioglu ZA. Zinc in the eye. *Surv Ophthalmol* 1982;**27**:114–22.

11. Warth JA, Prasad AS, Zwas F, Frank RN. Abnormal dark adaptation in sickle cell anemia. *J Lab Clin Med* 1981;**98**:189–94.

12. Cossack ZT, Prasad A, Koniuch D. Effect of zinc supplementation on retinal reductase in zinc deficient rats. *Nutr Rep Int* 1982;**26**: 841–8.

13. Mahajan SK, Prasad A, Brewer GJ, Zwas F, Lee DY. Effect of changes in dietary zinc intake on taste acuity and dark adaptation in normal human subjects. *J Trace Elem Exp Med* 1992;**5**:33–45.

14. Morrison SA, Russell RM, Carney EA, Oaks EV. Zinc deficiency: a cause of abnormal dark adaptation in cirrhotics. *Am J Clin Nutr* 1978;**31**:276–81.

15. Prasad AS. Impact of the discovery of human zinc deficiency on health. *J Am Coll Nutr* 2009;**28**:257–65.

16. Mochizuki K, Murase H, Imose M, Kawakami H, Sawada A. Improvement of scotopic electroretinograms and night blindness with recovery of serum zinc levels. *Jpn J Ophthalmol* 2006;**50**: 532–6.

17. Afridi HI, Kazi TG, Kazi N, Kandhro GA, Baig JA, Shah AQ, et al. Evaluation of status of zinc, copper, and iron levels in biological samples of normal children and children with night blindness with age groups of 3–7 and 8–12 years. *Biol Trace Elem Res* 2011;**142**: 323–34.

18. Tinley CG, Withers NJ, Sheldon CD, Quinn AG, Jackson AA. Zinc therapy for night blindness in cystic fibrosis. *J Cyst Fibros* 2008;**7**:333–5.

19. Smith JE, Brown ED, Smith Jr JC. The effect of zinc deficiency on the metabolism of retinol-binding protein in the rat. *J Lab Clin Med* 1974;**84**:692–7.

20. Dorea JG, Olson JA. The rate of rhodopsin regeneration in the bleached eyes of zinc-deficient rats in the dark. *J Nutr* 1986;**116**: 121–7.

21. Sturman JA, Wen GY, Wisniewski HM, Hayes KC. Histochemical localization of zinc in the feline tapetum. Effect of taurine depletion. *Histochemistry* 1981;**72**:341–50.

22. Hirayama Y. Histochemical localization of zinc and copper in rat ocular tissues. *Acta Histochem* 1990;**89**:107–11.

23. Redenti S, Ripps H, Chappell RL. Zinc release at the synaptic terminals of rod photoreceptors. *Exp Eye Res* 2007;**85**:580–4.

24. Ugarte M, Osborne NN. Zinc in the retina. *Prog Neurobiol* 2001;**64**:219–49.

25. Wyszynski RE, Bruner WE, Cano DB, Morgan KM, Davis CB, Sternberg P. A donor-age-dependent change in the activity of alpha-mannosidase in human cultured RPE cells. *Invest Ophthalmol Vis Sci* 1989;**30**:2341–7.

26. Ding X, Patel M, Chan CC. Molecular pathology of age-related macular degeneration. *Prog Retin Eye Res* 2009;**28**:1–18.

27. Pasantes-Morales H, Cruz C. Protective effect of taurine and zinc on peroxidation-induced damage in photoreceptor outer segments. *J Neurosci Res* 1984;**11**:303–11.

28. Wills NK, Ramanujam VM, Kalariya N, Lewis JR, van Kuijk FJ. Copper and zinc distribution in the human retina: relationship to cadmium accumulation, age, and gender. *Exp Eye Res* 2008;**87**: 80–8.

29. Lengyel I, Flinn JM, Peto T, Linkous DH, Cano K, Bird AC, et al. High concentration of zinc in sub-retinal pigment epithelial deposits. *Exp Eye Res* 2007;**84**:772–80.

30. Age-Related Eye Disease Study Research Group. The Age-Related Eye Disease Study (AREDS): design implications. AREDS report no. 1. *Control Clin Trials* 1999;**20**:573–600.

31. Newsome DA, Swartz M, Leone NC, Elston RC, Miller E. Oral zinc in macular degeneration. *Arch Ophthalmol* 1988;**106**:192–8.

32. Age-Related Eye Disease Study Research Group. A randomized, placebo-controlled, clinical trial of high-dose supplementation with vitamins C and E, beta carotene, and zinc for age-related macular degeneration and vision loss: AREDS report no. 8. *Arch Ophthalmol* 2001;**119**:1417–36.

33. van Leeuwen R, Boekhoorn S, Vingerling JR, Witteman JC, Klaver CC, Hofman A, et al. Dietary intake of antioxidants and risk of age-related macular degeneration. *J Am Med Assoc* 2005;**294**: 3101–7.

34. Tan JS, Wang JJ, Flood V, Rochtchina E, Smith W, Mitchell P. Dietary antioxidants and the long-term incidence of age-related macular degeneration: the Blue Mountains Eye Study. *Ophthalmology* 2008;**115**:334–41.

35. VandenLangenberg GM, Mares-Perlman JA, Klein R, Klein BE, Brady WE, Palta M. Associations between antioxidant and zinc intake and the 5-year incidence of early age-related maculopathy in the Beaver Dam Eye Study. *Am J Epidemiol* 1998;**148**:204–14.

36. Stur M, Tittl M, Reitner A, Meisinger V. Oral zinc and the second eye in age-related macular degeneration. *Invest Ophthalmol Vis Sci* 1996;**37**:1225–35.

37. Cho E, Seddon JM, Rosner B, Willett WC, Hankinson SE. Prospective study of intake of fruits, vegetables, vitamins, and carotenoids and risk of age-related maculopathy. *Arch Ophthalmol* 2004;**122**:883–92.

38. Seddon JM, Ajani UA, Sperduto RD, Hiller R, Blair N, Burton TC, et al. Dietary carotenoids, vitamins A, C, and E, and advanced age-related macular degeneration. Eye Disease Case–Control Study Group. *J Am Med Assoc* 1994;**272**:1413–20.

39. Tarwadi K, Agte V. Potential of commonly consumed green leafy vegetables for their antioxidant capacity and its linkage with the micronutrient profile. *Int J Food Sci Nutr* 2003;**54**:417–25.

40. Barash H, Poston HA, Rumsey GL. Differentiation of soluble proteins in cataracts caused by deficiencies of methionine, riboflavin or zinc in diets fed to Atlantic salmon, *Salmo salar*, rainbow trout, *Salmo gairdneri*, and lake trout. *Salvelinus namaycush Cornell Vet* 1982;**72**:361–71.

41. Cameron JD, McClain CJ. Ocular histopathology of acrodermatitis enteropathica. *Br J Ophthalmol* 1986;**70**:662–7.

42. Agte V, Tarwadi K. The importance of nutrition in the prevention of ocular disease with special reference to cataract. *Ophthalmic Res* 2010;**44**:166–72.

43. Ghosh KS, Pande A, Pande J. Binding of gamma-crystallin substrate prevents the binding of copper and zinc ions to the molecular chaperone alpha-crystallin. *Biochemistry* 2011;**50**:3279–81.

44. Beebe DC, Holekamp NM, Shui YB. Oxidative damage and the prevention of age-related cataracts. *Ophthalmic Res* 2010;**44**:155–65.

45. Age-Related Eye Disease Study Research Group. A randomized, placebo-controlled, clinical trial of high-dose supplementation with vitamins C and E and beta carotene for age-related cataract and vision loss: AREDS report no. 9. *Arch Ophthalmol* 2001;**119**:1439–52.

46. Sperduto RD, Hu TS, Milton RC, Zhao JL, Everett DF, Cheng QF, et al. The Linxian cataract studies. Two nutrition intervention trials. *Arch Ophthalmol* 1993;**111**:1246–53.

47. Robertson BT, Burns MJ. Zinc metabolism and the zinc-deficiency syndrome in the dog. *Am J Vet Res* 1963;**24**:997–1002.

48. Leure-Dupree AE. Vascularization of the rat cornea after prolonged zinc deficiency. *Anat Rec* 1986;**216**:27–32.

49. Aggett P. Severe zinc deficiency. In: Mills C, editor. *Zinc in Human Biology*. London: International Life Sciences Institute; 1989. p. 260.

50. Pati SK, Mukherji R. Serum zinc in corneal ulcer – a preliminary report. *Indian J Ophthalmol* 1991;**39**:134–5.

38

Impact of Impaired Maternal Vitamin A Status on Infant Eyes

David Coman[1], Glen Gole[2]

[1]Department of Metabolic Medicine, The Royal Children's Hospital, Brisbane, Queensland, Australia, [2]Paediatrics and Child Health, University of Queensland, Queensland, Australia

INTRODUCTION

Vitamin deficiency occurring in newborns and resulting in ocular defects appears to be quite rare. Except for vitamin A deficiency, no reports were found of ocular malformations occurring as a result of neonatal deficiencies in vitamins D or K or iron.

Neonatal vitamin A deficiency is usually seen in the context of maternal vitamin A deficiency, which in developed economies is seen most frequently after gastric bypass surgery for morbid obesity. Intrinsic disorders of vitamin A metabolism can also produce a variety of ocular malformations in infants.

Vitamin A 'metabolism' is a highly regulated and an incompletely understood process. The pleotropic effects of vitamin A reflect its important role in cell cycle fate, embryogenesis, and organogenesis.

Murine and human morphologic abnormalities have been demonstrated in vitamin A deficiency and excess alike.[1] Genetic heterogeneity is likely to exist among novel primary diseases in vitamin A synthesis, degradation, and signaling.

Vitamin A plays a crucial role in ocular embryogenesis and retinal function; vitamin A deficiency syndromes have provided insight into the molecular roles of vitamin A in the causation of the malformations as identified in murine models and in human patients,[1] where the ocular manifestations produce considerable morbidity.

Vitamin A metabolism plays a central role in the normal visual cycle and the normal function of retinal photoreceptors.

This chapter will briefly discuss vitamin A metabolism, clinical multisystem malformation syndromes associated with perturbations of vitamin A homeostasis and biosynthesis, and the role of vitamin A in ocular development. It will focus on the neonatal manifestations of maternal hypovitaminosis A.

SYSTEMIC VITAMIN A REGULATION

Dietary Sources and Intestinal Absorption

Retinoids are derived from dietary sources of vitamin A. Their main role is involvement in cell signaling mechanisms in such processes as cell proliferation, differentiation, and apoptosis. Dietary retinoids are taken up from the intestinal lumen into mucosa cells as the free alcohol retinol and enzymatically converted to retinaldehyde and subsequently to retinol within the cell. Within enterocytes, retinol is esterified with long-chain fatty acyl groups to form retinyl esters, which are packaged along with dietary lipids into chylomicrons.[2]

Vitamin A Metabolism

The active metabolite of vitamin A, retinoic acid (RA), is a powerful regulator of gene transcription, which circulates in the blood as retinol bound to retinol binding protein (RBP). Retinol is converted to retinaldehyde and then to all-*trans* retinoic acid via two oxidation steps (Fig. 38.1). All-*trans* retinoic acid then binds to the nuclear retinoic acid receptors (RARs).[1]

Once inside the cell, retinol is first converted to retinal by the retinol dehydrogenases (RDHs) and then to RA by the retinaldehyde dehydrogenases (RALDHs) (Fig. 38.1). The latter are thought to be the rate-limiting step in the synthesis of all-*trans*-RA.[3]

The two active isomers of RA, all-*trans*-RA and 9-*cis*-RA, act via different nuclear receptors: RARs and retinoic

---

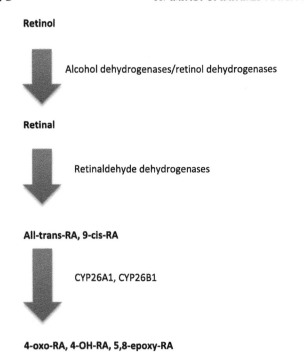

Retinol

Alcohol dehydrogenases/retinol dehydrogenases

Retinal

Retinaldehyde dehydrogenases

All-trans-RA, 9-cis-RA

CYP26A1, CYP26B1

4-oxo-RA, 4-OH-RA, 5,8-epoxy-RA

FIGURE 38.1 Overview of vitamin A synthesis and catabolism. This metabolic pathway, involving three enzyme systems, is responsible for the conversion of vitamin A (retinol) into the various active forms of retinoic acid (RA).

X receptors (RXRs). Subsequent downstream regulation of multiple genes occurs associated with such processes as neuronal differentiation and both anteroposterior and dorsoventral patterning.

Oxidative metabolism of RA is carried out by members of the cytochrome P450 (CYP) superfamily – CYP26A1, CYP26B1, and CYP26C1 – with CYP26A1 playing a major role in RA clearance[4] (Fig. 38.1). The *Cyp26a1*[−/−] mouse mimics the state of excess RA administration and is rescued by the heterozygous knockdown of *Raldh2*,[5] which provides compelling evidence that the CYP26 family plays an essential role in protecting tissues from excess exposure to all-*trans*-RA.

Storage of cellular vitamin A is dependent on lecithin:retinol acyl transferase (LRAT) and cellular retinol binding proteins (CRBP1, 2, and 3).

Normal Vitamin A Levels

Two biochemical indicators are currently used to determine vitamin A deficiency: serum retinol and serum RBP.[6] A cutoff for serum retinol concentration has been proposed at less than $0.70\,\mu mol/L$.[6] A cutoff cannot be reliably specified for RBP because available data are too few and too variable, but serum RBP concentration correlates well with serum retinol concentration.[6]

VITAMIN A AND OCULAR EMBRYOGENESIS

Retinol plays an essential role in embryogenesis and during pregnancy is transported by RBP from the maternal blood supply to the embryo through receptor-facilitated transport.[7] Within the embryonic tissues, retinol is metabolized to RA by a series of intracellular processes involving aldehyde enzymes. RA functions as a ligand for nuclear receptors, which regulate gene expression. It plays a vital role in the differentiation and development of ocular epithelia. RA signaling stimulates the transition of the optic cup from the optic vesicle, the closure of the choroid fissure, and the formation of the cornea and eyelids through apoptosis of the perioptic mesenchyme.[8] The first 8 weeks of the embryonic period are the most crucial concerning the development of congenital malformations. Before week 3 of gestation, miscarriage is more likely than gross structural defects. Organogenesis takes place between 3 and 8 weeks of gestation.[9]

VITAMIN A AND NORMAL OPHTHALMOLOGIC OCULAR FUNCTION

The retinoid cycle is essential for regeneration of rod and cone visual pigment and maintenance of continuous vision. The isomerization of 11-*cis*-retinal to all-*trans*-retinal in photoreceptors is the first step in vision. For photoreceptors to function in constant light, the all-*trans*-retinal must be converted back to 11-*cis*-retinal via the enzymatic steps of the visual cycle. Within the visual cycle, all-*trans*-retinal is reduced to all-*trans*-retinol in photoreceptors and transported to the retinal pigment epithelium (RPE). In the RPE, all-*trans*-retinol is converted to 11-*cis*-retinol, and in the final enzymatic step, 11-*cis*-retinol is oxidized to 11-*cis*-retinal. The first and last steps of the classical visual cycle are reduction and oxidation reactions, respectively, that utilize RDH enzymes.

VITAMIN A AND HUMAN OCULAR DISEASE

Pathogenic mutations, in knockout mice and human patients, in genes encoding enzymes associated with the retinal visual cycle can lead to blinding diseases.[10] Examples of such disease include:

- retinal-specific ATP-binding cassette transporter (*ABCA4*)
- Stargart disease

TABLE 38.1 Known Human Disease Phenotypes Consequent to Primary Defects in the Vitamin A Regulation Pathway

Gene	Clinical Phenotype	Biochemical Phenotype	Reference
N/A	Maternal dietary deficiency		
N/A	Maternal hypovitaminosis A after biliopancreatic diversion	Plasma vitamin A <0.1 mg/L (normal range: 0.3–0.9 mg/L)	11
N/A	Maternal vitamin A excess		
Neural tube defects			12
N/A	Maternal teratogens, i.e., isoretanion		
SAR1B	Anderson's disease/chylomicron retention disease	Decreased vitamin A with hypobetalipoproteinemia	13
STRA6	Microphthalmia, anophthalmia, coloboma, anophthalmia, diaphragmatic hernia, complex congenital heart disease, alveolar capillary dysplasia, intellectual impairment, Matthew–Wood syndrome, PDAC syndrome, and MCOPS9	Not reported	14–17
CYP26A1 and CYP26C1	Optic nerve hypoplasia	Not reported	18
CYP26B1	Craniosynostosis and skeletal dysplasia (radiohumeral synostosis (similar to what is seen in vitamin A excess))	Not reported	19
LRAT	Leber congenital amaurosis	Not reported	20
N/A	RBP deficiency secondary to renal disease, e.g., Dent's disease; night vision and xerophthalmia	Retinol decreased, plasma RBP undetectable, elevated urinary RBP	21
RBP4	Isolated ocular manifestations of vitamin A deficiency	Retinol decreased, plasma RBP undetectable	22

N/A: not applicable; RBP: retinol binding protein.

- autosomal recessive cone–rod dystrophy
- autosomal recessive retinitis pigmentosa
- macular degeneration
- retinol dehydrogenase 12 (RDH12)
- autosomal recessive retinal dystrophy
- lecithin retinal acyltransferase (LRAT)
- retinal dystrophy
- retinal pigment epithelium-specific protein (LCA2)
- Leber congenital amaurosis
- autosomal recessive retinitis pigmentosa
- retinal G protein-coupled receptor (RGR)
- autosomal dominant and autosomal recessive retinitis pigmentosa
- retinaldehyde binding protein (RLBP1)
- autosomal recessive retinitis punctate albescens
- autosomal recessive retinitis pigmentosa
- retinol dehydrogenase 5 (RDH5)
- autosomal recessive fundus albipunctatus.

VITAMIN A AND MULTISYSTEM DISEASE PHENOTYPES

The ophthalmologic manifestations of vitamin A deficiency or excess can be profoundly important clinical clues in discovering the primary underlying defect in vitamin A homeostasis (Table 38.1).[11–22]

Prolonged dietary deficiency of vitamin A results in xerophthalmia, Bitot spots, conjunctival xerosis, corneal scarring, and night blindness. The teratogenic effects of vitamin A excess include neural tube defects, craniofacial clefting, limb reduction defects, and internal organogenesis defects, such as intestinal atresia and congenital cardiac malformations.[12] Primary genetic defects in RA metabolism and transport have been reported most frequently associated with mutations in the stimulated by RA-6 gene (STRA6), which encodes a transmembrane receptor for the RBP and is responsible for mediating vitamin A uptake from the circulation to target organs.[14] Multiorgan manifestations of pathogenic mutations in the STRA6 gene include diaphragmatic hernia, complex congenital heart disease, alveolar capillary dysplasia, and intellectual impairment.

The ophthalmologic malformations in multisystem syndromes associated with aberrant vitamin A synthesis or transport can impose a profound disease burden on the patient and their family and include microphthalmia, anophthalmia, coloboma, optic nerve hypoplasia, and anophthalmia (Table 38.1).[10]

NEONATAL VITAMIN A DEFICIENCY SYNDROMES CONSEQUENT TO MATERNAL HYPOVITAMINOSIS A

The neonatal manifestations of maternal hypovitaminosis A tend to occur in two main maternal groups: in mothers with dietary deficiencies of vitamin A reflective of lower socioeconomic status[23] and in mothers who have become vitamin A deficient consequent to surgical procedures aimed at reducing morbid obesity, such as biliopancreatic diversion surgery.[24]

Biliopancreatic diversion surgery has been advocated as a treatment for morbid obesity for over 30 years.[25] The surgery involves bypassing the duodenum by creating a diversion between the stomach and also creating an anastomosis between the duodenum and distal ileum, causing biliopancreatic secretions to be released into the distal ileum.[25] The main benefit is the reduction of health risks associated with morbid obesity. The main complication is the malabsorption of dietary protein, minerals, and vitamins, including vitamin A, which is further exacerbated during pregnancy.[26]

The vast majority (approximately 80%) of patients who undergo gastric bypass surgery are females, and the majority of these are in their reproductive years.[11] A retrospective study of the outcome of pregnancies following biliopancreatic diversion advocated obesity surgery as an indirect treatment for infertility by restoring the patient to a healthy weight range and reducing the pregnancy risks associated with morbid obesity.[27] However, this may be counterbalanced by the detrimental effects of malabsorption caused by biliopancreatic diversion on the unborn fetus. A survey of vitamin and mineral utilization in patients following gastric bypass surgery found that 26% were iron deficient, 11% vitamin B_{12} deficient, 21% anemic, 18% protein deficient, 16% calcium deficient, and 6% deficient in fat-soluble vitamins.[28] These deficiencies, and the potentially deleterious effect they have on the fetus, are not limited to the immediate perioperative period associated with dramatic weight loss, as originally believed.[25]

Vitamin A supplementation during the first trimester of pregnancy in women who have undergone biliopancreatic diversion may be beneficial in preventing fetal ocular malformations. However, vitamin A supplementation cannot correct an already established fetal defect[29] and would need to be taken before becoming pregnant to ensure adequate retinol for early fetal development.

There is a causal relationship between maternal gastric bypass and neonatal hypovitaminosis A.[26] Ocular abnormalities, including microphthalmia, have been described in neonates as a result of maternal biliopancreatic diversion.[30] Anterior segment malformations may occur from perinatal infections, genetic mutations, such as in the key regulatory genes for eye development

(*PAX2* and *PAX6*), and nutritional deficiencies, such as vitamin A. Vitamin A metabolites are vital to the normal function and maintenance of the visual chromophore, the integrity of the corneal surface, and ocular morphogenesis.[31,32] Downstream gene regulation by vitamin A metabolites is mediated by RARs and RXRs. The transcriptional regulation of target genes involves proper assembly of RAR–RXR heterodimers into short promoter regions of the target genes known as retinoic acid responsive elements (RAREs).[1,31,32] These heterodimers play an important role in eye embryogenesis via regulation of apoptosis by *EYA2*[33] and via regulation of *PTIX2* and *FOXC1*.[32]

One of the present authors (GG) previously reported a case of maternal hypovitaminosis A caused by biliopancreatic bypass surgery, which produced severe ocular malformations in the infant.[24] Despite treatment, the mother's vitamin A levels had remained low throughout the pregnancy. The affected neonate presented with bilateral microcorneas, inferior corneal scarring (possibly caused by intrauterine corneal perforation), and malformed irides, which appeared to be pulled down inferiorly behind the corneal scar tissue in both eyes (Fig. 38.2A, B). Dilated fundus examination revealed bilateral hypoplastic foveas and optic discs (Fig. 38.2C, D). At 3 days of age, serum vitamin A was undetectable in the neonate. The infant underwent bilateral sector iridectomies at 8 weeks of age to allow some visual input. An electroretinogram at 9 months of age showed normal photopic tracings but absent scotopic tracings, suggesting rod dysfunction. At 10 months of age, there was jerk-type nystagmus and he appeared very light sensitive (possibly due to the sector iridectomies). Cycloplegic refraction revealed −4.25 diopter sphere in both eyes. He had poor vision. Based on this case, the authors advise measuring serum vitamin A levels before and during pregnancy in women who have undergone biliopancreatic diversion surgery.

World Health Organization guidelines recommend that vitamin A-deficient pregnant women take 200,000 IU/day for 2 weeks, followed by additional supplementation as required. In severe deficiencies and particularly in malabsorption syndromes, such as in women after biliopancreatic bypass surgery, oral vitamin A supplementation is likely to be ineffective and intramuscular injections may be necessary.[23] While vitamin A-deficient mothers do not develop blinding xerophthalmia, like their vitamin A-deficient infants, they do have higher rates of night blindness,[23] which reflects the importance of vitamin A in normal rod–cone function in the visual cycle.

Because the health system impacts of obesity are escalating in developed economies, maternal hypovitaminosis A secondary to morbid obesity surgical procedures may be an under-recognized entity and one in which

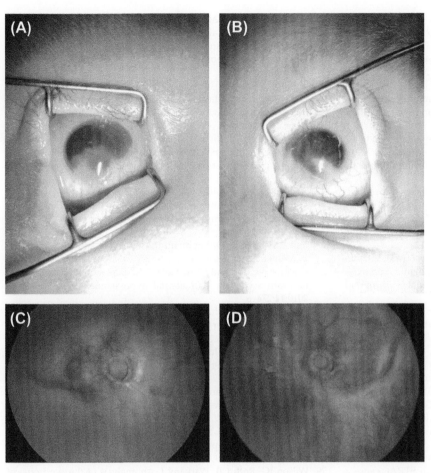

FIGURE 38.2 Ocular findings in a 2-day-old male with vitamin A deficiency. (A, B) Anterior segment photographs show microcorneas, inferior corneal scars, and down-drawn irides: (A) right eye, (B) left eye. (C, D) Fundus photographs obtained at 8 weeks of age following sector iridectomies showing optic disc and foveal hypoplasia: (C) right eye, (D) left eye. *Source: Gilchrist H, Taranath DA, Gole GA. Ocular malformation in a newborn secondary to maternal hypovitaminosis A. J AAPOS. 2010;14:274–6[24] with permission.* (See color plate section)

obstetricians, pediatricians, and ophthalmologists need to have a working knowledge of the manifestations and prevention.

aberrations in these pathways is vital not just for novel disease discovery but also for developing appropriate treatments.

CONCLUSIONS AND FUTURE PERSPECTIVES

Maternal hypovitaminosis A carries a potential for a significant neonatal disease burden. While obesity is traditionally considered to be a disorder seen in the developing world, the advent of surgical procedures to treat morbid obesity is creating yet another health system impact of obesity in developed economies.

Identification of knockout murine models and naturally occurring human pathogenic mutations in retinoid metabolism have provided great insight into the pleotropic roles of these compounds in embryogenesis and in later life. Much is still to be learned about RA synthesis and its molecular gene signaling targets, in which novel inherited defects are likely to be identified in time. The appropriate metabolism of retinols in the normal retinal visual cycle function is essential to normal visual function. Improved understanding of the retinol metabolic pathways and clinical disorders consequent to

TAKE-HOME MESSAGES

- Bitot spots and anterior segment malformations are prime clinical clues to the presence of neonatal hypovitaminosis A.
- Human disease phenotypes secondary to primary or secondary defects in vitamin A metabolism are being delineated.
- Maternal hypovitaminosis A secondary to morbid obesity surgery may be an underdiagnosed clinical entity that can lead to severe effects on the fetus.

References

1. Niederreither K, Dolle P. Retinoic acid in development: towards an integrated view. *Nat Rev Genet* 2008;**9**:541–53.
2. Wongsiriroj N, Piantedosi R, Palczewski K, Goldberg IJ, Johnston TP, Li E, et al. The molecular basis of retinoid absorption: a genetic dissection. *J Biol Chem* 2008;**283**:13510–9.
3. Perlmann T. Retinoid metabolism: a balancing act. *Nat Genet* 2002;**31**:7–8.

4. Ross AC, Zolfaghari R. Cytochrome P450s in the regulation of cellular retinoic acid metabolism. *Annu Rev Nutr* 2011;**31**:65–87.

5. Niederreither K, Vermot J, Fraulob V, Chambon P, Dolle P. Retinaldehyde dehydrogenase 2 (RALDH2)-independent patterns of retinoic acid synthesis in the mouse embryo. *Proc Natl Acad Sci USA* 2002;**99**:16111–6.

6. de Pee S, Dary O. Biochemical indicators of vitamin A deficiency: serum retinol and serum retinol binding protein. *J Nutr* 2002;**132** (Suppl. 9):2895–901. S.

7. Morriss-Kay GM, Sokolova N. Embryonic development and pattern formation. *FASEB J* 1996;**10**:961–8.

8. Duester G. Keeping an eye on retinoic acid signaling during eye development. *Chem Biol Interact* 2009;**178**:178–81.

9. Sadler TW. *Langman's Medical Embryology*. 10th ed. Lippincott: Williams & Wilkins; 2006.

10. Travis GH, Golczak M, Moise AR, Palczewski K. Diseases caused by defects in the visual cycle: retinoids as potential therapeutic agents. *Annu Rev Pharmacol Toxicol* 2007;**47**:469–512.

11. Huerta S, Rogers LM, Li Z, Heber D, Liu C, Livingston EH. Vitamin A deficiency in a newborn resulting from maternal hypovitaminosis A after biliopancreatic diversion for the treatment of morbid obesity. *Am J Clin Nutr* 2002;**76**:426–9.

12. Rothman KJ, Moore LL, Singer MR, Nguyen US, Mannino S, Milunsky A. Teratogenicity of high vitamin A intake. *N Engl J Med* 1995;**333**:1369–73.

13. Georges A, Bonneau J, Bonnefont-Rousselot D, Champigneulle J, Rabes JP, Abifadel M, et al. Molecular analysis and intestinal expression of SAR1 genes and proteins in Anderson's disease (chylomicron retention disease). *Orphanet J Rare Dis* 2011;**6**:1.

14. Casey J, Kawaguchi R, Morrissey M, Sun H, McGettigan P, Nielsen JE, et al. First implication of STRA6 mutations in isolated anophthalmia, microphthalmia, and coloboma: a new dimension to the STRA6 phenotype. *Hum Mutat* 2011;**32**:1417–26.

15. Pasutto F, Sticht H, Hammersen G, Gillessen-Kaesbach G, Fitzpatrick DR, Nurnberg G, et al. Mutations in STRA6 cause a broad spectrum of malformations including anophthalmia, congenital heart defects, diaphragmatic hernia, alveolar capillary dysplasia, lung hypoplasia, and mental retardation. *Am J Hum Genet* 2007;**80**:550–60.

16. Chassaing N, Golzio C, Odent S, Lequeux L, Vigouroux A, Martinovic-Bouriel J, et al. Phenotypic spectrum of STRA6 mutations: from Matthew–Wood syndrome to non-lethal anophthalmia. *Hum Mutat* 2009;**30**(5):E673–81.

17. Gavrilova R, Babovic N, Lteif A, Eidem B, Kirmani S, Olson T, et al. Vitamin A deficiency in an infant with PAGOD syndrome. *Am J Med Genet A* 2009;**149A**:2241–7.

18. Meire F, Delpierre I, Brachet C, Roulez F, Van Nechel C, Depasse F, et al. Nonsyndromic bilateral and unilateral optic nerve aplasia: first familial occurrence and potential implication of CYP26A1 and CYP26C1 genes. *Mol Vis* 2011;**17**:2072–9.

19. Laue K, Pogoda HM, Daniel PB, van Haeringen A, Alanay Y, von Ameln S, et al. Craniosynostosis and multiple skeletal anomalies in humans and zebrafish result from a defect in the localized degradation of retinoic acid. *Am J Hum Genet* 2011;**89**:595–606.

20. Thompson DA, Li Y, McHenry CL, Carlson TJ, Ding X, Sieving PA, et al. Mutations in the gene encoding lecithin retinol acyltransferase are associated with early-onset severe retinal dystrophy. *Nat Genet* 2001;**28**:123–4.

21. Becker-Cohen R, Rinat C, Ben-Shalom E, Feinstein S, Ivgi H, Frishberg Y. Vitamin A deficiency associated with urinary retinol binding protein wasting in Dent's disease. *Pediatr Nephrol* 2012;**27**: 1097–102.

22. Seeliger MW, Biesalski HK, Wissinger B, Gollnick H, Gielen S, Frank J, et al. Phenotype in retinol deficiency due to a hereditary defect in retinol binding protein synthesis. *Invest Ophthalmol Vis Sci* 1999;**40**:3–11.

23. Ramakrishnan U, Darnton-Hill I. Assessment and control of vitamin A deficiency disorders. *J Nutr* 2002;**132**(Suppl. 9):2947–53. S.

24. Gilchrist H, Taranath DA, Gole GA. Ocular malformation in a newborn secondary to maternal hypovitaminosis A. *J AAPOS* 2010;**14**:274–6.

25. Scopinaro N, Gianetta E, Civalleri D, Bonalumi U, Bachi V. Biliopancreatic bypass for obesity: II. Initial experience in man. *Br J Surg* 1979;**66**:618–20.

26. Cools M, Duval EL, Jespers A. Adverse neonatal outcome after maternal biliopancreatic diversion operation: report of nine cases. *Eur J Pediatr* 2006;**165**:199–202.

27. Marceau P, Kaufman D, Biron S, Hould FS, Lebel S, Marceau S, et al. Outcome of pregnancies after biliopancreatic diversion. *Obes Surg* 2004;**14**:318–24.

28. Brolin RE, Leung M. Survey of vitamin and mineral supplementation after gastric bypass and biliopancreatic diversion for morbid obesity. *Obes Surg* 1999;**9**:150–4.

29. Wilson JG, Roth CB, Warkany J. An analysis of the syndrome of malformations induced by maternal vitamin A deficiency. Effects of restoration of vitamin A at various times during gestation. *Am J Anat* 1953;**92**:189–217.

30. Smets KJ, Barlow T, Vanhaesebrouck P. Maternal vitamin A deficiency and neonatal microphthalmia: complications of biliopancreatic diversion? *Eur J Pediatr* 2006;**165**:502–4.

31. Nezzar H, Chiambaretta F, Marceau G, Blanchon L, Faye B, Dechelotte P, et al. Molecular and metabolic retinoid pathways in the human ocular surface. *Mol Vis* 2007;**13**:1641–50.

32. Matt N, Dupe V, Garnier JM, Dennefeld C, Chambon P, Mark M, et al. Retinoic acid-dependent eye morphogenesis is orchestrated by neural crest cells. *Development* 2005;**132**:4789–800.

33. Zhou J, Kochhar DM. Regulation of AP-2 and apoptosis in developing eye in a vitamin A-deficiency model. *Birth Defects Res A Clin Mol Teratol* 2003;**67**:41–53.

Optic Neuropathies Caused by Micronutrient Deficiencies and Toxins

Marco Spinazzi

Laboratory for the Research of Neurodegenerative Diseases, Department of Human Genetics, University of Leuven, Belgium

INTRODUCTION

The optic nerve represents a protrusion of the brain with peculiar anatomical and functional features, which make it particularly susceptible to metabolic insults. It is formed by the projections of about 1.5 million retinal ganglion cells embedded in the inner retina. In the prelaminar region, myelination of the axons is virtually absent, and the energy required for the propagation of action potentials is high. Therefore, retinal ganglion cells require a considerable amount of energy to be produced through mitochondrial oxidative phosphorylation. These features make the retinal ganglion cells particularly sensitive to acquired or inherited diseases affecting energy metabolism. This chapter will focus on micronutrient deficiencies, as well as some toxicants, which can affect the optic nerves. Many micronutrients are fundamental cofactors of enzymes with critical cellular functions such as mitochondrial energy production. Other important roles include DNA synthesis, antioxidant defenses, and lipid metabolism.

EPIDEMIOLOGY AND ETIOPATHOGENESIS OF NUTRITIONAL OPTIC NEUROPATHIES

The prevalence of nutritional optic neuropathies is unknown. Several epidemics of nutritional optic neuropathies of probable nutritional etiology have been observed in different populations affected by socioeconomic contingencies leading to poor nutrition (see below, in the section 'Epidemic Nutritional Optic Neuropathies'). The etiopathogenesis of these epidemics is likely to be multifactorial, including not only a generalized malnutrition but also exposure to toxicants, such as alcohol and smoking. Sporadic nutritional optic neuropathies also commonly have a multifactorial pathogenesis. The simultaneous deficiency of several micronutrients is frequently the result of an underlying poor diet, the presence of malabsorption diseases, and metabolic interactions between different vitamins.[1] Concomitant exposure to certain toxicants can precipitate the clinical signs of a micronutrient deficiency by interfering with its absorption but also by worsening the metabolic injury. Some drugs have the intrinsic properties of inducing optic nerve damage by mitochondrial impairment (Table 39.1). The rarity of both nutritional and toxic optic neuropathies strongly suggests a 'double-hit' mechanism and the presence of a susceptibility factor in individual patients. As already mentioned, in many cases the optic nerve damage results from a combination of nutritional and toxic insults. Less commonly, nutritional deficits may unmask a previously unapparent genetic disorder affecting the optic nerve.

Therefore, the frequency of nutritional neurologic manifestations may vary significantly depending on the specific socioeconomic situation, lifestyle, and genetic background. While they are now considered uncommon in Western developed countries, many cases are still unrecognized. Among people older than 60 years in the UK and the USA, cobalamin deficiency has been reported to exceed 6%.[2] More surprisingly, the neuropathologic detection of Wernicke's encephalopathy, caused by thiamine deficiency, was reported to be as high as 2.8% in an autoptic survey of individuals older than 20 years in western Australia.[3]

In terms of mechanisms, micronutrient deficiencies could arise from decreased dietary intake, insufficient absorption, increased consumption, or defective metabolism. Specific subgroups of individuals are particularly

TABLE 39.1　Examples of Optic Nerve Toxicants

Toxicant	Use
Methanol	Contaminant of poorly distillated alcoholic beverages
Clioquinol	Antibiotic
Isoniazid	Antibiotic
Ethambutol	Antibiotic
Chloramphenicol	Antibiotic
Carbon monoxide	Gas poison
Toluene	Alkylbenzene present in glues and paints
Smoke	Smoking tobacco (especially from pipes and cigars)

TABLE 39.2　Clinical Manifestations of Micronutrient Deficiencies

System Affected	Manifestations
Hematologic	Anemia, leukopenia, thrombocytopenia
Neurologic	Dorsolateral myelopathy
	Encephalopathy
	Optic neuropathy
	Peripheral neuropathy
	Myopathy
	Hearing loss
	Dysautonomia
Cardiologic	Cardiomyopathy (vitamin B_1 deficiency)
Dermatologic	Dermatosis, cheilitis, hyperpigmentation, desquamative lesions

at risk of developing nutritional optic neuropathies: malnourished individuals, elderly people, alcoholics, patients fed by tube or by parenteral nutrition,[4] patients treated with drugs interfering with micronutrient metabolism or absorption, pregnant women, and patients affected by malabsorption diseases (e.g., celiac disease) or who have undergone surgical interventions on the gastrointestinal tract. Importantly, the underlying malnutrition is often qualitative, rather than quantitative. As a consequence, patients affected by nutritional optic neuropathies often do not necessarily have phenotypical and biochemical signs of a generalized malnutrition, and their body mass index could well be within or even over the normal range. Obese patients have a basal high prevalence of nutritional deficits[5] and therefore are particularly at risk of developing micronutrient deficiency-related neurologic complications after surgical treatments for the obesity.[6] Optic neuropathy was reported in eight of 96 patients affected by neurologic complications after bariatric surgery.[7]

The importance of the recognition of nutritional optic neuropathies should be emphasized despite their rarity because of the availability of an effective supplementation therapy capable of preventing irreversible optic nerve degeneration and visual disability.

Finally, while most commonly nutritional optic neuropathies have a multifactorial pathogenesis, in some rare cases the nutritional optic neuropathy seems to be mostly related to a single nutritional deficiency, providing the proof of principle that optic neuropathy can be caused by the deficiency of a specific micronutrient *per se*. However, substantial caution should be applied in the interpretation of some reports not providing complete biochemical support or a positive response to supplementation therapy. The pathophysiology and clinical manifestations of micronutrients currently known to cause human optic neuropathies are summarized in the next section.

CLINICAL MANIFESTATIONS OF NUTRITIONAL OPTIC NEUROPATHIES

In most, if not all, cases of nutritional optic neuropathies, visual signs are accompanied by other neurologic and extraneurologic manifestations. Nutritional deficiencies are multisystem diseases, owing to the ubiquitous requirements of micronutrients, and especially in tissues with high energy demands.

Micronutrient deficiencies affect most commonly the hematologic system and the central and peripheral nervous systems. Less frequently, the heart and the skin[8] can also be affected (Table 39.2). Thus, the suspicion of nutritional optic neuropathy warrants careful research for these additional abnormalities, which may be subclinical.

The optic nerve involvement in nutritional optic neuropathies is quite stereotypical. It is characterized by symmetrical, painless, bilateral visual loss with a central visual field defect and color discrimination deficits (dyschromatopsia). The progression of the ocular manifestations is usually subacute or chronic. The severity of the visual loss is variable, often mild to moderate, but optic neuropathy can also be subclinical. Complete blindness is rare and should lead to alternative diagnoses being considered.

DIAGNOSIS OF NUTRITIONAL OPTIC NEUROPATHIES

The presence of unexplained, painless, bilateral visual loss with a central scotoma should prompt investigations into a nutritional cause. The diagnosis of

TABLE 39.3 Differential Diagnosis of Nutritional Optic Neuropathies

Optic Neuritis	Noninflammatory Optic Neuropathies
Multiple sclerosis	Ischemic optic neuropathies
Neuromyelitis optica	Neoplastic optic neuropathies
Infectious optic neuritis	Traumatic optic neuropathies
Connectivitis-associated optic neuritis	Metabolic optic neuropathies
Paraneoplastic optic neuritis	Nutritional optic neuropathies
	Hereditary optic neuropathies
	Toxic optic neuropathies
	Endocrine optic neuropathies

a nutritional optic neuropathy is based on the following criteria:

- Biochemical detection of deficiencies of micronutrients known to cause optic neuropathy in a patient with a consistent clinical/instrumental presentation.
- Exclusion of alternative diagnoses (Table 39.3): the differential diagnosis is quite extensive and includes both inflammatory (optic neuritis) and noninflammatory optic neuropathies. Some conditions can closely mimic the typical clinical manifestations of nutritional optic neuropathies and can only be differentiated by specific tests.
- Positive response (clinical and instrumental) to successful supplementation therapy.

Fundus examination can show optic atrophy, sometimes more prominent in the temporal side, in the chronic phase, but it may also be normal. A swollen optic disc can be revealed in the acute phase of thiamine deficiency, requiring a differential diagnosis with Leber hereditary optic neuropathy (LHON). The pupillary light reflex and ocular movements are usually normal, except in thiamine deficiency. The ophthalmologic examination will also include an accurate evaluation of visual acuity, color discrimination, and contrast sensitivity. Visual field assessment will show a bilateral central visual field defect.

Visual evoked potentials are useful in exploring the overall electrophysiologic function of the visual pathways, especially in cases with a normal fundus appearance despite significant visual loss, which may erroneously lead the clinician to diagnose malingering. Brain magnetic resonance imaging (MRI) with coronal sections and fat suppression acquisitions is helpful to quantify the atrophy of the optic nerves in the chronic phase,[9] as well as to rule out other etiologies. Optical

coherence tomography (OCT) is a novel technique allowing precise *in vivo* quantification of optic nerve fiber loss through evaluation of the thickness of the retinal nerve fiber layer.[10]

A complete biochemical workup should always be performed, as concomitant deficiencies of several micronutrients are commonly encountered. Vitamin supplementation should ideally be preceded by blood sampling for biochemical tests, although therapy should not be halted in cases with a potentially fatal outcome, such as thiamine deficiency. Unfortunately, it is common for multivitamins to be empirically prescribed before the necessary tests have been carried out, precluding a meaningful micronutrient assessment.

Importantly, if a nutritional deficiency is found, the etiologic factors causing the deficit (i.e., dietary or malabsorption factors) should be investigated. Recognition and cessation of toxins damaging the optic nerves is also fundamental.

The prognosis is dependent on the timeliness of the supplementation before irreversible neurodegeneration occurs, highlighting the critical importance of early recognition.

VITAMIN B₁ (THIAMINE) DEFICIENCY

Pathophysiology of Thiamine Deficiency

Thiamine is abundant in whole-grain cereals and meats. Thiamine deficiency, also known as beriberi, was the first nutritional deficiency syndrome to be described, in individuals fed with polished rice. The condition has become much rarer since the introduction of thiamine-enriched cereals, but it still occurs in at-risk individuals and populations.

Thiamine deficiency impairs the function of three thiamine-dependent enzymes, of which thiamine pyrophosphate is a coenzyme (Fig. 39.1):

- transketolase: a cytosolic enzyme involved in the pentose–phosphate pathway, producing reducing equivalents and 5-ribose-phosphate for nucleotide synthesis.
- pyruvate dehydrogenase: a mitochondrial enzyme converting pyruvic acid, produced by glycolysis, into acetyl-coenzyme A (CoA), for use in the Krebs cycle.
- α-ketoglutarate dehydrogenase: a mitochondrial Krebs cycle enzyme converting α-ketoglutarate into succinate.

The pathophysiology of thiamine deficiency is probably due to insufficient mitochondrial energy production, and possibly increased oxidative stress, due to the functional impairment of the above-mentioned enzymes.

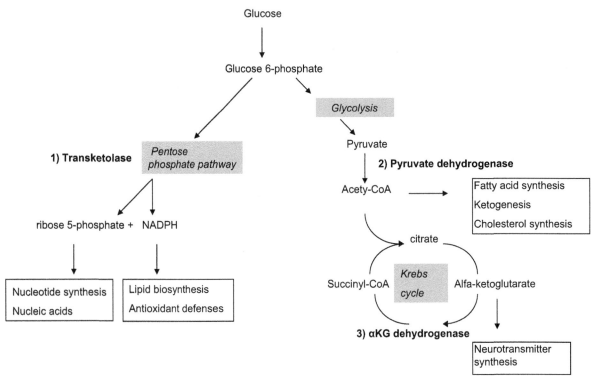

FIGURE 39.1   Thiamine-dependent enzymes. Pyruvate dehydrogenase and α-ketoglutarate dehydrogenase are mitochondrial enzymes involved in mitochondrial energy metabolism. Transketolase is a cytosolic enzyme involved in the pentose phosphate pathway, important for the generation of NADPH, maintenance of antioxidant defenses, and production of 5-ribose phosphate, needed for nucleotide and nucleic acid synthesis.

There are no systemic stores of thiamine in humans; therefore, thiamine depletion occurs within a few weeks of dietary deprivation. Frequent precipitating factors are alcoholism, previous gastrointestinal surgery, protracted vomiting, and increased intake of glucose, which consumes thiamine. Therefore, comatose patients admitted to the emergency room should always be supplemented with thiamine before intravenous glucose infusion to avoid provoking a potentially lethal Wernicke's encephalopathy.

Clinical Manifestations of Thiamine Deficiency

Thiamine deficiency leads to a multisystem clinical syndrome affecting the central and peripheral nervous systems and striated muscles. Clinical manifestations include:

- Wernicke's encephalopathy: a life-threatening encephalopathy characterized by acute cerebellar ataxia, confusion, and ophthalmoplegia. Dysautonomia is frequent. Neuropathologic features include degeneration of the thalami, hypothalamus, brainstem nuclei, and cerebellar vermis.[11]
- Polyneuropathy: an axonal polyneuropathy, usually with combined sensorimotor involvement leading

to muscle strength deficits, dysesthesias, and areflexia. Occasional patients show a predominantly motor phenotype with a rapidly progressive course mimicking Guillain–Barré syndrome. Conversely, other cases have a predominant sensory axonal polyneuropathy, resulting in sensory ataxia.[12]
- Cardiomyopathy: characterized by potentially lethal hyperdynamic shock and lactic acidosis.
- Skeletal muscle myopathy (rare).
- Optic neuropathy is a rare manifestation of thiamine deficiency, characterized by bilateral painless visual loss with central scotomas. The optic disc can be swollen in the acute phase and show peripapillary hemorrhages mimicking papillitis[13] or show a bilateral optic atrophy in the chronic phase (Fig. 39.2). However, fundus examination can be completely normal despite a severe visual defect.[14,15] The ocular manifestations are often associated with peripheral neuropathy or signs of Wernicke's encephalopathy, which can be limited to oculomotor deficits or nystagmus. These latter signs are not observed in other nutritional optic neuropathies and should be considered clinical red flags suggesting thiamine deficiency. Moreover, brain MRI may detect the typical radiologic findings of Wernicke's encephalopathy (see below) even in

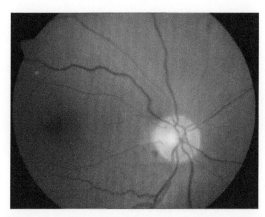

FIGURE 39.2 Optic atrophy in thiamine deficiency. Fundus photograph showing optic atrophy of a 71-year-old man affected by thiamine deficiency caused by a remote history of partial gastrectomy. He did not abuse alcohol and he did not smoke. He was admitted to the neurology emergency room with subacute onset of sensory ataxia, due to sensory axonal polyneuropathy, leading to the need for a wheelchair within 2 weeks. Despite the absence of subjective visual complaints, severe bilateral optic atrophy was evident. Visual evoked potentials were delayed bilaterally. Dramatic clinical and electrophysiologic improvements were recorded after vitamin supplementation. *Source: Spinazzi M, Angelini C, Patrini C. Subacute sensory ataxia and optic neuropathy with thiamine deficiency. Nat Rev Neurol. 2010;6:288–93.[12]*

the absence of the classical clinical triad. Reported risk factors predisposing to the development of thiamine deficiency optic neuropathy are alcoholism, bariatric surgery,[13] remote gastrectomy,[12] ketogenic diet,[14] parenteral hyperalimentation,[15] hyperemesis gravidarum,[16] prolonged vomiting,[17] intense physical activity, admission to an emergency room,[18] and rarely ingestion of food contaminated by thiaminases. The involvement of visual pathways has been demonstrated experimentally in thiamine-deficient rats.[19]

Patients harboring mutations in the *SLC19A2* gene, coding for the high-affinity thiamine transporter, which is mutated in thiamine-responsive megaloblastic anemia, do not reproduce the spectrum of manifestations of acquired thiamine deficiency but can develop optic atrophy.[20] A patient affected by pyruvate dehydrogenase deficiency caused by *PDHA1* gene mutations experienced thiamine-responsive peripheral neuropathy with optic atrophy.[21] A patient affected by Wolfram syndrome caused by a *WFS1* missense mutation was found to have a profound thiamine deficiency.[22] These observations demonstrate complex functional interactions between acquired nutritional factors and inherited disorders of metabolism.

Diagnosis and Therapy

The presence of bilateral optic neuropathy with optic disc edema, nystagmus, oculomotor palsy, or decreased consciousness should raise a high suspicion of thiamine deficiency, although these particular findings are not always present. Brain MRI can reveal typical diencephalic signal abnormalities in the mammillary bodies, dorsolateral thalami, and periaqueductal gray. The biochemical diagnosis is confirmed by decreased thiamine levels in whole blood, as determined by high-performance liquid chromatography,[23] or by decreased erythrocyte transketolase activity, which is rescued by the addition of thiamine diphosphate. Serum pyruvic acid levels can be increased owing to pyruvic dehydrogenase dysfunction.

Therapy comprises high-dose parenteral thiamine supplementation followed by chronic oral supplementation.

VITAMIN B$_{12}$ (COBALAMIN) DEFICIENCY

Pathophysiology of Cobalamin Deficiency

Vitamin B$_{12}$ is abundant in food derived from animals, such as meat, eggs, and milk. Therefore, vegans are particularly at risk of developing vitamin B$_{12}$ deficiency if they do not take appropriate supplements. Cobalamin is a critical cofactor of two enzymes (Fig. 39.3):

- Methionine synthase: catalyzes the conversion of homocysteine to methionine, which is converted to *S*-adenosylmethionine, required for methylation of different substrates, such as DNA, neurotransmitters, and lipids. The reaction leads to the conversion of 5-methyltetrahydrofolate into tetrahydrofolate, which is further converted into 5,10-methylenetetrahydrofolate, feeding both purine and thymidilate cycles, necessary for DNA synthesis. Therefore, vitamin B$_{12}$ deficiency results in a 'folate trap', a functional block of folate-dependent reactions.
- Methylmalonil-CoA mutase: a mitochondrial enzyme catalyzing the conversion of methylmalonil CoA into succinyl CoA, which can enter the Krebs cycle or the heme biosynthesis pathway.

Therefore, vitamin B$_{12}$ deficiency leads to an accumulation of the substrates homocysteine and methylmalonic acid. The pathophysiology of the disease may be due to neurotoxicity of these metabolites, but this has not been proven. An alternative, but not exclusive, hypothesis points to a deregulation of cytokine secretion.[24]

Vitamin B$_{12}$ deposits are sufficient to maintain vitamin B$_{12}$ levels for a number of years after dietary deprivation. Well-known conditions resulting in cobalamin deficiency include pernicious anemia, an autoimmune disorder associated with anti-intrinsic factor antibodies, malnutrition, 'veganism', malabsorption diseases, previous gastrointestinal surgery, and rarely exposure to

FIGURE 39.3 Vitamin B$_{12}$ and folate metabolism. The two vitamin B$_{12}$-dependent enzymes have different subcellular localization. Methymalonil coenzyme A (CoA) is a mitochondrial enzyme, while methionine synthase is cytosolic. 5-MTHF: 5-methyltetrahydrofolate; THF: tetrahydrofolate.

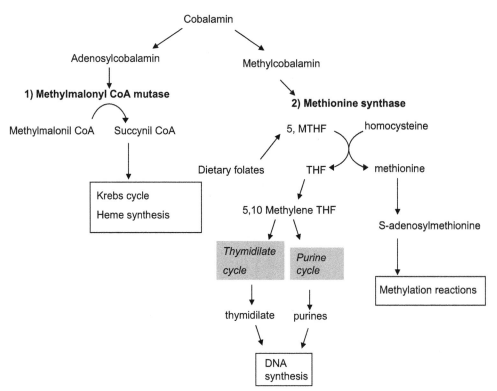

Clinical Manifestations of Cobalamin Deficiency

nitrous oxide, a gas used during surgical anesthesia that interferes with vitamin B$_{12}$ metabolism.

Clinical Manifestations of Cobalamin Deficiency

Vitamin B$_{12}$ has hematologic and neurologic clinical manifestations, which can occur simultaneously or independently. Anemia is absent in up to 30% of patients with neurologic involvement.[25] Therefore, it is an erroneous belief that the absence of megaloblastic anemia rules out a vitamin B$_{12}$ neurologic illness.

The different clinical signs of vitamin B$_{12}$ deficiency can occur in any combination:

- Megaloblastic anemia: it is fundamental to avoid inappropriate folate treatment of vitamin B$_{12}$-deficiency megaloblastic anemia. This intervention improves anemia through bypassing the folate metabolic trap but cannot prevent the neurologic deterioration.[26]
- Dorsolateral myelopathy, also called subacute combined sclerosis: characterized by a combined involvement of the pyramidal tracts and the dorsal columns of the spinal cord. Symptoms include sensory ataxia, paresthesias, and spasticity.
- Polyneuropathy: usually a sensory or sensorimotor axonal peripheral neuropathy.[27]
- Dementia.
- Optic neuropathy: symptomatic optic neuropathy due to isolated vitamin B$_{12}$ deficiency is very rare,

but subclinical involvement of the visual pathways is common and can be easily detected by visual evoked potentials. Prolonged P100 latencies, indicating delayed conduction velocities in the visual pathways, have been reported in up to 47.5% of patients affected by vitamin B$_{12}$ deficiency neurologic syndromes without any visual complaint.[28]

The first case of vitamin B$_{12}$ optic neuropathy was described by Bastianelli in 1897, in a patient affected by anemia and spastic paraplegia.[29] However, an etiopathologic link between vitamin B$_{12}$ deficiency and optic neuropathy was first postulated by Cohen in 1936. Further cases have since been reported,[29] although many of them lacked detailed biochemical studies ruling out other etiologies. However, in some of these cases, the reported dramatic response to vitamin B$_{12}$ supplementation strongly suggests a genuine etiopathogenic role of vitamin B$_{12}$ deficiency. In nine of 29 patients affected by vitamin B$_{12}$ deficiency optic neuropathy, the ophthalmic involvement was the first manifestation of the disease.[29] Moreover, in three autistic children with severe food selectivity, optic neuropathy was the only sign of cobalamin deficiency.[30]

The optic nerves are typically affected bilaterally and simultaneously, with central scotomas and dyschromatopsia. Fundus examination is normal or may show atrophy in the chronic phase.[29,31] Patients respond to vitamin B$_{12}$ supplementation.

Hereditary disorders of cobalamin metabolism can occasionally develop optic neuropathies,[32] demonstrating that a selective deficit of vitamin B_{12} can provoke optic nerve atrophy. Visual pathway degeneration with retinal ganglion cell loss has also been experimentally demonstrated in cobalamin-deficient monkeys.[33]

Diagnosis and Therapy

Low serum levels of vitamin B_{12} demonstrate a deficiency. Importantly, when cobalamin levels fall in the borderline range, the levels of homocysteine in the serum and methylmalonic acid in the 24-hour urine should be determined. Functional vitamin B_{12} deficiency is demonstrated by increased levels of these metabolites. Blood cell tests can detect a macrocytic anemia, and blood smear examination may reveal neutrophil hypersegmentation. In the presence of subacute combined sclerosis, spinal MRI will often, but not always, show a posterior column longitudinal abnormality in the spinal cord.

Visual evoked potentials will show a delay in the conduction velocities when optic nerve involvement is present. Importantly, the progression of the visual impairment despite appropriate supplementation, as well as significant asymmetry or complete blindness, should strongly suggest alternative diagnoses. In these cases, LHON should be ruled out using mitochondrial DNA analysis. Both conditions are characterized by degeneration of the papillomacular bundle. Moreover, vitamin B_{12} deficiency has been reported to precipitate the clinical signs in carriers of the LHON mutation.[34]

Treatment comprises cobalamin injections (1000 μg/day for 5 days) followed by chronic oral daily supplementation. The visual evoked potentials improve promptly with therapy.[28]

FOLATE DEFICIENCY

Pathophysiology of Folate Deficiency

Folate is a thermolabile B vitamin abundant in vegetables, fruit, legumes, and liver. It has a fundamental role in the methylation cycle necessary for nucleic acid and amino acid metabolism. Folate is inactivated by cooking food and its metabolism is hampered by several drugs, the most common of which are antiepileptic drugs (e.g., phenobarbital, phenytoin) and drugs for treating cancer (e.g., methotrexate, 5-fluorouracil). 5-Methyltetrahydrofolate provides the methyl group for the conversion of homocysteine to methionine, catalyzed by the cobalamin-dependent enzyme methionine synthase. Therefore, homocysteine levels are also elevated in folate deficiency. As already mentioned, folate and vitamin B_{12} metabolism are interdependent (Fig. 39.3), and vitamin B_{12} deficiency results in a folate trap, with an inability to convert folate in the chemical form that can be utilized for purine and thymidilate synthesis. The blocking of these metabolic pathways results in inhibition of DNA synthesis and, clinically, in megaloblastic anemia.

Clinical Manifestations of Folate Deficiency

Folate deficiency leads to both developmental defects and adult-onset clinical manifestations. Neurologic involvement is rare. Disorders related to folate deficiency include:

- Neural tube defects during embryogenesis.
- Macrocytic anemia.
- Dorsolateral myelopathy: this is a very rare complication of folate deficiency, mimicking subacute combined sclerosis.[35]
- Dementia.
- Optic neuropathy: pure folate deficiency optic neuropathy is extremely rare, and only a few cases have been reported. A 44-year-old smoking woman with poor dietary consumption of vegetables was reported to develop a folate-deficiency optic neuropathy and macrocytic anemia, which responded to a supplementation diet.[36] Six patients, habitual smokers and drinkers who developed bilateral optic neuropathy with folate deficiency, showed significant visual improvement after folate supplementation.[37] Methotrexate, an immunosuppressive agent interfering with folate metabolism, was reported to induce a folate-responsive optic neuropathy.[38]

The pathogenesis of folate deficiency optic neuropathy is unknown. Some authors have hypothesized that folate deficiency could lead to endogenous intoxication from formate,[37] which has been shown to be an inhibitor of cytochrome c oxidase of the mitochondrial respiratory chain.[39]

Surprisingly, to date, optic neuropathy does not appear to be a prominent sign of recently identified inherited disorders of folate metabolism leading to neurodegeneration, such as folate congenital malabsorption, due to *PCFT* mutations,[40] or cerebral folate transport deficit,[41] due to *FOLR1* mutations.

Diagnosis and Therapy

Determination of erythrocyte folate levels is more reliable than folate serum determination, which is significantly influenced by short-term dietary intakes. A positive response to folate supplementation therapy will support the diagnosis. Homocysteine levels are high during folate deficiency but return to normal after supplementation.

COPPER DEFICIENCY

Copper is an essential trace mineral present ubiquitously in foods but particularly in oysters, nuts, meat, and whole-grain products. The occurrence of copper deficiency in humans has been debated for decades,[42] but both hematologic and neurologic manifestations of copper deficiency have been clearly demonstrated.[43] Copper is an essential trace mineral, needed as a cofactor of several enzymes, including the antioxidant enzyme superoxide dismutase, cytochrome c oxidase of the mitochondrial respiratory chain, and the iron ferroxidases ceruloplasmin and hephaestin. Hematologic manifestations include anemia, variably associated with leukopenia and more rarely thrombocytopenia.

Causes of copper deficiency includes previous gastrointestinal surgery, malabsorption diseases, and zinc intoxication. Zinc can induce copper malabsorption by induction of metallothionein synthesis into the enterocytes. Intestinal metallothionein sequesters copper with high affinity and prevents its absorption into the portal circulation.

Deficiency of copper has recently been recognized to cause a spinal cord disease associated with myelopathy, which can be clinically identical to vitamin B_{12} deficiency subacute combined sclerosis.[44] The pathophysiology of copper deficiency myeloneuropathy involves impairment of copper-dependent proteins with critical functions in mitochondrial metabolism, antioxidant defenses, and iron metabolism.[45] Optic nerve involvement in copper deficiency has been described only very recently,[46–48] and optic nerve involvement was also demonstrated in copper-deficient rats.[49] Abnormal visual evoked potentials were detected in two of eight patients with copper deficiency where these data were available.[50] In one case, thinning of the retinal nerve fiber layer was demonstrated by OCT.[51]

Larger neurophthalmologic studies of copper-deficient patients are needed to better define the ocular involvement. Underlying causes of copper deficiency are variable and include previous gastric bypass[47] and surgery[46,51] and zinc intoxication from denture creams containing zinc.[46,48]

Diagnosis and Therapy

Diagnosis is based on the detection of reduced levels of serum copper and ceruloplasmin, a reliable marker of systemic stores of copper.

Oral copper treatment is effective in restoring normal copper levels in most cases, leading to a clear stop in the progression of disease. However, typically there is no significant regression of the neurologic damage, owing to underlying axonal degeneration.

EPIDEMIC NUTRITIONAL OPTIC NEUROPATHIES

Several epidemics of neurologic disturbances, variably including optic neuropathies, peripheral neuropathy, myelopathy, and hearing loss, have been reported in different populations during contingent generalized malnutrition, such as among World War II prisoners[52] and Cubans. Neuropathologic studies conducted in affected prisoners revealed papillomacular and fasciculus gracilis degeneration. Thousands of Cubans between 1991 and 1993 were affected by this illness, following sociopolitical events leading to food shortages. Highly prevalent deficiencies of thiamine, riboflavin, folate, and vitamin B_{12} were demonstrated.[53] Independent risk factors were tobacco smoking, high consumption of cassava (a root used as a carbohydrate source, containing cyanogenic glycosides), low serum levels of the antioxidant lycopene, low intakes of energy and methionine, and decreased availability of poultry.[54] Patients responded significantly to vitamin therapy,[55] demonstrating an etiologic role of micronutrient deficiencies. Moreover, the frequency of nutritional optic neuropathies dramatically decreased after the implementation in Cuba of a national campaign of multivitamin supplementation with B vitamins and retinol.

REDEFINITION OF TOBACCO–ALCOHOL AMBLYOPIA

The historic term 'tobacco–alcohol amblyopia' refers to the occurrence of bilateral optic neuropathy in smokers consuming alcoholic beverages. Pathologic features included retinal ganglion cell loss and papillomacular degeneration.[56] Patients were affected by subacute or chronic bilateral, painless visual loss of variable severity and color vision deficits. Fundus examination was normal or showed a bilateral optic atrophy. This terminology has recently been questioned, first because there is no real amblyopia and second because the condition is thought to be mainly of nutritional origin, rather than induced by alcohol abuse. Alcohol has no significant intrinsic toxicity to the retinal ganglion cells, but vitamin deficiencies are common in alcoholics affected by amblyopia.[57] Therefore, a more correct term for this condition, 'nutritional optic neuropathy', was proposed to replace 'tobacco–alcohol amblyopia'.[58] Consistent with this concept, patients dramatically improved with supplementation of vitamins of the B group, or sometimes even thiamine alone, despite uninterrupted drinking and smoking.[59] Moreover, the frequency of this condition, which was quite prevalent in the past, dropped dramatically after the introduction of fortified cereals in developed countries, despite little change in the consumption of alcohol and smoking. However, synergistic effects between smoking and vitamin deficiencies may also have been involved in the pathophysiology of this disease, as in the Cuban optic neuropathy outbreak. Many epidemic nutritional optic neuropathies (briefly described in the previous section) are likely to represent epidemic forms of 'tobacco–alcohol amblyopia' precipitated by unfavorable socioeconomic calamities. A recent case study reported significant remodeling of the retinal nerve fiber layer thickness evaluated by OCT and a remarkable response to vitamin B_{12} supplements and cessation of drinking and smoking.[60]

A different nosologic entity, termed tobacco optic neuropathy, refers to a now rare form of optic neuropathy caused by tabagism, which, for reasons unknown, was caused mainly by smoking pipes or cigars rather than cigarettes.[58] The pathophysiology is not known, but it has been proposed that it is related to impaired detoxification of cyanide released by smoking, although this would not explain the higher association with cigar smoking.[61] Cyanide is a powerful inhibitor of cytochrome c oxidase of the mitochondrial respiratory chain. However, the possibility of associated nutritional deficiencies as susceptibility factors cannot be ruled out. Vitamin B_{12} levels were found to be deficient in up to 26 of 65 patients affected by tobacco optic neuropathy.[62] Moreover, absorption of vitamin B_{12} was found to be reduced during smoking,[63] and treatment with hydroxycobalamin (a form of vitamin B_{12} used as an antidote for cyanide poisoning), but not cyanocobalamin, was often reported effective.

Currently, technologic medical innovations are enabling a much more precise diagnostic definition of nutritional optic neuropathies. Specific biochemical tests allow the underlying micronutrient deficiencies to be fully defined, sophisticated structural quantifications of the visual pathways to be performed, and many alternative diagnoses with a similar clinical presentation, which may have been missed in the past, to be ruled out. Hereditary optic neuropathies, among others (Table 39.3), should be considered in the differential diagnosis, owing to some clinical similarities. Intriguingly, smoking was found to be a significant risk factor for visual loss in patients harboring LHON mutations,[64] and four of 26 patients with 'tobacco alcohol amblyopia' were found to have LHON mutations,[65] although these mutations were not frequent in the Cuban population affected by optic neuropathy epidemics.[66] Future studies may be expected to discover several genetic variations as factors predisposing to the development of nutritional optic neuropathies.

TAKE-HOME MESSAGES

- Optic nerves are susceptible to mitochondrial dysfunction. Micronutrients implicated in nutritional optic neuropathies are fundamental cofactors of enzymes involved in mitochondrial metabolism.
- The presence of unexplained, painless, bilateral visual loss with central scotomas should prompt investigations into a nutritional cause.
- Nutritional optic neuropathies are commonly multifactorial diseases.
- Nutritional optic neuropathies are almost invariably multisystem diseases and may be life threatening.
- The diagnosis is supported by a consistent clinical presentation, appropriate biochemical tests, exclusion of alternative diagnoses, and positive response to supplementation therapy.
- Hereditary optic neuropathies rank high in the list of the differential diagnoses and could also be precipitated by nutritional deficiencies in some cases.
- Timely recognition is fundamental to providing appropriate therapy and preventing deleterious disease progression.

References

1. Dastur DK, Santhadevi N, Quadros EV, et al. The B-vitamins in malnutrition with alcoholism. A model of intervitamin relationships. *Br J Nutr* 1976;**36**:143–59.
2. Allen LH. How common is vitamin B-12 deficiency? *Am J Clin Nutr* 2009;**89**:693–6S.
3. Harper C. The incidence of Wernicke's encephalopathy in Australia – a neuropathological study of 131 cases. *J Neurol Neurosurg Psychiatry* 1983;**46**:593–8.
4. Francini-Pesenti F, Brocadello F, Manara R, Santelli L, Laroni A, Caregaro L. Wernicke's syndrome during parenteral feeding: not an unusual complication. *Nutrition* 2009;**25**:142–6.
5. Ernst B, Thurnheer M, Schmid SM, Schultes B. Evidence for the necessity to systematically assess micronutrient status prior to bariatric surgery. *Obes Surg* 2009;**19**:66–73.
6. Juhasz-Pocsine K, Rudnicki SA, Archer RL, Harik SI. Neurologic complications of gastric bypass surgery for morbid obesity. *Neurology* 2007;**68**:1843–50.
7. Koffman BM, Greenfield LJ, Ali II, Pirzada NA. Neurologic complications after surgery for obesity. *Muscle Nerve* 2006;**33**:166–76.
8. Ryan AS, Goldsmith LA. Nutrition and the skin. *Clin Dermatol* 1996;**14**:389–406.
9. Votruba M, Leary S, Losseff N, et al. MRI of the intraorbital optic nerve in patients with autosomal dominant optic atrophy. *Neuroradiology* 2000;**42**:180–3.
10. Lamirel C, Newman N, Biousse V. The use of optical coherence tomography in neurology. *Rev Neurol Dis* 2009;**6**:E105–20.
11. Sechi G, Serra A. Wernicke's encephalopathy: new clinical settings and recent advances in diagnosis and management. *Lancet Neurol* 2007;**6**:442–55.
12. Spinazzi M, Angelini C, Patrini C. Subacute sensory ataxia and optic neuropathy with thiamine deficiency. *Nat Rev Neurol* 2010;**6**:288–93.
13. Kulkarni S, Lee AG, Holstein SA, Warner JEA. You are what you eat. *Surv Ophthalmol* 2005;**50**:389–93.
14. Hoyt CS, Billson FA. Optic neuropathy in ketogenic diet. *Br J Clin Ophthalmol* 1979;**63**:191–4.

15. Suzuki S, Kumanomido T, Nagata E, Inoue J, Niikawa O. Optic neuropathy from thiamine deficiency. *Intern Med* 1997;**36**:532.

16. Tesfaye S, Achari V, Yang YC, Harding S, Bowden A, Vora JP. Pregnant, vomiting, and going blind. *Lancet* 1998;**352**:1594.

17. Cooke CA, Hicks E, Page AB, McKinstry S. An atypical presentation of Wernicke's encephalopathy in an 11-year-old child. *Eye (Lond)* 2006;**20**:1418–20.

18. Cruickshank AM, Telfer AB, Shenkin A. Thiamine deficiency in the critically ill. *Intens Care Med* 1988;**14**:384–7.

19. Rodger FC. Experimental thiamin deficiency as a cause of degeneration in the visual pathway of the rat. *Br J Clin Ophthalmol* 1953;**37**:11–29.

20. Borgna-Pignatti C, Azzalli M, Pedretti S. Thiamine-responsive megaloblastic anemia syndrome: long term follow-up. *J Pediatr* 2009;**155**:295–7.

21. Sedel F, Challe G, Mayer J-M, et al. Thiamine responsive pyruvate dehydrogenase deficiency in an adult with peripheral neuropathy and optic neuropathy. *J Neurol Neurosurg Psychiatry* 2008;**79**:846–7.

22. Lieber DS, Vafai SB, Horton LC, et al. Atypical case of Wolfram syndrome revealed through targeted exome sequencing in a patient with suspected mitochondrial disease. *BMC Med Genet* 2012;**13**:3.

23. Lu J, Frank EL. Rapid HPLC measurement of thiamine and its phosphate esters in whole blood. *Clin Chem* 2008;**54**:901–6.

24. Scalabrino G. Cobalamin (vitamin B(12)) in subacute combined degeneration and beyond: traditional interpretations and novel theories. *Exp Neurol* 2005;**192**:463–79.

25. Healton EB, Savage DG, Brust JC, Garrett TJ, Lindenbaum J. Neurologic aspects of cobalamin deficiency. *Medicine* 1991;**70**:229–45.

26. Goodman BP, Bosch EP, Ross MA, Hoffman-Snyder C, Dodick DD, Smith BE. Clinical and electrodiagnostic findings in copper deficiency myeloneuropathy. *J Neurol Neurosurg Psychiatry* 2009;**80**:524–7.

27. Fine EJ, Soria E, Paroski MW, Petryk D, Thomasula L. The neurophysiological profile of vitamin B_{12} deficiency. *Muscle Nerve* 1990;**13**:158–64.

28. Puri V, Chaudhry N, Goel S, Gulati P, Nehru R, Chowdhury D. Vitamin B_{12} deficiency: a clinical and electrophysiological profile. *Electromyogr Clin Neurophysiol* 2005;**45**:273–84.

29. Hamilton HE, Ellis PP, Sheets RF. Visual impairment due to optic neuropathy in pernicious anemia: report of case and review of the literature. *Blood* 1959;**14**:378–85.

30. Pineles SL, Avery RA, Liu GT. Vitamin B_{12} optic neuropathy in autism. *Pediatrics* 2010;**126**:967–70.

31. Enoksson P, Norden A. Vitamin B_{12} deficiency affecting the optic nerve. *Acta Med Scand* 1960;**167**:199–208.

32. Patton N, Beatty S, Lloyd IC, Wraith JE. Optic atrophy in association with cobalamin C (cblC) disease. *Ophthalmic Genet* 2000;**21**:151–4.

33. Chester EM, Agamanolis DP, Harris JW, Victor M, Hines JD, Kark JA. Optic atrophy in experimental vitamin B_{12} deficiency in monkeys. *Acta Neurol Scand* 1980;**61**:9–26.

34. Pott JWR, Wong KH. Leber's hereditary optic neuropathy and vitamin B_{12} deficiency. *Graefes Arch Clin Exp Ophthalmol* 2006;**244**:1357–9.

35. Manzoor M, Runcie J. Folate-responsive neuropathy: report of 10 cases. *BMJ* 1976;**1**:1176–8.

36. de Silva P, Jayamanne G, Bolton R. Folic acid deficiency optic neuropathy: a case report. *J Med Case Rep* 2008;**2**:299.

37. Golnik KC, Schaible ER. Folate-responsive optic neuropathy. *J Neuroophthalmol* 1994;**14**:163–9.

38. Clare G, Colley S, Kennett R, Elston JS. Reversible optic neuropathy associated with low-dose methotrexate therapy. *J Neuroophthalmol* 2005;**25**:109–12.

39. Nicholls P. Formate as an inhibitor of cytochrome c oxidase. *Biochem Biophys Res Commun* 1975;**67**:610–6.

40. Qiu A, Jansen M, Sakaris A, et al. Identification of an intestinal folate transporter and the molecular basis for hereditary folate malabsorption. *Cell* 2006;**127**:917–28.

41. Steinfeld R, Grapp M, Kraetzner R, et al. Folate receptor alpha defect causes cerebral folate transport deficiency: a treatable neurodegenerative disorder associated with disturbed myelin metabolism. *Am J Hum Genet* 2009;**85**:354–63.

42. Graham GG. Human copper deficiency. *N Engl J Med* 1971;**285**:857–8.

43. Jaiser SR, Winston GP. Copper deficiency myelopathy. *J Neurol* 2010;**257**:869–81.

44. Kumar N, Gross JB, Ahlskog JE. Copper deficiency myelopathy produces a clinical picture like subacute combined degeneration. *Neurology* 2004;**63**:33–9.

45. Spinazzi M, Sghirlanzoni A, Salviati L, Angelini A. Impaired copper and iron metabolism in blood cells and muscles of patients affected by copper deficiency myeloneuropathy. *Neuropathology and Applied Neurobiology* 2013. http://dx.doi.org/10.1111/nan.12111.

46. Spinazzi M, De Lazzari F, Tavolato B, Angelini C, Manara R, Armani M. Myelo-optico-neuropathy in copper deficiency occurring after partial gastrectomy. Do small bowel bacterial overgrowth syndrome and occult zinc ingestion tip the balance? *J Neurol* 2007;**254**:1012–7.

47. Naismith RT, Shepherd JB, Weihl CC, Tutlam NT, Cross AH. Acute and bilateral blindness due to optic neuropathy associated with copper deficiency. *Arch Neurol* 2009;**66**:1025–7.

48. Khaleeli Z, Healy DG, Briddon A, et al. Copper deficiency as a treatable cause of poor balance. *BMJ* 2010;**340**:c508.

49. Dake Y, Amemiya T. Electron microscopic study of the optic nerve in copper deficient rats. *Exp Eye Res* 1991;**52**:277–81.

50. Kumar N. Copper deficiency myelopathy (human swayback). *Mayo Clin Proc* 2006;**81**:1371–84.

51. Pineles SL, Wilson CA, Balcer LJ, Slater R, Galetta SL. Combined optic neuropathy and myelopathy secondary to copper deficiency. *Surv Ophthalmol* 2010;**55**:386–92.

52. Gill GV, Bell DR. Persisting nutritional neuropathy amongst former war prisoners. *J Neurol Neurosurg Psychiatry* 1982;**45**:861–5.

53. Arnaud J, Fleites-Mestre P, Chassagne M, et al. Vitamin B intake and status in healthy Havanan men, 2 years after the Cuban neuropathy epidemic. *Br J Nutr* 2001;**85**:741–8.

54. Cuba Neuropathy Field Investigation Team. Epidemic optic neuropathy in Cuba – clinical characterization and risk factors. *N Engl J Med* 1995;**333**:1176–82.

55. Sadun AA, Martone JF, Muci-Mendoza R, et al. Epidemic optic neuropathy in Cuba. Eye findings. *Arch Ophthalmol* 1994;**112**:691–9.

56. Victor M, Dreyfus PM. Tobacco–alcohol amblyopia. Further comments on its pathology. *Arch Ophthalmol* 1965;**74**:649–57.

57. Dreyfus PM. Blood transketolase levels in tobacco–alcohol amblyopia. *Arch Ophthalmol* 1965;**74**:617–20.

58. Grzybowski A, Holder GE. Tobacco optic neuropathy (TON) – the historical and present concept of the disease. *Acta Ophthalmol* 2011;**89**:495–9.

59. Carroll FD. Nutritional amblyopia. *Arch Ophthalmol* 1966;**76**:406–11.

60. Bhatnagar A, Sullivan C. Tobacco–alcohol amblyopia: can OCT predict the visual prognosis? *Eye (Lond)* 2009;**23**:1616–8.

61. Foulds WS, Chisholm IA, Pettigrew AR. The toxic optic neuropathies. *Br J Clin Ophthalmol* 1974;**58**:386–90.

62. Foulds WS, Chisholm IA, Brontë-Stewart J, Wilson TM. Vitamin B_{12} absorption in tobacco amblyopia. *Br J Clin Ophthalmol* 1969;**53**:393–7.

63. Watson-Williams EJ, Bottomley AC, Ainley RG, Phillips CI. Absorption of vitamin B_{12} in tobacco amblyopia. *Br J Clin Ophthalmol* 1969;**53**:549–52.

64. Kirkman MA, Yu-Wai-Man P, Korsten A, et al. Gene–environment interactions in Leber hereditary optic neuropathy. *Brain* 2009;**132**:2317–26.

65. Amaral-Fernandes MS, Marcondes AM. Miranda PM do AD, Maciel-Guerra AT, Sartorato EL. Mutations for Leber hereditary optic neuropathy in patients with alcohol and tobacco optic neuropathy. *Mol Vision* 2011;**17**:3175–9.

66. Newman NJ, Torroni A, Brown MD, Lott MT, Fernandez MM, Wallace DC. Epidemic neuropathy in Cuba not associated with mitochondrial DNA mutations found in Leber's hereditary optic neuropathy patients. Cuba Neuropathy Field Investigation Team. *Am J Ophthalmol* 1994;**118**:158–68.

Space Flight Ophthalmic Changes, Diet, and Vitamin Metabolism

S.R. Zwart[1], C.R. Gibson[2], S.M. Smith[3]

[1]Division of Space Life Sciences, Universities Space Research Association, Houston, Texas, USA, [2]Wyle Science, Technology and Engineering, Houston, Texas, USA, and Coastal Eye Associates, Webster, Texas, USA, [3]Biomedical Research and Environmental Sciences Division (MC SK3), NASA Johnson Space Center, Houston, Texas, USA

INTRODUCTION

The harsh space flight environment affects many physiologic systems, one of which is vision and ocular health. Some of the environmental aspects of space flight that can contribute to changes in ocular health are radiation exposure, cephalad shifts of body fluid, spacecraft cabin and spacesuit gas mixtures, and the space flight food system. These stressors contribute to an increased risk of cataracts among astronauts, and changes in vision accompanied by ophthalmic changes such as optic disc edema, globe flattening, hyperopic shifts, choroidal folds, and cotton wool spots have recently been observed in astronauts after long-duration space flight. These issues were first described in 2011, and the potential implications for exploration missions beyond low Earth orbit are significant. The etiology of the refractive and structural ophthalmic changes is not known and continues to be researched, but biochemical evidence indicates that the folate- and vitamin B_{12}-dependent one-carbon transfer pathway may be involved. This chapter reviews the available literature on this topic, along with planned research to understand and counteract these effects of space flight.

CATARACTS

Cataracts are opacities of the lens and have a multifactorial etiology. Diet, genetics, and environmental stressors can all play a role in initiating oxidative damage that can lead to cataract formation. Several studies have confirmed that astronauts and cosmonauts have an increased risk of cataract formation after space flight (Fig. 40.1).[1-4] In Figure 40.1, the low or high radiation dose classification was determined by flight duration, altitude, and destination of the mission.[4] Cucinotta et al.[2] identified an increased risk of all types of cataracts (including posterior subcapsular, cortical, and nuclear) among astronauts with higher exposure to radiation. Longitudinal follow-up studies have been conducted, and it was determined that progression of cortical cataracts, but not posterior subcapsular or nuclear cataracts, is related to space radiation exposure.[3,5] Although radiation exposure is a large driving force for the oxidative damage that leads to some types of cataracts, the longitudinal study provided evidence that intake of specific nutrients may provide some protective effects.[3] In the first report of the National Aeronautics and Space Administration (NASA) Study of Cataract in Astronauts, nutritional intake estimates were obtained from a questionnaire, and the data provided evidence that β-carotene and lycopene intake had a protective effect for some types of cataracts in astronauts.[3] As reviewed by Agte and Tarwadi, numerous ground-based studies have provided evidence for associations between micronutrients and antioxidants (either blood levels or estimated intakes) and cataracts.[6] It is not known whether altered micronutrient and antioxidant intake during space flight could minimize cataract incidence related to space radiation, as this requires further study and better estimates of in-flight nutrient intake and nutrient status assessments.

Handbook of Nutrition, Diet, and the Eye
http://dx.doi.org/10.1016/B978-0-12-401717-7.00040-X

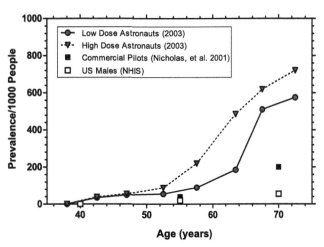

FIGURE 40.1 Prevalence of cataracts as a function of age in astronauts (exposed to a low or high dose of radiation), pilots, and healthy US males. The dose of radiation (low vs. high) was determined by the duration, altitude, and destination of the space flight mission. *Source: Reprinted from Jones et al. (2007)[4] with permission from* Aviation, Space, and Environmental Medicine.

VISION CHANGES AFTER LONG-DURATION SPACE FLIGHT

Long-term changes in vision among astronauts during and after long-duration space flight have recently been reported. Mader and colleagues described seven cases among long-duration crew members on the International Space Station (ISS) who had evidence of ophthalmic changes after flight, including optic disc edema, globe flattening, choroidal folds, and hyperopic shifts.[7] Myasnikov and Stepanova reported evidence of postflight edema of the optic nerve discs among cosmonauts and one case (out of 10) with signs of intracranial hypertension, although they note that the measurements were made before and after (not during) flight.[8] The etiology of the ophthalmic changes is not known, but microgravity-induced fluid shifts,[9,10] possible intraocular pressure (IOP) or intracranial pressure changes, and local intraorbital (choroidal and optic nerve sheath) changes have been suggested as possible contributing factors.

Evidence exists that IOP may be elevated during space flight. The evidence comes from parabolic flight, bed rest, and orbital flights.[11,12] As reviewed by Manuel and Mader, short-duration microgravity studies suggest that the initial rise in IOP may decline to roughly preflight values within several hours.[13] Data from long-duration missions are limited, so it is unknown how long the elevated pressure continues on those missions. The clinical significance of the hypothesized altered IOP during space flight is not yet known.

Fluid shifts and IOP may in turn be affected by increased carbon dioxide (CO_2) concentrations in the ISS atmosphere, and even potential effects of resistance exercise on headward fluid shifts. A large question that remains, however, is why some crew members experience changes and some do not, even though some flew on the same 6-month mission and any environmental factors would have been identical. Recent biochemical evidence that the folate- and vitamin B_{12}-dependent one-carbon transfer pathway may be involved could help to explain individual susceptibility to these ocular changes.

B-VITAMIN-DEPENDENT ONE-CARBON TRANSFER PATHWAY

Of the original seven cases described by Mader and colleagues,[7] five participated in a nutritional biochemistry surveillance study in which fasting blood samples were collected before, during, and after flight, and among other analytes, folate and intermediates of the folate and vitamin B_{12}-dependent one-carbon transfer pathway were measured.[14] Fifteen long-duration crew members who did not experience ophthalmic changes also participated in the study. Serum homocysteine, cystathionine, 2-methylcitric acid, and methylmalonic acid were significantly higher in the five crew members who had experienced ophthalmic changes during space flight than in the 15 crew members who had not experienced any changes (Fig. 40.2) and were higher before, during, and after flight. The mean values of all four biochemical intermediates in all crew members in the study were within the normal clinical range.

Homocysteine is an intermediate of a one-carbon transfer pathway in which methionine is formed in a methylation reaction catalyzed by methionine synthase (5-methyltetrahydrofolate homocysteine methyltransferase; EC 2.1.113) (Fig. 40.3). Vitamin B_{12} is a required cofactor, and folate (N_5-methyl-tetrahydrofolate) is a substrate in the same reaction. Homocysteine can also be condensed with cysteine and converted to cystathionine through the transsulfuration pathway, which is catalyzed by the enzyme cystathionine β-synthase (CBS; EC 4.2.1.22) and requires vitamin B_6 as a cofactor. Vitamin B_{12} is a required cofactor for methylmalonyl-coenzyme A (CoA) mutase (EC 5.4.99.2), which converts methylmalonyl CoA to succinyl CoA.

The intermediates homocysteine, cystathionine, 2-methylcitric acid, and methylmalonic acid can be elevated in conjunction with nutritional deficiencies, specifically folate, vitamin B_6, and vitamin B_{12} deficiencies.[16,17] Frank nutritional deficiencies were ruled out for all five crew members in the biochemical surveillance study who had ophthalmic changes, because all of them consumed a standardized menu during their 5–6-month mission, and the standardized menu for the ISS is known to contain adequate amounts of these B vitamins to meet dietary reference intake (DRI) requirements.[9] Serum

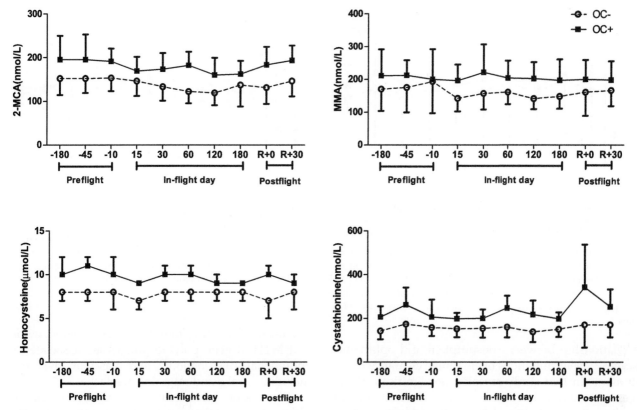

FIGURE 40.2 Serum concentrations of metabolites of the vitamin B_{12}-dependent one-carbon pathway before, during, and after long-duration space flight in crew members who had (OC+) or did not have (OC−) ophthalmic changes after flight. 2-MCA: 2-methylcitric acid; MMA: methylmalonic acid; R+0: landing day; R+30: 30 days after landing. 2-MCA, MMA, homocysteine, and cystathionine all had significant group effects ($P < 0.001$) in that metabolite concentrations were lower in the OC− group than in the OC+ group. 2-MCA also had a significant space flight effect (R+0 < preflight, $P < 0.05$; in-flight < preflight, $P < 0.05$). *Source: Data graphed from Table 1 in Zwart et al. (2012).*[14]

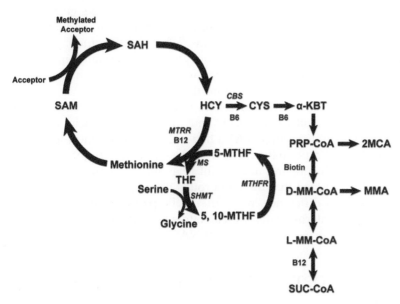

FIGURE 40.3 One-carbon transfer pathway. B6: vitamin B_6; B12: vitamin B_{12}; CBS: cystathionine β-synthase; CYS: cystathionine; HCY: homocysteine; α-KBT: α-ketobutyrate; MCA: methylcitric acid; MM-CoA: methylmalonyl CoA (D and L configurations); MMA: methylmalonic acid; MS: methionine synthase; 5-MTHF: 5-methyltetrahydrofolate; 5,10-MTHF: 5,10-methylenetetrahydrofolate; MTHFR: methylenetetrahydrofolate reductase; MTRR: 5-methyltetrahydrofolate homocysteine methyltransferase reductase; PRP-CoA: propionyl coenzyme A; SAH: S-adenosylhomocysteine; SAM: S-adenosylmethionine; SHMT: serine hydroxymethyltransferase; SUC-CoA: succinyl coenzyme A; THF: tetrahydrofolate. *Source: Adapted from Herrmann et al. (2001).*[15]

vitamin B_{12} was measured in conjunction with methylmalonic acid, and there were no differences between crew members with ophthalmic issues and those without them (Fig. 40.4). Furthermore, one of the five affected crew members reported taking a daily multivitamin supplement during the mission. Other dietary factors, such as coffee consumption, that could affect homocysteine concentration were investigated, and there was

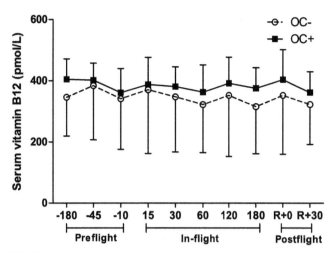

FIGURE 40.4 Serum vitamin B_{12} before, during, and after long-duration space flight in crew members who had (OC+) or did not have (OC−) ophthalmic changes after flight. Vitamin B_{12} was analyzed using a radioassay (Solid Phase No Boil Dualcount; Siemens Healthcare Diagnostics, Tarrytown, NY, USA). *Source: Data from Zwart et al. (2012).[14]*

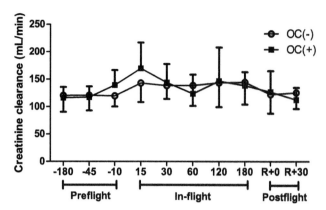

FIGURE 40.5 Creatinine clearance before, during, and after long-duration space flight in crew members who had (OC+) or did not have (OC−) ophthalmic changes after flight. A significant time effect was noted with a repeated-measures two-way analysis of variance, and a Bonferroni *post hoc* test revealed that creatinine clearance on flight day 15 was faster than at other time-points ($P < 0.05$). *Source: Data from Zwart et al. (2012).[14]*

no relationship between consumption and ophthalmic changes.[14] Determinations of creatinine clearance by the astronauts in that study revealed no significant differences in renal function between those who had ophthalmic changes after flight and those who did not (Fig. 40.5).

Besides diet and renal function, other factors can influence homocysteine, cystathionine, 2-methylcitric acid, and methylmalonic acid. Mild hyperhomocysteinemia can be produced by genetic variants of one or more of a number of enzymes involved in the pathway. These polymorphisms are common and result in a moderate elevation of homocysteine, whereas some rare mutations can cause severe inherited homocysteinemia. One of the most studied polymorphisms in this pathway is *MTHFR* 677C→T (a C→T substitution at base pair 677 of the gene, leading to a change from alanine to valine in the MTHFR enzyme protein), which has an allele frequency of about 30% in many ethnic groups.[18] Europeans and Russians have genetic gradients, with the T allele frequency ranging from 27% in Russians to 46% in some regions of Italy.[18] Heterozygotes (C/T) have a 30% decrease in MTHFR enzyme activity and homozygotes (T/T) a 60% decrease[19] in activity of this critical enzyme. The metabolic consequences of this reduced enzyme activity are most frequently observed in situations of moderately low folate nutritional status.[20] A higher intake of folate in individuals with this polymorphism can result in homocysteine values that are lower and within the normal clinical range.[19,20] Plasma homocysteine increased from $7.1 \pm 2.1\,\mu mol/L$ to $10.5 \pm 3.3\,\mu mol/L$ after 7 weeks of folate depletion in young women with the TT genotype for MTHFR.[21] Polymorphisms of enzymes involved in the one-carbon transfer pathway are associated with greater risk for several conditions of vascular dysfunction. Ground-based studies provide evidence that MTHFR gene polymorphisms are associated with endothelial dysfunction[22] and vascular dysfunction, including stroke and migraine headaches.[23–25]

When serum folate of ISS crew members who did or did not have ophthalmic changes was compared during flight (the preflight period was not analyzed because preflight diets were not standardized, and it is known that serum folate can be affected by recent folate intake), the five crew members who experienced ophthalmic changes had significantly lower serum folate than the 15 who did not have changes.[14] Some studies have suggested that folate requirements may be higher for individuals with MTHFR polymorphisms.[21,26] In the ISS study, preflight concentrations of homocysteine and cystathionine were significantly correlated with the refractive change in the right eye from before flight to after flight (Fig. 40.6).[14] Because the differences in homocysteine, cystathionine, 2-methylcitric acid, and methylmalonic acid between crew members who had ocular changes and those who did not have them were present before flight as well as during and after flight (Fig. 40.2), the authors hypothesized that the two groups of crew members may have a difference in the metabolic pathway involving these intermediates. Research to evaluate this in more detail is currently underway at NASA.

OPTIC NEUROPATHY AND B VITAMINS

Optic neuropathies caused by deficiencies in some B vitamins have been documented. Vitamin B_{12} or folate deficiencies, even those that are detected by screening of asymptomatic populations, can cause optic neuropathies such as optic disc swelling and vision impairment.[27–30]

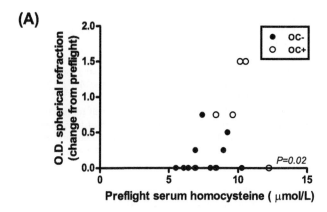

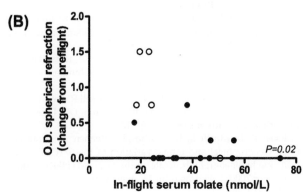

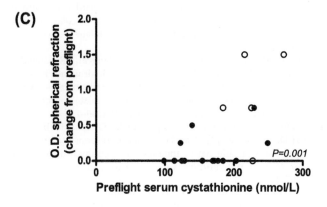

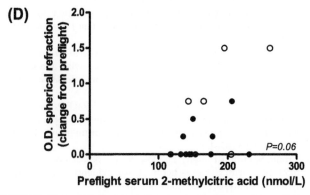

FIGURE 40.6 Associations before or during flight of serum homocysteine (A), folate (B), cystathionine (C), and 2-methylcitric acid (D) with absolute refractive change in the right eye after flight (relative to before flight) in crew members who had (OC+) or did not have (OC−) ophthalmic changes after flight. *Source: Reprinted from Zwart* et al. *(2012)[14] with permission from the* Journal of Nutrition.

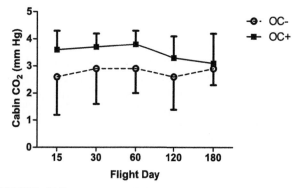

FIGURE 40.7 International Space Station cabin carbon dioxide (CO_2) levels during long-duration missions of crew members who had (OC+) or did not have (OC−) ophthalmic changes after flight. A significant group effect ($P < 0.05$) was noted on analysis by a repeated-measures two-way analysis of variance. *Source: Data graphed from Table 1 in Zwart* et al. *(2012)[14].*

Similarly, antifolate drugs such as methotrexate can cause side effects resulting in reversible optic neuropathy.[31,32] Patients taking methotrexate who have another insult to the metabolic pathway such as a dietary deficiency or polymorphism in the one-carbon transfer pathway have a decrease in serum folate.[33,34]

ENVIRONMENTAL FACTORS

The crew members who experienced ophthalmic changes were exposed to higher cabin CO_2 levels (Fig. 40.7), but it is important to note that overlap occurred, in that on some missions one crew member experienced ophthalmic changes and another on the same mission did not. Therefore, the cabin conditions cannot be the only causative factor. It is possible that metabolic or genetic differences in the one-carbon transfer pathway may influence anatomic or physiologic susceptibility to environmental stressors such as CO_2. Individuals with a polymorphism in an enzyme involved in the one-carbon transfer pathway may be more susceptible to vascular dysfunction if they are exposed to an environmental stressor. Another study in an extreme environment found this to be true. In divers with polymorphisms of the one-carbon transfer pathway discussed here, the response to increased atmospheric pressure was an increased risk for decompression sickness, also a vascular condition.[35]

TAKE-HOME MESSAGES

- Space flight can affect vision and risks to ophthalmic health in many ways.
- Cataract incidence is increased by space flight, secondary to radiation and oxygen exposure,

and this may be interactively affected by specific nutrients.

- Changes in vision and other ophthalmic changes, believed to be secondary to increased intracranial pressure and/or local intraorbital changes, have recently been identified in some astronauts after 4–6-month missions on the International Space Station.
- Alterations in folate and biochemical analytes associated with the vitamin B_{12}-dependent one-carbon transfer pathway have been noted in astronauts who had vision-related changes, and these differences were evident before as well as during and after flight.
- Polymorphisms of enzymes involved in folate metabolism and the B_{12}-dependent one-carbon transfer pathway have been associated with increased risk of stroke and migraine headaches, suggesting that they cause increased intracranial pressure. Furthermore, optic neuropathy has been associated with changes in folate and vitamin B_{12} metabolism.
- As with virtually all studies of human physiology in space flight, the implications for applying our understanding of space flight vision changes to Earth-based populations could be significant.

References

1. Rastegar N, Eckart P, Mertz M. Radiation-induced cataract in astronauts and cosmonauts. *Graefes Arch Clin Exp Ophthalmol* 2002;**240**:543–7.
2. Cucinotta FA, Manuel FK, Jones J, Iszard G, Murrey J, Djojonegro B, Wear M. Space radiation and cataracts in astronauts. *Radiat Res* 2001;**156**:460–6.
3. Chylack Jr LT, Peterson LE, Feiveson AH, Wear ML, Manuel FK, Tung WH, et al. NASA study of cataract in astronauts (NASCA). Report 1: Cross-sectional study of the relationship of exposure to space radiation and risk of lens opacity. *Radiat Res* 2009;**172**:10–20.
4. Jones JA, McCarten M, Manuel K, Djojonegoro B, Murray J, Feiversen A, Wear M. Cataract formation mechanisms and risk in aviation and space crews. *Aviat Space Environ Med* 2007;**78**(4 Suppl):A56–66.
5. Chylack Jr LT, Feiveson AH, Peterson LE, Tung WH, Wear ML, Marak LJ, et al. NASCA Report 2: Longitudinal study of relationship of exposure to space radiation and risk of lens opacity. *Radiat Res* 2012;**178**:25–32.
6. Agte V, Tarwadi K. The importance of nutrition in the prevention of ocular disease with special reference to cataract. *Ophthalmic Res* 2010;**44**:166–72.
7. Mader TH, Gibson CR, Pass AF, Kramer LA, Lee AG, Fogarty J, et al. Optic disc edema, globe flattening, choroidal folds, and hyperopic shifts observed in astronauts after long-duration space flight. *Ophthalmology* 2011;**118**:2058–69.
8. Mayasnikov VI, Stepanova SI. Features of cerebral hemodynamics in cosmonauts before and after flight on the MIR Orbital Station. *Orbital Station MIR Institute of Biomedical Problems: State Scientific Center of Russian Federation* 2008:300–5.
9. Smith SM, Zwart SR, Kloeris V, Heer M. *Nutritional Biochemistry of Space Flight*. New York: Nova Science Publishers; 2009.
10. Smith SM, Krauhs JM, Leach CS. Regulation of body fluid volume and electrolyte concentrations in spaceflight. *Adv Space Biol Med* 1997;**6**:123–65.
11. Mader TH, Gibson CR, Caputo M, Hunter N, Taylor G, Charles J, Meehan RT. Intraocular pressure and retinal vascular changes during transient exposure to microgravity. *Am J Ophthalmol* 1993;**115**:347–50.
12. Draeger J, Schwartz R, Groenhoff S, Stern C. Self-tonometry under microgravity conditions. *Aviat Space Environ Med* 1995;**66**:568–70.
13. Manuel FK, Mader TH. Ophthalmologic considerations. In: Barratt MR, Pool SL, editors. *Principles of Clinical Medicine for Space Flight*. New York: Springer; 2008. pp. 535–44.
14. Zwart SR, Gibson CR, Mader TH, Ericson K, Ploutz-Snyder R, Heer M, Smith SM. Vision changes after spaceflight are related to alterations in folate- and vitamin B_{12}-dependent one-carbon metabolism. *J Nutr* 2012;**142**:427–31.
15. Herrmann W, Schorr H, Purschwitz K, Rassoul F, Richter V. Total homocysteine, vitamin B(12), and total antioxidant status in vegetarians. *Clin Chem* 2001;**47**:1094–101.
16. Bailey LB, Gregory III JF. Folate. In: Bowman BA, Russell RM, editors. *Present Knowledge in Nutrition*. 9th ed. Washington, DC: International Life Sciences Institute; 2006. pp. 278–301.
17. Stabler SP. Vitamin B_{12}. In: Bowman BA, Russell RM, editors. *Present Knowledge in Nutrition*. 9th ed. Washington, DC: International Life Sciences Institute; 2006. pp. 302–13.
18. Wilcken B, Bamforth F, Li Z, Zhu H, Ritvanen A, Renlund M, et al. Geographical and ethnic variation of the 677C>T allele of 5,10 methylenetetrahydrofolate reductase (MTHFR): findings from over 7000 newborns from 16 areas world wide. *J Med Genet*. 2003;**40**:619–25.
19. Frosst P, Blom HJ, Milos R, Goyette P, Sheppard CA, Matthews RG, et al. A candidate genetic risk factor for vascular disease: a common mutation in methylenetetrahydrofolate reductase. *Nat Genet*. 1995;**10**:111–3.
20. Davis SR, Quinlivan EP, Shelnutt KP, Maneval DR, Ghandour H, Capdevila A, et al. The methylenetetrahydrofolate reductase 677C→T polymorphism and dietary folate restriction affect plasma one-carbon metabolites and red blood cell folate concentrations and distribution in women. *J Nutr* 2005;**135**:1040–4.
21. Shelnutt KP, Kauwell GP, Chapman CM, Gregory III JF, Maneval DR, Browdy AA, et al. Folate status response to controlled folate intake is affected by the methylenetetrahydrofolate reductase 677C→T polymorphism in young women. *J Nutr* 2003;**133**:4107–11.
22. Imamura A, Okumura K, Matsui H, Mizuno T, Ogawa Y, Imai H, et al. Endothelial nitric oxide synthase and methylenetetrahydrofolate reductase gene polymorphisms are associated with endothelial dysfunction in young, healthy men. *Can J Cardiol* 2004;**20**:1229–34.
23. Oterino A, Toriello M, Valle N, Castillo J, Alonso-Arranz A, Bravo Y, et al. The relationship between homocysteine and genes of folate-related enzymes in migraine patients. *Headache* 2010;**50**:99–168.
24. Kowa H, Yasui K, Takeshima T, Urakami K, Sakai F, Nakashima K. The homozygous C677T mutation in the methylenetetrahydrofolate reductase gene is a genetic risk factor for migraine. *Am J Med Genet* 2000;**96**:762–4.
25. Cronin S, Furie KL, Kelly PJ. Dose-related association of MTHFR 677T allele with risk of ischemic stroke: evidence from a cumulative meta-analysis. *Stroke* 2005;**36**:1581–7.
26. Solis C, Veenema K, Ivanov AA, Tran S, Li R, Wang W, et al. Folate intake at RDA levels is inadequate for Mexican American men with the methylenetetrahydrofolate reductase 677TT genotype. *J Nutr* 2008;**138**:67–72.
27. Chu C, Scanlon P. Vitamin B_{12} deficiency optic neuropathy detected by asymptomatic screening. *BMJ Case Rep* 2011.

28. Chavala SH, Kosmorsky GS, Lee MK, Lee MS. Optic neuropathy in vitamin B_{12} deficiency. *Eur J Intern Med* 2005;**16**:447–8.

29. de Silva P, Jayamanne G, Bolton R. Folic acid deficiency optic neuropathy: a case report. *J Med Case Rep* 2008;**2**:299.

30. Mojon DS, Kaufmann P, Odel JG, Lincoff NS, Marquez-Fernandez M, Santiesteban R, et al. Clinical course of a cohort in the Cuban epidemic optic and peripheral neuropathy. *Neurology* 1997;**48**:19–22.

31. Sbeity ZH, Baydoun L, Schmidt S, Loeffler KU. Visual field changes in methotrexate therapy. Case report and review of the literature. *J Med Liban* 2006;**54**:164–7.

32. Clare G, Colley S, Kennett R, Elston JS. Reversible optic neuropathy associated with low-dose methotrexate therapy. *J Neuroophthalmol* 2005;**25**:109–12.

33. Endresen GK, Husby G. Folate supplementation during methotrexate treatment of patients with rheumatoid arthritis. An update and proposals for guidelines. *Scand J Rheumatol* 2001;**30**:129–34.

34. van Ede AE, Laan RF, Blom HJ, Boers GH, Haagsma CJ, Thomas CM, et al. Homocysteine and folate status in methotrexate-treated patients with rheumatoid arthritis. *Rheumatology (Oxford)* 2002;**41**:658–65.

35. Candito M, Candito E, Chatel M, van Obberghen E, Dunac A. Homocysteinemia and thrombophilic factors in unexplained decompression sickness. *Rev Neurol (Paris)* 2006;**162**:840–4.

NUTRACEUTICALS

41

Flavonoids and Visual Function: Observations and Hypotheses

David T. Field, Lynne Bell, Sarah W. Mount, Claire M. Williams, Laurie T. Butler

School of Psychology and Clinical Language Sciences, University of Reading, Reading, UK

INTRODUCTION

Flavonoids are a class of organic polyphenol micronutrient found in high concentrations in plant-based foods such as berries, tea, and cocoa. There are several subclasses of flavonoid such as flavanols (including catechins), flavones (including flavonols), and anthocyanidins. The proportions of each vary across food types. Flavonoids have been linked to improvements in a number of different aspects of health through several different biologic and neurologic mechanisms. This chapter focuses on flavonoid-related effects within the human eye and visual system and brings together findings from diverse research areas. Visual function is considered in terms of three distinct subsystems: the eye, which focuses the image on the retina; the retina, which performs visual signal transduction and initial neural processing; and the brain, which receives the neural signal from the retina and processes it further, resulting in visual perception. Flavonoids have the potential to exert an influence on all three subsystems, and this complicates the task of pinpointing the mechanisms underlying observed improvements in behavioral measures of visual function. This chapter is divided into two main sections. The first section is a summary of published observations and hypotheses, including information on bioavailability in the eye, retina, and brain and the influence of flavonoids on various measures of vision. The second section contains new data from recent studies carried out by the authors specifically relating to contrast sensitivity and accommodation and eye-convergence testing.

REVIEW OF EXISTING EVIDENCE

Bioavailability in the Eye, Retina, and Brain

Vision begins with the eye, and so it is relevant to consider the transfer of the metabolites of ingested flavonoids from the bloodstream to ocular tissue. Animal studies, described below, have been used to ascertain the level of bioavailability. Eye and retinal tissues have molecular barriers similar to the blood–brain barrier that can prevent the crossing of large molecules; because of the efficiency of the blood–ocular barrier, pharmaceutical administration of medication for eye conditions generally favors topical application.[1] In the blood–brain barrier, endothelial cells along the capillaries prevent the diffusion of large molecules into the cerebrospinal fluid and provide active transport for essential metabolites across the blood–brain barrier. In the eye, the blood–ocular barrier is made up of two different barrier types.[2] The blood–aqueous barrier operates in ocular tissue in a similar way to the blood–brain barrier; in particular, molecule hydrophilicity and polarity as well as molecule size play a role in transbarrier transport.[3] The eye also contains a distinct blood–retinal barrier. This barrier combines an inner vascular epithelium (similar to the blood–aqueous barrier) with an outer retinal pigment epithelium. This pigmented layer shields the retina from excess light in addition to acting as a molecular barrier.[4]

Two studies have claimed to demonstrate the ocular bioavailability of anthocyanins (ACNs). First, Kalt et al.[5] fed pigs a blueberry-powder enriched *ad libitum* diet (1%, 2%, or 4% w/w feed) for 4 weeks, equivalent to an approximate dose range of 2.4–12mg ACN/kg body mass (BM)

Handbook of Nutrition, Diet, and the Eye
http://dx.doi.org/10.1016/B978-0-12-401717-7.00041-1

per day.[6] Prior to sample collection, the pigs were fasted for 18 hours. Tissue samples were rinsed with water to remove excess blood then freeze-dried for analysis. The mean ACN concentration found in the eye, across all test groups, was 0.709 ng/g fresh tissue weight. It is worth noting that ACNs were also found in the control group as trace amounts; ACNs were present in the feed owing to the inclusion of soy (0.0002% ACN w/w feed). Concurrent analysis of blood plasma samples, taken before the animals were killed, found no detectable ACNs, thus confirming the accumulative properties of ACNs within the ocular tissues. However it is not possible to state, based on this study, that ACNs were present in the retina itself because the ocular tissue samples analyzed were not specific to the retina. The study also demonstrated the presence of ACNs in the cerebral cortex, indicating that flavonoid compounds can cross the blood–brain barrier. Although it should be noted that some controversy remains about whether polyphenols enter the brain in biologically meaningful quantities, the bioactive nature of polyphenols *in vivo* is currently poorly understood.[7] In the second study, bioavailability was assessed using rodents (rats and rabbits), blackcurrant ACN, and various routes of administration.[8] In the first part of this study, an acute dose of 100 mg ACN/kg BM was orally administered to rats; after 30 minutes a maximum concentration of 115 ng/g was detectable in their ocular tissues. The same authors also assessed bioavailability using intravenous administration (20 mg/kg BM) in rabbits and found ACNs in the tissues of the ciliary body (2.04 μg/g) and retina (0.27 μg/g). Intraperitoneal injection in rats (500 mg/kg BM) resulted in 6.89 μg/g and 12.93 μg/g in the retina and in the ciliary body, respectively. Therefore, these experiments suggest that ACNs, and probably other flavonoids, can pass through the blood–retinal barrier. However, as with the previously cited concern over the concentration of polyphenols within the brain,[7] there is currently a lack of direct evidence that these ocular concentrations are sufficient to elicit effects. In addition, animal models may not be completely representative of bioavailability in humans.[9] Finally, it should be emphasized that whether or not flavonoids reach the eye in sufficient quantities to produce bioactivity, other less direct mechanisms may prove to be responsible for observed improvements in ocular function after flavonoid ingestion, for example, the general increase in blood flow to peripheral parts of the body caused by flavonoid consumption.[10, 11]

Observations of Flavonoid-Induced Changes in Visual Function

Protection Against Oxidative Stress and Disease Processes

Flavonoid compounds have well-documented antioxidant properties that protect against cell degeneration induced by free radicals *in vitro*. This is potentially important to vision, as the retina has the highest metabolic rate of any tissue and is highly vulnerable to oxidative injury.[6] An animal study by Youdim et al.[12] demonstrated the possible protective effects of ACNs; they were able to show that (blueberry) ACNs protected red blood cells (RBCs) from oxidative damage both *in vivo* and *in vitro*. They gave rats an acute oral dose of 100 mg ACN and measured RBC resistance to damage induced by reactive oxygen species after 1-, 6-, and 24-hour intervals. Human blood was used for the *in vitro* testing; RBCs were incubated with ACN at 0.5 mg/ml and with a dilute hydrogen peroxide solution (100 μM). Significant protection was afforded under both *in vivo* and *in vitro* conditions. Youdim et al. proposed that these antioxidant properties may help to maintain blood oxygen levels within tissues.[12] This is of particular relevance in aging, when blood oxygen levels are observed to decline following oxidative damage. The study was carried out on RBCs, but Youdim et al. proposed that ACN may also offer protection from reactive oxygen species in other cell systems.[12] However, it is uncertain whether tissue concentrations of flavonoids within the visual system are sufficient to produce meaningful antioxidative effects.[13] Below, several animal studies are briefly reviewed, which indicate that flavonoids can protect retinal function from oxidative insults.

Epigallocatechin gallate (EGCG) is a well-known catechin-base flavonoid found in green tea, which is believed to have protective antioxidant properties. This and its neuroprotective potential were investigated in a study conducted by Zhang et al.,[14] in which they examined the influence of EGCG on rat retinas. Rats were injected intraperitoneally, with either a single acute dose of 25 mg/kg BM EGCG or the same dose administered chronically, twice daily for 2 days. After treatment, the intraocular pressure of one eye was raised to induce ischemia. Damage to the photoreceptor cells and Müller cells in the retina as a result of ischemic restriction of blood flow was determined by electroretinography (ERG) and postmortem analysis. In control rats the ischemia caused a reduction in ERG wave amplitudes and a number of neurologic and biochemical changes within the retina. Significantly fewer of these changes were found in both acutely and chronically EGCG-treated rats. A separate *in vitro* experiment used cultured retinal ganglion cells. Cells previously exposed to EGCG showed a significant reduction in apoptosis on exposure to oxidative stress from hydrogen peroxide.

A similar study, conducted by Costa et al.,[15] addressed the question of whether or not orally administered EGCG in drinking water could reduce light-induced retinal damage in rats. Approximately 400 mg/kg BM per day was ingested for 2 days prior to light-induced retinal oxidative stress. Treatment was then continued for 4 days. The EGCG was found to attenuate light-induced damage to retinal photoreceptors in the test rats compared with controls. Reductions in ERG wave amplitudes in the EGCG rats were significantly less and changes to apoptotic proteins were significantly decreased. The study suggests that orally administered EGCG can reach the retina in sufficient concentration to have a physiologic effect in rats. On the basis of other *in vitro* studies, the authors postulate that once in the retina, the EGCG acts to indirectly facilitate endogenous protective mechanisms within the cells.

Flavonoids other than EGCG have also been observed to protect against light-induced retinal oxidative stress. Liu et al.[16] found positive results by supplementing pigmented rabbits with whole wild Chinese blueberries at doses of 1.2 or 4.9 g/kg BM/day, the main flavonoids present in blueberries being ACNs and flavanols. These blueberry treatments were administered chronically for 4 weeks and corresponded to approximate daily ACN intakes of 2 mg/kg BM and 9 mg/kg BM, respectively. In contrast to these findings, a similar study employing a daily dose of 2.8 mg blueberry ACN for 7 weeks in both albino and pigmented rats resulted in improvements only for the albino animals, and then only if the supplementation occurred before light-induced retinal damage. As an explanation for this effect, the authors propose that there may be a difference in the rhodopsin regeneration rate between brown and albino rats, and thus flavonoids may influence the rhodopsin cycle rather than offer protection against oxidative stress.[13]

Turning to human studies, Ohguro et al.[17] investigated the long-term effects of ingesting blackcurrant ACN on 38 glaucoma patients. This double-blind, randomized, controlled trial used an orally administered supplement of 50 mg/day for 24 months in addition to each participant's usual glaucoma medication. Differences in disease progression between the treatment and control groups were measured using several parameters including visual field mean deviation and ocular blood flow. After the 24-month test period, participants given blackcurrant ACN had significantly increased ocular blood flow and showed significantly less visual field deterioration compared with controls.

Both human and animal studies suggest that a diet high in flavonoids is beneficial to the long-term health of the retina.

Night Vision, Dark Adaptation, and the Rhodopsin Cycle

Flavonoid research has generally been motivated by observed links between health and the consumption of fruit and vegetables. In the case of visual function, interest has in part been driven by anecdotal evidence of a link between anthocyanidins and improvements in night vision. For example, World War II pilots are reported to have eaten flavonoid-rich bilberry jam before night missions. Scientific studies have investigated this possible bioactivity through measurement of dark adaptation.

To fully adapt the eyes to complete darkness takes approximately 30–40 minutes, at which point a visual sensation can be produced by an extremely dim test light (approximately 10^{-5} candela/m^2) that would still be undetectable even after 20 minutes of dark adaptation. Improvements in night vision due to flavonoid consumption would probably manifest themselves as a reduction in the time required to reach this maximum sensitivity to light during dark adaptation. Visual sensitivity at low light levels is provided by rods (cones are insensitive below about 0.03 candela/m^2). Within the rods, light is absorbed by the purple-colored rhodopsin chromophore, 11-*cis*-retinal, which results in its isomerization to *all-trans*-retinal. In this isomerized state rhodopsin is referred to as 'bleached' and can no longer absorb light. Before a rhodopsin molecule can once again perform the function of absorbing light, the photoisomerized *all-trans*-retinal must be converted back to 11-*cis*-retinal in a process called the retinoid cycle, which takes approximately 30 minutes. During dark adaptation, maximum sensitivity to light is achieved once 100% of rhodopsin has been regenerated. Given the strong causal link between the retinoid cycle and dark adaptation, if the latter were influenced by flavonoids then this would indicate that flavonoids can intervene in the retinoid cycle. Indeed, this is the explanation offered by Nakaishi et al.[18] They observed that acute supplementation with 50 mg blackcurrant anthocyanoside improved the level of sensitivity to light achieved after 30 minutes adapting to the dark, as well as at earlier time-points in the dark adaptation curve.

A number of other studies on the effect of ACNs on dark adaptation have been carried out with mixed results. For a review of these studies see Canter and Ernst,[19] who concluded that a negative outcome in such studies was correlated with the use of more rigorous methodology such as randomized control trials. However, Canter and Ernst[19] also point out that negative outcomes were correlated with low doses and with differences in the agricultural sources of the extract used. The use of extract can be problematic as overprocessing and prolonged storage lead to a loss of the more bioavailable monomers of ACNs.[6] In addition, studies did not always test participants at the same time of day, introducing confounds from diurnal effects related to physical and chemical changes in vision that occur throughout the day. Therefore, the issue of whether flavonoids can influence visual signal transduction via the retinoid cycle remains an open one, which would perhaps best be resolved by making direct measurements of retinal rhodopsin levels.

Contrast Sensitivity

Visual contrast sensitivity is a measure of the ability of the visual system to distinguish objects from the background. Tests usually involve the presentation of letters or digits that are adjusted to become increasingly similar in brightness to the background until they can no longer be seen (Fig. 41.1). Alternatively, gratings of parallel bars with different spatial frequencies are sometimes used in place of letters. The test is typically carried out under photopic conditions (daylight) but may also be carried under reduced (mesopic or scotopic) illumination in order to assess patients presenting with poor nocturnal visual function. Lee et al.[20] selected a sample of 60 such patients who were also myopic and used mesopic contrast sensitivity testing to ask whether their nocturnal visual function could be improved by supplementation with grape ACN (100 mg 85% anthocyanoside oligomer, or placebo tablet, twice daily for 4 weeks). Contrast sensitivity was improved compared with the control group at the end of the supplementation period, as were subjectively reported symptoms of visual fatigue. The authors compare their positive results with negative results from previous research and suggest two explanations for the difference: the larger dose given for longer in their study and the fact that they deliberately recruited participants whose contrast sensitivity was relatively poor. They go on to hypothesize that flavonoids may only be beneficial to subjects with a pre-existing visual deficiency. However, this has not proven to be the case; a recent experiment conducted by Field et al.[21] produced positive results of supplementation using participants with normal vision. In that study, 30 young participants took part

VRSKDR
NHCSOK
SCNOZV
CNHZOK
NODVHR

FIGURE 41.1 Measurement of contrast sensitivity. The Pelli–Robson contrast sensitivity chart is a standard method of assessing contrast sensitivity. A segment of the chart is reproduced here. Individuals are asked to read as many letters as possible. The letters are presented such that, when descending the page, the level of contrast between the letter and the background decreases until the letters are no longer legible.

in a crossover design comparing the effects of a single dose of dark chocolate containing 773 mg cocoa flavanol with a control condition. Contrast sensitivity was measured under photopic conditions 2 hours after consumption. While Lee et al.[20] demonstrated a chronic effect on individuals with poor vision, Field et al.[21] demonstrated an acute effect on individuals with normal vision.

The mechanism by which flavonoids improve contrast sensitivity is currently unknown. Rhodopsin regeneration, as discussed above in relation to dark adaptation, could play a role. However, no models currently exist of the relationship between visual contrast sensitivity and rhodopsin level. An alternative possibility is that the vasodilatory properties of flavonoids underlie these effects. Flavonoids are known to increase peripheral blood flow generally.[10,11] The resulting increase in the supply of oxygen, glucose, and other metabolites to the retina, as well as increased efficiency in the removal of the waste products of metabolism, could enhance retinal function, especially in light of the very high metabolic rate in retinal tissue. The plausibility of the blood flow hypothesis is increased by evidence from a nonflavonoid-based manipulation. Inducing hypercapnia by breathing air with a higher than normal carbon dioxide (CO_2) content, known to be vasodilatory, also enhances contrast sensitivity.[22] In this study, direct measurements of blood velocity in the central retinal artery were made and shown to change in response to hypercapnia. Finally, it must be acknowledged that while contrast sensitivity tests are effective at targeting retinal function, and in particular the function of the retinal ganglion cells, they do not do so perfectly; performance may be influenced by such factors as the motivation and arousal levels of participants or cortical visual processing. As all of these factors could potentially be influenced by flavonoid-induced blood flow changes in the brain they cannot at this stage be entirely ruled out. In support of this possibility it should be noted that Field et al.[21] also observed small performance improvements on a visual motion detection task, which is a task that relies on processing in the visual cortex of the brain. In order to distinguish between the three hypothetical mechanisms described here, a good starting point would be to measure the influence of flavonoids on retinal function using ERG.

Ocular Muscles

The studies reviewed above relate to effects of flavonoids on the retina, or possibly the retina and the brain in some cases. However, the initial process of vision is the formation by the eye of a focused image on the retina, with appropriate image correspondence across the two retinas. This process is controlled by muscles, the blood supply and function of which may be susceptible to influence from flavonoids. The first process considered here is ocular accommodation, which is the process by which the lens focuses a visual image on the retina. The lens is elastic, and by varying its shape it allows objects viewed over a range of distances to be brought into focus. This change in shape is achieved by contraction or dilation of the ciliary muscle surrounding the lens (Fig. 41.2). The second process considered here is ocular convergence (or 'vergence'), which is used to maintain image correspondence across the retinas, and therefore single binocular vision when viewing near objects. The medial rectus muscle, an extraocular muscle located within the eye orbit (Fig. 41.2), moves both eyes towards one another simultaneously in order to achieve correspondence when viewing near objects.

Tiring of both muscular systems is implicated in visual fatigue, such as that experienced as a result of prolonged computer use. With this in mind, Yoshihara et al.[23] investigated the effect of a peanut-derived polyphenol solution on visual fatigue. Office workers who spent high proportions of their time looking at computer monitors were given polyphenol-soaked cloths to place over their eyes for 5 minutes. Accommodation near point and accommodation time were measured at baseline and post-treatment after several hours at a computer screen (no measurements were made of the convergence process). The use of a polyphenol cloth was found to significantly reduce the accommodation near point and accommodation time compared with baseline, whereas no such effect was observed when a control moist cloth was used. In addition, use of the polyphenol significantly reduced subjective self-report measures of visual fatigue. In the next section of this chapter, the first studies on the effects of ingested flavonoids on accommodation and convergence are reported.

NEW FINDINGS

This section summarizes observations from three new studies carried out by the authors suggesting that flavonoid supplementation is beneficial for the function of the eye muscles and potentially also for the retina and/or cortical visual processing. Study 1 investigated the acute effect of cocoa flavanol ingestion on contrast sensitivity in older adults. Study 2 investigated the effects of cocoa flavanol ingestion on ocular accommodation. Study 3 investigated the effects of blueberry ACN on ocular accommodation and convergence.

Materials and Methods

In the first study, several improvements were made to the design used in the young adult study conducted by Field et al.[21] These included the introduction of baseline measurements, a double-blind test procedure, and the use of a control condition that matched the theobromine and caffeine content of the cocoa condition. Therefore, a positive result in this experiment would rule out potential confounding explanations for the findings of Field et al.[21] These include the influence of bioactive compounds other than flavanol and placebo effects. In this study, 33 healthy older adults (mean age 71 years, SD 6.5 years) participated in an acute, order-counterbalanced crossover design with 1-week washout periods (in addition, high-flavonoid foods were avoided for 24 hours before testing). Testing of binocular contrast sensitivity was performed by letter reading before and 2 hours after consumption of a milkshake containing 774 mg cocoa flavanol or a cocoa-flavored control matched for caffeine, theobromine, and calories. To guard against the possibility that participants would improve their performance by memorizing the

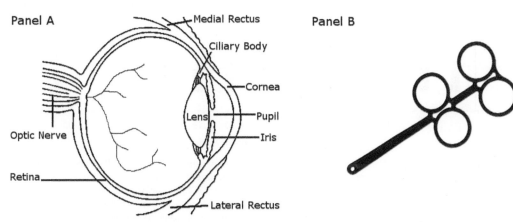

FIGURE 41.2 Ocular muscles: location and measurement of function. Panel A illustrates the location of the optical muscles in relation to other important eye structures. Ocular accommodation and convergence involve the ciliary and rectus muscles, respectively. Flipper arrangements of lenses or prisms, illustrated in Panel B, are used to fatigue these muscle systems. Optical lenses force the eye to focus on near and far objects successively using the ciliary muscle; accommodative function can be assessed by counting the number of cycles cleared. The rectus muscles are used to maintain single binocular vision when viewing near objects. Optical prisms induce double vision that must be counteracted using these muscles, allowing the measurement of convergence function.

sequences of 15 letters used at each contrast step, four separate versions of the test were created and their order of administration was counterbalanced. Two participants were referred to medical experts owing to very poor levels of contrast sensitivity suggestive of cataracts or eye disease. In addition to contrast sensitivity, the study measured visual acuity (the ability to resolve fine detail at high contrast) and visual attention using the useful field of view (UFOV) test,[24] a measure of the ability to allocate visual attention appropriately across the visual field.

In the second study, a group of nine young adult participants was supplemented daily with 35 g high-flavanol chocolate, equivalent to 774 mg flavanol. Supplementation was continued for 5 days in order to assess acute effects as well as any chronic changes that occurred as supplementation continued. A control group of eight participants was supplemented with 35 g low-flavanol chocolate matched to the active chocolate for caffeine and theobromine (two participants failed to complete the study). A test called 'accommodative facility',[25] which challenges the ciliary muscle, was used. Testing was carried out before supplementation, 2 hours after the first dose, and approximately 1 day after the final dose. The test involved focusing for 1 minute on a small target at a distance of 40 cm alternately through ±2 diopter lenses, the two lenses being interchanged each time the target appeared clearly in focus. The outcome measure was the number of cycles cleared per minute.

The third study used an acute, order-counterbalanced, crossover design with washout periods in which 25 young adult participants were tested 2 hours after consuming freeze-dried blueberry powder in drink form. Each participant consumed a high dose containing 235 mg ACN, a low (127 mg) dose, and a calorie-matched control drink, on three separate visits to the laboratory. In addition to accommodative facility (see Study 2), convergence facility was measured using 12 diopter base-out and 3 diopter base-in prisms instead of the ± lenses. During the test, prisms were interchanged once the target appeared single and clearly in focus, and, as for accommodative facility, the outcome measure was the number of cycles cleared per minute.

In Studies 2 and 3 attempts were made to measure contrast sensitivity using a letter-reading task, but

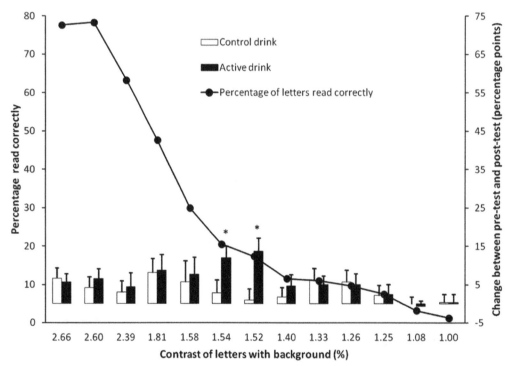

FIGURE 41.3 Effects of acute cocoa flavanol supplementation on visual contrast sensitivity in older adults. The solid line (primary y axis) plots the mean percentage of letters read correctly at each contrast step used in the experiment, averaged across all predrink and postdrink measures (four sets of measures per participant due to the crossover design). The authors expected 100% performance at the highest contrast levels used but failed to correctly anticipate the extent of visual decline in some of the participants. As an indication of the age-related decline in performance on this test relative to young adults, no older adult could read letters presented at 1% contrast, whereas in Field et al.,[21] young adult performance was on average over 90% correct at this contrast level. The hollow bars (secondary y axis, error bars show ±1 SE) plot the signed difference between the postdrink test scores and the predrink scores for the control drink and show a level of improvement that is probably explained by the effects of practice. The active drink (solid bars) produced greater improvement than the control drink at the two contrast levels, where overall performance level was at around 20%, but not when performance was high or when it was approaching chance levels. These differences were statistically significant by the t test, indicated by black stars. Therefore, the results show that cocoa flavanol improves the minimum level of visual contrast that the aged visual system can operate at. These results are similar to those of an earlier study in young adults, although in that study performance benefits occurred over a greater range of test difficulty levels, from approximately 75% correct performance down to approximately 20% correct performance. This difference is currently unexplained but may be accounted for by differences in methodology between the two studies.

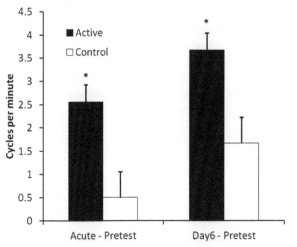

FIGURE 41.4 Effect of 5 days of flavanol supplementation on accommodative facility. For each participant, accommodative facility was measured before the first supplement was consumed, and measurements at subsequent time-points are plotted here as change relative to that baseline, in cycles cleared per minute. Hollow bars indicate the change from baseline at the acute dose measurement point for the control condition, at which performance did not differ reliably from baseline, and at the day 6 measurement time-point, at which performance was reliably higher than the baseline (error bars show ±1SE). The solid bars indicate that for the active condition there was a large improvement compared with baseline at the acute time-point and a further improvement at day 6. Therefore, the results show two separate influences: a practice effect that builds up owing to repeated testing and an acute effect of cocoa flavanol that is maintained over several days of supplementation. Practice effects were expected, as the test used is essentially a type of eye exercise. To place the size of the acute improvement in performance due to flavanol in context, the grand mean performance in the experiment was 7.7 cycles, and so an improvement of around two cycles is substantial.

technical problems with the visual display, discovered after the experiments were completed, rendered these data unusable.

Results

In Study 1, acute cocoa flavanol supplementation caused a significant improvement in the contrast sensitivity thresholds of older adults (Fig. 41.3). These findings suggest that supplementation could have a positive impact on quality of life for older adults, for whom poor contrast sensitivity often contributes to the decision to stop driving, as well as being a cause of falls.[26] Supplementation produced no improvements in visual acuity or visual attention.

In Study 2, acute flavanol supplementation caused a significant improvement in accommodative facility (Fig. 41.4). Continued supplementation over a 5-day period produced no additional improvement, although initial gains were maintained. There were also moderate practice effects in both active and control conditions attributable to repeated testing. These results suggest potential benefits for individuals with accommodative insufficiency.

In Study 3, acute ACN supplementation caused a small but significant improvement in convergence facility compared with controls (Fig. 41.5). However, there were no statistically reliable improvements in accommodative facility. There were no differences in the effect size between the low and high doses for convergence.

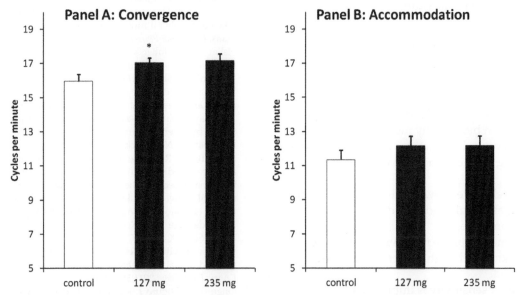

FIGURE 41.5 Acute effect of blueberry supplementation on accommodative facility and convergence facility. This study used a crossover design without pre-test measurements, and therefore the data plotted in the graph represent absolute performance at post-test rather than change relative to baseline as in the earlier figures. Hollow bars indicate the mean number of cycles per minute for the control drink, with solid bars indicating performance for the two acute anthocyanin (ACN) doses (error bars show ±1SE). Both doses of ACN produced a slight improvement in convergence facility compared with control (panel A). This was statistically significant by *t* test for the lower dose, but the size of the improvement was much smaller than that seen for cocoa flavanol in Studies 1 and 2. There was no reliable effect of ACN on accommodative facility (panel B). At this stage it is not clear why the effect of cocoa supplementation (see Fig. 41.4) was so much greater, although it may simply be that a larger dose of ACN than employed here would produce comparable effects.

CONCLUDING REMARKS

In this chapter the authors have reviewed promising evidence suggesting that dietary flavonoids are beneficial to the health and function of the visual system, as well as briefly reporting some new evidence from their laboratory. Several possible underlying mechanisms were outlined, which should not be considered mutually exclusive.

Ongoing research should focus on determining the specific mechanisms by which flavonoids are able to produce the kinds of improvements in visual test performance reported here. Because the determinants of performance on any behaviorally based measure are somewhat complex, rendering it difficult to determine underlying mechanisms, this research should probe the mechanisms more directly using ERG, other emerging retinal imaging techniques, or animal models.

Different flavonoid compounds may elicit varying levels of effect on the visual system, as tentatively suggested by comparing the outcomes of Studies 2 and 3 reported above. Future studies incorporating both manipulation of dose levels and direct comparison between the effects of flavanols and ACNs may establish variations in potency or even in mechanism of action.

Finally, while the results of existing laboratory tests suggest that flavonoid supplementation would be highly beneficial for the visual health and function of older adults, the best way to confirm this would be to perform large-scale longitudinal trials encompassing a broad range of outcome measures.

TAKE-HOME MESSAGES

- The evidence that dietary flavonoids can cross the blood–brain and blood–ocular barriers has been reviewed. It is also possible that without crossing these barriers in biologically meaningful quantities, flavonoids have indirect actions on the visual system by promoting a general increase in the circulation of the blood.
- Animal models have shown that flavonoids protect the retina against light-induced damage and raised intraocular pressure.
- Flavonoids may improve night vision and accelerate dark adaptation, although evidence regarding this point is mixed.
- Visual contrast sensitivity is improved by consumption of flavonoids; three potential mechanisms underlying this effect were described. New data are reported extending this recently observed effect of flavonoids to the older adult population.
- It was hypothesized that the known effects of flavonoids on the peripheral blood supply would prove beneficial for the functions of the ciliary muscle and extraocular muscles, leading to predictions of improved accommodation and convergence facility. Here, new data are reported confirming these predictions.

References

1. Stjernschantz J, Astin M. Anatomy and physiology of the eye. Physiological aspects of ocular drug therapy. In: Edman P, editor. *Biopharmaceutics of Ocular Drug Delivery*. Boca Raton, FL: CRC Press; 1993. pp. 1–25.
2. Cunha-Vaz J. The blood–ocular barriers. *Surv Ophthalmol* 1979;**23**: 279–96.
3. Tomi M, Hosoya K. The role of blood–ocular barrier transporters in retinal drug disposition: an overview. *Expert Opin Drug Metabol Toxicol* 2010;**6**:1111–24.
4. Runkle EA, Antonetti DA. The blood–retinal barrier: structure and functional significance. *Methods Mol Biol* 2011;**686**:133–48.
5. Kalt W, Blumberg JB, McDonald JE, et al. Identification of anthocyanins in the liver, eye, and brain of blueberry-fed pigs. *J Agric Food Chem* 2008;**56**:705–12.
6. Kalt W, Hanneken A, Milbury P, Tremblay F. Recent research on polyphenolics in vision and eye health. *J Agric Food Chem* 2010;**58**:4001–7.
7. Schaffer S, Halliwell B. Do polyphenols enter the brain and does it matter? Some theoretical and practical considerations. *Genes Nutr* 2012;**7**:99–109.
8. Matsumoto H, Nakamura Y, Iida H, Ito K, Ohguro H. Comparative assessment of distribution of blackcurrant anthocyanins in rabbit and rat ocular tissues. *Exp Eye Res* 2006;**83**:348–56.
9. Shanks N, Greek R, Greek J. Are animal models predictive for humans? *Philos Ethics Humanit Med* 2009;**4**:2.
10. Matsumoto H, Takenami E, Iwasaki-Kurashige K, Osada T, Katsumura T, Hamaoka T. Effects of blackcurrant anthocyanin intake on peripheral muscle circulation during typing work in humans. *Eur J Appl Physiol* 2005;**94**:36–45.
11. Fisher NDL, Hollenberg NK. Aging and vascular responses to flavanol-rich cocoa. *J Hypertens* 2006;**24**:1575–80.
12. Youdim KA, Shukitt-Hale B, MacKinnon S, Kalt W, Joseph JA. Polyphenolics enhance red blood cell resistance to oxidative stress: *in vitro* and *in vivo*. *Biochim Biophys Acta* 2000;**1523**:117–22.
13. Tremblay F, Waterhouse J, Nason J, Kalt W. Prophylactic neuroprotection by blueberry-enriched diet in a rat model of light-induced retinopathy. *J Nutr Biochem* 2013;**24**:647–55.
14. Zhang B, Safa R, Rusciano D, Osborne NN. Epigallocatechin gallate, an active ingredient from green tea, attenuates damaging influences to the retina caused by ischemia/reperfusion. *Brain Res* 2007;**1159**:40–53.
15. Costa BL, Fawcett R, Li G-Y, Safa R, Osborne NN. Orally administered epigallocatechin gallate attenuates light-induced photoreceptor damage. *Brain Res Bull* 2008;**76**:412–23.
16. Liu Y, Song X, Han Y, et al. Identification of anthocyanin components of wild Chinese blueberries and amelioration of light-induced retinal damage in pigmented rabbit using whole berries. *J Agric Food Chem* 2011;**59**:356–63.
17. Ohguro H, Ohguro I, Katai M, Tanaka S. Two-year randomized, placebo-controlled study of black currant anthocyanins on visual field in glaucoma. *Ophthalmologica* 2012;**228**:26–35.
18. Nakaishi H, Matsumoto H, Tominaga S, Hirayama M. Effects of black currant anthocyanoside intake on dark adaptation and VDT work-induced transient refractive alteration in healthy humans. *Altern Med Rev* 2000;**5**:553–62.

19. Canter PH, Ernst E. Anthocyanosides of *Vaccinium myrtillus* (bilberry) for night vision – a systematic review of placebo-controlled trials. *Surv Ophthalmol* 2004;**49**:38–50.

20. Lee J, Lee HK, Kim CY, et al. Purified high-dose anthocyanoside oligomer administration improves nocturnal vision and clinical symptoms in myopia subjects. *Br J Nutr* 2005;**93**:895.

21. Field DT, Williams CM, Butler LT. Consumption of cocoa flavanols results in an acute improvement in visual and cognitive functions. *Physiol Behav* 2011;**103**:255–60.

22. Huber KK, Adams H, Remky A, Arend KO. Retrobulbar haemodynamics and contrast sensitivity improvements after CO_2 breathing. *Acta Ophthalmol Scand* 2006;**84**:481–7.

23. Yoshihara A, Yamanaka K, Kawakami M. Effects of polyphenol on visual fatigue caused by VDT work. *Int J Occup Saf Ergon* 2009;**15**:339–43.

24. Ball K, Owsley C. The useful field of view test: a new technique for evaluating age-related declines in visual function. *J Am Optom Assoc* 1993;**64**:71–9.

25. Zellers JA, Alpert TL, Rouse MW. A review of the literature and a normative study of accommodative facility. *J Am Optom Assoc* 1984;**55**:31–7.

26. Lord SR, Clark RD, Webster IW. Visual acuity and contrast sensitivity in relation to falls in an elderly population. *Age Ageing* 1991;**20**:175–81.

Polyphenols in Vision and Eye Health

K. Srinivasan

Department of Biochemistry and Nutrition, CSIR – Central Food Technological Research Institute, Mysore, India

INTRODUCTION

Polyphenols are plant secondary metabolites with a large variety of chemical structures. These phytochemicals are of interest for many reasons, such as their role as antioxidants and in browning reactions, their properties of astringency and bitterness, and their color. They include simple phenols, hydroxycinnamates, and flavonoids (Fig. 42.1). With the exception of carotenes, the antioxidants in foods are phenolic compounds. Phenolic compounds have a wide range of biologic properties; of particular note are their antioxidant, anti-inflammatory potential and antitumoral properties, and hence they can potentially prevent coronary heart disease and cancer. Flavonoids are a group of polyphenolic compounds that impart color to fruits and vegetables; these include flavones, flavonols, flavanones, flavanols, anthocyanins, and isoflavones.[1] In recent years, flavonoids have gained importance because of their beneficial effects on human health. They are understood to play a dominant role in the prevention of cancer and heart diseases. Several biologic actions of flavonoids could also be useful in the prevention or treatment of ocular diseases responsible for vision loss, such as diabetic retinopathy, macular degeneration, and cataract.[2]

The etiology of most ocular diseases involves free-radical mediated oxidative damage, hypoxia, decreased blood supply to ocular tissues, and, in certain conditions, angiogenesis, increased vascular permeability, and leakage of vascular contents.[3,4] Flavonoids possess antioxidant, anti-inflammatory, and antiangiogenic activity and are also capable of reducing fluid retention and strengthening capillary walls. Recent research suggests that flavonoids may be involved in two major aspects in vision physiology and eye health. Flavonoids may function in their well-established role as antioxidants, which are particularly important in the eye, where oxidative stress is implicated in a number of vision pathologies, including macular degeneration.[5] Flavonoids may also function in visual signal transduction in ways that are not well understood. Thus, flavonoids may be effective in the prevention or treatment of ocular diseases (e.g., diabetic retinopathy, macular degeneration, and cataract) that lead to vision loss if left untreated.

BENEFICIAL EFFECTS OF FLAVONOIDS ON VISUAL SIGNAL TRANSDUCTION

Visual signal transduction involves the light activation of rhodopsin, leading to a change in its conformation and the creation of a visual signal, and subsequent regeneration of rhodopsin to its original conformation. Rhodopsin is composed of the chromophore retinal and the protein opsin. Rhodopsin is embedded in the membranes of discs in the outer segment of rod photoreceptors. The generation of a visual signal is initiated by the absorption of light by the rhodopsin chromophore, 11-*cis*-retinal, which results in its isomerization to all-*trans*-retinal. Isomerization to the all-*trans*-retinal causes the molecule to straighten, making it fit less favorably, energetically, in the opsin protein. This higher energy form, called metarhodopsin II, activates the G-protein transducin. Transducin, in turn, activates a cyclic guanosine monophosphate (cGMP) phosphodiesterase, which hydrolyzes cGMP, causing the closure of ion channels in the photoreceptor membrane and the hyperpolarization of the photoreceptor cell. The potential difference in the outer segment of the photoreceptor cell is transferred along this cell to its synaptic terminal and then transferred to second order neurons, including bipolar cells, amacrine and horizontal cells, and then to retinal ganglion cells.

Anthocyanins have been examined with respect to different steps in the visual signal transduction process described above. Cyanidin-3-glucoside has recently been reported to inhibit the activation of the G-protein transducin by metarhodopsin II.[6] The second major process

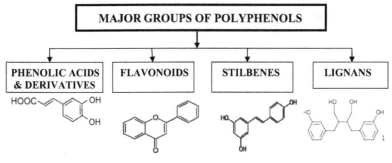

Subtypes of Flavonoids	Basic skeleton	Example
Flavones		Apigenin, Luteolin, etc.
Flavonols		Quercetin, Kaempferol, etc.
Flavanones		Naringenin, Hesperidin, etc.
Flavanonols		Dihydroquercetin, Dihydrokaempferol, etc.
Flavan-3-ols, Flavan-4-ols, Flavan-3,4-diols		Catechin, Gallocatechin, epicatechin, etc.
Anthocyanidins		Cyanidin, Delphindin, Pelargonidin, etc.
Isoflavones		Genistein, Daidzein, etc.

FIGURE 42.1   Classification of polyphenols.

involved in visual signal transduction is the regeneration of rhodopsin in a process called the retinoid cycle. Matsumoto *et al.*[7] report that specific blackcurrant anthocyanins stimulate the regeneration of rhodopsin.

BENEFICIAL ACTIONS OF FLAVONOIDS ON OCULAR DISEASES AND DISORDERS

Oxidative stress in the eye lens also plays a major role in the initiation and progression of cataracts, including diabetic cataract. Oxidative stress is higher in diabetes because of multiple factors, glucose auto-oxidation being the major source of free radicals.[8] The nonenzymatic glycation of proteins,[9] interaction between glycated products and their receptors,[10] and the polyol pathway[11,12] are other potential sources of hyperglycemia-induced oxidative stress in diabetes. Other factors include a reduction in antioxidant defenses, both antioxidant molecules and antioxidant enzymes that dispose of free radicals.

Antioxidant Effects of Flavonoids in the Retina

The retina is the most metabolically active tissue in the body[13] and hence has a high oxygen demand and vulnerability to oxidative stress.[5] The retina is also vulnerable to oxidative stress because of the high proportion of polyunsaturated fatty acids (PUFAs) in photoreceptor membranes and chronic exposure to light. Age-related changes within the retinal pigment epithelium (RPE), including an accumulation of lipofuscin, represent the earliest changes that ultimately lead to age-related macular degeneration (AMD).[14,15]

Effect of Flavonoids on Oxidative Damage to Retinal Cells In Vitro

The ability of flavonoids and other polyphenols to protect retinal cells *in vitro* through antioxidant effects has recently been studied using models and ocular cell types that are relevant to vision processes and pathologies. It is understood that flavonoids can protect both retinal cell types – RPE and retinal ganglion cells (RGCs) – through different mechanisms including direct scavenging of reactive oxygen species (ROS), antiapoptotic activity, and phase 2 induction.[16,17] The relative potencies of the flavonoids are similar in RPE[18] and RGCs,[16,17] suggesting that certain structural features are key to their effects. Structural features that increase the antioxidant potency of flavonoids include the presence of vicinal hydroxyl groups, unsaturation of the C ring, and a high degree of hydrophobicity.[19]

In Vivo *Antioxidant Activity of Polyphenols*

As the most external ocular tissue, the cornea is subject to the greatest irradiation and is most vulnerable to mechanical damage. Studies suggest that nonflavonoid phenolics may be protective against ultraviolet radiation in animals.[20] Several flavonoids have been documented as potential antioxidants.[21,22] These compounds exhibit their antioxidant activity through different mechanisms:

- By scavenging free radicals such as superoxide, peroxyl, alkoxyl, and hydroxyl directly.[21,22]
- By inhibiting nitric oxide (NO) production. The constitutive production of NO in cells including endothelial cells and macrophages is necessary to maintain the dilation of blood vessels. However, inducible nitric oxide synthase (iNOS) is responsible for the production of higher concentrations of NO during oxidative damage. NO reacts with free radicals and generates the highly reactive and damaging peroxynitrite. Flavonoids, through their free radical scavenging properties, can prevent the generation of peroxynitrite.[22] Flavonoids are capable of inhibiting iNOS directly and thus decrease production of NO.[23,24]
- By inhibiting certain enzymes responsible for the production of superoxide anions (xanthine oxidase and protein kinase C) and inhibiting those involved in ROS generation (cyclo-oxygenase, lipoxygenase, microsomal mono-oxygenase, glutathione-*S*-transferase, mitochondrial succinoxidase, and reduced nicotinamide adenine dinucleotide (NADH) oxidase).[21]

Effect on Ocular Blood Flow

A decrease in ocular blood flow can lead to diseases such as glaucoma, diabetic retinopathy, and macular degeneration. The literature suggests a significant increasing effect of bioflavonoids on ocular blood flow.[25,26]

Anti-Inflammatory Activity

Some flavonoids have been reported to inhibit several mediators that are activated in certain inflammatory conditions, such as NO, prostanoids, leukotrienes, cytokines, and adhesion molecules.[27] Flavonoids can act on multiple pathways in the inflammation process.

Aldose Reductase Inhibitory Activity

Aldose reductase, a key enzyme in the polyol pathway, has been identified as the primary factor responsible for the pathologic alterations by osmotic stress in diabetic cataract. In the diabetic situation, hyperglycemia is also observed in the aqueous humor. Aldose reductase catalyzes the conversion of glucose to sorbitol. This polyol, generated at high levels, cannot diffuse out of the lens passively and either accumulates or is converted to fructose. Therefore, an osmotic gradient is generated, inducing diffusion of water into the lens. The resultant swelling and electrolyte imbalance lead to cataract formation.[28] Several flavonoids are reported to inhibit the enzyme aldose reductase. Quercetin is the most promising aldose reductase inhibitor and is used as a positive control in many studies.[28]

BENEFICIAL INFLUENCES OF POLYPHENOLS ON OCULAR HEALTH EVIDENCED IN ANIMAL MODELS

Curcumin: The Bioactive Phytochemical of Turmeric (*Curcuma longa*)

Oxidative stress and inflammation are implicated in the pathogenesis of retinopathy in diabetes. Curcumin (Fig. 42.2a), a naturally occurring bioactive polyphenolic compound of the spice turmeric with known anti-inflammatory and antioxidative properties, has been examined by several investigators for its beneficial effects on eye disorders in rat models. Curcumin cotreatment prevented the oxidative damage and thus delayed the development of cataract in Wistar rat pups caused by selenium and showed no opacities in the lens.[29] The lipid peroxidation and xanthine oxidase enzyme activity in the lenses of curcumin- and selenium-cotreated animals were significantly lower than in selenium-treated animals. The activities of superoxide dismutase and catalase in curcumin- and selenium-cotreated animal lenses were higher.

A diet supplemented with 0.05% curcumin for 6 weeks soon after the induction of diabetes prevented diabetes-induced oxidative stress and proinflammatory markers in the retina.[30] Curcumin administration prevented the diabetes-induced decrease in antioxidant capacity and increase in oxidatively modified DNA 8-hydroxydeoxyguanosine and nitrotyrosine. Curcumin also inhibited

FIGURE 42.2 Structures of (a) curcumin and (b) nigerloxin explored for beneficial effects on eye health.

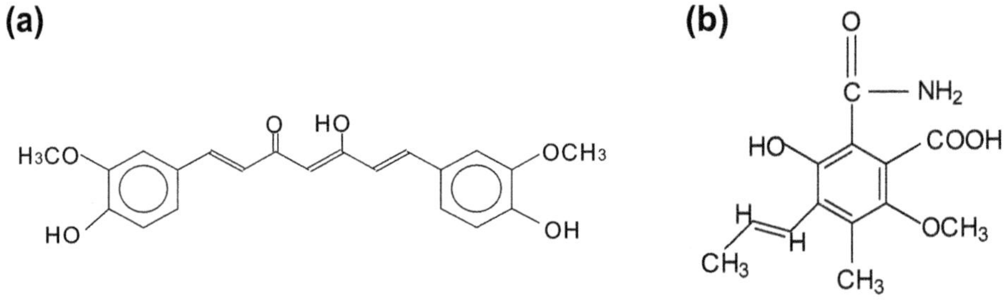

(a)

(b)

the diabetes-induced elevation in levels of interleukin-1β (IL-1β), vascular endothelial growth factor (VEGF), and nuclear factor-κB (NF-κB). These beneficial effects of curcumin on the metabolic abnormalities important in the development of diabetic retinopathy suggest that curcumin could have potential benefits in inhibiting the development of retinopathy in diabetes.

The effect of curcumin and its source, turmeric, has been investigated on streptozotocin-induced diabetic cataract in rats.[31] Diabetic animals received diets with 0.002% and 0.01% curcumin and 0.5% turmeric, respectively, for a period of 8 weeks. At the end of 8 weeks, oxidative stress, polyol pathway, and alterations in the crystallin profile in the lens involved in the pathogenesis of cataract were investigated. Both curcumin and turmeric delayed the progression and maturation of cataract, which was monitored by a slit-lamp biomicroscope. Curcumin and turmeric treatment countered the hyperglycemia-induced oxidative stress and minimized osmotic stress, as assessed by polyol pathway enzymes. Most importantly, aggregation and insolubilization of lens proteins due to hyperglycemia were prevented by curcumin and turmeric.

Chaperone-like activity of eye lens α-crystallin is declined in diabetic conditions. An animal study has shown that curcumin can manipulate the chaperone-like activity of α-crystallin in diabetic rat lens.[32] αH- and αL-crystallins isolated from streptozotocin-diabetic rats that received curcumin 0.002% or 0.01% in the diet showed improved chaperone-like activity compared with these proteins isolated from untreated diabetic rat lens. The decline in chaperone-like activity due to hyperglycemia was associated with reduced hydrophobicity and altered secondary and tertiary structure of αH- and αL-crystallins. Feeding of curcumin prevented the alterations in hydrophobicity and structural changes due to streptozotocin-induced hyperglycemia. Loss of chaperone activity of α-crystallin, particularly αL-crystallin, in diabetic rat lens could be attributed at least partly to increased oxidative stress. Curcumin, by virtue of its antioxidant activity, prevented the loss of α-crystallin chaperone activity and delayed the progression and maturation of diabetic cataract.

AMD is a complex disease that may involve inflammatory and oxidative stress-related pathways in its pathogenesis. Curcumin is understood to be an effective nutraceutical for the preventive and augmentative therapy of AMD.[33] These authors examined curcumin in

a rat model of light-induced retinal degeneration (LIRD) and in retina-derived cell lines. Significant retinal neuroprotection in rats fed diets supplemented with curcumin (0.2% in the diet) for 2 weeks was observed. The mechanism of retinal protection from LIRD by curcumin involved inhibition of NF-κB activation and downregulation of cellular inflammatory genes. Pretreatment with curcumin protected retina-derived cell lines (661W and ARPE-19) from hydrogen peroxide-induced cell death by upregulating cellular protective enzymes.

Neovascularization stimulated by hyperglycemia-mediated induction of VEGF has been implicated in the pathogenesis of diabetic retinopathy. Curcumin was investigated for its ability to inhibit angiogenesis, particularly VEGF expression, in streptozotocin-induced diabetic rat retina.[34] VEGF expression after feeding a diet containing 0.002% or 0.01% curcumin for a period of 8 weeks was analyzed by both real-time polymerase chain reaction and immunoblotting. There was an increase in VEGF expression in diabetic retina compared with control retina at both the transcript and the protein level. Notably, feeding of curcumin (0.002% or 0.01% in the diet) to diabetic rats inhibited expression of VEGF. This study indicated the potential of curcumin in the prevention and/or treatment of diabetic retinopathy.

The therapeutic potential of oral curcumin (1 g/kg body weight) in the prevention and treatment of streptozotocin-induced diabetic retinopathy was evaluated in Wistar rats.[35] Treatment with curcumin for 16 weeks showed significant increases in retinal glutathione levels and antioxidant enzymes, which were compromised in the diabetic condition. The elevation in proinflammatory cytokines, tumor necrosis factor-α, and VEGF in the diabetic retina was prevented by curcumin. Transmission electron microscopy showed degeneration of endothelial cell organelles and an increase in capillary basement membrane thickness in the diabetic retina, but curcumin prevented the structural degeneration and increase in capillary basement membrane thickness in the diabetic rat retina.

Polyphenols of Finger Millet (*Eleusine coracana*) Seed Coat

Polyphenols of finger millet (*Eleusine coracana*) (Fig. 42.3a) are antidiabetic and antioxidant components

(a) **(b)**

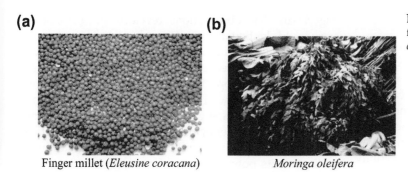

Finger millet (*Eleusine coracana*)   *Moringa oleifera*

FIGURE 42.3 Polyphenol-rich food materials explored for beneficial effects on eye health: (a) finger millet (*Eleusine coracana*); (b) *Moringa oleifera*. (See color plate section)

whose aldose reductase inhibiting activity has been evaluated.[36] Phenolic constituents in finger millet polyphenols such as gallic, protocatechuic, *p*-hydroxy benzoic, *p*-coumaric, vanillic, syringic, ferulic, and *trans*-cinnamic acids and quercetin inhibited the enzyme activity in cataract eye lens effectively; the latter was most potent, with an IC_{50} of 14.8 nM. Structure–function analysis revealed that for phenolics, having an OH group at the fourth position was important for the aldose reductase inhibitory property. The presence of a neighboring *O*-methyl group in phenolics denatured the aldose reductase activity. Finger millet seed coat polyphenols (SCPs) have been found to inhibit aldose reductase reversibly by noncompetitive inhibition.

In groups of Wistar rats rendered diabetic with streptozotocin administration and maintained on 5% finger millet seed coat matter (SCM) for 6 weeks, slit-lamp examination of the eyes was conducted to observe the cataract-associated changes in the eye lens.[37] The incidence of mature cataract was considerably lower in the diabetic animals maintained on the experimental diet, with only 10% of the diabetic experimental diet group of animals developing mature cataract compared with 90% of the diabetic control group at the end of the experimental period. The observations revealed the presence of very mild lenticular opacity and posterior subcapsular cataract (immature cataract) in the diabetic experimental group, in contrast to the significant lenticular opacity (mature cataract) and corneal vacuolization observed in the diabetic control group. No lenticular opacity was observed in the nondiabetic animals. These observations indicated that the finger millet SCM constituents delay cataractogenesis in diabetic animals. To gain further insight into the mechanism of anticataract action of the millet SCM, the activity of aldose reductase, a key enzyme involved in the etiology of complications in diabetes, was determined in the eye lenses. The activity of this enzyme was lower by 25% in the diabetic experimental group of animals than in the diabetic control group.

Glycation of lens protein crystallin may cause conformational changes resulting in exposure of thiol groups to oxidation and cross-link formation.[38] The lens crystallins readily accumulate as advanced glycation end-products (AGEs), which in turn cause aggregation of the lens crystallins producing the high molecular weight material responsible for opacification.[39] The millet SCM, being a rich source of phenolic compounds with antioxidant properties, may delay cataractogenesis in the diabetic rats. The aldose reductase inhibitory property of the millet SCPs *in vitro* has been previously reported.[36] Thus, the enzyme inhibitory property of the millet SCPs may also contribute to the anticataractogenic property. Lowered levels of AGEs, glycosylated hemoglobin (Hb_{A1c}), and aldose reductase activity, lesser lenticular opacity, and the absence of mature cataract in the diabetic experimental diet group of animals indicate that the millet SCM prevents protein glycation and delays cataractogenesis in diabetic animals. Reduced rat tail tendon collagen glycation has been reported in diabetic rats maintained on a millet diet.[40] Hence, the anticataractogenic properties of the millet SCM may partly be attributed to the enzyme inhibitory as well as to the antiprotein glycation properties of the millet SCP compounds.

Polyphenols of *Moringa oleifera* Leaves

The protective effects of the flavonoid fraction of *Moringa oleifera* leaves (FMO) (Fig. 42.3b) on selenite cataract was investigated *in vivo*[41] in Sprague–Dawley rat pups. The development of cataract was assessed and rat lenses were analyzed for the activities of antioxidant enzymes, generation of ROS, reduced glutathione, protein oxidation, and lipid peroxidation. The *M. oleifera* leaf extract, which contained a total phenolic content of 4.4 mg of catechin equivalent per gram of dried material, showed remarkable activity in *in vitro* antioxidant assays. FMO effectively prevented the morphologic changes and oxidative damage in the lens. FMO maintained the activities of antioxidant enzymes and sulfhydryl content and prevented ROS generation and lipid peroxidation. FMO was effective in preventing cataractogenesis in a selenite model by enhancing the activities of antioxidant enzyme, reducing the intensity of lipid peroxidation, and inhibiting free radical generation.

Fungal Metabolite Nigerloxin

Osmotic and oxidative stress has been implicated in the pathogenesis of diabetic cataract. Aldose reductase inhibitors and some dietary antioxidants are expected to have a beneficial influence in diabetic cataract. Nigerloxin (2-amido-3-hydroxy-6-methoxy-5-methyl-4-(prop-1'-enyl) benzoic acid) (Fig. 42.2b), obtained from solid-state fermentation of *Aspergillus niger*, has been shown to exhibit inhibitory activity on the partially purified rat lens aldose reductase *in vitro*.[42] Uncontrolled diabetes in rats is associated with enhanced activity of polyol pathway enzymes and compromised antioxidant defense system. This fungal metabolite nigerloxin has recently been evaluated for its aldose reductase inhibitory potential *in vivo* and hence beneficial modulation of eye lens injuries in streptozotocin-induced diabetes in experimental rats.[43] Nigerloxin administration at 100 mg/kg body weight for 30 days significantly countered the elevated activities of the polyol pathway enzymes aldose reductase and sorbitol dehydrogenase in the lenses of diabetic animals.

Administration of nigerloxin significantly countered the increase in lens lipid peroxides and advanced glycation end-products in diabetic rats. Hyperglycemia-induced glycation of lens proteins leads to the formation of advanced glycation end-products as a major source of the generation of superoxide radicals (O^{2-}).[44] Nigerloxin treatment significantly inhibited the formation of AGEs in the eye lens of diabetic animals. Lens glutathione and antioxidant enzyme activities were significantly elevated in nigerloxin-treated diabetic rats.[43] Examination of rat eyes indicated that nigerloxin delayed cataractogenesis in diabetic rats. The incidence of mature cataract was considerably lower in diabetic animals administered nigerloxin. The observations revealed the presence of very mild lenticular opacity and posterior subcapsular cataract (immature cataract) in the diabetic group of animals administered nigerloxin, in contrast to the significant lenticular opacity (mature cataract) and corneal vacuolization observed in the diabetic control group of animals. The results suggested the beneficial countering of polyol pathway enzymes and potentiation of antioxidant defense system by nigerloxin in diabetic animals, indicating its potential in ameliorating diabetic cataract (Fig. 42.4).

In another animal study, the beneficial influence of nigerloxin was investigated in galactose-induced juvenile cataract in rats.[45] Cataract was induced in Wistar rats by feeding 30% galactose in the diet. Groups of galactose-fed rats were orally administered with nigerloxin (100 mg/kg body weight/day) for 24 days. Lens aldose reductase activity was increased significantly in galactose-fed animals. Lens lipid peroxides and advanced glycation end-products were also significantly increased. Antioxidant molecule, reduced glutathione, total thiols, and activities of the antioxidant enzymes superoxide dismutase and glutathione peroxidase were decreased in the lens of galactose-fed animals. Oral administration of nigerloxin once a day for 24 days, at a dose of 100 mg/kg body weight, significantly decreased lens lipid peroxides and AGEs in galactose-fed rats. Lens aldose reductase activity was reduced and lens antioxidant molecules and antioxidant enzyme activities were elevated significantly by nigerloxin administration. The results suggest that alterations in the polyol pathway and antioxidant defense system were countered by nigerloxin in the lens of galactose-fed animals, suggesting the potential of nigerloxin in ameliorating the development of galactose-induced cataract in experimental animals.

FIGURE 42.4 Lens antioxidant status in nigerloxin-treated (100 mg/kg/day) diabetic rats. 1: Lipid peroxides; 2: superoxide dismutase (SOD) activity; 3: glutathione-*S*-transferase (GST) activity; 4: glutathione peroxidase (GPX) activity.

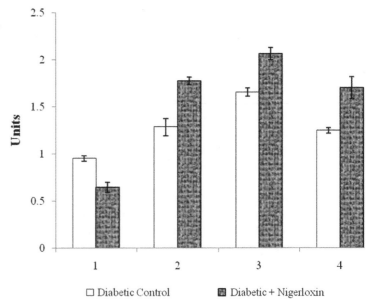

The mechanism of the beneficial effects of polyphenols on the eye lens is shown in Figure 42.5, and the influences of polyphenols on ocular health are summarized in Table 42.1.

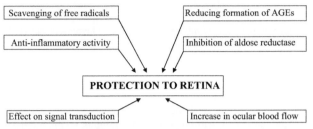

FIGURE 42.5 Mechanism of beneficial effects of polyphenols in eye lens. AGE: advanced glycation end-product.

TAKE-HOME MESSAGES

- Polyphenols play a role in vision physiology and antioxidant protection of the eye, which has only been characterized, for the most part, *in vitro*.
- Flavonoids are understood to act on various mechanisms or etiologic factors responsible for the development of ocular diseases: cataract, diabetic retinopathy, age-related macular degeneration, glaucoma, and dry eye syndrome.
- Recent *in vitro* studies demonstrate that anthocyanins and other flavonoids interact directly with rhodopsin and modulate visual pigment function. Additional *in vitro* studies show that flavonoids protect a variety of retinal cell types from oxidative stress-induced cell death, a neuroprotective property of significance

TABLE 42.1 Beneficial Influences of Polyphenols on Ocular Health Evidenced in Animal Models

Polyphenol Investigated	Effect Demonstrated	Reference
Curcumin	Curcumin cotreatment prevented the oxidative damage and thus delayed the development of cataract in Wistar rat pups by selenium and showed no opacities in the lens	29
	Dietary curcumin (0.5% for 6 weeks) inhibited diabetes-induced elevation in the levels of IL-1β, VEGF, and NF-κB, which are implicated in the development of diabetic retinopathy, suggesting that curcumin could have potential benefits in inhibiting the development of diabetic retinopathy	30
	Curcumin treatment countered hyperglycemia-induced oxidative stress and minimized osmotic stress, as assessed by polyol pathway enzymes. Aggregation and insolubilization of lens proteins due to hyperglycemia was prevented by curcumin	31
	Dietary curcumin (0.1%), by virtue of its antioxidant activity, prevented the loss of α-crystallin chaperone activity and delayed the progression and maturation of diabetic cataract in streptozotocin-induced diabetic rats	32
	Significant retinal neuroprotection in rats fed diets supplemented with curcumin (0.2% in the diet for 2 weeks) was observed. The mechanism of retinal protection from LIRD by curcumin involved inhibition of NF-κB activation and downregulation of cellular inflammatory genes	33
	Feeding of curcumin (0.002% or 0.01% in the diet for 8 weeks) to diabetic rats inhibited neovascularization stimulated by hyperglycemia-mediated induction of VEGF, which is implicated in the pathogenesis of diabetic retinopathy	34
	Elevation in proinflammatory cytokines, TNF-α and VEGF, in the diabetic retina, was prevented by dietary curcumin (1 g/kg for 16 weeks) in streptozotocin diabetic rats. Curcumin prevented the structural degeneration and increase in capillary basement membrane thickness in the diabetic rat retina	35
Nigerloxin (fungal metabolite)	Daily oral administration of nigerloxin delayed cataractogenesis in diabetic rats. Nigerloxin decreased lens lipid peroxides and AGEs in diabetic rats and countered the increase in lens aldose reductase and sorbitol dehydrogenase activities. Lens glutathione and antioxidant enzyme activities were elevated in nigerloxin-treated diabetic rats	43
	Increase in polyol pathway enzymes and diminished antioxidant defense system were countered by orally administered nigerloxin in the lens of galactose-fed animals, suggesting the potential of nigerloxin in ameliorating the development of galactose-induced cataract in experimental animals	45
Polyphenols of finger millet seed coat	Eye lenses of finger millet seed coat-fed diabetic rats had immature subcapsular cataract with mild lenticular opacity compared with the mature cataract with significant lenticular opacity and corneal vascularization in the diabetic control group	36
Polyphenols of *Moringa* leaves	*Moringa oleifera* extract was effective in preventing cataractogenesis in selenite-induced cataract model in rats by enhancing the activities of antioxidant enzyme, reducing the intensity of lipid peroxidation, and inhibiting free radical generation	41

IL-1β: interleukin-1β; VEGF: vascular endothelial growth factor; NF-κB: nuclear factor κB; LIRD: light-induced retinal degeneration; TNF-α: tumor necrosis factor-α; AGE: advanced glycation end-product.

because the retina has the highest metabolic rate of any tissue and is particularly vulnerable to oxidative injury.

- Age-related vision pathologies related to oxidative damage are well documented, and possible mitigation of such damage by a variety of nonanthocyanin flavonoids is supported by recent *in vitro* studies. This suggests that further investigation is warranted on nonanthocyanin flavonoids.
- There is also a need to explore exhaustively the potential for improved vision health through the intake of dietary polyphenols. There is a great opportunity to apply modern experimental and clinical approaches to expand our knowledge of the effects of flavonoids on vision physiology and eye health.
- Based on several studies, curcumin may have potential benefits in the prevention of retinopathy in diabetes.

References

1. Crozier A, Jaganath IB, Clifford MN. Dietary phenolics: chemistry, bioavailability and effects on health. *Nat Prod Rep* 2009;**26**:1001–43.
2. Kalt W, Hanneken A, Milbury P, Tremblay F. Recent research on polyphenolics in vision and eye health. *J Agric Food Chem* 2010;**58**:4001–7.
3. Ohia SE, Opere CA, Leday AM. Pharmacological consequences of oxidative stress in ocular tissues. *Mutat Res* 2005;**579**:22–36.
4. Erickson KK, Sundstrom JM, Antonetti DA. Vascular permeability in ocular disease and the role of tight junctions. *Angiogenesis* 2007;**10**:103–17.
5. Rhone M, Basu A. Phytochemicals and age-related eye diseases. *Nutr Rev* 2008;**66**:465–72.
6. Tirupula K, Balem F, Yanamala N, Klein-Seetharaman J. pH dependent interaction of rhodopsin with cyanidin-3-glucoside. 2. Functional aspects. *Photochem Photobiol* 2009;**85**:463–70.
7. Matsumoto H, Nakamura Y, Tachibanaki S, Kawamura S, Hirayama M. Stimulatory effect of cyanidin 3-glycosides on the regeneration of rhodopsin. *J Agric Food Chem* 2003;**51**:3560–3.
8. Wolff SP, Dean RT. Glucose autoxidation and protein modification. The potential role of 'autoxidative glycosylation' in diabetes. *Biochem J* 1987;**245**:243–50.
9. Mullarkey C, Edelstein D, Brownlee L. Free radical generation by early glycation products: a mechanism for accelerated atherogenesis in diabetes. *Biochem Biophys Res Commun* 1990;**173**:932–9.
10. Schmidt AM, Hori O, Brett J, Yan SD, Wautier JL, Stern D. Cellular receptors for advanced glycation end products. Implications for induction of oxidant stress and cellular dysfunction in the pathogenesis of vascular lesions. *Arterioscler Thromb* 1994;**14**:1521–8.
11. Cheng HM, Gonzalez RG. The effect of high glucose and oxidative stress on lens metabolism, aldose reductase, and senile cataractogenesis. *Metab Clin Exp* 1986;**35**:10–4.
12. Brownlee M. Biochemistry and molecular cell biology of diabetic complications. *Nature* 2001;**414**:813–20.
13. Lamb TD, Pugh Jr EN. Dark adaptation and the retinoid cycle of vision. *Prog Retin Eye Res* 2004;**23**:307–80.
14. Lu L, Hackett S, Mincey A, Lai H, Campochiaro P. Effects of different types of oxidative stress in RPE cells. *J Cell Physiol* 2006;**206**:119–25.
15. Suzuki M, Kamei M, Itabe H, Yoneda K, Bando H, Kume N, Tano Y. Oxidized phospholipids in the macula increase with age and in eyes with age-related macular degeneration. *Mol Vis* 2007;**13**:772–8.
16. Maher P, Hanneken A. Flavonoids protect retinal ganglion cells from oxidative stress-induced death. *Invest Ophthalmol Vis Sci* 2005;**46**:4796–803.
17. Maher P, Hanneken A. Flavonoids protect retinal ganglion cells from ischemia *in vitro*. *Exp Eye Res* 2008;**86**:366–74.
18. Hanneken A, Lin F, Johnson J, Maher P. Flavonoids protect human retinal pigment epithelial cells from oxidative-stress-induced death. *Invest Ophthalmol Vis Sci* 2006;**47**:3164–77.
19. Ishige K, Schubert D, Sagara Y. Flavonoids protect neuronal cells from oxidative stress by three distinct mechanisms. *Free Radic Biol Med* 2001;**30**:433–46.
20. Larrosa M, Lodovici M, Morbidelli L, Dolara P. Hydrocaffeic and *p*-coumaric acids, natural phenolic compounds, inhibit UV-B damage in WKD human conjunctival cells *in vitro* and rabbit eye *in vivo*. *Free Radical Res* 2008;**42**:903–10.
21. Pietta PG. Flavonoids as antioxidants. *J Nat Prod* 2000;**63**:1035–42.
22. Nijveldt RJ, van Nood E, van Hoorn DE, Boelens PG, van Norren K, van Leeuwen PA. Flavonoids: a review of probable mechanisms of action and potential applications. *Am J Clin Nutr* 2001;**74**:418–25.
23. Kim BH, Cho SM, Reddy AM, Kim YS, Min KR, Kim Y. Down-regulatory effect of quercitrin gallate on nuclear factor-kappa B-dependent inducible nitric oxide synthase expression in lipo-polysaccharide-stimulated macrophages RAW 264.7. *Biochem Pharmacol* 2005;**69**:1577–83.
24. Martinez-Florez S, Gutiérrez-Fernández B, Sánchez-Campos S, González-Gallego J, Tuñón MJ. Quercetin attenuates nuclear factorkappaB activation and nitric oxide production in interleukin-1beta-activated rat hepatocytes. *J Nutr* 2005;**135**:1359–65.
25. Park YH, Chiou GC. Structure–activity relationship between some natural flavonoids and ocular blood flow in the rabbit. *J Ocul Pharmacol Ther* 2004;**20**:34–41.
26. Xu XR, Park YH, Chiou CYG. Effects of dihydrogenation of flavones and number of hydroxy groups in the molecules on ocular blood flow in rabbits and retinal function recovery in rats. *J Ocul Pharmacol Ther* 2004;**20**:311–20.
27. Tunon MJ, Garcia-Mediavilla M, Sanchez-Campos S, Gonzalez-Gallego J. Potential of flavonoids as anti-inflammatory agents: modulation of pro-inflammatory gene expression and signal transduction pathways. *Curr Drug Metab* 2009;**10**:256–71.
28. Kawanishi K, Ueda H, Moriasu M. Aldose reductase inhibitors from the nature. *Curr Med Chem* 2003;**10**:1353–74.
29. Padmaja S, Raju TN. Antioxidant effect of curcumin in selenium induced cataract of Wistar rats. *Indian J Exp Biol* 2004;**42**:601–3.
30. Kowluru RA, Kanwar M. Effects of curcumin on retinal oxidative stress and inflammation in diabetes. *Nutr Metab (Lond)* 2007;**4**:8–9.
31. Suryanarayana P, Saraswat M, Mrudula T, Krishna TP, Krishnaswamy K, Reddy GB. Curcumin and turmeric delays treptozotocin-induced diabetic cataract in rats. *Invest Ophthalmol Vis Sci* 2005;**46**:2092–9.
32. Kumar PA, Suryanarayana P, Reddy PY, Reddy GB. Modulation of alpha-crystallin chaperone activity in diabetic rat lens by curcumin. *Mol Vis* 2005;**11**:561–8.
33. Mandal MN, Patiolla JM, Zheng L, Agbaga MP, Tran JT, Wicker L, et al. Curcumin protects retinal cells from light- and oxidant stress-induced cell death. *Free Radic Biol Med* 2009;**46**:672–9.
34. Mrudula T, Suryanarayana P, Srinivas PN, Reddy GB. Effect of curcumin on hyperglycemia-induced vascular endothelial growth factor expression in streptozotocin-induced diabetic rat retina. *Biochem Biophys Res Commun* 2007;**361**:528–32.

35. Gupta SK, Kumar B, Nag TC, Agrawal SS, Agrawal R, Agrawal P, et al. Curcumin prevents experimental diabetic retinopathy in rats through its hypoglycemic, antioxidant, and anti-inflammatory mechanisms. *J Ocul Pharmacol Ther* 2011;**27**:123–30.

36. Chethan S, Dharmesh SM, Malleshi NG. Inhibition of aldose reductase from cataracted eye lenses by finger millet (*Eleusine coracana*) polyphenols. *Bioorg Med Chem* 2008;**16**:10085–90.

37. Shobana S, Harsha MR, Platel K, Srinivasan K, Malleshi NG. Amelioration of hyperglycemia and its associated complications by finger millet (*Eleusine coracana*) seed coat matter in streptozotocin-induced diabetic rats. *Br J Nutr* 2010;**104**:1787–95.

38. Ansari NH, Awasthi YC, Srivastava SK. Role of glycosylation in protein disulphide formation and cataractogenesis. *Exp Eye Res* 1980;**31**:9–19.

39. Perry RE, Swamy MS, Abraham EC. Progressive changes in lens crystallin glycation and high molecular weight aggregate formation leading to cataract development in streptozotocin-diabetic rats. *Exp Eye Res* 1987;**44**:269–82.

40. Hegde PS, Rajasekaran NS, Chandra TS. Effects of the antioxidant properties of millet species on oxidative stress and glycemic status in alloxan-induced diabetic rats. *Nutr Res* 2005;**25**:1109–20.

41. Sasikala V, Rooban BN, Priya SG, Sahasranamam V, Abraham A. *Moringa oleifera* prevents selenite-induced cataractogenesis in rat pups. *J Ocul Pharmacol Ther* 2010;**26**:441–7.

42. Rao KCS, Divakar S, Babu KN, Rao AGA, Karanth NG, Sattur AP. Nigerloxin, a novel inhibitor of aldose reductase and lipoxygenase with free radical scavenging activity from *Aspergillus niger* CFR-W-105. *J Antibiot (Tokyo)* 2002;**55**:789–93.

43. Suresha BS, Sattur AP, Srinivasan K. Beneficial influence of fungal metabolite nigerloxin on eye lens abnormalities in experimental diabetes. *Can J Physiol Pharmacol* 2012;**90**:387–94.

44. Stitt W. The Maillard reaction in eye diseases. *Ann N Y Acad Sci* 2005;**1043**:582–97.

45. Suresha BS, Srinivasan K. Antioxidant potential of fungal metabolite nigerloxin during eye lens abnormalities in galactose-fed rats. *Curr Eye Res* 2013;**38**:1064–71.

Natural Products and Retinal Ganglion Cells: Protective Roles of Edible Wild Vegetables Against Oxidative Stress in Retinal Ganglion Cells

Sang Hoon Jung

Functional Food Center, Korea Institute of Science and Technology (KIST), Gangneung Institute, Gangneung, Republic of Korea

INTRODUCTION

Glaucoma is a chronic disease that causes progressive damage to the optic nerve and to retinal ganglion cells (RGCs), resulting in visual field damage from the side to the center vision.[1] This condition is expected to affect up to 80 million individuals worldwide by 2020, with at least 6–8 million individuals becoming blind in both eyes.[2,3]

Several risk factors have been suggested to play a role in the pathogenesis of glaucoma, including intraocular pressure (IOP) and mechanical damage to RGC axons,[4] ischemic insults to RGCs axons in the optic nerve head,[5,6] and genetic risks.[7]

Glaucoma has been classified into specific types, including primary open-angle glaucoma and closed-angle glaucoma. These types are specified by the contact between the iris and the trabecular meshwork, which in turn obstruct the aqueous humor from the eye. These types cause increased IOP in the eye, which is the most common risk factor for glaucoma.[4] Increased IOP (normal range 10–20 mm Hg) causes damage to the optic nerve head through the mechanical compression of RGC axons, and causes axonal transport blockade, thereby altering the appropriate nutritional requirements for the survival of RGCs.[8]

Thus, the current therapy for glaucoma is focused on the reduction of IOP.[9] Topical medications that reduce IOP have been used for treatment, including cholinergic agonists, cholinesterase inhibitors, carbonic anhydrase inhibitors, adrenergic agonists, β-blockers, and prostaglandin analogues.[3] Some epidemiologic studies have shown that reducing IOP is effective in slowing the progression of glaucoma in approximately 90% of cases.[4]

Although reduction of IOP is effective in most cases, not all patients with glaucoma are protected from disease progression with such a treatment.[10,11] This is because many glaucoma patients have normal IOP, but exhibit a decreased blood supply to or a weakness in the optic nerve; moreover, not all ocular hypertensive patients develop glaucoma.[11]

According to epidemiologic studies conducted in Korea, the prevalence of glaucoma among the population of rural Sangju in South Korea was 3.4%. Notably, open-angle glaucoma with low IOP was found to be the most common form, accounting for 94.4% of the total number of cases.[12] Greater knowledge of the underlying pathogenesis of glaucoma has resulted in new therapeutic approaches, and the prevention of RGC degeneration rather than IOP reduction is the goal of current treatment strategies.[13–15]

Several putative compounds with neuroprotective effects exist for the potential treatment of glaucoma, including memantine and brimonidine, which is currently in application for clinical trials.[3] Memantine failed its randomized phase III clinical trials in patients with glaucoma as it did not show significant efficacy that was similar to prior studies.[16] This lack of reported efficacy may be due to the various modes of action for neuroprotection in glaucoma, and the difficulty in detecting effects with present methodologies.[17]

Natural products that possess multiple properties, including antioxidant capacity, may provide neuroprotection in glaucoma and may potentially be used to prevent or treat such neurodegenerative diseases.[18] In this chapter, the neuroprotective role of natural products, particularly edible wild vegetables, in RGCs is investigated. One of these putative wild vegetables, *Gymnaster koraiensis*, was shown to have antiapoptotic and antioxidant properties in transformed RGCs, and protective effects on RGCs in an *N*-methyl-D-aspartate (NMDA)-induced excitotoxicity model and a partial optic nerve crush (PONC) model *in vivo*. The potential value of natural products and the possible application of edible wild vegetables as neuroprotective agents for glaucoma are summarized and discussed.

NEUROPROTECTION OF RETINAL GANGLION CELLS

RGC axons are the output of the eye and integrate information from the outer retina to the brain where they synapse at the thalamus, the hypothalamus, and the superior colliculus.[19] Neuroprotection in glaucoma is defined by the preservation and functional maintenance of damaged RGCs through interference with injury and death pathways.[20]

The reasons why and the processes involved in how RGCs actually die in glaucoma are still being investigated, although many studies have been carried out to identify the relevant mechanisms of action. The death of RGCs in glaucoma was shown to occur by apoptosis in various experimental models.[21–24]

Advances in the area of RGC neuroprotection can be achieved using knowledge derived from laboratory studies to predict clinical applications. Several approaches for the neuroprotection of RGCs have been proposed, including stopping or preventing apoptosis, preventing tumor necrosis factor activation, stabilizing Ca^{2+} homeostasis, blocking glutamate excitotoxicity, normalizing mitochondrial dysfunction, inhibiting nitric oxide production, modulating the expression of heat shock proteins, supplying neurotrophins, and improving blood flow to the optic nerve.[14,25–27]

There is a significant amount of evidence implicating the role of oxidative stress in glaucoma.[28] The mechanisms that have been proposed in RGCs, including excitotoxicity, IOP, and ischemia, can induce oxidative stress by various insults, resulting in RGC death.[29]

Reduced antioxidative capacity has been shown in the aqueous humor of glaucoma cases, and reduced glutathione (GSH) and increased malondialdehyde (MDA) plasma levels have also been found in the plasma of glaucoma patients.[30,31] In a clinical correlation study in patients with primary open-angle glaucoma, IOP elevation was shown to cause visual field damage by DNA oxidative damage in the human trabecular meshwork.[32] This finding may explain why oxidative stress is an important factor in the pathogenesis of glaucoma.

Therefore, it would be desirable for neuroprotective substances for RGCs to also possess antioxidant properties. Natural products, with their antioxidative properties, have long been shown to have beneficial effects in humans and can be tolerated when taken regularly. Thus, natural products may be beneficial in the treatment of a chronic disease such as glaucoma.[33]

USE OF NATURAL PRODUCTS TO PROTECT RETINAL GANGLION CELLS

There have been numerous reports of natural products, mostly from plants, with protective effects on RGCs, as summarized in Table 43.1. One of the natural products in clinical trials is *Ginkgo biloba*. *Ginkgo biloba* is a well-known dietary supplement containing many different flavonoids, terpenoids, and other polyphenols that have proven strong antioxidative properties.[73] Two well-known pharmacologic applications of *G. biloba* are in the treatment of blood disorders and the enhancement of memory.[74] Laboratories studies have shown that *G. biloba* enhances microcirculation by improving endothelium-dependent vasodilation,[75] which contributes to *G. biloba* acting as a putative substance for the treatment of glaucoma.

Notably, *G. biloba* extracts can increase ocular blood flow velocity and have various activities, including neuroprotection in RGCs, as shown by clinical studies in patients with normal tension glaucoma (Table 43.1).[54–61]

Green tea is an ancient tea from *Camellia sinensis*, and is widely used worldwide. Green tea contains high amounts of polyphenols, including (+)-epicatechin, (−)-epigallocatechin, (−)-epicatechin gallate, and (−)-epigallocatechin gallate (EGCG).[76] Green tea and its main constituents have various beneficial actions on conditions such as inflammation and thrombosis, as well as age-related neurodegeneration, by increasing learning and memory with a particular focus on the hippocampal formation.[77,78]

EGCG, the main constituent of green tea, has been shown to have beneficial effects on the neuroprotection of RGCs. Orally administered EGCG has been shown to attenuate retinal neuronal death in rats, and inhibit apoptosis caused by various insults, such as ischemia, oxidative stress, and light in transformed RGCs (RGC-5 cells) (Table 43.1).[44–48] EGCG tablets (Epinerve, Sifi, Italy) have been prescribed by ophthalmologists to treat glaucoma in Italy.

TABLE 43.1 Evidence Suggesting Protective Effects of Natural Substances in Retinal Ganglion Cells

Natural Product (Raw Materials)	Biologically Active Compounds	Methods	References
Thuja orientalis	Isoquercitrin	Oxidative stress-induced RGC-5 cell death	Jung et al. (2010)[33]
Carotenoid	Zeaxanthin, astaxanthin, lutein	Oxidative stress-induced RGC-5 cell death	Li and Lo (2010), Nakajima et al. (2008, 2009)[34–36]
Olive oil	Squalene	Alcohol damage in the chick embryo retina	Aguilera et al. (2005)[37]
Scutellaria baicalensis	Baicalin	Ischemia–reperfusion injury in retina, oxidative stress-induced RGC-5 cell death	Jung et al. (2008)[38]
Garlic	S-Allyl-L-cysteine (SAC)	Ischemia–reperfusion injury in retina, hypoxia-induced RGC-5 cell death	Chen et al. (2012)[39]
Phyllostachys nigra	Luteolin 6-C-(6″-O-trans-caffeoylglucoside)	Oxidative stress-induced RGC-5 cell death	Lee et al. (2010)[40]
Lycium barbarum	Polysaccharide	Ocular hypertension in rat, neuroprotection in primary cultured RGCs	Chan et al. (2007), Chiu et al. (2009), Yang et al. (2011)[41–43]
Green tea	Epigallocatechin gallate (EGCG), epicatechin (EC)	Rotenone-induced RGC-5 cell death, light-induced cell death in RGC-5 cell, ischemia–reperfusion injury in retina	Kamalden et al. (2012), Osborne et al. (2010), Zhang and Osborne (2006), Zhang et al. (2007, 2008)[44–48]
Brazilian green propolis	Dicaffeoylquinic acid	Oxygen–glucose deprivation-induced RGC-5 cell death, NMDA in rat	Inokuchi et al. (2006), Nakajima et al. (2007, 2009)[49–51]
Morus alba	–	Diabetes and hypercholesterolemia in mother rat	El-Sayyad et al. (2011)[52]
Rhus verniciflua	Fustin, fisetin, sulfuretin, butein	Oxidative stress-induced RGC-5 cell death	Choi et al. (2012)[53]
Ginkgo biloba	–	Clinical study for normal tension glaucoma, chronic glaucoma model by IOP elevation in cautery of three episcleral vessels in rats, optic nerve crush in rat	Chung et al. (1999), Hirooka et al. (2004), Ma et al. (2009, 2010), Park et al. (2011), Quaranta et al. (2003), Shim et al. (2012), Wang et al. (2011)[54–61]
Zuogui pill	–	Optic nerve crush in rat	Wang et al. (2011)[62]
Annatto (*Bixa orellana*)	Bixin	Endoplasmic reticulum stress-induced retinal damage, tunicamycin-induced cell death in rat	Tsuruma et al. (2012)[63]
Bilberry (*Vaccinium myrtillus*)	Anthocyanins	Oxidative stress-induced RGC-5 cell death, NMDA-induced cell death in rat	Matsunaga et al. (2009)[64]
Crocus sativus, *Gardenia jasminoides*	Crocetin	Light-induced retinal damage in rat	Yamauchi et al. (2011)[65]
Curcuma longa	Curcumin	Staurosporine-induced retinal damage in mice	Burugula et al. (2011)[66]
	Diosmin	Ischemia–reperfusion injury in retina	Tong et al. (2012)[67]
Eisenia bicyclis	Phlorotannins	Oxidative stress-induced RGC-5 cell death, NMDA-induced cell death in rat	Kim et al. (2012)[68]
Grape	Resveratrol	Endoplasmic reticulum stress-induced retinal damage in RGC-5 cells, ischemia–reperfusion injury in mice	Li et al. (2012), Yang et al. (2012)[69,70]
–	Flavonoids	Oxidative stress-induced RGC-5 cell death	Maher and Hanneken (2005)[71]
Aegle marmelos fruit	–	Water loading and steroid-induced models	Agarwal et al. (2009)[72]

RGC-5: transformed retinal ganglion cells; NMDA: N-methyl-D-aspartate; IOP: intraocular pressure.

The bioavailability of green tea is poor and small molecules, including epicatechin and epigallocatechin, have higher bioavailability than large molecules, such as EGCG.[79] Therefore, increasing the bioavailability of EGCG as the active compound in green tea is very important for the development of pharmaceuticals or dietary supplements.

Carotenoids are tetraterpenoids and can be classified into two specific classes: xanthophylls and carotenes. The chemical structures of carotenoids are shown in Figure 43.1. These molecules can absorb blue light, which can protect the eye, and particularly the macula lutea, from damaging blue and near-ultraviolet (UV) light. In this regard, oxygen-containing carotenoids (xanthophylls) such as lutein and zeaxanthin have been shown to play an important role in the prevention of age-related macular degeneration.[80] Given that humans are not able to synthesize xanthophylls, these must be consumed from natural products, including certain fruits, vegetables, and eggs.[81] Several studies have shown that lutein and zeaxanthin have beneficial antioxidative properties for neuroprotection against oxidative stress-induced RGC damage (Table 43.1).[34–36]

The well-known natural antioxidants resveratrol and anthocyanins, which are found in grapes and berries, respectively, have been shown to have protective effects on endoplasmic reticulum stress-induced retinal damage (Table 43.1).[64,69,70] Propolis has also been suggested as a putative natural substance for RGC protection, with effects shown both *in vitro* and *in vivo* (Table 43.1).[49–51]

Previous studies have shown that several natural products and their isolated compounds have protective effects on oxidative stress-induced RGC death. Flavones, such as isoquercitrin isolated from *Thuja orientalis*, were found to be effective at blunting the negative influence of oxidative insults to RGCs (Table 43.1).[33]

There are many reports on the biologic effects of *Scutellaria baicalensis*. Its active compound, baicalin, was shown to have neuroprotective effects on ischemia–reperfusion in the hippocampi of gerbils via antioxidative and antiapoptotic pathways.[82] Moreover, baicalin was found to attenuate inflammatory reactions and cerebral ischemia injury in rats, involving toll-like receptor 2 and 4 (TLR2/4) and the downstream nuclear factor-κB (NF-κB).[83] The neuroprotective effect of baicalin was also demonstrated by its inhibition

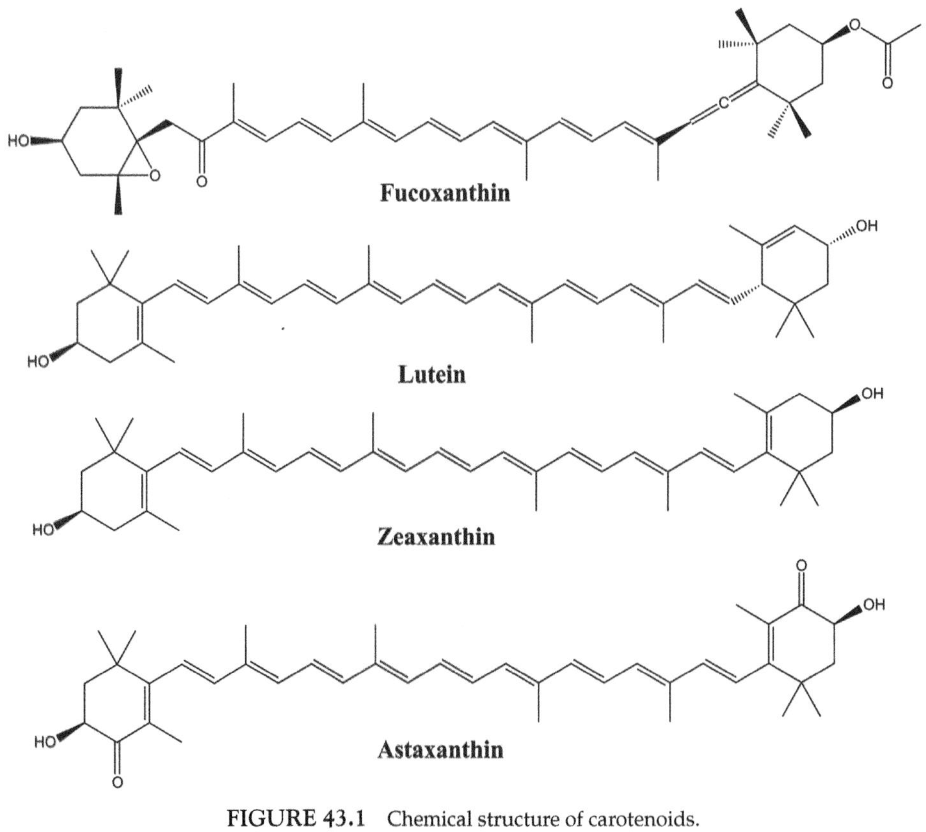

FIGURE 43.1 Chemical structure of carotenoids.

of NMDA receptor-mediated 5-lipoxygenase activation in rat cortical neurons.[84] Studies have shown that baicalin is a powerful neuroprotectant in RGCs. Baicalin attenuated ischemia–reperfusion injury in rats and oxidative stress-induced transformed RGC (RGC-5) death.[38]

Black bamboo (*Phyllostachys nigra*) and *Rhus verniciflua* have also been suggested as possible natural products for the treatment of glaucoma (Table 43.1).[40,53]

It has been reported that the hydroxyl groups of flavonoids at the catechol positions are effective in preventing microsomal lipid peroxidation and enable increased antioxidative properties.[85] Many flavonoids have catechol moieties and are potentially able to scavenge free radicals, including peroxynitrite.[86] This ability may be due to their strong reactivity with free radicals, and their efficient metal chelation and electron delocalization.[87,88]

The neuroprotective effects of isoquercitrin isolated from *T. orientalis*, baicalin isolated from *S. baicalensis*, and luteolin (6-C-6″-O-*trans*-caffeoylglucoside) isolated from *P. nigra* appear to be primarily related to the antioxidant characteristics of these substances. These compounds each have a catechol group, which may underlie the protective effects of the above compounds on RGCs.

In a recent study investigating the effects of seaweed on RGCs, the ethanol extract of *Eisenia bicyclis* was found to protect RGCs from NMDA-induced excitotoxicity in rat retinas and oxidative stress-induced RGC death. These protective effects on RGCs are due to the antioxidative properties of the extract, with phlorotannins being the likely active compounds (Table 43.1).[68] The beneficial effects of *E. bicyclis* may also be due in part to fucoxanthin (Fig. 43.1), a derivative of carotenoids that was isolated from *E. bicyclis* by centrifugal partition chromatography.[89]

Other suggested natural products that have been demonstrated to be effective in various experimental models are shown in Table 43.1.

USE OF WILD VEGETABLES TO PROTECT RETINAL GANGLION CELLS

Wild vegetables grown in Korea are consumed as healthy food, and several species have also been used as fermented food, particularly Kimchi. It is known that wild vegetables contains higher levels of minerals and vitamins than cultivated vegetables.[90] Moreover, it has also been reported that wild vegetables have higher carotene and calcium concentrations than locally cultivated species.[91] This may be because wild vegetables are under the constant stress of harsh conditions, including competition with other plants and with insects.

In recent years, the multiple roles of wild vegetables as both food and medicinal sources have been widely documented. Despite several publications on the importance of edible wild vegetables used in Korea, data have not been collected systematically and detailed biologic studies have not been conducted.

Research has investigated whether edible wild vegetables grown in Pyeongchang, Korea, exhibit any biologic activities (Daegwanryeong project), including neuroprotective effects in transformed RGCs (RGC-5). In particular, studies have focused on the family of Compositae; the list of vegetables belonging to this family is shown in Table 43.2. In total, 50 wild vegetables from the Compositae family consumed in Korea were tested for protective effects on oxidative stress-induced RGCs. Among the vegetables investigated, *Gymnaster koraiensis* N. showed the highest protective effect, and additional studies were carried out to further clarify this effect.

Protective Effect of *Gymnaster koraiensis* on Oxidative Stress-Induced Retinal Ganglion Cells

Gymnaster koraiensis N. (Compositae) (Fig. 43.2) is a wild vegetable that grows in Korea and has been used as a source of food in this country for many years. The reported biologic activities of *G. koraiensis* include inhibitory activities on AKR1B10 to prevent cancer,[92] on osteoclast formation by suppressing NFATc1 and DC-STAMP expression,[93] on aldose reductase and advanced glycation end-products,[94] and on nuclear factor of activated T cells (NFAT) transcription factor,[95] as well as hepatoprotective effects.[96]

The chemical constituents of *G. koraiensis* have been reported to include polyacetylenes, benzofurans, terpenes, squalenes, and several flavonoids [14,94,97-99] (Fig. 43.3), as well as dicaffeoylquinic acids. Brazilian green propolis also contains high amounts of dicaffeoylquinic acid, which has been shown to have significant neuroprotective effects on oxidative stress-induced retinal damage.[49] Therefore, dicaffeoylquinic acid may be the chemical constituent underlying the neuroprotective effects of *G. koraiensis*.

Many glaucoma risk factors lead to the overproduction of reactive oxygen species (ROS), which may play a major role in the demise of RGCs, as discussed in the 'Introduction' section.[100] It is clear that oxidative stress distorts intercellular homeostasis and that reducing oxidative stress in target tissues is a strategy worthy of consideration for the treatment of glaucoma. *Gymnaster koraiensis* has antioxidative properties that were shown to significantly reduce the production of ROS caused by different sources of oxidative stress, such as hydrogen peroxide (H_2O_2), $O_2{}^{\cdot-}$, or $\cdot OH$.[99] These findings indicate that *G. koraiensis* is effective in reducing ROS in RGCs.

TABLE 43.2   List of Compositae

No.	Code	Scientific Name	Parts
1	D-002	*Saussurea pulchella* (Fisch.) Fisch.	Whole plant
2	D-011	*Gymnaster koraiensis* (Nakai) Kitam.	Leaves, stem, and flower
3	D-014	*Cirsium nipponicum* (Maxim.) Makino	Leaves and stem
4	D-015	*Aster glehni* var. *hondoensis* kitamura	Leaves
5	D-016	*Solidago virgaurea subsp. gigantea* (Nakai) kitam	Leaves and stem
6	D-021	*Atractylodes ovata* (Thunb.) DC.	Leaves and stem
7	D-026	*Achillea alpina* (Ledeb)	Leaves and stem
8	D-027	*Lactuca indica* L. var. *laciniata* (O.Kuntze) Hara	Leaves, stem, and flower
9	D-037	*Lactuca indica* L. var. *laciniata* Hara	Leaves and stem
10	D-039	*Adenocaulon himalaicum* Edgew.	Whole plant
11	D-041	*Youngia sonchifolia* Maxim.	Leaves, stem, and flower
12	D-043	*Youngia denticulata* Kitamura	Whole plant
13	D-044	*Taraxacum officinale* Weber	Whole plant
14	D-050	*Artemisia dubia* Wall.	Leaves
15	D-058	*Ligularia fischeri* (Ledeb.) Turcz.	Leaves and stem
16	D-060	*Ligularia fischeri* var. *spiciformis* Nakai	Leaves and stem
17	D-061	*Ligularia fischeri* (Ledeb.) Turcz.	Leaves and stem
18	D-062	*Cirsium setidens* (Dunn) Nakai	Leaves and stem
19	D-066	*Lactuca indica* L. var. *laciniata* (O.Kuntze) Hara	Leaves
20	D-067	*Ligularia fischeri* (Ledeb.) Turcz.	Leaves, stem, and flower
21	D-068	*Aster scaber* Thunberg	Leaves, stem, and flower
22	D-076	*Taraxacum coreanum* Nakai	Leaves
23	D-080	*Carpesium abrotanoides* L.	Whole plant
24	D-082	*Artemisia gmelini* Weber ex Stechm	Leaves, stem, and flower
25	D-083	*Aster tataricus* L.f.	Whole plant
26	D-084	*Sonchus brachyotus* DC.	Whole plant
27	D-090	*Cirsium japonicum* var. *maackii Cirsium japonicum* var. *ussuriensis* (Kitam)	
28	D-094	*Synurus deltoides* (Aiton) Nakai.	
29	D-099	*Cirsium pendulum* Fisch. Ex DC.	
30	D-102	*Syneilesis palmata* (Thunb.) Maxim	
31	D-103	*Crepidiastrum chelidoniifolium* (Makino)	
32	D-110	*Erigeron canadensis* L.	
33	D-111	*Erigeron canadensis* L.	
34	D-112	*Erigeron annuus* L.	
35	D-122	*Solidago virgaaurea* var. *asiatica* Nakai	Leaves and shoot
36	D-135	*Breea segeta* Kitam.	Leaves and shoot
37	D-136	*Carpesium macrocephalum* Fr.	Leaves

TABLE 43.2 List of Compositae—cont'd

No.	Code	Scientific Name	Parts
38	D-142	*Parasenecio auriculata* var. *kamtschatica* (Maxim.) H.Koyama	Leaves and shoot
39	D-147	*Saussurea macrolepis* (Nakai) Kitam.	Leaves and shoot
40	D-148	*Saussurea seoulensis* Nakai	Leaves and shoot
41	D-153	*Synurus excelsus* (Makino) Kitam.	Leaves
42	D-156	*Lactuca triangulata* Maxim.	Leaves and stem
43	D-164	*Carpesium cernuum* L.	Leaves and stem
44	D-167	*Parasenecio auriculata* var. *kamtschatica* (Maxim.) H.Koyama	Leaves and stem
45	D-168	*Syneilesis palmata* (Thunb.) Maxim.	Leaves and stem
46	D-169	*Ixeris repens* (L.) A.Gray	Leaves, stem, and root
47	D-175	*Xanthium strumarium* L.	Stem and fruit
48	D-176	*Picris hieracioides* var. *koreana* Kitam.	Leaves, stem, and flower
49	D-187	*Helianthus tuberosus* L.	Leaves and stem
50	D-189	*Sigesbeckia glabrescens* (Makino) Makino	Leaves, stem, and flower

Fifty wild vegetables of Compositae were collected from 2007 to 2010 in the vicinity of Gangneung, Korea, and the voucher specimens were deposited at the Herbarium of KIST Gangneung, Korea.

FIGURE 43.2 *Gymnaster koraiensis.* The *G. koraiensis* was collected in 2009 from the vicinity of Gangneung, Korea, and the voucher specimen (voucher no. D-011) was deposited at the Herbarium of KIST Gangneung, Korea.

Studies by Wei *et al.* showed that nuclear factor erythroid 2-related factor 2 (Nrf2) is cytoprotective against neuronal and capillary degeneration in a model of retinal ischemia–reperfusion injury. The authors suggested that Nrf2 could be a new therapeutic target for retinal diseases.[101]

The author's previous studies investigated the cytoprotective effects of isolated compounds from *G. koraiensis* on oxidative stress-induced cytotoxicity in HepG2 cells.[96] Gymnasterkoreayne B, a polyacetylene derivative isolated from *G. koraiensis*, was shown to induce detoxification enzymes through the nuclear translocation of Nrf2, which is known to regulate antioxidant responsive element (ARE)-driven phase II detoxification genes.[102] Thus, these data suggest that the neuroprotective effects of *G. koraiensis* in RGCs may be due to both direct and indirect antioxidant mechanisms.

Online high-performance liquid chromatography (HPLC)–2,2'-azino-bis(3-ethylbenzothiazoline-6-sulfonic acid) (ABTS+), an online rapid screening method, was used to identify the antioxidative compound in *G. koraiensis* (Fig. 43.4). The detection of antioxidative compounds from complex extracts is possible by the online coupling of HPLC separation. The HPLC separation eluate is mixed with a stabilized solution of ABTS+ radicals from different HPLC pumps, and the mixed solution is detected with a UV/Vis detector monitoring absorbance at 734 nm. The ABTS+ radicals scavenging detection chromatogram is detected as a negative peak on the absorbance profile. The more rapidly the absorbance decreases, the more potent the antioxidant activity of the compound in terms of hydrogen-donating ability.[103]

Figure 43.4 shows the online HPLC–ABTS+ analysis of an extract of *G. koraiensis*. From the combined UV (positive signals) and ABTS+ quenching (negative signals) chromatograms, one major compound was identified as showing the highest free radical scavenging activity. The major compound was isolated from the ethyl acetate fraction of *G. koraiensis* (EAGK) using Diaion HP-20 and preparative HPLC, and the chemical structure was elucidated by spectral analysis as 3,5-di-*O*-caffeoylquinic acid (3,5-DCQA).[99] Both the EAGK and the isolated 3,5-DCQA

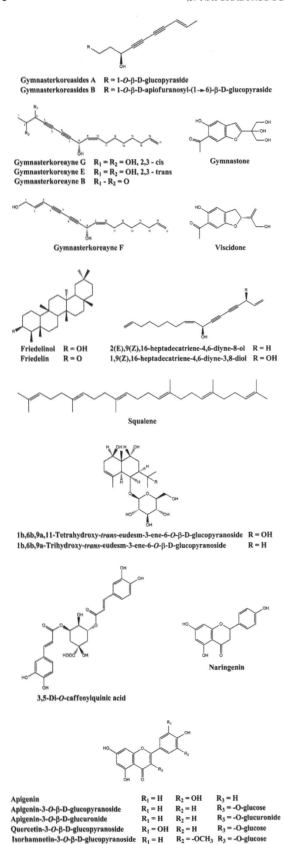

Gymnasterkoreasides A   R = 1-*O*-β-D-glucopyraside
Gymnasterkoreasides B   R = 1-*O*-β-D-apiofuranosyl-(1→6)-β-D-glucopyraside

Gymnasterkoreayne G   R₁ = R₂ = OH, 2,3 - cis
Gymnasterkoreayne E   R₁ = R₂ = OH, 2,3 - trans
Gymnasterkoreayne B   R₁ - R₂ = O

Gymnastone

Gymnasterkoreayne F

Viscidone

Friedelinol   R = OH
Friedelin   R = O

2(E),9(Z),16-heptadecatriene-4,6-diyne-8-ol   R = H
1,9(Z),16-heptadecatriene-4,6-diyne-3,8-diol   R = OH

Squalene

1b,6b,9a,11-Tetrahydroxy-*trans*-eudesm-3-ene-6-*O*-β-D-glucopyranoside   R = OH
1b,6b,9a-Trihydroxy-*trans*-eudesm-3-ene-6-*O*-β-D-glucopyranoside   R = H

Naringenin

3,5-Di-*O*-caffeoylquinic acid

	R₁	R₂	R₃
Apigenin	R₁ = H	R₂ = OH	R₃ = H
Apigenin-3-*O*-β-D-glucopyranoside	R₁ = H	R₂ = H	R₃ = -O-glucose
Apigenin-3-*O*-β-D-glucuronide	R₁ = H	R₂ = H	R₃ = -O-glucuronide
Quercetin-3-*O*-β-D-glucopyranoside	R₁ = OH	R₂ = H	R₃ = -O-glucose
Isorhamnetin-3-*O*-β-D-glucopyranoside	R₁ = H	R₂ = -OCH₃	R₃ = -O-glucose

FIGURE 43.3 Chemical composition of *Gymnaster koraiensis*. Several compound derivatives have been isolated from *G. koraiensis*, including polyacetylenes, terpenes, squalenes, caffeoylquinic acids, and flavonoids.

were found to protect RGC-5 cells against H_2O_2-induced oxidative stress in a concentration-dependent manner (Fig. 43.5A-C).[99]

H_2O_2-induced cell death in RGC-5 cells involves changes in the expression of various apoptotic proteins: cleaved poly (ADP-ribose) polymerase (PARP), apoptosis-inducing factor (AIF), and cleaved caspase-3 (Fig. 43.5D). EAGK and 3,5-DCQA were shown to attenuate the upregulation of cleaved PARP and cleaved caspase-3 proteins caused by exposing cultures to H_2O_2.[99] Moreover, H_2O_2 is known to cause AIF nuclear translocation. However, EAGK and 3,5-DCQA were found to significantly reduce the upregulation of AIF proteins in a concentration-dependent manner.[99]

Glutamate neurotoxicity caused by excessive excitatory neurotransmitter glutamate has been shown to cause glaucoma.[104] Excessive glutamate binds to the NMDA receptor and to other receptor subtypes, and is thought to activate an intracellular Ca^{2+} influx that increases ROS production and causes neurotoxicity.[105] Therefore, blocking glutamate neurotoxicity by inhibiting the NMDA receptor and/or decreasing glutamate release is a target strategy for neuroprotection in glaucoma.

Exposure to NMDA causes thinning of retinal thickness in the inner plexiform layer (IPL) of the retina and the apoptosis of RGCs, as shown by terminal deoxynucleotidyl transferase dUTP nick-end labeling (TUNEL) in the ganglion cell layer (GCL) (Fig. 43.6). EAGK and 3,5-DCQA were found to have protective effects on the thinning of the IPL and an increased number of TUNEL-positive cells in NMDA-induced rat retinas.[99] The neuroprotective effects of EAGK and 3,5-DCQA may be due not only to their direct antioxidative-mediated antiapoptotic effects, but also to their role in modulating the downstream signaling pathways involved in NMDA-induced excitotoxicity.

PONC is commonly used in the study of glaucoma and has been shown to cause chronic glaucoma that has been well characterized by optic nerve transections. PONC can be carried out by clamping the optic nerve for several minutes after surgically exposing the optic nerve.[106] PONC is known to induce the generation of ROS, causing the slow, chronic, and synchronous death of RGCs.[107]

To determine whether the EAGK has a neuroprotective effect in the retina, the PONC model with retrograde labeling using Fluoro-Gold into the superior colliculus was used (Fig. 43.7). Fluorescence in the Fluoro-Gold-labeled RGCs was observed 1 week after PONC. As shown in Figure 43.7, the average density of RGCs was significantly decreased by PONC; however, pretreatment with EAGK significantly decreased the loss of RGCs.

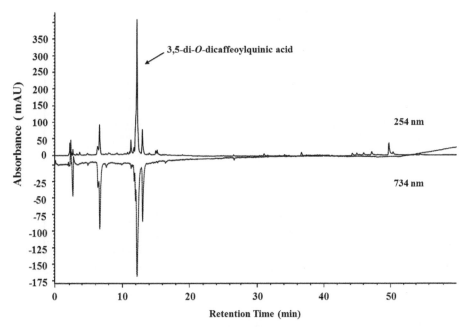

FIGURE 43.4 Online high-performance liquid chromatography (HPLC) 2,2′-azino-bis(3-ethylbenzothiazoline-6-sulfonic acid) (ABTS⁺) analysis of the ethyl acetate fraction of *Gymnaster koraiensis*. The column conditions were as follows: an Eclipse SB-C_{18} Rapid Resolution column (150 mm × 4.6 mm inner diameter, 3.5 μm, Agilent) was used; the column temperature was maintained at 25°C; and the mobile phase consisted of 0.1% trifluoroacetic acid (TFA; solvent A) and acetonitrile (solvent B) with a flow rate of 0.7 mL/minute. The gradient program conditions were as follows: 0–5 minutes, 10% to 20% B in A; 5–15 minutes, 20% B; 15–30 minutes, 20% to 35% B; 30–35 minutes, 35% B; 35–43 minutes, 35% to 100% B; 43–45 minutes, 100% to 10% B; 45–55 minutes, equilibration at 10% B; diode array detector at 254 nm (positive trace) prior to reaction with ABTS⁺ radicals and the analysis of antioxidant potential at 734 nm (negative trace); 5 mg/mL sample concentration; and 10 μL of injection volume. *Source: Unpublished data.*

CONCLUSIONS

The neuroprotective effects were investigated of edible wild vegetables collected from Pyeongchang, Korea, in transformed RGCs. Among them, *G. koraiensis* and 3,5-DCQA were found to possess potent neuroprotective effects both *in vitro* and *in vivo*. These effects may be due to their antiapoptotic properties by preventing the activation of caspase-3, PARP, and AIF (Fig. 43.8). The neuroprotective effects of *G. koraiensis* and 3,5-DCQA may also result from their antioxidant properties and involve the replenishment of the intracellular reduced GSH (rGSH):oxidized GSH (GSSG) ratio and the detoxification of H_2O_2 by stimulating antioxidant enzymes such as catalase and Gpx-1 (Fig. 43.8).[99]

It has been reported that excessive NMDA induces neurotoxicity in RGCs. A clinical study has shown that the level of glutamate in the vitreous body of glaucoma patients is two-fold higher than that observed in a control population of patients with only cataract.[108]

The NMDA receptor has a relatively high permeability to Ca^{2+} ions, which increase the production of ROS. Moreover, calcium overload changes mitochondrial membrane potential and induces apoptosis in the retina.[109,110] Intravitreal injection of NMDA caused a decrease in the levels of antioxidant proteins, such as Cu/Zn superoxide dismutase (SOD-1), catalase, and glutathione peroxidase-1.[99] However, the decreased level of antioxidant proteins was found to be attenuated by EAGK and 3,5-DCQA,[99] which may be explained by their ability to block the NMDA receptor or by indirectly reducing ROS production caused by the excessive activation of NMDA (Fig. 43.8).

Gymnaster koraiensis has long been used as a source of food and medicine in East Asia. Therefore, a major advantage of its use is that it is known to be well tolerated, even when considerable amounts are consumed. Edible wild vegetables, including *G. koraiensis*, may have a potential role in the prevention of glaucoma by inhibiting the overproduction of ROS. Most active compounds in natural products are phenolic compounds because of their antioxidant properties. 3,5-DCQA, the active compound in *G. koraiensis*, is also a phenolic compound and may serve as a biologically active compound template for standardization in the development of natural product medicines.

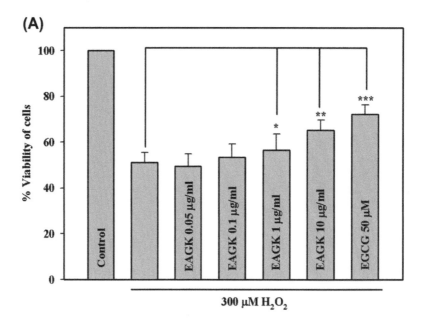

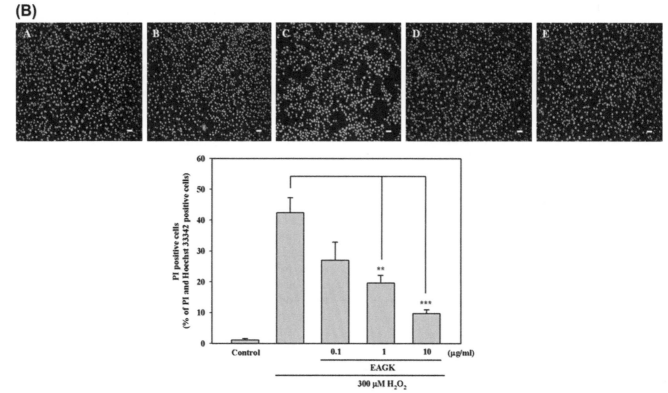

FIGURE 43.5 (A) Cell viability assay of the transformed retinal ganglion cells (RGC-5) exposed to 300 μM hydrogen peroxide (H_2O_2). The viability of the RGC-5 cells that were exposed to H_2O_2 was concentration dependent, with 300 μM H_2O_2 causing death in approximately 50% of the cells. The ethyl acetate fraction of *Gymnaster koraiensis* (EAGK) at 1 μg/mL and 10 μg/mL (as well as 50 μM epigallocatechin 3-gallate (EGCG)) significantly attenuated H_2O_2-induced cell death. Experimental values are expressed as the percentage of viable cells, with error bars indicating ±SEM from four independent experiments. *$P < 0.05$, **$P < 0.01$, and ***$P < 0.001$ indicate statistically significant differences compared with the cells exposed to H_2O_2 alone. (B) Microscopic analysis of RGC-5 cells using propidium iodide (PI) and Hoechst 33342 double staining. The RGC-5 cells were incubated with EAGK before exposure to 300 μM H_2O_2 for 24 hours, and then examined using fluorescence microscopy after staining with Hoechst 33342 and PI. Early apoptotic and necrotic cells were stained with PI (red), whereas apoptotic cells, having condensed and fragmented nuclei, were stained with Hoechst 33342 (blue). Control cells (A), and cells treated with 300 μM H_2O_2 (B), 300 μM H_2O_2 plus EAGK (0.1 μg/mL) (C), 300 μM H_2O_2 plus EAGK (1 μg/mL) (D), and 300 μM H_2O_2 plus EAGK (10 μg/mL) (E) clearly showed that EAGK blunted the negative effect of 300 μM H_2O_2, resulting in fewer red-stained cells. PI-positive cells were counted using a cell counter under a fluorescence microscope at 100× magnification, and four representative images were used to estimate the percent of PI-positive cells out of the total cell numbers (a minimum of 200 cells/well were counted), as shown in (F). The scale bar represents 50 μm. Experimental values are expressed as PI-positive cells, with error bars indicating mean±SEM from three independent experiments (**$P < 0.01$, ***$P < 0.001$).

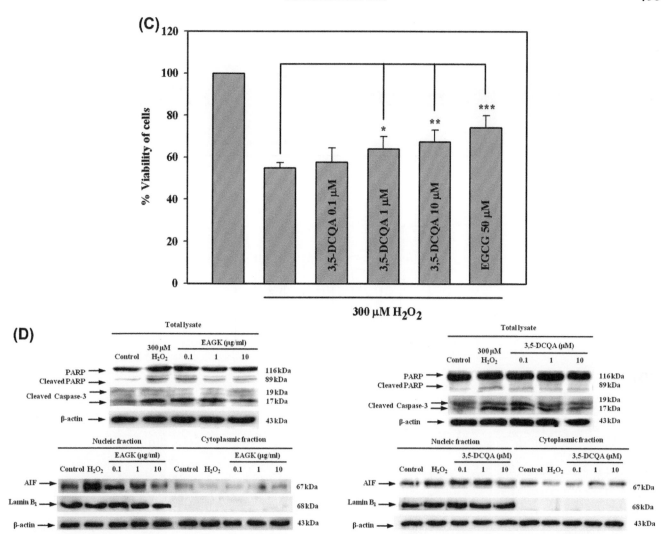

FIGURE 43.5 Cont'd (C) Effect of 3,5-di-*O*-caffeoylquinic acid (3,5-DCQA) isolated from EAGK on the viability of RGC-5 cells exposed to 300 μM H_2O_2 for 24 hours in culture, as measured by the MTT assay. Clearly, 1 μM and 10 μM 3,5-DCQA (as well as 50 μM EGCG) dose-dependently attenuated H_2O_2-induced cell death. Experimental values are expressed as a percentage of viable cells, with error bars indicating ±SEM from four independent experiments. *P<0.05, **P<0.01, and ***P<0.001 indicate statistically significant differences in the cells exposed to H_2O_2. (d) Evaluation of the levels of various apoptotic proteins (poly (ADP-ribose) polymerase (PARP), apoptosis-inducing factor (AIF), and cleaved caspase-3) in RGC-5 cells exposed to 300 μM H_2O_2, with or without either EAGK or 3,5-DCQA treatment. Representative Western blot shows the expression levels of the apoptotic proteins in the total lysates, nucleic fractions, and cytoplasmic fractions isolated from the RGC-5 cells exposed to 300 μM H_2O_2, with or without either EAGK or 3,5-DCQA treatment. The Western blot densitometric analysis of apoptotic protein levels in the RGC-5 cells is shown. Protein level values are expressed as mean ±SEM from three independent experiments. *Source: Data from* Kim et al. *(2011).*[99] *Reproduced with permission from Elsevier.*

TAKE-HOME MESSAGES

- Natural products may be putative substances for the prevention and treatment of glaucoma by enabling retinal ganglion cell (RGC) neuroprotection.
- Numerous reports have been published demonstrating the protective effects of natural products on RGCs.
- *Gymnaster koraiensis* prevents RGC death caused by oxidative stress.

- *Gymnaster koraiensis* has potent direct and/or indirect antioxidant activities.
- *Gymnaster koraiensis* protects against retinal degeneration caused by N-methyl-D-aspartate and optic nerve crush *in vivo*.
- 3,5-Di-*O*-caffeoylquinic acid is the active compound in *G. koraiensis*.
- Edible wild vegetables may have a potential role in the prevention of glaucoma through inhibition of the overproduction of reactive oxygen species.

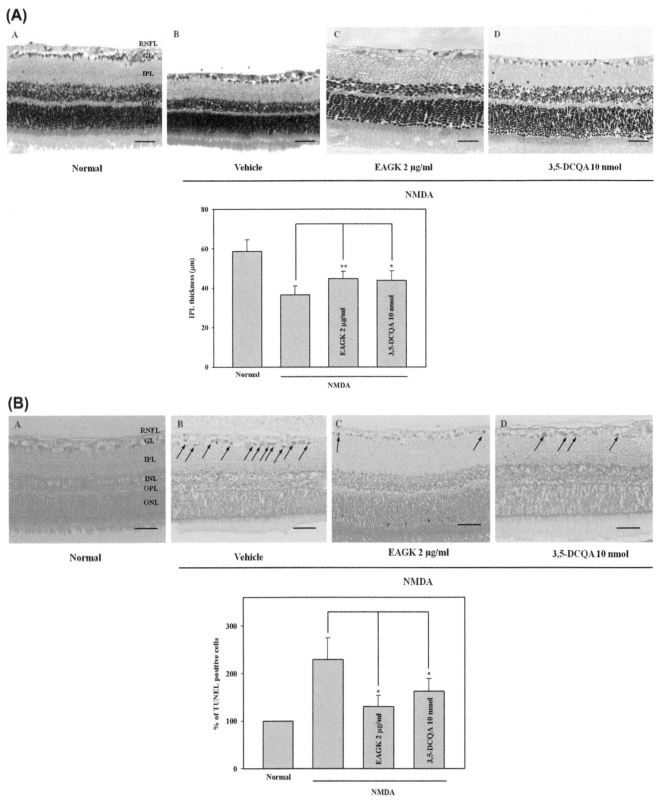

FIGURE 43.6 (A) Representative photomicrographs showing the histologic appearance of the retina induced by an intravitreous injection of *N*-methyl-D-aspartate (NMDA; hematoxylin and eosin staining, 400×). Non-treated (A), NMDA-treated (B), NMDA (5 nmol) plus the ethyl acetate fraction of *Gymnaster koraiensis* (EAGK, 2 μg/mL)-treated (C), and NMDA (5 nmol) plus 3,5-di-*O*-caffeoylquinic acid (3,5-DCQA; 10 nmol)-treated (D) retinal cross-sections after 7 days with or without NMDA are shown. The scale bar represents 50 μm. (F) shows the thickness of the inner plexiform layer (IPL). The results shown are the mean values with error bars indicating mean ± SEM from six independent experiments (*P < 0.05, **P < 0.01). (B) Antiapoptotic effect of EAGK or 3,5-DCQA demonstrated by the terminal deoxynucleotidyl transferase dUTP nick-end labeling (TUNEL) assay (400×). Retinal damage was induced with an intravitreous injection of NMDA. Non-treated (A), NMDA-treated (B), NMDA (5 nmol) plus EAGK (2 μg/mL)-treated (C), and NMDA (5 nmol) plus 3,5-DCQA (10 nmol)-treated (D) retinal cross-sections after 24 hours with or without NMDA are shown. The arrows indicate TUNEL-positive cells (brown stain). The number of TUNEL-positive cells increased after NMDA injection, but NMDA (5 nmol) plus EAGK at 2 μg/mL and 3,5-DCQA at 10 nmol decreased the NMDA-induced retinal damage. The scale bar represents 50 μm. (F) shows the TUNEL-positive cells. The results are the mean values with error bars indicating mean ± SEM from six independent experiments (*P < 0.05). *Source: Data from Kim et al. (2011).[99] Reproduced with permission from Elsevier.* (See color plate section)

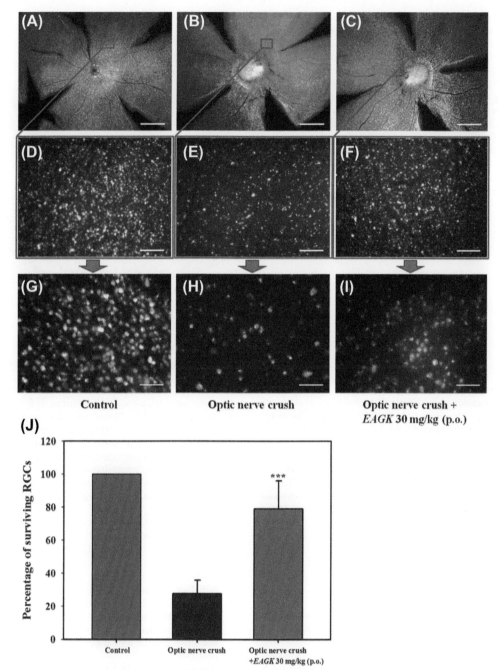

Control **Optic nerve crush** **Optic nerve crush +**
EAGK **30 mg/kg (p.o.)**

FIGURE 43.7 Fluoro-Gold-labeled retinal ganglion cells (RGCs) in the mouse 1 week after partial optic nerve crush (PONC). Retrograde-labeled RGCs of mice with uninjured and injured optic nerves are shown. RGCs were labeled by injecting 3% Fluoro-Gold into the superior colliculi of the brain. The figure shows representative micrographs of normal retina (A) and damaged retina 12 days after PONC with (C) or without the ethyl acetate fraction of *Gymnaster koraiensis* (EAGK, B). The scale bars in (A), (B), and (C) represent 500 μm. Low-magnification (200×) images of labeled RGCs designated by the dark red boxes in (A), (B), and (C) are shown in (D), (E), and (F). The scale bars in (D), (E), and (F) represent 100 μm. High-magnification (400×) images of (A), (B), and (C) are shown in (G), (H), and (I). The scale bars in (G), (H), and (I) represent 50 μm. The results showed a significant increase in RGC survival after EAGK treatment compared with vehicle treatment. (J) Experimental values are expressed as the percentage of surviving RGCs with error bars indicating mean ± SEM from three independent experiments (***P < 0.001). *Source: Unpublished data.*

FIGURE 43.8 Proposed mechanisms of action underlying the protective effects of *Gymnaster koraiensis* on oxidative stress-induced retinal damage. *Gymnaster koraiensis* can attenuate oxidative-stress induced retinal ganglion cells by inhibiting apoptotic proteins such as caspase-3, poly (ADP-ribose) polymerase (PARP) and apoptosis-inducing factor (AIF), and by inhibiting radical species and lipid peroxidation. Bcl-2: B-cell lymphoma 2; CAT: catalase; EAGK: ethyl acetate fraction of *Gymnaster koraiensis*; GPx: glutathione peroxidase; GSH: glutathione; GSSG: oxidized glutathione disulfide; NMDA: *N*-methyl-D-aspartate; NO: nitric oxide; NOS: nitric oxide synthase; RGCs: retinal ganglion cells; ROS: reactive oxygen species; SOD: superoxide dismutase. (See color plate section)

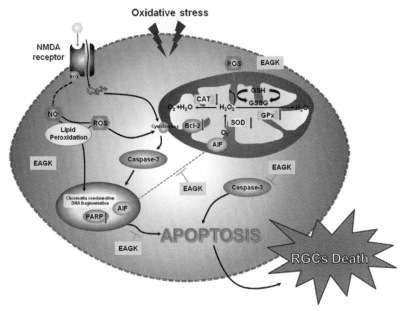

References

1. McKinnon SJ. Glaucoma, apoptosis, and neuroprotection. *Curr Opin Ophthalmol* 1997;**8**:28–37.
2. Quigley HA. Glaucoma. *Lancet* 2011;**377**:1367–77.
3. Zhang K, Zhang L, Weinreb RN. Ophthalmic drug discovery: novel targets and mechanisms for retinal diseases and glaucoma. *Nat Rev Drug Discov* 2012;**11**:541–59.
4. AGIS Investigators. The Advanced Glaucoma Intervention Study (AGIS): 7. The relationship between control of intraocular pressure and visual field deterioration. *Am J Ophthalmol* 2000;**130**:429–40.
5. Morgan JE. Optic nerve head structure in glaucoma: astrocytes as mediators of axonal damage. *Eye (Lond)* 2000;**14**(Pt 3B):437–44.
6. Flammer J, Orgul S, Costa VP, Orzalesi N, Krieglstein GK, Serra LM, et al. The impact of ocular blood flow in glaucoma. *Prog Retin Eye Res* 2002;**21**:359–93.
7. Wolfs RC, Klaver CC, Ramrattan RS, van Duijn CM, Hofman A, de Jong PT. Genetic risk of primary open-angle glaucoma. Population-based familial aggregation study. *Arch Ophthalmol* 1998;**116**:1640–5.
8. Anderson DR, Hendrickson A. Effect of intraocular pressure on rapid axoplasmic transport in monkey optic nerve. *Invest Ophthalmol Vis Sci* 1974;**13**:771–83.
9. Rahman KM, Sugie S, Watanabe T, Tanaka T, Mori H. Chemopreventive effects of melatonin on diethylnitrosamine and phenobarbital-induced hepatocarcinogenesis in male F344 rats. *Nutr Cancer* 2003;**47**:148–55.
10. Leske MC, Heijl A, Hussein M, Bengtsson B, Hyman L, Komaroff E. Factors for glaucoma progression and the effect of treatment: the early manifest glaucoma trial. *Arch Ophthalmol* 2003;**121**:48–56.
11. Tielsch JM, Katz J, Singh K, Quigley HA, Gottsch JD, Javitt J. Sommer AA population-based evaluation of glaucoma screening: the Baltimore Eye Survey. *Am J Epidemiol* 1991;**134**:1102–10.
12. Kim JH, Kang SY, Kim NR, Lee ES, Hong S, Seong GJ, et al. Prevalence and characteristics of glaucoma among Korean adults. *Korean J Ophthalmol* 2011;**25**:110–5.
13. Levin LA, Peeples P. History of neuroprotection and rationale as a therapy for glaucoma. *Am J Manag Care* 2008;**14**:S11–4.
14. Dat NT, Van Kiem P, Cai XF, Shen Q, Bae K, Kim YH. Gymnastone, a new benzofuran derivative from *Gymnaster koraiensis*. *Arch Pharm Res* 2004;**27**:1106–8.
15. Levin LA. Retinal ganglion cells and neuroprotection for glaucoma. *Surv Ophthalmol* 2003;**48**(Suppl. 1):S21–4.
16. Danesh-Meyer HV. Neuroprotection in glaucoma: recent and future directions. *Curr Opin Ophthalmol* 2011;**22**:78–86.
17. Osborne NN. Recent clinical findings with memantine should not mean that the idea of neuroprotection in glaucoma is abandoned. *Acta Ophthalmol* 2009;**87**:450–4.
18. Mozaffarieh M, Grieshaber MC, Orgul S, Flammer J. The potential value of natural antioxidative treatment in glaucoma. *Surv Ophthalmol* 2008;**53**:479–505.
19. Rodieck RW. Maintained activity of cat retinal ganglion cells. *J Neurophysiol* 1967;**30**:1043–71.
20. Danesh-Meyer HV, Levin LA. Neuroprotection: extrapolating from neurologic diseases to the eye. *Am J Ophthalmol* 2009;**148**:186–91. e182.
21. Berkelaar M, Clarke DB, Wang YC, Bray GM, Aguayo AJ. Axotomy results in delayed death and apoptosis of retinal ganglion cells in adult rats. *J Neurosci* 1994;**14**:4368–74.
22. Garcia-Valenzuela E, Gorczyca W, Darzynkiewicz Z, Sharma SC. Apoptosis in adult retinal ganglion cells after axotomy. *J Neurobiol* 1994;**25**:431–8.
23. Quigley HA, Nickells RW, Kerrigan LA, Pease ME, Thibault DJ, Zack DJ. Retinal ganglion cell death in experimental glaucoma and after axotomy occurs by apoptosis. *Invest Ophthalmol Vis Sci* 1995;**36**:774–86.
24. Kerrigan LA, Zack DJ, Quigley HA, Smith SD, Pease ME. TUNEL-positive ganglion cells in human primary open-angle glaucoma. *Arch Ophthalmol* 1997;**115**:1031–5.
25. Marcic TS, Belyea DA, Katz B. Neuroprotection in glaucoma: a model for neuroprotection in optic neuropathies. *Curr Opin Ophthalmol* 2003;**14**:353–6.
26. Cheung W, Guo L, Cordeiro MF. Neuroprotection in glaucoma: drug-based approaches. *Optom Vis Sci* 2008;**85**:406–16.
27. Baltmr A, Duggan J, Nizari S, Salt TE, Cordeiro MF. Neuroprotection in glaucoma – is there a future role? *Exp Eye Res* 2010;**91**:554–66.

28. Levkovitch-Verbin H, Martin KR, Quigley HA, Baumrind LA, Pease ME, Valenta D. Measurement of amino acid levels in the vitreous humor of rats after chronic intraocular pressure elevation or optic nerve transection. *J Glaucoma* 2002;**11**:396–405.

29. Kumar DM, Agarwal N. Oxidative stress in glaucoma: a burden of evidence. *J Glaucoma* 2007;**16**:334–43.

30. Ferreira SM, Lerner SF, Brunzini R, Evelson PA, Llesuy SF. Oxidative stress markers in aqueous humor of glaucoma patients. *Am J Ophthalmol* 2004;**137**:62–9.

31. Gherghel D, Griffiths HR, Hilton EJ, Cunliffe IA, Hosking SL. Systemic reduction in glutathione levels occurs in patients with primary open-angle glaucoma. *Invest Ophthalmol Vis Sci* 2005;**46**:877–83.

32. Sacca SC, Pascotto A, Camicione P, Capris P, Izzotti A. Oxidative DNA damage in the human trabecular meshwork: clinical correlation in patients with primary open-angle glaucoma. *Arch Ophthalmol* 2005;**123**:458–63.

33. Jung SH, Kim BJ, Lee EH, Osborne NN. Isoquercitrin is the most effective antioxidant in the plant *Thuja orientalis* and able to counteract oxidative-induced damage to a transformed cell line (RGC-5 cells). *Neurochem Int* 2010;**57**:713–21.

34. Li SY, Lo AC. Lutein protects RGC-5 cells against hypoxia and oxidative stress. *Int J Mol Sci* 2010;**11**:2109–17.

35. Nakajima Y, Inokuchi Y, Shimazawa M, Otsubo K, Ishibashi T, Hara H. Astaxanthin, a dietary carotenoid, protects retinal cells against oxidative stress *in vitro* and in mice *in vivo*. *J Pharm Pharmacol* 2008;**60**:1365–74.

36. Nakajima Y, Shimazawa M, Otsubo K, Ishibashi T, Hara H. Zeaxanthin, a retinal carotenoid, protects retinal cells against oxidative stress. *Curr Eye Res* 2009;**34**:311–8.

37. Aguilera Y, Dorado ME, Prada FA, Martínez JJ, Quesada A, Ruiz-Gutiérrez V. The protective role of squalene in alcohol damage in the chick embryo retina. *Exp Eye Res* 2005;**80**:535–43.

38. Jung SH, Kang KD, Ji D, Fawcett RJ, Safa R, Kamalden TA, Osborne NN. The flavonoid baicalin counteracts ischemic and oxidative insults to retinal cells and lipid peroxidation to brain membranes. *Neurochem Int* 2008;**53**:325–37.

39. Chen YQ, Pan WH, Liu JH, Chen MM, Liu CM, Yeh MY, et al. The effects and underlying mechanisms of S-allyl-l-cysteine treatment of the retina after ischemia/reperfusion. *J Ocul Pharmacol Ther* 2012;**28**:110–7.

40. Lee HJ, Kim KA, Kang KD, Lee EH, Kim CY, Um BH, Jung SH. The compound isolated from the leaves of *Phyllostachys nigra* protects oxidative stress-induced retinal ganglion cells death. *Food Chem Toxicol* 2010;**48**:1721–7.

41. Chan HC, Chang RC, Koon-Ching Ip A, Chiu K, Yuen WH, Zee SY, So KF. Neuroprotective effects of *Lycium barbarum* Lynn on protecting retinal ganglion cells in an ocular hypertension model of glaucoma. *Exp Neurol* 2007;**203**:269–73.

42. Chiu K, Chan HC, Yeung SC, Yuen WH, Zee SY, Chang RC, So KF. Modulation of microglia by wolfberry on the survival of retinal ganglion cells in a rat ocular hypertension model. *J Ocul Biol* 2009;**2**:47–56.

43. Yang M, Gao N, Zhao Y, Liu L-X, Lu X- J. Protective effect of *Lycium barbarum* polysaccharide on retinal ganglion cells *in vitro*. *Int J Ophthalmol* 2011;**4**:377–9.

44. Kamalden TA, Ji D, Osborne NN. Rotenone-induced death of RGC-5 cells is caspase independent, involves the JNK and p38 pathways and is attenuated by specific green tea flavonoids. *Neurochem Res* 2012;**37**:1091–101.

45. Osborne NN, Ji D, Abdul Majid AS, Fawcett RJ, Sparatore A, Del Soldato P. ACS67, a hydrogen sulfide-releasing derivative of latanoprost acid, attenuates retinal ischemia and oxidative stress to RGC-5 cells in culture. *Invest Ophthalmol Vis Sci* 2010;**51**:284–94.

46. Zhang B, Osborne NN. Oxidative-induced retinal degeneration is attenuated by epigallocatechin gallate. *Brain Res* 2006;**1124**:176–7.

47. Zhang B, Rusciano D, Osborne NN. Orally administered epigallocatechin gallate attenuates retinal neuronal death *in vivo* and light-induced apoptosis *in vitro*. *Brain Res* 2008;**1198**:141–52.

48. Zhang B, Safa R, Rusciano D, Osborne NN. Epigallocatechin gallate, an active ingredient from green tea, attenuates damaging influences to the retina caused by ischemia/reperfusion. *Brain Res* 2007;**1159**:40–53.

49. Inokuchi Y, Shimazawa M, Nakajima Y, Suemori S, Mishima S, Hara H. Brazilian green propolis protects against retinal damage *in vitro* and *in vivo*. *Evid Based Complement Alternat Med* 2006;**3**:71–7.

50. Nakajima Y, Shimazawa M, Mishima S, Hara H. Water extract of propolis and its main constituents, caffeoylquinic acid derivatives, exert neuroprotective effects via antioxidant actions. *Life Sci* 2007;**80**:370–7.

51. Nakajima Y, Shimazawa M, Mishima S, Hara H. Neuroprotective effects of Brazilian green propolis and its main constituents against oxygen–glucose deprivation stress, with a gene-expression analysis. *Phytother Res* 2009;**23**:1431–8.

52. El-Sayyad HI, El-Sherbiny MA, Sobh MA, Abou-El-Naga AM, Ibrahim MA, Mousa SA. Protective effects of *Morus alba* leaves extract on ocular functions of pups from diabetic and hypercholesterolemic mother rats. *Int J Biol Sci* 2011;**7**:715–28.

53. Choi SJ, Lee MY, Jo H, Lim SS, Jung SH. Preparative isolation and purification of neuroprotective compounds from *Rhus verniciflua* by high speed counter-current chromatography. *Biol Pharm Bull* 2012;**35**:559–67.

54. Chung HS, Harris A, Kristinsson JK, Ciulla TA, Kagemann C, Ritch R. *Ginkgo biloba* extract increases ocular blood flow velocity. *J Ocul Pharmacol Ther* 1999;**15**:233–40.

55. Hirooka K, Tokuda M, Miyamoto O, Itano T, Baba T, Shiraga F. The *Ginkgo biloba* extract (EGb 761) provides a neuroprotective effect on retinal ganglion cells in a rat model of chronic glaucoma. *Curr Eye Res* 2004;**28**:153–7.

56. Ma K, Xu L, Zhan H, Zhang S, Pu M, Jonas JB. Dosage dependence of the effect of *Ginkgo biloba* on the rat retinal ganglion cell survival after optic nerve crush. *Eye* 2009;**23**:1598–604.

57. Ma K, Xu L, Zhang H, Zhang S, Pu M, Jonas JB. The effect of *Ginkgo biloba* on the rat retinal ganglion cell survival in the optic nerve crush model. *Acta Ophthalmol* 2010;**88**:553–7.

58. Park JW, Kwon HJ, Chung WS, Kim CY, Seong GJ. Short-term effects of *Ginkgo biloba* extract on peripapillary retinal blood flow in normal tension glaucoma. *Korean J Ophthalmol* 2011;**25**:323–8.

59. Quaranta L, Bettelli S, Uva MG, Semeraro F, Turano R, Gandolfo E. Effect of *Ginkgo biloba* extract on preexisting visual field damage in normal tension glaucoma. *Ophthalmology* 2003;**110**:359–62. discussion 362–4.

60. Shim SH, Kim JM, Choi CY, Kim CY, Park KH. *Ginkgo biloba* extract and bilberry anthocyanins improve visual function in patients with normal tension glaucoma. *J Med Food* 2012;**15**:818–23.

61. Wang YS, Xu L, Ma K, Wang JJ. The protective effects of *Ginkgo biloba* extract on cultured human retinal ganglion cells. *Chin J Ophthalmol* 2011;**47**:824–8.

62. Wang YQ, Li XF, Zhou X, Liu XQ, Wang WP. Protective effects of Chinese herbal medicine Zuogui Pill on retina ganglion cells after optical nerve clipping injury in rats. *J Chin Integr Med* 2011;**9**:991–7.

63. Tsuruma K, Shimazaki H, Nakashima K, Yamauchi M, Sugitani S, Shimazawa M, et al. Annatto prevents retinal degeneration induced by endoplasmic reticulum stress *in vitro* and *in vivo*. *Mol Nutr Food Res* 2012;**56**:713–24.

64. Matsunaga N, Imai S, Inokuchi Y, Shimazawa M, Yokota S, Araki Y, Hara H. Bilberry and its main constituents have neuroprotective effects against retinal neuronal damage *in vitro* and *in vivo*. *Mol Nutr Food Res* 2009;**53**:869–77.

65. Yamauchi M, Tsuruma K, Imai S, Nakanishi T, Umigai N, Shimazawa M, Hara H. Crocetin prevents retinal degeneration induced by oxidative and endoplasmic reticulum stresses via inhibition of caspase activity. *Eur J Pharmacol* 2011;**650**:110–9.

66. Burugula B, Ganesh BS, Chintala SK. Curcumin attenuates staurosporine-mediated death of retinal ganglion cells. *Invest Ophthalmol Vis Sci* 2011;**52**:4263–73.

67. Tong N, Zhang Z, Gong Y, Yin L, Wu X. Diosmin protects rat retina from ischemia/reperfusion injury. *J Ocul Pharmacol Ther* 2012;**28**:459–66.

68. Kim KA, Kim SM, Kang SW, Jeon SI, Um BH, Jung SH. Edible seaweed, *Eisenia bicyclis*, protects retinal ganglion cells death caused by oxidative stress. *Mar Biotechnol* 2012;**14**:383–95.

69. Li C, Wang L, Huang K, Zheng L. Endoplasmic reticulum stress in retinal vascular degeneration: protective role of resveratrol. *Invest Ophthalmol Vis Sci* 2012;**53**:3241–9.

70. Yang H, Lee BK, Kook KH, Jung YS, Ahn J. Protective effect of grape seed extract against oxidative stress-induced cell death in a staurosporine-differentiated retinal ganglion cell line. *Curr Eye Res* 2012;**37**:339–44.

71. Maher P, Hanneken A. Flavonoids protect retinal ganglion cells from oxidative stress-induced death. *Invest Ophthalmol Vis Sci* 2005;**46**:4796–803.

72. Agarwal R, Gupta SK, Srivastava S, Saxena R, Agrawal SS. Intraocular pressure-lowering activity of topical application of *Aegle marmelos* fruit extract in experimental animal models. *Ophthalmic Res* 2009;**42**:112–6.

73. Ou HC, Lee WJ, Lee IT, Chiu TH, Tsai KL, Lin CY, Sheu WH. *Ginkgo biloba* extract attenuates oxLDL-induced oxidative functional damages in endothelial cells. *J Appl Physiol* 2009;**106**:1674–85.

74. Mahadevan S, Park Y. Multifaceted therapeutic benefits of *Ginkgo biloba* L.: chemistry, efficacy, safety, and uses. *J Food Sci* 2008;**73**:R14–9.

75. Wu Y, Li S, Cui W, Zu X, Du J, Wang F. *Ginkgo biloba* extract improves coronary blood flow in healthy elderly adults: role of endothelium-dependent vasodilation. *Phytomedicine* 2008;**15**:164–9.

76. Kanwar J, Taskeen M, Mohammad I, Huo C, Chan TH, Dou QP. Recent advances on tea polyphenols. *Front Biosci (Elite Ed)* 2012;**4**:111–31.

77. Andrade JP, Assuncao M. Protective effects of chronic green tea consumption on age-related neurodegeneration. *Curr Pharm Des* 2012;**18**:4–14.

78. Tedeschi E, Suzuki H, Menegazzi M. Antiinflammatory action of EGCG, the main component of green tea, through STAT-1 inhibition. *Ann N Y Acad Sci* 2002;**973**:435–7.

79. Lee MJ, Maliakal P, Chen L, Meng X, Bondoc FY, Prabhu S, et al. Pharmacokinetics of tea catechins after ingestion of green tea and (−)-epigallocatechin-3-gallate by humans: formation of different metabolites and individual variability. *Cancer Epidemiol Biomarkers Prev* 2002;**11**:1025–32.

80. Kijlstra A, Tian Y, Kelly ER, Berendschot TT. Lutein: more than just a filter for blue light. *Prog Retin Eye Res* 2012;**31**:303–15.

81. Calvo MM. Lutein: a valuable ingredient of fruit and vegetables. *Crit Rev Food Sci Nutr* 2005;**45**:671–96.

82. Cao Y, Mao X, Sun C, Zheng P, Gao J, Wang X, et al. Baicalin attenuates global cerebral ischemia/reperfusion injury in gerbils via anti-oxidative and anti-apoptotic pathways. *Brain Res Bull* 2011;**85**:396–402.

83. Tu XK, Yang WZ, Shi SS, Chen Y, Wang CH, Chen CM, Chen Z. Baicalin inhibits TLR2/4 signaling pathway in rat brain following permanent cerebral ischemia. *Inflammation* 2011;**34**:463–70.

84. Ge QF, Hu X, Ma ZQ, Liu JR, Zhang WP, Chen Z, Wei EQ. Baicalin attenuates oxygen-glucose deprivation-induced injury via inhibiting NMDA receptor-mediated 5-lipoxygenase activation in rat cortical neurons. *Pharmacol Res* 2007;**55**:148–57.

85. Heijnen CG, Haenen GR, Vekemans JA, Bast A. Peroxynitrite scavenging of flavonoids: structure activity relationship. *Environ Toxicol Pharmacol* 2001;**10**:199–206.

86. Heijnen CG, Haenen GR, van Acker FA, van der Vijgh WJ, Bast A. Flavonoids as peroxynitrite scavengers: the role of the hydroxyl groups. *Toxicol In Vitro* 2001;**15**:3–6.

87. Wang L, Tu YC, Lian TW, Hung JT, Yen JH, Wu MJ. Distinctive antioxidant and antiinflammatory effects of flavonols. *J Agric Food Chem* 2006;**54**:9798–804.

88. Yang B, Kotani A, Arai K, Kusu F. Estimation of the antioxidant activities of flavonoids from their oxidation potentials. *Anal Sci* 2001;**17**:599–604.

89. Kim SM, Shang YF, Um BH. A preparative method for isolation of fucoxanthin from *Eisenia bicyclis* by centrifugal partition chromatography. *Phytochem Anal* 2011;**22**:322–9.

90. Rajyalakshmi P, Geervani P. Nutritive value of the foods cultivated and consumed by the tribals of south India. *Plant Foods Hum Nutr* 1994;**46**:53–61.

91. Ogle BM, Johansson M, Tuyet HT, Johannesson L. Evaluation of the significance of dietary folate from wild vegetables in Vietnam. *Asia Pac J Clin Nutr* 2001;**10**:216–21.

92. Lee JY, Song D, Lee EH, Jung SH, Nho CW, Cha KH, Pan C. Inhibitory effects of 3,5-*O*-dicaffeoyl-epi-quinic acid from *Gymnaster koraiensis* on AKR1B10. *J Korean Soc Appl Biol Chem* 2009;**52**:731–4.

93. Kim HJ, Hong J, Jung JW, Kim TH, Kim JA, Kim YH, Kim SY. Gymnasterkoreayne F inhibits osteoclast formation by suppressing NFATc1 and DC-STAMP expression. *Int Immunopharmacol* 2010;**10**:1440–7.

94. Lee J, Lee YM, Lee BW, Kim JH, Kim JS. Chemical constituents from the aerial parts of *Aster koraiensis* with protein glycation and aldose reductase inhibitory activities. *J Nat Prod* 2012;**75**:267–70.

95. Dat NT, Cai XF, Shen Q, Lee IS, Lee EJ, Park YK, et al. Gymnasterkoreayne G, a new inhibitory polyacetylene against NFAT transcription factor from *Gymnaster koraiensis*. *Chem Pharm Bull (Tokyo)* 2005;**53**:1194–6.

96. Lee SB, Kang K, Oidovsambuu S, Jho EH, Yun JH, Yoo JH, et al. A polyacetylene from *Gymnaster koraiensis* exerts hepatoprotective effects *in vivo* and *in vitro*. *Food Chem Toxicol* 2010;**48**:3035–41.

97. Jung HJ, Min BS, Park JY, Kim YH, Lee HK, Bae KH. Gymnasterkoreaynes A-F, cytotoxic polyacetylenes from *Gymnaster koraiensis*. *J Nat Prod* 2002;**65**:897–901.

98. Park J, Min B, Jung H, Kim Y, Lee H, Bae K. Polyacetylene glycosides from *Gymnaster koraiensis*. *Chem Pharm Bull (Tokyo)* 2002;**50**:685–7.

99. Kim KA, Kang KD, Lee EH, Nho CW, Jung SH. Edible wild vegetable, *Gymnaster koraiensis* protects retinal ganglion cells against oxidative stress. *Food Chem Toxicol* 2011;**49**:2131–43.

100. Aslan M, Cort A, Yucel I. Oxidative and nitrative stress markers in glaucoma. *Free Radic Biol Med* 2008;**45**:367–76.

101. Wei Y, Gong J, Yoshida T, Eberhart CG, Xu Z, Kombairaju P, et al. Nrf2 has a protective role against neuronal and capillary degeneration in retinal ischemia–reperfusion injury. *Free Radic Biol Med* 2011;**51**:216–24.

102. Harvey CJ, Thimmulappa RK, Singh A, Blake DJ, Ling G, Wakabayashi N, et al. Nrf2-regulated glutathione recycling independent of biosynthesis is critical for cell survival during oxidative stress. *Free Radic Biol Med* 2009;**46**:443–53.

103. Wu JH, Huang CY, Tung YT, Chang ST. Online RP-HPLC-DPPH screening method for detection of radical-scavenging phytochemicals from flowers of *Acacia confusa*. *J Agric Food Chem* 2008;**56**:328–32.

104. Osborne NN, Ugarte M, Chao M, Chidlow G, Bae JH, Wood JP, Nash MS. Neuroprotection in relation to retinal ischemia and relevance to glaucoma. *Surv Ophthalmol* 1999;**43**(Suppl. 1):S102–28.

105. Yu SP, Yeh C, Strasser U, Tian M, Choi DW. NMDA receptor-mediated K$^+$ efflux and neuronal apoptosis. *Science* 1999;**284**: 336–9.

106. Leung CK, Lindsey JD, Crowston JG, Lijia C, Chiang S, Weinreb RN. Longitudinal profile of retinal ganglion cell damage after optic nerve crush with blue-light confocal scanning laser ophthalmoscopy. *Invest Ophthalmol Vis Sci* 2008;**49**:4898–902.

107. Steinsapir KD, Goldberg RA. Traumatic optic neuropathy. *Surv Ophthalmol* 1994;**38**:487–518.

108. Dreyer EB, Zurakowski D, Schumer RA, Podos SM, Lipton SA. Elevated glutamate levels in the vitreous body of humans and monkeys with glaucoma. *Arch Ophthalmol* 1996;**114**:299–305.

109. Schinder AF, Olson EC, Spitzer NC, Montal M. Mitochondrial dysfunction is a primary event in glutamate neurotoxicity. *J Neurosci* 1996;**16**:6125–33.

110. Kwong JM, Lam TT. *N*-Methyl-D-aspartate (NMDA) induced apoptosis in adult rabbit retinas. *Exp Eye Res* 2000;**71**:437–44.

44

Plant Stanol and Sterol Esters and Macular Pigment Optical Density

Tos T.J.M. Berendschot[1], Jogchum Plat[2]

[1]University Eye Clinic Maastricht, Maastricht, The Netherlands, [2]Department of Human Biology, Maastricht University Medical Center, Maastricht, The Netherlands

AGE-RELATED MACULAR DEGENERATION

Age-related macular degeneration (AMD) is a degenerative disease of the retina and in the developed world the leading cause of irreversible blindness in people 50 years and older.[1-3] The disease affects the macula, which is of utmost importance to human vision. The macula has the highest density of cone photoreceptor cells and is responsible for high-acuity vision. Dysfunction of the macula results in the inability to see fine detail, read, and recognize faces. AMD has a great impact on the quality of life and large negative financial and economic consequences.[4-9] Mild AMD causes a 17% decrement in the quality of life of the average patient, similar to that encountered with moderate cardiac angina or symptomatic human immunodeficiency virus syndrome. Moderate AMD causes a 40% decrease in the average patient's quality of life, similar to that associated with severe cardiac angina or renal dialysis. Very severe AMD causes a 63% decrease in the average patient's quality of life, similar to that encountered with end-stage prostatic cancer or a catastrophic stroke that leaves a person bedridden, incontinent, and requiring constant nursing care.

The prevalence of AMD is likely to rise as a consequence of increasing longevity. Although new treatments have emerged, they are suitable only for the small proportion of people with neovascular AMD. No treatments are available for the geographic atrophy stage of AMD. Therefore, it is very important to identify preventive factors for the development of AMD.

Genetic predisposition and environmental factors such as smoking play an important role in the development of this disease.[10] Recently, it has been shown that immunologic factors and inflammation also play an important role in the pathogenesis of AMD. The finding that drusen in the retina contain complement products[11,12] and the fact that one of the strongest genetic associations with AMD has been found in genes encoding for a protein of this system (CFH gene) have pointed to an important role for complement in the pathogenesis of AMD.[13-15] This was confirmed by recent findings showing an increase in the level of various complement components in the circulation of AMD patients, which also provided evidence for a systemic inflammatory component to the disease pathogenesis.[16-20]

From a dietary perspective, antioxidants are considered of particular importance in the prevention of AMD.[21] Recent epidemiologic studies have shown that individuals with a genetic predisposition for AMD can decrease their disease risk by eating a healthy diet rich in antioxidants.[22]

CAROTENOIDS

Carotenoids are colored pigments synthesized by flora and microbes. They perform critical functions in photosynthesis and photoprotection. Humans cannot synthesize carotenoids; they ingest them in their diet and subsequently utilize them for important physiologic functions. Humans typically consume a wide variety of carotenoids in the diet; β-carotene, along with lycopene, lutein, zeaxanthin, β-cryptoxanthin, and α-carotene, account for 90% of circulating carotenoids. Once in the circulation, the absorbed carotenoids are bound to lipoproteins and then targeted to a variety of tissues such as liver, macula, lung, adipose, brain, prostate, and skin. Some tissues such as skin are relatively nonselective. Their carotenoid levels and compositions largely reflect those of the serum.

MACULAR PIGMENT

The retina of the human eye has an extraordinary distribution of carotenoids. Figure 44.1 shows dissected eye globes from two donors before fixation.[23] The specific yellow color is caused by just three carotenoids: the stereoisomers lutein and zeaxanthin[24,25] and the intermediate meso-zeaxanthin.[26] The concentration of these macular pigments (MPs) reaches its peak in the center of the macula, which is therefore known as the macula lutea or yellow spot (see also Fig. 44.2).

The MP distribution can be quantified *in vivo* using fundus reflectance or autofluorescence.[27–33] Figure 44.2 shows reflectance and autofluorescence maps acquired at 488 nm and 514 nm using a scanning laser ophthalmoscope. Since the lens and the MP are the only absorbers in this wavelength region, digital subtraction of the log of the resulting gray-level maps at the two wavelengths provides density maps of the sum of both absorbers. The macular pigment optical density (MPOD) is assumed to be negligible at a peripheral site. If this site is then used to provide an estimate for the lens density, the mean MPOD at the fovea can be calculated,[31,32] and its spatial distribution can be imaged easily and with a high resolution, as illustrated in the lower part of Figure 44.3. Figure 44.4 shows the MPOD as a function of eccentricity obtained from similar maps by Berendschot and van Norren.[29] The distribution can be modeled with an exponentially decaying density as a function of eccentricity, in combination with a Gaussian distributed ring pattern that was visible in about 50% of all subjects.[29]

As already mentioned, humans are unable to synthesize carotenoids and therefore serum levels of lutein and zeaxanthin are dependent upon dietary intake. Lutein serum levels and the amount of MP correlate, in particular in men.[34–38] The amount of MP can be increased by dietary modification[39–42] or by supplements[32,35,43–56] in healthy subjects as well as in subjects with a diseased macula.[57–60]

Serum levels associated with the normal diet are far below the maximal levels achieved by supplementation. The normal Western diet contains 1.3–3 mg/day of lutein and zeaxanthin combined.[39] The highest concentration of lutein is found in food sources with yellow color, such as corn and egg yolk. Dark, leafy green vegetables, such as spinach and kale, are also good sources of lutein. The same is true for orange peppers, followed by egg yolk, corn, and orange juice for zeaxanthin.[61–64] In general, fruit and vegetables contain seven to 10 times more lutein than zeaxanthin. Meso-zeaxanthin is virtually nonexistent in food sources originating from plants; it is primarily formed in the retina following conversion from lutein.[26,65]

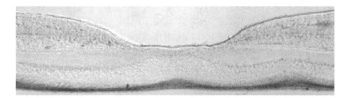

FIGURE 44.2 Unstained retina. *Source: Courtesy of MM Snodderly.* (See color plate section)

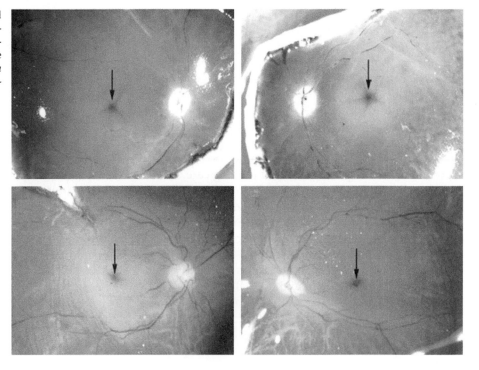

FIGURE 44.1 Photographs of dissected eye globes from two donors before fixation showing that the highest concentration of macular pigment is found in the fovea (black arrows). *Source: Reprinted from Powner et al. (2010)[23] with permission of Ophthalmology.* (See color plate section)

Figure 44.5 shows the extinction coefficients of lutein and zeaxanthin,[66, 67] demonstrating their specific absorption in the bluish wavelength region. Because of this specific absorption profile, many potential functions are attributed to MP, such as reducing the consequences of chromatic aberration,[68–70] minimizing stray light,[71] preservation of visual sensitivity[72] in older subjects, and improving glare disability and photostress recovery.[73,74]

In 1994 Seddon *et al.* observed an inverse association between a diet with a high content of the carotenoids lutein and zeaxanthin and the prevalence of AMD.[75] This pointed to another potential role of MP in preventing AMD. There are several biologically plausible mechanisms to support this hypothesis. First, MPs are efficient antioxidants that can quench singlet oxygen, reactive oxygen species, and various other free radicals that are by-products of metabolic process in cells.[76] Second, because of its ability to filter high-energy blue light, MP may limit the development of AMD by decreasing the chances for photochemical light damage.[77] Finally, lutein is capable of suppressing inflammation,[78,79] which plays a prominent role in AMD.[13–20]

Lipid-rich matrices result in a relatively high xanthophyll bioavailability to humans, compared with other dietary sources containing lutein in the same concentrations such as spinach.[42] Although low-density lipoproteins (LDLs) are capable of transporting lutein and zeaxanthin in plasma, high-density lipoproteins (HDLs) seem to be the main transporters for these carotenoids.[80] The LDL to HDL ratio of lutein and zeaxanthin is as high as 1:2, although the total surface lipid content of LDL in serum is about twice as high as HDL.[81,82] A study in Wisconsin hypoalpha mutant (WHAM) chickens, which have a mutation causing very low levels of HDL, showed that HDL is essential for the transport of lutein and zeaxanthin into plasma and more importantly into the retina.[83] Therefore, it is possible to increase plasma lutein concentrations. But what is known of ingredients that act in the other direction? In other words, are there also ingredients that may lower serum and macular lutein concentrations?

PLANT STEROLS AND STANOLS

Plant sterols and stanols are components that are naturally present in plants. Like cholesterol, they exist in a free and an esterified form. When incorporated as functional food ingredients, plant sterols and stanols are frequently esterified with a fatty acid ester to increase the solubility in the food matrix.[84] These functional foods enriched with plant sterol or stanol esters are well

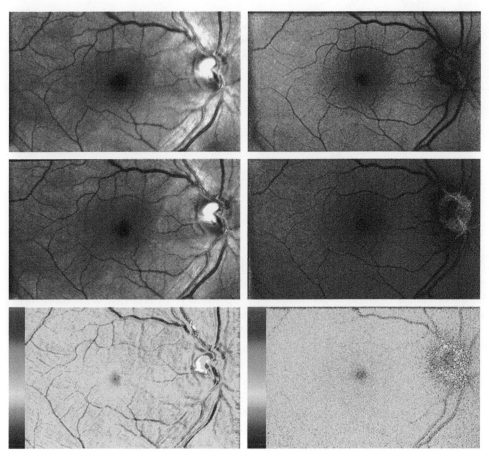

FIGURE 44.3 Reflectance (left) and autofluorescence maps (right) at a wavelength of 488 nm (top) and 514 nm (middle). The lower images show the corresponding color-coded macular pigment optical density maps. *Source: Reprinted from Berendschot and van Norren (2006)[29] with permission of* Investigative Ophthalmology and Visual Science. (See color plate section)

FIGURE 44.4 Macular pigment optical density (MPOD) as a function of eccentricity from reflectance (black) and autofluorescence (red) maps. The corresponding maps are shown as insets (reflectance maps on the left and autofluorescence maps on the right). Solid lines are the results of a model fit, explained in the text. *Source: Reprinted from Berendschot and van Norren (2006)[29] with permission of* Investigative Ophthalmology and Visual Science. (See color plate section)

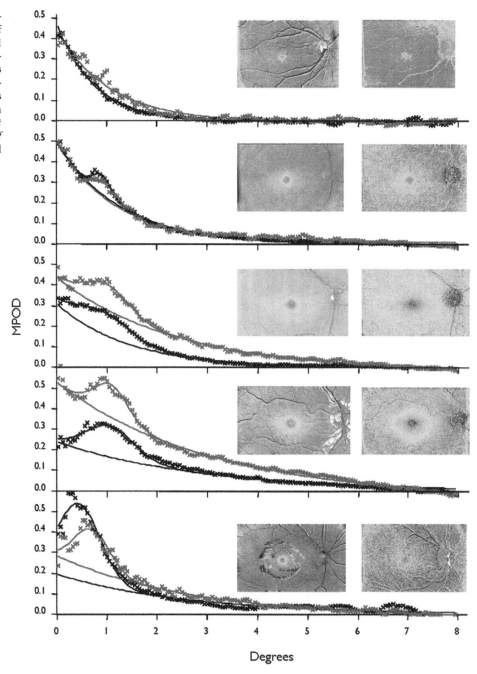

known for their serum LDL-cholesterol-lowering effect (Fig. 44.6).[85–87] This effect is not transient, as shown in an 85-week intervention study.[88] The effectiveness of these compounds resulted in their incorporation into national and international guidelines such as the National Cholesterol Education Program guidelines, which encourage a daily incorporation of 2 g plant sterols or stanols into a healthy diet low in saturated fatty acids. This aims to reduce the risk of cardiovascular disease for people with elevated LDL-C concentrations. Different meta-analyses have shown significant reductions in LDL-C concentrations after consumption of foods enriched with plant sterol or stanol esters.[89–91] On average, the addition of plant sterols and stanols can lower serum LDL-C concentrations by up to 10%. In contrast with these nonlinear dose–response curves, Mensink *et al.*[92] found a clear linear relationship between plant stanol intake and reductions in LDL-C up to 9 g/day. Compared with the control group, the reductions in serum LDL-C concentrations after a daily consumption of 3 g, 6 g, and 9 g were 7.5%, 12%, and 17.4%, respectively. Comparable findings were reported by Gylling *et al.*,[93] where a 17.4% reduction in serum LDL-C was found after a daily consumption of 8.8 g plant stanols provided as their fatty

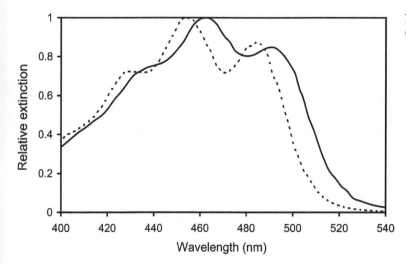

FIGURE 44.5 Relative extinction coefficient of lutein (dashed line) and zeaxanthin (solid line).[66,67]

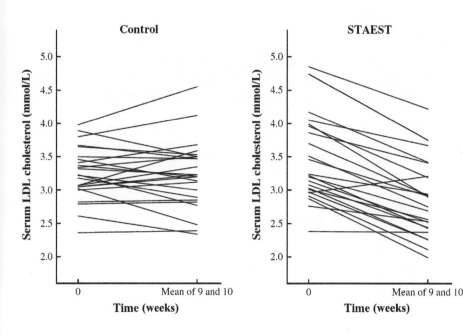

FIGURE 44.6 Absolute individual changes over time of low-density lipoprotein (LDL)-cholesterol concentration (mmol/L) in a plant stanol ester (STAEST) and a control group. *Source: Reprinted Gylling* et al. *(2010)[87] with permission of* Clinical Nutrition.

acid esters for a period of 10 weeks. In this respect, the most recent meta-analysis from Musa-Veloso *et al.* [86] suggested that consumption of plant stanols above the currently recommended 2 g/day is associated with an additional and dose-dependent reduction in serum LDL-C concentration. They included 113 publications and one unpublished study report and found that the maximal reduction in LDL-C was 16.4% after consumption of plant stanols and 8.3% after consumption of plant sterols at daily doses ranging from 0.8 g to 8.8 g and 0.19 g to 9 g, respectively.

Since functional foods containing plant sterols or stanols lower serum cholesterol values, they may also lower serum concentrations of the fat-soluble carotenoids lutein and zeaxanthin. Plat and Mensink studied the effect of a diet enriched with plant stanol ester on lutein and zeaxanthin concentrations and showed that

consumption of both vegetable oil- and wood-based plant stanol ester mixtures lowered the plasma carotenoid concentrations, although they remained within the normal range.[94] In investigating the effect on cholesterol metabolism of consuming a low-fat National Cholesterol Education Program diet, including a low-fat plant sterol ester-enriched spread, Colgan *et al.* found no effect on lutein and zeaxanthin concentrations.[95] However, this may be caused by their dietary advice to increase daily fruit and vegetable consumption. In a double-blind placebo-controlled human intervention trial with a duration of 18 months, Berendschot *et al.* evaluated possible effects of plant stanol and sterol esters on serum carotenoid concentrations and MP.[96] Compared with a control group, the cholesterol concentration lowered in both the sterol group and the stanol ester group.[88] The sterol group also showed a reduction in serum lutein concentration.[96] Mensink *et al.* also studied the relation between plant stanols provided as plant stanol esters on changes in serum concentrations of LDL-cholesterol and fat-soluble antioxidants in a dose-dependent manner.[92] They observed a statistically significant linear trend between stanol intake and changes in lutein and zeaxanthin concentrations. In contrast, Raeini-Sarjaz *et al.* found no effect of consumption of esterified plant sterols or stanols on serum fat-soluble carotenoid concentrations compared with a control diet.[97]

Apart from plant sterols and stanols, the sucrose polyester olestra has also been used to lower cholesterol levels. Olestra, a nonabsorbable, noncaloric fat substitute, is neither hydrolyzed by gastrointestinal enzymes nor absorbed, but it has the capability to interfere with the absorption of fat-soluble nutrients. Broekmans *et al.*[98] reported that compared with a control group, olestra consumption in an olestra chips and olestra spread group resulted in an 18% reduction of the serum lutein concentration and a 13% reduction in the serum zeaxanthin concentration, both lipid standardized. Neuhouser *et al.* studied whether customary consumption of olestra-containing savory snacks was associated with changes in serum fat-soluble carotenoid concentrations.[99] There were no significant associations of olestra with lutein and zeaxanthin concentrations, although they found fairly consistent reductions in other serum carotenoid concentrations associated with increasing olestra intake.

Yet another way to lower cholesterol levels is through the use of statins. Recently, Rydén *et al.* showed a reduction in lutein and zeaxanthin levels in a group using simvastatin for 6 weeks.[100] Also recently, Renzi *et al.* studied the relation between serum lipids and serum lutein and zeaxanthin concentrations in both healthy subjects and subjects using statins.[101] Unfortunately, they did not report on possible differences in lutein concentrations between statin and nonstatin users, although they showed that subjects using statins had significantly lower cholesterol levels and that lutein and zeaxanthin concentrations showed a significant correlation with HDL concentration.

Lowering lutein and zeaxanthin serum concentrations may, in turn, directly affect MP.[32,39,44,57,58] As argued above, this may lead to an increased susceptibility for AMD, although the literature is ambiguous on this point.[102–108] In evaluating possible effects of plant stanol and sterol esters, Berendschot *et al.* also studied MP.[96] Although they found significantly lowered carotenoid concentrations, this did not result in a decrease in the MPOD. Similar results were found by Cooper *et al.*[109] and Broekmans *et al.*,[98] who both reported that consumption of a dietary fat replacer was not associated with reduced MPOD. Renzi *et al.* also found that the MPOD was not lower in statin users compared with matched nonstatin

users.[101] However, they found that the MPOD decreased significantly with increased duration of statin use.

TAKE-HOME MESSAGES

- Macular pigment (MP) constitutes the stereoisomers lutein and zeaxanthin and the intermediate meso-zeaxanthin. The absorption peaks at the center of the fovea and decreases rapidly with eccentricity.
- Humans are unable to synthesize lutein and zeaxanthin. Thus, the MP optical density depends on dietary intake.
- There is some evidence that lutein intake and/or MP protects against or ameliorates the clinical course of age-related macular degeneration, the most common cause of irreversible blindness in the industrialized world.
- Plant sterols and stanols are components that are naturally present in plants. Functional foods enriched with plant sterol or stanol esters are well known for their serum low-density lipoprotein cholesterol-lowering effect.
- Serum concentrations of the fat-soluble carotenoids lutein and zeaxanthin decrease when using functional foods containing plant sterol or stanol. However, MP seems unaffected.

References

1. van Leeuwen R, Klaver CCW, Vingerling JR, Hofman A, de Jong PTVM. Epidemiology of age-related maculopathy: a review. *Eur J Epidemiol* 2003;**18**:845–54.
2. Klaver CCW, Wolfs RC, Vingerling JR, Hofman A, de Jong PTVM. Age-specific prevalence and causes of blindness and visual impairment in an older population: the Rotterdam Study. *Arch Ophthalmol* 1998;**116**:653–8.
3. Attebo K, Mitchell P, Smith W. Visual acuity and the causes of visual loss in Australia. The Blue Mountains Eye Study. *Ophthalmology* 1996;**103**:357–64.
4. Soubrane G, Cruess A, Lotery A, Pauleikhoff D, Mones J, Xu X, et al. Burden and health care resource utilization in neovascular age-related macular degeneration: findings of a multicountry study. *Arch Ophthalmol* 2007;**125**:1249–54.
5. Bandello F, Lafuma A, Berdeaux G. Public health impact of neovascular age-related macular degeneration treatments extrapolated from visual acuity. *Invest Ophthalmol Vis Sci* 2007;**48**:96–103.
6. Cruess A, Zlateva G, Xu X, Rochon S. Burden of illness of neovascular age-related macular degeneration in Canada. *Can J Ophthalmol* 2007;**42**:836–43.
7. Marback RF, Maia Junior OD, Morais FB, Takahashi WY. Quality of life in patients with age-related macular degeneration with monocular and binocular legal blindness. *Clinics (Sao Paulo)* 2007;**62**:573–8.
8. Gupta OP, Brown GC, Brown MM. Age-related macular degeneration: the costs to society and the patient. *Curr Opin Ophthalmol* 2007;**18**:201–5.

9. Covert D, Berdeaux G, Mitchell J, Bradley C, Barnes R. Quality of life and health economic assessments of age-related macular degeneration. *Surv Ophthalmol* 2007;**52**(Suppl. 1):S20–5.

10. de Jong PTVM. Age-related macular degeneration. *N Engl J Med* 2006;**355**:1474–85.

11. van der Schaft TL, Mooy CM, de Bruijn WC, de Jong PTVM. Early stages of age-related macular degeneration: an immunofluorescence and electron microscopy study. *Br J Ophthalmol* 1993;**77**:657–61.

12. Hageman GS, Luthert PJ, Chong NHV, Johnson LV, Anderson DH, Mullins RF. An integrated hypothesis that considers drusen as biomarkers of immune-mediated processes at the RPE–Bruch's membrane interface in aging and age-related macular degeneration. *Prog Retin Eye Res* 2001;**20**:705–32.

13. Khandhadia S, Cipriani V, Yates JRW, Lotery AJ. Age-related macular degeneration and the complement system. *Immunobiology* 2012;**217**:127–46.

14. Anderson DH, Radeke MJ, Gallo NB, Chapin EA, Johnson PT, Curletti CR, et al. The pivotal role of the complement system in aging and age-related macular degeneration: hypothesis revisited. *Prog Retin Eye Res* 2010;**29**:95–112.

15. Buschini E, Piras A, Nuzzi R, Vercelli A. Age related macular degeneration and drusen: neuroinflammation in the retina. *Prog Neurobiol* 2011;**15**(95):14–25.

16. Scholl HP, Charbel Issa P, Walier M, Janzer S, Pollok-Kopp B, Borncke F, et al. Systemic complement activation in age-related macular degeneration. *PLoS ONE* 2008;**3**(7):e2593.

17. Reynolds R, Hartnett ME, Atkinson JP, Giclas PC, Rosner B, Seddon JM. Plasma complement components and activation fragments: associations with age-related macular degeneration genotypes and phenotypes. *Invest Ophthalmol Vis Sci* 2009;**50**:5818–27.

18. Smailhodzic D, Klaver CC, Klevering BJ, Boon CJ, Groenewoud JM, Kirchhof B, et al. Risk alleles in CFH and ARMS2 are independently associated with systemic complement activation in age-related macular degeneration. *Ophthalmology* 2012;**119**:339–46.

19. Stanton CM, Yates JRW, den Hollander AI, Seddon JM, Swaroop A, Stambolian D, et al. Complement factor D in age-related macular degeneration. *Invest Ophthalmol Vis Sci* 2011;**52**:8828–34.

20. Hecker LA, Edwards AO, Ryu E, Tosakulwong N, Baratz KH, Brown WL, et al. Genetic control of the alternative pathway of complement in humans and age-related macular degeneration. *Hum Mol Genet* 2010;**19**:209–15.

21. Weikel KA, Chiu CJ, Taylor A. Nutritional modulation of age-related macular degeneration. *Mol Aspects Med* 2012;**33**:318–75.

22. Ho L, van Leeuwen R, Witteman JCM, van Duijn CM, Uitterlinden AG, Hofman A, et al. Reducing the genetic risk of age-related macular degeneration with dietary antioxidants, zinc, and ω-3 fatty acids: the Rotterdam Study. *Arch Ophthalmol* 2011;**129**:758–66.

23. Powner MB, Gillies MC, Tretiach M, Scott A, Guymer RH, Hageman GS, et al. Perifoveal Müller cell depletion in a case of macular telangiectasia type 2. *Ophthalmology* 2010;**117**:2407–16.

24. Bone RA, Landrum JT, Tarsis SL. Preliminary identification of the human macular pigment. *Vision Res* 1985;**25**:1531–5.

25. Bone RA, Landrum JT, Fernandez L, Tarsis SL. Analysis of the macular pigment by HPLC: retinal distribution and age study. *Invest Ophthalmol Vis Sci* 1988;**29**:843–9.

26. Bone RA, Landrum JT, Hime GW, Cains A, Zamor J. Stereochemistry of the human macular carotenoids. *Invest Ophthalmol Vis Sci* 1993;**34**:2033–40.

27. Berendschot TTJM, van Norren D. Objective determination of the macular pigment optical density using fundus reflectance spectroscopy. *Arch Biochem Biophys* 2004;**430**:149–55.

28. Ossewaarde-Van Norel J, van den Biesen PR, van de Kraats J, Berendschot TTJM, van Norren D. Comparison of fluorescence of sodium fluorescein in retinal angiography with measurements *in vitro*. *J Biom Optics* 2002;**7**:190–8.

29. Berendschot TTJM, van Norren D. Macular pigment shows ring-like structures. *Invest Ophthalmol Vis Sci* 2006;**47**:709–14.

30. Helb HM, Charbel Issa P, van der Veen RLP, Berendschot TTJM, Scholl HP, Holz FG. Abnormal macular pigment distribution in type 2 idiopathic macular telangiectasia. *Retina* 2008;**28**:808–16.

31. Elsner AE, Burns SA, Delori FC, Webb RH. Quantitative reflectometry with the SLO. In: Naseman JE, Burk ROW, editors. *Laser Scanning Ophthalmoscopy and Tomography*. Munich: Quintessenz-Verlag; 1990. pp. 109–21.

32. Berendschot TTJM, Goldbohm RA, Klöpping WA, van de Kraats J, van Norel J, van Norren D. Influence of lutein supplementation on macular pigment, assessed with two objective techniques. *Invest Ophthalmol Vis Sci* 2000;**41**:3322–6.

33. Wüstemeyer H, Mößner A, Jahn C, Wolf S. Macular pigment density in healthy subjects quantified with a modified confocal scanning laser ophthalmoscope. *Graefes Arch Clin Exp Ophthalmol* 2003;**241**:647–51.

34. Broekmans WMR, Berendschot TTJM, Klöpping WA, de Vries AJ, Goldbohm RA, Tijssen CC, et al. Macular pigment density in relation to serum and adipose tissue concentrations of lutein and serum concentrations of zeaxanthin. *Am J Clin Nutr* 2002;**76**:595–603.

35. Johnson EJ, Neuringer M, Russell RM, Schalch W, Snodderly DM. Nutritional manipulation of primate retinas, III: Effects of lutein or zeaxanthin supplementation on adipose tissue and retina of xanthophyll-free monkeys. *Invest Ophthalmol Vis Sci* 2005;**46**:692–702.

36. Johnson EJ, Hammond BR, Yeum KJ, Qin J, Wang XD, Castaneda C, et al. Relation among serum and tissue concentrations of lutein and zeaxanthin and macular pigment density. *Am J Clin Nutr* 2000;**71**:1555–62.

37. Landrum JT, Bone RA, Chen Y, Herrero C, Llerena CM, Twarowska E. Carotenoids in the human retina. *Pure Appl Chem* 1999;**71**:2237–44.

38. Loane E, Beatty S, Nolan JM. The relationship between lutein, zeaxanthin, serum lipoproteins and macular pigment optical density. *Invest Ophthalmol Vis Sci* 2009;**50**:1710.

39. Hammond BR, Johnson EJ, Russell RM, Krinsky NI, Yeum KJ, Edwards RB, et al. Dietary modification of human macular pigment density. *Invest Ophthalmol Vis Sci* 1997;**38**:1795–801.

40. Wenzel AJ, Gerweck C, Barbato D, Nicolosi RJ, Handelman GJ, Curran-Celentano JA. 12-wk egg intervention increases serum zeaxanthin and macular pigment optical density in women. *J Nutr* 2006;**136**:2568–73.

41. Burke JD, Curran-Celentano Wenzel AJ. Diet and serum carotenoid concentrations affect macular pigment optical density in adults 45 years and older. *J Nutr* 2005;**135**:1208–14.

42. Chung HY, Rasmussen HM, Johnson EJ. Lutein bioavailability is higher from lutein-enriched eggs than from supplements and spinach in men. *J Nutr* 2004;**134**:1887–93.

43. Landrum JT, Bone RA, Joa H, Kilburn MD, Moore LL, Sprague KE. A one year study of the macular pigment: the effect of 140 days of a lutein supplement. *Exp Eye Res* 1997;**65**:57–62.

44. Bone RA, Landrum JT, Guerra LH, Ruiz CA. Lutein and zeaxanthin dietary supplements raise macular pigment density and serum concentrations of these carotenoids in humans. *J Nutr* 2003;**133**:992–8.

45. Snodderly DM, Chung HC, Caldarella SM, Johnson EJ. The influence of supplemental lutein and docosahexaenoic acid on their serum levels and on macular pigment. *Invest Ophthalmol Vis Sci* 2005;**46**:1766.

46. Tanito M, Obana A, Okazaki S, Ohira A, Gellermann W. Change of macular pigment density quantified with resonance Raman spectrophotometry and autofluorescence imaging in normal subjects supplemented with oral lutein or zeaxanthin. *Invest Ophthalmol Vis Sci* 2009;**50**:1716.

47. Bhosale P. Zhao dY, Bernstein PS. HPLC measurement of ocular carotenoid levels in human donor eyes in the lutein supplementation era. *Invest Ophthalmol Vis Sci* 2007;**48**:543–9.

48. Bone RA, Landrum JT, Cao Y, Howard AN, varez-Calderon F. Macular pigment response to a supplement containing meso-zeaxanthin, lutein and zeaxanthin. *Nutr Metab (Lond)* 2007;**4**:12.

49. Zeimer M, Hense HW, Heimes B, Austermann U, Fobker M, Pauleikhoff D. The macular pigment: short- and intermediate-term changes of macular pigment optical density following supplementation with lutein and zeaxanthin and co-antioxidants. The LUNA Study. *Ophthalmologe* 2009;**106**:29–36.

50. Köpcke W, Schalch W. LUXEA-Study Group. Changes in macular pigment optical density following repeated dosing with lutein, zeaxanthin, or their combination in healthy volunteers – results of the LUXEA-Study. *Invest Ophthalmol Vis Sci* 2005;**46**:1768.

51. Richer S, Devenport J, Lang JC. LAST II: differential temporal responses of macular pigment optical density in patients with atrophic age-related macular degeneration to dietary supplementation with xanthophylls. *Optometry* 2007;**78**:213–9.

52. Rougier MB, Delyfer MN, Korobelnik JF. Measuring macular pigment *in vivo*. *J Fr Ophtalmol* 2008;**31**:445–53.

53. Johnson EJ, Chung HY, Caldarella SM, Snodderly DM. The influence of supplemental lutein and docosahexaenoic acid on serum, lipoproteins, and macular pigmentation. *Am J Clin Nutr* 2008;**87**:1521–9.

54. Neuringer M, Sandstrom MM, Johnson EJ, Snodderly DM. Nutritional manipulation of primate retinas, I: effects of lutein or zeaxanthin supplements on serum and macular pigment in xanthophyll-free rhesus monkeys. *Invest Ophthalmol Vis Sci* 2004;**45**:3234–43.

55. Schalch W, Cohn W, Barker FM, Kopcke W, Mellerio J, Bird AC, et al. Xanthophyll accumulation in the human retina during supplementation with lutein or zeaxanthin – the LUXEA (LUtein Xanthophyll Eye Accumulation) study. *Arch Biochem Biophys* 2007;**458**:128–35.

56. Connolly EE, Beatty S, Loughman J, Howard AN, Louw MS, Nolan JM. Supplementation with all three macular carotenoids: response, stability, and safety. *Invest Ophthalmol Vis Sci* 2011;**52**:9207–17.

57. Koh HH, Murray IJ, Nolan D, Carden D, Feather J, Beatty S. Plasma and macular responses to lutein supplement in subjects with and without age-related maculopathy: a pilot study. *Exp Eye Res* 2004;**79**:21–7.

58. Trieschmann M, Beatty S, Nolan JM, Hense HW, Heimes B, Austermann U, et al. Changes in macular pigment optical density and serum concentrations of its constituent carotenoids following supplemental lutein and zeaxanthin: the LUNA study. *Exp Eye Res* 2007;**84**:718–28.

59. Richer SP, Stiles W, Graham-Hoffman K, Levin M, Ruskin D, Wrobel J, et al. Randomized, double-blind, placebo-controlled study of zeaxanthin and visual function in patients with atrophic age-related macular degeneration: the Zeaxanthin and Visual Function Study (ZVF) FDA IND #78, 973. *Optometry* 2011;**82**:667–80.

60. Weigert G, Kaya S, Pemp B, Sacu S, Lasta M, Werkmeister RM, et al. Effects of lutein supplementation on macular pigment optical density and visual acuity in patients with age-related macular degeneration. *Invest Ophthalmol Vis Sci* 2011;**52**:8174–8.

61. Sommerburg O, Keunen JEE, Bird AC, van Kuijk FJGM. Fruits and vegetables that are sources for lutein and zeaxanthin: the macular pigment in human eyes. *Br J Ophthalmol* 1998;**82**:907–10.

62. Granado F, Olmedilla B, Herrero C, Perez-Sacristan B, Blanco I, Blazquez S. Bioavailability of carotenoids and tocopherols from broccoli: *in vivo* and *in vitro* assessment. *Exp Biol Med* 2006;**231**:1733–8.

63. Maiani G, Periago Caston MJ, Catasta G, Toti E, Cambrodon IG, Bysted A, et al. Carotenoids: actual knowledge on food sources, intakes, stability and bioavailability and their protective role in humans. *Mol Nutr Food Res* 2009;**53**(Suppl. 2):S194–218.

64. Thurnham DI. Macular zeaxanthins and lutein? A review of dietary sources and bioavailability and some relationships with macular pigment optical density and age-related macular disease. *Nutr Res Rev* 2007;**20**:163–79.

65. Bone RA, Landrum JT, Friedes LM, Gomez CM, Kilburn MD, Menendez E, et al. Distribution of lutein and zeaxanthin stereo-isomers in the human retina. *Exp Eye Res* 1997;**64**:211–8.

66. Handelman GJ, Snodderly DM, Krinsky NI, Russett MD, Adler AJ. Biological control of primate macular pigment. Biochemical and densitometric studies. *Invest Ophthalmol Vis Sci* 1991;**32**:257–67.

67. van de Kraats J, Kanis MJ, Genders SW, van Norren D. Lutein and zeaxanthin measured separately in the living human retina with fundus reflectometry. *Invest Ophthalmol Vis Sci* 2008;**49**:5568–73.

68. Schultze M. Zur Anatomie und Physiologie der Retina. *Arch Mickrosk Anat* 1866;**2**:165–286.

69. Engles M, Wooten BR, Hammond B. Macular pigment: a test of the acuity hypothesis. *Invest Ophthalmol Vis Sci* 2007;**48**:2922–31.

70. McLellan JS, Marcos S, Prieto PM, Burns SA. Imperfect optics may be the eye's defence against chromatic blur. *Nature* 2002;**417**:174–6.

71. Wooten BR, Hammond BR. Macular pigment: influences on visual acuity and visibility. *Prog Retin Eye Res* 2002;**21**:225–40.

72. Hammond BR, Wooten BR, Snodderly DM. Preservation of visual sensitivity of older subjects: association with macular pigment density. *Invest Ophthalmol Vis Sci* 1998;**39**:397–406.

73. Stringham JM, Hammond Jr BR. The glare hypothesis of macular pigment function. *Optom Vis Sci* 2007;**84**:859–64.

74. Stringham JM, Hammond BR. Macular pigment and visual performance under glare conditions. *Optom Vis Sci* 2008;**85**:82–8.

75. Seddon JM, Ajani UA, Sperduto RD, Hiller R, Blair N, Burton TC, et al. Dietary carotenoids, vitamins A, C, and E, and advanced age-related macular degeneration. Eye Disease Case–Control Study Group. *JAMA* 1994;**272**:1413–20.

76. Khachik F, Bernstein PS, Garland DL. Identification of lutein and zeaxanthin oxidation products in human and monkey retinas. *Invest Ophthalmol Vis Sci* 1997;**38**:1802–11.

77. Landrum JT, Bone RA, Kilburn MD. The macular pigment: a possible role in protection from age-related macular degeneration. *Adv Pharmacol* 1997;**38**:537–56.

78. Izumi-Nagai K, Nagai N, Ohgami K, Satofuka S, Ozawa Y, Tsubota K, et al. Macular pigment lutein is antiinflammatory in preventing choroidal neovascularization. *Arterioscler Thromb Vasc Biol* 2007;**27**:2555–62.

79. Kijlstra A, Tian Y, Kelly ER, Berendschot TT. Lutein: more than just a filter for blue light. *Prog Retin Eye Res* 2012;**31**:303–15.

80. Wang W, Connor SL, Johnson EJ, Klein ML, Hughes S, Connor WE. Effect of dietary lutein and zeaxanthin on plasma carotenoids and their transport in lipoproteins in age-related macular degeneration. *Am J Clin Nutr* 2007;**85**:762–9.

81. Parker RS. Carotenoids. 4. Absorption, metabolism, and transport of carotenoids. *FASEB J* 1996;**10**:542–51.

82. Krinsky NI, Cronwell DG, Oncley JL. The transport of vitamin A and carotenoids in human plasma. *Arch Biochem Biophys* 1958;**73**:233–46.

83. Connor WE, Duell PB, Kean R, Wang Y. The prime role of HDL to transport lutein into the retina: evidence from HDL-deficient WHAM chicks having a mutant ABCA1 transporter. *Invest Ophthalmol Vis Sci* 2007;**48**:4226–31.

84. Devaraj S, Jialal I. The role of dietary supplementation with plant sterols and stanols in the prevention of cardiovascular disease. *Nutr Rev* 2006;**64**(7 Pt 1):348–54.

85. Demonty I, Ras RT, van der Knaap HC, Duchateau GS, Meijer L, Zock PL, et al. Continuous dose–response relationship of the LDL-cholesterol-lowering effect of phytosterol intake. *J Nutr* 2009;**139**:271–84.

86. Musa-Veloso K, Poon TH, Elliot JA, Chung C. A comparison of the LDL-cholesterol lowering efficacy of plant stanols and plant sterols over a continuous dose range: results of a meta-analysis of randomized, placebo-controlled trials. *Prostaglandins Leukot Essent Fatty Acids* 2011;**85**:9–28.

87. Gylling H, Hallikainen M, Nissinen MJ, Miettinen TA. The effect of a very high daily plant stanol ester intake on serum lipids, carotenoids, and fat-soluble vitamins. *Clin Nutr* 2010;**29**:112–8.

88. de Jong A, Plat J, Lutjohann D, Mensink RP. Effects of long-term plant sterol or stanol ester consumption on lipid and lipoprotein metabolism in subjects on statin treatment. *Br J Nutr* 2008;**100**:937–41.

89. Law M. Plant sterol and stanol margarines and health. *BMJ* 2000;**320**:861–4.

90. Katan MB, Grundy SM, Jones P, Law M, Miettinen T, Paoletti R. Efficacy and safety of plant stanols and sterols in the management of blood cholesterol levels. *Mayo Clin Proc* 2003;**78**:965–78.

91. AbuMweis SS, Barake R, Jones PJ. Plant sterols/stanols as cholesterol lowering agents: a meta-analysis of randomized controlled trials. *Food Nutr Res* 2008:52.

92. Mensink RP, de Jong A, Lutjohann D, Haenen GR, Plat J. Plant stanols dose-dependently decrease LDL-cholesterol concentrations, but not cholesterol-standardized fat-soluble antioxidant concentrations, at intakes up to 9g/d. *Am J Clin Nutr* 2010;**92**:24–33.

93. Gylling H, Hallikainen M, Nissinen MJ, Simonen P, Miettinen TA. Very high plant stanol intake and serum plant stanols and non-cholesterol sterols. *Eur J Nutr* 2010;**49**:111–7.

94. Plat J, Mensink RP. Effects of diets enriched with two different plant stanol ester mixtures on plasma ubiquinol-10 and fat-soluble antioxidant concentrations. *Metab Clin Exp* 2001;**50**:520–9.

95. Colgan HA, Floyd S, Noone EJ, Gibney MJ, Roche HM. Increased intake of fruit and vegetables and a low-fat diet, with and without low-fat plant sterol-enriched spread consumption: effects on plasma lipoprotein and carotenoid metabolism. *J Hum Nutr Diet* 2004;**17**:561–9.

96. Berendschot TTJM, Plat J, de Jong A, Mensink RP. Long-term plant stanol and sterol ester-enriched functional food consumption, serum lutein/zeaxanthin concentration and macular pigment optical density. *Br J Nutr* 2009;**101**:1607–10.

97. Raeini-Sarjaz M, Ntanios FY, Vanstone CA, Jones PJH. No changes in serum fat-soluble vitamin and carotenoid concentrations with the intake of plant sterol/stanol esters in the context of a controlled diet. *Metabolism* 2002;**51**:652–6.

98. Broekmans WMR, Klopping-Ketelaars IA, Weststrate JA, Tijburg LB, van Poppel G, Vink AA, et al. Decreased carotenoid concentrations due to dietary sucrose polyesters do not affect possible markers of disease risk in humans. *J Nutr* 2003;**133**:720–6.

99. Neuhouser ML, Rock CL, Kristal AR, Patterson RE, Neumark-Sztainer D, Cheskin LJ, et al. Olestra is associated with slight reductions in serum carotenoids but does not markedly influence serum fat-soluble vitamin concentrations. *Am J Clin Nutr* 2006;**83**:624–31.

100. Rydén M, Leanderson P, Kastbom KO, Jonasson L. Effects of simvastatin on carotenoid status in plasma. *Nutr Metab Cardiovasc Dis* 2012;**22**:66–71.

101. Renzi L, Hammond B, Dengler M, Roberts R. The relation between serum lipids and lutein and zeaxanthin in the serum and retina: results from cross-sectional, case–control and case study designs. *Lipids Health Dis* 2012;**11**:33.

102. Mares-Perlman JA, Millen AE, Ficek TL, Hankinson SE. The body of evidence to support a protective role for lutein and zeaxanthin in delaying chronic disease. Overview. *J Nutr* 2002;**132**:518–24. S.

103. Loane E, Kelliher C, Beatty S, Nolan JM. The rationale and evidence base for a protective role of macular pigment in age-related maculopathy. *Br J Ophthalmol* 2008;**92**:1163–8.

104. Chong EW, Wong TY, Kreis AJ, Simpson JA, Guymer RH. Dietary antioxidants and primary prevention of age related macular degeneration: systematic review and meta-analysis. *BMJ* 2007;**335**:755.

105. Mares-Perlman JA. Too soon for lutein supplements. *Am J Clin Nutr* 1999;**70**:431–2.

106. O'Connell ED, Nolan JM, Stack J, Greenberg D, Kyle J, Maddock L, et al. Diet and risk factors for age-related maculopathy. *Am J Clin Nutr* 2008;**87**:712–22.

107. Carpentier S, Knaus M, Suh M. Associations between lutein, zeaxanthin, and age-related macular degeneration: an overview. *Crit Rev Food Sci Nutr* 2009;**49**:313–26.

108. Trumbo PR, Ellwood KC. Lutein and zeaxanthin intakes and risk of age-related macular degeneration and cataracts: an evaluation using the Food and Drug Administration's evidence-based review system for health claims. *Am J Clin Nutr* 2006;**84**:971–4.

109. Cooper DA, Curran-Celentano, Ciulla TA, Hammond BR, Danis RB, Pratt LM, et al. Olestra consumption is not associated with macular pigment optical density in a cross-sectional volunteer sample in Indianapolis. *J Nutr* 2000;**130**:642–7.

45

Seeds of *Cornus officinalis* and Diabetic Cataracts

Jin Sook Kim

Herbal Medicine Research Division, Korea Institute of Oriental Medicine, Daejeon, Republic of Korea

INTRODUCTION

Diabetes mellitus is a major cause of mortality and morbidity worldwide, and its prevalence is increasing at an alarming rate. Long-term hyperglycemia, which is the most common feature of diabetes mellitus, accelerates the irreversible formation of advanced glycation end-products (AGEs) through Amadori-type compounds from proteins and reducing sugars and the induction of aldose reductase (AR), an enzyme that plays important roles in the pathogenesis of diabetic complications.

The harmful effects of AGEs are due to structural and functional modifications in plasma and extracellular matrix proteins; protein cross-linking and the interaction of AGEs with their receptors and/or binding proteins lead to the enhanced formation of nuclear factor-κB and the release of proinflammatory cytokines, growth factors, and adhesion molecules.[1]

One of the enzymes in the polyol pathway is AR, which reduces excess D-glucose to D-sorbitol using reduced nicotinamide adenine dinucleotide phosphate (NADPH) as a cofactor. The increased accumulation of sorbitol has been linked to cellular damage.[2]

The mechanism of cataractogenesis is unclear, but irreversible AGE formation, polyol (e.g., sorbitol) accumulation, and oxidative stress within the eye lens have been regarded as major causes of diabetic cataract. Therefore, inhibitors of AGEs and AR could be a potential therapeutic strategy in the prevention of diabetic complications such as diabetic cataracts.[1]

Cataracts, which are characterized by cloudiness or opacification of the eye lens, are the leading cause of blindness worldwide. Diabetic cataracts are a major complication of diabetes and one of the most common causes of visual impairment in diabetic patients. There are few medications available to treat diabetic cataracts. Laser surgery can generally repair (or stop) the pathologic changes. Advances in cataract surgical techniques and instrumentation have generally improved the outcomes; however, surgery may not be safe or effective in individuals with pre-existing retinal pathology or limited visual potential. Natural foods, herbal extracts, and compounds isolated from natural resources have been proven to be effective in the prevention of diabetic complications such as diabetic cataracts. Clinical evidence has suggested that the appropriate use of traditional herbal medicine and/or natural functional foods with modern Western medicines, or mainstream antidiabetic drugs, can prevent or ameliorate the development of diabetic cataracts.

Aminoguanidine (AG), an anhydrazine-like molecule, was the first AGE inhibitor tested in clinical trials. However, this compound was not approved for commercial production because of side effects observed in diabetic patients during a phase III clinical trial.[3,4]

ALT-711 (Alagebrium), a highly potent AGE-cross-link breaker, has the ability to reverse already formed AGE cross-links.[5]

Epalrestat (EP) is a commercial synthetic AR inhibitor that was approved in Japan for the treatment and prevention of diabetic peripheral neuropathy.[6]

CORNUS OFFICINALIS SIEB. ET ZUCC.

Cornus officinalis Sieb. et Zucc. (corni fructus, Coraceacae), which is grown in Korea, China, and Japan, is a well-known traditional medicine owing to its tonic, analgesic, diuretic, and diabetic properties. Corni fructus has a slightly sweet and fruity taste and a slightly spicy and medicinal odor.

Corni fructus exerts its antidiabetic effects via various mechanisms: inhibition of α-glucosidase and sucrose tolerance,[7,8] increase in glucose transporter type 4 (GLUT4) expression,[9] antioxidant effect,[10] enhancement of insulin secretion and protection of pancreatic β-cells,[11] and protection against AGE-mediated renal injury in streptozotocin (STZ)-induced diabetic rats.[12]

The components of *C. officinalis* fruits with the seeds removed (i.e., pericarp) consist of galloyl glucosides and tannins,[13–16] iridoids,[12,15,17,18] flavonoids,[18] phenols,[18,19] terpenoids,[18,20] furan derivatives,[19,21] volatile flavor (organic acid) compounds,[12,18,21,22] and amino acids,[23]

The following compounds have been isolated from the fruits (pericarp) of *C. officinalis*: galloyl glucosides and tannins including 3-*O*-galloyl-β-D-glucose, 2,3-di-*O*-galloyl-β-D-glucose, 1,2,3-tri-*O*-galloyl-β-D-glucose, 1,2,6-tri-*O*-galloyl-β-D-glucose, 1,2,3,6-tetra-*O*-galloyl-β-D-glucose, 7-*O*-galloyl-D-sedoheptulose, tellimagrandin I, tellimagrandin II, germin D, cornusiin A, cornusiin B, cornusiin C, cornusiin G, and oenothein C; iridoid glucosides including loganin, loganic acid, morroniside, (7R)-*n*-butylmorronoside, and 7-*O*-methylmorronoside; secoiridoid glucosides such as sweroside and cornuside; flavonoids including quercetin-3-*O*-β-D-glucuronide, quercetin-3-*O*-β-D-(6″-*n*-butylglucuronide), and (−)-epicatechin-*O*-gallate; phenols including caffeic acid, caffeic acid monomethyl ester, and gallic acid; terpenoids including ursolic acid, oleanolic acid, and β-sitosterol; furan derivatives such as dimethyltetrahydrofuran *cis*-2,5-dicarboxylate and 5-hydroxyfurfural; and volatile flavor compounds such as malic acid, dimethylmalate, 2-butoxybutanedioic acid, palmitic acid, mevaloside, and benzyl cinnamate.

Morroniside and loganin are major compounds of the fruits (pericarp) of *C. officinalis*. Morroniside and loganin have been reported to prevent diabetic nephropathy through several mechanisms including the inhibition of connective tissue growth factor (CTGF) expression in serum and renal tissue,[24] reduction in AGE accumulation,[2,12,25] and inhibition of mesangial cell proliferation by reducing reactive oxygen species formation, increasing the activities of superoxide dismutase and glutathione peroxidase in type 1 and/or 2 diabetes models.[26] The neuroprotective action of morroniside against focal cerebral ischemia in rat was also reported.[27] The total triterpene acids of corni fruits have been reported to attenuate diabetic cardiomyopathy.[28]

In addition to the secondary metabolites mentioned above, primary metabolites isolated from *C. officinalis* have been reported. Compared with the pericarp (fruit with seeds removed) of *C. officinalis*, the seeds contain higher amounts of crude ash, crude protein, crude fat, and crude fiber and lower moisture and reducing sugar levels. The main organic acids in the pericarp and seeds are malic acid and citric acid, respectively. The contents of these acids are higher in the pericarp than in the seeds. Eighteen amino acids were analyzed in the pericarp and seeds. The total amino acid concentrations were 230.41 mg/100 g pericarp and 883.81 mg/100 g seed, respectively, and the essential amino acid concentrations were 124.44 mg/100 g pericarp and 375.53 mg/100 g seed, respectively. The main amino acids detected in the pericarp are glycine, leucine, and histidine; and those in the seeds are arginine, glycine, and leucine. Mineral component analysis revealed that pericarp and seeds both contain K, Ca, Mg, P, Na, Fe, Zn, Cu, and Mn. The most predominant minerals in both portions are K, Ca, and Mg. Linoleic, oleic, palmitic, and stearic acids have also been detected. Linoleic acid and linolenic acid are the main fatty acids in the pericarp, but linoleic acid is the principal fatty acid in the seeds of *C. officinalis*.[23]

The effects on eye-related disease of *C. officinalis* and/or compounds obtained from this plant have not been investigated before the study by the present group. This group evaluated the effects of *C. officinalis* seeds on diabetic complications by isolating compounds from the seeds and studying their effects on AGE formation, AR, and cataractogenesis.[23,29–31] It has been reported that the seeds of *C. officinalis* have a hypoglycemic action[32]; however, few studies have focused on the effects of *C. officinalis* seeds and their constituents.[19,23,30–32]

EXTRACTION AND ISOLATION OF COMPOUNDS FROM THE SEEDS OF *CORNUS OFFICINALIS*

The seeds of *C. officinalis* Sieb. et Zucc., which were cultivated in Gurye-gun, Jeolla Nam-do, Republic of Korea, were pulverized and extracted with alcohol (methanol or ethanol) at room temperature. The extract was concentrated *in vacuo* at 40°C to produce an alcohol extract. The alcohol extract was suspended in water and successively fractionated with *n*-hexane, ethyl acetate (EtOAc), and *n*-butanol (*n*-BuOH) to produce *n*-hexane-, EtOAc-, n-BuOH-, and water-soluble fractions, respectively.

The EtOAc-soluble extract was chromatographed over silica gel, resulting in 10 fractions (F01–F010). Fraction F06 was subjected to Sephadex column chromatography with MeOH, yielding nine subfractions (F0601–F0609). Gallic acid-4-*O*-β-D-(6′-*O*-galloyl)-glucoside, gallic acid-4-*O*-β-D-glucose, and 1,2,3-tri-*O*-galloylβ-D-glucose were purified from subfractions F0604, F0605, and F0607 by repeated column chromatography, respectively. Fraction F07 was subjected to a Diaion HP-20 column with a MeOH/water gradient as the stationary phase, yielding five subfractions (F0701–F0705). From the F0701 subfraction, 1,2,6-tri-*O*-galloyl-β-D-glucose, 1,2,3,6-tetra-*O*-galloyl-β-D-glucose, 1,2,4,6-tetra-*O*-galloyl-β-D-glucose, and 1,2,3,4,6-penta-*O*-galloyl-β-D-glucose were isolated. Tellimagrandin II was obtained from the F0705 subfraction.[30]

The structures of all compounds isolated from the *C. officinalis* seeds were assessed by spectroscopy and by comparing them to the published literature.

Betulinic acid, ursolic acid, 5-hydroxymethyl furfural, (+)-pinoresinol, (−)-balanophonin, gallicin, 4-hydroxybenzaldehyde, vanillin, coniferaldehyde, malic acid, kaempferol, and quercetin have also been isolated from the seeds of *C. officinalis*.[29]

Among these isolated compounds, ursolic acid, 1,2,3-tri-*O*-galloyl-β-D-glucose, 1,2,6-tri-*O*-galloyl-β-D-glucose, 1,2,3,6-tetra-*O*-galloyl-β-D-glucose, tellimagrandin II, 5-hydroxymethyl furfural, some organic acids such as malic acid, and some amino acids were isolated from both the fruits and the seeds of *C. officinalis*.

INHIBITORY EFFECT ON ADVANCED GLYCATION END-PRODUCT FORMATION

A reaction mixture (bovine serum albumin (BSA) in phosphate buffer at pH 7.4 with 0.02% sodium azide) was added to fructose and glucose. This mixture was mixed with the extract, solvent-soluble fractions, and the isolated compounds or AG. After incubating the mixture at 37°C for 14 days, an AGE assay was performed using a spectrofluorometric detector (Synergy HT; BioTek, Vermont, USA; excitation wavelength, 350 nm; emission wavelength, 450 nm).

The seeds of *C. officinalis* had a stronger inhibitory effect on AGE formation (half-maximal inhibitory concentration (IC_{50}): 1.13 μg/mL) than the pericarp (IC_{50}: 38.64 μg/mL) and fruits (pericarp plus seeds) (IC_{50}: 3.63 μg/mL) (Table 45.1). The *n*-hexane (IC_{50}: 17.46 μg/mL)-, EtOAc(IC_{50}: 1.52 μg/mL)-, *n*-BuOH (IC_{50}: 1.24 μg/mL)-, and water (IC_{50}: 3.27 μg/mL)-soluble fractions also inhibited AGE formation. However, the alcoholic extract of the seeds of *C. officinalis* had a stronger inhibitory effect on AGE formation (IC_{50}: 1.13 μg/mL) than the solvent-soluble fractions (Table 45.2).[31]

The compounds isolated from the seeds of *C. officinalis*, namely, 1,2,3-tri-*O*-galloyl-β-D-glucose (1), 1,2,6-tri-*O*-galloyl-β-D-glucose (2), 1,2,3,6-tetra-*O*-galloyl-β-D-glucose (3), 1,2,4,6-tetra-*O*-galloyl-β-D-glucose (4), 1,2,3,4,6-penta-*O*-galloyl-β-D-glucose (5), tellimagrandin II (6), gallic acid-4-*O*-β-D-glucose (7), and gallic acid-4-*O*-β-D-(6′-*O*-galloyl)-glucoside (8), were evaluated for their potential to inhibit AGE formation and rat lens aldose reductase (RLAR) activity. Among these compounds, compounds 1–6 dramatically inhibited AGE formation (Table 45.3).[30]

The compounds (−)-balanophonin (IC_{50}: 27.81 μM) and gallicin (IC_{50}: 18.04 μM) also had stronger inhibitory effects on AGE formation than AG (IC_{50}: 974.59 μM), the standard control.[29]

TABLE 45.1 Inhibitory Effect of the Fruit, Pericarp, and Seeds of 80% Ethanol Extract from *Cornus officinalis* on Advanced Glycation End-Product Formation

Part Used	Concentration (μg/mL)	Inhibitory Effect*	
		Inhibitory Effect (%)	IC_{50} (μg/mL)
Fruit (pericarp + seed)	1.25	−0.59 ± 2.31	3.63
	2.5	30.21 ± 4.49	
	5.0	76.97 ± 0.54	
Pericarp	10.0	−7.28 ± 1.65	38.64
	25.0	28.09 ± 2.77	
	50.0	70.16 ± 1.82	
Seed	1.0	14.18 ± 3.44	1.13
	1.25	27.66 ± 1.75	
	2.5	59.79 ± 1.22	
Aminoguanidine	18.5	12.70 ± 6.12	
	55.5	39.46 ± 1.91	
	74.0	50.68 ± 0.51	

*Expressed as the mean ± SD of three replicates for each sample. Half-maximal inhibitory concentration (IC_{50}) values were calculated from the dose–inhibition curve.
Source: Kim et al. (2008),[31] with permission from Korean Journal of Pharmacognosy.

TABLE 45.2 Inhibitory Effect of the Solvent Fractions of the 80% Ethanol Extract of the Seeds of *Cornus officinalis* on Advanced Glycation End-Product Formation

Solvent Fraction	Concentration (μg/mL)	Inhibitory Effect*	
		Inhibitory Effect (%)	IC_{50} (μg/mL)
n-Hexane	5.0	0.35 ± 0.83	17.64
	10.0	18.78 ± 0.77	
	25.0	79.52 ± 0.32	
EtOAc	0.5	17.46 ± 0.32	1.52
	0.75	23.84 ± 0.56	
	2.5	82.40 ± 0.44	
n-BuOH	0.75	17.49 ± 0.59	1.24
	1.0	30.63 ± 0.39	
	1.25	52.46 ± 0.49	
Water	1.25	10.03 ± 0.56	3.27
	2.5	43.08 ± 0.49	
	5.0	78.14 ± 0.49	

EtOAc: ethyl acetate; *n*-BuOH: *n*-butanol.
*Expressed as the mean ± SD of three replicates for each sample. Half-maximal inhibitory concentration (IC_{50}) values were calculated from the dose–inhibition curve.
Source: Kim et al. (2008),[31] with permission from Korean Journal of Pharmacognosy.

TABLE 45.3 Inhibitory Effects of Compounds 1–8 Extracted from the Seeds of *Cornus officinalis* on Advanced Glycation End-Product (AGE) Formation and Rat Lens Aldose Reductase (RLAR) Activity

	Inhibitory Effect (IC$_{50}$, μM)*	
Compound	AGE Formation†	RLAR
1	2.81 ± 0.03	2.35 ± 0.27
2	1.89 ± 0.24	4.01 ± 0.12
3	1.99 ± 0.07	0.70 ± 0.12
4	0.80 ± 0.02	0.76 ± 0.07
5	0.87 ± 0.01	1.93 ± 0.01
6	0.90 ± 0.01	0.90 ± 0.08
7	>150.6	82.74 ± 8.49
8	42.16 ± 0.82	11.77 ± 1.23
Aminoguanidine‡	961.2 ± 29.5	–
Epalrestat‡	–	0.067 ± 0.009

1: 1,2,3-tri-*O*-galloyl-β-D-glucose; 2: 1,2,6-tri-*O*-galloyl-β-D-glucose; 3: 1,2,3,6-tetra-*O*-galloyl-β-D-glucose; 4: 1,2,4,6-tetra-*O*-galloyl-β-D-glucose; 5: 1,2,3,4,6-penta-*O*-galloyl-β-D-glucose; 6: tellimagrandin II; 7: gallic acid-4-*O*-β-D-glucose; 8: gallic acid-4-*O*-β-D-(6′-*O*-galloyl)-glucose.

Expressed as the mean ± SD of triplicate experiments. Half-maximal inhibitory concentration (IC$_{50}$) values were calculated from the dose–inhibition curve.

†*After incubation for 14 days, the fluorescent reaction products were assayed on a spectrofluorometric detector.*

‡*Aminoguanidine and epalrestat were used as positive controls.*

Source: Lee et al. (2011),[30] with permission from Biological and Pharmaceutical Bulletin.

INDIRECT ENZYME-LINKED IMMUNOSORBENT ASSAY OF THE INHIBITORY EFFECT ON AGE FORMATION BY INDIRECT ENZYME-LINKED IMMUNOSORBENT ASSAY

Ninety-six-well plates were coated overnight with the reaction mixture in 50 mM sodium carbonate buffer, pH 9.6. After coating, the wells were blocked for 2 hours at 37°C with phosphate-buffered saline (PBS) containing 1% BSA. Anti-AGE antibodies (6D12) or anti-CML antibodies (NF-1G) were diluted at a titer of 1:1000 in PBS, incubated for 1 hour at 37°C, and washed. A horseradish peroxidase (HRP)-linked goat antimouse antibody (Santa Cruz, Biotechnology, CA, USA) was then added as the secondary antibody at a titer of 1:1000 in PBS, incubated for 1 hour at 37°C, and washed again. The wells were developed with 3,3′,5,5′-tetramethylbenzidine (TMB; Sigma, St Louis, MO, USA). The reaction was terminated by adding 1 M H$_2$SO$_4$, and absorbance at 450 nm was read on a microplate reader (Synergy HT; BioTek).

Compared with AG, the 80% alcohol extract and the n-EtOAc- and n-BuOH-soluble fractions of the seeds of *C. officinalis* had a more potent inhibitory effect on AGE–BSA formation in an enzyme-linked immunosorbent assay (ELISA) (Fig. 45.1).[31]

INHIBITORY EFFECT ON ADVANCED GLYCATION END-PRODUCT–BOVINE SERUM ALBUMIN CROSS-LINKING TO COLLAGEN

A mixture of AGE–BSA and either the extract, solvent-soluble fractions, or AG were added to each well of collagen-coated microtiter plates and then incubated for 4 hours at 37°C. The formation of the collagen–AGE–BSA complex was measured using an anti-AGE–BSA monoclonal antibody, a horseradish peroxidase-linked goat antimouse immunoglobulin G (IgG) antibody, and a hydrogen peroxide (H$_2$O$_2$) substrate containing a 2,2′-azinobis-(3-ethylbenzothiazoline-6-sulfonicacid)(ABTS)chromogen. The optical density (OD) was measured with an ELISA reader with a sample wavelength of 410 nm. The inhibition of AGE cross-linking was expressed as the OD decrease (%) when AGE–BSA and collagen were incubated in the presence of the seed compounds.

The alcoholic extract and its solvent-soluble fractions as well as galloyl glucose compounds (1–6) were further evaluated *in vitro* for their inhibitory effect on collagen–AGE–BSA cross-linking.

The alcoholic extract and its solvent soluble fractions showed stronger inhibitory effects on collagen–AGE–BSA cross-linking than AG, the positive control (Fig. 45.2).[31]

These galloyl glucosides also dramatically reduced collagen–AGE–BSA cross-linking in a dose-dependent manner. In addition, all of these compounds exhibited stronger inhibitory activity than AG (Fig. 45.3).[30]

Consequently, the alcoholic extract, solvent-soluble fractions, and these galloyl glucosides inhibited both AGE formation and collagen–AGE cross-linking.

BREAKING EFFECT ON ADVANCED GLYCATION END-PRODUCT–BOVINE SERUM ALBUMIN CROSS-LINKS FORMED *IN VITRO*

To ascertain the ability to break AGE cross-links formed *in vitro*, AGE cross-links were prepared from AGE–BSA and rat-tail tendon collagen *in vitro*. In brief, glycated BSA was allowed to react with collagen to form AGE–BSA–collagen complexes in collagen-coated 96-well plates for 4 hours at 37°C. AGE–BSA–collagen complexes were incubated with the extracts or ALT-711 (positive control) for 16 hours at 37°C. After incubation with the extracts or ALT-711, the amount of BSA remaining attached to collagen was quantified by ELISA using a mouse monoclonal anti-AGE–BSA antibody, an HRP-linked goat antimouse IgG antibody, and an H$_2$O$_2$ substrate containing an ABTS chromogen.

The alcoholic extract and EtOAc-, n-BuOH-, and water-soluble fractions of the seeds of *C. officinalis* had a potent and dose-dependent breaking activity against

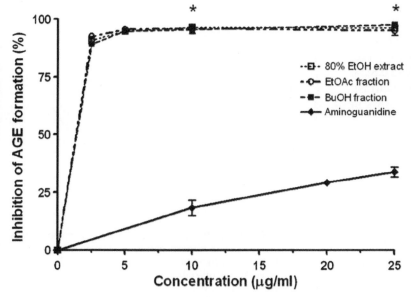

FIGURE 45.1 Percentage inhibition of the 80% ethanol (EtOH) extract and the solvent fractions of the seeds of *Cornus officinalis* on advanced glycation end-product (AGE) formation. AGE formation was determined using enzyme-linked immunosorbent assay with mouse anti-AGE antibodies (6D12) in the presence and absence of the extract, ethyl acetate (EtOAC)-, and *n*-butanol (*n*-BuOH)-soluble fractions and aminoguanidine (AG). Data are expressed as the mean±SD of four replicates for each sample. *P<0.01 vs. sample incubated with AG. *Source: Kim et al. (2008),[31] with permission from* Korean Journal of Pharmacognosy.

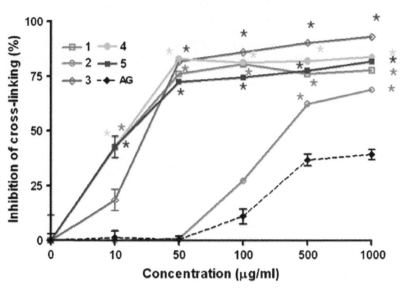

FIGURE 45.2 Inhibitory effect of the 80% ethanol (EtOH) extract and solvent fractions of the seeds of *Cornus officinalis* on collagen–advanced glycation end-product (AGE)–bovine serum albumin (BSA) cross-linking. AGE–BSA was cross-linked to collagen in the presence and absence of the alcoholic extract, solvent-soluble fractions of the seeds of *C. officinalis*, or amino-guanidine (AG); AGE–BSA cross-linked to collagen was determined by enzyme-linked immunosorbent assay using mouse anti-AGE antibodies (6D12). 1: 80% ethanol (EtOH) extract; 2: *n*-hexane-soluble fraction; 3: ethyl acetate (EtOAc)-soluble fraction; 4: *n*-butanol (*n*-BuOH)-soluble fraction; 5: water-soluble fraction. Data are expressed as the mean±SD of four replicates for each sample. *P<0.01 vs. sample incubated with AG. *Source: Kim et al. (2008),[31] with permission from* Korean Journal of Pharmacognosy.

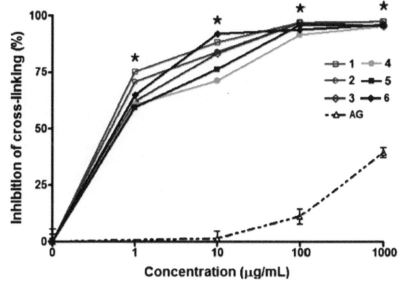

FIGURE 45.3 Inhibition (%) of advanced glycation end-product (AGE)–bovine serum albumin (BSA) formation by seed compounds 1–6 as measured by enzyme-linked immunosorbent assay using rabbit anti-AGE antibodies (6D12). 1: 1,2,3-tri-*O*-galloyl-β-D-glucose; 2: 1,2,6-tri-*O*-galloyl-β-D-glucose; 3: 1,2,3,6-tetra-*O*-galloyl-β-D-glucose; 4: 1,2,4,6-tetra-*O*-galloyl-β-D-glucose; 5: 1,2,3,4,6-penta-*O*-galloyl-β-D-glucose; 6: tellimagrandin II. Data are expressed as the mean±SD of four replicates for each sample. *P<0.01 vs. sample incubated with amino-guanidine (AG). *Source: Lee et al. (2011),[30] with permission from* Biological and Pharmaceutical Bulletin.

preformed AGE–BSA cross-linking. Furthermore, their effects were more potent than that of ALT-711, the positive control. The breaking effects of the alcoholic extract, EtOAc-, *n*-BuOH-, and water-soluble fractions at 1000 μg/mL were 81.26 ± 1.51%, 64.98 ± 0.81%, 85.0 ± 0.82%, and 59.8 ± 0.81%, respectively. In contrast, ALT-711 at 1000 μg/mL broke down to 33.38 ± 6.48% in the cross-linking to collagen (Fig. 45.4).[31]

RAT LENS ALDOSE REDUCTASE INHIBITION ASSAY

Crude RLAR was obtained from lenses removed from the eyes of 7–8-week-old Sprague–Dawley (SD) rats. The protein content of the enzyme was determined by the bicinchoninic assay using a BSA standard. RLAR activity was determined by measuring the amount of NADP produced per enzyme unit from NADPH at 37°C and pH 7.0. The incubation mixture contained Na,K-phosphate buffer, lithium sulfate, NADPH, DL-glyceraldehyde, and RLAR with or without the seed compounds and positive control, in a total volume of 1.0 mL. The reaction was initiated by adding NADPH at 37°C and stopped by adding HCl. Then, NaOH containing imidazole was added, and the mixture was incubated at 60°C for 10 minutes to convert NADP to a fluorescent product. The fluorescent product was measured on a spectrofluorophotometer (excitation wavelength, 360 nm; emission wavelength, 460 nm).

In the RLAR assay, 1,2,3,6-tetra-*O*-galloyl-β-D-glucose (3) and 1,2,4,6-tetra-*O*-galloyl-β-D-glucose (4) had the highest inhibitory activity; however, they were less inhibitory than EP, the positive control. The other galloyl glucose compounds, namely, 1,2,3-tri-*O*-galloyl-β-D-glucose (1), 1,2,6-tri-*O*-galloyl-β-D-glucose (2), and 1,2,3,4,6-penta-*O*-galloyl-β-D-glucose (5), also had potent inhibitory effect on RLAR activity, whereas the two gallic acid derivatives, gallic acid-4-*O*-β-D-glucose (7) and gallic acid-4-*O*-β-D-(6′-*O*-galloyl)-glucose (8), exhibited mild inhibitory activity (Table 45.3).[30]

RAT LENS ORGAN CULTURE AND ANALYSIS OF LENS OPACITY

The lenses of SD rats (6-week-old males) were carefully removed using a posterior approach. Lenses were cultured in 24-well dishes. Each well contained 2 mL of modified TC-199 medium supplemented with antibiotics. The lenses were allowed to equilibrate in the incubator (95% air, 5% carbon dioxide, 37°C) for 2 days. Sugar cataracts were induced by adding 20 mM D-(+)-xylose to the medium. The medium was changed every day and supplemented with the seed compounds or EP (positive control) and xylose.

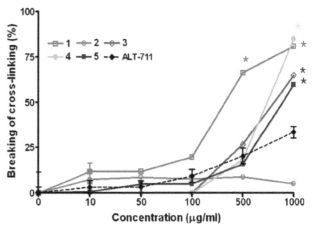

FIGURE 45.4 Effect of the 80% ethanol (EtOH) extract and solvent fractions of seeds of *Cornus officinalis* as breakers of advanced glycation end-product (AGE)–bovine serum albumin (BSA) cross-linking to collagen. Preformed AGE–BSA–collagen complex was incubated with the extract, solvent-soluble fractions, or ALT-711, then the remaining AGE–BSA attached to collagen was determined by enzyme-linked immunosorbent assay with mouse anti-AGE antibodies (6D12). 1: 80% EtOH extract; 2: *n*-hexane-soluble fraction; 3: ethyl acetate (EtOAc)-soluble fraction; 4: *n*-butanol (*n*-BuOH)-soluble fraction; 5: water-soluble fraction. Data are expressed as the mean ± SD of four replicates for each sample. *P < 0.01 vs. sample incubated with ALT-711. *Source: Kim et al. (2008),[31] with permission from Korean Journal of Pharmacognosy.*

Lenses were photographed under an optical microscope equipped with a charge-coupled device (CCD) camera. The opaque area of the lens was analyzed using an imaging system program (Image Analysis 42D 3D measuring software, TDI Scope Eyemedia 3.0, Olysia, Japan). Data were expressed as a percentage of the opaque lens area relative to the total lens area.

The compound 1,2,3,6-tetra-*O*-galloyl-β-D-glucose (3), which had the highest inhibitory effect on the RLAR activity, was further analyzed *ex vivo* on diabetic cataractogenesis. The xylose-treated lens had a significant opaque density. Treatment with 1,2,3,6-tetra-*O*-galloyl-β-D-glucose (3) (40 μM and 80 μM) significantly prevented opacification (P < 0.001) (Fig. 45.5).[30]

CONCLUSIONS

Most studies on *C. officinalis* have focused on the fruits with the seeds removed (i.e., the pericarp). The effects on eye-related disease of *C. officinalis* and/or compounds constituted have not been reported previously. Recent studies by this group have reported inhibitory effects of the seeds of *C. officinalis* and galloyl glucosides and lignans isolated from the seeds of *C. officinalis* on AGE formation, collagen–AGE cross-linking, aldose reductase activity, xylose-induced cataractogenesis, and breaking action on AGE–protein cross-links. The pericarp (fruits without seeds) had a lower inhibitory effect on AGE formation than the seeds and the pericarp containing

(A)

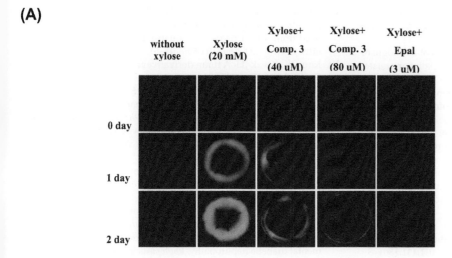

(B)

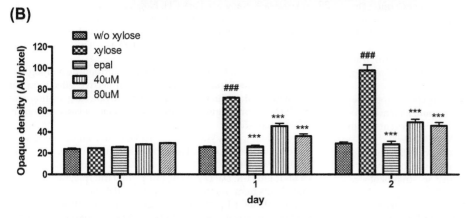

FIGURE 45.5 Preventive effect of compound 3 (1,2,3,6-tetra-*O*-galloyl-β-D-glucose) on lens opacification in rat lenses treated with 20 mM xylose. (a) Representative image of rat lenses incubated for different times; (b) opacity of each lens. Lens opacity was analyzed in each lens. Data are expressed as the mean ± SD of four replicates for 1,2,3,6-tetra-*O*-galloyl-β-D-glucose. $^{\#\#\#}P < 0.001$ vs. without xylose; $^{***}P < 0.01$ vs. xylose. *Source: Lee* et al. *(2011),[30] with permission from* Biological and Pharmaceutical Bulletin.

seeds. The seed extract and solvent-soluble fractions also showed a potent breaking effect on preformed AGE–BSA–collagen complexes. These findings suggest that the seeds of *C. officinalis* could be used to treat diabetic complications such as cataracts.

The seeds of *C. officinalis* are a potential candidate as a functional food or as medication for the prevention and treatment of diabetic cataracts.

TAKE-HOME MESSAGES

- The fruits without the seeds (pericarps) of *Cornus officinalis* have traditionally been used as a food and herbal medicine for the treatment of various diseases.
- Inhibitory effects of the alcoholic extract, solvent-soluble fractions, and constituents of *C. officinalis* seeds on advanced glycation end-product (AGE) formation, collagen–AGE cross-linking, aldose reductase (AR) activity, xylose-induced cataractogenesis, and breaking action on AGE–bovine serum albumin–protein complexes have been reported.
- Among the seeds, the fruits with seeds, and the pericarps (fruits with seeds removed), the seeds had the strongest inhibitory effects on AGE formation,

whereas the pericarps exhibited the weakest inhibitory effects on AGE formation.
- Some compounds such as galloyl glycosides that were isolated from both the seeds and pericarp exhibited inhibitory effects on AGEs, AR, and anticataractogenesis. However, other galloyl glycosides isolated from the seeds showed a tendency to be more effective against AGE formation.
- These studies demonstrated that the seeds of *C. officinalis* may be a potential candidate as a functional food or a medication for the prevention and treatment of diabetic complications such as cataracts.

Acknowledgment

This research was supported by grants (L07010, K09030, K10040, K1140) from the Korea Institute of Oriental Medicine, Korea.

References

1. Kyselova Z, Stefek M, Bauer V. Pharmacological prevention of diabetic cataract. *J Diabetes Complications* 2004;**18**:129–40.
2. Yokozawa T, Yamabe N, Kim HY, Kang KS, Hur JM, Park CH, Tanaka T. Protective effects of morroniside isolated from corni fructus against renal damage in streptozotocine-induced diabetic rats. *Biol Pharm Bull* 2008;**31**:1422–8.

3. Edelstein D, Brownlee M. Mechanistic studies of advanced glycosylation end product inhibition by aminoguanidin. *Diabetes* 1992;**41**:26–9.

4. Soulis T, Cooper ME, Bucala R, Jerums G. Effects of aminoguanidine in preventing experimental diabetic nephropathy are related to the duration of treatment. *Kidney Int* 1996;**50**:627–34.

5. Vasan S, Foiles P, Founds H. Therapeutic potential of breaker of advanced glycation end product–protein crosslinks. *Arch Biochem Biophys* 2003;**419**:89–96.

6. Schemmel KE, Padiyara RS, D'Souza JJ. Aldose reductase inhibitors in the treatment of diabetic peripheral neuropathy: a review. *J Diabetes Complications* 2010;**24**:354–60.

7. Park CH, Noh JS, Tanaka T, Uebaba K, Cho EJ, Yokozawa T. The effects of corni fructus extract and its fractions against α-glucosidase inhibitory activities *in vitro* and sucrose tolerance in normal rats. *Am J Chin Med* 2011;**39**:367–80.

8. He L-H. Comparative study for α-glucosidase inhibitory effects of total glycosides in the crude products and the wine-processed products from *Cornus officinalis. Yakugaku Zasshi* 2011;**131**:1801–5.

9. Qian DS, Zhu YF, Zhu Q. Effect of alcohol extract of *Cornus officinalis* Sieb. et Zucc on GLUT4 expression in skeletal muscle in type 2 (non-insulin-dependent) diabetic mellitus rats. *Zhongguo Zhong Yao Za Zhi* 2001;**26**:859–62.

10. Kim OK. Antidiabetic and antioxidative effects of corni fructus in streptozotocin-induced diabetic rats. *J Korea Oil Chem Soc* 2005;**22**:157–67.

11. Chen CC, Hsu CY, Chen CY, Liu HK. Fructus corni suppresses hepatic gluconeogenesis related gene transcription, enhances glucose responsiveness of pancreatic beta-cells, and prevents toxin induced beta-cell death. *J Ethnopharmacol* 2008;**117**:483–90.

12. Yamabe NY, Kang KS, Goto E, Tanaka T, Yokozawa T. Beneficial effect of corni fructus, a constituent of hachimi-jio-gan, on advanced glycation end-product-mediated renal injury in streptozotocin-treated diabetic rats. *Biol Pharm Bull* 2007;**30**:520–6.

13. Okuda T, Hatano T, Ogawa N, Kira R, Matsuda M, Cornusiin A. a dimeric ellagitannin forming four tautomers, and accompanying new tannins in *Cornus officinalis. Chem Pharm Bull* 1984;**32**:4662–5.

14. Hatano T, Ogawa N, Kira R, Yasuhara T, Okuda T. Tannins of cornaceous plants. I. Cornusiins A, B and C, dimeric monomeric and trimeric hydrolyzable tannins from *Cornus officinalis*, and orientation of valoneoyl group in related tannins. *Chem Pharm Bull* 1989;**37**:2083–90.

15. Hatano T, Yasuhara T, Abe R, Okuda T. A galloylated monoterpene glucoside and a dimeric hydrolysable tannin from *Cornus officinalis. Phytochemistry* 1990;**29**:2975–8.

16. Yamabe N, Kang KS, Park CH, Tanaka T, Yokozawa T. 7-O-Galloyl-D-sedoheptulose is a novel therapeutic agent against oxidative stress and advanced glycation end product in the diabetic kidney. *Biol Pharm Bull* 2009;**32**:657–64.

17. Endo T, Taguchi H. Study on the constituents of *Cornus officinalis* Sieb et Zucc. *Yakugaku Zasshi* 1973;**93**:30–2.

18. Lin MH, Liu HK, Huang CC, Wu TH, Hsu FL. Evaluation of the potential hypoglycemic and beta-cell protective constituents isolated from corni fructus to tackle insulin-dependent diabetes mellitus. *Agric Food Chem* 2011;**59**:7743–51.

19. Kim DK, Kwak JH. A furan derivative from *Cornus officinalis. Arch Pharm Res* 1998;**21**:787–9.

20. Kim DK, Kwak JH, Ryu JH, Kwon HC, Song KW, Kang SS, et al. A component from *Cornus officinalis* enhances hydrogen peroxide generation from macrophages. *Korean J Pharmacogn* 1996;**27**:101–4.

21. Miyazawa M, Kameoka H. Volatile flavor components of corni fructus (*Cornus officinalis* Sieb. et Zucc.). *Agric Biol Chem* 1989;**53**: 3337–40.

22. Miyazawa M, Anzai J, Fujioka J, Isikawa Y. Insecticidal compounds against drosophila melanogaster from *Cornus officinalis* Sieb. et Zucc. *Nat Prod Res* 2003;**17**:337–9.

23. Kim YD, Kim HK, Kim KJ. Analysis of nutritional components of *Cornus officinalis. J Korean Soc Food Sci Nutr* 2003;**32**:785–9.

24. Jiang WL, Zhang SP, Hou J, Zhu HB. Effect of loganin on experimental diabetic nephropathy. *Phytomedicine* 2012;**19**:217–22.

25. Yokozawa T, Kang KS, Park CH, Nog JS, Yamabe N, Shibahara N, Tanaka T. Bioactive constituents of *Cornus fructus*: the therapeutic use of morroniside, loganin, and 7-O-galloyl-D-sedoheptulose as renoprotective agent in type 2 diabetes. *Drug Discov Ther* 2010;**4**:223–34.

26. Xu H, Shen J, Liu H, Shi Y, Li L, Wei M. Morroniside and loganin extracted from *Cornus officinalis* have protective effects on rat mesangial cell proliferation exposed to advanced glycation end products by preventing oxidative stress. *Can J Physiol Pharmacol* 2006;**84**:1267–73.

27. Wang W, Xu J, Li L, Wang P, Ji X, Ai H, et al. Neuroprotective effect of morronisied on focal cerebral ischemia in rats. *Brain Res Bull* 2010;**83**:196–201.

28. Qi M-Y, Liu H-R, Dai D-Z, Li N, Dai Y. Total triterpene acids, active ingredients from fructus corni, attenuate diabetic cardiomyopathy by normalizing ET pathway and expression of FKBP12.6 and SERCA2a in streptozotocine-rats. *J Pharm Pharmacol* 2008;**60**:1687–94.

29. Lee GY, Jang DS, Lee YM, Kim YS, Kim JS. Constituents of the seeds of *Cornus officinalis* with inhibitory activity on the formation of advanced glycation end products (AGEs). *J Korean Soc Appl Biol Chem* 2008;**51**:316–20.

30. Lee J, Jang DS, Kim NH, Lee YM, Kim JH, Kim JS. Galloyl glucoses from the seeds of *Cornus officinalis* with inhibitory activity against protein glycation, aldose reductase, and cataractogenesis *ex vivo. Biol Pharm Bull* 2011;**34**:443–6.

31. Kim CS, Jang DS, Kim JH, Lee GY, Lee YM, Kim YS, Kim JS. Inhibitory effects of the seeds of *Cornus officinalis* on AGEs formation and AGEs-induced protein cross-linking. *Korean J Pharmacogn* 2008;**39**:249–54.

32. Park YK, Whang WK, Kim HI. The antidiabetic effects of extract from *Cornus officinalis* seed. *Chung-Ang J Pharm Sci* 1995;**9**:5–11.

Lutein and the Retinopathy of Prematurity

Costantino Romagnoli, Simonetta Costa, Carmen Giannantonio

Division of Neonatology, Catholic University of Rome, Rome, Italy

INTRODUCTION

Retinopathy of prematurity (ROP) is one of the main morbidities affecting surviving preterm infants and still represents a major cause of visual impairment and blindness, although its incidence has decreased over time with improvements in neonatal care.

The pathogenesis of ROP is multifactorial and, although the most relevant risk factors are premature birth and low birth weight, oxygen therapy has been identified as a factor that may have an important role in the pathogenic process.

Prolonged respiratory support and the use of high settings on pulse oximetry monitors for oxygen therapy control have been associated with an increased risk of ROP,[1] whereas the strict monitoring of oxygen therapy aimed at maintaining stable and relatively low oxygen saturation has been linked with a decreased incidence of ROP.[2]

Intrauterine and postnatal growth restrictions have been shown to be additional risk factors for ROP. It has been observed that the low serum insulin-like growth factor-1 levels detected in premature infants with extrauterine growth restriction have a role in the pathogenic process of ROP.[3]

Blood transfusions with adult red cells, which provide preterm infants with hemoglobin with a lower affinity for oxygen compared with fetal hemoglobin, are responsible for a higher oxygen delivery to the immature retina. For this reason blood transfusions have been demonstrated as a risk factor for the progression of ROP.[4]

Several other risk factors for the development and progression of ROP were identified by retrospective and observational studies involving large series of preterm infants. Bacterial and fungal sepsis,[5–7] hyperglycemia,[8] and postnatal steroid administration[9] are statistically associated with an increased risk of ROP or severe ROP. Unlike oxygen, the pathogenic mechanisms underlying these associations have only been hypothesized and are not clearly known but are likely to be epiphenomena of extreme prematurity or indicators of poor clinical status.

ROP is considered one of the oxygen radical diseases of prematurity. All clinical conditions and therapies that expose preterm infants to an oxidative stress have been identified as potential risk factors for the occurrence and the progression of ROP. Research has been directed towards clarifying the antiangiogenic effect of substances with antioxidant properties in the prevention and development of ROP. The prophylactic use of vitamin E has yielded conflicting results about the prevention of ROP, although some evidence suggests that this strategy could be beneficial in reducing the risk of stage 3+ ROP.[10] Similar results were observed with the enteral administration of D-penicillamine, a chelator of pro-oxidant heavy metals, but a recent randomized and placebo-controlled trial did not confirm its protective role.[11]

More recently, attention has been focused on carotenoids, mainly lutein, for their specific antioxidant properties in the retina.

RETINOPATHY OF PREMATURITY: PATHOGENESIS

In humans, fetal retinal vascular development begins during the fourth month of gestation and is completed *in utero* so that the retinal vasculature is almost complete in term infants.[12] Preterm infants at birth have incomplete retinal vascularization characterized by a peripheral zone of avascular retina (Fig. 46.1), the extent of which is inversely proportional to gestational age. Preterm birth is therefore the fundamental prerequisite for the development of ROP.

The pathogenesis of ROP is characterized by an abnormal vascular development of the retina that proceeds in two distinct phases.[13]

The infant born preterm is exposed to a state of hyperoxygenation due to the relative hyperoxia of the

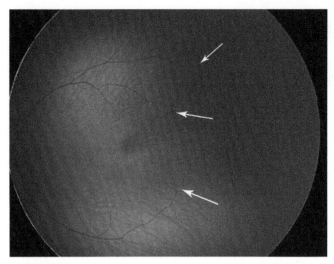

FIGURE 46.1 The avascular retina. The photograph shows the fundus of a preterm infant. Arrows show the incomplete vascularization of the immature retina. All photographs (Figs. 46.1–46.6) were taken using a digital imaging system (Retcam) by a team of neonatologists and ophthalmologists under Professor Romagnoli.

extrauterine environment. The hyperoxygenation is exacerbated by exposure to supplemental oxygen eventually needed for the treatment of a respiratory disease. Hyperoxia also occurs in the retinal tissue, where it influences the growth of retinal vessels that have already formed or are still developing (phase I of ROP)

The trait of phase I is delayed retinal vascular growth due to the hyperoxia that suppresses the expression of vascular endothelial growth factor (VEGF) and induces the cessation of normal retinal vessel growth and the regression of existing vessels.

The metabolic demand of the growing neural retina and the presence of avascular areas create a relative hypoxic microenvironment, in which the VEGF accumulates on the retinal plane and in the vitreous (phase II of ROP). Phase II is therefore characterized by an uncontrolled proliferative growth of retinal blood vessels due to overproduction of VEGF following oxygen deficit in the avascular retina.[14]

The role exerted by oxygen in the pathogenesis of ROP is also expressed, at least in part, through the oxidative damage mediated by the generation of free radicals and reactive oxygen species (ROS).[15] Several studies carried out on animal models showed that the oxygen fluctuations and the state of retinal tissue hypoxia typical of the two phases of ROP are responsible for the overproduction of ROS.[16,17] It has also been demonstrated that ROS activate nicotinamide adenine dinucleotide phosphate (NADPH)-oxidase, a major enzyme responsible for the production of superoxide radicals from macrophages. In retinal tissue under oxidative injury, NADPH-oxidase is able to direct the signaling pathways in the different directions of endothelial apoptosis or endothelial proliferation and

neoangiogenesis, so becoming a relevant factor in the development of ROP.[18]

RETINOPATHY OF PREMATURITY: STAGING

The International Classification of Retinopathy of Prematurity identified the degree of severity of acute ROP through four observations: (1) the stage or severity of retinopathy at the junction of the vascularized and avascular retina; (2) the location of retinal involvement by zone; (3) the extent of retinal involvement by clock hour; and (4) the presence or absence of plus disease, which is dilated and tortuous vessels of the posterior pole of the retina.[19] Shortly afterwards, this classification was extended to include the late stages of ROP such as retinal detachment.[20]

Stage 1 of ROP is a distinct white or grayish demarcation line between vascular and avascular retina. The demarcation line is both thin and flat and is located on the plane of the retina (Fig. 46.2a). The main characteristic of stage 2 of ROP is the presence of a 'ridge', which is an abnormal white to pink tissue that arises in the region of the demarcation line, extending above the plane of the retina (Fig. 46.2b). In stage 3 of ROP there is an extraretinal fibrovascular proliferation that extends from the ridge to the vitreous (Fig. 46.2c). Stage 4 of ROP is a partial detachment of the retina: stage 4A does not involve the macula whereas stage 4B involves the macula (Fig. 46.3a, b). Stage 5 is a total detachment of the retina (Fig. 46.3c).

Three concentric zones on the retina have been described for the purpose of defining the location of the ROP. Each zone is centered on the optic disc because normal retinal vasculature proceeds outward from the center of the optic disc towards the ora serrata.

Zone 1 is the most posterior zone centered on the optic nerve and with a radius of twice the distance from the optic nerve to the macula. Zone 2 is outside zone 1 but within a circle defined by a radius of the distance from the optic nerve to the nasal ora serrata. Zone 3 is the remaining zone of the retina that is external to zone 2.

The extension of the lesion of ROP on the retina is indicated by an hourly distribution. Three o'clock is located on the nasal retina of the right eye and on the temporal retina of the left eye.

A sign of more serious and rapidly progressive ROP is the presence of dilated veins and tortuous arterioles in the posterior pole of the retina that may later increase in severity to include iris vascular engorgement, poor pupillary dilatation, and vitreous haze. The presence of these characteristics indicates plus disease (Fig. 46.4).

The International Classification of Retinopathy of Prematurity was revised in 2005 and introduced the concept of a severe form of retinopathy that was observed

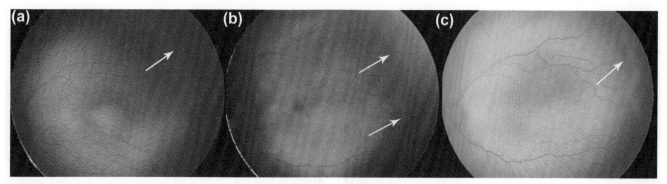

FIGURE 46.2 The stages of acute retinopathy of prematurity (ROP). The photographs show the lesions that characterize stages 1–3 of ROP: (a) the demarcation line (arrow) between vascular and avascular retina seen in stage 1 of ROP; (b) the ridge (arrows) that arises in the region of the demarcation line, extending above the plane of the retina; (c) the fibrovascular proliferation (arrow) that extends from the ridge to the vitreous in stage 3 of ROP.

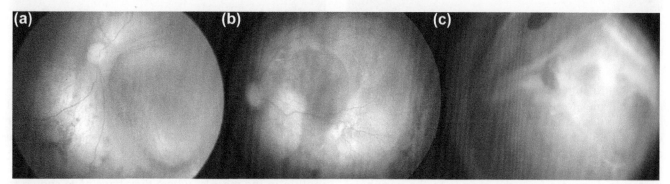

FIGURE 46.3 The late stages of retinopathy of prematurity (ROP). The photographs show the late stages of ROP: (a) partial detachment of the retina not involving the macula; (b) partial detachment involving the macula; (c) total detachment of the retina.

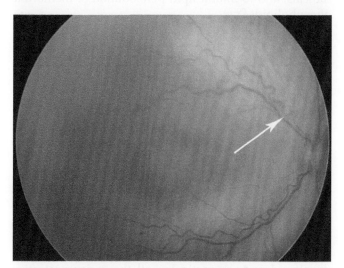

FIGURE 46.4 The plus disease of retinopathy of prematurity (ROP). The photograph shows the dilated veins and the tortuous arterioles (arrow) in the posterior pole of the retina that characterize the plus disease.

in the more premature infants: aggressive, posterior retinopathy of prematurity (AP-ROP). The revision also described an intermediate level of plus disease called pre-plus, which is a condition between normal posterior pole vessels and actual plus disease.[21]

The AP-ROP is an uncommon, rapidly progressing form of retinopathy that usually progresses to stage 5 ROP if untreated. The characteristic features of this type of ROP, previously referred to as type II ROP and Rush disease, are its posterior location and the severity of plus disease.

MACULAR CAROTENOIDS: RETINAL DISTRIBUTION, METABOLISM, AND FUNCTION

The macula is an elliptical area of about 5–6 mm located on the posterior pole of the retina, responsible for high spatial resolution and color vision.[22] Lutein, zeaxanthin, and meso-zeaxanthin are the main carotenoids of the macula, and for this reason they are referred to as macular pigments. Macular carotenoids are also known as macular xanthophylls because of their typical yellow appearance. Lutein and zeaxanthin are of dietary origin and are commonly found in carrots, corn, citrus fruits, and dark green leafy vegetables such as spinach and broccoli. Meso-zeaxanthin is not normally found in the human diet, and it is generated at the retina following lutein isomerization.[23,24]

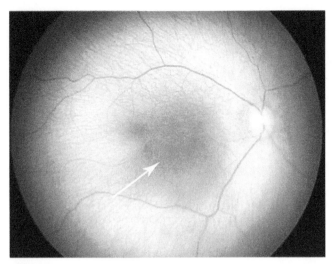

FIGURE 46.5   The fundus oculi of a term neonate. The photograph shows the retina of a term neonate. The macular spot (arrow) is weak because of the small amount of carotenoids.

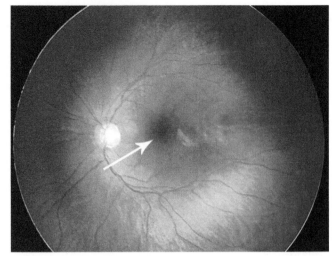

FIGURE 46.6   The fundus oculi of a child. The photograph shows the retina of a 1-year-old child. The macular spot (arrow) is more perceptible than in the neonatal retina.

Macular carotenoids are located mainly on the outer and inner plexiform layers of the retina.

Distribution studies, carried out on retinal tissue from donors, showed that the total mass of pigments per unit area and the ratio of lutein to zeaxanthin change with the distance from the macula and age. The total mass of pigments decreases from the macula to the peripheral retina by a factor of nearly 300; the lutein to zeaxanthin ratio is approximately 1:2.4 within 0.25 mm of the fovea, whereas at the periphery of the retina this relationship is reversed and is 2:1.[25] In relation to age, the two pigments are already present in retinal tissues of the fetus at 20 weeks of gestational age, but there is not enough of a quantity to appear as a yellow spot. In the first year of life, the total mass of macular pigments is generally much lower than that of older subjects so that the yellowish coloration of the infant macula is barely perceptible (Figs. 46.5 and 46.6). The lutein to zeaxanthin ratio changes during the first decade of life; the lutein is the predominant constituent of the macular pigment only in the first 2 years of life. This ratio decreases from a value of 1.59 at birth to 0.87 at 9 years of life.[25]

Carotenoids are absorbed from the gut in the form of chylomicrons,[26] and the efficacy of their absorption depends on the formulation or source from which they are derived. It has been demonstrated that a diester formulation of lutein is more bioavailable than free lutein owing to its dissolution property; that is, the ability to solubilize and form micelles. The degree of dissolution is the determining condition for absorption.[27] After absorption, the macular carotenoids are delivered to the liver via the portal circulation, and there they are bound to apolipoprotein E (ApoE) associated with high-density lipoproteins (HDLs) for subsequent release into the systemic circulation.[28] These ApoE/HDL complexes are then transported to the retina.

Macular carotenoids are a short-wavelength light filter with an important function of protection for retinal tissue. These carotenoids, possessing in their molecular structure a series of unconjugated double bonds, are believed to be very effective antioxidants. Several studies reported the antioxidant properties of lutein and zeaxanthin in the macula, where they have been shown to help quench singlet oxygen, to scavenge reactive free radicals, and to inhibit lipid peroxidation of membrane phospholipids.[29–31]

LUTEIN AND THE RETINOPATHY OF PREMATURITY

Oxidative stress may have a role in pathologic angiogenesis of the retina and thus in the pathogenesis of ROP.

The human retinal tissue is particularly vulnerable to oxidative damage because of its structural characteristics, such as the abundance of polyunsaturated fatty acids in cellular membranes and its high metabolic and functional properties, and also because of its exposure to light-induced free radicals. The retina of a preterm infant is much more vulnerable than that of adults because the antioxidant system of preterm infants is not completely developed and, in particular, all the major protective antioxidant systems are deficient in the immature retina.

Human milk is the only dietary source of lutein for newborn infants until weaning occurs[32,33]; as a result, it is thought that a deficiency of lutein can affect infants who receive formula without lutein or breast milk from mothers whose diet is low in lutein content. Nutritional studies indicated that newborn infants are able to absorb lutein when it is administered through supplemented formula milk[34] or through nutritional supplements mainly containing lutein.[35] Because of the results of the nutritional studies, some researchers[36–38] carried out clinical trials to assess whether lutein supplementation can reduce the occurrence or severity of ROP.

In a randomized, double-blind, controlled study by Romagnoli et al.,[36] 31 preterm infants with gestational age 32 weeks or less received orally 0.5 mg/kg/day of lutein and 0.02 mg/kg/day of zeaxanthin from the 7th day of life until 42 weeks of postmenstrual age. The occurrence and severity of ROP in these 31 infants were compared with those of 32 preterm newborns receiving a placebo. The authors did not find any difference in the incidence and severity of ROP between the two groups.[36]

A study conducted by Dani et al.[37] agrees with these findings. In their study, 114 preterm infants with gestational age 32 weeks or less were randomized to receive orally 0.14 mg/day of lutein and 0.05 mg/day of zeaxanthin or placebo from the first week of life until discharge from the hospital.[37]

The largest group of preterm infants at risk of ROP who have been studied and received oral lutein supplementation was described by Manzoni et al.[38] In this multicentric randomized trial, 228 preterm infants with gestational age 32 weeks or less were randomized to receive the same dose of lutein/zeaxanthin as administered by Dani et al.[37] or placebo. The purpose of the study was to evaluate the influence of carotenoids on oxidative stress-induced diseases of prematurity, namely ROP, necrotizing enterocolitis, and bronchopulmonary dysplasia. No significant beneficial effect on ROP was detected except for a statistically nonsignificant tendency towards a lower rate of progression from the early stages of ROP to the threshold stage.[38]

Despite the pathophysiologically rational use of lutein in the prevention of ROP, the results of these clinical trials did not provide evidence on the benefits of lutein supplementation in premature infants. There may be various reasons for this. First, while some evidence[39] suggests that in adults a serum lutein concentration between 0.6 and 1.05 mmol/L seems to be a safe and dietarily achievable target potentially associated with a beneficial impact on visual function and on the development of chronic diseases, no evidence exists regarding the ideal serum lutein concentration that could have a benefit for newborn infants. Consequently, the ideal dosage and form for lutein supplementation, such as with formula milk enriched with lutein or with nutritional supplement containing lutein, are unknown. Second, only one of the trials available has an adequate sample size to test the hypothesis that lutein affects ROP.[38] Third, considering that xanthophylls are mainly concentrated in the macula and that this region is primarily responsible for visual acuity, supplementation with lutein probably influences functional rather than structural vision outcomes in preterm infants.

Furthermore, one trial demonstrated that lutein supplementation of preterm infants of gestational age less than 33 weeks with formula milk containing 211 µg/L of lutein improved rod photoreceptor function.[40] This result was surprising because it was achieved with a dose of lutein much lower than that used by the other researchers. Assuming that preterm infants in full enteral feeding take about 150–160 mL/kg/day of formula, it can be deduced that in the study by Rubin et al. infants received a final supplementation of about 0.03 µg/kg/day of lutein.[40]

In light of the available studies, it still remains unproven whether lutein supplementation can decrease the incidence or severity of ROP in at-risk preterm infants.

TAKE-HOME MESSAGES

- Retinopathy of prematurity (ROP) is one of the main morbidities affecting surviving preterm infants and still represents a major cause of visual impairment and blindness.
- ROP is a vasoproliferative retinal disorder, and oxidative stress may play a role in the pathologic angiogenesis of the retina and then in the pathogenesis of retinopathy.
- As preterm infants are susceptible to oxidative damage owing to high metabolic demands and low levels of antioxidant enzymes, efforts have been made to identify protective strategies that will enhance their antioxidant capacities.
- Lutein, zeaxanthin, and meso-zeaxanthin, also known as macular xanthophylls or macular pigments, are the main carotenoids of the macula. They function as a short-wavelength light filter with an important role in protecting retinal tissue against light-induced oxidative damage.
- Since lutein is well absorbed from the gut when given to neonates through formula milk enriched with lutein or nutritional supplements mainly containing lutein, several studies have been designed to assess whether lutein supplementation could be an effective strategy in the prevention of the occurrence or progression of ROP.
- Available evidence does not prove the efficacy of lutein in the prevention of ROP.

References

1. Chen ML, Guo L, Smith LE, Dammann CE, Dammann O. High or low oxygen saturation and severe retinopathy of prematurity: a meta-analysis. *Pediatrics* 2010;**125**:e1483–92.
2. Chow LC, Wright KW, Sola A. Can changes in clinical practice decrease the incidence of severe retinopathy of prematurity in very low birth weight infants? *Pediatrics* 2003;**111**:339–45.

3. Hellstrom A, Engstrom E, Hard AL, Albertsson-Wikland K, Carlsson B, Niklasson A, et al. Postnatal serum insulin-like growth factor I deficiency is associated with retinopathy of prematurity and other complications of premature birth. *Pediatrics* 2003;**112**:1016–20.

4. Giannantonio C, Papacci P, Cota F, Vento G, Tesfagabir MG, Purcaro V, et al. Analysis of risk factors for progression to treatment-requiring ROP in a single neonatal intensive care unit: is the exposure time relevant? *J Matern Fetal Neonatal Med* 2012;**25**:471–7.

5. Liu PM, Fang PC, Huang CB, Kou HK, Chung MY, Yang YH, Chung CH. Risk factors of retinopathy of prematurity in premature infants weighing less than 1600 g. *Am J Perinatol* 2005;**22**:115–20.

6. Bourla DH, Gonzales CR, Valijan S, Yu F, Mango CW, Schwartz SD. Association of systemic risk factors with the progression of laser-treated retinopathy of prematurity to retinal detachment. *Retina* 2008;**28**:S58–64.

7. Bharwani SK, Dhanireddy R. Systemic fungal infection is associated with the development of retinopathy of prematurity in very low birth weight infants: a meta-review. *J Perinatol* 2008;**28**:61–6.

8. Garg R, Agthe AG, Donohue PK, Lehmann CU. Hyperglycemia and retinopathy of prematurity in very low birth weight infants. *J Perinatol* 2003;**23**:186–94.

9. Smolkin T, Steinberg M, Sujov P, Mezer E, Tamir A, Makhoul IR. Late postnatal systemic steroids predispose to retinopathy of prematurity in very-low-birth-weight infants: a comparative study. *Acta Paediatr* 2008;**97**:322–6.

10. Raju TN, Langenberg P, Bhutani V, Quinn GE. Vitamin E prophylaxis to reduce retinopathy of prematurity: a reappraisal of published trials. *J Pediatr* 1997;**131**:844–50.

11. Tandon M, Dutta S, Dogra MR, Gupta A. Oral D-penicillamine for the prevention of retinopathy of prematurity in very low birth weight infants: a randomized, placebo-controlled trial. *Acta Paediatr* 2010;**99**:1324–8.

12. Roth AM. Retinal vascular development in premature infants. *Am J Ophthalmol* 1977;**84**:636–40.

13. Smith LE. Pathogenesis of retinopathy of prematurity. *Acta Paediatr Suppl* 2002;**437**:26–8.

14. Chen J, Smith LE. Retinopathy of prematurity. *Angiogenesis* 2007;**10**:133–40.

15. Sapieha P, Joyal JS, Rivera JC, Kermorvant-Duchemin E, Sennlaub F, Hardy P, et al. Retinopathy of prematurity: understanding ischemic retinal vasculopathies at an extreme of life. *J Clin Invest* 2010;**120**:3022–32.

16. Hartnett ME. The effects of oxygen stresses on the development of features of severe retinopathy of prematurity: knowledge from the 50/10 OIR model. *Doc Ophthalmol* 2010;**120**:25–39.

17. Li SY, Fu ZJ, Lo ACY. Hypoxia-induced oxidative stress in ischemic retinopathy. *Oxid Med Cell Longev* 2012;**2012**:426769. Epub October 17.

18. Saito Y, Uppal A, Byfield G, Budd S, Hartnett ME. Activated NAD(P)H oxidase from supplemental oxygen induces neovascularization independent of VEGF in retinopathy of prematurity model. *Invest Ophthalmol Vis Sci* 2008;**49**:1591–8.

19. International Committee for the Classification of Retinopathy of Prematurity. An international classification of retinopathy of prematurity. *Arch Ophthalmol* 1984;**102**:1130–4.

20. International Committee for the Classification of Retinopathy of Prematurity. An international classification of retinopathy of prematurity. II. The classification of retinal detachment. *Arch Ophthalmol* 1987;**105**:906–12.

21. International Committee for the Classification of Retinopathy of Prematurity. The International Classification of Retinopathy of Prematurity revisited. *Arch Ophthalmol* 2005;**123**:991–9.

22. Hirsch J, Curcio CA. The spatial resolution capacity of human foveal retina. *Vis Res* 1989;**29**:1095–101.

23. Bone RA, Landrum JT, Hime GW, Cains A, Zamor J. Stereochemistry of the human macular carotenoids. *Invest Ophthalmol Vis Sci* 1993;**34**:2033–40.

24. Neuringer M, Sandstrom MM, Johnson EJ, Snodderly DM. Nutritional manipulation of primate retinas, I: effects of lutein or zeaxanthin supplements on serum and macular pigment in xanthophyll-free rhesus monkeys. *Invest Ophthalmol Vis Sci* 2004;**45**:3234–43.

25. Bone RA, Landrum JT, Fernandez L, Tarsis SL. Analysis of the macular pigment by HPLC: retinal distribution and age study. *Invest Ophthalmol Vis Sci* 1988;**29**:843–9.

26. Gärtner C, Stahl W, Sies H. Preferential increase in chylomicron levels of the xanthophylls lutein and zeaxanthin compared to beta-carotene in the human. *Int J Vitam Nutr Res* 1996;**66**:119–25.

27. Bowen PE, Herbst-Espinosa SM, Hussain EA, Stacewicz-Sapuntzakis M. Esterification does not impair lutein bioavailability in humans. *J Nutr* 2002;**132**:3668–73.

28. Clevidence BA, Bieri JG. Association of carotenoids with human plasma lipoproteins. *Methods Enzymol* 1993;**214**:33–46.

29. Ma L, Lin XM. Effects of lutein and zeaxanthin on aspects of eye health. *J Sci Food Agric* 2010;**90**:2–12.

30. Rapp LM, Maple SS, Choi JH. Lutein and zeaxanthin concentrations in rod outer segment membranes from peri-foveal and peripheral human retina. *Invest Ophthalmol Vis Sci* 2000;**41**:1200–9.

31. Stahl W, Sies H. Physical quenching of singlet oxygen and *cis–trans* isomerization of carotenoids. *Ann N Y Acad Sci* 1993;**691**:10–9.

32. Yeum KJ, Ferland G, Patry J, Russell RM. Relationship of plasma carotenoids, retinol and tocopherols in mothers and newborn infants. *J Am Coll Nutr* 1998;**17**:442–7.

33. Canfield LM, Clandinin MT, Davies DP, Fernandez MC, Jackson J, Hawkes J, et al. Multinational study of major breast milk carotenoids of healthy mothers. *Eur J Nutr* 2003;**42**:133–41.

34. Bettler J, Zimmer JP, Neuringer M, De Russo PA. Serum lutein concentrations in healthy term infants fed human milk or infant formula with lutein. *Eur J Nutr* 2010;**49**:45–51.

35. Romagnoli C, Tirone C, Persichilli S, Gervasoni J, Zuppi C, Barone G, Zecca E. Lutein absorption in premature infants. *Eur J Nutr* 2010;**64**:760–1.

36. Romagnoli C, Giannantonio C, Cota F, Papacci P, Vento G, Valente E, et al. A prospective, randomized, double blind study comparing lutein to placebo for reducing occurrence and severity of retinopathy of prematurity. *J Matern Fetal Neonatal Med* 2011;**24**(Suppl. 1.):147–50.

37. Dani C, Lori I, Favelli F, Frosini S, Messner H, Wanker P, et al. Lutein and zeaxanthin supplementation in preterm infants to prevent retinopathy of prematurity: a randomized controlled trial. *J Matern Fetal Neonatal Med* 2012;**25**:523–7.

38. Manzoni P, Guardione R, Bonetti P, Priolo C, Maestri A, Mansoldo C, et al. Lutein and zeaxanthin supplementation in preterm very low-birth weight neonates in neonatal intensive care units: a multicenter randomized controlled trial. *Am J Perinatol* 2013;**30**:25–32.

39. Delcourt C, Carrière I, Delage M, Barberger-Gateau P, Schalch W. POLA Study Group. Plasma lutein and zeaxanthin and other carotenoids as modifiable risk factors for age-related maculopathy and cataract: the POLA Study. *Invest Ophthalmol Vis Sci* 2006;**47**:2329–35.

40. Rubin LP, Chan GM, Barrett-Reis BM, Fulton AB, Hansen RM, Ashmeade TL, et al. Effect of carotenoid supplementation on plasma carotenoids, inflammation and visual development in preterm infants. *J Perinatol* 2012;**32**:418–24.

Dietary Wolfberry and Retinal Degeneration

Hua Ji[1,2], Hui He[3], Dingbo Lin[1]

[1]Department of Nutritional Sciences, Oklahoma State University, Stillwater, Oklahoma, [2]Institute of Genetics and Physiology, Hebei Academy of Agriculture and Forestry Sciences, Shijiazhuang, PR China, [3]Department of Diagnostic Medicine/Pathobiology, Kansas State University, Manhattan, Kansas, USA

INTRODUCTION

The retina is the light-sensitive layer of tissue lining the inside of the eye where light signals are converted to visual messages that are sent through the optic nerve to the brain. Any damage to this complicated network will cause retinal degeneration and lead to visual impairment or blindness in mammals. Retinal degeneration is an irreversible progressive neurologic disorder with no cure[1]; thus, dietary prevention or delay of the onset of the disease has received much attention in the field of retinoprotection.

Wolfberries are the fruits of the perennial plant *Lycium*, which is native to Asia and southeastern Europe, and they are also consumed in the USA.[2–4] Wolfberries boast uniquely high contents of polysaccharides, betaine, taurine, and diester forms of lutein and zeaxanthin.[5,6] Consumption of wolfberries has been claimed to improve vision, attenuate inflammation, potentiate the immune response, and prevent cancer in humans.[3–5] Recent research on the health benefits of wolfberry has addressed the fruit's ability to help prevent retinal degeneration, including retinitis pigmentosa (RP), loss of retinal ganglion cells (RGCs), glaucoma, age-related macular degeneration, and diabetic retinopathy. This chapter reviews this research and discusses the underlying mechanisms of potential bioactive components of wolfberry.

BOTANICAL ASPECTS OF WOLFBERRY

Wolfberry, or goji berry, is the common name for the fruit of two closely related species, *Lycium barbarum* L. and *Lycium chinense* Mill. Both species are members of the Solanaceae family. Varieties and related species include *Lycium barbarum* L. var. auranticarpum, K.F. Ching var. nov., and *L. chinense* Mill. var. potaninii (Pojark.) A.M. Lu. Wolfberry leaves are lanceolate to ovate. The light purple flowers grow in groups of one to three in the leaf axils (Fig. 47.1A). In the northern hemisphere, the plant blooms from June to September and the fruit matures from August to October. The oblong, orange to dark red fruits (Fig. 47.1B) can be consumed fresh or can be sun-dried (Fig. 47.1C) for longer shelf-life. *Lycium barbarum* L. is a deciduous shrub 1–3 m high, whereas *L. chinense* Mill is smaller. In wolfberry plantations, shrubs are trimmed to no taller than 1.8 m (Fig. 47.1D) to improve yield and save labor in harvesting. The majority of commercially produced wolfberries come from plantations of *L. barbarum* L. in the Ningxia Hui Autonomous Region and the Xinjiang Uyghur Autonomous Region, China.

BIOACTIVE CONSTITUENTS OF WOLFBERRY

Wolfberries are mainly used as fresh fruits, as tea, or as an ingredient in bread, wine, stews and soups, porridges, and juices. Dried wolfberries are chewy, like raisins. Although wolfberries have been used for hundreds of years in China as a general tonic to protect the liver, improve vision, and promote longevity, the bioactive constituents of wolfberry have been well identified and characterized only in the past two decades with the development of modern analytic methods, such as high-performance liquid chromatography (HPLC), nuclear magnetic resonance, and mass spectrometry. Compositions of the fruits of *L. barbarum* L. and *L. chinense* Mill are similar. Wolfberries contain numerous bioactive constituents including, but not

FIGURE 47.1 Wolfberry in Zhongning County, Ningxia, China. Wolfberry flowers (A) and fruits (B); sun-drying wolfberry fruits (C); and trimmed wolfberry trees in a plantation in Zhongning, Ningxia, China (D). (See color plate section)

limited to, polysaccharides, carotenoids, flavonoids, betaine, taurine, scopoletin, β-sitosterol, ϱ-coumaric acid, and daucosterol.[5,7–13] Vitamin C content is significantly higher in the fresh (raw) fruits than in most other fruits and vegetables. Similarly to most other dried fruits, vitamin C content varies widely in dried wolfberries.

Polysaccharides are the predominant bioactive constituents in wolfberries. According to the literature, the content varies significantly, from 16% to 23% based on the dried fruit weight.[7,11] Chang and coworkers reported the polysaccharide content to be as high as 40%.[14] In most cases, the glycosidic portion of the polysaccharides accounts for 90–95% of the mass and consists of arabinose, glucose, galactose, galactuoric acid, mannose, rhamnose, and xylose.[5] Polysaccharides are naturally conjugated with proteins to form the bioactive unit.

Carotenoids are another major bioactive constituent group in wolfberries. Total carotenoid contents vary from about 2 to 4 mg/g in dried fruits.[6,7] HPLC showed zeaxanthin to be the most abundant carotenoid (about 95%), followed by lutein and β-cryptoxanthin (Fig. 47.2). Chen's group further characterized wolfberry zeaxanthin as being highly diesterified by palmitate (1143.7 μg/g dried fruit),[8] which makes wolfberry unique among foods in that the diester form of zeaxanthin is more bioavailable than the free form of this carotenoid.[15]

In addition, wolfberries have a total phenolic content comparable to blueberries and significantly higher than strawberries (Fig. 47.3). Of the various flavonols, quercetin and its glucosides are most enriched in wolfberries.[9] About 10% of total phenolics were found to be rutin in wolfberries.[16] Wolfberries are also rich in betaine, at 0.15–0.2% (dried fruit),[12] and taurine, at about 3 mg/g (dried fruit).[13] Both are proposed to have enhanced antioxidant activity.

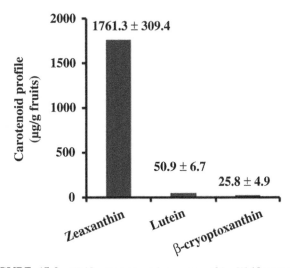

FIGURE 47.2 Wolfberries are rich in zeaxanthin. Wolfberry sun-dried fruits were subjected to carotenoid profiling by high-performance liquid chromatography. Wolfberries contained high levels of zeaxanthin, lutein, and β-cryptoxanthin. Data are shown as mean±SD, expressed as μg/g fruit (n = 3).

BIOAVAILABILITY OF WOLFBERRY

Wolfberries have gained popularity in the USA in the past two decades because of their antioxidant properties and potential health benefits; however, bioavailability data are largely from *in vitro* and animal studies. Human data are very limited.

Zeaxanthin and lutein are carotenoids that are prevalent in wolfberry. Overall, zeaxanthin and lutein bioavailability is low, particularly in neurons. The first bioavailability study in wolfberry was presented by Benzie *et al.*[17] The authors demonstrated that homogenizing wolfberries in hot skim milk (80 °C) enhanced the bioavailability of zeaxanthin threefold compared with homogenization in hot water (80 °C) and/or warm skim milk (40 °C) in subjects at an average of 28 years of age.

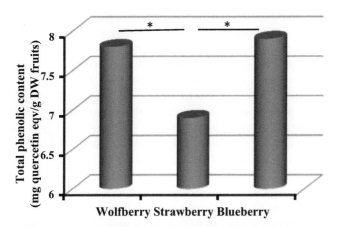

FIGURE 47.3 Wolfberries contain high total phenolics. The amount of total phenolics in fruits of wolfberry, strawberry, and blueberry was determined according to the Folin–Ciocalteu method as described previously.[16] Data are shown as mean±SD and expressed as mg quercetin equivalent/g dry weight (DW) fruit (n=3). Statistical significance at P<0.05 (*).

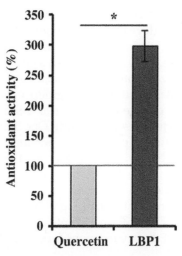

FIGURE 47.4 Wolfberry polysaccharides have an enhanced antioxidant activity. Wolfberry polysaccharides (LBP) were extracted as described previously,[20] and fraction 1 (LBP1) was subjected to the antioxidant activity assay using a colorimetric microplate assay kit (catalog no. TA02; Oxford Biomedical Research, Oxford, MI, USA). Total antioxidant power was normalized to quercetin (%). Data are shown as mean±SD (n=3). Statistical significance at P<0.05 (*).

This formulation of wolfberries was named lacto-wolfberry (LWB). More recent studies showed that LWB at 13.7g/day significantly increased plasma zeaxanthin levels and antioxidant capacity, by 26% and 57%, respectively, in healthy elderly people (65–70 years old) in a randomized, double-blind, placebo-controlled trial.[18] Studies suggest that the plasma response to wolfberry zeaxanthin depends on the subject's age, the baseline level of plasma zeaxanthin, and the amount of wolfberry taken. Individuals with low daily consumption of food containing zeaxanthin may have a higher response to wolfberry.

The ingestion, uptake, and metabolism of wolfberry polysaccharides in humans and mice are unknown, so it is difficult to conclude whether antioxidant efficacies are directly or indirectly caused by wolfberry in humans. Polysaccharide bioavailability in mammals has not been reported.[5] The antioxidant capability and physiologic impact of polysaccharide-standardized wolfberry juice in healthy adults have been investigated recently.[19] It was found that wolfberry polysaccharide fraction-1 (LBP1) enhanced *in vitro* antioxidant activity by up to 300% compared with quercetin (Fig. 47.4).[20] Consumption of wolfberry juice (120mL/day) improved serum antioxidant biomarkers, superoxide dismutase, and glutathione peroxidase and decreased lipid peroxidation, as indicated by decreased levels of malondialdehyde.[19] The human data suggest that drinking wolfberry juice may help to prevent or reduce oxidative stress, which was confirmed by a parallel study using a mouse model under ultraviolet-radiation induced skin stress conditions.[21] However, these observational studies have not elucidated the underlying mechanisms of polysaccharides.

WOLFBERRY AND PREVENTION OF MACULAR DEGENERATION

Macular degeneration is a leading cause of central vision impairment and blindness in elderly people. Decreases in macular pigment density and the formation of soft drusen are features of the development of macular degeneration. Zeaxanthin and lutein are dominant retinal pigments preferentially accumulated in the macula in humans.[22,23] Epidemiologic studies suggest that levels of plasma zeaxanthin and lutein are inversely associated with the risk of macular degeneration.[24,25] Increased dietary intake of zeaxanthin and lutein has been proposed to lower the risk of this retinal disease by preventing the decline in macular pigment density.[24,25]

Wolfberries have been used traditionally in China to improve vision, as well as liver and kidney function, but no high-quality human studies have addressed the mechanism of action of the bioactive constituents of wolfberry. As described above, dietary wolfberry is the richest natural source of plant-derived zeaxanthin, mostly in diesterified form.

Bucheli *et al.* recently reported the preventive effect of wolfberry on macular hypopigmentation and soft drusen formation in healthy elderly people aged 65–70 years.[26] During the 90-day study period, individuals in the LWB group consumed LWB product at 13.7g/day (about 7g wolfberries containing 10mg/day zeaxanthin and 68.5mg/day wolfberry-derived vitamin C precursor). The placebo group used the product without wolfberry. Before and after the dietary supplementation treatment, subjects underwent a detailed ophthalmic examination and fasting blood tests. The authors observed that the placebo group developed hypopigmentation and an increase in soft drusen in the macula, but the LWB group remained stable. They also observed significant increases in both plasma zeaxanthin and antioxidant capacity in the LWB group but not in the placebo group; however, no evidence links the change in plasma zeaxanthin and altered macular characteristics in this study. One limitation of the study was the subject pool, which contained only healthy subjects. Revealing whether dietary wolfberry increases macular pigment density in patients in the early stages of macular degeneration and developing good animal models would enhance understanding of the pathogenesis of macular degeneration[27] and eventually may lead to the development of dietary regimens for disease prevention.

WOLFBERRY POLYSACCHARIDES AND RETINITIS PIGMENTOSA

RP is a group of inherited neurodegenerative diseases characterized by progressive photoreceptor degeneration. RP is caused by a number of mutations that result in rod cell death followed by gradual death of cone cells. The progression of the disease is potentiated by oxidative stress in the outer retina.[28] A mixture of antioxidants, including α-tocopherol, ascorbic acid, Mn(III) tetrakis (4-benzoic acid) porphyrin, and α-lipoic acid (ALA), markedly reduced cone oxidative damage and preserved function of cone cells in the Q344ter, rd1, and rd10 mouse models of RP.[29] Miranda et al. recently found that wolfberry-derived antioxidants rescued photoreceptors in rd1 mice.[30] In the study, rd1 mice from postnatal day 3 were given a daily oral infusion of a mixture of antioxidants including wolfberry (L. barbarum L.) polysaccharides (175 mg/kg body weight (BW)), zeaxanthin and lutein (both 0.67 mg/kg BW), and glutathione and ALA (both 10 mg/kg BW) for 1 week. The control rd1 group was given only the vehicle. Results revealed that survival of photoreceptors through the addition of antioxidants in the rd1 mouse retina was directly related to thiol contents and thiol-dependent peroxide metabolism. Although no data address whether antioxidants protect the retina from RP in the long term, protection from oxidative damage may be a broadly applicable treatment strategy in RP.

WOLFBERRY POLYSACCHARIDES AND GANGLION CELL SURVIVAL IN RETINAL ISCHEMIA AND GLAUCOMA

An RGC is a type of neuron located near the inner surface of the retina that transmits visual information from photoreceptors to the brain. Oxidative stress is one of the causative factors in RGC degeneration under various pathologic conditions, such as ischemia, glaucoma, and diabetes. Researchers have taken advantage of the enhanced antioxidant activity of wolfberry's bioactive constituents to investigate RGC protection by wolfberry in rodent models.

The first in vivo study demonstrating that wolfberry polysaccharides were retinoprotective was conducted in an ocular hypertension rat model of glaucoma.[31] Adult female Sprague–Dawley rats at 10–12 weeks of age were fed daily for 7 days with phosphate-buffered saline (PBS) and/or LBP at 0.01, 0.1, 1, 10, 100, or 1000 mg/kg BW before the induction of ocular hypertension (OH) by photocoagulation. Increases in intraocular pressure (IOP) and loss of RGC as determined by changes in RGC density were measured at days 3, 7, 14, and 28 after photocoagulation. Results showed that the IOP levels increased at day 3 and remained high until study termination at day 28 in OH mice with PBS. LBP did not reduce IOP. OH-treated rats fed with LBP showed a significant reduction in RGC loss compared with those fed PBS, and the loss worsened over time. LBP retinoprotection was dose dependent and appeared as a U-shaped curve. There was no significant difference in RGC loss between the groups fed with 1, 10, and 100 mg/kg BW. The authors suggested that upregulated crystallins by LBP could be critical to preventing RGC loss.[32] LBP was also retinoprotective in a mouse model of acute glaucoma.[33]

Protection by wolfberry polysaccharide against RGC degeneration induced by retinal ischemia–reperfusion injury was also reported recently by a group from the University of Hong Kong.[14] Male C57BL/6N mice at 10–12 weeks of age were given orally by gavage either PBS (the vehicle) or LBP at 1 mg/kg BW for 7 days. The mice were then subjected to ischemia for 2 hours followed by reperfusion for 22 hours. Retinal damage was evaluated by histology and immunohistochemistry. Ischemia–reperfusion caused severe retinal damage in the PBS group, including significant increases in apoptotic RGCs, loss of RGCs in the central and peripheral retina, and increases in the inner retinal thickness of the central retina. Preadministration of LBP for 7 days significantly protected the retinal structure from ischemia–reperfusion injury. The authors suggested that LBP attenuated oxidative stress and protected the integrity of the blood–retinal barrier in the ischemic retina.

WOLFBERRY AND DIABETIC RETINAL DEGENERATION

Wolfberry's ability to protect against diabetic retinal degeneration has been reported recently. The present group is investigating the bioavailability of wolfberry's bioactive components, zeaxanthin and lutein, and the regulation of gene expression and retinoprotection using C57BL/6J (B6) and db/db type 2 diabetic mouse models.

Bioavailability of Wolfberry in the Wild-Type B6 Mouse

The levels of zeaxanthin and lutein in the retina and circulation system are inversely associated with the risk of diabetic retinopathy,[34] the most common diabetic eye disease and a leading cause of blindness in US adults. Thus, it was hypothesized that wolfberry may be retinoprotective owing to its high zeaxanthin and lutein contents. First, it was demonstrated through HPLC analysis that 6-week-old B6 mice fed 1% (kcal) wolfberries for 8 weeks had significantly increased hepatic concentrations of zeaxanthin plus lutein (323%) compared with the control group (B6 on the control diet without addition of wolfberry) (Fig. 47.5). It was also shown that the concentration of retinal zeaxanthin plus lutein increased by 13.7% in B6 mice with 1% (kcal) wolfberry compared with the control group. Although zeaxanthin and lutein are bioavailable only at low levels in the retina of rodents, the data suggest that they are detectable when enough retinal tissues are pooled.

Upregulation of Carotenoid Metabolic Genes in the Diabetic Retina

As discussed above, wolfberries have high polysaccharide and zeaxanthin contents, but the metabolic mechanism of wolfberry polysaccharide remains unknown, both in humans and in rodent models. There is no evidence to show whether polysaccharides can pass through the blood–retinal barrier or what the targets are. The metabolism of lutein and zeaxanthin has recently

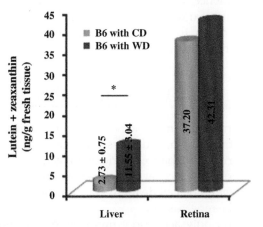

FIGURE 47.5 Dietary wolfberry increases overall lutein and zeaxanthin concentrations in the liver and retinal tissues of C57BL/6J mice. Six-week-old C57BL/6J (B6) mice were fed the control diet (CD) or wolfberry diet (control diet with 1% (kcal) wolfberry, WD) for 8 weeks. Lutein and zeaxanthin contents in the liver and retinal tissues were then detected by high-performance liquid chromatography. Data are shown as mean±SD (liver tissues, n=3) or just mean (from eight pooled retinal tissues), expressed as ng/g fresh tissue. Statistical significance at P<0.05 (*).

been revealed. Uptake, transport, and metabolism of zeaxanthin and lutein in healthy humans and mice have been well documented.[35,36] Some key proteins regulating homeostasis of zeaxanthin and lutein in mice have also been identified recently; for instance, scavenger receptor class B type I (SR-BI) is one of the uptake proteins in the retina.[37] Glutathione S-transferase Pi 1 (GSTP1) mediates the binding and transport of carotenoid within the retinal cells.[38] β,β-Carotene 15′,15′-monooxygenase-1 and β,β-carotene 9′,10′-oxygenase 2 (BCO2) are two enzymes that cleave carotenoids.[35,39] BCO2 cleaves lutein and zeaxanthin in the mitochondrion.[40] Six-week-old db/db type 2 diabetic mice were treated with 1% (kcal) wolfberries for 8 weeks, then gene expression was tested by quantitative real-time polymerase chain reaction and Western blot. Wolfberries at 1% (kcal) did not systemically affect hyperglycemia and hyperinsulinemia, but they ameliorated hypoxia.[6,41] Expression of BCO2, SR-BI, and GSTP1 transcripts and proteins decreased in db/db mice at 14 weeks of age and could be completely reversed by dietary wolfberries. These results suggest that wolfberries alter carotenoid metabolic gene expression in diabetes.

Attenuation of Retinal Hypoxia in db/db Type 2 Diabetic Mice

The retinal vascular system provides nutrients and oxygen to the inner retina, whereas the choroidal vasculature supplies the outer retina. In diabetes, elevated blood glucose and decreases in blood flow result in hyperglycemia and hypoxia in the retina.[42] The development of hypoxia was confirmed by determining the levels of hypoxia-inducible factor-1α (HIF-1α) and vascular endothelial growth factor (VEGF). Dietary wolfberries inhibited HIF-1α and VEGF in the retina of db/db mice.[41] The inhibition of hypoxia signaling may be beneficial to maintaining visual acuity and delaying the progression of diabetic retinodegeneration.

Enhancement of Retinal Mitochondrial Biogenesis in Diabetes

Mitochondrial dysfunction is the hallmark of retinal degeneration in diabetes.[43] Retinal mitochondrial apoptosis and mitochondrial DNA (mtDNA) damage are associated with changes in retinal structure and function. The mtDNA copy number and mitochondrial mass decreased significantly in the retina of db/db diabetic mice; mitochondrial function was also impaired significantly, as evidenced by decreased activity of citrate synthase and inhibited mitochondrial transcription factor A protein expression in mitochondria. Application of 1% (kcal) wolfberries for 8 weeks in db/db mice completely reversed these parameters to levels similar to

FIGURE 47.6　Wolfberry ameliorates dispersion of mitochondria and increased pigment granules in retinal pigment epithelium (RPE) cells of db/db diabetic mice retina. The wild-type and db/db diabetic mice at 6 weeks of age were fed the control diet (CD) or wolfberry diet (control diet with 1% (kcal) wolfberry, WD) for 8 weeks. At termination, whole eyeballs were fixed and subjected to transmission electron microscopy focusing on RPE. The distribution of mitochondria (mt) and pigment granules (pg) in RPE of db/db mice is shown. (A) db/db with CD; (B) db/db with WD. Scale bar: 1.5 μm.

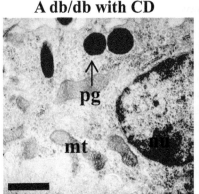

A db/db with CD

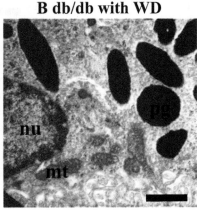

B db/db with WD

those in wild-type mice. In addition, expression of peroxisome proliferator-activated receptor-γ coactivator-1α was significantly inhibited by the onset of diabetes at both transcriptional and translational levels, which were completely reversed by wolfberry, suggesting that wolfberry enhanced mitochondrial biogenesis in the retina of db/db diabetic mice. It was also shown that wolfberries ameliorated the dispersion of mitochondria and increased pigment granules in the retinal pigment epithelia (RPE) of db/db diabetic mice (Fig. 47.6).[41]

Protection against Retinal Degeneration in the Early Stage of Diabetes

The present authors and others have reported that changes in the structure and function of the retina such as damage to retinal photoreceptor and inner nuclear layers, loss of RGCs, and altered retinal thickness occur in the early stages of diabetes, before the observation of clinical retinopathy in db/db type 2 diabetic mice.[6,44] The db/db mice fed 1% (kcal) wolfberries did not develop observable retinal structural abnormalities as tested by histologic analysis.[6] Thus, wolfberry and/or its bioactive components, including but not limited to zeaxanthin and lutein, exert neuroprotective qualities via regulation of gene expression, although their antioxidant capabilities cannot be excluded in the diabetic retina.

POTENTIAL WOLFBERRY–DRUG INTERACTIONS AND SIDE EFFECTS

A few cases have reported side effects of wolfberry. Three cases of a markedly elevated, indeterminate international normalized ratio after consumption of concentrated wolfberry tea and/or juice have been reported in patients receiving anticoagulant warfarin therapy.[45–47] One case report showed that wolfberries altered systemic photosensitivity in a 53-year-old man.[48] A 60-year-old woman who consumed three cups of wolfberry tea per day for 10 days (one handful of wolfberries in each cup) developed liver dysfunction, but her liver function recovered to normal 1 month after withdrawal of wolfberries.[49] Another case report showed the allergenic potential of wolfberry in high-risk individuals with food allergies.[50]

CONCLUSION

Wolfberry has been used for thousands of years as a health food and traditional herbal medicine, but it has not been subjected to rigorous clinical trials. Consumption of 10–15 g dried berries per day is commonly recommended for humans.[18] Despite the lack of substantial clinical evidence, wolfberry dried fruits, juice, and other products are marketed as a functional food to promote eye health owing to enriched bioactive components such as zeaxanthin, lutein, and polysaccharides.

TAKE-HOME MESSAGES

- Wolfberry is rich in bioactive constituents, such as zeaxanthin, lutein, and polysaccharides.
- Wolfberry has enhanced antioxidant activities.
- Wolfberry polysaccharides protect against retinal ganglion cell death in glaucoma and ischemia–reperfusion.
- Wolfberry polysaccharides protect photoreceptors from oxidative damage in retinitis pigmentosa.
- Dietary wolfberry prevents macular hypopigmentation and soft drusen formation in healthy elderly people.
- Dietary wolfberry protects against retinal degeneration in the early stage of type 2 diabetes.

Acknowledgments

We apologize to all investigators whose research work could not be appropriately cited owing to limitations of space. I extend great thanks to Drs. Medeiros, Wang, and Lindshield of Kansas State University for fruitful discussions and collaborations. A special thank you to my past and current students and coworkers for their wonderful contributions. The authors are grateful to Ms. Sarah Hancock for proofreading and editing. This work was supported by grants from NIH COBRE P20-RR-017686 and the NIH K-INBRE Major Starter Grant P20-RR-016475, by the K-INBRE Summer Scholarship, and by the K-State Cancer Center Research Initiative Award to D.L. Contribution no. 13-171-B from the Kansas Agricultural Experiment Station.

References

1. Schmidt KG, Bergert H, Funk RH. Neurodegenerative diseases of the retina and potential for protection and recovery. *Curr Neuropharmacol* 2008;**6**:164–78.
2. Carpenter TM, Steggerda M. The food of the present-day Navajo Indians of New Mexico and Arizona. *J Nutr* 1939;**18**:297–305.
3. Potterat O. Goji (*Lycium barbarum* and *L. chinense*): phytochemistry, pharmacology and safety in the perspective of traditional uses and recent popularity. *Planta Med* 2010;**76**:7–19.
4. Cassileth B. Complementary therapies, herbs, and other OTC agents. *Oncology* 2010;**22**:1353.
5. Chang RC, So KF. Use of anti-aging herbal medicine, *Lycium barbarum*, against aging-associated diseases. What do we know so far? *Cell Mol Neurobiol* 2008;**28**:643–52.
6. Tang L, Zhang Y, Jiang J, Willard L, Ortiz E, Wark L, et al. Dietary wolfberry ameliorates retinal structure abnormalities in db/db mice at the early stage of diabetes. *Exp Biol Med* 2011;**236**:1051–63.
7. Lin NC, Lin JC, Chen SH, Ho CT, Yeh AI. Effect of goji (*Lycium barbarum*) on expression of genes related to cell survival. *J Agric Food Chem* 2011;**59**:10088–96.
8. Inbaraj BS, Lu H, Hung CF, Wu WB, Lin CL, Chen BH. Determination of carotenoids and their esters in fruits of *Lycium barbarum* Linnaeus by HPLC-DAD-APCI-MS. *J Pharm Biomed Anal* 2008;**47**:812–8.
9. Inbaraj BS, Lu H, Kao TH, Chen BH. Simultaneous determination of phenolic acids and flavonoids in *Lycium barbarum* Linnaeus by HPLC-DAD-ESI-MS. *J Pharm Biomed Anal* 2010;**51**:549–56.
10. Xie C, Xu LZ, Li XM, Li KM, Zhao BH, Yang SL. Studies on chemical constituents in fruit of *Lycium barbarum* L. *Zhongguo Zhong Yao Za Zhi* 2001;**26**:323–4.
11. Mikulic-Petkovsek M, Schmitzer V, Slatnar A, Stampar F, Veberic R. Composition of sugars, organic acids, and total phenolics in 25 wild or cultivated berry species. *J Food Sci* 2012;**77**:C1064–70.
12. Shin YG, Cho KH, Kim JM, Park MK, Park JH. Determination of betaine in *Lycium chinense* fruits by liquid chromatography–electrospray ionization mass spectrometry. *J Chromatogr A* 1999;**857**:331–5.
13. Xie H, Zhang S. Determination of taurine in *Lycium barbarum* L. by high performance liquid chromatography with OPA–urea precolumn derivatization. *Se Pu* 1997;**15**:54–6.
14. Li SY, Yang D, Yeung CM, Yu WY, Chang RC, So KF, et al. *Lycium barbarum* polysaccharides reduce neuronal damage, blood–retinal barrier disruption and oxidative stress in retinal ischemia/reperfusion injury. *PLoS ONE* 2011;**6**:e16380.
15. Breithaupt DE, Weller P, Wolters M, Hahn A. Comparison of plasma responses in human subjects after the ingestion of 3R,3R′-zeaxanthin dipalmitate from wolfberry (*Lycium barbarum*) and non-esterified 3R,3R′-zeaxanthin using chiral high-performance liquid chromatography. *Br J Nutr* 2004;**91**:707–13.
16. Jiang Y, Zhang Y, Wark L, Ortiz E, Lim S, He H, et al. Wolfberry water soluble phytochemicals down-regulate ER stress biomarkers and modulate multiple signaling pathways leading to inhibition of proliferation and induction of apoptosis in Jurkat cells. *J Nutr Food Sci* 2011;**S2**. Pii:001.
17. Benzie IF, Chung WY, Wang J, Richelle M, Bucheli P. Enhanced bioavailability of zeaxanthin in a milk-based formulation of wolfberry (Gou Qi Zi; *Fructus barbarum* L.). *Br J Nutr* 2006;**96**:154–60.
18. Vidal K, Bucheli P, Gao Q, Moulin J, Shen LS, Wang J, et al. Immunomodulatory effects of dietary supplementation with a milk-based wolfberry formulation in healthy elderly: a randomized, double-blind, placebo-controlled trial. *Rejuvenation Res* 2012;**15**:89–97.
19. Amagase H, Sun B, Borek C. *Lycium barbarum* (goji) juice improves *in vivo* antioxidant biomarkers in serum of healthy adults. *Nutr Res* 2009;**29**:19–25.
20. Zou S, Zhang X, Yao W, Niu Y, Gao G. Structure characterization and hypoglycemic activity of a polysaccharide isolated from the fruit of *Lycium barbarum* L. *Carbohydr Polym* 2010;**80**:1161–7.
21. Reeve VE, Allanson M, Arun SJ, Domanski D, Painter N. Mice drinking goji berry juice (*Lycium barbarum*) are protected from UV radiation-induced skin damage via antioxidant pathways. *Photochem Photobiol Sci* 2010;**9**:601–7.
22. Hammond Jr BR. Possible role for dietary lutein and zeaxanthin in visual development. *Nutr Rev* 2008;**66**:695–702.
23. Sangiovanni JP, Neuringer M. The putative role of lutein and zeaxanthin as protective agents against age-related macular degeneration: promise of molecular genetics for guiding mechanistic and translational research in the field. *Am J Clin Nutr* 2012;**96**:1223–33S.
24. Berendschot TT, Plat J, de Jong A, Mensink RP. Long-term plant stanol and sterol ester-enriched functional food consumption, serum lutein/zeaxanthin concentration and macular pigment optical density. *Br J Nutr* 2009;**101**:1607–10.
25. Thurnham DI. Macular zeaxanthins and lutein – a review of dietary sources and bioavailability and some relationships with macular pigment optical density and age-related macular disease. *Nutr Res Rev* 2007;**20**:163–79.
26. Bucheli P, Vidal K, Shen L, Gu Z, Zhang C, Miller LE, Wang J. Goji berry effects on macular characteristics and plasma antioxidant levels. *Optom Vis Sci* 2011;**88**:257–62.
27. Tuo J, Bojanowski CM, Zhou M, Shen D, Ross RJ, Rosenberg KI, et al. Murine ccl2/cx3cr1 deficiency results in retinal lesions mimicking human age-related macular degeneration. *Invest Ophthalmol Vis Sci* 2007;**48**:3827–36.
28. Hodge WG, Barnes D, Schachter HM, Pan YI, Lowcock EC, Zhang L, et al. The evidence for efficacy of omega-3 fatty acids in preventing or slowing the progression of retinitis pigmentosa: a systematic review. *Can J Ophthalmol* 2006;**41**:481–90.
29. Komeima K, Rogers BS, Lu L, Campochiaro PA. Antioxidants reduce cone cell death in a model of retinitis pigmentosa. *Proc Natl Acad Sci U S A* 2006;**103**:11300–5.
30. Miranda M, Arnal E, Ahuja S, Alvarez-Nölting R, López-Pedrajas R, Ekström P, et al. Antioxidants rescue photoreceptors in rd1 mice: relationship with thiol metabolism. *Free Radic Biol Med* 2010;**48**:216–22.
31. Chan HC, Chang RC, Koon-Ching Ip A, Chiu K, Yuen WH, Zee SY, So KF. Neuroprotective effects of *Lycium barbarum* Lynn on protecting retinal ganglion cells in an ocular hypertension model of glaucoma. *Exp Neurol* 2007;**203**:269–73.
32. Chiu K, Zhou Y, Yeung SC, Lok CK, Chan OO, Chang RC, et al. Up-regulation of crystallins is involved in the neuroprotective effect of wolfberry on survival of retinal ganglion cells in rat ocular hypertension model. *J Cell Biochem* 2010;**110**:311–20.

33. Mi XS, Feng Q, Lo AC, Chang RC, Lin B, Chung SK, So KF. Protection of retinal ganglion cells and retinal vasculature by *Lycium barbarum* polysaccharides in a mouse model of acute ocular hypertension. *PLoS ONE* 2012;7:e45469.

34. Brazionis L, Rowley K, Itsiopoulos C, O'Dea K. Plasma carotenoids and diabetic retinopathy. *Br J Nutr* 2009;**101**:270–7.

35. von Lintig J. Colors with functions: elucidating the biochemical and molecular basis of carotenoid metabolism. *Annu Rev Nutr* 2010;**30**:35–56.

36. Borel P. Genetic variations involved in interindividual variability in carotenoid status. *Mol Nutr Food Res* 2012;**56**:228–40.

37. Provost AC, Vede L, Bigot K, Keller N, Tailleux A, Jaïs JP, et al. Morphologic and electroretinographic phenotype of SR-BI knockout mice after a long-term atherogenic diet. *Invest Ophthalmol Vis Sci* 2009;**50**:3931–42.

38. Bhosale P, Larson AJ, Frederick JM, Southwick K, Thulin CD, Bernstein PS. Identification and characterization of a Pi isoform of glutathione *S*-transferase (GSTP1) as a zeaxanthin-binding protein in the macula of the human eye. *J Biol Chem* 2004;**279**:49447–54.

39. Bhatti RA, Yu S, Boulanger A, Fariss RN, Guo Y, Bernstein SL, et al. Expression of beta-carotene 15,15′ monooxygenase in retina and RPE-choroid. *Invest Ophthalmol Vis Sci* 2003;**44**:44–9.

40. Amengual J, Lobo GP, Golczak M, Li HN, Klimova T, Hoppel CL, et al. A mitochondrial enzyme degrades carotenoids and protects against oxidative stress. *FASEB J* 2011;**25**:948–59.

41. Yu H, Han J, Wark L, Willard L, He H, Ortiz E, et al. Dietary wolfberry up-regulates carotenoid metabolic proteins and enhances mitochondrial biogenesis in the diabetic retina. *Mol Nutr Food Res* 2013;**57**:1158–69.

42. Arden GB, Sivaprasad S. Hypoxia and oxidative stress in the causation of diabetic retinopathy. *Curr Diabetes Rev* 2011;7: 291–304.

43. Kowluru RA. Diabetic retinopathy: mitochondrial dysfunction and retinal capillary cell death. *Antioxid Redox Signal* 2005;7:1581–7.

44. Ly A, Yee P, Vessey KA, Phipps JA, Jobling AI, Fletcher EL. Early inner retinal astrocyte dysfunction during diabetes and development of hypoxia, retinal stress, and neuronal functional loss. *Invest Ophthalmol Vis Sci* 2011;**52**:9316–26.

45. Lam AY, Elmer GW, Mohutsky MA. Possible interaction between warfarin and *Lycium barbarum* L. *Ann Pharmacother* 2001;**35**: 1199–201.

46. Leung H, Hung A, Hui AC, Chan TY. Warfarin overdose due to the possible effects of *Lycium barbarum* L. *Food Chem Toxicol* 2008;**46**:1860–2.

47. Rivera CA, Ferro CL, Bursua AJ, Gerber BS. Probable interaction between *Lycium barbarum* (goji) and warfarin. *Pharmacotherapy* 2012;**32**(3):e50–3.

48. Gómez-Bernal S, Rodríguez-Pazos L, Martínez FJ, Ginarte M, Rodríguez-Granados MT, Toribio J. Systemic photosensitivity due to goji berries. *Photodermatol Photoimmunol Photomed* 2011;**27**: 245–7.

49. Arroyo-Martinez Q, Sáenz MJ, Arias FA, Acosta MSJ. Lycium barbarum: a new hepatotoxic 'natural' agent? *Dig Liver Dis* 2011;**43**:749.

50. Larramendi CH, García-Abujeta JL, Vicario S, García-Endrino A, López-Matas MA, García-Sedeño MD, Carnés J. Goji berries (*Lycium barbarum*): risk of allergic reactions in individuals with food allergy. *J Invest Allergol Clin Immunol* 2012;**22**:345–50.

Sea Buckthorn, Dry Eye, and Vision

Petra S. Larmo[1], Riikka L. Järvinen[2], Baoru Yang[3], Heikki P. Kallio[3]

[1]Aromtech Ltd, Tornio, Finland, [2]Finnsusp Ltd, Lieto, Finland, [3]Food Chemistry and Food Development, Department of Biochemistry, University of Turku, Finland

INTRODUCTION

Sea buckthorn (*Hippophaë*) berries are widely used as traditional medicine in Asia. In Central Nepal sea buckthorn is among the medicinal plants with the widest range of uses. It is recommended for gastrointestinal disorders, coughs, and colds, as well as for menstrual disorders.[1] Sea buckthorn berries are listed in the *Chinese Pharmacopeia* as an ingredient for the treatment of cough and for improving blood circulation and digestion.[2] In Russia, sea buckthorn oil was tested as a treatment for eye disorders in the 1950s. The oil was reported to have beneficial effects on lesions of the cornea, dark adaptations, and visual acuity, among other things.[3]

Scientific research has found a rationale for the traditional uses of sea buckthorn based on the unique combination of bioactive compounds in the berries[4–11] and their effects on inflammatory markers,[12,13] dry eye,[14] platelet aggregation,[15] atopic dermatitis,[16] and circulating lipids[16] in humans. Sea buckthorn grows wild in the cold and low-rainfall regions of Asia and Europe.[17] It has been introduced to North America and is nowadays cultivated as a source of berries for food use in several countries. The main sea buckthorn resources are found in China, India, and Russia.[17] The objective of this chapter is to discuss the effects of sea buckthorn on vision-related health and the possible mechanisms of effect.

BIOACTIVE COMPOUNDS OF SEA BUCKTHORN BERRY

In comparison to most other berries, the soft parts of sea buckthorn berries have an exceptionally high oil content, approximately 2–3% of the fresh weight.[4] The oil content of sea buckthorn seeds varies from 7 to greater than 10% (by weight) depending on genetic background.[4,6,18] The main lipid class in both the berry and the seed oil is triacylglycerols, but the fatty acid composition of the oils differs (Table 48.1). Sea buckthorn pulp oil is characterized by its content of approximately 30% palmitoleic acid (16:1n-7), rare in such amounts in food sources. Consumption of high-palmitoleic acid oils has been shown to affect circulating total and low-density lipoprotein (LDL)-cholesterol beneficially.[20] Recent animal studies suggest lipokine effects and modulation of insulin resistance and hepatic lipid accumulation by palmitoleic acid.[21,22] The relevance of these findings for humans, however, remains unclear.[23]

The main fatty acids of sea buckthorn seed oil are the essential α-linolenic (18:3n-3) and linoleic (18:2n-6) acids, which are present in a nutritionally favorable ratio.[24] In humans, these are important as precursors for the metabolic pathways to eicosanoids (Fig. 48.1). They regulate several functions of the body, including inflammation,[25] which, according to current knowledge, is involved in the development of disorders affecting vision.[28–30] Epidemiologic studies indicate beneficial effects of n-3 fatty acids on dry eye[31] and age-related macular degeneration (AMD).[30] Oral intake of combined linoleic and γ-linolenic (18:3n-6) acids has been shown to relieve dry eye.[32–36] Adequate intake of n-3 fatty acids is important for the normal development of the retina.[37]

Sea buckthorn seed and pulp oils are rich sources of tocopherols and tocotrienols (Table 48.1). Typically, α-tocopherol is the main tocol in both seeds and pulp.[5,6] Carotenoids give sea buckthorn berry its orange color and are enriched in the oil from the soft part (Table 48.1). β-Carotene, β-cryptoxanthin, and zeaxanthin are among the main carotenoids reported.[38] Carotenoids, tocopherols, and tocotrienols are important for their function as antioxidants and in their anti-inflammatory potential. For vision, the function of carotenoids having a β-ionone ring as a precursor to vitamin A is of special importance.[39,40] Lutein and zeaxanthin, nonvitamin A carotenoids found in sea buckthorn oil, are present in

TABLE 48.1 Fatty Acids, Tocopherols and Tocotrienols, and Carotenoids in Sea Buckthorn Oils

	Oil from Soft Parts	Oil from Seeds	References
Fatty acids (%, by weight)			
Palmitoleic acid (16:1n-7)	27–33		4
Palmitic acid (16:0)	27–28	7–9	4
Vaccenic acid (18:1n-7)	8–9	2–3	4
Oleic acid (18:1n-9)	≈17	17–19	4
Linoleic acid (18:2n-6)	9–13	39–41	4
α-Linolenic acid (18:3n-3)	3–7	27–31	4
Tocopherols and tocotrienols (mg/100 g)	100–700	100–300	5,6
Carotenoids (mg/100 g)	70–820	24–28	6,19

the macula and lens of the human eye. Intake of lutein and zeaxanthin has been associated with a reduced risk of AMD and been reported to have a beneficial effect on the course of the disease.[41,42]

In addition to the oils, sea buckthorn is rich in more polar bioactive compounds. Even with the great variation that arises from species/subspecies, harvesting time, growth conditions, and processing, sea buckthorn berries are among the best food sources of vitamin C (Table 48.2). Flavonoids and phenolic acids are secondary metabolites produced by plants to defend against insects, pathogens, and ultraviolet radiation. The main flavonoids in sea buckthorn berries are flavonols and proanthocyanidins, which in combination with the vitamin C account for the high antioxidant activity of sea buckthorn juice *in vitro*.[7] The flavonols and flavonoid-rich fractions of sea buckthorn possess anti-inflammatory and immune-modulating activity in animals and *in vitro*.[43–47] Clinical interventions suggest that flavonoids and vitamin C in combination with other

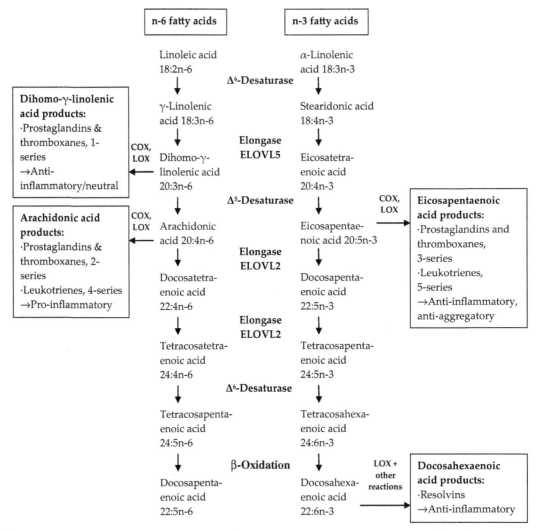

FIGURE 48.1 Synthesis of long-chain n-3 and n-6 fatty acids from α-linolenic and linoleic acids, respectively, and a simplified overview of the messengers derived from them in humans. *Source: Adapted from Schmitz and Ecker (2008),[25] Larmo (2011),[26] and Rosenberg and Asbell (2010).[27]*

antioxidants may beneficially affect components of dry eye.[48,49] Evidence as to the effect of noncarotenoid antioxidants on AMD is conflicting.[41,42] The vision-related bioactivities associated with compounds rich in sea buckthorn berries are summarized in Figure 48.2.

SEA BUCKTHORN OIL FOR DRY EYE

Dry eye is a disease affecting tears and the ocular surface. It causes symptoms of discomfort and visual disturbance. There are two main types of dry eye: the aqueous tear-deficient and the evaporative types. Both of them lead to hyperosmolarity of the tear film that protects the ocular surface. Hyperosmolarity provokes inflammatory cascades and may ultimately damage the ocular surface. Aqueous tear-deficient dry eye is characterized by insufficient supply of tears from the lacrimal gland to the tear film. In patients with Sjögren's syndrome, the lacrimal

TABLE 48.2 Hydrophilic Bioactive Compounds of Sea Buckthorn Berries

	Concentration	References
Flavonol-glycosides (mg/100 g FW)	27–130	11
Isorhamnetin (mg aglycone/100 g FW)	17–66	9,11
Flavanols (mg/100 ml juice)	2–3	7
Proanthocyanidins (mg/100 g FW)	<1–280	8,9
Vitamin C (mg/100 ml juice)	30–1300	5,10

FW: fresh weight.

glands are affected by an autoimmune process that leads to inflammatory changes and hyposecretion. Aqueous deficiency is not always associated with Sjögren's syndrome, and in addition to hyposecretion it may be due to obstruction of the lacrimal duct. In evaporative dry eye, the evaporation of aqueous tear film is excessive. This may be due to intrinsic factors, including deficiencies in the meibomian oil that forms the outermost lipid layer of the tear film. Extrinsic factors associated with evaporative dry eye include vitamin A deficiency, wearing of contact lenses, and allergies affecting the ocular surface.[29]

At worst, dry eye can significantly compromise experienced quality of life. The condition is more common in women than in men, and the risk increases with age. Low humidity of the environment, hormone therapy in menopause, and computer use are among the factors that are associated with dry eye. The prevalence of dry eye in epidemiologic studies varies with the population studied and the definition employed. Among those aged 40 years or older, prevalence figures of approximately 6% to over 30% have been reported in large epidemiologic studies using symptom-based criteria.[50] The usual treatments are ocular lubricants and artificial tears. They relieve the symptoms but do not cure the inflammation that maintains and encourages the dry eye.[51]

The effect of oral sea buckthorn oil on dry eye was investigated in a double-blind, randomized, placebo-controlled study with 100 participants.[14] Healthy women (85%) and men from 20 to 75 years of age experiencing symptoms of dry eye were recruited to the study, carried out at the University of Turku in Finland. Half of the participants were contact-lens wearers. During the study period of 3 months, the participants took 2 g daily of sea buckthorn or placebo oil in the form of two capsules twice a day. The sea

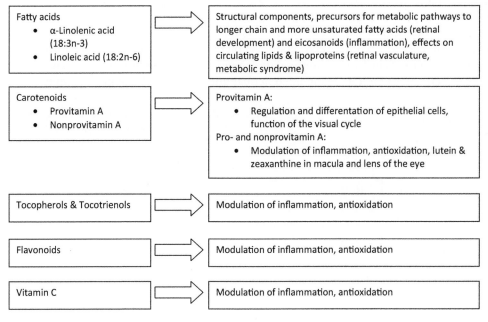

FIGURE 48.2 Summary of vision-related bioactivities associated with compounds rich in sea buckthorn berries.

buckthorn oil used in the study was a standardized composition of sea buckthorn pulp and seed oils produced through supercritical carbon dioxide extraction by Aromtech (Tornio, Finland). The main fatty acids in the standardized sea buckthorn oil were palmitoleic acid (345 mg/day), palmitic acid (340 mg/day), oleic acid (315 mg/day), linoleic acid (245 mg/day), and α-linolenic acid (150 mg/day). The daily dose contained 2 mg carotenoids, 6 mg α-tocopherol, and 0.8 mg γ-tocopherol. The placebo oil was composed of triacylglycerols of medium-chain fatty acids fractionated from coconut and palm kernel. It did not contain carotenoids or γ-tocopherol in quantifiable amounts. The amount of α-tocopherol was 0.2 mg/day.

The study included four study visits: at the beginning; after 1 month; after 3 months, when the intervention ended; and in addition, a postcheck 1–2 months after the end of the intervention period (Fig. 48.3). At each visit, dry eye was evaluated through measurement of the tear film osmolarity (mOsm/L) with an electrochemical osmolarity meter (TearLab™; Ocusense, San Diego, CA, USA), tear film stability (seconds until break-up of fluorescein tear film, TBUT), tear secretion (mm/5 minutes) via the Schirmer test without anesthesia, and symptoms of dry eye via the modified Ocular Surface Disease Index (mOSDI). The fatty acid composition of tear film was analyzed from Schirmer papers that had absorbed tears. The participants kept a daily log of their dry eye symptoms and compliance from the beginning of the study to the end of the capsule intake period, at 3 months. In the logbook, the severity of typical dry eye symptoms was evaluated on a four-point scale from 0 (none) to 3 (severe).

The results of the diagnostic tests for dry eye and the symptoms in the sea buckthorn and placebo groups were compared via suitable statistical methods. The primary analyses were performed including both the compliant participants and the noncompliant participants to simulate the real-life situation of supplement use. Participants who, according to the logbook follow-up, reported taking the study capsules for at least 80% of the study days were considered compliant. Analyses including only the compliant participants were performed as

well. A total of 86 participants completed the study and 81 were considered compliant (Fig. 48.3).

There was a general increase in tear film osmolarity in both the sea buckthorn and the placebo groups. This could be expected due to the timing of the observation. The intervention lasted from autumn to winter. From the beginning to the end of the study, the mean temperature in Turku dropped from +8°C to −5°C. In the colder months the air humidity is low both indoors and outdoors, which causes increased tear evaporation.[52,53] The dry eye symptoms are more common during the periods when indoor heating systems are used.[52,53] The increase in tear film osmolarity was significantly less in the sea buckthorn group (+8 mOsm/L) than in the placebo group (+12 mOsm/L) (P for group difference: for all participants P = 0.04; for compliant participants P = 0.02), suggesting a beneficial effect of sea buckthorn oil. High tear film osmolarity is directly involved in the mechanism of dry eye, associated with different types of dry eye, and has been suggested as the 'gold standard' of diagnosis.[29,54] The Schirmer test, TBUT, and mOSDI were not affected by sea buckthorn oil.

The symptoms of burning and redness of the eyes reported in the daily symptom logbooks were less severe in the sea buckthorn than in the placebo group. The group difference for redness was significant when all participants were included in the analysis (P = 0.04). A beneficial trend (nonsignificant, P = 0.11) was observed when only the compliant participants were included. A significant effect of sea buckthorn on burning of the eyes was observed among the compliant participants (P = 0.04). Among all participants (compliant and noncompliant) a tendency towards beneficial sea buckthorn effects was observed (P = 0.05). Other symptoms of dry eye were not affected by sea buckthorn oil.

The outermost lipid layer of the tear film is essential in preventing excessive evaporation of tears. The tear film lipid layer is thinner in people experiencing dry eye symptoms,[55] and abnormalities in the fatty acid and lipid class composition have been reported.[56,57] In addition, the carotenoids of meibum may be compromised,[58] emphasizing the potential importance of the nonprovitamin A and

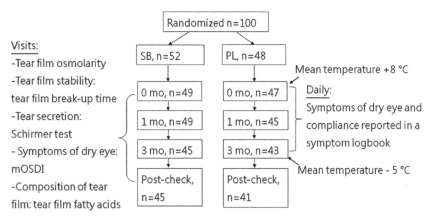

FIGURE 48.3 Summary of the flow of the study of sea buckthorn oil for dry eye.[14] SB: sea buckthorn oil; PL: placebo oil; mOSDI: modified Ocular Surface Disease Index.

provitamin A carotenoids in sea buckthorn oil. Regardless of the beneficial changes induced by sea buckthorn oil, in this study no effects on the fatty acid composition of the tear film were observed.[59] Beneficial effects of n-3 fatty acid supplementation without direct changes in meibum lipid composition have been reported.[60] Effects of intake of n-3 fatty acids or combined linoleic and γ-linolenic acids on eicosanoids and the inflammatory component of dry eye have been reported in clinical studies.[33,35] The effects of sea buckthorn oil on osmolarity and symptoms were probably mediated by modulation of the inflammation that induces and reinforces dry eye, whether due to deficiency in the production of aqueous tears or to excess tear film evaporation.[29]

In the study reviewed above, different types of dry eye were included. Yang and Erkkola[61] reported beneficial effects of sea buckthorn oil on dryness of eyes associated with Sjögren's syndrome. The effects of sea buckthorn oil were investigated in a double-blind, randomized, placebo-controlled crossover study of 25 women with Sjögren's syndrome. The main interest lay in the effects on the genital mucosa. During the study period of 3 months, the participants consumed 3 g of sea buckthorn or placebo oil daily. The sea buckthorn oil was a standardized composition of pulp and seed oils, produced by means of supercritical carbon dioxide extraction (Aromtech). The placebo oil was triacylglycerols of medium-chain fatty acids. The symptoms of Sjögren's syndrome were evaluated on a visual analogue scale and by verbal description at the beginning of the study, the end of first intervention period, and the end of the second intervention period. During the first 3 months of the study, a significantly higher proportion of participants reported improvement of overall symptoms of Sjögren's syndrome in the sea buckthorn group compared with the placebo group. Among the individual symptoms beneficially affected was the experienced dryness of the eyes. In the sea buckthorn group approximately 90% of participants reported improvement from baseline in the dryness of eyes, compared with around 70% of participants reporting improvement in the placebo group.

EFFECTS OF SEA BUCKTHORN ON COMPONENTS OF METABOLIC SYNDROME ASSOCIATED WITH RETINAL FUNCTION

Components of metabolic syndrome, including obesity, dyslipidemia, hypertension, and impaired fasting glucose, affect the retinal circulation and may ultimately lead to changes detrimental to vision.[28,62] Endothelial dysfunction and inflammation are likely to be the central mechanisms that mediate the effects of metabolic disorders on the retinal vascular system.[28] Clinical studies have shown sea buckthorn berries and their fractions affect several components of metabolic syndrome and inflammation beneficially (Table 48.3).

TABLE 48.3 Indications of Effects of Sea Buckthorn on Inflammation and Components of Metabolic Syndrome in Clinical Studies

Sea Buckthorn Berry Fraction	Dose and Duration	Study Design	Result	Reference
Combined pulp and seed oil, CO_2 extraction	6 g/day for 1 month	Randomized, open, crossover; 80 women	Reduction in circulating VCAM	13
Ethanol extract (phenolic extract: flavonoids and sugars)	14.6 g (1:1 extract: maltodextrin) for 1 month	Randomized, open, crossover; 80 women	Reduction in circulating ICAM	13
Frozen purée of whole berries	28 g/day for 3 months	Randomized, double-blind, parallel, placebo-controlled; 250 women and men	Reduction in circulating hs-CRP	12
Sea buckthorn pulp oil, CO_2 extraction	5 g/day for 4 months	Randomized, double-blind, parallel, placebo-controlled; 78 women	Increase in plasma HDL cholesterol	16
Sea buckthorn juice	300 ml/day for 8 weeks	Randomized, double-blind, parallel, placebo-controlled; 30 men	Decrease in the susceptibility of LDL to oxidation	63
Dried whole berries, CO_2-extracted oil-free berries, ethanol-extracted oil-free berries	One 20–40 g meal of each	Postprandial response, each subject as his own control; 10 men	Suppression of peak insulin and stabilization of hyperglycemia by ethanol-soluble components	64
Dried whole berries, CO_2-extracted oil-free berries, ethanol-extracted oil-free berries	One 36–80 g meal of each	Postprandial response, each subject as his own control; 25 men	Delayed lipemia by fiber, complementary effect of polyphenols	65

CO_2: carbon dioxide; VCAM: vascular cell adhesion molecule; ICAM: intracellular adhesion molecule; hs-CRP: high-sensitivity C-reactive protein; HDL: high-density lipoprotein; LDL: low-density lipoprotein.

In a study of 80 overweight women, intake of sea buckthorn oil and ethanol-extracted flavonoid-rich fraction of the berry for 1 month significantly reduced the circulating levels of vascular cell adhesion molecule (VCAM) and intercellular adhesion molecule (ICAM), respectively.[13] VCAM and ICAM are markers of endothelial dysfunction associated with the early stages of developing atherosclerosis. In a double-blind, randomized, placebo-controlled study of 250 healthy volunteers, daily intake of 28 g of sea buckthorn berries significantly reduced the circulating levels of serum high-sensitivity C-reactive protein (hs-CRP).[12] CRP is an inflammatory marker considered an independent predictor of increased coronary risk.[66] Klein and colleagues[67] found that, while ICAM alone was not associated with retinal venular or arteriolar diameter, the largest venular diameters were observed in people with the highest concentrations of markers of both inflammatory and endothelial dysfunction. They concluded that this supports the hypothesis of vascular endothelial damage by activated leukocytes being involved in the increase in venular diameter. In the same study, higher levels of hs-CRP were associated with larger retinal venular diameters.

In a randomized, double-blind, placebo-controlled study of 78 women, Yang et al.[16] found that intake of sea buckthorn pulp oil for 4 months increased the plasma levels of high-density lipoprotein (HDL)-cholesterol. Reduced HDL-cholesterol levels have been associated with retinal venular dilatation.[28] A beneficial tendency for increased resistance of LDL to oxidation was observed after consuming sea buckthorn juice for 2 months, even though in this study ICAM was unaffected.[63] Recently, beneficial effects of sea buckthorn polyphenols and fibers on postprandial glucose and insulin response and lipemia have been observed in humans.[64,65] The effects of sea buckthorn berries, oils, and hydrophilic fractions on inflammation and components of metabolic disorder indicate that sea buckthorn may affect other aspects of vision in addition to dry eye. The clinical findings on metabolic markers indicate that the effects of sea buckthorn on the retinal vascular system merit investigation.

TAKE-HOME MESSAGES

- Sea buckthorn berries and oil are rich in bioactive compounds that are associated with effects on vision-related health.
- Dry eye is a common condition that may seriously compromise a person's quality of life.
- Intake of 2 g of sea buckthorn oil per day had beneficial effects on dry eye (tear film osmolarity and symptoms of redness and burning) among volunteers experiencing dry eye symptoms.

- Sea buckthorn oil beneficially affected experienced dryness of the eyes in patients with Sjögren's syndrome.
- Components of metabolic syndrome have detrimental effects on the retinal vascular system.
- Clinical studies indicate beneficial effects of sea buckthorn's berries, oil, and flavonoid-rich fractions on inflammation and components of metabolic syndrome. The possible effects of sea buckthorn on the retinal vasculature merit investigation.

References

1. Uprety Y, Asselin H, Boon EK, Yadav S, Shrestha KK. Indigenous use and bio-efficacy of medicinal plants in the Rasuwa District, Central Nepal. *J Ethnobiol Ethnomed* 2010;**6**:3.
2. *Pharmacopeia of the People's Republic of China.* English edn, vol. 1. Beijing: Chemical Industry Press; 2000.
3. Gurevich SK. [Use of the *Hippophae rhamnoides* oil in ophthalmology.] (In Russian). *Vestn Oftalmol.* 1956;**69**:30–3.
4. Yang B, Kallio HP. Fatty acid composition of lipids in sea buckthorn (*Hippophaë rhamnoides* L.) berries of different origins. *J Agric Food Chem* 2001;**49**:1939–47.
5. Kallio H, Yang B, Peippo P. Effects of different origins and harvesting time on vitamin C, tocopherols, and tocotrienols in sea buckthorn (*Hippophaë rhamnoides*) berries. *J Agric Food Chem* 2002;**50**:6136–42.
6. St George SD, Cenkowski S. Influence of harvest time on the quality of oil-based compounds in sea buckthorn (*Hippophae rhamnoides* L. ssp. *sinensis*) seed and fruit. *J Agric Food Chem* 2007;**55**:8054–61.
7. Rösch D, Bergmann M, Knorr D, Kroh LW. Structure–antioxidant efficiency relationships of phenolic compounds and their contribution to the antioxidant activity of sea buckthorn juice. *J Agric Food Chem* 2003;**51**:4233–9.
8. Hosseinian FS, Li W, Hydamaka AW, Tsopmo A, Lowry L, Friel J, et al. Proanthocyanidin profile and ORAC values of Manitoba berries, chokecherries, and seabuckthorn. *J Agric Food Chem* 2007;**55**:6970–6.
9. Määttä-Riihinen KR, Kamal-Eldin A, Mattila PH, González-Paramás AM, Törrönen AR. Distribution and contents of phenolic compounds in eighteen Scandinavian berry samples. *J Agric Food Chem* 2004;**52**:4477–86.
10. Tiitinen KM, Yang B, Haraldsson GG, Jonsdottir S, Kallio HP. Fast analysis of sugars, fruit acids, and vitamin C in sea buckthorn (*Hippophaë rhamnoides* L.) varieties. *J Agric Food Chem* 2006;**54**:2508–13.
11. Yang B, Halttunen T, Raimo O, Price K, Kallio H. Flavonol glycosides in wild and cultivated berries of three major subspecies of *Hippophaë rhamnoides* and changes during harvesting period. *Food Chem* 2009;**115**:657–64.
12. Larmo P, Alin J, Salminen E, Kallio H, Tahvonen R. Effects of sea buckthorn berries on infections and inflammation: a double-blind, randomized, placebo-controlled trial. *Eur J Clin Nutr* 2008;**62**:1123–30.
13. Lehtonen H, Suomela J, Tahvonen R, Yang B, Venojärvi M, Viikari J, et al. Different berries and berry fractions have various but slightly positive effects on the associated variables of metabolic diseases on overweight and obese women. *Eur J Clin Nutr* 2011;**65**:394–401.
14. Larmo P, Järvinen R, Setälä N, Yang B, Viitanen M, Engblom J, et al. Oral sea buckthorn oil attenuates tear film osmolarity and symptoms in individuals with dry eye. *J Nutr* 2010;**140**:1462–8.

15. Johansson A, Laine T, Linna M, Kallio H. Variability in oil content and fatty acid composition in wild northern currants. *Eur Food Res Technol* 2002;**211**:277–83.

16. Yang B, Kalimo KO, Mattila LM, Kallio SE, Katajisto JK, Peltola OJ, et al. Effects of dietary supplementation with sea buckthorn (*Hippophaë rhamnoides*) seed and pulp oils on atopic dermatitis. *J Nutr Biochem* 1999;**10**:622–30.

17. Singh V. Geographical adaptation and distribution of seabuckthorn resources. In: Singh V, Kallio H, Sawhney RC, et al., editors. *Seabuckthorn (Hippophae L.): A Multipurpose Wonder Plant*. New Delhi: Indus; 2003. pp. 21–34.

18. Kallio H, Yang B, Peippo P, Tahvonen R, Pan R. Triacylglycerols, glycerophospholipids, tocopherols, and tocotrienols in berries and seeds of two subspecies (ssp. *sinensis* and *mongolica*) of sea buckthorn (*Hippophaë rhamnoides*). *J Agric Food Chem* 2002;**50**:3004–9.

19. Ranjith A, Kumar KS, Venugopalan VV, Arumughan C, Sawhney RC, Singh V. Fatty acids, tocols, and carotenoids in pulp oil of three sea buckthorn species (*Hippophae rhamnoides, H. salicifolia,* and *H. tibetana*) grown in the Indian Himalayas. *J Am Oil Chem Soc* 2006;**83**:359–64.

20. Garg M, Blake R, Wills R. Macadamia nut consumption lowers plasma total and LDL cholesterol levels in hypercholesterolemic men. *J Nutr* 2003;**133**:1060–3.

21. Yang ZH, Miyahara H, Hatanaka A. Chronic administration of palmitoleic acid reduces insulin resistance and hepatic lipid accumulation in KK-Ay mice with genetic type 2 diabetes. *Lipids Health Dis* 2011;**10**:120.

22. Cao H, Gerhold K, Mayers JR, Wiest MM, Watkins SM, Hotamisligil GS. Identification of a lipokine, a lipid hormone linking adipose tissue to systemic metabolism. *Cell* 2008;**134**:933–44.

23. Mozaffarian D, Cao H, King IB, Lemaitre RN, Song X, Siscovick DS, et al. Circulating palmitoleic acid and risk of metabolic abnormalities and new-onset diabetes. *Am J Clin Nutr* 2010;**92**:1350–8.

24. Simopoulos AP. The importance of the ratio of omega-6/omega-3 essential fatty acids. *Biomed Pharmacother* 2002;**56**:365–79.

25. Schmitz G, Ecker J. The opposing effects of n-3 and n-6 fatty acids. *Prog Lipid Res* 2008;**47**:147–55.

26. Larmo P. *The Health Effects of Sea Buckthorn Berries and Oil. PhD Thesis*. Department of Biochemistry and Food Chemistry, University of Turku; 2011.

27. Rosenberg ES, Asbell PA. Essential fatty acids in the treatment of dry eye. *Ocul Surf* 2010;**8**:18–28.

28. Nguyen TT, Wong TY. Retinal vascular manifestations of metabolic disorders. *Trends Endocrinol Metab* 2006;**17**:262–8.

29. The definition and classification of dry eye disease: report of the Definition and Classification Subcommittee of the International Dry Eye WorkShop (2007). *Ocul Surf* 2007;**5**:75–92.

30. Tan JS, Wang JJ, Flood V, Mitchell P. Dietary fatty acids and the 10-year incidence of age-related macular degeneration: the Blue Mountains Eye Study. *Arch Ophthalmol* 2009;**127**:656–65.

31. Miljanovic B, Trivedi KA, Dana MR, Gilbard JP, Buring JE, Schaumberg DA. Relation between dietary n-3 and n-6 fatty acids and clinically diagnosed dry eye syndrome in women. *Am J Clin Nutr* 2005;**82**:887–93.

32. Kokke KH, Morris JA, Lawrenson JG. Oral omega-6 essential fatty acid treatment in contact lens associated dry eye. *Cont Lens Anterior Eye* 2008;**31**:141–6.

33. Aragona P, Bucolo C, Spinella R, Giuffrida S, Ferreri G. Systemic omega-6 essential fatty acid treatment and PGE1 tear content in Sjögren's syndrome patients. *Invest Ophthalmol Vis Sci* 2005;**46**:4474–9.

34. Macri A, Giuffrida S, Amico V, Iester M, Traverso CE. Effect of linoleic acid and gamma-linolenic acid on tear production, tear clearance and on the ocular surface after photorefractive keratectomy. *Graefes Arch Clin Exp Ophthalmol* 2003;**241**:561–6.

35. Barabino S, Rolando M, Camicione P, Ravera G, Zanardi S, Giuffrida S, et al. Systemic linoleic and gamma-linolenic acid therapy in dry eye syndrome with an inflammatory component. *Cornea* 2003;**22**:97–101.

36. Pinna A, Piccinini P, Carta F. Effect of oral linoleic and gamma-linolenic acid on meibomian gland dysfunction. *Cornea* 2007;**26**:260–4.

37. Heinemann KM, Waldron MK, Bigley KE, Lees GE, Bauer JE. Long-chain (n-3) polyunsaturated fatty acids are more efficient than alpha-linolenic acid in improving electroretinogram responses of puppies exposed during gestation, lactation, and weaning. *J Nutr* 2005;**135**:1960–6.

38. Raffo A, Paoletti F, Antonelli M. Changes in sugar, organic acid, flavonol and carotenoid composition during ripening of berries of three seabuckthorn (*Hippophae rhamnoides* L.) cultivars. *Eur Food Res Technol* 2004;**219**:360–8.

39. Traber M. Vitamin E. In: Shils ME, Shike M, Ross CA, Cabarello B, Cousins R, editors. *Modern Nutrition in Health and Disease*. 10th ed. Philadelphia, PA: Lippincott Williams & Wilkins; 2006. pp. 396–411.

40. Ross CA. Vitamin A and carotenoids. In: Shils ME, Shike M, Ross CA, Cabarello B, Cousins R, editors. *Modern Nutrition in Health and Disease*. 10th ed. Philadelphia, PA: Lippincott Williams & Wilkins; 2006. pp. 351–75.

41. Age-Related Eye Disease Study Research Group, SanGiovanni JP, Chew EY, Clemons TE, Ferris III FL, Gensler G, et al. The relationship of dietary carotenoid and vitamin A, E, and C intake with age-related macular degeneration in a case–control study: AREDS Report No. 22. *Arch Ophthalmol* 2007;**125**:1225–32.

42. Weigert G, Kaya S, Pemp B, Sacu S, Lasta M, Werkmeister RM, et al. Effects of lutein supplementation on macular pigment optical density and visual acuity in patients with age-related macular degeneration. *Invest Ophthalmol Vis Sci* 2011;**52**:8174–8.

43. Hämäläinen M, Nieminen R, Vuorela P, Heinonen M, Moilanen E. Anti-inflammatory effects of flavonoids: genistein, kaempferol, quercetin, and daidzein inhibit STAT-1 and NF-kappaB activations, whereas flavone, isorhamnetin, naringenin, and pelargonidin inhibit only NF-kappaB activation along with their inhibitory effect on iNOS expression and NO production in activated macrophages. *Mediators Inflamm* 2007;**2007**:456–73.

44. Boivin D, Blanchette M, Barrette S, Moghrabi A, Beliveau R. Inhibition of cancer cell proliferation and suppression of TNF-induced activation of NFkappaB by edible berry juice. *Anticancer Res* 2007;**27**:937–48.

45. Bao M, Lou Y. Isorhamnetin prevent endothelial cell injuries from oxidized LDL via activation of p38MAPK. *Eur J Pharmacol* 2006;**547**:22–30.

46. Mishra KP, Chanda S, Karan D, Ganju L, Sawhney RC. Effect of seabuckthorn (*Hippophae rhamnoides*) flavone on immune system: an *in-vitro* approach. *Phytother Res* 2008;**22**:1490–5.

47. Gupta A, Kumar R, Pal K, Singh V, Banerjee PK, Sawhney RC. Influence of sea buckthorn (*Hippophae rhamnoides* L.) flavone on dermal wound healing in rats. *Mol Cell Biochem* 2006;**290**:193–8.

48. Blades KJ, Patel S, Aidoo KE. Oral antioxidant therapy for marginal dry eye. *Eur J Clin Nutr* 2001;**55**:589–97.

49. Peponis V, Papathanasiou M, Kapranou A, Magkou C, Tyligada A, Melidonis A, et al. Protective role of oral antioxidant supplementation in ocular surface of diabetic patients. *Br J Ophthalmol* 2002;**86**:1369–73.

50. The epidemiology of dry eye disease: report of the Epidemiology Subcommittee of the International Dry Eye WorkShop (2007). *Ocul Surf* 2007;**5**:93–107.

51. Management and therapy of dry eye disease: report of the Management and Therapy Subcommittee of the International Dry Eye WorkShop (2007). *Ocul Surf* 2007;**5**:163–78.

52. Uchiyama E, Aronowicz JD, Butovich IA, McCulley JP. Increased evaporative rates in laboratory testing conditions simulating airplane

cabin relative humidity: an important factor for dry eye syndrome. *Eye Contact Lens* 2007;**33**:174–6.

53. Moss SE, Klein R, Klein BE. Prevalence of and risk factors for dry eye syndrome. *Arch Ophthalmol* 2000;**118**:1264–8.

54. Methodologies to diagnose and monitor dry eye disease: report of the Diagnostic Methodology Subcommittee of the International Dry Eye WorkShop (2007). *Ocul Surf*. 2007;**5**:108–23.

55. Blackie CA, Solomon JD, Scaffidi RC, Greiner JV, Lemp MA, Korb DR. The relationship between dry eye symptoms and lipid layer thickness. *Cornea* 2009;**28**:789–94.

56. Joffre C, Souchier M, Gregoire S, Viau S, Bretillon L, Acar N, et al. Differences in meibomian fatty acid composition in patients with meibomian gland dysfunction and aqueous-deficient dry eye. *Br J Ophthalmol* 2008;**92**:116–9.

57. Shine WE, McCulley JP. Keratoconjunctivitis sicca associated with meibomian secretion polar lipid abnormality. *Arch Ophthalmol* 1998;**116**:849–52.

58. Oshima Y, Sato H, Zaghloul A, Foulks GN, Yappert MC, Borchman D. Characterization of human meibum lipid using Raman spectroscopy. *Curr Eye Res* 2009;**34**:824–35.

59. Järvinen RL, Larmo PS, Setala NL, Yang B, Engblom JRK, Viitanen MH, et al. Effects of oral sea buckthorn oil on tear film fatty acids in individuals with dry eye. *Cornea* 2011;**30**:1013–9.

60. Wojtowicz JC, Butovich I, Uchiyama E, Aronowicz J, Agee S, McCulley JP. Pilot, prospective, randomized, double-masked, placebo-controlled clinical trial of an omega-3 supplement for dry eye. *Cornea* 2011;**30**:308–14.

61. Yang B, Erkkola R. Sea buckthorn oils, mucous membranes and Sjögren's syndrome with special reference to latest studies.

In: Singh V, et al., editors. *Seabuckthorn (Hippophae L.). A Multipurpose Wonder Plant. Advances in Research and Development*, vol. III. New Delhi: Dya; 2006. pp. 254–67.

62. Kawasaki R, Tielsch JM, Wang JJ, Wong TY, Mitchell P, Tano Y, et al. The metabolic syndrome and retinal microvascular signs in a Japanese population: the Funagata study. *Br J Ophthalmol* 2008;**92**:161–6.

63. Eccleston C, Baoru Y, Tahvonen R, Kallio H, Rimbach GH, Minihane AM. Effects of an antioxidant-rich juice (sea buckthorn) on risk factors for coronary heart disease in humans. *J Nutr Biochem* 2002;**13**:346–54.

64. Lehtonen HM, Järvinen R, Linderborg K, Viitanen M, Venojärvi M, Alanko H, et al. Postprandial hyperglycemia and insulin response are affected by sea buckthorn (*Hippophaë rhamnoides* ssp. *turkestanica*) berry and its ethanol-soluble metabolites. *Eur J Clin Nutr* 2010;**64**:1465–71.

65. Linderborg KM, Lehtonen HM, Järvinen R, Viitanen M, Kallio H. The fibres and polyphenols in sea buckthorn (*Hippophaë rhamnoides*) extraction residues delay postprandial lipemia. *Int J Food Sci Nutr* 2012;**63**:483–90.

66. Pearson TA, Mensah GA, Alexander RW, Anderson JL, Cannon III RO, Criqui M, et al. Markers of inflammation and cardiovascular disease: application to clinical and public health practice: a statement for healthcare professionals from the Centers for Disease Control and Prevention and the American Heart Association. *Circulation* 2003;**107**:499–511.

67. Klein R, Klein BE, Knudtson MD, Wong TY, Tsai MY. Are inflammatory factors related to retinal vessel caliber? The Beaver Dam Eye Study. *Arch Ophthalmol* 2006;**124**:87–94.

Resveratrol and the Human Retina

Adela Mariana Pintea, Dumitriţa Olivia Rugină

University of Agricultural Sciences and Veterinary Medicine, Cluj-Napoca, Romania

INTRODUCTION

Resveratrol (3,5,4′-trihydroxystilbene) is a polyphenolic compound belonging to the class of stilbenes. The first report on resveratrol dates back to 1940, when the compound was isolated from the roots of white hellebore (*Veratrum grandiflorum* O. LOES). Resveratrol was later isolated from the roots of a medicinal plant, Japanese knotweed (*Polygonum cuspidatum*).[1] Stilbenes are phytoalexins produced by *Vitis vinifera* and other plants in response to injuries, stress, ultraviolet radiation, or fungal infection.[2] The presence of resveratrol in wine has led to the hypothesis that resveratrol is the active ingredient in wine's lipid-lowering and cardioprotective effects. High wine consumption may be responsible for the so-called 'French paradox', defined as the low death rates from coronary heart diseases despite a high intake of dietary cholesterol and saturated fat.[1] In recent decades, many *in vitro* and *in vivo* studies have demonstrated the ability of resveratrol to target intracellular molecules and processes. In addition to its cardioprotective effects, resveratrol may possess antioxidant, anti-inflammatory, antiaging, anticarcinogenic, and neuroprotective properties.[1,3] The specific mechanisms involved in the ability of resveratrol to protect against eye diseases are presented.

STRUCTURE AND PROPERTIES OF RESVERATROL AND ITS DERIVATIVES

Resveratrol is a white powder that is fat soluble and practically insoluble in water but exhibits high membrane permeability.[4] The compound exists in both the *cis* and *trans* forms, but the *trans* isomer is more abundant in nature and more biologically active. Piceid (resveratrol-3-*O*-beta-glucoside) is the major derivative found in grape juices, but several other compounds such as pterostilbene (a dimethoxylated analogue of resveratrol) and oxyresveratrol have been identified.[5] Both the skin of red grapes and red wines are rich sources of resveratrol (up to 14.3 mg/L), but the compound has also been found in dozens of plant species such as blueberries, bilberries, cranberries, and rhubarb root.[1] Some biologically active oligomers of resveratrol, including ε-viniferin (a dimer of resveratrol), vaticanol B (a tetramer of resveratrol), gnetin H, and suffruticosol, have also been isolated and characterized from other plant families[6,7] (Fig. 49.1).

BIOAVAILABILITY AND SAFETY OF RESVERATROL

Many *in vitro* and *in vivo* studies have demonstrated the beneficial effects of resveratrol, most of which emerged from cell cultures and animal models. Relatively few studies have been performed on human subjects, with few convincing results.[8,9] The discrepancies between the results obtained from *in vitro* studies and those obtained from animal or human experiments are likely to be related to the low bioavailability of resveratrol, which was extensively reviewed by Walle[9] and Cottart *et al.*[10] Resveratrol is easily absorbed in humans but rapidly metabolized into sulfo- or gluco-conjugates or hydrogenated derivatives, which, in turn, are excreted in urine. When ^{14}C-labeled resveratrol was administered as a single oral dose (25 mg = 110 μmol), the absorption was at least 70%, with the resveratrol and metabolite concentrations reaching the peak at approximately 2 μM after 1 hour. Only traces of unmetabolized free resveratrol could be found in plasma, at concentrations less than 5 ng/mL.[11] Owing to its lipophilic nature, resveratrol could be bound to the cellular fraction in blood, leading to an underestimation of its concentration.[10] Some important questions remain unsolved with respect to the bioavailability of resveratrol. Resveratrol conjugates, which reach significantly higher concentrations in plasma, could also be biologically active in tissues. However, resveratrol may also be released from

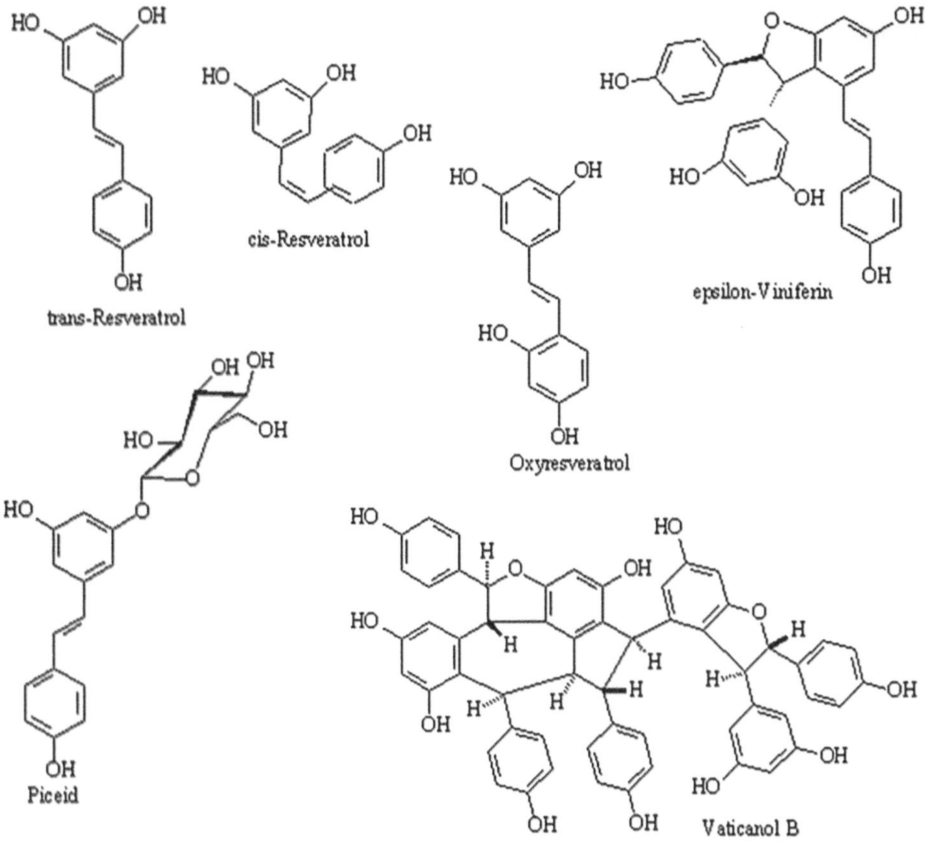

FIGURE 49.1 Chemical structure of resveratrol and of some of its derivatives. *Source: Original contribution.*

the glucuronides by tissue β-glucuronidases. Little is known about the accumulation of resveratrol in the tissues. Some evidence from animal studies (rats and mice) has indicated a preferential accumulation of resveratrol in liver, kidney, and intestinal cells. Cell culture studies have also shown a high accumulation level in Caco-2 colonic human cells.[9,10] The levels of resveratrol and resveratrol metabolites were measured in tissue samples obtained before biopsy and after surgery from patients with colorectal cancer who had been given daily supplements of 0.5–1 g resveratrol over 8 days prior to resection. The concentration of free resveratrol reached a maximum of 674 nmol/g tissue, reducing the tumor cell proliferation by 5%.[12]

The limited bioavailability of resveratrol could be solved using more bioavailable analogues, such as pterostilbene, 3,5,4'-trimethoxy-*trans*-stilbene; 3,4,5,4'-tetramethoxystilbene.[10] Increasing the solubility, stability, and release of resveratrol via improved formulations such as liposomes, β-cyclodextrin nanosponges, lipid-core nanocapsules, or calcium-pectinate beads could provide another solution.[4]

Most toxicity studies have been performed on animals, while few data are available for humans. Animal studies have indicated that resveratrol is well tolerated.[10] A 90-day study on rats and rabbits showed that high-purity resveratrol did not cause any adverse effects

at up to 700 mg/kg/day; it did not induce embryo–fetal toxicity and was nonirritating to the eyes and skin of the rabbits.[13] Several human clinical studies indicated that resveratrol is well tolerated, producing only mild adverse effects.[12] Most data in the literature have been obtained from *in vivo* studies, leading to the conclusion that no valid data exist on the toxicity of chronic resveratrol intake in human subjects.[8]

ANTIOXIDANT PROPERTIES OF RESVERATROL IN EXPERIMENTAL EYE DISEASE MODELS

Oxidative stress is defined as a serious imbalance between the production of reactive species and antioxidant defenses. The most important reactive oxygen species (ROS) produced by physiologic processes are superoxide anion, hydroxyl radical, hydrogen peroxide (H_2O_2), lipid peroxyl radicals, nitric oxide, and peroxynitrite. ROS can cause oxidative damage to lipids, proteins, and nucleic acids.[14] Enzymatic and nonenzymatic antioxidants are involved in protection against free radicals or oxidizing agents. The most important endogenous antioxidants at the retinal level are superoxide dismutase (SOD), catalase (CAT), peroxidase (GPx), glutathione (GSH), and tocopherols.[15]

The retina is a tissue with a high metabolism and a high rate of oxygen consumption: thus, maintaining the balance between the retinal oxygen supply and oxygen consumption is essential for retinal homeostasis.[15] Owing to the presence of polyunsaturated fatty acids (in the membranes of photoreceptors), photosensitizers (e.g., lipofuscin), and exposure to light, the retina is prone to the production of ROS.[15] The retina possesses endogenous enzymatic and nonenzymatic antioxidants as a mechanism for ROS detoxification, but this antioxidant defense decreases with age. The imbalance between the production and neutralization of ROS in the retina is important to the development of several eye diseases, such as age-related macular degeneration (AMD),[15] diabetic retinopathy,[16] retinitis pigmentosa,[17] glaucoma,[18–20] uveitis,[21] and cataract.[22]

During recent years, interest has significantly increased in using antioxidant phytochemicals for the prevention of or as complementary therapies for eye diseases. A recent extensive review described the results of human clinical trials and laboratory tests on the beneficial effects of resveratrol including its antioxidant potential.[23] A National Health Nutrition and Examination Survey has indicated that moderate wine consumption can be associated with a decreased risk of AMD development.[24]

An antioxidant is defined as any substance that delays, prevents, or removes oxidative damage to a target molecule.[14] Resveratrol is a nonflavonoid polyphenol that exhibits remarkable antioxidant properties. Resveratrol is an effective scavenger of hydroxyl, superoxide, and metal-induced radicals and exhibits a protective effect against lipid peroxidation and DNA damage by ROS.[1,25] The para-(4′)-hydroxyl group and the *trans*-configuration are structural determinants of the antioxidant activity of resveratrol.[26] In addition to its intrinsic antioxidant properties, the effect of resveratrol could be due to the upregulation of antioxidant enzymes.[1]

Because there are a large number of publications on the health benefits of resveratrol, this section focuses only on the antioxidant effects as they relate to the retina and eye diseases.

Diabetic retinopathy is the most severe ocular complication associated with diabetes and is a leading cause of acquired blindness in developed countries. Diabetes, characterized by sustained hyperglycemia, is a major risk factor for developing diabetic retinopathy, cataract, and glaucoma. The more severe form, proliferative diabetic retinopathy, affects up to 50% of patients with type 1 diabetes and approximately 10% of patients with type 2 diabetes who have had the disease for 15 years. The disease affects the blood vessels of the retina, first causing microaneurysms and intraretinal hemorrhaging, leading to neovascularization in the later stages. Macular edema also occurs and can cause a loss of central vision. The molecular mechanism of diabetic retinopathy is complex and involves several pathways. However, oxidative stress, nitric oxide, and vascular endothelial growth factor (VEGF) production play pivotal roles in the development of diabetic retinopathy. Oxidative stress induces the expression of VEGF, which in turn induces angiogenesis and neovascularization in the retina.[16]

Resveratrol (10 mg/kg/day) was administered for 4 weeks to Wistar albino male rats with streptozotocin-induced diabetes mellitus. The diabetic group showed increased nitrite–nitrate and protein carbonyl levels and decreased GSH concentrations, as evidence of oxidative stress. The expression of endothelial nitric oxide synthase (eNOS) messenger RNA (mRNA) was increased in the diabetic and resveratrol-treated groups but decreased in the diabetic rats supplemented with resveratrol.[27] Resveratrol was found to reduce both the oxidative stress (decreasing the level of lipid peroxidation and peroxynitrite) and DNA fragmentation in the diabetic neuropathy experimentally induced in the rats.[28]

Retinopathy of prematurity (ROP) is a complex disease of the immature retina in premature infants, with a higher risk for babies born weighing less than 1000 g. Resveratrol was tested for its potential protective effect via nitric oxide (NO)-modulating actions in an animal model of *in vivo* oxygen-induced retinopathy and in retinal primary cell cultures obtained from neonatal rats. The ROP was induced by exposing newborn Sprague–Dawley rats for 7 days to cyclic hyperoxia. The resveratrol was then injected intravitreally. In primary cell culture, the resveratrol was delivered to the cell culture (5 µg/mL), and the cells were exposed to hyperoxia for 6 hours. The NOS catalyzed the formation of NO by converting the L-arginine to L-citrulline. Under the conditions of these experiments (both *in vivo* and *in vitro*), the expression of the inducible nitric oxide synthase (iNOS) antibody and mRNA increased, while the expression of eNOS and neuronal nitric oxide synthase (nNOS) decreased in the resveratrol-treated group.[29]

Glaucoma is a major cause of irreversible blindness, affecting more than 70 million people.[18] Glaucoma is characterized by the accelerated death of the retinal ganglion cells (RGCs) and their axons, leading to progressive visual field loss and eventual blindness.

Elevated intraocular pressure (IOP) is a major risk factor for glaucoma, but other risk factors such as age, ethnicity, and oxidative stress are also involved. Human aqueous humor samples obtained during surgery showed a decreased total reactive antioxidant potential and elevated activities of SOD, GPx, and malondialdehyde (MDA).[30] However, more recent studies have found a reduced expression of SOD, glutamine synthase (GS), glutathione-*S*-transferase (GST), and iNOS in the aqueous humor of patients with glaucoma.[31]

A transformed rat retinal ganglion cell line (RGC-5) was exposed to hydrostatic pressure (0, 30, 60, and 100 mm Hg over atmospheric pressure) for 2 hours, conditions that mimic those observed in glaucoma patients with elevated IOP. Increased levels were found in the adduct 4-hydroxy-2-nonenal (HNE) proteins and heme oxygenase (HO-1) expression.[18] Similar results were obtained from the same experiment on retina samples procured from mice and exposed for 1 hour to the same hydrostatic pressure. The HNE is produced via lipid peroxidation in response to oxidative stress and forms irreversible adducts with proteins but can also induce cell death by apoptosis. Both HNE–protein adducts and HO-1 levels are considered good biomarkers for oxidative stress conditions. The pretreatment of RGCs with resveratrol at 60 mm Hg was better at reducing the HNE–protein adducts than quercetin. These results demonstrated the involvement of oxidative stress as an early event in the IOP-induced neuronal damage and the protective role of resveratrol.[18] Chemical ischemia induced with iodoacetic acid was used as a model for retinal ischemia in the same cell line, RGC-5. In this model, resveratrol significantly increased the cell viability when administered during both ischemia and recovery but provided modest protection when administered only during recovery.[19]

Trabecular meshwork (TM) is a specialized tissue responsible for draining most of the aqueous humor from the anterior chamber of the eye, thereby controlling the IOP. The TM cells are essential in maintaining the outflow system homeostasis, and their damage is involved in the pathogenesis of glaucoma. Oxidative DNA damage to the TM in patients with glaucoma was found to correlate significantly with visual field damage and intraocular pressure.[32] To evaluate the role of oxidative stress, cultured primary porcine TM cells were exposed to 40% oxygen (O_2) and then treated with resveratrol (15 days). The resveratrol was associated with a significant (five-fold) decrease in the intracellular ROS production during chronic exposure at a level comparable to that of unstressed cells (5% O_2).[33]

AMD is the leading cause of blindness in patients over 65 years of age in developed countries.[15,34] It affects the macula at the center of the eye, resulting in the loss of central vision. Two forms of AMD exist, dry and wet. The dry form (atrophic) affects the largest number of subjects (85%) and is characterized by elevated autofluorescence due to the formation of lipofuscin and the accumulation of drusen between the retinal pigment epithelium (RPE) and choroid. This accumulation event results in RPE cell death, which in turn determines the degeneration of the overlying photoreceptors leading to atrophy of the choroidal capillaries.[15,34] The wet form of AMD is the most severe and is characterized by subretinal neovascularization, which produces rapid and severe

vision loss. The dry form of AMD can progress into the wet form because of the impairment of VEGF diffusion from RPE. Currently, the wet form of AMD is treated with antiangiogenic agents,[35] but no effective therapy exists for the dry form. The multifactorial pathogenesis of AMD involves genetic, metabolic, and environmental factors. However, evidence that the oxidative damage to the RPE cells is primarily responsible is increasing.[15] Mitochondrial dysfunction, phagocytosis, nicotinamide adenine dinucleotide phosphate (NADPH) oxidase, and the presence of photosensitizers are the principal cellular sources of ROS in the retina; in addition, moderate smoking, radiation, and drugs represent external sources of ROS.[15]

Pyrydinium bysretinoid (A2E) is a major lipofuscin fluorophore that accumulates in the RPE cells with aging and results from light-related vitamin A cycling in the retina. Under illumination with blue light, A2E generates singlet oxygen, which in turn oxidizes the A2E to A2E epoxides. The resveratrol treatment of A2E-laden RPE cells was found to inhibit the epoxidation of A2E in irradiated cells through the quenching of the singlet oxygen. In the presence of resveratrol, only three of the seven A2E epoxides were formed. When A2E was exposed to singlet oxygen generated by the decomposition of the endoperoxide of 1,4-dimethylnaphtalene, resveratrol reduced the A2E loss from 50 to 5%.[36]

In another model, the RPE cells were exposed to H_2O_2 to induce oxidative stress. Resveratrol treatment reduced the intracellular ROS generation by 20% for cells in basal conditions and 15% for cells exposed to H_2O_2. In addition, resveratrol treatment reduced the RPE proliferation in a dose-dependent manner, independent of the resveratrol-induced cytotoxicity.[37] Resveratrol also demonstrated a protective effect against H_2O_2-induced oxidative stress at lower concentrations (25 and 50 μM) without exhibiting a cytotoxic effect up to 100 μM. In addition to the direct antioxidant effects of quenching the intracellular ROS, resveratrol increased the activity of the oxidant enzymes – SOD, GPx, and CAT – and the concentration of the essential endogenous antioxidant GSH.[38] Acrolein, a component of cigarette smoke, inhibited the phagocytic function of RPE cells in acute and chronic exposures. Pretreatment with resveratrol alleviated the inhibition of phagocytosis, acting synergistically with zeaxanthin and meclofenamic acid.[39] Resveratrol reversed the production of ROS and reactive nitrogen species (RNS) in RPE treated with another component of cigarette smoke: benzo(e)pyrene.[40] The β-integrin knockout mice lack the diurnal burst of RPE phagocytosis and lose visual function owing to the accumulation of lipofuscin. The effects of the oxidative stress were demonstrated in the high level of HNE–protein adducts in the RPE cells of β5$^{-/-}$ mice. A diet rich in antioxidants (grapes or marigold extracts) administered for up to 13 months prevented RPE oxidation, cytoskeletal damage, and vision loss, suggesting that the loss of photoreceptors is due to oxidative damage to the RPE.[41] One case study reported that daily supplementation for 5 months with a polyphenolic mixture containing resveratrol improved the visual function and caused a visible clearing of RPE lipofuscin in an 80-year-old man.[42]

Cataract development is accelerated by ROS generation and the subsequent damage to the epithelial and fiber cells of the lens.[34] Resveratrol was found to reduce the oxidative stress (increased levels of GSH and decreased levels of lipid peroxidation) and cataract formation in a sodium selenite-induced experimental cataract model in rats.[43] Resveratrol pretreatment improved the cell survival rate and inhibited the apoptosis of HLEB-3 lens epithelial cells exposed to H_2O_2. In addition, the expression of the antioxidant enzymes SOD, CAT, and HO-1 was upregulated, and the ROS generation was reduced after cell treatment with resveratrol.[44] Resveratrol treatment together with grape seed proanthocyanidin extract reduced the level of ROS generation in a primary canine epithelial cell line.[45] Primary human and porcine lens epithelial cells exposed to acute H_2O_2 (600 μM) oxidative stress under normal and chronic hyperoxic 40% oxygen conditions were protected by resveratrol (25 μM daily for 8 days), which significantly reduced (69%) the intracellular ROS production and increased the mitochondrial membrane potential.[46] The protective effect of resveratrol in this model appears to be mediated by the upregulation of forkhead box-O transcription factors (FoxOs), known for promoting oxidative stress resistance.

Endotoxin-induced uveitis (EIU) is an established model for ocular inflammation exhibiting elevated oxidative biomarkers. Resveratrol was tested as an oral supplement (5–200 mg/kg) on an EIU model in male mice (C57/B6). The antioxidant protection was demonstrated by the inhibition of 8-hydroxy-2′-deoxyguanosine (8-OHdG) generation and by the suppression of nuclear factor-κB (NF-κB) p65 translocation[47] (Tables 49.1 and 49.2).

PROAPOPTOTIC AND ANTIAPOPTOTIC PROPERTIES OF RESVERATROL FROM *IN VITRO* EXPERIMENTS

Retinoblastoma is the most common intraocular malignancy in infants and children and is characterized by the loss or mutation of both alleles in the retinoblastoma tumor-suppressor gene (*RB1*). Anticancer agents arrest tumor proliferation, modulating the cell cycle and leading to apoptosis. Apoptosis is a genetically programmed mechanism whose final event is the cell committing suicide. The extrinsic pathway of the apoptotic process is initiated by activating death receptors TNFR, Fas, and TRAIL from the cell membrane, causing the activation of initiator caspases (caspases-8 and -10) and then the activation of an effector caspase (caspase-3). Caspase-3, which is responsible for cleavage of some death substrates, leads to the characteristic hallmarks of an apoptotic cell including DNA fragmentation, nuclear fragmentation, and membrane blebbing.[48] The intrinsic pathway involves mitochondria-dependent processes that result in the release of cytochrome *c* from the mitochondria and the activation of caspase-9. In tumor cells, resveratrol prevents cell proliferation and induces apoptosis,[49] but normal cells are protected by resveratrol from apoptotic induced death.[50] In human Y79 retinoblastoma cells, resveratrol triggered the release of cytochrome *c* from the mitochondria after 24 hours of treatment. Forty-eight hours following resveratrol addition, an accumulation of cytochrome *c* was observed in the cytosolic fraction. Resveratrol treatment also induced the activation of caspase-9 and caspase-3. Hence, the resveratrol inhibited the Y79 cell proliferation, induced an S-phase growth arrest, and stimulated apoptosis through the activation of a mitochondrial intrinsic apoptotic pathway, indicating that resveratrol could serve as adjuvant to conventional anticancer therapies for retinoblastoma.[49] Resveratrol suppressed Bax in the cytoplasm and therefore protected the normal retina cells E1A-NR.3 from apoptotic death induced by antiretinal antibodies[50] (Fig. 49.2).

ANTIPROLIFERATIVE POTENTIAL OF RESVERATROL *IN VITRO* EXPERIMENTS

The mitogen-activated protein kinases (MAPKs) are serine/threonine protein kinases that follow a signal transduction pathway from the cell membrane to the nucleus. The MAPKs regulate a variety of cellular functions from cell proliferation to cell death. They are activated through a cascade of phosphorylation events. The MAPK kinase (MAP3K) is the first to be activated and is responsible for the phosphorylation and activation of the MAPK kinase (MAP2K), which in turn stimulates MAPK activation.[51] The MAPK subfamilies include three major enzymes such as extracellular signal-regulated kinases (ERKs, including the ERK-1 and ERK-2 isoforms), p38 (including p38-α, p38-β, p38-γ, and p38-δ isoforms), and c-Jun N-terminal kinase (JNKs, including JNK-1, JNK-2, and JNK-3 isoforms), which is also involved in cell growth and death. The ERK cascade activated by MAP/ERK kinase (MEK) is activated by Raf and is the most intensely studied of the MAPK pathways. Raf, which is an MAP3K, is indirectly activated by the epidermal growth factor (EGF) receptor.[52] Resveratrol reduced the H_2O_2-induced activation of both ERK-1/2 and its upstream activator, MEK, thereby inhibiting the RPE cell proliferation[37] (Fig. 49.2).

TABLE 49.1 Effect of Resveratrol on Oxidative Stress-Induced Models of Eye Diseases: *In Vitro* Evidence

Cell Type	Inductor of Oxidative Stress	Resveratrol Dose	Effect of Resveratrol	Reference
RGC (rat)	Hydrostatic pressure	20, 40 µM	↓ HNE–protein adducts	[18]
RGC (rat)	Chemical ischemia (iodoacetic acid)	25 µM	↑ Cell survival	[19]
TM (primary culture porcine)	Oxygen (40%)	25 µM	↓ ROS generation	[33]
			↓ Carbonylated proteins	
			↓ Autofluorescence	
			↓ Inflammatory markers	
RPE cells (human)	A2E and light	100 µM	↓ A2E epoxidation	[36]
	Singlet oxygen	40 mM	↓ Singlet oxygen	
RPE cells (human)	H_2O_2	100 µM	↓ ROS generation	[37]
RPE cells (human)	Benzo(e)pyrene	30 µM	↓ ROS and RNS generation	[40]
RPE cells (human)	H_2O_2	25–100 µM	↑ SOD, Cat, GPx, GSH	[38]
			↓ ROS generation	
RPE cells (primary culture β5$^{-/-}$ mice)	HNE	30 µM	No effect on: HNE adduct formation	[41]
			HNE-induced destabilization of actin	
LEC (primary culture canine)	*tert*-Butyl-hydroperoxide	100 mM	↓ ROS generation	[45]
LEC (human)	H_2O_2	2.5–20 µM	↑ SOD, CAT, HO-1	[44]
			↓ ROS	
LEC (primary culture human and porcine)	Hyperoxia and H_2O_2	25 µM	↓ ROS generation	[46]
			↑ Mitochondrial membrane potential	
Retinal primary cells (rat)	Hyperoxia	5 µg/mL	↑ iNOS	[29]
			↓ eNOS and nNOS	

RGC: retinal ganglion cell; TM: trabecular meshwork; RPE: retinal pigment epithelium; LEC: lens epithelial cell; A2E: pyrydinium bysretinoid; H_2O_2: hydrogen peroxide; HNE: 4-hydroxy-2-nonenal; ROS: reactive oxygen species; RNS: reactive nitrogen species; SOD: superoxide dismutase; Cat: catalase; GPx: glutathione peroxidase; GSH: glutathione *S*-transferase; HO-1: heme oxygenase-1.

ANTIANGIOGENIC PROPERTIES OF RESVERATROL IN EXPERIMENTAL RETINA MODELS

Angiogenesis is the process of new blood vessel formation. The process is not only important for the normal development of the human body but also occurs in tumorigenesis, inflammatory disorders, and ocular neovascularization diseases. One of the most important cytokines in the promotion of angiogenesis is VEGF.[53] The VEGF human proteins contain five members – VEGF-A, VEGF-B, VEGF-C, VEGF-D, and placental growth factor (PlGF), which belong to the heparin-binding growth factor family. The most intensely studied thus far, the VEGF-A protein, is secreted in response to inflammatory stimuli by different

retinal cells, such as glial cells, vascular endothelial cells, RPE cells, and ganglion cells, which modulate the development and growth of blood vessels.[54] The VEGF-A isoform differs in its interactions with the receptor and in its function, binding to the VEGFR1 and VEGFR2 receptors. The binding of VEGF-A to the extracellular portion of the receptor induce the activation of the intracellular signaling proteins, such as protein kinase C (PKC), VEGF receptor-associated protein (VRAP), and MAPK. The VEGF coreceptors neuropilin-1 and -2 (NRP-1, NRP-2) and the heparin sulfate proteoglycans (HSPGs) exist near the receptors, on the cell surface, and are responsible for the enhanced binding and a prolonged signal transduction.[55]

Resveratrol can protect the RPE cells against hyperglycemia-induced conditions by inhibiting the inflammation

TABLE 49.2 Effect of Resveratrol on Oxidative Stress-Induced Models of Eye Diseases: *In Vivo* Evidence

Animal Model	Inductor of Oxidative Stress	Resveratrol Dose	Effect of Resveratrol	Reference
Rats (lenses)	Na-selenite	40 mg/kg	↑ GSH-l	43
			↓ MDA	
Rats	Diabetes neuropathy (streptozotocin)	10–20 mg/kg	↓ MDA in plasma	28
			↑ CAT	
Male mice (C57/B6)	Intraperitoneal injection of lipopolysaccharide	5–200 mg/kg	↓ 8-OHdG	47
Neonatal rats	Hyperoxia	30 mg/kg	↑ iNOS (retina)	29
			↓ eNOS	
			↓ nNOS	
Wistar albino rats	Diabetes (streptozotocin)	10 mg/kg	↓ eNOS	27
β5⁻/⁻ mice (RPE cells)	Impaired phagocytosis in RPE cells	6.7 g fresh grapes/kg	↓ Formation of HNE adducts	41
	Accumulation of lipofuscin		↓ Lipofuscin accumulation	
			↓ Loss of photoreceptors	

RPE: retinal pigment epithelium; GSH: glutathione *S*-transferase-1; MDA: malondialdehyde; CAT: catalase; 8-OHdG: 8-hydroxy-2′-deoxyguanosine; iNOS: inducible nitric oxide synthase; eNOS: endothelial nitric oxide synthase; HNE: 4-hydroxy-2-nonenal.

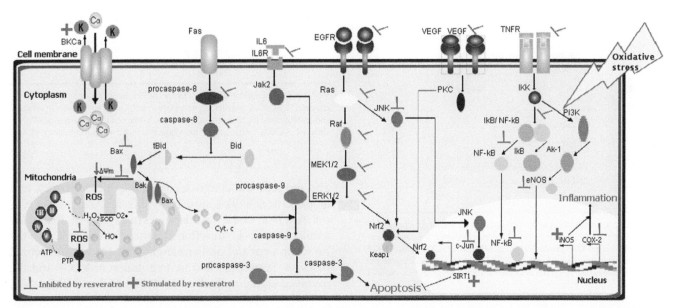

FIGURE 49.2 A proposed molecular mechanism of resveratrol protection against cellular stress induced in normal retinal cells. Ak-1: adenylate kinase-1; ATP: adenosine-5′-triphosphate; BK_{Ca}: large-conductance calcium-activated potassium; COX-2: cyclooxygenase-2; Cyt. c: cytochrome *c*; EGFR: epidermal growth factor receptor; eNOS: endothelial nitric oxide synthase; ERK: extracellular signal-regulated kinase; H_2O_2: hydrogen peroxide; IKK: IκB kinase; IL6: interleukin-6; IL6R: interleukin-6 receptor; iNOS: inducible nitric oxide synthase; Jak2: janus kinase-2; JNK: c-Jun N-terminal kinase; Keap1: Kelch-like ECH-associated protein-1; MEK: MAP/ERK kinase; NF-kB: nuclear factor-κB; Nrf2: nuclear factor erythroid 2-related factor-2; PI3K: phosphatidylinositol 3-kinase; PKC: protein kinase C; PTP: protein tyrosine phosphatase; ROS: reactive oxygen species; SIRT1: sirtuin-1; SOD: superoxide dismutase; TNFR: tumor necrosis factor receptor; VEGF: vascular endothelial growth factor. *Source: Original contribution.*

biomarkers interleukin-6 (IL-6) and interleukin-8 (IL-8), by transforming growth factor-β_1 (TGF-β_1), cyclooxygenase-2 (COX-2), and through VEGF accumulation.[56] Chronic treatment with resveratrol induced the mRNA expression of the inflammatory markers IL-1α, IL-6, IL-8, and endothelium leukocyte adhesion molecule-1 (ELAM-1) under oxidative stress conditions in primary porcine TM cells. No significant effects of the resveratrol treatment were identified on DNA damage, cell proliferation, or proteasomal activity.[33]

Angiotensin II (Ang II) stimulates the formation of new retinal blood vessels by VEGF-induced angiogenic

TABLE 49.3 Molecular Targets of Resveratrol in Retinal Cellular Signaling Pathways: *In Vitro* Evidence

Cell Line	Molecular Target	Resveratrol Dose	References
ARPE-19	↓ VEGF	10 μM	[56]
	↓ TGF-β₁		
	↓ COX-2		
	↓ IL-6		
	↓ IL-8		
	↑ PKCβ		
ARPE-19	↓ MEK	100 μmol/L	[37]
	↓ ERK-1/2		
Retinal stem cells	↑ SIRT1 mRNA levels	15 μM	[61]
Y79 retinoblastoma (human)	↑ Cytochrome *c*	50, 100 μM	[49]
	↑ Caspase 3, 9		
	↓ Mitochondrial membrane potential		
RPE R-50	Open BK$_{Ca}$ channels	10 μM/20 min	[62,63]
	↓ Oxidative damage		
RPE R-50	↑ Inhibitory effects on acrolein	10 μM	[39]
E1A.NR3 retinal cells	↓ Bax	40 μM	[50]
	↓ Caspase 3		
	↑ SIRT1		
ARPE-19	↓ Oxysterol-induced cell death	1 μM	[60]
	↓ VEGF		

RPE: retinal pigment epithelium; VEGF: vascular endothelial growth factor; TGF-β₁: transforming growth factor β₁; COX-2: cyclooxygenase-2; IL-6: interleukin-6; IL-8: interleukin-8; PKCβ: protein kinase Cβ; MEK: MAP/ERK kinase; ERK-1/2: extracellular signal-regulated kinase-1/2; SIRT1: sirtuin-1; mRNA: messenger RNA; BK$_{Ca}$ channel: large-conductance calcium-activated potassium channel.

TABLE 49.4 Molecular Targets of Resveratrol in Retinal Signaling Pathways: Evidence from Animal Models

Animals	Molecular Target/ Biologic Event	Resveratrol Dose	References
Mice	↓ Corneal vascularization	0.4 μg/ml	[64]
	↓ Vessel density		
Wistar albino rats (streptozotocin-induced diabetes)	↓ VEGF	10 mg/kg	[27]
	↓ ACE		
	↓ MMP-9 mRNA levels		
	↓ eNOS		
	↓ NO		
	↑ GSH		
C57BL/6 mice (streptozotocin-induced diabetes)	↓ Increased vessel leakage	20 mg/kg	[59]
	↓ Pericyte loss		
	↓ VEGF		
C57BL/6 mice	↑ SIRT1 activity	5–200 mg/kg	[47,65]
	↓ 8-OHdG generation		
	↓ NF-κB		

VEGF: vascular endothelial growth factor; ACE: angiotensin-converting enzyme; MMP-9: matrix metalloproteinase-9; mRNA: messenger RNA; eNOS: endothelial nitric oxide synthase; NO: nitric oxide; GSH: glutathione; SIRT1: sirtuin-1; 8-OHdG: 8-hydroxy-2′-deoxyguanosine; NF-κB: nuclear factor-κB.

and reduced the cytotoxic, oxidative, inflammatory, and angiogenic activities induced by oxysterols in human retinal ARPE-19 cells.[60] Antiangiogenic synthetic drugs are available only as injections. Hence, resveratrol fills the need for a natural oral antiangiogenic therapy for retinal diseases, targeting the downregulation of VEGF and the involved signaling pathways (Tables 49.3 and 49.4).

RESVERATROL AS A SIRTUIN ACTIVATOR

Sirtuins (silent information regulator two proteins) in mammals are a group of seven proteins containing nicotinamide adenine dinucleotide (NAD$^+$)-dependent deacetylases, which catalyze the deacetylation of histones and nonhistone proteins (e.g., p53, peroxisome proliferator-activated receptor-γ (PPARγ), NF-κB, or FoxOs). Sirtuins are involved in the cell cycle, the regulation of apoptosis, DNA repair, the regulation of energy

activity in the retina.[57] Ang I is converted to Ang II by angiotensin-converting enzyme (ACE), which induces a vascular contraction.[57] The inhibition of ACE was found to reduce the accumulation of VEGF mRNA in diabetic rat retinas.[58] Moreover, the mRNA expression of ACE was decreased in resveratrol-treated diabetic rats compared with that in nontreated diabetic rats.[27] Resveratrol decreased the VEGF vascular lesions and the VEGF induction in the retinas of C57BL/6 mice with early diabetes over those in nondiabetic mice.[59] Resveratrol inhibited the expression of VEGF in human retinal vascular endothelial cells induced by CoCl$_2$ in a dose-dependent manner[53]

metabolism, and several other processes.[1,66] Resveratrol is known to be a potent activator of sirtuins. The SIRT1 activating compounds promote longevity in different species and provide protective effects against acute or chronic neurodegenerative diseases, including retinal injury.[1,66] The products of lipid oxidation (aldehydes, e.g., 4-hydroxynonenal) and the oxidants inhibit sirtuins by reacting with cysteine residues.[66]

Sirtuins, especially SIRT1, are highly expressed in the central nervous system, and a SIRT-1 deficiency in mutant mice is associated with developmental defects and abnormal retinal histology.[67] In addition, in rd10 mice models of retinitis pigmentosa, SIRT1 exhibited an antiapoptotic and neuroprotective effect resulting from its involvement both in double DNA strand-break repair mechanisms and in maintaining energy homeostasis in the photoreceptor cells.[68] The RGC loss observed in the experimental optic neuritis induced by immunization was inhibited by intravitreal injections of two SIRT1 activators, one of which is a formulation for resveratrol.[69] The antiapoptotic and protective effects of resveratrol on the antibody-induced apoptosis of retinal cells were modulated through the upregulation of SIRT1 and Ku70 (a DNA repair protein present in the nucleus in its deacetylated form). Pretreatment with resveratrol in E1A.NR3 cells yielded a five-fold increase in SIRT1 expression.[50] The efficacy of resveratrol as a SIRT1 activator in the RPE choroid was demonstrated in mice with EIU.[47] Resveratrol increased the viability of rat retinal stem cells, protecting them against induced oxidative stress via the activation of SIRT1.[61] In the same experiment, resveratrol promoted the activity of SIRT1 and enhanced the self-renewal abilities to increase the number of neurospheres in the retinal stem cells, especially in older rats (12 months). Recently, resveratrol demonstrated neuroprotective properties by increasing the number of RGCs in an animal model of multiple sclerosis (experimental autoimmune encephalomyelitis). Pure resveratrol (100 and 250 mg/kg) and a pharmaceutical formulation (250 mg/kg) were compared by administering an oral gavage for 30 days. Resveratrol did not prevent or reduce the inflammation in the spinal cords or optic nerves but prevented neuronal loss and delayed the visual decline[70] (Tables 49.3 and 49.4).

RESVERATROL AS A LARGE-CONDUCTANCE CALCIUM-ACTIVATED POTASSIUM CHANNEL MODULATOR

Large-conductance calcium-activated potassium channels (BK_{Ca} channels) are found in RPE cells and are responsible for the dark adaptation of photoreceptor activity, transepithelial transport, phagocytosis, growth factor secretion, and differentiation.[71] At the cellular level, BK_{Ca} channels provide K^+ transport and control Ca^{2+} influx and Ca^{2+}-dependent physiologic processes.[63] A cell membrane depolarization or an increase in the intracellular free Ca^{2+} concentration can open the BK_{Ca} channels.[72] The increase in intracellular Ca^{2+} concentrations is determined by adenosine-5′-triphosphate (ATP), through the activation of purinergic receptors, in the RPE cells.[73] In human RPE cells, the BK_{Ca} channel activity could be suppressed by oxidizing agents,[74] but resveratrol could have a reverse action, as a BK_{Ca} opener.[75]

The mechanisms of ultraviolet irradiation and H_2O_2 damage on RPE phagocytosis may involve BK_{Ca} channel activation by meclofenamic acid and resveratrol.[62,63]

Therefore, resveratrol is a potential pharmacologic agent with an ability to modulate the BK_{Ca} channel activity and could be a target for the treatment or prevention of ocular diseases (Fig. 49.2).

TAKE-HOME MESSAGES

- Resveratrol is a polyphenolic compound belonging to the class of stilbenes. The best sources of resveratrol are red grapes and red wines, but other plants including peanuts, blueberries, and cranberries contain significant quantities.
- Resveratrol is well absorbed in humans, but it is rapidly metabolized. Hence, the plasma concentration of the free, unmetabolized form is low. Resveratrol is generally well tolerated in humans, but further data are needed on the toxicity of chronic intake.
- In recent years, a large number of *in vitro* and *in vivo* studies have demonstrated the protective effects of resveratrol on several pathologies including eye diseases.
- Oxidative stress is involved in several eye diseases such as age-related macular degeneration (AMD), glaucoma, diabetic retinopathy, and cataract. Resveratrol acts as an antioxidant by directly quenching free radicals, enhancing the activity of antioxidant enzymes, upregulating forkhead box O transcription factors, or acting as a sirtuin activator.
- The antiangiogenic potential of resveratrol, demonstrated by the inhibition of inflammatory markers such as vascular endothelial growth factor, can be exploited to prevent neovascularization in diabetic retinopathy or wet AMD.
- Resveratrol is an inhibitor of mitogen-activated protein kinase pathways and modulates the large-conductance calcium-activated potassium channels.
- Resveratrol exhibits an antiapoptotic effect on normal retinal pigment epithelial cells but acts as a proapoptotic factor in retinoblastoma cells.
- The ability of resveratrol to target many intracellular molecules and processes renders it applicable to clinical studies on eye diseases.

References

1. Baur JA, Sinclair DA. Therapeutic potential of resveratrol: the in vivo evidence. Nat Rev Drug Discov 2006;5:493–506.

2. Langcake P, Pryce RJ. The production of resveratrol by Vitis vinifera and other members of the Vitaceae as a response to infection or injury. Physiol Plant Pathol 1976;9:77–86.

3. Marques FZ, Markus MA, Morris BJ. Resveratrol: cellular actions of a potent natural chemical that confers a diversity of health benefits. Int J Biochem Cell Biol 2009;41:2125–8.

4. Amri A, Chaumeil JC, Sfar S, Charrueau C. Administration of resveratrol: what formulation solutions to bioavailability limitations? J Control Release 2012;158:182–93.

5. Romero-Pérez AI, Ibern-Gómez M, Lamuela-Raventós RM, de la Torre-Boronat MC. Piceid, the major resveratrol derivative in grape juices. J Agric Food Chem 1999;47:1533–6.

6. Kim HJ, Chang EJ, Cho SH, Chung SK, Park HD, Choi SW. Antioxidative activity of resveratrol and its derivatives isolated from seeds of Paeonia lactiflora. Biosci Biotechnol Biochem 2002;66:1990–3.

7. Tabata Y, Takano K, Ito T, Iinuma M, Yoshimoto T, Miura H, et al. Vaticanol B, a resveratrol tetramer, regulates endoplasmic reticulum stress and inflammation. Am J Physiol Cell Physiol 2007;293:C411–8.

8. Vang O, Ahmad N, Baile CA, Baur JA, Brown K, Csiszar A, et al. What is new for an old molecule? Systematic review and recommendations on the use of resveratrol. PLoS ONE 2011;6:e19881.

9. Walle T. Bioavailability of resveratrol. Ann N Y Acad Sci 2011;1215:9–15.

10. Cottart CH, Nivet-Antoine V, Laguillier-Morizot C, Beaudeux JL. Resveratrol bioavailability and toxicity in humans. Mol Nutr Food Res 2010;54:7–16.

11. Walle T, Hsieh F, DeLegge MH, Oatis Jr JE, Walle UK. High absorption but very low bioavailability of oral resveratrol in humans. Drug Metab Dispos 2004;32:1377–82.

12. Patel KR, Brown VA, Jones DJ, Britton RG, Hemingway D, Miller AS, et al. Clinical pharmacology of resveratrol and its metabolites in colorectal cancer patients. Cancer Res 2010;70:7392–9.

13. Williams LD, Burdock GA, Edwards JA, Beck M, Bausch J. Safety studies conducted on high-purity trans-resveratrol in experimental animals. Food Chem Toxicol 2009;47:2170–82.

14. Halliwell B, Gutteridge JM. Free Radicals in Biology and Medicine. New York: Oxford University Press; 1999.

15. Jarrett SG, Boulton ME. Consequences of oxidative stress in age-related macular degeneration. Mol Aspects Med 2012;33:399–417.

16. Kowluru RA, Chan PS. Oxidative stress and diabetic retinopathy. Exp Diabetes Res 2007;2007:43603.

17. Obolensky A, Berenshtein E, Lederman M, Bulvik B, Alper-Pinus R, Yaul R, et al. Zinc-desferrioxamine attenuates retinal degeneration in the rd10 mouse model of retinitis pigmentosa. Free Radic Biol Med 2011;51:1482–91.

18. Liu Q, Ju WK, Crowston JG, Xie F, Perry G, Smith MA, et al. Oxidative stress is an early event in hydrostatic pressure induced retinal ganglion cell damage. Invest Ophthalmol Visual Sci 2007;48:4580–9.

19. Maher P, Hanneken A. Flavonoids protect retinal ganglion cells from ischemia in vitro. Exp Eye Res 2008;86:366–74.

20. Tezel G. Oxidative stress in glaucomatous neurodegeneration: mechanisms and consequences. Prog Retin Eye Res 2006;25:490–513.

21. Ohia SE, Opere CA, Leday AM. Pharmacological consequences of oxidative stress in ocular tissues. Mutat Res 2005;579:22–36.

22. Beebe DC, Holekamp NM, Shui YB. Oxidative damage and the prevention of age-related cataracts. Ophthalmic Res 2010;44:155–65.

23. Smoliga JM, Baur JA, Hausenblas HA. Resveratrol and health – a comprehensive review of human clinical trials. Mol Nutr Food Res 2011;55:1129–41.

24. Obisesan TO, Hirsch R, Kosoko O, Carlson L, Parrott M. Moderate wine consumption is associated with decreased odds of developing age-related macular degeneration in NHANES-1. J Am Geriatr Soc 1998;46:1–7.

25. Leonard SS, Xia C, Jiang BH, Stinefelt B, Klandorf H, Harris GK, Shi X. Resveratrol scavenges reactive oxygen species and effects radical-induced cellular responses. Biochem Biophys Res Commun 2003;309:1017–26.

26. Cheng JC, Fang JG, Chen WF, Zhou B, Yang L, Liu ZL. Structure–activity relationship studies of resveratrol and its analogues by the reaction kinetics of low density lipoprotein peroxidation. Bioorg Chem 2006;34:142–57.

27. Yar AS, Menevse S, Dogan I, Alp E, Ergin V, Cumaoglu A, et al. Investigation of ocular neovascularization-related genes and oxidative stress in diabetic rat eye tissues after resveratrol treatment. J Med Food 2012;15:391–8.

28. Kumar A, Kaundal RK, Iyer S, Sharma SS. Effects of resveratrol on nerve functions, oxidative stress and DNA fragmentation in experimental diabetic neuropathy. Life Sci 2007;80:1236–44.

29. Kim WT, Suh ES. Retinal protective effects of resveratrol via modulation of nitric oxide synthase on oxygen-induced retinopathy. Korean J Ophthalmol 2010;24:108–18.

30. Ghanem AA, Arafa LF, El-Baz A. Oxidative stress markers in patients with primary open-angle glaucoma. Curr Eye Res 2010;35:295–301.

31. Bagnis A, Izzotti A, Centofanti M, Sacca SC. Aqueous humor oxidative stress proteomic levels in primary open angle glaucoma. Exp Eye Res 2012;103:55–62.

32. Sacca SC, Pascotto A, Camicione P, Capris P, Izzotti A. Oxidative DNA damage in the human trabecular meshwork: clinical correlation in patients with primary open-angle glaucoma. Arch Ophthalmol 2005;123:458–63.

33. Luna C, Li G, Liton PB, Qiu J, Epstein DL, Challa P, Gonzalez P. Resveratrol prevents the expression of glaucoma markers induced by chronic oxidative stress in trabecular meshwork cells. Food Chem Toxicol 2009;47:198–204.

34. Fletcher AE. Free radicals, antioxidants and eye diseases: evidence from epidemiological studies on cataract and age-related macular degeneration. Ophthalmic Res 2010;44:191–8.

35. Andreoli CM, Miller JW. Anti-vascular endothelial growth factor therapy for ocular neovascular disease. Curr Opin Ophthalmol 2007;18:502–8.

36. Sparrow JR, Vollmer-Snarr HR, Zhou J, Jang YP, Jockusch S, Itagaki Y, Nakanishi K. A2E-epoxides damage DNA in retinal pigment epithelial cells. Vitamin E and other antioxidants inhibit A2E-epoxide formation. J Biol Chem 2003;278:18207–13.

37. King RE, Kent KD, Bomser JA. Resveratrol reduces oxidation and proliferation of human retinal pigment epithelial cells via extracellular signal-regulated kinase inhibition. Chem Biol Interact 2005;151:143–9.

38. Pintea A, Rugina D, Pop R, Bunea A, Socaciu C, Diehl HA. Antioxidant effect of trans-resveratrol in cultured human retinal pigment epithelial cells. J Ocul Pharmacol Ther 2011;27:315–21.

39. Sheu SJ, Liu NC, Chen JL. Resveratrol protects human retinal pigment epithelial cells from acrolein-induced damage. J Ocul Pharmacol Ther 2010;26:231–6.

40. Mansoor S, Gupta N, Patil AJ, Estrago-Franco MF, Ramirez C, Migon R, et al. Inhibition of apoptosis in human retinal pigment epithelial cells treated with benzo(e)pyrene, a toxic component of cigarette smoke. Invest Ophthalmol Vis Sci 2010;51:2601–7.

41. Yu CC, Nandrot EF, Dun Y, Finnemann SC. Dietary antioxidants prevent age-related retinal pigment epithelium actin damage and blindness in mice lacking alphavbeta5 integrin. Free Radic Biol Med 2012;52:660–70.

42. Richer S, Stiles W, Thomas C. Molecular medicine in ophthalmic care. *Optometry* 2009;**80**:695–701.

43. Doganay S, Borazan M, Iraz M, Cigremis Y. The effect of resveratrol in experimental cataract model formed by sodium selenite. *Curr Eye Res* 2006;**31**:147–53.

44. Zheng Y, Liu Y, Ge J, Wang X, Liu L, Bu Z, Liu P. Resveratrol protects human lens epithelial cells against H_2O_2-induced oxidative stress by increasing catalase, SOD-1, and HO-1 expression. *Mol Vis* 2010;**16**:1467–74.

45. Barden CA, Chandler HL, Lu P, Bomser JA, Colitz CM. Effect of grape polyphenols on oxidative stress in canine lens epithelial cells. *Am J Vet Res* 2008;**69**:94–100.

46. Li G, Luna C, Navarro ID, Epstein DL, Huang W, Gonzalez P, Challa P. Resveratrol prevention of oxidative stress damage to lens epithelial cell cultures is mediated by forkhead box O activity. *Invest Ophthalmol Vis Sci* 2011;**52**:4395–401.

47. Kubota S, Kurihara T, Mochimaru H, Satofuka S, Noda K, Ozawa Y, et al. Prevention of ocular inflammation in endotoxin-induced uveitis with resveratrol by inhibiting oxidative damage and nuclear factor-kappaB activation. *Invest Ophthalmol Vis Sci* 2009;**50**: 3512–9.

48. Portt L, Norman G, Clapp C, Greenwood M, Greenwood MT. Anti-apoptosis and cell survival: a review. *Biochim Biophys Acta* 2010;**1813**:238–59.

49. Sareen D, van Ginkel PR, Takach JC, Mohiuddin A, Darjatmoko SR, Albert DM, Polans AS. Mitochondria as the primary target of resveratrol-induced apoptosis in human retinoblastoma cells. *Invest Ophthalmol Vis Sci* 2006;**47**:3708–16.

50. Anekonda TS, Adamus G. Resveratrol prevents antibody-induced apoptotic death of retinal cells through upregulation of Sirt1 and Ku70. *BMC Res Notes* 2008;**1**:122.

51. Brown MD, Sacks DB. Protein scaffolds in MAP kinase signalling. *Cell Signal* 2009;**21**:462–9.

52. Ramos JW. The regulation of extracellular signal-regulated kinase (ERK) in mammalian cells. *Int J Biochem Cell Biol* 2008;**40**: 2707–19.

53. Li W-L, Zhang L, Zhang Y-L. Effects of resveratrol on proliferation of retinal vascular endothelial cells and expression of VEGF. *Int J Ophthalmol* 2008;**1**:237–40.

54. Qazi Y, Maddula S, Ambati BK. Mediators of ocular angiogenesis. *J Genet* 2009;**88**:495–515.

55. Mataftsi A, Dimitrakos SA, Adams GG. Mediators involved in retinopathy of prematurity and emerging therapeutic targets. *Early Hum Dev* 2011;**87**:683–90.

56. Losso JN, Truax RE, Richard G. *trans*-Resveratrol inhibits hyperglycemia-induced inflammation and connexin downregulation in retinal pigment epithelial cells. *J Agric Food Chem* 2010;**58**: 8246–52.

57. Sjolie AK, Chaturvedi N. The retinal renin–angiotensin system: implications for therapy in diabetic retinopathy. *J Hum Hypertens* 2002;**16**(Suppl 3):S42–6.

58. Gilbert RE, Kelly DJ, Cox AJ, Wilkinson-Berka JL, Rumble JR, Osicka T, et al. Angiotensin converting enzyme inhibition reduces retinal overexpression of vascular endothelial growth factor and hyperpermeability in experimental diabetes. *Diabetologia* 2000;**43**:1360–7.

59. Kim YH, Kim YS, Roh GS, Choi WS, Cho GJ. Resveratrol blocks diabetes-induced early vascular lesions and vascular endothelial growth factor induction in mouse retinas. *Acta Ophthalmol* 2012;**90**:e31–7.

60. Dugas B, Charbonnier S, Baarine M, Ragot K, Delmas D, Menetrier F, et al. Effects of oxysterols on cell viability, inflammatory cytokines, VEGF, and reactive oxygen species production on human retinal cells: cytoprotective effects and prevention of VEGF secretion by resveratrol. *Eur J Nutr* 2010;**49**:435–46.

61. Peng CH, Chang YL, Kao CL, Tseng LM, Wu CC, Chen YC, et al. SirT1 – a sensor for monitoring self-renewal and aging process in retinal stem cells. *Sensors Basel* 2010;**10**:6172–94.

62. Sheu SJ, Wu TT. Resveratrol protects against ultraviolet A-mediated inhibition of the phagocytic function of human retinal pigment epithelial cells via large-conductance calcium-activated potassium channels. *Kaohsiung J Med Sci* 2009;**25**:381–8.

63. Sheu SJ, Bee YS, Chen CH. Resveratrol and large-conductance calcium-activated potassium channels in the protection of human retinal pigment epithelial cells. *J Ocul Pharmacol Ther* 2008;**24**:551–5.

64. Brakenhielm E, Cao R, Cao Y. Suppression of angiogenesis, tumor growth, and wound healing by resveratrol, a natural compound in red wine and grapes. *FASEB J* 2001;**15**:1798–800.

65. Kubota S, Ozawa Y, Kurihara T, Sasaki M, Yuki K, Miyake S, et al. Roles of AMP-activated protein kinase in diabetes-induced retinal inflammation. *Invest Ophthalmol Vis Sci* 2011;**52**:9142–8.

66. Zhang F, Wang S, Gan L, Vosler PS, Gao Y, Zigmond MJ, Chen J. Protective effects and mechanisms of sirtuins in the nervous system. *Prog Neurobiol* 2011;**95**:373–95.

67. Cheng HL, Mostoslavsky R, Saito S, Manis JP, Gu Y, Patel P, et al. Developmental defects and p53 hyperacetylation in Sir2 homolog (SIRT1)-deficient mice. *Proc Natl Acad Sci U S A* 2003;**100**:10794–9.

68. Jaliffa C, Ameqrane I, Dansault A, Leemput J, Vieira V, Lacassagne E, et al. Sirt1 involvement in rd10 mouse retinal degeneration. *Invest Ophthalmol Vis Sci* 2009;**50**:3562–72.

69. Shindler KS, Ventura E, Rex TS, Elliott P, Rostami A. SIRT1 activation confers neuroprotection in experimental optic neuritis. *Invest Ophthalmol Vis Sci* 2007;**48**:3602–9.

70. Fonseca-Kelly Z, Nassrallah M, Uribe J, Khan RS, Dine K, Dutt M, Shindler KS. Resveratrol neuroprotection in a chronic mouse model of multiple sclerosis. *Front Neurol* 2012;**3**:84.

71. Wimmers S, Karl MO, Strauss O. Ion channels in the RPE. *Prog Retin Eye Res* 2007;**26**:263–301.

72. Wimmers S, Halsband C, Seyler S, Milenkovic V, Strauss O. Voltage-dependent Ca^{2+} channels, not ryanodine receptors, activate Ca^{2+}-dependent BK potassium channels in human retinal pigment epithelial cells. *Mol Vis* 2008;**14**:2340–8.

73. Ryan JS, Baldridge WH, Kelly ME. Purinergic regulation of cation conductances and intracellular Ca^{2+} in cultured rat retinal pigment epithelial cells. *J Physiol* 1999;**520**:745–59.

74. Sheu SJ, Wu SN. Mechanism of inhibitory actions of oxidizing agents on calcium-activated potassium current in cultured pigment epithelial cells of the human retina. *Invest Ophthalmol Vis Sci* 2003;**44**:1237–44.

75. Li HF, Chen SA, Wu SN. Evidence for the stimulatory effect of resveratrol on Ca^{2+}-activated K^+ current in vascular endothelial cells. *Cardiovasc Res* 2000;**45**:1035–45.

Acetyl-L-Carnitine as a Nutraceutical Agent in Preventing Selenite-Induced Cataract

Pitchairaj Geraldine[1], Arumugam R. Muralidharan[1], Rajan Elanchezhian[1], P. Archana Teresa[2], Philip A. Thomas[2]

[1]Department of Animal Science, School of Life Sciences, Bharathidasan University, Tiruchirappalli, Tamilnadu, India; [2]Institute of Ophthalmology, Joseph Eye Hospital, Tiruchirappalli, Tamilnadu, India

INTRODUCTION

Cataract, characterized by opacification of the crystalline lens, is the most common cause of preventable blindness worldwide.[1] Currently, surgery is the most effective treatment for cataract. However, the need for highly trained personnel and the cost of surgery pose a significant economic problem. It has also been suggested that a delay of 10 years in cataract onset would halve the number of individuals needing surgery, thereby reducing the economic burden.[2] Thus, the need for preventing or delaying the progression of cataract formation by alternative therapeutic modality is essential. Acetyl-L-carnitine (ALCAR) is a quaternary amine, synthesized endogenously in human brain, liver, and kidneys by the acetyl carnitine transferase enzyme.[3] ALCAR has been shown to act as a scavenger of free radicals in mammalian tissues and to reverse age-related alterations in fatty acid profiles.[4,5] In this chapter, the potential efficacy of ALCAR as a nutraceutical protecting agent in selenite-induced cataract is reviewed.

OXIDATIVE STRESS AND CATARACT FORMATION

Oxidative stress arises owing to an imbalance between the rate of production and rate of degradation of oxidants.[6] The complete reduction of the oxygen molecule to form water occurs within the mitochondria. However, a partial reduction of oxygen produces superoxide and various reactive oxidative intermediates. In addition to these endogenous oxidants, there are exogenous sources of oxidants, including food, air pollutants, ionizing radiation, infrared radiation, and sunlight.[7] In spite of cells being equipped with a well-organized antioxidant defense system (ADS) comprising many substances such as enzymes and vitamins, an imbalance between generation of free radicals and their disposal by antioxidants occurs with advancing age, resulting in numerous degenerative changes and senescence.[8] Oxidative stress plays a significant role in the degradation, oxidation, cross-linking, and aggregation of lenticular proteins, thereby initiating the formation of cataract.[9] The major enzyme systems involved in the degradation and detoxification of reactive oxygen species (ROS) in the lens include superoxide dismutase (SOD)[10] and catalase (CAT),[11] which constitute the primary line of defense against superoxide anion (O_2^-) and hydrogen peroxide (H_2O_2), respectively. Glutathione peroxidases (GPx), which reduce small organic peroxides through a catalytic mechanism, are another class of antioxidant enzyme involved in the defense mechanism.[12] Augmentation of the antioxidant defense of the lens has been reported to prevent or delay experimental cataracts.[13]

SELENITE CATARACT

Ostadalova *et al.*[14] first described the potential of selenite to cause cataract. Selenite cataract is usually produced by a single subcutaneous injection of 19–30 mmoles/kg body weight of sodium selenite (Na_2SeO_3) into 10–14-day-old suckling rats. Selenite is cataractogenic only when administered to young rats before completion of the critical maturation period of the lens (approximately 16 days of age). Severe, bilateral nuclear cataracts are produced within 4–6 days.

Precursor stages include: posterior subcapsular cataract (day 1), swollen fibers (day 2–3), and perinuclear refractile ring (day 3). Although the model has been used extensively as a model for nuclear cataract, a transient cortical cataract also forms 15–30 days after injection.[15]

Mechanism of Selenite-Induced Cataract

Selenite cataract is characterized by several biochemical processes in the lens, which include altered metabolism of epithelial cells, accumulation of calcium, calpain-induced proteolysis, precipitation of crystallins, phase transition, and loss of cytoskeletal proteins.[16] Shearer et al.[15] also hypothesized that selenite oxidizes critical sulfhydryl groups in the lenticular epithelium, resulting in altered biochemical processes in the lens followed by limited proteolysis of crystallins, especially β-crystallin polypeptides, thereby leading to abnormal interaction of crystallins and insolubilization of proteins.

Selenite Cataract and Human Senile Cataract

Selenite cataract shows a number of general similarities to human senile cataract (in addition to vesicle formation), such as increased levels of calcium, insoluble protein, and proteolysis and decreased levels of water-soluble protein and reduced glutathione (GSH). In contrast to human senile cataract, selenite cataract is dominated by calpain-induced proteolysis, whereas human senile cataract appears to be mainly caused by prolonged oxidative stress. Despite these differences, selenite cataract is an extremely rapid and convenient model of senile nuclear cataract. The reliability and extreme characterization of selenite cataract make it an effective rodent model for rapid screening of anticataract agents.[16]

INFLUENCE OF NUTRITION SUPPLEMENTATION AND CATARACT PREVENTION

The development of cataract is an age-related phenomenon, in which a concomitant decrease in the activity of antioxidant enzymes is noted.[17] Prevention of cataract by nutritional and metabolic antioxidant supplementation has gained much attention. Several experimental studies have highlighted the potential role of antioxidants in the amelioration of cataract; the various antioxidants thus evaluated include ALCAR,[18,19] lycopene,[20] α-ketoglutarate,[21] resveratrol,[22] rutin,[23] and ellagic acid.[24] Although various compounds possessing antioxidant activity have been evaluated in experimental models, only a few compounds have reached the stage of clinical trials. A recent report by Ugboaja et al.[25] indicated the effectiveness of nutritional supplementation in preventing cataract. Table 50.1 summarizes the salient aspects of supplementation with antioxidants, multivitamins, and minerals and various clinical trials.

ACETYL-L-CARNITINE

L-Carnitine and its short-chain derivatives are essential cofactors for mitochondrial transport and oxidation of long-chain fatty acids. Carnitine is a small water-soluble molecule present in cells and tissues in relatively high concentrations either as free carnitine or as acylcarnitines, including ALCAR.[38] Carnitine is taken up into cells by a stereospecific transport system, resulting in a 10–100-fold gradient between intracellular and extracellular concentrations.[39] L-Carnitine has been reported to act as a scavenger of oxygen free radicals in mammalian tissues.[5] ALCAR, a quaternary amine, is a naturally occurring, short-chain derivative of L-carnitine (Fig. 50.1), which is synthesized endogenously in the human brain, liver, and kidneys by the enzyme acetyl carnitine transferase or obtained from dietary sources.[3] ALCAR, produced in the mitochondria, facilitates the uptake of acetyl-coenzyme A (CoA) into the mitochondria during fatty acid oxidation, thereby enhancing production of acetylcholine, and stimulates synthesis of protein and membrane phospholipids. ALCAR counteracts oxidative stress by inhibiting the increase in lipid hydroperoxidation.[40]

The antioxidant potential of ALCAR has been highlighted by studies showing that ALCAR can inhibit xanthine oxidase-induced damage to rat skeletal muscle[41] and lipid peroxidation in the periphery of the brain.[41,42] Boerrigter et al.[43] reported that ALCAR inhibits oxidant-induced DNA single-strand breaks in human peripheral blood lymphocytes. In a related study, Liu et al.[44] observed that ALCAR decreased lipid peroxidation and the extent of the oxidized molecules oxo8dG/oxo8G in the brain tissue of old rats, while Calabrese et al.[45] observed that ALCAR-induced heme oxygenase in rat astrocytes protects against oxidative stress. Hagen et al.[4] reported that a diet supplemented with ALCAR reversed age-related alterations in fatty acid profiles, metabolic rate, and cardiolipin levels. ALCAR has been reported to effectively suppress oxidative stress in and around mitochondria, thereby preventing the mitochondrial signaling pathway that leads to apoptosis.[46] ALCAR is known to enhance mitochondrial metabolism via the carnitine/acetyl carnitine-mediated transfer of the acetyl group that plays an important role in signal transduction pathways and gene expression in neuronal cells.[47]

The possible role of ALCAR in preventing diabetic cataractogenesis as well as lenticular opacification induced by exposure to H_2O_2 was suggested by studies

TABLE 50.1 Clinical Trials Evaluating Nutritional Supplementation in Prevention of Cataract Formation in Humans

No.	Study Setting (Investigators) [Reference]	Age (years), Population Size (N)	Nutrition Supplements and Frequency	Conclusion
1.	Linxian Study, China[26]	45–75, N = 4000	Multivitamin or mineral preparation	Reported 43% reduction in nuclear cataract formation
2.	[27]	65–74, N = 2141	Daily multivitamin or vitamin:	36% reduction in nuclear cataract formation observed with riboflavin and niacin combination
			Retinol (5000 IU) and zinc (22 mg)	
			Riboflavin (3 mg) and niacin (40 mg)	
			Vitamin C (120 mg) and molybdenum (30 mg)	
			Vitamin E (30 mg), β-carotene (15 mg), and selenium (50 mg)	
3.	Age-Related Eye Disease Study (AREDS)[28]	55–80, N = 3640	Daily supplementation with antioxidant combination:	Protective effect against nuclear cataract
			Vitamin E (400 IU)	
			Vitamin C (500 mg)	
			β-Carotene (15 mg)	
			Zinc oxide (80 mg)	
			Copper II oxide (2 mg)	
			Duration: average of 6.3 years	
4.	Roche European–American Cataract Trial (REACT)[29]	N = 445	Daily multivitamin supplementation:	Significant decrease in cataract progression in US participants but not in UK participants
			Vitamin E (600 mg)	
			Vitamin C (750 mg)	
			β-Carotene (18 mg)	
			Duration: 3 years	
5.	Nurses' Health Studies Nutrition and Vision Project[30]	Women, 53–73, N = 492	Vitamin C (362 mg/day)	57% lower incidence of nuclear cataract
6.	[31]	Women, ≥45, N = 47,152	Daily vitamin C	28% lower incidence of nuclear cataract
7.	Melbourne Visual Impairment Project[32]	≥40, N = 3271	Fish lutein and zeaxanthin	Inverse correlation between lutein and zeaxanthin intake and risk of nuclear cataracts
8.	Physicians' Health Study (PHS)[33]	Men, N = 22,000	Daily β-carotene (50 mg) with low-dose aspirin	≈21,000 participants had no incidence of cataract at the beginning. After 13.2 years, 2017 developed cataract, of which 998 in the β-carotene group exhibited no positive or negative effect
9.	Vitamin E, Cataract and Age-related Maculopathy Trial (VECAT)[34]	55–80, N = 1204	Vitamin E (500 mg/day)	Vitamin E had no significant effect on incidence or progression of cataracts
			Duration: 4 years	

(Continued)

TABLE 50.1   Clinical Trials Evaluating Nutritional Supplementation in Prevention of Cataract Formation in Humans—cont'd

No.	Study Setting (Investigators) [Reference]	Age (years), Population Size (N)	Nutrition Supplements and Frequency	Conclusion
10.	[35]	Men, ≥50, N = 11,545	Vitamin E (400 IU) Vitamin C (500 mg) Duration: 8 years	No significant effect on incidence of cataract
11.	[36]	55–80, N = 1193	Daily supplementation of vitamin E (500 IU) Duration: 4 years	Failed to reduce progression of cataract
12.	Cohort study[37]	Women, 49–83, N = 24,593	Vitamin C supplementation ≈1000 mg and multivitamins containing vitamin C ≈60 mg Duration: 8.2 years	Greater incidence of age-related cataract

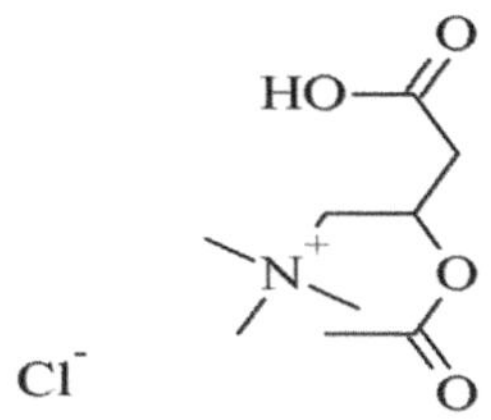

FIGURE 50.1 Two-dimensional chemical structure of acetyl-L-carnitine (ALCAR).

performed around 2001.[48] ALCAR appears to prevent the oxidative insult to the lens caused by exposure to selenite, under *in vitro* as well as *in vivo* conditions, by restoring the antioxidant status,[19] preventing modulation of connexin,[49] and maintaining the calcium level, thereby preventing activation of calpain and alterations to crystalline proteins (Fig. 50.2).[50]

The biologic activity of ALCAR is mainly due to the presence of the acetyl-group moiety; in particular, the acetyl moiety of ALCAR serves as an essential precursor in the synthesis of acetyl-CoA, which is involved in the redox reaction that eliminates ROS.[51] The acetyl group present in ALCAR plays a vital role in the maintenance of high intracellular concentrations of acetyl-CoA and the acetylation of amino and carboxyl groups of the side-chains of amino acids. Such modifications caused by ALCAR may alter the structure, function, and turnover of proteins, therein conferring a molecular chaperone property on ALCAR.[38]

Potential Efficacy of Acetyl-L-Carnitine on Antioxidant and Redox System

The antioxidant system is a potentially important system for ocular tissues since it protects against various pathologic processes, including cataract formation,

in the eye.[52] The balance between the production and catabolism of oxidants by cells and tissue is critical for maintenance of the biologic integrity of the tissue. Ocular tissues contain antioxidants that prevent damage from excessive oxygen metabolites; these act by either decomposing peroxide or trapping the free radicals, leading to interference with the chain of oxidation reactions.[53] The major protective enzymes of the antioxidant system in the lens are SOD, catalase (CAT),[54] and GPx.[55] In addition, there are several nonenzymatic free radical scavengers, such as glutathione[56] and vitamin E.[57]

SOD, a chain-breaking antioxidant, was described by McCord and Fridovich[58] in red blood cells. Bhuyan and Bhuyan[59] first described its occurrence in the lenses of different species. SOD converts superoxide to H_2O_2. The enzyme exists in two forms, one containing Mn^{2+}, restricted to the mitochondria, and a cytosolic form containing Zn^{2+} and Cu^{2+}. The occurrence of GPx in the lens was first shown by Pirie.[60] GPx is required to check lipid peroxidation initiated by superoxide in the phospholipid bilayer for maintenance of membrane integrity. CAT is a hemeprotein that requires nicotinamide adenine dinucleotide phosphate (NADPH) for regeneration to its active form. The presence of CAT in the lens has been well demonstrated.[61] Both GPx and CAT catalyze the transformation of H_2O_2 within the cell to harmless by-products, thereby curtailing the quantity of cellular destruction inflicted by products of lipid peroxidation.[62]

A high concentration of GSH has been found to protect the lens from oxidative damage and toxic chemicals[63]; thus, depletion of GSH seriously affects GSH-dependent enzymes such as GPx, glutathione reductase (GR), and glutathione S-transferase (GST) and also leukotriene C4 synthetase, and the glutaredoxin system, rendering the cells susceptible to a toxic challenge.[64] A progressive decrease in GSH in the lens was found to be associated

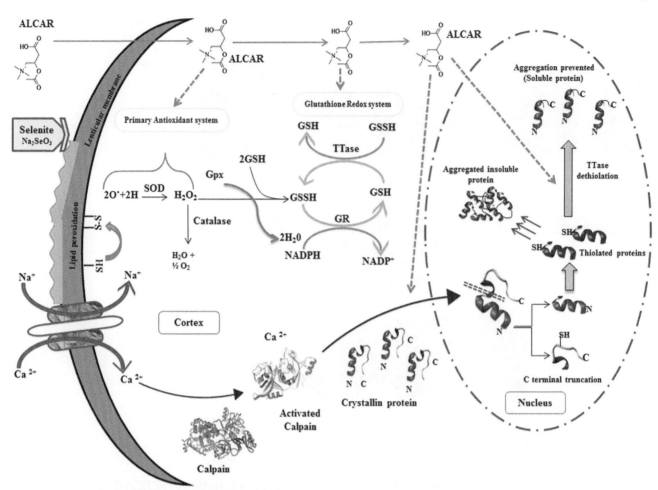

FIGURE 50.2 Hypothetical schematic representation of acetyl-L-carnitine (ALCAR) in preventing selenite cataract. Gpx: glutathione peroxidase; GR: glutathione reductase; GSH: reduced glutathione; GSSH: oxidized glutathione; NADP$^+$: nicotinamide adenine dinucleotide; NADPH: nicotinamide adenine dinucleotide phosphate; SOD: superoxide dismutase; TTase: thiotransferase.

with experimental cataract formation or human senile cataract formation.[65] A depletion of GSH content in postmitotic tissues has also been found to occur during aging,[66] possibly owing to enhanced oxidative damage due to free radicals. GR maintains the intracellular level of GSH by preserving the integrity of cell membranes and by stabilizing the sulfhydryl groups of proteins. Administration of carnitine to aged rats has been found to increase the activity of GR by increasing the levels of GSH and reducing equivalent NADPH.[67]

Lipid peroxidation is considered to be the basic mechanism of cellular damage caused by free radicals.[68] Phospholipids in the lenticular membrane are affected by oxidative stress, with a consequent increase in lipid peroxidation, which is manifested by a rise in the level of malondialdehyde (MDA).[21] In a noteworthy study, Elanchezhian et al.[18] reported higher mean activities and levels of the antioxidant enzymes and redox system components and a lower mean level of MDA in selenite-challenged, ALCAR-treated rat lenses than those in selenite-challenged, untreated rat lenses; these observations

were believed to highlight the antiradical effects of ALCAR and its ability to partially prevent the reduction in lenticular antioxidant enzyme activities wrought by exposure of rat pups to selenite. This protective effect was also manifested morphologically as a decreased frequency and intensity of lenticular opacification (Fig. 50.3). Thus, the protective effect of ALCAR can be attributed to several factors, including inhibition of free radical formation by decreasing the concentration of cytosolic ion, which plays an important role in the formation of ROS, and by protecting DNA from alkylating injury.[43]

Modulatory Effect of Acetyl-L-Carnitine on Connexin-Mediated Lenticular Homeostasis and Calpain Activity in Selenite-Induced Cataractogenesis

Precise regulation of calcium homeostasis is essential to the viability of most cells.[69] In all human cataractous lenses, total calcium is elevated 3000-fold, compared with normal human lenses. The opacification of the

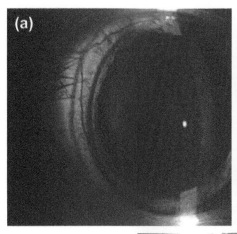

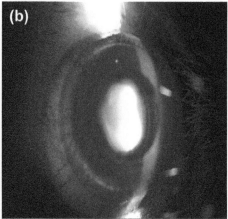

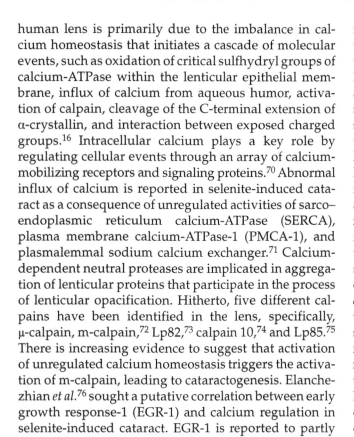

FIGURE 50.3   Slit-lamp appearance of Wistar rat lenses: (a) normal, (b) selenite challenged-untreated, and (c) selenite challenged-acetyl-L-carnitine-treated eyes.

human lens is primarily due to the imbalance in calcium homeostasis that initiates a cascade of molecular events, such as oxidation of critical sulfhydryl groups of calcium-ATPase within the lenticular epithelial membrane, influx of calcium from aqueous humor, activation of calpain, cleavage of the C-terminal extension of α-crystallin, and interaction between exposed charged groups.[16] Intracellular calcium plays a key role by regulating cellular events through an array of calcium-mobilizing receptors and signaling proteins.[70] Abnormal influx of calcium is reported in selenite-induced cataract as a consequence of unregulated activities of sarco–endoplasmic reticulum calcium-ATPase (SERCA), plasma membrane calcium-ATPase-1 (PMCA-1), and plasmalemmal sodium calcium exchanger.[71] Calcium-dependent neutral proteases are implicated in aggregation of lenticular proteins that participate in the process of lenticular opacification. Hitherto, five different calpains have been identified in the lens, specifically, μ-calpain, m-calpain,[72] Lp82,[73] calpain 10,[74] and Lp85.[75] There is increasing evidence to suggest that activation of unregulated calcium homeostasis triggers the activation of m-calpain, leading to cataractogenesis. Elanchezhian *et al.*[76] sought a putative correlation between early growth response-1 (EGR-1) and calcium regulation in selenite-induced cataract. EGR-1 is reported to partly

inhibit the transcription of SERCA-2, which is reported to actively transport calcium ions from the cytosol into the endoplasmic reticulum and to restore intracellular calcium pools in myocytes.[77] Elanchezhian *et al.*[76] also reported that the messenger RNA (mRNA) transcripts of EGR-1 were upregulated as a consequence of selenite injection. However, ALCAR treatment maintained the expression of EGR-1 at near normal levels. In selenite cataractogenesis, oxidative stress-induced lipid peroxidative damage to lenticular membranes is reported to contribute to the loss of Ca^{2+}-ATPase activity, with subsequent accumulation of Ca^{2+}.[16] In a noteworthy study, the administration of ALCAR was found to prevent the autolysis of calpain Lp82 proteins; moreover, an increase in m-calpain mRNA transcript levels was demonstrated in selenite-induced cataractogenesis.[50] In a recent study, Muralidharan *et al.*[49] observed that ALCAR is capable of modulating the activities of calcium-ATPase and sodium/potassium-ATPase in selenite-induced cataracts *in vivo*; the molecular mechanism underlying the anticataractogeneic effect of ALCAR was hypothesized to be its protection against oxidation of sulfhydryl groups in the lenticular epithelium, thereby maintaining the activity of these ATPases. Li *et al.*[78] reported that maintenance of lenticular homeostasis and transparency depends on an extensive network of gap junctions. Connexins (connexin 43, connexin 46, and connexin 50) constitute the major component of lenticular gap junctions. In an earlier study, Fleschner[79] showed that a disruption of the functioning of gap junctions was possibly an important factor in the mechanism of sodium selenite-induced cataractogenesis. Muralidharan et al.[49] have recently demonstrated the modulatory effect of ALCAR in maintaining calcium homeostasis mediated by connexin protein; the gap junctional proteins were found to be upregulated at both the transcriptional and translational levels following treatment with ALCAR. These researchers also predicted the phosphorylation sites of gap junctional proteins at specific amino acids, which would have presumably influenced the inactivation of the gap junctional proteins connexin 46 and connexin 50. The encouraging outcome of this study clearly highlights the regulation of protein expression at the molecular level by ALCAR treatment.

Regulatory Effect of Acetyl-L-Carnitine on Expression of Apoptotic Genes in Selenite-Induced Cataractogenesis

Lenticular cells are derived from epithelial cells, where cell division and fiber differentiation are highly localized. Identifying the processes controlling normal cellular function in the lens is of great importance, as modification of these highly regulated mechanisms results in lenticular pathologies, namely cataract.[80] Earlier studies have shown that death of lenticular epithelial cells can disturb normal lenticular homeostasis and transparency of lenticular fiber cells, thereby initiating cataractogenesis.[81,82] Several notable studies have reported the association of apoptosis in lenticular epithelial cells with the development of cataracts induced by exposure to H_2O_2,[13] ultraviolet radiation,[82] or selenite.[83] Maintaining the integrity of epithelium and survival of epithelial cells during adverse conditions is an extremely important function of the lens. Consistent with the findings of Li and Spector,[82] there have been several studies where apoptosis has been noted in capsular epithelial cells of cataractous human lenses or mouse lenses.[84,85] Varying levels of apoptosis in lenticular epithelial cells have also been detected in capsulorhexis specimens from patients with cataract or from murine cataractous lenses.[84,86]

Apoptosis in lenticular epithelial cells has been suggested as an early event in selenite cataractogenesis,[87] which is accompanied by activation of caspase and calpain.[83] Tamada et al.[83] reported that in selenite cataract, apoptosis occurred in the lenticular epithelium 1 day after the selenite injection, while the number of apoptotic cells substantially increased on the second day. The apoptotic component of selenite-induced cell death can be confirmed by electron microscopy, activation of caspase-3, and cleavage of poly(ADP-ribose) polymerase (PARP). Decreased mRNA transcript level of cytochrome c oxidase subunit I (COX-I) has also been noted in a selenite-exposed, mouse lenticular epithelial cell line (α-TN4), which also showed leakage of cytochrome c from mitochondria into the cytosol.[87] A majority of the downregulated genes were mitochondrial genes, which suggests that the loss of integrity of lenticular epithelial cells, following exposure to selenite, occurs through the mitochondria-mediated intrinsic death pathway.[87] COX-I is associated with mitochondrial membranes and is the terminal enzyme of the electron transport chain; the electron donor is cytochrome c, and the reaction results in the translocation of protons across the mitochondrial membrane.[46]

The caspase cascade has been identified as playing a key role in apoptosis in several tissues.[88] PARP nuclear protein is often found to be degraded in apoptotic cells. In many examples of programmed cell death, caspase-3 has been found to cleave intact 116kDa PARP into 85kDa (carboxyl terminus) and 25kDa fragments.[89,90] The occurrence of apoptosis in lenses undergoing selenite-induced cataractogenesis can be inferred from the presence of proteolysis of intact PARP and accumulation of 85kDa and 25kDa PARP fragments. The increased activity of caspase noted during apoptosis in selenite-cataractous lenses was hypothesized to arise at the enzyme level rather than being due to upregulation of mRNA.[83] Expression of the EGR-1 gene has also been linked to cell death.[89] Enhanced expression of EGR-1 is associated with damage to lenticular epithelial cells. Wang et al.[91] reported induction of mRNA for EGR-1 in the lenticular epithelium following injections of selenite into young rats.

In an earlier study, ALCAR was reported to strongly inhibit mitochondria-dependent apoptosis in mouse fibroblast culture.[92] More recently, ALCAR was found to prevent alterations in expression of apoptosis-related genes such as COX-I, caspase-3, and EGR-1 (Fig. 50.4) that occur during selenite-induced cataractogenesis,[76] possibly by means of improved energy metabolism, inhibition of electron leakage from mitochondrial electron transport systems,[93] and enhanced repair of oxidized membrane/lipid bilayers.[94]

Protective Role of Acetyl-L-Carnitine on Lenticular Proteins in Selenite-Induced Cataractogenesis

Transparency is the fundamental characteristic of the normal lens that distinguishes cells of the lens from all other mammalian cells.[95] The transparency and refractive power of the ocular lens depend on an even distribution of the protein within lenticular cells on the scale of the

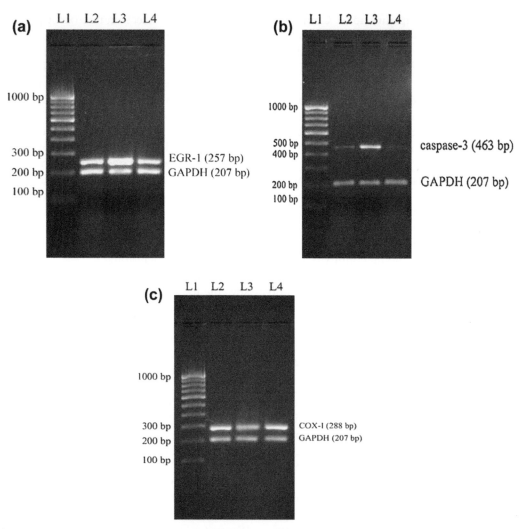

FIGURE 50.4 Semi-quantitative reverse transcription–polymerase chain reaction (RT-PCR) analysis: (a) early growth response protein-1 (EGR-1), (b) caspase-3, and (c) cytochrome *c* oxidase subunit I (COX-I) messenger RNA (mRNA) in rat lenses visualized on an ethidium bromide-stained agarose gel. L1: 100 bp DNA ladder; L2: normal; L3: selenite challenged-untreated; L4: selenite challenged-acetyl-L-carnitine-treated. The results depicted are normalized to levels of glyceraldehyde-3-phosphate dehydrogenase (GAPDH) mRNA. *Source: Adapted from Elanchezhian et al. (2010),[64] with permission from Elsevier.*

wavelength of visible light.[96] Dense opacification results when the proteins form large insoluble aggregates that approach or exceed the dimensions of the wavelength of light and produce large fluctuations in the index of refraction that result in increased light scattering.[95,97]

The mammalian lens is principally composed of proteins, which account for over 30% of its total weight, and contains three major structural proteins known as α-, β-, and γ-crystallin, which constitute about 90% of the total lenticular proteins.[98] A variety of cytoskeletal proteins that constitute intermediate filaments, microfilaments, and microtubules may have important roles in the development and maintenance of the transparent structure of lenticular cells.[99] Typically, the content of cytoskeletal proteins in the lens decreases with age, particularly in the nucleus. Loss of these cytoskeletal proteins, including vimentin and α-actin, has been observed

to accelerate the development of senile cataract in the human lens.[100]

Several biochemical alterations in lenticular proteins have been recorded in human cataract: an abnormal increase in the ratio of insoluble to soluble protein,[97] proteolysis of lenticular crystallins,[98] and decreased content of cytoskeletal proteins.[100] Pioneering studies have suggested a functional role for the cytoskeletal proteins, such as α-actin,[101] vimentin,[102] and spectrin,[100] in the maintenance of the normal structure of transparent lenticular cells. Furthermore, the development of senile nuclear cataract is associated with an increase in insoluble proteins.[98] The proportion of water-insoluble proteins increases in human lenses with aging and also during development of cataract.[103] Therefore, changes in the water-soluble, high molecular weight proteins leading to water-insoluble proteins of cataractous lenses may

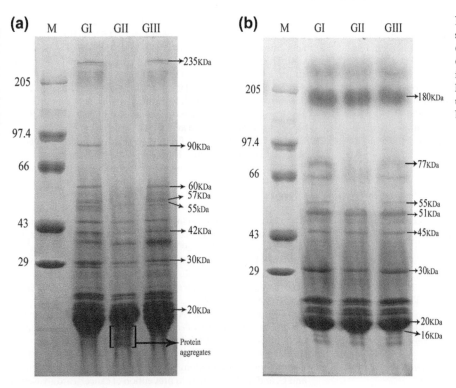

FIGURE 50.5 4–20% sodium dodecyl sulfate–polyacrylamide gel electrophoresis (SDS-PAGE) pattern of Wistar rat lenses: (a) lens-soluble and (b) lens-insoluble proteins in the lenses of experimental groups of rats. M: marker; GI: normal; GII: selenite challenged-untreated; GIII: selenite challenged-acetyl-L-carnitine-treated (unpublished data).

play a critical role in producing cataract-specific aggregates and cross-linked species.

The β- and γ-crystallins share a common polypeptide chain fold, have conserved sequences, and together form a superfamily of βγ-crystallins.[104] In contrast, the α-crystallins form a separate family of proteins related to the small heat-shock proteins.[105] Mutations in βγ-crystallin genes can lead to nonspecific aggregation of crystallins, resulting in cataract formation.[106] It has previously been shown that crystallins in the nucleus of the normal lens undergo partial proteolysis and insolubilization from birth to 4 months of age.[107] Changes in protein composition during normal lenticular differentiation may occur owing to interactions between the cytoskeleton and crystallins; these interactions may also be important in the proteolysis and insolubilization of crystallins and cytoskeletal proteins observed in the selenite cataract.[108,109] The loss of cytoskeletal proteins that has been found to follow selenite injection[110] is hypothesized to be due to the activity of calpain.[111] Glutamic acid and aspartic acid are presumably needed for the synthesis of α-, β-, and γ-crystallins; loss of aspartic acid and glutamic acid is one of the reasons for the aggregation of crystalline proteins.[112] Unpublished data (Fig. 50.5) suggest that these alterations in lenticular proteins in selenite-challenged rats can be prevented by treatment with ALCAR, possibly by maintaining the levels of aspartic acid, glutamic acid, and glycine. It has also been shown that ALCAR is able to maintain the normal interaction between the crystallin and cytoskeletal proteins by inhibiting calpains.[50]

TAKE-HOME MESSAGES

- Acetyl-L-carnitine appears to prevent selenite-induced cataractogenesis by maintaining lenticular antioxidant and redox system components at near normal levels and by preventing excessive lipid peroxidation.
- Acetyl-L-carnitine also appears to possess a novel paradigm in maintaining lenticular homeostasis through connexin proteins, thereby preventing elevated calcium influx in the lenticular environment and subsequent activation of calpain.
- Acetyl-L-carnitine has been demonstrated to prevent accelerated expression of apoptosis-related genes.
- Acetyl-L-carnitine also appears to completely inhibit loss of cytoskeletal proteins and to partially inhibit loss of αA-crystallin proteins.
- These original observations suggest a novel use for acetyl-L-carnitine as a possible cataract-preventing drug. However, further study is required to extrapolate these results to the use of acetyl-L-carnitine in humans for the prophylaxis of cataractogenesis in humans.

References

1. Resnikoff S, Pascolini D, Etya'ale D, Kocur I, Pararajasegaram R, Pokharel GP, Mariotti SP. Global data on visual impairment in the year 2002. *Bull World Health Organ* 2004;**82**:844–51.
2. Brian G, Taylor H. Cataract blindness – challenges for the 21st century. *Bull World Health Organ* 2001;**79**:249–56.

3. Goa KL, Brogden RN. L-Carnitine. A preliminary review of its pharmacokinetics, and its therapeutic use in ischaemic cardiac disease and primary and secondary carnitine deficiencies in relationship to its role in fatty acid metabolism. *Drugs* 1987;**34**:1–24.

4. Hagen TM, Liu J, Lykkesfeldt J, Wehr CM, Ingersoll RT, Vinarsky V, et al. Feeding acetyl-L-carnitine and lipoic acid to old rats significantly improves metabolic function while decreasing oxidative stress. *Proc Natl Acad Sci USA* 2002;**99**:1870–5.

5. Izgut-Uysal VN, Agac A, Derin N. Effect of carnitine on stress-induced lipid peroxidation in rat gastric mucosa. *J Gastroenterol* 2001;**36**:231–6.

6. Sies H. Oxidative stress: oxidants and antioxidants. *Exp Physiol* 1997;**82**:291–5.

7. Vinson JA. Oxidative stress in cataracts. *Pathophysiology* 2006;**13**:151–62.

8. Reiter RJ. Oxygen radical detoxification processes during aging: the functional importance of melatonin. *Aging (Milano)* 1995;**7**:340–51.

9. Lou MF. Redox regulation in the lens. *Prog Retin Eye Res* 2003;**22**:657–82.

10. Behndig A, Svensson B, Marklund SL, Karlsson K. Superoxide dismutase isoenzymes in the human eye. *Invest Ophthalmol Vis Sci* 1998;**39**:471–5.

11. Yang Y, Spector A, Ma W, Wang RR, Larsen K, Kleiman NJ. The effect of catalase amplification on immortal lens epithelial cell lines. *Exp Eye Res* 1998;**67**:647–56.

12. Brigelius-Flohe R. Tissue-specific functions of individual glutathione peroxidases. *Free Radic Biol Med* 1999;**27**:951–65.

13. Spector A. Oxidative stress-induced cataract: mechanism of action. *FASEB J* 1995;**9**:1173–82.

14. Ostadalova I, Babicky A, Obenberger J. Cataract induced by administration of a single dose of sodium selenite to suckling rats. *Experientia* 1978;**34**:222–3.

15. Shearer TR, David LL, Anderson RS, Azuma M. Review of selenite cataract. *Curr Eye Res* 1992;**11**:357–69.

16. Shearer TR, Ma H, Fukiage C, Azuma M. Selenite nuclear cataract: review of the model. *Mol Vis* 1997;**3**:8.

17. Berman ER. *Biochemistry of the Eye*. New York: Plenum Press; 1991. pp. 223–224.

18. Elanchezhian R, Ramesh E, Sakthivel M, Isai M, Geraldine P, Rajamohan M, et al. Acetyl-L-carnitine prevents selenite-induced cataractogenesis in an experimental animal model. *Curr Eye Res* 2007;**32**:961–71.

19. Geraldine P, Sneha BB, Elanchezhian R, Ramesh E, Kalavathy CM, Kaliamurthy J, Thomas PA. Prevention of selenite-induced cataractogenesis by acetyl-L-carnitine: an experimental study. *Exp Eye Res* 2006;**83**:1340–9.

20. Gupta SK, Trivedi D, Srivastava S, Joshi S, Halder N, Verma SD. Lycopene attenuates oxidative stress induced experimental cataract development: an *in vitro* and *in vivo* study. *Nutrition* 2003;**19**:794–9.

21. Varma SD, Hegde KR. Effect of alpha-ketoglutarate against selenite cataract formation. *Exp Eye Res* 2004;**79**:913–8.

22. Doganay S, Borazan M, Iraz M, Cigremis Y. The effect of resveratrol in experimental cataract model formed by sodium selenite. *Curr Eye Res* 2006;**31**:147–53.

23. Isai M, Sakthivel M, Ramesh E, Thomas PA, Geraldine P. Prevention of selenite-induced cataractogenesis by rutin in Wistar rats. *Mol Vis* 2009;**15**:2570–7.

24. Sakthivel M, Elanchezhian R, Ramesh E, Isai M, Jesudasan CN, Thomas PA, Geraldine P. Prevention of selenite-induced cataractogenesis in Wistar rats by the polyphenol, ellagic acid. *Exp Eye Res* 2008;**86**:251–9.

25. Ugboaja OC, Bielory L, Bielory BP, Ehiorobo ES. Antioxidant vitamins, minerals and cataract: current opinion. *Curr Opin Allergy Clin Immunol* 2012;**12**:517–22.

26. Bunce GE. Evaluation of the impact of nutrition intervention on cataract prevalence in China. *Nutr Rev* 1994;**52**:99–101.

27. Sperduto RD, Hu TS, Milton RC, Zhao JL, Everett DF, Cheng QF, et al. The Linxian cataract studies. Two nutrition intervention trials. *Arch Ophthalmol* 1993;**111**:1246–53.

28. AREDS. A randomized, placebo-controlled, clinical trial of high-dose supplementation with vitamins C and E, beta carotene, and zinc for age-related macular degeneration and vision loss: AREDS report no. 8. *Arch Ophthalmol* 2001;**119**:1417–36.

29. Chylack Jr LT, Brown NP, Bron A, Hurst M, Kopcke W, Thien U, Schalch W. The Roche European American Cataract Trial (REACT): a randomized clinical trial to investigate the efficacy of an oral antioxidant micronutrient mixture to slow progression of age-related cataract. *Ophthalmic Epidemiol* 2002;**9**:49–80.

30. Taylor A, Jacques PF, Chylack Jr LT, Hankinson SE, Khu PM, Rogers G, et al. Long-term intake of vitamins and carotenoids and odds of early age-related cortical and posterior subcapsular lens opacities. *Am J Clin Nutr* 2002;**75**:540–9.

31. Chasan-Taber L, Willett WC, Seddon JM, Stampfer MJ, Rosner B, Colditz GA, Hankinson SE. A prospective study of vitamin supplement intake and cataract extraction among US women. *Epidemiology* 1999;**10**:679–84.

32. Vu HT, Robman L, Hodge A, McCarty CA, Taylor HR. Lutein and zeaxanthin and the risk of cataract: the Melbourne visual impairment project. *Invest Ophthalmol Vis Sci* 2006;**47**:3783–6.

33. Christen WG, Gaziano JM, Hennekens CH. Design of Physicians' Health Study II – a randomized trial of beta-carotene, vitamins E and C, and multivitamins, in prevention of cancer, cardiovascular disease, and eye disease, and review of results of completed trials. *Ann Epidemiol* 2000;**10**:125–34.

34. Garrett SK, Mcneil JJ, Silagy C, Sinclair M, Thomas AP, Robman LP, et al. Methodology of the VECAT study: vitamin E intervention in cataract and age-related maculopathy. *Ophthalmic Epidemiol* 1999;**6**:195–208.

35. Christen WG, Glynn RJ, Sesso HD, Kurth T, Macfadyen J, Bubes V, et al. Age-related cataract in a randomized trial of vitamins E and C in men. *Arch Ophthalmol* 2010;**128**:1397–405.

36. Mcneil JJ, Robman L, Tikellis G, Sinclair MI, Mccarty CA, Taylor HR. Vitamin E supplementation and cataract: randomized controlled trial. *Ophthalmology* 2004;**111**:75–84.

37. Rautiainen S, Lindblad BE, Morgenstern R, Wolk A. Vitamin C supplements and the risk of age-related cataract: a population-based prospective cohort study in women. *Am J Clin Nutr* 2010;**91**:487–93.

38. Pettegrew JW, Levine J, Mcclure RJ. Acetyl-L-carnitine physical–chemical, metabolic, and therapeutic properties: relevance for its mode of action in Alzheimer's disease and geriatric depression. *Mol Psychiatry* 2000;**5**:616–32.

39. Broquist HP. Carnitine. In: Shils ME, Olsen J, Shike M, editors. *Modern Nutrition in Health and Disease*. Philadelphia, PA: Lea & Febiger; 1994. pp. 459–65.

40. Yasui F, Matsugo S, Ishibashi M, Kajita T, Ezashi Y, Oomura Y, et al. Effects of chronic acetyl-L-carnitine treatment on brain lipid hydroperoxide level and passive avoidance learning in senescence-accelerated mice. *Neurosci Lett* 2002;**334**:177–80.

41. Di Giacomo C, Latteri F, Fichera C, Sorrenti V, Campisi A, Castorina C, et al. Effect of acetyl-L-carnitine on lipid peroxidation and xanthine oxidase activity in rat skeletal muscle. *Neurochem Res* 1993;**18**:1157–62.

42. Kaur J, Sharma D, Singh R. Acetyl-L-carnitine enhances Na(+), K(+)-ATPase glutathione-S-transferase and multiple unit activity and reduces lipid peroxidation and lipofuscin concentration in aged rat brain regions. *Neurosci Lett* 2001;**301**:1–4.

43. Boerrigter ME, Franceschi C, Arrigoni-Martelli E, Wei JY, Vijg J. The effect of L-carnitine and acetyl-L-carnitine on the disappearance of DNA single-strand breaks in human peripheral blood lymphocytes. *Carcinogenesis* 1993;**14**:2131–6.

44. Liu J, Head E, Kuratsune H, Cotman CW, Ames BN. Comparison of the effects of L-carnitine and acetyl-L-carnitine on carnitine levels, ambulatory activity, and oxidative stress biomarkers in the brain of old rats. *Ann NY Acad Sci* 2004;**1033**:117–31.

45. Calabrese V, Ravagna A, Colombrita C, Scapagnini G, Guagliano E, Calvani M, et al. Acetylcarnitine induces heme oxygenase in rat astrocytes and protects against oxidative stress: involvement of the transcription factor Nrf2. *J Neurosci Res* 2005;**79**:509–21.

46. Zhu X, Sato EF, Wang Y, Nakamura H, Yodoi J, Inoue M. Acetyl-L-carnitine suppresses apoptosis of thioredoxin 2-deficient DT40 cells. *Arch Biochem Biophys* 2008;**478**:154–60.

47. Lazebnik YA, Kaufmann SH, Desnoyers S, Poirier GG, Earnshaw WC. Cleavage of poly(ADP-ribose) polymerase by a proteinase with properties like ICE. *Nature* 1994;**371**:346–7.

48. Peluso G, Petillo O, Barbarisi A, Melone MA, Reda E, Nicolai R, Calvani M. Carnitine protects the molecular chaperone activity of lens alpha-crystallin and decreases the post-translational protein modifications induced by oxidative stress. *FASEB J* 2001;**15**:1604–6.

49. Muralidharan AR, Leema G, Annadurai T, Anitha TS, Thomas PA, Geraldine P. Deciphering the potential efficacy of acetyl-L-carnitine (ALCAR) in maintaining connexin-mediated lenticular homeostasis. *Mol Vis* 2012;**18**:2076–86.

50. Elanchezhian R, Sakthivel M, Geraldine P, Thomas PA. The effect of acetyl-L-carnitine on lenticular calpain activity in prevention of selenite-induced cataractogenesis. *Exp Eye Res* 2009;**88**:938–44.

51. Chang B, Nishikawa M, Nishiguchi S, Inoue M. L-carnitine inhibits hepatocarcinogenesis via protection of mitochondria. *Int J Cancer* 2005;**113**:719–29.

52. Bhuyan KC, Bhuyan DK. Molecular mechanism of cataractogenesis: III. Toxic metabolites of oxygen as initiators of lipid peroxidation and cataract. *Curr Eye Res* 1984;**3**:67–81.

53. Rao NA, Thaete LG, Delmage JM, Sevanian A. Superoxide dismutase in ocular structures. *Invest Ophthalmol Vis Sci* 1985;**26**:1778–81.

54. Fecondo JV, Augusteyn RC. Superoxide dismutase, catalase and glutathione peroxidase in the human cataractous lens. *Exp Eye Res* 1983;**36**:15–23.

55. Dwivedi RS, Pratap VB. Alteration in glutathione metabolism during cataract progression. *Ophthalmic Res* 1987;**19**:41–4.

56. Ross WM, Creighton MO, Trevithick JR, Stewart-Dehaan PJ, Sanwal M. Modelling cortical cataractogenesis: VI. Induction by glucose *in vitro* or in diabetic rats: prevention and reversal by glutathione. *Exp Eye Res* 1983;**37**:559–73.

57. Trevithick JR, Linklater HA, Mitton KP, Dzialoszynski T, Sanford SE. Modeling cortical cataractogenesis: IX. Activity of vitamin E and esters in preventing cataracts and gamma-crystallin leakage from lenses in diabetic rats. *Ann N Y Acad Sci* 1989;**570**:358–71.

58. McCord JM, Fridovich I. Superoxide dismutase. An enzymic function for erythrocuprein (hemocuprein). *J Biol Chem* 1969;**244**:6049–55.

59. Bhuyan KC, Bhuyan DK. Superoxide dismutase of the eye: relative functions of superoxide dismutase and catalase in protecting the ocular lens from oxidative damage. *Biochim Biophys Acta* 1978;**542**:28–38.

60. Pirie A. Glutathione peroxidase in lens and a source of hydrogen peroxide in aqueous humour. *Biochem J* 1965;**96**:244–53.

61. Bhuyan KC, Bhuyan DK. Catalase in ocular tissue and its intracellular distribution in corneal epithelium. *Am J Ophthalmol* 1970;**69**:147–53.

62. Santini SA, Marra G, Giardina B, Cotroneo P, Mordente A, Martorana GE, et al. Defective plasma antioxidant defenses and enhanced susceptibility to lipid peroxidation in uncomplicated IDDM. *Diabetes* 1997;**46**:1853–8.

63. Hightower KR, McCready JP. Effect of selenite on epithelium of cultured rabbit lens. *Invest Ophthalmol Vis Sci* 1991;**32**:406–9.

64. Kumaran S, Savitha S. Anusuya Devi M, Panneerselvam C. L-Carnitine and DL-alpha-lipoic acid reverse the age-related deficit in glutathione redox state in skeletal muscle and heart tissues. *Mech Ageing Dev* 2004;**125**:507–12.

65. Harding JJ. Free and protein-bound glutathione in normal and cataractous human lenses. *Biochem J* 1970;**117**:957–60.

66. Suh JH, Wang H, Liu RM, Liu J, Hagen TM. (R)-alpha-lipoic acid reverses the age-related loss in GSH redox status in post-mitotic tissues: evidence for increased cysteine requirement for GSH synthesis. *Arch Biochem Biophys* 2004;**423**:126–35.

67. Rani PJ, Panneerselvam C. Effect of L-carnitine on brain lipid peroxidation and antioxidant enzymes in old rats. *J Gerontol A Biol Sci Med Sci* 2002;**57**:B134–7.

68. Esterbauer H, Schaur RJ, Zollner H. Chemistry and biochemistry of 4-hydroxynonenal, malonaldehyde and related aldehydes. *Free Radic Biol Med* 1991;**11**:81–128.

69. Berridge MJ, Bootman MD, Lipp P. Calcium – a life and death signal. *Nature* 1998;**395**:645–8.

70. Missiaen L, Robberecht W, Van Den Bosch L, Callewaert G, Parys JB, Wuytack F, et al. Abnormal intracellular Ca(2+) homeostasis and disease. *Cell Calcium* 2000;**28**:1–21.

71. Rhodes JD, Sanderson J. The mechanisms of calcium homeostasis and signalling in the lens. *Exp Eye Res* 2009;**88**:226–34.

72. Andersson M, Sjostrand J, Andersson AK, Andersen B, Karlsson JO. Calpains in lens epithelium from patients with cataract. *Exp Eye Res* 1994;**59**:359–64.

73. Ma H, Hata I, Shih M, Fukiage C, Nakamura Y, Azuma M, Shearer TR. Lp82 is the dominant form of calpain in young mouse lens. *Exp Eye Res* 1999;**68**:447–56.

74. Ma H, Fukiage C, Kim YH, Duncan MK, Reed NA, Shih M, et al. Characterization and expression of calpain 10. A novel ubiquitous calpain with nuclear localization. *J Biol Chem* 2001;**276**:28525–31.

75. Shih M, Ma H, Nakajima E, David LL, Azuma M, Shearer TR. Biochemical properties of lens-specific calpain Lp85. *Exp Eye Res* 2006;**82**:146–52.

76. Elanchezhian R, Sakthivel M, Geraldine P, Thomas PA. Regulatory effect of acetyl-L-carnitine on expression of lenticular antioxidant and apoptotic genes in selenite-induced cataract. *Chem Biol Interact* 2010;**184**:346–51.

77. Arai M, Yoguchi A, Takizawa T, Yokoyama T, Kanda T, Kurabayashi M, Nagai R. Mechanism of doxorubicin-induced inhibition of sarcoplasmic reticulum Ca(2+)-ATPase gene transcription. *Circ Res* 2000;**86**:8–14.

78. Li L, Cheng C, Xia CH, White TW, Fletcher DA, Gong X. Connexin mediated cataract prevention in mice. *PLoS ONE* 2010;5.

79. Fleschner CR. Connexin 46 and connexin 50 in selenite cataract. *Ophthalmic Res* 2006;**38**:24–8.

80. Wormstone IM, Collison DJ, Hansom SP, Duncan G. A focus on the human lens in vitro. *Environ Toxicol Pharmacol* 2006;**21**:215–21.

81. Li DW, Spector A. Hydrogen peroxide-induced expression of the proto-oncogenes, c-jun, c-fos and c-myc in rabbit lens epithelial cells. *Mol Cell Biochem* 1997;**173**:59–69.

82. Li WC, Spector A. Lens epithelial cell apoptosis is an early event in the development of UVB-induced cataract. *Free Radic Biol Med* 1996;**20**:301–11.

83. Tamada Y, Fukiage C, Nakamura Y, Azuma M, Kim YH, Shearer TR. Evidence for apoptosis in the selenite rat model of cataract. *Biochem Biophys Res Commun* 2000;**275**:300–6.

84. Lee EH, Wan XH, Song J, Kang JJ, Cho JW, Seo KY, Lee JH. Lens epithelial cell death and reduction of anti-apoptotic protein Bcl-2 in human anterior polar cataracts. *Mol Vis* 2002;**8**:235–40.

85. Majima K, Itonaga K, Yamamoto N, Marunouchi T. Localization of cell apoptosis in the opaque portion of anterior polar cataract and anterior capsulotomy margin. *Ophthalmologica* 2003;**217**:215–8.

86. Okamura N, Ito Y, Shibata MA, Ikeda T, Otsuki Y. Fas-mediated apoptosis in human lens epithelial cells of cataracts associated with diabetic retinopathy. *Med Electron Microsc* 2002;**35**:234–41.

87. Belusko PB, Nakajima T, Azuma M, Shearer TR. Expression changes in mRNAs and mitochondrial damage in lens epithelial cells with selenite. *Biochim Biophys Acta* 2003;**1623**:135–42.

88. Fan TJ, Han LH, Cong RS, Liang J. Caspase family proteases and apoptosis. *Acta Biochim Biophys Sin (Shanghai)* 2005;**37**:719–27.

89. Miura M, Zhu H, Rotello R, Hartwieg EA, Yuan J. Induction of apoptosis in fibroblasts by IL-1 beta-converting enzyme, a mammalian homolog of the *C. elegans* cell death gene ced-3. *Cell* 1993;**75**:653–60.

90. Nicholson DW, Ali A, Thornberry NA, Vaillancourt JP, Ding CK, Gallant M, et al. Identification and inhibition of the ICE/CED-3 protease necessary for mammalian apoptosis. *Nature* 1995;**376**:37–43.

91. Wang Z, Bunce GE, Hess JL. Selenite and Ca^{2+} homeostasis in the rat lens: effect on Ca-ATPase and passive Ca^{2+} transport. *Curr Eye Res* 1993;**12**:213–8.

92. Pillich RT, Scarsella G, Risuleo G. Reduction of apoptosis through the mitochondrial pathway by the administration of acetyl-L-carnitine to mouse fibroblasts in culture. *Exp Cell Res* 2005;**306**:1–8.

93. Trush MA, Kensler TW. An overview of the relationship between oxidative stress and chemical carcinogenesis. *Free Radic Biol Med* 1991;**10**:201–9.

94. Liu J, Atamna H, Kuratsune H, Ames BN. Delaying brain mitochondrial decay and aging with mitochondrial antioxidants and metabolites. *Ann N Y Acad Sci* 2002;**959**:133–66.

95. Benedek GB. Theory of transparency of the eye. *Appl Opt* 1971;**10**:459–73.

96. Delaye M, Tardieu A. Short-range order of crystallin proteins accounts for eye lens transparency. *Nature* 1983;**302**:415–7.

97. Spector A. Oxidation and cataract. *Ciba Found Symp* 1984;**106**:48–64.

98. Harding JJ. Post-translational modification of lens proteins in cataract. *Lens Eye Toxic Res* 1991;**8**:245–50.

99. Clark JI, Matsushima H, David LL, Clark JM. Lens cytoskeleton and transparency: a model. *Eye (Lond)* 1999;**13**:417–24.

100. Tagliavini J, Gandolfi SA, Maraini G. Cytoskeleton abnormalities in human senile cataract. *Curr Eye Res* 1986;**5**:903–10.

101. Mousa GY, Trevithick JR. Actin in the lens: changes in actin during differentiation of lens epithelial cells *in vivo*. *Exp Eye Res* 1979;**29**:71–81.

102. Ireland M, Maisel H. A cytoskeletal protein unique to lens fiber cell differentiation. *Exp Eye Res* 1984;**38**:637–45.

103. Clark R, Zigman S, Lerman S. Studies on the structural proteins of the human lens. *Exp Eye Res* 1969;**8**:172–82.

104. Lubsen NH, Aarts HJ, Schoenmakers JG. The evolution of lenticular proteins: the beta- and gamma-crystallin super gene family. *Prog Biophys Mol Biol* 1988;**51**:47–76.

105. Caspers GJ, Leunissen JA, De Jong WW. The expanding small heat-shock protein family, and structure predictions of the conserved 'alpha-crystallin domain'. *J Mol Evol* 1995;**40**:238–48.

106. Chambers C, Russell P. Deletion mutation in an eye lens beta-crystallin. An animal model for inherited cataracts. *J Biol Chem* 1991;**266**:6742–6.

107. David LL, Azuma M, Shearer TR. Cataract and the acceleration of calpain-induced beta-crystallin insolubilization occurring during normal maturation of rat lens. *Invest Ophthalmol Vis Sci* 1994;**35**:785–93.

108. Carter JM, Hutcheson AM, Quinlan RA. *In vitro* studies on the assembly properties of the lens proteins CP49:CP115: coassembly with alpha-crystallin but not with vimentin. *Exp Eye Res* 1995;**60**:181–92.

109. Nicholl ID, Quinlan RA. Chaperone activity of alpha-crystallins modulates intermediate filament assembly. *EMBO J* 1994;**13**:945–53.

110. Sakthivel M, Geraldine P, Thomas PA. Alterations in the lenticular protein profile in experimental selenite-induced cataractogenesis and prevention by ellagic acid. *Graefes Arch Clin Exp Ophthalmol* 2011;**249**:1201–10.

111. Matsushima H, David LL, Hiraoka T, Clark JI. Loss of cytoskeletal proteins and lens cell opacification in the selenite cataract model. *Exp Eye Res* 1997;**64**:387–95.

112. Mitton KP, Hess JL, Bunce GE. Causes of decreased phase transition temperature in selenite cataract model. *Invest Ophthalmol Vis Sci* 1995;**36**:914–24.

51

Taurine Deficiency and the Eye

Nicolas Froger[1,2,3], José-Alain Sahel[1,2,3,4,5,6,7], Serge Picaud[1,2,3,6]

[1]INSERM, U968, Institut de la Vision, Paris, France; [2]Sorbonne Universités, UPMC Univ Paris 06, UMR_S 968 Paris, France; [3]CNRS, UMR 7210, Institut de la Vision, Paris, France; [4]Centre Hospitalier National d'Ophtalmologie des Quinze-Vingts, Paris, France; [5]Institute of Ophthalmology, University College of London, UK; [6]Fondation Ophtalmologique Adolphe de Rothschild, Paris, France; [7]French Academy of Sciences, Paris, France

INTRODUCTION

Distribution of Taurine

Taurine or 2-aminoethane-sulfonic acid is a free amino acid, which is found almost exclusively in the animal kingdom. It occurs in traces in the plant kingdom, representing less than 10 nmol/g fresh tissue. By contrast, in mammals, taurine is the most abundant intracellular free amino acid, supplied ubiquitously to all tissues.[1] However, some differences between species exist, such that overall taurine levels are lower in primates than in rodents or in rabbits (Table 51.1), probably owing to their reduced ability to synthesize taurine (see 'Metabolism of Taurine' section). High concentrations of taurine are found in skeletal muscles, which constitute the most important pool of taurine in the adult body.[1] All vital organs, including the central nervous system (CNS), heart, liver, and kidney, also contain elevated concentrations of taurine.[2] High taurine amounts are found in the retina, where they exceed those observed in brain from all adult species examined,[6] going up to 50 μmol/g tissue in rats (Table 51.1). Specifically, in this neural part of the eye, taurine represents the most abundant amino acid after glutamate.[7] The taurine concentration is even greater in the developing CNS tissues, with a maximum in the developing retina, where it can reach more than 100 μmol/g tissue in mice.[8] During development, taurine levels are very high in mammalian neonates, falling progressively until adulthood to one-third of the birth level.[8] In the retina of rat neonates, Macaione and colleagues[7] reported that taurine levels progressively reach a peak of 50 μmol/mg protein at 30 days old then decrease and stabilize at 40 μmol/mg protein after 45 days old. The taurine increase correlates with the formation of the photoreceptor layer. Quantitative measurements showed

that a significant proportion of this taurine was provided to pups from the mother's milk.[8] Indeed, milk is highly enriched in taurine in mammals (≈600 μM in gerbil, ≈300 μM in cat, and ≈40 μM in human[2]). Accordingly, feeding infant monkeys with vegetal milk, such as soy milk, can induce a significant taurine depletion.[9] As a consequence, taurine has been introduced into artificial milk for human babies.

Metabolism of Taurine

One source of taurine in mammals relies on endogenous synthesis occurring in different tissues (for reviews see Huxtable[1] and Hayes and Sturman[2]). The major pathway for this taurine endogenous formation is from L-methionine and/or L-cysteine (Fig. 51.1); despite this, other theoretical biosynthesis processes can follow several other enzymatic pathways linked to the metabolism of sulfur amino acids. In the first part of this major metabolic pathway, the precursor L-methionine

TABLE 51.1 Taurine Levels in Plasma and Tissues from Rodents and Primates

Species	Plasma	Tissues (μmol/g Wet Weight)	
	(μM)	Brain	Heart
Rat	222–450	3–9	27
Mouse	740–770	9	43
Monkey	100–190	2–4	11
Human	80	2	6

Values are taurine concentrations measured in adult tissues, except for the value for the monkey retina, which was obtained in 5-month-old animals.
Source: Data in tissues provided from Hayes and Sturman (1981)[2]; taurine concentrations in plasma from rats and mice are ranges described in Gaucher et al. (2012), Jammoul et al. (2009), and Froger et al. (2012).[3–5]

FIGURE 51.1 Metabolic pathway of sulfur amino acids leading to taurine biosynthesis in mammalian liver and brain.

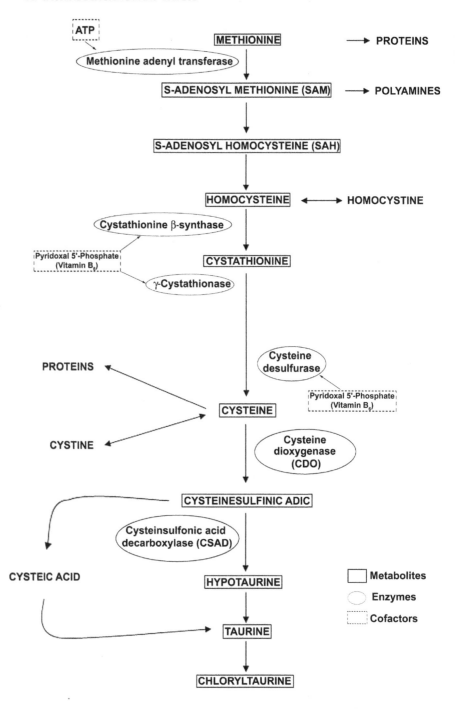

is converted into L-cysteine, after four intermediate steps involving the formation of *S*-adenosyl methionine (SAM), *S*-adenosyl homocysteine (SAH), homocysteine, and cystathionine (Fig. 51.1). The formation of cystathionine results from the condensation of one homocysteine with one L-serine, and this reaction is catalyzed by a cystathionine β-synthase, indicating that the serine level can theoretically regulate taurine biosynthesis. Finally, cystathionine produces one L-cysteine after the successive intervention of two enzymes, γ-cystathionase and cysteine desulfurase (Fig. 51.1).

In the second part of this metabolic pathway, L-cysteine, resulting either from L-methionine catabolism or from the diet, is transformed to cysteinesulfonic acid by the activity of cysteinedioxygenase (CDO) (Fig. 51.1). The product is then changed in hypotaurine following the action of cysteinesulfonic decarboxylase (CSAD). The concentrations of these enzymes in tissue constitute a limiting factor for endogenous taurine synthesis. Three enzymes involved in this taurine biosynthesis pathway, cystathionine synthase, γ-cystathionase, and CSAD, require pyridoxal 5′-phosphate (vitamin B_6) as a cofactor to exert their full enzymatic activities (Fig. 51.1). This means that dietary deficiency in vitamin B_6 can also result in reduced taurine synthesis, leading to taurine depletion.[10]

Taurine anabolism was first characterized in the liver[11] and subsequently generalized in brain tissues[12] (for a review see Huxtable[1]). Evidence that taurine synthesis can occur in the brain was first provided by incubating radioactive [^{35}S]methionine or [^{35}S]cysteine in rat cortical slices to produce radioactive labeled taurine.[12] The two enzymes for taurine synthesis from cysteine, CDO and CSAD, were then identified in rat brain.[13] However, taurine synthesis was restricted to cell bodies and dendrites of neurons since no CSAD was found in axons.[14] Dominy et al.[15] suggested more recently that taurine synthesis in brain does not have a purely neuronal origin but could result from metabolic interactions between neurons and astroglia.

The capacity for taurine biosynthesis is different between hepatic and cerebral tissue and also depends on age and species. Both liver and brain of young animals from all species studied have a reduced capacity for taurine synthesis compared with adults.[16] This observation is explained by a limited activity of CSAD and accounts for the need to provide taurine intake in baby food supplementation.[17] For adults, species including cats, monkeys, and humans have extremely low activity of CSAD in liver, thus limiting the endogenous synthesis of taurine and causing dependence on nutritional taurine intake. By contrast, rats and dogs have a higher concentration of this enzyme, thus conferring a weaker dependence on nutritional exogenous taurine intake. For example, 80% of cysteine is catabolized to taurine in rats, whereas in cats, only 20% of cysteine goes to this metabolic pathway. However, in all species, taurine biosynthesis cannot explain the physiologic taurine concentrations in tissues, meaning that taurine intake provided by nutrition is essential to maintain taurine homeostasis.

Taurine in the Diet

As stated above, taurine endogenous synthesis cannot account for physiologic taurine content in tissue. Accordingly, exogenous taurine should be provided to tissues through nutrition. Because taurine is exclusively found in the animal kingdom, it is highly enriched in meat, seafood, and fish (Table 51.2) (see also Zhao[18]). For example, taurine content reaches up to $30\,\mu mol/g$ (wet weight) in beef and $9\,\mu mol/g$ (wet weight) in lamb. In fish, some reported taurine concentrations are remarkably high: up to $83\,\mu mol/g$ (wet weight) in yellowtail (Japanese amberjack) and $78\,\mu mol/g$ (wet weight) in mackerel. Taurine is also present in high concentrations in milk and eggs. By contrast, taurine is absent, or occurs in traces, in vegetables and mushrooms.[1] Accordingly, people following a vegan diet can develop taurine deficiency,[20,21] further demonstrating that in humans, endogenous taurine

TABLE 51.2 Taurine Content in Different Foods

Seafood and Fish	Taurine Content	Meat	Taurine Content
	(mg/100 g)		(mg/100 g)
Hairtail fish	56	Beef muscle	50–100
Carp	90	Beef liver	42
Crab	278	Chicken	378
Octopus	390	Uncooked lamb muscle	310
Prawn	143	Cooked lamb muscle	171
Shrimp	115	Pork muscle	118
Sole	256	Pork kidney	120
Tuna	332	Pork liver	42
Yellow fish	88	Quail	9.5–28

Values, expressed as mg of taurine per 100 g food, represent the taurine concentrations measured in each corresponding food.
Source: Data from Zhao (1994) and Purchas et al. (2004).[18,19]

biosynthesis alone is not sufficient to sustain *in situ* physiologic amounts of taurine.

Taurine and Nutrition

In cats, taurine is considered an essential amino acid, while in primates (including humans), taurine is classed as a conditioned essential amino acid, owing to its crucial role in development and a limited capacity for endogenous synthesis. Taurine tissue intake from nutrition is highly dependent on taurine transporter (TauT) expression. Taurine from our diet has to cross from the intestine to the blood by taurine transport mechanisms. There are two intestinal transport mechanisms that mediate taurine movements across the brush-border membrane of the human intestine.[22] Intestinal taurine transport relies on a high-affinity, low-capacity Na^+/Cl^--dependent TauT (*SLC6A6* gene). TauT has been isolated from a number of species and tissues including rat, mouse, and human. Nevertheless, previous studies suggested that the major absorptive mechanism for amino acids such as taurine was a low-affinity transport.[23] An H^+-coupled transporter of a broad range of amino acids was identified at the brush-border membrane of human intestinal cells, named the proton-coupled amino acid transporter-1 (PAT1).[24] PAT1 is found in rat, mouse, and human tissues but is absent from the small intestine of rabbit and guinea pig, in which the high-affinity TauT remains the only transport mechanism for taurine uptake from nutrients.[23,25] A recent study assessed the differentiated functions between PAT1 and TauT and suggested that PAT1 may be responsible for the bulk of taurine uptake during meals, whereas TauT may be involved in active taurine uptake into the intestinal epithelium between meals, even at low concentrations.[22]

TAURINE DEPLETION AND PHOTORECEPTOR DEGENERATION

Taurine-Free Diet

Taurine is the most abundant amino acid in the mammalian retina, both during development and in adulthood, and the retina is the organ containing the highest taurine concentrations (Table 51.1). The distribution of TauT in the retina[26] suggests that plasma taurine is sequestered either by the retinal pigment epithelium (RPE) in the outer retina to be diffused to photoreceptors or by capillary endothelium to be provided to the inner retina and retinal ganglion cell (RGC) layer. Despite the huge retinal concentration of this amino acid having been known about for a long time, its functional role remains enigmatic. However, the importance of taurine to the retina was demonstrated by the observation of retinal damage with massive photoreceptor loss during prolonged taurine depletion.[27] Such retinal damage was first described in cats on a taurine-free diet[27]; as previously mentioned (see 'Metabolism of Taurine' section), cats have a low endogenous production of taurine. Since the original study in 1975, the consequences of taurine depletion have been well documented, first showing a focal lesion at the area centralis (feline central retinal degeneration), then progressing to form a horizontal band of retinal degeneration, and finally extending to the whole retina.[28] The retinal damage primarily affects the outer layers, with a structural disorganization of photoreceptor outer segments leading to cell death, suggesting that photoreceptors are the more vulnerable retinal neurons to taurine deficiency. These retinal lesions were associated with depressed (or absent) electroretinograms (ERGs),[29] reflecting functional alterations prior to any morphologic changes in the retinal tissue. At later stages, histologic examination revealed a surprising displacement of photoreceptor cell bodies into the subretinal space. Such tissue damage was seen as soon as taurine plasma concentrations had been decreased by a factor of two or more.[29] In baby primates (cebus monkey and cynomolgus monkey), taurine-free milk produced similar taurine depletion,[30] inducing a loss of visual acuity associated with morphologic changes in central foveal cones and even retinal degeneration.[31]

Depletion Induced by Taurine Transporter Inhibition or Suppression

In addition to feeding a taurine- or vitamin B_6-free diet, taurine deficiency can be achieved following prolonged pharmacologic treatments in animals. The pharmacologic agents first used to induce taurine deficiency were the TauT blockers β-alanine and guanidoethanesulfonate (GES). These agents limit both exogenous taurine assimilation from the diet and taurine uptake within cells and tissues. Measurements of [³H]taurine uptake on isolated heart and retinal tissues demonstrated that β-amino acids can exert specific and competitive inhibition of taurine uptake, whereas α-amino acids exhibit no affinity for TauT. Among them, L-glutamate and L-cysteine were active at millimolar concentrations, but L-β-alanine exerted a more efficient action on TauT inhibition. GES, a structural analogue of taurine discovered by Huxtable's team, also generates a powerful inhibition of TauT. The authors reported that administration of GES for 4 weeks induced an 80% decrease in taurine content in the heart, 76% in the liver, and 67% in the cerebellum. In the retina, taurine depletion induced by chronic GES treatment, as well as β-alanine in rats, has been shown to induce morphologic alterations in photoreceptors specifically located at the outer segments, similar to the findings initially described in cats fed a taurine-free diet. These results with GES were recently confirmed in a study showing that cone photoreceptors were much more sensitive to taurine depletion than rods.[3] The role of TauT was also indicated by the photoreceptor degeneration observed in taurine transporter knockout (TauT KO) mice.[32]

TAURINE DEPLETION AND RETINAL GANGLION CELL DEGENERATION

Taurine Depletion Induced by Pharmacologic Agents

A recent study demonstrated that taurine depletion is the likely cause of the retinal toxicity of vigabatrin, an irreversible inhibitor of γ-aminobutyric acid (GABA) transaminase.[4] Despite this retinal toxicity, the drug remains a third-line treatment for refractory complex partial seizures in adults and the first-line treatment in infantile spasms. This retinal toxicity was first demonstrated in rats in 1987 by Butler and colleagues[33] before commercialization of the drug. They described a disorganization of the photoreceptor nuclear layer, with displaced photoreceptor nuclei moving towards the RPE and sclera. In humans, chronic vigabatrin treatments are associated with a strong risk of retinopathy, evidenced by an irreversible peripheral visual field constriction. This visual field constriction, first evoked in 1997 by Eke et al.,[34] can start after 4 months of treatment and is established after 9 months of treatment in adults and 11 months in children. One postmortem histologic study indicated retinal damage in both the photoreceptor layer and the RGC layer but with an apparent primary site of lesion in RGCs.[35] Damage in RGCs was confirmed by in vivo imaging techniques in both infants and adults.[36,37]

An experimental animal study indicated a photopic ERG decrease with a sequence of events in the photoreceptor layer from cone damage with a glial response to rod cell layer disorganization and bipolar cell plasticity (Fig. 51.2).[38,39] Measurements of amino acid concentrations in the plasma of vigabatrin-treated rats revealed a decrease in the taurine concentration.[4] This observation of vigabatrin-induced taurine depletion was then extended to vigabatrin-treated children.[4] If taurine supplementation can limit all the vigabatrin-induced damage in the rat photoreceptor layer,[4] these observations do not explain the RGC degeneration reported in patients.

To investigate whether taurine depletion could also trigger RGC degeneration, the authors investigated whether RGCs were degenerating in vigabatrin-treated animals.[40] This study not only showed a decrease in RGCs but also demonstrated that taurine supplementation prevented this RGC loss.[40] To confirm that taurine depletion could trigger RGC degeneration, GES-treated mice were re-examined to determine whether these taurine-depleted animals would also exhibit RGC loss. This study confirmed that taurine depletion could induce both cone and RGC loss.[3]

Taurine Depletion in Retinal Diseases with Retinal Ganglion Cell Degeneration

Having discovered that taurine depletion can induce RGC degeneration as in the retinal toxicity of vigabatrin, the authors wondered whether this mechanism could

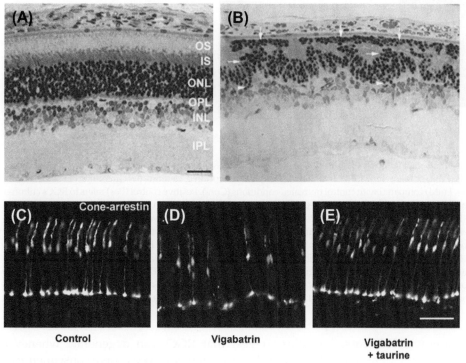

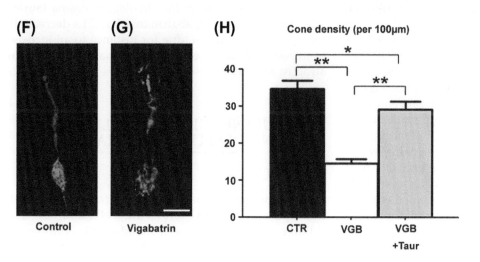

FIGURE 51.2 Photoreceptor damage in taurine-depleted animals. (A,B) Effects of chronic administration of vigabatrin on retinal histology. Retinal section from vigabatrin-treated rat presented histologic disorganization (B) compared with that from control rats (A). (C–H) Prevention of vigabatrin-induced cone toxicity by taurine supplementation in neonate rats. Immunostaining of cones with the specific immune marker cone-arrestin shows that vigabatrin treatment alters the cone density (D), compared with control untreated animals (C), which can be recovered by simultaneous taurine supplementation (E). Enlarged views of cone-arrestin immunolabeled photoreceptor showing morphologic alterations in outer segment of cones from vigabatrin-treated rats (G) compared with control untreated rats (F). The counting of cone density (H) corroborates the observations in (C)–(D). *Source: (A–B) From Duboc et al. (2004),[38] with the permission of the publisher; (C–H) from Jammoul et al. (2010),[40] with the permission of the publisher.*

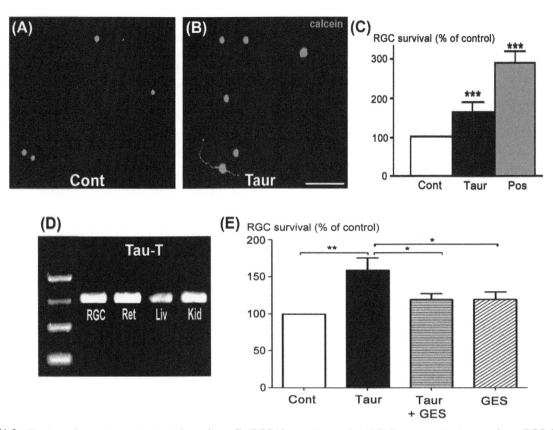

FIGURE 51.3 Taurine-enhanced survival of retinal ganglion cells (RGCs) by taurine uptake. (A,B) Representative images of pure RGCs labeled with calcein after 6 days *in vitro*, in control conditions (Cont) or following taurine incubation (Taur) into the cultured medium. (C) Quantification of the density of surviving RGC after taurine incubation (Taur, 1mM) compared with control untreated conditions (Cont). Positive control (Pos) refers to RGCs cultured in an enriched medium (B27 supplement). Data represent mean±SEM of 21 independent experiments in which each RGC density, counted in triplicate for each condition, was normalized to that of the control condition (Cont) at 6 days *in vitro*. ***$P < 0.001$, one-way analysis of variance followed by Dunn's *post hoc* test. (D) Gene amplification (polymerase chain reaction) of the taurine transporter (TauT) performed on total complementary DNA obtained from total RNA extracted from freshly purified rat RGCs, rat full retina (Ret), rat liver (Liv), or rat kidney (Kid). TauT gene amplification revealed high expression in pure retinal ganglion cells (A). (E) Implication of TauT in the taurine-induced increase in RGC survival: quantification of RGC densities of alive RGCs, treated with taurine alone (Taur), taurine plus the TauT blocker guanidinoethanesulfonate (Taur+GES), and guanidinoethanesulfonate (GES), compared with control conditions (Cont). Data represent mean ± SEM of 9-11 independent experiments. *$p<0.05$ and **$p<0.01$ as compared to indicated groups. One-way analysis of variance followed by Dunn's post hoc test. *Source: Adapted from Froger* et al. *(2012)[5] (online, free access).*

have medical relevance in retinal diseases with RGC loss, such as glaucoma.[5] They first tested whether taurine could directly affect RGC survival using a pure adult rat RGC culture. Taurine incubation at a millimolar concentration significantly improved RGC survival after 6 days *in vitro* (Fig. 51.3).[5] This effect was dependent on taurine uptake function, since the blockade of TauT suppressed taurine-induced enhanced RGC survival in culture (Fig. 51.3). Similarly, taurine significantly prevented the glutamate-induced excitotoxicity of ganglion cell death on retinal explants cultured for 4 days.[5] When taurine supplementation was provided in the drinking water, two models of glaucoma, DBA/2J mice and rats with epicleral vein occlusion, showed an increase in RGC survival.[5] Finally, a similar taurine supplementation increased RGC survival in a model of retinitis pigmentosa, P23H rats, displaying secondary RGC degeneration following the primary photoreceptor loss.[5] These

studies suggest that RGCs can degenerate whenever local retinal taurine depletion occurs as a consequence of (1) a general decrease in the circulating plasma taurine concentration, as in vigabatrin toxicity; (2) a decrease in retinal blood perfusion, due for instance to an increase in intraocular pressure as in glaucoma; or (3) vascular atrophy, as in the late stages of retinitis pigmentosa.

TAURINE-INDUCED MOLECULAR AND CELLULAR MECHANISMS

Taurine Uptake

In many cells, taurine is taken up from the extracellular space by TauT to exert its intracellular action. The cellular content of taurine is based on the equilibrium between its active transport and its passive release from the cell, regulated by osmotic pressure. Active taurine transport is mediated by a selective TauT, which is dependent on Na^+ and Cl^- ions. The taurine concentration in the intracellular space can reach 1 mM,[1] whereas plasma concentrations are between 80 and 700 μM, varying according to species (Table 51.1). This difference in taurine concentration between the intracellular compartment and plasma suggests that TauT plays a crucial role in maintaining the intracellular taurine content, which results in the homeostasis between the active taurine uptake and the passive taurine diffusion due to osmotic pressure. The disruption of TauT function by either pharmacologic agents or its genetic invalidation (TauT KO mice) induces severe taurine depletion in the plasma leading to undetectable taurine concentrations in some tissues. This effect is probably a consequence of taurine uptake inhibition at the intestinal level. In TauT KO mice, taurine depletion resulted in alterations in the development of the kidney, heart, and brain, as well as a progressive and early degeneration of photoreceptor loss.[41] In the retina, TauT is expressed not only by cells generating the hematoretinal barrier (endothelial cells and RPE cells) but also in photoreceptors and RGCs. This cellular localization of TauT suggests that its functional role could involve intracellular actions, such as osmoregulation and reduction in oxidative stress.

Antioxidant Action

Taurine plays a major role in cell physiology owing to its antioxidant properties. The sulfonate group in the molecule confers zwitterionic properties to taurine in the physiologic pH range. This sulfur group enables taurine to directly counteract the production of reactive oxygen species (ROS). For example, taurine can neutralize hypochlorous acid, nitric oxide, and hydrogen peroxide. Thus, taurine inserted into membranes could prevent membrane permeabilization induced by ROS production. In addition, by inhibiting the activity of methyltransferase, taurine affects membrane fluidity, improving its effects in preventing oxidative stress. This protective role of taurine against ROS production was established under light exposure and in aging (for a review see Schaffer et al.[42]).

Nonselective Agonist at γ-Aminobutyric Acid, Glycine, and 5-Hydroxytryptamine Receptors

In the retina and more generally in the CNS, taurine is considered as an agonist for all GABA receptors, even $GABA_C$ ionotropic and $GABA_B$ metabotropic receptors.[43,44] Activation of strychnine-sensitive glycine ionotropic receptors by taurine was also evoked in RGCs and

cone photoreceptors.[45] A recent study showed taurine-induced 5-hydroxytryptamine receptor activation in retinal neurons.[46] Although some studies have suggested the existence of specific taurine receptors, they have never been clearly identified.

OTHER EYE STRUCTURES

Taurine was found in large amount in the other eye structures where it constitutes the most abundant free amino acid, as in the retina.[47] Thus, taurine is present in cornea (10 μM/g dry weight), iris–ciliary body (9.2 μM/g dry weight), and lens (13.0 μM/g dry weight). Taurine was also detected in the vitreous at a concentration of 1.72 μmol/ml of vitreous fluid.[47] Precursor amino acids involved in taurine biosynthesis (i.e., methionine and cysteine) were also present in these different structures. While methionine was found at concentrations ranging between 0.09 and 0.37 μM, cysteine was measured as traces.[47] However, few data are available describing the functional role in these eye structures. In the lens, taurine uptake was weak, suggesting a potential endogenous biosynthesis rather than exogenous intake from nutrition. The involvement of taurine in cataract was evoked owing to the decreased taurine contents observed in human senile cataractous lenses, as well as in rat lenses with galactose-elicited cataract.[48] By contrast, taurine incubation can reduce lens damage in a model of diabetic cataract in vitro.[49] In the cornea, taurine uptake is also weak, suggesting endogenous production to reach a level of 10 μM. The role of taurine in osmoregulation of corneal epithelium was demonstrated such that taurine was proposed to prevent hypertonicity-induced ocular dry eye.[50] The vitreous probably acts as an exchange medium between the anterior segment and the retina and serves as the lymphatic drainage route. Taurine may participate in these processes, although its exact role in the vitreous is still unknown.

TAKE-HOME MESSAGES

- Taurine is the most abundant amino acid in the eye, particularly in retina, but its ocular functions are still unclear.
- Although endogenous synthesis occurs in liver and brain from mammals, the main taurine content is provided from nutrition.
- Because taurine is exclusively found in the animal kingdom, its nutritive sources have to be found in meat, fish, seafood, eggs, and milk.
- A specific taurine transporter is involved in the taurine uptake from nutrients to blood and from blood to specific tissues and their constitutive cells.

- Taurine depletion can be induced by taurine-free diet in species with reduced endogenous taurine synthesis, such as cats and primates.
- Depletion of taurine was first reported to induce retinal degeneration characterized by photoreceptor loss.
- Retinal ganglion cells and cone photoreceptors are the two primary sites of retinal damage under taurine deficiency.
- Taurine supplementation can provide neuroprotection to retinal ganglion cells in animal models of glaucoma.
- In other eye structures, taurine may protect from cataract and dry eye.

Acknowledgments

This work was supported by Institut National de la Santé et de la Recherche Médicale (INSERM), Pierre et Marie Curie University (UPMC), La Fondation Ophtalmologique A. de Rothschild (Paris), Agence Nationale pour la Recherche (ANR: GLAUCOME), the European Community contract TREATRUSH (no. HEALTH-F2-2010-242013), the Fédération des Aveugles de France, IRRP, the city of Paris, the Regional Council of Ile-de-France, and the French State program 'Investissements d'Avenir', managed by the Agence Nationale de la Recherche (LIFESENSES: ANR-10-LABX-65). Nicolas Froger received postdoctoral fellowships from the Fondation pour la Recherche Médicale and from the Fondation Rolland Bailly.

References

1. Huxtable RJ. Physiological actions of taurine. *Physiol Rev* 1992;**72**:101–63.
2. Hayes KC, Sturman JA. Taurine in metabolism. *Annu Rev Nutr* 1981;**1**:401–25.
3. Gaucher D, Arnault E, Husson Z, Froger N, Dubus E, Gondouin P, et al. Taurine deficiency damages retinal neurones: cone photoreceptors and retinal ganglion cells. *Amino Acids* 2012;**43**:1979–93.
4. Jammoul F, Wang Q, Nabbout R, Coriat C, Duboc A, Simonutti M, et al. Taurine deficiency is a cause of vigabatrin-induced retinal phototoxicity. *Ann Neurol* 2009;**65**:98–107.
5. Froger N, Cadetti L, Lorach H, Martins J, Bemelmans AP, Dubus E, et al. Taurine provides neuroprotection against retinal ganglion cell degeneration. *PLoS ONE* 2012;**7**:e42017.
6. Pasantes-Morales H. Taurine function in excitable tissues: the retina as a model. In: Moroan MM, editor. *Retinal Transmitters and Modulators: Models for the Brain*. Boca Raton, FL: CRC Press; 1985. pp. 33–62.
7. Macaione S, Ruggeri P, De Luca F, Tucci G. Free amino acids in developing rat retina. *J Neurochem* 1974;**22**:887–91.
8. Sturman JA. Taurine in development. *J Nutr* 1988;**118**:1169–76.
9. Sturman JA, Messing JM, Rossi SS, Hofmann AF, Neuringer MD. Tissue taurine content and conjugated bile acid composition of rhesus monkey infants fed a human infant soy-protein formula with or without taurine supplementation for 3 months. *Neurochem Res* 1988;**13**:311–6.
10. Yamaguchi K, Shigehisa S, Sakakibara S, Hosokawa Y, Ueda I. Cysteine metabolism *in vivo* of vitamin B$_6$-deficient rats. *Biochim Biophys Acta* 1975;**381**:1–8.
11. Lombardini JB, Singer TP, Boyer PD. Cystein oxygenase. II. Studies on the mechanism of the reaction with ^{18}oxygen. *J Biol Chem* 1969;**244**:1172–5.

12. Tappaz M, Almarghini K, Legay F, Remy A. Taurine biosynthesis enzyme cysteine sulfinate decarboxylase (CSD) from brain: the long and tricky trail to identification. *Neurochem Res* 1992;**17**:849–59.
13. Remy A, Henry S, Tappaz M. Specific antiserum and monoclonal antibodies against the taurine biosynthesis enzyme cysteine sulfinate decarboxylase: identity of brain and liver enzyme. *J Neurochem* 1990;**54**:870–9.
14. Sturman JA. Cysteinesulfinic acid decarboxylase activity in the mammalian nervous system: absence from axons. *J Neurochem* 1981;**36**:304–6.
15. Dominy J, Eller S, Dawson Jr R. Building biosynthetic schools: reviewing compartmentation of CNS taurine synthesis. *Neurochem Res* 2004;**29**:97–103.
16. Loriette C, Chatagner F. Cysteine oxidase and cysteine sulfinic acid decarboxylase in developing rat liver. *Experientia* 1978;**34**:981–2.
17. Gaull GE. Taurine in pediatric nutrition: review and update. *Pediatrics* 1989;**83**:433–42.
18. Zhao X-H. Dietary protein, amino acids and their relation to health. *Asia Pac J Clin Nutr* 1994;**3**:131–4.
19. Purchas RW, Rutherfurd SM, Pearce PD, Vather R, Wilkinson BH. Concentrations in beef and lamb of taurine, carnosine, coenzyme Q(10), and creatine. *Meat Sci* 2004;**66**:629–37.
20. Laidlaw SA, Shultz TD, Cecchino JT, Kopple JD. Plasma and urine taurine levels in vegans. *Am J Clin Nutr* 1988;**47**:660–3.
21. Rana SK, Sanders TA. Taurine concentrations in the diet, plasma, urine and breast milk of vegans compared with omnivores. *Br J Nutr* 1986;**56**:17–27.
22. Anderson CM, Howard A, Walters JR, Ganapathy V, Thwaites DT. Taurine uptake across the human intestinal brush-border membrane is via two transporters: H+-coupled PAT1 (SLC36A1) and Na+- and Cl(−)-dependent TauT (SLC6A6). *J Physiol* 2009;**587**:731–44.
23. Munck LK, Munck BG. Distinction between chloride-dependent transport systems for taurine and beta-alanine in rabbit ileum. *Am J Physiol* 1992;**262**:G609–15.
24. Thwaites DT, McEwan GT, Simmons NL. The role of the proton electrochemical gradient in the transepithelial absorption of amino acids by human intestinal Caco-2 cell monolayers. *J Membr Biol* 1995;**145**:245–56.
25. Munck LK, Munck BG. Chloride-dependent intestinal transport of imino and beta-amino acids in the guinea pig and rat. *Am J Physiol* 1994;**266**:R997–1007.
26. Vinnakota S, Qian X, Egal H, Sarthy V, Sarkar HK. Molecular characterization and in situ localization of a mouse retinal taurine transporter. *J Neurochem* 1997;**69**:2238–50.
27. Hayes KC, Carey RE, Schmidt SY. Retinal degeneration associated with taurine deficiency in the cat. *Science* 1975;**188**:949–51.
28. Leon A, Levick WR, Sarossy MG. Lesion topography and new histological features in feline taurine deficiency retinopathy. *Exp Eye Res* 1995;**61**:731–41.
29. Schmidt SY, Berson EL, Watson G, Huang C. Retinal degeneration in cats fed casein. III. Taurine deficiency and ERG amplitudes. *Invest Ophthalmol Vis Sci* 1977;**16**:673–8.
30. Hayes KC, Stephan ZF, Sturman JA. Growth depression in taurine-depleted infant monkeys. *J Nutr* 1980;**110**:2058–64.
31. Imaki H, Moretz R, Wisniewski H, Neuringer M, Sturman J. Retinal degeneration in 3-month-old rhesus monkey infants fed a taurine-free human infant formula. *J Neurosci Res* 1987;**18**:602–14.
32. Rascher K, Servos G, Berthold G, Hartwig HG, Warskulat U, Heller-Stilb B, Haussinger D. Light deprivation slows but does not prevent the loss of photoreceptors in taurine transporter knockout mice. *Vis Res* 2004;**44**:2091–100.
33. Butler WH, Ford GP, Newberne JW. A study of the effects of vigabatrin on the central nervous system and retina of Sprague Dawley and Lister-Hooded rats. *Toxicol Pathol* 1987;**15**:143–8.

34. Eke T, Talbot JF, Lawden MC. Severe persistent visual field constriction associated with vigabatrin. *BMJ* 1997;**314**:180–1.

35. Ravindran J, Blumbergs P, Crompton J, Pietris G, Waddy H. Visual field loss associated with vigabatrin: pathological correlations. *J Neurol Neurosurg Psychiatry* 2001;**70**:787–9.

36. Frisen L, Malmgren K. Characterization of vigabatrin-associated optic atrophy. *Acta Ophthalmol Scand* 2003;**81**:466–73.

37. Wild JM, Robson CR, Jones AL, Cunliffe IA, Smith PE. Detecting vigabatrin toxicity by imaging of the retinal nerve fiber layer. *Invest Ophthalmol Vis Sci* 2006;**47**:917–24.

38. Duboc A, Hanoteau N, Simonutti M, Rudolf G, Nehlig A, Sahel JA, Picaud S. Vigabatrin, the GABA-transaminase inhibitor, damages cone photoreceptors in rats. *Ann Neurol* 2004;**55**:695–705.

39. Wang QP, Jammoul F, Duboc A, Gong J, Simonutti M, Dubus E, et al. Treatment of epilepsy: the GABA-transaminase inhibitor, vigabatrin, induces neuronal plasticity in the mouse retina. *Eur J Neurosci* 2008;**27**:2177–87.

40. Jammoul F, Degardin J, Pain D, Gondouin P, Simonutti M, Dubus E, et al. Taurine deficiency damages photoreceptors and retinal ganglion cells in vigabatrin-treated neonatal rats. *Mol Cell Neurosci* 2010;**43**:414–21.

41. Heller-Stilb B, van Roeyen C, Rascher K, Hartwig HG, Huth A, Seeliger MW, et al. Disruption of the taurine transporter gene (taut) leads to retinal degeneration in mice. *FASEB J* 2002;**16**:231–3.

42. Schaffer SW, Azuma J, Mozaffari M. Role of antioxidant activity of taurine in diabetes. *Can J Physiol Pharmacol* 2009;**87**:91–9.

43. Albrecht J, Schousboe A. Taurine interaction with neurotransmitter receptors in the CNS: an update. *Neurochem Res* 2005;**30**:1615–21.

44. Jones SM, Palmer MJ. Activation of the tonic GABAC receptor current in retinal bipolar cell terminals by nonvesicular GABA release. *J Neurophysiol* 2009;**102**:691–9.

45. Balse E, Tessier LH, Forster V, Roux MJ, Sahel JA, Picaud S. Glycine receptors in a population of adult mammalian cones. *J Physiol* 2006;**571**:391–401.

46. Bulley S, Liu Y, Ripps H, Shen W. Taurine activates delayed rectifier KV channels via a metabotropic pathway in retinal neurons. *J Physiol* 2013;**591**:123–32.

47. Heinamaki AA, Muhonen AS, Piha RS. Taurine and other free amino acids in the retina, vitreous, lens, iris–ciliary body, and cornea of the rat eye. *Neurochem Res* 1986;**11**:535–42.

48. Gupta K, Mathur RL. Distribution of taurine in the crystalline lens of vertebrate species and in cataractogenesis. *Exp Eye Res* 1983;**37**:379–84.

49. Kilic F, Bhardwaj R, Caulfeild J, Trevithick JR. Modelling cortical cataractogenesis 22: is *in vitro* reduction of damage in model diabetic rat cataract by taurine due to its antioxidant activity? *Exp Eye Res* 1999;**69**:291–300.

50. Shioda R, Reinach PS, Hisatsune T, Miyamoto Y. Osmosensitive taurine transporter expression and activity in human corneal epithelial cells. *Invest Ophthalmol Vis Sci* 2002;**43**:2916–22.

NUTRIGENOMICS AND MOLECULAR BIOLOGY OF EYE DISEASE

52

Gene Expression and the Impact of Antioxidant Supplements in the Cataractous Lens

Rijo Hayashi

Department of Ophthalmology, Koshigaya Hospital, Dokkyo University, School of Medicine, Koshigaya, Saitama, Japan

INTRODUCTION

Cataracts have been reported to be responsible for 51% of cases of blindness in the world, which represents approximately 20 million people, according to the statistics of World Health Organization (WHO). Cataracts are not only one of the most important causes of blindness in developing countries but also make up a significant portion of medical expenses in developed countries. According to statistics from the Japanese Ministry of Health, Labor and Welfare in 2010, nearly 40% of ophthalmologic medical costs were spent on treating cataracts in patients over 65 years old. The prevention of cataracts is one of the most important public health issues in the 21st century.

Cataracts develop owing to the interaction of multiple factors such as aging, ultraviolet radiation, and diabetes. Oxidation has been reported to be an important mechanism of cataract formation. The correlations between decreases in antioxidant substances such as superoxide dismutase (SOD), reduced glutathione (GSH), and L-ascorbic acid and the incidence and progression of cataract formation have been investigated.[1–3] Inhibiting the decrease in these antioxidant substances could be the key to delaying cataract formation.

ANTIOXIDANTS AND PREVENTION OF CATARACT PROGRESSION

Antioxidant substances are presumed to decrease oxidative damage and are therefore thought to be effective in the prevention of cataract progression. Results of animal studies support that antioxidants such as vitamins C and E can protect against oxidative damage and inhibit the progression of cataract formation.[4,5] Several epidemiologic studies have reported a decrease in cataracts correlated with dietary vitamin C[6–8] and vitamin E.[9] The intake of multivitamin supplements was also reported to decrease the risk of cataracts.[10,11]

Ocuvite+Lutein®, an antioxidant supplement, is reported to be effective in decreasing the risk of age-related macular degeneration.[12] Some laboratory reports support the antioxidative effects of lutein. Lutein has been reported to decrease intracellular hydrogen peroxide (H_2O_2) accumulation by scavenging superoxide and H_2O_2.[13] It has also been reported that lutein can effectively block H_2O_2-induced protein oxidation, lipid peroxidation, and DNA damage in lens epithelial cells.[14] In addition, the lutein included in Ocuvite+Lutein® has been reported to decrease the risk of cataract formation in epidemiologic studies.[15,16]

ROLE OF NUTRITION IN CATARACT PREVENTION: EPIDEMIOLOGIC STUDIES

Although *in vitro* studies indicated that lutein possesses antioxidant, cataract-preventing effects in the lens, epidemiologic studies demonstrated inconsistent results on the effects of lutein on human cataract prevention. Several epidemiologic studies indicated that high dietary intake of lutein was associated with a decreased risk of cataract formation.[16–19] There were also reports that indicated an association between serum lutein levels and a decrease in cataract severity.[16,17] However, other studies did not reveal a significant association between lutein and cataract prevention.[20–23]

Reviewing epidemiologic reports that did not support an effect of lutein on cataract prevention revealed that some studies were cross-sectional.[21,22] These studies investigated serum lutein levels and cataract severity simultaneously, which could make any causal relationship unclear. The amount of intake was also an important factor. The study of Vu et al.[23] utilized an intake amount less than one-third of other studies.[15,19] Importantly, most of the previous studies investigated the correlation between cataract grade and either the intake amount or serum level of antioxidant nutrients from different individuals, which could introduce confounding factors and interfere with the results. Measuring the changes before and after intake in the same person could be more accurate for assessing the effects of antioxidant supplementation. Confounding factors of personal differences, such as genetic factors and ultraviolet exposure, can be controlled by measuring samples from the same patient who has binocular cataracts of the same grade. In addition, several factors including age, body composition, and gender were reported to affect the absorption of lutein.[24,25] Investigating the same patient before and after supplementation can control for these factors. Moreover, instead of measuring the grades of cataracts, a more accurate method could be to investigate the changes in antioxidative enzyme levels and the production of antioxidative enzymes.

Serial studies were conducted that investigated the same patient before and after consuming consistent doses of an antioxidant supplement. One of these studies measured the changes in expression of glucose-6-phosphate dehydrogenase (G6PDH) and 18S ribosomal RNA (18S rRNA) in the lenticular anterior capsules of cataractous patients after intake of an antioxidant supplement.[26] The details of this investigation will be described later in this chapter.

POSSIBLE INDICATORS OF THE EFFECTS OF ANTIOXIDANT SUPPLEMENTS IN THE LENS

G6PDH

The antioxidative capacities of cells ultimately depend on the production of reduced nicotinamide adenine dinucleotide phosphate (NADPH).[27] G6PDH, the rate-limiting enzyme of the pentose phosphate cycle, determines the production of NADPH by controlling the metabolism of glucose. G6PDH is regulated by the ratio of NADPH to nicotinamide adenine dinucleotide phosphate (NADP) and is also regulated by hormones, nutrients, and oxidative stress.[28] As an effect of antioxidation, G6PDH was reported to decrease H_2O_2-induced cell death.[29] Since another product of the pentose phosphate cycle, ribose-5-phosphate (R5P), is essential for nucleic acid synthesis, G6PDH may also play an important role in transcription and protein synthesis and is required for the maintenance of basic cellular function.

The lens has an anaerobic status, and therefore the pentose phosphate cycle is important for energy production from glucose. As described above, G6PDH is the rate-limiting enzyme of the pentose phosphate cycle and

NADPH is one of the products of the pentose phosphate cycle. In the lens, G6PDH diverts over 10% of the tissue glucose to the hexose monophosphate shunt pathway[30] and supports reductive biosynthesis. It is an important enzyme of NADPH production and for maintaining a reducing environment and preventing oxidative damage. G6PDH could be an ideal indicator of the oxidative status and could be useful for measuring the effects of antioxidant supplements in the lens.

18S rRNA

18S rRNA is the active center of protein synthesis in the 40S ribosomal subunit. Increased numbers of ribosomes, which lead to increases in the amount of RNA transcription and protein synthesis, are presumed to be proportional to increases in 18S rRNA. It is well known that nucleic acids and proteins are damaged by oxidation. The expression level of 18S rRNA has been reported to decrease as markers of oxidative damage increase in arteriosclerotic plaque.[31] 18S rRNA could also be an ideal indicator for measuring the synthesis of proteins, including antioxidative enzymes, and for further investigating the effects of antioxidant supplements.

GENE EXPRESSION AS AN INDICATOR OF ANTIOXIDANT ENZYME PRODUCTION

Measuring changes in the synthesis of antioxidative enzymes after taking a supplement may be important for verifying the effectiveness of antioxidant supplementation in inhibiting cataract progression. Because the first stage of protein synthesis is initiated by messenger RNA (mRNA) transcription, measuring changes in mRNA expression may indicate changes in the synthesis of certain proteins such as antioxidative enzymes. Therefore, measuring changes in the mRNA expression levels of antioxidative enzymes in the lenticular anterior capsule, which mostly occur in lens epithelial cells, may be important for verifying the effects of antioxidant supplementation on the prevention of cataract progression.

G6PDH AND 18S rRNA GENE EXPRESSION IN LENTICULAR ANTERIOR CAPSULE AFTER OCUVITE+LUTEIN®

Changes in mRNA expression of G6PDH and 18S rRNA in the lenticular anterior capsule were measured by investigating the same patient before and after intake of an antioxidant supplement, Ocuvite+Lutein®.[26] The antioxidant nutrients were assimilated in a fixed amount

TABLE 52.1 Composition of Ocuvite + Lutein®

Substance	Amount
Lutein	6.0 mg
Vitamin C	300.0 mg
Vitamin E	60.0 mg
Vitamin B$_2$	3.0 mg
β-Carotene	1200.0 μg
Niacin	12.0 mg
Zinc	9.0 mg
Selenium	45.0 μg
Copper	0.6 mg
Manganese	1.5 mg

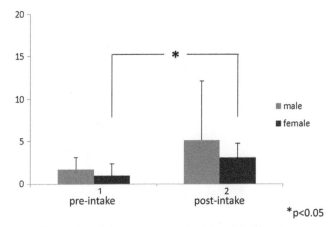

FIGURE 52.1 Gene expression levels of glucose-6-phosphate dehydrogenase (G6PDH) in pre- and post-intake samples. The messenger RNA expression levels of G6PDH were higher in the post-intake samples than in the pre-intake samples, and a significant difference was observed among the female patients (P < 0.05). G6PDH expression increased by more than four times in the post-intake samples compared with those in the pre-intake samples from female patients. There were no significant differences between genders in the pre- or post-intake samples.

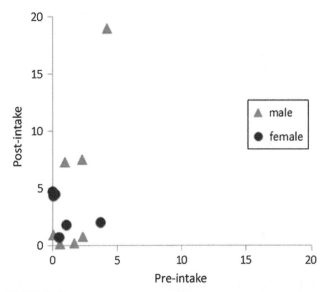

FIGURE 52.2 Correlation between gene expression levels of glucose-6-phosphate dehydrogenase (G6PDH) in pre- and post-intake samples. No significant correlations were observed between the expression levels of G6PDH in the pre- and post-intake samples.

by administering the supplement at a definite daily dosage. Confounding factors of differences between people, such as genetic factors, ultraviolet exposure, and factors affecting the absorption of nutrients, can be controlled by measuring samples from the same patient who has binocular cataracts of the same grade. Changes in the mRNA expression levels of antioxidative enzymes in the lenticular anterior capsule, which mostly occur in lens epithelial cells, may be good indicators of the antioxidative status of lens and verify the effects of antioxidant supplementation.

The composition of the antioxidant supplement used in this study, Ocuvite + Lutein®, is described in Table 52.1. Patients who had undergone bilateral cataract surgery were included in this study after providing informed consent according to the tenets of the Declaration of Helsinki. Approval from the institutional human experimentation committee was also granted. All of the patients had the same WHO classification and the same type of lens opacity in both eyes. Patients with ocular complications or systemic diseases were excluded from this study.

Pre-intake samples consisted of central regions of the anterior capsule measuring 5 mm in diameter that were collected from one eye of each patient via continuous curvilinear capsulorhexis during cataract surgery. Three tablets (the recommended daily dosage) of Ocuvite + Lutein®, the antioxidant supplement, were administered orally each day beginning the day after surgery. Six weeks after first surgery, post-intake samples of the anterior capsule were collected during cataract surgery on the opposite eye by the same method used for the pre-intake samples. Both the pre- and post-intake samples were filled with nitrogen to suppress oxidation and frozen immediately after collection. The samples were stored at −80 °C until processed. The samples were processed with the methods described in 'Methodologic Considerations', below.

After supplementation for 6 weeks, the mRNA expression levels of G6PDH increased. Among the female patients, G6PDH expression increased by more than four times in the post-intake samples (Fig. 52.1). There were no significant differences between genders in the pre- or post-intake samples, and no significant correlations were observed between the expression levels of G6PDH in the pre- and post-intake samples (Fig. 52.2).

The expression levels of 18S rRNA were also increased after supplementation. Among the female patients, 18S rRNA expression almost doubled in the post-intake samples (Fig. 52.3). There were no differences between

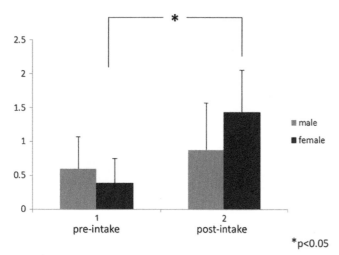

FIGURE 52.3 Expression levels of 18S ribosomal RNA (18S rRNA) in pre- and post-intake samples. The expression levels of 18S rRNA were higher in the post-intake samples compared with expression levels in the pre-intake samples, and a significant difference existed in samples from the female patients (P < 0.05). 18S rRNA expression almost doubled in the post-intake samples compared with the expression levels in the pre-intake samples from female patients. There were no differences between genders in pre- or post-intake samples.

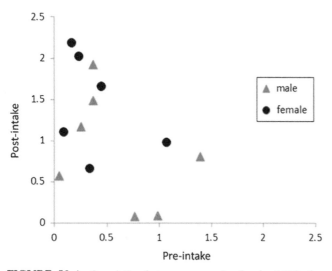

FIGURE 52.4 Correlation between expression levels of 18S ribosomal RNA (18S rRNA) in pre- and post-intake samples. There were no significant correlations between the 18S rRNA expression levels in the pre- and post-intake samples. However, there were tendencies towards lower expression of 18S rRNA in the pre-intake samples and higher expression of 18S rRNA in the post-intake samples.

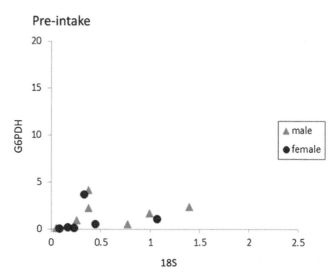

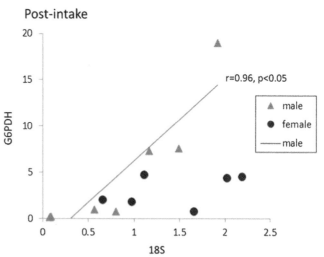

FIGURE 52.5 Correlation between gene expression levels of glucose-6-phosphate dehydrogenase (G6PDH) and 18S ribosomal RNA (18S rRNA). Positive correlations between G6PDH and 18S rRNA expression levels were observed in both the pre-intake and the post-intake samples, and a significant positive correlation existed between the expression levels of G6PDH and 18S rRNA in the post-intake samples from the male patients (r = 0.96, P < 0.05).

genders in pre- or post-intake samples, and there were no significant correlations between the 18S rRNA expression levels in the pre- and post-intake samples. However, there were tendencies towards lower expression of 18S rRNA in the pre-intake samples and higher expression of 18S rRNA in the post-intake samples (Fig. 52.4).

Positive correlations between G6PDH and 18S rRNA expression levels were observed in both the pre-intake and the post-intake samples, and a significant positive correlation existed between the expression levels of G6PDH and 18S rRNA in the post-intake samples from the male patients (Fig. 52.5).

METHODOLOGIC CONSIDERATIONS OF MEASURING MESSENGER RNA EXPRESSION OF G6PDH AND 18S rRNA IN LENTICULAR ANTERIOR CAPSULE

The anterior capsule pieces collected during cataract surgery were homogenized and proteolyzed using MagNa Lyser Green Beads (Roche). The RNA was purified using the MagNa Pure Compact RNA Isolation Kit

TABLE 52.2 Probes and Primers of the Messenger RNAs of Glucose-6-Phosphate Dehydrogenase and 18S Ribosomal RNA

ProbeFinder has designed an optimal real-time PCR assay for:
NM_004285.3 Homo sapiens hexose-6-phosphate dehydrogenase (glucose 1-dehydrogenase) (H6PD), mRNA

Assay rank 1
Use probe **#89** (cat. no. 04689143001)

Primer	Length	Position	Tm	%GC	Sequence
Left	23	983 - 1005	60	43	tggagatcatcatgaaagagacc
Right	20	1037 - 1056	60	55	gcgaatgacaccgtactcct

Amplicon (74 nt)

tggagatcatcatgaaagagaccgtggatgctgaaggccgcaccagcttctatgaggagtacggtgtcattcgc

X03205.1|X03205:EMBL|X03205:SILVA-SSU Human 18S ribosomal RNA

Assay rank 1
Use probe **#48** (cat. no. 04688082001)

Primer	Length	Position	Tm	%GC	Sequence
Left	20	1617 - 1636	60	45	gcaattattccccatgaacg
Right	20	1665 - 1684	59	50	gggacttaatcaacgcaagc

Amplicon (68 nt)

gcaattattccccatgaacgaggaattcccagtaagtgcgggtcataagcttgcgttgattaagtccc

(Roche). Using a Transcriptor High Fidelity cDNA Synthesis Kit (Roche), complementary DNA (cDNA) was reverse transcribed by the Gene Amp PCR system 9700 (AB). The cDNA of the control RNA (positive and negative control RNA included with the kit) was also reverse transcribed at the same time. Using the cDNA, Universal® Probes (Roche), and primers (Nihon Gene Research), as shown in Table 52.2, the mRNAs of G6PDH and 18S rRNA were quantified using the Light Cycler® LC480 (Roche) for reverse transcription–polymerase chain reaction (RT-PCR).

The expression levels of each mRNA were represented as a Cp value, where Cp was defined as the threshold cycle of RT-PCR at which the amplified product was detected. The Cp values were calculated based on a standard curve that was determined from an internal standard and the control RNA for each substance. The mRNA levels of each substance were indicated relative to internal standards and control RNA.

POSSIBLE ROLE OF ANTIOXIDANT SUPPLEMENT IN INCREASED G6PDH AND 18S rRNA EXPRESSION

According to the results described above, the expression levels of G6PDH and 18S rRNA were significantly increased after supplementation, especially among female patients. Positive correlations between the expression levels of G6PDH and 18S rRNA were observed,

especially from the male patients following supplementation. These data suggest that mRNA expression of G6PDH increased proportionally with the expression of 18S rRNA in cataractous lenses after supplementation.

The observed increase in expression of G6PDH in the post-intake samples may indicate activation of the pentose phosphate cycle following supplement intake (Fig. 52.6). The activation of the pentose phosphate cycle may be followed by an increase in NADPH. Glutathione reductase (GR) is activated owing to an increase in the level of coenzyme NADPH, leading to an increase in the level of GSH. Increased mRNA expression of GR and glutathione peroxidase (GPx) has also been observed (unpublished data). Therefore, increasing G6PDH after supplement intake further increases the levels of GSH, GR, and GPx with activation of the glutathione redox cycle, which in turn increases the scavenging activity of peroxides.

G6PDH showed significantly higher expression levels in the post-intake samples from the female patients in this study. Antioxidative abilities decrease in postmenopausal female patients as a result of hormonal imbalance. Estrogens have antioxidant effects[32,33] due to the scavenging of free radicals[34] and upregulation of antioxidative enzymes.[35] Therefore, oxidation increases following dramatically decreased estrogen levels in postmenopausal female patients. These observations could potentially explain the greater prevalence and progression of cataracts in female patients.[36,37] Moreover, NADPH from the pentose phosphate cycle is a

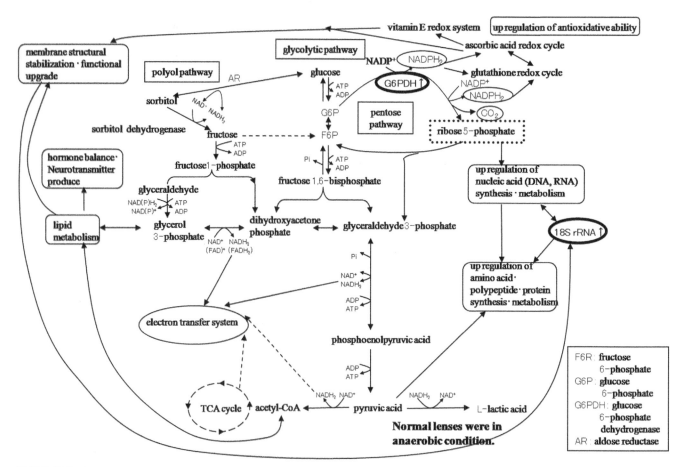

FIGURE 52.6 Presumed mechanisms of increases in gene expression levels of glucose-6-phosphate dehydrogenase (G6PDH) and 18S ribosomal RNA (18S rRNA) after supplement intake. ADP: adenosine diphosphate; AR: aldose reductase; ATP: adenosine triphosphate; F6P: fructose 6-phosphate; FAD: flavin adenine dinucleotide; FADH$_2$: reduced flavin adenine dinucleotide; G6P: glucose 6-phosphate; NAD(P)$^+$: nicotinamide adenine dinucleotide (phosphate); NAD(P)H$_2$: reduced nicotinamide adenine dinucleotide (phosphate); NAD$^+$: nicotinamide adenine dinucleotide; NADH$_2$: reduced nicotinamide adenine dinucleotide; Pi: inorganic phosphate; TCA: tricarboxylic acid. *Source: Hayashi et al. (2012),[26] with permission from Elsevier.*

coenzyme in the estrogen antioxidation cycle, which continuously regenerates the antioxidative potential of estrogens.[38] An increase in G6PDH expression is thought to increase NADPH to compensate for the decrease in estrogen among the postmenopausal female patients in this study. This could explain why the expression of G6PDH increased significantly among female patients rather than male patients after supplement intake.

Copper, one of the components of Ocuvite+Lutein®, combined with GSH (GSH–Cu complex) has been reported to be a strong superoxide scavenger and is found in cataractous lenses of diabetic patients.[39] GSH–Cu complex levels are thought to increase to compensate for the decrease in SOD in diabetic cataractous lenses, which thereby reduces oxidative stress. These complexes could have the same action in the aging cataractous lens.

Ocuvite+Lutein®, the antioxidant supplement used in this study, also contains vitamins B$_2$, C, and E, niacin, and selenium. Niacin is the base component of

NADP, vitamin B$_2$ is the coenzyme of NADPH dehydrogenase, and selenium is a key factor in the activation of GPx. The production of NADPH and the activity of GPx are presumed to increase after taking this supplement. Consequently, G6PDH expression increases with the increasing synthesis of NADPH. Intake of vitamin E combined with selenium has been reported to prevent a glucocorticoid-induced decrease in G6PDH.[40] In addition, L-ascorbic acid (reduced vitamin C) with vitamin E synergistically strengthens the antioxidative effects and stabilizes the cell membranes. Moreover, GSH, GPx, and NADPH play roles in converting dehydroascorbic acid (oxidized vitamin C) to L-ascorbic acid (reduced vitamin C). Although GSH, GPx, and NADPH are not included in the supplement used in this study, as previously mentioned, these factors are presumed to increase after intake of the supplement. In this study, the observed increase in the expression of G6PDH is presumed to be a result of the synergistic effects of selenium, vitamins E, C, and B$_2$, and niacin, contained in the supplement Ocuvite+Lutein®.

Lutein and zeaxanthin are the only carotenoids proven to be found in the human lens.[41] An inverse association between high dietary lutein intake and the prevalence of cataracts has been reported,[17] especially in women.[19] Lutein has also been reported to decrease intracellular H_2O_2 accumulation by scavenging superoxide and H_2O_2.[13] Lutein removes free radicals and is regenerated during oxidation by coupling to nonradical reducing systems such as glutathione–glutathione disulfide, NADPH–NADP$^+$, and NADH–NAD$^+$.[42] The effects of this supplement on the mechanisms leading to the increased G6PDH expression reported in this study are possibly due to the cooperation of lutein with vitamins E, C, and B_2 and niacin, which are included in the supplement.

The 18S rRNA is the active center of protein synthesis in the 40S ribosomal subunit. Increased numbers of ribosomes lead to increases in RNA transcription and protein synthesis, which are presumed to be proportional to increases in the expression levels of 18S rRNA. The increased expression levels of 18S rRNA in post-intake samples suggest an increase in the transcription and synthesis of proteins, including enzymes involved in the pentose phosphate cycle. Furthermore, as previously mentioned, G6PDH expression increased after supplement intake. The amount of R5P, a product of the pentose phosphate cycle that is related to G6PDH, may increase with G6PDH expression levels. As R5P is essential for nucleic acid synthesis, the production of RNA, including ribosomal RNA such as 18S rRNA, may be activated and increased with the increase in G6PDH expression levels. This is a possible mechanism for the proportional increase in 18S rRNA expression relative to the increase in G6PDH expression after taking the supplement.

The significant increase in expression of 18S rRNA in female patients suggests that metabolism and protein synthesis increased after taking the antioxidant supplement. Estrogen has been reported to upregulate antioxidative enzymes,[35] and the dramatically decreased levels of estrogen following menopause are presumed to affect transcriptional regulation. Zinc, which was present in the supplement used in this study, is an important component of RNA and DNA polymerases and regulates the metabolism of nucleic acids and proteins. The transcriptional signal is initiated when zinc associated with RNA polymerase binds to a zinc ligand in the DNA-binding domain of the nuclear receptor of a steroid hormone (such as estrogen).[43] Therefore, the zinc in the supplement may also regulate transcription and increase 18S rRNA expression in the female patients. The increasing 18S rRNA expression after supplement intake is possibly due to the cooperation of nutrients included in the supplement and the interaction with G6PDH.

CONCLUSIONS

The increases in the expression levels of G6PDH and 18S rRNA after antioxidant supplementation may be related to the activation of the pentose phosphate cycle. This suggests that intake of an antioxidant supplement can upregulate the pentose phosphate cycle and may further regulate antioxidation in the lens. Gene expression of G6PDH and 18S rRNA is a good indicator for evaluating the status of cellular functions, such as antioxidation, and the effects of antioxidants.

TAKE-HOME MESSAGES

- An antioxidant supplement is expected to have a preventive effect on cataract formation.
- As glucose-6-phosphate dehydrogenase (G6PDH) is the key enzyme for antioxidation, gene expression of G6PDH is a good indicator of oxidative status and antioxidative effect.
- As 18S ribosomal RNA (18S rRNA) is the active center of protein synthesis in the 40S ribosomal subunit, expression of 18S rRNA could be a good indicator of protein synthesis, including production of antioxidative enzymes.
- The increases in the expression levels of G6PDH and 18S rRNA suggest that there is an antioxidative effect of Ocuvite + Lutein® in the lens that may be effective in preventing the progression of cataract formation.
- The effects of antioxidant nutrients could differ between male and female patients.

References

1. Obara Y. The oxidative stress in the cataract formation. *Nihon Ganka Gakkai Zassi* 1995;**99**:1303–41.
2. Lin J. The association between copper ions and peroxidative reaction in diabetic cataract. *J Jpn Ophthalmol Soc* 1996;**100**:672–9.
3. Truscott RJ. Age-related nuclear cataract-oxidation is the key. *Exp Eye Res* 2005;**80**:701–25.
4. Blondin J, Baragi V, Schwartz E, Sadowski JA, Taylor A. Delay of UV-induced eye lens protein damage in guinea pigs by dietary ascorbate. *J Free Radic Biol Med* 1986;**24**:275–81.
5. Varma SD, Devamanoharan PS, Mansour S, Teter B. Studies on Emory mouse cataracts: oxidative factors. *Ophthalmic Res* 1994;**26**:141–8.
6. Taylor A, Jacques PF, Chylack Jr LT, Hankinson SE, Khu PM, Rogers G, et al. Long-term intake of vitamins and carotenoids and odds of early age-related cortical and posterior subcapsular lens opacities. *Am J Clin Nutr* 2002;**7**:540–9.
7. Yoshida M, Takashima Y, Inoue M, Iwasaki M, Otani T, Sasaki S, Tsugane S. JPHC Study Group. Prospective study showing that dietary vitamin C reduced the risk of age-related cataracts in a middle-aged Japanese population. *Eur J Nutr* 2007;**46**:118–24.
8. Tan AG, Mitchell P, Flood VM, Burlutsky G, Rochtchina E, Cumming RG, Wang JJ. Antioxidant nutrient intake and the long-term incidence of age-related cataract: the Blue Mountain Eye Study. *Am J Clin Nutr* 2008;**87**:1899–905.

9. Jacques PF, Taylor A, Moeller S, Hankinson SE, Rogers G, Tung W, et al. Long-term nutrient intake and 5-year change in nuclear lens opacities. *Arch Ophthalmol* 2005;**123**:517–26.

10. Leske MC, Wu SY, Connell AM, Hyman L, Schachat AP. Lens opacities, demographic factors and nutritional supplements in the Barbados Eye Study. *Int J Epidemiol* 1997;**26**:1314–22.

11. Mares-Perlman JA, Lyle BJ, Klein R, Fisher AI, Brady WE, Vanden-Langenberg GM, et al. Vitamin supplement use and incident cataracts in a population-based study. *Arch Ophthalmol* 2000;**118**:1556–63.

12. Age-Related Eye Disease Study Group. A randomized, placebo-controlled, clinical trial of high dose supplementation with vitamins C and E, beta carotene and zinc for age-related macular degeneration and vision loss. AREDS report no. 8. *Arch Ophthalmol* 2001;**119**:1417–36.

13. Kim JH, Na HJ, Kim CK, Kim JY, Ha KS, Chung HT, et al. The non-provitamin A carotenoid, lutein, inhibits HF-kappaB-dependent gene expression through redox-based regulation of the phosphatidylinositol 3-kinase/PTEN/Akt and NF-kappaB-inducing kinase pathway: role of H_2O_2 in NF-kappaB activation. *Free Radic Biol Med* 2008;**45**:885–96.

14. Gao S, Qin T, Liu Z, Caceres MA, Ronchi CF, Chen CY, et al. Lutein and zeaxanthin supplementation reduces H_2O_2-induced oxidative damage in human lens epithelial cells. *Mol Vis* 2011;**17**:3180–90.

15. Chasan-Taber L, Willett WC, Seddon JM, Stampfer MJ, Rosner B, Colditz GA, et al. A prospective study of carotenoid and vitamin A intake and risk of cataract extraction. *Am J Clin Nutr* 1999;**70**:509–16.

16. Moeller SM, Voland R, Tinker L, Blodi BA, Klein ML, Gehrs KM, et al. CAREDS Study Group, Women's Health Initiative. Association between age-related nuclear cataract and lutein and zeaxanthin in the diet and serum in the Carotenoid in the Age-Related Eye Disease Study (CAREDS), an ancillary study of the Women's Health Initiative. *Arch Ophthalmol* 2008;**126**:354–64.

17. Lyle BJ, Mares-Perlman JA, Klein BE, Klein R, Greger JL. Antioxidant intake and risk of incident age-related nuclear cataracts in the Beaver Dam Eye Study. *Am J Epidemiol* 1999;**149**:801–9.

18. Jacques PF, Chylack Jr LT, Hankinson SE, Khu PM, Rogers G, Friend J, et al. Long-term nutrient intake and early age-related nuclear lens opacities. *Arch Ophthalmol* 2001;**119**:1009–19.

19. Christen WG, Liu S, Glynn RJ, Gaziano JM, Buring JE. Dietary carotenoids, vitamins C and D, and risk of cataract in women: a prospective study. *Arch Ophthalmol* 2008;**126**:102–9.

20. Brown L, Rimm EB, Seddon JM, Giovannucci EL, Chasan-Taber L, Spiegelman D, et al. A prospective study of carotenoid intake and risk of cataract extraction in US men. *Am J Clin Nutr* 1999;**70**:517–24.

21. Gale CR, Hall NF, Phillips DIW, Martyn CN. Plasma antioxidant vitamins and carotenoids and age-related cataract. *Ophthalmology* 2001;**108**:1992–8.

22. Delcourt C, Carriere I, Delage M, Barberger-Gateau P, Schalch W. POLA Study Group. Plasma lutein and zeaxanthin and other carotenoids as modifiable risk factors for age-related maculopathy and cataract: the POLA study. *Invest Ophthalmol Vis Sci* 2006;**47**:2329–35.

23. Vu HTV, Robman L, Hodge A, McCarty CA, Taylor HR. Lutein and zeaxanthin and the risk of cataract: the Melbourne Visual Impairment Project. *Invest Ophthalmol Vis Sci* 2006;**47**:3783–6.

24. Brady WE, Mares-Perlman JA, Bowen P, Stacewicz-Sapuntzakis M. Human serum carotenoid concentrations are related to physiologic and lifestyle factors. *J Nutr* 1996;**126**:129–37.

25. Williams AW, Boileau TW, Erdman Jr JW. Factors influencing the uptake and absorption of carotenoids. *Proc Soc Exp Biol Med* 1998;**218**:106–8.

26. Hayashi R, Hayashi S, Arai K, Chikuda M, Obara Y. Effects of antioxidant supplementation on mRNA expression of glucose-6-phosphate dehydrogenase, β-actin and 18S rRNA in the anterior capsule of the lens in cataract patients. *Exp Eye Res* 2012;**96**:48–54.

27. Martini G, Ursini MV. A new lease of life for an old enzyme. *Bioessays* 1996;**18**:631–7.

28. Kletzien RF, Harris PKW, Foellmi LA. Glucose-6-phosphate dehydrogenase: a 'housekeeping' enzyme subject to tissue-specific regulation by hormone, nutrients and oxidant stress. *FASEB J* 1994;**8**:174–81.

29. Tian WN, Braunstein LD, Apse K, Pang J, Rose M, Tain X, Stanton RC. Importance of glucose-6-phosphate dehydrogenase activity in cell death. *Am J Physiol Cell Physiol* 1999;**276**:C1121–31.

30. Zhao W, Devamanoharan PS, Varma SD. Fructose induced deactivation of glucose-6-phosphate dehydrogenase activity and its prevention by pyruvate: implication in cataract prevention. *Free Radic Res* 1998;**29**:315–20.

31. Martinet W, de Meyer GR, Herman AG, Kockx MM. Reactive oxygen species induce RNA damage in human atherosclerosis. *Eur J Clin Invest* 2004;**34**:323–7.

32. Mooradian AD. Antioxidant properties of steroids. *J Steroid Biochem Mol Biol* 1993;**45**:509–11.

33. Gomez-Zubeldia MA, Arbues JJ, Hinchado G, Nogales AG, Millan JC. Influence of estrogen replacement therapy on plasma lipid peroxidation. *Menopause* 2001;**8**:274–80.

34. Ruiz-Larrea MB, Martin C, Martinez R, Navarro R, Lacort M, Miller NJ. Antioxidant activities of estrogens against aqueous and lipophilic radicals: differences between phenol and catechol estrogen. *Chem Phys Lipids* 2000;**105**:179–88.

35. Borras C, Gambini J, Gomez-Cabrera MC, Sastre J, Pallardo FV, Mann GE, Vina J. 17β-Oestradiol up-regulates longevity-related, antioxidant enzyme expression via the ERK1 and ERK2(MAPK)/NFκB cascade. *Aging Cell* 2005;**4**:113–8.

36. McCarty CA, Mukesh BN, Fu CL, Taylor HR. The epidemiology of cataract in Australia. *Am J Ophthalmol* 1999;**128**:446–65.

37. Klein BE, Klein R, Lee KE. Incidence of age-related cataract. The Beaver Dam Eye Study. *Arch Ophthalmol* 1998;**116**:219–25.

38. Prokai L, Prokai-Tatrai K, Perjesi P, Zharikova AD, Simpkins JW. Quinol-based bioreversible metabolic cycle for estrogens in rat liver microsomes. *Drug Metab Dispos* 2003;**31**:701–4.

39. Hayashi S, Tanaka Y, Arai K, Hayashi L, Arai K, Chikuda M, Obara Y. Glutathion–Cu complex in cataractous lenses of diabetics. *Biomed Res Trace Elements* 1998;**9**:69–70.

40. Yilmaz S, Beytut E, Erisir M, Ozan S, Aksakal M. Effects of additional vitamin E and selenium supply on G6PDH activity in rats treated with high dose of glucorticoid. *Neurosci Lett* 2006;**393**:85–9.

41. Yeum KJ, Taylor A, Tang G, Russell RM. Measurement of carotenoid, retinoid, and tocopherols in human lenses. *Invest Ophthalmol Vis Sci* 1995;**36**:2756–61.

42. Sies H, Stahl W, Vitamin E, C, β-carotene, and other carotenoids as antioxidants. *Am J Clin Nutr* 1995;**62**: 1315-21S.

43. Werner F, Weinzierl ROJ. Direct modulation of RNA polymerase core function by basal transcription factor. *Mol Cell Biol* 2005;**25**:8344–55.

The Adenosine A$_{2a}$ Receptor and Diabetic Retinopathy

Bashira A. Charles

Center for Research on Genomic and Global Health, National Human Genome Research Institute, Bethesda, Maryland, USA

INTRODUCTION

The pathology of the two primary causes of diabetes is marked by drastic impairment or total lack of insulin production due to pancreatic islet beta-cell destruction caused by autoimmune processes or by impaired insulin sensitivity or secretion, both of which result in chronic hyperglycemia. The prevalence of diabetes in the global population was approximately 7% (366 million) in 2011 and is projected to increase to 8.3% (552 million) by 2030,[1] placing affected individuals at an increased risk for diabetic retinopathy (DR), a microvascular complication of diabetes. The Meta-Analysis for Eye Disease (META-EYE) Study Group investigated 22,896 individuals diagnosed with diabetes for prevalence of any DR, proliferative diabetic retinopathy (PDR), and vision-threatening diabetic retinopathy[2] and demonstrated prevalences of 34.6%, 6.9%, and 10.2%, respectively. These investigators have estimated the global prevalence of DR, PDR, and vision-threatening retinopathy to be 93 million, 17 million, and 28 million, respectively.[2] DR may cause visual impairment and is the most prevalent cause of new-onset blindness in individuals 20–75 years of age in Western countries.[3] In the USA, the prevalence of DR among adults is estimated to be 4.2 million (28.5%), of whom 0.7 million have been diagnosed with PDR.[3]

The pathogenesis of DR has four stages.[4–6] Before the first stage of DR, chronic hyperglycemia induces pericyte cell death.[7] In the first stage, mild nonproliferative diabetic retinopathy, the weakening of segments of the vessel wall and attendant endothelial cell proliferation following pericyte cell death result in the formation of microaneurysms in the retinal microvasculature.[4] The second stage of DR, moderate nonproliferative

diabetic retinopathy, is associated with impairment in the blood supply of vessels of the retina caused by clot formation.[4,8] The third stage of DR, severe nonproliferative diabetic retinopathy, is marked by increased impaired blood flow through the vessels supplying the retina and vascular neogenesis (formation of new vessels) with weak and malformed vessels.[8] The fourth and most severe stage of DR is proliferative retinopathy, which is marked by increasing vascular neogenesis near the vitreous humor. These vascular changes may result in blindness by vitreous hemorrhage or by retinal detachment caused by fibrosis.[7] Many of these changes result in debilitating visual impairment or blindness (Fig. 53.1).[4,9]

HYPERGLYCEMIA

A complex series of metabolic reactions utilizing different biologic pathways occurs in the presence of hyperglycemia. Metabolic pathways that are activated during hyperglycemia influence the cellular environment and expose the cells to many toxic substances. Although this is a process that occurs in all individuals experiencing hyperglycemia, the nondiabetic individual will respond by producing enough insulin to mediate glucose uptake by the peripheral cells and tissues to return serum glucose levels to the euglycemic state, thereby decreasing their exposure to hyperglycemia and the pathophysiologic responses that include production of substances toxic to the cells. In contrast to the response of nondiabetic individuals, serum glucose levels are elevated more frequently and for more extended periods in the individual with diabetes. The toxic substances and states that the cells are exposed to include sorbitol (increasing

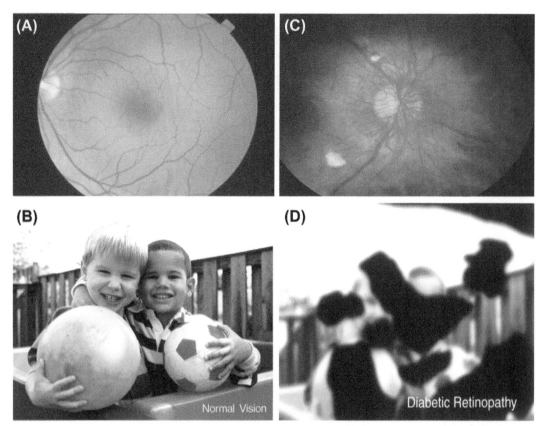

FIGURE 53.1 Photographs of the normal retinal and the retina affected by proliferative diabetic retinopathy paired with an example of the vision observed in normal and pathologic conditions. (A) Fundus photograph, normal retina; (B) normal vision; (C) retinal fundus proliferative retinopathy, an advanced form of diabetic retinopathy, occurs when abnormal new blood vessels and scar tissue form on the surface of the retina; (D) scene as it might be viewed by a person with diabetic retinopathy. *Source: National Eye Institute, National Institutes of Health: (A) Ref. EDA06; (B) EDS01; (C) EDA01; (D) EDS04.* (See color plate section)

osmotic pressure), hydrogen peroxide, hypoxia, lactate, toxic levels of glucose, reactive oxygen species (ROS), and advanced glycation end-products. When the cells are exposed to hypoxia and to this proinflammatory environment for extended periods, the cells become injured and dysfunctional and subsequently die.[7,10–13]

Hypoxia-associated medical conditions such as diabetes result in an exocytotic release of large amounts of adenosine triphosphate (ATP) (from damaged cells) into the extracellular environment.[14,15] During states of hyperglycemia retinal cells exhibit an increase in the extracellular release and catabolism of ATP.[16] The increase in extracellular ATP levels may be as high as 60% above those observed in retinal cells that have not been exposed to hyperglycemia.[16] In addition, cells cultured under hyperglycemic conditions have lower rates of catabolism of ATP than cells cultured under normal–glycemic conditions. The catabolism and subsequent release of adenosine have significant implications for individuals affected by diabetes. The physiologic effects of adenosine in diabetes and hyperglycemic conditions will be discussed in greater detail later in this chapter.

ADENOSINE AND ITS RECEPTORS

Adenosine is an endogenous purine nucleoside that plays a role in many biochemical processes. It is produced during the metabolism of ATP. Adenosine phosphorylation results in the production of ATP, and the final stage of ATP hydrolysis results in the production of adenosine[17] following its conversion to adenosine diphosphate (ADP) and adenosine monophosphate (AMP). The metabolic cascade associated with ATP and purine metabolism is noted in Figure 53.2. This metabolic cascade promotes energy generation and transfer as well as extracellular signaling and receptor activation in neuronal and non-neuronal tissues.[18,19] Jacobson and Gao describe adenosine as serving a cytoprotective role *in vivo*, 'increasing the ratio of oxygen supply and demand, protecting against ischemic damage by cell conditioning, triggering anti-inflammatory responses and promotion of angiogenesis'.[20] Elevated levels of adenosine are associated with vasodilatation, bronchospasm, and feelings of fatigue.[21] Adenosine also plays a role in cell proliferation, vascular tone, neurotransmission, platelet function, renal function, and amelioration

Adenosine Triphosphate and Purine Metabolism

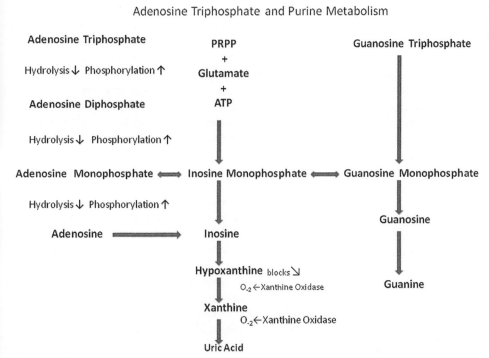

FIGURE 53.2 Adenosine triphosphate (ATP) and purine metabolism. PRPP: phosphoribosyl pyrophosphate.

of oxidative stress. In the ocular system adenosine has been associated with regulation of aqueous humor secretion and intraocular pressure. In general, adenosine receptors either inhibit or stimulate adenylate cyclase activity.[20] Signaling of adenosine is due to P1 receptors and P2 receptors.[19] The P1 receptors are G-coupled protein receptors. They consist of the adenosine A$_1$ receptor (ADORA$_1$), the adenosine A$_{2A}$ receptor (ADORA$_{2A}$), the adenosine A$_{2B}$ receptor (ADORA$_{2B}$), and the adenosine A$_3$ receptor (ADORA$_3$). The P2 receptors consist of ADP and uridine 5′-triphosphate. These receptors exert their effects via guanosine triphosphate proteins binding with ion channels, phospholipases, and adenylate cyclase.[22] This chapter will focus primarily on the P1 receptors and specifically on ADORA$_{2A}$.

ADORA$_{2A}$ and ADORA$_{2B}$ stimulate adenylate cyclase activity, whereas ADORA$_1$ and ADORA$_3$ inhibit adenylate cyclase activity.[21] All four receptors are ubiquitously expressed; however, there is tissue-specific expression for each receptor.[20,21] For example, *ADORA$_1$*, located on chromosome 1q32.1, is more highly expressed in tissues of the central nervous system. *ADORA$_{2B}$*, located on chromosome 17p12, is expressed in the alimentary track, with the highest expression identified in the esophagus, appendix, colon, and gastric atrium. *ADORA$_3$* is located on chromosome 1q21-p13 and is highly expressed in vascular smooth muscle.[21] All four P1 receptors have similar expression levels in the ocular ciliary body but have differential levels of expression in the ciliary epithelium, with *ADORA$_3$* expressed in the nonpigmented ciliary epithelium. *ADORA$_{2A}$* is located on chromosome 22q11.23 and

is highly expressed in tissues of the vasculature, basal ganglia, and platelets. ADORA$_{2A}$ ameliorates and/or arrests tissue-specific and systemic inflammation. The human ADORA$_{2A}$ is a larger protein than ADORA$_1$, ADORA$_{2B}$, or ADORA$_3$.[23] Olah and Stiles also noted that ADORA$_{2A}$ has 80–90 more amino acids than the other receptors, has a 93% protein identity, and is more rapidly deactivated than ADORA$_1$, possibly owing to an increased number of phosphorylation sites. All of these features have implications for potential therapeutic interventions.

OXIDATIVE STRESS AND THE ADENOSINE A$_{2A}$ RECEPTOR

The production of ROS in cells is a normal component of oxidative metabolism. Approximately 95–99.9% of the oxygen that enters oxidative metabolism is used for metabolic processes, while the remaining 0.1–5% results in the production of ROS.[24] Under nonpathologic conditions, the ROS that are produced are scavenged by vitamin E and C or endogenous antioxidants such as glutathione peroxidase and superoxide dismutase (SOD).[24,25] Elevated levels of oxidative stress occur when the production of ROS exceeds the capacity of the available endogenous antioxidants to scavenge them. Both acute and chronic hyperglycemia result in excessive production of ROS,[26] contributing to the development of complications of diabetes including diabetic neuropathy and DR. Activation of ADORA$_{2A}$ inhibits SOD production and ameliorates inflammation.[27]

VASCULAR CELLS, HYPERGLYCEMIA, AND ADENOSINE

Over time, diabetes has numerous effects on the ocular system, including the microenvironment, basement membrane, vascular cells, and vessels, all of which contribute to the development of microvascular complications of diabetes. The basement membrane is a fine sheet of fibrous tissue lining the vascular endothelium. It plays a role in angiogenesis, cell differentiation, growth, repair, and migration. When retinal capillaries are occluded or closed as a result of diabetes-associated pathologic changes, proteolytic enzymes increase and neoangiogenesis is initiated at the site of local basement membrane degradation in the area of the nonperfused capillary.[28] This process is associated with a concomitant increase in growth factors including vascular endothelial growth factor (VEGF), which is also associated with increased expression in the retinas of human[29] and animal models[30] with DR. Increased VEGF levels have been observed in retinal pericytes, retinal endothelial cells,[31,32] Müller cells,[30] retinal pigment epithelial cells,[33] and in the vitreous fluid[34] in ischemic environments. These observations are consistent with the elevated VEGF levels observed in individuals with PDR, and increased VEGF levels are responsible at least in part for the neoangiogenesis associated with this microvascular complication of diabetes.

Investigators have suggested that increased VEGF levels during states of hypoxia may be attributed to adenosine, which is also increased in an ischemic cellular environment.[35] Follow-up studies identified ADORA$_{2A}$ as the receptor mediating adenosine-associated elevations in VEGF.[36] Specifically, using bovine retinal capillary endothelial cells and microvascular pericytes, the ADORA$_{2A}$ agonists N^6-[2-(3,5-dimethoxpheynl)-2-(2-methylphenyl)-ethyl] adenosine (DPMA), 5'-N-ethylcarboxamidoadenosine (NECA), and 5'-N-ethylcarboxamidoadenosine (CGS21680) in the absence of hypoxia increased VEGF messenger RNA (mRNA) by up to 2.3-fold (P < 0.0001), 2.8-fold (P = 0.016), and 2.2-fold (P = 0.025), respectively. The ADORA$_1$ agonist N^6-cyclopentyladenosine (CPA) failed to increase VEGF mRNA expression, whereas the ADORA$_{2A}$ antagonist 8-(3-chlorostyryl)-caffeine (CSC) resulted in a significantly reduced (68%) stimulatory response of VEGF mRNA following exposure of the cells to hypoxia. In addition, VEGF production was increased 4.7-fold in cells exposed to both hypoxia and NECA. The work of these investigators provides evidence that vascular endothelial cells and pericyte cells exposed to hypoxia have an increased production of adenosine, which, via the action of the ADORA$_{2A}$ and the cyclic adenosine monophosphate (cAMP) pathway, upregulates expression of VEGF. Increased levels of VEGF, in turn, increase vascular cell permeability and neovascularization.

Pericytes are microvascular contractile cells that support the endothelial cell lining of retinal capillary walls and play a role in maintaining vessel tone. Chronic hyperglycemia associated with diabetes and the resultant cascade of biologic events creating a toxic cellular environment for the vessels of the eye are associated with pericyte cell death. The increased production of adenosine in pericytes exposed to hypoxia[35] has been shown to inhibit pericyte cell proliferation mediated by ADORA$_{2A}$. Increased pericyte cell apoptosis and subsequent inhibition of pericyte cell proliferation mediated by ADORA$_{2A}$ are likely to play a role in the disruption of the pericyte to endothelial cell ratio. This ratio is 1:1 in the healthy eye but has been found to be reduced to as low as 1:4 in individuals with diabetes.[37] The death of pericytes in the absence of compensatory repair and replacement results in the formation of aneurysms. Over time, these aneurysms rupture,[7,10–13] producing increased hypoxia and vascular impairment, stimulating a physiologic response that attempts to repair the injury to the vascular system of the eye by accelerating the production of endothelial cells, and causing narrowing of the capillary lumen and further ischemia. As mentioned above, in nondiabetic individuals this feedback mechanism is not activated as frequently or for extended periods, and therefore this pathologic growth of new vessels is not activated in nondiabetic individuals.[10]

Retinal endothelial cells play a major role in preventing large molecules such as glucose from entering the retinal microvasculature. This process helps to maintain retinal vascular cell glucose homeostasis. Hypoxia, which can be induced by the cascade of biologic processes associated with hyperglycemia, is associated with increased glucose transport into the retinal vasculature. Takagi and colleagues, using a bovine model, demonstrated that adenosine via ADORA$_{2A}$ plays a major role in hypoxia-induced glucose transport into the retinal vasculature.[38] Increased intracellular glucose in endothelial vascular cells is a prominent factor in the development of DR. Hypoxia results in an increase in adenosine via reduced initiation of the metabolic cascade that converts adenosine to AMP by adenosine kinase.[38] Increased adenosine mediated by ADORA$_{2A}$, in turn, increases the expression of VEGF and kinase insert domain-containing receptor (KDR).[39] These investigators provide evidence that hypoxia increased adenosine expression, which, in turn, increased GLUT1 mRNA expression twofold. Ischemia-induced glucose transport into bovine retinal epithelial cells is another process mediated by adenosine, ADORA$_{2A}$, and the cAMP–protein kinase A pathway, which contributes to the pathophysiology of DR. The increased GLUT1 expression mediated by ADORA$_{2A}$ was arrested when the cells were exposed to the ADORA$_{2A}$ antagonist and stimulated when exposed to the ADORA$_{2A}$ agonists 2-p(2-carboxyethyl)-phenethyl-amino-5'-N-ethylcarboxamidoadenosine and NECA. These investigators also demonstrated that increased *GLUT1* expression was not initiated when cells were exposed to the ADORA$_1$ agonist CPA, providing additional evidence for adenosine and ADORA$_{2A}$ in the pathophysiology of DR.

NERVE CELLS

The retina consists of neurons and glial cells. Retinal neurons (photoreceptors, amacrine, ganglion cells, etc.) and glial cells (Müller cells and astrocytes) (http://webvision.med.utah.edu) make up 95% of the retina; however, they are transparent and defects are not visible during routine clinical exams.[40] Müller cells begin at the surface of the vitreous and cover up to 70% of the retinal depth.[41] Müller cells and astrocytes support the neurons and blood vessels.[42] They play a role in cellular transport including transport of ions, neurotransmitter, and metabolites. These cells generate energy from glycolysis and production of ATP.[41] Müller cells and astrocytes have minimal oxygen and glucose requirements compared to retinal neurons. The reduced oxygen and glucose requirements render Müller and astrocyte cells less vulnerable to the ischemia and hypoglycemia that may occur in individuals with diabetes due to hypoxia and/or vascular nonperfusion resulting from ischemia, vessel occlusion, and/or rupture. Retinal neurons have higher oxygen requirements and are more susceptible to ischemic cellular injury. The toxic cellular environment of diabetes leads to disruption of the blood-retinal barrier (BRB) and apoptosis of retinal neurons.[43,44] Microglial cells are derived from the hematpoietic system[45] and migrate to the retina as a part of the response to inflammation and infection. They play a role in generation of cytokines, growth factors, and ROS resulting from chronic hyperglycemia.

Vascular changes are the first clinically detectable sign of DR[46,47]; however, there is some controversy regarding whether vascular damage or neuronal damage is the earliest pathologic manifestation of DR.[48,49] Individuals with preclinical DR are typically asymptomatic, and routine examination reveals a normal-appearing retina; however, more sensitive tests may reveal decreased blue-yellow color perception, visual field defects, decreased oscillatory potential amplitudes on electroretinogram (ERG), increased blood-retinal barrier (BRB) permeability on vitreous fluorometry,[40] and impaired contrast sensitivity.[50] During this preclinical phase histopathology testing may uncover neural apoptosis, decreased depth of the nerve fiber layers, glial cell dysfunction, and microglial cell activation including production of cytokines such as VEGF and TNFα. During preclinical diabetes tight junctions are loosened and basement membrane thickening occurs in vascular cells. Patients experiencing NPDR may be asymptomatic or may experience blurred vision and/ or problems with glare. On exam the clinician may observe microaneurysms, cotton-wool spots, retinal hemorrhage, and/or depression. The evidence noted above suggests that neurovascular changes of DR likely occur in concert.

Nerve cells exposed to hypoxic conditions respond by catabolizing ATP to adenosine and releasing glutamate and other biochemicals associated with excitotoxicity. ROS, extracellular signal regulated kinase (ERK), and other regulators of cytokine production result in release of inflammatory cytokines including TNFα.[51] Acting through its receptors, adenosine regulates the resultant inflammation. ADORA$_1$ receptors are primarily expressed in the inner retina, while ADORA$_{2A}$ receptors are expressed in the outer retina.[52] ADORA$_1$ stimulation causes a reduction in nerve cell excitability and metabolism. Activation of ADORA$_1$ thereby reduces oxygen demand, protecting neuronal tissue from ischemic injury.[53,54] ADORA$_{2A}$ relaxes the blood vessels, increasing the flow of oxygenated blood to the tissues. It also stimulates glycogenolysis in Müller cells.[55] Acting via microglial cells, ADORA$_{2A}$ binds with adenosine, increasing adenylate cyclase production, which inhibits signaling of biochemicals such as MEK and ERK, preventing generating proinflammatory cytokines and thereby ameliorating retinal inflammation, nerve cell damage, and apoptosis.[56]

The role the receptors play during states of inflammation is time and tissue dependent. The work of Bing *et al.* demonstrated that administration of an ADORA$_1$ antagonist 5–30 minutes before initiation of an ischemic event decreased a- and b-wave recovery time by more than 50% and that this effect persisted for a week. They also found that administration of CSC, the ADORA$_{2A}$ antagonist, before initiation of a 5-minute ischemic event decreased a- and b-wave recovery for 2 hours but had a protective effect when ischemia was initiated for longer than 30 minutes. These data suggest the receptors may work in a complementary manner to enhance the beneficial effects of adenosine,[57] with ADORA$_1$ working as a first responder during acute inflammation and ADORA$_{2A}$ being activated during more chronic inflammatory states. The effects of both ADORA$_1$ and ADORA$_{2A}$ *in vivo* warrant further investigation to elucidate their complex roles in living organisms. Treatment of DR is limited to intermittent injection with anti-VEGF or steroids, photocoagulation, and vitrectomy. None of these interventions result in permanent reversal of disease or complete restoration of vision and all have adverse effects. The biologic properties associated with the adenosine receptors suggests pharmacologic therapies using adenosine receptor agonists and antagonists warrant further exploration for use in neuronal and vascular tissues.

VARIANTS OF THE ADENOSINE A$_{2A}$ RECEPTOR AND PROLIFERATIVE DIABETIC RETINOPATHY

Adenosine's association with hyperglycemia and hypoxia, coupled with the role it plays in vasodilatation, glucose transport, and VEGF expression acting via ADORA$_{2A}$, renders variants of this receptor biologically plausible candidate genes for investigating genetic factors contributing to the development of diabetes complications. In 2011 investigators reported variants of the ADORA$_{2A}$ gene to be associated with protection from development of PDR.[58] These investigators used data and banked DNA from the participants of the Pittsburgh Epidemiology of Diabetes Complications (EDC) study to conduct a study investigating the genetic basis of DR. The EDC is prospective study of the natural history of diabetes complications.[59,60] EDC participants comprise a well-phenotyped cohort of individuals with type 1 diabetes for longer than 25 years. Participants (n = 658) underwent biennial physical examinations including collection of blood pressure measurements, high-density lipoprotein, low-density lipoprotein, triglycerides, glycosylated hemoglobin, body mass index, banked DNA, and stereoscopic retinal examinations.[61] Ocular examinations were conducted at baseline and then biennially for 10 years and consisted of dilated eye examinations and stereoscopic imaging. Using the Early Treatment of Diabetic Retinopathy Study (ETDRS) guidelines, fields 1, 2, and 4 were captured using the Zeiss fundus camera (Carl Zeiss, Oberkochen, Germany).[62] Diagnosis and DR severity staging was determined by the Fundus Photography Reading Center at the University of Wisconsin using the Arlie House system. The sensitivity and reliability of using the three images mentioned above instead of the standard seven were established previously.[59,63] Participants in the EDC study with banked DNA available (75.4%) were recruited into the Genetic Basis of Diabetic Retinopathy (GBDR) study.

The vast majority of individuals with diabetes will develop some form of DR during the course of their lives; however, a greater proportion of individuals with type 1 disease progress to develop PDR than do individuals with type 2 disease: 50% versus 10%, respectively.[58,64,65] Individuals with grades of 60 or higher in one eye, or grade less than 60 but with pan-retinal chorioretinal scars consistent with laser therapy, were categorized as having PDR.[58] GBDR participants with PDR at baseline were categorized as having prevalent PDR. GBDR participants who developed PDR during the course of follow-up were defined as having incident PDR.

Two tag single nucleotide polymorphisms (rs2236624-C/T and rs4822489-G/T) were identified for the HapMap CEU population using an r$_2$ of 0.8 and a minor allele frequency of 0.2. Clinical characteristics for the GBDR study participants based on incidence prevalence or any PDR are noted in Table 53.1. Controlling for covariates identified in Table 53.1, logistic regression analysis and Cox proportional hazards regression analysis revealed that homozygous carriers of rs2236624 (T) and rs4822489 (T) were associated with protection from development of PDR (Table 53.2). The strongest association was identified for individuals homozygous for the T allele of rs2236624. These individuals had a decreased risk of developing incident PDR (hazard ratio = 0.156, P-value (P) = 0.009), prevalent PDR (odds ratio (OR) = 0.36, P = 0.04), and any PDR (OR = 0.23, P = 0.001). Homozygous carriers of the T allele of rs4822489 also displayed protection from development of PDR, although the effect was not as striking (OR = 0.55, P = 0.04). The Kaplan–Meier curve provides compelling evidence supporting a role for ADORA$_{2A}$ in protection from the development of PDR (Fig. 53.3). In fact, among those homozygous for the T allele of rs2236624, 11% progressed to PDR compared with 49% of carriers of any copy of the C allele.

The GBDR sample is a European ancestry population. There are population-specific differences in the allele frequencies of these variants. Based on 1000 Genomes data, the T allele has a frequency of 5%, 23%, 32%, and 28% in the African, admixed American, Asian, and European ancestry populations, respectively.[66] The minor allele differs between populations for the rs4822489 variant; specifically, the G allele is the minor allele in the African population, with a frequency of 24%, while the T allele is the minor allele for the admixed American, Asian, and European populations, with frequencies of 46%, 49%, and 41%, respectively (Fig. 53.4). Based on 1000 Genomes data, African ancestry populations have a lower frequency of the T allele for the rs2236624 variant; however, the frequency of the T allele is almost 80% for the rs4822489 variant. Asian ancestry populations have higher frequencies of the T allele for both variants of the gene. It is unclear what effect differing allele frequencies for these variants has on the risk of developing DR or whether allele dosing will have an impact on the response to genetic-based therapies that may be developed based on mounting evidence that ADORA$_{2A}$ ameliorates the hypoxia and oxidative damage associated with chronic hyperglycemia.

THERAPIES BASED ON THE ADENOSINE A$_{2A}$ RECEPTOR

Throughout the lifespan the human body is in a constant battle to maintain homeostasis. It must guard the individual against pathogens as well as environmental assaults. Decades ago, the major factors contributing to disease development were biologic pathogens that are the source of communicable diseases such as malaria, smallpox, and poliomyelitis. Over the past several

TABLE 53.1 Patient Characteristics

(a) Patients Who Developed Incident PDR vs. Patients Who Were PDR Free at End of Follow-Up

	Incident PDR	No PDR during Follow-Up	P-Value
Age (years)	25.8 ± 6.9	24.8 ± 8.1	0.1992
Duration (years)	17.0 ± 6.3	16.3 ± 7.1	0.7675
Male/female (n)	79/82	92/91	0.9132
Glycosylated Hb	10.78 ± 1.9	9.9 ± 1.7	<0.0001
Hypertension (n)	15 (9.3)	7 (3.8)	0.0435
HDL (mg/dL)	54.7 ± 12.1	55.3 ± 12.0	0.6155
LDL (mg/dL)	114.3 ± 31.4	102.2 ± 26.6	0.0004
Triglycerides (mg/dL)	106.6 ± 87.0	81.1 ± 50.1	0.0012
Body mass index	23.7 ± 3.0	22.9 ± 3.4	0.0466
Ever-smoker	46 (28.8)	61 (34.7)	0.1602

(b) Patients Who Had PDR at Baseline vs. Patients Who Were PDR Free at Baseline

	Prevalent PDR	No PDR at Baseline	P-Value
Age (years)	37.2 ± 5.8	24.8 ± 8.1	<0.0001
Duration (years)	25.1 ± 6.1	16.3 ± 7.1	<0.0001
Duration (years)	79/63	92/91	0.4562
Male/female (n)	10.3 ± 1.8	9.9 ± 1.7	0.0020
Glycosylated Hb	35.9	3.8	<0.0001
HDL (mg/dL)	51.7 ± 11.3	55.3 ± 12.0	0.0058
LDL (mg/dL)	128.9 ± 38.0	102.2 ± 26.6	<0.0001
Triglycerides (mg/dL)	120.8 ± 71.2	81.1 ± 50.1	<0.0001
Body mass index	24.2 ± 3.3	22.9 ± 3.4	0.0051
Ever-smoker	66 (47.5)	61 (34.7)	0.9456

(c) Patients Who Developed PDR (Incident or Prevalent) vs. Patients Who Were PDR Free at End of Follow-Up

	Any PDR	No PDR	P-Value
Age (years)	29.0 ± 7.3	24.7 ± 7.9	<0.375
Duration Years	20.8 ± 7.8	16.2 ± 7.0	<0.009
Duration (years)	159/147	94/91	0.456
Male/female (n)	10.5 ± 1.8	9.9 ± 1.7	<0.001
Glycosylated Hb	68 (90.7)	7 (9.3)	<0.001
HDL (mg/dL)	53.4 ± 12.1	55.2 ± 11.9	0.145
LDL (mg/dL)	120.9 ± 35.2	102.5 ± 26.7	<0.001
Triglycerides (mg/dL)	113.2 ± 79.5	81.7 ± 50.0	<0.001
Body mass index	23.9 ± 3.1	22.9 ± 3.4	0.016
Ever-smoker	113 (64.6)	62 (35.4)	0.434

Data are shown as mean ± SD unless otherwise specified. Figures in parentheses are percentages.
PDR: proliferative diabetic retinopathy; Hb: hemoglobin; HDL: high-density lipoprotein; LDL: low-density lipoprotein.
Source: Charles et al. Ophthalmic Res 2011; 46:1–8,[58] with permission from S. Karger AG, Basel.

TABLE 53.2 Association of Adenosine Receptor Single Nucleotide Polymorphisms (SNPs) with All Cases of Proliferative Diabetic Retinopathy under Three Genetic Models

SNP (Reference Allele)	OR	95% CI	P-Value
rs2236624 T			
Recessive adjusted	0.18	0.06 to 0.59	0.004
Dominant adjusted	1.24	0.78 to 1.95	0.363
Additive adjusted	0.93	0.64 to 1.34	0.890
rs4822489			
Recessive adjusted	0.55	0.31 to 0.97	0.040
Dominant adjusted	0.97	0.62 to 1.49	0.873
Additive adjusted	0.84	0.62 to 1.13	0.240

Adjusted for duration of type 1 disease, low-density lipoprotein, hypertension, and glycosylated hemoglobin.
OR: odds ratio; CI: confidence interval.
Source: Charles et al. Ophthalmic Res 2011;46:1–8,[58] with permission from S. Karger AG, Basel.

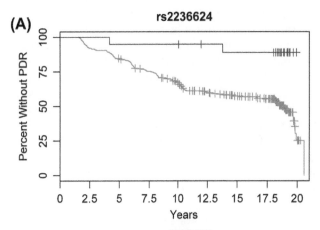

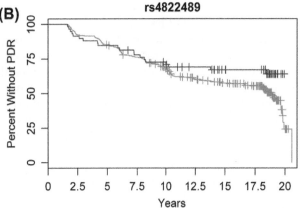

FIGURE 53.3 Survival analysis for proliferative diabetic retinopathy (PDR) based on adenosine A$_{2A}$ (ADORA$_{2A}$) allele carrier status. Kaplan–Meier curves showing follow-up time for development of PDR in 2.5-year increments by (A) rs2236624 and (B) rs4822489 genotype. (A) Blue line: T allele; red line: C allele; (B) blue line: T allele; red line: G allele.

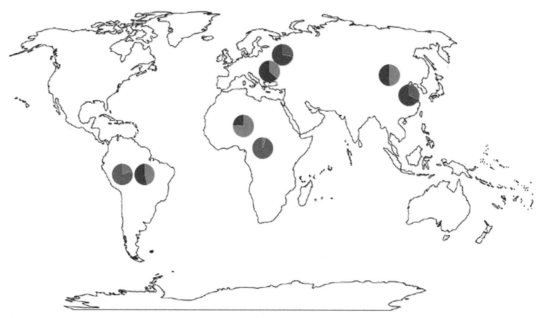

FIGURE 53.4   Population-based allele frequencies for the rs2236624 and rs4822489 variants in the 1000 Genomes populations. rs2236624: the T allele is represented by royal blue and the C allele by red; rs4822489: the G allele is represented by dark blue and the T allele by yellow.

decades, noncommunicable diseases such as diabetes have become a major source of morbidity and mortality in industrialized nations, and the epidemic has begun to spread to developing countries where resources for diagnosing and treating the populace are often scarce. To combat communicable diseases, large-scale vaccination has been implemented in many countries, virtually eliminating their threat to the populations of industrialized nations. Noncommunicable diseases such as diabetes are more challenging to address; they often require drastic changes in lifestyle for individuals with type 2 diabetes, whereas for individuals with type 1 diabetes the goal of maintaining euglycemic control while avoiding extreme hypoglycemia has not been mastered. In fact, as mentioned above, most individuals with diabetes will develop some form of DR during the course of their lives.

To date, strategies for preventing diabetes complications have focused on maintaining euglycemic control. It is not until after PDR has developed that physicians implement therapy with corticosteroids or photocoagulation therapy, or both. Corticosteroid injections have an unfavorable side-effect profile, increasing intraocular pressure and the development of cataracts.[67] In recent years, genetic-based therapeutic interventions have been instituted. One such therapy is the use of anti-VEGF injections to prevent the progression of PDR. Intravitreal injections of bevacizumab and ranibizumab, as well as administration of these two anti-VEGF therapies in combination with photocoagulation, are the gold standard for the treatment of PDR.

The role of inflammation in the development of DR has been established, with many investigators documenting that retinal inflammation plays a role in development of DR, and many cytokines including interleukin-1, interleukin-6, and tumor necrosis factor-α (TNF-α) have been implicated in this process.[68] Using a rodent model, Liou et al. demonstrated that the ADORA$_{2A}$ agonists NECA and CGS21680 inhibit TNF-α expression in rodent microglial cells. They also suggest that cannabidiol is able to block diabetes-induced retinal damage by enhancing the effects of ADORA$_{2A}$. Ibrahim et al. also demonstrated that ADORA$_{2A}$ agonists significantly decreased hyperglycemia-induced TNF-α production and retinal apoptosis.[69] These findings provide evidence supporting a role for ADORA$_{2A}$ in the development of future therapies for DR.

TAKE-HOME MESSAGES

- Global prevalences of diabetic retinopathy (DR), proliferative diabetic retinopathy (PDR), and vision-threatening retinopathy are estimated to be 93 million, 17 million, and 28 million, respectively.
- DR may cause visual impairment and is the most prevalent cause of new-onset blindness in individuals 20–75 years of age in Western countries.
- The final stage of adenosine triphosphate hydrolysis results in the production of adenosine.
- Acting via the adenosine A$_{2A}$ receptor (ADORA$_{2A}$), adenosine plays a role in vascular cell proliferation, vascular tone, neurotransmission, and amelioration of oxidative stress and inflammation in nerve and vascular tissues.
- ADORA$_{2A}$ genetic variants are associated with protection from the development of PDR in individuals with type 1 diabetes.

- Population allele frequencies for the $ADORA_{2A}$ gene differ between populations of African, admixed American, Asian, and European ancestry.
- Agonists and antagonists of $ADORA_{2A}$ are under investigation for the treatment of DR.

Acknowledgment

This research was supported in part by the Intramural Research Program of the National Human Genome Research Program of the National Human Genome Research Institute, National Institutes of Health, in the Center for Research on Genomics and Global Health (Z01GH2003362). The funders had no role in the preparation of this manuscript.

References

1. International Diabetes Federation. *IDF Diabetes Atlas: The Global Burden.* 5th ed ; 2011. IDF.
2. Yau JW, Rogers SL, Kawasaki R, Lamoureaux EL, Kawalski JW, Bek T, et al. Global prevalence and risk factors of diabetic retinopathy. *Diabetes Care* 2012;**35**:556–64.
3. American Diabetes Association. Diabetes Basics: Diabetes Statistics. *American Diabetes Association* 2012.
4. National Eye Institute. *Facts About Diabetic Retinopathy.* Bethesda, MD: NEI; 2012.
5. American Academy of Ophthalmology Retina Panel. *Guidelines.* San Francisco, CA: American Academy of Ophthalmology; 2012.
6. Chew E, Ferris FI. *Nonproliferative Diabetic Retinopathy.* Philadelphia, PA: Elsevier/Mosby; 2006.
7. Sheetz MJ, King GL. Molecular understanding of hyperglycemia's adverse effects for diabetes complications. *JAMA* 2002;**288**:2579–88.
8. Petrovic MG, Hawlina M, Peterlin B, Petrovic D. Bg/II gene polymorphism of the A2B1 integrin gene is a risk factor for diabetic retinopathy in caucasians with type 2 diabetes. *J Hum Genet* 2003;**48**:457–60.
9. National Eye Institute. *Search NEI Photos, Images, and Videos. Diabetic Retinopathy and Normal Eye.* Bethesda, MD: NEI; 2012.
10. Boehm BO, Lang G, Feldman B, Kurkhaus A, Rosinger S, Volper O, et al. Proliferative diabetic retinopathy is associated with a low level of the natural anti-angiogenic agent pigment epithelium-derived factor (PEDF) in aqueous humor: a pilot study. *Horm Metab Res* 2003;**50**:382–6.
11. Whikert DR. *Biochemistry of the Eye.* Philadelphia: Butterworth-Heinemann, Philadelphia, PA; 2003.
12. Chowdhury TA, Kumar S, Barnett AH, Bain SC. Nephropathy in diabetes. *Diabet Med* 1995;**12**:1059–67.
13. Kumaramanickavel G, Sripriya S, Ramprasad VL, Upadyay NK, Paul PG, Sharma T. Z-2Aldose reductase allele and diabetic retinopathy in India. *Ophthalmic Genet* 2003;**24**:41–8.
14. Santos P, Carmelo O, Carvalho A, Duarte C. Characterization of ATP release from cultures enriched in cholinergic amacrine-like neurons. *J Neurobiol* 1999;**41**:340–8.
15. Volonte C, Amadio S, Cavaliere F, D'Ambrosi N, Vacca F, Bernardi G. Extracellular ATP and neurodegeneration. *Curr Drug Targets* 2003;**2**:403–12.
16. Costa G, Pereira T, Neto A, Cristóvão A, Ambrósio A, Santos P. High glucose changes extracellular adenosine triphosphate levels in rat retinal cultures. *J Neurosci Res* 2009;**87**:1375–80.
17. Wilson K, Walker J, editors. *Principles and Techniques of Biochemistry and Molecular Biology.* Cambridge: Cambridge University Press; 2008.
18. Burnstock G. Purinergic signaling. *Br J Pharmacol* 2006;**147**:S172–81.
19. Burnstock G, Knight G. Cellular distribution and functions of P2 receptor subtypes in different systems. *Int Rev Cytol* 2004;**240**:31–04.
20. Jacobson K, Gao Z-G. Adenosine receptors as therapeutic targets. *Nat Rev.* 2006;**5**:247–64.
21. Online Mendelian Inheritance in Man. *OMIM.* McKusick-Nathans Institute of Genetic Medicine, 2005
22. Linden F. Structure and function of A_1 adenosine receptors. *FASEB J* 1991;**5**:2668–76.
23. Olah ME, Stiles GL. Adenosine receptor subtypes: characterization and therapeutic regulation. *Annu Rev Toxicol* 1995;**35**:581–606.
24. Kowluru RA, Chan P-S. Oxidative stress and diabetic retinopathy. *Exp Diabetes Res* 2007;**2007**(Article ID 43603):12.
25. Maiese K, Chong ZZ, Shang YC. Mechanistic insights into diabetes mellitus and oxidative stress. *Curr Med Chem* 2007;**14**:1729–38.
26. Yano M, Hasegawa G, Ishii M, Yamasaki M, Fukui M, Nakamura N, et al. Short-term exposure of high glucose concentration induces generation of reactive oxygen species in endothelial cells: implication for the oxidative stress associated with postprandial hyperglycemia. *Redox Rep* 2004;**9**:111–6.
27. Cronstein B. Adenosine and endogenous anti-inflammatory agent. *J Appl Physiol* 1994;**76**:5–13.
28. Roy S, Trudeau K, Roy S, Behl Y, Dhar S, Chronopoulos A. New insights into hyperglycemia-induced molecular changes in microvascular cells. *J Dent Res* 2009;**89**:116–27.
29. Pe'er J, Folberg R, Gnessin H, Hemo I, Keshet E. Upregulated expression of vascular endothelial growth factor in proliferative diabetic retinopathy. *Br J Ophthalmol* 1996;**80**:241–5.
30. Pierce E, Avery R, Foley E, Aiello LP, Smith L. Vascular endothelial growth factor/vascular permeability factor expression in a mouse model of retinal neovascularization. *Proc Natl Acad Sci U S A* 1995;**92**:905–9.
31. Aiello LP, Northrup J, Keyt B, Takagi H, Iwamoto M. Hypoxic regulation of vascular endothelial growth factor in retinal cells. *Arch Ophthalmol* 1995;**113**:1538–44.
32. Simorre P, Guerrin M, Chollet P, Penary M, Clamens S, Malacaze F, et al. Vasculotropin-VEGF stimulates retinal capillary endothelial cells through an autocrine pathway. *Invest Ophthalmol Vis Sci* 1994;**35**:3393–400.
33. Adamis A, Shima D, Yeo K, Keo T, Brown L, Berse B, et al. Synthesis and secretion of vascular permeability factor/vascular endothelial growth factor by human retinal pigment epithelial cells. *Biochem Biophys Res Commun* 1993;**193**:631–8.
34. Aiello LP, Avery R, Arrigg P, Keyt B, Jampel H, Shah S, et al. Vascular endothelial growth factor in ocular fluid of patients with diabetic retinopathy and other retinal disorders. *N Engl J Med* 1994;**331**:1480–7.
35. Bontemps F, Mimouni M, Van den Berghe G. Mechanisms of elevation of adenosine levels in anoxic hepatocytes. *J Biochem* 1993;**290**:671–7.
36. Takagi H, King GL, Ferrara N, Aiello LP. Hypoxia regulates vascular endothelial growth factor receptor KDR/Flk gene expression through adenosine A_2 receptor in retinal capillary endothelial cells. *Invest Ophthalmol Vis Sci* 1996;**37**:1311–21.
37. Robinson Jr WG, Kador PF, Kinoshita JH. Early retinal microangiopathy: prevention with aldose reductase inhibitors. *Diabet Med* 1985;**2**:196–9.
38. Takagi H, King G, Aiello LP. Hypoxia upregulates glucose transport activity through an adenosine-mediated increase of GLUT1 expression in retinal capillary endothelial cells. *Diabetes* 1998;**47**:1480–7.
39. Takagi H, King GL, Robinson G, Ferrara N, Aiello LP. Hypoxic-induction of vascular endothelial growth factor (VEGF) is mediated by adenosine, adenosine A_2 receptors, and cAMP-dependent protein kinase A in retinal microvascular pericytes and vascular endothelial cells. *Invest Ophthalmol Vis Sci* 1996;**37**:2165–76.
40. Gardner TW, Aiello LP. Pathogenesis of diabetic retinopathy. In: Scott IU, Flynn Jr H, Smiddy WE, editors. *Diabetes and Ocular Disease: Past, Present and FutureTherapies.* Oxford: Oxford University Press; 2010. pp. 49–70.
41. Winkler BS, et al. Energy metabolism in human retinal Müller cells. *Invest Ophthalmol Vis Sci* 2000;**41**(10):3183–90.

42. Stone J, Dreher Z. Relationship between astrocytes, ganglion cells and vasculature of the retina. *J Comp Neurol* 1987;**255**(1): 35–49.

43. Rungger-Brandle E, Dosso AA, Leuenberger PM. Glial reactivity, an early feature of diabetic retinopathy. *Invest Ophthalmol Vis Sci* 2000;**41**(7):1971–80.

44. Zeng XX, Ng YK, Ling EA. Neuronal and microglial response in the retina of streptozotocin-induced diabetic rats. *Vis Neurosci* 2000;**17**(3):463–71.

45. Chen L, Yang P, Kijlstra A. Distribution, markers, and functions of retinal microglia. *Ocul Immunol Inflamm* 2002;**10**(1):27–39.

46. Bursell SE, et al. Retinal blood flow changes in patients with insulin-dependent diabetes mellitus and no diabetic retinopathy. *Invest Ophthalmol Vis Sci* 1996;**37**(5):886–97.

47. Cogan DG, Toussaint D, Kuwabara T. Retinal vascular patterns. IV. Diabetic retinopathy. *Arch Ophthalmol* 1961;**66**:366–78.

48. Verma A, et al. Does neuronal damage precede vascular damage in subjects with type 2 diabetes mellitus and having no clinical diabetic retinopathy? *Ophthalmol Res* 2012;**47**(4):202–7.

49. Ozawa Y, et al. Neural degeneration in the retina of the streptozotocin-induced type 1 diabetes model. *Exp Diabetes Res* 2011; **2011**:7. pgs.

50. Sokol S, et al. Contrast sensitivity in diabetics with and without background retinopathy. *Arch Ophthalmol* 1985;**103**(1):51–4.

51. Liou GI, et al. Role of adenosine in diabetic retinopathy. *J Ocul Biol Dis Infor* 2011;**4**(1–2):19–24.

52. Blazynski C. Discrete distributions of adenosine receptors in mammalian retina. *J Neurochem* 1990;**54**(2):648–55.

53. Rudolphi KA, Schubert P. Modulation of neuronal and glial cell function by adenosine and neuroprotection in vascular dementia. *Behav Brain Res* 1997;**83**(1–2):123–8.

54. Schubert P, et al. Modulation of nerve and glial function by adenosine–role in the development of ischemic damage. *Int J Biochem* 1994;**26**(10–11):1227–36.

55. Osborne NN. [(3)H]Glycogen hydrolysis elicited by adenosine in rabbit retina: involvement of A(2)-receptors. *Neurochem Int* 1989;**14**(4):419–22.

56. Liao SD, Puro DG. NAD+induced vasotoxicity in the pericyte-containing microvasculature of the rat retina: effect of diabetes. *Invest Ophthalmol Vis Sci* 2006;**47**(11):5032–8.

57. Li B, et al. Differing roles of adenosine receptor subtypes in retinal ischemia-reperfusion injury in the rat. *Exp Eye Res* 1999;**68**(1):9–17.

58. Charles B, Conley Y, Chen G, Miller R, Dorman J, Gorin M, et al. Variants of the adenosine A$_{2A}$ receptor gene are protective against proliferative diabetic retinopathy in patients with type 1 diabetes. *Ophthalmic Res* 2011;**46**:1–8.

59. Kostraba J, Klein R, Dorman J, Becker D, Drash A, Maser R, et al. The epidemiology of diabetes complications study. *Am J Epidemiol* 1991;**133**:381–91.

60. Costacou T, Chang Y, Ferrell R, Orchard T. Identifying genetic susceptibilities to diabetes-related complications among individuals at low risk of complications: an application of tree structures survival analysis. *Am J Epidemiol* 2006;**164**:862–72.

61. LLoyd C, Klein R, Maser R, Kuller L, Becker D, Orchard T. The progression of retinopathy over 2 years: the Pittsburgh Epidemiology of Diabetes Complications (EDC) Study. *Diabetes Complications* 1995;**9**:140–8.

62. Diabetic Retinopathy Study Group. A modification of the Arlie House classification of diabetic retinopathy: Diabetic Retinopathy Report No. 7. *Invest Ophthalmol Vis Sci* 1981;**21**:210–26.

63. Moss S, Meuer S, Klein R, Hubbard L, Brothers R, Klein BEK. Are seven standard photographic fields necessary for classification of diabetic retinopathy? *Invest Ophthalmol Vis Sci* 1989;**30**:823–8.

64. Frank R. Diabetic retinopathy. *N Engl J Med* 2004;**350**:48–58.

65. American Diabetes Association. All about Diabetes. *American Diabetes Association* 2008.

66. 1000 Genomes Project. *1000 Genomes: A Deep Catalog of Human Genetic Variation*. http://browser.1000genomes.org/Homo_sapiens/Info/Index; 2012.

67. Kumar B, Gupta SK, Saxena R, Srivastava S. Current trends in the pharmacotherapy of diabetic retinopathy. *J Postgrad Med* 2012;**58**:132–9.

68. Liou GI, Auchampach JA, Hillard CJ, Zhu G, Yousufzai B, Mian S, et al. Mediation of cannabidiol anti-inflammation in the retina by equilibrative nucleoside transporter and A$_{2A}$ adenosine receptor. *Invest Ophthalmol Vis Sci* 2008;**49**:5526–31.

69. Ibrahim AS, El-shishtawy MM, Zhang W, Caldwell RB, Liou GI. A$_{2A}$ adenosine receptor (A2AAR) as a therapeutic target in diabetic retinopathy. *Am J Pathol* 2011;**178**:2136–45.

54

Effects of Environmental, Genetic, and Epigenetic Factors on Platelet Glycoproteins and the Development of Diabetic Retinopathy

Daniel Petrovič

Institute of Histology and Embryology, Medical Faculty Ljubljana, University of Ljubljana, Ljubljana, Slovenia

INTRODUCTION

Changes in dietary habits and lifestyles associated with rapid economic growth have dramatically increased the incidence of diabetes, obesity, and related vascular complications.[1,2] Both type 1 and type 2 diabetes are associated with hyperglycemia, oxidant stress, and inflammation and a significantly increased risk of macrovascular and microvascular complications (Table 54.1).[1,2]

Diabetic retinopathy (DR) is associated with both environmental and genetic factors. Several metabolic abnormalities are implicated in its pathogenesis; however, the exact mechanism remains to be determined. While several studies have been devoted to the evaluation of genetic factors related to diabetes and its complications (including DR), much less is known about environmental factors, nutrition, and epigenetic mechanisms related to DR.[3–5] So far, several genes and their polymorphisms have been implicated in the pathogenesis (Table 54.2).[5–7]

PATHOGENESIS OF DIABETIC RETINOPATHY

Epidemiologic and experimental studies indicate that besides the duration of diabetes and glycemic control, several other factors are involved in the development of DR (Table 54.3).[8,9]

DR is characterized by increased vascular permeability, hemostatic abnormalities, increased tissue ischemia, and neoangiogenesis.[2] The pathogenetic mechanisms of DR are very complex. Although many hyperglycemia-induced metabolic abnormalities are implicated in its pathogenesis, the exact mechanism of the development of retinopathy remains elusive. Alteration in retinal blood flow, metabolic changes, hemostatic abnormalities, increased oxidative stress, increased polyol pathway flux, activation of protein kinase C isoforms, increased hexosamine pathway flux, and increased advanced glycation end-product formation, nonenzymatic glycosylation of collagen, and other tissue proteins are observed during long-term hyperglycemia.[2,9] It is also important to know that many of the systemic abnormalities in both type 1 and type 2 diabetic patients are prothrombotic: hyperreactive platelets, decreased vascular prostacyclin production, endothelial dysfunction resulting in increased circulating levels of von Willebrand factor and leukocyte adhesion molecules, hypercoagulability, and decreased fibrinolysis. Increased levels of intercellular adhesion molecule-1 and plasminogen activator inhibitor-1 messenger RNA (mRNA), as well as decreased levels of tissue plasminogen activator, are specifically found in retinal vessels of diabetic compared with nondiabetic individuals.[10,11]

TABLE 54.1   Chronic Complications of Diabetes

Microvascular	Macrovascular
Retinopathy	Coronary heart disease
Nephropathy	Cerebrovascular disease
Neuropathy	Peripheral arterial obstructive disease

TABLE 54.2 Pathways and Genes Implicated in the Pathogenesis of Diabetic Retinopathy

Pathway/System	Gene
Polyol pathway	Aldose reductase
Renin–angiotensin system	Renin
	Angiotensinogen
	Angiotensin-1 converting enzyme
	Aldosterone
Advanced glycation end-products	Receptor for advanced glycation end-products
Growth factors	Vascular endothelial growth factor
	Basic fibroblast growth factor
	Insulin-like growth factor
Peroxisome proliferator-activated receptor	Peroxisome proliferator-activated receptor
	Coactivator for peroxisome proliferator-activated receptor
Inflammatory genes	Interleukins
	Tumor necrosis factor
Thrombotic system	Fibrinogen
Platelet function	Integrin
Oxidative system	Manganese superoxide dismutase
	Catalase
	Myeloperoxidase
	Glutathione S-transferase
	NADPH oxidase
	Endothelial nitric oxide synthase
	Inducible nitric oxide synthase
Adhesion molecules	Intercellular adhesion molecule
	Vascular cell adhesion protein
	Platelet endothelial cell adhesion molecule
Extracellular matrix homeostasis	Matrix homeostasis genes
	Matrix metalloproteinase
Hormones/vitamins	Growth hormone
	Vitamin D
Undefined	Glucose transporter-1
	Growth hormone

NADPH: reduced nicotinamide adenine dinucleotide.

TABLE 54.3 Risk Factors for Diabetic Retinopathy

Modifiable	Nonmodifiable
Cigarette smoking	Genetic factors
Blood pressure	Duration of diabetes
Plasma cholesterol	Age of onset of diabetes
Fasting glycemia	Positive family history
Hb_{A1c}	
Proteinuria	
Body mass index	
Waist-to-hip ratio	
Insulin treatment	
Vitamin B_{12} deficiency	

Hb_{A1c}: glycosylated hemoglobin.

PLATELETS AND DIABETIC RETINOPATHY

Platelets are thought to be involved in the pathogenesis of DR.[12] Platelets from diabetic patients may interact with exposed subendothelial fibrillar collagens, major components of the subendothelial matrix. Moreover, the amount of nonenzymatically glycosylated collagen, which is prone to interact with platelets, was demonstrated to be higher in diabetic patients than in nondiabetic controls.[10] Platelets activated by contact with collagens can trigger thrombus formation and small vessel occlusion. Platelets from diabetic patients are hyperreactive to aggregating agents, such as collagen, thrombin, and adenosine diphosphate (ADP).[12] The platelet membrane glycoprotein Ia/IIa, $\alpha_2\beta_1$ integrin, serves as a platelet receptor for collagen.[13] Some genetic variations of the glycoprotein involved in platelet adhesion, aggregation, and activation in favor of thrombogenesis may be considered as risk factors for thrombotic events.[13]

C825T POLYMORPHISM OF THE G-PROTEIN-COUPLED RECEPTOR GENE

Platelet aggregation that varies among individuals and genetic factors may alter platelet activation through G-protein-coupled receptors.[14] Recently, Dusse and co-workers demonstrated that the C825T polymorphism of the gene *GNB3* encoding the G-protein β_3 subunit influences platelet aggregation. In human whole blood, the *GNB3* 825CC genotype has been reported to be associated with enhanced platelet aggregation.[14] The study evaluating the effect of this polymorphism related to DR development has not been reported yet.

GENETIC FACTORS AND α₂β₁ INTEGRIN

Genetic variations in $\alpha_2\beta_1$ integrin were reported to affect the density of $\alpha_2\beta_1$ receptors on the platelet surface.[10] Namely, an association between genetic variations in $\alpha_2\beta_1$ and the density of $\alpha_2\beta_1$ receptors on the platelet surface was reported.[10] Kunicki et al.[10] demonstrated that the density of $\alpha_2\beta_1$ receptors on the platelet surface affects platelet adhesion to collagen, contributing to an increased risk of thrombosis.[10]

Matsubara et al.[15] demonstrated for the first time in 2000 the association between genetic variations in $\alpha_2\beta_1$ integrin and DR in the Japanese population. This finding was confirmed a few years later in a Caucasian population by Petrovic and co-workers.[7] In both studies, the BglII (+/+) genotype of the gene polymorphism of the $\alpha_2\beta_1$ integrin gene was reported to be an independent risk factor for DR in Caucasians with type 2 diabetes (Table 54.4).

Moreover, Matsubara et al.[15] demonstrated an association between genetic variations in $\alpha_2\beta_1$ integrin and another microvascular complication of diabetes, diabetic nephropathy, whereas Tsai et al.[16] failed to find such an association with diabetic nephropathy in a Chinese population with type 2 diabetes.

GENETIC FACTORS AND GLYCOPROTEIN IIB-IIIA

Glycoprotein (GP) IIb-IIIa ($\alpha_{IIb}\beta_3$-integrin), also known as integrin β_3 or antigen CD61, is the central receptor of platelet aggregation. GPIIIa is a surface protein found in various tissues, participating in cell adhesion and cell-surface mediated signaling. It builds glycoprotein IIb/IIIa in the platelet membrane and constitutes a fibrinogen receptor, which also exhibits an ability to bind the von Willebrand factor, fibronectin, and thrombospondin.[17] Activated GPIIb-IIIa binds fibrinogen or von Willebrand factor, which forms molecular bridges between aggregating platelets.

GPIIb-IIIa is a calcium-dependent heterodimer that is expressed in megakaryocytes, platelets, and mast cells.[18]

On the surface of resting platelets, GPIIb-IIIa is exhibited in a low-affinity conformation, in which the ligand-binding site is not exposed. Following the activation of platelets by agonists such as ADP, thrombin, or collagen, inside–out 3 signaling occurs, giving rise to the exposure of the ligand binding site of GPIIb-IIIa.[19] The binding of fibrinogen to the active form of GPIIb-IIIa is the final step leading to platelet aggregation and the formation of thrombus.[19,20]

Both GPIIb and GPIIIa are known to bear a number of single amino acid substitutions affecting conformational changes and the ligand binding function with little or no effect on platelet function. A platelet-specific antigen (PlA1/A2) polymorphism has been by far the most investigated GPIIIa gene polymorphism. A substitution of cytosine for thymidine at position 1565 in exon 2 of the GPIIIa gene leads to an amino acid difference at position 33: a leucine (A1 allele) or a proline (A2 allele).[17] The mentioned polymorphism can influence both platelet activation and aggregation[21] and affect postoccupancy signaling by the platelet fibrinogen receptor IIb/IIIa.[22] The presence of one or two PlA2 alleles is associated with an increased binding affinity to fibrinogen as well as with platelet aggregability in response to epinephrine, ADP, and collagen in vitro.[23] It has also been suggested that the PlA2 allelic variant causes an increased sensitivity to platelet aggregation by various agonists and an altered sensitivity to aspirin[24–26] (Table 54.5).

Platelet receptor polymorphism has also been implicated in the pathogenesis of glucose metabolism. It has been suggested that the function of platelet GPIIIa is modulated by a cysteine protease, calpain 10, which is not characteristic of platelets. Calpain 10 was reported to influence glucose metabolism, insulin secretion, and insulin action.[27] Studies on the association of the PlA1/A2 polymorphism of glycoprotein IIIa and type 2 diabetes have reported conflicting results[17,28,29] (Table 54.6).

Although there have been many studies on this polymorphism, conflicting results on its association with stroke, coronary artery disease, and myocardial infarction have been reported in a number of case-controlled clinical studies[30–38] (Table 54.7). A meta-analysis of 12 epidemiologic studies showed that there was an association between the PlA2 variant and an increased risk of

TABLE 54.4 Reported Studies of Polymorphism of $\alpha_2\beta_1$ Integrin and Diabetic Retinopathy (DR) in Subjects with Type 2 Diabetes

Population	Cases with DR* (n)	Control Group† (n)	Bgl II (+/+) Genotype: Cases (%)	Bgl II (+/+) Genotype: Controls* (%)	OR (P Value)‡	Reference
Japanese	119	108	23.5	11.2	2.1 (0.04)	Matsubara et al.[15]
Caucasians	163	95	19.2	7.4	2.4 (<0.05)	Petrovic et al.[6]

*Diabetic retinopathy.
†Subjects with duration of type 2 diabetes of more than 10 years.
‡Odds ratio and P value in logistic regression analysis.

TABLE 54.5 Reported Polymorphisms of Glycoprotein IIIa (GPIIIa) and Their Effect on Platelet Function

Gene	Polymorphic Site	Effect on Platelet Function	Reference
GPIIIa	Arg143Gln	No effect	Bajt and Lotus[26]
GPIIIa	Asp119	No effect	Bajt and Lotus[26]
GPIIIa	Ser121	No effect	Bajt and Lotus[26]
GPIIIa	Ser123	No effect	Bajt and Lotus[26]
GPIIIa	Ser130	No effect	Bajt and Lotus[26]
GPIIIa	Asp126	No effect	Bajt and Lotus[26]
GPIIIa	Asp127	No effect	Bajt and Lotus[26]
GPIIIa	−400 C/A*	No available information	Kozierdska et al.[17]
GPIIIa	−425 A/C*	No available information	Kozierdska et al.[17]
GPIIIa	−468 G/A*	No available information	Kozierdska et al.[17]
GPIIIa	1565 C/T	Influence on platelet activation and aggregation	Feng et al.[21]

Polymorphic site on the promoter of the gene.

coronary heart disease.[36] The studies were not directly comparable because they differed greatly in their patient pool and also in the way they were analyzed.[36] The epigenetic effect was analyzed in only one study. Oksala and co-workers reported that smokers with stable coronary artery disease and carriers of the *PlA2* allele were at an increased risk of subsequent cardiac events in comparison with nonsmokers with the *PlA2* allele.[39]

The *PlA1/PlA2* polymorphism of the *GPIIIa* gene was also implicated in the pathogenesis of DR (Table 54.8). Recently, the *A2A2* genotype of the *GPIIIa PlA1/A2* polymorphism has been reported to be a protective factor in the development of DR in Slovene Caucasians with type 2 diabetes. Contrary to this study, Pucci and co-workers[40] failed to demonstrate an important contribution of the *PlA1/PlA2* polymorphism of the *GPIIIa* gene on either nephropathy or retinopathy development in type 2 diabetes. This study recruited 605 subjects with type 2 diabetes. The *PlA1/PlA2* polymorphism of the *GPIIIa* gene failed to contribute to the development of nephropathy or retinopathy in type 1 and type 2 diabetes.

Although there have been many studies on this polymorphism, conflicting results on its association with stroke, myocardial infarction, or microvascular complications of diabetes have been reported in a number

TABLE 54.6 Reported Studies of the PlA1/A2 Polymorphism of Glycoprotein IIIa and Type 2 Diabetes

Population	Cases* (n)	Control Group* (n)	A2 Allele: Cases* (%)	A2 Allele: Controls† (%)	OR (P Value)‡	Reference
Caucasians	112	59	34.8	14.6	3.1 (P < 0.01)	Tschoepe et al.[28]
Caucasians	113	95	20.4	21.5	1.0 (ns)	Kozieradzka et al.[17]
Caucasians	1051	2247	15.6	15.6	1.0 (ns)	Marz et al.[29]

Subjects with type 2 diabetes.
†Subjects without diabetes.
‡odds ratio and P value; ns: not statistically significant.

TABLE 54.7 Reported Studies of Polymorphism of $\alpha_2\beta_1$ Integrin and Myocardial Infarction

Population	Cases with MI (n)	Control Group (n)	Bgl II (+/+) Genotype: Cases (%)	Bgl II (+/+) Genotype: Controls (%)	OR (P Value)*	Reference
Caucasians	71	68	39	19	2.7 (0.04)	Weiss et al.[30]
Men, USA†	375	14,212	13.5	14.8	0.9 (0.4)	Ridker et al.[31]
Caucasians	242	209	1.2	2.4	0.9 (0.6)	Samani et al.[32]
Caucasians	156	216	§	§	1.66 (0.007)	Carter et al.[33]
Caucasians	1066	512	§	§	1.1 (0.9)	Gardemann et al.[34]
Caucasians	225	170	33.8	26.9	1.47 (0.06)	Anderson et al.[35]
Caucasians	529	1191	3.6	2.9	1.4 (0.01)	Grove et al.[37]
Caucasians‡	229	325	11.6	14.1	1.0 (0.9)	Starčević and Petrović[38]

Odds ratio and P value in logistic regression analysis.
†Men participating in the prospective Physicians' Health Study.
‡Subjects with type 2 diabetes; §data not available.

of case-controlled clinical studies. Genetic association studies are prone to beta-statistical error and population-specific genotype effects, all of which make the results difficult to reproduce. One could speculate that discrepancies between different studies may occur owing to other factors and mechanisms, such as differences in environmental factors (e.g., nutrition) and epigenetic mechanisms. Because of this, further studies enrolling larger numbers of patients from different populations with consideration of various environmental and epigenetic factors are needed to confirm the results of genetic studies published so far on the importance of platelet function in DR.

EPIGENETIC FACTORS AND DIABETIC RETINOPATHY

Microthrombosis due to platelet dysfunction is implicated in the pathogenesis of DR via complex interactions of environmental and genetic factors, none of which can be fully and solely responsible for the higher risk of developing DR and for DR progression.

Good glycemic control, if started in the initial stage of diabetes, prevents the development of retinopathy but, if reinstituted after a period of poor control, fails to halt its development, suggesting a metabolic memory phenomenon.[4] Patients in the conventional treatment regimen during the Diabetes Complications and Control Trial had a higher incidence of complications several years after switching to intensive therapy than patients in the intensive control group.[41,42] Studies in rats have demonstrated that the retina continues to experience oxidative stress, manganese superoxide dismutase (MnSOD) remains compromised, and nuclear factor-κB is activated for at least 6 months after reinstitution of good glycemic control that has followed 6 months of poor control. This phenomenon has recently been reported to be due to the global acetylation of retinal histone H3.[4] The epigenetic regulation has been demonstrated in the *MnSOD* gene,[4] whereas it has not been studied in any integrin gene yet. A similar mechanism (i.e., global acetylation of retinal histone H3) has been proposed in the pathogenesis of the progression of DR.

Epigenetic changes occur without alterations in the DNA sequence and can affect gene transcription in response to environmental changes and nutrition. Switching between the active and inactive state of chromatin is the central mechanism of gene regulation, and this is defined as epigenetic factor. Several pathways may be involved in epigenetic regulation, such as DNA methylation, histone acetylation, and noncoding RNAs or microRNAs.[3,4]

Modulation of epigenetic changes by pharmaceutical means may provide a potential strategy to retard the progression of DR. Besides intense medical management, these strategies include dietary measures and the introduction of epigenetic drugs, such as inhibitors of DNA methylation and histone demethylases.

ENVIRONMENTAL FACTORS, NUTRITION, AND DIABETIC RETINOPATHY

The impact of nutrition on the manifestation and progression of DR and other retinal disease has become an important and controversial topic in recent years. Awareness of this topic in the general population has increased partly as a result of commercial advertisements for supplements and diets. Although well-designed clinical trials have demonstrated the importance of some environmental factors (e.g., smoking, arterial hypertension), the significance of other factors (e.g., nutritional factors, vitamins) remains to be confirmed in well-designed clinical trials.[43,44] So far, predominantly cross-sectional and interventional studies have been performed to address this question.

Several environmental factors have been reported to be associated with DR and the progression of DR.[41,43–46]

Cigarette smoking was reported to be significantly associated with the incidence of DR and the rate of progression of DR.[43,44]

The consumption of fatty acids and dietary fiber was reported to be significantly associated with the rate of progression of DR.[41]

In particular, studies on patients with DR have implicated an impact of higher cholesterol levels on the progression of the disease.[44] Fibrates, like statins, may act directly to decrease the progression of diabetic complications through their lipid-lowering effects

TABLE 54.8 Reported Studies of the PlA1/PlA2 Polymorphism of the *GPIIIa* Gene and Diabetic Retinopathy in Subjects with Type 2 Diabetes

Population	Cases with DR (n)	Control Group* (n)	A2A2 Genotype: Cases (%)	A2A2 Genotype: Controls* (%)	OR (P Value)†	Reference
Caucasians, Italy	339	266	1.0	4.0	0.9 (0.6)	Pucci *et al.*[40]
Caucasians, Slovenia	222	120	12.6	22.2	0.61 (0.037)	Nikolajević *et al.*[7]

Subjects with duration of type 2 diabetes of more than 10 years.
†*Odds ratio and P value in logistic regression analysis.*

but may also go beyond that via pleiotropic effects [44] (Table 54.9).

Data on some chronic disorders (e.g., asthma, chronic obstructive pulmonary disease, diabetes) indicate that an increase in vegetable and fruit consumption may contribute to the prevention of these diseases, whereas there is insufficient evidence for DR regarding an association with the consumption of fruits and vegetables.[47]

Vitamin B$_{12}$ deficiency has been recently demonstrated to be an independent risk factor for the development of DR,[48] although more data are needed on the role of vitamins. Nutritional supplementation is receiving increasing interest with regard to DR via different mechanisms (e.g., antioxidants, platelet function).

Vitamin C was reported to affect retinal blood flow in animal models and human supplementation trials.[49] Vitamin C reduces platelet aggregation[50] and protects against oxidative stress caused by nonenzymatic glycation and metabolic stress in people with diabetes.[51] Vitamin E was reported to reduce oxidative stress at levels of 200 mg/day or more.[51]

Epidemiologic evidence does not support a relationship between antioxidant intake and reduced risk for DR; however, intervention trials indicate that further research is needed to clarify the role of several nutritional factors. Pycnogenol was reported to protect vitamin E from oxidation and prevent platelet activity in humans without increasing bleeding time.[52]

So far, only a few studies on agents affecting platelet function have been reported.[50] The platelet proteomics approach revealed novel insights into the regulation of cellular biomarkers of atherogenic and thrombotic pathways by a dietary 80:20 cis9,trans11-conjugated linoleic acid blend.[46]

Recently, the effect of delphinidin-3-glucoside (Dp-3-g), one of the predominant bioactive compounds of anthocyanins in many plant foods, on platelet function has been demonstrated.[53] This study found that Dp-3-g significantly inhibited human and murine platelet aggregation. Platelet activation markers were examined via flow cytometry, and Dp-3-g significantly inhibited the expression of P-selectin, CD63, and CD40L, which reflect platelet α- and δ-granule release, and cytosol protein secretion, respectively. They further demonstrated that Dp-3-g downregulated the expression of active integrin αIIbβ$_3$ on platelets and attenuated fibrinogen binding to platelets following agonist treatment without interfering with the direct interaction between fibrinogen and integrin αIIbβ$_3$. Thus, Dp-3-g significantly inhibits platelet activation, and this may have a favorable effect on the progression of DR.[53]

Optimizing the medical management of DR should address the control of glycemia, blood pressure, and lipids, and based on recent trials, specific therapies using fenofibrate with a statin and candesartan should be considered. Results from experimental studies indicate that further prospective clinical studies are warranted on the prevention and inhibition of the progression of DR.

CONCLUSIONS

Changes in dietary habits and lifestyles associated with rapid economic growth have dramatically increased the incidence of diabetes and related macrovascular and microvascular complications. Several factors, such as hyperglycemia, oxidative stress, inflammation, and platelet dysfunction, are implicated in the pathogenesis of DR. So far, several studies have demonstrated the importance of several environmental and genetic factors, whereas much less is known about gene–environment interactions and epigenetic changes.

Alarming estimates indicate that the rate of diabetes and associated complications (including DR) are rapidly increasing; therefore, additional strategies to arrest these trends are needed. Besides intense medical management, these strategies include dietary measures and the introduction of epigenetic drugs, such as inhibitors of DNA methylation and histone demethylases.

Finally, based on current knowledge, optimizing the medical management of DR should address the control of glycemia, blood pressure, and lipids, and specific therapies using fenofibrate with a statin and candesartan should be considered. To conclude, the impact of nutritional factors is still insufficiently understood for patients with DR, and well-designed prospective randomized clinical trials are needed to address the role of nutritional factors.

TABLE 54.9 Agents and Drugs Used to Prevent or Slow the Progression of Diabetic Retinopathy

Agent/Drug	Mechanism of Action
Statins	Lipid-lowering effect plus pleiotropic effect
Vitamin B$_{12}$	Antioxidants, platelet function
Vitamin C	Reduces platelet aggregation and protects against oxidative stress
Vitamin E	Reduces oxidative stress
Delphinidin-3-glucoside	Inhibits platelet activation

TAKE-HOME MESSAGES

- Changes in dietary habits and lifestyles associated with rapid economic growth have dramatically increased the incidence of diabetes and related macrovascular and microvascular complications.

- Several biochemical mechanisms and genetic factors have been implicated in the pathogenesis of diabetic retinopathy (DR).
- Owing to the influences of gene–environmental interactions, epigenetic mechanisms regulate at least some of the pathologic mechanisms in the development of DR.
- Epigenetic changes occur without alterations in the DNA sequence and can affect gene transcription in response to environmental changes and nutrition.
- Optimal medical management of DR should address the control of glycemia, blood pressure, and lipids, and specific therapies using fenofibrate with a statin and candesartan should be considered.
- The impact of nutritional factors is still insufficiently understood for patients with DR.
- Well-designed prospective randomized clinical trials are needed to address the role of nutritional factors.

Acknowledgment

The authors thank Ms Visam Bajt, BA, for revising the English.

References

1. Klein R, Klein BEK. Visual disorders in diabetes: diabetes in America. In: Harris CI, Cowie CC, Stern MP, Boyko EJ, Reiber GE, Bennett PH, editors. *Report of National Institutes of Diabetes and Digestive and Kidney Diseases*. Bethesda, MD: National Institutes of Health; 1995. pp. 293–338.
2. Brownlee M. The pathobiology of diabetic complications: a unifying mechanism. *Diabetes* 2005;**54**:1615–25.
3. Reddy MA, Natarajan R. Epigenetic mechanisms in diabetic vascular complications. *Cardiovasc Res* 2011;**90**:421–9.
4. Zhong Q, Kowluru RA. Role of histone acetylation in the development of diabetic retinopathy and the metabolic memory phenomenon. *J Cell Biochem* 2010;**110**:1306–13.
5. Uhlmann K, Kovacs P, Boettcher Y, Hammes HP, Paschke R. Genetics of diabetic retinopathy. *Exp Clin Endocrinol Diab* 2006;**114**:275–94.
6. Petrovic MG, Hawlina M, Peterlin B, Petrovic D. BglII gene polymorphism of the alpha2beta1 integrin gene is a risk factor for diabetic retinopathy in Caucasians with type 2 diabetes. *J Hum Genet* 2003;**48**:457–60.
7. Nikolajević-Starčević J, Petrovic MG, Petrovic D. A1/A2 polymorphism of the glycoprotein IIIa gene and diabetic retinopathy in Caucasians with type 2 diabetes. *Clin Exp Ophthalmol* 2011;**39**:665–72.
8. Engerman RL, Kern TS. Progression of incipient diabetic retinopathy during good glycemic control. *Diabetes* 1987;**36**:808–12.
9. Porta M, Sjoelie AK, Chaturvedi N, Stevens L, Rottiers R, Veglio M, Fuller JH. Risk factors for progression to proliferative diabetic retinopathy in the EURODIAB Prospective Complications Study. *Diabetologia* 2001;**44**:2203–9.
10. Kunicki TJ, Kritzik M, Annis DS, Nugent DJ. Hereditary variation in platelet integrin a2b1 density is associated with two silent polymorphisms in the alpha 2 gene coding sequence. *Blood* 1997;**89**:1939–43.
11. McLeod DS, Lefer DJ, Merges C, Lutty GA. Enhanced expression of intracellular adhesion molecule-1 and P-selectin in the diabetic human retina and choroid. *Am J Pathol* 1995;**147**:642–53.
12. Barnett AH. Pathogenesis of diabetic microangiopathy: an overview. *Am J Med* 1991;**90**:67–73S.
13. Santoro SA, Zutter MM. The a2b1 integrin: a collagen receptor on platelets and other cells. *Thromb Haemost* 2001;**74**:813–21.
14. Dusse F, Frey UH, Bilalic A, Dirkmann D, Görlinger K, Siffert W, Peters J. The GNB3 C825T polymorphism influences platelet aggregation in human whole blood. *Pharmacogenet Genomics* 2012;**22**:43–9.
15. Matsubara Y, Murata M, Maruyama T, Handa M, Yamagata N, Watanabe G, et al. Association between diabetic retinopathy and genetic variations in a2b1 integrin, a platelet receptor for collagen. *Blood* 2000;**95**:1560–4.
16. Tsai DH, Jiang YD, Wu KD, Tai TY, Chuang LM. Platelet collagen receptor a2b1 integrin and glycoprotein IIIa Pl(A1/A2) polymorphisms are not associated with nephropathy in type 2 diabetes. *Am J Kidney Dis* 2001;**38**:1185–90.
17. Kozieradzka A, Kamiński K, Pepiński W, Janica J, Korecki J, Szepietowska B, Musiał WJ. The association between type 2 diabetes mellitus and A1/A2 polymorphism of glycoprotein IIIa gene. *Acta Diabetol* 2007;**44**:30–3.
18. Coller BS, Shattil SJ. The GPIIb/IIIa (integrin alphaIIbbeta3) odyssey: a technology-driven saga of a receptor with twists, turns, and even a bend. *Blood* 2008;**112**:3011–25.
19. Shattil SJ, Newman PJ. Integrins: dynamic scaffolds for adhesion and signaling in platelets. *Blood* 2004;**104**:1606–15.
20. Du X, Ginsberg MH. Integrin alpha(IIb)beta(3) and platelet function. *Thromb Haemost* 1997;**78**:96–100.
21. Feng D, Lindpaintner K, Larson MG, Rao VS, O'Donnell CJ, Lipinska I, et al. Increased platelet aggregability associated with platelet GpIIIa PlA2 polymorphism. The Framingham Offspring Study. *Arterioscler Thromb Vasc Biol* 1999;**19**:1142–7.
22. Vijayan KV, Liu Y, Dong JF, Bray PF. Enhanced activation of mitogen-activated protein kinase and myosin light chain kinase by the Pro33 polymorphism of integrin β3. *Biol Chem* 2003;**278**:3860–7.
23. Michelson AD, Furman MI, Goldschmidt-Clermont P, Mascelli MA, Hendrix C, Coleman L, et al. Platelet GP IIIa Pl(A) polymorphisms display different sensitivities to agonists. *Circulation* 2000;**101**:1013–8.
24. Cooke GE, Liu-Stratton Y, Ferketich AK, Moeschberger ML, Frid DJ, Magorien RD, et al. Effect of platelet antigen polymorphism on platelet inhibition by aspirin, clopidogrel, or their combination. *J Am Coll Cardiol* 2006;**47**:541–6.
25. Szczeklik A, Undas A, Sanak M, Frolow M, Wegrzyn W. Relationship between bleeding time, aspirin, and the PlA1/A2 polymorphism of platelet glycoprotein IIIa. *Br J Haematol* 2000;**110**:965–7.
26. Bajt ML, Loftus JC. Mutation of a ligand binding domain of beta 3 integrin. Integral role of oxygenated residues in alpha IIb beta 3 (GPIIb-IIIa) receptor function. *J Biol Chem* 1994;**269**:20913–9.
27. Harris F, Chatfield L, Singh J, Phoenix DA. Role of calpains in diabetes mellitus: a mini review. *Mol Cell Biochem* 2004;**261**:161–7.
28. Tschoepe D, Menart B, Ferber P, Altmann C, Haude M, Haastert B, Roesen P. Genetic variation of the platelet-surface integrin GPIIb-IIIa (PIA1/A2-SNP) shows a high association with type 2 diabetes mellitus. *Diabetologia* 2003;**46**:984–9.
29. März W, Boehm BO, Winkelmann BR, Hoffmann MM. The PlA1/A2 polymorphism of platelet glycoprotein IIIa is not associated with the risk of type 2 diabetes. The Ludwigshafen Risk and Cardiovascular Health study. *Diabetologia* 2004;**47**:1969–73.
30. Weiss EJ, Bray PF, Tayback M, Schulman SP, Kickler TS, Becker LC, et al. A polymorphism of a platelet glycoprotein receptor as an inherited risk factor for coronary thrombosis. *N Engl J Med* 1996;**334**:1090–4.
31. Ridker PM, Hennekens CH, Schmitz C, Stampfer MJ, Lindpaintner K. PIA1/A2 polymorphism of platelet glycoprotein IIIa and risks of myocardial infarction, stroke, and venous thrombosis. *Lancet* 1997;**349**:385–8.

32. Samani NJ, Lodwick D. Glycoprotein IIIa polymorphism and risk of myocardial infarction. *Cardiovasc Res* 1997;**33**:693–7.

33. Carter AM, Ossei-Gerning N, Wilson IJ, Grant PJ. Association of the platelet Pl(A) polymorphism of glycoprotein IIb/IIIa and the fibrinogen Bbeta 448 polymorphism with myocardial infarction and extent of coronary artery disease. *Circulation* 1997;**96**:1424–31.

34. Gardemann A, Humme J, Stricker J, Nguyen QD, Katz N, Philipp M, et al. Association of the platelet glycoprotein IIIa PlA1/A2 gene polymorphism to coronary artery disease but not to nonfatal myocardial infarction in low risk patients. *Thromb Haemost* 1998;**80**:214–7.

35. Anderson JL, King GJ, Bair TL, Elmer SP, Muhlestein JB, Habashi J, Carlquist JF. Associations between a polymorphism in the gene encoding glycoprotein IIIa and myocardial infarction or coronary artery disease. *J Am Coll Cardiol* 1999;**33**:727–33.

36. Burr D, Doss H, Cooke GE, Goldschmidt-Clermont PJ. A meta-analysis of studies on the association of the platelet PlA polymorphism of glycoprotein IIIa and risk of coronary heart disease. *Statist Med* 2003;**22**:1741–60.

37. Grove EL, Ørntoft TF, Lassen JF, Jensen HK, Kristensen SD. The platelet polymorphism PlA2 is a genetic risk factor for myocardial infarction. *J Intern Med* 2004;**255**:637–44.

38. Nikolajević-Starčević J, Petrovič D. The a1/a2 polymorphism of the glycoprotein IIIa gene and myocardial infarction in Caucasians with type 2 diabetes. *Mol Biol Rep* 2013;**40**:2077–81.

39. Oksala NK, Heikkinen M, Mikkelsson J, Pohjasvaara T, Kaste M, Erkinjuntti T, Karhunen PJ. Smoking and the platelet fibrinogen receptor glycoprotein IIb/IIIA PlA1/A2 polymorphism interact in the risk of lacunar stroke and midterm survival. *Stroke* 2007;**38**:50–5.

40. Pucci L, Lucchesi D, Fotino C, Grupillo M, Miccoli R, Penno G, Del Prato S. Integrin Beta 3 PlA1/PlA2 polymorphism does not contribute to complications in both type 1 and type 2 diabetes. *G Ital Nefrol* 2003;**20**:461–9.

41. Cundiff DK, Nigg CR. Diet and diabetic retinopathy: insights from the Diabetes Control and Complications Trial (DCCT). *MedGenMed* 2005;**7**:3.

42. Writing Team for the Diabetes Control and Complications Trial/ Epidemiology of Diabetes Interventions and Complications Research Group. Sustained effect of intensive treatment of type 1 diabetes mellitus on development and progression of diabetic nephropathy: the Epidemiology of Diabetes Interventions and Complications (EDIC) study. *JAMA* 2003;**290**:2159–67.

43. Matthews DR, Stratton IM, Aldington SJ, Holman RR, Kohner EM. UK Prospective Diabetes Study Group. Risks of progression of retinopathy and vision loss related to tight blood pressure control in type 2 diabetes mellitus: UKPDS 69. *Arch Ophthalmol* 2004;**122**:1631–40.

44. Stratton IM, Kohner EM, Aldington SJ, Turner RC, Holman RR, Manley SE, Matthews DR. UKPDS 50: risk factors for incidence and progression of retinopathy in type II diabetes over 6 years from diagnosis. *Diabetologia* 2001;**44**:156–63.

45. Lee CT, Gayton EL, Beulens JW, Flanagan DW, Adler AI. Micronutrients and diabetic retinopathy: a systematic review. *Ophthalmology* 2010;**117**:71–8.

46. Bachmair EM, Bots ML, Mennen LI, Kelder T, Evelo CT, Horgan GW, et al. Effect of supplementation with an 80:20 *cis*9,*trans*11 conjugated linoleic acid blend on the human platelet proteome. *Mol Nutr Food Res* 2012;**56**:1148–59.

47. Boeing H, Bechthold A, Bub A, Ellinger S, Haller D, Kroke A, et al. Critical review: vegetables and fruit in the prevention of chronic diseases. *Eur J Nutr* 2012;**51**:637–63.

48. Satyanarayana A, Balakrishna N, Pitla S, Reddy PY, Mudili S, Lopamudra P, et al. Status of B-vitamins and homocysteine in diabetic retinopathy: association with vitamin-B_{12} deficiency and hyperhomocysteinemia. *PLoS ONE* 2011;**6**:e26747.

49. Cunningham JJ, Mearkle PL, Brown RG, Vitamin C: an aldose reductase inhibitor that normalizes erythrocyte sorbitol in insulin-dependent diabetes mellitus. *J Am Coll Nutr* 1994;**13**:344–50.

50. Wilkinson IB, Megson IL, MacCallum H, Sogo N, Cockcroft JR, Webb DJ. Oral vitamin C reduces arterial stiffness and platelet aggregation in humans. *J Cardiovasc Pharmacol* 1999;**34**:690–3.

51. Baynes JW, Thorpe SR. Role of oxidative stress in diabetic complications: a new perspective on an old paradigm. *Diabetes* 1999;**48**:1–9.

52. Pütter M, Grotemeyer KH, Würthwein G, Araghi-Niknam M, Watson RR, Hosseini S, Rohdewald P. Inhibition of smoking-induced platelet aggregation by aspirin and pycnogenol. *Thromb Res* 1999;**95**:155–61.

53. Yang Y, Shi Z, Reheman A, Jin JW, Li C, Wang Y, et al. Plant food delphinidin-3-glucoside significantly inhibits platelet activation and thrombosis: novel protective roles against cardiovascular diseases. *PLoS ONE* 2012;**7**:e37323.

Haptoglobin Genotype and Diabetic Retinopathy

N. Goldenberg-Cohen[1,2]

[1]Pediatric Ophthalmology Unit, Schneider Children's Medical Center of Israel, Petach Tikva, Israel, [2]The Krieger Eye
Research Laboratory, Felsenstein Medical Research Center, Sackler School of Medicine, Tel Aviv University, Tel Aviv, Israel

INTRODUCTION

Diabetes mellitus is a major cause of avoidable blindness in both the developing and the developed countries.[1] Diabetic retinopathy (DR) has been observed in around a quarter of diabetic patients and can result in poor vision and even blindness. The mechanism suggested is the weakening of the retinal capillary wall by high glucose levels, leading to leakage of blood into the surrounding space.[2]

The duration of diabetes is the most significant predictor of visual impairment. Overall, diabetic microvascular complications are caused by prolonged exposure to high glucose levels. Other associated factors are treatment type,[3] low metabolic control,[4-6] and genetic background. The extent of diabetic tissue damage is determined by genetic factors that influence individual susceptibility. Atherosclerosis, and the presence of independent accelerating factors such as hypertension and dyslipidemia, may contribute to the severity of the disease. Some evidence of a modifying genetic effect on the onset and severity of DR has been reported.[7,8] However, most of the involved genes remain unknown and are yet to be explored.

The role of hyperglycemia has been established by large-scale prospective studies for both type 1 and type 2 diabetes, the Diabetes Control and Complications Trial (DCCT/EDIC),[9] and the UK Prospective Diabetes Study (UKPDS).[10] Similar data have been reported by the Steno-2 study.[11] However, further analysis of the DCCT data shows that although intensive therapy reduced the risk of sustained retinopathy progression by 73% compared with standard treatment (glycosylated hemoglobin (Hb_{A1c})), the duration of diabetes (glycemic exposure) explained only 11% of the variation in retinopathy risk for the entire study population. These facts suggest that the remaining 89% of the variation in risk is presumably explained by aspects of glycemia not captured by Hb_{A1c}.[11]

Oxidative Stress

Oxidative stress plays a pivotal role in the development of complications of diabetes, both microvascular and cardiovascular. The metabolic abnormalities of diabetes cause mitochondrial superoxide overproduction in endothelial cells of both large and small vessels.[12] Increased intracellular reactive oxygen species (ROS) cause defective angiogenesis in response to ischemia, activate a number of proinflammatory pathways, and lead to long-lasting epigenetic changes that drive persistent expression of proinflammatory genes after glycemia is normalized (Fig. 55.1).

Protection by Haptoglobin Against Hemoglobin Damage

Extracorpuscular hemoglobin, also known as free hemoglobin, is a potent oxidant protein[14,15] produced as a by-product of red blood cell hemolysis. Free hemoglobin can potentially enter the vessel wall and mediate low-density lipoprotein oxidation, thereby promoting the development and progression of atherosclerosis. This process is prevented by haptoglobin (Hp), which binds to hemoglobin during its transport and also helps to facilitate the removal of hemoglobin from the extravascular compartment via the CD163 macrophage scavenger receptor.[14]

Haptoglobin is encoded by two different alleles, whose products, Hp1-1, Hp2-1, and Hp2-2, differ

Handbook of Nutrition, Diet, and the Eye
http://dx.doi.org/10.1016/B978-0-12-401717-7.00055-1

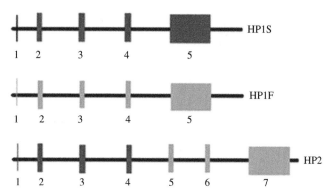

FIGURE 55.1 Schematic representation of the organization of the haptoglobin gene. The boxes indicate exons. *Source: Adapted from Wobeto VPA, Zaccariotto TR, Sonati MF. Polymorphism of human haptoglobin and its clinical importance.* Genet Mol Biol. [online] 2008;31:602–20,[13] *with permission from the Brazilian Society of Genetics (© BSG).*

markedly in their antioxidant ability.[16] Patients who are homozygous for the 1 allele (Hp1-1) form dimeric haptoglobin molecules, whereas patients who are heterozygous (Hp1-2) or homozygous (Hp2-2) for the 2-allele form multimers of two or more haptoglobin molecules. Haptoglobin multidimers are sterically hindered from binding to hemoglobin and have less access to the extravascular space; their haptoglobin/hemoglobin complexes have a lower affinity for CD163.[17,18]

COMPLICATIONS OF DIABETES

Atherosclerosis and cardiomyopathy in type 2 diabetes are caused in part by pathway-selective insulin resistance, which increases mitochondrial ROS production from free fatty acids and by inactivation of antiatherosclerosis enzymes by ROS. There are no sufficient data regarding the mechanisms of DR, but some oxidative stress may cause retinal ischemia and neovascularization. Experimental mechanisms of protection, such as the overexpression of superoxide dismutase (SOD) in transgenic diabetic mice, prevented DR, nephropathy, and cardiomyopathy.[12]

HAPTOGLOBIN AND RISK OF DIABETIC RETINOPATHY

Studies in diabetic patients have shown that the development of DR is related to hyperglycemia-induced oxidative stress and that individuals with the Hp2-2 genotype appear to be at significantly higher risk of microvascular and macrovascular complications.[18]

Genetic factors underlying DR have been investigated for many years.[5] The possible effect of haptoglobin genotype on the onset of DR in patients with type 2 diabetes, controlling for other potential factors and treatment

risks, was recently examined.[19] The findings show no role for the haptoglobin genotype in the development or the deterioration of DR.[19] This study did not find any increased risk of proliferative diabetic retinopathy (PDR) for Hp2-2 patients.[19] The haptoglobin genotype apparently played no role in the development or severity of proliferative retinopathy in type 2 diabetes. However, Hp1-1 was associated with a delay in the onset of diabetes, while Hp2-2 posed a microvascular risk.[19]

Chandra *et al.*[20] compared 180 healthy individuals with 81 diabetic patients without retinopathy and 122 cases with DR, matched for age and gender, for haptoglobin phenotype. A significant decrease in Hp2-1 frequency was found in the diabetes group, suggesting the existence of protection for heterozygotes in both diabetes and DR (with a relative risk of about 0.31). As an acute-phase reactant, haptoglobin may be functionally involved in the etiology of diabetes and DR, which are associated with immunologic and inflammatory processes, respectively. No significant differences were found with respect to gender, age at onset, duration of DR, types of diabetes and DR, or family history.[20]

HAPTOGLOBIN AND ENDOTHELIAL DYSFUNCTION

Hp2-2 Genotype and Diabetes

Haptoglobin polymorphism has been found to correlate with diabetic vascular complications, owing to the molecular basis behind this interaction.[21] Endothelial function is impaired in patients with diabetes owing to an increase in oxidative stress. Hp2-2 diabetic patients had a worse endothelial function than controls and Hp1-1 diabetic subjects.[22] Individuals with the Hp2-2 genotype and diabetes mellitus appear to be at a significantly higher risk of microvascular and macrovascular complications.[23]

Mechanisms of Endothelial Dysfunction

Endothelial dysfunction and pericyte loss in the diabetic retina have been explained by several mechanisms.[24] Hyperglycemia persistently activates protein kinase C (PKC) and p38α mitogen-activated protein kinase (MAPK) to increase the expression of a previously unknown target of PKC signaling, SHP-1 (Src homology-2 domain-containing phosphatase-1), a protein tyrosine phosphatase. This signaling cascade leads to platelet-derived growth factor (PDGF) receptor-β dephosphorylation and a reduction in downstream signaling from this receptor, resulting in pericyte apoptosis.[24] The same pathway, activated by increased fatty acid oxidation in insulin-resistant arterial endothelial cells and heart, may play an equally important role in diabetic atherosclerosis and cardiomyopathy.

Therefore, the role of haptoglobin was investigated in atherosclerosis, cardiac myopathies, and infarctions, as well as diabetic nephropathy and retinopathy.[22,23,25-35] Haptoglobin was found to play a role in diabetes microvasculopathy in multiple organs, including heart, kidney, and retina.[22,23,25-35]

HAPTOGLOBIN GENOTYPE AND TYPE OF DIABETES

The combination of diabetes and Hp2-2 has been shown to increase the risk of the development of diabetes complications, particularly diabetic cardiovascular disease,[32] and accelerated coronary atherosclerosis in subjects with type 1 diabetes.[31] In the northern Chinese Han population, Hp2-2 was reported to be associated with an increased risk of type 2 diabetes,[30] while in Japanese diabetic patients, haptoglobin genotype was associated with diabetic microangiopathies.[36]

Another study reported a higher risk of carotid plaque iron deposition in male diabetic patients (not clearly defining type 1 or type 2 diabetes) with increased plasma levels of homocysteine and the Hp2-2 genotype.[29] Current evidence and pathophysiologic considerations suggest that the increased intraplaque iron deposition may be associated with increased oxidative stress, affecting the stability of the carotid plaque.[29]

Earlier studies on the potential role of the haptoglobin genotype in type 1 diabetes suggested that, through its antioxidative and hemoglobin-binding capacity, Hp-1 exerts a protective effect against both diabetic nephropathy and restenosis following percutaneous transluminal coronary angioplasty, whereas the presence of Hp-2 is linked to diabetes-related cardiovascular disease, in addition to other conditions.[17,23,27,31,37-41] Others have shown both nephropathy[42] and cardiovascular risk[43] to be associated with the haptoglobin genotype. Furthermore, ophthalmologic studies focusing on DR in murine models,[44] rats,[45] and humans[46] suggest that the Hp-2 genotype plays an accelerating role.

Unlike the findings for the haptoglobin genotype in type 1 diabetes,[16] and despite the increased risk of cardiovascular disease in patients with retinopathy (both non-PDR and PDR),[41] a group study[19] failed to show an association of the haptoglobin genotype with the existence or severity of DR in type 2 diabetes, either between or within the patient groups. The haptoglobin genotypes were equally distributed in the PDR group and in the no-retinopathy type 2 diabetes group. Although the Hp2-2 genotype was represented at a higher rate than the Hp1-2 and Hp1-1 genotypes, patients showed no tendency towards an increased risk

of progression to PDR. Of note, in both groups, patients with the Hp1-1 genotype were diagnosed with type 2 diabetes at a later age than patients with the Hp-2 genotype (owing to the small size of the PDR group, this finding was significant only for the no-retinopathy group). Barbosa et al.[47] reported no correlation between haptoglobin and diabetes occurrence or linkage with a secondary susceptibility locus, thus supporting the above-mentioned findings.[47,48]

Similarly, serum haptoglobin levels and lipid profile were normal in children with type 1 diabetes with persistent microalbuminuria but without DR.[49]

ANIMAL MODELS OF DIABETIC RETINOPATHY

The pathogenesis of DR has been investigated using several animal models of diabetes. These models have been generated by pharmacologic induction, feeding a galactose diet, and spontaneously by selective inbreeding or genetic modification. Among the available animal models, rodents have been studied most extensively owing to their short generation time and the inherited hyperglycemia and/or obesity that affect certain strains. In particular, mice have proven useful for studying DR and evaluating novel therapies because of their amenability to genetic manipulation. Mouse models suitable for replicating the early, nonproliferative stages of the retinopathy have been characterized, but no animal model has yet been found to demonstrate all of the vascular and neural complications that are associated with the advanced, proliferative stages of DR that occur in humans. The common animal models of DR use various *in vivo* imaging techniques to characterize DR. By highlighting the ocular pathologic findings, clinical implications, advantages, and disadvantages of these models, essential information was provided for planning experimental studies of DR that led to new strategies for its prevention and treatment.[50]

Gene Expression Changes Associated with Diabetic Retinopathy in a Mouse Model

Recently, the advanced technology of next generation sequencing was used to define gene expression changes associated with DR in a mouse model and showed that diabetes alters many transcripts in the retina. A high-throughput RNA-sequencing strategy using Illumina GAIIx was applied to characterize the entire retinal transcriptome from nondiabetic and streptozotocin-treated mice 32 weeks after the induction of diabetes. Some of the diabetic mice were treated with inhibitors of receptor for advanced glycation end-products (RAGE) and p38 MAPK, which have previously been shown to inhibit

DR in rodent models. The transcripts and alternatively spliced variants were determined in all experimental groups. Next generation sequencing-based RNA-sequencing profiles provided comprehensive signatures of transcripts that are altered in early stages of DR. These transcripts encoded proteins involved in distinct yet physiologically relevant disease-associated pathways such as inflammation, microvasculature formation, apoptosis, glucose metabolism, Wnt signaling, xenobiotic metabolism, and photoreceptor biology. The significant upregulation of crystallin transcripts was observed in diabetic animals, and the diabetes-induced upregulation of these transcripts was inhibited in diabetic animals treated with inhibitors of either RAGE or p38 MAPK.[51]

Diabetic Mice Genetically Modified at the Haptoglobin Locus

Individuals with diabetes mellitus homozygous for the Hp-1 allele are at decreased risk of retinopathy compared with diabetic individuals with the Hp-2 allele.[44] These findings were assessed in diabetic mice genetically modified at the haptoglobin locus. An early morphologic characteristic of the microangiopathy seen in diabetic retinal disease is retinal capillary basement membrane (RCBM) thickening. A highly significant increase in RCBM thickness was observed in diabetic mice with the Hp-2 genotype. These data provide important support for human association studies showing an increased prevalence of DR in individuals with the Hp-2 genotype.[44]

Haptoglobin and Free Iron Serum Levels in Diabetic Rats

Haptoglobin is a hemoglobin-binding acute-phase protein that possesses anti-inflammatory and antioxidative properties.[28] Changes in protein expression of rat haptoglobin under diabetes-related inflammatory and oxidative stress were measured.[28,45] An inverse correlation was observed between the haptoglobin and free iron serum levels in diabetic rats.[28] The higher levels of haptoglobin during the first 2 weeks were accompanied by a lower level of free iron.[28] Haptoglobin possibly decreases the oxidative stress during the early stage of diabetic conditions induced by an intraperitoneal injection of streptozotocin.[28]

rs10811661 Predisposes to Type 2 Diabetes and to Nephropathy in Type 1 Diabetes

A single nucleotide polymorphism (SNP) predisposing to type 2 diabetes, rs10811661 near CDKN2A/B, is associated with diabetic nephropathy in patients with type 1 diabetes.[52]

The same SNP was also associated with severe retinopathy, but the association did not remain after Bonferroni statistical correction. None of the other selected SNPs examined was associated with nephropathy, severe retinopathy, or cardiovascular disease.

PREVALENCE OF MILD DIABETIC RETINOPATHY

Epidemiologic studies show that the worldwide prevalence of mild DR is about 30–36%[53,54] and that of PDR is 7–9%.[55,56] About 75% of all cases of DR are mild. The Anglo–Danish–Dutch study of Intensive Treatment in People with Screen-detected Diabetes in Primary Care (ADDITION) documented only a 6% rate of severe DR.[55] Of the 670 diagnosed patients,[48] only 6.8% had any retinopathy, mainly minimal.[48] Reported rates of PDR ranged from 19.7% in France to 31.5% in the UK.[57] This is in line with the reported decrease in the overall incidence of retinopathy,[53] especially PDR.[58]

The haptoglobin phenotype distribution and the association between haptoglobin polymorphism and type 2 diabetes were studied in a Jordanian population of 265 diabetic and 618 nondiabetic patients.[54] However, their retinopathy level was not documented.[54] In both groups, Hp2-2 was the predominant genotype, followed by Hp2-1 (nondiabetic: Hp2-2, 0.529, Hp2-1, 0.387; diabetic: Hp2-2, 0.540, Hp2-1, 0.381), with no statistically significant variation, similar to other findings in the (larger) no-retinopathy group.[19] Overall, the Hp-2 allele occurred at a frequency of 0.722 in nondiabetic patients and 0.730 in diabetic patients.[54] These results suggest that type 2 diabetes may be independent of haptoglobin phenotype.[54]

TAKE-HOME MESSAGES

- Diabetic retinopathy (DR) is a progressive disease that results from vascular injury due to chronic hyperglycemia.
- Hyperglycemia induces weakening of the retinal capillary wall, leading to leakage of blood into the surrounding space and to DR.
- Free hemoglobin is a potent oxidant protein. Haptoglobin (Hp) protects from potential oxidative damage by hemoglobin. Haptoglobin binds to hemoglobin during its transport and facilitates its removal from the extravascular compartment.
- Hp1-1, Hp2-1, and Hp2-2 differ markedly in their antioxidant ability.
- Patients who are homozygous for the 1-allele (Hp1-1) form dimeric haptoglobin molecules, whereas patients who are heterozygous (Hp1-2) or

homozygous (Hp2-2) for the 2-allele form multimers of two or more haptoglobin molecules.

- Haptoglobin multidimers are sterically hindered from binding to hemoglobin and have less access to the extravascular space.
- Hp2-2 genotype and diabetes mellitus appear to carry a higher risk of microvascular and macrovascular complications.
- A higher prevalence of Hp1-1 in older patients suggests its possible role in delaying the onset of retinopathy.
- Heterozygotes (Hp2-1) are protected from both diabetes and DR, but other studies showed that haptoglobin genotype apparently plays no role in the development or severity of proliferative retinopathy in type 2 diabetes.
- Type 2 diabetes may be haptoglobin phenotype independent.

References

1. Singh R, Ramasamy K, Abraham C, Gupta V, Gupta A. Diabetic retinopathy: an update. *Indian J Ophthalmol* 2008;**56**:178–88.
2. Chen YH, Chou HC, Lin ST, Chen YW, Lo YW, Chan HL. Effect of high glucose on secreted proteome in cultured retinal pigmented epithelium cells: its possible relevance to clinical diabetic retinopathy. *J Proteomics* 2012;**77**:111–28.
3. Yam JC, Kwok AK. Update on the treatment of diabetic retinopathy. *Hong Kong Med J* 2007;**13**:46–60.
4. Xie X, Xu L, Yang H, Wang S, Jonas JB. Frequency of diabetic retinopathy in the adult population in China: the Beijing Eye Study 2001. *Int Ophthalmol* 2009;**29**:485–93.
5. Xie XW, Xu L, Jonas JB, Wang YX. Prevalence of diabetic retinopathy among subjects with known diabetes in China: the Beijing Eye Study. *Eur J Ophthalmol* 2009;**19**:91–9.
6. Xie XW, Xu L, Wang YX, Jonas JB. Prevalence and associated factors of diabetic retinopathy. The Beijing Eye Study 2006. *Graefes Arch Clin Exp Ophthalmol* 2008;**246**:1519–26.
7. Abhary S, Hewitt AW, Burdon KP, Craig JE. A systematic meta-analysis of genetic association studies for diabetic retinopathy. *Diabetes* 2009;**58**:2137–47.
8. Sobrin L, Green T, Sim X, Jensen RA, Tai ES, Tay WT, et al. Candidate gene association study for diabetic retinopathy in persons with type 2 diabetes: the Candidate gene Association Resource (CARe). *Invest Ophthalmol Vis Sci* 2011;**52**:7593–602.
9. Diabetes Control and Complications Trial Research Group. The effect of intensive treatment of diabetes on the development and progression of long-term complications in insulin-dependent diabetes mellitus. *N Engl J Med* 1993;**329**:977–86.
10. UK Prospective Diabetes Study (UKPDS) Group. Intensive blood-glucose control with sulphonylureas or insulin compared with conventional treatment and risk of complications in patients with type 2 diabetes (UKPDS 33). *Lancet* 1998;**352**:837–53.
11. Lachin JM, Genuth S, Nathan DM, Zinman B, Rutledge BN. Effect of glycemic exposure on the risk of microvascular complications in the diabetes control and complications trial – revisited. *Diabetes* 2008;**57**:995–1001.
12. Giacco F, Brownlee M. Oxidative stress and diabetic complications. *Circ Res* 2010;**107**:1058–70.
13. Wobeto VPA, Zaccariotto TR, Sonati MF. Polymorphism of human haptoglobin and its clinical importance. *Genet Mol Biol* 2008;**31**:602–20.
14. Melamed-Frank M, Lache O, Enav BI, Szafranek T, Levy NS, Ricklis RM, Levy AP. Structure–function analysis of the antioxidant properties of haptoglobin. *Blood* 2001;**98**:3693–8.
15. Tseng CF, Lin CC, Huang HY, Liu HC, Mao SJ. Antioxidant role of human haptoglobin. *Proteomics* 2004;**4**:2221–8.
16. Nakhoul FM, Marsh S, Hochberg I, Leibu R, Miller BP, Levy AP. Haptoglobin genotype as a risk factor for diabetic retinopathy. *JAMA* 2000;**284**:1244–5.
17. Langlois MR, Delanghe JR. Biological and clinical significance of haptoglobin polymorphism in humans. *Clin Chem* 1996;**42**:1589–600.
18. Asleh R, Guetta J, Kalet-Litman S, Miller-Lotan R, Levy AP. Haptoglobin genotype- and diabetes-dependent differences in iron-mediated oxidative stress *in vitro* and in vivo. *Circ Res* 2005;**96**:435–41.
19. Goldenberg-Cohen N, Gabbay M, Dratviman-Storobinsky O, Reich E, Axer-Siegel R, Weinberger D, Gabbay U. Does haptoglobin genotype affect early onset of diabetic retinopathy in patients with type 2 diabetes? *Retina* 2011;**31**:1574–80.
20. Chandra T, Lakshmi CN, Padma T, Vidyavathi M, Satapathy M. Haptoglobin phenotypes in diabetes mellitus and diabetic retinopathy. *Hum Hered* 1991;**41**:347–50.
21. Asleh R, Levy AP. *In vivo* and *in vitro* studies establishing haptoglobin as a major susceptibility gene for diabetic vascular disease. *Vasc Health Risk Manag* 2005;**1**:19–28.
22. Dayan L, Levy AP, Blum S, Miller-Lotan R, Melman U, Alshiek J, Jacob G. Haptoglobin genotype and endothelial function in diabetes mellitus: a pilot study. *Eur J Appl Physiol* 2009;**106**:639–44.
23. Levy AP, Asleh R, Blum S, Levy NS, Miller-Lotan R, Kalet-Litman S, et al. Haptoglobin: basic and clinical aspects. *Antioxid Redox Signal* 2010;**12**:293–304.
24. Geraldes P, Hiraoka-Yamamoto J, Matsumoto M, Clermont A, Leitges M, Marette A, et al. Activation of PKC-delta and SHP-1 by hyperglycemia causes vascular cell apoptosis and diabetic retinopathy. *Nat Med* 2009;**15**:1298–306.
25. Costacou T, Levy AP. Haptoglobin genotype and its role in diabetic cardiovascular disease. *J Cardiovasc Transl Res* 2012;**5**:423–35.
26. Farbstein D, Levy AP. The genetics of vascular complications in diabetes mellitus. *Cardiol Clin* 2010;**28**:477–96.
27. Goldenstein H, Levy NS, Levy AP. Haptoglobin genotype and its role in determining heme-iron mediated vascular disease. *Pharmacol Res* 2012;**66**:1–6.
28. Jelena A, Mirjana M, Desanka B, Svetlana IM, Aleksandra U, Goran P, Ilijana G. Haptoglobin and the inflammatory and oxidative status in experimental diabetic rats: antioxidant role of haptoglobin. *J Physiol Biochem* 2013;**69**:45–58.
29. Lioupis C, Barbatis C, Drougou A, Koliaraki V, Mamalaki A, Klonaris C, et al. Association of haptoglobin genotype and common cardiovascular risk factors with the amount of iron in atherosclerotic carotid plaques. *Atherosclerosis* 2011;**216**:131–8.
30. Shi X, Sun L, Wang L, Jin F, Sun J, Zhu X, et al. Haptoglobin 2-2 genotype is associated with increased risk of type 2 diabetes mellitus in northern Chinese. *Genet Test Mol Biomarkers* 2012;**16**:563–8.
31. Simpson M, Snell-Bergeon JK, Kinney GL, Lache O, Miller-Lotan R, Anbinder Y, et al. Haptoglobin genotype predicts development of coronary artery calcification in a prospective cohort of patients with type 1 diabetes. *Cardiovasc Diabetol* 2011;**10**:99.
32. Vardi M, Levy AP. Is it time to screen for the haptoglobin genotype to assess the cardiovascular risk profile and vitamin E therapy responsiveness in patients with diabetes? *Curr Diab Rep* 2012;**12**:274–9.
33. Viener HL, Levy AP. Haptoglobin genotype and the iron hypothesis of atherosclerosis. *Atherosclerosis* 2011;**216**:17–8.
34. Wobeto VP, Garcia PM, Zaccariotto TR, Sonati Mde F. Haptoglobin polymorphism and diabetic nephropathy in Brazilian diabetic patients. *Ann Hum Biol* 2009;**36**:437–41.
35. Wobeto VP, Pinho Pda C, Souza JR, Zaccariotto TR, Zonati Mde F. Haptoglobin genotypes and refractory hypertension in type 2 diabetes mellitus patients. *Arq Bras Cardiol* 2011;**97**:338–45.

36. Koda Y, Soejima M, Yamagishi S, Amano S, Okamoto T, Inagaki Y, et al. Haptoglobin genotype and diabetic microangiopathies in Japanese diabetic patients. *Diabetologia* 2002;**45**:1039–40.

37. Chandy A, Pawar B, John M, Isaac R. Association between diabetic nephropathy and other diabetic microvascular and macrovascular complications. *Saudi J Kidney Dis Transpl* 2008;**19**:924–8.

38. Jeganathan VS, Cheung N, Wong Y. Diabetic retinopathy is associated with an increased incidence of cardiovascular events in type 2 diabetic patients. *Diabet Med* 2008;**25**:882. author reply 882–3.

39. Levy AP, Roguin A, Marsh S. Haptoglobin phenotype and vascular complications in patients with diabetes. *N Engl J Med* 2000;**343**:969–70.

40. Roguin A, Ribichini F, Ferrero V, Matullo G, Herer P, Wijns W, Levy AP. Haptoglobin phenotype and the risk of restenosis after coronary artery stent implantation. *Am J Cardiol* 2002;**89**:806–10.

41. Gimeno-Orna JA, Faure-Nogueras E, Castro-Alonso FJ, Boned-Juliani B. Ability of retinopathy to predict cardiovascular disease in patients with type 2 diabetes mellitus. *Am J Cardiol* 2009;**103**:1364–7.

42. Conway BR, Savage DA, Brady HR, Maxwell AP. Association between haptoglobin gene variants and diabetic nephropathy: haptoglobin polymorphism in nephropathy susceptibility. *Nephron Exp Nephrol* 2007;**105**:e75–9.

43. Blum S, Asaf R, Guetta J, Miller-Lotan R, Asleh R, Kremer R, et al. Haptoglobin genotype determines myocardial infarct size in diabetic mice. *J Am Coll Cardiol* 2007;**49**:82–7.

44. Miller-Lotan R, Miller B, Nakhoul F, Aronson D, Asaf R, Levy AP. Retinal capillary basement membrane thickness in diabetic mice genetically modified at the haptoglobin locus. *Diabetes Metab Res Rev* 2007;**23**:152–6.

45. Obrosova IG, Minchenko AG, Marinescu V, Fathallah L, Kennedy A, Stockert CM, et al. Antioxidants attenuate early up regulation of retinal vascular endothelial growth factor in streptozotocin-diabetic rats. *Diabetologia* 2001;**44**:1102–10.

46. Wobeto VP, Rosim ET, Melo MB, Calliari LE, Sonati Mde F. Haptoglobin polymorphism and diabetic retinopathy in Brazilian patients. *Diabetes Res Clin Pract* 2007;**77**:385–8.

47. Barbosa J, Segall M, Rich S. Genetic relationships between type I and type II diabetes mellitus. *Adv Exp Med Biol.* 1988;**246**:127–30.

48. Rich SS, Panter SS, Goetz FC, Hedlund B, Barbosa J. Shared genetic susceptibility of type 1 (insulin-dependent) and type 2 (non-insulin-dependent) diabetes mellitus: contributions of HLA and haptoglobin. *Diabetologia* 1991;**34**:350–5.

49. Cam H, Pusuroglu K, Aydin A, Ercan M. Effects of hemorheological factors on the development of hypertension in diabetic children. *J Trop Pediatr* 2003;**49**:164–7.

50. Robinson R, Barathi VA, Chaurasia SS, Wong TY, Kern TS. Update on animal models of diabetic retinopathy: from molecular approaches to mice and higher mammals. *Dis Model Mech* 2012;**5**:444–56.

51. Kandpal RP, Rajasimha HK, Brooks MJ, Nellissery J, Wan J, Qian J, et al. Transcriptome analysis using next generation sequencing reveals molecular signatures of diabetic retinopathy and efficacy of candidate drugs. *Mol Vis* 2012;**18**:1123–46.

52. Fagerholm E, Ahlqvist E, Forsblom C, Sandholm N, Syreeni A, Parkkonen M, on behalf of the FinnDiane Study Group, et al. SNP in the genome-wide association study hotspot on chromosome 9p21 confers susceptibility to diabetic nephropathy in type 1 diabetes. *Diabetologia* 2012;**55**:2386–93.

53. Sloan FA, Belsky D, Ruiz Jr D, Lee P. Changes in incidence of diabetes mellitus-related eye disease among US elderly persons, 1994–2005. *Arch Ophthalmol* 2008;**126**:1548–53.

54. Awadallah S, Hamad M. The prevalence of type II diabetes mellitus is haptoglobin phenotype-independent. *Cytobios* 2000;**101**:145–50.

55. Bek T, Lund-Andersen H, Hansen AB, Johnsen KB, Sandbaek A, Lauritzen T. The prevalence of diabetic retinopathy in patients with screen-detected type 2 diabetes in Denmark: the ADDITION study. *Acta Ophthalmol* 2009;**87**:270–4.

56. Goh PP. Status of diabetic retinopathy among diabetics registered to the Diabetic Eye Registry, National Eye Database, 2007. *Med J Malaysia* 2008;**63**(Suppl C.):24–8.

57. Rubino A, Rousculp MD, Davis K, Wang J, Girach A. Diagnosed diabetic retinopathy in France, Italy, Spain, and the United Kingdom. *Prim Care Diabetes* 2007;**1**:75–80.

58. Aylward GW. Progressive changes in diabetics and their management. *Eye (Lond)* 2005;**19**:1115–8.

SLC23A2 Gene Variation, Vitamin C Levels, and Glaucoma

Vicente Zanon-Moreno[1,2], Jose J. Garcia-Medina[3,4], Pedro Sanz-Solana[5], Maria D. Pinazo-Duran[6,7], Dolores Corella[1,2]

[1]Genetic and Molecular Epidemiology Unit, Department of Preventive Medicine and Public Health, School of Medicine, University of Valencia, Valencia, Spain, [2]CIBER Fisiopatología de la Obesidad y Nutrición, ISCIII, Valencia, Spain, [3]Department of Ophthalmology, University General Hospital Reina Sofía, Murcia, Spain, [4]Department of Ophthalmology, School of Medicine, University of Murcia, Murcia, Spain, [5]Department of Ophthalmology, Dr. Peset University Hospital, Valencia, Spain, [6]Ophthalmology Research Unit 'Santiago Grisolia', Dr. Peset University Hospital, Valencia, Spain, [7]Department of Surgery, School of Medicine, University of Valencia, Valencia, Spain

INTRODUCTION

Glaucoma is a group of diseases characterized by progressive loss of nerve fibers of the optic nerve. When the atrophy of optic nerve tissue is significant, visual field loss starts.[1] Glaucoma is the leading cause of irreversible blindness in the world (2.4% prevalence), and there are approximately 60 million people with glaucoma worldwide, of whom 8.4 million are bilaterally blind. In Spain there are about 300,000 people diagnosed with glaucoma and 700,000 people with the subclinical form.[2,3]

Despite all this, glaucomatous blindness is avoidable, according to the World Health Organization (WHO), as a specific hypotensive treatment can be effective in certain types of glaucoma when the disease is detected in its early stages. The objective of ophthalmologic societies, therefore, should be the early diagnosis of the disease, before visual impairment is too severe, to address it with the most appropriate and effective treatment and to change the fact that an avoidable disease remains a leading cause of blindness worldwide.

There are many types of glaucoma (Table 56.1), with primary open-angle glaucoma (POAG) being the most common form, representing between 60% and 70% of all cases. POAG is characterized mechanically by a high intraocular pressure (IOP), morphologically by an alteration in the optic nerve head, and functionally by a progressive loss of visual field and vision. There are no symptoms in the early stages of POAG. Initially, the peripheral visual field is gradually affected, leaving the central visual field apparently unharmed. Therefore, visual acuity is not significantly affected until later stages of the disease. That is why early diagnosis of this optic neuropathy is vital to halt the progression of the disease to irreversible blindness.

Although the relationship between IOP and optic nerve damage is undisputed, the existence of normotensive glaucoma (glaucomatous optic nerve damage with normal IOP) indicates that other factors must influence the pathogenesis of glaucoma. Therefore, POAG is a multifactorial disease involving both environmental and genetic factors.

ENVIRONMENTAL FACTORS IN GLAUCOMA

Among all the environmental factors that may be associated with this optic neuropathy, those related to nutrition are highlighted here.

The relationship between nutrition and health has been known for a long time.[4–7] It is well known that poor nutrition adversely affects cardiovascular health.[8] Foods free of or with low contents of saturated fats, cholesterol, and sodium are recommended for people to reduce the risk of cardiovascular diseases.[9] Other diseases, such as dental caries, anorexia, cancer, and obesity, are also influenced by malnutrition.[10–13]

TABLE 56.1   Classification of Glaucomas

Amplitude of the Iridocorneal Angle	
Open-angle glaucoma	The most common form of glaucoma
	Normal iridocorneal angle
	Chronic
	Asymptomatic in its early stages
	A special type of open-angle glaucoma is normal tension glaucoma
	Main risk factors: age, ethnicity, family history, ocular hypertension
Closed-angle glaucoma	Iridocorneal angle completely or partly blocked
	Acute
	Symptoms: severe pain, blurred vision, red eye, nausea and vomiting
	Main risk factors: age, family history, hyperopia, stress
Origin	
Primary glaucoma	Not related to other pathologies, trauma, or medical treatments
	The iridocorneal angle can be open or closed
Secondary glaucoma	Caused by other ocular pathologies, trauma, or medical treatments
	The iridocorneal angle can be open or closed
	– neovascular glaucoma
	– pseudoexfoliative glaucoma
	– traumatic glaucoma
	– inflammatory glaucoma
Age at Diagnosis	
Congenital glaucoma	Infrequent (1 in 10,000 people)
	From birth until 3 years old
	Abnormally large corneas, tearing and sensitivity to light
Juvenile glaucoma	Infrequent (1 in 50,000 people)
	Affects children, teenagers, and young adults
	Frequently asymptomatic, but in some cases may have blurred vision
Adult glaucoma	The most common form (after 40 years old, 1% of people in the world)
	Symptoms depend on whether it is chronic or acute glaucoma
Location of Obstruction of Aqueous Humor	
Pretrabecular glaucoma	Obstruction is located in front of trabeculum
Trabecular glaucoma	Obstruction is located in the trabeculum
Post-trabecular glaucoma	Blockage is caused by increased episcleral venous pressure

However, the relationship between nutritional status and eye health, and the fact that nutritional status affects visual function, is less well known among the general population.[14] Many people are not aware that what you eat is very important for your eyes. For example, docosahexaenoic acid is an omega-3 polyunsaturated fatty acid (PUFA), present in large amounts in fish oil, which protects retinal ganglion cells from oxidative damage caused by free radicals.[15] Vitamins and carotenoids are also very important in preventing ocular diseases,[16,17] but they are not synthesized by human cells, and therefore people have to consume these nutrients in the diet. A deficiency of these essential nutrients can cause health problems. Furthermore, not only a deficiency but also an excess in some of these essential nutrients has been associated with various pathologies. Chang *et al.* reported a 44-year-old woman with peripheral corneal rings who admitted to taking more than 40 different vitamin supplement pills daily. On examination, only carotenoid serum concentrations were above the normal range, and the woman was diagnosed with hypercarotenemia.[18]

GENETIC FACTORS IN GLAUCOMA

In addition to environmental factors, genetic factors are also related to this optic neuropathy. Many genes have been linked to glaucoma (Table 56.2), and new studies associating genetic variants with different types of glaucoma are published almost daily. Research groups generally focus their resources and efforts on the study of the possible association between genetic polymorphisms and POAG, so that people who are most likely to develop POAG can be identified at an early stage. The main hurdle to achieving this objective is that POAG is a polygenic disease (caused by two or more genes), so these genetic studies are difficult to carry out because it is necessary to investigate the influence of polymorphisms of all of the genes on the disease and to analyze the interactions between the genes.

Moreover, it is important to determine not only the individual effect of all these genetic factors and the gene–gene interactions in the pathogenesis of POAG but also the interactions between all the factors related to this disease and how these interactions influence the emergence and development of glaucomatous optic neuropathy (Fig. 56.1).

NUTRITIONAL GENOMICS

It is not yet possible to study all the factors involved in POAG, but the interaction between nutritional and genetic factors can be studied by means of nutritional genomics, a relatively young discipline whose aim is to

TABLE 56.2 Genes Associated with Different Types of Glaucoma

Gene	OMIM	Chromosomal Location	Glaucoma	References
MYOC	601652	1q23–q24	POAG	Sheffield (1993), Stone (1997), Michels-Rautenstrauss (2002), Mengkegale (2008)[19–22]
OPTN	602432	10p13	NTG	Sarfarazi (1998), Rezaie (2002), Fuse (2004), Cheng (2010)[23–26]
WDR36	609669	5q22	POAG, NTG	Monemi (2005), Miyazawa (2007), Weisschuh (2007), Blanco-Marchite (2011)[27–30]
CYP1B1	601771	2p22	POAG	Stoilov (1997), Vasilou (2008), Choudhary (2009), Su (2012)[31–34]
LOXL1	153456	15q23	ExG	Thorleifsson (2007), Fan (2011)[35,36]
TMCO1	614123	1q22–q25	POAG	Burdon (2011), van Koolwijk (2012), Sharma (2012)[37–39]
CAV1	601047	7q31	POAG	Thorleifsson (2010), Wiggs (2011), Surgucheva (2011)[40–42]
CAV2	601048	7q31	POAG	Thorleifsson (2010), Wiggs (2011), Surgucheva (2011)[35,41,42]
CDKN2B	600431	9p21	POAG, NTG	Fan (2011), Takamoto (2012)[43,44]
SIX6	606326	14q23	POAG	Osman (2012)[45]
MMP9	120361	20q11–q13	PACG	Cong (2009), Mossböck (2010), Awadalla (2011)[46–48]
TLR4	603030	9q33	POAG, NTG, ExG	Shibuya (2008), Chen (2012), Takano (2012)[49–51]
iNOS	163730	17q11–q12	POAG	Motallebipour (2005)[52]
SRBD1	–	2p21	NTG	Meguro (2010)[53]
ELOVL5	611805	6p12	NTG	Meguro (2010)[53]

OMIM: Online Mendelian Inheritance in Man; POAG: primary open-angle glaucoma; NTG: normal tension glaucoma; ExG: exfoliative glaucoma.

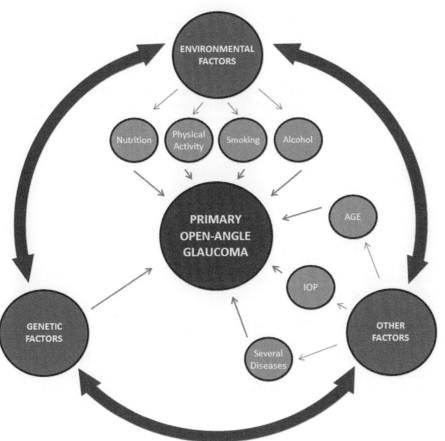

FIGURE 56.1 Diagram of the factors that influence the development and progression of primary open-angle glaucoma. IOP: intraocular pressure.

gain enough knowledge to adjust the diet of people to their genotype. In other words, it aims to achieve personalized diets with which the onset of many diseases could be avoided or prevented.[54]

The ALIENOR study (antioxidants, essential lipids, nutrition and ocular disease) investigates the influence of nutritional factors (antioxidants, lipids, and PUFA omega-3) in age-related eye diseases, such as age-related macular degeneration (AMD), cataracts, and glaucoma,[55] although the articles published up to now refer only to AMD.[56–58]

There are no studies on nutrigenomics and POAG in the reference list, but the relationship between nutrition and glaucoma has been widely studied.[59–61]

VITAMIN C

One of the most important nutritional factors for eye health is vitamin C.[62–65] It is present in various ocular structures, with a higher concentration in aqueous humor, lens, and retina than in plasma.[65–68] This vitamin is involved in the regulation of the trabecular outflow by means of increasing nitric oxide production in trabecular meshwork cells[69] and also in counteracting the oxidative damage caused by free radicals.[70]

As humans lack the enzyme necessary for the synthesis of vitamin C (L-gulonolactone oxidase), this antioxidant has to be incorporated in the diet (Table 56.3), absorbed through the buccal mucosa, stomach, and small intestine, transported to the liver, and, from this organ, distributed to the tissues where it is needed.[71] The concentration of vitamin C is higher in the cells and organs than in plasma, suggesting the active transport of ascorbate into cells. The transporter of vitamin C was named sodium-dependent vitamin C transporter by Tsukaguchi and colleagues,[72] and two isoforms were identified: sodium-dependent vitamin C transporter-1 (SVCT1), encoded by the *SLC23A1* gene, and sodium-dependent vitamin C transporter-2 (SVCT2), encoded by the *SLC23A2* gene.

These two isoforms have 65% identity in humans, but they have different functions, as they have different tissue localization. *SVCT1* is expressed mainly in epithelial cells of the intestine and kidney, whereas the highest expression of *SVCT2* is in brain, lung, and bone.[73] The *SLC23A2* gene is also expressed in the eye, and its expression and the activity of SVCT2 are essential to the uptake of ascorbic acid in the peripheral nervous system.[74] A study carried out at the University of Southern California's Keck School of Medicine (Los Angeles, USA) showed that the sodium-dependent transport of vitamin C in normal lens epithelium is likely to be mediated by SVCT2 rather than by SVCT1.[75]

TABLE 56.3 Food Rich in Vitamin C

Food	Vitamin C (mg/100 g)
Guava	273
Red pepper	225
Parsley	200
Blackcurrant	160
Green pepper	120
Brussels sprout	110
Broccoli	110
Papaya	64
Kale	62
Chive	60
Watercress	60
Lychee	60
Cauliflower	47
Mango	44
Lime	42
Tomato	38
Lemon	37
Mint	31
Soy	29
Spinach	26

Source: Spanish database of food composition (www.bedca.net).

TABLE 56.4 General Characteristics of Primary Open-Angle Glaucoma and Control Subjects

	Glaucoma (n=150)	Control (n=150)	P
Females (%)	59.3	59.3	1.000
Age (years)	68±9	68±8	0.973
BMI (kg/m²)	27.5±4.7	27.4±4.9	0.801
Cup:disc ratio	0.7	0.3	<0.001*
Smokers (%)	28	22	0.230
Alcohol consumers (%)	74	68	0.252

BMI: body mass index.

Taking these findings into account, the present research group carried out a case-control study in subjects with POAG and healthy volunteers, matched by age, gender, and body mass index, all from a Mediterranean population (see the characteristics of the subjects in Table 56.4) to study some single nucleotide polymorphisms (SNPs) in several genes related to vitamins A (*RBP1* gene) and

C (*SLC23A1* and *SLC23A2* genes). The aim was to investigate the possible association between selected polymorphisms in key proteins related to plasma vitamin C and vitamin A concentrations and POAG. The findings demonstrated that the GG genotype of rs1279683 polymorphism in *SLC23A2* gene is associated with a high risk of POAG (odds ratio = 1.67, 95% confidence interval 1.03 to 2.71, P = 0.038). An analysis adjusting for potential confounders (tobacco smoking and alcohol consumption) was conducted, and the POAG risk associated with this SLC23A2 genotype reached statistical significance (P = 0.010). The other SNPs studied were not associated with POAG.

SLC23A2 GENE VARIATION

Variations in the *SLC23A2* gene have been related to various pathologies, including several types of cancer. Chen *et al.* found a higher risk of head and neck squamous cell carcinoma (HNSCC) associated with human papillomavirus type 16 (HPV16)-positive serology in carriers of the wild-type allele of rs4987219 polymorphism in the *SLC23A2* gene, while in homozygous carriers of the variant the risk of HNSCC associated with HPV16 was diminished. Moreover, the authors observed an increased risk of HNSCC in subjects with a wild-type rs4987219 *SLC23A2* allele, HPV16-positive serology, and high citrus intake.[76]

Researchers from 10 countries participating in the European Prospective Investigation into Cancer and Nutrition (EPIC) demonstrated that high plasma levels of vitamin C are associated with a low risk of gastric cancer.[77] Related to that, Wright and colleagues reported that subjects with the AA genotype of rs12479919 in the *SLC23A2* gene had 41% lower risk of gastric cancer

than homozygous carriers for the G allele in a Polish population.[78]

Variations in the *SLC23A2* gene have also been associated with a higher risk of preterm birth,[79] but, to the authors' knowledge, this research group reported for the first time an association between this gene and an ocular disease, POAG, as well as publishing one of the first papers about nutritional genomics in POAG.[80]

In addition to the genetic study of vitamin-related genes in POAG, the plasma levels of vitamin C were determined. Significantly lower concentrations of this vitamin were found in glaucomatous subjects compared with healthy subjects (Fig. 56.2). PUFAs, carotenoids, and some vitamins have antioxidant properties and play a key role in the defense against oxidative damage caused by reactive oxygen species. Oxidative stress has been strongly associated with glaucoma, and it has been demonstrated that antioxidant concentration is decreased in subjects with POAG,[81] in agreement with the authors' results. It remains to be elucidated whether oxidative stress is a cause or consequence of POAG, but the relationship between them is unquestionable.

The next step in the study was to analyze whether the SNP in the *SLC23A2* gene had some effect on plasma levels of vitamin C. Homozygous carriers of the G allele had significantly lower plasma levels of this vitamin than heterozygous and homozygous carriers of the A allele in both POAG and control groups (Fig. 56.3). There are no studies in the reference list on the effect of *SLC23A2* genotype on vitamin C levels in glaucomatous subjects (except for the one by the present authors), but the *SLC23A2* genotype has previously been linked to serum ascorbic acid concentration. Cahill and El-Sohemy reported that *SLC23A2* genotypes (rs6139591 and rs2681116 polymorphisms)

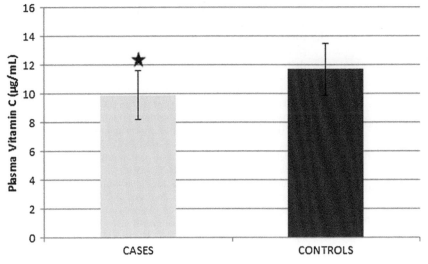

FIGURE 56.2 Plasma vitamin C levels between glaucomatous subjects and controls. The concentration of plasma vitamin C in the glaucoma group was significantly lower than in the control group (*P < 0.05).

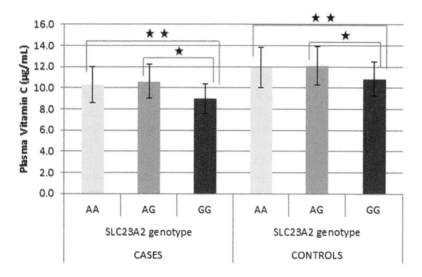

FIGURE 56.3 Plasma vitamin C levels by SLC23A2 genotype in cases and control subjects. The GG genotype had a lower plasma vitamin C concentration compared with the other two genotypes in both glaucoma and control groups. *Statistically significant differences between GG and AG genotypes (P<0.05); **statistically significant differences between GG and AA genotypes (P<0.05).

modify the strength of the correlation between dietary vitamin C intake and serum levels of the vitamin.[82] The current group's findings demonstrated that the *SLC23A2* genotype not only modifies the serum levels of vitamin C but is also associated with an increased risk of POAG, thus highlighting the importance of nutrition and genetics in the etiopathogenesis of POAG.

CONCLUSIONS

In summary, there are many factors involved in glaucomatous optic neuropathy, both genetic and environmental. It is a daunting task to discover all these factors, but the scientific community is gradually progressing in the understanding of the etiopathogenic mechanisms of glaucoma, with a firm commitment to prevent the blindness caused by this neurodegenerative eye disease. Someday, it may be possible to reverse the vision loss caused by glaucoma.

TAKE-HOME MESSAGES

- Good nutrition is essential for maintaining eye health.
- The plasma concentration of vitamin C is reduced in primary open-angle glaucoma (POAG).
- The *SLC23A2* genotype modifies the plasma vitamin C concentration.
- Variation in *SLC23A2* gene is associated with a high risk of POAG.
- Intake of foods rich in vitamin C may be beneficial for the development and progression of POAG.

References

1. Alward W. *Glaucoma: Los Requisitos en Oftalmología*. Madrid: Ediciones Harcourt; 2001; 11–12.
2. Quigley HA. Glaucoma. *Lancet* 2011;**377**:1367–77.
3. Carretero M. Tratamiento del glaucoma. *Av Farmacol* 2002; **21**:172–4.
4. Mellanby E. Diet and nutrition: proper feeding and good health. *Can Med Assoc J* 1939;**40**:597–9.
5. Baxter DM. Food, nutrition and good health. *Manit Med Rev* 1966;**46**:269–71.
6. Winckler I. The importance of good nutrition. *Nurs Times* 1976;**72**:1890–1.
7. Krehl WA. The role of nutrition in maintaining health and preventing disease. *Health Values* 1983;**7**:9–13.
8. Mayer J. Nutrition, exercise and cardiovascular disease. *Fed Proc* 1967;**26**:1768–71.
9. Shay CM, Stamler J, Dyer AR, Brown IJ, Chan Q, Elliott P, et al. Nutrient and food intakes of middle-aged adults at low risk of cardiovascular disease: the international study of macro-/micronutrients and blood pressure (INTERMAP). *Eur J Nutr* 2012;**51**:917–26.
10. Psoter WJ, Reid BC, Katz RV. Malnutrition and dental caries: a review of the literature. *Caries Res* 2005;**39**:441–7.
11. Mattar L, Godart N, Melchior JC, Pichard C. Anorexia nervosa and nutritional assessment: contribution of body composition measurements. *Nutr Res Rev* 2011;**8**:1–7.
12. Ames BN, Wakimoto P. Are vitamin and mineral deficiencies a major cancer risk? *Nat Rev Cancer* 2002;**2**:694–704.
13. Astrup A. Dietary management of obesity. *JPEN J Parenter Enteral Nutr* 2008;**32**:575–7.
14. Lien EL, Hammond BR. Nutritional influences on visual development and function. *Prog Retin Eye Res* 2011;**30**:188–203.
15. Shimazawa M, Nakajima Y, Mashima Y, Hara H. Docosahexaenoic acid (DHA) has neuroprotective effects against oxidative stress in retinal ganglion cells. *Brain Res* 2009;**1251**:269–75.
16. Bartlett H, Eperjesi F. An ideal ocular nutritional supplement? *Ophthalmic Physiol Opt* 2004;**24**:339–49.
17. Hammond Jr BR. Possible role for dietary lutein and zeaxanthin in visual development. *Nutr Rev* 2008;**66**:695–702.
18. Chang JS, Oellers P, Karp CL. Peripheral corneal ring due to hypercarotenaemia in a case of nutritional supplement abuse. *Br J Ophthalmol* 2012;**96**:605.

19. Sheffield VC, Stone EM, Alward WL, Drack AV, Johnson AT, Streb LM, et al. Genetic linkage of familial open angle glaucoma to chromosome 1q21-q31. *Nat Genet* 1993;**4**:47–50.

20. Stone EM, Fingert JH, Alward WL, Nguyen TD, Polansky JR, Sunden SL, et al. Identification of a gene that causes primary open angle glaucoma. *Science* 1997;**275**:668–70.

21. Michels-Rautenstrauss K, Mardin C, Wakili N, Jünemann AM, Villalobos L, Mejia C, et al. Novel mutations in the MYOC/GLC1A gene in a large group of glaucoma patients. *Hum Mutat* 2002;**20**:479–80.

22. Mengkegale M, Fuse N, Miyazawa A, Takahashi K, Seimiya M, Yasui T, et al. Presence of myocilin sequence variants in Japanese patients with open-angle glaucoma. *Mol Vis* 2008;**14**:413–7.

23. Sarfarazi M, Child A, Stoilova D, Brice G, Desai T, Trifan OC, et al. Localization of the fourth locus (GLC1E) for adult-onset primary open-angle glaucoma to the 10p15-p14 region. *Am J Hum Genet* 1998;**62**:641–52.

24. Rezaie T, Child A, Hitchings R, Brice G, Miller L, Coca-Prados M, et al. Adult-onset primary open-angle glaucoma caused by mutations in optineurin. *Science* 2002;**295**:1077–9.

25. Fuse N, Takahashi K, Akiyama H, Nakazawa T, Seimiya M, Kuwahara S, et al. Molecular genetic analysis of optineurin gene for primary open-angle and normal tension glaucoma in the Japanese population. *J Glaucoma* 2004;**13**:299–303.

26. Cheng JW, Li P, Wei RL. Meta-analysis of association between optineurin gene and primary open-angle glaucoma. *Med Sci Monit* 2010;**16**:369–77.

27. Monemi S, Spaeth G, DaSilva A, Popinchalk S, Ilitchev E, Liebmann J, et al. Identification of a novel adult-onset primary open-angle glaucoma (POAG) gene on 5q22.1. *Hum Mol Genet* 2005;**14**:725–33.

28. Miyazawa A, Fuse N, Mengkegale M, Ryu M, Seimiya M, Wada Y, et al. Association between primary open-angle glaucoma and WDR36 DNA sequence variants in Japanese. *Mol Vis* 2007;**13**:1912–9.

29. Weisschuh N, Wolf C, Wissinger B, Gramer E. Variations in the WDR36 gene in German patients with normal tension glaucoma. *Mol Vis* 2007;**13**:724–9.

30. Blanco-Marchite C, Sánchez-Sánchez F, López-Garrido MP, Iñigez-de-Onzoño M, López-Martínez F, López-Sánchez E, et al. WDR36 and P53 gene variants and susceptibility to primary open-angle glaucoma: analysis of gene–gene interactions. *Invest Ophthalmol Vis Sci* 2011;**52**:8467–78.

31. Stoilov I, Akarsu AN, Sarfarazi M. Identification of three different truncating mutations in cytochrome P4501B1 (CYP1B1) as the principal cause of primary congenital glaucoma (buphthalmos) in families linked to the GLC3A locus on chromosome 2p21. *Hum Mol Genet* 1997;**6**:641–7.

32. Vasiliou V, Gonzalez FJ. Role of CYP1B1 in glaucoma. *Annu Rev Pharmacol Toxicol* 2008;**48**:333–54.

33. Choudhary D, Jansson I, Schenkman JB. CYP1B1, a developmental gene with a potential role in glaucoma therapy. *Xenobiotica* 2009;**39**:606–15.

34. Su CC, Liu YF, Li SY, Yang JJ, Yen YC. Mutations in the CYP1B1 gene may contribute to juvenile-onset open-angle glaucoma. *Eye (Lond)* 2012;**26**:1369–77.

35. Thorleifsson G, Magnusson KP, Sulem P, Walters GB, Gudbjartsson DF, Stefansson H, et al. Common sequence variants in the LOXL1 gene confer susceptibility to exfoliation glaucoma. *Science* 2007;**317**:1397–400.

36. Fan BJ, Pasquale LR, Rhee D, Li T, Haines JL, Wiggs JL. LOXL1 promoter haplotypes are associated with exfoliation syndrome in a U.S. Caucasian population. *Invest Ophthalmol Vis Sci* 2011;**52**:2372–8.

37. Burdon KP, Macgregor S, Hewitt AW, Sharma S, Chidlow G, Mills RA, et al. Genome-wide association study identifies susceptibility loci for open angle glaucoma at TMCO1 and CDKN2B-AS1. *Nat Genet* 2011;**43**:574–8.

38. van Koolwijk LM, Ramdas WD, Ikram MK, Jansonius NM, Pasutto F, Hysi PG, et al. Common genetic determinants of intraocular pressure and primary open-angle glaucoma. *PLoS Genet* 2012;**8**:e1002611.

39. Sharma S, Burdon KP, Chidlow G, Klebe S, Crawford A, Dimasi DP, et al. Association of genetic variants in the TMCO1 gene with clinical parameters related to glaucoma and characterization of the protein in the eye. *Invest Ophthalmol Vis Sci* 2012;**53**:4917–25.

40. Thorleifsson G, Walters GB, Hewitt AW, Masson G, Helgason A, DeWan A, et al. Common variants near CAV1 and CAV2 are associated with primary open-angle glaucoma. *Nat Genet* 2010;**42**:906–9.

41. Wiggs JL, Kang JH, Yaspan BL, Mirel DB, Laurie C, Crenshaw A, et al. Common variants near CAV1 and CAV2 are associated with primary open-angle glaucoma in Caucasians from the USA. *Hum Mol Genet* 2011;**20**:4707–13.

42. Surgucheva I, Surguchov A. Expression of caveolin in trabecular meshwork cells and its possible implication in pathogenesis of primary open angle glaucoma. *Mol Vis* 2011;**17**:2878–88.

43. Fan BJ, Wang DY, Pasquale LR, Haines JL, Wiggs JL. Genetic variants associated with optic nerve vertical cup-to-disc ratio are risk factors for primary open angle glaucoma in a US Caucasian population. *Invest Ophthalmol Vis Sci* 2011;**52**:1788–92.

44. Takamoto M, Kaburaki T, Mabuchi A, Araie M, Amano S, Aihara M, et al. Common variants on chromosome 9p21 are associated with normal tension glaucoma. *PLoS ONE* 2012;**7**:e40107.

45. Osman W, Low SK, Takahashi A, Kubo M, Nakamura Y. A genome-wide association study in the Japanese population confirms 9p21 and 14q23 as susceptibility loci for primary open angle glaucoma. *Hum Mol Genet* 2012;**21**:2836–42.

46. Cong Y, Guo X, Liu X, Cao D, Jia X, Xiao X, et al. Association of the single nucleotide polymorphisms in the extracellular matrix metalloprotease-9 gene with PACG in southern China. *Mol Vis* 2009;**15**:1412–7.

47. Mossböck G, Weger M, Faschinger C, Zimmermann C, Schmut O, Renner W, et al. Role of functional single nucleotide polymorphisms of MMP1, MMP2, and MMP9 in open angle glaucomas. *Mol Vis* 2010;**16**:1764–70.

48. Awadalla MS, Burdon KP, Kuot A, Hewitt AW, Craig JE. Matrix metalloproteinase-9 genetic variation and primary angle closure glaucoma in a Caucasian population. *Mol Vis* 2011;**17**:1420–4.

49. Shibuya E, Meguro A, Ota M, Kashiwagi K, Mabuchi F, Iijima H, et al. Association of Toll-like receptor 4 gene polymorphisms with normal tension glaucoma. *Invest Ophthalmol Vis Sci* 2008;**49**:4453–7.

50. Chen LJ, Tam PO, Leung DY, Fan AH, Zhang M, Tham CC, et al. SNP rs1533428 at 2p16.3 as a marker for late-onset primary open-angle glaucoma. *Mol Vis* 2012;**18**:1629–39.

51. Takano Y, Shi D, Shimizu A, Funayama T, Mashima Y, Yasuda N, et al. Association of Toll-like receptor 4 gene polymorphisms in Japanese subjects with primary open-angle, normal-tension, and exfoliation glaucoma. *Am J Ophthalmol* 2012;**154**:825–32. e1.

52. Motallebipour M, Rada-Iglesias A, Jansson M, Wadelius C. The promoter of inducible nitric oxide synthase implicated in glaucoma based on genetic analysis and nuclear factor binding. *Mol Vis* 2005;**11**:950–7.

53. Meguro A, Inoko H, Ota M, Mizuki N, Bahram S. Genome-wide association study of normal tension glaucoma: common variants in SRBD1 and ELOVL5 contribute to disease susceptibility. *Ophthalmology* 2010;**117**:1331–8. e5.

54. Palou A. From nutrigenomics to personalised nutrition. *Genes Nutr* 2007;**2**:5–7.

55. Delcourt C, Korobelnik JF, Barberger-Gateau P, Delyfer MN, Rougier MB, Le Goff M, et al. Nutrition and age-related eye dis-

eases: the Alienor (Antioxydants, Lipides Essentiels, Nutrition et maladies OculaiRes) Study. *J Nutr Health Aging* 2010;**14**:854–61.

56. Delcourt C, Delyfer MN, Rougier MB, Lambert JC, Amouyel P, Colin J, et al. ARMS2 A69S polymorphism and the risk for age-related maculopathy: the ALIENOR study. *Arch Ophthalmol* 2012;**130**:1077–8.

57. Merle B, Delyfer MN, Korobelnik JF, Rougier MB, Colin J, Malet F, et al. Dietary omega-3 fatty acids and the risk for age-related maculopathy: the Alienor Study. *Invest Ophthalmol Vis Sci* 2011;**52**:6004–11.

58. Delcourt C, Delyfer MN, Rougier MB, Amouyel P, Colin J, Le Goff M, et al. Associations of complement factor H and smoking with early age-related macular degeneration: the ALIENOR study. *Invest Ophthalmol Vis Sci* 2011;**52**:5955–62.

59. Stewart WC. The effect of lifestyle on the relative risk to develop open-angle glaucoma. *Curr Opin Ophthalmol* 1995;**6**:3–9.

60. Kang JH, Pasquale LR, Willett WC, Rosner BA, Egan KM, Faberowski N, Hankinson SE. Dietary fat consumption and primary open-angle glaucoma. *Am J Clin Nutr* 2004;**79**:755–64.

61. Pasquale LR, Kang JH. Lifestyle, nutrition, and glaucoma. *J Glaucoma* 2009;**18**:423–8.

62. Birch TW, Dann WJ. Ascorbic acid in the eye-lens and aqueous humour of the ox. *Biochem J* 1964;**28**:638–41.

63. Pirie A. Ascorbic acid content of cornea. *Biochem J* 1946;**40**:96–100.

64. Purcell EF, Lerner LH, Kinsey VE. Ascorbic acid in aqueous humor and serum of patients with and without cataract; physiologic significance of relative concentrations. *AMA Arch Ophthalmol* 1954;**51**:1–6.

65. Virno M, Bucci MG, Pecori-Giraldi J, Cantore GP. Sodium ascorbate as an osmotic agent in glaucoma. *Boll Ocul* 1965;**44**:542–50.

66. Varma SD. Ascorbic acid and the eye with special reference to the lens. *Ann N Y Acad Sci* 1987;**498**:280–306.

67. Helbig H, Korbmacher C, Wiederholt M. Mechanism of ascorbic acid transport in the aqueous humor. *Fortschr Ophthalmol* 1990;**87**:421–4.

68. Garland DL. Ascorbic acid and the eye. *Am J Clin Nutr* 1991;**54**:1198–202.

69. Kim JW. Ascorbic acid enhances nitric oxide production in trabecular meshwork cells. *Korean J Ophthalmol* 2005;**19**:227–32.

70. Yin J, Thomas F, Lang JC, Chaum E. Modulation of oxidative stress responses in the human retinal pigment epithelium following treatment with vitamin C. *J Cell Physiol* 2011;**226**:2025–32.

71. Naidu KA. Vitamin C in human health and disease is still a mystery? An overview. *Nutr J* 2003;**2**:7.

72. Tsukaguchi H, Tokui T, Mackenzie B, Berger UV, Chen XZ, Wang Y, et al. A family of mammalian Na+-dependent L-ascorbic acid transporters. *Nature* 1999;**399**:70–5.

73. May JM. The SLC23 family of ascorbate transporters: ensuring that you get and keep your daily dose of vitamin C. *Br J Pharmacol* 2011;**164**:1793–801.

74. Gess B, Lohmann C, Halfter H, Young P. Sodium-dependent vitamin C transporter 2 (SVCT2) is necessary for the uptake of L-ascorbic acid into Schwann cells. *Glia* 2010;**54**:287–99.

75. Kannan R, Stolz A, Ji Q, Prasad PD, Ganapathy V. Vitamin C transport in human lens epithelial cells: evidence for the presence of SVCT2. *Exp Eye Res* 2001;**73**:159–65.

76. Chen AA, Marsit CJ, Christensen BC, Houseman EA, McClean MD, Smith JF, et al. Genetic variation in the vitamin C transporter, SLC23A2, modifies the risk of HPV16-associated head and neck cancer. *Carcinogenesis* 2009;**30**:977–81.

77. Jenab M, Riboli E, Ferrari P, Sabate J, Slimani N, Norat T, et al. Plasma and dietary vitamin C levels and risk of gastric cancer in the European Prospective Investigation into Cancer and Nutrition (EPIC-EURGAST). *Carcinogenesis* 2006;**27**:2250–7.

78. Wright ME, Andreotti G, Lissowska J, Yeager M, Zatonski W, Chanock SJ, et al. Genetic variation in sodium-dependent ascorbic acid transporters and risk of gastric cancer in Poland. *Eur J Cancer* 2009;**45**:1824–30.

79. Erichsen HC, Engel SA, Eck PK, Welch R, Yeager M, Levine M, et al. Genetic variation in the sodium-dependent vitamin C transporters, SLC23A1, and SLC23A2 and risk for preterm delivery. *Am J Epidemiol* 2006;**163**:245–54.

80. Zanon-Moreno V, Ciancotti-Olivares L, Asencio J, Sanz P, Ortega-Azorin C, Pinazo-Duran MD, Corella D. Association between a *SLC23A2* gene variation, plasma vitamin C levels, and risk of glaucoma in a Mediterranean population. *Mol Vis* 2011;**17**:2997–3004.

81. Zanon-Moreno V, Marco-Ventura P, Lleo-Perez A, Pons-Vazquez S, Garcia-Medina JJ, Vinuesa-Silva I, et al. Oxidative stress in primary open-angle glaucoma. *J Glaucoma* 2008;**17**:263–8.

82. Cahill LE, El-Sohemy A. Vitamin C transporter gene polymorphisms, dietary vitamin C and serum ascorbic acid. *J Nutrigenet Nutrigenomics* 2009;**2**:292–301.

Molecular Pathways, Green Tea Extract, (−)-Epigallocatechin Gallate, and Ocular Tissue

Yao Jin[1], Chen Xi[1], Jiang Qin[1], Victor R. Preedy[2], Ji Yong[3]

[1]Nanjing Medical University Eye Hospital, Nanjing, Jiangsu Province, China, [2]Diabetes and Nutritional Sciences, School of Medicine, King's College London, London, UK, [3]Department of Pathophysiology, Nanjing Medical University, Nanjing, Jiangsu Province, China

INTRODUCTION

Tea, from the evergreen plant *Carmellia sinensis*, is an ancient drink, which contains large amounts of various flavanoids. Catechins are one of the major classes of flavanoid and include epigallocatechin gallate (EGCG), epicatechin (EC), epigallocatechin (EGC), and epicatechin-3-gallate (ECG). EGCG is the most abundant of the catechins and has been demonstrated to have both anti-inflammatory and antioxidant properties in multiple cell types.[1–3] EGCG can bind to proteins and nucleic acids more tightly than other green tea extracts, and this binding is attributed to its polyphenolic structure[4] (Fig. 57.1). EGCG also acts as a strong metal ion chelator.[2]

Because of its structural and functional properties, EGCG may have therapeutic benefits in numerous inflammatory diseases such as atherosclerosis and arthritis and, in relation to this book, dry eye disease (DED) and many kinds of ocular disease.[3] Recent studies have also shown that EGCG has antivasodilatory, anticarcinogenic, and neuroprotective properties both *in vivo* and *in vitro*.[5,6]

MOLECULAR PATHWAY OF GREEN TEA AND (−)-EPIGALLOCATECHIN GALLATE

Inhibition of Reactive Oxygen Species

Reactive oxygen species (ROS) include superoxide and hydroxyl radicals, nitric oxide, singlet oxygen, nitrogen dioxide, and peroxynitrite. Many reports have shown that EGCG inhibits the formation of or damage caused by ROS. For example, EGCG can block the reduced nicotinamide adenine dinucleotide phosphate (NADPH)–cytochrome P450 production of ROS.[7]

Induction of Cell-cycle Arrest and Apoptosis

EGCG can induce cell-cycle arrest at the G2/M phase and increase the sub-G1 phase of cells. It achieves this by both reducing the expression of cyclin D1 and increasing cyclin-dependent kinase (CDK) inhibitors. This deactivates CDKs, causing the arrest of cells.[2]

It has been observed that EGCG can inhibit the expression of antiapoptotic proteins Bcl-2 and Bcl-XL and at the same time increase the expression of Bax and Bak proapoptotic proteins. It exerts these effects via three processes: activating caspase-3 and caspase-9, regulating the functions of the mitochondria, and cleaving the activity of apoptosis poly(ADP-ribose) polymerase.[2] EGCG decreases the activation of not only extracellular signal-regulated kinases (ERKs) but also the downstream transcription factors fos and jun. Together with proapoptotic Bax upregulation, as well as antiapoptotic protein Bcl2 downregulation, EGCG leads to apoptotic cell death. In addition, EGCG decreases the levels of glutathione, which results in the increasing production of ROS. Thus, EGCG can induce cell apoptosis through various pathways.[8]

Effect on the Ras/Mitogen-Activated Protein Kinase Pathway

Ras is a small GTPase that plays a role in diverse signaling pathways. Mitogen-activated protein kinases (MAPKs) are a group of serine/threonine kinases. While

Handbook of Nutrition, Diet, and the Eye
http://dx.doi.org/10.1016/B978-0-12-401717-7.00057-5

activated by different kinds of stimulus out of the cell, MAPK could act as a signal transducer of cell proliferation. MAPKs mainly consist of three subtypes: c-jun N-terminal kinases (JNKs), ERK, and p38 MAPK. It has been widely reported that EGCG regulates many cytokines and agents through this pathway, and thus EGCG produces marked effect on cell activities.[2]

Effect on the Phosphoinositide-3 Kinase/Akt Pathway

Active Akt induces phosphorylation of apoptotic proteins such as Bad and caspase-9 and results in the inhibition of apoptosis. Activation of Akt is accompanied by resistance to apoptosis and increased proliferation. It has been reported that by inhibiting tyrosine kinase phosphorylation of the platelet-derived growth factor (PDGF) receptor, EGCG is able to inhibit the downstream signaling and activation of the phosphoinositide-3 kinase (PI3K)/Akt pathway.[2]

Effect on Nuclear Factor-κB and Activator Protein-1

Nuclear factor-κB (NF-κB) acts as a sequence-specific transcription factor and is sensitive to oxidative stress. NF-κB regulates I-κB phosphorylation by inducing kinase/I-κB kinase. Phosphorylation of I-κB releases active NF-κB, which then translocates to the nucleus, and this effect also induces the expression of over 200 genes. Many of these expressed genes can then act to inhibit apoptosis and/or induce cell proliferation. EGCG can inhibit the activity of another transcription factor, activator protein-1 (AP-1), through the suppression of MAPK.[2]

Apart from all the pathways introduced above, EGCG can affect the human body through many other pathways (Table 57.1), including the insulin-like growth factor-1

pathway, cyclooxygenases, vascular endothelial growth factor (VEGF), and proteasomes. All of these effects of EGCG could act on the ocular tissues and may be beneficial in treating diseases associated with the visual senses.

The next part of this chapter will introduce recent studies on the beneficial effects of green tea extract and EGCG on ocular tissues and detail the possible molecular pathways.

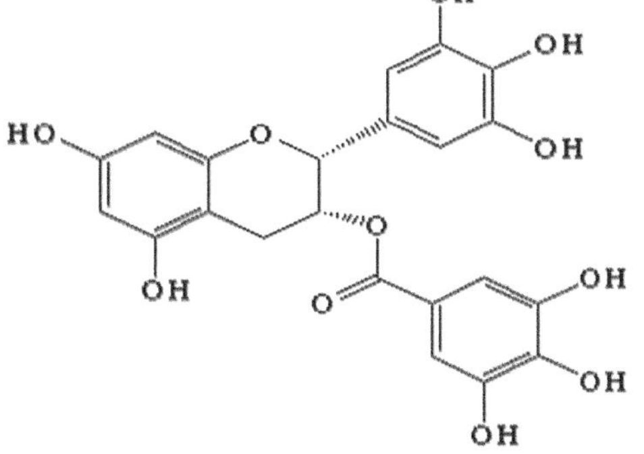

BENEFICIAL EFFECTS OF GREEN TEA EXTRACT AND (−)-EPIGALLOCATECHIN GALLATE ON THE CORNEAL EPITHELIUM

Anti-inflammatory and Antioxidant Effects on the Corneal Epithelium

The corneal epithelium acts as an initial physical barrier to injury and infection and also plays an important role in the ocular immune response system by producing inflammatory cytokines.[9] Many factors can

TABLE 57.1 Mechanisms of Action of (−)-Epigallocatechin Gallate (EGCG)

Action of EGCG	Pathway and Factors	
	↑	↓
Cell-cycle proteins	P21, P21, pRb	CDK, cyclin D, PgP, BCRP
Pro- and antiapoptosis	Bax & Bak, Cas3 & Cas9, PARP, Cas8 & Trail	Bcl2, Bclxl, ID2
Transcription factors	p53, IκB	AP-1, NF-κB, c-jun & c-fos, STAT1, STAT3, STAT5, β-catenin
Metastasis	TIMP, COX2	MMP1, MMP3, MMP4, MMP9, MMP13, IL-8, VEGF
Protein kinases	Erk1/2	HER2, JAK2, JNK, PKA, PKC, PI3K/Akt, Ras/ MAPK
Growth factor pathway		TNF, PDGF, TGFα/β, IGF1, IGFR, IGFBP3, INFγ

Cas: caspase; PARP: poly(ADP-ribose) polymerase; Trail: TNF-related apoptosis-inducing ligand; TIMP: tissue inhibitor of MMP; COX: cyclooxygenase; Erk: extracellular signal-regulated kinase; CDK: cyclin-dependent kinase; PgP: P-glycoprotein; BCRP: breast cancer resistance protein; ID: inhibitor of differentiation; AP: activator protein; NF: nuclear factor; STAT: signal transducer and activator of transcription; MMP: matrix metalloproteinase; IL: interleukin; VEGF: vascular endothelial growth factor; HER: human epidermal growth factor receptor; JAK: janus kinase; JNK: c-Jun N-terminal kinase; PK: protein kinase; PI3K: phosphoinositide-3 kinase; MAPK: mitogen-activated protein kinase; TNF: tumor necrosis factor; PDGF: platelet-derived growth factor; TGF: transforming growth factor; IGF: insulin-like growth factor; IGFR: insulin-like growth factor receptor; IGFBP3: insulin-like growth factor binding protein; INF: interferon.

EGCG

FIGURE 57.1 Chemical structure of EGCG: (2R,3R)-2-(3,4,5-trihydroxy-phenyl)-3,4-dihydro-1(2H)-benzopyran-3,5,7-triol 3-(3,4, 5-trihydroxybenzoate), (−)-epigallocatechin gallate, (−)-epigallocatechin gallate 3-O-gallate (molecular formula: C22H18O11).

induce inflammation of ocular surface, including DED, pathogens, and allergic reactions. Acute inflammation is usually beneficial, as it promotes the process of healing. But chronic inflammation may result in cell damage or even cell death.[10] Thus, in simple terms, when short-term inflammation turns into long-term inflammation, the compensatory processes become pathologic with consequential cellular damage.

The damage to the epithelium in many corneal diseases, such as DED, is commonly related to ocular surface inflammation. DED is one of the most common ophthalmic pathologies in the world.[11] The exact mechanism of the appearance and development of DED is still unknown.[11,12] Luo et al.,[12] in experiments on mice, found altered expression and production of interleukin-1β (IL-1β), tumor necrosis factor-α (TNF-α), and matrix metalloproteinase-9 (MMP-9) and activation of MAPK signaling pathways on the ocular surface in DED. MAPKs are known to stimulate the production of inflammatory cytokines and MMPs,[13] and these together could play an important role in the induction of those factors that have been implicated in the pathogenesis of DED.[12]

The hyperosmolarity of tears is an important mediator in the inflammation of the ocular surface, and this inflammation plays a role in DED.[14] Many factors are involved in the appearance and development of DED, such as activator protein-1 (AP-1), NF-κB, IL-1, IL-6, IL-8, and TNF-α. These factors can be activated by hyperosmolarity. Furthermore, oxidative stress is a contributor to the cellular damage that occurs during inflammation. Besides these factors, there is increased lipid peroxide and myeloperoxidase activity in the tears of patients with DED.[15] ROS also play a role in corneal inflammatory diseases.[16]

EGCG is proposed to act as an anti-inflammatory and antioxidant agent in human corneal epithelium cells (HCEpiCs).[17] This study confirmed that 3–30 μM EGCG could significantly inhibit the phosphorylation of the MAPKs p38 and JNK and NF-κB and AP-1 transcriptional activities. These, in turn, inhibit the IL-1β-induced release of inflammatory cytokines, including colony-stimulating factor (CSF), granulocyte–macrophage colony-stimulating factor (GM-CSF), IL-6, IL-8, and monocyte chemotactic protein-1 (MCP-1) and the hyperosmolarity-induced release of IL-6 and MCP-1. At the same time, EGCG can inhibit glucose oxidase-induced ROS in HCEpiCs. From this experiment, it was suggested that EGCG may have therapeutic potential for ocular inflammation such as in DED.[17]

Another experiment confirmed that topical EGCG administration can reduce the clinical signs and inflammatory changes in DED.[18] In this study mice received topical application of 1% atropine sulfate and subcutaneous administration of scopolamine to maximize ocular dryness.[18] Apart from these treatments,

mice with experimental DED were maintained in a controlled environment chamber. Corneal fluorescein staining was used to confirm the introduction of DED. Mice were divided into treatment groups (treated with different concentrations of EGCG) and a control group. After optical EGCG administration, all the corneal staining scores of the treatment groups decreased compared with the control group.[18] The number of central corneal CD11b+ cells (also called CD11b+ inflammatory cells) in the EGCG treatment groups was found to be significantly decreased compared with the untreated group. However, the authors did not confirm the activity of these CD11b+ cells. Inflammatory factors such as IL-1β, TNF-α, and chemokine (C-C motif) ligand 2 (CCL2) were detected in the experiment. Both the expression of IL-1β and CCL2 were reduced in the treatment group compared with the untreated one. However, there seemed to be no association between TNF-α transcription and topical EGCG administration. Proangiogenic and prolymphangiogenic growth factors (VEGF-A, VEGF-C, and VEGF-D) were also detected in the cornea of these mice.[18] The levels and the area of VEGF were lower and smaller in the treatment groups than in the control group, but the differences were not statistically significant. The number of apoptotic cells was significantly increased in dry eye corneas compared with normal controls, while EGCG treatment also decreased the number of apoptotic cells of the corneal epithelium.[18]

One may speculate on the kinds of agent that may affect DED via the administration of EGCG. Inflammatory mediators can induce immune-mediated inflammation, such as the activation of antigen-presenting cells (APCs), which as a result can exacerbate ocular surface cell damage. Besides corneal infiltration by CD11b+, APCs are increased in dry eye, and IL-1β induces the loss of corneal epithelial barrier function associated with ocular inflammation.[18] It has been demonstrated that CCL2 plays an important role in the regulation of monocytes and macrophages and can depress the inflammatory reactions in DED.[19] Finally, Lee et al.[18] suggested that direct EGCG administration on the eye could decrease inflammatory mediators including IL-1β and CCL2, the number of central corneal CD11b+ cells, and the number of apoptotic cells.

From the above, EGCG may be a potential therapeutic agent for the treatment of DED. Further studies are required to understand the mechanisms more clearly, and there is a need to conduct experiments in the clinical context.

Inhibition of Corneal Neovascularization

Corneal neovascularization has been reported to be associated with many ocular conditions including infectious keratitis, corneal ischemia, corneal trauma, and degeneration. It is thought that angiogenic factors play an important role in corneal neovascularization. VEGF is an angiogenic factor involved in the development of corneal neovascularization.[15] MMPs are considered to be other key mediators of angiogenesis. MMPs are a family of zinc-dependent enzymes that are involved in the degradation of the extracellular matrix and vascular basement membrane during angiogenesis. IL-1β, a multipotent inflammatory cytokine, also induces angiogenic activity mediated by NF-κB.[16] These mediators play key roles in corneal neovascularization. Neuhaus et al.[6] favored the hypothesis that EGCG, enriched in the plasma membrane, may prevent the binding of growth factors to their respective receptors, thereby suppressing tyrosine phosphorylation of these receptor types. Thus, EGCG acts against angiogenesis, and the effect is concentration dependent.[6]

An interesting phenomenon has been observed, in which drinking tea (concentration 708 mg/mL EGCG; plasma concentration 0.1–0.3 μmol/L) instead of water could significantly inhibit corneal neovascularization induced by VEGF.[20] Sánchez-Huerta et al.[21] suggested that EGCG may inhibit corneal neovascularization through inhibiting factors such as VEGF, IL-1, and NF-κB, but this demands further study. If tea is effective for the inhibition of corneal neovascularization via EGCG, there could be a mode of treatment for corneal diseases such as alkali or acid burns, infections, trauma, inflammation, and corneal graft rejection. One study reported that a nutrient mixture containing green tea extract affects umbilical vein endothelial cells via a number of routes including inhibition of MMP-2 secretion, reduction of endothelial cell migration, and inhibition of capillary tube formation.[22] The study demonstrated that a nutrient mixture (NM) containing green tea extract, proline, ascorbic acid, and lysine could diminish the concentration of cells that secrete VEGF and MMPs. Moreover, the NM was shown to have effects on inhibition of endothelial cell migration and capillary tube formation as well.[22] In conclusion, the authors suggested that topical application of NM is useful for the suppression of corneal neovascularization and could promote the restoration of corneal clarity. However, the NM should be investigated in various animal models of corneal neovascularization before investigations on humans and clinical application can commence.[22]

EGCG's high solubility in water and production of hydrogen peroxide (H_2O_2) necessitate further investigation, as well as testing the safety of its long-term administration, before the therapeutic potential of the drug can be realized. Toxicity aspects can easily be overcome by decreasing the concentration of EGCG, coupled with the use of buffered diluents close to the neutral pH.

ANTIOXIDANT EFFECTS OF (−)-EPIGALLOCATECHIN GALLATE ON HUMAN LENS EPITHELIAL CELLS

The most common pathologic change in human lens epithelial cells (HLECs) is opacification, which results in a multiple and common disease of ophthalmology named cataract.

Cataracts are defined as any opacification of the lens and are often considered to be an unavoidable consequence of aging. They may be induced by a series of factors including aging, light injury, metabolism, toxins, trauma, and smoking. Oxidative damage is widely thought to be a major cause of cortical and nuclear cataracts, the most common types of cataracts.[23] The present authors have shown that the migration of HLECs plays a crucial role in the remodeling of lens capsule and cataract formation.[24] Oxidative damage is a prominent feature of the mechanisms for nuclear and cortical cataracts, while absorption of ultraviolet (UV) light can generate free radicals, leading to increased oxidative damage, finally resulting in the formation of cataracts.

Epidemiologic and experimental studies in both humans and animals have shown that cataract may be caused by UV radiation.[24] UV radiation can induce the generation of ROS in HLECs. UV radiation-induced cataracts are believed to be associated with direct damage to lens epithelium and degradation of cellular components. Therefore, antioxidants may be applied for protection against UV radiation-induced cataract.

EGCG is a powerful antioxidant, which is considered to be effective in protecting HLECs from the oxidative damage induced by UV radiation. EGCG gives protection against UV-induced DNA damage to HLECs before and after UV exposure.[25] Both the total number and the viability of cultured HLECs after UV irradiation are enhanced by EGCG.[26] The enzyme catalase in the lens is quite sensitive to inactivation due to ultraviolet A (UVA) exposure. Zigman et al.[27] confirmed that the inactivation of catalase due to UVA exposure was significantly reduced by the antioxidant effect of EGCG.

In a study by this group, it was found that EGCG could also inhibit the injury to HLECs induced by ultraviolet B (UVB) radiation.[24] The data showed that UVB induces NADPH oxidase activity in a dose-dependent manner. EGCG treatment inhibited the activity of NADPH oxidase compared with the control group.[24] In these studies HLECs were preincubated with diphenylene iodonium (DPI, 10 μM) or EGCG (100 μg/mL) for 1 hour. After UVB (30 mJ/cm²) irradiation, cells were loaded with dihydrorhodamine (DHR, 20 μM) for 1 hour and then collected and fixed at 30 minutes post-treatment for fluorescence-activated cell sorting (FACS) analysis. FACS analysis showed that UVB induced ROS production, while DPI and EGCG reduced ROS production (Fig. 57.2). This suggested that UVB induces ROS generation partly through the NADPH oxidase pathway. This study also showed that EGCG treatment significantly reduced intracellular ROS, suggesting that EGCG not only is an antioxidant but also acts as an ROS scavenger. MMP-2 and MMP-9 mRNA and their activities were measured. Treatment with EGCG decreased MMP-2 and MMP-9 activation induced by UVB (Fig. 57.3). Furthermore, reverse transcription–polymerase chain reaction (RT-PCR) showed that UVB treatment increased the MMP-2 and MMP-9 mRNA levels, which had been partly blocked by EGCG (Fig. 57.4). Finally, immunofluorescence microscopy and Western blotting showed that NF-κB translocates to the nucleus in response to UVB irradiation. Treatment with EGCG significantly inhibited UVB-induced NF-κB activation. In conclusion, it is proposed that UVB radiation activates the redox-sensitive transcription factor NF-κB through the NADPH oxidase-mediated generation of ROS and therefore increases the expression of MMPs and cell migration in HLECs as a consequence. The NADPH oxidase activity of cultured HLECs decreased and the increase in ROS level induced by UVB was reversed after pretreatment with EGCG. The mechanisms may be related to the inhibition of activity and mRNA level of MMP.[28]

In another experiment involving HLECs exposed to UV, the protective effect of antioxidants such as vitamin C, taurine, superoxide dismutase (SOD), and EGCG were analyzed.[25] All four antioxidants decreased the DNA damage to HLECs: SOD was the most efficient and EGCG the second most efficient antioxidant. Thus, EGCG is considered to be a promising antioxidant against UV-induced cataracts.

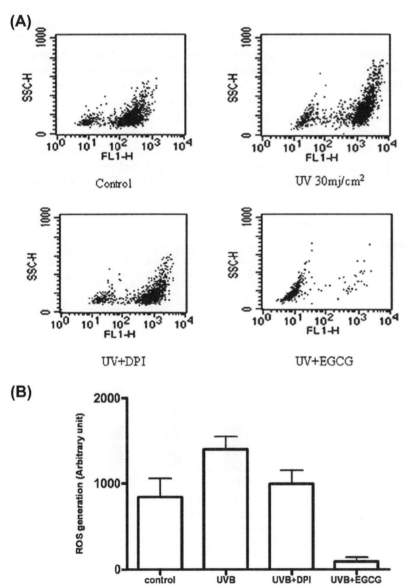

FIGURE 57.2 Reactive oxygen species (ROS) generation induced by ultraviolet B (UVB) is partly mediated by NADPH oxidase. (A) Induction of ROS generation is expressed in arbitrary units (B) as the mean ± SD of triplicate experiments. UV: ultraviolet; DPI: diphenylene iodonium; EGCG: (−)-epigallocatechin gallate. *Source: From Liu et al. (2011).[28]*

Apoptosis could be another process by which EGCG protects against the appearance or development of cataracts. Huang et al.[29] showed that the proliferation of cultured rabbit lens epithelial cells was inhibited by EGCG in a dose- and time-dependent manner. The cells became smaller and more rounded, with their nuclei condensed and broken. Apoptotic bodies were also seen under the electron microscope after EGCG was added to the medium. DNA ladders and evident apoptosis peaks in flow cytometry strengthened the notion that EGCG is proapoptotic.

Another study showed that EGCG could protect HLECs from cell death induced by H_2O_2.[30] EGCG reduced the generation of ROS, the loss of mitochondrial membrane potential, and the release of cytochrome c from the mitochondria into the cytosol of cells exposed to H_2O_2. In addition, EGCG inhibited the increased expression of caspase-9 and caspase-3, decreased the Bcl-2/Bax ratio, and attenuated the H_2O_2-induced reduced activation and expression of ERK, p38 MAPK, and Akt. Thus, these studies suggest another pathway by which EGCG protects against damage to HLECs.[30]

As UV is not the only way to induce cataracts, other models are needed for the study of lens epithelial cells. In a study of bovine lens, a tumor promoter, 10(-7)M12-O-tetradecanoylphorbol-13-acetate (TPA), was used to induce lens opacification. The H_2O_2 concentration in the whole lens was elevated by TPA in a dose-dependent manner. The study showed that preincubation of the lens with EGCG stopped the TPA-mediated opacification process and suppressed elevation of measured H_2O_2.[31]

All of these studies on the mechanisms by which EGCG delays or even reverses the pathologic changes of HLECs are significant for potential clinical applications. Further studies to confirm the protective effects of EGCG on HLECs are warranted.

PROTECTIVE EFFECTS OF (−)-EPIGALLOCATECHIN GALLATE ON RETINAL PIGMENT EPITHELIUM CELLS

Human retinal pigment epithelium (RPE) cells perform important functions in the visual process, and thus the dysfunction of RPE cells may be related to various retinal diseases.

Experiments have shown that EGCG plays a role in protecting irradiated cells against free-radical DNA damage.[32] The protective effect of green tea polyphenols against UVB-induced oxidant damage to RPE cells has been found in vitro.[33] As mentioned earlier, green tea polyphenols include catechins, of which EGCG is the main component. In this study the authors used green tea polyphenols to treat RPE cells before and after exposure to UVB. Viability, ultrastructure, and survivin gene expression were examined.[33] The results showed that green tea polyphenols protect the RPE cells against UVB-induced damage and reduce the UVB-depressed RPE expression. Green tea polyphenols also reduced UVB-induced DNA fragmentation. It was suggested that green tea polyphenols could protect RPE cells from UVB-induced damage

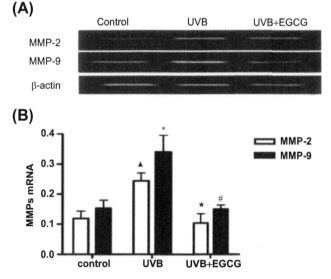

FIGURE 57.3 (−)-Epigallocatechin gallate (EGCG) could reverse the effect of ultraviolet B (UVB)-induced activity of matrix metalloproteinase (MMP). (A) Glatin zymography; (B) blank bar vs. control group, ▲P<0.05; blank bar UVB treatment group vs. UVB+EGCG group, ★P<0.05; black bar vs. with control group, *P<0.05; black bar UVB treatment group vs. UVB+EGCG group, #P<0.05. Source: From Liu et al. (2011).[28]

FIGURE 57.4 (−)-Epigallocatechin gallate (EGCG) could decrease the messenger RNA (mRNA) of ultraviolet B (UVB)-induced matrix metalloproteinase (MMP). (A) mRNA of MMP; (B) blank bar vs. control group, ▲P<0.05; blank bar UVB treatment group vs. UVB+EGCG group, ★P<0.05; black bar vs. control group, *P<0.05; black bar UVB treatment group vs. UVB+EGCG group, #P<0.05. Source: From Liu et al. (2011).[28]

through suppressing the decrease in survivin expression level and alleviating mitochondrial dysfunction.

Based on the above experiments, green tea polyphenols may have protective effects for some age-related diseases, such as age-related macular degeneration (AMD), but further studies are demanded to test and verify this point of view.

Chan et al.[34] suggest that EGCG is an effective inhibitor of RPE cell migration and adhesion to fibronectin and, therefore, may prevent epiretinal membrane formation. Their experimental results showed that VEGF could inhibit PDGF-BB (a subtype of PDGF)-induced RPE migration and RPE cell adhesion to fibronectin. However, EGCG did not directly bind to PDGF-BB in dot binding assays. As well as the inhibition effects in REP cells, EGCG was shown to alter actin cytoskeleton organization during cell adhesion. Finally, they found that EGCG (at or above $3\,\mu M$) could inhibit PDGF-BB-induced PDGF receptor-β (PDGFR-β), PI3K/Akt, and MAPK phosphorylation. However, JNK phosphorylation was not significantly affected by EGCG. This was detected by the extent of phosphorylation of PDGFR-β and its downstream components. It is already known that PDEF can evoke activation of signaling including PI3K, ERK1/2, and P38 in RPE cells.[34] Adult RPE cells are usually quiescent and differentiated and reside in the Go phase of the cell cycle. However, by re-entering the cell cycle, RPE cells can initiate proliferation and migration and induce the secretion of extracellular matrix proteins in diseases such as proliferative diabetic retinopathy, proliferative vitreoretinopathy, and AMD.[34] All the diseases mentioned above are common diseases that affect the human eye and have no effective methods of treatment, even by surgery. This study reveals a new direction for the treatment of these diseases.

EFFECTS ON OCULAR HYPERTENSION AND GLAUCOMA

Glaucoma is a progressive, multifactorial neurodegeneration whose primary effective site locates to the retinal ganglion cells (RGCs), particularly their axons.[35] Elevated intraocular pressure (IOP) is considered to be an important risk factor for glaucoma. Some academics have hypothesized that there are other contributing factors as well, such as ischemia, genetic factors, and failure of trophic support.[36] Thus, IOP control is very important for RGC protection.

Epigallocatechin and EGCG were confirmed to lower IOP in some animal studies. IOP could be lowered below control levels with 1 mg of epigallocatechin and EGCG.[37] In this study only phenolic antioxidants containing a pyrogallol B-ring system and nonaromatic C-ring (epigallocatechin, EGCG) were active in lowering IOP.[37]

In addition to animal studies, protective effects of EGCG on retinal neurons have been found in clinical trials. Falsini et al.[36] randomly assigned 18 patients with ocular hypertension (OHT) and 18 with open-angle glaucoma (OAG) to consume oral placebo or EGCG in a randomized, placebo-controlled, double-blind, crossover design clinical trial over a 3-month period. The pattern-evoked electroretinograms (PERGs) and perimetry of these patients were assessed at study entry and after 3 months of placebo or EGCG. The PERG amplitudes were distinctly increased in the OAG patient group after 3 months of EGCG. However, there was no obvious change in perimetry in either the OHT or the OAG patient group. The study showed that EGCG may protect the function of the inner retina, although the observed effect was small. Further long-term studies on EGCG are important to evaluate its full potential.

From the above studies, it can be inferred that green tea extract and EGCG may ameliorate the pathologic changes of glaucoma by lowering IOP and via antioxidative, anti-ischemic, and antineurodegenerative processes.

BENEFITS FOR RETINAL ISCHEMIA–REPERFUSION INJURY, OPHTHALMIC ARTERY, AND OPTIC NERVE

Ischemic retinal injury can result from various factors including ischemic optic neuropathy, central retinal artery occlusion, diabetic retinopathy, retinopathy of prematurity, glaucoma, energy-dependent dysfunction, tissue edema, and RGCs.[38] Oxidative stress is considered to play a key role in ischemic retinal injuries. Reoxygenation during the reperfusion period can cause excessive production of both ROS and reactive nitrogen species. The free radicals may contribute to the damage to cellular components both directly and indirectly by modulating the receptor–postreceptor signal pathways.

In the experiments of Peng et al.,[38] rats were divided into four groups: a normal control group, an EGCG with sham operation group, an ischemic retinal group, and an EGCG with ischemic retina group. Ischemic retinal injury was induced by increasing the IOP in the rats' eyes 150 mm Hg for 60 minutes. The IOP was increased by inserting saline into the anterior chamber of the eye after corneal anesthesia with one drop of proparacaine hydrochloride. Retinal ischemia was signaled by the whitening of the iris and the loss of the red reflex, while perfusion of the retinal vasculature was confirmed by examination of the fundus. RGC death from ischemic retinal injury was reduced by approximately 10% 3 days later in those animals pretreated with EGCG. The ischemic retina-induced glial fibrillary acidic protein expression was significantly downregulated by EGCG. EGCG treatment also reduced terminal deoxynucleotidyl transferase dUTP nick-end labeling (TUNEL)-positive cells in the inner retina after ischemic retinal injury as well as ischemic retina-induced lipid peroxidation. Histology showed that EGCG reduced neuronal nitric oxide synthase (NOS) and nicotinamide adenine dinucleotide phosphate diaphorase-positive cells in the retina. Thus, EGCG acts as a cellular and molecular protector in ischemic retinal injury via the pathways mentioned above.

Zhang et al.[39] demonstrated that orally administered EGCG attenuates injury to the retina caused by ischemia where caspases had been activated. Their studies showed that white light (1000 lx, 48 hour)-induced apoptosis is caspase independent and could be blunted by EGCG in vitro.

Romano and Lograno[40] determined that EGCG could relax the isolated bovine ophthalmic artery. This study showed that EGCG produces a relaxant responses on bovine ophthalmic arterial rings precontracted with 5-hydroxytryptamine. They considered the PI3K/Akt/nitric oxide/cyclic GMP signaling pathway to be the major upstream activator for NOS and that nitric oxide and cGMP are required for the vasorelaxant response. The authors subsequently discussed the potential use of EGCG as a new treatment for ocular neurodegeneration and glaucoma.[40]

Xie et al.[41] investigated the neuroprotective effects of EGCG in an optic nerve crush (ONC) model in rats. They divided the animals into four groups at random: normal control (group A), sham operation + EGCG (group B), ONC + vehicle (group C), and ONC + EGCG (group D). ONC injury was achieved using a micro-optic nerve clipper with 40 g power at approximately 2 mm from the optic nerve head for 60 seconds. A progressive loss of RGCs was observed after ONC in group C, while a significantly higher density of RGCs was observed in group D compared with group C. The expression of neurofilament triplet L (NF-L) protein (a viable measure of RGC viability) was much higher in group D than in group C in both immunohistochemical and Western blotting analyses. These findings suggest that EGCG has protective effects on RGCs after ONC. The authors concluded that EGCG was a potential therapeutic agent for optic nerve diseases.

These studies suggest that EGCG may be effective in preventing or ameliorating retinal, optical nerve, ophthalmic artery, and retinal vascular damage.

OTHER EFFECTS

EGCG inhibits gelatinase activity of some bacterial isolates from ocular infection and limits their invasion through gelatine.[42] In addition, EGCG inhibits biofilm formation by ocular staphylococcal isolates via interference with the polysaccharides that form the glycocalyx, disrupting their interactions either reciprocally or with the cell wall and thus reducing the amount of slime that accumulates. EGCG breaks the integrity of the bacterial cell wall by binding to the peptidoglycan and interferes with the initial docking phase of biofilm formation.[43]

A study by Siu et al.[44] showed that catechin can protect against glutamate-induced retinal lipid and protein damage.

HOW TO SUPPLY GREEN TEA EXTRACT AND EPIGALLOCATECHIN GALLATE

One bag of green tea contains 80–100 mg of polyphenols, of which EGCG accounts for about 25–30 mg.[45] An experiment on rats showed that the plasma level of EGCG was 0.1% (v/v) after intragastric administration of decaffeinated green tea[46]

The maximum concentration of green tea polyphenols in human plasma is 4400 pmol/mL.[47] This concentration of EGCG would be sufficient to act as an antioxidant and have other biologic activities in the blood.[2] It is generally considered that green tea contains more catechins than tea obtained from other leaf-processing methods (Fig. 57.5),

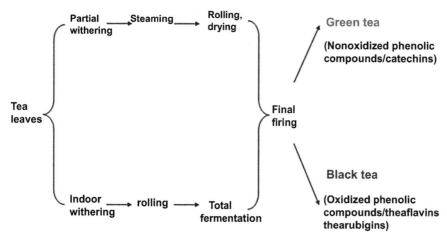

FIGURE 57.5 Principal differences between green and black tea processing and its influence on the final polyphenol content. Compared with black tea, green tea contains more catechins and more effective polyphenols, including (−)-epigallocatechin gallate (EGCG).

and it is usually considered that green tea contains more effective polyphenols, including EGCG, compared with black tea, red tea, and oolong tea.[5]

Smith *et al.*[48] investigated the ability of one preformulation method to improve the oral bioavailability of EGCG. They found that forming nanolipidic EGCG particles could more than double the oral bioavailability of EGCG *in vivo* compared with free EGCG. It has also been found that EGCG can retain its biologic efficacy when encapsulated in polylactic acid–polyethylene glycol nanoparticles with a greater than 10-fold dose advantage both in cell culture and *in vivo* compared with non-encapsulated EGCG.[49]

Drinking tea, especially green tea, would be a good way to supplement the intake of EGCG. Topical EGCG administration in eyedrops may be useful in the treatment of ophthalmologic diseases.

CONCLUSIONS

In conclusion, EGCG acts on a variety of cell types including those of the cornea, lens, and retina. The mechanisms through which EGCG acts are not known precisely, but there is evidence to support the involvement of molecular processes. These range from receptor-mediated processes to signaling and altered gene expression. Potentially, green tea extracts and EGCG could be effectively applied to prevent and/or treat various ophthalmologic diseases, although further studies are needed.

TAKE-HOME MESSAGES

- The green tea extract (−)-epigallocatechin gallate (EGCG) is produced from *Carmellia sinensis* and is the major polyphenol found in green tea.
- EGCG has anti-inflammatory and antioxidant properties in many cell types.
- Recent studies have shown that EGCG may exert its effects via various molecular pathways including reactive oxygen species, cell-cycle arrest and apoptosis, the Ras/mitogen-activated protein kinase and phosphoinositide-3 kinase/Akt pathways, nuclear factor-κB, and activator protein-1.
- Many studies have shown the benefits of EGCG for ocular tissues such as corneal epithelium cells, human lens epithelial cells, retinal pigment epithelium cells, and retinal ganglion cells.
- EGCG may act as a protector in many eye diseases.
- Drinking green tea to enhance EGCG intake is a suitable and practical approach compared with supplementation with purified extracts.

References

1. Zheng Y, Toborek M, Hennig B. Epigallocatechin gallate-mediated protection against tumor necrosis factor-alpha-induced monocyte chemoattractant protein-1 expression is heme oxygenase-1 dependent. *Metabolism* 2010;**59**:35–1528.
2. Kanwar J, Taskeen M, Mohammad I, Huo C, Chan TH, Dou QP. Recent advances on tea polyphenols. *Front Biosci (Elite Ed)* 2012;**4**:31–111.
3. Wu D, Wang J, Pae M, Meydani SN. Green tea EGCG, T cells, and T cell-mediated autoimmune diseases. *Mol Aspects Med* 2012;**33**:18–07.
4. Chen D, Milacic V, Chen MS, Wan SB, Lam WH, Huo C, et al. Tea polyphenols, their biological effects and potential molecular targets. *Histol Histopathol* 2008;**23**:96–487.
5. Cabrera C, Artacho R, Gimenez R. Beneficial effects of green tea – a review. *J Am Coll Nutr* 2006;**25**:79–99.
6. Neuhaus T, Pabst S, Stier S, Weber AA, Schror K, Sachinidis A, et al. Inhibition of the vascular-endothelial growth factor-induced intracellular signaling and mitogenesis of human endothelial cells by epigallocatechin-3 gallate. *Eur J Pharmacol* 2004;**483**:7–223.
7. Surh Y. Molecular mechanisms of chemopreventive effects of selected dietary and medicinal phenolic substances. *Mutat Res* 1999;**428**:27–305.
8. Manohar M, Fatima I, Saxena R, Chandra V, Sankhwar PL, Dwivedi A. (−)-Epigallocatechin-3-gallate induces apoptosis in human endometrial adenocarcinoma cells via ROS generation and p38 MAP kinase activation. *J Nutr Biochem* 2013;**24**:7–940.
9. Ueta M, Kinoshita S. Innate immunity of the ocular surface. *Brain Res Bull* 2009;**81**:28–19.
10. Wells CA, Ravasi T, Hume DA. Inflammation suppressor genes: please switch out all the lights. *J Leukoc Biol* 2005;**78**:9–13.
11. Pflugfelder SC, de Paiva CS, Li DQ, Stern ME. Epithelial–immune cell interaction in dry eye. *Cornea* 2008;**27**(Suppl 1.):S9–11.
12. Luo L, Li DQ, Doshi A, Farley W, Corrales RM, Pflugfelder SC. Experimental dry eye stimulates production of inflammatory cytokines and MMP-9 and activates MAPK signaling pathways on the ocular surface. *Invest Ophthalmol Vis Sci* 2004;**45**:301–4293.
13. Zhang Y, Liu J, Kou J, Yu J, Yu B. DT-13 suppresses MDA-MB-435 cell adhesion and invasion by inhibiting MMP-2/9 via the p38 MAPK pathway. *Mol Med Rep* 2012;**6**:5–1121.
14. Pflugfelder SC, Stern ME. Immunoregulation on the ocular surface: 2nd Cullen Symposium. *Ocul Surf* 2009;**7**:67–77.
15. Azar DT. Corneal angiogenic privilege: angiogenic and antiangiogenic factors in corneal avascularity, vasculogenesis, and wound healing (an American Ophthalmological Society thesis). *Trans Am Ophthalmol Soc* 2006;**104**:264–302.
16. Kondo Y, Fukuda K, Adachi T, Nishida T. Inhibition by a selective IkappaB kinase-2 inhibitor of interleukin-1-induced collagen degradation by corneal fibroblasts in three-dimensional culture. *Invest Ophthalmol Vis Sci* 2008;**49**:7–4850.
17. Cavet ME, Harrington KL, Vollmer TR, Ward KW, Zhang JZ. Anti-inflammatory and anti-oxidative effects of the green tea polyphenol epigallocatechin gallate in human corneal epithelial cells. *Mol Vis* 2011;**17**:42–533.
18. Lee HS, Chauhan SK, Okanobo A, Nallasamy N, Dana R. Therapeutic efficacy of topical epigallocatechin gallate in murine dry eye. *Cornea* 2011;**30**:1465–72.
19. Melgarejo E, Medina MA, Sanchez-Jimenez F, Botana LM, Dominguez M, Escribano L, et al. (−)-Epigallocatechin-3-gallate interferes with mast cell adhesiveness, migration and its potential to recruit monocytes. *Cell Mol Life Sci* 2007;**64**:2690–701.
20. Cao Y, Cao R. Angiogenesis inhibited by drinking tea. *Nature* 1999;**398**:381.
21. Sánchez-Huerta V, Gutierrez-Sanchez L, Flores-Estrada J. (−)-Epigallocatechin 3-gallate (EGCG) at the ocular surface inhibits corneal neovascularization. *Med Hypotheses* 2010;**76**:311–3.

22. Shakiba Y, Mostafaie A. Inhibition of corneal neovascularization with a nutrient mixture containing lysine, proline, ascorbic acid, and green tea extract. *Arch Med Res* 2007;**38**:789–91.

23. Beebe DC, Holekamp NM, Shui YB. Oxidative damage and the prevention of age-related cataracts. *Ophthalmic Res* 2010;**44**:155–65.

24. Yao J, Liu Y, Wang X, Shen Y, Yuan S, Wan Y, et al. UVB radiation induces human lens epithelial cell migration via NADPH oxidase-mediated generation of reactive oxygen species and up-regulation of matrix metalloproteinases. *Int J Mol Med* 2009;**24**:153–9.

25. Wu ZH, Wang MR, Yan QC, Pu W, Zhang JS. UV-induced DNA damage and protective effects of antioxidants on DNA damage in human lens epithelial cells studied with comet assay. *Zhonghua Yan Ke Za Zhi* 2006;**42**:1002–7.

26. Heo J, Lee BR, Koh JW. Protective effects of epigallocatechin gallate after UV irradiation of cultured human lens epithelial cells. *Korean J Ophthalmol* 2008;**22**:183–6.

27. Zigman S, Rafferty NS, Rafferty KA, Lewis N. Effects of green tea polyphenols on lens photooxidative stress. *Biol Bull* 1999;**197**:285–6.

28. Liu Y-Y, Chen M-R, Yao J, Jiang Q. Experimental study of epigallocatechin gallate for inhibiting lens epithelial cell injury induced by UVB. *Rec Adv Ophthalmol* 2011;**31**:516–23.

29. Huang W, Li S, Zeng J, Liu Y, Wu M, Zhang M. Growth inhibition, induction of apoptosis by green tea constituent (−)-epigallocatechin-3-gallate in cultured rabbit lens epithelial cells. *Yan Ke Xue Bao* 2000;**16**:194–8.

30. Yao K, Ye P, Zhang L, Tan J, Tang X, Zhang Y. Epigallocatechin gallate protects against oxidative stress-induced mitochondria-dependent apoptosis in human lens epithelial cells. *Mol Vis* 2008;**14**:217–23.

31. Ye JJ, Frenkel K, Zadunaisky JA. Lens opacification and H2O2 elevation induced by a tumor promoter. *Lens Eye Toxic Res* 1992;**9**:37–48.

32. Paul B, Hayes CS, Kim A, Athar M, Gilmour SK. Elevated polyamines lead to selective induction of apoptosis and inhibition of tumorigenesis by (−)-epigallocatechin-3-gallate (EGCG) in ODC/Ras transgenic mice. *Carcinogenesis* 2005;**26**:119–24.

33. Xu JY, Wu LY, Zheng XQ, Lu JL, Wu MY, Liang YR. Green tea polyphenols attenuating ultraviolet B-induced damage to human retinal pigment epithelial cells *in vitro*. *Invest Ophthalmol Vis Sci* 2011;**51**:6665–70.

34. Chan CM, Huang JH, Chiang HS, Wu WB, Lin HH, Hong JY, et al. Effects of (−)-epigallocatechin gallate on RPE cell migration and adhesion. *Mol Vis* 2010;**16**:586–95.

35. Casson RJ, Chidlow G, Ebneter A, Wood JP, Crowston J, Goldberg I. Translational neuroprotection research in glaucoma: a review of definitions and principles. *Clin Exp Ophthalmol* 2012;**40**:350–7.

36. Falsini B, Marangoni D, Salgarello T, Stifano G, Montrone L, Di Landro S, et al. Effect of epigallocatechin-gallate on inner retinal function in ocular hypertension and glaucoma: a short-term study by pattern electroretinogram. *Graefes Arch Clin Exp Ophthalmol* 2009;**247**:1223–33.

37. Hodges LC, Kearse CE, Green K. Intraocular pressure-lowering activity of phenolic antioxidants in normotensive rabbits. *Curr Eye Res* 1999;**19**:234–40.

38. Peng PH, Ko ML, Chen CF. Epigallocatechin-3-gallate reduces retinal ischemia/reperfusion injury by attenuating neuronal nitric oxide synthase expression and activity. *Exp Eye Res* 2008;**86**:637–46.

39. Zhang B, Rusciano D, Osborne NN. Orally administered epigallocatechin gallate attenuates retinal neuronal death *in vivo* and light-induced apoptosis *in vitro*. *Brain Res* 2008;**1198**:141–52.

40. Romano MR, Lograno MD. Epigallocatechin-3-gallate relaxes the isolated bovine ophthalmic artery: involvement of phosphoinositide 3-kinase-Akt-nitric oxide/cGMP signalling pathway. *Eur J Pharmacol* 2009;**608**:48–53.

41. Xie J, Jiang L, Zhang T, Jin Y, Yang D, Chen F. Neuroprotective effects of Epigallocatechin-3-gallate (EGCG) in optic nerve crush model in rats. *Neurosci Lett* 2010;**479**:26–30.

42. Blanco AR, La Terra Mule S, Babini G, Garbisa S, Enea V, Rusciano D. (−)Epigallocatechin-3-gallate inhibits gelatinase activity of some bacterial isolates from ocular infection, and limits their invasion through gelatine. *Biochim Biophys Acta* 2003;**1620**:273–81.

43. Blanco AR, Sudano-Roccaro A, Spoto GC, Nostro A, Rusciano D. Epigallocatechin gallate inhibits biofilm formation by ocular staphylococcal isolates. *Antimicrob Agents Chemother* 2005;**49**:4339–43.

44. Siu AW, Lau MK, Cheng JS, Chow CK, Tam WC, Li KK, et al. Glutamate-induced retinal lipid and protein damage: the protective effects of catechin. *Neurosci Lett* 2008;**432**:193–7.

45. Kanwar J, Taskeen M, Mohammad I, Huo C, Chan TH, Dou QP. Recent advances on tea polyphenols. *Front Biosci (Elite Ed)* 2012;**111**:31.

46. Okushio K, Suzuki M, Matsumoto N, Nanjo F, Hara Y. Identification of (−)-epicatechin metabolites and their metabolic fate in the rat. *Drug Metab Dispos* 1999;**27**:309–16.

47. Lee MJ, Wang ZY, Li H, Chen L, Sun Y, Gobbo S, et al. Analysis of plasma and urinary tea polyphenols in human subjects. *Cancer Epidemiol Biomarkers Prev* 1995;**4**:393–9.

48. Smith A, Giunta B, Bickford PC, Fountain M, Tan J, Shytle RD. Nanolipidic particles improve the bioavailability and alpha-secretase inducing ability of epigallocatechin-3-gallate (EGCG) for the treatment of Alzheimer's disease. *Int J Pharm* 2010;**389**:207–12.

49. Siddiqui IA, Mukhtar H. Nanochemoprevention by bioactive food components: a perspective. *Pharm Res* 2010;**27**:1054–60.

Dietary Antioxidants, αvβ5 Integrin, and Ocular Protection: Long-Term Consequences of Arrhythmic Retinal Pigment Epithelium Phagocytosis

Saumil Sethna, Silvia C. Finnemann

Department of Biological Sciences, Center for Cancer, Genetic Diseases, and Gene Regulation, Fordham University, Bronx, New York, USA

INTRODUCTION

Retinal Pigment Epithelial Cells

The retinal pigment epithelium (RPE) forms the outermost layer of the vertebrate retina. The tight junctions and adherence junctions of the RPE constitute the outer blood–retinal barrier. Basolaterally, RPE cells adhere to a complex, pentalaminar basement membrane, called Bruch's membrane, which faces the highly vascularized choroidal connective tissue.[1] Apically, in sharp contrast, RPE cells face the avascular subretinal space of the neural retina where they interact with secreted proteins of the interphotoreceptor matrix and possibly with the outer segment portions of photoreceptor rods and cones. RPE cells extend thin, highly elongated microvilli from their apical plasma membrane that ensheath outer segments. Surface receptor proteins localizing to the RPE's microvilli mediate mechanically robust retinal adhesion. In the adult mammalian eye, both photoreceptors and RPE cells are postmitotic such that they act as a single functional unit in the retina for life. While the barrier function of the RPE cells maintains the integrity of the neural retina in general,[2] RPE cells perform numerous other tasks that support function and integrity specifically of photoreceptor rods and cones. First, RPE cells use numerous cell surface transporter and channel proteins to maintain the volume of the subretinal space and control its ionic composition.[2,3] Second, RPE cells shuttle nutrients towards the neural retina and metabolic waste towards the choroid. Third, RPE cells secrete a number of growth and survival factors that act on photoreceptors to maintain their viability and that act on endothelial cells controlling growth and permeability of choroidal vessels.[4,5] Fourth, RPE cells possess a large number of melanosome organelles filled with melanin pigment that absorbs stray light and combats oxidative stress.[6,7] As reviewed by Boulton and Dayhaw-Barker, RPE cells also prevent photo-oxidative damage using enzymatic antioxidants such as superoxide dismutase and catalase, in addition to nonenzymatic antioxidants such as β-carotene, α-tocopherol, lutein and zeaxanthin, and ascorbate.[8] Fifth, RPE cells participate in the visual cycle. The RPE-specific isomerase enzyme, RPE65, catalyzes the conversion of bleached all-*trans* retinal to 11-*cis* retinal that is then transported back to photoreceptors for reuse.[9,10] Finally, RPE cells participate in the continuous renewal of the outer segment portions of photoreceptors. A complex, multiprotein phagocytic mechanism serves to clear shed outer segment debris from the retina by prompting its engulfment and subsequent phagolysosomal digestion.

Retinal Pigment Epithelium Phagocytosis in the Mammalian Eye

Photoreceptor rods and cones are permanent cells in the mature mammalian retina and do not divide or turn over. Photoisomerization of 11-*cis* retinal to all-*trans* retinal stimulates phototransduction by activating the G-protein transducin and its downstream signaling pathways, which ultimately reduce synaptic glutamate neurotransmitter release towards retinal interneurons.[11,12] In both rods and cones, light detection and phototransduction are restricted to the outer segment portion of the cell, which is filled with tightly packed stacks of membranes called outer segment discs. In rods, the individual discs are separated from each other and from the outer segment plasma membrane, while in cones, the discs are contiguous.[13] Membrane lipid and protein constituents of outer segment discs are subjected to high levels of photo-oxidative stress. It is thought that the process of outer segment renewal evolved to enable long-term functionality and survival of photoreceptors despite this damaging influence.

A series of elegant experiments by Young and colleagues first established in the late 1960s that the outer segments of rods and cones are dynamic structures that are continuously renewed.[14–16] These investigators also found that production of new discs containing rhodopsin in rods is balanced by the diurnal removal of the oldest, distal tips of outer segments by a process termed shedding. Shed photoreceptor outer segment fragments (POS) are engulfed and digested by adjacent RPE cells.[17] In mammals, the process of outer segment renewal is circadian in nature and entrained by the daily light cycle.[18,19] In all species tested thus far, shedding and phagocytosis of rod photoreceptor POS peak shortly after light onset. While shedding and uptake of cone photoreceptor POS also exhibits a diurnal rhythm, the timing of its peak varies depending on species.[18,20–22] By far the majority of photoreceptors (approximately 97%) are rods in rodent as well as in human retina.[23] Hence, the early morning uptake of shed rod POS constitutes the primary burden for the phagocytic uptake and digestive machinery of RPE cells. It is, however, important to bear in mind that nothing is known to date about shedding or RPE phagocytosis of those POS shed by the specialized cones that are located in a tightly packed mosaic in the central foveal region of the primate retina, which is virtually devoid of rods.

ROLE OF INTEGRIN ADHESION RECEPTOR αVβ5 IN PHAGOCYTOSIS

The process of phagocytosis is broadly divided into three separate but interconnected phases: recognition and binding of POS to RPE apical surface receptors, engulfment of surface-tethered POS, and digestion of engulfed POS in phagolysosomal organelles of the RPE. The strict rhythmicity of rod outer segment renewal and the increasing availability of genetically modified rat or mouse animal models have allowed the process of POS phagocytosis to be studied *in vivo*.[24] Yet,

the different phases of phagocytosis cannot be studied separately by analyzing RPE *in situ*. It is therefore difficult to assign specific functions to individual components of the phagocytic machinery based on animal experiments alone. However, RPE cells retain their phagocytic activity even if isolated from the eye and maintained as cell culture. Synchronized (pulse-chase) phagocytosis experiments using such RPE cell cultures allow the recognition, engulfment, and digestion phases of phagocytosis to be investigated largely in isolation. Studies exploring POS phagocytosis by RPE *in vivo* and in cell culture have identified important aspects of POS shedding and phagocytic functions of RPE cells: at light onset, distal tips of photoreceptor rods externalize the anionic phospholipid phosphatidylserine that is hidden in the internal leaflet of the rod plasma membrane at other times.[25] The secreted glycoprotein milk fat globule EGF-like protein 8 (MFG-E8) is present in the subretinal space and binds to exposed phosphatidylserine via a specialized motif.[26] MFG-E8 acts as opsonin for shedding or shed POS because it also possesses a binding motif recognized by integrin receptors.[27] While RPE cells express numerous adhesion receptors of the integrin family, the αvβ5 heterodimer is the only integrin receptor that localizes to the apical, phagocytic surface of the RPE.[28,29] Cell culture studies have shown that αvβ5 integrin is stabilized at the RPE surface by complex formation with the tetraspanin CD81.[30] Thus, CD81 alone does not influence POS uptake, unlike other RPE surface receptors, such as the scavenger receptor CD36, which acts in POS clearance independently of αvβ5 integrin.[31] Active αvβ5 integrin receptors are required for recognition of POS by RPE cells in culture.[26–29,32–34]

In addition to acting as a tethering receptor, αvβ5 integrin, upon binding of POS, initiates at least two distinct signaling pathways in RPE cells, both of which are required for engulfment of bound POS. The first pathway hinges on the cytosolic tyrosine kinase p125 focal adhesion kinase (p125FAK), which associates either directly or indirectly with apical αvβ5 integrin receptors in resting, nonphagocytic RPE cells.[35] Binding of MFG-E8-opsonized POS to αvβ5 integrin leads to swift autophosphorylation and activation of p125FAK. Active p125FAK then dissociates from the apical integrin complex and relocalizes, ultimately causing increased tyrosine phosphorylation and hence activity of another RPE surface receptor, Mer tyrosine kinase (MerTK).[34,35] MerTK activity is indispensable for POS engulfment. MerTK deficiency abolishes engulfment by RPE cells in culture[36,37] and leads to rapid accumulation of POS debris in the subretinal space, causing photoreceptor death by young adulthood in a spontaneous rat model of hereditary blindness, the Royal College of Surgeons rat (commonly known as the RCS rat),[38,39] and in genetically engineered mutant mice[40] and rat chimeras.[41]

Importantly, mutations in MerTK have been demonstrated to cause severe forms of retinitis pigmentosa.[42] Despite its enormous importance for the phagocytic clearance of POS by RPE cells, it is still largely unknown how exactly MerTK promotes engulfment. Recent evidence suggests a possible functional relationship of MerTK with the F-actin cytoskeletal motor protein myosin II, but details of MerTK–myosin II interactions remain to be established.[43]

In a second signaling pathway that is entirely independent of p125FAK and MerTK, αvβ5 integrin receptor ligation by MFG-E8-opsonized POS increases activity of the small Rho family GTPase Rac1. Rac1 activation, in turn, promotes F-actin polymerization and recruitment beneath surface-bound POS.[44] Efficient POS engulfment requires F-actin assembly into a defined structure called a phagocytic cup. F-actin in phagocytic cups also recruits additional cytosolic proteins including myosin II and the actin regulatory protein annexin A2.[36,43–45]

The digestion phase is the least well understood of the three phases of POS phagocytosis. Electron micrographs suggest a multistep process of phagolysosomal maturation that includes both complete fusion of lysosomal vesicles with the POS phagosome and temporary contacts of lysosomal vesicles with POS phagosomes, presumably to deliver soluble content.[46] Lysosomal acidification by vacuolar adenosine triphosphate (ATP)-driven proton pumps is essential for phagolysosomal digestion, as in other forms of phagocytosis. Degradation of opsin, which is by far the most abundant protein of POS, requires the ubiquitously expressed aspartyl lysosomal protease, cathepsin D.[46–49] A recent study has further shown a role in lysosome-mediated POS digestion for melanoregulin (MREG). MREG is a small, highly charged protein thought to be involved in lysosome-related organelle biogenesis; however, its exact function remains to be identified.[50] Myosin VIIa is required for proper phagosome movement within the RPE, and phagosome–lysosome fusion is delayed in myosin VIIa knockout RPE, causing delayed clearance of ingested POS.[51] The specific molecules and conditions known to date to be important for the different phases of POS phagocytosis by RPE cells are summarized in Table 58.1.

PRIMARY RETINAL PIGMENT EPITHELIUM DEFECTS IN MICE LACKING αVβ5 INTEGRIN

Comparison of phagosome size and abundance at defined times after light onset in the RPE of age- and background-matched wild-type and β5 integrin knockout mice has shown that β5 integrin knockout RPE lacks the peak of phagosome burden after light onset that is characteristic of the RPE in wild-type mice. The POS phagosome content of wild-type RPE is greatest at 1 and 2 hours after light onset (Fig. 58.1, black bars). At time-points later than 3 hours after light onset, very few POS phagosomes reside in wild-type RPE, indicating that engulfment has ceased and that POS engulfed after light onset have been digested. In contrast, the experiment reveals no discernible peak of POS phagosome load at any time of day in the RPE of age- and strain-matched β5 integrin knockout mice (Fig. 58.1, gray bars). These data show that αvβ5 integrin receptors are essential for synchronizing the rhythm of POS phagocytosis. Notably, β5 integrin knockout RPE cells are capable of engulfing POS as they maintain a POS phagosome load that is above normal at any time-point other than immediately after light onset (Fig. 58.1).

Quantifying the amount of melanin pigment associated with detached retina in retinal isolation assays allows relative quantification of the strength of retinal adhesion.[52] The quantity of melanin in retinal samples directly correlates with RPE attachment to the outer retina. Melanin levels can further be correlated with the presence of RPE-specific proteins in retinal tissue fractions. Such experiments established that significantly less RPE melanin coisolates with β5 integrin knockout neural retina than with wild-type retina.[53] Retinal adhesion in both wild-type and β5 integrin knockout retina significantly decreases with age. However, retinal adhesion in mice lacking αvβ5 integrin was impaired to the same extent regardless of whether mice were studied at 2.5 months of age (in young adulthood) or at 12 months of age (in old age). These results suggest that deficiency of αvβ5 integrin impairs retinal adhesion acutely and directly rather than secondary to arrhythmic phagocytosis. Notably, weakened retinal adhesion does not cause visual impairment in wild-type mice at 12 months of age. Taken together, these studies imply that αvβ5 integrin receptors perform two independent functions at the apical surface of the RPE: synchronizing diurnal phagocytosis of shed POS and contributing to permanent retinal adhesion to intact outer segments.

TABLE 58.1 Molecules and Conditions Involved in Specific Phases of the Diurnal Phagocytosis of Photoreceptor Outer Segment Fragments (POS) by Retinal Pigment Epithelium Cells

Phase of POS Phagocytosis	Molecules/Conditions
Recognition and binding	αvβ5 integrin receptor,[28,31,24] integrin ligand MFG-E8,[26] CD81,[30] externalized phosphatidylserine[25]
Internalization	p125FAK,[35] MerTK,[36,37,39,41] annexin A2,[46] myosin II,[43] Rac1,[44] CD36 [31]
Digestion	Myosin VIIa,[51] cathepsin D,[46,47,49] lysosomal acidification,[46,49] melanoregulin[50]

Age-Related Accumulation of Pro-Oxidant Lipofuscin in Retinal Pigment Epithelium and Loss of Vision in Mice Lacking αVβ5 Integrin

Despite their two functional defects, neural retina and RPE in β5 integrin knockout mice up to 12 months of age do not reveal significant morphological differences from the same tissues in age-matched wild-type mice.[54] Careful morphometry and histological analysis do not reveal significant differences in outer or inner segment length or signs of cell death in β5 integrin knockout mice. Furthermore, scotopic electroretinograms (ERGs) reveal normal retinal responses to both high- and low-intensity white light flash, indicating normal function of photoreceptor rods and cones as well as retinal bipolar cells.[34,54] However, between 6 and 12 months of age the a-waves of both light- and dark-adapted ERGs dramatically decrease in β5 integrin knockout mice, indicating an age-dependent loss of function of both rod and cone photoreceptors in β5 integrin knockout mice.[34,54]

With age, postmitotic tissues such as the RPE tend to accumulate autofluorescent compounds in inclusion bodies that are contiguous with the cells' lysosomal organelles.[55] These compounds are commonly termed lipofuscin. However, it is important to bear in mind that RPE lipofuscin differs considerably from lipofuscin in other cell types. Extensive chemical analysis of RPE lipofuscin has revealed that it is comprised of a complex mix of retinoid derivatives and lipids that are oxidized to varying extent but very little protein.[56] This unique ocular lipofuscin forms and accumulates in the acidic environment of RPE lysosomes from incompletely digested phagocytosed POS.[57] Many RPE lipofuscin constituents are products of oxidative stress, and lipofuscin acts as a photosensitizer. This leads to a gradually and

continuously increasing oxidative burden on the RPE in the aging eye.[58] In β5 integrin knockout mice, detectable accumulation of lipofuscin starting around 6 months of age parallels the loss of photoreceptor function. By 12 months of age, both lipofuscin granule abundance per RPE cell and brightness of individual autofluorescent granules are dramatically elevated in β5 integrin knockout mice compared with age-matched wild-type mice (Fig. 58.2). The presence of the pyridinium bis-retinoid commonly called A2E indicates that the lipofuscin forming in aging β5 integrin knockout RPE stems from undigested POS and is highly similar to the lipofuscin accumulating in the aging human eye.[59] A2E, its derivatives, and oxidative degradation products have proven to be toxic to RPE cells in culture in numerous ways. For instance, even low levels of A2E decrease resistance to light stress of RPE cells in culture.[60,61] Lipofuscin may also contribute to local and/or systemic inflammation, although the cellular mechanisms involved remain largely obscure.[62] Furthermore, accumulation of A2E at physiologic levels in lysosomes of RPE cells in culture specifically slows digestion of the lipid component of phagocytosed POS, while opsin protein digestion, POS recognition, and engulfment remain unaffected.[63] Finally, A2E impairs the activity of the mitochondrial respiratory chain and thus the efficiency of ATP synthesis by increasing cytoplasmic oxidative stress independently of photic stress.[64]

Effects of Antioxidant-Enriched Diets

RPE lipofuscin in general and A2E specifically are both products and producers of oxidative stress, exacerbating oxidative burden on cells in a positive feedback loop. Elevated levels of lipofuscin in the RPE thus provide evidence of the increased oxidative burden on the RPE and retina

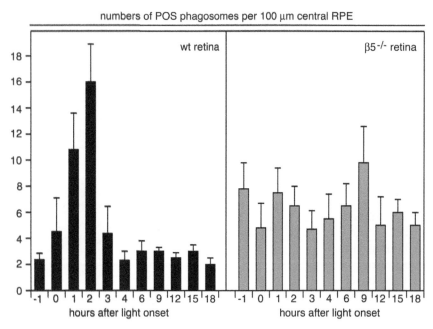

numbers of POS phagosomes per 100 μm central RPE

wt retina

β5⁻/⁻ retina

hours after light onset

hours after light onset

FIGURE 58.1 β5 Integrin knockout retinal pigment epithelium (RPE) *in situ* lacks the diurnal peak of photoreceptor outer segment (POS) clearance. Bars show the average number of phagosomes (±SD) detected by electron microscopy in sections obtained from RPE of 2-month-old wild-type (wt) mice (black bars) or age- and strain-matched β5 integrin knockout (β5⁻/⁻) mice (gray bars) killed at different time-points before and after light onset, as indicated. *Source:* © *Nandrot et al. J Exp Med. 2004;200:1539–45,*[34] *with permission.*

of aging β5 integrin knockout mice. The present authors recently completed a research project that directly tested the effects of dietary supplementation with antioxidants on RPE lipofuscin content and on retinal and RPE function in β5 integrin knockout mice. Two distinct dietary regimens were designed to provide increased antioxidant content as part of the regular mouse chow. One regimen compared consumption of standard rodent chow #5053 supplemented with 1% w/w FloraGLO® natural marigold extract (lutein diet) with consumption of standard rodent chow #5053 (control diet). FloraGLO contains 5% w/w lutein and 0.2% w/w zeaxanthin, the two xanthophylls that are highly and specifically enriched in the human macula.[65] Macular lutein and zeaxanthin protect retinal functionality by acting as antioxidants.[66] These compounds also protect against light damage and contribute to cellular membrane organization and signal transduction.[67] Mice fed the lutein diet consumed on average 52mg lutein and 2mg zeaxanthin/kg body weight. The other regimen compared consumption of standard rodent chow enriched with 1.5% w/w freeze-dried grapes (grape diet) and standard rodent chow enriched in glucose and fructose to reach an equal sugar content to the grape diet (sugar diet). Freeze-dried grapes retain the mix of natural antioxidants such as resveratrol, flavans, flavonols, anthocyanins, and simple phenolics that are present in fresh grapes. *In vitro* and *in vivo* studies have revealed that catechins and anthocyanins can inhibit lipid peroxidation and may scavenge superoxide and hydroxyl radicals in cells.[68–70] Furthermore, resveratrol, a stilbene antioxidant abundant in grapes, acts on numerous tissues to combat aging, reduces the risk of cardiovascular disease, and improves immune cell functions.[71–73] The production of diet pellets avoided prolonged or excessive heat or irradiation to ensure the full potency of additives. Details of diet preparation are provided in Yu *et al.*[54] All diets were fed *ad libitum* to mice starting at 3 months of age, which is equivalent to young adulthood.

To determine the effects of the antioxidant-enriched diets on oxidative burden specifically of the RPE, lipofuscin content was measured in RPE tissue of mice fed with the four diets continuously from 3 to 12 months of age. Consumption of the grape diet and, to a lesser extent, the lutein diet significantly reduced the accumulation of autofluorescent compounds in the RPE of β5 integrin knockout mice (Fig. 58.3A). Furthermore, the level of the toxic A2E component of lipofuscin was reduced to the level accumulating in wild-type mice if

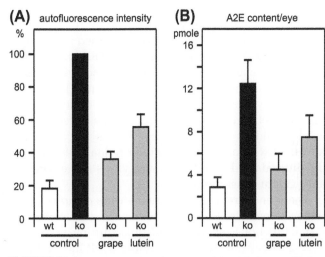

FIGURE 58.3 Consumption of an antioxidant-rich diet is sufficient to prevent lipofuscin accumulation in β5 integrin knockout retinal pigment epithelium (RPE). Bars show relative autofluorescence (A) and absolute levels of the RPE lipofuscin component A2E (B) of the RPE *in situ* of 12-month-old wild-type (wt) mice fed with control diet (white bar) or β5 integrin knockout (ko) mice fed with control diet (black bar) or with diets enriched in natural lutein/zeaxanthin or grapes (gray bars). Mice consumed respective diets *ad libitum* continuously from 3 to 12 months of age. Bars show averages ±SD. Either antioxidant-enriched diet significantly reduced autofluorescence and A2E content (P<0.05). *Source:* © *Yu* et al. Free Rad Biol Med. 2012;52:660–70,[54] *with permission.*

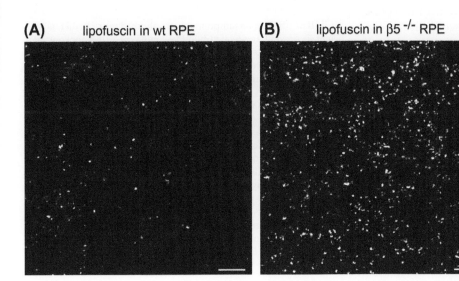

FIGURE 58.2 β5 Integrin knockout retinal pigment epithelium (RPE) *in situ* accumulates excess numbers of autofluorescent lipofuscin granules. Images show cytoplasmic autofluorescent granules (excitation 488nm, emission 580–670nm) in representative fields of whole mounts of RPE obtained from 1-year-old wild-type (wt) (A) or β5 integrin (β5−/−) knockout mice (B). *Source:* © *Yu* et al. Free Rad Biol Med. 2012;52:660–70,[54] *with permission.*

β5 integrin knockout mice consumed the grape diet and was also significantly reduced if β5 integrin knockout mice consumed the lutein diet (Fig. 58.3B).

An initial study of visual function recorded monthly ERGs from 6 to 12 months of age of the four cohorts of age-matched β5 integrin knockout mice fed the different diets. As expected, mice consuming the regular diet or sugar diet progressively lost photoreceptor function as recorded by scotopic ERGs. Twelve-month-old β5 integrin knockout mice fed the control diet exhibited severely impaired cone and rod photoreceptor responses to high-intensity light flashes, causing a dramatic decline in a- and b-wave amplitudes (Fig. 58.4). At the same age, mice that consumed the lutein diet retained robust light responses, with a distinct a-wave (Fig. 58.4A). Similarly, mice that consumed the grape diet retained visual function, whereas those on the sugar diet did not (Fig. 58.4B). Importantly, the efficacy of antioxidant enrichment on visual function of the mutant mice as measured by ERG is independent of the primary defects in β5 integrin knockout RPE or retina, as neither retinal adhesion nor the diurnal rhythm of POS phagocytosis are restored by antioxidant consumption.[54]

These encouraging experiments were complemented with a second study asking whether a grape diet benefits visual function even if only consumed during a restricted period rather than during the entire adult life. To better discern effects on functionality of rod versus cone photoreceptors, scotopic ERGs were recorded with dim white light flashes, which specifically activate rods, and photopic ERGs with high-intensity white light flashes, which specifically activate cones. β5 integrin knockout mice received grape diets throughout adulthood as before (from 3 to 12 months of age), or only during young adulthood (from 3 to 6 months of age), mid-adulthood but before onset of vision loss (from 6 to 9 months of age), or during old age after the onset of vision loss (from 9 to 12 months of age). Comparison of cone a-wave amplitudes recorded from mice of all feeding groups at 12 months of age shows that grape consumption only during young and mid-adulthood significantly improved cone photoreceptor activity in old age, even if the diet was discontinued in old age (Fig. 58.4C). Similarly, mice consuming the grape diet in young or mid-adulthood only retained markedly greater rod activity than mice consuming the sugar diet or grape diet during old age only (Fig. 58.4D). However, grape consumption only during old age after deterioration of vision had no effects on cone or rod function (Fig. 58.4C, D). It is important to note, however, that continuous grape consumption from 3 to 12 months of age yielded the strongest protective effect for both rods and cones (Fig. 58.4C, D). Taken together, the results show that consumption of an antioxidant-rich diet is sufficient to significantly defer the age-dependent impairment of photoreceptor activity in β5 integrin

knockout mice even if the protective diet is consumed only during early or mid-adulthood, before the onset of vision loss. In contrast, a protective diet does not benefit visual function if consumed only during old age.

ACTIN CYTOSKELETAL DAMAGE IN RETINAL PIGMENT EPITHELIUM CELLS

Increased oxidative stress due to accumulation of A2E-containing lipofuscin is likely to cause oxidative post-translational modifications of both lipids and protein

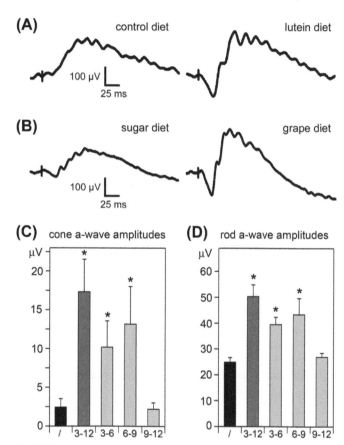

FIGURE 58.4 Consumption of an antioxidant-rich diet is sufficient to prevent age-related blindness in β5 integrin knockout mice even if antioxidants are only consumed during either young or mid-adulthood and discontinued in old age. Scotopic and photopic electroretinograms were recorded from 12-month-old β5 integrin knockout mice. Representative scotopic individual responses to high-intensity white light flashes are shown for mice fed from 3 to 12 months of age with control diet and lutein-/zeaxanthin-enriched diet (A) or with diet enriched in sugar as contained in grapes or diet enriched with grapes (B). Mean photopic (C) and scotopic (D) a-wave amplitudes (±SD) quantified from groups of mice fed with sugar diet (/, black bar) or with a grape-enriched diet either continuously from 3 to 12 months of age (3–12, dark gray bar), or for a period of 3 months only (light gray bars), from 3 to 6 (3–6), 6 to 9 (6–9), or 9 to 12 (9–12) months of age. Mice receiving the grape-enriched diet for 3 months were fed sugar control diet at all other times. Asterisks indicate significant improvement in light responses by mice fed throughout with a sugar diet. *Source: © Yu et al. Free Rad Biol Med. 2012;52:660–70,[54] with permission.*

components in aging retina or RPE. Such protein damage has been well documented for the RPE and retina in the aging human eye, and it is particularly prominent in human RPE and retina from donors with age-related macular degeneration (AMD).[60,74] 4-Hydroxynonenal (HNE) is an α,β-unsaturated hydroxyalkenal that is a product of peroxidation of cellular lipids in response to oxidative stress. HNE is highly reactive and covalently modifies proteins, yielding HNE adducts. Quantification of HNE adducts in homogenates obtained from wild-type and β5 integrin knockout mouse eyes detects only low levels of modified proteins in mice at 6 months of age regardless of genotype. However, the HNE adduct content in eyes of 12-month-old β5 integrin knockout mice is 3.5-fold higher than in eyes of age-matched wild-type mice. Notably, HNE adducts form in the RPE but not in the neural retina.[54]

It is generally assumed that oxidative modifications lead to dysfunction of cellular macromolecules and cause cell death if excessive. Yet, relatively little is known about lipids or proteins that are particularly vulnerable to sublethal oxidation causing specific cellular dysfunction prior to cell death. Directed mobility of organelles in RPE cells in culture is highly sensitive to low levels of oxidative stress, suggesting that cytoskeletal functionality may be reduced.[75] Indeed, actin, the building block of the F-actin (microfilament) cytoskeleton, is prominently modified by HNE in the RPE of aged β5 integrin knockout mice but not in the RPE of young mice or in aged wild-type mice.[54] HNE modification does not alter total cellular actin protein content in aging β5 integrin knockout RPE but significantly decreases its stability (Fig. 58.5). HNE–actin forms and the actin cytoskeleton is destabilized acutely upon incubation with sublethal concentrations of HNE, which generates HNE modified proteins in RPE cells in culture at the same level as that measured in the aging

β5 integrin knockout RPE.[54] This damage due to protein oxidation in β5 integrin knockout RPE is specific to the F-actin cytoskeleton as modification of tubulin, the building block of the microtubule cytoskeletal system, remains below levels of detection.[54] Likewise, all other proteins tested, the integrin subunit αv, CD36, CD81, the essential POS engulfment receptor MerTK, and the visual cycle isomerase enzyme RPE65, do not harbor detectable levels of HNE modification.[54] Taken together, formation or maintenance of F-actin that yields the essential microfilament cytoskeleton is specifically and significantly destabilized by HNE adduct formation in aging β5 integrin knockout RPE.

Like vision loss and lipofuscin accumulation, age-related formation of HNE adducts is significantly reduced in the RPE of β5 integrin knockout mice that consume an antioxidant-enriched diet supplemented with either lutein/zeaxanthin or grapes.[54] The same holds true for levels of protein carbonylation, a different type of post-translational, covalent protein modification that is also caused by increased levels of cellular oxidative stress.[54] Consuming an antioxidant-enriched diet is also sufficient to prevent actin modification by HNE and F-actin microfilament weakening (Fig. 58.5). Notably, providing RPE cells in culture during or immediately preceding HNE incubation with either the vitamin E derivative trolox or vitamin C (ascorbate) fails to protect actin from modification by synthetic HNE or from destabilization.[54] These findings further strengthen the causal chain from oxidative stress, through formation of reactive HNE and HNE–actin adduct formation, to F-actin cytoskeleton damage. They also clarify that the presence of antioxidants does not affect the reactivity of HNE in a scenario where reactive HNE is formed or supplied experimentally. However, consumption of a diet rich in antioxidants is clearly effective in reducing

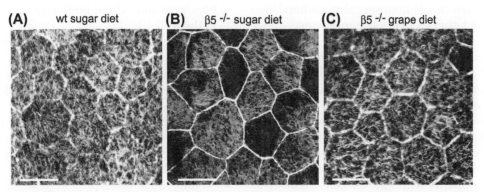

FIGURE 58.5 Consumption of an antioxidant-enriched diet is sufficient to prevent cytoskeletal destabilization in β5 integrin knockout retinal pigment epithelium (RPE) cells. Images show fluorescent phalloidin labeling of stable F-actin cytoskeleton in RPE whole-mount samples obtained from 12-month-old wild-type (wt) mice on sugar control diet (A), β5 integrin knockout (β5−/−) mice on sugar control diet (B), and β5 integrin knockout mice on diet enriched with grapes (C). Mice consumed respective diets *ad libitum* continuously from 3 to 12 months of age. Images show representative three-dimensional projections of image stacks. Scale bars: 20 μm. *Source: © Yu et al. Free Rad Biol Med. 2012;52:660–70,[54] with permission.*

oxidative stress at the cellular level and as a result formation of reactive HNE, which in turn lowers the incidence of HNE adduct formation.

CONCLUSIONS

It has long been hypothesized that antioxidant consumption may defer age-related human diseases that are associated with increased oxidative stress. Over the past several years, a number of clinical studies have tested whether dietary antioxidants may be beneficial for human patients suffering from age-related diseases involving oxidative stress, including neurodegenerative diseases. Among the largest of these, the Age-Related Eye Disease Study (AREDS) was sponsored by the US National Institutes of Health to determine in a multicenter clinical trial whether controlled antioxidant intake affects the course or onset of age-related eye disease.[76] AREDS focused specifically on cataracts and on AMD. High doses of vitamins C and E and zinc were found to result in a moderately decreased risk of progression to severe vision loss among patients afflicted with at least intermediate-stage AMD but showed no effect on disease onset or progression of early forms.[76] Advances in understanding of the cell biology and physiology of the human macula and retina suggest that different types of antioxidant compound may exert more powerful benefits than the original AREDS formulation. Therefore, follow-up clinical trials, most prominently AREDS-2, are underway.[77] AREDS-2 seeks to determine the effects of intake of dietary supplements containing lutein and zeaxanthin, among others, on the progression to advanced AMD. AREDS-2 is currently following thousands of patients enrolled at age 50 and older over a period of 5–6 years. The first results were recently published.[78] Smaller studies with similar objectives have also focused on providing dietary supplements to elderly people and reported modest effects of lutein supplementation.[79]

Human subject studies like those mentioned above are tremendously costly and difficult to conduct and to interpret. In contrast, dietary interventions can be rapidly tested in appropriate, well-controlled experimental animal models that represent relevant aspects of human blinding disease. The β5 integrin knockout mouse provides an experimental model for late-onset, progressive loss of photoreceptor function due to elevated oxidative stress in the RPE. While rod and cone photoreceptors in β5 integrin knockout retina lose functionality with age, the cells themselves largely survive, which is an important prerequisite for the design of rescue strategies. The age-related, slowly progressing loss of function and pro-oxidant lipofuscin accumulation in β5 integrin knockout retina presents an opportunity to test preventive as well as restorative therapies in a well-controlled model system that exhibits two of the cardinal features of human retina afflicted by AMD.

Enrichment of a standard diet with lutein/zeaxanthin or with grapes throughout adult life has proven to be largely sufficient to prevent the dramatic loss of cone and rod photoreceptor function in β5 integrin knockout mice. Mice consumed antioxidants as part of their regular diet rather than as a single daily supplement as commonly tested in human subject trials. The clear efficacy shown in this mouse study emphasizes the benefits specifically for vision of a healthy, naturally antioxidant-rich diet. Moreover, the visual function of β5 integrin knockout mice benefits even if mice consume an antioxidant-rich diet only during young- or mid-adulthood and discontinue antioxidant consumption in old age. Yet, vision does not benefit if mice commence antioxidant consumption only late in life when lipofuscin and oxidative damage to the RPE are already apparent. Directly comparing long-term and short-term effects of healthy diets in a similar study of human patients would be fraught with difficulty. There are significant differences in physiology between mice and humans. However, the β5 integrin knockout mouse study suggests that natural antioxidants consumed as part of the normal diet throughout one's lifetime can, in principle, prevent oxidative damage to the RPE and subsequent loss of vision.

TAKE-HOME MESSAGES

- Retinal pigment epithelial (RPE) cells use the integrin receptor αvβ5 to stimulate the diurnal burst of outer-segment phagocytosis and to strengthen retinal adhesion.
- Secondary to loss of the phagocytic rhythm, mice lacking αvβ5 integrin develop age-related blindness and exhibit lipofuscin accumulation in the RPE.
- An antioxidant-enriched diet is efficacious in preventing age-related RPE damage and blindness in mice lacking αvβ5 integrin, even if consumed only during young or mid-adulthood and discontinued in old age.
- Consumption of an antioxidant-enriched diet has no effect on age-related blindness in mice lacking αvβ5 integrin if consumed only immediately before and during old age.
- At the cellular level, protein oxidation is more pronounced in the RPE than in the neural retina in aging mice lacking αvβ5 integrin.
- A major target of oxidative damage in the RPE of mice lacking αvβ5 integrin is the F-actin cytoskeleton, which is destabilized by HNE modification.

References

1. Guymer R, Luthert P, Bird A. Changes in Bruch's membrane and related structures with age. *Prog Ret Eye Res* 1999;**18**:59–90.

2. Wimmers S, Karl MO, Strauss O. Ion channels in the RPE. *Prog Ret Eye Res* 2007;**26**:263–301.

3. Gundersen D, Orlowski J, Rodriguez-Boulan E. Apical polarity of Na, K-ATPase in retinal pigment epithelium is linked to a reversal of the ankyrin–fodrin submembrane cytoskeleton. *J Cell Biol* 1991;**112**:863–72.

4. Philp NJ, Wang D, Yoon H, Hjelmeland LM. Polarized expression of monocarboxylate transporters in human retinal pigment epithelium and ARPE-19 cells. *Invest Ophthalmol Vis Sci* 2003;**44**:1716–21.

5. Strauss O. The retinal pigment epithelium in visual function. *Physiol Rev* 2005;**85**:845–81.

6. Rózanowski B, Burke JM, Boulton ME, Sarna T, Rózanowska M. Human RPE melanosomes protect from photosensitized and iron-mediated oxidation but become pro-oxidant in the presence of iron upon photodegradation. *Invest Ophthalmol Vis Sci* 2008;**49**:2838–47.

7. Burke JM, Kaczara P, Skumatz CM, Zareba M, Raciti MW, Sarna T. Dynamic analyses reveal cytoprotection by RPE melanosomes against non-photic stress. *Mol Vis* 2011;**17**:2864–77.

8. Boulton M, Dayhaw-Barker P. The role of the retinal pigment epithelium: topographical variation and ageing changes. *Eye* 2001;**15**:384–9.

9. Moiseyev G, Chen Y, Takahashi Y, Wu BX, Ma JX. RPE65 is the isomerohydrolase in the retinoid visual cycle. *Proc Natl Acad Sci U S A* 2005;**102**:12413–8.

10. Redmond TM, Yu S, Lee E, Bok D, Hamasaki D, Chen N, et al. RPE65 is necessary for production of 11-*cis*-vitamin A in the retinal visual cycle. *Nat Genet* 1998;**20**:344–51.

11. Hargrave PA, McDowell JH. Rhodopsin and phototransduction: a model system for G protein-linked receptors. *FASEB J* 1992;**6**:2323–31.

12. Archer S. Molecular biology of visual pigments. In: Djamgoz MBA, Archer SN, Vallerga S, editors. *Neurobiology and Clinical Aspects of the Outer Retina*. London: Chapman & Hall; 1995. pp. 74–104.

13. Steinberg RH, Fisher SK, Anderson DH. Disc morphogenesis in vertebrate photoreceptors. *J Comp Neurol* 1980;**190**:501–18.

14. Young RW. The renewal of photoreceptor cell outer segments. *J Cell Biol* 1967;**33**:61–72.

15. Young RW. The daily rhythm of shedding and degradation of rod and cone outer segment membranes in the chick retina. *Invest Ophthalmol Vis Sci* 1978;**17**:105–16.

16. Anderson DH, Fisher SK, Steinberg RH. Mammalian cones: disc shedding, phagocytosis, and renewal. *Invest Ophthalmol Vis Sci* 1978;**17**:117–33.

17. Young R, Bok D. Participation of the retinal pigment epithelium in the rod outer segment renewal process. *J Cell Biol* 1969;**42**:392–403.

18. LaVail MM. Rod outer segment disc shedding in relation to cyclic lighting. *Exp Eye Res* 1976;**23**:277–80.

19. LaVail MM. Circadian nature of rod outer segment disc shedding in the rat. *Invest Ophthalmol Vis Sci* 1980;**19**:407–11.

20. Steinberg RH, Wood I, Hogan MJ. Pigment epithelial ensheathment and phagocytosis of extrafoveal cones in human retina. *Philos Trans R Soc Lond B Biol Sci* 1977;**277**:459–74.

21. Young RW. An hypothesis to account for a basic distinction between rods and cones. *Vision Res* 1971;**11**:1–5.

22. Young RW. The renewal of rod and cone outer segments in the rhesus monkey. *J Cell Biol* 1971;**49**:303–18.

23. Jeon CJ, Strettoi E, Masland RH. The major cell populations of the mouse retina. *J Neurosci* 1998;**18**:8936–46.

24. Sethna S, Finnemann SC. Analysis of photoreceptor rod outer segment phagocytosis by RPE cells *in situ*. In: Weber BHF, Langmann T, editors. *Retinal Degeneration: Methods and Protocols*. New York: Humana Press; 2013. pp. 245–54.

25. Ruggiero L, Connor MP, Chen J, Langen R, Finnemann SC. Diurnal, localized exposure of phosphatidylserine by rod outer segment tips in wild-type but not Itgb5$^{-/-}$ or Mfge8$^{-/-}$ mouse retina. *Proc Natl Acad Sci U S A* 2012;**109**:8145–8.

26. Nandrot EF, Anand M, Almeida D, Atabai K, Sheppard D, Finnemann SC. Essential role for MFG-E8 as ligand for αvβ5 integrin in diurnal retinal phagocytosis. *Proc Natl Acad Sci U S A* 2007;**104**:12005–10.

27. Hanayama R, Tanaka M, Miwa K, Shinohara A, Iwamatsu A, Nagata S. Identification of a factor that links apoptotic cells to phagocytes. *Nature* 2002;**417**:182–7.

28. Finnemann SC, Bonilha VL, Marmorstein AD, Rodriguez-Boulan E. Phagocytosis of rod outer segments by retinal pigment epithelial cells requires αvβ5 integrin for binding but not for internalization. *Proc Natl Acad Sci U S A* 1997;**94**:12932–7.

29. Anderson DH, Johnson LV, Hageman GS. Vitronectin receptor expression and distribution at the photoreceptor–retinal pigment epithelial interface. *J Comp Neurol* 1995;**360**:1–16.

30. Chang Y, Finnemann SC. Tetraspanin CD81 is required for the αvβ5 integrin-dependent particle-binding step of RPE phagocytosis. *J Cell Sci* 2007;**120**:3053–63.

31. Finnemann SC, Silverstein RL. Differential roles of CD36 and αvβ5 integrin in photoreceptor phagocytosis by the retinal pigment epithelium. *J Exp Med* 2001;**194**:1289–98.

32. Lin H, Clegg D. Integrin αvβ5 participates in the binding of photoreceptor rod outer segments during phagocytosis by cultured human retinal pigment epithelium. *Invest Ophthalmol Vis Sci* 1998;**39**:1703–12.

33. Miceli MV, Newsome DA, Tate Jr DJ. Vitronectin is responsible for serum-stimulated uptake of rod outer segments by cultured retinal pigment epithelial cells. *Invest Ophthalmol Vis Sci* 1997;**38**:1588–97.

34. Nandrot EF, Kim Y, Brodie SE, Huang X, Sheppard D, Finnemann SC. Loss of synchronized retinal phagocytosis and age-related blindness in mice lacking αvβ5 integrin. *J Exp Med* 2004;**200**:1539–45.

35. Finnemann SC. Focal adhesion kinase signaling promotes phagocytosis of integrin-bound photoreceptors. *EMBO J* 2003;**22**:4143–54.

36. Chaitin MH, Hall MO. The distribution of actin in cultured normal and dystrophic rat pigment epithelial cells during the phagocytosis of rod outer segments. *Invest Ophthalmol Vis Sci* 1983;**24**:821–31.

37. Nandrot E, Dufour EM, Provost AC, Péquignot MO, Bonnel S, Gogat K, et al. Homozygous deletion in the coding sequence of the c-mer gene in RCS rats unravels general mechanisms of physiological cell adhesion and apoptosis. *Neurobiol Dis* 2000;**7**:586–99.

38. Bourne MC, Campbell DA, Tansley K. Hereditary degeneration of the rat retina. *Br J Ophthalmol* 1938;**22**:613–23.

39. D'Cruz PM, Yasumura D, Weir J, Matthes MT, Abderrahim H, LaVail MM, Vollrath D. Mutation of the receptor tyrosine kinase gene Mertk in the retinal dystrophic RCS rat. *Hum Mol Genet* 2000;**9**:645–51.

40. Duncan JL, LaVail MM, Yasumura D, Matthes MT, Yang H, Trautmann N, et al. An RCS-like retinal dystrophy phenotype in Mer knockout mice. *Invest Ophthalmol Vis Sci* 2003;**44**:826–38.

41. Mullen R, LaVail M. Inherited retinal dystrophy: primary defect in pigment epithelium determined with experimental rat chimeras. *Science* 1976;**192**:799–801.

42. Gal A, Li Y, Thompson DA, Weir J, Orth U, Jacobson SG, et al. Mutations in MerTK, the human orthologue of the RCS rat retinal dystrophy gene, cause retinitis pigmentosa. *Nat Genet* 2000;**26**:270–1.

43. Strick DJ, Feng W, Vollrath D. Mertk drives myosin II redistribution during retinal pigment epithelial phagocytosis. *Invest Ophthalmol Vis Sci* 2009;**50**:2427–35.

44. Mao Y, Finnemann SC. Essential diurnal Rac1 activation during retinal phagocytosis requires αvβ5 integrin but not tyrosine kinases focal adhesion kinase or Mer tyrosine kinase. *Mol Biol Cell* 2012;**23**:1104–14.

45. Law AL, Ling Q, Hajjar KA, Futter CE, Greenwood J, Adamson P, et al. Annexin A2 regulates phagocytosis of photoreceptor outer segments in the mouse retina. *Mol Biol Cell* 2009;**20**:3896–904.

46. Bosch E, Horwitz J, Bok D. Phagocytosis of outer segments by retinal pigment epithelium: phagosome–lysosome interaction. *J Histochem Cytochem* 1993;**41**:253–63.

47. Rakoczy PE, Lai CM, Baines M, Di Grandi S, Fitton JH, Constable IJ. Modulation of cathepsin D activity in retinal pigment epithelial cells. *Biochem J* 1997;**324**:935–40.

48. Deguchi J, Yamamoto A, Yoshimori T, Sugasawa K, Moriyama Y, Futai M, et al. Acidification of phagosomes and degradation of rod outer segments in rat retinal pigment epithelium. *Invest Ophthalmol Vis Sci* 1994;**35**:568–79.

49. Regan CM, de Grip WJ, Daemen FJM, Bonting SL. Degradation of rhodopsin by a lysosomal fraction of retinal pigment epithelium: biochemical aspects of the visual process. XLI. *Exp Eye Res* 1980;**30**:183–91.

50. Damek-Poprawa M, Diemer T, Lopes VS, Lillo C, Harper DC, Marks MS, et al. Melanoregulin (MREG) modulates lysosome function in pigment epithelial cells. *J Biol Chem* 2009;**284**:10877–89.

51. Gibbs D, Kitamoto J, Williams DS. Abnormal phagocytosis by retinal pigmented epithelium that lacks myosin VIIa, the Usher syndrome 1B protein. *Proc Natl Acad Sci U S A* 2003;**100**:6481–6.

52. Endo EG, Yao XY, Marmor MF. Pigment adherence as a measure of retinal adhesion: dependence on temperature. *Invest Ophthalmol Vis Sci* 1988;**29**:1390–6.

53. Nandrot EF, Anand M, Sircar M, Finnemann SC. Novel role for αvβ5 integrin in retinal adhesion and its diurnal peak. *Am J Physiol Cell Physiol* 2006;**290**:C1256–62.

54. Yu C-C, Nandrot EF, Dun Y, Finnemann SC. Dietary antioxidants prevent age-related retinal pigment epithelium actin damage and blindness in mice lacking αvβ5 integrin. *Free Rad Biol Med* 2012;**52**:660–70.

55. Chen H, Lukas T, Du N, Suyeoka G, Neufield A. Dysfunction of retinal pigment epithelium with age: increased iron decreases phagocytosis and lysosomal activity. *Invest Ophthalmol Vis Sci* 2009;**50**:1895–902.

56. Ng KP, Gugiu B, Renganathan K, Davies MW, Gu X, Crabb JS, et al. Retinal pigment epithelium lipofuscin proteomics. *Mol Cell Proteomics* 2008;**7**:1397–405.

57. Brunk UT, Terman A. Lipofuscin: mechanisms of age-related accumulation and influence on cell function. *Free Rad Biol Med* 2002;**33**:611–9.

58. Sparrow JR, Gregory-Roberts E, Yamamoto K, Blonska A, Ghosh SK, Ueda K, Zhou J. The bisretinoids of retinal pigment epithelium. *Prog Ret Eye Res* 2012;**31**:121–35.

59. Liu J, Itagaki Y, Ben-Shabat S, Nakanishi K, Sparrow JR. The biosynthesis of A2E, a fluorophore of aging retina, involves the formation of the precursor, A2-PE, in the photoreceptor outer segment membrane. *J Biol Chem* 2000;**275**:29354–60.

60. Sparrow J, Boulton M. RPE lipofuscin and its role in retinal pathobiology. *Exp Eye Res* 2005;**80**:595–606.

61. Roberts JE, Kukielczak BM, Hu DN, Miller DS, Bilski P, Sik RH, et al. The role of A2E in prevention or enhancement of light damage in human retinal pigment epithelial cells. *Photochem Photobiol* 2002;**75**:184–90.

62. Zhou J, Jang YP, Kim SR, Sparrow JR. Complement activation by photooxidation products of A2E, a lipofuscin constituent of the retinal pigment epithelium. *Proc Natl Acad Sci U S A* 2006;**103**:16182–7.

63. Finnemann S, Leung L, Rodriguez-Boulan E. The lipofuscin component of A2E selectively inhibits phagolysosomal degradation of photoreceptor phospholipid by retinal pigment epithelium. *Proc Natl Acad Sci U S A* 2002;**99**:3842–7.

64. Vives-Bauza C, Anand M, Shirazi AK, Magrane J, Gao J, Vollmer-Snarr HR, et al. The age lipid A2E and mitochondrial dysfunction synergistically impair phagocytosis by retinal pigment epithelial cells. *J Biol Chem* 2008;**283**:24770–80.

65. Bone RA, Landrum JT, Tarsis SL. Preliminary identification of the human macular pigment. *Vision Res* 1985;**25**:1531–5.

66. Sommer A, Vyas KS. A global clinical view on vitamin A and carotenoids. *Am J Clin Nutr* 2012;**96**:1204–6S.

67. Lien EL, Hammond BR. Nutritional influences on visual development and function. *Prog Retin Eye Res* 2011;**30**:188–203.

68. Chidambara Murthy KN, Singh RP, Jayaprakasha GK. Antioxidant activities of grape (*Vitis vinifera*) pomace extracts. *J Agric Food Chem* 2002;**50**:5909–14.

69. Shafiee M, Carbonneau M-A, Urban N, Descomps B, Leger CL. Grape and grape seed extract capacities at protecting LDL against oxidation generated by Cu^{2+}, AAPH or SIN-1 and at decreasing superoxide THP-1 cell production. A comparison to other extracts or compounds. *Free Rad Res* 2003;**37**:573–84.

70. Ueda T, Armstrong D. Preventive effect of natural and synthetic antioxidants on lipid peroxidation in the mammalian eye. *Ophthalmic Res* 1996;**28**:198–202.

71. Falchetti R, Fuggetta M, Lanzilli G, Tricarico M, Ravagnan G. Effects of resveratrol on human immune cell function. *Life Sci* 2001;**70**:81–96.

72. Folts J. Potential health benefits from the flavonoids in grape products on vascular disease. *Adv Exp Med Biol* 2002;**505**:95–111.

73. Martin AR, Villegas I, La Casa C, de la Lastra CA. Resveratrol, a polyphenol found in grapes, suppresses oxidative damage and stimulates apoptosis during early colonic inflammation in rats. *Biochem Pharmacol* 2004;**67**:1399–410.

74. Wu Y, Yanase E, Feng X, Siegel MM, Sparrow JR. Structural characterization of bisretinoid A2E photocleavage products and implications for age-related macular degeneration. *Proc Natl Acad Sci U S A* 2010;**107**:7275–80.

75. Burke JM, Zareba M. Sublethal photic stress and the motility of RPE phagosomes and melanosomes. *Invest Ophthalmol Vis Sci* 2009;**50**:1940–7.

76. The AREDS group. A randomized, placebo-controlled, clinical trial of high-dose supplementation with vitamins C and E, beta carotene, and zinc for age-related macular degeneration and vision loss: AREDS report no. 8. *Arch Ophthalmol* 2001;**119**:1417–36.

77. The AREDS group. The Age-Related Eye Disease Study 2 (AREDS2): study design and baseline characteristics. AREDS2 Report No. 1. *Ophthalmology* 2012;**26**:26.

78. Age-Related Eye Disease Study 2 Research Group. Lutein + zeaxanthin and omega-3 fatty acids for age-related macular degeneration: the Age-Related Eye Disease Study 2 (AREDS2) randomized clinical trial. *JAMA* 2013;**309**:2005–15.

79. Berrow EJ, Bartlett HE, Eperjesi F, Gibson JM. The effects of a lutein-based supplement on objective and subjective measures of retinal and visual function in eyes with age-related maculopathy – a randomised controlled trial. *Br J Nutr* 2013;**109**:2008–14.

Nutrition, Diet, the Eye, and Vision: Molecular Aspects of Vitamin A Binding Proteins and Their Importance in Vision

H.A. Tajmir-Riahi, Daniel Agudelo, Philippe Bourassa

Department of Chemistry–Biology, University of Québec at Trois-Rivières, Trois-Rivières, Québec, Canada

INTRODUCTION

Vitamin A is ingested from dietary sources as a retinyl ester or synthesized from β-carotene and is stored in the liver as a retinyl ester until it is mobilized for delivery to various target tissues. Retinol (Scheme 59.1) is one of the forms of vitamin A obtained from foods of animal origin. Retinal (retinaldehyde), the aldehyde derived from retinol, is essential for vision, while retinoic acid (Scheme 59.1) is essential for skin health and bone growth.[1,2]

Retinol and retinoic acid are related to vitamin A, which is an essential micronutrient that plays a key role in vision, cell growth, and differentiation.[1,2] *In vivo*, retinoids must bind with specific proteins to perform their necessary functions. Plasma retinol-binding protein and epididymal retinoic acid-binding protein carry retinoids in bodily fluids, while cellular retinol-binding proteins and cellular retinoic acid-binding proteins carry retinoids within cells.[3–7] Even though all of these transport proteins possess similar structures, the modes of binding for the different retinoids with their carrier proteins are different.[2] This chapter reviews the binding sites of retinol and retinoic acid transport proteins such as human and bovine serum albumins (HSA and BSA), milk β-lactoglobulin, and α- and β-caseins, using different spectroscopic methods and molecular modeling. The results show that in the same family of proteins and subcellular location, the orientation of a retinoid molecule within a binding protein is the same, whereas when different families of proteins are considered, the orientation of the bound retinoid is completely different. In addition, none of

the amino acid residues involved in ligand binding are conserved between the transport proteins. However, for each specific binding protein, the amino acids involved in the ligand binding are conserved. The results of this study allow a possible transport model for retinoids to be proposed, which plays a major role in eye function and vision.

ANALYTICAL METHODS

Fourier Transform Infrared Spectroscopy

Retinoid binding modes to different amino acids with hydrophobic and hydrophilic contacts are determined using infrared spectroscopy. Infrared spectroscopy with its derivative methods is also used to quantify the protein conformational changes upon ligand interaction.[8–12]

Circular Dichroism Spectroscopy

Circular dichroism (CD) spectroscopy is a powerful tool for the analysis of protein conformational changes upon ligand complexation.[13,14] CD spectroscopy is applied to quantify the protein secondary structural changes such as α-helix, β-sheet, turn, and random coil structures.[15]

Fluorescence Spectroscopy

Fluorescence quenching is an important technique for measuring binding affinity between ligands and proteins. Fluorescence quenching is the decrease in the quantum yield of fluorescence from a fluorophore,

Handbook of Nutrition, Diet, and the Eye
http://dx.doi.org/10.1016/B978-0-12-401717-7.00059-9

induced by a variety of molecular interactions with quencher molecule(s).[16]

Assuming that there are (n) substantive binding sites for quencher (Q) on protein (B_0), the quenching reaction can be shown as follows:

$$nQ + B \Leftrightarrow Q_nB \qquad (59.1)$$

The binding constant (K_A), can be calculated as:

$$K_A = [Q_nB] / [Q]^n [B] \qquad (59.2)$$

where [Q] and [B] are the quencher and retinoid concentration, respectively, [Q_nB] is the concentration of nonfluorescent fluorophore–quencher complex, and [B_0] gives total retinoid concentration:

$$[Q_nB] = [B_0] - [B] \qquad (59.3)$$

$$K_A = ([B_0] - [B]) / [Q]^n [B] \qquad (59.4)$$

The fluorescence intensity is proportional to the retinoid concentration as described:

$$[B] / [B_0] \propto F/F_0 \qquad (59.5)$$

Results from fluorescence measurements can be used to estimate the binding constant of retinoid–protein complex from Equation (59.4):

$$log [(F_0 - F) / F] = logK_A + nlog [Q] \qquad (59.6)$$

The accessible fluorophore fraction (f) can be calculated by the modified Stern–Volmer equation[16]:

$$F_0 / (F_0 - F) = 1/fK [Q] + 1/f \qquad (59.7)$$

where F_0 is the initial fluorescence intensity and F is the fluorescence intensity in the presence of quenching agent (or interacting molecule). K is the Stern–Volmer quenching constant, [Q] is the molar concentration of quencher, and f is the fraction of accessible fluorophore to a polar quencher, which indicates the fractional fluorescence contribution of the total emission for an interaction with a hydrophobic quencher.[16,17] The plot of $F_0/(F_0 - F)$ versus $1/[Q]$ yields f^{-1} as the intercept on the y-axis and $(f K)^{-1}$ as the slope. Thus, the ratio of the ordinate and the slope gives K.

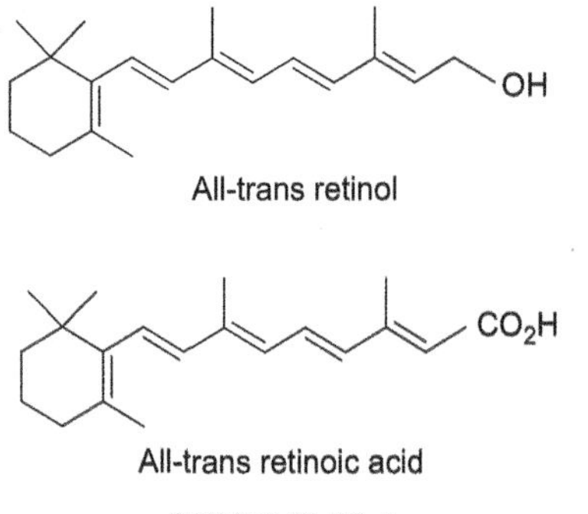

All-trans retinol

All-trans retinoic acid

SCHEME 59.1

Molecular Modeling

The docking studies were carried out with ArgusLab 4.0.1 software (Mark A. Thompson, Planaria Software LLC, Seattle, WA, USA; http://www.arguslab.com). The protein structures were obtained from literature reports,[18–20] and the retinoid three-dimensional structures were generated from PM3 semiempirical calculations using Chem3D Ultra 11.0.

STRUCTURAL ANALYSIS

Fourier Transform Infrared Spectra of Retinoid–Protein Complexes

Fourier transform infrared (FTIR) spectroscopy and its derivative methods are widely used to characterize the nature of retinoid binding to different proteins.[10,11,21] Figures 59.1 and 59.2 present the infrared spectra and difference spectra of several protein complexes with retinol and retinoic acid. The retinoid–protein interaction was characterized using spectral changes (shifting and intensity variations) of protein amide I band at 1657–1654 cm^{-1} (mainly C=O stretch) and amide II band at 1546–1540 cm^{-1} (C–N stretching coupled with N–H bending modes)[12,22] upon retinoid complexation. The difference spectra [(Protein solution + Retinoid solution) − (Protein solution)] were obtained in order to monitor the intensity variations of these vibrations and the results are shown in Figures 59.1 and 59.2. Similarly, the infrared self-deconvolution with second derivative resolution enhancement and curve-fitting procedures[23] were used to determine the protein secondary structures in the presence of retinoids (Table 59.1).

At low retinoid concentration (0.125 mM), an increase in intensity was reported for the protein amide I at 1657–1654 cm^{-1} and amide II at 1546–1540 cm^{-1} in the difference spectra of the retinol–protein and retinoic acid–protein complexes. Positive features are observed in the difference spectra for amide I and II bands (spectra not shown). These positive features are related to the increase in intensity of the amide I and amide II bands upon retinoid complexation. The increase in intensity of the amide I and amide II bands is due to retinoid binding to protein C=O, C–N, and N–H groups (hydrophilic interaction).[10,11,21]

As retinoid concentration increased to 0.5 mM, increases in the intensity of the amide I and amide II bands were observed with positive features at 1657 and 1542 cm^{-1} (retinol) and 1656 and 1542 cm^{-1} (retinoic acid), in the difference spectra of retinoid–HSA complexes (Fig. 59.1A, B) (diff. 0.5 mM), where negative features appeared at 1658 and 1538 (retinol) and 1660 and 1538 cm⁻¹ (retinoic acid), and in the difference spectra of retinoid–BSA complexes (Fig. 59.1B) (diff. 0.5 mM). The observed increase in intensity of the amide I band in the spectra of the retinoid–HSA complexes suggests a major increase in protein α-helical structure at high retinoid concentrations, while the loss of intensity of amide I band in the spectra of retinoid–BSA complexes indicates a major reduction in protein α-helical structure (Fig. 59.1A, B). Similar infrared spectral changes were observed for

protein amide I band in several ligand–protein complexes, where major protein conformational changes occurred.[24] The spectral differences showed different binding modes for retinoid complexes with HSA and BSA.[11] However, it has been demonstrated that the addition of a high concentration of BSA controls the uptake of 13-*cis* and all-*trans*-retinoic acid in cells, reducing significantly the isomerization of all-*cis*-retinoic acid to all-*trans*-retinoic acid, which is different from that of retinoid–HSA complexation.[10] Retinol

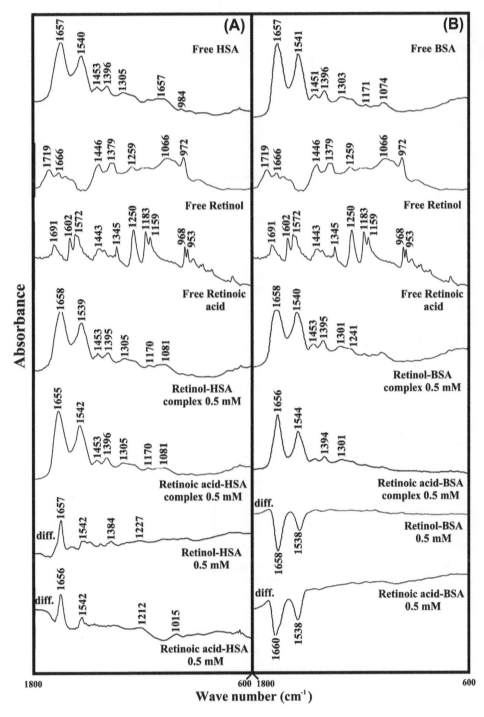

FIGURE 59.1 Fourier transform infrared spectra and difference spectra [(Protein solution + Retinoid solution) − (Protein solution)] (diff.) in the region of 1800–600 cm⁻¹ for (A) free human serum albumin (HSA) and (B) free bovine serum albumin (BSA) and their retinoid complexes in aqueous solution at pH 7.4 with various retinoid concentrations and constant protein concentration (0.25 mM).

and retinoic acid share similar binding sites in the central calyx of β-lactoglobulin, with similar stability of retinoid–protein complexes.[21] Similar infrared difference spectra were observed for retinol and retinoic acid complexes with β-lactoglobulin (Fig. 59.2A) (diff. 0.5 mM). The negative features at 1636 and 1546 cm^{-1} (retinol) and 1639 and 1536 cm^{-1} (retinoic acid) result from the loss of intensity of the amide I and amide

II bands, indicating partial protein conformational changes upon retinoid binding (Fig. 59.2A). Different binding modes were observed for retinoids with α- and β-caseins.[25] The positive features observed in the difference spectra of retinoid–α-casein for amide I and amide II bands at 1656 and 1544 (retinol) and 1649 and 1559 cm^{-1} (retinoic acid) are indicative of no major alterations in protein conformation upon

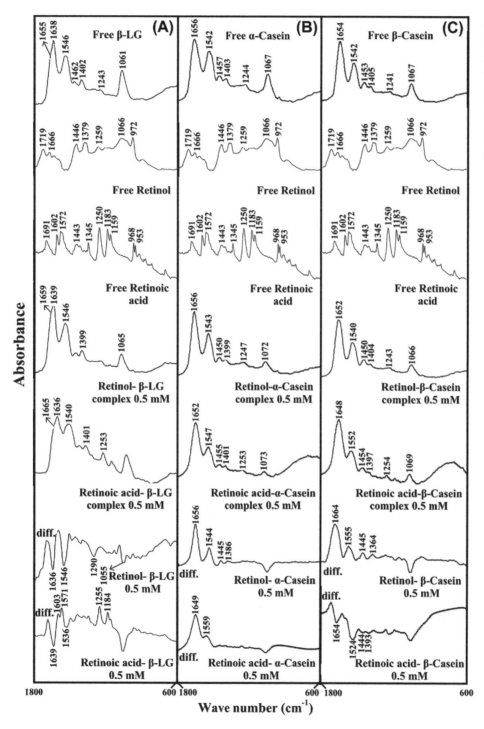

FIGURE 59.2 Fourier transform infrared spectra and difference spectra [(Protein solution + Retinoid solution) − (Protein solution)] (diff.) in the region of 1800–600 cm^{-1} for (A) free β-lactoglobulin (β-LG), (B) free α-casein, and (C) and free β-casein and their retinoid complexes in aqueous solution at pH 7.4 with various retinoid concentrations and constant protein concentration (0.25 mM).

retinoid–protein complexation (Fig. 59.2B). However, retinol binding did not alter β-casein structure with positive features at 1664 and 1555 cm⁻¹, while retinoic acid interaction altered β-casein conformation with negative features at 1654 and 1524 cm⁻¹ in the difference spectra of retinoid–β-casein complexes (Fig. 59.2C). These spectral changes showed that while retinol and retinoic acid complexation did not alter α-casein conformation, retinoic acid induced some alterations in β-casein secondary structure.[25]

A quantitative analysis of the protein secondary structures for the free HSA, BSA, β-lactoglobulin, α- and β-caseins, and their retinoid adducts was reported (Table 59.1). The serum albumins mainly have an α-helical structure[26,27] with HSA 55% (α-helix) and BSA 61% (α-helix) in the native states (Table 59.1).[10,11] Upon retinoid interaction, a minor increase in α-helix from 55% free HSA to 58% in retinoid–HSA and a major decrease in α-helix from 61% free BSA to 36–26% in retinoid–BSA complexes were observed (Table 59.1).[10,11] The differences were attributed to different modes of retinoid binding to HSA and BSA in these retinoid–protein complexes. Retinoid induced structural stabilization for HSA, while a partial protein unfolding occurred for BSA upon retinoid complexation.[10,11] Milk

β-lactoglobulin is a member of the lipocalin superfamily of transporters for small hydrophobic molecules such as retinoids. There are two potential binding sites in β-lactoglobulin for small hydrophobic molecules. One is located in the interior cavity and the other on the surface cleft of the β-lactoglobulin structure.[19,28] On retinoid binding, β-lactoglobulin did not show major conformational changes from 59% β-sheet in the free protein to 52–51% in the retinoid–β-lactoglobulin complexes (Table 59.1), indicating no protein structural destabilization.[21] Similarly, no major protein conformational changes were reported for milk α- and β-caseins upon retinoid complexation (Table 59.1). The α-casein with 35% α-helix and β-casein with 32% α-helix[20] showed no major alterations upon retinol and retinoic acid complex formation (Table 59.1). These results showed no major alterations in protein secondary structures for milk β-lactoglobulin, α- and β casein complexes with retinol, and retinoic acid.[21,25]

Circular Dichroism Spectra

The conformational changes observed from infrared results for retinoid–protein complexes were also characterized by the CD spectroscopic method

TABLE 59.1 Secondary Structure Analysis (Infrared Spectra) of Free Proteins and Their Retinoid Complexes in Hydrated Film at pH 7.4 with Protein Concentration of 0.25 mM and Retinoid 0.5 mM

		Amide I Components (cm⁻¹)				
		β-Anti (±1%)	Turn (±2%)	α-Helix (±2%)	Random Coil (±4%)	β-Sheet (±2%)
Protein	Complexes	1692–1680	1680–1660	1660–1650	1648–1641	1640–1610
HSA	Free	12	11	55	3	22
	Retinol	12	11	58	14	19
	Retinoic acid	11	12	58	13	19
BSA	Free	4	11	61	7	17
	Retinol	10	15	36	20	19
	Retinoic acid	7	27	26	25	15
β-LG	Free	17	13	11	–	59
	Retinol	14	22	12	–	52
	Retinoic acid	15	24	10	–	51
α-Casein	Free	5	23	35	14	24
	Retinol	9	16	37	19	19
	Retinoic acid	12	17	36	15	19
β-Casein	Free	7	24	32	17	20
	Retinol	7	20	33	13	27
	Retinoic acid	10	27	27	12	24

HSA: human serum albumin; BSA: bovine serum albumin; β-LG: β-lactoglobulin.

(Table 59.2). The CD results were consistent with infrared data showing free HSA with 59% α-helix and BSA with 61% α-helix (Table 59.2), consistent with an earlier report.[33] Upon retinoid complexation, no changes for the HSA α-helix structure were observed, while BSA showed major alterations in α-helix from 61% to 52–46% on retinoid complexation (Table 59.2). This was related to stabilization of HSA conformation and destabilization of BSA secondary structure on retinoid interaction.[10,11] Free β-lactoglobulin with 58% β-sheet[29] showed a minor reduction to 50–49% in the retinoid–protein complexes (Table 59.2). This was due to a minor perturbation of protein structure by retinoid interaction.[21] However, no major conformational changes were observed for α- and β-caseins[30] on retinoid complexation (Table 59.2).

Stability of Retinoid–Protein Complexes by Fluorescence Spectroscopy

HSA contains a single polypeptide of 585 amino acids with only one tryptophan (Trp-214) located in

subdomain IIA. BSA contains two tryptophan residues, Trp-134 and Trp-212, located in the first and second domains of protein hydrophobic regions.[17] Milk β-lactoglobulin has two tryptophan residues, Trp-19 and Trp-61. Trp-19 is in an apolar environment and contributes to 80% of total fluorescence, while Trp-61 is partly exposed to aqueous solvent and makes a minor contribution to tryptophan fluorescence.[31,32] Milk α-casein (mixture of α_{s1}-casein and α_{s2}-casein) has two tryptophan residues, Trp-66 and Trp-37 (in α_{s1}-casein) and Trp-109 and Trp-193 (in α_{s2}-casein), while β-casein contains one tryptophan, Trp-143, with intrinsic fluorescence.[33,34] These tryptophan residues are located in the protein surfaces. Tryptophan emission dominates casein fluorescence spectra in the ultraviolet region. When other molecules interact with casein, tryptophan fluorescence may change depending on the impact of such interaction on the protein conformation.[31,32] The decrease in fluorescence intensity of proteins has been monitored at 330–350 nm for retinoid–protein systems. The plot of $F_0/(F_0 - F)$ versus $1/[\text{retinoid}]$ is used to calculate the binding constants for retinoid–protein complexes.[35] Assuming that the observed changes in fluorescence come from the interaction between retinoid and protein, the quenching constant can be taken as the binding constant of the complex formation. The K values are presented in Table 59.3. The fluorescence quenching constants (K_q), estimated according to the Stern–Volmer equation[36]:

$$F_0/F = 1 + k_Q t_0 [Q] = 1 + K_D [Q] \tag{59.8}$$

TABLE 59.2 Secondary Structure Analysis (Circular Dichroism Spectra) of Free Proteins (12.5 μM) and Their Retinoid Complexes (0.5 mM) at pH 7.4 Calculated by CDSSTR Software

Protein	Complexes	Amide I Components			
		α-Helix (±3%)	β-Sheet (±2%)	Turn (±2%)	Random (±2%)
HSA	Free	59	10	10	21
	Retinol	59	8	11	22
	Retinoic acid	58	9	10	23
BSA	Free	62	13	12	13
	Retinol	52	16	14	18
	Retinoic acid	44	18	18	20
β-LG	Free	12	58	10	20
	Retinol	11	49	16	24
	Retinoic acid	11	50	18	21
α-Casein	Free	35	12	20	33
	Retinol	35	14	20	31
	Retinoic acid	34	13	21	32
β-Casein	Free	33	16	17	34
	Retinol	32	13	20	35
	Retinoic acid	28	14	22	36

HSA: human serum albumin; BSA: bovine serum albumin; β-LG: β-lactoglobulin.

TABLE 59.3 Binding Constant (K), Fluorescence Quenching Constant (K_q), and Number of Bound Retinoid Molecules per Protein (n) for Retinoid–Protein Complexes

Protein	Complexes	K (M⁻¹)	n	K_q (M⁻¹s⁻¹)
HSA	Retinol	1.3×10^5	1.0	2.5×10^{13}
	Retinoic acid	3.3×10^5	1.1	1.5×10^{13}
BSA	Retinol	5.3×10^6	1.2	4.5×10^{13}
	Retinoic acid	2.3×10^6	1.8	2.5×10^{13}
β-LG	Retinol	6.4×10^6	1.1	1.9×10^{14}
	Retinoic acid	3.3×10^6	1.5	1.9×10^{14}
α-Casein	Retinol	1.2×10^5	1.5	5.8×10^{13}
	Retinoic acid	6.2×10^4	1.0	1.6×10^{13}
β-Casein	Retinol	1.1×10^5	1.0	2.4×10^{13}
	Retinoic acid	6.3×10^4	1.0	2.3×10^{13}

HSA: human serum albumin; BSA: bovine serum albumin; β-LG: β-lactoglobulin.

showed that quenching is predominant in the retinoid–protein complexes (Table 59.3). The number of retinoid molecules bound per protein (n) was also calculated from $log\,[(F_0 - F)\,/F] = logK_S + nlog$ [retinoid] for the static quenching.[8,15] The n values showed that between one and two retinoid molecules are bound per protein (Table 59.3). The stability of the complexes showed that retinol binds protein more strongly than retinoic acid, with major hydrophobic contacts.

Docking Studies

Based on the spectroscopic data, docking has been reported for several retinoid–protein complexes.[11,21,25] The docking results presented in Figures 59.3 and 59.4 and Table 59.4 show detailed participation of amino acids in retinoid–protein binding. The binding sites of retinol and retinoic acids are not similar on HSA, whereas there are common binding sites on BSA,

β-lactoglobulin, and α- and β-caseins (Table 59.4). In the retinol–HSA adduct, the amino acids surrounding retinol show more hydrophobic characters, Asp-451, *Cys-448 (hydrogen (H)-bonding), Leu-198, Leu-347, Leu-481, Lys-195, Ser-454, Trp-214, Tyr-452, Val-343, Val-344, Val-455, Val-482, than those in the vicinity of retinoic acid, Ala-126, Arg-117, Asn-130, Leu-115, Leu-182, Lys-137, Met-123, *Phe-134 (H-bonding), Phe-165, Thr-133, Tyr-138, Tyr-161 (Fig. 59.3 and Table 59.4). However, retinol and retinoic acid binding sites are very similar on BSA with retinol, *Arg-209 (H-bonding), Asn-185, Asp-153, Glu-154, Glu-206, Ile-165, Leu-162, Phe-150, Phe-157, Trp-134, Tyr-161, and for retinoic acid, *Arg-209 (H-bonding), Asp-153, Glu-154, Ile-165, Leu-162, Phe-150, Phe-157, Trp-134, Tyr-161 (Fig. 59.3 and Table 59.4). Retinol and retinoic acid share similar binding sites in the central calyx of β-lactoglobulin with similar stability of retinoid–protein complexes for retinol, Ala-86, Asn-88, Asn-90, Asp-85,

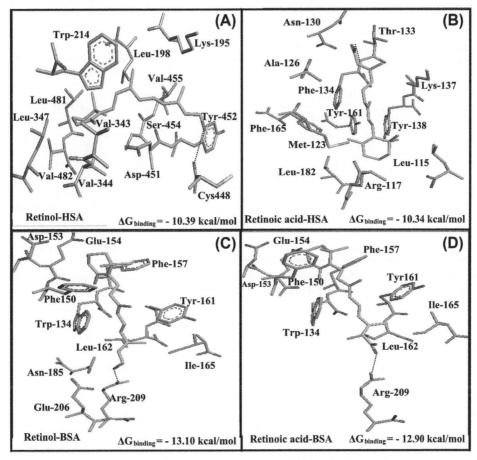

FIGURE 59.3 Best docked conformations of retinoid–bovine serum albumin (BSA) and retinoid–human serum albumin (HSA) complexes: (A) retinol complexed to HSA, (B) retinoic acid complexed to HSA, (C) retinol complexed to BSA, and (D) retinoic acid complexed to BSA. $\Delta G_{binding}$: free binding energy.

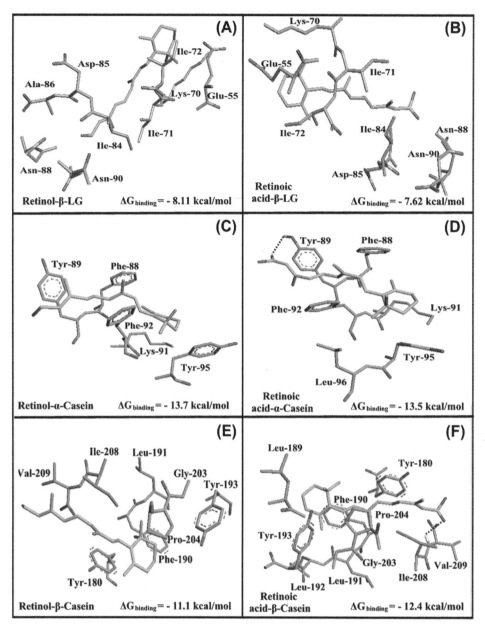

FIGURE 59.4 Best docked conformations of retinoid–protein complexes: (A) retinol complexed to β-lactoglobulin (β-LG), (B) retinoic acid complexed to β-LG, (C) retinol complexed to α-casein, (D) retinoic acid complexed to α-casein, (E) retinol complexed to β-casein, and (F) retinoic acid complexed to β-casein. ΔG$_{binding}$: free binding energy.

Glu-55, Ile-71, Ile-72, Ile-84, Lys-70, and retinoic acid, Asn-88, Asn-90, Asp-85, Glu-55, Ile-71, Ile-72, Ile-84, Lys-70 (Fig. 59.4 and Table 59.4). Retinoids also share similar binding sites with α-casein for retinol, Phe-88, Tyr-89, Lys-91, Phe-92, Tyr-95, and for retinoic acid, Phe-88, Tyr-89* (H-bonding), Lys-91, Phe-92, Tyr-95, Leu-96 (Fig. 59.4 and Table 59.4). Similar binding sites are also observed for retinoids on β-casein with retinol, Tyr-180, Phe-190, Leu-191, Tyr-193, Gly-203, Pro-204, Ile-208, Val-209, and for retinoic acid, Tyr-180, Leu-189, Phe-190, Leu-191, Leu-192, Tyr-193, Gly-203, Pro-204, Ile-208, Val-209* (H-bonding) (Fig. 59.4 and Table 59.4). The free binding energy shows that stronger complexes are formed with retinol than retinoic acid, except for β-casein, which has an additional H-bonding system (Table 59.4). Similar molecular modeling was reported for other ligands docked to HSA, BSA, β-lactoglobulin, and caseins.[37–44]

TABLE 59.4 Amino Acid Residues Involved in Retinoid–Protein Complexes with Free Binding Energy ($\Delta G_{binding}$) for the Best Selected Docking Positions

Protein	Complexes	Amino Acids in the Vicinity of Retinoids	$\Delta G_{binding}$ (kcal/mol)
HSA	Retinol	Asp-451, *Cys-448, Leu-198, Leu-347, Leu-481, Lys-195, Ser-454, Trp-214, Tyr-452, Val-343, Val-344, Val-455, Val-482	−10.39
	Retinoic acid	Ala-126, Arg-117, Asn-130, Leu-115, Leu-182, Lys-137, Met-123, *Phe-134, Phe-165, Thr-133, Tyr-138, Tyr-161	−10.34
BSA	Retinol	*Arg-209, Asn-185, Asp-153, Glu-154, Glu-206, Ile-165, Leu-162, Phe-150, Phe-157, Trp-134, Tyr-161	−13.10
	Retinoic acid	*Arg-209, Asp-153, Glu-154, Ile-165, Leu-162, Phe-150, Phe-157, Trp-134, Tyr-161	−12.90
β-LG	Retinol	Ala-86, Asn-88, Asn-90, Asp-85, Glu-55, Ile-71, Ile-72, Ile-84, Lys-70	−8.11
	Retinoic acid	Asn-88, Asn-90, Asp-85, Glu-55, Ile-71, Ile-72, Ile-84, Lys-70	−7.62
α-Casein	Retinol	Phe-88, Tyr-89, Lys-91, Phe-92, Tyr-95	−13.7
	Retinoic acid	Phe-88, Tyr-89* Lys-91, Phe-92, Tyr-95, Leu-96	−13.5
β-Casein	Retinol	Tyr-180, Phe-190, Leu-191, Tyr-193, Gly-203, Pro-204, Ile-208, Val-209	−11.1
	Retinoic acid	Tyr-180, Leu-189, Phe-190, Leu-191, Leu-192, Tyr-193, Gly-203, Pro-204, Ile-208, Val-209*	−12.4

Hydrogen bonding observed with this amino acid residue.

TAKE-HOME MESSAGES

- Retinol and retinoic acid binding sites include several amino acids with hydrophilic and hydrophobic characters.
- Stronger retinoid–protein complexes formed with retinol than with retinoic acid, except for β-casein.
- The binding sites of retinol and retinoic acids are not similar on HSA, while there are common binding sites shared on bovine serum albumin (BSA), β-lactoglobulin, and α- and β-caseins.
- Retinoid–protein interaction stabilizes human serum albumin (HSA) conformation while inducing a partial protein structural destabilization for BSA, β-lactoglobulin, and α- and β-caseins.
- BSA forms more stable complexes than HSA, while milk β-lactoglobulin forms weaker complexes than those of α- and β-caseins.
- HSA and BSA, as well as milk protein, can transport retinoids *in vitro*.
- Models of retinoid–protein complexes presented here provide useful information regarding the binding of retinoids to different receptors, which is essential for eye and vision.

Acknowledgment

This work is supported by grant from Natural Sciences and Engineering Research Council of Canada (NSERC).

References

1. Newcomer ME. Retinoid-binding proteins: structural determinants important for function. *FASEB J* 1995;**9**:229–39.
2. Zhang YR, Zhao YQ, Huang JF. Retinoid-binding proteins: similar protein architectures bind similar ligands via completely different ways. *PLoS ONE* 2012;**7**(5):e36772. 1–8.
3. Folli C, Calderone V, Ottonello S, Bolchi A, Zanotti G, Stoppini M, Berni R. Identification, retinoid binding, and X-ray analysis of human retinol binding protein. *Proc Natl Acad Sci USA* 2001;**98**:3710–5.
4. Calderone V, Bern R, Zanotti G. High-resolution structures of retinol-binding protein in complex with retinol: pH-induced protein structural changes in the crystal state. *J Mol Biol* 2003;**329**:841–50.
5. Zanotti G, Ottonello S, Berni R, Monaco HL. Crystal structure of the trigonal form of human plasma retinol-binding proteins at 2.5 Å resolution. *J Mol Biol* 1993;**230**:613–24.
6. Newcomer ME. Structure of the epididmyal retinoic acid binding protein at 2.1 Å resolution. *Structure* 1993;**1**:7–18.
7. Tsukada M, Schroder H, Seltmann CE, Orfanos CC, Zouboutis CC. High albumin levels restrict the kinetics of 13-*cis* retinoic acid uptake and intercellular isomerization of all-*trans* retinoic acid and inhibit its anti-proliferative effect of SZ95 sebocytes. *J Invest Dermatol* 2002;**19**:182–5.
8. Mandeville JS, Tajmir-Riahi HA. Complexes of dendrimers with bovine serum albumin. *Biomacromolecules* 2010;**11**:465–72.
9. Dubeau S, Bourassa P, Thomas TJ, Tajmir-Riahi HA. Biogenic and synthetic polyamines bind bovine serum albumin. *Biomacromolecules* 2010;**11**:1507–15.
10. N'soukpoé-Kossi CN, Sedaghat-Herati R, Ragi C, Hotchandani S, Tajmir-Riahi HA. Retinol and retinoic acid bind human serum albumin: stability and structural features. *Int J Biol Macromol* 2007;**40**:484–90.
11. Belatik A, Hotchandani S, Bariyanga J, Tajmir–Riahi HA. Binding sites of retinol and retinoic acid with serum albumins. *Eur J Med Chem* 2012;**48**:114–23.
12. Dousseau F, Therrien M, Pezolet M. On the spectral subtraction of water from the FT-IR spectra of aqueous solutions of proteins. *Appl Spectrosc* 1989;**43**:538–42.
13. Johnson WC. Analyzing protein circular dichroism spectra for accurate secondary structure. *Proteins Struct Funct Genet* 1999;**35**:307–12.

14. Sreerama N, Woddy RW. Estimation of protein secondary structure from circular dichroism spectra: comparison of CONTIN, SELCON and CDSST methods with an expanded reference set. *Anal Biochem* 2000;**287**:252–60.

15. Charbonneau D, Beauregard M, Tajmir-Riahi HA. Structural analysis of human serum albumin complexes with cationic lipids. *J Phys Chem B* 2009;**113**:1777–84.

16. Lakowicz JR. *Principles of Fluorescence Spectroscopy*. 3rd ed. New York: Springer; 2006.

17. Tayeh N, Rungassamy T, Albani JR. Fluorescence spectral resolution of tryptophan residues in bovine and human serum albumins. *J Pharm Biomed Anal* 2009;**50**:107–16.

18. Sugio S, Kashima A, Mochizuki S, Noda M, Kobayashi K. Crystal structure of human serum albumin at 2.5 Å resolution. *Protein Eng* 1999;**12**:439–46.

19. Qin BY, Bewley MC, Creamer LK, Baker HM, Baker EN, Jameson JB. Structural basis of the Tanford transition of bovine β-lactoglobulin. *Biochemistry* 1998;**37**:14014–23.

20. Kumosinski TF, Brown EM, Farell Jr HM. Three-dimensional molecular modeling of bovine caseins: a refined, energy-minimized beta-casein structure. *J Dairy Sci* 1993;**76**:931–45.

21. Belatik A, Kanakis DC, Hotchandani S, Tarantilis P, Polissiou MG, Tajmir-Riahi HA. Locating the binding sites of retinol and retinoic acid with milk β-lactoglobulin. *J Biomol Struct Dyn* 2012;**30**:437–47.

22. Krimm S, Bandekar J. Vibrational spectroscopy and conformation of peptides, polypeptides, and proteins. *Adv Protein Chem* 1986;**38**:181–364.

23. Byler DM, Susi H. Examination of the secondary structure of proteins by deconvoluted FTIR spectra. *Biopolymers* 1986;**25**:469–87.

24. Ahmed Ouameur A, Diamantoglou S, Sedaghat-Herati MR, Sh Nafisi, Carpentier R, Tajmir-Riahi HA. An overview of drug binding to human serum albumin: protein folding and unfolding. *Cell Biochem Biophys* 2006;**45**:203–13.

25. Bourassa P, N'soukpoé-Kossi CN, Tajmir-Riahi HA. Binding of vitamin A with milk α- and β-caseins. *Food Chem* 2013;**138**:444–53.

26. Carter DC, Ho JX. Structure of serum albumin. *Adv Protein Chem* 1994;**45**:153–203.

27. Peters T. Serum albumin. *Adv Protein Chem* 1985;**37**:161–245.

28. Brownlow S, Cabral JM, Cooper R, Flower DR, Yewdall SJ, Polikarpov I, et al. Bovine β-lactoglobulin at 1.8 Å resolution – still an enigmatic lipocalin. *Structure* 1997;**5**:481–95.

29. Bourassa P, Tajmir-Riahi HA. Locating the binding sites of folic acid with milk α- and β-caseins. *J Phys Chem B* 2012;**116**:513–9.

30. Chakraborty A, Basak S. Effect of surfactants on casein structure: a spectroscopic study. *Colloids Surf B Biointerfaces* 2008;**63**:83–90.

31. Liang L, Subirade M. β-Lactoglobulin/folic acid complexes: formation, characterization, and biological implication. *J Phys Chem B* 2010;**114**:6707–12.

32. Liang L, Tajmir-Riahi HA, Subirade M. Interaction of β-lactoglobulin with resveratrol and its biological implications. *Biomacromolecules* 2008;**9**:50–5.

33. Chakraborty A, Basak S. pH-induced structural transitions of casiens. *J Photochem Photobiol B* 2007;**87**:191–9.

34. Zhang X, Fu X, Zhang H, Liu C, Jiao W, Chang Z. Chaperone-like activity of β-casein. *Int J Biochem Cell Biol* 2005;**37**:1232–40.

35. Dufour C, Dangles O. Flavonoid-serum albumin complexation: determination of binding constants and binding sites by fluorescence spectroscopy. *Biochim Biophys Acta* 2005;**1721**:164–73.

36. Zhang G, Que Q, Pan J, Guo J. Study of the interaction between icarin and human serum albumin by fluorescence spectroscopy. *J Mol Struct* 2008;**881**:132–8.

37. Bourassa P, Hasni I, Tajmir-Riahi HA. Folic acid complexes with human and bovine serum albumins. *Food Chem* 2011;**129**:1148–55.

38. Bourassa P, Tajmir-Riahi HA. Locating the binding sites of folic acid with milk α- and β-caseins. *J Phys Chem B* 2012;**116**:513–9.

39. Hasni I, Bourassa P, Hamdani S, Samson G, Carpentier R, Tajmir-Riahi HA. Interaction of milk α- and β-caseins with tea polyphenols. *Food Chem* 2011;**126**:630–9.

40. Hasni I, Tajmir-Riahi HA. Binding of cationic lipids to milk β-lactoglobulin. *J Phys Chem B* 2011;**115**:6683–90.

41. Kanakis CD, Hasni I, Bourassa P, Tarantilis P, Polissiou MG, Tajmir-Riahi HA. Milk β-lactoglobulin complexes with tea polyphenols. *Food Chem* 2011;**127**:1046–55.

42. Froehlich E, Jennings CJ, Sedaghat-Herati MR, Tajmir-Riahi HA. Dendrimers bind human serum albumin. *J Phys Chem B* 2009;**113**:6986–93.

43. Essemine J, Hasni I, Carpentier R, Thomas TJ, Tajmir-Riahi HA. Binding of biogenic and synthetic polyamines to β-lactoglobulin. *Int J Biol Macromol* 2011;**49**:201–11.

44. Agudelo D, Bourassa P, Bruneau J, Bérubé G, Asselin E, Tajmir-Riahi HA. Probing the binding sites of antibiotic drugs doxorubicin and N-(trifluoroacetyl) doxorubicin with human and bovine serum albumins. *PLoS ONE* 2012;**7**(8):1–13. e43814.

Lycopene and Retinal Pigment Epithelial Cells: Molecular Aspects

Chi-Ming Chan[1,2], Chi-Feng Hung[1]

[1]School of Medicine, Fu Jen Catholic University, New Taipei City, Taiwan, [2]Department of Ophthalmology, Cardinal Tien Hospital, New Taipei City, Taiwan

INTRODUCTION

Lycopene, a naturally occurring red carotenoid pigment found in tomatoes, pink grapefruit, watermelon, papaya, guava, and other fruits, is associated with a decreased risk of chronic diseases such as cardiovascular disease[1] and cancer.[2] Lycopene belongs to a large family of plant pigments known as carotenoids. Carotenoids produce colors ranging from the yellow color of squash, to the orange color of pumpkins, to the red color of tomatoes. Carotenoids also contribute to the aroma of some plant foods.[3] In humans, carotenoids function primarily as dietary sources of provitamin A. Thus, the carotenoids can be categorized as follows: vitamin A precursors that do not pigment, such as β-carotene; pigments with partial vitamin A activity, such as cryptoxanthin, β-apo-8′-carotenoic acid ethyl ester; nonvitamin A precursors that do not pigment or pigment poorly, such as violaxanthin and neoxanthin; and nonvitamin A precursors that pigment, such as lycopene, lutein, zeaxanthin, and canthaxanthin. Lycopene lacks the β-ionone ring structure required to form vitamin A and has no provitamin A activity.[4] Moreover, there are two primary types of carotenoids: hydrocarbon carotenoids and xanthophylls. Hydrocarbon carotenoids, which include lycopene, are composed entirely of hydrogen and carbon. In contrast, xanthophylls, such as lutein, contain oxygen in addition to carbon and hydrogen.[5] Carotenoids typically contain 40 carbons. Apo-carotenoids contain fewer than 40 carbons. The prefix 'apo' is used to identify carotenoids that have been shortened in length by one or more carbons. Regardless of the number of carbons, all carotenoids possess an isoprenoid backbone.[6]

The chemical formula for lycopene is $C_{40}H_{56}$. The 11 conjugated and two unconjugated double bonds present in lycopene allow for extensive isomerization, resulting in 1056 theoretical *cis–trans* configurations. In nature, lycopene exists in all-*trans* form, and seven of these bonds can isomerize from the *trans* form to the mono or poly-*cis* form under the influence of heat, light, or certain chemical reactions.[7] The thermodynamic stabilities of the common lycopene isomers have been determined relative to the all-*trans* isomer. The 5-*cis* isomer is the most stable, followed by all-*trans*, 9-*cis*, 13-*cis*, 15-*cis*, 7-*cis*, and 11-*cis*[7] (Fig. 60.1). The bioavailability of lycopene from tomatoes has been found to increase with processing, and its uptake in humans is higher (two-fold increase) from processed tomato juice than from unprocessed juice. This is attributed to the isomerization of the *trans* to *cis* form of lycopene on heating and its release from ruptured cells as a result of thermal processing.[8] The *cis* isomers of lycopene are better absorbed and more bioavailable than their *trans* counterparts.[9] The lycopene isomers found in human blood plasma, breast milk, and human tissues are mainly of the *cis*-isomer type.[9–11] The most abundant isomeric forms in human tissue and blood are the all-*trans* and 5-*cis* isomers, with the sum of *cis*-isomers accounting for the majority (58–88%) of the total lycopene.[9] The pigment character of carotenoids is imparted to the colorless basal C_{40} structure of phytoene by introducing additional double bonds. Multiple desaturation steps lead to the red pigment lycopene (13 double bonds), which confers its color to ripe tomatoes. Additional desaturation up to 15 fully conjugated double bonds has been achieved artificially by molecular breeding, resulting in the pink 3,4,30,40-tetradehydro-lycopene.[12]

This chapter focuses on *in vitro* studies of the molecular mechanisms behind the bioactivities of lycopene and its derivatives in humans. The beneficial effects of lycopene against the diseases age-related macular

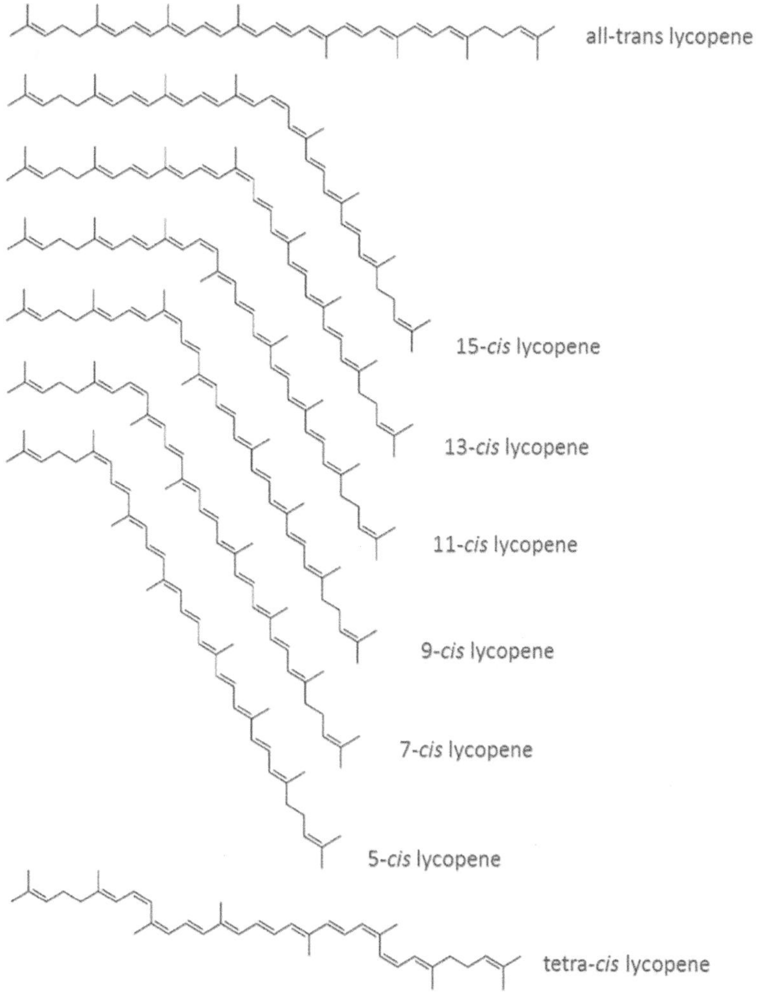

FIGURE 60.1 Structure of lycopene (all-*trans*) and several of the (*cis-*) isomers.

all-trans lycopene

15-*cis* lycopene

13-*cis* lycopene

11-*cis* lycopene

9-*cis* lycopene

7-*cis* lycopene

5-*cis* lycopene

tetra-*cis* lycopene

degeneration (AMD) and proliferative vitreoretinopathy (PVR) are discussed, along with proposed mechanisms of action (Figs. 60.2 and 60.3).

LYCOPENE AND ITS METABOLITES IN HUMAN OCULAR TISSUES

Carotenoid extracts from retinal pigment epithelium (RPE) cells and ocular tissues have been analyzed with high-performance liquid chromatography.[13,14] For the endogenous carotenoids in adult retinal pigment epithelial cell-19 (ARPE-19) cells, the concentration of β-carotene is 0.06 ng/μg DNA while concentrations of lycopene and lutein are below detectable levels.[15] Uveal structures (iris, ciliary body, and RPE-choroid) account for approximately 50% of the carotenoids and approximately 30% of the lutein and zeaxanthin in the eye. The concentration of lycopene in pooled extracts of human ciliary body is 7.8 ng/0.2 g of tissue and the concentration in the human RPE-choroid is 8.64 ng/0.2 g of tissue.[14] It has been suggested that these pigments are likely to play a role in

filtering out phototoxic short-wavelength visible light in the iris while more likely to act as antioxidants in the ciliary body. Both mechanisms, light screening and antioxidative action, may be operative in the RPE-choroid to a possible function of this tissue in the transport of carotenoids from the circulating blood to the retina. Another study has detected no lycopene in the lens, retina, and macula.[13] The presence of *all-E*-lycopene and its 5Z isomer has been shown in the RPE-choroid, ciliary body, and iris. The ratio of *all-E*-lycopene to 5Z-lycopene in human RPE-choroid and ciliary body is almost consistent with that in plasma, whereas in the iris, the *all-E* isomer predominates.[13]

The structure of metabolites of lycopene are shown in Fig. 60.4.

OXIDATIVE STRESS IN THE RETINAL PIGMENT EPITHELIUM

The RPE is a single layer of pigment-containing epithelial cells in the retina, situated close to the outer

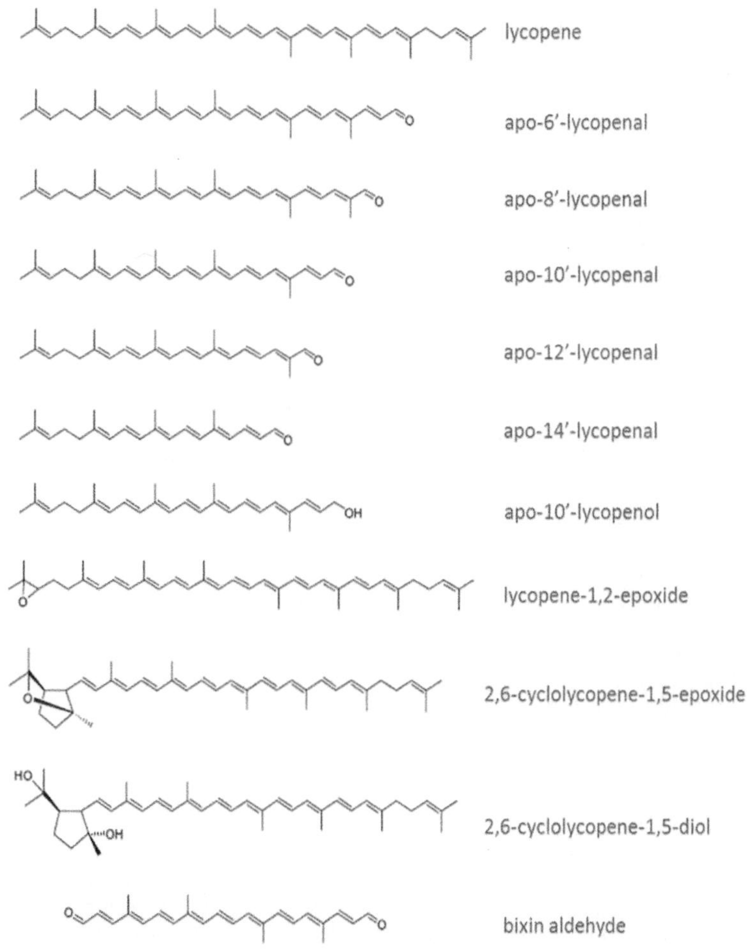

FIGURE 60.2 Structure of various lycopenoids.

segments of the retinal photoreceptor cells. These epithelial cells perform essential functions for the photoreceptor cells, such as nutrient transport, phagocytosis of the shed photoreceptor membranes, and ensuring retinal attachment. RPE cells exhibit robust metabolic activity to meet their high energy needs. This amplified oxidative phosphorylation produces large amounts of adenosine triphosphate (ATP) and in the process also generates high local concentrations of reactive oxygen species (ROS). As the RPE ages, the capacity to utilize and/or neutralize these mitochondrial-derived ROS probably diminishes. The result is increased collisions of free radicals with DNA, proteins, and lipids within the mitochondria and in the cytoplasm. The damage caused by these free radicals can wreak havoc on mitochondrial function and integrity as well as affect cytoplasmic processes and overall cellular health.

The retina has a uniquely high metabolic demand for oxygen that is normally met by a highly efficient vascular supply.[16] Oxygen plays an essential role in oxidative phosphorylation as an electron acceptor in the mitochondrial respiratory chain for the synthesis of ATP required to support the metabolic demand, including that of the visual cycle. Maintenance of normal retinal function depends on a continuous supply of oxygen and on the capability to detect and respond rapidly to local oxygen deficiency (hypoxia).[17]

Ultraviolet (UV) light with a wavelength less than 400 nm is a very important source of free radical damage in tissues such as skin and cornea, which are directly exposed to this stress. Although the anterior structures of the eye absorb much of the UV component of the optical radiation spectrum, a portion of the UVA band (315–400 nm) penetrates into the retina. Natural sources, such as the sun, emit energetic UV photons that typically do not result in energy confinement in the retina, and thus do not produce thermal or mechanical damage, but are capable of inducing photochemical damage. Photochemical damage in the retina proceeds through type 1 (direct reactions involving proton or electron transfers) and type 2 (reactions involving ROS) mechanisms.[18] Visible light also has significant effects on the RPE. Rhodopsin absorbs light at 500 nm, and light toxicity at that wavelength is associated with damage to rod cells.[19,20] It has also been demonstrated that blue light of 430 nm can have a major impact on photoreceptor and RPE function, inducing photochemical damage and apoptotic cell death.[21] The mechanism of this damage is not clear,

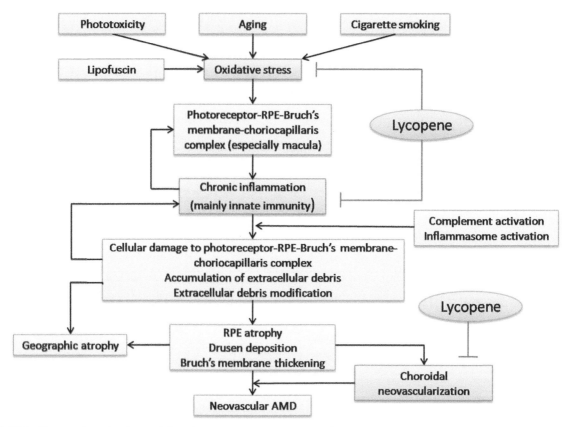

FIGURE 60.3 Proposed mechanisms of the inhibition of age-related macular degeneration (AMD) development by lycopene. In the pathogenesis of AMD, phototoxicity, cigarette smoke, aging, and lipofuscin deposits induce oxidative stress in the retinal pigment epithelium (RPE) cells, which in turn causes the inflammation of the photoreceptor–RPE–Bruch's membrane–choriocapillaris complex. The resulting RPE atrophy enhances the choriocapillaris invasion to the subretinal space and leads to the development of neovascular AMD. Thus, the antioxidative, anti-inflammatory, and antiangiogenic effects of lycopene make it a promising nutrient against the progression of AMD.

but one possibility is a free radical mechanism involving $O_2\bullet$ and the Fe^{3+} of cytochrome a3. Light can initiate lipid peroxidation by singlet oxygen.[22,23] In particular, light-emitting diodes and incandescent light have been shown to promote the production of ROS, raise the lipid peroxidation level, lower the activity of the antioxidant key enzymes in mammalian cells, and eventually cause a number of cell deaths.[24]

Another source of intracellular oxidative stress in RPE is the age-dependent accumulation of lipofuscin, also known as aging pigment.[25] This heterogeneous, autofluorescent complex of lipid–protein aggregates is derived largely from photoreceptor outer segments that are phagocytosed by RPE cells and presumably also contain remnants of triaged RPE organelles.[26,27] The ability to avoid the accumulation of cellular cytotoxic protein aggregates is reduced in senescent RPE cells, which may evoke lipofuscin accumulation in the lysosomes of postmitotic RPE cells. This lipofuscin accumulation decreases lysosomal enzyme activity, impairs autophagic clearance of damaged proteins, and produces free radicals.[28] In addition, some components of lipofuscin can be photo-oxidized and become

toxic to the RPE. Among these components, the reaction of two molecules of all-*trans*-retinal with ethanolamine forms bis-retinoid pyridinium compound A2E, which has gained much attention and resulted in a variety of studies to understand its potential role in the phototoxicity of lipofuscin. When oxidized by 430 nm light, various A2E epoxides are generated that are potentially harmful to RPE health and function. A2E epoxides have pleiotropic effects, ranging from destabilizing mitochondrial and/or lysosomal membrane integrity to reducing the capacity of RPE cells to process phagocytosed photoreceptor outer segment in cultured RPE cells.[29,30] Photodegradation of A2E can generate the advanced glycation end-product dicarbonyls glyoxal and methylglyoxal.[31] In addition, as the aerobic photoreactivity of A2E is not a major contributor to the phototoxicity of lipofuscin, other hydrophobic components of lipofuscin have been suggested as major contributors to the photoexcitatory-induced free radical production and the subsequent damage to RPE cells.[32] Taken together, photoexcitation of these outer segment-derived bis-retinoids can lead to singlet oxygen, superoxide anion, and hydroxyl

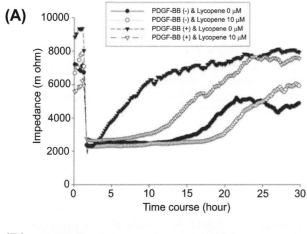

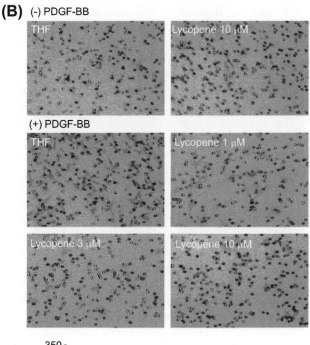

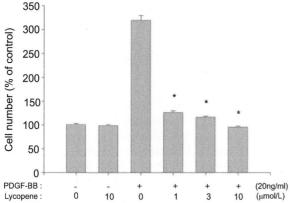

radical generation, all of which can initiate a chain reaction of lipid peroxidations.[33]

Cigarette smoke is a complex mixture of more than 5000 chemicals, and several individual components contained within cigarette smoke have been shown to induce oxidative stress in RPE cells.[34] Chronic exposure to cigarette smoke has led to features of early AMD in mice.[35] In one study, mice exposed to cigarette smoke developed oxidative damage and ultrastructural degeneration of the RPE and Bruch's membrane, as well as RPE cell apoptosis.[36] Moreover, in 16-month-old C57Bl6 mice, cigarette smoke at a higher concentration required even shorter duration of exposure to induce ultrastructural changes to the Bruch's membrane and choriocapillaris endothelium that are compatible with early AMD.[37] The exposure of human RPE cells to cigarette smoke extract or hydroquinone, a component of cigarette smoke, results in oxidative stress and cell death, characterized by a reduction in cell size and nuclear condensation. Oxidative damage, including increased lipid peroxidation and mitochondrial superoxide production, as well as a decrease in intracellular glutathione concentration, adds to the observed effects of cigarette smoke on human RPE cells.[38]

The generation of ROS has long been considered to have harmful consequences and has been thought to be a major factor in the development of AMD. Oxidative damage in the retina and macula can cause local alterations in cellular functions. The compensatory local antioxidant activity within the retina appears to increase in AMD patients. Nevertheless, ROS activity may exceed the antioxidant capacity within the retina and macula, with subsequent progression of AMD.[39]

Antioxidative Effects of Lycopene

Carotenoids have gained much attention as natural antioxidants. The roles and actions of carotenoids as

which represents the number of migrated cells, increased sharply during the first 10 hours. By contrast, the impedance in the well containing both PDGF-BB and lycopene increased slowly during the same time. Compared with these two wells, the impedance in the well containing no PDGF-BB increased mildly and gradually regardless of whether it contained lycopene. (B) In the Transwell migration assay, ARPE19 cells (5×10^4 cells in 200 μl) were seeded in the upper chambers in the absence or presence of lycopene. The Transwell inserts coated with 15 μg/ml fibronectin were assembled in the lower chambers, which were filled with 600 μl serum-free ((−)PDGF-BB) (upper panel) or 20 ng/ml PDGF-BB-containing medium ((+)PDGF-BB) (middle panel) and preincubated with various concentrations of lycopene for 30 minutes at 37 °C. After incubating for 5 hours at 37 °C, ARPE19 cells that had migrated to the underside of the filter membranes were photographed (upper and middle panels) and counted with a phase-contrast light microscope under high power field (magnification 100 ×) (lower panel). *Significantly different from cells stimulated with PDGF-BB alone (P < 0.05). *Source: Reprinted from Chan* et al. Biochem Biophys Res Commun. *2009;388:172–6,*[85] *with permission.*

FIGURE 60.4 Inhibitory effects of platelet-derived growth factor (PDGF)-BB-induced ARPE19 cell migration by lycopene in ECIS® and Transwell® migration assays. (A) In ECIS migration assay, the ARPE19 cells were treated with different combinations of PDGF-BB (20 ng/ml) and lycopene (0 μM indicates that it contained only tetrahydrofuran) before being preincubated together at 37 °C for 30 minutes. Cell migration was then assessed by continuous resistance measurements for 30 hours. In the well containing PDGF-BB but no lycopene, the impedance,

singlet oxygen quenchers due to their chemical structures have been well demonstrated.[40] Lycopene consists of a tetraterpene hydrocarbon polyene chain with 11 conjugated and two unconjugated double bonds that can easily be attacked by electrophilic reagents, resulting in an extreme reactivity towards oxygen and free radicals.[41] In terms of oxidation potential, lycopene is the easiest carotenoid to oxidize to its radical cation and astaxanthin is the most difficult.[42] The radical cations arising from the three carotenoids in the macular (lutein, zeaxanthin, and mesozeaxanthin) are all effectively removed by lycopene but not by β-carotene. A reduced level of serum lycopene is correlated with AMD, even though lycopene does not accumulate in the macula.[43] The repair of the radical cations from dietary lutein and zeaxanthin by lycopene allows these two xanthophylls, upon reaching the macula, to protect the eye.[42]

Carotenoids in general have been suggested to react with free radicals in three major ways, which are electron transfer, hydrogen abstraction, and radical addition.[44] For the reactions of hydrocarbon carotenoids with peroxyl radicals, over 98% are predicted to proceed, in both polar and nonpolar environments, via the radical addition mechanism. Lycopene and torulene are more reactive than β-carotene towards peroxyl radicals, and for lycopene the C5 position is the main –OOH addition site.[45] The peroxyl radical scavenging capacity of β-carotene and lycopene has been shown to be about one-tenth of that observed for α-tocopherol, and their efficacy in lipid peroxidation inhibition is much smaller.[46] Lycopene has also been reported to trap peroxynitrite, an important biologic oxidant.[47] The protective effect of lycopene against oxidative stress in the cells could be due to the scavenging and reduction of intracellular ROS and reactive nitrogen species (RNS) by lycopene. Furthermore, with a demonstrated ability to reduce the δ-tocopheryl radical, lycopene may enhance the cellular antioxidant defense system by regenerating the nonenzymatic antioxidants vitamins E and C from their radicals.[48] The resulting lycopene radical cations would react with each other to form stable products in the absence of tocopherols.[49] Lycopene exerts its maximal antioxidant activity in the cellular membranes and interacts with the lipid components owing to its highly lipophilic nature even though it has only low efficacy in lipid peroxidation inhibition.[50] In addition, the ability of lycopene to scavenge nitrogen dioxide and thiyl and sulfonyl radicals has been shown.[49] Compared with β-carotene, lycopene is more efficient in preventing nitrogen dioxide-induced oxidation of lipid membranes and subsequent cell deaths.[51]

When RPE cells and photoreceptor outer segments are exposed to normobaric hyperoxia (40% O_2), lipofuscin accumulates within the cells. Addition of lycopene, lutein, zeaxanthin, or α-tocopherol has been shown to reduce the amounts of lipofuscin formation in RPE cells.[52] Moreover, after incubation with tomato extract, ARPE-19 cells have been observed to preferentially accumulate β-carotene and lutein, rather than lycopene, and exhibit a reduced level of nitrotyrosine, protein carbonyls, and thiobarbituric acid-reactive substance formation after hydrogen peroxide (H_2O_2) treatment.[15]

INFLAMMATION IN RETINAL PIGMENT EPITHELIUM

Oxidative stress and inflammation are known to be associated with AMD. RPE cells play a principal role in the immune defense of the macula, and their dysfunction is a crucial development leading up to the clinically relevant changes seen in AMD. Aging in its various manifestations induces a vicious circle of stress, inflammation, and lipofuscin accumulation in RPE cells.[53,54] Drusen deposits in atrophic AMD contain evidence of chronic low-grade inflammation, supporting the hypothesis that inflammation is central to the pathogenesis of AMD.[55] Moreover, several genetic risk factors associated with AMD are involved in the activation and regulation of the complement pathway. In particular, single nucleotide polymorphisms (SNPs) from complement factor H (CFH), complement component 2 (C2), complement component 3 (C3), and complement factor B (CFB) have been implicated in the pathogenesis of AMD.[56] Inflammatory mediators also govern the progression from atrophic to neovascular AMD. Histopathology of AMD has demonstrated that all stages of the disease, including drusen formation, geographic atrophy, and choroidal neovascularization (CNV), are associated with inflammatory cells, especially macrophages.[57] Depletion of macrophages in a laser-induced animal model of CNV has been shown to inhibit the immune response and subsequent neovascularization.[58] Moreover, oxidative stress can activate the NLR family, pyrin domain-containing 3 (NLRP3) inflammasomes in RPE cells which occupy the center stage in the development of AMD.[59]

The formation of extracellular drusen involves various inflammatory proteins such as vitronectin, apolipoprotein B and E, α-crystallin, complement components, LOC387715(ARMS2)/HtrA1 (high temperature requirement factor A-1), and lipids.[60] In addition, the presence of many exosome proteins and the autophagy marker Atg5 in drusen has been found.[61] Local inflammatory and immune-mediated events are involved in the development of drusen.[62] A variety of inflammatory cells in the choroid and in isolated CNV membranes of eyes with advanced AMD emphasizes their important contribution to AMD pathogenesis.[63]

Anti-Inflammatory Effects of Lycopene

Lycopene has been shown to inhibit nuclear factor-κB (NF-κB) activity and modulate inflammatory pathways.[64,65] Moreover, lycopene downregulates the cell surface expression of three types of receptor, namely the toll-like receptor (TLR)-2 and -4 and receptor for advanced glycation end-products (RAGE), which are known to bind high-mobility group box 1 (HMGB1) protein to initiate proinflammatory responses in endothelial cells.[66] Lycopene prevents the oxysterol-induced increase in proinflammatory cytokine interleukin-1β (IL-1β), IL-6, IL-8, and tumor necrosis factor-α (TNF-α) secretion and expression, which is accompanied by inhibition of oxysterol-induced ROS production, mitogen-activated protein kinase phosphorylation, and NF-κB activation. The anti-inflammatory effects of peroxisome proliferator-activated receptor-γ (PPAR-γ) activators have been shown to inhibit cytokine production by preventing the activation and translocation of NF-κB.[67] Lycopene also increases levels of PPAR-γ in human macrophages.[68]

ROS and oxidative stress activate NF-κB signaling, and hence all antioxidants, including phytochemicals, can suppress NF-κB-dependent signaling.[69] Moreover, the inflammatory signaling induced by LPS and TNF cytokines is mediated via ROS-dependent signaling.[69] Lycopene reduces NF-κB, interferon regulatory factor-1, and signal transducer and activator of transcription-1α (STAT-1α)/DNA binding activity, which has been observed to correlate with a suppression of ROS generation as well as reduced macrophage activation.[70] This correlation can be explained by the inhibition of ROS-mediated NF-κB signaling by lycopene.

ANGIOGENESIS IN RETINAL PIGMENT EPITHELIUM

There is strong evidence that ROS and hypoxia inducible factor (HIF) are directly involved in stimulating angiogenesis, both in tumors and in the retina.[71,72] In particular, intermittent hypoxia, followed by reoxygenation, has been shown to affect the production of ROS, which increases HIF protein expression and prevents the hydroxylation of HIF-1α protein.[73] These molecular responses ultimately lead to the increased expression of vascular endothelial growth factor (VEGF), which is the major growth factor triggering CNV in the exudative AMD form.[74] One study showed that immunohistochemical analysis has revealed areas within the CNV membranes that are predominantly immunopositive for CD68 and cytokeratin, demonstrating the presence of RPE and/or macrophages, and that these cells strongly colocalize with the presence of HIF and VEGF. This localization of HIF expression supports the idea that hypoxia is a major stimulus for the development of submacular wound healing and CNV.[75]

Antiangiogenic Effects of Lycopene

HIF-1α can be induced in the presence of free radicals and suppressed if these free radicals are removed by antioxidants. Lycopene can directly remove the free radicals that make the survival of HIF-1α difficult and bind at the active site of the protein to inhibit its activity.[76] The inhibitory effects of lycopene on the migration of endothelial cells give it its antiangiogenic ability. Lycopene reduces network branching of the junction numbers, the number of tubules, and the tubule length in both human umbilical vein endothelial cells (HUVECs) and rat aortic rings at physiologic concentrations.[77] The inhibitory effect of lycopene on angiogenesis has been shown to be independent of the presence of proangiogenic agents, VEGF, and TNF-α. Thus, the antiangiogenic effects of lycopene have been demonstrated at a concentration that is achievable through dietary intake.[77] Lycopene has been shown to inhibit tube formation, invasion, and migration in HUVECs, and such action is accompanied by reduced activity of matrix metalloproteinase-2 (MMP-2), urokinase-type plasminogen activator, and protein expression of Rac1 and enhanced protein expression of tissue inhibitors of MMP-2 and plasminogen activator inhibitor-1. Moreover, lycopene has been shown to attenuate VEGF receptor-2 (VEGFR2)-mediated phosphorylation of extracellular signal-regulated kinase (ERK), p38, and Akt, as well as protein expression of PI3K.[78] In a study on human peripheral blood mononuclear cells (MNCs), conditioned medium of MNCs stimulated with lycopene resulted in the inhibition of endothelial cell proliferation and migration. Lycopene showed antiangiogenic effects by immunomodulation with upregulation of IL-12 and interferon-γ (IFN-γ) in MNCs. Furthermore, pretreatment of HUVECs with dexamethasone, an IL-12 inhibitor, blocked the antiangiogenic effects of lycopene-stimulated MNC conditioned medium in parallel with inhibition of IL-12 and IFN-γ.[79]

CELL PROLIFERATION AND MIGRATION OF RETINAL PIGMENT EPITHELIAL CELLS

In ocular diseases such as AMD, PVR, and proliferative diabetic retinopathy (PDR), RPE cell proliferation and migration result in severe visual impairment.[80] The pathologic process of PVR begins with a number of growth factors, such as platelet-derived growth factor (PDGF), transforming growth factor-β, heparin-binding epidermal growth factor, hepatocyte growth factor, connective tissue growth factor, fibroblast growth factor, and

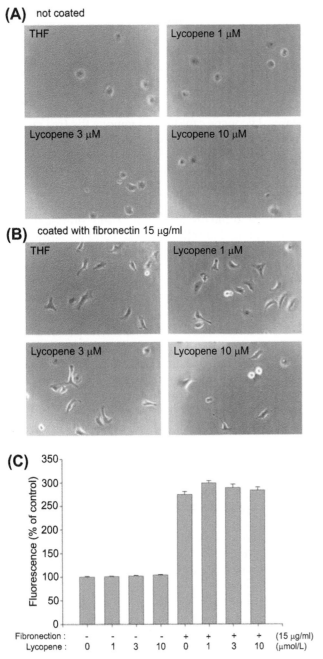

FIGURE 60.5 Adult retinal pigment epithelial cell-19 (ARPE19) cell adhesion was not affected following lycopene treatment. BCECF-labeled cells were treated with tetrahydrofuran (THF) or lycopene for 30 minutes and then seeded and allowed to undergo adhesion on plates (A) with or (B) without fibronectin (15 μg/ml) precoated at 37°C for 1 hour. Cells photographed under a phase-contrast microscope were used for morphologic analysis. (C) Fluorescence was measured using excitation and emission wavelength at 485 and 535 nm, respectively. *Source: Reprinted from Chan et al. Biochem Biophys Res Commun. 2009;388:172–6,[85] with permission.*

epidermal growth factor, which are released into the vitreous cavity.[81] Moreover, several cytokines, for example, TNF-α, TNF-β, IL-3, IL-6, and intercellular adhesion molecule-1, are upregulated.[82,83] These factors can disturb the local ocular cells, including the RPE and glial cells, causing them to migrate and proliferate in the vitreous cavity and form the PVR membrane.[84] RPE cells are considered a key element in the development of PVR,[84] and the diverse growth factors can be shown to trigger a variety of PVR-related phenotypical changes of RPE.[85]

Effects of Lycopene on Cell Proliferation and Migration

In many types of cancer cell, lycopene has shown antiproliferative effects. Lycopene suppresses the insulin-like growth factor-1 (IGF-1) pathway,[86] increases the level of IGF binding protein-3 (IGFBP-3), and decreases IGF type-I receptor (IGF-IR) expression in cells treated with IGF-1.[87] Moreover, lycopene inactivates Ras signaling, as evidenced by its translocation from cell membranes to cytosol, by inhibiting the expression of 3-hydroxy-3-methylglutaryl-coenzyme A (HMG-CoA) reductase expression and thereby modulating the mevalonate pathway in prostate, colon, and lung cancer cells.[88] Lycopene inhibits the transcriptional activity of MMP-9 by preventing the binding of the transcription factors NF-κB and stimulatory protein (Sp1) to the regulatory sequences of the MMP-9 gene, rather than inhibiting the enzyme directly.[89] In PVR, RPE cells are stimulated by chemotactic factors, such as PDGF, to migrate away from a monolayer into a provisional extracellular matrix, where they participate in epiretinal membrane formation.[90] Studies have shown that PDGF is a strong chemotactic factor for RPE cells in the presence of fibronectin,[91] and lycopene can significantly suppress PDGF-BB-induced RPE cell migration without cytotoxicity (Fig. 60.5). On the other hand, RPE cell adhesion and morphologic change on fibronectin are not affected by lycopene (Fig. 60.6). Preincubation of PDGF-BB with lycopene resulted in a marked inhibitory effect on the signaling of this growth factor in RPE cells, including phosphorylation of PDGFR-β. Lycopene binds to PDGF-BB, subsequently blocks the interaction of PDGF-BB with its receptors, and inactivates PDGF-BB functions. Furthermore, the inhibition of PDGF-BB-induced cell migration has been shown to coincide with reduced activation of PDGF-BB-induced Akt, ERK, and p38.[85]

CONCLUSIONS

Naturally present in tomatoes and other common fruits and vegetables, lycopene is associated with many beneficial effects on human health. Lycopene is not found in the human ARPE-19 cells, but the presence of lycopene has been shown in the RPE-choroid, ciliary body, and iris. It shows suppressive effects on oxidation,

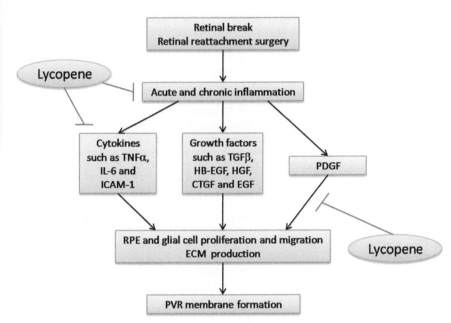

FIGURE 60.6 Proposed mechanisms of the suppression of proliferative vitreoretinopathy (PVR) membrane formation by lycopene. In PVR, retinal breaks and retinal reattachment surgery cause the ocular inflammation that promotes the proliferation and migration of retinal pigment epithelium (RPE) and glial cells. The suppression of inflammatory cascade and the inhibition of platelet-derived growth factor (PDGF)-BB-induced RPE cell migration are potential beneficial effects of lycopene in the prevention of PVR. TNFα: tumor necrosis factor-α; IL-6: interleukin-6; ICAM-1: intercellular adhesion molecule-1; TGFβ: transforming growth factor-β; HB-EGF: heparin-binding epidermal growth factor; HGF: hepatocyte growth factor; CTGF: connective tissue growth factor; EGF: epidermal growth factor; ECM: extracellular matrix.

inflammation, angiogenesis, and processes that cause cell proliferation and migration in many types of cells. This chapter has summarized the effects of lycopene and its derivatives *in vitro*. In RPE cells, lycopene shows antioxidative effects. Tomato extract reduces nitrotyrosine, protein carbonyls, and thiobarbituric acid-reactive substance formation in ARPE19 cells after H_2O_2 treatment. In the pathogenesis of AMD, phototoxicity, cigarette smoke, aging, and lipofuscin deposits induce oxidative stress in the RPE cells, which causes inflammation of the photoreceptor–RPE–Bruch's membrane–choriocapillaris complex. The resulting RPE atrophy enhances the choriocapillaris invasion to the subretinal space and leads to the development of neovascular AMD. Thus, the antioxidative, anti-inflammatory, and antiangiogenic effects of lycopene make it a promising nutrient against the progression of AMD (Fig. 60.5). In PVR, retinal breaks and retinal reattachment surgery cause the ocular inflammation that promotes the proliferation and migration of RPE and glial cells. The suppression of the inflammatory cascade and the inhibition of PDGF-BB-induced RPE cell migration are potential beneficial effects of lycopene in the preventions of PVR (Fig. 60.6). With its demonstrated antioxidant and antimigratory effects on RPE cells, lycopene is a strong candidate for the prevention and treatment of AMD and PVR.

TAKE-HOME MESSAGES

- Lycopene is not detectable in human retinal pigment epithelium (RPE) cells.
- The presence of lycopene has been shown in RPE-choroid, ciliary body, and iris.

- Lycopene shows inhibitory effects of oxidation, inflammation, angiogenesis, proliferation, and migration in different cell types.
- Lycopene demonstrates antioxidant and antimigratory effects in RPE cells.
- The beneficial effects of lycopene on the development of age-related macular degeneration are proposed through its antioxidative, anti-inflammatory, and antiangiogenic effects.
- The beneficial effects of lycopene on the progression of proliferative vitroretinopathy are suggested through its anti-inflammatory, antiproliferative, and antimigratory effects.

References

1. Mordente A, Guantario B, Meucci E, Silvestrini A, Lombardi E, Martorana GE, et al. Lycopene and cardiovascular diseases: an update. *Curr Med Chem* 2011;**18**:1146–63.
2. Tanaka T, Shnimizu M, Moriwaki H. Cancer chemoprevention by carotenoids. *Molecules* 2012;**17**:3202–42.
3. Rodriguez-Bustamante E, Sanchez S. Microbial production of C13-norisoprenoids and other aroma compounds via carotenoid cleavage. *Crit Rev Microbiol* 2007;**33**:211–30.
4. Rao AV, Rao LG. Carotenoids and human health. *Pharmacol Res* 2007;**55**:207–16.
5. Paiva SA, Russell RM. Beta-carotene and other carotenoids as antioxidants. *J Am Coll Nutr* 1999;**18**:426–33.
6. Britton G. Structure and properties of carotenoids in relation to function. *FASEB J* 1995;**9**:1551–8.
7. Shi J, Le Maguer M. Lycopene in tomatoes: chemical and physical properties affected by food processing. *Crit Rev Food Sci Nutr* 2000;**40**:1–42.
8. Unlu NZ, Bohn T, Francis DM, Nagaraja HN, Clinton SK, Schwartz SJ. Lycopene from heat-induced cis-isomer-rich tomato sauce is more bioavailable than from all-trans-rich tomato sauce in human subjects. *Br J Nutr* 2007;**98**:140–6.

9. Boileau TW, Boileau AC, Erdman Jr JW. Bioavailability of all-*trans* and *cis*-isomers of lycopene. *Exp Biol Med (Maywood)* 2002;**227**: 914–9.

10. Alien CM, Smith AM, Clinton SK, Schwartz SJ. Tomato consumption increases lycopene isomer concentrations in breast milk and plasma of lactating women. *J Am Diet Assoc* 2002;**102**:1257–62.

11. Hadley CW, Clinton SK, Schwartz SJ. The consumption of processed tomato products enhances plasma lycopene concentrations in association with a reduced lipoprotein sensitivity to oxidative damage. *J Nutr* 2003;**133**:727–32.

12. Schmidt-Dannert C, Umeno D, Arnold FH. Molecular breeding of carotenoid biosynthetic pathways. *Nat Biotechnol* 2000;**18**:750–3.

13. Khachik F, de Moura FF, Zhao DY, Aebischer CP, Bernstein PS. Transformations of selected carotenoids in plasma, liver, and ocular tissues of humans and in nonprimate animal models. *Invest Ophthalmol Vis Sci* 2002;**43**:3383–92.

14. Bernstein PS, Khachik F, Carvalho LS, Muir GJ, Zhao DY, Katz NB. Identification and quantitation of carotenoids and their metabolites in the tissues of the human eye. *Exp Eye Res* 2001;**72**: 215–23.

15. Chichili GR, Nohr D, Frank J, Flaccus A, Fraser PD, Enfissi EM, Biesalski HK. Protective effects of tomato extract with elevated beta-carotene levels on oxidative stress in ARPE-19 cells. *Br J Nutr* 2006;**96**:643–9.

16. Yu DY, Cringle SJ. Oxygen distribution and consumption within the retina in vascularised and avascular retinas and in animal models of retinal disease. *Prog Retin Eye Res* 2001;**20**:175–208.

17. Lange CA, Bainbridge JW. Oxygen sensing in retinal health and disease. *Ophthalmologica* 2011;**227**:115–31.

18. Glickman RD. Ultraviolet phototoxicity to the retina. *Eye Contact Lens* 2011;**37**:196–205.

19. Demontis GC, Longoni B, Marchiafava PL. Molecular steps involved in light-induced oxidative damage to retinal rods. *Invest Ophthalmol Vis Sci* 2002;**43**:2421–7.

20. Organisciak DT, Vaughan DK. Retinal light damage: mechanisms and protection. *Prog Retin Eye Res* 2010;**29**:113–34.

21. Algvere PV, Marshall J, Seregard S. Age-related maculopathy and the impact of blue light hazard. *Acta Ophthalmol Scand* 2006;**84**: 4–15.

22. Girotti AW, Kriska T. Role of lipid hydroperoxides in photo-oxidative stress signaling. *Antioxid Redox Signal* 2004;**6**:301–10.

23. Girotti AW. Photosensitized oxidation of membrane lipids: reaction pathways, cytotoxic effects, and cytoprotective mechanisms. *J Photochem Photobiol B* 2001;**63**:103–13.

24. Song J, Gao T, Ye M, Bi H, Liu G. The photocytotoxicity of different lights on mammalian cells in interior lighting system. *J Photochem Photobiol B* 2012;**117C**:13–8.

25. Sparrow JR, Gregory-Roberts E, Yamamoto K, Blonska A, Ghosh SK, Ueda K, Zhou J. The bisretinoids of retinal pigment epithelium. *Prog Retin Eye Res* 2012;**31**:121–35.

26. Spaide R. Autofluorescence from the outer retina and subretinal space: hypothesis and review. *Retina* 2008;**28**:5–35.

27. Boulton M, Rozanowska M, Rozanowski B. Retinal photodamage. *J Photochem Photobiol B* 2001;**64**:144–61.

28. Kaarniranta K, Hyttinen J, Ryhanen T, Viiri J, Paimela T, Toropainen E, et al. Mechanisms of protein aggregation in the retinal pigment epithelial cells. *Front Biosci (Elite Ed)* 2010;**2**:1374–84.

29. Bergmann M, Schutt F, Holz FG, Kopitz J. Inhibition of the ATP-driven proton pump in RPE lysosomes by the major lipofuscin fluorophore A2-E may contribute to the pathogenesis of age-related macular degeneration. *FASEB J* 2004;**18**:562–4.

30. Vives-Bauza C, Anand M, Shirazi AK, Magrane J, Gao J, Vollmer-Snarr HR, et al. The age lipid A2E and mitochondrial dysfunction synergistically impair phagocytosis by retinal pigment epithelial cells. *J Biol Chem* 2008;**283**:24770–80.

31. Yoon KD, Yamamoto K, Ueda K, Zhou J, Sparrow JR. A novel source of methylglyoxal and glyoxal in retina: implications for age-related macular degeneration. *PLoS ONE* 2012;**7**:e41309.

32. Pawlak A, Rozanowska M, Zareba M, Lamb LE, Simon JD, Sarna T. Action spectra for the photoconsumption of oxygen by human ocular lipofuscin and lipofuscin extracts. *Arch Biochem Biophys* 2002;**403**:59–62.

33. Kim SR, Jockusch S, Itagaki Y, Turro NJ, Sparrow JR. Mechanisms involved in A2E oxidation. *Exp Eye Res* 2008;**86**:975–82.

34. Jia L, Liu Z, Sun L, Miller SS, Ames BN, Cotman CW, Liu J. Acrolein, a toxicant in cigarette smoke, causes oxidative damage and mitochondrial dysfunction in RPE cells: protection by (R)-alpha-lipoic acid. *Invest Ophthalmol Vis Sci* 2007;**48**:339–48.

35. Fujihara M, Nagai N, Sussan TE, Biswal S, Handa JT. Chronic cigarette smoke causes oxidative damage and apoptosis to retinal pigmented epithelial cells in mice. *PLoS ONE* 2008;**3**:e3119.

36. Cano M, Thimmalappula R, Fujihara M, Nagai N, Sporn M, Wang AL, et al. Cigarette smoking, oxidative stress, the anti-oxidant response through Nrf2 signaling, and age-related macular degeneration. *Vision Res* 2010;**50**:652–64.

37. Espinosa-Heidmann DG, Suner IJ, Catanuto P, Hernandez EP, Marin-Castano ME, Cousins SW. Cigarette smoke-related oxidants and the development of sub-RPE deposits in an experimental animal model of dry AMD. *Invest Ophthalmol Vis Sci* 2006;**47**:729–37.

38. Bertram KM, Baglole CJ, Phipps RP, Libby RT. Molecular regulation of cigarette smoke induced-oxidative stress in human retinal pigment epithelial cells: implications for age-related macular degeneration. *Am J Physiol Cell Physiol* 2009;**297**:C1200–10.

39. Khandhadia S, Lotery A. Oxidation and age-related macular degeneration: insights from molecular biology. *Expert Rev Mol Med* 2010;**12**:e34.

40. Bohm F, Edge R, Truscott G. Interactions of dietary carotenoids with activated (singlet) oxygen and free radicals: potential effects for human health. *Mol Nutr Food Res* 2012;**56**:205–16.

41. Krinsky NI. The antioxidant and biological properties of the carotenoids. *Ann N Y Acad Sci* 1998;**854**:443–7.

42. Bohm F, Edge R, Truscott TG. Interactions of dietary carotenoids with singlet oxygen ($1O_2$) and free radicals: potential effects for human health. *Acta Biochim Pol* 2012;**59**:27–30.

43. Cardinault N, Abalain JH, Sairafi B, Coudray C, Grolier P, Rambeau M, et al. Lycopene but not lutein nor zeaxanthin decreases in serum and lipoproteins in age-related macular degeneration patients. *Clin Chim Acta* 2005;**357**:34–42.

44. Young AJ, Lowe GM. Antioxidant and prooxidant properties of carotenoids. *Arch Biochem Biophys* 2001;**385**:20–7.

45. Galano A, Francisco-Marquez M. Reactions of OOH radical with beta-carotene, lycopene, and torulene: hydrogen atom transfer and adduct formation mechanisms. *J Phys Chem B* 2009;**113**:11338–45.

46. Takashima M, Shichiri M, Hagihara Y, Yoshida Y, Niki E. Capacity of peroxyl radical scavenging and inhibition of lipid peroxidation by beta-carotene, lycopene, and commercial tomato juice. *Food Funct* 2012;**3**:1153–60.

47. Muzandu K, Ishizuka M, Sakamoto KQ, Shaban Z, El Bohi K, Kazusaka A, Fujita S. Effect of lycopene and beta-carotene on peroxynitrite-mediated cellular modifications. *Toxicol Appl Pharmacol* 2006;**215**:330–40.

48. Bast A, Haenen GR, van den Berg R, van den Berg H. Antioxidant effects of carotenoids. *Int J Vitam Nutr Res* 1998;**68**:399–403.

49. Mortensen A, Skibsted LH. Relative stability of carotenoid radical cations and homologue tocopheroxyl radicals. A real time kinetic study of antioxidant hierarchy. *FEBS Lett* 1997;**417**:261–6.

50. Rao AV, Agarwal S. Role of antioxidant lycopene in cancer and heart disease. *J Am Coll Nutr* 2000;**19**:563–9.

51. Bohm F, Tinkler JH, Truscott TG. Carotenoids protect against cell membrane damage by the nitrogen dioxide radical. *Nat Med* 1995;**1**:98–9.

52. Sundelin SP, Nilsson SE. Lipofuscin-formation in retinal pigment epithelial cells is reduced by antioxidants. *Free Radic Biol Med* 2001;**31**:217–25.

53. Ehrlich R, Harris A, Kheradiya NS, Winston DM, Ciulla TA, Wirostko B. Age-related macular degeneration and the aging eye. *Clin Interv Aging* 2008;**3**:473–82.

54. Beatty S, Koh H, Phil M, Henson D, Boulton M. The role of oxidative stress in the pathogenesis of age-related macular degeneration. *Surv Ophthalmol* 2000;**45**:115–34.

55. Anderson DH, Mullins RF, Hageman GS, Johnson LV. A role for local inflammation in the formation of drusen in the aging eye. *Am J Ophthalmol* 2002;**134**:411–31.

56. Bergeron-Sawitzke J, Gold B, Olsh A, Schlotterbeck S, Lemon K, Visvanathan K, et al. Multilocus analysis of age-related macular degeneration. *Eur J Hum Genet* 2009;**17**:1190–9.

57. Grossniklaus HE, Ling JX, Wallace TM, Dithmar S, Lawson DH, Cohen C, et al. Macrophage and retinal pigment epithelium expression of angiogenic cytokines in choroidal neovascularization. *Mol Vis* 2002;**8**:119–26.

58. Espinosa-Heidmann DG, Suner IJ, Hernandez EP, Monroy D, Csaky KG, Cousins SW. Macrophage depletion diminishes lesion size and severity in experimental choroidal neovascularization. *Invest Ophthalmol Vis Sci* 2003;**44**:3586–92.

59. Kauppinen A, Niskanen H, Suuronen T, Kinnunen K, Salminen A, Kaarniranta K. Oxidative stress activates NLRP3 inflammasomes in ARPE-19 cells – implications for age-related macular degeneration (AMD). *Immunol Lett* 2012;**147**:29–33.

60. Coleman HR, Chan CC, Ferris III FL, Chew EY. Age-related macular degeneration. *Lancet* 2008;**372**:1835–45.

61. Wang AL, Lukas TJ, Yuan M, Du N, Tso MO, Neufeld AH. Autophagy and exosomes in the aged retinal pigment epithelium: possible relevance to drusen formation and age-related macular degeneration. *PLoS ONE* 2009;**4**:e4160.

62. Donoso LA, Kim D, Frost A, Callahan A, Hageman G. The role of inflammation in the pathogenesis of age-related macular degeneration. *Surv Ophthalmol* 2006;**51**:137–52.

63. Oh H, Takagi H, Takagi C, Suzuma K, Otani A, Ishida K, et al. The potential angiogenic role of macrophages in the formation of choroidal neovascular membranes. *Invest Ophthalmol Vis Sci* 1999;**40**:1891–8.

64. Kim GY, Kim JH, Ahn SC, Lee HJ, Moon DO, Lee CM, Park YM. Lycopene suppresses the lipopolysaccharide-induced phenotypic and functional maturation of murine dendritic cells through inhibition of mitogen-activated protein kinases and nuclear factor-kappaB. *Immunology* 2004;**113**:203–11.

65. Feng D, Ling WH, Duan RD. Lycopene suppresses LPS-induced NO and IL-6 production by inhibiting the activation of ERK, p38MAPK, and NF-kappaB in macrophages. *Inflamm Res* 2010;**59**:115–21.

66. Park JS, Svetkauskaite D, He Q, Kim JY, Strassheim D, Ishizaka A, Abraham E. Involvement of toll-like receptors 2 and 4 in cellular activation by high mobility group box 1 protein. *J Biol Chem* 2004;**279**:7370–7.

67. Touyz RM, Schiffrin EL. Peroxisome proliferator-activated receptors in vascular biology – molecular mechanisms and clinical implications. *Vascul Pharmacol* 2006;**45**:19–28.

68. Palozza P, Simone R, Catalano A, Monego G, Barini A, Mele MC, et al. Lycopene prevention of oxysterol-induced proinflammatory cytokine cascade in human macrophages: inhibition of NF-kappaB nuclear binding and increase in PPARgamma expression. *J Nutr Biochem* 2011;**22**:259–68.

69. Surh J, Kundu JK, Na HK, Lee JS. Redox-sensitive transcription factors as prime targets for chemoprevention with anti-inflammatory and antioxidative phytochemicals. *J Nutr* 2005;**135**:2993–3001S.

70. De Stefano D, Maiuri MC, Simeon V, Grassia G, Soscia A, Cinelli MP, Carnuccio R. Lycopene, quercetin and tyrosol prevent macrophage activation induced by gliadin and IFN-gamma. *Eur J Pharmacol* 2007;**566**:192–9.

71. Al-Shabrawey M, Rojas M, Sanders T, Behzadian A, El-Remessy A, Bartoli M, et al. Role of NADPH oxidase in retinal vascular inflammation. *Invest Ophthalmol Vis Sci* 2008;**49**:3239–44.

72. Ushio-Fukai M, Nakamura Y. Reactive oxygen species and angiogenesis: NADPH oxidase as target for cancer therapy. *Cancer Lett* 2008;**266**:37–52.

73. Yuan G, Nanduri J, Khan S, Semenza GL, Prabhakar NR. Induction of HIF-1alpha expression by intermittent hypoxia: involvement of NADPH oxidase, Ca^{2+} signaling, prolyl hydroxylases, and mTOR. *J Cell Physiol* 2008;**217**:674–85.

74. Schlingemann RO. Role of growth factors and the wound healing response in age-related macular degeneration. *Graefes Arch Clin Exp Ophthalmol* 2004;**242**:91–101.

75. Sheridan CM, Pate S, Hiscott P, Wong D, Pattwell DM, Kent D. Expression of hypoxia-inducible factor-1alpha and -2alpha in human choroidal neovascular membranes. *Graefes Arch Clin Exp Ophthalmol* 2009;**247**:1361–7.

76. Upadhyay J, Kesharwani RK, Misra K. Comparative study of antioxidants as cancer preventives through inhibition of HIF-1 alpha activity. *Bioinformation* 2009;**4**:233–6.

77. Elgass S, Cooper A, Chopra M. Lycopene inhibits angiogenesis in human umbilical vein endothelial cells and rat aortic rings. *Br J Nutr* 2012;**108**:431–9.

78. Chen ML, Lin YH, Yang CM, Hu ML. Lycopene inhibits angiogenesis both *in vitro* and *in vivo* by inhibiting MMP-2/uPA system through VEGFR2-mediated PI3K-Akt and ERK/p38 signaling pathways. *Mol Nutr Food Res* 2012;**56**:889–99.

79. Huang CS, Chuang CH, Lo TF, Hu ML. Anti-angiogenic effects of lycopene through immunomodulation of cytokine secretion in human peripheral blood mononuclear cells. *J Nutr Biochem* 2013;**24**:428–34.

80. de Silva DJ, Kwan A, Bunce C, Bainbridge J. Predicting visual outcome following retinectomy for retinal detachment. *Br J Ophthalmol* 2008;**92**:954–8.

81. Chen YJ, Tsai RK, Wu WC, He MS, Kao YH, Wu WS. Enhanced PKCdelta and ERK signaling mediate cell migration of retinal pigment epithelial cells synergistically induced by HGF and EGF. *PLoS ONE* 2012;**7**:e44937.

82. Ricker LJ, Kijlstra A, Kessels AG, de Jager W, Liem AT, Hendrikse F, La Heij EC. Interleukin and growth factor levels in subretinal fluid in rhegmatogenous, retinal detachment: a case–control study. *PLoS ONE* 2011;**6**:e19141.

83. Sadaka A, Giuliari GP. Proliferative vitreoretinopathy: current and emerging treatments. *Clin Ophthalmol* 2012;**6**:1325–33.

84. Pastor JC, de la Rua ER, Martin F. Proliferative vitreoretinopathy: risk factors and pathobiology. *Prog Retin Eye Res* 2002;**21**:127–44.

85. Chan CM, Fang JY, Lin HH, Yang CY, Hung CF. Lycopene inhibits PDGF-BB-induced retinal pigment epithelial cell migration by suppression of PI3K/Akt and MAPK pathways. *Biochem Biophys Res Commun* 2009;**388**:172–6.

86. Karas M, Amir H, Fishman D, Danilenko M, Segal S, Nahum A, et al. Lycopene interferes with cell cycle progression and insulin-like growth factor I signaling in mammary cancer cells. *Nutr Cancer* 2000;**36**:101–11.

87. Kanagaraj P, Vijayababu MR, Ravisankar B, Anbalagan J, Aruldhas MM, Arunakaran J. Effect of lycopene on insulin-like growth factor-I, IGF binding protein-3 and IGF type-I receptor in prostate cancer cells. *J Cancer Res Clin Oncol* 2007;**133**:351–39.

88. Palozza P, Colangelo M, Simone R, Catalano A, Boninsegna A, Lanza P, et al. Lycopene induces cell growth inhibition by altering mevalonate pathway and Ras signaling in cancer cell lines. *Carcinogenesis* 2010;**31**:1813–21.

89. Huang CS, Fan YE, Lin CY, Hu ML. Lycopene inhibits matrix metalloproteinase-9 expression and down-regulates the binding activity of nuclear factor-kappa B and stimulatory protein-1. *J Nutr Biochem* 2007;**18**:449–56.

90. Casaroli Marano RP, Vilaro S. The role of fibronectin, laminin, vitronectin and their receptors on cellular adhesion in proliferative vitreoretinopathy. *Invest Ophthalmol Vis Sci* 1994;**35**: 2791–803.

91. Hinton DR, He S, Graf K, Yang D, Hsueh WA, Ryan SJ, Law RE. Mitogen-activated protein kinase activation mediates PDGF-directed migration of RPE cells. *Exp Cell Res* 1998;**239**: 11–5.

Ascorbate Transport in Retinal Cells and Its Relationship with the Nitric Oxide System

C.C. Portugal[1], R. Socodato[1], T.G. Encarnação[1], I.C.L. Domith[1], M. Cossenza[1, 3], R. Paes-de-Carvalho[1, 2]

[1]Program of Neurosciences, Fluminense Federal University, Niterói, Brazil, [2]Department of Neurobiology, Institute of Biology, Fluminense Federal University, Niterói, Brazil, [3]Departament of Physiology and Pharmacology, Biomedical Institute, Fluminense Federal University, Niterói, Brazil

INTRODUCTION

The visual system is important for the perception of environmental stimuli, and for this perception many neurochemical signals are needed. Vitamin C, which is composed of the reduced form ascorbate and the oxidized form dehydroascorbate, is an important factor for this system, regulating various functions such as neuronal differentiation and survival. This chapter will explore the bioavailability of vitamin C (measured by the regulation of its transport) in the central nervous system (CNS), using the retina as a model, modulated by important cellular messengers such as L-arginine and nitric oxide.

VITAMIN C

Brief Overview

Ascorbate is an important antioxidant molecule with historical importance in the prevention of scurvy, a disease that affected sailors in the past and was caused by a lack of vitamin C on long sea journeys. The major natural source of vitamin C is citrus fruits, where it can be found in reduced or oxidized forms, ascorbate or dehydroascorbate, respectively. Ascorbate is capable of acting in various biologic processes such as myelin shaft formation,[1] prevention of dopamine oxidation,[2] and regulation of collagen synthesis,[3] and has antioxidant/pro-oxidant properties.[4]

Ascorbate and dehydroascorbate are hydrophilic molecules, and their diffusion through the plasma membrane is insignificant. Thus, membrane machinery is needed for their transport across cell membranes.

There are two such transport systems: sodium-dependent transport performed by vitamin C cotransporters (SVCTs), which transport ascorbate stereospecifically, and glucose transporters (GLUTs), which transport dehydroascorbate. For certain species the only source of vitamin C is from food. Most species are capable of synthesizing ascorbate from glucose, through the enzyme α-gulono-gamma-lactone oxidase, but in humans, guinea-pigs, and some bat species, which lack a functional gene to synthesize the enzyme, the biosynthetic pathway cannot be used for ascorbate synthesis.[5]

There are four SVCT isoforms, but only two of them, SVCT-1 and SVCT-2, are capable of transporting ascorbate. Those transporters are distributed in several tissues differing in their localization and affinity for ascorbate. SVCT-1 has a higher maximum activity (V_{max}) and lower affinity for ascorbate, whereas SVCT-2 has a higher affinity for ascorbate.[6] SVCT-1 is mainly expressed in epithelial organs such as the intestine and skin, while SVCT-2 is abundant in tissues with high metabolic activity such as the heart and retina.[6] GLUT isoforms 1, 2, 3, and 4 are capable of transporting dehydroascorbate through a facilitated diffusion mechanism, with higher affinity for dehydroascorbate than for glucose.[5,7,8,79]

Ascorbate Transport

Ascorbate transport is carried out in a stoichiometric ratio, with two sodium (Na^+) ions being transported for each ascorbate taken up.[7,9] Therefore, the transport mechanism relies on cooperation between Na^+ and ascorbate. In SVCT-2 there are two Na^+ binding sites. For ascorbate uptake to occur, first one Na^+ ion must bind to the transporter, increasing its affinity for ascorbate. Then, ascorbate binding cooperatively promotes the binding of another Na+ as well as ascorbate, which, in this way, is carried into the cell. Ouabain, a specific inhibitor of Na+/K+-ATPase, inhibits the uptake of ascorbate, showing the requirement for a constant content of Na+ to create an electrochemical gradient for the transport of ascorbate by SVCT-2.[10]

There are three hypotheses to explain ascorbate release from nerve cells (Fig. 61.1). The heteroexchange hypothesis states that there is a transporter capable of performing a heteroexchange of glutamate/ascorbate. In this way, the transporter is able to conduct glutamate uptake and ascorbate release[11] (Fig. 61.1A), and this phenomenon has been shown to be blocked by inhibitors of glutamate transporters.[12] The volume-sensitive organic anion channels hypothesis predicts that glutamate uptake by an ordinary glutamate transporter, which is Na+ dependent, causes swelling in astrocytes, activating volume-sensitive organic anion channels and allowing ascorbate to passively diffuse according to its concentration gradient (Fig. 61.1B). The third and most recent hypothesis predicts that glutamate, in a ligand-gated ion channel-dependent manner, causes SVCT-2 reversion, allowing the efflux of ascorbate[13] (Fig. 61.1C). This phenomenon is Na+ dependent, blocked by sulfinpyrazone, an SVCT inhibitor, and independent of glutamate transporters.[13]

Ascorbate Absorption and the Role of SVCT-2

After food ingestion, vitamin C can be absorbed in the intestine by both SVCT, with a better absorption in distal segments of the human intestine, and GLUT, through the entire length of the human small intestine.[7] Epithelial cells from the intestine have a polarized distribution of SVCT-1, with a heterogeneous localization in intracellular organelles, such as trafficking vesicles and endosomes, in addition to its localization at the apical membrane of polarized cells.[14] Export from the endoplasmic reticulum to the apical brush-border membrane is dependent on the C-terminal region of the transporter since amino acid modifications in this region result in a loss of apical targeting.[14] Moreover, ascorbate uptake in endothelial cells from SVCT-2+/− mice is reduced.[15] SVCT-2 protein localization is preferentially basolateral in endothelial cells, suggesting that SVCT-2 does not contribute to ascorbate absorption but could contribute, in organisms that synthesize ascorbate, in the transport of ascorbate from the blood to enterocytes through the basolateral membrane.[15] Vitamin C in the bloodstream can be excreted and reabsorbed in the kidney, which expresses SVCT-1.[16] In the retina, vitamin C transport occurs through the blood–retinal barrier[17,18] (detailed information can be found in Chapter 32, 'Vitamin Transport across the Blood–Retinal Barrier: Focus on Vitamins C, E, and Biotin').

In the retina, ascorbate acts as an antioxidant molecule, regenerating α-tocopherol (vitamin E) from α-tocopherol radical, thus preventing vitamin E itself from acting as a pro-oxidant molecule, which in turn prevents harmful events in the cornea and conjunctiva caused by vitamin A deficiency.[19,20]

Ascorbate Modulation of Retinal Neurotransmission and Visual Processing

The retina is a specialized tissue of the CNS (Fig. 61.2A), which originates from the neural ectoderm. It includes both sensory neurons that respond to light and intricate neural circuits that perform the first stages of image processing; ultimately, an electrical message travels down the optic nerve into the superior colliculus and

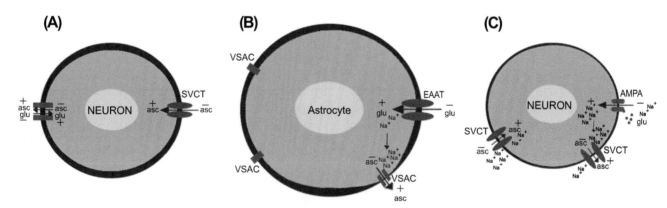

FIGURE 61.1 Scheme highlighting ascorbate release hypotheses. (A) The heteroexchange hypothesis predicts that there is a transporter capable of performing glutamate/ascorbate antiport, promoting ascorbate release. (B) The volume-sensitive organic anion channels hypothesis predicts that glutamate uptake, by a Na+-dependent glutamate transporter, causes a membrane microdomain swelling by increasing intracellular Na+ in astrocytes, activating volume-sensitive anion channels (VSACs), allowing ascorbate to diffuse according to its concentration gradient. EAAT: excitatory amino acid transporter. (C) The vitamin C transporter (SVCT-2) reversal hypothesis predicts that Na+ influx by activation of α-amino-3-hydroxy-5-methyl-4-isoazo-lepropionic acid (AMPA) receptors is capable of inducing SVCT-2 reversal, releasing ascorbate.

then to higher centers responsible for vision processing and to the visual cortex. Ascorbate is present at high concentrations in the retina,[21] and it has been detected in early stages of development in the interstitial matrix between the lens anlagen and the optic vesicle, suggesting its importance in eye formation.[22] As in other CNS regions, the retina contains neural circuits employing the excitatory transmitter glutamate, the inhibitory transmitter γ-aminobutyric acid (GABA), and dopamine, the major catecholamine in the vertebrate retina. There are two such circuits in the retina, one in the outer plexiform layer (OPL), involving glutamatergic photoreceptors, GABAergic horizontal cells, and dopaminergic interplexiform cells that converge on bipolar cells, and another in the inner plexiform layer (IPL), involving glutamatergic bipolar cells, GABAergic amacrine cells, and dopaminergic interplexiform cells that converge on ganglion cells.[23] Moreover, an ascorbate modulatory effect within this circuitry has been described.[24]

In the retina, release and uptake of neurotransmitters are related to membrane potential, which is determined by the activity of potassium (K^+) ion channels. Ascorbate, at physiologic concentrations (around 200 μM), inhibits voltage-gated K^+ currents ($I_{K(V)}$) of bipolar cells through a dopamine D_1 receptor/G-protein/ cyclic adenosine monophosphate (cAMP)-dependent protein kinase (PKA)-dependent mechanism[25] (Fig. 61.2B). Moreover, ascorbate released from retinal cells plays an important role in increasing dopamine half-life, consequently increasing the efficiency of dopaminergic neurotransmission in the retina.[26] In the dark, photoreceptors release glutamate that promotes GABA efflux from horizontal cells. However, the extracellular concentrations of retinal dopamine are increased when photoreceptors are stimulated by light, and this effect is recognized as a light adaptation signal by reducing photosensitivity. Ascorbate, like dopamine, inhibits dark-induced GABA efflux.[24] Moreover, $GABA_A$ receptors, located on the presynaptic terminals of bipolar cells, can be modulated by ascorbate, which potentiates $GABA_A$-mediated currents in the intact retinal tissue. Ascorbate can also modulate $GABA_C$ receptors, probably by increasing its affinity for GABA.[27] However, ascorbate can reduce metal ions, leading to the formation of free radicals such as ascorbate/Fe^{2+}. It has been shown that this complex is a nonenzymatic membrane peroxidation agent in biologic systems, causing oxidative stress and increasing GABA release through the reversal of the GABA transporter in the developing retina[28] (Fig. 61.2C).

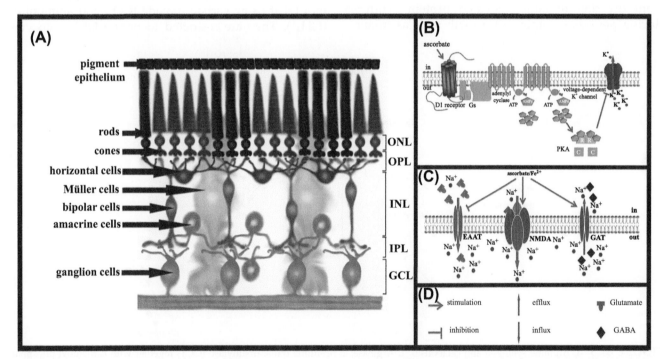

FIGURE 61.2 The retina and ascorbate signaling in bipolar cells. (A) The mature retina consists of well-defined tissue arranged in a laminar structure: outer nuclear layer (ONL), where the photoreceptors are found (rods and cones); outer plexiform layer (OPL), synapse region between photoreceptors, bipolar, and horizontal cells; inner nuclear layer (INL), which shows the cell bodies of horizontal, bipolar, Müller glia, and amacrine cells; inner plexiform layer (IPL), the region of synapse between INL cells and ganglion cell layer (GCL), consisting of ganglion cells and displaced amacrine. (B) Ascorbate at physiologic concentrations inhibits voltage-dependent K^+ currents ($I_{K(V)}$) in retinal bipolar cells through a D_1 dopamine receptor/G-protein/protein kinase A (PKA)-dependent mechanism. (C) Oxidative stress, caused by ascorbate/Fe^{2+} complex, increases N-methyl-D-aspartate (NMDA) receptor activity, further contributing to neuronal hyperexcitability through Na^+ entry. EAAT: excitatory amino acid transporter; GAT: γ-aminobutyric acid (GABA) transporter. (D) Symbols used in parts (A)–(C).

A redox site in *N*-methyl-D-aspartate (NMDA) receptors, probably localized in its extracellular region, has been implicated with its gating properties. Oxidative stress induced by ascorbate/Fe^{2+} increases NMDA receptor activity (Fig. 61.2C), further contributing to neuronal hyperexcitability through the entry of Na^+ and consequently triggering the release of GABA. Moreover, the increased concentration of Na^+ in retinal cells due to oxidative stress, mediated by ascorbate/Fe^{2+}, inhibits excitatory amino acid transporter (EAAT) systems in retinal cells (Fig. 61.2C), which are responsible for glutamate uptake.[29] On the other hand (as stated above), glutamate induces ascorbate release in retinal cells through a mechanism independent of EAAT but dependent on Na^+.[13] Thereby, visual processing in the retina involves a reciprocal modulation between ascorbate and classic neurotransmitters such as glutamate, GABA, and dopamine.

L-ARGININE AND NITRIC OXIDE

L-Arginine: An Outline

L-Arginine (L-Arg) has been classified as a 'semiessential' amino acid owing to its mandatory intake from the diet to attain elevated protein synthesis during periods of rapid cell growth such as infancy, childhood, illness, or trauma.[30] For its synthesis and catabolism, there is a complex network of enzymes located in different tissues. Moreover, there is no single organ or cell type expressing all enzymes for complete L-Arg recycling (Fig. 61.3). Extracellular traffic of L-Arg and its derivatives ensures the balance between production and consumption of these compounds. In this way, both intracellular metabolism and membrane transport features are pivotal for the physiologic or pathophysiologic roles played by L-Arg within organisms.[30]

L-Arginine Synthesis and Catabolism

In the first half of the 20th century, Krebs and Henseleit (1932) discovered the urea cycle (Fig. 61.3), which was the first biochemical cycle related to detoxification of nitrogen excess from the body using L-Arg, L-ornithine (L-Orn), and L-citrulline (L-Cit) as enzymatic substrates.[31] The enzymes of the urea cycle are essentially as follows: arginase; ornithine carbamoyltransferase (OCT), which works together with carbamoyl phosphate synthetase-I (CPS-I); argininosuccinate synthase (ASS); and argininosuccinate lyase (ASL) (Fig. 61.3).

OCT, together with CPS-I, is located in the mitochondrial matrix and catalyzes the transformation of L-Orn into L-Cit using carbon dioxide (CO_2) and ammonium (NH_4^+). They are restricted to periportal hepatocytes,

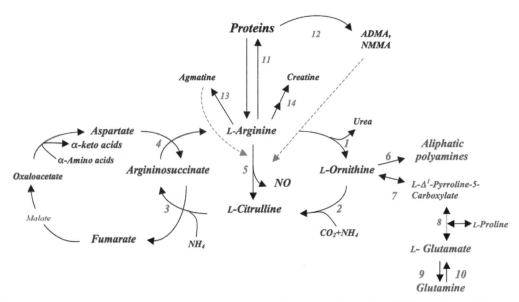

FIGURE 61.3 L-Arginine synthesis and catabolism. 1: Arginase; 2: ornithine transcarbamylase (OCT) together with carbamoyl phosphate synthetase-I (CPS-I); 3: argininosuccinate synthase (ASS); 4: argininosuccinate lyase (ASL); 5: nitric oxide synthase (NOS); 6: ornithine decarboxylase (ODC); 7: ornithine-δ-aminotransferase (OAT). 8: Three steps, two involving the enzyme Δ1-pyrroline-5-carboxylate synthetase (P5CS), with γ-glutamyl kinase and γ-glutamyl phosphate reductase activity and a subsequent spontaneous reaction with addition of L-proline; 9: glutamine synthetase; 10: glutaminase; 11: a process involving many enzymatic pathways, starting with the arginyl-tRNA synthetase; this same enzyme is also involved in post-translational *arginilation* in L-Asp residues or L-Glu N-terminal regions of proteins; 12: a process involving post-translational methylation in L-Arg residues by arginine *N*-methyltransferase family (PRMTs) and subsequent release of *N*-monomethylarginine (NMMA) and asymmetric dimethylarginine (ADMA) by protein degradation; 13: arginine decarboxylase (ADC); 14: L-arginine–glycine amidinotransferase. The headless arrows indicate inhibition.

epithelial cells of the small intestine mucosa, and, to a lesser extent, large intestine colonocytes.[30] In the retina, the urea cycle is incomplete, leading ornithine metabolism to other important enzymes such as ornithine-δ-aminotransferase (OAT). Another critical enzyme is ornithine decarboxylase (ODC), which catalyzes the decarboxylation of ornithine to form putrescine (Fig. 61.3). Putrescine is the first in a sequence of aliphatic polyamines produced in some tissues, and its levels are established by ODC activity. Other polyamines such as spermidine and spermine are synthesized sequentially by S-adenosyl-L-methionine decarboxylase (SAMdc), both putrescine availability (reflecting the activity of ODC) and SAMdc activity being rate-limiting steps for polyamine synthesis[32] (Fig. 61.3).

Sequentially in the urea cycle, ASS and ASL are both present in a wide variety of cells, including retinal neurons, and are responsible for synthesizing argininosuccinate and L-Arg, respectively. Because of the absence of OCT and CPS-I in the CNS, detoxification of surplus NH_4^+, which would be involved in this reaction, occurs, primarily through the glutamate/GABA–glutamine cycle between neuronal and glial cells[33] (Fig. 61.4). Specifically in the retina, there is no detailed information regarding detoxification of NH_4^+, but the absence of CPS-I and OCT is well characterized.[34]

The hydrolytic cleavage of L-Arg, to generate urea and L-Orn, is catalyzed by two isoforms of arginase (I and II) (Fig. 61.3). These isoforms are encoded by different genes and differ in their molecular properties, tissue distribution, subcellular location, and expression.[35] Substantial arginase I activity was found in periportal hepatocytes, which play an essential role in hepatic and intestinal NH_4^+ detoxication. Even though arginase II seems to be expressed in a variety of tissues, including the retina, its specific distribution in retinal cell types remains elusive.[36] Intracellular compartmentalization of arginases in endothelial cells has important implications for the production of L-Arg derivatives. Cytosolic colocalization of arginase I with ODC preferentially directs L-Orn towards the synthesis of aliphatic polyamines, whereas arginase II may preferentially direct ornithine towards either proline or glutamate production owing to its colocalization with OAT in the mitochondria.[35] ODC and SAMdc, as well as putrescine, spermidine, and spermine, have been described in retinal pigmented epithelium cells (RPEs), where they seem to be essential for retinal Müller glia migration,[34] while polyamines may act as endogenous modulators of the activity of inward rectifying K^+ channels in retinal cells.[37] Likewise, in the neonatal retina, discrete localization in photoreceptor outer segments and ganglion cells may also suggest an important role for these compounds during retinal tissue development.[32]

Relationship between Nitric Oxide Synthase and Arginase

Almost all mammalian cells contain one of the isoforms of nitric oxide synthase (NOS), which synthesizes nitric oxide (NO) and L-Cit from L-Arg (Fig. 61.3). There are three NOS isoforms, which were initially described according to their location. NOS-I is constitutively expressed in neurons, and its presence was extensively demonstrated in the retina.[38] NOS-II, the inducible isoform, has been attributed to various inflammatory effects when, for instance, microglia cells are stimulated with proinflammatory agents such as interleukin-1β (IL-1β), tumor necrosis factor-α (TNF-α), interferon-γ (IFN-γ), or lipopolysaccharide (LPS) that are strongly implicated in cellular defense mechanisms and neurodegenerative disorders.[39] NOS-III is mainly expressed in the vascular system.[40] All isoforms play a substantial role in both physiologic and pathophysiologic contexts within the CNS.

Both arginase and NOS use L-Arg as a substrate (Fig. 61.3), and therefore arginase was recently recognized as a critical regulator of NO production by competing with NOS for L-Arg. Arginase has been reported to inhibit endogenous NOS function, and its hyperfunction has been linked to several diseases, including atherosclerosis, hypertension, asthma, cystic fibrosis, arthritis, psoriasis, glomerulonephritis, and endothelial dysfunction in diabetes.[36,41,42] Moreover, increased arginase expression was also related to impaired NO synthesis and wound healing in diabetes.[36,41] Within the cell, competition between NOS and arginase for L-Arg is kinetically explained. Although NOS ($K_m \approx 5\,\mu M$) has a much higher affinity for L-Arg than arginase ($K_m \approx 5\,mM$), the maximum activity (V_{max}) of arginase is 1000-fold greater than NOS, indicating similar rates of substrate use at physiologic L-Arg concentrations.[42]

In the 1970s, Simell and Takki reported hyperornithinemia with gyrate atrophy of the choroid and retina, an autosomal recessive chorioretinal degeneration. Patients with this disease usually exhibit progressive night blindness and loss of vision.[43] It has been postulated that the initial site of insult in gyrate atrophy of the choroid and retina occurs within the RPE,[44] and some authors have reported that inactivation of OAT in human RPE cell lines leads to cell death by excessive spermine generation.[45]

Another important mechanism associated with the deleterious actions of arginase is related to L-Arg availability. Under normal physiologic conditions, NOS uses L-Arg to produce NO. However, if L-Arg supply is limited, NOS can use molecular oxygen to produce superoxide. Superoxide reacts rapidly with any residual NO to form peroxynitrite, creating an extremely noxious environment. In line with this, increased production of reactive oxygen species (ROS), including peroxynitrite, has been implicated in the pathogenesis of vascular inflammation, injury, and dysfunction.[41,42]

Neuronal Nitric Oxide Synthase Activity

Several groups have shown that NOS-I is the primary source of neuronal NO, and it is the most abundant NOS isoform expressed in the retina.[38] Cultured retinal cells have also been reported to express NOS-III.[46] However, it was argued that perinuclear distribution of NOS-III may not contribute to NO production in retinal cells under basal conditions.[46] The main regulator of NOS-I activity is cytosolic calcium (Ca^{2+}), through interaction with calmodulin (Fig. 61.5). Indeed, in the CNS, NO synthesis seems predominantly to be regulated by Ca^{2+} influx through ligand-gated channels, in particular following postsynaptic stimulation of NMDA receptors by glutamate.[47] Specifically in the retina, activation of calcium-permeable α-amino-3-hydroxy-5-methyl-4-isoazo-lepropionic acid (AMPA) receptors by glutamate may also trigger the activation of NOS-I with concomitant NO production, an effect implicated in retinal neuronal cell death.[48,49] Long-term exposure of retinal neurons to NO donors is able to protect from cell death.[49] Retinal NOS-I expression has been detected using immunohistochemistry, which revealed that this enzyme is predominantly expressed in the inner retina,

but ganglion cells may also express NOS-I in all vertebrate species.[38] Amacrine cells of different mammals express NOS-I to produce NO and were termed NO synthesizing amacrine cells.[38] Bipolar cells from both mouse and rat retina have also been shown to express NOS-I.[38] In the human retina, both outer and inner segments of photoreceptor cells were reported to contain NOS-I.[38]

In some preparations of CNS tissues, the NMDA–Ca^{2+}–calmodulin complex is responsible for activating the enzyme eEF2K (eukaryotic elongation factor-2 kinase), which phosphorylates the transcription factor involved in polypeptide chain elongation (eEF2 pathway) and inhibits protein synthesis. In a model of cultured retinal cells, this effect was primarily related to an increased availability of L-Arg for NO synthesis.[50] Indeed, in those preparations, glutamate, L-Arg, NO, or protein synthesis inhibitors enhanced AKT activation, an effect involved in cell resilience in an NO/cGMP-dependent protein kinase (PKG)-dependent manner.[51,52]

The interaction of Ca^{2+}–calmodulin, together with tetrahydrobiopterin and L-Arg (already bound in the heme group), creates a complex that promotes the dimerization of two NOS monomers.[53] In this configuration and in the presence of flavin mononucleotide (FMN) and flavin adenine dinucleotide (FAD) as cofactors, the enzyme changes its conformation, allowing an electron flux from the reductase domain to the heme-containing group in

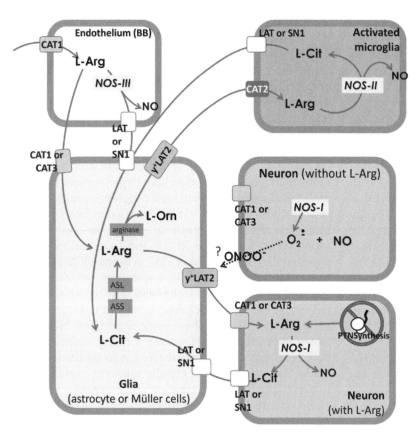

FIGURE 61.4 L-Arginine/nitric oxide (L-Arg/NO) cycle in retinal cells. Hypothetic model of the regulation of the L-Arg/NO/L-citrulline (L-Cit) cycle. L-Arg is taken up by transporters present in vascular endothelium and made available to the CNS, especially by glial cell accumulation. The local vascular blood flow is highly dependent on nitric oxide synthase-III (NOS-III). L-Arg can be released from glia cells by a transport system that can be regulated by depolarizing glutamatergic signaling (not shown) and taken up by adjacent neurons, where it serves the synthesis of NO. This glial–neuronal L-Arg transfer prevents the formation of free $O_2^{-•}$ (superoxide) by the NOS-I. If the supply of L-Arg is impaired, the NOS-I, which is activated mainly via stimulation of N-methyl-D-aspartate (NMDA) receptors, generates both NO and $O_2^{-•}$, then $ONOO^-$ (peroxinitrite). The oxidative stress seems to somehow provide the glial release of L-Arg. The inhibition of protein synthesis also can become L-Arg available for NO synthesis. In activated microglial cells, NO is generated by two immunologic-induced proteins, NOS-II enzyme and CAT2 L-Arg transporter, which enhance several times the NO availability. BB: blood–brain barrier or blood–retina barrier; Glu: glutamate; ASS: argininosuccinate synthetase; ASL: argininosuccinate lyase; PTN: protein.

the catalytic domain[54] (Fig. 61.5). NOS hydroxylates a guanidine nitrogen of L-Arg and then oxidizes it into N-hydroxy–L-arginine intermediate, producing NO and L-Cit.[53]

NOS isoforms have in their N-termini a region known as the lead sequence, which has multiple sites for interaction with cellular proteins. NOS-I exhibits in this region a PDZ domain that may interact with postsynaptic density protein-95 (PSD-95), a scaffold protein located at postsynaptic regions coupling NOS-I with NMDA receptors. Another scaffold protein important for NOS-I regulation is CAPON (carboxyl-terminal PDZ ligand of NOS). Upon binding with NOS-I PDZ domains, CAPON may force it to disassociate from the plasma membrane, and this regulates neuronal NO production at excitatory synapses.[55] Also related to protein–protein interaction for modulation of NOS-I activity, a protein known as protein inhibitor of nitric oxide synthase (PIN) was identified. PIN is a member of the cellular dynein light-chain family and interacts with the lead sequence of NOS-I,

destabilizing this dimeric structure and inhibiting NOS enzymatic activity.[56]

L-Arginine Transport and the Nitric Oxide/L-Citrulline Cycle: Role of Glial Cells

As mentioned above, L-Arg availability is strictly related to several processes that create a balance among molecules committed to cellular metabolism. Since enzymes from L-Arg metabolism are not necessarily expressed within the same cell type, L-Arg transport systems arise as pivotal elements in maintaining the balance. For instance, the synthesis of NO has been shown to be dependent on the availability of extracellular L-Arg despite its huge saturating intracellular concentration. Controversially, acute application of exogenous L-Arg elicits NO production, a phenomenon known as the 'arginine paradox'.[57] Thus, membrane transport of extracellular L-Arg represents one of the rate-limiting steps

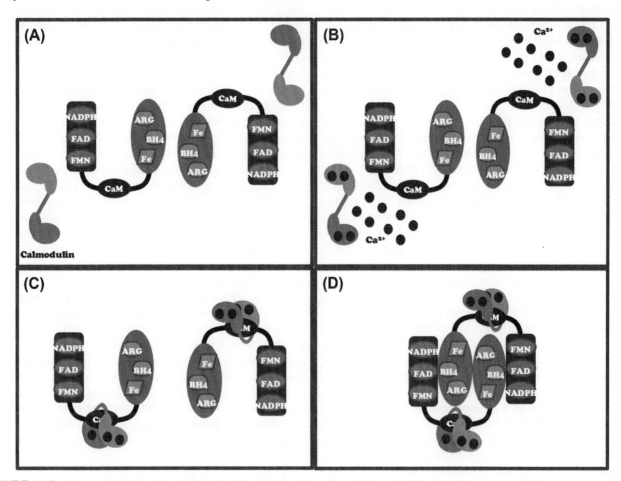

FIGURE 61.5 Scheme of nitric oxide synthase (NOS) activation. As a dimer (A), each NOS subunit is comprised of an oxygenase domain constituted by a catalytic site (ARG), a binding site for tetrahydrobiopterin (BH4), and a reductase domain comprising binding sites for flavin mononucleotide (FMN), flavin adenine dinucleotide (FAD), and nicotinamide adenine dinucleotide (NADPH). (B) Calmodulin activation by Ca²⁺, followed by (C) calmodulin binding to NOS (on its appropriate site) (CaM), (D) allows binding of cofactors in the oxidase domain and enzyme dimerization. Only in this configuration can electrons flow from NADPH to heme, converting L-arginine into L-citrulline and allowing NO release.

in establishing constitutive and compartmentalized NO levels within cells.

Most of cationic amino acid transport is mediated by the y^+ system, which facilitates the high-affinity transport of L-lysine, L-Orn, and L-Arg in a Na^+-independent manner, while transporting glutamine and homoserine in a Na^+-dependent fashion.[58-60] Other transport systems such as y^+L, $b^{0,+}$, which is sodium independent, and $B^{0,+}$, which is sodium dependent, also participate in cationic amino acids transport in some tissues.[61]

Recently, several transport proteins named cationic amino acid transporters (CATs) – CAT1 (Slc7a1), CAT2a (Slc7a2a), CAT2b (Slc7a2b), CAT3 (Slc7a3), and CAT4 – were cloned and identified as functionally similar to the y+ system. *Cat1* codes for the protein CAT1, which is constitutively expressed, and it has been shown to be involved in L-Arg uptake in TR-iBRB2 cells, an *in vitro* model of the inner blood–retinal barrier.[62] *Cat2* codes for two different proteins (CAT2a and CAT2b), which differ in amino acids segments 41–42.[63] Some authors have shown that inflammatory mediators such as IL-1β, TNF-α, and LPS can stimulate the expression of CAT2, but not CAT1, in rat vascular and astroglial cells.[64,65] *Cat3*, which codes for the CAT3 transporter, is exclusively expressed in the adult rat brain,[66] and *Cat4* was shown to mediate the expression of CAT4 transporter in the brain, testis, and human placenta[30].

Plasma membrane proteins that form heterodimers (with heavy and light chains) represent the second class of L-Arg carriers. Disulfide binding may form a transporter composed of the glycoproteins rBAT and 4F2hc (the dimer heavy chain) and seven other proteins that represent the light chain. For L-Arg transport, the proteins y^+LAT1 and y^+LAT2 represent two isoforms of the y^+L system when linked to 4F2hc glycoprotein ($4F2hc/y^+LAT$), and $b^{0,+}AT$ when associated to rBAT ($b^{0,+}AT$) represents the $b^{0,+}$ system. Expression of the heterodimer $4F2hc/y^+LAT$ has been found in the brain and in cultured neurons and astrocytes, transporting either neutral amino acids (in a Na^+-dependent manner) or cationic amino acids (in a Na^+-independent manner).[67,68]

Very little information has been provided regarding the $B^{0,+}$ ($ATB^{0,+}$) system, but it is well established that it transports, in a Na^+-dependent way, both cationic and neutral amino acids.[67] A Na^+-dependent transport system was identified in the rabbit conjunctiva, in which the inhibition pattern by other amino acids appears to be in accordance with a default $B^{0,+}$ system.[69] In chicken retinal cultures, the participation of at least two transport components, one Na^+ dependent and the other Na^+ independent, was also shown.[70]

In relation to L-Arg recycling and transport systems, the existence of transcellular cooperation involving neuronal and glial cells for L-Arg/NO/L-Cit turnover has been reported[30] (Fig. 61.4). Pow demonstrated immunolabeling for L-Cit in neurons and for L-Arg primarily in glial cells in the neurohypophysis.[71] Cossenza and Paes-de-Carvalho, using an autoradiographic approach for detecting L-Arg and immunolabeling for L-Cit in cultured retinal cells, corroborate Pow's data, showing that L-Cit was present in neurons and L-Arg in glial cells.[70] Studies on the distribution of NOS and ASS have shown that ASS is absent from many neurons that synthesize NO.[72] Similarly, some works showed a low immunoreactivity for ASS in glial cells.[73] Therefore, glial cells may have little ability to synthesize argininosuccinate. However, these cells display robust immunoreactivity for ASL.[73] Some authors also showed that glial cells have a great capacity to take up L-Arg and accumulate it in available intracellular pools[50,70] (Fig. 61.4). L-Arg release can be achieved in cultured neuronal cells by using depolarizing agents, such as high K^+ concentrations[70] or by non-NMDA receptor agonists, and then neurons can take up L-Arg for NO generation.[70,74,75] These data strongly suggest a cooperative cycle involving glial cells (providing L-Arg) and neurons (synthesizing NO and L-Cit) (Fig. 61.4). Stoichiometrically produced L-Cit is therefore released from neurons to glial cells and restores L-Arg levels using a transextracellular communication[70] (Fig. 61.4). In line with these pieces of evidence, excitatory stimulation of cultured retinal cells using glutamate has been demonstrated to induce NO production in retinal neurons but not glial cells.[76] However, neuronally produced NO may directly reach Müller glial cells and regulate the activation of cAMP response element-binding protein (CREB) transcription factor in an extracellular signal-regulated kinase (ERK) and mitogen-activated protein kinase (MAPK)-dependent manner.[76] Moreover, exogenous NO has been demonstrated to directly decrease the proliferation of developing Müller glial cells in culture.[77] This phenomenon was shown to be independent of the canonical NO signaling pathway, namely soluble guanylyl cyclase/cGMP/protein kinase G.[77] In this particular case, S-nitrosylation was proposed to be the main regulator of this NO-mediated decrease in proliferation responses in retinal glial cells.[77] Furthermore, it was recently demonstrated that NO also couples death-signaling cascades within retinal neurons. Stimulation of calcium-permeable AMPA receptors led to NOS-I phosphorylation/activation and NO production in retinal neurons.[48] NO, in that way, triggers the activation of the cytoplasmic protein tyrosine kinase Src and regulates excitatory retinal neuronal cell death.[48]

L-Arginine Supplementation

The artificial irrelevant intake of L-Arg should be regarded with extreme caution. It seems to be clear that exogenous L-Arg intake induces both arginase I and arginase II expression.[78] As mentioned above, hyperfunction of arginases correlates with the developing of several disorders by creating an imbalance between arginase and NOS activities. Thus, eventually, if excessive L-Arg has been provided, its rate of consumption would be increased. In line with this, an uncoupled NOS function might be established and reduce physiologic NO formation, providing a pro-oxidative environment. The formation of L-Orn in this noxious environment would also increase and could be detrimental to the development of chronic diseases, including vascular and avascular retinal degeneration.[41]

NITRIC OXIDE REGULATION OF ASCORBATE SYSTEM

Regulation of Ascorbate Uptake by Nitric Oxide

Recently, it was demonstrated that endogenous levels of NO were capable of modulating SVCT-2 expression and therefore ascorbate transport in cultured retinal cells[46] (Fig. 61.6). Endogenous NO production was induced in this paradigm by applying L-Arg to cultured neuronal cells. Furthermore, exogenous NO application, using the NO donors S-nitroso-N-acetyl-DL-penicillamine (SNAP) and NOC5, was also capable of increasing ascorbate transport and SVCT-2 expression in retinal cultures.[46] In line with this, quantitative reverse transcription polymerase chain reaction (qRT-PCR) data demonstrated that NO increased SVCT-2 expression by regulating the transcription of the gene *slc23a2* in these cells.[46] Moreover, kinetic studies of ascorbate transport using L-Arg and SNAP showed a regulation of the transport capacity of SVCT-2 for ascorbate but not its

affinity.[46] This suggests that NO regulates the *de novo* synthesis of SVCT-2 to regulate ascorbate transport at the plasma membrane level in cultured retinal neurons (Fig. 61.6). In addition, it has been previously described that NOS-I expression in retinal cells[70] is restricted to cultured neurons, and these cells are prone to producing NO in response to excitatory stimulation.[48,76] Taking into account that SVCT-2 is preferentially expressed in retinal neurons,[13] it may therefore be suggested that neuronally produced NO regulates ascorbate transport in retinal neuronal cells.

SVCT-2 Transcription: Modulation by the Canonical Nitric Oxide Pathway

Within the SVCT-2 gene, there are promoter regions responsive to both activator protein-1 (AP-1) and nuclear factor-κB (NF-κB) family of transcription factors.[6] In cultured retinal neurons, it was demonstrated that NO regulates the phosphorylation of endogenous repressors of the NF-κB system, suggesting that NO may regulate NF-κB-dependent transcription in cultured retinal cells.[46] Moreover, NO-induced SVCT-2 expression in these cells was demonstrated to depend on the activity of NF-κB, since pharmacologic blockade of NF-κB pathway prevented both NO-induced ascorbate uptake and increased SVCT-2 expression.[46] Furthermore, data in retinal cells suggested that NO exerted its effects on NF-κB regulation in a cGMP-PKG-dependent fashion (Fig. 61.6), which points towards the canonical NO pathway as an endogenous regulator of NF-κB-dependent SVCT-2 expression in retinal neurons.[46] In addition to NF-κB, AP-1 family transcription factors may also regulate

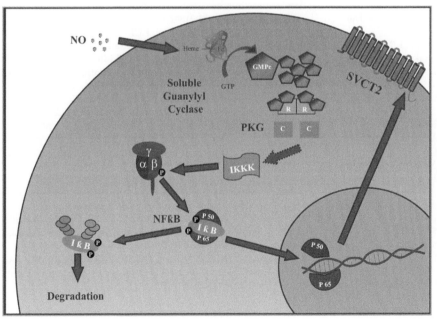

FIGURE 61.6 Nitric oxide (NO) signaling to control ascorbate uptake in retinal neurons. NO interaction with the heme domain of soluble guanylyl cyclase activates this enzyme, which catalyzes the production of cyclic guanosine monophosphate (cGMP) from guanosine triphosphate (GTP). Increased levels of cGMP activate cGMP-dependent protein kinase (PKG), leading to IκK kinase (IKKK) activation. IκB kinase (IKK) activation, by IKKK, induces IκB phosphorylation. When phosphorylated, IKB is ubiquitinated and degraded, releasing nuclear factor-κB (NF-κB) subunits, p50 and p65, which are allowed to migrate to the nucleus, acting as transcription factors. NF-κB promotes sodium-dependent vitamin C transporter-2 (SVCT2) gene transcription, increasing SVCT2 expression at the plasma membrane, which is responsible for ascorbate uptake.

SVCT-2 transcription. In line with this, ERKs are classically described as cellular modulators of AP-1 members' activity, such as c-fos and c-jun. In cultured retinal cells, NO has been described to couple ERK2 activation in a PKG-dependent manner.[76] Furthermore, recent data[46] suggest that a MEK inhibitor (the upstream activator of ERK2 in neurons) blocked the SNAP-induced ascorbate uptake increase in these cultures. This may also implicate ERK2, through the NO pathway, as a regulator of SVCT-2 expression and consequently ascorbate transport in retinal cells.

TAKE-HOME MESSAGES

- Ascorbate is an important bioactive molecule in the retina.
- L-Arginine is an important dietary supplementation for nitric oxide (NO) production.
- NO is an important regulatory molecule in the retina.
- NO regulates SVCT-2 expression by means of nuclear factor-κB stimulation.
- NO modulates ascorbate bioavailability in the retina.

References

1. Eldridge CF, Bunge MB, Bunge RP, Wood PM. Differentiation of axon-related Schwann cells *in vitro*. I. Ascorbic acid regulates basal lamina assembly and myelin formation. *J Cell Biol* 1987;**105**: 1023–34.
2. Neal MJ, Cunningham JR, Matthews KL. Release of endogenous ascorbic acid preserves extracellular dopamine in the mammalian retina. *Invest Ophthalmol Vis Sci* 1999;**40**:2983–7.
3. Englard S, Seifter S. The biochemical functions of ascorbic acid. *Annu Rev Nutr* 1986;**6**:365–406.
4. Poulsen HE, Weimann A, Salonen JT, Nyyssonen K, Loft S, Cadet J, et al. Does vitamin C have a pro-oxidant effect? *Nature* 1998;**395**:231–2.
5. Wilson JX. Regulation of vitamin C transport. *Annu Rev Nutr* 2005;**25**:105–25.
6. Savini I, Rossi A, Pierro C, Avigliano L, Catani M. SVCT1 and SVCT2: key proteins for vitamin C uptake. *Amino Acids* 2008;**34**:347–55.
7. Malo C, Wilson JX. Glucose modulates vitamin C transport in adult human small intestinal brush border membrane vesicles. *J Nutr* 2000;**130**:63–9.
8. Rumsey SC, Kwon O, Xu GW, Burant CF, Simpson I, Levine M. Glucose transporter isoforms GLUT1 and GLUT3 transport dehydroascorbic acid. *J Biol Chem* 1997;**272**:18982–9.
9. Tsukaguchi H, Tokui T, Mackenzie B, Berger UV, Chen XZ, Wang Y, et al. A family of mammalian Na+-dependent L-ascorbic acid transporters. *Nature* 1999;**399**:70–5.
10. Garcia ML, Salazar K, Millan C, Rodriguez F, Montecinos H, Caprile T, et al. Sodium vitamin C cotransporter SVCT2 is expressed in hypothalamic glial cells. *Glia* 2005;**50**:32–47.
11. Grunewald RA, Fillenz M. Release of ascorbate from a synaptosomal fraction of rat brain. *Neurochem Int* 1984;**6**:491–500.
12. Cammack J, Ghasemzadeh B, Adams RN. The pharmacological profile of glutamate-evoked ascorbic acid efflux measured by *in vivo* electrochemistry. *Brain Res* 1991;**565**:17–22.
13. Portugal CC, Miya VS, Calaza KC, Santos RAM, Paes-de-Carvalho R. Glutamate receptors modulate sodium-dependent and calcium-independent vitamin C bidirectional transport in cultured avian retinal cells. *J Neurochem* 2009;**108**:507–20.
14. Subramanian VS, Marchant JS, Boulware MJ, Said HMA. C-terminal region dictates the apical plasma membrane targeting of the human sodium-dependent vitamin C transporter-1 in polarized epithelia. *J Biol Chem* 2004;**279**:27719–28.
15. Boyer JC, Campbell CE, Sigurdson WJ, Kuo SM. Polarized localization of vitamin C transporters, SVCT1 and SVCT2, in epithelial cells. *Biochem Biophys Res Commun* 2005;**334**:150–6.
16. Takanaga H, Mackenzie B, Hediger MA. Sodium-dependent ascorbic acid transporter family SLC23. *Pflugers Archiv* 2004;**447**: 677–82.
17. Hosoya K, Tomi M. Advances in the cell biology of transport via the inner blood–retinal barrier: establishment of cell lines and transport functions. *Biol Pharm Bull* 2005;**28**:1–8.
18. Hosoya K, Minamizono A, Katayama K, Terasaki T, Tomi M. Vitamin C transport in oxidized form across the rat blood–retinal barrier. *Invest Ophthalmol Vis Sci* 2004;**45**:1232–9.
19. Fujikawa A, Gong H, Amemiya T. Vitamin E prevents changes in the cornea and conjunctiva due to vitamin A deficiency. *Graefes Arch Clin Exp Ophthalmol* 2003;**241**:287–97.
20. Lien EL, Hammond BR. Nutritional influences on visual development and function. *Progr Retin Eye Res* 2011;**30**:188–203.
21. Woodford BJ, Tso MO, Lam KW. Reduced and oxidized ascorbates in guinea pig retina under normal and light-exposed conditions. *Invest Ophthalmol Vis Sci* 1983;**24**:862–7.
22. Lam KW, Zwaan J, Garcia A, Shields C. Detection of ascorbic acid in the eye of the early chicken embryo by silver staining. *Exp Eye Res* 1993;**56**:601–4.
23. Kolb H. How the retina works. *Am Sci* 2003;**91**:28–35.
24. Yazulla S. Evoked efflux of [3H]GABA from goldfish retina in the dark. *Brain Res* 1985;**325**:171–80.
25. Fan SF, Yazulla S. Modulation of voltage-dependent K+ currents (IK(V)) in retinal bipolar cells by ascorbate is mediated by dopamine D1 receptors. *Vis Neurosci* 1999;**16**:923–31.
26. Neal MJ, Cunningham JR, Matthews KL. Release of endogenous ascorbic acid preserves extracellular dopamine in the mammalian retina. *Invest Ophthalmol Vis Sci* 1999;**40**:2983–7.
27. Calero CI, Vickers E, Cid GM, Aguayo LG, von Gersdorff H, Calvo DJ. Allosteric modulation of retinal GABA receptors by ascorbic acid. *J Neurosci* 2011;**31**:9672–82.
28. Agostinho P, Duarte CB, Carvalho AP, Oliveira CR. Effect of oxidative stress on the release of [3H]GABA in cultured chick retina cells. *Brain Res* 1994;**655**:213–21.
29. Agostinho P, Duarte CB, Oliveira CR. Impairment of excitatory amino acid transporter activity by oxidative stress conditions in retinal cells: effect of antioxidants. *FASEB J* 1997;**11**:154–63.
30. Wiesinger H. Arginine metabolism and the synthesis of nitric oxide in the nervous system. *Prog Neurobiol* 2001;**64**:365–91.
31. Krebs HA, Henseleit K. Untersuchungen uber die Harnstoffbildung im Tierkörper. *Hoppe Seyler Z Physiol Chem* 1932;**210**:33–66.
32. Withrow C, Ashraf S, O'Leary T, Johnson LR, Fitzgerald MEC, Johnson DA. Effect of polyamine depletion on cone photoreceptors of the developing rabbit retina. *Invest Ophthalmol Vis Sci* 2002;**43**:3081–90.
33. Bak LK, Schousboe A, Waagepetersen HS. The glutamate/GABA-glutamine cycle: aspects of transport, neurotransmitter homeostasis and ammonia transfer. *J Neurochem* 2006;**98**:641–53.
34. Johnson DA, Fields C, Fallon A, Fitzgerald MEC, Viar MJ, Johnson LR. Polyamine-dependent migration of retinal pigment epithelial cells. *Invest Ophthalmol Vis Sci* 2002;**43**:1228–33.
35. Li Q, Verma A, Han P, Nakagawa T, Johnson RJ, Grant MB, et al. Diabetic eNOS-knockout mice develop accelerated retinopathy. *Invest Ophthalmol Vis Sci* 2010;**51**:5240–6.
36. Zhang W, Baban B, Rojas M, Tofigh S, Virmani SK, Patel C, et al. Arginase activity mediates retinal inflammation in endotoxin-induced uveitis. *Am J Pathol* 2009;**175**:891–902.

37. Biedermann B, Skatchkov SN, Brunk I, Bringmann A, Pannicke T, Bernstein HG, et al. Spermine/spermidine is expressed by retinal glial (Müller) cells and controls distinct K+ channels of their membrane. *Glia* 1998;**23**:209–20.

38. Vielma A, Retamal M, Schmachtenberg O. Nitric oxide signaling in the retina: what have we learned in two decades? *Brain Res* 2012;**1430**:112–25.

39. Moncada S, Bolaños JP. Nitric oxide, cell bioenergetics and neurodegeneration. *J Neurochem* 2006;**97**:1676–89.

40. Ignarro LJ, Cirino G, Casini A, Napoli C. Nitric oxide as a signaling molecule in the vascular system: an overview. *J Cardiovasc Pharmacol* 1999;**34**:879–86.

41. Durante W, Johnson FK, Johnson RA. Arginase: a critical regulator of nitric oxide synthesis and vascular function. *Clin Exp Pharmacol Physiol* 2007;**34**:906–11.

42. Caldwell RB, Zhang W, Romero MJ, Caldwell RW. Vascular dysfunction in retinopathy – an emerging role for arginase. *Brain Res Bull* 2010;**81**:303–9.

43. Simell O, Takki K. Raised plasma-ornithine and gyrate atrophy of the choroid and retina. *Lancet* 1973;**301**:1031–3.

44. Wang T, Milam A, Steel G, Valle D. A mouse model of gyrate atrophy of the choroid and retina. Early retinal pigment epithelium damage and progressive retinal degeneration. *J Clin Invest* 1996;**97**:2753–62.

45. Kaneko S, Ueda-Yamada M, Ando A, Matsumura S, Okuda-Ashitaka E, Matsumura M, et al. Cytotoxic effect of spermine on retinal pigment epithelial cells. *Invest Ophthalmol Vis Sci* 2007;**48**:455–63.

46. Portugal CC, da Encarnação TG, Socodato R, Moreira SR, Brudzewsky D, Ambrósio AF, Paes-de-Carvalho R. Nitric oxide modulates sodium vitamin C transporter 2 (SVCT-2) protein expression via protein kinase G (PKG) and nuclear factor-kappaB (NF-kB). *J Biol Chem* 2012;**287**:3860–72.

47. Garthwaite J. Concepts of neural nitric oxide mediated transmission. *Eur J Neurosci* 2008;**27**:2783–802.

48. Socodato R, Santiago FN, Portugal CC, Domingues AF, Santiago AR, Relvas JB, et al. Calcium-permeable AMPA receptors trigger neuronal NOS activation to promote nerve cell death in an Src kinase-dependent fashion. *J Biol Chem* 2012;**287**:38680–94.

49. Mejía-García T, Paes-de-Carvalho R. Nitric oxide regulates cell survival in purified cultures of avian retinal neurons: involvement of multiple transduction pathways. *J Neurochem* 2007;**100**:382–94.

50. Cossenza M, Cadilhe DV, Coutinho RN, Paes-de-Carvalho R. Inhibition of protein synthesis by activation of NMDA receptors in cultured retinal cells: a new mechanism for the regulation of nitric oxide production. *J Neurochem* 2006;**97**:1481–513.

51. Cossenza M, Mejía-García TA, Lima W, Paes-de-Carvalho R. Inhibition of protein synthesis increases nitric oxide production and activation of downstream signalling pathways in avian retina. *J Neurochem* 2011;**118**(Suppl. 1):73.

52. Paes-de-Carvalho R, Mejía-García TA, Portugal CC, Encarnação TG. Nitric oxide regulates AKT phosphorylation and nuclear translocation in cultured avian retinal cells (WE07-07). *Cell Signal* 2013;**25**:2424–39.

53. Stuehr DJ, Santolini J, Wang ZQ, Wei CC, Adak S. Update on mechanism and catalytic regulation in the NO synthases. *J Biol Chem* 2004;**279**:36167–70.

54. Boehning D, Snyder SH. Novel neural modulators. *Annu Rev Neurosci* 2003;**26**:105–31.

55. Jaffrey SR, Snowman AM, Eliasson MJL, Cohen NA, Snyder SH. CAPON: a protein associated with neuronal nitric oxide synthase that regulates its interactions with PSD95. *Neuron* 1998;**20**:115–24.

56. Jaffrey S, Snyder S. PIN: an associated protein inhibitor of neuronal nitric oxide synthase. *Science* 1996;**274**:774–7.

57. Kurz S, Harrison D. Insulin and the arginine paradox. *J Clin Invest* 1997;**99**:369–70.

58. White MF. The transport of cationic amino acids across the plasma membrane of mammalian cells. *Biochim Biophys Acta* 1985;**822**:355–74.

59. Macleod CL, Finley K, Kakuda DK. y(+)-type cationic amino acid transport: expression and regulation of the mCAT genes. *J Exp Biol* 1994;**196**:109–21.

60. Kakuda DK, MacLeod CL. Na(+)-independent transport (uniport) of amino acids and glucose in mammalian cells. *J Exp Biol* 1994;**196**:93–108.

61. MacLeod C, Kakuda D. Regulation of CAT: cationic amino acid transporter gene expression. *Amino Acids* 1996;**11**:171–91.

62. Tomi M, Kitade N, Hirose S, Yokota N, Akanuma S, Tachikawa M, Hosoya K. Cationic amino acid transporter 1-mediated l-arginine transport at the inner blood–retinal barrier. *J Neurochem* 2009;**111**:716–25.

63. Kakuda DK, Finley KD, Maruyama M, MacLeod CL. Stress differentially induces cationic amino acid transporter gene expression. *Biochim Biophys Acta* 1998;**1414**:75–84.

64. Gill DJ, Low BC, Grigor MR. Interleukin-1 and tumor necrosis factor-stimulate the cat-2 gene of the L-arginine transporter in cultured vascular smooth muscle cells. *J Biol Chem* 1996;**271**:11280–3.

65. Stevens BR, Kakuda DK, Yu K, Waters M, Vo CB, Raizada MK. Induced nitric oxide synthesis is dependent on induced alternatively spliced CAT-2 encoding L-arginine transport in brain astrocytes. *J Biol Chem* 1996;**271**:24017–22.

66. Ito K, Groudine M. A new member of the cationic amino acid transporter family is preferentially expressed in adult mouse brain. *J Biol Chem* 1997;**272**:26780–6.

67. Deves R, Boyd C. Transporters for cationic amino acids in animal cells: discovery, structure, and function. *Physiol Rev* 1998;**78**:487–545.

68. Bröer A, Wagner C, Lang F, Bröer S. The heterodimeric amino acid transporter 4F2hc/y+ LAT2 mediates arginine efflux in exchange with glutamine. *Biochem J* 2000;**349**:787–95.

69. Hosoya K, Horibe Y, Kim K, Lee V. Na(+)-dependent L-arginine transport in the pigmented rabbit conjunctiva. *Exp Eye Res* 1997;**65**:547–53.

70. Cossenza M, Paes-de-Carvalho R. L-Arginine uptake and release by cultured avian retinal cells: differential cellular localization in relation to nitric oxide synthase. *J Neurochem* 2000;**74**:1885–94.

71. Pow DV. Immunocytochemical evidence for a glial localisation of arginine, and a neuronal localisation of citrulline in the rat neurohypophysis: implications for nitrergic transmission. *Neurosci Lett* 1994;**181**:141–4.

72. Arnt-Ramos L, O'Brien W, Vincent S. Immunohistochemical localization of argininosuccinate synthetase in the rat brain in relation to nitric oxide synthase-containing neurons. *Neuroscience* 1992;**51**:773–89.

73. Nakamura H, Yada T, Saheki T, Noda T, Nakagawa S. L-Argininosuccinate modulates L-glutamate response in acutely isolated cerebellar neurons of immature rat. *Brain Res* 1991;**539**:312–5.

74. Grima G, Benz B, Do KQ. Glial-derived arginine, the nitric oxide precursor, protects neurons from NMDA-induced excitotoxicity. *Eur J Neurosci* 2001;**14**:1762–70.

75. Grima G, Benz B, Do KQ. Glutamate-induced release of the nitric oxide precursor, arginine, from glial cells. *Eur J Neurosci* 2006;**9**:2248–58.

76. Socodato RES, Magalhães CR, Paes-de-Carvalho R. Glutamate and nitric oxide modulate ERK and CREB phosphorylation in the avian retina: evidence for direct signaling from neurons to Müller glial cells. *J Neurochem* 2009;**108**:417–29.

77. Magalhães CR, Socodato RES, Paes-de-Carvalho R. Nitric oxide regulates the proliferation of chick embryo retina cells by a cyclic GMP-independent mechanism. *Int J Dev Neurosci* 2006;**24**:53–60.

78. Dioguardi FS. To give or not to give? Lessons from the arginine paradox. *J Nutrigenet Nutrigenomics* 2011;**4**:90–8.

79. Mardones L, Ormazabal V, Romo X, Jaña C, Binder C, Peña E, Vergara M, Zúñiga FA. The glucose transporter-2 (GLUT2) is a low affinity dehydroascorbic acid transporter. *Biochem Biophys Res Comm* 2011;**410**:7–12.

ADVERSE EFFECTS AND REACTIONS

62

Dietary Hyperlipidemia and Retinal Microaneurysms

Maria Cristina de Oliveira Izar, Tatiana Helfenstein, Francisco Antonio Helfenstein Fonseca

Cardiology Division, Department of Medicine, Federal University of São Paulo, São Paulo, Brazil

INTRODUCTION

The link between diet, cardiovascular risk factors, and retinal vascular abnormalities is discussed in this chapter. Dyslipidemia, impaired glucose levels, high blood pressure, metabolic syndrome, and obesity affect endothelial function and arterial health, leading to retinal vascular damage. A well-balanced diet is important for controlling modifiable risk factors and preventing organ damage.

CARDIOVASCULAR RISK FACTORS, ENDOTHELIAL DYSFUNCTION, AND INFLAMMATION

Cardiovascular disease is the main cause of death and disability in both the developed and developing world.[1–3] The European guideline on cardiovascular disease prevention in clinical practice[3] mentions that hyperlipidemia (especially elevated levels of low-density lipoprotein (LDL)-cholesterol), smoking, hypertension, age, obesity, and diabetes mellitus type 1 or 2 are considered risk factors for coronary heart disease.

All these risk factors have a close relationship with vascular damage and inflammatory status (Fig. 62.1). Endothelial dysfunction is responsible for alteration of mediators (E-selectin), surface proteins, and autacoids (nitric oxide, endothelin) involved in vasoregulation, coagulation, and inflammation, and in obese subjects the accumulation of macrophages in the adipose tissue produces proinflammatory factors and promotes systemic oxidative stress.[4] These alterations may also occur in patients with insulin resistance, diabetes, or obesity and in first-degree relatives of people with insulin resistance.[5–8] Two studies (with more than 6000 people) demonstrated the association between inflammatory factors and endothelial dysfunctions and retinal venular dilatation.[9,10] Furthermore, there are many studies relating diabetes, obesity, hypertension, dyslipidemia, and metabolic syndrome to retinal vascular abnormalities, such as retinopathy, venular dilatation, arteriolar narrowing, arteriovenous ratio, retinal artery emboli, vein occlusion, hard exudates, and macular edema in humans.[11]

HYPERCHOLESTEROLEMIA AND RETINAL VASCULAR LESIONS

In humans there are no specific data associating isolated hypercholesterolemia with retinal vascular lesions, but many experimental studies have been conducted to assess a link between isolated dyslipidemia and eye disease. These vascular changes usually appear to involve inflammatory factors and endothelial dysfunction. Monkeys fed a diet high in cholesterol and saturated fat presented a reduced amount of polyunsaturated fatty acids, especially 22:6, in retinal lipids; this alteration can lead to impairments in the visual process, especially with regard to the electroretinographic a-wave.[12] In cholesterol-fed apolipoprotein E (apoE)-deficient mice, abnormalities such as retinal neuronal cell dropout and cell layer thinning have been described,[13] as well as ultrastructural changes in Bruch's membrane with a pattern that resembles age-related maculopathy:[14] the electroretinographic a- and b-wave implicit times were lengthened, indicating retinal neuronal dysfunction, and there was a reduction in cell number and cell layer thickness, suggesting degeneration of the retina.[15] In C57BL/6 mice fed a high-fat diet other retinal alterations were found, including lipid-like droplet accumulation and an increased number and size of autophagocytic and empty cytoplasmic vacuoles in the retinal pigment epithelium (RPE) and thickening and fragmentation of the elastic lamina in Bruch's membrane.[16] A study in which two groups of miniature pigs were fed a high-cholesterol diet showed higher retinal oxidative stress, greater release of superoxide anion, and the development of pyknosis, irregular nuclear membranes, and cytoplasmic accumulation of lipids and autophagocytic vacuoles in RPE cells.[17] However, in the group receiving a high-cholesterol diet supplemented with vitamins C and E these biochemical and structural events were mostly prevented. Regarding retinal vascular changes, caliber irregularity and narrowed capillaries in the inner and outer layers of the capillary network were found in rats with inherited hypercholesterolemia.[18] In addition, rats fed a high-cholesterol diet presented greater entrapment of leukocytes in the retinal microcirculation,[19] leading to vascular dysfunction. The present group has recently assessed whether New Zealand white rabbits fed a high-fat/high-cholesterol diet would develop more retinal microaneurysms (Fig. 62.2), and other target organ lesions, compared with normal animals (Fig. 62.3). There were no differences between normal and hypercholesterolemic rabbits regarding the presence of microaneurysms at 12 or 24 weeks.[20]

In addition to these animal experiments, the Atherosclerosis Risk in the Communities (ARIC)[21] and Rotterdam[9] studies revealed the relationship between low levels of high-density lipoprotein (HDL)-cholesterol and high levels of total cholesterol and triglycerides with retinal venular dilatation. The Fenofibrate Intervention and Event Lowering in Diabetes (FIELD) study[22] showed that lipid control with fenofibrate in patients with type 2 diabetes reduced retinopathy needing laser treatment compared with placebo-treated subjects.

OBESITY, METABOLIC SYNDROME, AND SIGNS OF RETINOPATHY

According to the National Cholesterol Education Program Adult Treatment Panel III (NCEP/ATPIII) definition, metabolic syndrome can be diagnosed based on three of the five following characteristics:[23]

- waist circumference (>102 cm for men and >88 cm for women)
- hypertriglyceridemia (≥150 mg/dL)
- low HDL-cholesterol (<40 mg/dL for men and <50 mg/dL for women)
- high blood pressure (≥130/85 mm Hg or use of medication for hypertension)

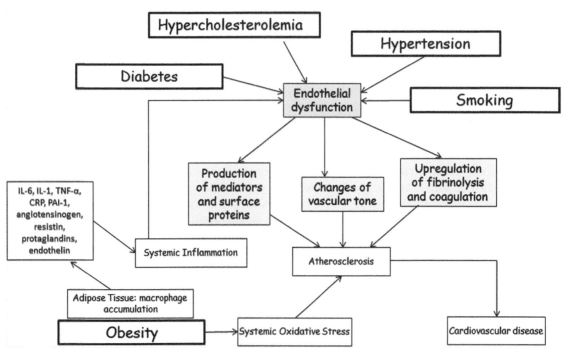

FIGURE 62.1 Links between cardiovascular risk factors, inflammation, and endothelial dysfunction. Endothelial dysfunction is responsible for alterations in mediators, surface proteins, and in autacoids involved in vasomotion, coagulation, and inflammation. In adipose tissue there is an accumulation of macrophages in obese subjects producing factors as interleukin-1 (IL-1), interleukin-6 (IL-6), tumor necrosis factor-α (TNF-α), resistin, prostaglandin E$_2$, angiotensinogen, endothelin, plasminogen activator inhibitor-1 (PAI-1), and C-reactive protein (CRP), leading to systemic inflammation. All these factors can also increase oxidative stress.

- impaired fasting glucose (IFG) (≥100 mg/dL or use of medication for hyperglycemia).

These factors are also related to endothelial dysfunction and inflammation, reinforcing the concept that microvascular disease may be a marker of underlying systemic vascular diseases and metabolic syndrome.

The Multiethnic Study of Atherosclerosis (MESA)[24] revealed retinal microvascular changes in metabolic syndrome, showing associations between venular dilatation and obesity, whereas hypertriglyceridemia, low HDL-cholesterol levels with hyperglycemia, and arteriolar narrowing were associated with hypertension.

The ARIC study showed that patients with metabolic syndrome tended to present signs of retinopathy, characteristically of diabetic retinopathy, and retinal venular dilatation and that the higher the number of components of metabolic syndrome, the higher the prevalence of retinal microvascular signs.[21]

Visceral fat implies lower expression of phosphatidylinositol 3-kinase (PI3K) in different tissues and increased hepatic gluconeogenesis.[25] Higher levels of free fatty acids or their minor intracellular metabolism lead to greater levels of diacylglycerols and other metabolites, which activate the serine/threonine kinase pathway,[26] thus diminishing the action of serine/threonine protein kinase Akt and reducing endothelial nitric oxide production in endothelial cells.[27] In addition, the augmentation of tumor necrosis factor-α and interleukin-6 stimulates the c-Jun N-terminal kinase and 3-phosphoinositide-dependent protein kinase-1-mediated IκB kinase-β pathways. This results in the upregulation of proinflammatory mediators and greater release of monocyte chemotactic protein-1, leading to higher macrophage entrapment in endothelial and adipose tissues[28,29] (Fig. 62.4).

According to these studies, inflammation and endothelial dysfunction may be the mechanisms behind retinal vascular abnormalities in obese people, who have increased caliber retinal veins.[9, 10] In the ARIC study,[21]

there was a positive correlation between central obesity and arteriovenous nicking, retinal arteriolar narrowing, and retinal venular dilatation. In addition, many clinical trials have related obesity to retinopathy in type 1[31–34] and type 2 diabetes.[35–37] Furthermore, in an experimental study with New Zealand white rabbits fed a high-fat/high-sucrose and cholesterol-enriched diet for 24 weeks, the presence of retinal microaneurysms was evident at 12 weeks and progressed during the 24-week follow-up (Fig. 62.5).[20]

GLYCEMIA AND DIABETIC RETINOPATHY

Hemorrhages and microaneurysms, signs of diabetic retinopathy, are usually seen in subjects before the diagnosis of diabetes. These retinal abnormalities can be seen 4–7 years before clinical diabetes is detected[38] and also in nondiabetic subjects over 40 years of age.[39,40] In people with IFG or glucose intolerance,[41] eye vascular damage was present in almost 7% of an Australian study population, and this sign was also positively correlated with clinical stroke, suggesting that retinal microvascular abnormalities could be considered as a possible marker of systemic vascular disease. In the ARIC study,[21] impaired fasting glucose was associated with retinopathy and venular dilatation.

In the USA, diabetic retinopathy is the main cause of blindness among people aged 20–64 years.[42]

BLOOD PRESSURE AND RETINAL VASCULAR LESIONS

Hypertensive retinopathy is characterized by generalized and focal arteriolar narrowing, flame- and blot-shaped retinal hemorrhages, arteriovenous nicking, optic disc swelling, and cotton-wool spots.[24] In the

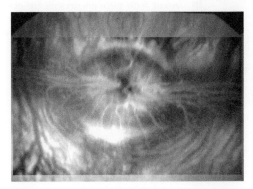

FIGURE 62.2 Angiofluoresceinography of a normal New Zealand white rabbit on a 24-week normal diet. Few white hyperfluorescent dots corresponding to microaneurysms are observed.

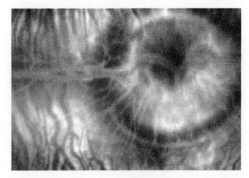

FIGURE 62.3 Angiofluoresceinography of a New Zealand white rabbit on a 12-week high-fat (cholesterol and fat) diet. Retinal vessels with few white hyperfluorescent dots corresponding to microaneurysms are seen.

ARIC study,[21] hypertension was found to be related to retinopathy, arteriovenous nicking, and focal arteriolar narrowing. In another study with 711 nondiabetic subjects, microaneurysms were associated with hypertension and obesity, although the study did not find further evidence that microvascular retinopathy in patients without diabetes was a consequence of past hyperglycemia.[43]

Some studies have shown greater expression of angiotensin II type 1 receptor (AT1R) in the hyperinsulinemic state.[44–47] In this scenario, endothelial dysfunction occurs with minor endothelial nitric oxide synthase phosphorylation, diminishing glucose uptake and contributing to higher blood pressure levels. Insulin has an important vasodilatory action via production of nitric oxide; this effect is impaired in an insulin-resistant state, contributing to the association between hypertension and insulin resistance, which is reflected in changes to the retinal microvasculature.[48]

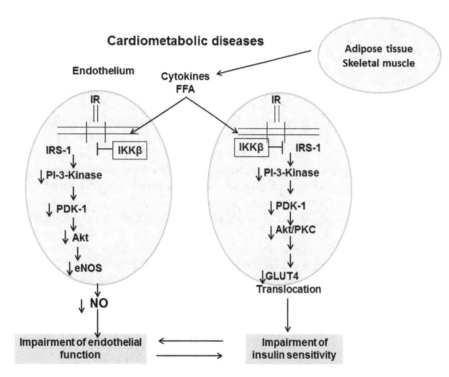

FIGURE 62.4 Cardiometabolic disease and insulin resistance: crosslinking of insulin receptor pathways in the insulin resistance state on the endothelium, adipose tissue, and skeletal muscle cells. Akt: serine/threonine protein kinase Akt; eNOS: endothelial nitric oxide synthase; FFA: free fatty acid; GLUT4: glucose transporter type 4; IKKβ: 3-phosphoinositide-dependent protein kinase-1-mediated IκB kinase-β; IR: insulin receptor; IRS-1: insulin receptor substrate 1; NO: nitric oxide; PDK-1: phosphoinositide-dependent kinase-1; PI-3-kinase: phosphatidylinositol 3-kinase; PKC: protein kinase C.

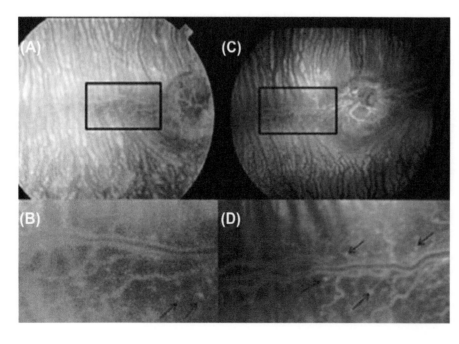

FIGURE 62.5 Angiofluoresceinography representative of rabbits' retinas from high-fat/high-sucrose and cholesterol-enriched diet at (A, B) 12 weeks and (C, D) 24 weeks. Magnification of microaneurysms is shown in (B) and (D) (arrows). White hyperfluorescent dots consistent with microaneurysms are observed in (A) and (B). Progression of the number and size of microaneurysms can be seen in (C) and (D).

HOW TO KEEP HEALTHY

According to the European guideline on cardiovascular disease prevention in clinical practice,[3] it is important to eat two to three portions of fruits and vegetables daily, stop smoking, take regular exercise, maintain a body mass index between 20 and 25 kg/m^2, and have a waist circumference below 102 cm for men and 88 cm for women, blood pressure below 140/90 mmHg, a lipid profile with triglycerides under 140 mg/dL, HDL-cholesterol above 50 mg/dL, and LDL-cholesterol under 130 mg/dL (or less, depending on risk stratification for cardiovascular disease), glucose below 100 mg/dL, and glycosylated hemoglobin (Hb$_{A1c}$) under 6.5%.

In summary, dyslipidemia, impaired glucose levels, high blood pressure, metabolic syndrome, and obesity clearly affect endothelial function and arterial health, promoting retinal vascular damage.

Therefore, it is important to consume a well-balanced diet to control modifiable risk factors and prevent cardiovascular disease and damage to target organs.

TAKE-HOME MESSAGES

- Cardiovascular risk factors have a close relationship with vascular damage and inflammatory status. There are many studies relating diabetes, obesity, hypertension, dyslipidemia, and metabolic syndrome to retinal vascular abnormalities, such as retinopathy, venular dilatation, arteriolar narrowing, arteriovenous ratio, retinal artery emboli, vein occlusion, hard exudates, and macular edema in humans.
- In humans, there are no specific data associating isolated hypercholesterolemia and retinal vascular lesions, but many experimental studies have been conducted to assess a link between isolated dyslipidemia and eye disease.
- Inflammation and endothelial dysfunction may be the mechanisms of retinal vascular abnormalities in obese people, which present larger retinal caliber veins.
- Hemorrhages and microaneurysms, signs of diabetic retinopathy, are usually seen in subjects 4–7 years before a diagnosis of diabetes, and diabetic retinopathy is the main cause of blindness among people aged 20–64 years in the USA.
- Hypertensive retinopathy is characterized by generalized and focal arteriolar narrowing, flame- and blot-shaped retinal hemorrhages, arteriovenous nicking, optic disc swelling, and cotton-wool spots.
- Greater expression of angiotensin II type 1 receptor (AT1R) in the hyperinsulinemic state has been reported, and endothelial dysfunction occurs with minor endothelial nitric oxide synthase phosphorylation and diminishing glucose uptake contributing to higher blood pressure levels.
- Insulin has an important vasodilatory action via nitric oxide production, and in the insulin-resistant state this effect is impaired, contributing to the association between hypertension and insulin resistance reflected in the retinal microvasculature changes.
- Dyslipidemia, impaired glucose levels, high blood pressure, metabolic syndrome, and obesity clearly affect endothelial function and arterial health, promoting retinal vascular damage.
- It is important to follow a well-balanced diet to control modifiable risk factors and prevent cardiovascular disease and target organ lesions.

References

1. World Health Organization. Global burden of coronary heart disease. *WHO* 2002. Available from: http://www.who.int/cardiovascular_diseases/en/cvd_atlas_13_coronaryHD.pdf. [cited 25 September 2012].
2. World Health Organization. Deaths from coronary heart disease. *WHO* 2002. Available from: http://www.who.int/cardiovascular_diseases/en/cvd_atlas_14_deathHD.pdf. [cited 25 September 2012].
3. Perk J, De Backer G, Gohlke H, Graham I, Reiner Z, Verschuren M, et al. European Guidelines on cardiovascular disease prevention in clinical practice (version 2012). The Fifth Joint Task Force of the European Society of Cardiology and Other Societies on Cardiovascular Disease Prevention in Clinical Practice (constituted by representatives of nine societies and by invited experts). Developed with the special contribution of the European Association for Cardiovascular Prevention & Rehabilitation (EACPR). *Eur Heart J* 2012;**33**:1635–701.
4. Wang Z, Nakayama T. Inflammation, a link between obesity and cardiovascular disease. *Mediators Inflamm* 2010:535918.
5. Caballero AE. Metabolic and vascular abnormalities in subjects at risk for type 2 diabetes: the early start of a dangerous situation. *Arch Med Res* 2005;**36**:241–9.
6. Couillard C, Ruel G, Archer WR, Pomerleau S, Bergeron J, Couture P, et al. Circulating levels of oxidative stress markers and endothelial adhesion molecules in men with abdominal obesity. *J Clin Endocrinol Metab* 2005;**90**:6454–9.
7. Festa A, D'Agostino Jr R, Howard G, Mykkanen L, Tracy RP, Haffner SM. Chronic subclinical inflammation as part of the insulin resistance syndrome: the Insulin Resistance Atherosclerosis Study (IRAS). *Circulation* 2000;**102**:42–7.
8. van Hecke MV, Dekker JM, Nijpels G, Moll AC, Heine RJ, Bouter LM, et al. Inflammation and endothelial dysfunction are associated with retinopathy: the Hoorn Study. *Diabetologia* 2005;**48**:1300–6.
9. Ikram MK, de Jong FJ, Vingerling JR, Witteman JC, Hofman A, Breteler MM, de Jong PT. Are retinal arteriolar or venular diameters associated with markers for cardiovascular disorders? The Rotterdam Study. *Invest Ophthalmol Vis Sci* 2004;**45**:2129–34.
10. Klein R, Klein BE, Knudtson MD, Wong TY, Tsai MY. Are inflammatory factors related to retinal vessel caliber? The Beaver Dam Eye Study. *Arch Ophthalmol* 2006;**124**:87–94.
11. Nguyen TT, Wong TY. Retinal vascular manifestations of metabolic disorders. *Trends Endocrinol Metab* 2006;**17**:262–8.

12. Hyman BT, Haimann MH, Armstrong ML, Spector AA. Fatty acid and lipid composition of the monkey retina in diet-induced hypercholesterolemia. *Atherosclerosis* 1981;**40**:321–8.

13. Ong JM, Rosenberg SE, Zorapapel NC, Rajeev B, Nesburn AB, Kenney MC. Ocular cytopathy in apolipoprotein E-deficient mice [ARVO Abstract]. *Invest Ophthalmol Vis Sci* 1999;**40**:S920. abstract no. 4851.

14. Dithmar S, Curcio CA, Le NA, Brown S, Grossniklaus HE. Ultrastructural changes in Bruch's membrane of apolipoprotein E-deficient mice. *Invest Ophthalmol Vis Sci* 2000;**41**:2035–42.

15. Ong JM, Zorapapel NC, Rich KA, Wagstaff RE, Lambert RW, Rosenberg SE, et al. Effects of cholesterol and apolipoprotein E on retinal abnormalities in ApoE-deficient mice. *Invest Ophthalmol Vis Sci* 2001;**42**:1891–900.

16. Miceli MV, Newsome DA, Tate Jr DJ, Sarphie TG. Pathologic changes in the retinal pigment epithelium and Bruch's membrane of fat-fed atherogenic mice. *Curr Eye Res* 2000;**20**:8–16.

17. Fernandez-Robredo P, Moya D, Rodriguez JA, Garcia-Layana A. Vitamins C and E reduce retinal oxidative stress and nitric oxide metabolites and prevent ultrastructural alterations in porcine hypercholesterolemia. *Invest Ophthalmol Vis Sci* 2005;**46**:1140–6.

18. Yamakawa K, Bhutto IA, Lu Z, Watanabe Y, Amemiya T. Retinal vascular changes in rats with inherited hypercholesterolemia – corrosion cast demonstration. *Curr Eye Res* 2001;**22**:258–65.

19. Tomida K, Tamai K, Matsuda Y, Matsubara A, Ogura Y. Hypercholesterolemia induces leukocyte entrapment in the retinal microcirculation of rats. *Curr Eye Res* 2001;**23**:38–43.

20. Helfenstein T, Fonseca FA, Ihara SS, Bottos JM, Moreira FT, Pott Jr H, et al. Impaired glucose tolerance plus hyperlipidaemia induced by diet promotes retina microaneurysms in New Zealand rabbits. *Int J Exp Pathol* 2011;**92**:40–9.

21. Wong TY, Duncan BB, Golden SH, Klein R, Couper DJ, Klein BE, et al. Associations between the metabolic syndrome and retinal microvascular signs: the Atherosclerosis Risk in Communities study. *Invest Ophthalmol Vis Sci* 2004;**45**:2949–54.

22. Keech A, Simes RJ, Barter P, Best J, Scott R, Taskinen MR, et al. FIELD Study Investigators. Effects of long-term fenofibrate therapy on cardiovascular events in 9795 people with type 2 diabetes mellitus (the FIELD study): randomised controlled trial. *Lancet* 2005;**366**:1849–61.

23. Executive Summary of the Third Report of the National Cholesterol Education Program (NCEP) Expert Panel on Detection. Evaluation, and Treatment of High Blood Cholesterol in Adults (Adult Treatment Panel III). *JAMA* 2001;**285**:2486–97.

24. Wong TY, Mohamed Q, Klein R, Couper DJ. Do retinopathy signs in nondiabetic individuals predict the subsequent risk of diabetes? *Br J Ophthalmol* 2006;**90**:301–3.

25. Ferroni P, Basili S, Falco A, Davi G. Inflammation, insulin resistance, and obesity. *Curr Atheroscler Rep* 2004;**6**:424–31.

26. Shulman GI. Cellular mechanisms of insulin resistance. *J Clin Invest* 2000;**106**:171–6.

27. Zdychova J, Komers R. Emerging role of Akt kinase/protein kinase B signaling in pathophysiology of diabetes and its complications. *Physiol Res* 2005;**54**:1–16.

28. Fain JN, Madan AK, Hiler ML, Cheema P, Bahouth SW. Comparison of the release of adipokines by adipose tissue, adipose tissue matrix, and adipocytes from visceral and subcutaneous abdominal adipose tissues of obese humans. *Endocrinology* 2004;**145**:2273–82.

29. Andreozzi F, Laratta E, Sciacqua A, Perticone F, Sesti G. Angiotensin II impairs the insulin signaling pathway promoting production of nitric oxide by inducing phosphorylation of insulin receptor substrate-1 on Ser312 and Ser616 in human umbilical vein endothelial cells. *Circ Res* 2004;**94**:1211–8.

30. Kim JA, Montagnani M, Koh KK, Quon MJ. Reciprocal relationships between insulin resistance and endothelial dysfunction: molecular and pathophysiological mechanisms. *Circulation* 2006;**113**:1888–904.

31. Chaturvedi N, Sjoelie AK, Porta M, Aldington SJ, Fuller JH, Songini M, Kohner EM. EURODIAB Prospective Complications Study. Markers of insulin resistance are strong risk factors for retinopathy incidence in type 1 diabetes. *Diabetes Care* 2001;**24**:284–9.

32. Henricsson M, Nystrom L, Blohme G, Ostman J, Kullberg C, Svensson M, et al. The incidence of retinopathy 10 years after diagnosis in young adult people with diabetes: results from the nationwide population-based Diabetes Incidence Study in Sweden (DISS). *Diabetes Care* 2003;**26**:349–54.

33. Keen H, Lee ET, Russell D, Miki E, Bennett PH, Lu M. The appearance of retinopathy and progression to proliferative retinopathy: the WHO Multinational Study of Vascular Disease in Diabetes. *Diabetologia* 2001;**44**:S22–30.

34. Zhang L, Krzentowski G, Albert A, Lefebvre PJ. Risk of developing retinopathy in Diabetes Control and Complications Trial type 1 diabetic patients with good or poor metabolic control. *Diabetes Care* 2001;**24**:1275–9.

35. UK Prospective Diabetes Study Group. Tight blood pressure control and risk of macrovascular and microvascular complications in type 2 diabetes: UKPDS 38. *BMJ* 1998;**317**:703–13.

36. UK Prospective Diabetes Study (UKPDS) Group. Effect of intensive blood-glucose control with metformin on complications in overweight patients with type 2 diabetes (UKPDS 34). *Lancet* 1998;**352**:854–65.

37. van Leiden HA, Dekker JM, Moll AC, Nijpels G, Heine RJ, Bouter LM, et al. Blood pressure, lipids, and obesity are associated with retinopathy: the Hoorn Study. *Diabetes Care* 2002;**25**:1320–5.

38. Harris MI, Klein R, Welborn TA, Knuiman MW. Onset of NIDDM occurs at least 4–7 yr before clinical diagnosis. *Diabetes Care* 1992;**15**:815–9.

39. Cugati S, Cikamatana L, Wang JJ, Kifley A, Liew G, Mitchell P. Five-year incidence and progression of vascular retinopathy in persons without diabetes: the Blue Mountains Eye Study. *Eye (Lond)* 2006;**20**:1239–45.

40. Klein R, Klein BE, Moss SE. The relation of systemic hypertension to changes in the retinal vasculature: the Beaver Dam Eye Study. *Trans Am Ophthalmol Soc* 1997;**95**:329–48. discussion 48–50.

41. Wong TY, Barr EL, Tapp RJ, Harper CA, Taylor HR, Zimmet PZ, Shaw JE. Retinopathy in persons with impaired glucose metabolism: the Australian Diabetes Obesity and Lifestyle (AusDiab) study. *Am J Ophthalmol* 2005;**140**:1157–9.

42. Congdon NG, Friedman DS, Lietman T. Important causes of visual impairment in the world today. *JAMA* 2003;**290**:2057–60.

43. Munch IC, Kessel L, Borch-Johnsen K, Glümer C, Lund-Andersen H, Larsen M. Microvascular retinopathy in subjects without diabetes: the Inter99 Eye Study. *Acta Ophthalmol* 2012;**90**:613–9.

44. Samuelsson AM, Bollano E, Mobini R, Larsson BM, Omerovic E, Fu M, et al. Hyperinsulinemia: effect on cardiac mass/function, angiotensin II receptor expression, and insulin signaling pathways. *Am J Physiol Heart Circ Physiol* 2006;**291**:H787–96.

45. Banday AA, Siddiqui AH, Menezes MM, Hussain T. Insulin treatment enhances AT1 receptor function in OK cells. *Am J Physiol Renal Physiol* 2005;**288**:F1213–9.

46. Golovchenko I, Goalstone ML, Watson P, Brownlee M, Draznin B. Hyperinsulinemia enhances transcriptional activity of nuclear factor-kappaB induced by angiotensin II, hyperglycemia, and advanced glycosylation end products in vascular smooth muscle cells. *Circ Res* 2000;**87**:746–52.

47. Nickenig G, Roling J, Strehlow K, Schnabel P, Bohm M. Insulin induces upregulation of vascular AT1 receptor gene expression by posttranscriptional mechanisms. *Circulation* 1998;**98**:2453–60.

48. Kuboki K, Jiang ZY, Takahara N, Ha SW, Igarashi M, Yamauchi T, et al. Regulation of endothelial constitutive nitric oxide synthase gene expression in endothelial cells and *in vivo*: a specific vascular action of insulin. *Circulation* 2000;**101**:676–81.

63

Iron-Induced Retinal Damage

David Dunaief[1], Alyssa Cwanger[2], Joshua L. Dunaief[2]

[1]Medical Compass, MD Private Practice, New York, USA; [2]FM Kirby Center for Molecular Ophthalmology, Scheie Eye Institute, University of Pennsylvania, Philadelphia, Pennsylvania, USA

INTRODUCTION

When we think of iron, the associations that usually come to mind are positive, including effects that improve our strength and energy and even the Ironman triathlon. In other words, we have a perception that iron is good for us and that the more iron we get in our diets and from supplements, the better.

It is true that iron is important in many biologic processes in the body, such as DNA production and adenosine triphosphate (ATP) formation.[1,2] Photoreceptors in the eye depend on enzymes containing iron to help produce lipids used to develop disc membranes.[3] Retinal pigment epithelium (RPE) cells require iron to regenerate 11-*cis*-retinal for the visual cycle.[4]

Many people perceive that iron is always valuable, that they should get as much as possible. Is this perception correct, or is it a myth? It depends on the demographics: gender and age. Children, women of reproductive age, and those who are anemic may benefit from more iron. Men and postmenopausal women, however, may increase their risks of systemic diseases, such as cancer and heart disease, and may also increase their risk of eye diseases, such as age-related macular degeneration, glaucoma, and cataracts, with higher levels of iron intake.

The focus of this chapter is on iron's potential detrimental effect on eye disease, specifically retinal diseases. The problem develops when there is free iron, excess amounts that are not bound. Free iron is involved in the Fenton reaction, which converts hydrogen peroxide into reactive oxygen species (ROS), hydroxyl radicals. Ultimately, these radicals may cause breakdown of biomolecules, lipid peroxidation, and nicks in the strands of the DNA double helix.[5] Thus, iron's damaging effects to the retina may be due to its reactive nature, which leads to oxidative stress and then inflammation.

Does this mean that men or postmenopausal women should avoid iron intake? Interestingly, not all iron is created equal. Iron that comes from red meat in the form of heme is absorbed much more readily than iron from plant-based and iron-fortified foods. Ultimately, it depends on the population and type of iron intake.

This chapter will investigate the eye diseases associated with iron overload; how iron is imported, exported, and stored in the eye; the impact of nutrition on retinal iron; and finally preventive and therapeutic approaches to iron overload in the retina.

SYSTEMIC AND RETINAL DISEASES WITH EXCESS IRON

There are numerous hereditary and acquired diseases that cause iron overload in the retina and other organs. These diseases include aceruloplasminemia, Alzheimer's disease, and Parkinson's disease (Table 63.1).

Iron homeostasis is critically important to the body, but when the process goes awry there is an accumulation of iron, which has been associated with neurodegenerative diseases, such as Parkinson's.[6] Since iron is imported by neurons, in part to serve as a cofactor in the production of neurotransmitters, iron dysregulation with excess uptake or impaired export can lead to neurodegeneration.

In a study to determine whether iron may play a role in Parkinson's disease, neonatal mice were fed excessive amounts of iron.[7] The iron dose used for the neonates was similar to that used in first year iron-fortified human infant formulas. In old mice, there was significant loss of dopaminergic neurons in substantia nigra that was not seen in younger mice. This study was preceded by one showing that iron supplementation in neonatal mice caused elevated iron levels in the substantia nigra of adults.[8]

In younger mice, at the age of 12 months, there was a substantial reduction in striatum dopamine, which continued to the old age of 24 months. The effects on dopaminergic neurons are potentially relatable to human Parkinson's disease. The implication of this study is that dietary iron not only was transported through the blood–brain barrier, but the excess iron caused neurodegeneration. Its ability to pass through the blood–brain barrier may indicate its ability to also pass through the blood–retinal barrier and cause retinal degeneration. Studies are needed to assess the ability of dietary iron to cross the blood–retinal barrier at different ages.

Alzheimer's disease has also been associated with elevated brain iron. *Post mortem* brains from Alzheimer's patients have increased iron in the hippocampus. Redox-active iron is associated with the senile plaques and neurofibrillary tangles seen in the disease.[9]

Studies in mice and humans have suggested that iron chelation may be therapeutic for neurodegeneration. It is still too early to tell whether this approach will be effective and safe, but it does generate excitement about a novel therapeutic approach to chronic neurodegenerative and retinal degenerative diseases for which new therapies are needed.

Iron overload may play a significant role in a number of retinal disorders and diseases. These include ocular siderosis, age-related macular degeneration (AMD), and hereditary diseases, such as Friedreich's ataxia, pantothenate kinase-associated neurodegeneration, and aceruloplasminemia.

Aceruloplasminemia is a disorder that, as the name indicates, is caused by a deficiency in ceruloplasmin resulting from a mutation of the gene at chromosome 3q. It is an autosomal recessive disease that occurs in adults.

TABLE 63.1 Systemic and Ocular Diseases Associated with Iron Overload

Systemic	Ocular
Hereditary hemochromatosis (caused by mutations in HFE, ferroportin, hemojuvelin, or hepcidin)	Age-related macular degeneration
Thalassemia/sickle cell anemia with transfusional iron overload	Diabetic retinopathy
Diabetes	Aceruloplasminemia
Heart disease	Friedreich's ataxia
Alzheimer's disease	Pantothenate kinase-associated neurodegeneration
Parkinson's disease	Traumatic iron foreign body (siderosis)
	Intraocular hemorrhage
	Glaucoma
	Cataract

Patients with this rare disease may have three clinical disorders: retinal degeneration, diabetes mellitus, and dementia.[10] The importance of ceruloplasmin is that it is involved in facilitating iron export from cells. This is achieved by causing the oxidation of ferrous to ferric iron, which then is bound to transferrin.[11] Thus, ceruloplasmin helps to regulate the homeostasis of iron. In mice without ceruloplasmin and its homolog hephaestin, iron overload causes damage to the RPE and photoreceptors.[12] Early-onset macular degeneration was found in a Caucasian patient with aceruloplasminemia.[13] The findings of subretinal white lesions in the macula paralleled those seen in AMD with drusen. On histopathology, drusen, RPE hypertrophy, and RPE pigmentary changes were found.[14]

This brings us to the next disease that is potentially affected by iron, AMD. With AMD, the dry, or nonexudative, form produces drusen in its early stages, which may progress to significant amounts of geographic atrophy. In its wet, or exudative, form, choroidal neovascularization occurs, leading to the number one cause of blindness in patients who are at least 65 years old.[15,16] The development of AMD may be attributable to ROS and the resultant damage.[17,18] In the Age-Related Eye Disease Study (AREDS), antioxidants in supplement form, including vitamins β-carotene, C, and E and zinc, were shown to reduce the progression of the early dry form of AMD to late-stage geographic atrophy, or wet AMD.[19]

Although oxidative stress may be involved, it is not known where the oxidants originate. Iron is an attractive candidate. In a histopathology study involving 19 *post mortem* eyes, 10 with AMD and nine without AMD, maculas were stained with Perls' Prussian blue and then images were analyzed with computer-assisted densitometry. The results showed that there was a greater amount of stain, representing more iron, in the macula of AMD patients.[20] Both the RPE and Bruch's membrane contained iron, regardless of the stage of AMD, early or advanced.

Increased transferrin levels also suggest the presence of more iron. In *post mortem* retinas of AMD patients compared with controls, those with AMD had a doubling of transferrin levels, with dry AMD retinas showing a greater effect than wet AMD.[21]

Again, it is too early in the investigative process to say definitively that iron is a cause of AMD. In fact, iron is not likely to be the principal cause of the disease but rather increases oxidative stress, fueling the flames of inflammation, in part by directly activating the complement cascade.[22] Thus, the question remains: which came first, the iron or the AMD? Results have indicated that in mice denuded of vital iron homeostasis proteins, there was iron overload and resultant retinal damage, similar to that seen in AMD.[12] Furthermore, as people age, there is an accumulation of retinal and systemic iron.[23] This makes sense, since humans do not have a natural way to purge excess iron. These findings support the hypothesis that iron may be a cause rather than merely a consequence of AMD.

AMD is also associated with cardiovascular disease and Alzheimer's disease, both of which may potentially be exacerbated by iron.[24,25] Ultimately, inflammation may be the underlying factor that connects all three diseases. Inflammation is strongly implicated in AMD because complement cascade factors have polymorphism linked to the disease and both complement proteins and leukocytes have been detected in AMD retinas.[26–32]

In siderosis, the RPE and photoreceptors are damaged by ROS catalyzed by iron. Intraocular iron deposits are introduced into the eye by iron-containing foreign bodies that penetrate the eye during trauma.[33] After the initial insult, the time frame in which sight may be affected is as short as 18 days and as long as 8 years. Ocular siderosis may result in a range of complications, such as retinal detachment, cataract development, RPE atrophy and clumping, pupillary mydriasis, and iris heterochromia. If the Schlemm's canal and the trabecular meshwork are affected, then secondary glaucoma can occur.[33–35]

One of the best known diseases involving iron overload is hemochromatosis, which is divided into two types: primary, or hereditary, and secondary. Hereditary hemochromatosis is the predominant type. It is most commonly caused by a mutation in the histocompatibility leukocyte antigen class I-like gene called *HFE*. Mutation results in elevated iron uptake by the gut and iron accumulation in some organs, especially the liver.[36]

The *HFE* gene may play a role in retinal iron regulation since both the mouse RPE and neural retina express the gene.[37,38] To reinforce this point, without the *HFE* gene, retinal degeneration occurs in mice.[39] In three areas of the eye – the sclera, the ciliary epithelium, and the peripapillary RPE – several patients with hereditary hemochromatosis had accumulations of iron. These patients also had RPE pigmentary abnormalities.[40] Further studies on the association between hereditary hemochromatosis as well as acquired iron overload and retinal disease will be of interest.

IMPORTATION, STORAGE, AND EXPORTATION OF IRON IN THE EYE

Importation of Iron

Transferrin is a much-needed protein that binds to iron in the bloodstream. One transferrin molecule accepts two iron atoms following their export from cells.[41] The iron–transferrin complex can also be taken into cells following binding to the transferrin receptor on the cell surface and endocytosis bringing it across the cell membrane into the cell. Within the endosome, the acidic environment facilitates separation of iron from the transferrin. Then, the iron can be released from the endosome into the cytoplasm of the cell, where it is either bound to ferritin for storage, incorporated into iron-dependent enzymes, or exported from the cell.

Where are the transporters found and produced in the eye? The transferrin receptor has been identified in multiple cell layers in the eye including the RPE, the inner segments of photoreceptors, the outer plexiform layer, the inner nuclear layer, and the ganglion cell layer (Fig. 63.1).[42] Transferrin is produced by the RPE and neural retina.

Why is iron transport into the cell layers of the eye so intricate? The eye, like the brain, has a barrier. In the central nervous system (CNS), it is the blood–brain barrier, whereas in the eye, it is referred to as the blood–retinal barrier. These barriers are meant to separate the CNS and eye from the less filtered circulation of the body.

It is not just the cell layers of the eye, however, that contain the protein transferrin but also the vitreous and aqueous humor.[43–45] This implies that these substances may also play a role in iron transport within the eye.

Divalent metal transporter-1 (DMT1) is an important transporter of dietary free iron from the luminal surface of the gut into the epithelial cells of the small intestine, the enterocytes. First, the iron must be reduced from

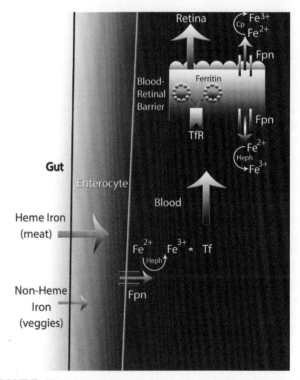

FIGURE 63.1 Importation, storage, and exportation of iron in the eye. Cp: ceruloplasmin; Fe^{2+}: ferrous iron; Fe^{3+}: ferric iron; Fpn: ferroportin; Heph: hephaestin; Tf: transferrin; TfR: transferrin receptor.

ferric (Fe^{3+}) to ferrous (Fe^{2+}) before DMT1 can transport it. In the rat brain, there are significant amounts of DMT1 detected by immunochemistry in ependymal and vascular cells.[46] DMT1 is also found in the retina, where its role in retinal iron uptake is a topic of current investigation.

Storage of Iron

Ferritin is the primary site of iron storage in the cell. Not surprisingly, iron and ferritin are found in relatively the same distribution in the eye. This was determined by looking at adult rat retinas using immunohistochemistry to indicate ferritin levels and proton-induced X-ray emissions to demonstrate iron levels. There was significant overlap in the choroid, the inner nuclear layer, the inner segments of photoreceptors, the RPE, and the ganglion cell layer.[47]

Ferritin has a high capacity for holding iron while in its ferric state (Fe^{3+}). Ferritin can carry up to 4500 iron molecules in its core. The ferritin complex contains two types of chains, heavy and light.[48] The heavy chains gather iron quickly and retain it through enzymatic ferroxidase activity, while light chains lack this enzyme activity but facilitate iron storage.[49]

Using immunohistochemistry, mitochondrial ferritin was found throughout the inner retina and inner segments of photoreceptors.[50] This form of ferritin closely resembles heavy ferritin[51] and is likely to be important for mitochondrial iron storage.

Exportation of Iron

Several proteins are involved in iron export from the cell. These include the only known iron exporter, ferroportin, which works in cooperation with ferroxidases such as ceruloplasmin and hephaestin, and the iron regulatory hormone hepcidin.

Like heavy ferritin, ceruloplasmin and hephaestin have ferroxidase activity. Ceruloplasmin protects against oxidative stress when it oxidizes ferrous iron (Fe^{2+}) to its ferric state, thus avoiding the Fenton reaction, which, as stated above, can produce ROS.[52] There are two different forms of ceruloplasmin. One, the membrane-anchored glycosylphosphatidylinositol-linked form, is generated in the brain, and the other, secreted form, is synthesized systemically in the liver.[53,54] Both forms were found in the mouse retina using reverse transcription–polymerase chain reaction (RT-PCR).[55]

In the healthy human eye, ceruloplasmin is found in the retina and in the aqueous and vitreous humor by Western analysis. Ceruloplasmin may play a significant role in defending against oxidative stress; in the diabetic rat retina, there is a significant increase in ceruloplasmin in response to oxidative stress.[56] Ceruloplasmin

messenger RNA levels are also increased in response to light damage[55] and in human AMD compared with age-matched normal retinas.[57]

Hephaestin is very similar to ceruloplasmin, identical in half the amino acids. Both are multicopper ferroxidases.[58] Hephaestin is found in the RPE cells of humans and mice.[12] RT-PCR was used to make this discovery. Just like with ceruloplasmin, hephaestin helps to export iron across the cell membrane using ferroxidase activity to oxidize iron to its ferric form so that it binds to transferrin.

Another role for hephaestin is that it facilitates the absorption of iron from the intestine to the bloodstream. The sequelae of ceruloplasmin and hephaestin insufficiency are retinal iron overload and age-dependent RPE degeneration associated with oxidative stress.[12,59]

Ferroportin is a transmembrane protein that transports iron out of the cell, as first demonstrated in *Xenopus* oocytes.[60] Hephaestin and ceruloplasmin are ferroxidases that function along with ferroportin to carry ferrous iron out of the cell and then oxidize it so that iron can bind to apotransferrin. When ceruloplasmin is added to tissue culture medium, it makes ferroportin more effective for iron export.[61]

Expression of ferroportin occurs in many locations, including the retina, brain, hepatocytes, intestine, placenta, and reticuloendothelial macrophages.[50,60–65] When mutation occurs in the human ferroportin gene, ferroportin disease can occur, which is also known as type IV hereditary hemochromatosis. This is an autosomal dominant disease.

Hepcidin is an iron regulatory hormone located in the urine and plasma of humans. It is a 20–25 amino acid peptide. Synthesis of hepcidin takes place primarily in the liver.[66,67] Secreted hepcidin binds ferroportin, triggering its degradation, thus preventing iron export from cells. This protein may also play a role in immunology and oxidative stress. In the environment of inflammation, infection, and iron overload, hepcidin's transcription is upregulated.[68,69] Hepcidin is also produced by the retina, upregulated by retinal iron, and its systemic deficiency leads to retinal iron overload and degeneration in the mouse retina.[70] These findings suggest that hepcidin plays a role in iron regulation locally, within the retina.

THE ROLE OF NUTRITION AND RETINAL IRON

Iron is absorbed through the gut by the epithelium of the duodenum. At this site, enterocyte cells regulate the absorption of iron into the bloodstream. Enterocytes use DMT1 and heme transporters to absorb iron. These cells typically allow 10% of nonheme dietary iron and 50% of dietary heme iron (from meat) to be absorbed. Ferroportin and hephaestin are proteins needed to help the enterocyte to export iron into the bloodstream, where the iron becomes primarily bound to transferrin.

Transferrin is a protein that enables iron to be transported throughout bodily fluids, such as vitreous and aqueous humor, as well as lymph and serum. As long as iron is bound to transferrin, it cannot potentiate free radicals causing oxidative stress.[71]

It is unclear, however, how iron is absorbed in the CNS because of the blood–brain barrier, and in particular into the eye, because of the blood–retinal barrier. Therefore, even though iron consumed in the diet may be absorbed by the body, it may not be directly absorbed by the eye. It is known that the retina expresses transferrin receptor, DMT1, ferroportin, ceruloplasmin, hephaestin, and hepcidin and that disruption of ceruloplasmin, hephaestin, or hepcidin in mice leads to retinal iron overload and degeneration, but the full details of retinal iron regulation await further investigation.

NUTRITIONAL IRON AND EYE DISEASE

There are numerous eye diseases potentially associated with high iron levels from diet and/or supplements. These diseases include diabetic retinopathy, AMD, glaucoma, and cataracts. The deleterious effects of iron probably result from oxidative stress and inflammation.[72–74]

In the Melbourne Collaborative Cohort Study, iron intake from red meat was shown to potentially increase the risk of early AMD, but it was unclear whether it affected advanced AMD since only a small number of participants developed the advanced type.[75] The results showed that people who ate red meat at least 10 times per week were 47% more likely to develop AMD than those who consumed red meat five or fewer times per week. These results suggest that systemic iron intake may influence the risk of AMD.

In a cross-sectional study, iron supplementation of at least 18 mg/day increased the risk of glaucoma by 3.8-fold compared with not taking iron supplements.[76] There were even more dramatic results, a 7.2-fold increased risk of developing glaucoma, when participants took both calcium (≥800 mg/day) and iron supplements. The database used for this study was from the National Health and Nutrition Examination Survey (NHANES). There were 3848 participants involved in this study, and approximately 10% of participants developed normal tension glaucoma.

In a subsequent study using the NHANES data, researchers investigated the impact of dietary iron and calcium on glaucoma. The results showed a decreased risk of glaucoma as dietary iron intake was decreased in a linear manner. The participants in the fifth quintile, having the least amount of iron in their diet, showed the most reduction, 69%, in the risk of developing glaucoma. For dietary calcium, the opposite was true: the more calcium consumed, the greater the reduction in risk of glaucoma.[77] These results were based on 24-hour recall food frequency questionnaires. The negative effects of iron, whether through supplementation or food, may result from oxidative damage to the retina. These preliminary studies involving glaucoma suggest that lower levels of iron, in both supplemental and dietary forms, decrease the risk of glaucoma.

Higher levels of iron appear to have detrimental effects in yet another disease affecting the eye: diabetic retinopathy. Studies have shown that the levels of iron are 150% higher in proliferative diabetic retinopathy than in the normal retina.[78] *In vitro* studies indicate that hyperglycemia may be responsible for the degradation of heme from both myoglobin and hemoglobin, thus increasing the levels of free iron in the eye and the risk of oxidation.[79,80]

PREVENTION AND THERAPEUTICS

In order to potentially reduce the impact that high levels of iron and its subsequent oxidative stress have on the retina, especially the RPE, both preventive measures and therapeutic options should be considered. Preventive measures may reduce the risk of all-too-common chronic retinal degenerative diseases, such as AMD, glaucoma, and diabetic retinopathy. Prevention is probably the most powerful weapon in the fight against retinal degeneration caused by iron overload. So, how can iron overload be prevented? One of the problems is that the standard American diet perpetuates consuming more iron than is excreted. Thus, there is a build up of iron that may contribute to retinal degeneration and other age-associated disorders.[81]

Heme iron is absorbed much more readily than nonheme iron. Since red meat is the primary source of heme iron, reducing consumption may be the first step in reducing iron levels in the blood and potentially in the blood–retinal barrier. Another preventive measure could be to donate blood on a quarterly basis, which would help to lower iron stores. Decreased iron stores appear to be beneficial for increasing lifespan in fruit fly, *Drosophila*. As the level of systemic iron is reduced, lifespan is increased.[82,83] It is not clear, however, whether giving blood will decrease levels of iron in the retina.

Therapeutic options include the use of iron chelators and possibly phlebotomy. Using iron chelators is complicated; the most effective chelator would be relatively specific for iron so that it did not also reduce levels of other metals such as zinc, which may have beneficial effects.[84]

An oral chelator also has to be able to pass through the gastrointestinal tract and then penetrate the blood–retinal barrier in large enough amounts to have an impact on iron levels in the retina. This means that, in order to pass through the barriers, the chelator should be relatively lipophilic, with a neutral charge and low molecular weight.[85,86]

Two potential iron chelators look promising: oral deferiprone and salicylaldehyde isonicotinoyl hydrazone (SIH). Like most drugs, deferiprone is not without side effects, but with monitoring, these may be limited or avoided. Thus far, deferiprone has shown efficacy in mouse studies. It has reduced the amount of iron in the retina and, subsequently, the amount of oxidative damage. These results are encouraging because the mice were deficient in both ceruloplasmin and hephaestin, proteins that export iron. Thus, they were experiencing iron overload before treatment.[87] Deferiprone can also protect the mouse retina against degeneration caused by light, the oxidant sodium iodate, or the *rd6* mutation.[88]

SIH has the potential to diminish oxygen radical species that are produced by iron's ability to catabolize hydrogen peroxide into reactive hydroxyl ions. This could potentially prevent cell death in the RPE and, ultimately, prevent retinal neurodegenerative disease, such as AMD.[89,90] When applied to RPE cells in tissue culture, SIH prevents degradation of this layer of the eye by harmful stimuli, such as blue light, staurosporine, and Fas pathway signaling, all of which produce free radicals.

The options for the present and future prevention and treatment of chronic diseases that cause retinal damage are encouraging. The results with iron-specific chelators are exciting potential treatment options worthy of future clinical trials for the treatment of age-related neurodegenerative diseases. While the future treatment options have tremendous potential, they are just that – in the future. Right now, prevention of these neurodegenerative retinal diseases may be achieved through lifestyle modifications, including reducing the amount of red meat consumed and possibly donating blood on a regular basis. With additional study, there is strong potential to reduce the risk of retinal disease by limiting iron-induced damage.

TAKE-HOME MESSAGES

- Iron has the potential to be highly toxic to the retina.
- Hereditary iron dysregulation can cause retinopathy in aceruloplasminemia, Friedreich's ataxia, and pantothenate kinase-associated neurodegeneration.
- Acquired retinal iron accumulation is associated with age-related macular degeneration (AMD).
- Iron chelation provides protection against mouse retinal degenerations of diverse etiology.

- While it is unknown whether high oral iron intake results in increased retinal iron, epidemiologic studies suggest that there may be an increased risk of AMD and glaucoma.
- Given the current state of knowledge, it is prudent to avoid excess systemic iron accumulation by limiting red meat intake and limiting the use of iron supplements to patients with iron deficiency.

References

1. Poss KD, Tonegawa S. Heme oxygenase 1 is required for mammalian iron reutilization. *Proc Natl Acad Sci U S A* 1997;**94**:10919–24.
2. Wigglesworth J, Baum H. Iron-dependent enzymes in the brain. In: Youdim M, editor. *Topics in Neurochemistry and Neuropharmacology*. London: Taylor & Francis; 1988. pp. 25–66.
3. Schichi H. Microsomal electron transport system of bovine retinal pigment epithelium. *Exp Eye Res* 1969;**8**:60–8.
4. Moiseyev G, Takahashi Y, Chen Y, Gentleman S, Redmond TM, Crouch RK, Ma JX. RPE65 is an iron(II)-dependent isomerohydrolase in the retinoid visual cycle. *J Biol Chem* 2006;**281**:2835–40.
5. Halliwell B, Gutteridge JM. Oxygen toxicity, oxygen radicals, transition metals and disease. *Biochem J* 1984;**219**:1–14.
6. Youdim MBH. Neuropharmacological and neurobiochemical aspects of iron deficiency. In: Dobbing J, editor. *Brain, Behavior, and Iron in the Infant Diet*. London: Springer; 1990. pp. 83–106.
7. Kaur D, Peng J, Chinta SJ, Rajagopalan S, Di Monte DA, Cherny RA, Andersen JK. Increased murine neonatal iron intake results in Parkinson-like neurodegeneration with age. *Neurobiol Aging* 2007;**28**:907–13.
8. Fredriksson A, Schroder N, Eriksson P, Izquierdo I, Archer T. Neonatal iron exposure induces neurobehavioural dysfunctions in adult mice. *Toxicol Appl Pharmacol* 1999;**159**:25–30.
9. Smith MA, Harris PL, Sayre LM, Perry G. Iron accumulation in Alzheimer disease is a source of redox-generated free radicals. *Proc Natl Acad Sci U S A* 1997;**94**:9866–8.
10. Yamaguchi K, Takahashi S, Kawanami T, Kato T, Sasaki H. Retinal degeneration in hereditary ceruloplasmin deficiency. *Ophthalmologica* 1998;**212**:11–4.
11. Jeong SY, David S. Glycosylphosphatidylinositol-anchored ceruloplasmin is required for iron efflux from cells in the central nervous system. *J Biol Chem* 2003;**278**:27144–8.
12. Hahn P, Qian Y, Dentchev T, Chen L, Beard J, Harris ZL, Dunaief JL. Disruption of ceruloplasmin and hephaestin in mice causes retinal iron overload and retinal degeneration with features of age-related macular degeneration. *Proc Natl Acad Sci U S A* 2004;**101**:13850–5.
13. Dunaief JL, Richa C, Franks EP, Schultze RL, Aleman TS, Schenck JF, et al. Macular degeneration in a patient with aceruloplasminemia, a disease associated with retinal iron overload. *Ophthalmology* 2005;**112**:1062–5.
14. Wolkow N, Song Y, Wu TD, Qian J, Guerquin-Kern JL, Dunaief JL. Aceruloplasminemia: retinal histopathologic manifestations and iron-mediated melanosome degradation. *Arch Ophthalmol* 2011;**129**:1466–74.
15. Klein R, Wang Q, Klein BE, Moss SE, Meuer SM. The relationship of age-related maculopathy, cataract, and glaucoma to visual acuity. *Invest Ophthalmol Vis Sci* 1995;**36**:182–91.
16. Leibowitz HM, Krueger DE, Maunder LR, Milton RC, Kini MM, Kahn HA, et al. The Framingham Eye Study monograph: an ophthalmological and epidemiological study of cataract, glaucoma, diabetic retinopathy, macular degeneration, and visual acuity in a general population of 2631 adults, 1973–1975. *Surv Ophthalmol* 1980;**24**(Suppl.):335–610.

17. Beatty S, Koh H, Phil M, Henson D, Boulton M. The role of oxidative stress in the pathogenesis of age-related macular degeneration. *Surv Ophthalmol* 2000;**45**:115–34.

18. Zarbin MA. Current concepts in the pathogenesis of age-related macular degeneration. *Arch Ophthalmol* 2004;**122**:598–614.

19. Age-Related Eye Disease Study Research Group. A randomized, placebo-controlled, clinical trial of high-dose supplementation with vitamins C and E, beta carotene, and zinc for age-related macular degeneration and vision loss: AREDS report no. 8. *Arch Ophthalmol* 2001;**119**:1417–36.

20. Hahn P, Milam AH, Dunaief JL. Maculas affected by age-related macular degeneration contain increased chelatable iron in the retinal pigment epithelium and Bruch's membrane. *Arch Ophthalmol* 2003;**121**:1099–105.

21. Chowers I, Wong R, Dentchev T, Farkas RH, Iacovelli J, Gunatilaka TL, et al. The iron carrier transferrin is upregulated in retinas from patients with age-related macular degeneration. *Invest Ophthalmol Vis Sci* 2006;**47**:2135–40.

22. Vogt W, Nolte R, Brunahl D. Binding of iron to the 5th component of human complement directs oxygen radical-mediated conversion to specific sites and causes nonenzymic activation. *Complement Inflamm* 1991;**8**:313–9.

23. Hahn P, Ying GS, Beard J, Dunaief JL. Iron levels in human retina: sex difference and increase with age. *Neuroreport* 2006;**17**:1803–6.

24. Lee DW, Andersen JK, Kaur D. Iron dysregulation and neurodegeneration: the molecular connection. *Mol Interv* 2006;**6**:89–97.

25. Zacharski LR, Gerhard GS. Atherosclerosis: a manifestation of chronic iron toxicity? *Vasc Med* 2003;**8**:153–5.

26. Edwards AO, Ritter III R, Abel KJ, Manning A, Panhuysen C, Farrer LA. Complement factor H polymorphism and age-related macular degeneration. *Science* 2005;**308**:421–4.

27. Glatt H, Machemer R. Experimental subretinal hemorrhage in rabbits. *Am J Ophthalmol* 1982;**94**:762–73.

28. Gold B, Merriam JE, Zernant J, Hancox LS, Taiber AJ, Gehrs K, et al. Variation in factor B (BF) and complement component 2 (C2) genes is associated with age-related macular degeneration. *Nat Genet* 2006;**38**:458–62.

29. Hageman GS, Anderson DH, Johnson LV, Hancox LS, Taiber AJ, Hardisty LI, et al. A common haplotype in the complement regulatory gene factor H (HF1/CFH) predisposes individuals to age-related macular degeneration. *Proc Natl Acad Sci U S A* 2005;**102**:7227–32.

30. Haines JL, Hauser MA, Schmidt S, Scott WK, Olson LM, Gallins P, et al. Complement factor H variant increases the risk of age-related macular degeneration. *Science* 2005;**308**:419–21.

31. Klein RJ, Zeiss C, Chew EY, Tsai JY, Sackler RS, Haynes C, et al. Complement factor H polymorphism in age-related macular degeneration. *Science* 2005;**308**:385–9.

32. Magnusson KP, Duan S, Sigurdsson H, Petursson H, Yang Z, Zhao Y, et al. CFH Y402H confers similar risk of soft drusen and both forms of advanced AMD. *PLoS Med* 2006;**3**: e5.

33. Cibis PA, Yamashita T, Rodriguez F. Clinical aspects of ocular siderosis and hemosiderosis. *AMA Arch Ophthalmol* 1959;**62**:180–7.

34. Sneed SR. Ocular siderosis. *Arch Ophthalmol* 1988;**106**:997.

35. Talamo JH, Topping TM, Maumenee AE, Green WR. Ultrastructural studies of cornea, iris and lens in a case of siderosis bulbi. *Ophthalmology* 1985;**92**:1675–80.

36. Feder JN, Penny DM, Irrinki A, Lee VK, Lebron JA, Watson N, et al. The hemochromatosis gene product complexes with the transferrin receptor and lowers its affinity for ligand binding. *Proc Natl Acad Sci U S A* 1998;**95**:1472–7.

37. Dunaief JL. Iron induced oxidative damage as a potential factor in age-related macular degeneration: the Cogan Lecture. *Invest Ophthalmol Vis Sci* 2006;**47**:4660–4.

38. Martin PM, Gnana-Prakasam JP, Roon P, Smith RG, Smith SB, Ganapathy V. Expression and polarized localization of the hemo-

39. Gnana-Prakasam JP, Thangaraju M, Liu K, Ha Y, Martin PM, Smith SB, Ganapathy V. Absence of iron-regulatory protein Hfe results in hyperproliferation of retinal pigment epithelium: role of cystine/glutamate exchanger. *Biochem J* 2009;**424**:243–52.

40. Roth AM, Foos RY. Ocular pathologic changes in primary hemochromatosis. *Arch Ophthalmol* 1972;**87**:507–14.

41. Baker E, Morgan EH. Iron transport. In: Brock J, Halliday JH, Pippard MH, Powell LW, editors. *Iron Metabolism in Health and Disease*. Philadelphia, PA: WB Saunders; 1994. pp. 63–95.

42. Yefimova MG, Jeanny JC, Keller N, Sergeant C, Guillonneau X, Beaumont C, Courtois Y. Impaired retinal iron homeostasis associated with defective phagocytosis in Royal College of Surgeons rats. *Invest Ophthalmol Vis Sci* 2002;**43**:537–45.

43. Hawkins KN. Contribution of plasma proteins to the vitreous of the rat. *Curr Eye Res* 1986;**5**:655–63.

44. Tripathi RC, Millard CB, Tripathi BJ, Chailertborisuth NS, Neely KA, Ernest JT. Aqueous humor of cat contains fibroblast growth factor and transferrin similar to those in man. *Exp Eye Res* 1990;**50**:109–12.

45. Yu TC, Okamura R. Quantitative study of characteristic aqueous humor transferrin, serum transferrin and desialized serum transferrin in aqueous humor. *Jpn J Ophthalmol* 1988;**32**:268–74.

46. Burdo JR, Antonetti DA, Wolpert EB, Connor JR. Mechanisms and regulation of transferrin and iron transport in a model blood–brain barrier system. *Neuroscience* 2003;**121**:883–90.

47. Yefimova MG, Jeanny JC, Guillonneau X, Keller N, Nguyen-Legros J, Sergeant C, et al. Iron, ferritin, transferrin, and transferrin receptor in the adult rat retina. *Invest Ophthalmol Vis Sci* 2000;**41**:2343–51.

48. Aisen P, Enns C, Wessling-Resnick M. Chemistry and biology of eukaryotic iron metabolism. *Int J Biochem Cell Biol* 2001;**33**:940–59.

49. Levi S, Santambrogio P, Cozzi A, Rovida E, Corsi B, Tamborini E, et al. The role of the L-chain in ferritin iron incorporation. Studies of homo and heteropolymers. *J Mol Biol* 1994;**238**:649–54.

50. Hahn P, Dentchev T, Qian Y, Rouault T, Harris ZL, Dunaief JL. Immunolocalization and regulation of iron handling proteins ferritin and ferroportin in the retina. *Mol Vis* 2004;**10**:598–607.

51. Levi S, Corsi B, Bosisio M, Invernizzi R, Volz A, Sanford D, et al. A human mitochondrial ferritin encoded by an intronless gene. *J Biol Chem* 2001;**276**:24437–40.

52. Osaki S, Johnson DA, Frieden E. The possible significance of the ferrous oxidase activity of ceruloplasmin in normal human serum. *J Biol Chem* 1966;**241**:2746–51.

53. Patel BN, David S. A novel glycosylphosphatidylinositol-anchored form of ceruloplasmin is expressed by mammalian astrocytes. *J Biol Chem* 1997;**272**:20185–90.

54. Patel BN, Dunn RJ, David S. Alternative RNA splicing generates a glycosylphosphatidylinositol-anchored form of ceruloplasmin in mammalian brain. *J Biol Chem* 2000;**275**:4305–10.

55. Chen L, Dentchev T, Wong R, Hahn P, Wen R, Bennett J, Dunaief JL. Increased expression of ceruloplasmin in the retina following photic injury. *Mol Vis* 2003;**9**:151–8.

56. Gerhardinger C, Costa MB, Coulombe MC, Toth I, Hoehn T, Grosu P. Expression of acute-phase response proteins in retinal Muller cells in diabetes. *Invest Ophthalmol Vis Sci* 2005;**46**:349–57.

57. Newman AM, Gallo NB, Hancox LS, Miller NJ, Radeke CM, Maloney MA, et al. Systems-level analysis of age-related macular degeneration reveals global biomarkers and phenotype-specific functional networks. *Genome Med* 2012;**4**:16.

58. Vulpe CD, Kuo YM, Murphy TL, Cowley L, Askwith C, Libina N, et al. Hephaestin, a ceruloplasmin homologue implicated in intestinal iron transport, is defective in the sla mouse. *Nat Genet* 1999;**21**:195–9.

59. Hadziahmetovic M, Dentchev T, Song Y, Haddad N, He X, Hahn P, et al. Ceruloplasmin/hephaestin knockout mice model morpho-

logic and molecular features of AMD. *Invest Ophthalmol Vis Sci* 2008;**49**:2728–36.

60. Donovan A, Brownlie A, Zhou Y, Shepard J, Pratt SJ, Moynihan J, et al. Positional cloning of zebrafish ferroportin1 identifies a conserved vertebrate iron exporter. *Nature* 2000;**403**:776–81.

61. McKie AT, Marciani P, Rolfs A, Brennan K, Wehr K, Barrow D, et al. A novel duodenal iron-regulated transporter, IREG1, implicated in the basolateral transfer of iron to the circulation. *Mol Cell* 2000;**5**:299–309.

62. Abboud S, Haile DJ. A novel mammalian iron-regulated protein involved in intracellular iron metabolism. *J Biol Chem* 2000;**275**:19906–12.

63. Burdo JR, Menzies SL, Simpson IA, Garrick LM, Garrick MD, Dolan KG, et al. Distribution of divalent metal transporter 1 and metal transport protein 1 in the normal and Belgrade rat. *J Neurosci Res* 2001;**66**:1198–207.

64. Dentchev T, Hahn P, Dunaief JL. Strong labeling for iron and the iron-handling proteins ferritin and ferroportin in the photoreceptor layer in age-related macular degeneration. *Arch Ophthalmol* 2005;**123**:1745–6.

65. Yang F, Wang X, Haile DJ, Piantadosi CA, Ghio AJ. Iron increases expression of iron-export protein MTP1 in lung cells. *Am J Physiol Lung Cell Mol Physiol* 2002;**283**:L932–9.

66. Krause A, Neitz S, Magert HJ, Schulz A, Forssmann WG, Schulz-Knappe P, Adermann K. LEAP-1, a novel highly disulfide-bonded human peptide, exhibits antimicrobial activity. *FEBS Lett* 2000;**480**:147–50.

67. Park CH, Valore EV, Waring AJ, Ganz T. Hepcidin, a urinary antimicrobial peptide synthesized in the liver. *J Biol Chem* 2001;**276**:7806–10.

68. Nicolas G, Chauvet C, Viatte L, Danan JL, Bigard X, Devaux I, et al. The gene encoding the iron regulatory peptide hepcidin is regulated by anemia, hypoxia, and inflammation. *J Clin Invest* 2002;**110**:1037–44.

69. Pigeon C, Ilyin G, Courselaud B, Leroyer P, Turlin B, Brissot P, Loreal O. A new mouse liver-specific gene, encoding a protein homologous to human antimicrobial peptide hepcidin, is overexpressed during iron overload. *J Biol Chem* 2001;**276**:7811–9.

70. Hadziahmetovic M, Song Y, Ponnuru P, Iacovelli J, Hunter A, Haddad N, et al. Age-dependent retinal iron accumulation and degeneration in hepcidin knockout mice. *Invest Ophthalmol Vis Sci* 2011;**52**:109–18.

71. Picard E, Fontaine I, Jonet L, Guillou F, Behar-Cohen F, Courtois Y, Jeanny JC. The protective role of transferrin in Muller glial cells after iron-induced toxicity. *Mol Vis* 2008;**14**:928–41.

72. Honda K, Casadesus G, Petersen RB, Perry G, Smith MA. Oxidative stress and redox-active iron in Alzheimer's disease. *Ann N Y Acad Sci* 2004;**1012**:179–82.

73. Wolozin B, Golts N. Iron and Parkinson's disease. *Neuroscientist* 2002;**8**:22–32.

74. Wong RW, Richa DC, Hahn P, Green WR, Dunaief JL. Iron toxicity as a potential factor in AMD. *Retina* 2007;**27**:997–1003.

75. Chong EW, Simpson JA, Robman LD, Hodge AM, Aung KZ, English DR, et al. Red meat and chicken consumption and its association with age-related macular degeneration. *Am J Epidemiol* 2009;**169**:867–76.

76. Wang SY, Singh K, Lin SC. The association between glaucoma prevalence and supplementation with the oxidants calcium and iron. *Invest Ophthalmol Vis Sci* 2012;**53**:725–31.

77. Wang SYSK, Lin SC. *Glaucoma and intake of the oxidants calcium and iron in a US population sample.* Chicago: American Academy of Ophthalmology Annual Meeting; 2012.

78. Konerirajapuram NS, Coral K, Punitham R, Sharma T, Kasinathan N, Sivaramakrishnan R. Trace elements iron, copper and zinc in vitreous of patients with various vitreoretinal diseases. *Indian J Ophthalmol* 2004;**52**:145–8.

79. Belcher JD, Beckman JD, Balla G, Balla J, Vercellotti G. Heme degradation and vascular injury. *Antioxid Redox Signal* 2010;**12**:233–48.

80. Cussimanio BL, Booth AA, Todd P, Hudson BG, Khalifah RG. Unusual susceptibility of heme proteins to damage by glucose during non-enzymatic glycation. *Biophys Chem* 2003;**105**:743–55.

81. Sullivan JL. Is stored iron safe? *J Lab Clin Med* 2004;**144**:280–4.

82. Massie HR, Aiello VR, Williams TR. Iron accumulation during development and ageing of *Drosophila*. *Mech Ageing Dev* 1985;**29**:215–20.

83. Massie HR, Aiello VR, Williams TR. Inhibition of iron absorption prolongs the life span of *Drosophila*. *Mech Ageing Dev* 1993;**67**:227–37.

84. Liu ZD, Hider RC. Design of clinically useful iron(III)-selective chelators. *Med Res Rev* 2002;**22**:26–64.

85. Kalinowski DS, Richardson DR. The evolution of iron chelators for the treatment of iron overload disease and cancer. *Pharmacol Rev* 2005;**57**:547–83.

86. Maxton DG, Bjarnason I, Reynolds AP, Catt SD, Peters TJ, Menzies IS. Lactulose, 51Cr-labelled ethylenediaminetetra-acetate, L-rhamnose and polyethyleneglycol 400 [corrected] as probe markers for assessment *in vivo* of human intestinal permeability. *Clin Sci (Lond)* 1986;**71**:71–80.

87. Hadziahmetovic M, Song Y, Wolkow N, Iacovelli J, Grieco S, Lee J, et al. The oral iron chelator deferiprone protects against iron overload-induced retinal degeneration. *Invest Ophthalmol Vis Sci* 2011;**52**:959–68.

88. Hadziahmetovic M, Pajic M, Grieco S, Song Y, Song D, Li Y, et al. The oral iron chelator deferiprone protects against retinal degeneration induced through diverse mechanisms. *Trans Vis Sci Tech* 2012;**1**(3).

89. Kurz T, Karlsson M, Brunk UT, Nilsson SE, Frennesson C. ARPE-19 retinal pigment epithelial cells are highly resistant to oxidative stress and exercise strict control over their lysosomal redox-active iron. *Autophagy* 2009;**5**:494–501.

90. Lukinova N, Iacovelli J, Dentchev T, Wolkow N, Hunter A, Amado D, et al. Iron chelation protects the retinal pigment epithelial cell line ARPE-19 against cell death triggered by diverse stimuli. *Invest Ophthalmol Vis Sci* 2009;**50**:1440–7.

Hypoglycemia and Retinal Cell Death

R. Roduit[1, 2], *D. Balmer*[1], *M. Ibberson*[3], *D.F. Schorderet*[1, 2, 4]

[1]IRO, Institute for Research in Ophthalmology, Sion, Switzerland; [2]Department of Ophthalmology, University of Lausanne, Lausanne, Switzerland; [3]Vital-IT Group, Swiss Institute of Bioinformatics, Lausanne, Switzerland; [4]Faculty of Life Sciences, Ecole Polytechnique Fédérale de Lausanne, Lausanne, Switzerland

INTRODUCTION

Glucose is the most important metabolic substrate of the retina, and maintenance of normoglycemia is an essential challenge for diabetic patients. This chapter describes the processes of retinal glucose metabolism, hypoglycemia, and the pathways involved in retinal cell death induced by hypoglycemia.

GLUCOSE METABOLISM IN THE RETINA

Energy supply is obtained from glucose, proteins, and lipids by different metabolic pathways. Carbohydrates, the most important fuel molecules, are metabolized by nearly all known organisms and are an essential metabolic substrate for all mammalian cells, particularly neural tissues, brain and retina, which require a higher rate of respiration and glucose oxidation than any other tissues. Indeed, both tissues need a constant supply of blood glucose to sustain their functions.

Energy demand and use differ depending on light conditions and cell types. In the dark, photoreceptors (cones and rods) expend substantial energy to regulate Ca^{2+} flux for synaptic transmission and to maintain circulating dark current.[1] Adenosine triphosphate (ATP) is required to sustain the dark current due to strong pumping of Na^+, which is probably needed at the inner segment of photoreceptors to preserve its electrochemical gradient. Thus, to restore ATP to the levels needed for proper cell function, a large amount of oxygen is required.[2] In light conditions, cells demand a significant amount of energy to perform visual signal transduction, which takes place in the outer segment of photoreceptors.

Hence, retinal glucose metabolism begins with the transport of glucose across the blood–retinal barrier (BRB).[3] Glucose has to pass through two barriers: the retinal capillary endothelial cells of the inner BRB and the retinal pigment epithelium (RPE), which is located between the photoreceptors and the choroid and is known as the outer BRB.[4] Both barriers prevent (or strongly limit) the passive diffusion of glucose into the retinal interstitial space. A specific facilitated transport process involving members of the sodium-independent glucose transporter family, the GLUT family, mainly the Glut1 transporter, has two roles within the glucose transport process. The first one is the transport of glucose from the choroidal vasculature to the outer retina across the BRB formed by the RPE cells expressing Glut1 at both the apical and basolateral membranes.[5] The second role is to mediate glucose uptake by the photoreceptors themselves, expressing the glucose transporter Glut1 mainly in the inner segment with a few in the outer part.[1] This idea is reinforced by the fact that hexokinase, an enzyme involved in the first step of glucose metabolism that transforms glucose to glucose-6-phosphate (G6P), is confined to the inner segment of the photoreceptors and is required to hold the glucose inside the cell.

Since the photoreceptor outer segments are devoid of mitochondria, G6P resources may be supplied by diffusion from the inner segments via the connecting cilium. When present in the outer segments, G6P can produce, via glycolysis, the ATP mandatory for visual function. Moreover, it can enter the pentose phosphate pathway to produce the anabolic coenzyme nicotinamide adenine dinucleotide phosphate oxidase (NADPH), which is required in the regeneration of all-*trans*-retinal used during the visual cycle.[1] Lastly, Glut1 is also expressed in the ganglion cell layer, in the neuronal processes of both plexiform layers, and in the cell bodies of the inner nuclear layer.[1] In addition to Glut1, Glut3 is localized in the inner and the outer plexiform layers, where it can be efficient in glucose transport at low glucose concentration because of its very low Michaelis–Menten constant.[6] More recent studies also suggest a

Handbook of Nutrition, Diet, and the Eye
http://dx.doi.org/10.1016/B978-0-12-401717-7.00064-2

role for Glut2 in retinal glucose transport. Glut2 is present at the apical ends of Müller cells that face the interphotoreceptor space. It may control glucose homeostasis within the retina by performing anterior and posterior transport of glucose.[7] Further studies are needed to elucidate the mechanism of glucose within the retina and to clearly define its specific roles in maintaining visual function.

Glucose metabolism activity takes place mainly in the photoreceptors and Müller cells. They form the 'cell complex' and act in a symbiotic way: both cell types are needed and they are interconnected. It has been considered that photoreceptors metabolize lactate obtained from Müller cells as their major substrate, but this hypothesis is highly controversial.[8] In fact, glucose seems to be the major energy source in the retina in normal conditions. However, photoreceptors can also take up and metabolize lactate, in case of hypoglycemia, for instance.[1] Müller cells exhibit a high rate of aerobic glycolysis, which results in the production of lactate and pyruvate. These molecules can be secreted into the extracellular space and/or can be used by the photoreceptors. All retinal cells use glucose as a first substrate, but during periods of starvation or intense neuronal activity (as in dark conditions), photoreceptors can metabolize lactate or pyruvate as additional fuel for their oxidative energy metabolism. These monocarboxylates come from Müller cells or are formed in the photoreceptors themselves.

Müller cells are quite resistant to injury, since they mainly depend on glycolysis (aerobic and anaerobic). They can switch from aerobic to anaerobic mode when there is a lack of oxygen, thus continuing to supply the retina with its required energy. Moreover, Müller cells can resist glucose limitation, as they use other substrates such as lactate or pyruvate to produce energy substrate by the tricarboxylic acid cycle.[9] These cells are the major glycogen storage sites and therefore can, in stress conditions (e.g., hypoglycemia), break down glycogen to restore sufficient glucose molecules.[10] Glycogen content is regulated by glycogen synthase (GS) and glycogen phosphorylase (GP), which are mainly localized in Müller cells.[11] Few studies have analyzed in detail the regulation of retinal glycogen metabolism in normal or pathologic conditions, but very recently Osorio-Paz et al. showed the importance of glycogen as an energy supply during hypoglycemia.[12]

Glucose is important since electrical activity (needed for visual functions) depends on its availability. Furthermore, glucose metabolism is primordial because it supports several intermediate metabolites or key activities essential for photoreceptor survival and for visual functions. ATP production is an important product of this metabolism because it supplies energy requirements. Glucose also contributes to the synthesis of N-acetylglucosamine, which is required for asparagine formation and for the production of cytosolic NADPH. This latter compound is needed to inhibit caspase-mediated apoptosis, support anabolic activity, and help to maintain appropriate levels of reactive oxygen species (ROS).

As glucose is so critical, the following sections will describe the roles of these pathways in hypoglycemia-induced cell death.

HYPOGLYCEMIA

As described above, neural tissue, including retina, is totally dependent on glucose for normal metabolic activity. In addition, the level of glucose storage is negligible compared with the eye's glucose demand. Therefore, the retina is dependent on blood glucose delivery. Glucose is the most important metabolic substrate of the retina, and maintenance of normoglycemia is an essential challenge for diabetic patients. Glycemic excursions, in both type 1 and type 2 diabetes, can lead to cardiovascular diseases, nephropathy, neuropathy, and retinopathy.[13] Although diabetes-related diseases are generally linked to hyperglycemia,[14] hypoglycemia could play an important role in the onset of diabetic retinopathy. There are several different causes of hypoglycemia, including iatrogenic hypoglycemia, inappropriate diet, and irregular physical activity. Indeed, iatrogenic hypoglycemia, due to inappropriate insulin dosage, causes morbidity in most people with type 1 diabetes and in many with advanced type 2 diabetes.[15] Hypoglycemia could also result from other causes such as low glucose intake due to inappropriate choices regarding food or the use of too much glucose linked to irregular and unpredictable physical activity. A 2005 report, written by the American Diabetes Association Workgroup on hypoglycemia,[16] defined hypoglycemia in diabetes as 'all episodes of an abnormally low plasma glucose concentration that expose the individual to potential harm'. But in regard to this definition, we can separate symptomatic hypoglycemia (thousands of hypoglycemic episodes over diabetes lifetime; blood glucose about 3.9 mM) and severe hypoglycemia (one or two episodes per year; blood glucose below 2 mM) (for an exhaustive review see Cryer[15]). Patients with type 1 diabetes suffer up to two severe hypoglycemic episodes, whereas type 2 diabetic patients experience on average one episode over 5 years.[17] In addition, retinopathy of prematurity may be the consequence of hypoglycemic periods that alter brain (and retina) development.[18] Although this is still controversial because of the small number of studies in this field, the belief is that hypoglycemia could also be detrimental in preterm infant.

Diabetic retinopathy is the result of microvascular retinal changes promoted by hyperglycemia through the formation of advanced glycation end-products, resulting in weakening and blockage of blood vessels through upregulation and secretion of vascular endothelium growth factor (VEGF).[13,19] Moreover, hyperglycemia has been linked to ROS production, which alters mitochondrial metabolism leading to cell death. The role of hyperglycemia in vascular complications has been extensively studied in a large number of *in vivo* and *ex vivo* models.[20,21] On the opposite side, few studies exist that implicate hypoglycemia as a key factor in visual disorders, while hypoglycemia has been reported to have an important role in brain dysfunction and neuronal cell death (for reviews see Suh *et al.*[22,23]). In studies related to hypoglycemia and retina, the majority of data collected has focused on *in vitro* or *ex vivo* analyses. These analyses have shown that conditions of low glucose reduced viability of all retinal cell types in a mixed primary cell culture[24] and demonstrated the sensitivity of isolated chick retinas to *in vitro* aglycemic conditions.[25] Recently, one *in vivo* study showed a decrease in retinal function and visual acuity that correlated directly with the degree of hypoglycemia.[26] The authors showed that chronic moderate hypoglycemia in mice led to retinal degeneration

(Fig. 64.1A, B) and loss of vision (Fig. 64.1C). Another study suggested that cone death in diverse mouse models of retinitis pigmentosa could be, at least in part, the result of the starvation of cones via the insulin/mammalian target of rapamycin (mTOR) pathway.[27] Khan *et al.* also reported a decrease in central vision in humans during hypoglycemia that was reversed with restoration of normal glycemia.[28] A recent study by the present authors showed that insulin-induced hypoglycemia led to retinal cell death in mice. As a paradigm, hyperinsulinemic/hypoglycemic clamps were used (Fig. 64.2), with hyperinsulinemic/euglycemic clamps or sham-operated mice as controls. This acute hypoglycemic model provided some important new results. Figure 64.3 shows that the number of terminal deoxynucleotidyl transferase dUTP nick-end labeling (TUNEL)-positive cells was much higher in mice that underwent 5 hours of hypoglycemia (blood glucose concentration about 2 mM) than in their specific

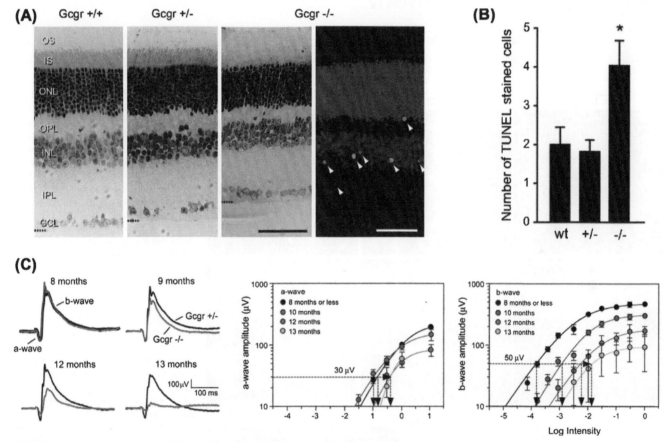

FIGURE 64.1 Retinal cell death and age-related changes in the electroretinogram (ERG) in chronic hypoglycemic mouse models. (A) Sections of retinas from age-matched adult $Gcgr^{+/+}$, $Gcgr^{+/-}$, and moderately hypoglycemic $Gcgr^{-/-}$ mice. Shown are photoreceptors (outer segments, OS; inner segments, IS) at the top and the ganglion cell layer (GCL) at the bottom. Visible in between are the outer plexiform layer (OPL) and inner plexiform layer (IPL) and the densely stained cell bodies in the outer nuclear layer (ONL) and inner nuclear layer (INL). The oblique cryosection on the right from a $Gcgr^{-/-}$ mouse shows terminal deoxynucleotidyl transferase dUTP nick-end labeling (TUNEL) staining primarily in the INL. (B) Average number of TUNEL-positive cells is significantly greater for Gcgr-deficient mice than for heterozygous and wild-type (wt) mice. (C) Average ERGs recorded from a group of moderately hypoglycemic $Gcgr^{-/-}$ mice (red traces) and euglycemic $Gcgr^{+/+}$ mice (black traces) in response to 10 ms flashes. Intensity–response functions of a-waves and b-waves recorded from the same ERGs. Dashed horizontal lines and vertical intercepts on the abscissa show the method for determining threshold light intensities. Responses were recorded at 8, 10, 12, and 13 months. *Source: Adapted from Umino et al. Proc Natl Acad Sci U S A. 2006;103:19541–5,[26] with the permission of PNAS.* (See color plate section)

(A)

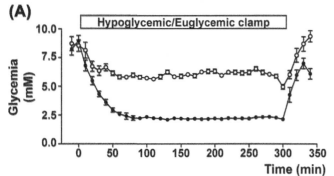

(B)

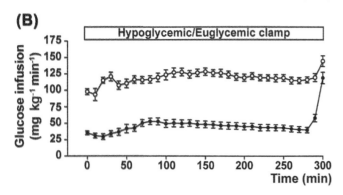

FIGURE 64.2 Paradigm used for studying the hypoglycemic effect on retinal cells. An indwelling catheter was inserted in the femoral vein of isoflurane-anesthetized mice, which were allowed to recover for 4–7 days. After a 5-hour fasting period, awake and freely moving mice were subjected to 5 hours of either a hyperinsulinemic/hypoglycemic or hyperinsulinemic/euglycemic clamp. (A) Graphic representation of plasma glucose levels; (B) glucose infusion rates during the hyperinsulinemic/hypoglycemic clamp (black circle) and the control hyperinsulinemic/euglycemic clamp (white circle). *Source: From Emery* et al. PLoS ONE 2011;6:e21586,[29] *with permission of* PLoS ONE.

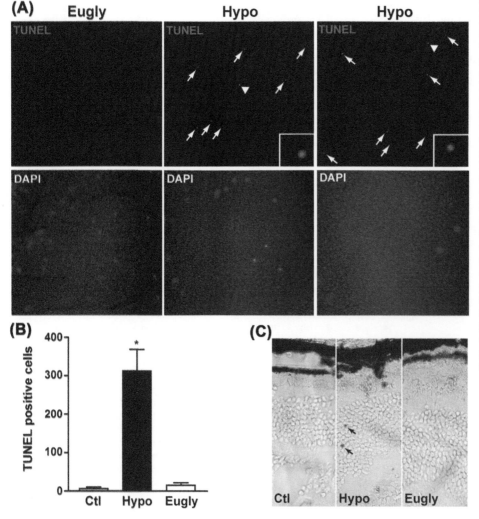

FIGURE 64.3 Acute insulin-induced hypoglycemia leads to retinal cell death in mice. (A) Flat-mounted retinas were isolated 48 hours after the clamp, stained for cell death by terminal deoxynucleotidyl transferase dUTP nick-end labeling (TUNEL) assay, and DAPI counterstained. White arrows show TUNEL-positive cells in hypoglycemic condition. (B) Quantification of TUNEL-positive cells was performed under a fluorescence microscope on retinal flat-mounts. Results are expressed as mean±SEM of three different retinas for each group, *P < 0.006 hypoglycemic (Hypo) vs. euglycemic (Eugly). (C) Ten-micrometer embedded frozen sections of enucleated eyes isolated from control (Ctl), hypoglycemic (Hypo), and euglycemic (Eugly) animals were stained for cell death by a colorimetric TUNEL system. Using this procedure, apoptotic nuclei are stained dark brown (black arrows). A representative region of three different isolated retinas is shown. *Source: From Emery* et al. PLoS ONE *2011;6:e21586,[29] with permission of* PLoS ONE. (See color plate section)

euglycemic controls. The mechanisms involved may be acting through an increase in the caspase-3 pathway and ROS production, as well as activation of the B-cell lymphoma-2 protein (Bcl-2)/Bcl-2-associated X protein (Bax) pathway and a reduction in glutathione (GSH) content[29] (see below for more details).

MICROARRAY ANALYSIS AND MAJOR PATHWAYS INVOLVED

To underscore and more precisely define which pathways are altered by an acute hypoglycemic event, a microarray analysis was performed with retina isolated from the hypoglycemic and the euglycemic groups (manuscript in preparation). Statistical analysis of the results highlights diverse clusters of genes involved in many different pathways, including lysosomes, apoptosis, and GSH metabolism. Gene set enrichment analysis (GSEA)[30] confirmed that genes involved in GSH metabolism may play a key role in hypoglycemia-induced cell death (Fig. 64.4A). Among all these genes, two key enzymes were

found that were further characterized.[29] Glutathione peroxidase-3 (GPX3) and glutathione transferase (GSTo1) were both increased in hypoglycemic conditions (Fig. 64.4B), as already published.[29] Similar regulation was also observed occurring in normal isolated mouse retina cultured in low glucose conditions, and these results correlated with a decrease in the GSH content in the group of mice that underwent hypoglycemia (Fig. 64.4C) and in 661W cells cultured in low glucose conditions (Fig. 64.4D).[29] The implication of the lysosomal pathway is currently being evaluated, and the apoptotic pathways are partially described below.

APOPTOSIS AND CELL DEATH

Key Role of Glutathione

GSH (γ-L-glutamyl-L-cystein-glycine) is a scavenger and antioxidant protein involved in many cellular functions including regulation of DNA and protein synthesis, signal transduction, and cell cycle regulation, as well as

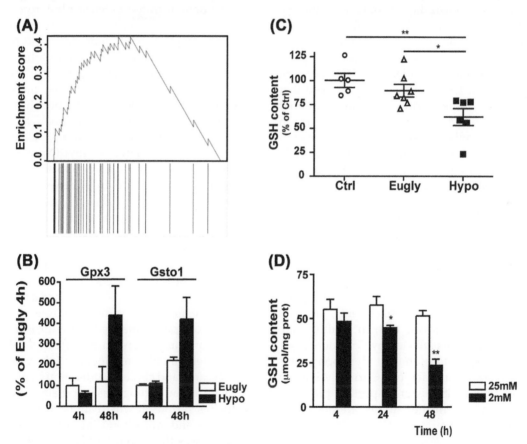

FIGURE 64.4 Glutathione (GSH) metabolism plays a central role in the hypoglycemic process. (A) Results of gene set enrichment analysis (GSEA) for glutathione metabolism pathway (KEGG).[30] The positions of the pathway gene in the gene list (ranked by log 2-fold change) are indicated as vertical bars under the plot. This pathway is enriched for genes that are upregulated under hypoglycemia. (B) Two genes (*GPX3* and *GSTo1*, implicated in GSH depletion) were identified by microarray analysis (Affymetrix) in the hypoglycemic (Hypo, black box) and euglycemic (Eugly, white box) groups of animals 48 hours after the clamp. (C) GSH content was measured in protein lysates obtained from retina of control (Ctrl, white circle), euglycemic (Eugly, white triangle), and hypoglycemic (Hypo, black square) mice.[29] (D) 661W cells were cultured at low (2 mM, black box) and high (25 mM, white box) glucose conditions and GSH content was measured at diverse periods. *Source: From Emery* et al. PLoS ONE *2011;6:e21586,*[29] *with permission of PLoS ONE, and from Ibberson* et al. 2013; *in preparation.*[31]

maintaining a stable thiol redox state.[32] GSH metabolism is complex; the level of GSH is increased by direct uptake of the compound or by *de novo* synthesis through a two-step reaction utilizing L-glutamate, L-cysteine, glycine, ATP, and both glutamate–cysteine ligase (GCL) and glutathione synthase (GS) (for reviews see references[32] and[33]). A decrease in cellular GSH (the reduced glutathione) occurs essentially in reactions in which glutathione peroxidase reduces hydrogen peroxide (H_2O_2) by producing the oxidized glutathione (GSSG) and H_2O. NADPH is used to restore GSH from GSSG with the help of glutathione reductase (GR) (for exhaustive reviews on cellular GSH regulation transport and functions, see references[33–35]). Glutathione transferase serves as a protective enzyme by adding GSH to proteins, targeting them for export from the cell. This mechanism, called glutathionylation, may be involved in the metabolic adaptation of cells to stress stimuli by modifying protein activity or degradation. Glutathionylation of proteins has been described in multiple studies.[36,37] Without giving a complete list of enzymes modified by glutathionylation, caspase-3,[38] nuclear factor-κB (NF-κB),[39] c-Jun[40] protein kinase C,[41] and thioredoxin (Trx)[42] may be mentioned.

A few studies have shown the role of GSH in the eye and analyzed GR activity and GSH levels in diseases involving oxidative damage of cells (for review see Ganea and Harding[32]). To the authors' knowledge, no study to date has related a potential role of GSH during a hypoglycemic event. This group recently showed that retinal GSH level was modified by low glucose, either *in vivo* in insulin-induced hyploglycemia or *in vitro* in 661W photoreceptor cells exposed to low glucose conditions (Fig. 64.4C, D). Moreover, modulation of GSH level in 661W cells modified the cellular response to poor nutrient stimuli. Restoration of GSH level, by the addition of exogenous GSH ethyl ester, blocked the low glucose-induced cell death, while inhibition of GSH synthesis, using buthionine sulfoximine (BSO) at high glucose concentration, induced the decrease of GSH and apoptosis of photoreceptor cells.[29]

The key role of GSH in the retina is highlighted by the retinal dystrophy observed in two sisters with glutathione synthase deficiency.[43] In these patients, the level of GSH content was 80% lower than in healthy subjects. About 25% of patients affected by GSH deficiency develop retinitis pigmentosa, retinal dystrophy, lens opacities, or decreased visual acuity.[44] Currently, it is not known whether retinal GSH is an important player in diabetic retinopathy, and further studies are needed to clarify this important point.

Oxidative Stress and Reactive Oxygen Species Production

Oxidative stress (essentially ROS production) is considered as a critical mediator of pathways implicated in diabetic retinopathy. In fact, oxidative metabolism, mediated by glucose availability and thus glycolysis, can produce oxygen species. This can disrupt the RPE cell junctions and the integrity of the BRB, as well as the redox state of cells, thus triggering cell death and generating the retinal abnormalities characteristic of diabetes.[45] Most, if not all, studies have made a direct link between oxidative stress and hyperglycemia.[13,46–49] Very few analyses have tried to understand the role of oxidative stress induced by hypoglycemic events, and these studies were mainly looking at the brain[22,23,50] rather than the retina. A few analyses showed induction of oxidative stress by hypoglycemia in cellular models,[29,51–54] while others studied *in vivo* interactions.[55,56] *In vitro* studies principally showed that hypoglycemia induced endoplasmic reticulum stress and unfold protein response in retinal pericytes.[53] Cotter's group demonstrated a prosurvival role of NADPH oxidases, Nox2 and Nox4, in retina-derived cells, suggesting a protective role of ROS in the retina.[51,52] Recently, the same group showed that rods and cones generated ROS in reaction to serum deprivation through NADPH oxidase proteins.[54] Mitochondrial superoxide production has been demonstrated in 661W photoreceptor cells cultured in low glucose conditions,[29] and it was postulated that, following nutrient deprivation, ROS production was principally involved in the apoptotic pathway. The authors are currently analyzing ROS production in the insulin-induced hypoglycemic mouse model. In hyperglycemic conditions, ROS production has been linked to a downregulation of the GSH pool, via activation of aldolase[13] and activation of caspase-3, leading to neuronal dysfunction through N-methyl-D-aspartate (NMDA) receptors.[47] These results are supported by several studies showing an increase in superoxide production in hippocampal neurons cultured under hypoglycemic conditions[57] and in neurons after glucose deprivation.[58] Moreover, *in vivo* studies have recently shown production of mitochondrial ROS in the newborn brain during acute hypoglycemia[56] and in insulin-induced hypoglycemic stress in healthy subjects.[55]

Very recently, Santos *et al.* published a review focusing on diabetic retinopathy and the role of antioxidants in mitochondrial damages.[49] Kowluru and his group worked for many years on the effect of hyperglycemia in the development of retinopathy. It is well known from basic research and clinical studies that hyperglycemia is a key factor in the pathogenesis of diabetic complications. It activates various downstream factors, increases oxidative stress and mitochondrial superoxide production, and decreases the activity and expression of scavenger proteins. Among others, Kowluru *et al.* showed that overexpression of mitochondrial superoxide dismutase protected the retina of mice from diabetes-induced oxidative stress.[59] The challenge is to counteract oxidative stress, stabilize mitochondrial homeostasis, and prevent superoxide production by increasing antioxidant defense. There are few studies showing the benefits of antioxidant-complemented diets in preventing retinopathy in diabetic animals.[49] A similar approach in delaying the onset of diabetic side effects could be considered in the case of hypoglycemia.

The Bcl2-/Bax Pathway

The Bcl-2 family of proteins is a key factor in the apoptotic process. This family is composed of more than 20 different homologues, which are either proapoptotic (e.g., Bax, Bak, Bok, Bcl-XS, Bad, Bid) or antiapoptotic (e.g., Bcl-2, Bcl-XL, McL-1, BcL-W).[60] These proteins bind to one another to form homodimers or heterodimers and modulate cell survival. They also are able to form ionic channels that permeabilize the mitochondrial membrane to allow cytochrome *c*, apoptosis-inducing factor, second mitochondria-derived activator of caspases, and direct IAP-binding protein with low PI (Smac/Diablo) release from the mitochondrial inner space, leading to activation of caspases.[61] The Bcl-2/Bax complex has been extensively studied in multiple and different cell types and stress-induced apoptosis models. Proapoptotic proteins Bak and Bax are involved in developmental photoreceptor apoptosis. Indeed, double Bak- and Bax-deficient mice showed an absence of TUNEL-positive cells at postnatal day 7 (p7), which normally corresponds to the peak of developmental apoptosis.[62] In addition, the retinas of these mice are more resistant to light-induced photoreceptor cell death,[63] and retinas from Bax$^{-/-}$ mice are less sensitive to cell death after retinal detachment.[64] In rats, Bax expression was increased in ischemia when compared with nonoperated animals[65] and translocated to the mitochondria in 8-month streptozotocin-induced diabetes.[66] A decrease in the Bcl-2/Bax ratio was described

in a murine glaucoma model[67] and in RPE65$^{-/-}$ mice representing a retinitis pigmentosa model.[68] These results argue for an implication of proapoptotic Bax proteins in retinal apoptosis. However, other reports failed to replicate these studies, and in a similar rat ischemia experiment, a reduction in Bax expression was found compared with sham-operated retinas.[69] Similarly, the implication of antiapoptotic proteins, such as Bcl-2 proteins, is controversial. Diverse studies showed that overexpression of Bcl-2 is able to reduce photoreceptor cell death in different retinal degeneration models,[70,71] while others reported a lack of protection by the antiapoptotic protein.[72] A recent study also suggested a role of a Bcl-2-related protein, Bad, in the modulation of the counter-regulatory hormonal response to hypoglycemia. Indeed, Bad deficiency was associated with impaired glucoprivic feeding.[73]

From all of these studies, it seems that the ratio of antiapoptotic to proapoptotic proteins is implicated in deciding the death or survival of cells. In 661W photoreceptor cells, low glucose induced a strong decrease in the antiapoptotic Bcl-2 at messenger RNA[74] and protein levels (Fig. 64.5A), while expression of the proapoptotic Bax was not modified (Fig. 64.5A) but the level of free Bax was increased (Fig. 64.5B). The hypothesis was that low glucose induced the decrease in Bcl-2 protein, thus freeing Bax, which could exert its apoptotic effect and induce retinal cell death.[74] Further analyses are needed to confirm this hypothesis. It is also necessary to clarify the *in vivo* role of the Bcl-2/Bax ratio during hypoglycemia.

As described above, activation of the Bcl-2 family of proteins led to the activation of caspases, including caspase-3. This enzyme was activated by ROS in long-term diabetic rats,[66] mice, and patients with type 2 diabetes.[75] Caspase-3 activation has been implicated

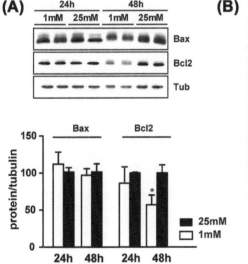

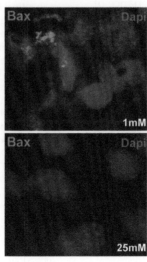

FIGURE 64.5 Hypoglycemia-induced cell death is Bcl2/Bax dependent. (A) Culture of 661W photoreceptor cells at low glucose induces a decrease in Bcl-2 protein expression. (B) The decrease in *Bcl-2* gene expression releases the active proapoptotic Bax that is recognized by a specific antibody. *Source: Adapted from Balmer et al.* PLoS ONE *2013;10.1371/journal.pone.0074162.*[74] (See color plate section)

in several models of neuronal apoptosis. Its expression was increased in cerebral cortex of hypoglycemic rats,[76] in oxygen–glucose deprivation of cerebrocortical cultures,[77] and in rat cortical neurons cultured at low glucose.[78] It was clearly shown that caspase-3 was activated during insulin-induced hypoglycemia in mice and in 661W photoreceptor cells cultured at low glucose. Moreover, inhibition of caspase-3, by Z-VAD-FMK inhibitor, blocked low glucose-induced cell death.[29] A recent study showed that in the hypoglycemic chicken retina, spontaneous episodes of spreading depression could occur, resulting in irreversible lesions in the macula, possibly via activation of caspase-3.[79]

CONCLUSIONS

Hypoglycemic conditions cause very similar patterns to those observed during hyperglycemic conditions (mitochondrial superoxide production, caspase-3 activation, decrease in the scavenger GSH protein leading to cell death).

One hypothesis is that sensitivity of retinal cells to stress (in this case low glucose) is directly dependent on the GSH level. Enzymes involved in GSH homeostasis try to counteract apoptosis by adaptation to harmful environmental conditions. This leads to a strong decrease in GSH that is deleterious for retinal cells. Modulation of the GSH/GSSG ratio is crucial to maintaining normal redox state and cell function. An increase in GSH will protect cells against oxidative stress, while low glucose-induced depletion of GSH will induce apoptosis. However, other pathways playing a role in survival or apoptosis need to be investigated.

Hypoglycemia plays an important role in the establishment of diabetic retinopathy. This reinforces the idea that glycemic excursions are not harmless. A strict control of glycemia, to prevent not only hyperglycemia but also hypoglycemia, is critical for the maintenance of good retinal function and vision. Hypoglycemic events induce superoxide production, decrease scavenger protein GSH, and induce cell death. Treatment of diabetic patients with antioxidant compounds may reduce some negative effects of hypoglycemia by maintaining stable redox within the cells and reducing retinal cell death. New ways are being investigated of preventing retinal degeneration and decreased eyesight in diabetic patients.

TAKE-HOME MESSAGES

- Hypoglycemia, as well as hyperglycemia, is involved in the development of diabetic retinopathy, since acute hypoglycemia induces retinal cell death.

- Microarray analysis shows that multiple pathways, including lysosomal and apoptotic pathways, as well as glutathione metabolism, are involved in the process.
- The Bcl2/Bax pathway plays a role not only in apoptosis but also in autophagy defects.
- Mitochondrial reactive oxygen species production and low levels of the antioxidant scavenger protein glutathione are implicated in cell death.
- Glycemic excursions are dangerous, and strict control of blood glucose is important to avoid the secondary effects of diabetes.

References

1. Gospe S, Baker S, Arshavsky V. Facilitative glucose transporter Glut1 is actively excluded from rod outer segments. *J Cell Sci* 2010;**123**:3639–44.
2. Tsacopoulos M, Poitry-Yamate C, MacLeish P, Poitry S. Trafficking of molecules and metabolic signals in the retina. *Prog Retin Eye Res* 1998;**17**:429–42.
3. Ola M, Berkich D, Xu Y, King M, Gardner T, Simpson I, LaNoue K. Analysis of glucose metabolism in diabetic rat retinas. *Am J Physiol Endocrinol Metab* 2006;**290**:E1057–67.
4. Kumagai A. Glucose transport in brain and retina: implications in the management and complications of diabetes. *Diabetes Metab Res Rev* 1999;**15**:261.
5. Takata K, Kasahara T, Kasahara M, Ezaki O, Hirano H. Ultracytochemical localization of the erythrocyte/HepG2-type glucose transporter (GLUT1) in cells of the blood–retinal barrier in the rat. *Invest Ophthalmol Vis Sci* 1992;**33**:377–83.
6. Watanabe T, Matsushima S, Okazaki M, Hirosawa S, Uchimura H, Nakahara K. Localization and ontogeny of GLUT3 expression in the rat retina. *Dev Brain Res* 1996;**94**:60–6.
7. Watanabe T, Mio Y, Hoshino F, Nagamatsu S, Hirosawa K, Nakahara K. GLUT2 expression in the rat retina: localization at the apical ends of Müller cells. *Brain Res* 1994;**655**:128–34.
8. Winkler B. Cultured retinal neuronal cells and Müller cells both show net production of lactate. *Neurochem Int* 2004;**45**:311–20.
9. Reichenbach A, Bringmann A. *Müller Cells in the Healthy and Diseased Retina*. New York: Springer; 2010.
10. Winkler B, Arnold J, Brassell M, Puro D. Energy metabolism in human retinal Müller cells. *Invest Ophthalmol Vis Sci* 2000;**41**:3183–90.
11. Pfeiffer-Guglielmi B, Francke M, Reichenbach A, Fleckenstein B, Jung G, Hamprecht B. Glycogen phosphorylase isozyme pattern in mammalian retinal Müller (glial) cells and in astrocytes of retina and optic nerve. *Glia* 2005;**49**:84–95.
12. Osorio-Paz I, Sanchez-Chavez G, Salceda R. Control of glycogen content in retina: allosteric regulation of glycogen synthase. *PLoS ONE* 2012;**7**(2):e30822.
13. Brownlee M. Biochemistry and molecular cell biology of diabetic complications. *Nature* 2001;**414**:813–20.
14. Nguyen T, Wong T. Retinal vascular changes and diabetic retinopathy. *Curr Diab Rep* 2009;**9**:277–83.
15. Cryer P. The barrier of hypoglycemia in diabetes. *Diabetes* 2008;**57**:3169–76.
16. Association AD. Defining and reporting hypoglycemia in diabetes: a report from the American Diabetes Association Workgroup on Hypoglycemia. *Diabetes Care* 2005;**28**:1245–9.
17. Perlmuter L, Flanagan B, Shah P, Singh S. Glycemic control and hypoglycemia: is the loser the winner? *Diabetes Care* 2008;**31**:2072–6.

18. Lucas A, Morley R, Cole T. Adverse neurodevelopmental outcome of moderate neonatal hypoglycaemia. *BMJ* 1988;**297**:1304–8.

19. Hirata C, Nakano K, Nakamura N, Kitagawa Y, Shigeta H, Hasegawa G, et al. Advanced glycation end products induce expression of vascular endothelial growth factor by retinal Müller cells. *Biochem Biophys Res Commun* 1997;**236**:712–5.

20. Crawford T, Alfaro D, Kerrison J, Jablon E. Diabetic retinopathy and angiogenesis. *Curr Diabetes Rev* 2009;**5**:8–13.

21. Geraldes P, Hiraoka-Yamamoto J, Matsumoto M, Clermont A, Leitges M, Marette A, et al. Activation of PKC-delta and SHP-1 by hyperglycemia causes vascular cell apoptosis and diabetic retinopathy. *Nat Med* 2009;**15**:1298–306.

22. Suh S, Gum E, Hamby A, Chan P, Swanson R. Hypoglycemic neuronal death is triggered by glucose reperfusion and activation of neuronal NADPH oxidase. *J Clin Invest* 2007;**117**:910–8.

23. Suh S, Hamby A, Swanson R. Hypoglycemia, brain energetics, and hypoglycemic neuronal death. *Glia* 2007;**55**:1280–6.

24. Luo X, Lambrou G, Sahel J, Hicks D. Hypoglycemia induces general neuronal death, whereas hypoxia and glutamate transport blockade lead to selective retinal ganglion cell death *in vitro*. *Invest Ophthalmol Vis Sci* 2001;**42**:2695–705.

25. Zeevalk G, Nicklas W. Lactate prevents the alterations in tissue amino acids, decline in ATP, and cell damage due to aglycemia in retina. *J Neurochem* 2000;**75**:1027–34.

26. Umino Y, Everhart D, Solessio E, Cusato K, Pan J, Nguyen T, et al. Hypoglycemia leads to age-related loss of vision. *Proc Natl Acad Sci U S A* 2006;**103**:19541–5.

27. Punzo C, Kornacker K, Cepko C. Stimulation of the insulin/mTOR pathway delays cone death in a mouse model of retinitis pigmentosa. *Nat Neurosci* 2009;**12**:44–52.

28. Khan M, Barlow R, Weinstock R. Acute hypoglycemia decreases central retinal function in the human eye. *Vis Res* 2011;**51**:1623–6.

29. Emery M, Schorderet D, Roduit R. Acute hypoglycemia induces retinal cell death in mouse. *PLoS ONE* 2011;**6**:e21586.

30. Subramanian A, Tamayo P, Mootha V, Mukherjee S, Ebert B, Gillette M, et al. Gene set enrichment analysis: a knowledge-based approach for interpreting genome-wide expression profiles. *Proc Natl Acad Sci U S A* 2005;**102**:15545–50.

31. Ibberson M, Balmer D, Schorderet D, Roduit R. *Microarray analysis reveal diverse pathways in insulin-induced hypoglycemia in mouse*; 2013. in preparation.

32. Ganea E, Harding J. Glutathione-related enzymes and the eye. *Curr Eye Res* 2006;**31**:1–11.

33. Deneke S, Fanburg B. Regulation of cellular glutathione. *Am J Physiol* 1989;**257**:L163–73.

34. Aoyama K, Watabe M, Nakaki T. Regulation of neuronal glutathione synthesis. *J Pharmacol Sci* 2008;**108**:227–38.

35. Lushchak V. Glutathione homeostasis and functions: potential targets for medical interventions. *J Amino Acids* 2012;**2012**:1–26.

36. Dalle-Donne I, Colombo G, Gagliano N, Colombo R, Giustarini D, Rossi R, Milzani A. S-glutathiolation in life and death decisions of the cell. *Free Radic Res* 2011;**45**:3–15.

37. Dalle-Donne I, Rossi R, Giustarini D, Colombo R, Milzani A. S-glutathionylation in protein redox regulation. *Free Radic Biol Med* 2007;**43**:883–98.

38. Huang Z, Pinto J, Deng H, Richie J. Inhibition of caspase-3 activity and activation by protein glutathionylation. *Biochem Pharmacol* 2008;**75**:2234–44.

39. Pineda-Molina E, Klatt P, Vazquez J, Marina A, Garcia de Lacoba M, Perez-Sala D, Lamas S. Glutathionylation of the p50 subunit of NF-kappaB: a mechanism for redox-induced inhibition of DNA binding. *Biochemistry* 2001;**40**:14134–42.

40. Klatt P, Lamas S. c-Jun regulation by S-glutathionylation. *Methods Enzymol* 2002;**348**:157–74.

41. Ward N, Stewart J, Ioannides C, O'Brian C. Oxidant-induced S-glutathiolation inactivates protein kinase C-alpha (PKC-alpha): a potential mechanism of PKC isozyme regulation. *Biochemistry* 2000;**39**:10319–29.

42. Casagrande S, Bonetto V, Fratelli M, Gianazza E, Eberini I, Massignan T, et al. Glutathionylation of human thioredoxin: a possible crosstalk between the glutathione and thioredoxin systems. *Proc Natl Acad Sci U S A* 2002;**99**:9745–9.

43. Ristoff E, Burstedt M, Larsson A, Wachtmeister L. Progressive retinal dystrophy in two sisters with glutathione synthetase (GS) deficiency. *J Inherit Metab Dis* 2007;**30**:102.

44. Ristoff E, Mayatepek E, Larsson A. Long-term clinical outcome in patients with glutathione synthetase deficiency. *J Pediatr* 2001;**139**:79–84.

45. Coffe V, Carbajal R, Salceda R. Glucose metabolism in rat retinal pigment epithelium. *Neurochem Res* 2006;**31**:103–8.

46. Wieloch T. Hypoglycemia-induced neuronal damage prevented by an N-methyl-D-aspartate antagonist. *Science* 1985;**230**:681–3.

47. Russell J, Golovoy D, Vincent A, Mahendru P, Olzmann J, Mentzer A, Feldman E. High glucose-induced oxidative stress and mitochondrial dysfunction in neurons. *FASEB J* 2002;**16**:1738–48.

48. Singh P, Jain A, Kaur G. Impact of hypoglycemia and diabetes on CNS: correlation of mitochondrial oxidative stress with DNA damage. *Mol Cell Biochem* 2004;**260**:153–9.

49. Santos J, Mohammad G, Zhong Q, Kowluru R. Diabetic retinopathy, superoxide damage and antioxidants. *Curr Pharm Biotechnol* 2011;**12**:352–61.

50. Suh S, Aoyama K, Chen Y, Garnier P, Matsumori Y, Gum E, et al. Hypoglycemic neuronal death and cognitive impairment are prevented by poly(ADP-ribose) polymerase inhibitors administered after hypoglycemia. *J Neurosci* 2003;**23**:10681–90.

51. Groeger G, Mackey A, Pettigrew C, Bhatt L, Cotter T. Stress-induced activation of Nox contributes to cell survival signalling via production of hydrogen peroxide. *J Neurochem* 2009;**109**:1544–54.

52. Mackey A, Sanvicens N, Groeger G, Doonan F, Wallace D, Cotter T. Redox survival signalling in retina-derived 661W cells. *Cell Death Differ* 2008;**15**:1291–303.

53. Ikesugi K, Mulhern M, Madson C, Hosoya K, Terasaki T, Kador P, Shinohara T. Induction of endoplasmic reticulum stress in retinal pericytes by glucose deprivation. *Curr Eye Res* 2006;**31**:947–53.

54. Bhatt L, Groeger G, McDermott K, Cotter T. Rod and cone photoreceptor cells produce ROS in response to stress in a live retinal explant system. *Mol Vis* 2010;**16**:283–93.

55. Razavi Nematollahi L, Kitabchi A, Stentz F, Wan J, Larijani B, Tehrani M, et al. Proinflammatory cytokines in response to insulin-induced hypoglycemic stress in healthy subjects. *Metabolism* 2009;**58**:443–8.

56. McGowan J, Chen L, Gao D, Trush M, Wei C. Increased mitochondrial reactive oxygen species production in newborn brain during hypoglycemia. *Neurosci Lett* 2006;**399**:111–4.

57. Hernandez-Fonseca K, Cardenas-Rodriguez N, Pedraza-Chaverri J, Massieu L. Calcium-dependent production of reactive oxygen species is involved in neuronal damage induced during glycolysis inhibition in cultured hippocampal neurons. *J Neurosci Res* 2008;**86**:1768–80.

58. Paramo B, Hernandez-Fonseca K, Estrada-Sanchez A, Jiminez N, Hernandez-Cruz A, Massieu L. Pathways involved in the generation of reactive oxygen and nitrogen species during glucose deprivation and its role on the death of cultured hippocampal neurons. *Neuroscience* 2010;**167**:1057–69.

59. Kowluru R, Kowluru V, Xiong Y, Ho Y. Overexpression of mitochondrial superoxide dismutase in mice protects the retina from diabetes-induced oxidative stress. *Free Radic Biol Med* 2006;**41**:1191–6.

60. Reed J. Mechanisms of apoptosis. *Am J Pathol* 2000;**157**:1415–30.

61. Willis S, Day C, Hinds M, Huang D. The Bcl-2-regulated apoptotic pathway. *J Cell Sci* 2003;**116**:4053–6.

62. Hahn P, Lindsten T, Ying G-S, Bennett J, Milam A, Thompson C, Dunaief J. Proapoptotic Bcl-2 family members, Bax and Bak, are essential for developmental photoreceptor apoptosis. *Invest Ophthalmol Vis Sci* 2003;**44**:3598–605.

63. Hahn P, Lindsten T, Lyubarsky A, Ying G, Pugh E, Thompson C, Dunaief J. Deficiency of Bax and Bak protects photoreceptors from light damage *in vivo*. *Cell Death Differ* 2004;**11**:1192–7.

64. Yang L, Bula D, Arroyo J, Chen D. Preventing retinal detachment-associated photoreceptor cell loss in Bax-deficient mice. *Invest Ophthalmol Vis Sci* 2004;**45**:648–54.

65. Kaneda K, Kashii S, Kurosawa T, Kaneko S, Akaike A, Honda Y, et al. Apoptotic DNA fragmentation and upregulation of Bax induced by transient ischemia of the rat retina. *Brain Res* 1999;**815**:11–20.

66. Kowluru R, Abbas S. Diabetes-induced mitochondrial dysfunction in the retina. *Invest Ophthalmol Vis Sci* 2003;**44**:5327–34.

67. Ji J, Chang P, Pennesi M, Yang Z, Zhang J, Li D, et al. Effects of elevated intraocular pressure on mouse retinal ganglion cells. *Vis Res* 2005;**45**:169–79.

68. Cottet S, Schorderet D. Triggering of Bcl-2-related pathway is associated with apoptosis of photoreceptors in Rpe65$^{-/-}$ mouse model of Leber's congenital amaurosis. *Apoptosis* 2008;**13**: 329–42.

69. Produit-Zengaffinen N, Pournaras C, Schorderet D. Retinal ischemia-induced apoptosis is associated with alteration in Bax and Bcl-x(L) expression rather than modifications in Bak and Bcl-2. *Mol Vis* 2009;**15**:2101–10.

70. Chen J, Flannery J, LaVail M, Steinberg R, Xu J, Simon M. bcl-2 overexpression reduces apoptotic photoreceptor cell death in three different retinal degenerations. *Proc Natl Acad Sci U S A* 1996;**93**:7042–7.

71. Nir I, Kedzierski W, Chen J, Travis G. Expression of Bcl-2 protects against photoreceptor degeneration in retinal degeneration slow (rds) mice. *J Neurosci* 2000;**20**:2150–4.

72. Quiambao A, Tan E, Chang S, Komori N, Naash M, Peachey N, et al. Transgenic Bcl-2 expressed in photoreceptor cells confers both death-sparing and death-inducing effects. *Exp Eye Res* 2001;**73**:711–21.

73. Osundiji M, Godes M, Evans M, Danial N. BAD modulates counterregulatory responses to hypoglycemia and protective glucoprivic feeding. *PLoS ONE* 2011;**6**(12):e28016.

74. Balmer D, Emery M, Andreux P, Auwerx J, Ginet V, Puyal J, et al. Autophagy defect is associated with low glucose-induced apoptosis in 661W photoreceptor cells. *PLoS ONE* 2013. http://dx.doi.org/10.1371/journal.pone.0074162.

75. Mohr S, Xi X, Tang J, Kern T. Caspase activation in retinas of diabetic and galactosemic mice and diabetic patients. *Diabetes* 2002;**51**:1172–9.

76. Rao R, Sperr D, Ennis K, Tran P. Postnatal age influences hypoglycemia-induced poly(ADP-ribose) polymerase-1 activation in the brain regions of rats. *Pediatr Res* 2009;**66**:642–7.

77. Nath R, Probert A, McGinnis K, Wang K. Evidence for activation of caspase-3-like protease in excitotoxin- and hypoxia/hypoglycemia-injured neurons. *J Neurochem* 1998;**71**:186–95.

78. Turner C, Blackburn M, Rivkees S. A1 adenosine receptors mediate hypoglycemia-induced neuronal injury. *J Mol Endocrinol* 2004;**32**:129–44.

79. Yu Y, Santos L, Mattiace L, Costa M, Ferreira L, Benabou K, et al. Reentrant spiral waves of spreading depression cause macular degeneration in hypoglycemic chicken retina. *Proc Natl Acad Sci U S A* 2012;**109**:2585–9.

Yellow Corneal Rings, Age-Related Macular Degeneration, and Carotenoid Supplementation and Metabolism

Ian R. Gorovoy[1], Andrew W. Eller[2]

[1]University of California, San Francisco, CA; [2]Retina Service, UPMC Eye Center, University of Pittsburgh School of Medicine, and The Eye and Ear Institute, Pittsburgh, PA

AGE-RELATED MACULAR DEGENERATION, AREDS1, AND AREDS2

Age-related macular degeneration (AMD) is a multifactorial disease likely caused by a genetic predisposition coupled with a lifelong exposure to free radicals and environmental toxins. It is the most common cause of severe vision loss in people over 50 years of age, especially in patients over 70 years of age.[1,2] Fortunately, the most visually significant form of AMD, neovascular AMD, accounts for only 10% of AMD cases. Of the 90% of remaining patients with the dry form of AMD, geographic atrophy (GA), comprises 35% of all cases of end-stage dry AMD and about 20% of legal blindness in AMD patients.[3]

Currently, the only evidence-based therapy for dry AMD is the Age-Related Eye Disease Study (AREDS)-based vitamin supplements. Theoretically, this supplementation works vis-à-vis the oxidative stress and depletion of essential micronutrients, which are important factors in AMD progression. AREDS1 (Bausch & Lomb Pharmaceuticals Inc., USA) was the first multicenter, randomized clinical trial to evaluate the effect of high-dose antioxidants (vitamin E, vitamin C, and β-carotene) and zinc on the progression of AMD.[4] Unfortunately, AREDS1 vitamins were unable to prevent vision loss in the majority of patients. However, it benefited patients at high risk of developing advanced AMD by lowering their risk by about 25% when treated with the micronutrients combination. In the same high-risk group, which included people with intermediate AMD or advanced AMD in one eye but not the other eye, the nutrients reduced the risk of vision loss by about 19%.

For patients who had early AMD, the nutrients did not provide an apparent benefit.[4]

AREDS2 was designed to evaluate the effects of oral vitamin supplementation of macular xanthophylls (lutein and zeaxanthin) and/or long-chain omega-3 fatty acids (docosahexaenoic acid (DHA) and eicosapentaenoic acid (EPA)) on the progression to advanced AMD. These micronutrients are believed to function as antioxidants, anti-inflammatory, and anti-angiogenic agents. An additional goal of the study is to assess whether forms of the AREDS nutritional supplement with reduced zinc and/or no β-carotene works as well as the original supplement in reducing the risk of progression to advanced AMD. Enrollment was concluded in 2008 and participants are being followed for approximately 5 years (ClinicalTrials.gov number, NCT00345176).

Besides high-dose vitamin supplementation, the only other US Food and Drug Administration (FDA)-approved treatment for AMD is anti-vascular endothelial growth factor (VEGF) medications, which have proven to be useful in neovascular AMD but also carry the cost of repeated intraocular injections. Other ongoing studies have targeted putative breakdown points in the healthy maintenance of normal photoreceptors and retinal pigment epithelial cells found in AMD. Visual cycle modulators theoretically may reduce the activity of photoreceptors that normally cause the toxic accumulation of fluorophores and lipofuscin in AMD. The discovery of complement mutations in patients with AMD has suggested that targeted therapy against these molecules may be beneficial. Additionally, there is hope that neuroprotective drugs, such as ciliary neurotrophic factor, brimonidine tartrate, tandospirone, and anti-amyloid β antibodies, will be effective. Historically, the field of ophthalmology has been extremely successful in borrowing medications used in other fields of medicine such as steroids, β-blockers, carbonic anhydrase inhibitors, and anti-VEGF medications, and perhaps future innovations for AMD will follow suit.

LIMBAL CIRCULATION, YELLOW RINGS, AND OTHER PERIPHERAL CORNEAL RINGS

Although the cornea is a relatively avascular structure, the transparency and proximity of the limbal circulation predisposes the peripheral cornea to the deposition of pigment from the systemic circulation. Characteristic depth and color of these deposited substances can help to determine their etiology. Heavy metals in particular have a propensity for accumulation in the peripheral cornea. Greenish-gold Kayser-Fleischer rings, found in the majority of patients with Wilson disease, are the result of abnormal copper deposition at the level of Descemet's membrane.[5,6] Iron from intraocular foreign bodies in the anterior chamber may leave rust-colored deposits in the posterior corneal stroma and anterior lens capsule in a form of localized siderosis. In many cases, corneal deposition is accompanied by visible and often irreversible changes in the discoloration of the conjunctiva and/or coloration as seen in agyrosis (silver), chyrsiasis (gold), hydragyria (mercury), and chlorpromazine.[7–9]

These rings are best understood in Wilson disease, where an inability to excrete copper causes cirrhosis and neuropsychiatric problems from the elevated levels of copper in the blood. Interestingly, Kayser-Fleischer rings are more prevalent in Wilson patients with neuropsychiatric manifestations, which may be secondary to the higher levels of copper reaching the cerebral circulation. After correction of the hypercupremia via copper chelation or liver transplantation, Kayser-Fleischer rings wash out centripetally in the opposite sequence in which they were deposited.[10,11]

Besides heavy metals and medications, other molecules can also deposit in the corneal periphery. Pigmented rings in the peripheral cornea secondary to hyperbilirubinemia in advanced liver disease have also been described, especially in the setting of scleral icterus.[12] Arcus senilis occurs as a milky white to gray peripheral ring secondary to cholesterol deposits. In middle-aged individuals, it generally is of little clinical relevance as it occurs most commonly in the setting of normal lipid levels. One prospective cohort study of 12,745 Danish people with mean follow-up of 22 years found that arcus did not portend a greater risk of cardiovascular disease.[13] In contrast, unilateral arcus is clinically relevant as it can be a sign of decreased blood flow to the unaffected eye due to carotid artery disease or ocular hypotony. In individuals under 40 years of age, arcus is referred to as arcus juvenilis, where it often accompanies a disturbance in lipid metabolism such as familial hypercholesterolemia, hyperlipoproteinemia, or hyperlipidemia.

Yellow peripheral corneal rings have recently been described in the setting of increased levels of carotenoids secondary to nutritional supplements for AMD (including the AREDS1 and AREDS2 formulations and generic forms) or for general health maintenance purposes and excessive consumption of foods rich in carotenoids.[14–16] We believe that these yellow rings are more common than generally noted, as they are often faint, misdiagnosed as arcus senilis, or masked by topical fluorescein dye. These rings may in fact represent carotenoid staining of older arcus senilis. They were not previously reported in the AREDS1 trial or in other clinical trials with β-carotene supplementation (Fig. 65.1; Fig. 65.2).[4,17,18]

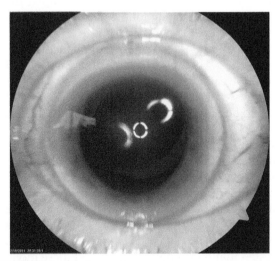

FIGURE 65.1 Slit lamp photograph demonstrating an orange–yellow corneal ring from carotenoids used as part of the AREDS1 formula near the limbus and extending toward the pupil in a 72-year-old female. (See color plate section)

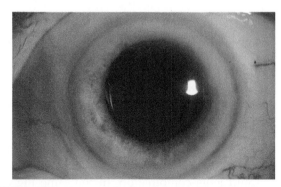

FIGURE 65.2 An example of yellow peripheral corneal ring secondary to carotenoids from a generic form of AREDS supplementation that was marigold flower-based in an 89-year-old female. (See color plate section)

The etiology of the rings was deduced by characteristics consistent with carotenoids and by a process of elimination. The patients' history was negative for intraocular foreign bodies or other heavy metal exposure, exposure to medications such as phenothiazines, which are known to cause ocular pigmentation, or liver disease. Several of these patients had a similar, albeit more subtle, reddish-orange pigmentation of the skin, especially the palms and soles, all of which is suggestive of carotenoid deposition.

Several patients with excessive consumption of dietary carotenoids had associated retinal paramacular crystal deposition or diffuse yellow–brown discoloration of the anterior lens capsule.[19] This was not noted in the AMD patients on vitamin supplementation, but it is possible that drusen may have obfuscated this finding. Retinal paramacular carotenoid crystals have been described in patients using canthaxantine, which is another carotenoid used in artificial tanning, and high-dose β-carotene for the treatment of retinitis pigmentosa.[20,21] These crystals are almost universally observed in patients who have used cumulative doses of canathaxine greater than 60 g, slowly diminish with discontinuation of the drug, and are asymptomatic, although they cause small localized defects on formal perimetry.[22]

It is still unclear how patients with these yellow rings should be counseled, as their significance is still not understood. They do not appear to cause or be a marker for corneal dysfunction. Perhaps the rings indicate medication compliance in patients who are taking AREDS-based vitamin supplementation with carotenoids and their presence is suggestive of sufficient carotenoid levels over sufficient time to be beneficial in AMD. As every individual metabolizes carotenoids at different rates yet receives the same dose in the AREDS trials, perhaps it is more appropriate to dose patients until these rings occur. It should be noted that not all patients with yellow rings had high levels of serum carotenoids, although serum testing has many inherent problems such as molecule breakdown from light, the inability to test all carotenoids simultaneously, or changing levels with red blood cell destruction. The breakdown of carotenoids induced by light may also explain the prominence of these rings superiorly and inferiorly where the eyelids cover the limbus.[23]

An alternative hypothesis is that these rings occur in the setting of an underlying systemic disease that alters carotenoid metabolism such as hypothyroidism, hyperlipidemia, diabetes mellitus, nephrotic syndrome, or hepatic disease. These diseases cause hypercarotenemia through the impaired conversion of β-carotene into retinol with an associated increase in serum lipids.[24] Currently, only a small number of patients with these rings have been described, and no obvious association is apparent. Any correlation, however, may not be very robust as it is also known that both the absorption and conversion of carotenoids to vitamin A between individuals is variable, suggesting a more benign etiology.[25]

We believe that the rings are clinically insignificant due to the downregulation of carotenoid conversion to vitamin A when levels are too high. Nonetheless, we should question whether or not this form of vitamin supplementation is safe. Evidence of these rings illustrates the widespread deposition in a relatively avascular structure such as the cornea. The use of megadoses of vitamins may ultimately prove to be harmful to patients or select subsets of patients.

CAROTENOID AND VITAMIN A METABOLISM AND ITS CLINICAL SIGNIFICANCE

Vitamin A is a fat-soluble molecule derived from two food sources: 1) preformed vitamin A from animal sources and 2) provitamin A from plant sources in the form of carotenoids. In the eye, vitamin A is critical in the phototransduction cascade and in the prevention of xerophthalmia. Preformed vitamin A is the more active form and includes retinol, retinal, retinoic acid, and retinyl esters. Carotenoids such as β-carotene, β-cryptoxanthin, lycopene, lutein, and zeaxanthin are found mostly in green, yellow, and orange vegetables.[25] The initial steps in the metabolism of β-carotene to vitamin A are subject to feedback regulation, which depends on vitamin A status.

Carotenoids have been postulated to serve many roles in the retina including limiting chromatic aberration at the fovea and quenching free radical oxygen molecules. Histologic studies have confirmed that the yellow pigmentation of the fovea is derived from carotenoids.[26] Interestingly, the retina contains very little β-carotene. In humans, zeanthin is found primarily in the fovea whereas lutein is found throughout the retina. The retinal pigment epithelium is essentially devoid of carotenoids.[27]

Preformed vitamin A absorption occurs with its breakdown on the small intestine mucosal brush border by retinyl estery hydrolases and is not well regulated. Because it is stored in the liver and adipose tissues, serum levels may not adequately reflect whole body stores. Due to the efficiency of its absorption and storage, preformed vitamin A can cause toxicity if excessive quantities are ingested. Hypervitaminosis A is associated with birth defects, liver failure, yellowing and xerosis of the skin, osteoporosis, and raised intracranial pressure.

β-Carotene is also absorbed in the small bowel and is stored in liver and adipose tissue. However, early steps in its metabolism are highly regulated, which prevents the transformation of excessive β-carotene into biologically active vitamin A.[28] Furthermore, in the setting of high serum vitamin A levels, the conversion of β-carotene to vitamin A is also downregulated, suggesting that high doses of β-carotene are safe even in the setting of high levels of vitamin A.[29] This mechanism has not been well described with the other carotenoids, but likely a comparable process exists.

Despite these metabolic regulatory steps, the Institute of Medicine does not advise carotenoid supplementation for the general population.[28] The recommended daily intake for males and females and different age groups is outlined in Table 65.1.[28] The FDA, however, has approved the use of carotenoids as a dietary supplement and dye additive.[30] Carotenoids have been studied in a wide variety of clinical trials to treat many health conditions. Individuals who consume a diet rich in carotenoids have lower mortality rates from chronic disease and a decreased incidence of some forms of cancer.[31,32] However, other trials have paradoxically found that carotenoid supplementation may increase the risk for lung cancer in male smokers and may harm the cardiovascular system.[17,33] Animal models have also demonstrated exacerbation of alcoholic steatosis with high β-carotene diets.[34]

High levels of carotenoids can create a clinical syndrome called carotenoderma (also called xanthoderma), which is a benign and reversible subtle yellow or orange skin discoloration.[35] The skin discoloration is more prominent in areas of skin with more sweat glands such as the palms, soles, and nasolabial folds. Unlike in patients with jaundice, the sclera is spared. Yellowing of the skin can have other etiologies including endogenous sources such as hyperbilirubinemia and riboflavinemia or the ingestion of saffron, quinacrine, or fluorescein. Carotenoid pigmentation has also been described in the liver, reflecting the systemic nature of pigment deposition.[36] Corneal carotenoid pigmentation demonstrates that even poorly vascularized organs such as the cornea are susceptible to carotenoid deposition. Carotenoderma is especially common in infants who often eat baby food fortified with high levels of carotenoids. It also can be found in the treatment of erythropoietic protoporphyria where high levels of oral β-carotene are used.[37] Lycopene,

a red carotenoid pigment, can rarely cause a deep orange discoloration of the skin.

In summary, although carotenoid supplementation is beneficial in some groups of patients with AMD, a common misconception is that it is recommended for the general population. Peripheral rings in the poorly vascularized cornea secondary to carotenoids highlight the fact that systemic deposition occurs with high doses and can remain for years after the discontinuation of high doses. The significance of these peripheral rings is unclear and merits further study.

References

1. Ambati J, Ambati BK, Yoo SH, Ianchulev S, Adamis AP. Age-related macular degeneration: etiology, pathogenesis, and therapeutic strategies. *Surv Ophthalmol* 2003;**48**(3):257–93.
2. van Leeuwen R, Klaver CC, Vingerling JR, Hofman A, de Jong PT. Epidemiology of age-related maculopathy: a review. *Eur J Epidemiol* 2003;**18**(9):845–54.
3. Klein R, Klein BE, Knudtson MD, Meuer SM, Swift M, Gangnon RE. Fifteen-year cumulative incidence of age-related macular degeneration: the Beaver Dam Eye Study. *Ophthalmology* 2007;**114**(2):253–62.
4. Age-Related Eye Disease Study Research Group. A randomized, placebo-controlled, clinical trial of high-dose supplementation with vitamins C and E, beta carotene, and zinc for age-related macular degeneration and vision loss: AREDS report no 8. *Arch Ophthalmol* 2001;**119**(10):1417–36 [erratum in: *Arch Ophthalmol* 2008;126(9):1251].
5. Roberts EA, Schilsky ML. Diagnosis and treatment of Wilson disease: an update. *Hepatology* 2008;**47**:2089–111.
6. Ala A, Walker AP, Ashkan K, Dooley JS, Schilsky ML. Wilson's disease. *Lancet* 2007;**369**(9559):397–408.
7. Webber SK, Domniz Y, Sutton GL, Rogers CM, Lawless MA. Corneal deposition after high-dose chlorpromazine hydrochloride therapy. *Cornea* 2001;**20**(2):217–9.
8. Lopez JD, del Castillo JM, Lopez CD, Sánchez JG. Confocal microscopy of ocular chrysiasis. *Cornea* 2003;**22**(6):573–5.
9. Palamar M, Midilli R, Egrilmez S, Akalin T, Yagci A. Black tears (melanodacryorrhea) from agyrosis. *Arch Ophthalmol* 2010;**128**(4):503–5.
10. Aggarwal A, Bhatt M. Eye sign in a 18 year old man with psychosis. *BMJ* 2009;**9**:339.
11. Schoenberger M, Ellis PP. Disappearance of Kayser-Fleischer rings after liver transplantation. *Arch Ophthalmol* 1979;**97**:1914–5.
12. Fleming CR, Dickson ER, Wahner HW, Hollenhorst RW, McCall JT. Pigmented corneal rings in non-Wilsonian liver disease. *Ann Intern Med* 1977;**86**:285–8.
13. Christoffersen M, Frikke-Schmidt R, Schnohr P, Jensen GB, Nordestgaard BG, Tybjærg-Hansen A. Xanthelasmata, arcus corneae, and ischaemic vascular disease and death in general population: prospective cohort study. *BMJ* 2011;**343**:d5497.
14. Rahmani B, Jampol LM, Feder RS. Peripheral pigmented corneal ring: a new finding in hypercarotenemia. *Arch Ophthalmol* 2003;**121**:403–7.
15. Eller AW, Gorovoy IR, Mayercik VM. Yellow corneal ring associated with vitamin supplementation for age-related macular degeneration. *Ophthalmology* 2012;**119**(5):1011–6.
16. Chang JS, Oellers P, Karp CL. Peripheral corneal ring due to hypercarotenaemia in a case of nutritional supplement abuse. *Cr J Ophthalmol* 2012;**96**:605.

TABLE 65.1 Recommended Daily Allowances for Vitamin A by Age

Age (years)				
1–3	4–8	9–13	14–18	> 19
1000 IU	1320 IU	2000 IU	3000 IU (males); 2310 IU (females)	3000 IU (males); 2310 IU (females)

IU: International Unit.

17. The Alpha-Tocopherol, Beta Carotene Cancer Prevention Study Group. The effect of vitamin E and beta carotene on the incidence of lung cancer and other cancers in male smokers. *New Engl J Med* 1994;**330**:1029–103.

18. Yoser DSL, Heckenlively JR. The appearance of retinal crystals in retinitis pigmentosa patients using beta carotene. *Invest Ophthalmol Vis Sci* 1989;**30**:305.

19. Rahmani B, Jampol LM, Feder RS. Peripheral pigmented corneal ring: a new finding in hypercarotenemia. *Arch Ophthalmol* 2003;**121**:403–7.

20. Poh-Fitzpatrick MB, Babera LG. Absence of crystalline retinopathy after long-term therapy with b-carotene. *J Am Acad Dermatol* 1984;**11**:111–3.

21. Rousseau A. Canthaxantine deposits in the eye. *J Am Acad Dermatol* 1983;**8**:123–4.

22. Ros AM, Leyon H, Wennersten G. Crystalline retinopathy in patients taking an oral drug containing canthaxanthinge. *Photodermatology* 1985;**2**:183–5.

23. Pryor WA, Stahl W, Rock CL. Beta carotene: from biochemistry to clinical trials. *Nutr Rev* 2000;**58**:39–53.

24. Leung AK. Carotenemia. *Adv Pediatr* 1987;**34**:223–48.

25. Van Arnum SD. Vitamin A. In: Kirk-Othmer, editor. *Encyclopedia of Chemical Technology*. New York: Wiley; 1998. p. 99–107.

26. Khachik F, Bertstein PS, Garland DL. Identification of lutein and zeanthin oxidation products in human and monkey retinas. *Invest Ophthalmol Vis Sci* 1997;**38**(9):1802–11.

27. Mayne ST. Beta carotene, carotenoids, and disease prevention in humans. *FASEB J* 1996;**10**(7):690–701.

28. Institute of Medicine. Food and Nutrition Board. In: *Dietary Reference Intakes for Vitamin A, Vitamin K, Arsenic, Boron, Chromium, Copper, Iodine, Iron, Manganese, Molybdenum, Nickel, Silicon, Vanadium, and Zinc*. Washington, DC: National Academy Press; 2001.

29. Bendich A, Langseth L. Safety of vitamin A. *Am J Clin Nutr* 1989;**49**:358–71.

30. US Food and Drug Administration. Ingredients, Packaging & Labeling. http://www.fda.gov/Food/IngredientsPackagingLabeling/default.htm [last accessed 28.01.14].

31. Diplock AT, Charleux JL, Crozier-Willi G, Kok FJ, Rice-Evans C, Roberfroid M, et al. Functional food science and defence against reactive oxidative species. *Br J Nutr* 1998;**80**(Suppl. 1): S77–112.

32. Ziegler RG. A review of epidemiologic evidence that carotenoids reduce the risk of cancer. *J Nutr* 1989:116–22.

33. Omenn GS, Goodman GE, Thornquist MD, Balmes J, Cullen MR, Glass A, et al. Effects of a combination of beta carotene and vitamin A on lung cancer and cardiovascular disease. *New Engl J Med* 1996;**334**:1150–5.

34. Leo MA, Aleynik SI, Aleynik MK, Leiber CS. β-Carotene beadlets potentiate hepatotoxicity of alcohol. *Am J Clin Nutr* 1997;**66**(6): 1461–9.

35. Monk BE. Metabolic carotenemia. *Br J Dermatol* 1982;**106**:485–8.

36. Nishimura T. A correlation between carotenemia and biliary dyskinesia. *J Dermatol* 1993;**20**(5):287–92.

37. Kuhlwein A, Beykirch W. β-carotene as the therapeutic agent for erythropoetic protoporphyria, not the last resort. *A Hautkr* 1980;**55**(12):817–23.

Index

Note: Page numbers with "*f*" denote figures; "*t*" tables and "*b*" boxes.

Color Plates

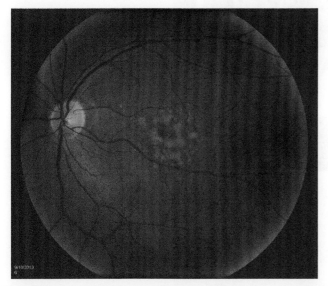

FIGURE 2.1 A fundus photograph of non-neovascular age-related macular degeneration. *Source: Wills Eye Hospital Diagnostic Testing Center. Reproduced with permission.*

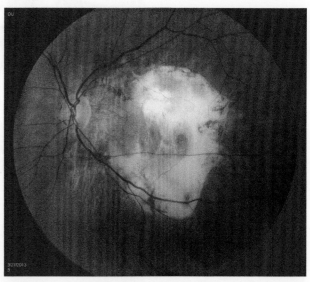

FIGURE 2.2 A fundus photo of neovascular age-related macular degeneration. *Source: Wills Eye Hospital Diagnostic Testing Center. Reproduced with permission.*

Early AMD	Intermediate AMD	Advanced non-neovascular AMD	Advanced neovascular AMD
Several small drusen or few medium-sized drusen	Many medium-sized drusen or at least one large drusen or geographic atrophy not extending into macula	Many drusen and geographic atrophy into macula	Choroidal neovascularization

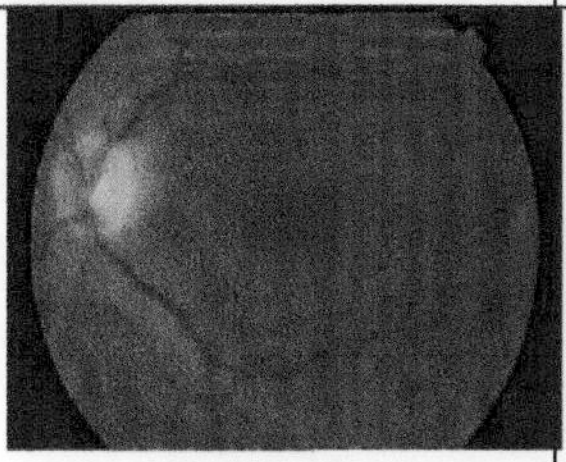

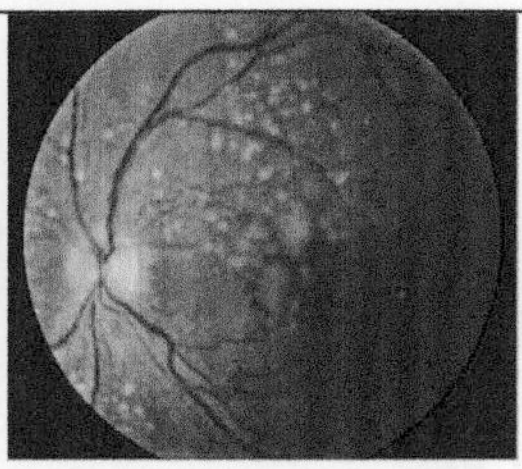

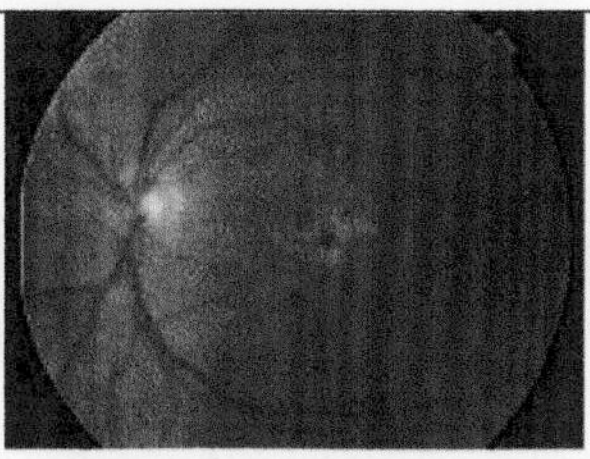

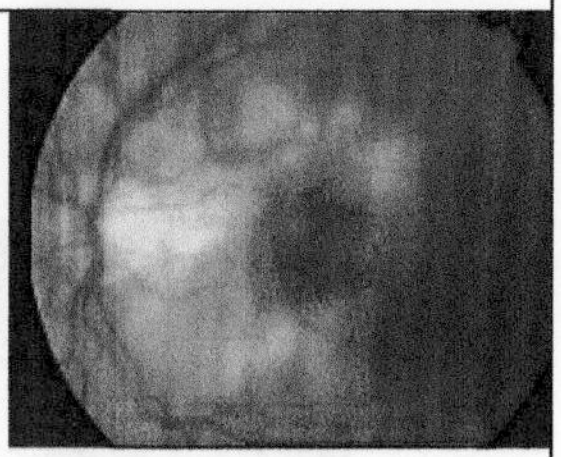

FIGURE 2.3 Fundus photos and descriptions of the Age-Related Eye Disease Study (AREDS) classification system for age-related macular degeneration (AMD). *Sources: Coleman HR, Chan C, Ferris F, Chew E. Age-related macular degeneration. Lancet 2008;372:1835–45.[1] Stringham J, Hammond B, Nolan J, et al. The utility of using customized heterochromatic flicker photometry (cHFP) to measure macular pigment in patients with age-related macular degeneration. Exp Eye Res. 2008;87:445–53.[41] Reproduced with permission from Elsevier via the Copyright Clearance Centre.*

FIGURE 2.4 Vision loss due to age-related macular degeneration. Upper photograph: normal vision; lower photograph: vision loss with age-related macular degeneration. *Source: National Eye Institute, National Institutes of Health.*

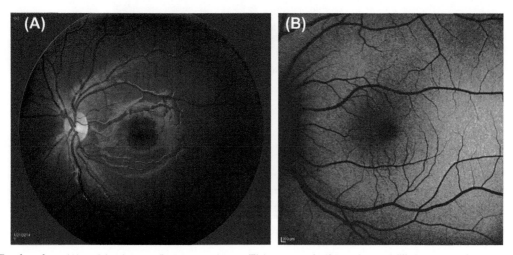

FIGURE 2.7 Fundus photo (A) and fundus autofluorescence image (B) in a normal subject. *Source: Wills Eye Hospital Retina Service. Reproduced with permission.*

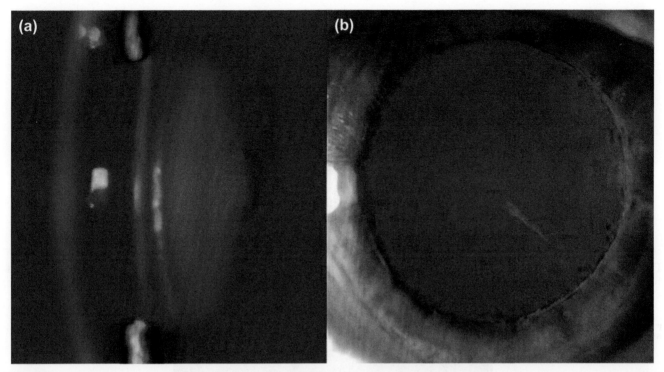

FIGURE 3.1 Appearance of age-related cataracts through a dilated pupil on slit-lamp biomicroscopy; a. nuclear sclerotic cataract; b. cortical cataract.

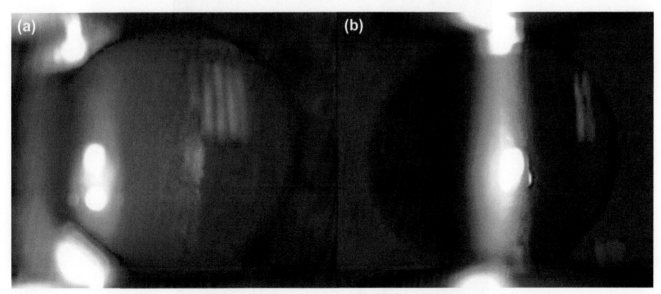

FIGURE 3.2 Appearance of posterior subcapsular cataract on slit-lamp biomicroscopy; a. subcapsular opacity is located in the cortex near the posterior capsule; b. the same eye under retroillumination.

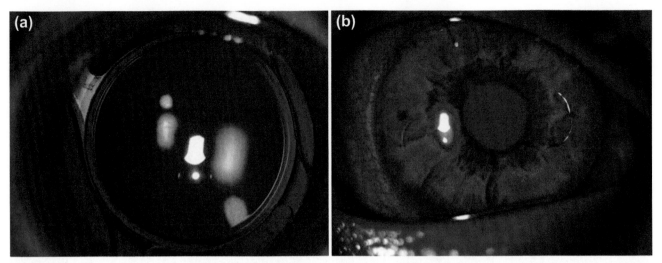

FIGURE 3.5 a. Appearance of four haptic posterior chamber intraocular lenses (Acreos, Bausch & Lomb) 6 months postoperative after uneventful phacoemulsification. b. iris claw anterior chamber intraocular lens in a patient with previous aphakia due to complicated cataract surgery.

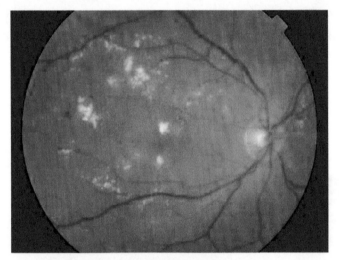

FIGURE 5.1 Nonproliferative diabetic retinopathy with diabetic macular edema.

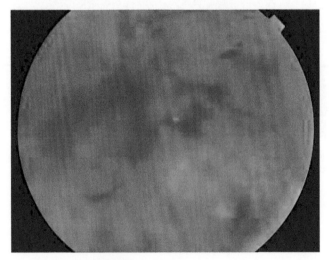

FIGURE 5.2 Proliferative diabetic retinopathy with vitreous hemorrhage.

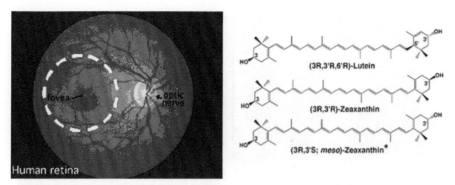

FIGURE 8.1 Ophthalmoscopic view of a human retina and the chemical structures of the major macular pigment carotenoids are shown on the right.[6] *Reproduced by permission of The Royal Society of Chemistry (RSC) on behalf of the European Society for Photobiology, the European Photochemistry Association, and RSC.*

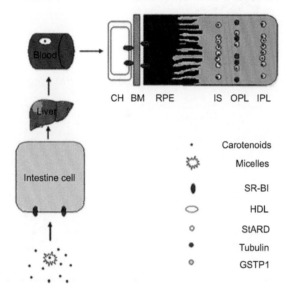

FIGURE 8.3 Possible pathway for macular pigment carotenoid uptake, transport, and accumulation in the human retina.[6] BM: Bruch's membrane; CH: choriocapillaris; HDL: high-density lipoprotein; IPL: inner plexiform layer; IS: inner segments; OPL: outer plexiform layer; RPE: retinal pigment epithelium; SR-BI: scavenger receptor class BI. *Reproduced by permission of The Royal Society of Chemistry (RSC) on behalf of the European Society for Photobiology, the European Photochemistry Association, and RSC.*

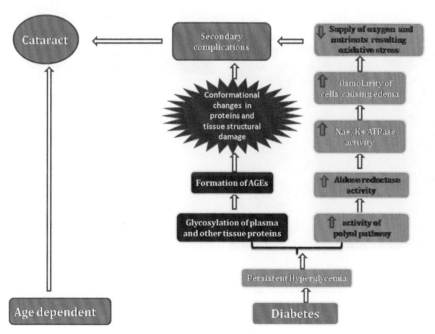

FIGURE 14.1 Pathogenesis of diabetic complications.

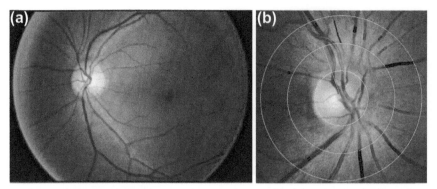

FIGURE 20.1 Retina allows for a noninvasive visualization of human microcirculation. Computer-assisted program for the measurement of retinal vascular caliber to quantify structural vascular microcirculatory changes. Zone B is marked in IVAN software by 0.5–1.0 optic disc diameter away from the margin of optic disc. The biggest eight retinal vascular arterioles and venules were located and assessed within zone B.

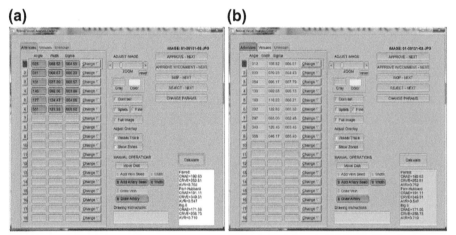

FIGURE 20.2 Vessels are assessed by clicking on the seed point (the proximal end of the vessel trace) on the image display or by clicking on the vessel number button on the data table. Retinal arteriolar analysis control (a) and retinal venular analysis control (b) show the visual vessel tracing and the identified data (angle, width, Sigma) for each vessel.

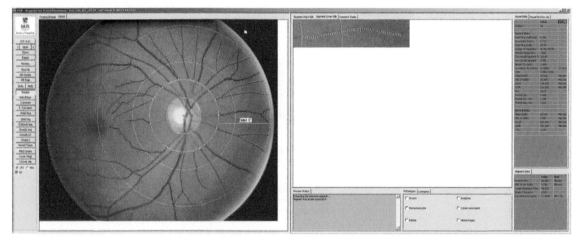

FIGURE 20.3 A screenshot of a computer-assisted program for measurement of new geometrical retinal vascular parameters from a retinal fundus photograph. Zone B and zone C are marked in IVAN software by 0.5–1.0 and 0.5–2.0 optic disc diameter away from the margin of optic disc, respectively. All retinal arterioles and venules larger than 25 μm are marked and assessed within zone B and zone C.

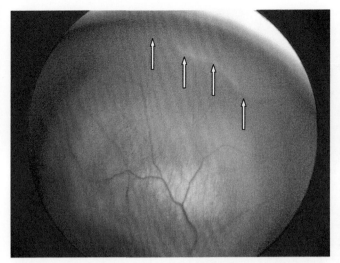

FIGURE 23.1 Retinopathy of prematurity (ROP) – vascular ridge with peripheral avascular zone. The second stage of retinopathy of prematurity: arrows indicate vascular ridge and the beginning of avascular zone. *Source: author's collection.*

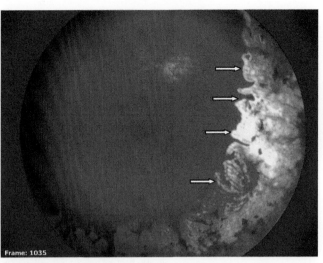

FIGURE 23.2 Retina – the effect of laser photocoagulation. The arrows indicate scars of the retina – the effect of laser photocoagulation. *Source: author's collection.*

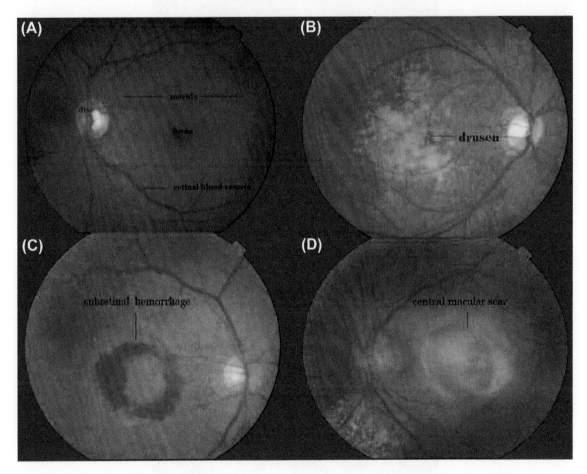

FIGURE 34.3 Retinal photographs: (A) normal; (B) dry age-related macular degeneration (AMD); (C) wet AMD; (D) end-stage macular scar. *Source: Nanjing Medical University Affiliated Eye Hospital.*

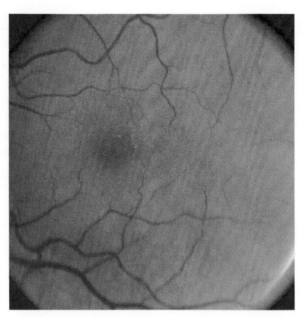

FIGURE 37.1 Small and intermediate drusen (yellow deposits) in the worse eye of a patient who meets criteria for Age-Related Eye Disease Study (AREDS) category 2 (early age-related macular degeneration).

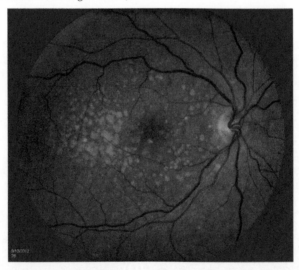

FIGURE 37.2 Extensive large drusen (yellow deposits) in the worse eye of a patient who meets criteria for Age-Related Eye Disease Study (AREDS) category 3 (intermediate age-related macular degeneration).

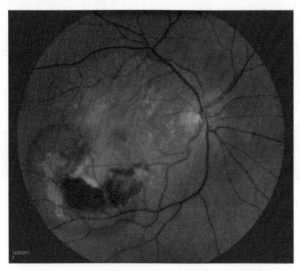

FIGURE 37.3 Retinal hemorrhage, a sign of choroidal neovascularization and advanced age-related macular degeneration (AMD), and large drusen in the worse eye of a patient who meets criteria for Age-Related Eye Disease Study (AREDS) category 4 (advanced AMD).

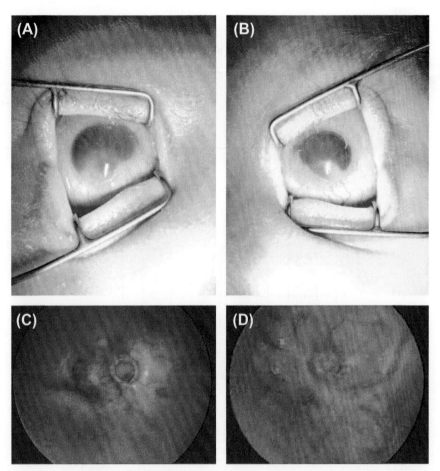

FIGURE 38.2 Ocular findings in a 2-day-old male with vitamin A deficiency. (A, B) Anterior segment photographs show microcorneas, inferior corneal scars, and down-drawn irides: (A) right eye, (B) left eye. (C, D) Fundus photographs obtained at 8 weeks of age following sector iridectomies showing optic disc and foveal hypoplasia: (C) right eye, (D) left eye. *Source: Gilchrist H, Taranath DA, Gole GA. Ocular malformation in a newborn secondary to maternal hypovitaminosis A. J AAPOS. 2010;14:274–6[24] with permission.*

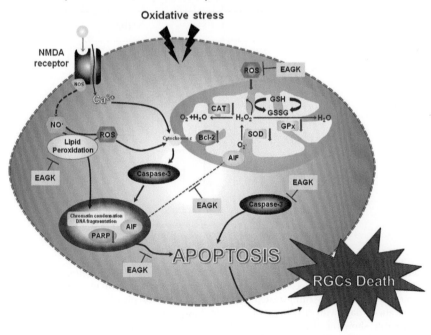

FIGURE 43.8 Proposed mechanisms of action underlying the protective effects of *Gymnaster koraiensis* on oxidative stress-induced retinal damage. *Gymnaster koraiensis* can attenuate oxidative-stress induced retinal ganglion cells by inhibiting apoptotic proteins such as caspase-3, poly (ADP-ribose) polymerase (PARP) and apoptosis-inducing factor (AIF), and by inhibiting radical species and lipid peroxidation. Bcl-2: B-cell lymphoma 2; CAT: catalase; EAGK: ethyl acetate fraction of *Gymnaster koraiensis*; GPx: glutathione peroxidase; GSH: glutathione; GSSG: oxidized glutathione disulfide; NMDA: *N*-methyl- D-aspartate; NO: nitric oxide; NOS: nitric oxide synthase; RGCs: retinal ganglion cells; ROS: reactive oxygen species; SOD: superoxide dismutase.

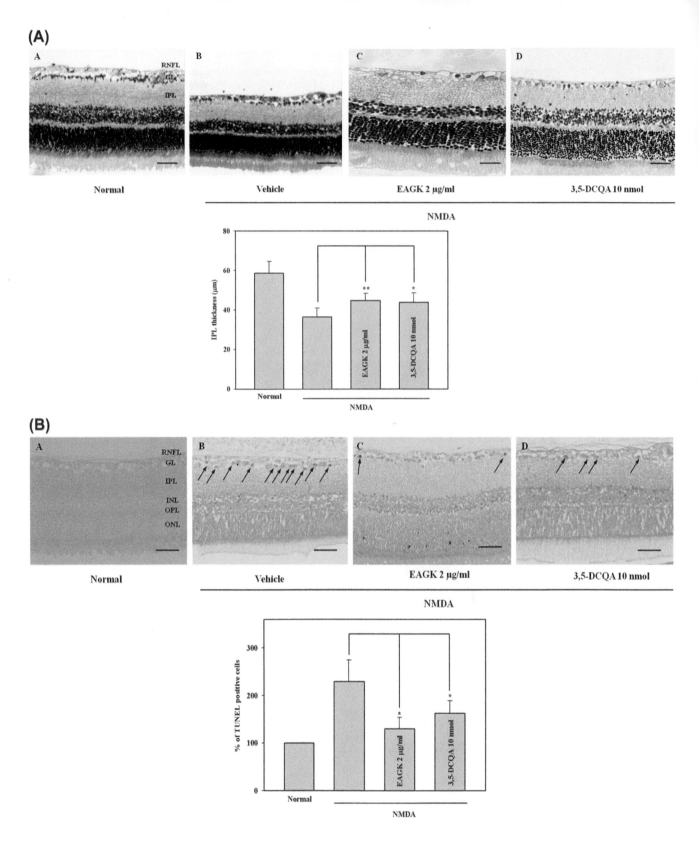

FIGURE 43.6 (A) Representative photomicrographs showing the histologic appearance of the retina induced by an intravitreous injection of *N*-methyl-D-aspartate (NMDA; hematoxylin and eosin staining, 400×). Non-treated (A), NMDA-treated (B), NMDA (5 nmol) plus the ethyl acetate fraction of *Gymnaster koraiensis* (EAGK, 2 μg/mL)-treated (C), and NMDA (5 nmol) plus 3,5-di-*O*-caffeoylquinic acid (3,5-DCQA; 10 nmol)-treated (D) retinal cross-sections after 7 days with or without NMDA are shown. The scale bar represents 50 μm. (F) shows the thickness of the inner plexiform layer (IPL). The results shown are the mean values with error bars indicating mean ± SEM from six independent experiments (*P < 0.05, **P < 0.01). (B) Antiapoptotic effect of EAGK or 3,5-DCQA demonstrated by the terminal deoxynucleotidyl transferase dUTP nick-end labeling (TUNEL) assay (400×). Retinal damage was induced with an intravitreous injection of NMDA. Non-treated (A), NMDA-treated (B), NMDA (5 nmol) plus EAGK (2 μg/mL)-treated (C), and NMDA (5 nmol) plus 3,5-DCQA (10 nmol)-treated (D) retinal cross-sections after 24 hours with or without NMDA are shown. The arrows indicate TUNEL-positive cells (brown stain). The number of TUNEL-positive cells increased after NMDA injection, but NMDA (5 nmol) plus EAGK at 2 μg/mL and 3,5-DCQA at 10 nmol decreased the NMDA-induced retinal damage. The scale bar represents 50 μm. (F) shows the TUNEL-positive cells. The results are the mean values with error bars indicating mean ± SEM from six independent experiments (*P < 0.05). *Source: Data from Kim et al. (2011).[99] Reproduced with permission from Elsevier.*

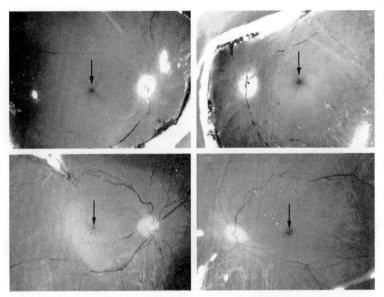

FIGURE 44.1 Photographs of dissected eye globes from two donors before fixation showing that the highest concentration of macular pigment is found in the fovea (black arrows). *Source: Reprinted from Powner* et al. *(2010)[23] with permission of Ophthalmology.*

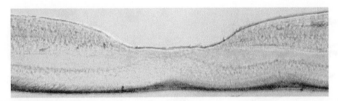

FIGURE 44.2 Unstained retina. *Source: Courtesy of MM Snodderly.*

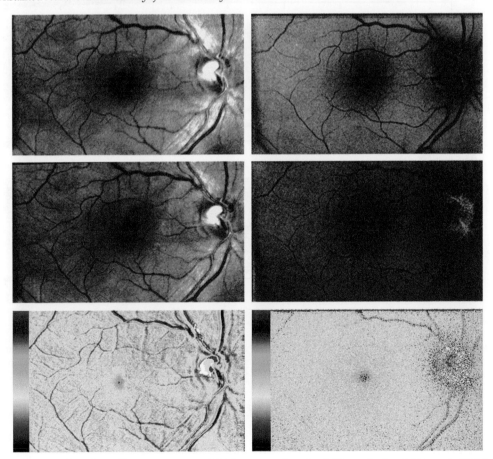

FIGURE 44.3 Reflectance (left) and autofluorescence maps (right) at a wavelength of 488 nm (top) and 514 nm (middle). The lower images show the corresponding color-coded macular pigment optical density maps. *Source: Reprinted from Berendschot and van Norren (2006)[29] with permission of Investigative Ophthalmology and Visual Science.*

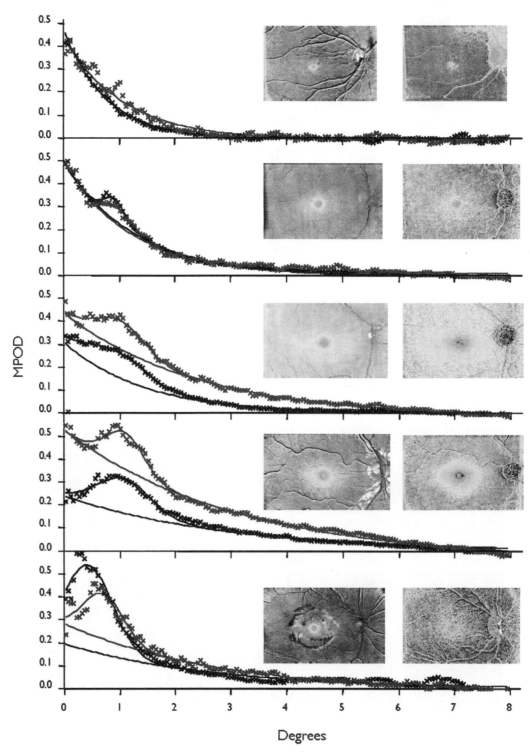

FIGURE 44.4 Macular pigment optical density (MPOD) as a function of eccentricity from reflectance (black) and autofluorescence (red) maps. The corresponding maps are shown as insets (reflectance maps on the left and autofluorescence maps on the right). Solid lines are the results of a model fit, explained in the text. *Source: Reprinted from Berendschot and van Norren (2006)[29] with permission of* Investigative Ophthalmology and Visual Science.

FIGURE 47.1 Wolfberry in Zhongning County, Ningxia, China. Wolfberry flowers (A) and fruits (B); sun-drying wolfberry fruits (C); and trimmed wolfberry trees in a plantation in Zhongning, Ningxia, China (D).

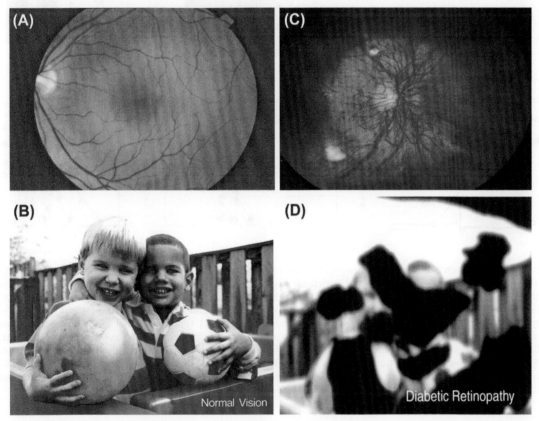

FIGURE 53.1 Photographs of the normal retinal and the retina affected by proliferative diabetic retinopathy paired with an example of the vision observed in normal and pathologic conditions. (A) Fundus photograph, normal retina; (B) normal vision; (C) retinal fundus proliferative retinopathy, an advanced form of diabetic retinopathy, occurs when abnormal new blood vessels and scar tissue form on the surface of the retina; (D) scene as it might be viewed by a person with diabetic retinopathy. *Source: National Eye Institute, National Institutes of Health: (A) Ref. EDA06; (B) EDS01; (C) EDA01; (D) EDS04.*

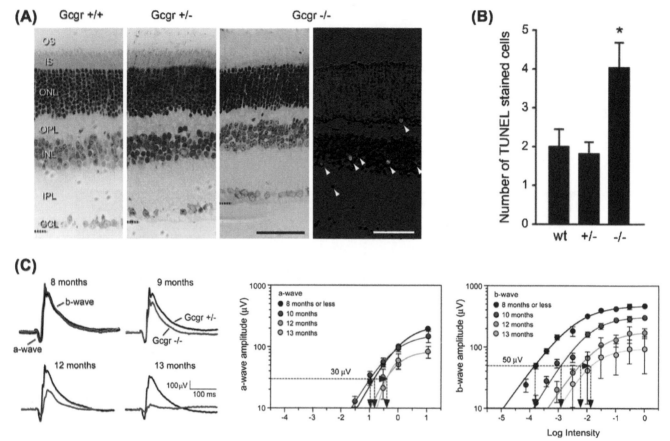

FIGURE 64.1 Retinal cell death and age-related changes in the electroretinogram (ERG) in chronic hypoglycemic mouse models. (A) Sections of retinas from age-matched adult Gcgr$^{+/+}$, Gcgr$^{+/-}$, and moderately hypoglycemic Gcgr$^{-/-}$ mice. Shown are photoreceptors (outer segments, OS; inner segments, IS) at the top and the ganglion cell layer (GCL) at the bottom. Visible in between are the outer plexiform layer (OPL) and inner plexiform layer (IPL) and the densely stained cell bodies in the outer nuclear layer (ONL) and inner nuclear layer (INL). The oblique cryosection on the right from a Gcgr$^{-/-}$ mouse shows terminal deoxynucleotidyl transferase dUTP nick-end labeling (TUNEL) staining primarily in the INL. (B) Average number of TUNEL-positive cells is significantly greater for Gcgr-deficient mice than for heterozygous and wild-type (wt) mice. (C) Average ERGs recorded from a group of moderately hypoglycemic Gcgr$^{-/-}$ mice (red traces) and euglycemic Gcgr$^{+/+}$ mice (black traces) in response to 10 ms flashes. Intensity–response functions of a-waves and b-waves recorded from the same ERGs. Dashed horizontal lines and vertical intercepts on the abscissa show the method for determining threshold light intensities. Responses were recorded at 8, 10, 12, and 13 months. *Source: Adapted from Umino* et al. Proc Natl Acad Sci U S A. *2006;103:19541–5,[26] with the permission of* PNAS.

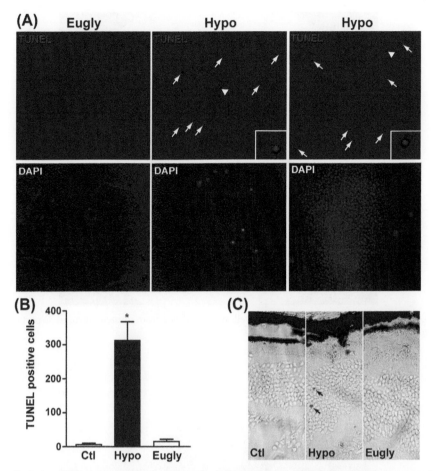

FIGURE 64.3 Acute insulin-induced hypoglycemia leads to retinal cell death in mice. (A) Flat-mounted retinas were isolated 48 hours after the clamp, stained for cell death by terminal deoxynucleotidyl transferase dUTP nick-end labeling (TUNEL) assay, and DAPI counterstained. White arrows show TUNEL-positive cells in hypoglycemic condition. (B) Quantification of TUNEL-positive cells was performed under a fluorescence microscope on retinal flat-mounts. Results are expressed as mean ± SEM of three different retinas for each group, *P < 0.006 hypoglycemic (Hypo) vs. euglycemic (Eugly). (C) Ten-micrometer embedded frozen sections of enucleated eyes isolated from control (Ctl), hypoglycemic (Hypo), and euglycemic (Eugly) animals were stained for cell death by a colorimetric TUNEL system. Using this procedure, apoptotic nuclei are stained dark brown (black arrows). A representative region of three different isolated retinas is shown. *Source: From Emery* et al. PLoS ONE *2011;6:e21586,*[29] *with permission of* PLoS ONE.

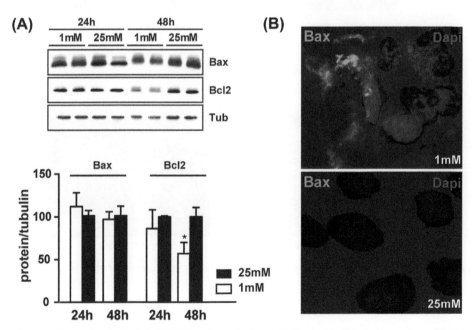

FIGURE 64.5 Hypoglycemia-induced cell death is Bcl2/Bax dependent. (A) Culture of 661W photoreceptor cells at low glucose induces a decrease in Bcl-2 protein expression. (B) The decrease in *Bcl-2* gene expression releases the active proapoptotic Bax that is recognized by a specific antibody. *Source: Adapted from Balmer* et al. PLoS ONE *2013;10.1371/journal.pone.0074162.*[74]

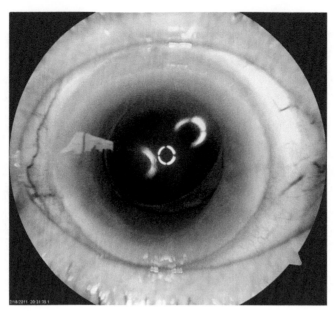

FIGURE 65.1 Slit lamp photograph demonstrating an orange–yellow corneal ring from carotenoids used as part of the AREDS1 formula near the limbus and extending toward the pupil in a 72-year-old female.

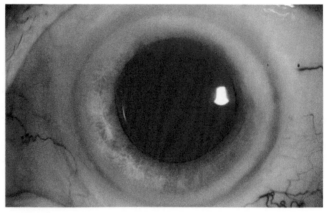

FIGURE 65.2 An example of yellow peripheral corneal ring secondary to carotenoids from a generic form of AREDS supplementation that was marigold flower-based in an 89-year-old female.

(a)   (b)

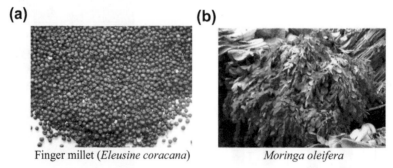

Finger millet (*Eleusine coracana*)   *Moringa oleifera*

FIGURE 42.3 Polyphenol-rich food materials explored for beneficial effects on eye health: (a) finger millet (*Eleusine coracana*); (b) *Moringa oleifera*.

Printed and bound by CPI Group (UK) Ltd, Croydon, CR0 4YY

08/05/2025

01865034-0007